More Than Just a Textbook

Log on to algebra1.com to...

- access your Online Student Edition from home so you don't need to bring your textbook home each night.
- link to Student Workbooks and Online Study Tools.

See mathematical concepts come to life

- Personal Tutor
- Concepts in Motion: BrainPOP® Movies
- Concepts in Motion: Interactive Labs
- Concepts in Motion: Animations

Practice what you've learned

- Chapter Readiness
- Extra Examples
- Self-Check Quizzes
- Reading in the Content Area
- Graphing Calculator Keystrokes
- Vocabulary Review
- Chapter Tests
- Texas Test Practice

Try these other fun activities

- Cross-Curricular Projects
- Real-World Careers
- GameZone Games

The Cheecks LOL

5209♡

Algebra 1

Authors

Holliday • Luchin • Marks • Day
Cuevas • Carter • Casey • Hayek

New York, New York Columbus, Ohio Chicago, Illinois Woodland Hills, California

About the Cover

Reaching a dizzying height of 420 feet and a top speed of 120 miles per hour, the Top Thrill Dragster in Sandusky, Ohio, is the tallest and fastest roller coaster in the world. Riders of the Top Thrill Dragster accelerate to top speed in approximately four seconds and then zoom straight up on a track that rotates 90 degrees. In Chapter 4, you will learn about rates of change like speeds.

About the Graphics

Twisted torus. Created with *Mathematica.*
A torus with rose-shaped cross section is constructed. Then the cross section is rotated around its center as it moves along a circle to form a twisted torus. For more information, and for programs to construct such graphics, see: www.wolfram.com

The McGraw·Hill Companies

Send all inquiries to:
Glencoe/McGraw-Hill
8787 Orion Place
Columbus, OH 43240-4027

ISBN: 978-0-07-873822-7
MHID: 0-07-873822-9

Printed in the United States of America.

3 4 5 6 7 8 9 10 043/058 15 14 13 12 11 10 09 08 07

Contents in Brief

Authors

Berchie Holliday, Ed.D.
National Mathematics Consultant
Silver Spring, MD

Gilbert J. Cuevas, Ph.D.
Professor of Mathematics Education
University of Miami
Miami, FL

Beatrice Luchin
Mathematics Consultant
League City, TX

John A. Carter, Ph.D.
Director of Mathematics
Adlai E. Stevenson High School
Lincolnshire, IL

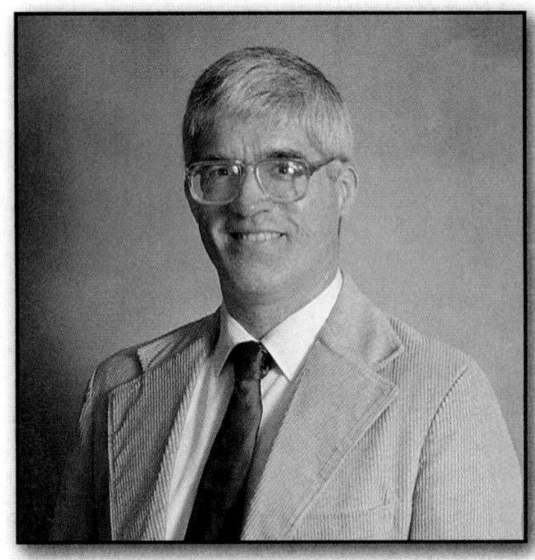

Daniel Marks, Ed.D
Professor Emeritus of Mathematics
Auburn University at Montgomery
Montgomery, AL

Roger Day, Ph.D.
Mathematics Department Chairperson
Pontiac Township High School
Pontiac, IL

Ruth M. Casey
Mathematics Teacher
Department Chair
Anderson County High School
Lawrenceburg, KY

Linda M. Hayek
Mathematics Teacher
Ralston Public Schools
Omaha, NE

Contributing Authors

Carol E. Malloy, Ph.D
Associate Professor
University of North Carolina at Chapel Hill
Chapel Hill, NC

Viken Hovsepian
Professor of Mathematics
Rio Hondo College
Whittier, CA

FOLDABLES **Dinah Zike**
Educational Consultant, Dinah-Might Activities, Inc.
San Antonio, TX

Consultants

Glencoe/McGraw-Hill wishes to thank the following professionals for their feedback. They were instrumental in providing valuable input toward the development of this program in these specific areas.

Mathematical Content

Viken Hovsepian
Professor of Mathematics
Rio Hondo College
Whittier, California

Bob McCollum
Associate Principal
Curriculum and Instruction
Glenbrook South High School
Glenview, Illinois

Differentiated Instruction

Nancy Frey, Ph.D.
Associate Professor of Literacy
San Diego State University
San Diego, California

English Language Learners

Mary Avalos, Ph.D.
Assistant Chair, Teaching and Learning
Assistant Research Professor
University of Miami, School of Education
Coral Gables, Florida

Jana Echevarria, Ph.D.
Professor, College of Education
California State University, Long Beach
Long Beach, California

Josefina V. Tinajero, Ph.D.
Dean, College of Educatifon
The University of Texas at El Paso
El Paso, Texas

Gifted and Talented

Ed Zaccaro
Author
Mathematics and science books for gifted children
Bellevue, Iowa

Graphing Calculator

Ruth M. Casey
Mathematics Teacher
Department Chair
Anderson County High School
Lawrenceburg, Kentucky

Jerry Cummins
Past President
National Council of Supervisors of Mathematics
Western Springs, Illinois

Learning Disabilities

Kate Garnett, Ph.D.
Chairperson, Coordinator Learning Disabilities
School of Education
Department of Special Education
Hunter College, CUNY
New York, New York

Mathematical Fluency

Jason Mutford
Mathematics Instructor
Coxsackie-Athens Central School District
Coxsackie, New York

Pre-AP

Dixie Ross
AP Calculus Teacher
Pflugerville High School
Pflugerville, Texas

Reading and Vocabulary

Douglas Fisher, Ph.D.
Director of Professional Development and Professor
City Heights Educational Collaborative
San Diego State University
San Diego, California

Lynn T. Havens
Director of Project CRISS
Kalispell School District
Kalispell, Montana

Teacher Reviewers

Each Reviewer reviewed at least two chapters of the Student Edition, giving feedback and suggestions for improving the effectiveness of the mathematics instruction.

Chrissy Aldridge
Teacher
Charlotte Latin School
Charlotte, North Carolina

Rick Ardary
Mathematics Department Chair
Eastern York High School
Wrightsville, Pennsylvania

Harriette Neely Baker
Mathematics Teacher
South Mecklenburg High School
Charlotte, North Carolina

Danny L. Barnes, NBCT
Mathematics Teacher
Speight Middle School
Stantonsburg, North Carolina

Aimee Barrette
Special Education Teacher
Sedgefield Middle School
Charlotte, North Carolina

Karen J. Blackert
Mathematics Teacher
Myers Park High School
Charlotte, North Carolina

Patricia R. Blackwell
Mathematics Department Chair
East Mecklenburg High School
Charlotte, North Carolina

Rebecca B. Caison
Mathematics Teacher
Walter M. Williams High School
Burlington, North Carolina

Myra Cannon
Mathematics Department Chair
East Davidson High School
Thomasville, North Carolina

Peter K. Christensen
Mathematics/AP Teacher
Central High School
Macon, Georgia

Rebecca Claiborne
Mathematics Department Chairperson
George Washington Carver High School
Columbus, Georgia

Laura Crook
Mathematics Department Chair
Middle Creek High School
Apex, North Carolina

Dayl F. Cutts
Teacher
Northwest Guilford High School
Greensboro, North Carolina

Angela S. Davis
Mathematics Teacher
Bishop Spaugh Community Academy
Charlotte, North Carolina

Sheri Dunn-Ulm
Teacher
Bainbridge High School
Bainbridge, Georgia

Susan M. Fritsch
Mathematics Teacher, NBCT
David W. Butler High School
Matthews, North Carolina

Dr. Jesse R. Gassaway
Teacher
Northwest Guilford Middle School
Greensboro, North Carolina

Tina Gleason
8th Grade Algebra Teacher
Samuel Morse Middle School
Milwaukee, Wisconsin

Matt Gowdy
Mathematics Teacher
Grimsley High School
Greensboro, North Carolina

Wendy Hancuff
Teacher
Jack Britt High School
Fayetteville, North Carolina

Ernest A. Hoke Jr.
Mathematics Teacher
E. B. Aycock Middle School
Greenville, North Carolina

Carol B. Huss
Mathematics Teacher
Independence High School
Charlotte, North Carolina

Deborah Ivy
Mathematics Teacher
Marie G. Davis Middle School
Charlotte, North Carolina

Lynda B. (Lucy) Kay
Mathematics Department Chair
Martin Middle School
Raleigh, North Carolina

Julia Kolb
Mathematics Teacher/Department Chair
Leesville Road High School
Raleigh, North Carolina

M. Kathleen Kroh
Mathematics Teacher
Z. B. Vance High School
Charlotte, North Carolina

Tosha S. Lamar
Mathematics Instructor
Phoenix High School
Lawrenceville, Georgia

Kay S. Laster
8th Grade Pre-Algebra/Algebra Teacher
Rockingham County Middle School
Reidsville, North Carolina

Joyce M. Lee
Lead Mathematics Teacher
National Teachers Teaching with Technology Instructor
George Washington Carver High School
Columbus, Georgia

Susan Marshall
Mathematics Chairperson
Kernodle Middle School
Greensboro, North Carolina

Alice D. McLean
Mathematics Coach
West Charlotte High School
Charlotte, North Carolina

Portia Mouton
Mathematics Teacher
Westside High School
Macon, Georgia

Elaine Pappas
Mathematics Department Chair
Cedar Shoals High School
Athens, Georgia

Susan M. Peeples
Retired 8th Grade Mathematics Teacher
Richland School District Two
Columbia, South Carolina

Carolyn G. Randolph
Mathematics Department Chair
Kendrick High School
Columbus, Georgia

Mary Roden
Math Curriculum Facilitator
Centennial ISD #12
Circle Pines, Minnesota

Tracey Shaw
Mathematics Teacher
Chatham Central High School
Bear Creek, North Carolina

Marjorie Smith
Mathematics Teacher
Eastern Randolph High School
Ramseur, North Carolina

McCoy Smith, III
Mathematics Department Chair
Sedgefield Middle School
Charlotte, North Carolina

Bridget Sullivan
8th Grade Mathematics Teacher
Northeast Middle School
Charlotte, North Carolina

Marilyn R. Thompson
Geometry/Mathematics Vertical Team Consultant
Charlotte-Mecklenburg Schools
Charlotte, North Carolina

Gwen Turner
Mathematics Teacher
Clarke Central High School
Athens, Georgia

Elizabeth Webb
Mathematics Department Chair
Myers Park High School
Charlotte, North Carolina

Jack Whittemore
C & I Resource Teacher
Charlotte-Mecklenburg Schools
Charlotte, North Carolina

Angela Whittington
Mathematics Teacher
North Forsyth High School
Winston-Salem, North Carolina

Kentucky Consultants

Amy Adams Cash
Mathematics Educator/Department Chair
Bowling Green High School
Bowling Green, Kentucky

Susan Hack, NBCT
Mathematics Teacher
Oldham County High School
Buckner, Kentucky

Kimberly L. Henderson Hockney
Mathematics Educator
Larry A. Ryle High School
Union, Kentucky

Unit 1

Foundations for Functions

The Language and Tools of Algebra

Prerequisite Skills

Reading and Writing Mathematics

Standardized Test Practice

H.O.T. Problems
Higher Order Thinking

Solving Linear Equations

Unit 2

Linear Functions

Functions and Patterns

Analyzing Linear Equations

Prerequisite Skills

Reading and Writing Mathematics

Standardized Test Practice

H.O.T. Problems
Higher Order Thinking

CHAPTER 5 Solving Systems of Linear Equations

Prerequisite Skills

Reading and Writing Mathematics

Standardized Test Practice

H.O.T. Problems
Higher Order Thinking

Solving Linear Inequalities

Unit 3

Polynomials and Nonlinear Functions

Polynomials

Prerequisite Skills

Reading and Writing Mathematics

Standardized Test Practice

H.O.T. Problems
Higher Order Thinking

CHAPTER 8 Factoring

CHAPTER 9

Quadratic and Exponential Functions

Unit 4

Advanced Expressions and Data Analysis

Chapter 10 Radical Expressions and Triangles

CHAPTER 11

Rational Expressions and Equations

Prerequisite Skills

Reading and Writing Mathematics

Standardized Test Practice

H.O.T. Problems
Higher Order Thinking

Statistics and Probability

Student Handbook

Built-In Workbooks

Reference

Prerequisite Skills

Reading and Writing Mathematics

Standardized Test Practice

H.O.T. Problems
Higher Order Thinking

UNIT 1

Foundations for Functions

Focus

Use symbols to express relationships and solve real-world problems.

CHAPTER 1

The Language and Tools of Algebra

BIG Idea Identify and use the arithmetic properties of subsets of integers and rational, irrational, and real numbers.

BIG Idea Use properties of numbers to demonstrate whether assertions are true or false and to construct simple, valid arguments (direct or indirect) for, or formulate counterexamples to claimed assertions.

CHAPTER 2

Solving Linear Equations

BIG Idea Simplify expressions before solving linear equations and inequalities in one variable.

BIG Idea Solve multistep problems, including word problems, involving linear equations and linear inequalities in one variable.

BIG Idea Apply algebraic techniques to solve rate problems, work problems, and percent mixture problems.

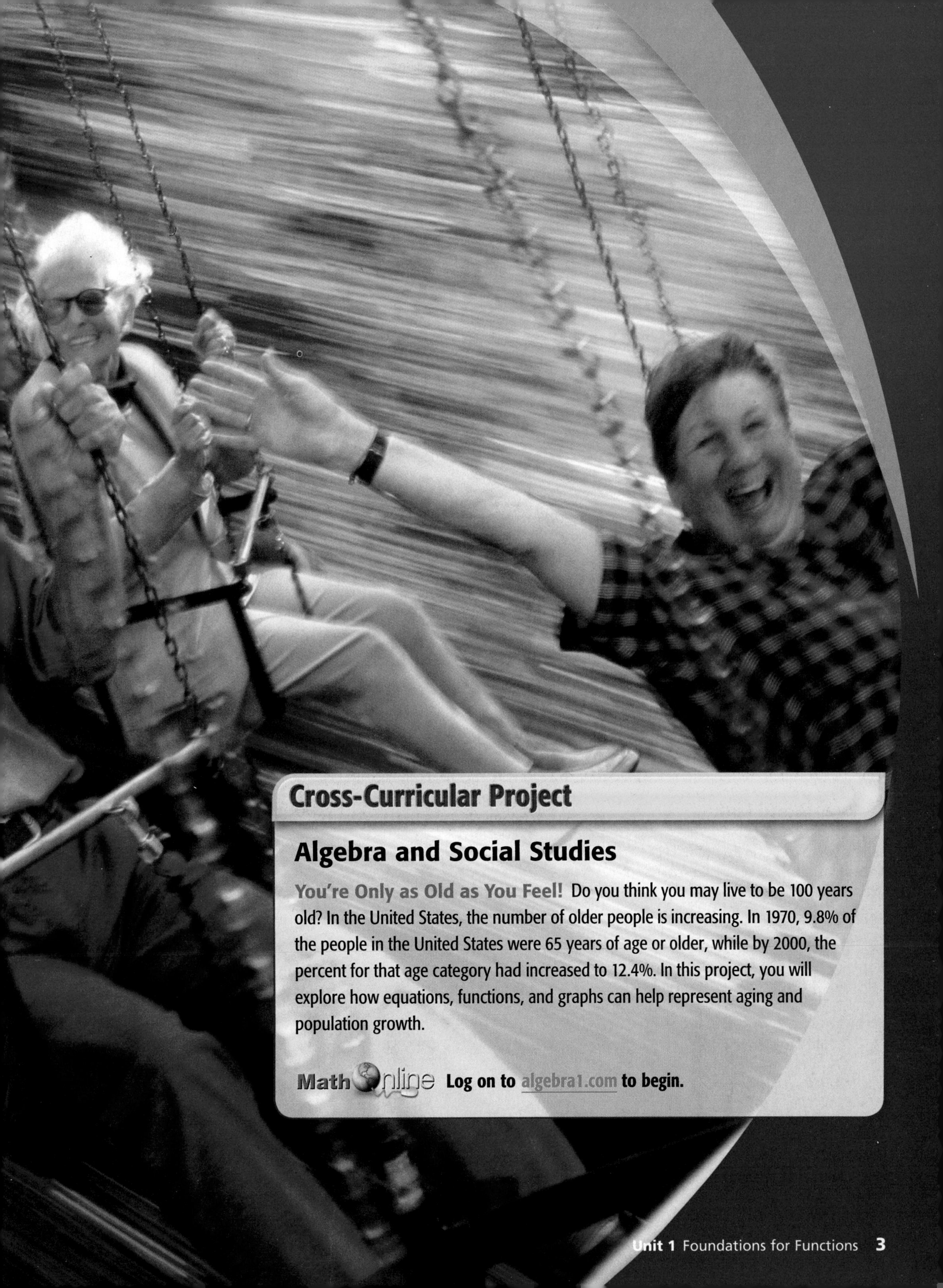

Cross-Curricular Project

Algebra and Social Studies

You're Only as Old as You Feel! Do you think you may live to be 100 years old? In the United States, the number of older people is increasing. In 1970, 9.8% of the people in the United States were 65 years of age or older, while by 2000, the percent for that age category had increased to 12.4%. In this project, you will explore how equations, functions, and graphs can help represent aging and population growth.

Math Online **Log on to algebra1.com to begin.**

The Language and Tools of Algebra

BIG Ideas

- Write algebraic expressions.
- Evaluate expressions and solve open sentences.
- Use algebraic properties of identity and equality.
- Use conditional statements and counterexamples.

Key Vocabulary

algebraic expression (p. 6)

coefficient (p. 29)

equation (p. 15)

function (p. 53)

Real-World Link

Architecture Architects can use algebraic expressions to describe the shapes of the structures they design. A few of the shapes these buildings can resemble are a rectangle, a triangle, or even a pyramid.

FOLDABLES™ Study Organizer

The Language and Tools of Algebra Make this Foldable to help you organize information about algebraic properties. Begin with a sheet of notebook paper.

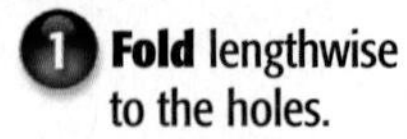

1. **Fold** lengthwise to the holes.

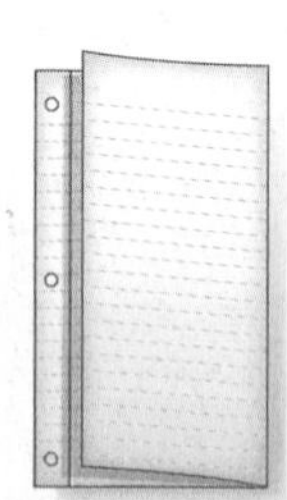

2. **Cut** along the top line and then cut 10 tabs.

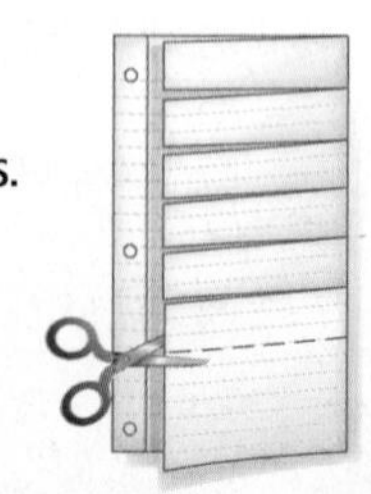

3. **Label** the tabs using the lesson numbers and concepts.

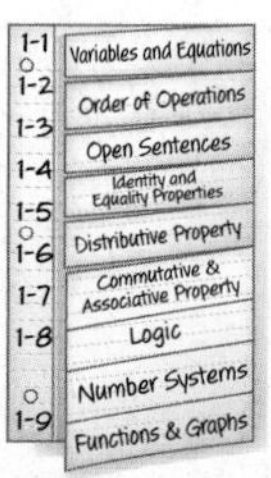

GET READY for Chapter 1

Diagnose Readiness You have two options for checking Prerequisite Skills.

Option 2

Math Online Take the Online Readiness Quiz at algebra1.com.

Option 1

Take the Quick Check below. Refer to the Quick Review for help.

QUICKCheck

Write each fraction in simplest form. If the fraction is already in simplest form, write *simplest form*. (Prerequisite Skill)

1. $\frac{52}{13}$ **2.** $\frac{6}{18}$ **3.** $\frac{9}{15}$ **4.** $\frac{13}{25}$

5. $\frac{26}{100}$ **6.** $\frac{3}{81}$ **7.** $\frac{17}{1}$ **8.** $\frac{15}{75}$

9. SURVEY Thirty-three out of 198 students surveyed said that they preferred hockey to all other sports. What fraction of students surveyed is this?

Find the perimeter of each figure.
(Prerequisite Skill)

10.

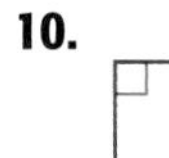

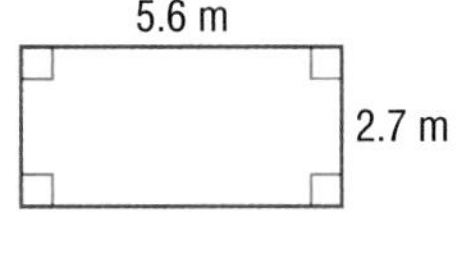

11.

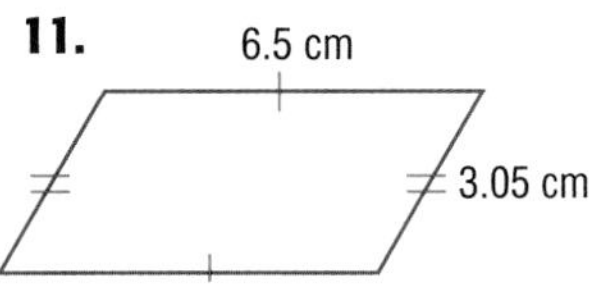

12.

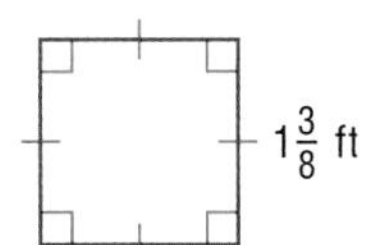

13.

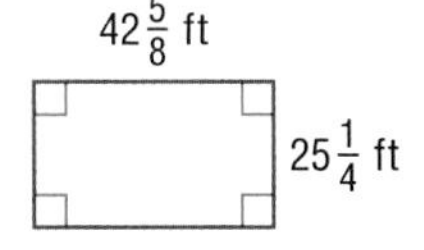

14. HOMES The dimensions of a rectangular backyard are 45 feet by 84 feet. What is its perimeter?

Find each product or quotient.
(Prerequisite Skill)

15. $6 \cdot 1.12$ **16.** $0.5 \cdot 3.9$

17. $3.24 \div 1.8$ **18.** $10.64 \div 1.4$

19. $\frac{3}{4} \cdot 12$ **20.** $1\frac{2}{3} \cdot \frac{3}{4}$

21. $\frac{5}{16} \div \frac{9}{12}$ **22.** $\frac{5}{6} \div \frac{2}{3}$

QUICKReview

EXAMPLE 1

Write $\frac{30}{36}$ in simplest form.

Find the greatest common factor (GCF) of 30 and 36.

factors of 30: 1, 2, 3, 5, (6), 10, 15, 30

factors of 36: 1, 2, 3, 4, (6), 9, 12, 18, 36

The GCF of 30 and 36 is 6.

$\frac{30 \div 6}{36 \div 6} = \frac{5}{6}$ Divide the numerator and denominator by their GCF, 6.

EXAMPLE 2

Find the perimeter of the figure.

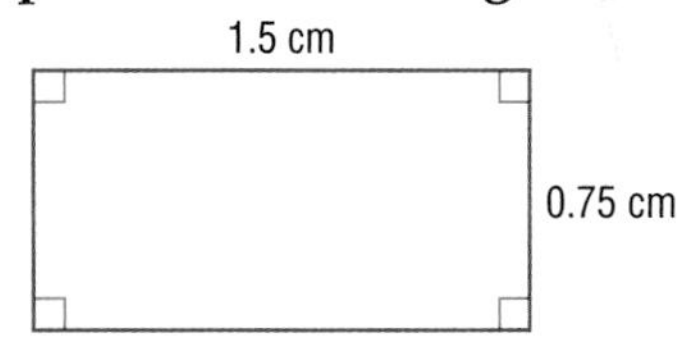

$P = 2\ell + 2w$

$P = 2(1.5) + 2(0.75)$ $\ell = 1.5$ and $w = 0.75$

$P = 3 + 1.5$ or 4.5 Simplify.

The perimeter is 4.5 centimeters.

EXAMPLE 3

Find $\frac{4}{5} \div \frac{12}{15}$.

$\frac{4}{5} \div \frac{12}{15} = \frac{4}{5}\left(\frac{15}{12}\right)$ Multiply $\frac{4}{5}$ by $\frac{15}{12}$, the reciprocal of $\frac{12}{15}$.

$= \frac{4(15)}{5(12)}$ Multiply the numerators and the denominators.

$= \frac{60}{60}$ or 1 Simplify.

1-1 Variables and Expressions

Main Ideas

- Write mathematical expressions for verbal expressions.
- Write verbal expressions for mathematical expressions.

New Vocabulary

variables
algebraic expression
factors
product
power
base
exponent
evaluate

GET READY for the Lesson

A baseball infield is a square with a base at each corner. Each base lies the same distance from the next one. Suppose s represents the length of each side. Since the infield is a square, you can use the expression 4 times s, or $4s$, to find the perimeter.

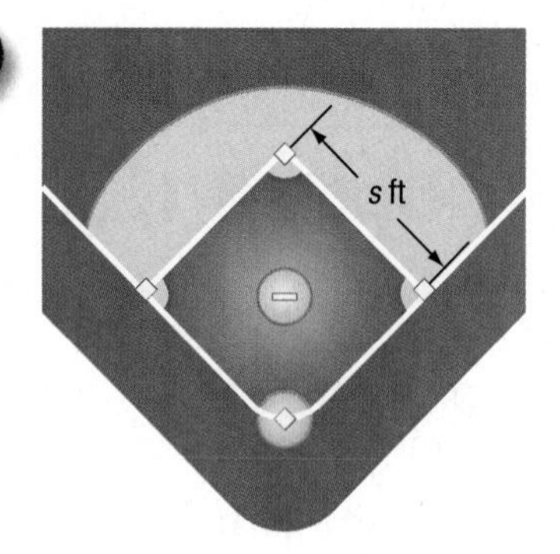

Write Mathematical Expressions In the algebraic expression $4s$, the letter s is called a variable. In algebra, **variables** are symbols used to represent unspecified numbers or values. Any letter may be used as a variable. *The letter s was used above because it is the first letter of the word side.*

An **algebraic expression** consists of one or more numbers and variables along with one or more arithmetic operations. Here are some examples of algebraic expressions.

$$5x \qquad 3x - 7 \qquad 4 + \frac{p}{q} \qquad m \times 5n \qquad 3ab \div 5cd$$

In algebraic expressions, a raised dot or parentheses are often used to indicate multiplication as the symbol $\times$ can be easily mistaken for the letter x. Here are several ways to represent the product of x and y.

$$xy \qquad x \cdot y \qquad x(y) \qquad (x)y \qquad (x)(y)$$

In each expression, the quantities being multiplied are called **factors**, and the result is called the **product**.

An expression like x^n is raised is called a **power**. The variable x is called the **base**, and n is called the **exponent**. The word *power* can also refer to the exponent. The exponent indicates the number of times the base is used as a factor. The expression x^n is read "x to the nth power."

Reading Math

First Power
When no exponent is shown, it is understood to be 1. For example, $a = a^1$.

Symbols	Words	Meaning
3^1	3 to the first power	3
3^2	3 to the second power or 3 squared	$3 \cdot 3$
3^3	3 to the third power or 3 cubed	$3 \cdot 3 \cdot 3$
3^4	3 to the fourth power	$3 \cdot 3 \cdot 3 \cdot 3$
$2b^6$	2 times b to the sixth power	$2 \cdot b \cdot b \cdot b \cdot b \cdot b \cdot b$
x^n	x to the nth power	$\underbrace{x \cdot x \cdot x \cdot \ldots \cdot x}_{n \text{ factors}}$

By definition, $x^0 = 1$ for any nonzero number x.

It is often necessary to translate verbal expressions into algebraic expressions.

EXAMPLE 1 Write Algebraic Expressions

1 Write an algebraic expression for each verbal expression.

a. eight more than a number

The words *more than* suggest addition. Let n represent the number.

Thus, the algebraic expression is $n + 8$.

b. 7 less the product of 4 and a number x

Less implies subtract, and *product* implies multiply. So the expression can be written as $7 - 4x$.

c. one third of the original area a

The word *of* implies multiply. The expression can be written as $\frac{1}{3}a$ or $\frac{a}{3}$.

d. the product of 7 and m to the fifth power

$7m^5$

CHECK Your Progress

1A. 13 less than a number

1B. 9 more than the quotient of b and 5

1C. three-fourths of the perimeter p

1D. n cubed divided by 2

Online Personal Tutor at algebra1.com

Reading Math

Subtraction

5 less x is $5 - x$.
5 less than x is $x - 5$.

To **evaluate** an expression means to find its value.

EXAMPLE 2 Evaluate Powers

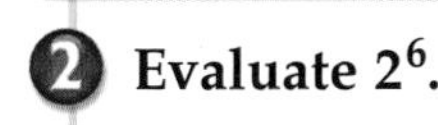

2 Evaluate 2^6.

$2^6 = 2 \cdot 2 \cdot 2 \cdot 2 \cdot 2 \cdot 2$ Use 2 as a factor 6 times.

$= 64$ Multiply.

CHECK Your Progress

2. Evaluate 4^3.

Study Tip

Multiple Translations

There may be more than one way to translate an algebraic expression into a verbal expression.

$4m^3 \rightarrow$ 4 times m cubed

$c^2 + 21d \rightarrow$ c squared plus the product of 21 and d

Write Verbal Expressions Another important skill is translating algebraic expressions into verbal expressions.

EXAMPLE 3 Write Verbal Expressions

3 Write a verbal expression for each algebraic expression.

a. $4m^3$
4 times m to the third power

b. $c^2 + 21d$
the sum of c squared and 21 times d

CHECK Your Progress

3. Write a verbal expression for $x^4 - \frac{y}{9}$.

Math Online Extra Examples at algebra1.com

Example 1 (p. 7)

Write an algebraic expression for each verbal expression.

1. the sum of a number and 14

2. 6 less a number t

3. 24 less than 3 times a number

4. 1 minus the quotient of r and 7

5. two-fifths of a number j squared

6. n cubed increased by 5

7. MONEY Lorenzo bought a bag of peanuts that cost p dollars, and he gave the cashier a $20 bill. Write an expression for the amount of change that he will receive.

Example 2 (p. 7)

Evaluate each expression.

8. 9^2

9. 4^4

Example 3 (p. 7)

Write a verbal expression for each algebraic expression.

10. $2m$

11. $\frac{1}{2}n^3$

12. $a^2 - 18b$

Exercises

HOMEWORK HELP	
For Exercises	See Examples
13–25	1
26–29	2
30–37	3

Write an algebraic expression for each verbal expression.

13. x more than 7

14. a number less 35

15. 5 times a number

16. one third of a number

17. f divided by 10

18. the quotient of 45 and r

19. 49 increased by twice a number

20. 18 decreased by 3 times d

21. k squared minus 11

22. 20 divided by t to the fifth power

23. GEOMETRY The area of a circle is the number π times the square of the radius. Write an expression that represents the area of a circle with a radius r.

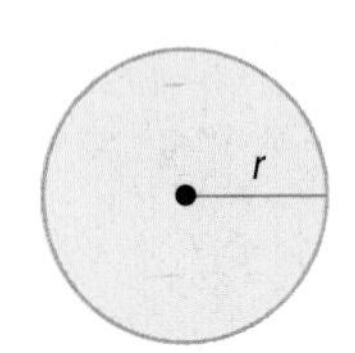

RECYCLING For Exercises 24 and 25, use the following information.

Each person in the United States produces about 3.5 pounds of trash a day.

24. Write an expression to describe the pounds of trash produced per day by a family with m members.

25. Use the expression you wrote to predict the amount of trash produced by a family of four each day.

Evaluate each expression.

26. 8^2

27. 10^6

28. 3^5

29. 15^3

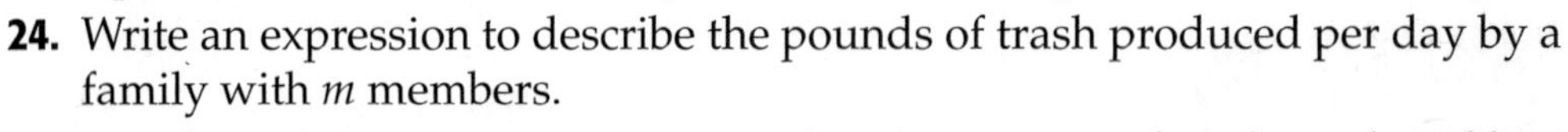

Write a verbal expression for each algebraic expression.

30. $7p$

31. $\frac{1}{8}y$

32. $15 + r$

33. $w - 24$

34. $3x^2$

35. $\frac{r^4}{9}$

36. $2a + 6$

37. $n^3 \cdot p^5$

EXTRA PRACTICE
See pages 714, 744.

Math Online
Self-Check Quiz at algebra1.com

Write a verbal expression for each algebraic expression.

38. $17 - 4m^5$

39. $\frac{12z^2}{5}$

40. $3x^2 - 2x$

41. **SAVINGS** Kendra is saving to buy a new computer. Write an expression to represent the total amount of money she will have if she has s dollars saved and she adds d dollars per week for the next 12 weeks.

H.O.T. Problems

42. **MUSIC** Mario has 55 CDs. Write an expression to represent the total number of CDs he will have after 18 months if he buys x CDs per month.

43. **REASONING** Determine whether the product given by the expression $-3a$ is *always, sometimes,* or *never* a negative number. Explain.

44. **CHALLENGE** In the square, x represents a positive whole number. Find the value of x such that the area and the perimeter of the square have the same value.

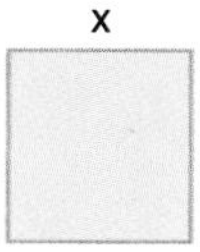

45. ***Writing in Math*** Use the data about baseball found on page 6 to explain how expressions can be used to find the perimeter of a baseball diamond. Include two different verbal expressions and an algebraic expression other than 4s to represent the perimeter of a square.

STANDARDIZED TEST PRACTICE

46. Which expression best represents the perimeter of the rectangle?

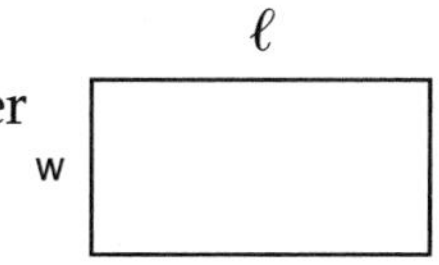

A $2\ell w$ C $2\ell + 2w$

B $\ell + w$ D $4(\ell + w)$

47. The yards of fabric needed to make curtains is 3 times the width of a window in inches, divided by 36. Which expression best represents the yards of fabric needed in terms of the width of the window w?

F $\frac{3 + w}{36}$ H $\frac{3}{36w}$

G $\frac{3w}{36}$ J $3w(36)$

48. **REVIEW** Which expression best represents the volume of the cube?

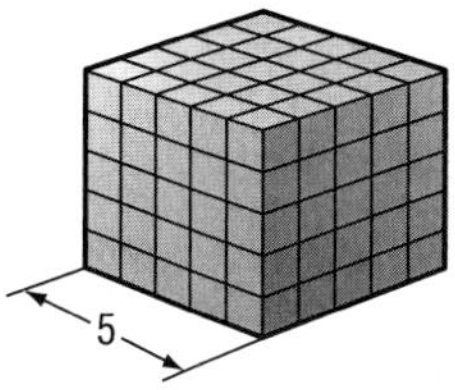

A the product of five and three

B three to the fifth power

C five squared

D five cubed

GET READY for the Next Lesson

PREREQUISITE SKILL Evaluate each expression. (Pages 696–697)

49. $10 - 3.24$

50. 1.04×4.3

51. $15.36 \div 4.8$

52. $\frac{1}{3} + \frac{2}{5}$

53. $\frac{3}{8} \times \frac{4}{9}$

54. $\frac{7}{10} \div \frac{3}{5}$

1-2 Order of Operations

Main Ideas

- Evaluate numerical expressions by using the order of operations.
- Evaluate algebraic expressions by using the order of operations.

New Vocabulary

order of operations

GET READY for the Lesson

Nicole's Internet service provider charges \$4.95 a month, which includes 100 hours of access. If she is online more than 100 hours, she pays an additional \$0.99 per hour. Suppose Nicole is online 117 hours this month. The expression below represents what she must pay for the month.

$4.95 + 0.99(117 - 100)$

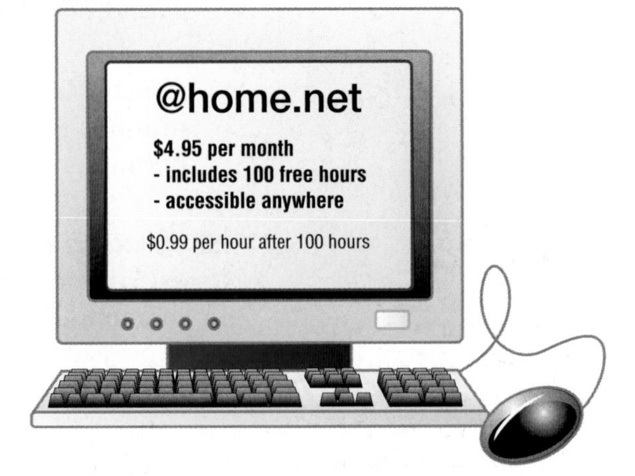

Evaluate Numerical Expressions Numerical expressions often contain more than one operation. A rule is needed to let you know which operation to perform first. This rule is called the **order of operations**.

KEY CONCEPT — *Order of Operations*

Step 1 Evaluate the expressions inside grouping symbols.

Step 2 Evaluate all powers.

Step 3 Multiply and/or divide in order from left to right.

Step 4 Add and/or subtract in order from left to right.

EXAMPLE Evaluate Expressions

1 **Evaluate $15 \div 3 \cdot 6 - 4^2$.**

$15 \div 3 \cdot 6 - 4^2 = 15 \div 3 \cdot 6 - 16$	Evaluate power.
$= 5 \cdot 6 - 16$	Divide 15 by 3.
$= 30 - 16$	Multiply 5 by 6.
$= 14$	Subtract 16 from 30.

CHECK Your Progress

Evaluate each expression.

1A. $8 - 6 \cdot 4 \div 3$

1B. $32 + 7^2 - 5 \cdot 2$

Grouping symbols such as parentheses ( ), brackets [ ], and braces { } are used to clarify or change the order of operations. They indicate that the expression within the grouping symbol is to be evaluated first. A fraction bar also acts as a grouping symbol. It indicates that the numerator and denominator should each be treated as a single value.

EXAMPLE Grouping Symbols

2 **Evaluate each expression.**

a. $2(5) + 3(4 + 3)$

$$2(5) + 3(4 + 3) = 2(5) + 3(7) \quad \text{Evaluate inside parentheses.}$$
$$= 10 + 21 \quad \text{Multiply expressions left to right.}$$
$$= 31 \quad \text{Add 10 and 21.}$$

Study Tip

Grouping Symbols

When more than one grouping symbol is used, start evaluating within the innermost grouping symbols.

b. $2[5 + (30 \div 6)^2]$

$$2[5 + (30 \div 6)^2] = 2[5 + (5)^2] \quad \text{Evaluate innermost expression first.}$$
$$= 2[5 + 25] \quad \text{Evaluate power.}$$
$$= 2[30] \quad \text{Evaluate expression inside grouping symbols.}$$
$$= 60 \quad \text{Multiply.}$$

c. $\frac{6 + 4}{3^2 \cdot 4}$

$$\frac{6 + 4}{3^2 \cdot 4} = \frac{10}{3^2 \cdot 4} \quad \text{Add 6 and 4 in the numerator.}$$
$$= \frac{10}{9 \cdot 4} \quad \text{Evaluate the power in the denominator.}$$
$$= \frac{10}{36} \text{ or } \frac{5}{18} \quad \text{Multiply 9 and 4 in the denominator. Then simplify.}$$

CHECK Your Progress

2A. $(15 - 9) + 3 \cdot 6$ **2B.** $45 + [(1 + 1)^3 \div 4]$ **2C.** $\frac{6^2 - 8}{4(3 + 7)}$

Evaluate Algebraic Expressions To evaluate an algebraic expression, replace the variables with their values. Then, find the value of the numerical expression using the order of operations.

EXAMPLE Evaluate an Algebraic Expression

3 **Evaluate $a^2 - (b^3 - 4c)$ if $a = 7$, $b = 3$, and $c = 5$.**

$$a^2 - (b^3 - 4c) = 7^2 - (3^3 - 4 \cdot 5) \quad \text{Replace } a \text{ with 7, } b \text{ with 3, and } c \text{ with 5.}$$
$$= 49 - (27 - 4 \cdot 5) \quad \text{Evaluate } 7^2 \text{ and } 3^3.$$
$$= 49 - (27 - 20) \quad \text{Multiply 4 and 5.}$$
$$= 49 - 7 \quad \text{Subtract 20 from 27.}$$
$$= 42 \quad \text{Subtract.}$$

CHECK Your Progress

3. Evaluate $x(y^3 + 8) \div 12$ if $x = 3$ and $y = 4$.

Real-World Career...
Architect

Architects use math to find measures and calculate areas. They must consider the function, safety, and appearance when they design buildings.

For more information, go to algebra1.com.

EXAMPLE

4 ARCHITECTURE The Pyramid Arena in Memphis, Tennessee, is the third largest pyramid in the world. The area of its base is 360,000 square feet, and it is 321 feet high. The volume of a pyramid is one third of the product of the area of the base B and its height h.

a. Write an expression that represents the volume of a pyramid.

Words	one third	of	the product of area of base and height
Variables	B = area of base and h = height		
Expression	$\frac{1}{3}$	$\times$	$(B \cdot h)$ or $\frac{1}{3}Bh$

b. Find the volume of the Pyramid Arena.

$V = \frac{1}{3}(Bh)$ — Volume of a pyramid

$= \frac{1}{3}(360{,}000 \cdot 321)$ — Replace B with 360,000 and h with 321.

$= \frac{1}{3}(115{,}560{,}000)$ — Multiply 360,000 by 321.

$= 38{,}520{,}000$ — Multiply $\frac{1}{3}$ by 115,560,000.

The volume of the Pyramid Arena is 38,520,000 cubic feet.

CHECK Your Progress

According to market research, the average consumer spends \$78 per trip to the mall on weekends and only \$67 per trip during the week.

4A. Write an algebraic expression to represent how much the average consumer spends at the mall in x weekend trips and y weekday trips.

4B. Evaluate the expression to find what the average consumer spends after going to the mall twice during the week and 5 times on the weekends.

Online **Personal Tutor at** algebra1.com

CHECK Your Understanding

Examples 1, 2 (pp. 10–11)

Evaluate each expression.

1. $30 - 14 \div 2$
2. $5 \cdot 5 - 1 \cdot 3$
3. $6^2 + 8 \cdot 3 + 7$
4. $(4 + 6)7$
5. $50 - (15 + 9)$
6. $[8(2) - 4^2] + 7(4)$
7. $\frac{11 - 8}{1 + 7 \cdot 2}$
8. $\frac{(4 \cdot 3)^2}{9 + 3}$
9. $\frac{3 + 2^3}{5^2(6)}$

Example 3 (p. 11)

Evaluate each expression if $a = 4$, $b = 6$, and $c = 8$.

10. $8b - a$
11. $2a + (b^2 \div 3)$
12. $\frac{b(9 - c)}{a^2}$

Example 4 (p. 12)

13. GEOMETRY Write an algebraic expression to represent the area of the rectangle. Then evaluate it to find the area when $n = 4$ centimeters.

Exercises

HOMEWORK HELP

For Exercises	See Examples
14–17	1
18–25	2
26–31	3
32–33	4

Evaluate each expression.

14. $22 + 3 \cdot 7$

15. $18 \div 9 + 2 \cdot 6$

16. $10 + 8^3 \div 16$

17. $12 \div 3 \cdot 5 - 4^2$

18. $(11 \cdot 7) - 9 \cdot 8$

19. $29 - 3(9 - 4)$

20. $(12 - 6) \cdot 5^2$

21. $3^5 - (1 + 10^2)$

22. $108 \div [3(9 + 3^2)]$

23. $[(6^3 - 9) \div 23]4$

24. $\frac{8 + 3^3}{12 - 7}$

25. $\frac{(1 + 6)9}{5^2 - 4}$

Evaluate each expression if $r = 2$, $s = 3$, and $t = 11$.

26. $r + 6t$

27. $7 - rs$

28. $(2t + 3r) \div 4$

29. $s^2 + (r^3 - 8)5$

30. $t^2 + 8st + r^2$

31. $3r(r + s)^2 - 1$

32. BOOKS At a bookstore, Luna bought one new book for \$20 and three used books for \$4.95 each. Write and evaluate an expression to find how much money the books cost, not including sales tax.

33. ENTERTAINMENT Derrick sold tickets for the school musical. He sold 50 tickets for floor seats and 90 tickets for balcony seats. Write and evaluate an expression to find how much money Derrick collected.

School Musical Tickets

Type of Seat	Cost per Ticket (\$)
floor	7.50
balcony	5.00

Evaluate each expression.

34. $\frac{2 \cdot 8^2 - 2^2 \cdot 8}{2 \cdot 8}$

35. $6 - \left[\frac{2 + 7}{3} - (2 \cdot 3 - 5)\right]$

36. $7^3 - \frac{2}{3}(13 \cdot 6 + 9)4$

Evaluate each expression if $x = 12$, $y = 8$, and $z = 3$.

37. $\frac{2xy - z^3}{z}$

38. $\left(\frac{x}{y}\right)^2 - \frac{z}{(x - y)^2}$

39. $\frac{x - z^2}{xy} + \frac{2y - x}{y^2}$

40. BIOLOGY The cells of a certain type of bacteria double in number every 20 minutes. Suppose 100 of these cells are in one culture dish and 250 cells are in another culture dish. Write and evaluate an expression to find the total number of bacteria cells in both dishes after 20 minutes.

BUSINESS For Exercises 41 and 42, use the following information.
A sales representative receives an annual salary s, an average commission each month c, and a bonus b for each sales goal that she reaches.

41. Write an algebraic expression to represent her total earnings in one year if she receives four equal bonuses.

42. Suppose her annual salary is \$52,000 and her average commission is \$1225 per month. If each bonus is \$1150, how much does she earn in a year?

EXTRA PRACTICE
See pages 717, 744.

Self-Check Quiz at
algebra1.com

H.O.T. Problems

43. FIND THE ERROR Leonora and Chase are evaluating $3[4 + (27 \div 3)]^2$. Who is correct? Explain your reasoning.

Leonora

$3[4 + (27 \div 3)]^2 = 3(4 + 9^2)$
$= 3(4 + 81)$
$= 3(85)$
$= 255$

Chase

$3[4 + (27 \div 3)]^2 = 3(4 + 9)^2$
$= 3(13)^2$
$= 3(169)$
$= 507$

44. OPEN ENDED Write a numerical expression involving division in which the first step in evaluating the expression is addition. Discuss why addition rather than division is the first step.

45. CHALLENGE Choose three numbers from 1 to 6. Using each of the numbers exactly once in each expression, write five expressions that have different results when they are evaluated. Justify your choices.

46. ***Writing in Math*** Use the information about the Internet on page 10 to explain how expressions can be used to determine the monthly cost of Internet service. Include an expression for the cost of service if Nicole has a coupon for $25 off her base rate for her first six months.

STANDARDIZED TEST PRACTICE

47. REVIEW What is the perimeter of the triangle if $a = 9$ and $b = 10$?

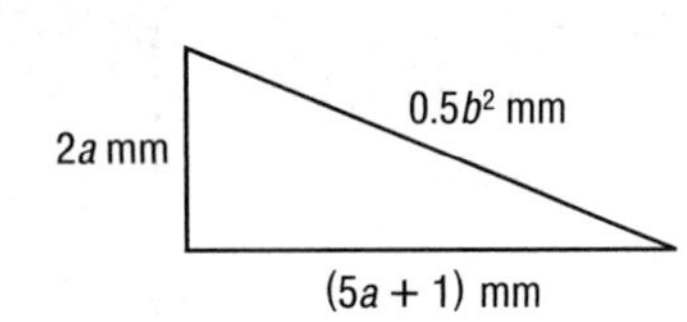

A 164 mm

B 114 mm

C 28 mm

D 4 mm

48. Simplify: $[10 + 15(2^3)] \div [7(2^2) - 2]$

Step 1 $[10 + 15(8)] \div [7(4) - 2]$

Step 2 $[10 + 120] \div [28 - 2]$

Step 3 $130 \div 26$

Step 4 $\frac{1}{5}$

Which is the first *incorrect* step?

F Step 1

G Step 2

H Step 3

J Step 4

Spiral Review

Write an algebraic expression for each verbal expression. (Lesson 1-1)

49. the product of 13 and p

50. one eighth of a number b

51. 20 increased by twice a number

52. 6 less than the square of y

53. TRAVEL Sari's car has 23,500 miles on the odometer. She takes a trip and drives an average of m miles each day for two weeks. Write an expression that represents the mileage on Sari's odometer after her trip. (Lesson 1-1)

Write a verbal expression for each algebraic expression. (Lesson 1-1)

54. $5 + \frac{n}{2}$

55. $q^2 - 12$

56. $\frac{x^3}{9}$

GET READY for the Next Lesson

PREREQUISITE SKILL Find the value of each expression. (pages 696–697, 700–701)

57. $0.5 - 0.075$

58. $5.6 + 1.612$

59. $2.4(6.425)$

60. $4\frac{1}{8} - 1\frac{1}{2}$

61. $\frac{3}{5} + 2\frac{5}{7}$

62. $8 \div \frac{2}{9}$

1-3 Open Sentences

Main Ideas

- Solve open sentence equations.
- Solve open sentence inequalities.

New Vocabulary

open sentence
solving the open sentence
solution
equation
replacement set
set
element
solution set
inequality

GET READY for the Lesson

The Daily News sells garage sale ads and kits. Spring Creek residents are planning a community garage sale, and their budget for advertising is \$135. The expression $15.50 + 5n$ represents the cost of an ad and n kits. The open sentence below can be used to ensure that the budget is met.

$$15.50 + 5n \leq 135$$

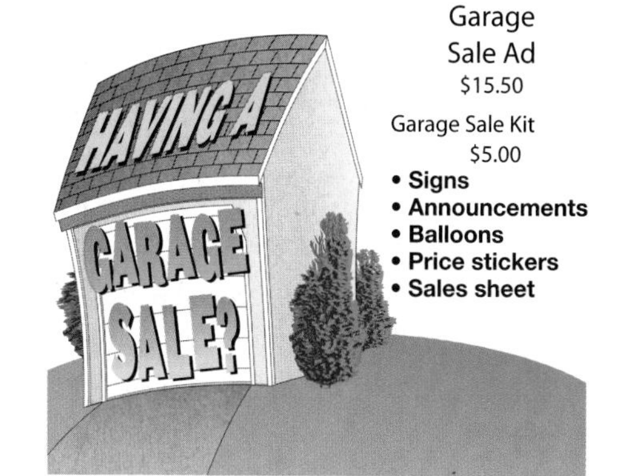

Solve Equations A mathematical statement with one or more variables is called an **open sentence**. An open sentence is neither true nor false until the variables have been replaced by specific values. The process of finding a value for a variable that results in a true sentence is called **solving the open sentence**. This replacement value is called a **solution**. A sentence that contains an equals sign, =, is called an **equation.**

A set of numbers from which replacements for a variable may be chosen is called a **replacement set.** A **set** is a collection of objects or numbers. It is often shown using braces, { }. Each number in the set is called an **element**, or member. The **solution set** of an open sentence is the set of elements from the replacement set that make the open sentence true.

EXAMPLE Use a Replacement Set to Solve an Equation

1 Find the solution set for each equation if the replacement set is {3, 4, 5, 6, 7}.

a. $6n + 7 = 37$

Replace n in $6n + 7 = 37$ with each value in the replacement set.

n	$6n + 7 = 37$	True or False?
3	$6(3) + 7 \stackrel{?}{=} 37 \rightarrow 25 \neq 37$	false
4	$6(4) + 7 \stackrel{?}{=} 37 \rightarrow 31 \neq 37$	false
5	$6(5) + 7 \stackrel{?}{=} 37 \rightarrow 37 = 37$	true ✓
6	$6(6) + 7 \stackrel{?}{=} 37 \rightarrow 43 \neq 37$	false
7	$6(7) + 7 \stackrel{?}{=} 37 \rightarrow 49 \neq 37$	false

Since $n = 5$ makes the equation true, the solution of $6n + 7 = 37$ is 5. The solution set is {5}.

b. $5(x + 2) = 40$

Replace x in $5(x + 2) = 40$ with each value in the replacement set.

x	$5(x + 2) = 40$	True or False?
3	$5(3 + 2) \stackrel{?}{=} 40 \rightarrow 25 \neq 40$	false
4	$5(4 + 2) \stackrel{?}{=} 40 \rightarrow 30 \neq 40$	false
5	$5(5 + 2) \stackrel{?}{=} 40 \rightarrow 35 \neq 40$	false
6	$5(6 + 2) \stackrel{?}{=} 40 \rightarrow 40 = 40$	true ✓
7	$5(7 + 2) \stackrel{?}{=} 40 \rightarrow 45 \neq 40$	false

The solution of $5(x + 2) = 40$ is 6. The solution set is $\{6\}$.

CHECK Your Progress

Find the solution set for each equation if the replacement set is {0, 1, 2, 3}.

1A. $8m - 7 = 17$

1B. $28 = 4(1 + 3d)$

You can often solve an equation by applying the order of operations.

EXAMPLE Use Order of Operations to Solve an Equation

2 **Solve** $\dfrac{13 + 2(4)}{3(5 - 4)} = q$.

$\dfrac{13 + 2(4)}{3(5 - 4)} = q$ Original equation

$\dfrac{13 + 8}{3(1)} = q$ Multiply 2 and 4 in the numerator. Subtract 4 from the 5 in the denominator.

$\dfrac{21}{3} = q$ Simplify.

$7 = q$ Divide. The solution is 7.

CHECK Your Progress

Solve each equation.

2A. $t = 9^2 \div (2 + 1)$

2B. $x = \dfrac{3^2 - (7 - 5)}{3(4) + (2 + 1)}$

Reading Math

Symbols Inequality symbols are read as follows.

$<$ *is less than*

$\leq$ *is less than or equal to*

$>$ *is greater than*

$\geq$ *is greater than or equal to*

Solve Inequalities An open sentence that contains the symbol $<$, $\leq$, $>$, or $\geq$ is called an **inequality**. Inequalities can be solved in the same way as equations.

Real-World EXAMPLE

3 **SHOPPING Meagan had $18. After she went to the used book store, she had less than $10 left. Could she have spent $8, $9, $10, or $11? Find the solution set for** $18 - y < 10$ **if the replacement set is {8, 9, 10, 11}.**

Replace y in $18 - y < 10$ with each value in the replacement set.

Math Online **Extra Examples at** algebra1.com

y	$18 - y < 10$	True or False?
8	$18 - 8 \stackrel{?}{<} 10 \rightarrow 10 \not< 10$	false
9	$18 - 9 \stackrel{?}{<} 10 \rightarrow 9 < 10$	true ✓
10	$18 - 10 \stackrel{?}{<} 10 \rightarrow 8 < 10$	true ✓
11	$18 - 11 \stackrel{?}{<} 10 \rightarrow 7 < 10$	true ✓

The solution set is {9, 10, 11}. So, Meagan could have spent \$9, \$10, or \$11.

CHECK Your Progress

Find the solution set for each inequality if the replacement set is {5, 6, 7, 8}.

3A. $30 + n \geq 37$

3B. $19 > 2y - 5$

Real-World EXAMPLE

Concepts in Motion
Interactive Lab
algebra1.com

4 **FUND-RAISING Refer to the application at the beginning of the lesson. If the residents buy an ad, what is the maximum number of garage sale kits they can buy and stay within their budget?**

Explore The residents can spend no more than \$135. Let n = the number of ads. The situation can be represented by the inequality $15.50 + 5n \leq 135$.

Plan Estimate to find reasonable values for the replacement set.

Solve Start by letting $n = 10$ and then adjust values as needed.

$15.50 + 5n \leq 135$ Original inequality

$15.50 + 5(10) \stackrel{?}{\leq} 135$ Replace n with 10.

$15.50 + 50 \stackrel{?}{\leq} 135$ Multiply 5 and 10.

$65.50 \leq 135$ Add 15.50 and 50.

The inequality is true, but the estimate is too low. Increase the value of n.

n	$15.50 + 5n \leq 135$	Reasonable?
20	$15.50 + 5(20) \stackrel{?}{\leq} 135 \rightarrow 115.50 \leq 135$	too low
25	$15.50 + 5(25) \stackrel{?}{\leq} 135 \rightarrow 140.50 \leq 135$	too high
23	$15.50 + 5(23) \stackrel{?}{\leq} 135 \rightarrow 130.50 \leq 135$	almost
24	$15.50 + 5(24) \stackrel{?}{\leq} 135 \rightarrow 135.50 \leq 135$	too high

Check The solution set is {0, 1, 2, 3, …, 21, 22, 23}. In addition to the ad, the residents can buy as many as 23 garage sale kits.

Study Tip

Ellipsis

In {1, 2, 3, 4, …}, the three dots are an *ellipsis*. In math, an ellipsis is used to indicate that numbers continue in the same pattern.

CHECK Your Progress

4. ENTERTAINMENT Trevor and his brother have a total of \$15. They plan to buy 2 movie tickets at \$6.50 each and then play video games in the arcade for \$0.50 each. Write and solve an inequality to find the greatest number of video games v that they can play.

Online **Personal Tutor at** algebra1.com

Math Online **Problem Solving Handbook at** algebra1.com

CHECK Your Understanding

Example 1 (p. 15)

Find the solution of each equation if the replacement set is {11, 12, 13, 14, 15}.

1. $n + 10 = 23$
2. $7 = \frac{c}{2}$
3. $29 = 3x - 7$
4. $(k - 8)12 = 84$

Find the solution of each equation using the given replacement set.

5. $36 = 18 + a$; {14, 16, 18, 20}
6. $\frac{d + 5}{11} = 2$; {4, 17, 23, 30, 45}

Example 2 (p. 16)

Solve each equation.

7. $x = 4(6) + 3$
8. $\frac{14 - 8}{2} = w$
9. $\frac{3(9) - 2}{1 + 4} = d$
10. $j = 15 \div 3 \cdot 5 - 4^2$

Example 3 (pp. 16–17)

Find the solution set of each inequality using the given replacement set.

11. $\frac{a}{5} \geq 2$; {5, 10, 15, 20, 25}
12. $24 - 2x \geq 13$; {0, 1, 2, 3, 4, 5, 6}

Example 4 (p. 17)

13. **ANALYZE TABLES** Suppose you have \$102.50 to buy sweaters from an online catalog. Using the information in the table, write and solve an inequality to find the maximum number of sweaters that you can purchase.

Online Catalog Prices	
Item	**Cost (\$)**
sweaters	39.00 each
shipping	10.95 per order

Exercises

HOMEWORK HELP	
For Exercises	**See Examples**
14–25	1
26–33	2
34–37	3
38–39	4

Find the solution of each equation if the replacement sets are a = {0, 3, 5, 8, 10} and b = {12, 17, 18, 21, 25}.

14. $b - 12 = 9$
15. $22 = 34 - b$
16. $\frac{15}{a} = 3$
17. $68 = 4b$
18. $31 = 3a + 7$
19. $5(a - 1) = 10$
20. $\frac{40}{a} - 4 = 0$
21. $27 = a^2 + 2$

Find the solution of each equation using the given replacement set.

22. $t - 13 = 7$; {10, 13, 17, 20}
23. $14(x + 5) = 126$; {3, 4, 5, 6, 7}
24. $22 = \frac{n}{3}$; {62, 64, 66, 68, 70}
25. $35 = \frac{g - 8}{2}$; {78, 79, 80, 81}

Solve each equation.

26. $a = 32 - 9(2)$
27. $w = 56 \div (2^2 + 3)$
28. $\frac{27 + 5}{16} = g$
29. $\frac{12 \cdot 5}{15 - 3} = y$
30. $r = \frac{9(6)}{(8 + 1)3}$
31. $a = \frac{4(14 - 1)}{3(6) - 5} + 7$

32. **FOOD** During a lifetime, the average American drinks 15,579 glasses of milk, 6220 glasses of juice, and 18,995 glasses of soda. Write and solve an equation to find g, the total number of glasses of milk, juice, and soda that the average American drinks in a lifetime.

33. **ENERGY** A small electric generator can power 3550 watts of electricity. Write and solve an equation to find the most 75-watt light bulbs one small generator could power.

Find the solution set for each inequality using the given replacement set.

34. $s - 2 < 6$; {6, 7, 8, 9, 10, 11}

35. $5a + 7 > 22$; {3, 4, 5, 6, 7}

36. $3 \geq \frac{25}{m}$; {1, 3, 5, 7, 9, 11}

37. $\frac{2a}{4} \leq 8$; {12, 14, 16, 18, 20, 22}

ENTERTAINMENT For Exercises 38 and 39, use the table.

38. Mr. and Mrs. Conkle are taking their three children to an amusement park. Write and solve an inequality to determine whether they can all go to the park for under $200. Describe what the variables in your inequality represent and explain your answer.

Amusement Park Admission Prices	
Person	**Cost ($)**
Adult	41.99
Child	26.99

39. Write and solve an inequality to find how many children can go with three adults if the budget is $300. Determine whether your answer is reasonable.

Find the solution of each equation or inequality using the given replacement set.

40. $x + \frac{2}{5} = 1\frac{3}{20}$; $\left\{\frac{1}{4}, \frac{1}{2}, \frac{3}{4}, 1, 11, 4)\right\}$

41. $\frac{2}{5}(x + 1) = \frac{8}{15}$; $\left\{\frac{1}{6}, \frac{1}{3}, \frac{1}{2}, \frac{2}{3}\right\}$

42. $2.7(x + 5) = 17.28$; {1.2, 1.3, 1.4, 1.5}

43. $16(x + 2) = 70.4$; {2.2, 2.4, 2.6, 2.8}

44. $4a - 3 \geq 10.6$; {3.2, 3.4, 3.6, 3.8, 4}

45. $3(12 - x) + 2 \leq 28$ {2.5, 3, 3.5, 4}

NUTRITION For Exercises 46 and 47, use the following information.

A person must burn 3500 Calories to lose one pound of weight.

46. Define a variable and write an equation for the number of Calories that a person would have to burn each day to lose four pounds in two weeks.

47. How many Calories would the person have to burn each day?

H.O.T. Problems

48. REASONING Describe the difference between an expression and an open sentence.

49. OPEN ENDED Write an inequality that has a solution set of {8, 9, 10, 11, ...}. Explain your reasoning.

50. REASONING Explain why an open sentence always has at least one variable.

51. CHALLENGE Describe the solution set for x if $3x \leq 1$.

52. ***Writing in Math*** Use the information about budgets on page 15 to explain how you can use open sentences when you have to stay within a budget. Also explain and give examples of real-world situations in which you would use inequalities and equations.

STANDARDIZED TEST PRACTICE

53. What is the solution set of the inequality $(5 + n^2) - n < 50$ if the replacement set is {5, 7, 9}?

A {5} C {7}

B {5, 7} D {7, 9}

54. $27 \div 3 + (12 - 4) =$

F $-\frac{11}{5}$ H 17

G $\frac{27}{11}$ J 25

55. REVIEW A box in the shape of a rectangular prism has a volume of 56 cubic inches. If the length of each side is multiplied by 2, what will be the approximate volume of the resulting box?

A 112 in^3 C 336 in^3

B 224 in^3 D 448 in^3

56. REVIEW Ms. Beal had 1 bran muffin, 16 ounces of orange juice, 3 ounces of sunflower seeds, 2 slices of turkey, and a half cup of spinach. According to the table, which equation best represents the total grams of protein that she consumed?

Protein Content	
Food	**Protein (g)**
bran muffin (1)	3
orange juice (8 oz)	2
sunflower seeds (1 oz)	6
turkey (1 slice)	12
spinach (1 c)	5

F $P = 3 + 2 + 6 + 12 + 5$

G $P = 3 + \frac{1}{2}(2) + \frac{1}{3}(6) + \frac{1}{2}(12) + 2(5)$

H $P = 3 + 16(2) + 3(6) + 2(12) + 2(5)$

J $P = 3 + 2(2) + 3(6) + 2(12) + \frac{1}{2}(5)$

Spiral Review

Evaluate each expression. (Lesson 1-2)

57. $5 + 3(4^2)$ **58.** $\frac{38 - 12}{2 \cdot 13}$ **59.** $[5(1 + 1)]^3 + 4$

60. RING TONES Andre downloaded three standard ringtones for $1.99 each and two premium ringtones at $3.49 each. Write and evaluate an expression to find how much the ringtones cost. (Lesson 1-2)

Write a verbal expression for each algebraic expression. (Lesson 1-1)

61. $n^5 - 8$ **62.** $r^2 + 3s$ **63.** $b \div 5a$.

Write an algebraic expression for each verbal expression.

64. two-thirds the square of a number

65. 6 increased by one half of a number *n*

66. one-half the cube of *x*

67. one fourth of the cube of a number

GET READY for the Next Lesson

PREREQUISITE SKILL **Find each product. Express answers in simplest form.** (pages 700–701)

68. $\frac{1}{6} \cdot \frac{2}{5}$ **69.** $\frac{4}{9} \cdot \frac{3}{7}$ **70.** $\frac{5}{6} \cdot \frac{15}{16}$

71. $\frac{6}{14} \cdot \frac{12}{18}$ **72.** $\frac{2}{5} \cdot \frac{3}{4}$ **73.** $\frac{11}{12} \cdot \frac{4}{5}$

Identity and Equality Properties

Main Ideas

- Recognize the properties of identity and equality.
- Use the properties of identity and equality.

New Vocabulary

additive identity
additive inverse
multiplicative identity
multiplicative inverse
reciprocal

GET READY for the Lesson

During the college football season, teams are ranked weekly. The table shows the last three rankings of the top five teams for the 2005 football season. The open sentence below represents the change in rank of Texas from Week 6 to Week 7.

College Football Team	Week 6	Week 7	Final Rank
Texas	2	2	1
Southern California	1	1	2
Penn State	16	5	3
Ohio State	6	15	4
West Virginia	34	26	5

Rank in Week 6	plus	change in rank	equals	rank for Week 7.
2	+	r	=	2

The solution of this equation is 0. Texas' rank changed by 0 from Week 6 to Week 7. In other words, $2 + 0 = 2$.

Identity and Equality Properties The sum of any number and 0 is equal to the number. Thus, 0 is called the **additive identity**.

> **KEY CONCEPT** — *Additive Identity*
>
> **Words** For any number a, the sum of a and 0 is a.
>
> **Symbols** $a + 0 = a, 0 + a = a$
>
> **Examples** $5 + 0 = 5, 0 + 5 = 5$

Two numbers with a sum of zero are called **additive inverses**. For example, $a + (-a) = 0$.

There are also special properties associated with multiplication. Consider the following equations.

$7 \cdot n = 7$ The solution of the equation is 1. Since the product of any number and 1 is equal to the number, 1 is called the **multiplicative identity**.

$9 \cdot m = 0$ The solution of the equation is 0. The product of any number and 0 is equal to 0. This is called the **Multiplicative Property of Zero**.

$\frac{1}{p} \cdot p = 1$ Two numbers whose product is 1 are called **multiplicative inverses** or **reciprocals**. Zero has no reciprocal because any number times 0 is 0.

The multiplicative properties are summarized in the following table.

Study Tip

Properties

These properties are true for all *real numbers*. You will learn more about real numbers in Lesson 1-8.

KEY CONCEPT *Multiplication Properties*

Property	Words	Symbols	Examples
Multiplicative Identity	For any number a, the product of a and 1 is a.	$a \cdot 1 = a$, $1 \cdot a = a$	$12 \cdot 1 = 12$, $1 \cdot 12 = 12$
Multiplicative Property of Zero	For any number a, the product of a and 0 is 0.	$a \cdot 0 = 0$, $0 \cdot a = 0$	$8 \cdot 0 = 0$, $0 \cdot 8 = 0$
Multiplicative Inverse	For every number $\frac{a}{b}$, where $a, b \neq 0$, there is exactly one number $\frac{b}{a}$ such that the product of $\frac{a}{b}$ and $\frac{b}{a}$ is 1.	$\frac{a}{b} \cdot \frac{b}{a} = 1$, $\frac{b}{a} \cdot \frac{a}{b} = 1$	$\frac{2}{3} \cdot \frac{3}{2} = \frac{6}{6} = 1$, $\frac{3}{2} \cdot \frac{2}{3} = \frac{6}{6} = 1$

EXAMPLE Identify Properties

1 Find the value of n in each equation. Then name the property that is used.

a. $42 \cdot n = 42$

$n = 1$, since $42 \cdot 1 = 42$. This is the Multiplicative Identity Property.

b. $n \cdot 9 = 1$

$n = \frac{1}{9}$, since $\frac{1}{9} \cdot 9 = 1$. This is the Multiplicative Inverse Property.

CHECK Your Progress

1. Find the value of n in the equation $28n = 0$. Then name the property that is used.

Several properties of equality are summarized below.

KEY CONCEPT *Properties of Equality*

Property	Words	Symbols	Examples
Reflexive	Any quantity is equal to itself.	For any number a, $a = a$.	$7 = 7$, $2 + 3 = 2 + 3$
Symmetric	If one quantity equals a second quantity, then the second quantity equals the first.	For any numbers a and b, if $a = b$, then $b = a$.	If $9 = 6 + 3$, then $6 + 3 = 9$.
Transitive	If one quantity equals a second quantity and the second quantity equals a third quantity, then the first quantity equals the third quantity.	For any numbers a, b, and c, if $a = b$, and $b = c$, then $a = c$.	If $5 + 7 = 8 + 4$, and $8 + 4 = 12$, then $5 + 7 = 12$.
Substitution	A quantity may be substituted for its equal in any expression.	If $a = b$, then a may be replaced by b in any expression.	If $n = 15$, then $3n = 3(15)$.

Math Online **Extra Examples at** algebra1.com

Use Identity and Equality Properties The properties of identity and equality can be used to justify each step when evaluating an expression.

EXAMPLE Evaluate Using Properties

2 **Evaluate $2(3 \cdot 2 - 5) + 3 \cdot \frac{1}{3}$. Name the property used in each step.**

$2(3 \cdot 2 - 5) + 3 \cdot \frac{1}{3} = 2(6 - 5) + 3 \cdot \frac{1}{3}$	Substitution; $3 \cdot 2 = 6$
$= 2(1) + 3 \cdot \frac{1}{3}$	Substitution; $6 - 5 = 1$
$= 2 + 3 \cdot \frac{1}{3}$	Multiplicative Identity; $2 \cdot 1 = 2$
$= 2 + 1$	Multiplicative Inverse; $3 \cdot \frac{1}{3} = 1$
$= 3$	Substitution; $2 + 1 = 3$

CHECK Your Progress

2. Evaluate $8 + (15 - 3 \cdot 5)$. Name the property used in each step.

CHECK Your Understanding

Example 1 (p. 22)

Find the value of n in each equation. Then name the property that is used.

1. $13n = 0$ **2.** $17 + 0 = n$ **3.** $1 = \frac{1}{6}n$

Example 2 (p. 23)

Evaluate each expression. Name the property used in each step.

4. $11 + 2(8 - 7)$ **5.** $6(12 - 48 \div 4)$ **6.** $\left(15 \cdot \frac{1}{15} + 8 \cdot 0\right) \cdot 12$

7. HISTORY Abraham Lincoln's Gettysburg Address began "Four score and seven years ago…." Since a score is equal to 20, the expression $4(20) + 7$ represents this quote. Evaluate the expression to find how many years Lincoln was referring to. Name the property used in each step.

Exercises

HOMEWORK HELP	
For Exercises	See Examples
8–17	1
18–25	2

Find the value of n in each equation. Then name the property that is used.

8. $12n = 12$ **9.** $n \cdot 1 = 5$

10. $8n = 1$ **11.** $1.5 = n + 1.5$

12. $6 = 6 + n$ **13.** $1 = 2n$

14. $n + 0 = \frac{1}{3}$ **15.** $4 \cdot \frac{1}{4} = n$

16. $4 - n = 0$ **17.** $n - \frac{1}{2} = 0$

Evaluate each expression. Name the property used in each step.

18. $(1 \div 5)5 \cdot 14$ **19.** $7 + (9 - 3^2)$ **20.** $\frac{3}{4}[4 \div (7 - 4)]$

21. $[3 \div (2 \cdot 1)]\,\frac{2}{3}$ **22.** $2(3 \cdot 2 - 5) + 3 \cdot \frac{1}{3}$ **23.** $6 \cdot \frac{1}{6} + 5(12 \div 4 - 3)$

24. **MILITARY PAY** An enlisted member of the military at grade E-2 earns \$1427.40 per month in the first year of service. After 5 years of service, a person at grade E-2 earns \$1427.40 per month. Write and solve an equation using addition that shows the change in pay from 1 year of service to 5 years. Name the property or identity used.

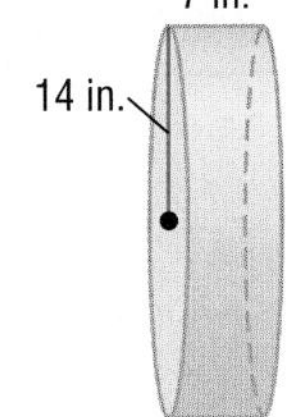

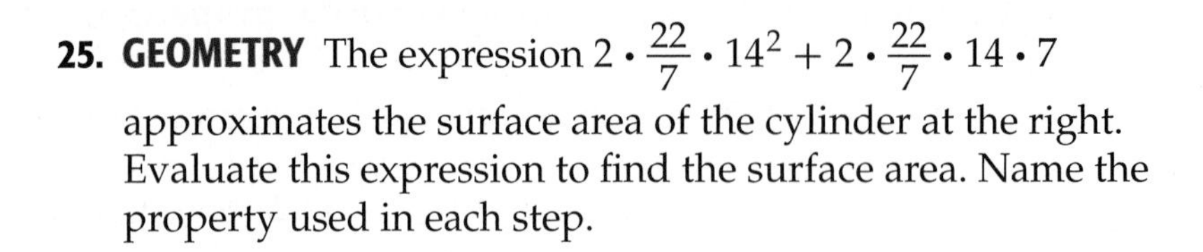

25. **GEOMETRY** The expression $2 \cdot \frac{22}{7} \cdot 14^2 + 2 \cdot \frac{22}{7} \cdot 14 \cdot 7$ approximates the surface area of the cylinder at the right. Evaluate this expression to find the surface area. Name the property used in each step.

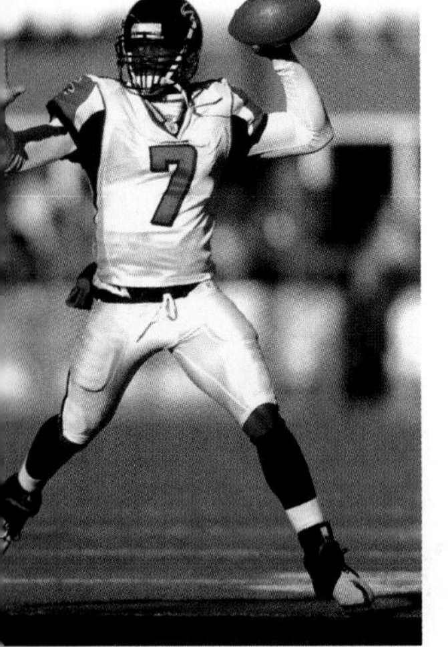

Find the value of n in each equation. Then name the property that is illustrated.

26. $1 = \frac{(9-5)}{3^2} n$

27. $n\left(5^2 \cdot \frac{1}{25}\right) = 3$

28. $6\left(\frac{1}{2} \cdot n\right) = 6$

Evaluate each expression. Name the property used in each step.

29. $3 + 5(4 - 2^2) - 1$

30. $7 - 8(9 - 3^2)$

31. $\left[\frac{5}{8}\left(1 + \frac{3}{8}\right)\right] \cdot 17$

32. **FOOTBALL** The table shows various bonus plans for the NFL in a recent year. Write an expression that could be used to determine what a team owner would pay in bonuses for the following:

- eight players who keep their weight below 240 pounds and have averaged 4.5 yards per carry, and
- three players who score 12 touchdowns and score 76 points.

Name the property used in each step.

NFL Bonuses

Goal	Bonuses(\$)
Average 4.5 yards per carry	50,000
12 touchdowns	50,000
76 points scored	50,000
Keep weight below 240 pounds	100,000

Source: *ESPN Sports Almanac*

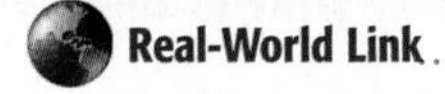

Real-World Link

Nationally organized football began in 1920 and originally included five teams. In 2005, there were 32 professional teams.

Source: www.infoplease.com

ANALYZE TABLES **For Exercises 33 and 34, use the following information.**

The spirit club at Marshall High School is selling school bumper stickers, buttons, and caps. The profit for each item is the difference between the selling price and the cost.

School Spirit items

Item	Cost (\$)	Selling Price (\$)
Bumper Sticker	0.30	2.00
Button	1.00	2.50
Cap	6.00	10.00

33. Write an expression that represents the profit for selling 25 bumper stickers, 80 buttons, and 40 caps.

34. Evaluate the expression, indicating the property used in each step.

EXTRA PRACTICE

See pages, 718, 744.

Math Online

Self-Check Quiz at algebra1.com

H.O.T. Problems

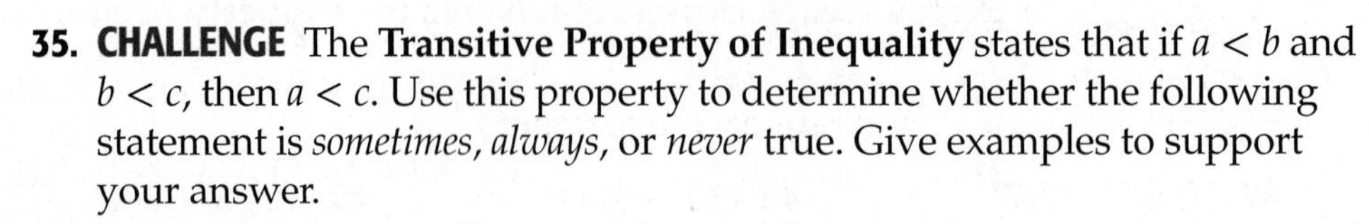

35. **CHALLENGE** The **Transitive Property of Inequality** states that if $a < b$ and $b < c$, then $a < c$. Use this property to determine whether the following statement is *sometimes*, *always*, or *never* true. Give examples to support your answer.

$$\text{If } x > y \text{ and } z > w, \text{ then } xz > yw.$$

36. **REASONING** Explain whether 1 can be an additive identity. Give an example to justify your answer.

37. **OPEN ENDED** Write two equations showing the Transitive Property of Equality. Justify your reasoning.

38. **REASONING** Explain why 0 has no multiplicative inverse.

39. ***Writing in Math*** Use the data about football on page 21 to explain how properties can be used to compare data. Include an example of the Transitive Property using three teams' rankings as an example.

STANDARDIZED TEST PRACTICE

40. Which illustrates the Symmetric Property of Equality?

A If $a = b$, then $b = a$.

B If $a = b$ and $b = c$, then $a = c$.

C If $a = b$, then $b = c$.

D If $a = a$, then $a + 0 = a$.

41. Which property is used below?

If $4xy^2 = 8y^2$ *and* $8y^2 = 72$, *then* $4xy^2 = 72$.

F Reflexive Property

G Substitution Property

H Symmetric Property

J Transitive Property

Spiral Review

Find the solution set for each inequality using the given replacement set. (Lesson 1-3)

42. $10 - x > 6$; {3, 5, 6, 8}

43. $4x + 2 < 58$; {11, 12, 13, 14, 15}

44. **EXERCISE** It takes about 2000 steps to walk one mile. Use the table to determine how many miles of walking it would take to burn all the Calories contained in a cheeseburger and two 12-ounce sodas. (Lesson 1-3)

Steps Needed to Burn Calories	
Food	**Number of Steps**
cheeseburger	7590
12 oz. soda	3450

45. **SHOPPING** In a recent year, the average U.S. household spent $213 on toys and games. In San Jose, California, the average spending was $59 less than twice this amount. Write and evaluate an expression to find the average spending on toys and games in San Jose during that year. (Lesson 1-2)

46. Write an algebraic expression for *the sum of twice a number squared and 7.* (Lesson 1-1)

GET READY for the Next Lesson

PREREQUISITE SKILL **Evaluate each expression.** (Lesson 1-2)

47. $10(6) + 10(2)$

48. $(15 - 6) \cdot 8$

49. $12(4) - 5(4)$

50. $3(4 + 2)$

1-5 The Distributive Property

Main Ideas

- Use the Distributive Property to evaluate expressions.
- Use the Distributive Property to simplify algebraic expressions.

New Vocabulary

term
like terms
equivalent expressions
simplest form
coefficient

GET READY for the Lesson

Instant Replay Video Games sells new and used games. During a sale, the first 8 customers each bought a bargain game and a new release. To calculate the total sales for these customers, you can use the Distributive Property.

Sale Prices	
Used Games	$9.95
Bargain Games	$14.95
Regular Games	$24.95
New Releases	$34.95

Evaluate Expressions There are two methods you could use to calculate the video game sales.

Method 1

sales of bargain games	plus	sales of new releases
$8(14.95)$	$+$	$8(34.95)$

$= 119.60 + 279.60$

$= 399.20$

Method 2

number of customers	times	each customer's purchase price
8	$\times$	$(14.95 + 34.95)$

$= 8(49.90)$

$= 399.20$

Either method gives total sales of $399.20 because the following is true.

$$8(14.95) + 8(34.95) = 8(14.95 + 34.95)$$

This is an example of the **Distributive Property.**

Vocabulary Link
Distribute
Everyday Use to divide among several or many
Distributive
Math Use property that allows you to multiply each number in a sum or difference by a number outside the parentheses

KEY CONCEPT *Distributive Property*

Symbols For any numbers a, b, and c,
$a(b + c) = ab + ac$ and $(b + c)a = ba + ca$ and
$a(b - c) = ab - ac$ and $(b - c)a = ba - ca$.

Examples

$3(2 + 5) = 3 \cdot 2 + 3 \cdot 5$
$3(7) = 6 + 15$
$21 = 21$ ✓

$4(9 - 7) = 4 \cdot 9 - 4 \cdot 7$
$4(2) = 36 - 28$
$8 = 8$ ✓

Notice that it does not matter whether a is placed on the right or the left of the expression in the parentheses. The Symmetric Property of Equality allows the Distributive Property to be written as follows.

If $a(b + c) = ab + ac$, then $ab + ac = a(b + c)$.

EXAMPLE Distribute Over Addition or Subtraction

Real-World Link

Ramona is America's oldest continuing outdoor drama. It has been performed since 1923 for more than 2 million people. Its location, called the *Ramona Bowl*, has been designated a California State Historical Landmark.

Source: ramonabowl.com

1 **Tickets for a play are $8. A group of 10 adults and 4 children are planning to go. Rewrite $8(10 + 4)$ using the Distributive Property. Then evaluate to find the total cost for the group.**

$8(10 + 4) = 8(10) + 8(4)$ Distributive Property

$= 80 + 32$ Multiply.

$= 112$ Add.

CHECK Your Progress

Rewrite each expression using the Distributive Property. Then evaluate.

1A. $(5 + 1)9$ **1B.** $3(11 - 8)$

Online Personal Tutor at algebra1.com

The Distributive Property can be used to simplify mental calculations involving multiplication. To use this method, rewrite one factor as a sum or difference. Then use the Distributive Property to multiply. Finally, find the sum or difference.

EXAMPLE The Distributive Property and Mental Math

2 **Use the Distributive Property to rewrite $15 \cdot 99$. Then evaluate.**

$15 \cdot 99 = 15(100 - 1)$ Think: $99 = 100 - 1$

$= 15(100) - 15(1)$ Distributive Property

$= 1500 - 15$ Multiply

$= 1485$ Subtract.

CHECK Your Progress

Use the Distributive Property to rewrite each expression. Then evaluate.

2A. $402(12)$ **2B.** $60 \cdot 7\frac{2}{3}$

Simplify Expressions You can use algebra tiles to investigate how the Distributive Property relates to algebraic expressions.

ALGEBRA LAB

The Distributive Property

Use a product mat and algebra tiles to model $3(x + 2)$ as the area of a rectangle with dimensions of 3 and $(x + 2)$.

Make a rectangle with algebra tiles that is 3 units wide and $x + 2$ units long. The rectangle has 3 x-tiles and 6 1-tiles. The area of the rectangle is $x + 1 + 1 + x + 1 + 1 + x + 1 + 1$ or $3x + 6$. Therefore, $3(x + 2) = 3x + 6$.

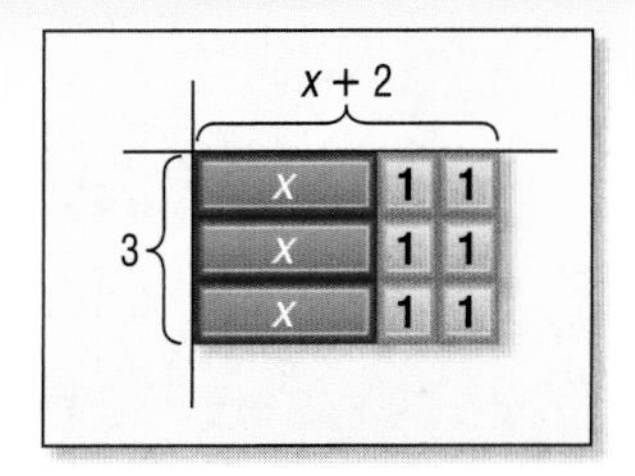

(continued on the next page)

MODEL AND ANALYZE

Find each product by using algebra tiles.

1. $2(x + 1)$ **2.** $5(x + 2)$ **3.** $2(2x + 1)$

Tell whether each statement is *true* or *false*. Justify your answer with algebra tiles and a drawing.

4. $3(x + 3) = 3x + 3$ **5.** $x(3 + 2) = 3x + 2x$ **6.** $2(2x + 1) = 2x + 2$

You can apply the Distributive Property to algebraic expressions.

EXAMPLE Algebraic Expressions

3 Rewrite each product using the Distributive Property. Then simplify.

a. $5(g - 9)$

$5(g - 9) = 5 \cdot g - 5 \cdot 9$ Distributive Property

$= 5g - 45$ Multiply.

b. $3(x^2 + x - 1)$

$3(x^2 + x - 1) = 3(x^2) + 3(x) - 3(1)$ Distributive Property

$= 3x^2 + 3x - 3$ Simplify.

CHECK Your Progress

3A. $2(8 + n)$ **3B.** $-6(r - s - t)$

Reading Math

Algebraic Expressions The expression $5(g - 9)$ is read *5 times the quantity g minus 9* or *5 times the difference of g and 9.*

A **term** is a number, a variable, or a product or quotient of numbers and variables. For example, y, p^3, $4a$, and $5g^2h$ are all terms. **Like terms** contain the same variables, with corresponding variables having the same power.

$2x^2 + 6x + 5$ — three terms

$3a^2 + 5a^2 + 2a$ — like terms ($3a^2$, $5a^2$); unlike terms ($5a^2$, $2a$)

Concepts in Motion
BrainPOP®
algebra1.com

The Distributive Property and the properties of equality can be used to show that $5n + 7n = 12n$. In this expression, $5n$ and $7n$ are like terms.

$5n + 7n = (5 + 7)n$ Distributive Property

$= 12n$ Substitution

The expressions $5n + 7n$ and $12n$ are called **equivalent expressions** because they denote the same number. An expression is in **simplest form** when it is replaced by an equivalent expression having no like terms or parentheses.

EXAMPLE Combine Like Terms

4 a. Simplify $15x + 18x$.

$15x + 18x = (15 + 18)x$ Distributive Property

$= 33x$ Substitution

Math Online Extra Examples at algebra1.com

b. **Simplify $10n + 3n^2 + 9n^2$.**

$10n + 3n^2 + 9n^2 = 10n + (3 + 9)n^2$ Distributive Property

$= 10n + 12n^2$ Substitution

CHECK Your Progress

Simplify each expression. If not possible, write *simplified*.

4A. $6t - 4t$ **4B.** $b^2 + 13b + 13$

Reading Math

Like terms may be defined as terms that are the same or vary only by the coefficient.

The **coefficient** of a term is the numerical factor. For example, in $17xy$, the coefficient is 17, and in $\frac{3y^2}{4}$, the coefficient is $\frac{3}{4}$. In the term m, the coefficient is 1 since $1 \cdot m = m$ by the Multiplicative Identity Property.

CHECK Your Understanding

Example 1 (p. 27)

Rewrite each expression using the Distributive Property. Then evaluate.

1. $6(12 - 3)$ **2.** $8(1 + 5)$ **3.** $(19 + 3)10$

4. COSMETOLOGY A hair stylist cut 12 customers' hair. She earned $29.95 for each haircut and received an average tip of $4 for each. Write and evaluate an expression to determine the total amount that she earned.

Example 2 (p. 27)

Use the Distributive Property to rewrite each expression. Then find the product.

5. $16(103)$ **6.** $\left(3\frac{1}{17}\right)(34)$

Example 3 (p. 28)

Rewrite each expression using the Distributive Property. Then simplify.

7. $2(4 + t)$ **8.** $(g - 9)5$

Example 4 (pp. 28–29)

Simplify each expression. If not possible, write *simplified*.

9. $13m + m$ **10.** $14a^2 + 13b^2 + 27$ **11.** $3(x + 2x)$

Exercises

HOMEWORK HELP	
For Exercises	See Examples
12–19	1
20–23	2
24–27	3
28–33	4

Rewrite each expression using the Distributive Property. Then evaluate.

12. $(5 + 7)8$ **13.** $7(13 + 12)$ **14.** $6(6 - 1)$

15. $(3 + 8)15$ **16.** $12(9 - 5)$ **17.** $(10 - 7)13$

18. COMMUNICATION A consultant keeps a log of all contacts she makes. In a typical week, she averages 5 hours using e-mail, 18 hours on the telephone, and 12 hours of meetings in person. Write and evaluate an expression to predict how many hours she will spend on these activities over the next 12 weeks.

19. OLYMPICS The table shows the average daily attendance for two venues at the 2004 Summer Olympics. Write and evaluate an expression to estimate the total number of people at these venues over a 4-day period.

Average Olympic Attendance

Venue	Number of People
Olympic Stadium	110,000
Aquatic Center	17,500

Source: www.olympic.org

Use the Distributive Property to rewrite each expression. Then find the product.

20. $5 \cdot 97$

21. $8(990)$

22. $18 \cdot 2\frac{1}{9}$

23. $\left(3\frac{1}{6}\right)48$

Rewrite each expression using the Distributive Property. Then simplify.

24. $2(x + 4)$

25. $3(5 + n)$

26. $8(4 - 3m)$

27. $-3(x - 6)$

Simplify each expression. If not possible, write *simplified*.

28. $2x + 9x$

29. $4b - 1 + 5b$

30. $5n^2 - 7n$

31. $3a^2 + a + 14a^2$

32. $12(4 + 3c)$

33. $15(3x - 5)$

ANALYZE TABLES **For Exercises 34 and 35, use the table that shows the monthly cost of a company health plan.**

Available Insurance Plans – Monthly Charges			
Coverage	Medical	Dental	Vision
Employee	\$78	\$20	\$12
Family (additional charge)	\$50	\$15	\$7

34. Write and evaluate an expression to calculate the total cost of medical, dental, and vision insurance for an employee for 6 months.

35. How much would an employee expect to pay for family medical and dental coverage per year?

Rewrite each expression using the Distributive Property. Then simplify.

36. $27\left(\frac{1}{3} - 2b\right)$

37. $4(p + q - r)$

38. $-6(2 - d^2 + d)$

39. $5(6m^3 + 4n - 3n)$

EXTRA PRACTICE
See pages 718, 744.

Math Online
Self-Check Quiz at algebra1.com

Simplify each expression. If not possible, write *simplified*.

40. $6x^2 + 14x - 9x$

41. $4y^3 + 3y^3 + y^4$

42. $a + \frac{a}{5} + \frac{2}{5}a$

H.O.T. Problems

43. REASONING Explain why the Distributive Property is sometimes called the Distributive Property of Multiplication Over Addition.

44. OPEN ENDED Write an expression that has five terms, three of which are like terms and one that has a coefficient of 1. Describe how to simplify the expression.

45. FIND THE ERROR Courtney and Che are simplifying $3(x + 4)$. Who is correct? Explain your reasoning.

Courtney
$3(x + 4) = 3(x) + 3(4)$
$= 3x + 12$

Che
$3(x + 4) = 3(x) + 4$
$= 3x + 4$

46. CHALLENGE The expression $2(\ell + w)$ can be used to find the perimeter of a rectangle with a length ℓ and width w. What are the length and width of a rectangle if the area is $13\frac{1}{2}$ square units and the length of one side is $\frac{1}{5}$ the measure of the perimeter? Explain your reasoning.

47. *Writing in Math* Use the data about video game prices on page 26 to explain how the Distributive Property can be used to calculate quickly. Also, compare and contrast the two methods of finding the total video game sales.

STANDARDIZED TEST PRACTICE

48. In three months, Mayuko had 108 minutes of incoming calls on her voice mail. What was the total cost of voice mail for those three months?

Voice Mail	
Item	**Cost**
service fee	\$4.95 per month
incoming calls	\$0.07 per minute

A \$7.61

B \$12.51

C \$22.41

D \$37.80

49. REVIEW If each dimension of the prism is tripled, which expression represents the new volume?

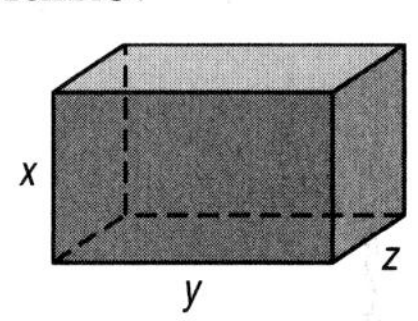

F $x^3y^3z^3$

G $3xyz$

H $3(x + y + z)$

J $27xyz$

Spiral Review

Name the property illustrated by each statement or equation. (Lesson 1-4)

50. If $7 \cdot 2 = 14$, then $14 = 7 \cdot 2$.

51. $mnp = 1mnp$

52. $\frac{3}{4} \cdot \frac{4}{3} = 1$

53. $32 + 21 = 32 + 21$

54. PHYSICAL SCIENCE Sound travels through air at an approximate speed of 344 meters per second. Write and solve an equation to find how far sound travels through air in 2 seconds. (Lesson 1-3)

Evaluate each expression if $a = 4$, $b = 6$, and $c = 3$. (Lesson 1-2)

55. $3b - c$

56. $8(a - c)^2 + 3$

57. $\frac{6ab}{2(c + 5)}$

GET READY for the Next Lesson

PREREQUISITE SKILL **Find the area of each figure.** (Pages 704–705)

58.

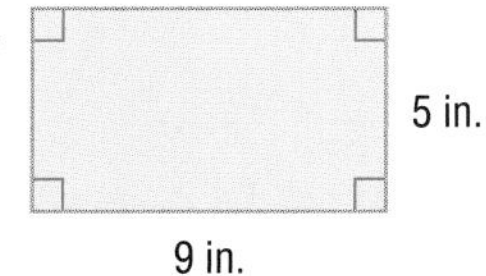

59.

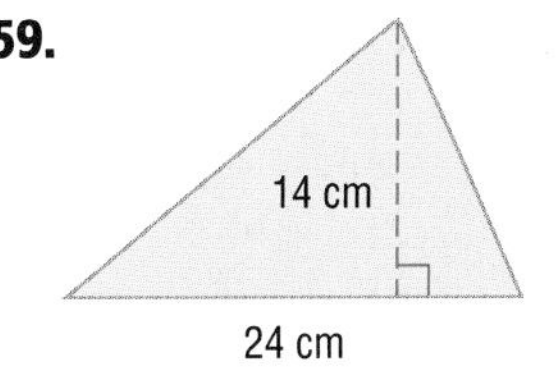

60.

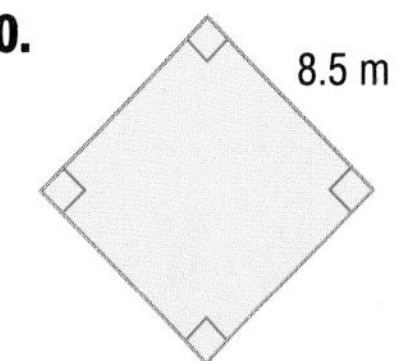

CHAPTER 1 Mid-Chapter Quiz

Lessons 1-1 through 1-5

Write an algebraic expression for each verbal expression. (Lesson 1-1)

1. the quotient of y and 3
2. 5 minus the product of 7 and t
3. x squared increased by 2

Write a verbal expression for each algebraic expression. (Lesson 1-1)

4. $5n + 2$
5. a^3

6. **GOLF** At a driving range, a small bucket of golf balls costs $6 and a large bucket costs $8. Write an expression for the total cost of buying s small and t large buckets. (Lesson 1-1)

7. **MULTIPLE CHOICE** Jasmine bought a satellite radio receiver and a subscription to satellite radio. What was her total cost after 7 months? (Lesson 1-2)

Satellite Radio	
Item	**Cost**
receiver	$78
subscription	$12.95 per month

A $90.95
B $168.65
C $558.95
D $636.65

Evaluate each expression. (Lesson 1-2)

8. $2 + 18 \div 9$
9. $6 \cdot 9 - 2(8 + 5)$
10. $9(3) - 4^2$
11. $\dfrac{(5 - 2)^2}{4 \times 2 - 7}$

12. Evaluate $\dfrac{5a^2 + c - 2}{6 + b}$ if $a = 4$, $b = 5$, and $c = 10$. (Lesson 1-2)

Find the solution of each equation if the replacement set is {10, 11, 12, 13}. (Lesson 1-3)

13. $x - 3 = 10$
14. $25 = 2r + 1$
15. $\dfrac{t}{5} = 2$
16. $4y - 9 = 35$

17. Find the solution set for $2n^2 + 3 \leq 75$ if the replacement set is {4, 5, 6, 7, 8, 9}. (Lesson 1-3)

18. **MULTIPLE CHOICE** Dion bought 1 pound of dried greens, 3 pounds of sesame seeds, and 2 pounds of flax seed to feed his birds. According to the table, which expression best represents the total cost? (Lesson 1-3)

Bird Food	Cost ($)
dried greens (0.5 lb)	4.95
sesame seeds (1 lb)	5.75
flax seed (2 lb)	2.75

F $4.95 + 5.75 + 2.75$
G $4.95 + 3(5.75) + 2(2.75)$
H $2(4.95) + 3(5.75) + 2.75$
J $0.5(4.95) + 5.75 + 2(2.75)$

Find the value of n in each equation. Then name the property that is used. (Lesson 1-4)

19. $n = 11 + 0$
20. $\dfrac{1}{3} \cdot 3 = n$

21. **GEOMETRY** The expression $\frac{1}{2}(7)(a + b)$ represents the area of the trapezoid. What is the area if $a = 22.4$ centimeters and $b = 10.8$ centimeters? (Lesson 1-5)

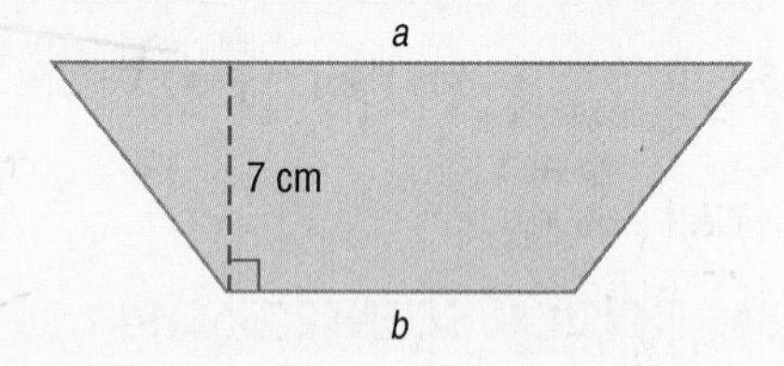

22. **MULTIPLE CHOICE** Which expression represents the second step of simplifying the algebraic expression? (Lesson 1-5)

Step 1 $9(x + 4y) + 5 + 2(x + 7)$
Step 2 [ ]
Step 3 $11x + 36y + 19$

A $9x + 36y + 7(x + 7)$
B $9x + 36y + 5 + 2x + 14$
C $9(x + 4y + 5) + 2x + 14$
D $9(x + 5) + 4y + 2x + 14$

Commutative and Associative Properties

Main Ideas

- Recognize the Commutative and Associative Properties.
- Use the Commutative and Associative Properties to simplify algebraic expressions.

GET READY for the Lesson

The South Line of the Atlanta subway leaves Five Points and heads for Garnett and then West End. The distance from Five Points to West End can be found by evaluating $0.4 + 1.5$. Likewise, the distance from West End to Five Points can be found by evaluating $1.5 + 0.4$.

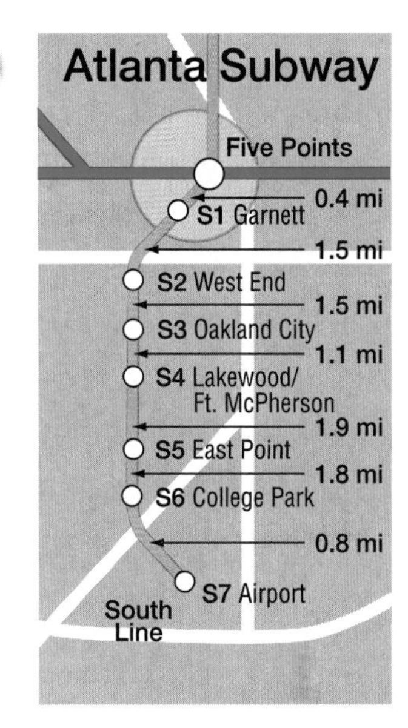

Commutative and Associative Properties In the situation above, the distance from Five Points to West End is the same as the distance from West End to Five Points.

The distance from Five Points to West End	equals	the distance from West End to Five Points.
$0.4 + 1.5$	$=$	$1.5 + 0.4$

This is an example of the **Commutative Property** for addition.

Vocabulary Link
Commute
Everyday Use to travel back and forth, as in commute to work
Commutative
Math Use property that allows you to change the order in which numbers are added or multiplied

KEY CONCEPT *Commutative Property*

Words The order in which you add or multiply numbers does not change their sum or product.

Symbols For any numbers a and b, $a + b = b + a$ and $a \cdot b = b \cdot a$.

Examples $5 + 6 = 6 + 5$, $3 \cdot 2 = 2 \cdot 3$

An easy way to find the sum or product of numbers is to group, or associate, the numbers using the **Associative Property**.

Vocabulary Link
Associate
Everyday Use to come together as in associate with friends
Associative
Math Use property that allows you to group three or more numbers when adding or multiplying

KEY CONCEPT *Associative Property*

Words The way you group three or more numbers when adding or multiplying does not change their sum or product.

Symbols For any numbers a, b, and c,
$(a + b) + c = a + (b + c)$ and $(ab)c = a(bc)$.

Examples $(2 + 4) + 6 = 2 + (4 + 6)$, $(3 \cdot 5) \cdot 4 = 3 \cdot (5 \cdot 4)$

Real-World EXAMPLE Use Addition Properties

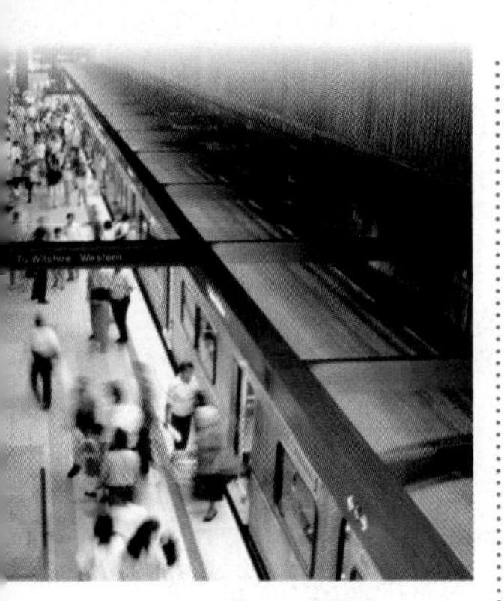

Real-World Link

New York City has the most extensive subway system, covering 842 miles of track and serving about 4.3 million passengers per day.

Source: *The Guinness Book of Records*

1 **TRANSPORTATION Refer to the beginning of the lesson. Find the distance between Five Points and Lakewood/Ft. McPherson.**

Five Points to Garnett		Garnett to West End		West End to Oakland City		Oakland City to Lakewood/Ft. McPherson
0.4	+	1.5	+	1.5	+	1.1

$0.4 + 1.5 + 1.5 + 1.1 = 0.4 + 1.1 + 1.5 + 1.5$ Commutative (+)

$= (0.4 + 1.1) + (1.5 + 1.5)$ Associative (+)

$= 1.5 + 3.0$ or 4.5 Add mentally.

Lakewood/Ft. McPherson is 4.5 miles from Five Points.

CHECK Your Progress

Evaluate each expression using properties of numbers. Name the property used in each step.

1A. $35 + 17 + 5 + 3$ **1B.** $8\frac{3}{4} + 12 + 5\frac{1}{4}$

EXAMPLE Use Multiplication Properties

2 **Evaluate $8 \cdot 2 \cdot 3 \cdot 5$ using properties of numbers. Name the property used in each step.**

$8 \cdot 2 \cdot 3 \cdot 5 = 8 \cdot 3 \cdot 2 \cdot 5$ Commutative (×)

$= (8 \cdot 3) \cdot (2 \cdot 5)$ Associative (×)

$= 24 \cdot 10$ or 240 Multiply mentally.

Concepts in Motion
BrainPOP®
algebra1.com

CHECK Your Progress

Evaluate each expression using properties of numbers. Name the property used in each step.

2A. $2.9 \cdot 4 \cdot 10$ **2B.** $\frac{5}{3} \cdot 25 \cdot 3 \cdot 2$

Simplify Expressions The Commutative and Associative Properties can be used with other properties when evaluating and simplifying expressions.

CONCEPT SUMMARY *Properties of Numbers*

The following properties are true for any numbers a, b, and c.

Properties	Addition	Multiplication
Commutative	$a + b = b + a$	$ab = ba$
Associative	$(a + b) + c = a + (b + c)$	$(ab)c = a(bc)$
Identity	0 is the identity. $a + 0 = 0 + a = a$	1 is the identity. $a \cdot 1 = 1 \cdot a = a$
Zero	—	$a \cdot 0 = 0 \cdot a = 0$
Distributive	$a(b + c) = ab + ac$ and $(b + c)a = ba + ca$	
Substitution	If $a = b$, then a may be substituted for b.	

EXAMPLE 3 Write and Simplify an Expression

Use the expression *four times the sum of a and b increased by twice the sum of a and 2b.*

a. Write an algebraic expression for the verbal expression.

Words	four times the sum of a and b increased by twice the sum of a and $2b$
Variables	Let a and b represent the numbers.
Expression	$4(a + b) + 2(a + 2b)$

Concepts in Motion
BrainPOP®
algebra1.com

b. Simplify the expression and indicate the properties used.

$4(a + b) + 2(a + 2b) = 4(a) + 4(b) + 2(a) + 2(2b)$	Distributive Property
$= 4a + 4b + 2a + 4b$	Multiply and Associative (×).
$= 4a + 2a + 4b + 4b$	Commutative (+)
$= (4a + 2a) + (4b + 4b)$	Associative (+)
$= (4 + 2)a + (4 + 4)b$	Distributive Property
$= 6a + 8b$	Substitution

CHECK Your Progress

3A. Write an algebraic expression for *5 times the difference of q squared and r plus 8 times the sum of 3q and 2r*. Then simplify the expression and indicate the properties used.

3B. Simplify $6(x - 2y) + 4(-3x + y)$ and indicate the properties used.

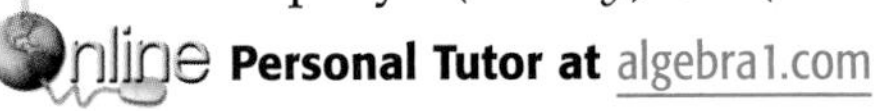

CHECK Your Understanding

Examples 1, 2 (p. 34)

Evaluate each expression using properties of numbers. Name the property used in each step.

1. $14 + 18 + 26$
2. $3\frac{1}{2} + 4 + 2\frac{1}{2}$
3. $5 \cdot 3 \cdot 6 \cdot 4$
4. $\frac{5}{6} \cdot 16 \cdot \frac{3}{4}$

Example 3 (p. 35)

Simplify each expression.

5. $4x + 5y + 6x$
6. $5a + 3b + 2a + 7b$
7. $3(4x + 2) + 2x$
8. $7(ac + 2b) + 2ac$

9. **GEOMETRY** Find the perimeter of the triangle.

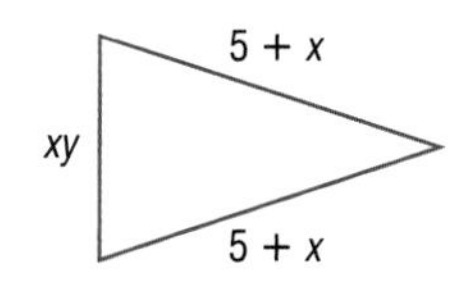

10. Write an algebraic expression for *half the sum of p and 2q increased by three-fourths q*. Then simplify, indicating the properties used.

Exercises

HOMEWORK HELP	
For Exercises	See Examples
11–16	1
17–22	2
23–34	3

Evaluate each expression using properties of numbers. Name the property used in each step.

11. $17 + 6 + 13 + 24$

12. $8 + 14 + 22 + 9$

13. $4.25 + 3.50 + 8.25$

14. $6.2 + 4.2 + 4.3 + 5.8$

15. $6\frac{1}{2} + 3 + \frac{1}{2} + 2$

16. $2\frac{3}{8} + 4 + 3\frac{3}{8}$

17. $5 \cdot 11 \cdot 4 \cdot 2$

18. $3 \cdot 10 \cdot 6 \cdot 3$

19. $0.5 \cdot 2.4 \cdot 4$

20. $8 \cdot 1.6 \cdot 2.5$

21. $3\frac{3}{7} \cdot 14 \cdot 1\frac{1}{4}$

22. $2\frac{5}{8} \cdot 24 \cdot 6\frac{2}{3}$

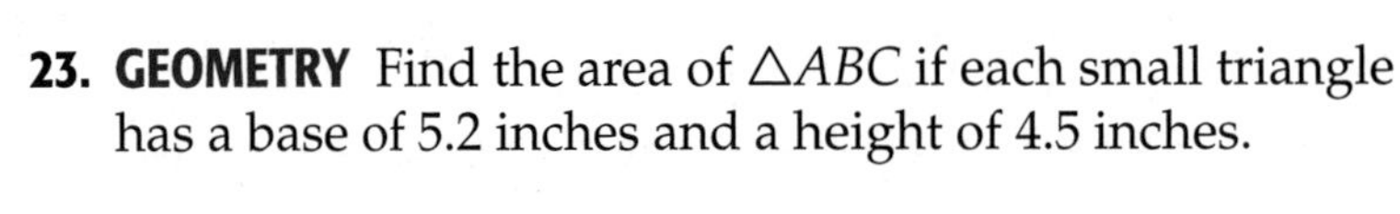

23. GEOMETRY Find the area of $\triangle ABC$ if each small triangle has a base of 5.2 inches and a height of 4.5 inches.

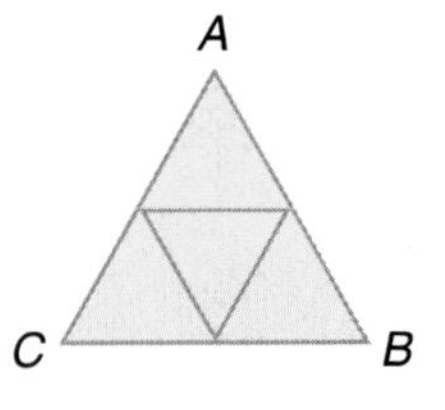

24. GEOMETRY A regular hexagon measures $(3x + 5)$ units on each side. What is the perimeter?

Simplify each expression.

25. $4a + 2b + a$

26. $2y + 2x + 8y$

27. $x^2 + 3x + 2x + 5x^2$

28. $4a^3 + 6a + 3a^3 + 8a$

29. $6x + 2(2x + 7)$

30. $4(3n + 9) + 5n$

Write an algebraic expression for each verbal expression. Then simplify, indicating the properties used.

31. twice the sum of s and t decreased by s

32. 5 times the product of x and y increased by $3xy$

33. the product of 6 and the square of z, increased by the sum of 7, z^2, and 6

34. 6 times the sum of x and y squared minus 3 times the sum of x and y squared

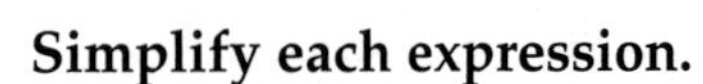

Simplify each expression.

35. $\frac{1}{4}q + 2q + 2\frac{3}{4}q$

36. $3.2(x + y) + 2.3(x + y) + 4x$

37. $3(4m + n) + 2m(4 + n)$

38. $6(0.4f + 0.2g) + 0.5f$

39. $\frac{3}{4} + \frac{2}{3}(s + 2t) + s$

40. $2p + \frac{3}{5}\left(\frac{1}{2}p + 2q\right) + \frac{2}{3}$

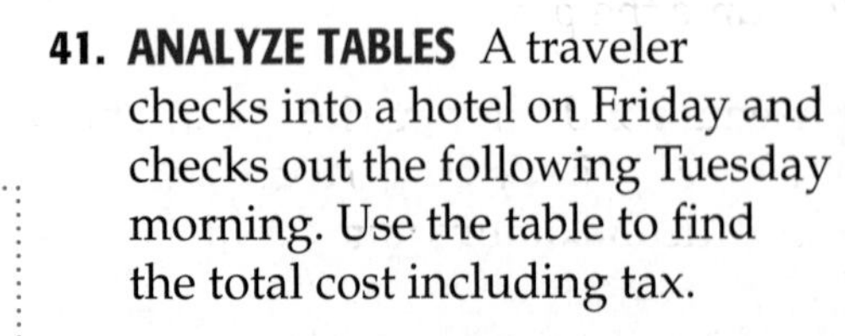

41. ANALYZE TABLES A traveler checks into a hotel on Friday and checks out the following Tuesday morning. Use the table to find the total cost including tax.

Hotel Rates Per Day		
Day	Room Charge	Sales Tax
Monday–Friday	$72	$5.40
Saturday–Sunday	$63	$5.10

Real-World Link

In 1943, Jacques Cousteau and Emile Gagnan invented the *aqualung*, or SCUBA (Self-Contained Underwater Breathing Apparatus).

Source: howstuffworks.com

SCUBA DIVING For Exercises 42 and 43, use the following information. A scuba diving store rents air tanks for $7.50, dive flags for $5.00, and wet suits for $10.95. The store also sells disposable underwater cameras for $18.99.

42. Write two expressions to represent the total sales after renting 2 wet suits, 3 air tanks, 2 dive flags, and selling 5 underwater cameras.

43. What are the total sales?

H.O.T. Problems

44. CHALLENGE Does the Commutative Property *sometimes*, *always*, or *never* hold for subtraction? Explain your reasoning.

45. Which One Doesn't Belong? Identify the sentence that does not belong with the other three. Explain your reasoning.

$x + 12 = 12 + x$	$7h = h \cdot 7$	$1 + a = a + 1$	$n \div 2 = 2 \div n$

EXTRA PRACTICE See pages 718, 744.

Math Online
Self-Check Quiz at algebra1.com

46. OPEN ENDED Write examples of the Commutative Property of Addition and the Associative Property of Multiplication using the numbers 1, 5, and 8 in each. Justify your examples.

47. *Writing in Math* Use the information about subways on page 33 to explain how the Commutative and Associative Properties are useful in performing calculations. Include an expression using the properties that could help you find the distance from the airport to Five Points Station.

STANDARDIZED TEST PRACTICE

48. Which expression is equivalent to $4 + 6(ac + 2b) + 2ac$?

A $4 + 8ac + 12b$

B $4 + 10ab + 2ac$

C $4 + 12abc + 2ac$

D $12ac + 20b$

49. REVIEW Daniel is buying a jacket that is regularly \$59.99 and is on sale for $\frac{1}{3}$ off. Which expression can he use to estimate the discount on the jacket?

F 0.0003×60

G 0.003×60

H 0.03×60

J 0.33×60

Spiral Review

Simplify each expression. If not possible, write *simplified*. (Lesson 1-5)

50. $5x - 8 + 7x$
51. $7m + 6n + 8$
52. $t^2 + 2t + 4t$
53. $3(5 + 2p)$
54. $(a + 2b)3 - 3a$
55. $(d + 5)8 + 2f$

56. Evaluate $3(5 - 5 \cdot 1^2) + 21 \div 7$. Name the property used in each step. (Lesson 1-4)

57. LAUNDRY Jonathan is meeting friends for dinner in 3 hours and he wants to do laundry beforehand. If it takes 50 minutes to do each load of laundry, write and use an inequality to find the maximum number of loads that he can finish. (Lesson 1-3)

GET READY for the Next Lesson

PREREQUISITE SKILL **Evaluate each expression.** (Lesson 1-2)

58. If $x = 4$, then $2x + 7 = \underline{?}$.

59. If $x = 8$, then $6x + 12 = \underline{?}$.

60. If $n = 6$, then $5n - 14 = \underline{?}$.

61. If $n = 7$, then $3n - 8 = \underline{?}$.

READING MATH

Arguments with Properties of Real Numbers

A lawyer who is presenting a case in court pays careful attention to make statements that are accurate, that follow a logical order, and that are justified.

In writing an argument that uses many steps, it is important to evaluate each step to check for errors. It is also important to provide the correct justification for each statement.

Example

Manuel has simplified the expression $3y + 5(x + y) - 3(y - x) + 2x$ and listed the properties used in each step. Evaluate each step. Determine whether the solution is accurate. If not, indicate the correct steps.

Step 1	$3y + 5(x + y) - 3(y - x) + 2x$	Original expression
Step 2	$3y + 5x + 5y - 3y + 3x + 2x$	Distributive Property
Step 3	$3y + 5y - 3y + 5x + 3x + 2x$	Commutative Property of Addition
Step 4	$(3y + 5y - 3y) + (5x + 3x + 2x)$	Commutative Property of Addition
Step 5	$(3 + 5 - 3)y + (5 + 3 + 2)x$	Distributive Property
Step 6	$5y + 10x$	Substitution

Manuel's solution is accurate until Step 4. The property used in this step is the Associative Property of Addition, not the Commutative Property of Addition. There are no other errors.

READING TO LEARN

Evaluate each step. Determine whether the solution is accurate. If not, indicate the correct steps.

1.	**Step 1**	$3(2 + x) + 4(x - 8) - 3x$	Original expression
	Step 2	$3(2) + 3(x) + 4(x) + 4(-8) - 3x$	Distributive Property
	Step 3	$6 + 3x + 4x - 32 - 3x$	Multiply.
	Step 4	$3x + 4x - 3x + 6 - 32$	Commutative Property of Addition
	Step 5	$(3x + 4x - 3x) + 6 - 32$	Associative Property of Addition
	Step 6	$(1 + 4 - 3)x + 6 - 32$	Distributive Property
	Step 7	$2x - 26$	Substitution
2.	**Step 1**	$4(3b - 2a) + 3(a + b) + b + 2(a - b)$	Original expression
	Step 2	$12b - 2a + 3a + 3b + b + 2a - 2b$	Distributive Property
	Step 3	$12b + 3b + b - 2b - 2a + 3a + 2a$	Commutative Property of Addition
	Step 4	$(12b + 3b + b - 2b) + (-2a + 3a + 2a)$	Associative Property of Addition
	Step 5	$(12 + 3 + 1 - 2)b + (-2 + 3 + 2)a$	Distributive Property
	Step 6	$14b + 3a$	Substitution

Logical Reasoning and Counterexamples

Main Ideas

- Identify the hypothesis and conclusion in a conditional statement.
- Use a counterexample to show that an assertion is false.

New Vocabulary

conditional statement
if-then statement
hypothesis
conclusion
deductive reasoning
counterexample

GET READY for the Lesson

The directions at the right can help you make perfect popcorn.

If the popcorn burns, then the heat was too high or the kernels heated unevenly.

Stovetop Popping

To pop popcorn on a stovetop, you need:

- A 3- to 4-quart pan with a loose lid that allows steam to escape
- Enough popcorn to cover the bottom of the pan, one kernel deep
- 1/4 cup of oil for every cup of kernels

Heat the oil to 400–460°F (if the oil smokes, it is too hot). Test the oil on a couple of kernels. When they pop, add the rest of the popcorn, cover the pan, and shake to spread the oil. When the popping begins to slow, remove the pan from the stovetop.

Source: Popcorn Board

Conditional Statements The statement *If the popcorn burns, then the heat was too high or the kernels heated unevenly* is called a conditional statement. **Conditional statements** can be written in the form *If A, then B*. Statements in this form are called **if-then statements**.

If	A,	then	B.
If	the popcorn burns,	then	the heat was too high or the kernels heated unevenly.

The part of the statement immediately following *if* is called the **hypothesis**.

The part of the statement immediately following *then* is called the **conclusion**.

EXAMPLE Identify Hypothesis and Conclusion

1 Identify the hypothesis and conclusion of each statement.

a. ENTERTAINMENT If it is Friday, then Ofelia and Miguel are going to the movies.

The hypothesis follows the word *if* and the conclusion follows the word *then*.

Hypothesis: it is Friday

Conclusion: Ofelia and Miguel are going to the movies

b. If $4x + 3 > 27$, then $x > 6$.

Hypothesis: $4x + 3 > 27$ Conclusion: $x > 6$

Reading Math

If-Then Statements

Note that "if" is not part of the hypothesis and "then" is not part of the conclusion.

CHECK Your Progress

1A. If it is warm this afternoon, then we will have the party outside.

1B. If $8w - 5 = 11$, then $w = 2$.

Sometimes a conditional statement is written without using the words *if* and *then*. But a conditional statement can always be rewritten in if-then form.

Real-World Link

In 2005, more than 74 million people attended a Major League baseball game.

Source: ballparksofbaseball.com

EXAMPLE Write a Conditional in If-Then Form

2 Identify the hypothesis and conclusion of each statement. Then write each statement in if-then form.

a. I will go to the ball game with you on Saturday.

Hypothesis: it is Saturday

Conclusion: I will go to the ball game with you

If it is Saturday, then I will go to the ball game with you.

b. For a number x such that $6x - 8 = 16$, $x = 4$.

Hypothesis: $6x - 8 = 16$

Conclusion: $x = 4$

If $6x - 8 = 16$, then $x = 4$.

CHECK Your Progress

2A. Brianna wears goggles when she is swimming.

2B. A rhombus with side lengths of $(x - y)$ units has a perimeter of $(4x - 4y)$ units.

Deductive Reasoning and Counterexamples **Deductive reasoning** is the process of using facts, rules, definitions, or properties to reach a valid conclusion. Suppose you have a true conditional and you know that the hypothesis is true for a given case. Deductive reasoning allows you to say that the conclusion is true for that case.

EXAMPLE Deductive Reasoning

3 Determine a valid conclusion that follows from the statement below for each condition. If a valid conclusion does not follow, write *no valid conclusion* and explain why.

"If two numbers are odd, then their sum is even."

a. The two numbers are 7 and 3.

7 and 3 are odd, so the hypothesis is true.

Conclusion: The sum of 7 and 3 is even.

CHECK $7 + 3 = 10$ ✓ The sum, 10, is even.

b. The sum of two numbers is 14.

The conclusion is true. If the numbers are 11 and 3, the hypothesis is true also. However, if the numbers are 8 and 6, the hypothesis is false. Therefore, there is no valid conclusion that can be drawn from the given conditional.

Study Tip

Common Misconception

Suppose the conclusion of a conditional is true. This does not mean that the hypothesis is true. Consider the conditional "If it rains, Annie will stay home." If Annie stays home, it does not necessarily mean that it is raining.

CHECK Your Progress

Determine a valid conclusion that follows from the statement "There will be a quiz every Wednesday."

3A. It is Wednesday.

3B. It is Tuesday.

To show that a conditional is false, we can use a counterexample. A **counterexample** is a specific case in which the hypothesis is true and the conclusion is false. For example, consider the conditional *if a triangle has a perimeter of 3 centimeters, then each side measures 1 centimeter.* A counterexample is a triangle with perimeter 3 and sides 0.9, 0.9, and 1.2 centimeters long. It takes only one counterexample to show that a statement is false.

STANDARDIZED TEST EXAMPLE

4 Rachel believes that if $x \div y = 1$, then x and y are whole numbers. José states that this theory is not always true. Which pair of values for x and y could José use to disprove Rachel's theory?

A $x = 2, y = 2$

B $x = 1.2, y = 0.6$

C $x = 0.25, y = 0.25$

D $x = 6, y = 3$

Read the Test Item

The question is asking for a counterexample. Find the values of x and y that make the statement false.

Solve the Test Item

Replace x and y in the equation $x \div y = 1$ with the given values.

A $x = 2, y = 2$

$2 \div 2 \stackrel{?}{=} 1$

$1 = 1$ ✓

The hypothesis is true, and both values are whole numbers. The statement is true.

B $x = 1.2, y = 0.6$

$1.2 \div 0.6 \stackrel{?}{=} 1$

$2 \neq 1$

The hypothesis is false, and the conclusion is false. This is not a counterexample.

C $x = 0.25, y = 0.25$

$0.25 \div 0.25 \stackrel{?}{=} 1$

$1 = 1$ ✓

The hypothesis is true, but 0.25 is not a whole number. Thus, the statement is false.

D $x = 6, y = 3$

$6 \div 3 \stackrel{?}{=} 1$

$2 \neq 1$

The hypothesis is false. Therefore, this is not a counterexample even though the conclusions are true.

The only values that prove the statement false are $x = 0.25$ and $y = 0.25$. So these numbers are counterexamples. The answer is C.

Test-Taking Tip

Checking Results Since choice C is the correct answer, you can check your results by testing the other values.

CHECK Your Progress

4. Which numbers disprove the statement below?

If $x + y > xy$, then $x > y$.

F $x = 1, y = 2$ **G** $x = 2, y = 3$ **H** $x = 4, y = 1$ **J** $x = 4, y = 2$

Online **Personal Tutor at** algebra1.com

CHECK Your Understanding

Example 1 (p. 39)

Identify the hypothesis and conclusion of each statement.

1. If it is April, then it might rain.
2. If you play tennis, then you run fast.
3. If $34 - 3x = 16$, then $x = 6$.

Example 2 (p. 40)

Identify the hypothesis and conclusion of each statement. Then write each statement in if-then form.

4. Colin watches television when he does not have homework.
5. A number that is divisible by 10 is also divisible by 5.
6. A rectangle is a quadrilateral with four right angles.

Example 3 (p. 40)

Determine a valid conclusion that follows from the statement below for each given condition. If a valid conclusion does not follow, write *no valid conclusion* and explain why.

If the last digit of a number is 2, then the number is divisible by 2.

7. The number is 10,452.
8. The number is divisible by 2.
9. The number is 946.

Example 4 (p. 41)

Find a counterexample for each conditional statement.

10. If Anna is in school, then she has a science class.
11. If you can read 8 pages in 30 minutes, then you can read a book in a day.
12. If a number x is squared, then $x^2 > x$.
13. If $3x + 7 \geq 52$, then $x > 15$.
14. **STANDARDIZED TEST PRACTICE** Which number disproves the statement $x < 2x$?

 A 0 **B** 1 **C** 2 **D** 4

Exercises

HOMEWORK HELP	
For Exercises	See Examples
15–18	1
19–24	2
25–28	3
29–38	4

Identify the hypothesis and conclusion of each statement.

15. If both parents have red hair, then their children have red hair.
16. If you are in Hawaii, then you are in the tropics.
17. If $2n - 7 > 25$, then $n > 16$.
18. If $a = b$ and $b = c$, then $a = c$.

Identify the hypothesis and conclusion of each statement. Then write each statement in if-then form.

19. The trash is picked up on Monday.
20. Vito will call after school.
21. For $x = 8$, $x^2 - 3x = 40$.
22. $4s + 6 > 42$ when $s > 9$.
23. A triangle with all sides congruent is an equilateral triangle.
24. The sum of the digits of a number is a multiple of 9 when the number is divisible by 9.

Determine whether a valid conclusion follows from the statement below for each given condition. If a valid conclusion does not follow, write *no valid conclusion* and explain why.

If a DVD box set costs less than $70, then Ian will buy one.

25. A DVD box set costs $59.

26. A DVD box set costs $89.

27. Ian will not buy a DVD box set.

28. Ian bought 2 DVD box sets.

Find a counterexample for each conditional statement.

29. If you were born in North Carolina, then you live in North Carolina.

30. If you are a professional basketball player, then you play in the United States.

31. If the product of two numbers is even, then both numbers must be even.

32. If two times a number is greater than 16, then the number must be greater than 7.

33. If $4n - 8 \geq 52$, then $n > 15$.

34. If $x \cdot y = 1$, then x or y must equal 1.

GEOMETRY **For Exercises 35 and 36, use the following information. If points *P*, *Q*, and *R* lie on the same line, then *Q* is between *P* and *R*.**

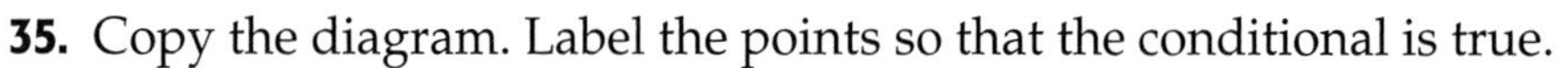

35. Copy the diagram. Label the points so that the conditional is true.

36. Copy the diagram. Provide a counterexample for the conditional.

Real-World Link

Groundhog Day has been celebrated in the United States since 1897. The most famous groundhog, Punxsutawney Phil, has seen his shadow about 85% of the time.

Source: www.infoplease.com

Determine whether a valid conclusion follows from the statement below for each given condition. If a valid conclusion does not follow, write *no valid conclusion* and explain why.

If the dimensions of rectangle ABCD are doubled, then the perimeter is doubled.

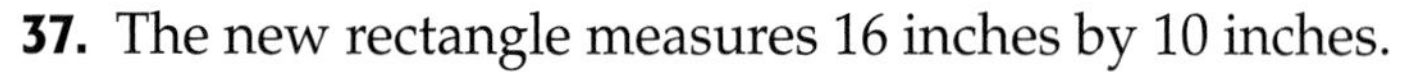

37. The new rectangle measures 16 inches by 10 inches.

38. The perimeter of the new rectangle is 52 inches.

39. RESEARCH On Groundhog Day (February 2) of each year, some people say that if a groundhog comes out of its hole and sees its shadow, then there will be 6 more weeks of winter weather. If it does not see its shadow, then there will be an early spring. Use the Internet or another resource to research the weather on Groundhog Day for your city for the past 10 years. Summarize your data as examples or counterexamples for this belief.

NUMBER THEORY **For Exercises 40–42, use the following information.**

Copy the Venn diagram and place the numbers 1 to 25 in the appropriate places on the diagram.

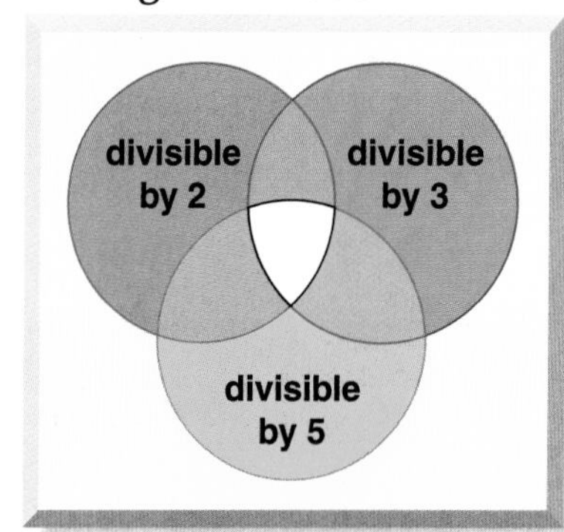

40. What conclusions can you make about the numbers and where they appear on the diagram?

41. What conclusions can you form about numbers that are divisible by both 2 and 3?

42. Provide a counterexample for the data you have collected if possible.

EXTRA PRACTICE
See pages 719, 744.

Math Online
Self-Check Quiz at
algebra1.com

H.O.T. Problems

43. **CHALLENGE** Determine whether the following statement is always true. If it is not, provide a counterexample.

If the mathematical operation $*$ is defined for all numbers a and b as $a * b = a + 2b$, then the operation $*$ is commutative.

44. **OPEN ENDED** Write a conditional statement and label the hypothesis and conclusion. Describe how conditional statements are used to solve problems.

45. **REASONING** Explain how deductive reasoning is used to show that a conditional is true or false.

46. ***Writing in Math*** Use the information about popcorn found on page 39 to explain how logical reasoning is helpful in cooking. Include in your answer the hypothesis and conclusion of the statement *If you have small, underpopped kernels, then you have not used enough oil in your pan.*

STANDARDIZED TEST PRACTICE

47. Which number serves as a counterexample to the statement below?

$$2x < 3x$$

A -2 B $\frac{1}{4}$ C $\frac{1}{2}$ D 2

48. **REVIEW** If $4a = a$, which of the following is true?

F $a > 4$ H $a = 1$

G $a = 4$ J $a = 0$

49. What value of n makes the following statement true?

If $14n - 12 \geq 100$, then $n \geq$ ______.

A 8

B 10

C 12

D 24

Spiral Review

Simplify each expression. (Lesson 1-6)

50. $2x + 5y + 9x$

51. $4(5mn + 6) + 3mn$

52. $2(3a + b) + 3b + 4$

53. **ENVIRONMENT** A typical family of four uses the water shown in the table. Write two expressions that represent the amount of water a typical family of four uses for these activities in d days. (Lesson 1-5)

Average Water Usage Per Day	
Activity	**Gallons Used**
flushing toilet	100
showering/bathing	80
using bathroom sink	8

Source: U.S. Environmental Protection Agency

Find the value of n in each expression. Then name the property used. (Lesson 1-4)

54. $1(n) = 64$

55. $12 + 7 = 12 + n$

56. $(9 - 7)5 = 2n$

57. $4n = 1$

58. $n + 18 = 18$

59. $36n = 0$

GET READY for the Next Lesson

PREREQUISITE SKILL **Evaluate each expression.** (Lesson 1-2)

60. 6^2

61. $(-8)^2$

62. 1.6^2

63. $(-11.5)^2$

Algebra Lab
Logic and Properties of Numbers

You can apply what you have learned about the properties of numbers to determine whether a mathematical statement is *always, sometimes,* or *never* true.

ACTIVITY 1

Determine whether $2y \leq 6$ is *always, sometimes,* or *never* true.

First determine the greatest value of y that satisfies the inequality.

$2(3) \leq 6$ Substitute 3 for y.

$6 \leq 6$ Simplify.

Next, substitute a value less than 3 for y and a value greater than 3.

$2(2) \leq 6$ Substitute 2 for y.

$4 \leq 6$ Simplify.

$2(4) \leq 6$ Substitute 4 for y.

$8 \not\leq 6$ Simplify.

Substituting 2 for y yields a true inequality. The inequality is true when $y \leq 3$. When $y > 3$, the inequality is not true. Therefore, the inequality is sometimes true.

ACTIVITY 2

Determine whether the following statement is *true* or *false*.
Use the properties of numbers to justify your answer.
The equation $y = 2(x + 4) - 3$ is negative when $x < 0$.

Substitute a negative value for x in the equation.

$y = 2(x + 4) - 3$ Original equation
$y = 2(-1 + 4) - 3$ Substitute -1 for x.
$y = -2 + 8 - 3$ Distributive Property
$y = 3$ Add.

Since substituting -1 for x into the equation yields a positive value for y, the statement is false.

Exercises

Determine whether each statement or inequality is *always, sometimes,* or *never* true.

1. $3t > -6$

2. $-2v < 4$

3. $3w + 4 > 0$

Determine whether each statement is *true* or *false*. Use the properties of numbers to justify your answer.

4. In the equation $y = 3(x + 2)$, y is positive when $x > 0$.

5. In the linear equation $y = -2x$, y is always positive.

1-8 Number Systems

Main Ideas

- Classify and graph real numbers.
- Find square roots and order real numbers.

New Vocabulary

positive number
negative number
natural number
whole number
integers
rational number
square root
perfect square
irrational numbers
real numbers
Closure Property
graph
coordinate
radical sign
principal square root
rational approximation

GET READY for the Lesson

In the 2000 Summer Olympics, Australian sprinter Cathy Freeman wore a special running suit that covered most of her body. The surface area of the human body may be approximated using the expression $\sqrt{\frac{\text{height} \times \text{weight}}{3600}}$, where height is in centimeters, weight is in kilograms, and surface area is in square meters. The symbol $\sqrt{}$ designates a nonnegative square root of a nonnegative number.

Classify and Graph Real Numbers A number line can be used to show the sets of natural numbers, whole numbers, and integers. Values greater than 0, or **positive numbers**, are listed to the right of 0, and values less than 0, or **negative numbers**, are listed to the left of 0.

natural numbers: 1, 2, 3, ...

whole numbers: 0, 1, 2, 3, ...

integers:
..., −3, −2, −1, 0, 1, 2, 3, ...

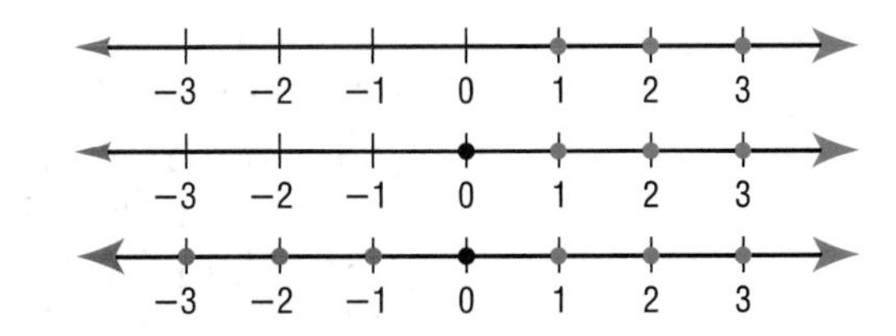

rational numbers: numbers that can be expressed in the form $\frac{a}{b}$, where a and b are integers and $b \neq 0$.

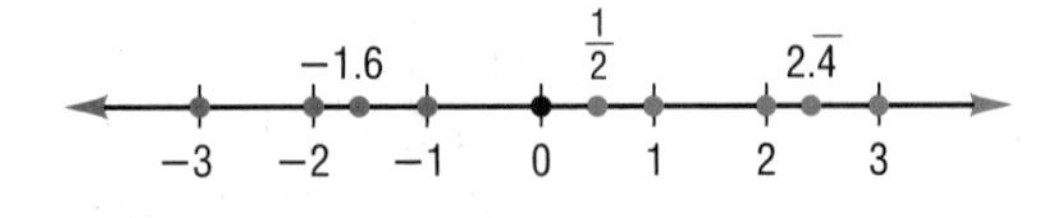

A rational number can also be expressed as a decimal that terminates, or as a decimal that repeats indefinitely.

A **square root** is one of two equal factors of a number. For example, one square root of 64, written as $\sqrt{64}$, is 8 since $8 \cdot 8$ or 8^2 is 64. Another square root of 64 is −8 since $(-8) \cdot (-8)$ or $(-8)^2$ is also 64. A number like 64, with a square root that is a rational number, is called a **perfect square**. The square roots of a perfect square are rational numbers.

A number such as $\sqrt{3}$ is the square root of a number that is *not* a perfect square. It cannot be expressed as a terminating or repeating decimal.

$\sqrt{3} = 1.73205080\ldots$

Numbers that cannot be expressed as terminating or repeating decimals, or in the form $\frac{a}{b}$, where a and b are integers and $b \neq 0$, are called **irrational numbers**. Irrational numbers and rational numbers together form the set of **real numbers**.

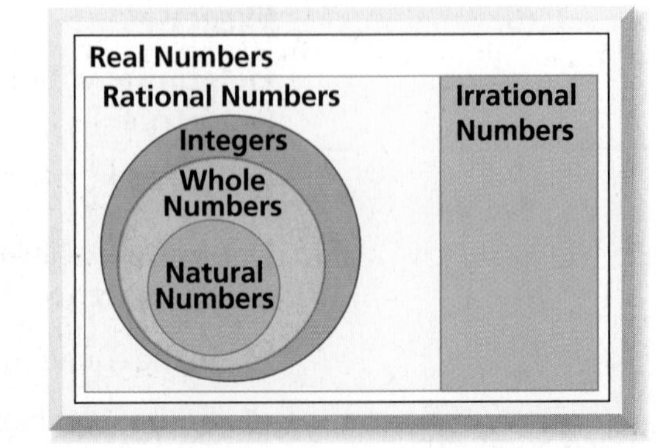

Concepts in Motion
Animation at algebra1.com

Study Tip

Common Misconception

Pay close attention to the placement of a negative sign when working with square roots. $\sqrt{-81}$ is undefined for real numbers since no real number multiplied by itself can result in a negative product.

EXAMPLE Classify Real Numbers

1 Name the set or sets of numbers to which each real number belongs.

a. $\frac{5}{22}$

Because 5 and 22 are integers and $5 \div 22 = 0.2272727\ldots$ or $0.2\overline{27}$, which is a repeating decimal, this number is a rational number.

b. $\sqrt{81}$

Because $\sqrt{81} = 9$, this number is a natural number, a whole number, an integer, and a rational number.

c. $\sqrt{56}$

Because $\sqrt{56} = 7.48331477\ldots$, which is not a repeating or terminating decimal, this number is irrational.

CHECK Your Progress

1A. $\frac{6}{11}$ **1B.** $-\sqrt{9.16}$

In Lesson 1-4, you learned about properties of real numbers. Another property of real numbers is the **Closure Property**. For example, the sum of any two whole numbers is a whole number. So, the set of whole numbers is said to be *closed* under addition.

EXAMPLE Closure Property

2 Determine whether each set of numbers is closed under the indicated operation.

a. whole numbers, multiplication

Select two different whole numbers and then determine whether the product is a whole number.

$0 \times 4 = 0$ $5 \times 2 = 10$ $1 \times 6 = 6$

Since the products of each pair of whole numbers are whole numbers, the set of whole numbers is closed under multiplication.

b. whole numbers, subtraction

We need to determine whether the difference of any two whole numbers is a whole number.

$3 - 4 = -1$

This is not a whole number, so the set of whole numbers is not closed under subtraction.

CHECK Your Progress

2A. integers, division **2B.** integers, addition

To **graph** a set of numbers means to draw, or plot, the points named by those numbers on a number line. The number that corresponds to a point on a number line is called the **coordinate** of that point. The rational numbers alone do not complete the number line. By including irrational numbers, the number line is complete.

EXAMPLE Graph Real Numbers

3 **Graph each set of numbers.**

a. $\left\{-\frac{4}{3}, -\frac{1}{3}, \frac{2}{3}, \frac{5}{3}\right\}$

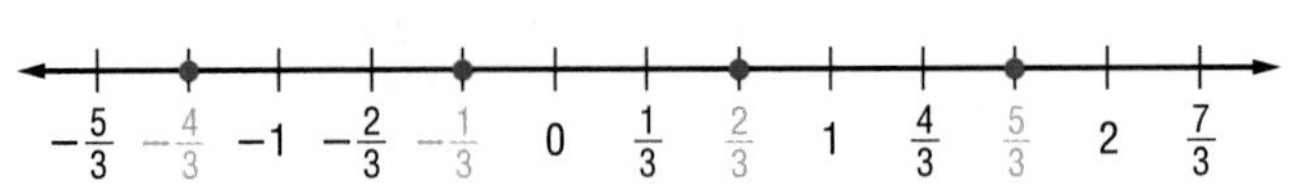

b. $x > -2$

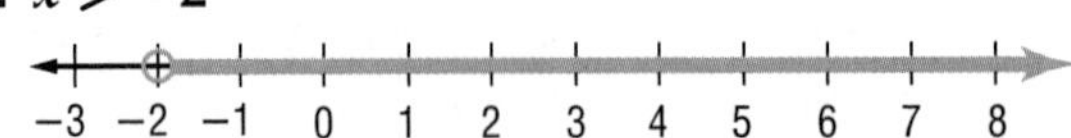

The heavy arrow indicates that all numbers to the right of −2 are included in the graph. Not only does this set include integers like 3 and −1, but it also includes rational numbers like $\frac{3}{8}$ and $-\frac{12}{13}$ and irrational numbers like $\sqrt{40}$ and π. The circle at −2 indicates −2 is *not* included in the graph.

c. $a \leq 4.5$

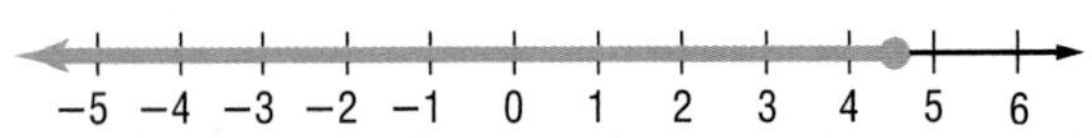

The heavy arrow indicates that all points to the left of 4.5 are included in the graph. The dot at 4.5 indicates that 4.5 *is* included in the graph.

CHECK Your Progress

3A. $\{-5, -4, -3, -2, \ldots\}$ **3B.** $x \leq 8$

Reading Math

Square Roots
$\pm\sqrt{64}$ is read *plus or minus the square root of 64.*
Exponents can also be used to indicate the square root. $9^{\frac{1}{2}}$ means the same thing as $\sqrt{9}$. $9^{\frac{1}{2}}$ is read *nine to the one-half power*. $9^{\frac{1}{2}} = 3$.

Square Roots and Ordering Real Numbers The symbol $\sqrt{\ }$, called a **radical sign**, is used to indicate a nonnegative or **principal square root** of the expression under the radical sign.

$\sqrt{64} = 8$ ← $\sqrt{64}$ indicates the *principal* square root of 64.

$-\sqrt{64} = -8$ ← $-\sqrt{64}$ indicates the *negative* square root of 64.

$\pm\sqrt{64} = \pm 8$ ← $\pm\sqrt{64}$ indicates *both* square roots of 64.

EXAMPLE Find Square Roots

4 **Find** $-\sqrt{\frac{49}{256}}$.

$-\sqrt{\frac{49}{256}}$ represents the negative square root of $\frac{49}{256}$.

$$\sqrt{\frac{49}{256}} = \left(\frac{7}{16}\right)^2 \longrightarrow -\sqrt{\frac{49}{256}} = -\frac{7}{16}$$

CHECK Your Progress

Find each square root.

4A. $\sqrt{\frac{4}{121}}$ **4B.** $\pm\sqrt{1.69}$

Real-World EXAMPLE

5 **SPORTS SCIENCE Refer to the application at the beginning of the lesson. Find the surface area of an athlete whose height is 192 centimeters and whose weight is 48 kilograms.**

$$\text{surface area} = \sqrt{\frac{\text{height} \times \text{weight}}{3600}}$$ Write the formula.

$$= \sqrt{\frac{192 \times 48}{3600}}$$ Replace height with 192 and weight with 48.

$$= \sqrt{\frac{9216}{3600}}$$ Simplify.

$$= \sqrt{\left(\frac{96}{60}\right)^2}$$ $\frac{9216}{3600} = \left(\frac{96}{60}\right)^2$

$$= \frac{96}{60} \text{ or } 1.6$$ Simplify.

The surface area of the athlete is 1.6 square meters.

Real-World Career
Scientist
Sports scientists use math to conduct experiments and interpret data. One of their jobs is to help athletes enhance their performance.

Math Online
For more information, go to algebra1.com.

CHECK Your Progress

5. Find the surface area of the athlete whose height is 200 centimeters and whose weight is 50 kilograms.

Online Personal Tutor at algebra1.com

To express irrational numbers as decimals, you need to use a rational approximation. A **rational approximation** of an irrational number is a rational number that is close to, but not equal to, the value of the irrational number. For example, a rational approximation of $\sqrt{2}$ is 1.41.

EXAMPLE Compare Real Numbers

6 **Replace each ● with <, >, or = to make each sentence true.**

a. $\sqrt{19}$ ● $3.\overline{8}$

Find two perfect squares closest to $\sqrt{19}$, and write an inequality.

$16 < 19 < 25$ 19 is between 16 and 25.

$\sqrt{16} < \sqrt{19} < \sqrt{25}$ Find the square root of each number.

$4 < \sqrt{19} < 5$ $\sqrt{19}$ is between 4 and 5.

Since $\sqrt{19}$ is between 4 and 5, it must be greater than $3.\overline{8}$.

So, $\sqrt{19} > 3.\overline{8}$.

b. $7.\overline{2}$ ● $\sqrt{52}$

You can use a calculator to find an approximation for $\sqrt{52}$.

$\sqrt{52} = 7.211102551\ldots$

$7.\overline{2} = 7.222\ldots$ Therefore, $7.\overline{2} > \sqrt{52}$.

CHECK Your Progress

6A. $2\frac{2}{3}$ ● $\sqrt{5}$

6B. $0.\overline{8}$ ● $\frac{8}{9}$

Study Tip

Ordering Real Numbers

When ordering real numbers, it is helpful to first simplify any square roots of perfect squares. For example, $\sqrt{\frac{25}{64}} = \frac{5}{8}$.

To order a set of real numbers from greatest to least or from least to greatest, find a decimal approximation for each number in the set and compare.

EXAMPLE Order Real Numbers

7 **Order $2.\overline{63}$, $-\sqrt{7}$, $\frac{8}{3}$, and $\frac{53}{-20}$ from least to greatest.**

$2.\overline{63} = 2.6363636\ldots$ or about 2.636

$-\sqrt{7} = -2.64575131\ldots$ or about -2.646

$\frac{8}{3} = 2.66666666\ldots$ or about 2.667

$\frac{53}{-20} = -2.65$

$-2.65 < -2.646 < 2.636 < 2.667$

The numbers arranged in order from least to greatest are $\frac{53}{-20}$, $-\sqrt{7}$, $2.\overline{63}$, $\frac{8}{3}$.

CHECK Your Progress

Order each set of numbers from greatest to least.

7A. $\sqrt{0.\overline{42}}, 0.\overline{63}, \sqrt{\frac{4}{9}}$

7B. $-1.\overline{46}, 0.2, \sqrt{2}, -\frac{1}{6}$

CHECK Your Understanding

Example 1 (p. 47)

Name the set or sets of numbers to which each real number belongs.

1. $-\sqrt{64}$ **2.** $\frac{8}{3}$ **3.** $\sqrt{28}$ **4.** $\frac{56}{7}$

Example 2 (p. 47)

Determine whether each set of numbers is closed under the indicated operation.

5. whole, division

6. rational, addition

7. rational, division

8. natural, subtraction

Example 3 (p. 48)

Graph each set of numbers.

9. $\{-4, -2, 1, 5, 7\}$ **10.** $x < -3.5$ **11.** $x \geq -7$

Example 4 (p. 48)

Find each square root.

12. $-\sqrt{25}$ **13.** $\sqrt{1.44}$ **14.** $\pm\sqrt{\frac{16}{49}}$ **15.** $\sqrt{361}$

Example 5 (p. 49)

16. **PHYSICAL SCIENCE** The time it takes a falling object to travel a distance d is given by $t = \sqrt{\frac{d}{16}}$, where t is in seconds and d is in feet. If Isabel drops a ball from 29.16 feet, how long will it take for it to reach the ground?

Example 6 (p. 49)

Replace each ● with <, >, or = to make each sentence true.

17. $\sqrt{17}$ ● $4\frac{1}{10}$ **18.** $\frac{2}{9}$ ● $0.\overline{2}$ **19.** $\frac{1}{6}$ ● $\sqrt{6}$

Example 7 (p. 50)

Order each set of numbers from least to greatest.

20. $\frac{1}{8}, \sqrt{\frac{1}{4}}, 0.\overline{15}, -15$

21. $\sqrt{30}, 5\frac{4}{9}, 13, \sqrt{\frac{1}{30}}$

Exercises

HOMEWORK HELP	
For Exercises	See Examples
22–27	1
28–37	2
38–41	3
42–47	4, 5
48–51	6
52–55	7

Name the set or sets of numbers to which each real number belongs.

22. $-\sqrt{22}$ **23.** $\frac{36}{6}$ **24.** $-\frac{5}{12}$

25. $\sqrt{10.24}$ **26.** $\frac{-54}{19}$ **27.** $\sqrt{\frac{82}{20}}$

Determine whether each set of numbers is closed under the indicated operation.

28. irrational, addition
29. irrational, subtraction
30. natural, addition
31. natural, multiplication
32. irrational, multiplication
33. irrational, division
34. integers, subtraction
35. rational, subtraction
36. rational, multiplication
37. integers, multiplication

Graph each set of numbers.

38. $\{-4, -2, -1, 1, 3\}$
39. $\{\ldots -2, 0, 2, 4, 6\}$
40. $x > -12$
41. $x \geq -10.2$

Find each square root.

42. $\sqrt{49}$ **43.** $\pm\sqrt{0.64}$ **44.** $\pm\sqrt{5.29}$

45. $-\sqrt{6.25}$ **46.** $\sqrt{\frac{169}{196}}$ **47.** $\sqrt{\frac{25}{324}}$

Replace each ● with $<$, $>$, or $=$ to make each sentence true.

48. $5.\overline{72}$ ● $\sqrt{5}$ **49.** $\sqrt{22}$ ● 4.7 **50.** $\sqrt{\frac{2}{3}}$ ● $\frac{2}{3}$ **51.** 8 ● $\sqrt{67}$

Order each set of numbers from least to greatest.

52. $\sqrt{0.06}, 0.\overline{24}, \sqrt{\frac{9}{144}}$
53. $0.6, \sqrt{\frac{16}{49}}, \frac{5}{9}$

54. $-4.\overline{83}, 0.4, \sqrt{8}, -\frac{3}{8}$
55. $-0.25, 0.\overline{14}, -\sqrt{\frac{5}{8}}, \sqrt{0.5}$

Evaluate each expression if $a = 4$, $b = 9$, and $c = 100$.

56. $2 \cdot \sqrt{a}$ **57.** $\sqrt{a} \cdot \sqrt{b}$ **58.** $\sqrt{a \cdot b}$ **59.** $\sqrt{a} + \sqrt{c}$

60. RIVERS The table shows the change in river depths for various rivers over a 24-hour period. Use a number line to graph these values and compare the changes in each river.

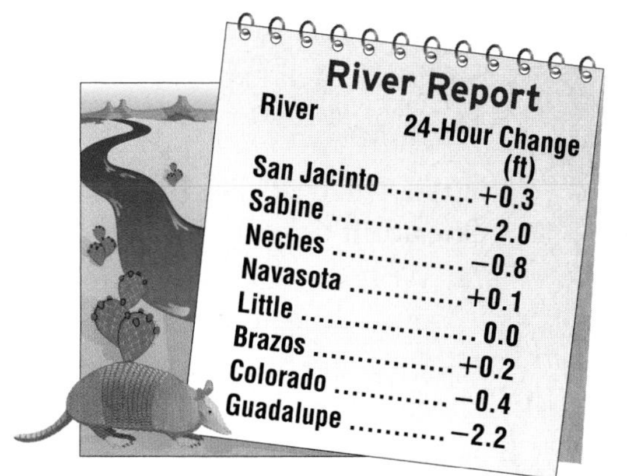

River Report

River	24-Hour Change (ft)
San Jacinto	+0.3
Sabine	−2.0
Neches	−0.8
Navasota	+0.1
Little	0.0
Brazos	+0.2
Colorado	−0.4
Guadalupe	−2.2

GEOMETRY For Exercises 61–63, consider squares with the following areas: 1 unit^2, 4 units^2, 9 units^2, 16 units^2, and 25 units^2.

61. Find the side length and perimeter of each square.
62. Describe the relationship between the lengths of the sides and the areas.
63. Write an expression to find the perimeter of a square with a area of a units^2.

H.O.T. Problems

64. **REASONING** Determine whether the following statement is *true* or *false*. Justify your answer. *Since the natural numbers are a subset of the whole numbers then for each operation that the whole numbers are closed, the natural numbers are closed also.*

65. **CHALLENGE** Determine whether the following statement is *true* or *false*. Include an example or a counterexample in your answer. *The average of two irrational numbers is an irrational number.*

66. **CHALLENGE** Determine when the following statements are all true for real numbers q and r.

 a. $q^2 > r^2$ **b.** $\frac{1}{q} < \frac{1}{r}$ **c.** $\sqrt{q} > \sqrt{r}$ **d.** $\frac{1}{\sqrt{q}} < \frac{1}{\sqrt{r}}$

67. **OPEN-ENDED** Give a real-life example in which numbers are ordered.

68. ***Writing in Math*** Use the information on page 46 to explain how square roots can be used to find the surface area of the human body.

STANDARDIZED TEST PRACTICE

69. **REVIEW** Which is an irrational number?

 A -6

 B $\frac{3}{2}$

 C $-\sqrt{8}$

 D $-\sqrt{4}$

70. For what value of a is $-\sqrt{a} < -\frac{1}{\sqrt{a}}$ true?

 F 2

 G $\frac{1}{3}$

 H 1

 J -4

Spiral Review

Find a counterexample for each statement. (Lesson 1-7)

71. If the sum of two numbers is even, then both numbers must be even.

72. If $x^2 < 1$, then $x = 0$.

Simplify each expression. (Lesson 1-6)

73. $8x + 2y + x$

74. $7(5a + 3b) - 4a$

75. $4[1 + 4(5x + 2y)]$

MOVIE THEATERS For Exercises 76 and 77, use the following information. (Lesson 1-2)
One adult ticket at a movie theater costs $8.50. One small popcorn costs $3.50.

76. Write an algebraic expression to represent how much could be spent at the movie theater.

77. Evaluate the expression to find the total cost if Julio and his three friends each bought a ticket and popcorn.

GET READY for the Next Lesson

PREREQUISITE SKILL

78. Refer to the table. If the pattern continues, what are the values for y if $x = 6$ and $x = 7$?

x	0	1	2	3	4	5
y	2	5	8	11	14	17

1-9 Functions and Graphs

Main Ideas

- Interpret graphs of functions.
- Draw graphs of functions.

New Vocabulary

function
coordinate system
y-axis
origin
x-axis
ordered pair
x-coordinate
y-coordinate
independent variable
dependent variable
relation
domain
range
discrete function
continuous function

GET READY for the Lesson

The graph shows that as the number of days after a concussion increases, the percent of blood flow increases.

The return of normal blood flow is said to be a function of the number of days since the concussion.

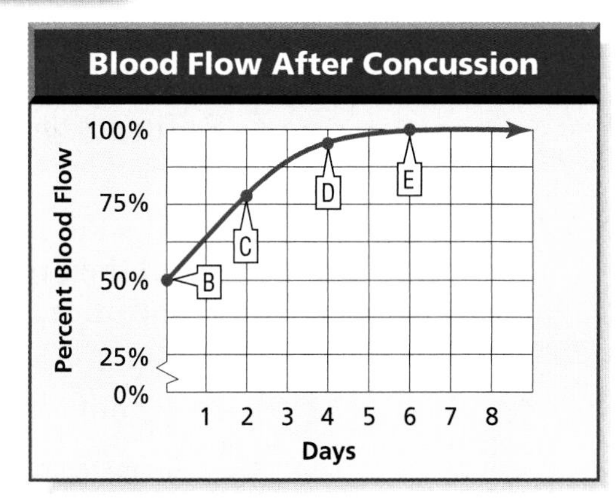

Source: *Scientific American*

Interpret Graphs A **function** is a relationship between input and output. In a function, the output depends on the input. There is exactly one output for each input. A function is graphed using a **coordinate system**, or *coordinate plane*. It is formed by the intersection of two number lines, the *horizontal axis* and the *vertical axis*.

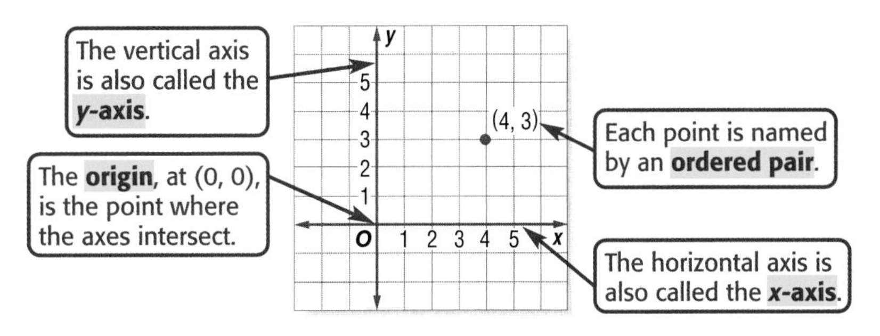

Each input x and its corresponding output y is graphed using an ordered pair in the form (x, y). The x-value, called the **x-coordinate** corresponds to the x-axis and the y-value, or **y-coordinate**, corresponds to the y-axis.

Real-World EXAMPLE Identify Coordinates

1 **MEDICINE Refer to the application above. Name the ordered pair at point *C* and explain what it represents.**

Point C is at 2 along the x-axis and about 80 along the y-axis. So, its ordered pair is (2, 80). This represents 80% normal blood flow 2 days after the injury.

CHECK Your Progress

1. Name the ordered pair at point E and explain what it represents.

Study Tip

Graphs of Functions

Usually the independent variable is graphed on the horizontal axis, and the dependent variable is graphed on the vertical axis.

In Example 1, the blood flow depends on the number of days from the injury. Therefore, the number of days from the injury is called the **independent variable** and the percent of normal blood flow is called the **dependent variable**.

EXAMPLE Independent and Dependent Variables

2 Identify the independent and dependent variables for each function.

a. In general, the average price of gasoline slowly and steadily increases throughout the year.

Time is the independent variable as it is unaffected by the price of gasoline, and the price is the dependent quantity as it is affected by time.

b. Art club members are drawing caricatures of students to raise money for their trip to New York City. The profit that they make increases as the price of their drawings increases.

In this case, price is the independent quantity. Profit is the dependent quantity as it is affected by the price.

CHECK Your Progress

2A. The distance a person runs increases with time.

2B. As the dimensions of a square decrease, so does the area.

Functions can be graphed without using a scale on either axis to show the general shape of the graph.

EXAMPLE Analyze Graphs

3 The graph at the right represents the speed of a school bus traveling along its morning route. Describe what is happening in the graph.

At the origin, the bus is stopped. It accelerates and maintains a constant speed. Then it begins to slow down, eventually stopping. After being stopped for a short time, the bus accelerates again. The process repeats continually.

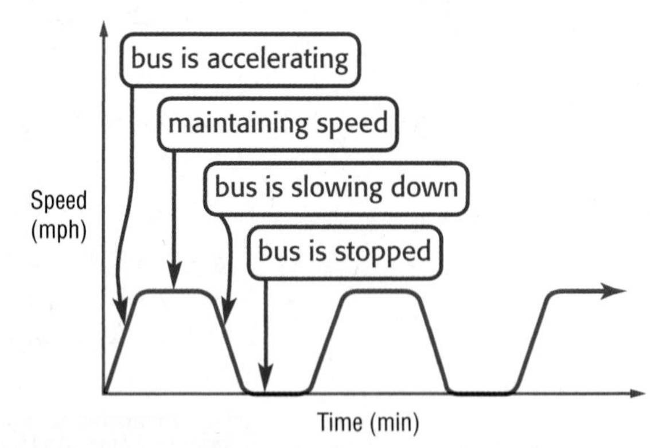

CHECK Your Progress

3. Identify the graph that represents the altitude of a space shuttle above Earth, from the moment it is launched until the moment it lands.

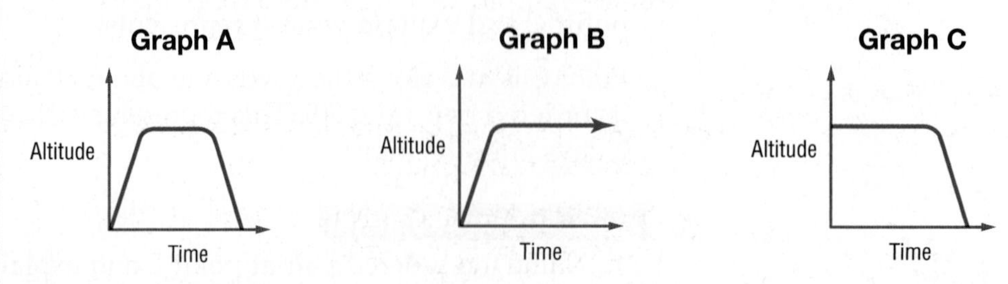

Draw Graphs Graphs can be used to represent many real-world situations.

Real-World EXAMPLE Draw Graphs

4 SHOPPING For every two pairs of earrings you buy at the regular price of $29 each, you get a third pair free.

a. Make a table showing the cost of buying 1 to 5 pairs of earrings.

Pairs of Earrings	1	2	3	4	5
Total Cost ($)	29	58	58	87	116

Study Tip

Different Representations

Example 4 illustrates some of the ways data can be represented—tables, ordered pairs, and graphs.

b. Write the data as a set of ordered pairs. Then graph the data.

Use the table. The number of pairs of earrings is the independent variable, and the cost is the dependent variable. So, the ordered pairs are (1, 29), (2, 58), (3, 58), (4, 87), and (5, 116).

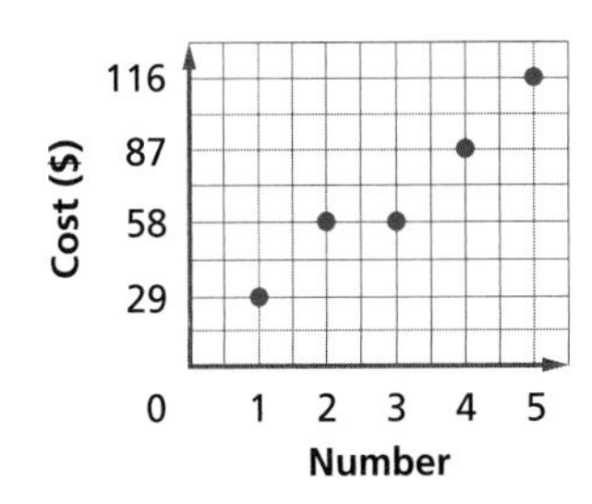

CHECK Your Progress

4A. Suppose you earn $0.25 for each book that you sell. Make a table showing how much you earn when you sell 1 to 5 books.

4B. Write the data as a set of ordered pairs. Then graph the data.

A set of ordered pairs, like those in Example 4, is called a **relation**. The set of the first numbers of the ordered pairs is the **domain**. The set of second numbers of the ordered pairs is the **range**. The function in Example 4 is a **discrete function** because its graph consists of points that are not connected. A function graphed with a line or smooth curve is a **continuous function.**

Study Tip

Domain and Range

The domain contains all values of the independent variable. The range contains all values of the dependent variable.

Real-World EXAMPLE Domain and Range

5 EXERCISE Rasha rides her bike an average of 0.25 mile per minute up to 36 miles each week. The distance that she travels each week is a function of the number of minutes that she rides.

a. Identify a reasonable domain and range for this situation.

The domain is the number of minutes that Rasha rides her bike. Since she rides up to 36 miles each week, she rides up to $36 \div 0.25$ or 144 minutes a week. A reasonable domain would be values from 0 to 144 minutes. The range is the weekly distance traveled. A reasonable range is 0 to 36 miles.

b. Draw a graph that shows the relationship.

Graph the ordered pairs (0, 0) and (144, 36). Since she rides any number of miles up to 36 miles, connect the two points with a line to include those points.

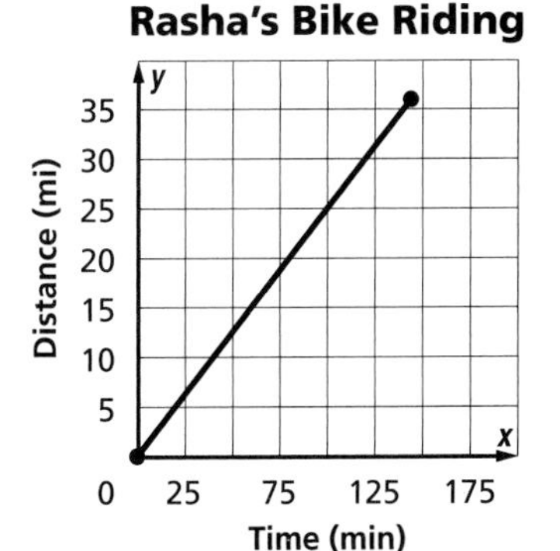

c. State whether the function is discrete or continuous. Explain.

Since the points are connected with a line, this function is continuous.

Cross-Curricular Project

Math Online A graph of the number of people over 65 in the U.S. for the years since 1900 will help you predict trends. Visit algebra1.com to continue work on your project.

CHECK Your Progress

5A. At Go-Cart World, each go-cart ride costs $4. For a birthday party, a group of 7 friends are each allowed to take 1 go-cart ride. The total cost is a function of the number of rides. Identify a reasonable domain and range for this situation.

5B. Draw a graph that shows the relationship between the number of rides and the total cost.

5C. State whether the function is discrete or continuous. Explain.

Online Personal Tutor at algebra1.com

CHECK Your Understanding

For Exercises 1–3, use the graph at the right.

Example 1 (p. 53)

1. Name the ordered pair at point *A* and explain what it represents.

2. Name the ordered pair at point *B* and explain what it represents.

Example 2 (p. 54)

3. Identify the independent and dependent variables for the function.

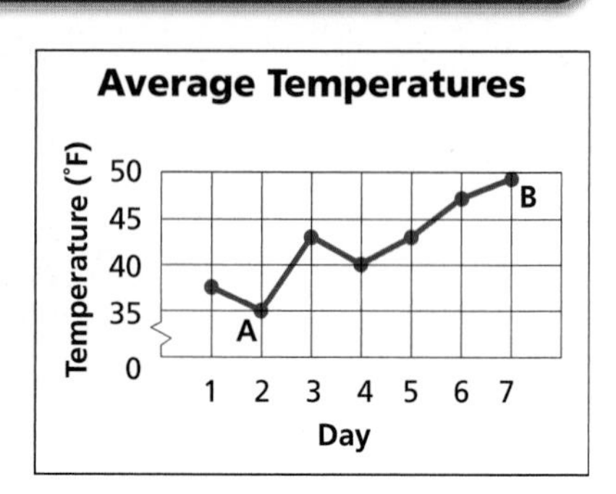

Example 3 (p. 54)

4. The graph at the right represents Alexi's speed as he rides his bike. Describe what is happening in the graph.

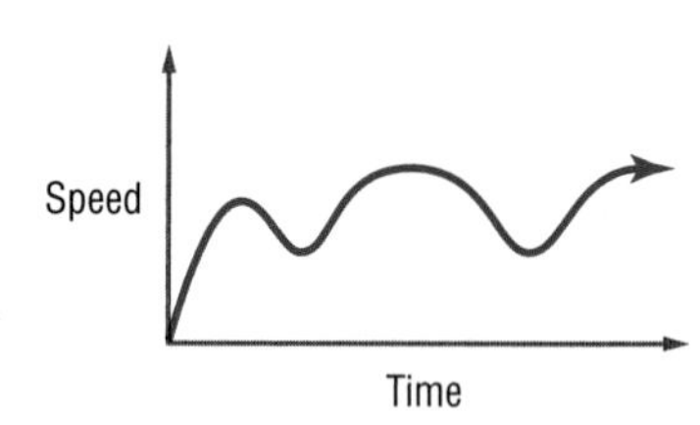

5. Identify the graph that represents the altitude of a skydiver just before she jumps from a plane until she lands.

Graph A

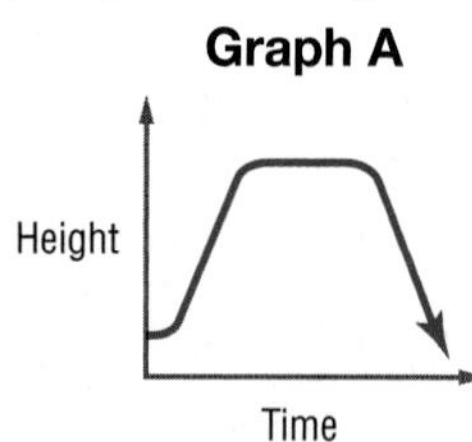

Graph B

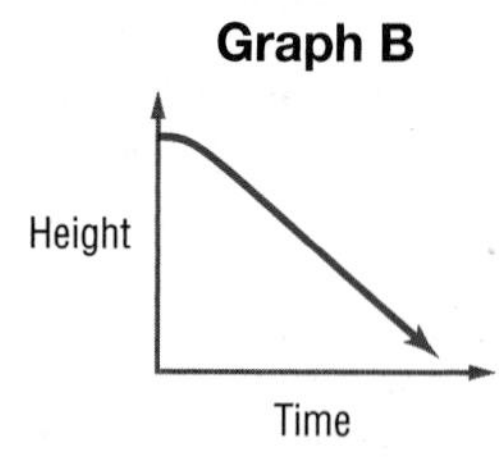

Graph C

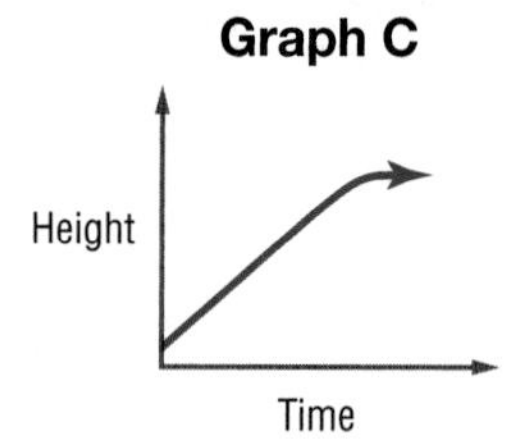

PHYSICAL SCIENCE For Exercises 6–9, use the table and the information. Ms. Blackwell's students recorded the height of an object above the ground at several intervals after it was dropped from a height of 5 meters.

Time (s)	0	0.2	0.4	0.6	0.8	1
Height (cm)	500	480	422	324	186	10

Example 4 (p. 55)

6. Write a set of ordered pairs representing the data in the table.

7. Draw a graph showing the relationship between the height of the falling object and time.

Example 5 (p. 55)

8. Identify the domain and range for this situation.

9. State whether the function is discrete or continuous. Explain.

Exercises

HOMEWORK HELP	
For Exercises	See Examples
10, 11, 13, 14	1
12, 15	2
16–19	3
20, 21	4
22–25	5

For Exercises 10–12, use the graph at the right.

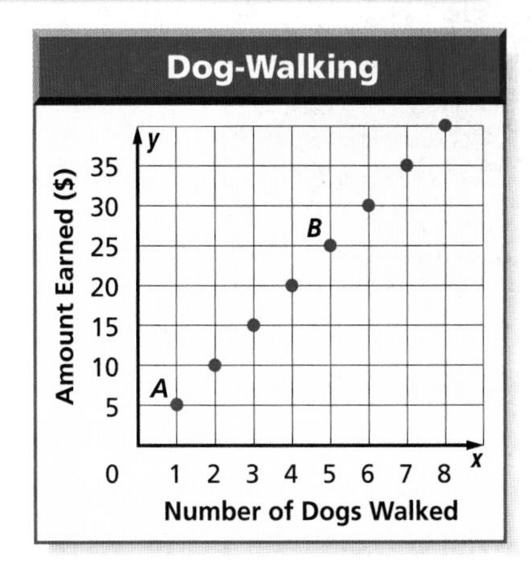

10. Name the ordered pair at point A and explain what it represents.

11. Name the ordered pair at point B and explain what it represents.

12. Identify the independent and dependent variables for the function.

For Exercises 13–15, use the graph at the right.

13. Name the ordered pair at point C and explain what it represents.

14. Name the ordered pair at point D and explain what it represents.

15. Identify the independent and dependent variables.

16. The graph below represents Teresa's temperature when she was sick. Describe what is happening in the graph.

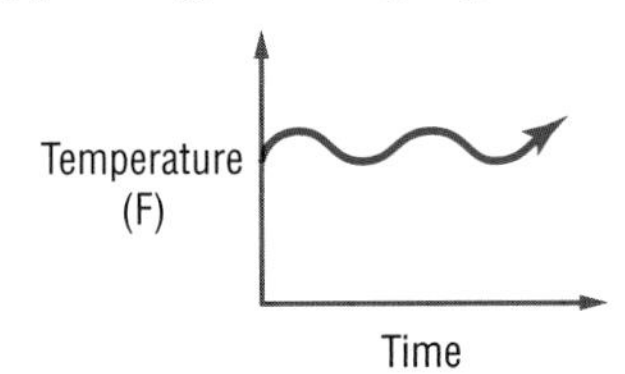

17. The graph below represents the altitude of a group of hikers. Describe what is happening in the graph.

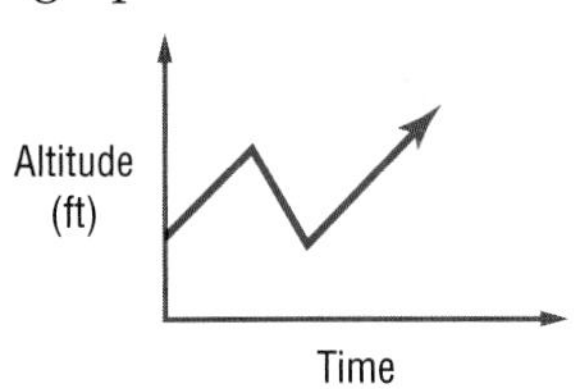

18. TOYS Identify the graph that displays the speed of a radio-controlled car as it moves along and then hits a wall.

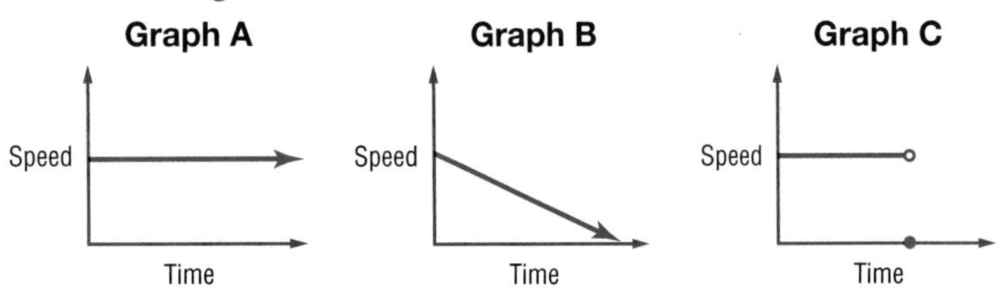

19. INCOME In general, as people get older, their incomes increase steadily until they retire. Which of the graphs below represents this?

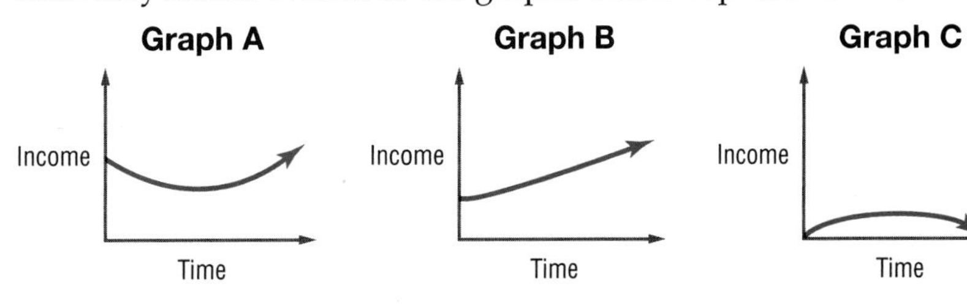

20. **CARS** Refer to the information at the left. A car was purchased new in 1990. The owner has taken excellent care of the car, and it has relatively low mileage. Draw a reasonable graph to show the value of the car from the time it was purchased to the present.

21. **CHEMISTRY** When ice is exposed to temperatures above 32°F, it begins to melt. Draw a reasonable graph showing the relationship between the temperature of a block of ice and the time after it is removed from a freezer.

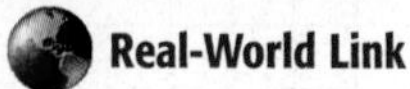

Real-World Link

Most new cars lose 15 to 30 percent of their value in the first year. After about 12 years, more popular cars tend to increase in value.

Source: *Consumer Guide*

GEOMETRY For Exercises 22–25, use the table below.

Polygon	triangle	quadrilateral	pentagon	hexagon	heptagon
Number of Sides	3	4	5	6	7
Interior Angle Sum	180	360	540	720	900

22. Identify the independent and dependent variables. Then graph the data.
23. Identify the domain and range for this situation.
24. State whether the function is discrete or continuous. Explain.
25. Predict the sum of the interior angles for an octagon, nonagon, and decagon.

H.O.T. Problems

26. **REASONING** Compare and contrast dependent and independent variables.

27. **OPEN ENDED** Give an example of a relation. Identify the domain and range.

28. **CHALLENGE** Eva is 23 years older than Lisa. Draw a graph showing Eva's age as a function of Lisa's age for the first 40 years of Lisa's life. Then find the point on the graph when Eva is twice as old as Lisa.

29. ***Writing in Math*** Use the data about concussions on page 53 to explain how real-world situations can be modeled using graphs and functions.

STANDARDIZED TEST PRACTICE

30. What is the range for the function $\{(1, 3), (5, 7), (9, 11)\}$?

    A $\{1, 5, 9\}$ C $\{3\}$

    B $\{3, 7, 11\}$ D $\varnothing$

31. **REVIEW** If $3x - 2y = 5$ and $x = 2$, what value of y makes the equation true?

    F 0.5 H 2

    G 1 J 5

Spiral Review

32. What is $\pm\sqrt{121}$? (Lesson 1-8)

Identify the hypothesis and conclusion of each statement. (Lesson 1-7)

33. You can send e-mail with a computer.
34. The express lane is for shoppers with 9 or fewer items.
35. Evaluate $ab(a + b)$ and name the property used in each step. (Lesson 1-6)

Write an algebraic expression for each verbal expression. (Lesson 1-1)

36. the product of 8 and a number x all raised to the fourth power
37. three times a number decreased by 10

Algebra Lab

Investigating Real-World Functions

ACTIVITY

The table shows the number of students enrolled in elementary and secondary schools for the given years.

Year	1900	1920	1940	1960	1970	1980	1990	2000
Enrollment (thousands)	15,503	21,578	25,434	36,807	45,550	41,651	40,543	46,857

Source: *The World Almanac*

- On **grid paper**, draw a vertical and horizontal axis as shown. Make your graph large enough to fill most of the sheet. Label the horizontal axis 0 to 120 and the vertical axis 1 to 50,000.
- To make graphing easier, let x represent the number of years since 1900. Write the eight ordered pairs using this method. The first will be (0, 15,503).
- Graph the ordered pairs on your grid paper.

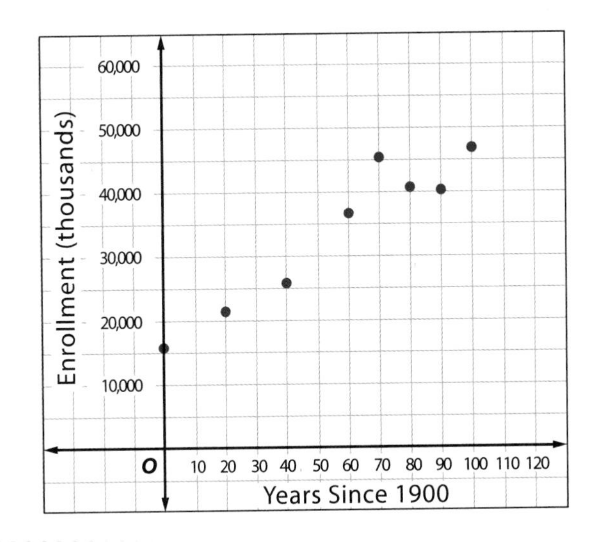

Analyze the Results

1. Describe the graph you made.
2. Use your graph to estimate the number of students in elementary and secondary school in 1910, 1975, and 2020.
3. Describe the method you used to make estimates for Exercise 2.
4. Do you think your prediction for 2020 will be accurate? Explain your reasoning.
5. Graph this set of data, which shows the number of students per computer in U.S. schools. Predict the number of students per computer in 2010. Explain how you made your prediction.

Year	Students per Computer	Year	Students per Computer	Year	Students per Computer
1984	125	1990	22	1996	10
1985	75	1991	20	1997	7.8
1986	50	1992	18	1998	6.1
1987	37	1993	16	1999	5.7
1988	32	1994	14		
1989	25	1995	10.5		

Source: *The World Almanac*

CHAPTER 1

Study Guide and Review

FOLDABLES™ Study Organizer — GET READY to Study

Be sure the following Key Concepts are noted in your Foldable.

Key Concepts

Order of Operations (Lesson 1-2)

Simplify the expression inside grouping symbols. Evaluate all powers. Multiply and divide then add and subtract in order from left to right.

Properties (Lessons 1-4, 1-5, and 1-6)

- Additive Identity: For any number a, the sum of a and 0 is a.
- Additive Inverse: For any number a, the sum of a and $-a$ is zero.
-

Multiplication Properties	
Identity	For any number a, $a \cdot 1 = a$
Property of Zero	For any number a, $a \cdot 0 = 0$
Inverse	For every number $\frac{a}{b}$, where a, $b \neq 0$, there is exactly one number $\frac{b}{a}$ such that $\frac{a}{b} \cdot \frac{b}{a} = 1$.

-

Properties of Equality	
Reflexive	For any numbers a, b, and c
Symmetric	$a = a$
Transitive	If $a = b$, then $b = a$.
Substitution	If $a = b$, then a may be replaced by b in any expression.
Distributive	$a(b + c) = ab + ac$ and $a(b - c) = ab - ac$
Commutative	$a + b = b + a$ and $ab = ba$
Associative	$(a + b) + c = a + (b + c)$ and $(ab)c = a(bc)$

Number Systems (Lesson 1-8)

- Real Numbers:
  - Natural Numbers $\{1, 2, 3, \ldots\}$
  - Whole Numbers $\{0, 1, 2, 3, \ldots\}$
  - Integers $\{\ldots, -2, -1, 0, 1, 2, \ldots\}$
  - Rational Numbers: $\frac{a}{b}$, where a and b are integers and $b \neq 0$.

Key Vocabulary

additive inverse (p. 21)
algebraic expression (p. 6)
coefficient (p. 29)
conditional statement (p. 39)
continuous function (p. 55)
coordinate system (p. 53)
counterexample (p. 41)
deductive reasoning (p. 40)
dependent variable (p. 54)
discrete function (p. 55)
domain (p. 55)
exponent (p. 6)
factors (p. 6)
function (p. 53)
independent variable (p. 54)
inequality (p. 16)
integers (p. 46)
irrational numbers (p. 46)
like terms (p. 28)
multiplicative inverses (p. 21)
open sentence (p. 15)
order of operations (p. 10)
perfect square (p. 46)
power (p. 6)
principal square root (p. 48)
range (p. 55)
rational approximation (p. 49)
rational number (p. 46)
real numbers (p. 46)
reciprocal (p. 21)
solution set (p. 15)
variable (p. 6)

Vocabulary Check

State whether each sentence is *true* or *false*. If *false*, replace the underlined word or number to make a true sentence.

1. The vertical axis is also called the y-axis.
2. Two numbers with a product of 1 are called elements.
3. A collection of objects or numbers is called a function.
4. A nonnegative square root is called a principal square root.
5. Irrational numbers and rational numbers together form the set of negative numbers.
6. The exponent indicates the number of times the base is used as a factor.
7. To evaluate an expression means to find its value.

Vocabulary Review at algebra1.com

Lesson-by-Lesson Review

1–1 Variables and Expressions (pp. 6–9)

Write an algebraic expression for each verbal expression.

8. a number to the fifth power

9. five times a number y squared

10. the sum of a number p and twenty-one

11. the difference of twice a number k and 8

Evaluate each expression.

12. 4^6 **13.** 2^5

Write a verbal expression for each algebraic expression.

14. $\frac{1}{2} + 7y$ **15.** $6p^2$ **16.** $3m^4 - 5$

17. FROGS A frog can jump twenty times the length of its body. If a frog's body length is b, write an algebraic expression to describe the length the frog could jump.

Example 1 Write an algebraic expression for *the sum of twice a number and five.*

Variable Let n represent the number.

Expression $2n + 5$

Example 2 Evaluate 7^5.

$7^5 = 7 \cdot 7 \cdot 7 \cdot 7 \cdot 7$ Use 7 as a factor 5 times.

$= 16{,}807$ Multiply.

Example 3 Write a verbal expression for $4x^2 - 11$.

four times a number x squared minus eleven

1–2 Order of Operations (pp. 10–14)

Evaluate each expression.

18. $3 + 16 \div 8 \cdot 5$ **19.** $4^2 \cdot 3 - 5(6 + 3)$

20. $288 \div [3(2^3 + 4)]$ **21.** $\frac{6(4^3 + 2^2)}{9 + 3}$

Evaluate each expression if $x = 3$, $t = 4$, and $y = 2$.

22. $t^2 + 3y$ **23.** xty^3

24. $\frac{6ty}{x}$ **25.** $8(x - y)^2 + 3t$

26. RUNNING Alan ran twice as many miles on Tuesday as he did on Monday and five more miles on Wednesday than he did on Monday. Write and evaluate an expression to find the total number of miles he ran if he ran 5 miles on Monday.

Example 4 Evaluate $x^2 - (y + 2)$ if $x = 4$ and $y = 3$.

$x^2 - (y + 2) = 4^2 - (3 + 2)$ Replace x with 4 and y with 3.

$= 4^2 - 5$ Add 3 and 2.

$= 16 - 5$ Evaluate 4^2.

$= 11$ Subtract 5 from 16.

Study Guide and Review

1-3 Open Sentences (pp. 15–20)

Find the solution of each equation or the solution set of each inequality if the replacement set is {4, 5, 6, 7, 8}.

27. $g + 9 = 16$ **28.** $10 - p < 7$

29. $\frac{x+1}{3} = 2$ **30.** $2a + 5 \geq 15$

Solve each equation.

31. $w = 4 + 3^2$ **32.** $d = \frac{7(4 \cdot 3)}{18 \div 3}$

33. $k = 5[2(4) - 1^3]$ **34.** $\frac{6(7) - 2(3)}{4^2 - 6(2)} = n$

35. HOMECOMING Tickets to the homecoming dance are $9 for one person and $15 for two people. If a group of seven students wish to go to the dance, write and solve an equation that would represent the least expensive price p of their tickets.

Example 5 **Find the solution set for $14 + 2h < 24$ if the replacement set is {1, 3, 5, 7}.**

Replace h in $14 + 2h < 24$ with each value in the replacement set.

h	$14 + 2h < 24$	True or False?
1	$14 + 2(1) < 24$	True
3	$14 + 2(3) < 24$	True
5	$14 + 2(5) < 24$	False
7	$14 + 2(7) < 24$	False

The solution set is {1, 3}.

Example 6 **Solve $5^2 - 3 = y$.**

$5^2 - 3 = y$ Original equation

$25 - 3 = y$ Evaluate 5^2.

$22 = y$ Subtract 3 from 25.

The solution is 22.

1-4 Identity and Equality Properties (pp. 21–25)

Find the value of n in each equation. Name the property that is used.

36. $4 + 0 = n$ **37.** $5n = 1$ **38.** $0 = 17n$

Evaluate each expression. Name the property used in each step.

39. $3(4 \div 4)^2 - \frac{1}{4}(4)$

40. $(7 - 7)(5) + 3 \cdot 1$

41. $\frac{1}{2} \cdot 2 + 2[2 \cdot 3 - 1]$

42. COOKIES Emilia promised her brother one half of the cookies that she made. If she did not make any cookies, determine how many she owes her brother and identify the property represented.

Example 7 **Find the value of n in $n \cdot 9 = 9$. Name the property that is used.**

$n = 1$, since $1 \cdot 9 = 9$. This is the Multiplicative Identity Property.

Example 8 **Evaluate $7 \cdot 1 + 5(2 - 2)$. Name the property used in each step.**

$7 \cdot 1 + 5(2 - 2) = 7 \cdot 1 + 5(0)$ Substitution

$= 7 + 5(0)$ Multiplicative Identity

$= 7 + 0$ Multiplicative Property of Zero

$= 7$ Substitution

Mixed Problem Solving
For mixed problem-solving practice, see page 744.

1-5 The Distributive Property (pp. 26–31)

Rewrite each expression using the Distributive Property. Then evaluate.

43. $8(15 - 6)$ **44.** $4(x + 1)$

Simplify each expression. If not possible, write *simplified*.

45. $3w - w + 4v + 3v$

46. $4np + 7mp$

47. EXPENSES Nikki's monthly expenses are \$550 for rent, \$225 for groceries, \$110 for transportation, and \$150 for utilities. Use the Distributive Property to write and evaluate an expression for her total expenses for nine months.

Example 9 **Rewrite $5(t + 3)$ using the Distributive Property. Then evaluate.**

$5(t + 3) = 5(t) + 5(3)$ Distributive Property

$= 5t + 15$ Multiply.

Example 10 **Simplify $2x^2 + 4x^2 + 7x$. If not possible, write *simplified*.**

$2x^2 + 4x^2 + 7x = (2 + 4)x^2 + 7x$ Distributive Property

$= 6x^2 + 7x$ Substitution

1-6 Commutative and Associative Properties (pp. 33–37)

Simplify each expression.

48. $7w^2 + w + 2w^2$

49. $3(2 + 3x) + 21x$

50. ZOO At a zoo, each adult admission costs \$9.75, and each child costs \$7.25. Find the cost of admission for two adults and four children.

Example 11 **Write an algebraic expression for *five times the sum of x and y increased by 2x*. Then simplify.**

$5(x + y) + 2x = 5x + 5y + 2x$

$= 5x + 2x + 5y$

$= (5 + 2)x + 5y$

$= 7x + 5y$

1-7 Logical Reasoning (pp. 39–44)

51. Identify the hypothesis and conclusion of the statement. Then write the statement in if-then form.
School begins at 7:30 A.M.

52. Find a counterexample for *if you have no umbrella, you will get wet.*

53. LIGHTNING It is said that lightning never strikes twice in the same place. Identify the hypothesis and conclusion of this statement and write it in if-then form.

Example 12 **Identify the hypothesis and conclusion of the statement *The trumpet player must audition to be in the band.* Then write the statement in if-then form.**

Hypothesis: a person is a trumpet player

Conclusion: the person must audition to be in the band

If a person is a trumpet player, then the person must audition to be in the band.

Study Guide and Review

1-8 Number Systems (pp. 46–52)

Name the set or sets of numbers to which each real number belongs.

54. $\frac{7}{15}$ **55.** $\sqrt{45}$

Graph each set of numbers.

56. $\{-3, -1, 0, 2, 5\}$ **57.** $x \leq 6.5$

Order each set of numbers from least to greatest.

58. $-\sqrt{4}, 3.5, \sqrt{11}, 3\frac{11}{20}$

59. $\sqrt{27}, 5\frac{1}{5}, -\sqrt{34}, -\frac{47}{9}$

60. ROOMS Belinda's square bedroom is $10\frac{41}{50}$ feet long. The area of Jarrod's bedroom is 115 square feet. Whose bedroom is larger? Explain.

Example 13 **Name the set or sets of numbers to which $\frac{48}{6}$ belongs.**

$\frac{48}{6} = 8$, so $\frac{48}{6}$ is a natural number, a whole number, an integer, and a rational number.

Example 14 **Graph $x > -1$.**

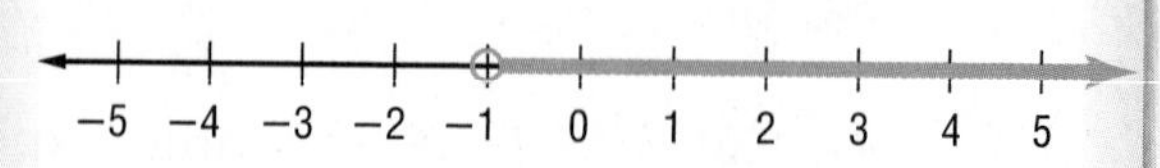

Example 15 **Order $-\frac{29}{4}$, $\sqrt{7}$, and -7.85 from least to greatest.**

$-\frac{29}{4} = -7.25$ Write each number as a decimal and compare.

$\sqrt{7} \approx 2.646$

So, the order is -7.85, $-\frac{29}{4}$, and $\sqrt{7}$.

1-9 Functions and Graphs (pp. 53–58)

For Exercises 61 and 62, use the graph in Example 16.

61. Name the ordered pair at point A and explain what it represents.

62. Identify the independent and dependent variables for the function. Explain.

63. ALTITUDE Identify the graph that represents the altitude of an airplane taking off, flying for a while, then landing.

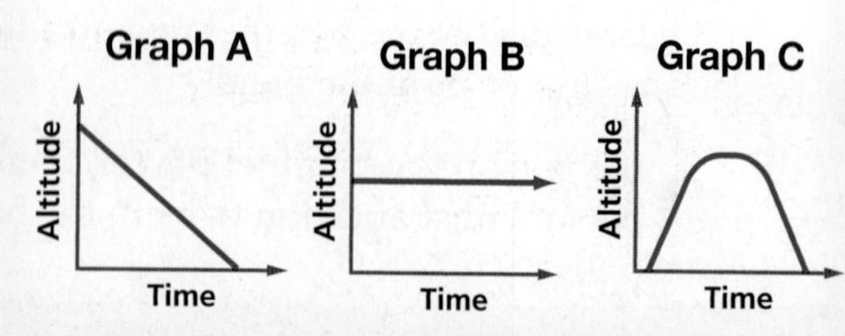

Example 16 **Name the ordered pair at point B and explain what it represents.**

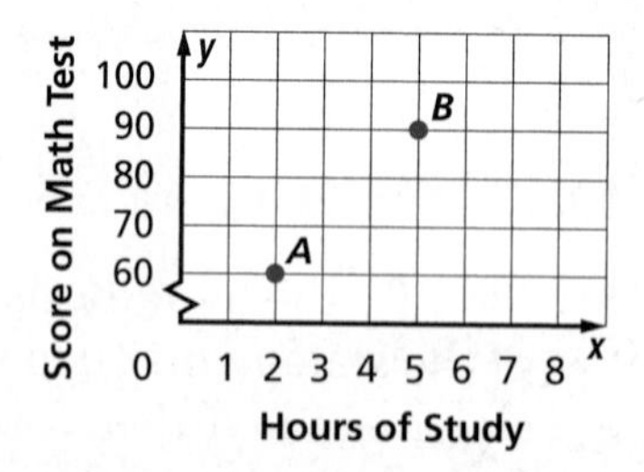

Point B is at 5 along the x-axis and 90 along the y-axis. So, its ordered pair is (5, 90). This represents a score of 90 on the math test with 5 hours of study.

CHAPTER 1

Practice Test

Write an algebraic expression for each verbal expression.

1. the sum of a number x and 13
2. 25 increased by the product of 2 and a number
3. 7 less a number y squared

Simplify each expression.

4. $5(9 + 3) - 3 \cdot 4$
5. $12 \cdot 6 \div 3 \cdot 2 \div 8$

Evaluate each expression if $a = 2$, $b = 5$, $c = 3$, and $d = 1$.

6. $a^2b + c$
7. $(cd)^3$
8. $(a + d)c$

9. **MULTIPLE CHOICE** Tia owns a dog grooming business. How much did she earn in one week if she groomed 14 small dogs, 11 medium-size dogs, and 3 large dogs?

Dog Grooming	
Size of Dog	**Cost ($)**
small	25.00
medium	27.50
large	36.50

A $117.00

B $636.65

C $700.00

D $762.00

Solve each equation.

10. $y = (4.5 + 0.8) - 3.2$
11. $4^2 - 3(4 - 2) = x$
12. Evaluate $3^2 - 2 + (2 - 2)$. Name the property used in each step.

Rewrite each expression in simplest form.

13. $4x + 2y - 2x + y$
14. $3(2a + b) - 5a + 4b$

Find a counterexample for each conditional statement.

15. **EXERCISE** If you run 15 minutes today, then you will be able to run a marathon.
16. If $x \leq 6$, then $2x - 3 < 9$.

Replace each ● with <, >, or = to make each sentence true.

17. $\sqrt{43}$ ● $6.\overline{5}$
18. $\frac{1}{10}$ ● $\sqrt{10}$
19. $-\sqrt{7}$ ● $-\frac{5}{2}$
20. $0.\overline{36}$ ● $\frac{4}{11}$

21. **MULTIPLE CHOICE** Which number serves as a counterexample to the statement below?

If n is a prime number, then n is odd.

F 5

G 4

H 3

J 2

22. **MULTIPLE CHOICE** Riders must be at least 52 inches tall to ride the newest ride at an amusement park. Which graph represents the heights of these riders?

A 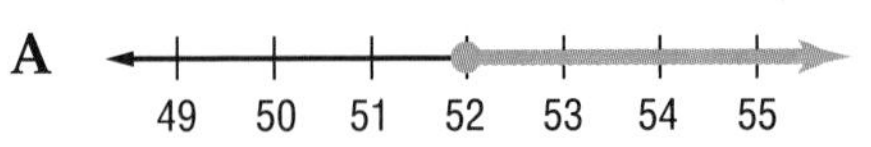

B
49 50 51 52 53 54 55

C
49 50 51 52 53 54 55

D
49 50 51 52 53 54 55

Sketch a reasonable graph for each situation.

23. A basketball is shot from the free throw line and falls through the net.
24. A nickel is dropped on a stack of pennies and bounces off.

CHAPTER 1

Standardized Test Practice

Chapter 1

Read each question. Then fill in the correct answer on the answer document provided by your teacher or on a sheet of paper.

1. Karla has a backyard play area that has the dimensions shown below. She is expanding the play area by adding x feet to the 60 foot side. The area of the addition is $50x$. How can the area of the new play area be expressed in terms of x?

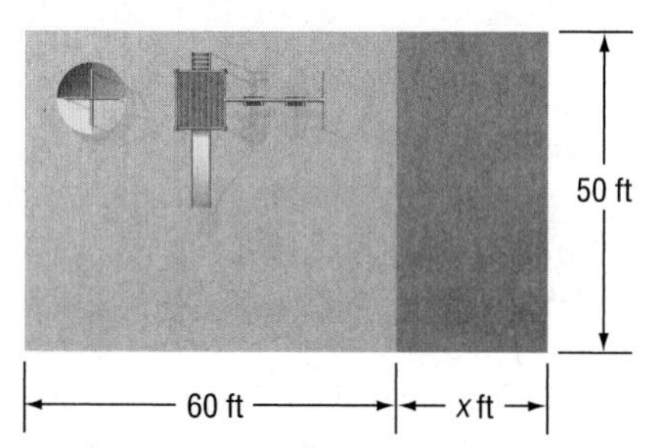

A $3000 + 50x$
B $3000 - 50x$
C $60 + 50x$
D $60 - 50x$

2. **GRIDDABLE** When $-3(4m^2 - 14)$ is simplified, what is the value of the constant term?

3. Laura makes \$8 an hour working at a day care center. 20% of her income is deducted for taxes. She needs to make more than \$288 each month to make her car payment. Solve the inequality $8x - 0.20(8x) > 288$ to determine how many hours per month she must work.

F more than 45 hours
G less than 45 hours
H more than 36 hours
J less than 36 hours

TEST-TAKING TIP

Question 3 Some multiple-choice questions ask you to solve an equation or inequality. You can check your solution by replacing the variable in the equation or inequality with your answer. The answer choice that results in a true statement is the correct answer.

4. The triangle in the graph is to be dilated by a scale factor of $\frac{1}{2}$.

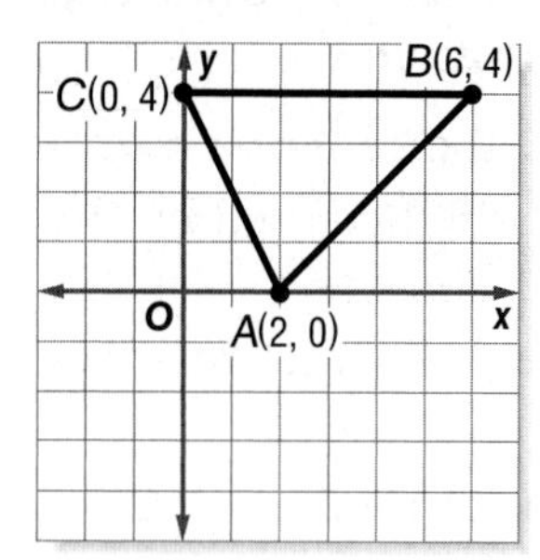

Which graph shows this transformation?

A

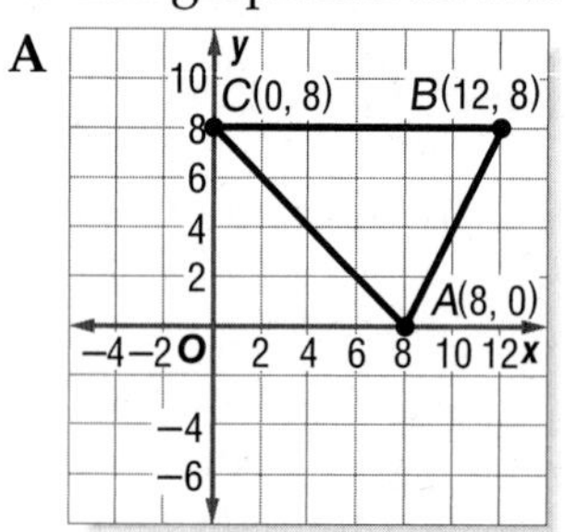

B

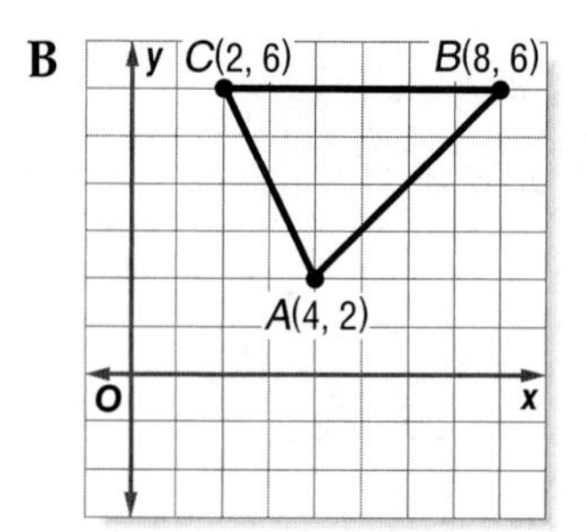

C

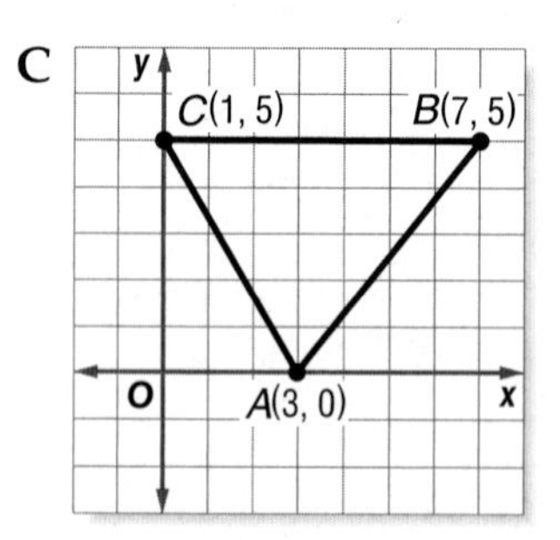

D

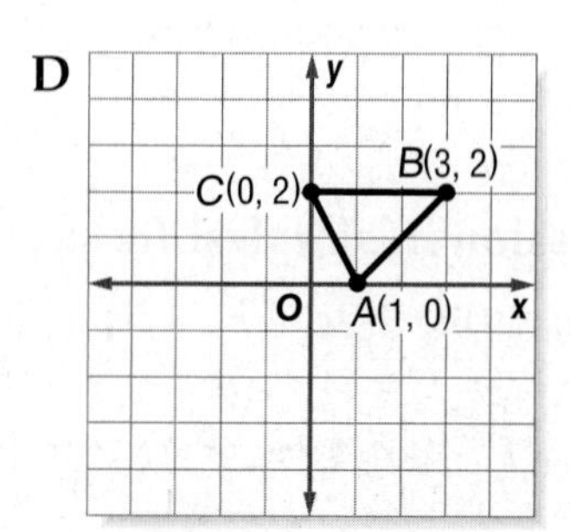

For test-taking strategies and more practice, see pages 756–773.

5. Connor is redecorating his living room. He has budgeted \$1200 for the project, and has already spent \$420 on paint. If he wants to spend the rest of the money on curtains to cover 5 windows, which expression represents the maximum amount w he can spend on each window?

F $1200 = 420 - 5w$

G $420 = 1200 + 5w$

H $1200 = 420 + 5w$

J $420 = 1200 + 5 - w$

6. Which expression below illustrates the Commutative Property?

A $(x + y) + 2 = x + (y + 2)$

B $-3(a + b) = -3a + -3b$

C $qrs = rqs$

D $2 + 0 = 2$

7. The net of a cube is shown below.

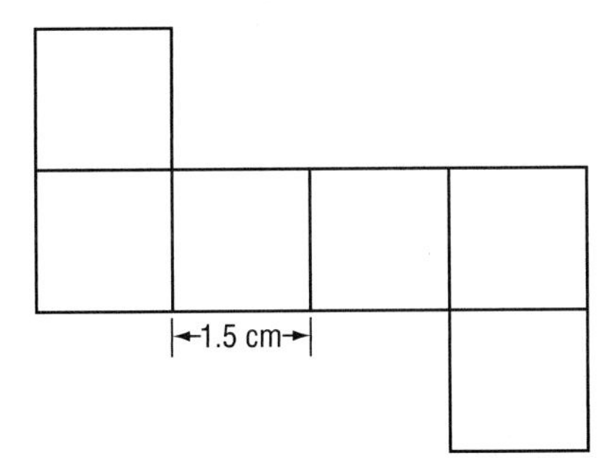

Which best represents the volume of this cube to the nearest cubic centimeter?

F 3 cm^3

G 8 cm^3

H 14 cm^3

J 42 cm^3

8. Millie's Market sells peanuts in two different containers. The first container is a rectangular prism that has a height of 6 inches and a square base with a side length of 3 inches. The other container is a cylinder with a radius of 1.5 inches and a height of 6 inches. Which best describes the relationship between the two containers?

A The prism has the greater volume.

B The cylinder has the greater volume.

C The volumes are equivalent.

D The volumes cannot be determined.

9. GRIDDABLE Hugo's deck is shaped like a rectangle with a width of 15 feet and a length of 20 feet. Staining the deck will cost \$1.25 per square foot. How much money will it cost to stain the deck?

Pre-AP

Record your answers on a sheet of paper. Show your work.

10. The Lee family is going to play miniature golf. The family is composed of two adults and four children.

	Greens Fees	
	Before 6 P.M.	After 6 P.M.
Adult (a)	\$5.00	\$6.50
Children (c)	\$3.00	\$4.50

a. Write an inequality to show the cost for the family to play miniature golf if they don't want to spend more than \$30.

b. How much will it cost the family to play after 6 P.M.?

c. How much will it cost the family to play before 6 P.M.?

NEED EXTRA HELP?										
If You Missed Question...	1	2	3	4	5	6	7	8	9	10
Go to Lesson or Page...	1-1	1-5	1-5	10-6	1-3	1-2	708	708	704	1-3

CHAPTER 2

Solving Linear Equations

BIG Ideas

- Translate verbal sentences into equations and equations into verbal sentences.
- Solve equations and proportions.
- Find percents of change.
- Solve equations for given variables.
- Solve mixture problems and uniform motion problems.

Key Vocabulary

equivalent equations (p. 79)

identity (p. 99)

proportion (p. 105)

percent of change (p. 111)

Real-World Link

Baseball Linear equations can be used to solve problems in every facet of life. For example, a baseball player's slugging percentage is found by using an equation based on a mixture, or weighted average, of five factors, including at bats and hits.

FOLDABLES™ Study Organizer

Solving Linear Equations Make this Foldable to help you organize information about solving linear equations. Begin with 5 sheets of plain $8\frac{1}{2}$" by 11" paper.

1. **Fold** in half along the width.
2. **Fold** the bottom to form a pocket. Glue edges.
3. **Repeat** four times and glue all five pieces together.

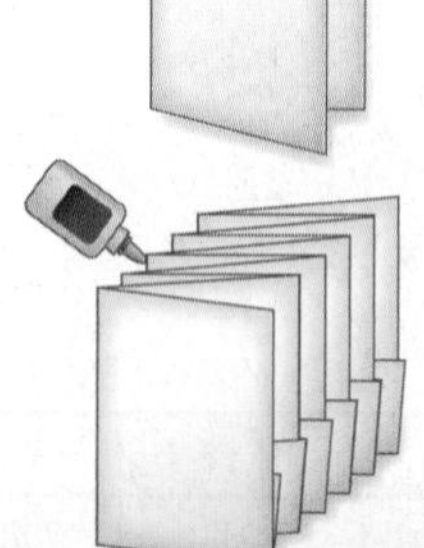

4. **Label** each pocket. Place an index card in each pocket.

GET READY for Chapter 2

Diagnose Readiness You have two options for checking Prerequisite Skills.

Option 2

Math Online Take the Online Readiness Quiz at algebra1.com

Option 1

Take the Quick Check below. Refer to the Quick Review for help.

QUICKCheck

Write an algebraic expression for each verbal expression. (Lesson 1-1)

1. five greater than half a number t
2. the product of seven and s divided by the product of eight and y
3. the sum of three times a and the square of b
4. w to the fifth power decreased by 37

Evaluate each expression. (Lesson 1-2)

5. $3 \cdot 6 - \frac{12}{4}$
6. $5(13 - 7) - 22$
7. $5(7 - 2) - 3^2$
8. $\frac{2 \cdot 6 - 4}{2}$
9. $(25 - 4)(2^2 - 1)$
10. $36 \div 4 - 2 + 3$
11. $\frac{19 - 5}{7} + 3$
12. $\frac{1}{4}(24) - \frac{1}{2}(12)$
13. **MONEY** For his birthday, Brad received $25 from each of his four aunts. After putting 10% of this money in the bank, Brad bought 5 CDs at $10 each. How much money does he have left?

Find each percent. (Prerequisite Skill)

14. Five is what percent of 20?
15. What percent of 5 is 15?
16. What percent of 300 is 21?
17. **ELECTIONS** In an exit poll, 456 of 900 voters said that they voted for Dunlap. Approximately what percentage of voters polled voted for Dunlap?

QUICKReview

EXAMPLE 1

Write an algebraic expression for the phrase *two more than five times a number t.*

two	more than	five	times	a number t
2	+	5	×	t

The expression is $2 + 5t$.

EXAMPLE 2

Evaluate $7 - \frac{6}{3}[5(17 - 6) + 2] + 108$.

$= 7 - \frac{6}{3}[5(17 - 6) + 2] + 108$

$= 7 - \frac{6}{3}[5(11) + 2] + 108$ Evaluate inside the parentheses.

$= 7 - \frac{6}{3}[55 + 2] + 108$ Multiply.

$= 7 - \frac{6}{3}[57] + 108$ Add inside the brackets.

$= 7 - 2[57] + 108$ Simplify.

$= 7 - 114 + 108$ Multiply.

$= 1$ Simplify.

EXAMPLE 3

What percent of 64 is 6?

$\frac{a}{b} = \frac{p}{100}$ Use the percent proportion.

$\frac{6}{64} = \frac{p}{100}$ Replace a with 6 and b with 64.

$6(100) = 64p$ Find the cross products.

$600 = 64p$ Multiply.

$9.375 = p$ Divide each side by 64.

6 is 9.375% of 64.

2-1 Writing Equations

Main Ideas

- Translate verbal sentences into equations.
- Translate equations into verbal sentences.

New Vocabulary

four-step problem-solving plan
defining a variable
formula

GET READY for the Lesson

The Statue of Liberty stands on a pedestal that is 154 feet high. The height of the pedestal and the statue is 305 feet. You can write an equation to represent this situation.

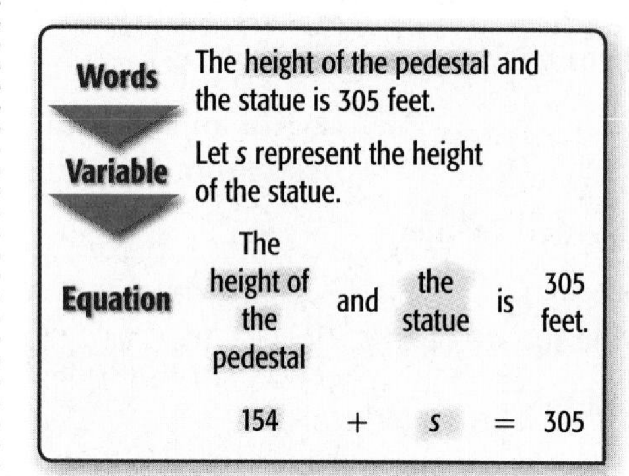

Words	The height of the pedestal and the statue is 305 feet.						
Variable	Let s represent the height of the statue.						
Equation	The height of the pedestal	and	the statue	is	305 feet.		
	154	$+$	s	$=$	305		

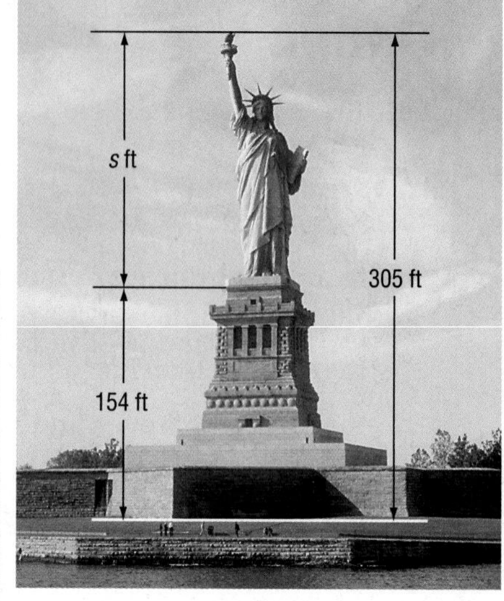

Source: *World Book Encyclopedia*

Write Equations When writing equations, use variables to represent the unspecified numbers or measures. Then write the verbal expressions as algebraic expressions. Some verbal expressions that suggest the *equals sign* are listed below.

is is as much as is the same as is identical to

Study Tip

Look Back

To review **translating verbal expressions to algebraic expressions**, see Lesson 1-1.

EXAMPLE Translate Sentences into Equations

1 Translate each sentence into an equation.

a. Five times the number a squared is three times the sum of b and c.

Five	times	a squared	is	three	times	the sum of b and c.
5	$\cdot$	a^2	$=$	3	$\cdot$	$(b + c)$

The equation is $5a^2 = 3(b + c)$.

b. Nine times a number subtracted from 95 equals 37.
Rewrite the sentence so it is easier to translate. Let $n =$ the number.

95	less	nine times n	equals	37.
95	$-$	$9n$	$=$	37

The equation is $95 - 9n = 37$.

CHECK Your Progress

1A. Two plus the quotient of a number and 8 is the same as 16.

1B. Twenty-seven times k is h squared decreased by 9.

Using the **four-step problem-solving plan** can help you solve any word problem.

KEY CONCEPT *Four-Step Problem-Solving Plan*

Step 1 Explore the problem.

Step 2 Plan the solution.

Step 3 Solve the problem.

Step 4 Check the solution.

Reading Math

Verbal Problems
In a verbal problem, the sentence that tells what you are asked to find usually contains *find, what, when,* or *how*.

Step 1 **Explore the Problem**
To solve a verbal problem, first read the problem carefully and explore what the problem is about.

- Identify what information is given.
- Identify what you are asked to find.

Step 2 **Plan the Solution**
One strategy you can use to solve a problem is to write an equation. Choose a variable to represent one of the unspecified numbers in the problem. This is called **defining a variable**. Then use the variable to write expressions for the other unspecified numbers.

Step 3 **Solve the Problem**
Use the strategy you chose in Step 2 to solve the problem.

Step 4 **Check the Solution**
Check your answer in the context of the original problem.

- Does your answer make sense?
- Does it fit the information in the problem?

Real-World Link
The first ice cream plant was established in 1851 by Jacob Fussell. Today, 2,000,000 gallons of ice cream are produced in the United States each day.

Source: *World Book Encyclopedia*

Real-World EXAMPLE **Use the Four-Step Plan**

2 **ICE CREAM Use the information at the left. In how many days can 40,000,000 gallons of ice cream be produced in the United States?**

Explore You know that 2,000,000 gallons of ice cream are produced in the United States each day. You want to know how many days it will take to produce 40,000,000 gallons of ice cream.

Plan Write an equation. Let d represent the number of days.

2,000,000	times	the number of days	equals	40,000,000.
2,000,000	$\cdot$	d	$=$	40,000,000

Solve $2{,}000{,}000d = 40{,}000{,}000$ Find d mentally by asking, "What number times 2,000,000 equals 40,000,000?"
$d = 20$

It will take 20 days to produce 40,000,000 gallons of ice cream.

Check If 2,000,000 gallons of ice cream are produced in one day, $2{,}000{,}000 \cdot 20$ or 40,000,000 gallons are produced in 20 days. The answer makes sense.

2. **GOVERNMENT** There are 50 members in the North Carolina Senate. This is 70 fewer than the number in the North Carolina House of Representatives. How many members are in the North Carolina House of Representatives?

A **formula** is an equation that states a rule for the relationship between certain quantities.

ALGEBRA LAB

Surface Area

- Mark each side of a rectangular box as the length ℓ, the width w, or the height h.
- Use scissors to cut the box so that each surface or face of the box is a separate piece.

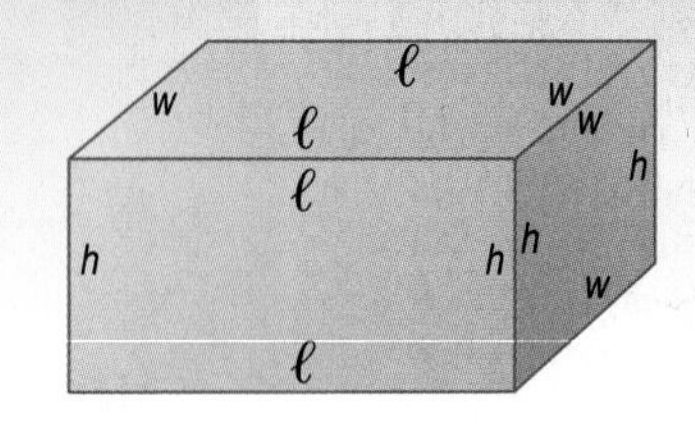

ANALYZE

Write an expression for the area of each side of the box.

1. front	**2.** back	**3.** left side
4. right side	**5.** top	**6.** bottom

7. The surface area S of a rectangular box is the sum of all the areas of the faces of the box. Write a formula for the surface area of a rectangular box.

MAKE A CONJECTURE

8. If s represents the length of the side of a cube, write a formula for the surface area S of a cube.

EXAMPLE Write a Formula

3 **Translate the sentence into a formula.** *The perimeter of a rectangle equals two times the length plus two times the width.*

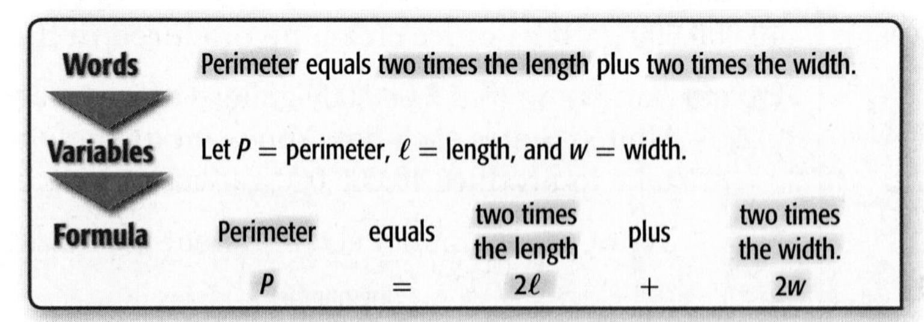

Words	Perimeter equals two times the length plus two times the width.				
Variables	Let P = perimeter, ℓ = length, and w = width.				
Formula	Perimeter	equals	two times the length	plus	two times the width.
	P	$=$	2ℓ	$+$	$2w$

The formula for the perimeter of a rectangle is $P = 2\ell + 2w$.

CHECK Your Progress

3. **GEOMETRY** Translate the sentence into a formula. *In a right triangle, the square of the measure of the hypotenuse c is equal to the sum of the squares of the measures of the legs, a and b.*

Online **Personal Tutor at** algebra1.com

Write Verbal Sentences You can also translate equations into verbal sentences or make up your own verbal problem if you are given an equation.

EXAMPLE Translate Equations into Sentences

4 **Translate each equation into a verbal sentence.**

a. $3m + 5 = 14$

$3m$	$+$	5	$=$	14
Three times m	plus	five	equals	fourteen.

b. $w + v = y^2$

$w + v$	$=$	y^2
The sum of w and v	equals	the square of y.

CHECK Your Progress

4A. $13 = 2 + 6t$

4B. $\frac{1}{4}n + 5 = n - 7$

EXAMPLE Write a Problem

5 **Write a problem based on the given information.**

a = Rafael's age **$a + 5$ = Tierra's age** **$a + 2(a + 5) = 46$**

The equation adds Rafael's age a plus twice Tierra's age $(a + 5)$ to get 46.

Sample problem:
Tierra is 5 years older than Rafael. The sum of Rafael's age and twice Tierra's age equals 46. How old is Rafael?

CHECK Your Progress

5. p = price of jeans $0.2p$ = discount $p - 0.2p = 31.20$

CHECK Your Understanding

Example 1 (p. 70)

Translate each sentence into an equation.

1. Twice a number t decreased by eight equals seventy.

2. Five times the sum of m and n is the same as seven times n.

3. Half of p is the same as p minus three.

4. A number squared is as much as twelve more than the number.

Example 2 (p. 71)

5. **SAVINGS** Misae has $1900 in the bank. She wishes to increase her account to a total of $2500 by depositing $30 per week from her paycheck. Write and use an equation to find how many weeks she needs to reach her goal.

Example 3 (p. 72)

Translate each sentence into a formula.

6. The area A of a triangle equals one half times the base b times the height h.

7. The circumference C of a circle equals the product of 2, π, and the radius r.

Example 4 (p. 72)

Translate each equation into a verbal sentence.

8. $1 = 7 + \frac{z}{5}$

9. $14 + d = 6d$

10. $\frac{1}{3}b - 4 = 2a$

Example 5 (p. 73)

Write a problem based on the given information.

11. c = cost of a suit
$c - 25 = 150$

12. p = price of a new backpack
$0.055p$ = tax; $p + 0.055p = 31.65$

Exercises

HOMEWORK HELP	
For Exercises	See Examples
13–20	1
21–22	2
23–26	3
27–32	4
33–34	5

Translate each sentence into an equation.

13. The sum of twice r and three times s is identical to thirteen.

14. The quotient of t and forty is the same as twelve minus half of s.

15. Two hundred minus three times a number is equal to nine.

16. The sum of one-third a number and 25 is as much as twice the number.

17. The square of m minus the cube of n is sixteen.

18. Two times z is equal to two times the sum of v and w.

19. GEOGRAPHY The Pacific Ocean covers about 46% of Earth. If P represents the area of the Pacific Ocean and E represents the area of Earth, write an equation for this situation.

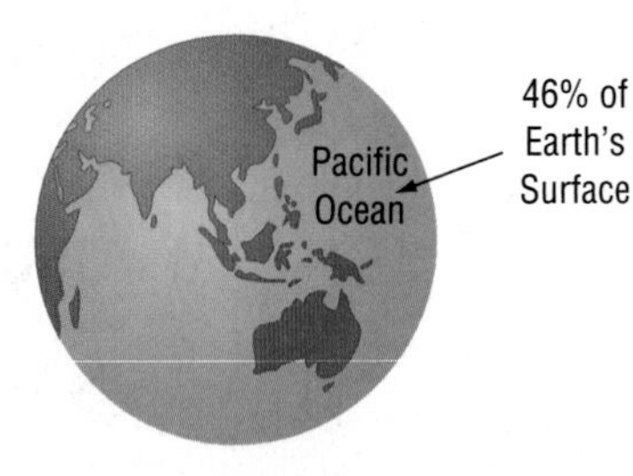

Source: *World Book Encyclopedia*

20. GARDENING Mrs. Patton is planning to place a fence around her vegetable garden. The fencing costs $1.75 per yard. She buys f yards of fencing and pays $3.50 in tax. If the total cost is $73.50, write an equation to represent the situation.

Real-World Link

More than 50 movies featuring Tarzan have been made. The first, *Tarzan of the Apes*, in 1918, was among the first movies to gross over $1 million.

Source: www.tarzan.org

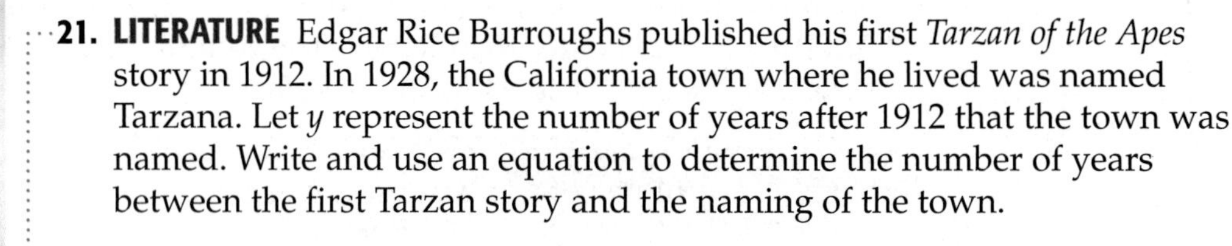

21. LITERATURE Edgar Rice Burroughs published his first *Tarzan of the Apes* story in 1912. In 1928, the California town where he lived was named Tarzana. Let y represent the number of years after 1912 that the town was named. Write and use an equation to determine the number of years between the first Tarzan story and the naming of the town.

22. WRESTLING Darius weighs 155 pounds. He wants to start the wrestling season weighing 160 pounds. If g represents the number of pounds he wants to gain, write an equation to represent this situation. Then use the equation to find the number of pounds Darius needs to gain.

Translate each sentence into a formula.

23. The area A of a parallelogram is the base b times the height h.

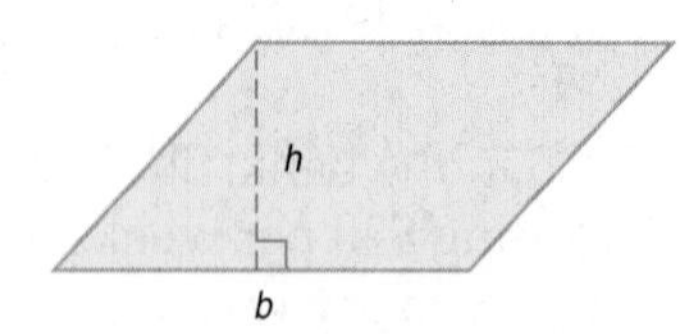

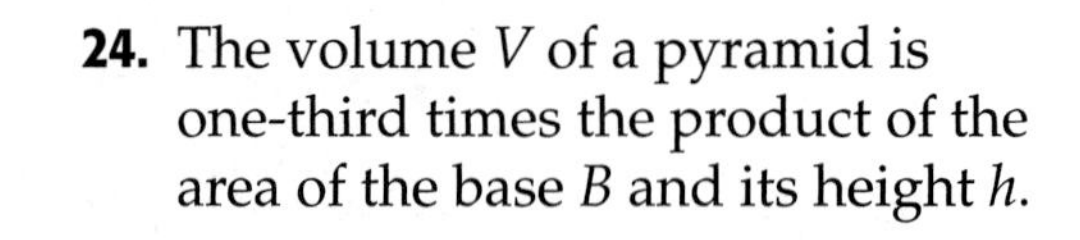

24. The volume V of a pyramid is one-third times the product of the area of the base B and its height h.

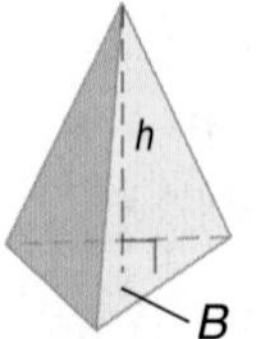

25. The perimeter P of a parallelogram is twice the sum of the lengths of the two adjacent sides, a and b.

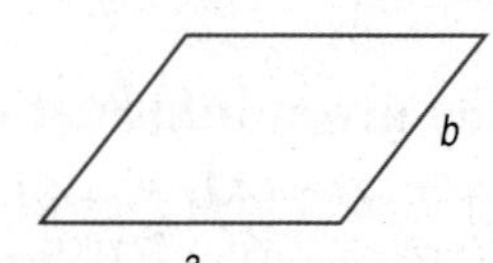

26. The volume V of a cylinder equals the product of π, the square of the radius r of the base, and the height.

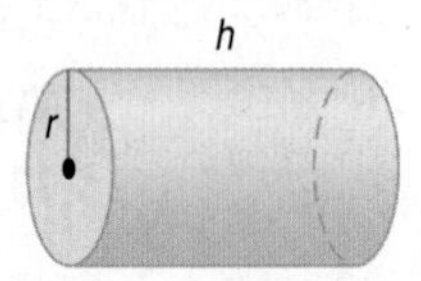

Translate each equation into a verbal sentence.

27. $d - 14 = 5$

28. $2f + 6 = 19$

29. $k^2 + 17 = 53 - j$

30. $2a = 7a - b$

31. $\frac{3}{4}p + \frac{1}{2} = p$

32. $\frac{2}{5}w = \frac{1}{2}w + 3$

Write a problem based on the given information.

33. y = Yolanda's height in inches
$y + 7$ = Lindsey's height in inches
$2y + (y + 7) = 193$

34. b = price of a book
$0.065b$ = tax
$2(b + 0.065b) = 42.49$

Real-World Link

A typical half-hour show runs for 22 minutes and has 8 minutes of commercials.

Source: gaebler.com

Translate each sentence into an equation.

35. Half the sum of nine and a number is the same as the number minus three.

36. The quotient of g and h is the same as seven more than twice the sum of g and h.

GEOMETRY For Exercises 37 and 38, use the following information.
The volume V of a cone equals one third times the product of π, the square of the radius r of the base, and the height h.

37. Write the formula for the volume of a cone.

38. Find the volume of a cone if r is 10 centimeters and h is 30 centimeters.

GEOMETRY For Exercises 39 and 40, use the following information.
The volume V of a sphere is four thirds times π times the radius r of the sphere cubed.

39. Write a formula for the volume of a sphere.

40. Find the volume of a sphere if r is 4 inches.

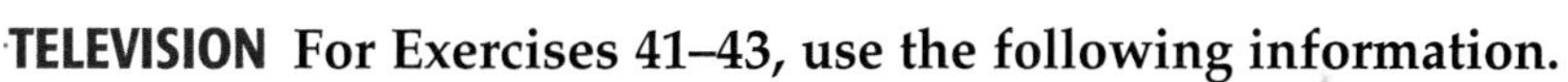

TELEVISION For Exercises 41–43, use the following information.
During a highly rated one-hour television show, the entertainment portion lasted 15 minutes longer than 4 times the advertising portion a.

41. Write an expression for the entertainment portion.

42. Write an equation to represent the situation.

43. Use the guess-and-check strategy to determine the number of minutes spent on advertising.

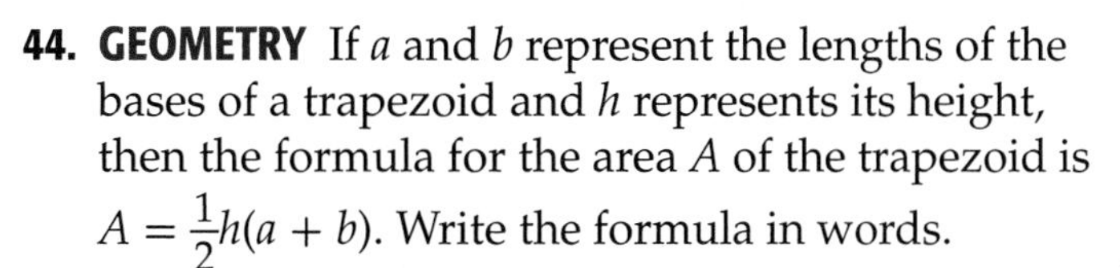

44. GEOMETRY If a and b represent the lengths of the bases of a trapezoid and h represents its height, then the formula for the area A of the trapezoid is $A = \frac{1}{2}h(a + b)$. Write the formula in words.

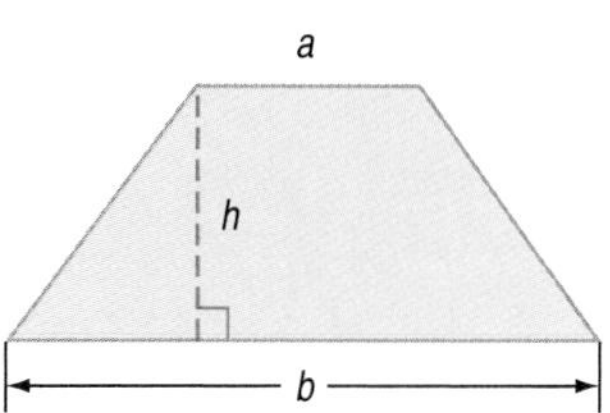

Translate each equation into a verbal sentence.

45. $4(t - s) = 5s + 12$

46. $7(x + y) = 35$

H.O.T. Problems

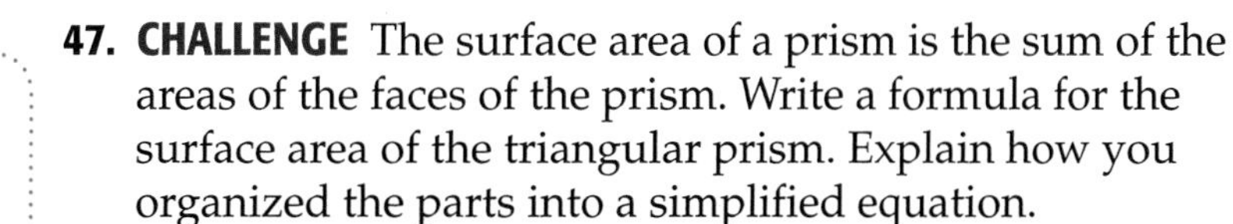

47. CHALLENGE The surface area of a prism is the sum of the areas of the faces of the prism. Write a formula for the surface area of the triangular prism. Explain how you organized the parts into a simplified equation.

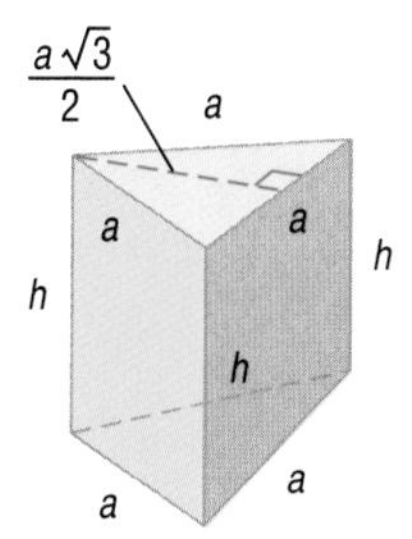

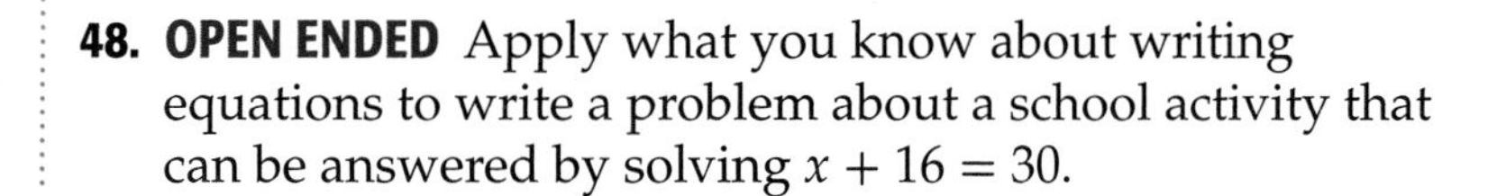

48. OPEN ENDED Apply what you know about writing equations to write a problem about a school activity that can be answered by solving $x + 16 = 30$.

49. ***Writing in Math*** Use the information about the Statue of Liberty on page 70 to explain how equations are used to describe heights. Include an equation relating the heights of the Sears Tower, which is 1454 feet tall, the two antenna towers on top of the building, which are a feet tall, and the total height, which is 1707 feet.

STANDARDIZED TEST PRACTICE

50. Which equation *best* represents the relationship between the number of hours an electrician works h and the total charges c?

Cost of Electrician	
emergency house call	$30 one-time fee
rate	$55/hour

A $c = 30 + 55$ **C** $c = 30 + 55h$

B $c = 30h + 55$ **D** $c = 30h + 55h$

51. REVIEW A car traveled at 55 miles per hour for 2.5 hours and then at 65 miles per hour for 3 hours. How far did the car travel in all?

F 300.5 mi

G 305 mi

H 330 mi

J 332.5 mi

Spiral Review

52. ENTERTAINMENT Juanita has the volume on her stereo turned up. When her telephone rings, she turns the volume down. After she gets off the phone, she returns the volume to its previous level. Identify which graph shows the volume of Juanita's stereo during this time. (Lesson 1-9)

Graph A

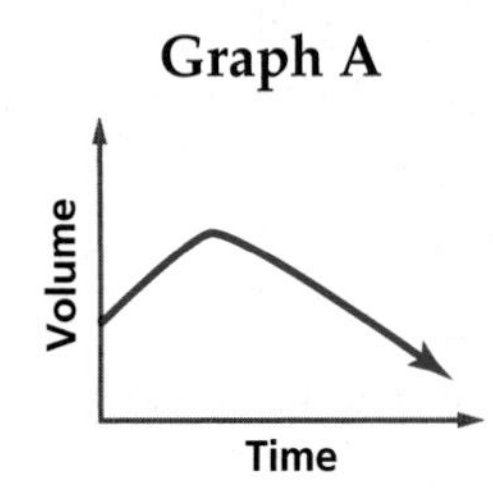

Graph B

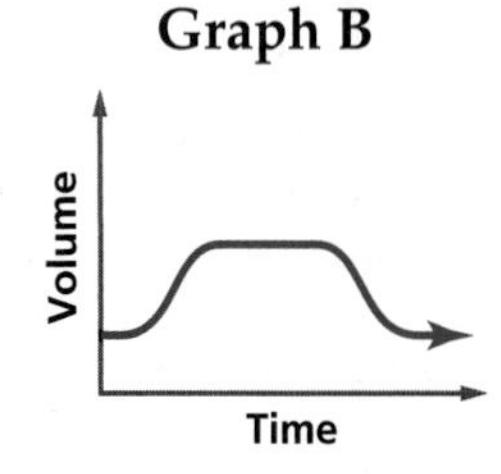

Graph C

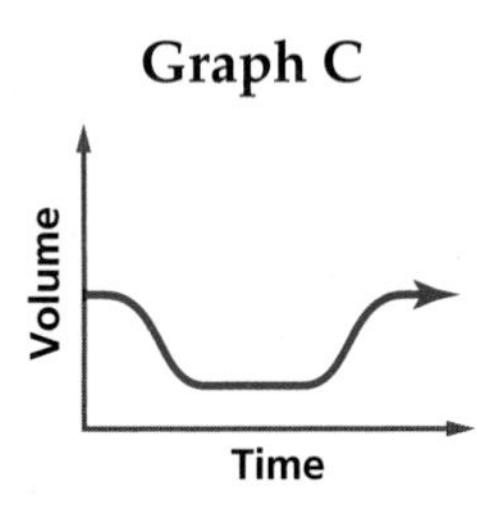

Find each square root. If necessary, round to the nearest hundredth. (Lesson 1-8)

53. $\sqrt{8100}$ **54.** $-\sqrt{\frac{25}{36}}$ **55.** $\sqrt{90}$ **56.** $-\sqrt{55}$

Simplify each expression. (Lesson 1-5)

57. $12d + 3 - 4d$ **58.** $7t^2 + t + 8t$ **59.** $3(a + 2b) + 5a$

Evaluate each expression. (Lesson 1-2)

60. $5(8 - 3) + 7 \cdot 2$ **61.** $6(4^3 + 2)$ **62.** $7(0.2 + 0.5) - 0.6$

GET READY for the Next Lesson

PREREQUISITE SKILL Find each sum or difference. (Pages 694–697)

63. $0.57 + 2.8$ **64.** $5.28 - 3.4$ **65.** $9 - 7.35$

66. $\frac{2}{3} + \frac{1}{5}$ **67.** $\frac{1}{6} + \frac{2}{3}$ **68.** $\frac{7}{9} - \frac{2}{3}$

EXPLORE 2-2

Algebra Lab
Solving Addition and Subtraction Equations

You can use algebra tiles to solve equations. To solve an equation means to find the value of the variable that makes the equation true. After you model the equation, the goal is to get the x-tile by itself on one side of the mat using the rules stated below.

Reading Math

Variables You *isolate* the variable by getting it by itself on one side of the equation.

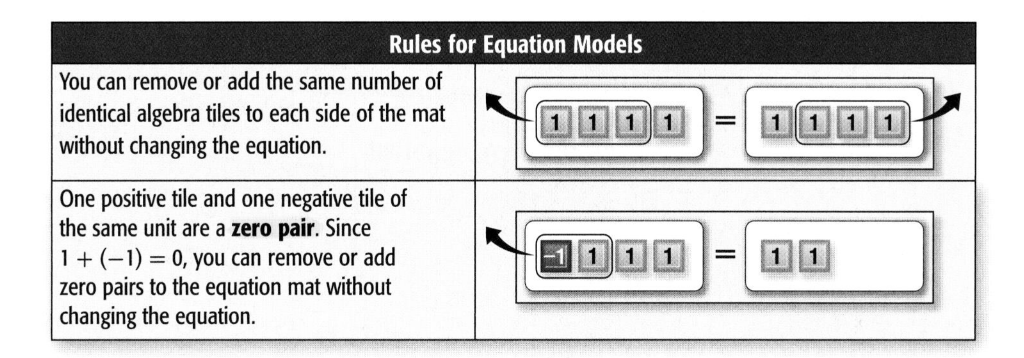

Rules for Equation Models	
You can remove or add the same number of identical algebra tiles to each side of the mat without changing the equation.	
One positive tile and one negative tile of the same unit are a **zero pair**. Since $1 + (-1) = 0$, you can remove or add zero pairs to the equation mat without changing the equation.	

Concepts in Motion
Animation algebra1.com

ACTIVITY

Use an equation model to solve $x - 3 = 2$.

Step 1 Model the equation.

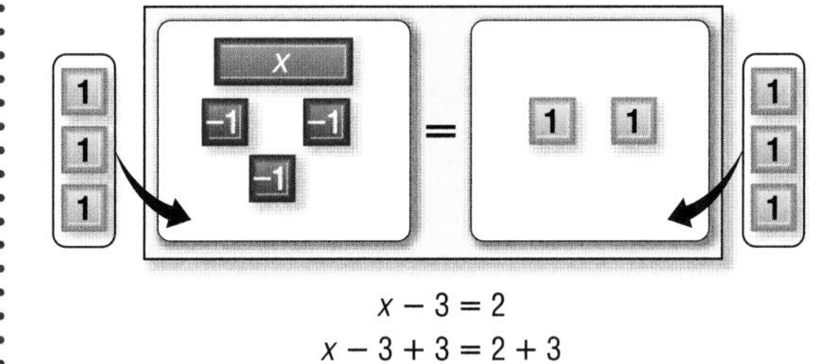

$x - 3 = 2$
$x - 3 + 3 = 2 + 3$

Place 1 x-tile and 3 negative 1-tiles on one side of the mat. Place 2 positive 1-tiles on the other side of the mat. Then add 3 positive 1-tiles to each side.

Step 2 Isolate the x-term.

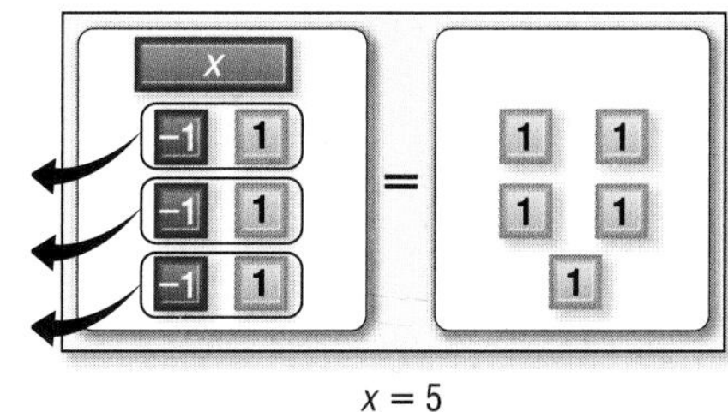

$x = 5$

Group the tiles to form zero pairs. Then remove all the zero pairs. The resulting equation is $x = 5$.

Model and Analyze

Use algebra tiles to solve each equation.

1. $x + 5 = 7$
2. $x + (-2) = 8$
3. $x + 4 = 5$
4. $x + (-3) = 4$
5. $x + 3 = -4$
6. $x + 7 = 2$

Make a Conjecture

7. If $a = b$, what can you say about $a + c$ and $b + c$?
8. If $a = b$, what can you say about $a - c$ and $b - c$?

2-2 Solving Equations by Using Addition and Subtraction

Main Ideas

- Solve equations by using addition.
- Solve equations by using subtraction.

New Vocabulary

equivalent equations
solve an equation

Study Tip

Look Back

To review **writing algebraic expressions,** see Lesson 1-1.

GET READY for the Lesson

The graph shows some of the fastest-growing occupations from 1992 to 2005.

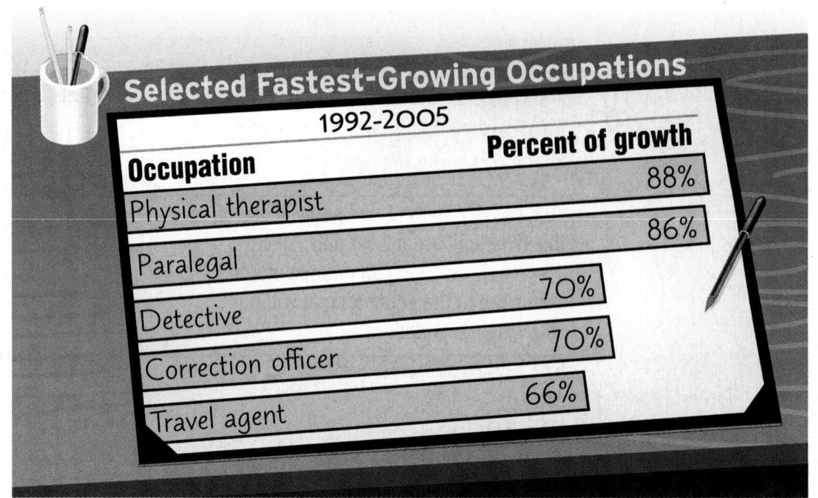

Source: Bureau of Labor Statistics

The percent of growth for travel agents is 5 less than the percent of growth for medical assistants. An equation can be used to find the percent of growth expected for medical assistants. If m is the percent of growth for medical assistants, then $66 = m - 5$. You can use a property of equality to find the value of m.

Solve Using Addition Suppose the boys' soccer team has 15 members and the girls' soccer team has 15 members. If each team adds 3 new players, the number of members on the boys' and girls' teams would still be equal.

$15 = 15$	Each team has 15 members before adding the new players.
$15 + 3 = 15 + 3$	Each team adds 3 new members.
$18 = 18$	Each team has 18 members after adding the new members.

This example illustrates the **Addition Property of Equality**.

KEY CONCEPT — *Addition Property of Equality*

Words If an equation is true and the same number is added to each side, the resulting equation is true.

Symbols For any numbers a, b, and c, if $a = b$, then $a + c = b + c$.

Examples

$$7 = 7 \qquad 14 = 14$$
$$7 + 3 = 7 + 3 \qquad 14 + (-6) = 14 + (-6)$$
$$10 = 10 \qquad 8 = 8$$

If the same number is added to each side of an equation, then the result is an equivalent equation. **Equivalent equations** have the same solution.

$t + 3 = 5$	The solution of this equation is 2.
$t + 3 + 4 = 5 + 4$	Using the Addition Property of Equality, add 4 to each side.
$t + 7 = 9$	The solution of this equation is also 2.

Study Tip

Coefficients
Remember that x means $1 \cdot x$. The coefficient of x is 1.

To **solve an equation** means to find all values of the variable that make the equation a true statement. One way to do this is to isolate the variable having a coefficient of 1 on one side of the equation. You can sometimes do this by using the Addition Property of Equality.

EXAMPLE Solve by Adding

1 Solve each equation. Check your solution.

a. $m - 48 = 29$

$m - 48 = 29$	Original equation
$m - 48 + 48 = 29 + 48$	Add 48 to each side.
$m = 77$	$-48 + 48 = 0$ and $29 + 48 = 77$

To check that 77 is the solution, substitute 77 for m in the original equation.

CHECK	$m - 48 = 29$	Original equation
	$77 - 48 \stackrel{?}{=} 29$	Substitute 77 for m.
	$29 = 29$ ✓	Subtract.

The solution 77 is correct.

b. $21 + q = -18$

$21 + q = -18$	Original equation
$21 + q + (-21) = -18 + (-21)$	Add -21 to each side.
$q = -39$	$21 + (-21) = 0$ and $-18 + (-21) = -39$

The solution is -39. To check, substitute -39 for q in the original equation.

CHECK Your Progress

1A. $32 = r - 8$

1B. $7 = 42 + t$

Solve Using Subtraction Similar to the Addition Property of Equality, the **Subtraction Property of Equality** can also be used to solve equations.

KEY CONCEPT *Subtraction Property of Equality*

Words If an equation is true and the same number is subtracted from each side, the resulting equation is true.

Symbols For any numbers a, b, and c, if $a = b$, then $a - c = b - c$.

Examples

$17 = 17$	$3 = 3$
$17 - 9 = 17 - 9$	$3 - 8 = 3 - 8$
$8 = 8$	$-5 = -5$

EXAMPLE Solve by Subtracting

2 **Solve $142 + d = 97$. Check your solution.**

$142 + d = 97$	Original equation
$142 + d - \mathbf{142} = 97 - \mathbf{142}$	Subtract 142 from each side.
$d = -45$	$142 - 142 = 0$ and $97 - 142 = -45$

The solution is -45. To check, substitute -45 for d in the original equation.

CHECK Your Progress **Solve each equation. Check your solution.**

2A. $27 + k = 30$ **2B.** $-12 = p + 16$

Remember that subtracting a number is the same as adding its inverse.

EXAMPLE Solve by Adding or Subtracting

3 **Solve $g + \frac{3}{4} = -\frac{1}{8}$ in two ways.**

Method 1 Use the Subtraction Property of Equality.

$g + \frac{3}{4} = -\frac{1}{8}$	Original equation
$g + \frac{3}{4} - \frac{3}{4} = -\frac{1}{8} - \frac{3}{4}$	Subtract $\frac{3}{4}$ from each side.
$g = -\frac{7}{8}$	$\frac{3}{4} - \frac{3}{4} = 0$ and $-\frac{1}{8} - \frac{3}{4} = -\frac{1}{8} - \frac{6}{8}$ or $-\frac{7}{8}$

Method 2 Use the Addition Property of Equality.

$g + \frac{3}{4} = -\frac{1}{8}$	Original equation
$g + \frac{3}{4} + \left(-\frac{3}{4}\right) = -\frac{1}{8} + \left(-\frac{3}{4}\right)$	Add $-\frac{3}{4}$ to each side.
$g = -\frac{7}{8}$	$\frac{3}{4} + \left(-\frac{3}{4}\right) = 0$ and $-\frac{1}{8} + \left(-\frac{3}{4}\right) = -\frac{1}{8} + \left(-\frac{6}{8}\right)$ or $-\frac{7}{8}$

CHECK Your Progress

3. Solve $t + 10 = 55$ in two ways.

EXAMPLE Write and Solve an Equation

Study Tip

Checking Solutions

You should always check your solution in the context of the original problem. For instance, in Example 4, is 37 increased by 5 equal to 42? The solution checks.

4 **Write an equation for the problem. Then solve the equation.**

A number increased by 5 is equal to 42. Find the number.

A number	increased by	5	is equal to	42.
n	$+$	5	$=$	42

$n + 5 = 42$	Original equation
$n + 5 - 5 = 42 - 5$	Subtract 5 from each side.
$n = 37$	

The solution is 37.

CHECK Your Progress

4. Twenty-five is 3 less than a number. Find the number.

5 **HISTORY In the fourteenth century, part of the Great Wall of China was repaired and the wall was extended. When the wall was completed, it was 2500 miles long. How much of the wall was added during the 1300s?**

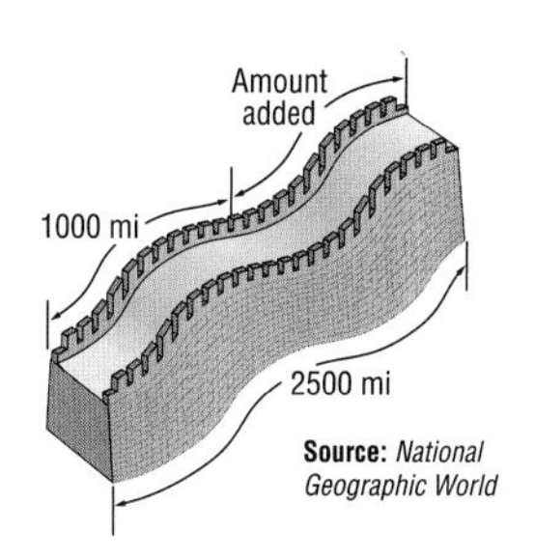

Source: *National Geographic World*

Words	The original length plus the additional length is 2500.				
Variable	Let a = the additional length.				
Equation	The original length	plus	the additional length	is	2500.
	1000	+	a	=	2500

$1000 + a = 2500$ Original equation

$1000 + a - 1000 = 2500 - 1000$ Subtract 1000 from each side.

$a = 1500$ $1000 - 1000 = 0$ and $2500 - 1000 = 1500$

The Great Wall of China was extended 1500 miles in the 1300s.

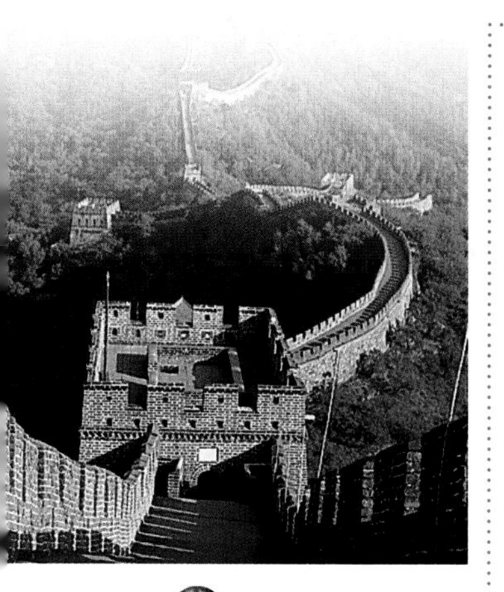

Real-World Link

The first emperor of China, Qui Shi Huangdi, ordered the building of the Great Wall of China to protect his people from nomadic tribes that attacked and looted villages. By 204 B.C., this wall guarded 1000 miles of China's border.

Source: *National Geographic World*

5. **DEER** In a recent year, 1286 female deer were born in Lewis County. That was 93 fewer than the number of male deer born. How many male deer were born that year?

Online Personal Tutor at algebra1.com

CHECK Your Understanding

Examples 1–3 (pp. 79–80)

Solve each equation. Check your solution.

1. $n - 20 = 5$
2. $104 = y - 67$
3. $-4 + t = -7$
4. $g + 5 = 33$
5. $19 + p = 6$
6. $15 = b - (-65)$
7. $h + \frac{2}{5} = \frac{7}{10}$
8. $-6 = \frac{1}{4} + m$
9. $\frac{2}{3} + w = 1\frac{1}{2}$

Example 4 (p. 80)

Write an equation for each problem. Then solve the equation and check your solution.

10. Twenty-one subtracted from a number is −8. Find the number.
11. A number increased by 91 is 37. Find the number.

Example 5 (p. 81)

12. **HISTORY** Over the years, the height of the Great Pyramid at Giza, Egypt, has decreased. Use the figure to write an equation to represent the situation. Then find the decrease in the height of the pyramid.

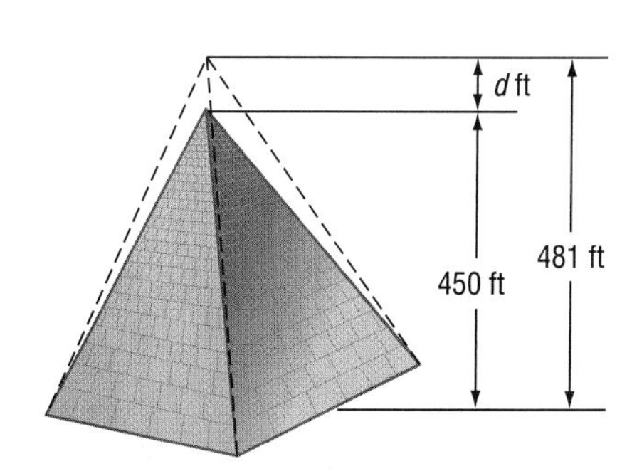

Source: *World Book Encyclopedia*

Exercises

HOMEWORK HELP	
For Exercises	See Examples
13–26	1–3
27–30	4
31, 32	5

Solve each equation. Check your solution.

13. $v - 9 = 14$

14. $44 = t - 72$

15. $-61 = d + (-18)$

16. $p + (-26) = 16$

17. $18 + z = 40$

18. $19 = c + 12$

19. $n + 23 = 4$

20. $-67 = 11 + k$

21. $18 - (-f) = 91$

22. $88 = 125 - (-u)$

23. $\frac{2}{3} + r = -\frac{4}{9}$

24. $\frac{3}{4} = w + \frac{2}{5}$

25. $-\frac{1}{2} + a = \frac{5}{8}$

26. $-\frac{7}{10} = y - \frac{3}{5}$

Write an equation for each problem. Then solve the equation and check your solution.

27. Eighteen subtracted from a number equals 31. Find the number.

28. What number decreased by 77 equals −18?

29. A number increased by −16 is −21. Find the number.

30. The sum of a number and −43 is 102. What is the number?

For Exercises 31 and 32, write an equation for each situation. Then solve the equation.

31. GAS MILEAGE A midsize car with a 4-cylinder engine goes 34 miles on a gallon of gasoline. This is 10 miles more than a luxury car with an 8-cylinder engine goes on a gallon of gasoline. How many miles does a luxury car travel on a gallon of gasoline?

32. IN THE MEDIA The world's biggest-ever passenger plane, the *Airbus A380*, was first used by Singapore Airlines in 2005. The following description appeared on a news Web site after the plane was introduced.

> "That airline will see the A380 transporting some 555 passengers, 139 more than a similarly set-up 747." **Source:** cnn.com

How many passengers does a similarly set-up 747 transport?

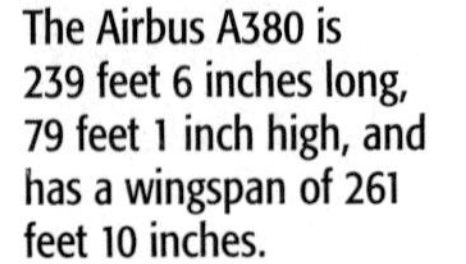

Real-World Link

The Airbus A380 is 239 feet 6 inches long, 79 feet 1 inch high, and has a wingspan of 261 feet 10 inches.

Source: nationalgeographic.com

Solve each equation. Then check your solution.

33. $k + 0.6 = -3.8$

34. $8.5 + t = 7.1$

35. $4.2 = q - 3.5$

36. $q - 2.78 = 4.2$

37. $6.2 = -4.83 + y$

38. $-6 = m + (-3.42)$

Write an equation for each problem. Then solve the equation and check your solution.

39. What number minus one half is equal to negative three fourths?

40. The sum of 19 and 42 and a number is equal to 87. What is the number?

41. **CARS** The average time t it takes to manufacture a car in the United States is 24.9 hours. This is 8.1 hours longer than the average time it takes to manufacture a car in Japan. Write and solve an addition equation to find the average time to manufacture a car in Japan.

42. If $x - 7 = 14$, what is the value of $x - 2$?

43. If $t + 8 = -12$, what is the value of $t + 1$?

ANALYZE GRAPHS For Exercises 44–47, use the graph at the right to write an equation for each situation. Then solve the equation.

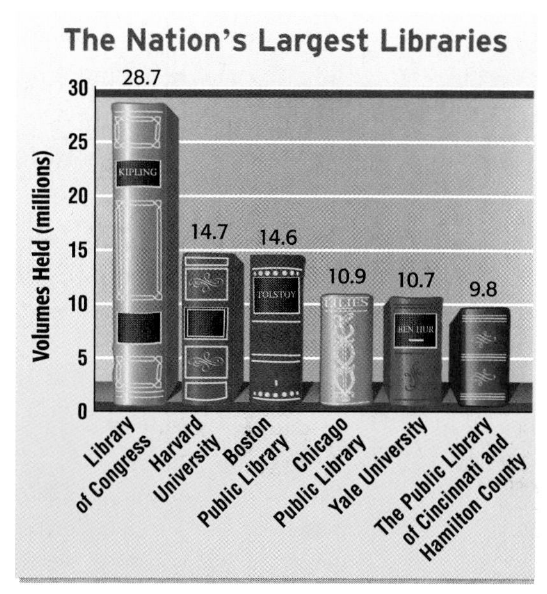

Source: American Library Association

44. How many more volumes does the Library of Congress have than the Harvard University Library?

45. How many more volumes does the Harvard University Library have than the Boston Public Library?

46. How many more volumes does the Library of Congress have than the Boston Public Library?

47. What is the total number of volumes in the three largest U.S. libraries?

EXTRA PRACTICE
See pages 720, 745.
Math Online
Self-Check Quiz at
algebra1.com

H.O.T. Problems

48. **Which One Doesn't Belong?** Identify the equation that does not belong with the other three. Explain your reasoning.

$n + 14 = 27$	$12 + n = 25$	$n - 16 = 29$	$n - 4 = 9$

49. **OPEN ENDED** Write an equation involving addition and demonstrate two ways to solve it.

50. **CHALLENGE** If $a - b = x$, what values of a, b, and x would make the equation $a + x = b + x$ true? Explain your reasoning.

51. **CHALLENGE** Determine whether each sentence is *sometimes*, *always*, or *never* true. Explain your reasoning.
   a. $x + x = x$
   b. $x + 0 = x$

52. ***Writing in Math*** Use the data about occupations on page 78 to explain how equations can be used to compare data. Include a sample problem and related equation using the information in the graph and an explanation of how to solve the equation.

STANDARDIZED TEST PRACTICE

53. Which problem is best represented by the equation $w - 15 = 33$?

A Jake added w ounces of water to his water bottle, which originally contained 33 ounces of water. How much water did he add?

B Jake added 15 ounces of water to his water bottle, for a total of 33 ounces of water. How much water w was originally in the bottle?

C Jake drank 15 ounces of water from his water bottle and 33 ounces were left. How much water w was originally in the bottle?

D Jake drank 15 ounces of water from his water bottle, which originally contained 33 ounces. How much water w was left?

54. REVIEW The table shows the results of a survey given to 500 international travelers. Based on the data, which statement is true?

Vacation Plans	
Destination	**Percent**
The Tropics	37
Europe	19
Asia	17
Other	17
No Vacation	10

F Fifty international travelers have no vacation plans.

G Fifteen international travelers are going to Asia.

H One third of international travelers are going to the tropics.

J One hundred international travelers are going to Europe.

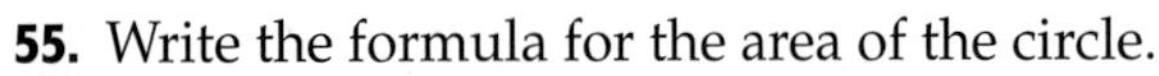

Spiral Review

GEOMETRY For Exercises 55 and 56, use the following information. (Lesson 2-1)
The area of a circle is the product of π times the radius r squared.

55. Write the formula for the area of the circle.

56. If a circle has a radius of 16 inches, find its area.

Replace each ● with >, <, or = to make the sentence true. (Lesson 1-8)

57. $\frac{1}{2} \bullet \sqrt{2}$　　**58.** $\frac{3}{4} \bullet \frac{2}{3}$　　**59.** $0.375 \bullet \frac{3}{8}$

Identify the hypothesis and conclusion of each statement. Then write the statement in if-then form. (Lesson 1-7)

60. There is a science quiz every Friday.　　**61.** For $y = 2$, $4y - 6 = 2$.

62. SHOPPING Shawnel bought 8 bagels at $0.95 each, 8 doughnuts at $0.80 each and 8 small cartons of milk for $1.00 each. Write and solve an expression to determine the total cost. (Lesson 1-5)

GET READY for the Next Lesson

PREREQUISITE SKILL Find each product or quotient. (Pages 698–701)

63. 6.5×2.8　　**64.** $17.8 \div 2.5$　　**65.** $\frac{2}{3} \times \frac{5}{8}$　　**66.** $\frac{8}{9} \div \frac{4}{15}$

2-3 Solving Equations by Using Multiplication and Division

Main Ideas

- Solve equations by using multiplication.
- Solve equations by using division.

GET READY for the Lesson

The diagram shows the distance between Earth and each star in the Big Dipper. Light travels at a rate of about 5,870 billion miles per year. The rate or speed at which something travels times the time equals the distance it travels. The following equation can be used to find the time it takes light from the closest star in the Big Dipper to reach Earth.

$$rt = d$$

$$5870t = 311{,}110$$

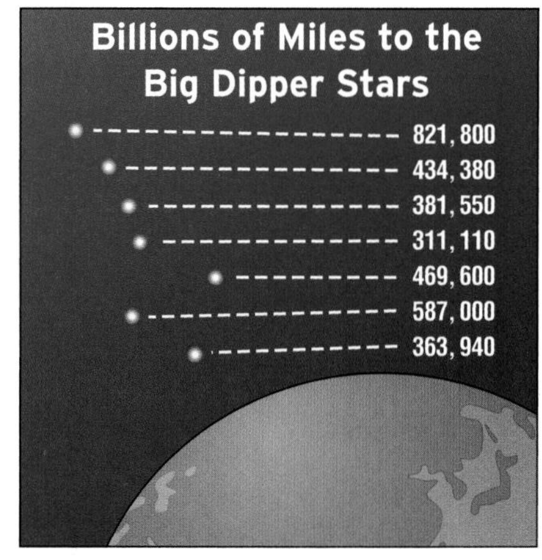

Source: *National Geographic World*

Solve Using Multiplication To solve equations such as the one above, you can use the **Multiplication Property of Equality**.

KEY CONCEPT *Multiplication Property of Equality*

Words If an equation is true and each side is multiplied by the same number, the resulting equation is true.

Symbols For any numbers a, b, and c, if $a = b$, then $ac = bc$.

Examples

$6 = 6$	$9 = 9$	$10 = 10$
$6 \times \mathbf{2} = 6 \times \mathbf{2}$	$9 \times \mathbf{(-3)} = 9 \times \mathbf{(-3)}$	$10 \times \frac{1}{2} = 10 \times \frac{1}{2}$
$12 = 12$	$-27 = -27$	$5 = 5$

EXAMPLE Solve Using Multiplication by a Positive Number

1 **Solve $\frac{t}{3} = 7$. Check your solution.**

$\frac{t}{3} = 7$	Original equation
$3\left(\frac{t}{3}\right) = 3(7)$	Multiply each side by 3.
$t = 21$	$\frac{t}{3}(3) = t$ and $7(3) = 21$

CHECK

$\frac{t}{3} = 7$	Original equation
$\frac{21}{3} \stackrel{?}{=} 7$	Substitute 21 for t.
$7 = 7$ ✓	The solution is 21.

CHECK Your Progress

Solve each equation. Check your solution.

1A. $18 = \frac{w}{2}$ **1B.** $\frac{n}{3} = -\frac{2}{5}$

EXAMPLE Solve Using Multiplication by a Fraction

2 **Solve each equation.**

a. $\left(2\frac{1}{4}\right)g = \frac{1}{2}$

$\left(2\frac{1}{4}\right)g = \frac{1}{2}$ Original equation

$\left(\frac{9}{4}\right)g = \frac{1}{2}$ Rewrite the mixed number as an improper fraction.

$\frac{4}{9}\left(\frac{9}{4}\right)g = \frac{4}{9}\left(\frac{1}{2}\right)$ Multiply each side by $\frac{4}{9}$, the reciprocal of $\frac{9}{4}$.

$g = \frac{2}{9}$ Check this result.

b. $42 = -6m$

$42 = -6m$ Original equation

$-\frac{1}{6}(42) = -\frac{1}{6}(-6m)$ Multiply each side by $-\frac{1}{6}$, the reciprocal of -6.

$-7 = m$ Check this result.

CHECK Your Progress

2A. $\frac{3}{5}k = 6$

2B. $-\frac{1}{4} = \frac{2}{3}b$

Real-World EXAMPLE

3 **SPACE TRAVEL** **Refer to the information at the left. An item's weight on the Moon is about one sixth its weight on Earth. What was the weight of Neil Armstrong's suit and life-support backpacks on Earth?**

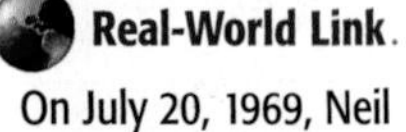

Real-World Link

On July 20, 1969, Neil Armstrong stepped on the surface of the Moon. On the Moon, his suit and life-support backpacks weighed about 33 pounds.

Source: NASA

Words	One sixth	times	the weight on Earth	equals	the weight on the Moon.
Variables	Let w = the weight on Earth.				
Equation	$\frac{1}{6}$	$\cdot$	w	$=$	33

$\frac{1}{6}w = 33$ Original equation

$6\left(\frac{1}{6}w\right) = 6(33)$ Multiply each side by 6.

$w = 198$ $\frac{1}{6}(6) = 1$ and $33(6) = 198$

Neil Armstrong's suit and backpacks were about 198 pounds on Earth.

CHECK Your Progress

3. SURVEYS In a recent survey of 13- to 15-year-old girls, 225, or about $\frac{9}{20}$ of those surveyed, said they talk on the telephone while they watch television. About how many girls were surveyed?

Personal Tutor at algebra1.com

Solve Using Division In Example 2b, the equation $42 = -6m$ was solved by multiplying each side by $-\frac{1}{6}$. The same result could have been obtained by dividing each side by -6. This method uses the **Division Property of Equality**.

KEY CONCEPT — *Division Property of Equality*

Words If an equation is true and each side is divided by the same non-zero number, the resulting equation is true.

Symbols For any numbers a, b, and c with $c \neq 0$, if $a = b$, then $\frac{a}{c} = \frac{b}{c}$.

Examples

$15 = 15$	$28 = 28$
$\frac{15}{3} = \frac{15}{3}$	$\frac{28}{-7} = \frac{28}{-7}$
$5 = 5$	$-4 = -4$

EXAMPLE Solve Using Division

4 **Solve each equation. Check your solution.**

a. $13s = 195$

$13s = 195$	Original equation
$\frac{13s}{13} = \frac{195}{13}$	Divide each side by 13.
$s = 15$	$\frac{13s}{13} = s$ and $\frac{195}{13} = 15$

CHECK	
$13s = 195$	Original equation
$13(15) \stackrel{?}{=} 195$	Substitute 15 for s.
$195 = 195$ ✓	Multiply.

b. $-3x = 12$

$-3x = 12$	Original equation
$\frac{-3x}{-3} = \frac{12}{-3}$	Divide each side by -3.
$x = -4$	$\frac{-3x}{-3} = x$ and $\frac{12}{-3} = -4$

CHECK	
$-3x = 12$	Original equation
$-3(-4) \stackrel{?}{=} 12$	Substitute -4 for x.
$12 = 12$ ✓	Multiply.

✓ CHECK Your Progress

4A. $84 = 3b$ **4B.** $-42 = -3s$

Study Tip

Alternative Method

You can also solve equations like those in Examples 4 and 5 by using the Multiplication Property of Equality. For instance, in Example 4b, you could multiply each side by $-\frac{1}{3}$.

EXAMPLE Write and Solve an Equation Using Division

5 **Write an equation for the problem below. Then solve the equation.**

Negative eighteen times a number equals −198.

Negative eighteen	times	a number	equals	−198.
-18	$\times$	n	$=$	-198

$-18n = -198$	Original equation
$\frac{-18n}{-18} = \frac{-198}{-18}$	Divide each side by -18.
$n = 11$	Check this result.

✓ CHECK Your Progress

5. Write an equation for the following problem. Then solve the equation.
Negative forty-two equals the product of six and a number.

✓CHECK Your Understanding

Examples 1 and 2 (pp. 85–86)

Solve each equation. Check your solution.

1. $\frac{t}{7} = -5$
2. $\frac{a}{36} = \frac{4}{9}$
3. $\frac{2}{3}n = 10$
4. $\frac{8}{9} = \frac{4}{5}k$
5. $12 = \frac{x}{-3}$
6. $-\frac{r}{4} = \frac{1}{7}$

Example 3 (p. 86)

7. **GEOGRAPHY** The *discharge* of a river equals the product of its width, its average depth, and its speed. At one location in St. Louis, the Mississippi River is 533 meters wide, its speed is 0.6 meter per second, and its discharge is 3198 cubic meters per second. How deep is the Mississippi River at this location?

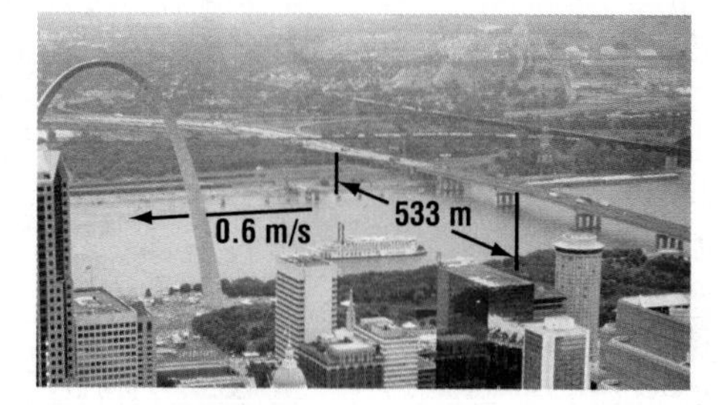

Example 4 (p. 87)

Solve each equation. Check your solution.

8. $8t = 72$
9. $20 = 4w$
10. $45 = -9a$
11. $-2g = -84$

Example 5 (p. 87)

Write an equation for each problem. Then solve the equation.

12. Five times a number is 120. What is the number?
13. One third equals negative seven times a number. What is the number?

Exercises

HOMEWORK HELP	
For Exercises	See Examples
14–25	1–2
26, 27	3
28–33	4
34–37	5

Solve each equation. Check your solution.

14. $\frac{x}{9} = 10$
15. $\frac{b}{7} = -11$
16. $\frac{3}{4} = \frac{c}{24}$
17. $\frac{2}{3} = \frac{1}{8}y$
18. $\frac{2}{3}n = 14$
19. $\frac{3}{5}g = -6$
20. $4\frac{1}{5} = 3p$
21. $-5 = 3\frac{1}{2}x$
22. $6 = -\frac{1}{2}n$
23. $-\frac{2}{5} = -\frac{z}{45}$
24. $-\frac{g}{24} = \frac{5}{12}$
25. $-\frac{v}{5} = -45$

26. **GENETICS** About two twenty-fifths of the male population in the world cannot distinguish red from green. If there are 14 boys in the ninth grade that cannot distinguish red from green, about how many ninth-grade boys are there in all? Write and solve an equation to find the answer.

27. **WORLD RECORDS** In 1998, Winchell's House of Donuts in Pasadena, California, created the world's largest doughnut. It weighed 2.5 tons and had a circumference of about 298.5 feet. What was its diameter? (*Hint*: $C = \pi d$)

Solve each equation. Check your solution.

28. $8d = 48$
29. $-65 = 13t$
30. $-5r = 55$
31. $-252 = 36s$
32. $-58 = -29h$
33. $-26a = -364$

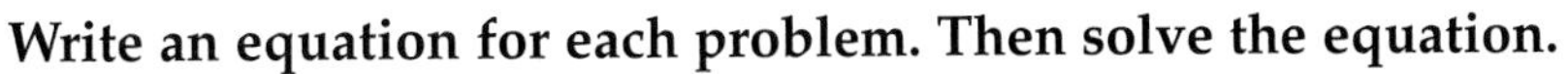

Write an equation for each problem. Then solve the equation.

34. Seven times a number equals –84. What is the number?

35. Two fifths of a number equals −24. Find the number.

36. Negative 117 is nine times a number. Find the number.

37. Twelve is one fifth of a number. What is the number?

Solve each equation. Check your solution.

38. $\left(3\frac{1}{4}\right)p = 2\frac{1}{2}$ **39.** $-5h = -3\frac{2}{3}$ **40.** $\left(-2\frac{3}{5}\right)t = -22$

41. $3.15 = 1.5y$ **42.** $-11.78 = 1.9f$ **43.** $-2.8m = 9.8$

Write an equation for each problem. Then solve the equation.

44. Negative three eighths times a number equals 12. What is the number?

45. Two and one half times a number equals one and one fifth. Find the number.

46. One and one third times a number is −4.82. What is the number?

Real-World Link

Nolan Ryan pitched a fastball that was officially clocked by the *Guinness Book of World Records* at 100.9 miles per hour in a game played on August 20, 1974.

Source: www.baseball-almanac.com

BASEBALL For Exercises 47 and 48, use the following information.

In baseball, if all other factors are the same, the speed of a four-seam fastball is faster than a two-seam fastball. The distance from the pitcher's mound to home plate is 60.5 feet.

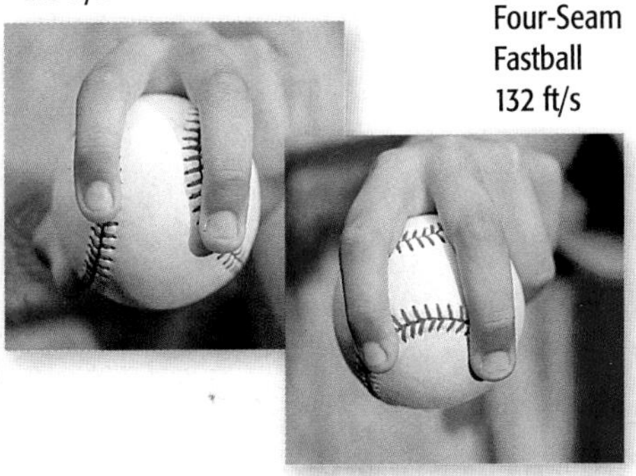

Source: *Baseball and Mathematics*

47. How long does it take a two-seam fastball to go from the pitcher's mound to home plate? Round to the nearest hundredth. (*Hint*: $d = rt$)

48. How much longer does it take for a two-seam fastball to reach home plate than a four-seam fastball?

PHYSICAL SCIENCE For Exercises 49–51, use the following information.

For every 8 grams of oxygen in water, there is 1 gram of hydrogen. In science lab, Ayame and her classmates are asked to determine how many grams of hydrogen and oxygen are in 477 grams of water.

49. If x represents the number of grams of hydrogen, write an expression to represent the number of grams of oxygen.

50. Write an equation to represent the situation.

51. How many grams of hydrogen and oxygen are in 477 grams of water?

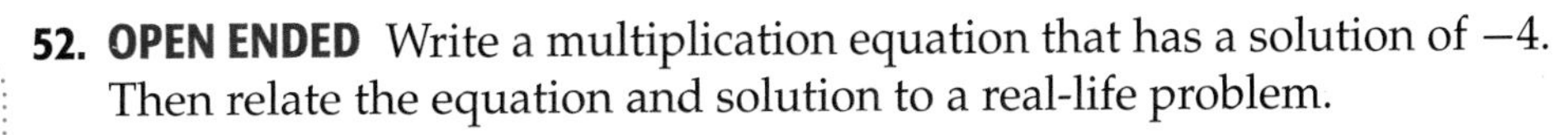

52. OPEN ENDED Write a multiplication equation that has a solution of −4. Then relate the equation and solution to a real-life problem.

53. REASONING Compare and contrast the Multiplication Property of Equality and the Division Property of Equality and explain why they can be considered the same property.

54. CHALLENGE Discuss how you can use $6y - 7 = 4$ to find the value of $18y - 21$. Then find the value of y.

55. FIND THE ERROR Casey and Camila are solving $8n = -72$. Who is correct? Explain your reasoning.

Casey	Camila
$8n = -72$	$8n = -72$
$8n(8) = -72(8)$	$\frac{8n}{8} = \frac{-72}{8}$
$n = -576$	$n = -9$

56. ***Writing in Math*** Use the data about light speed on page 85 to explain how equations can be used to find how long it takes light to reach Earth. Include an explanation of how to find how long it takes light to reach Earth from the closest star in the Big Dipper and an equation describing the situation for the star in the Big Dipper farthest from Earth.

STANDARDIZED TEST PRACTICE

57. REVIEW Which is the *best* estimate for the number of minutes on the calling card advertised below?

$10 Prepaid Calling Card
Only 5.4¢ per minute

A 10 min
B 20 min
C 50 min
D 200 min

58. REVIEW Mr. Morisson is draining his cylindrical, above ground pool. The pool has a radius of 10 feet and a standard height of 4.5 feet. If the pool water is pumped out at a constant rate of 5 gallons per minute, about how long will it take to drain the pool? $(1\ \text{ft}^3 = 7.5\ \text{gal})$

F 37.8 min
G 7 h
H 25.4 h
J 35.3 h

Spiral Review

Solve each equation. Check your solution. (Lesson 2-2)

59. $m + 14 = 81$

60. $d - 27 = -14$

61. $17 - (-w) = -55$

62. Translate the following sentence into an equation. (Lesson 2-1)
Ten times a number a is equal to 5 times the sum of b and c.

63. MUSIC Ryan practiced playing his violin 40 minutes on Monday and n minutes each on Tuesday, Wednesday, and Thursday. Write an expression for the total amount of time that he practiced during those four days. (Lesson 1-1)

GET READY for the Next Lesson

PREREQUISITE SKILL Use the order of operations to find each value. (Lesson 1-2)

64. $9 + 2 \times 8$

65. $24 \div 3 - 8$

66. $\frac{3}{8}(17 + 7)$

67. $\frac{15 - 9}{26 + 12}$

EXPLORE 2-4

Algebra Lab

Solving Multi-Step Equations

You can use an equation model to solve multi-step equations.

ACTIVITY

Concepts in Motion
Animation algebra1.com

Solve $3x + 5 = -7$.

Step 1 Model the equation.

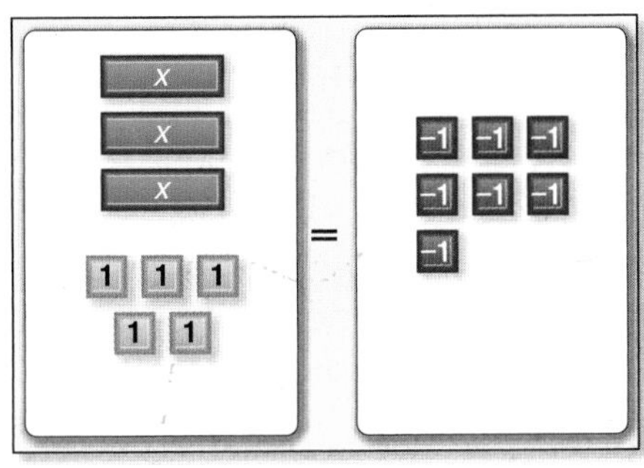

$3x + 5 = -7$

Place 3 x-tiles and 5 positive 1-tiles on one side of the mat. Place 7 negative 1-tiles on the other side of the mat.

Step 2 Isolate the x-term.

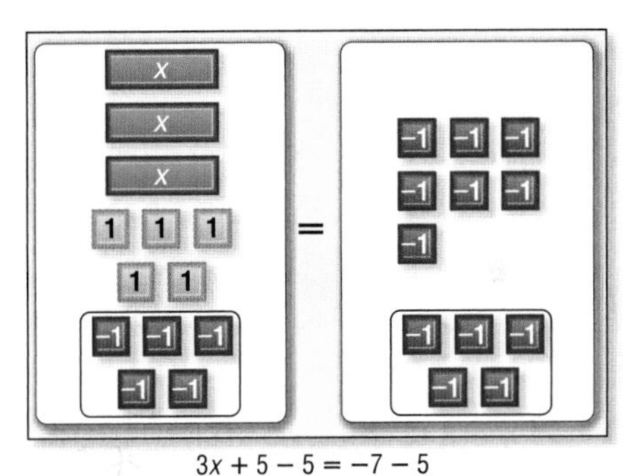

$3x + 5 - 5 = -7 - 5$

Since there are 5 positive 1-tiles with the x-tiles, add 5 negative 1-tiles to each side to form zero pairs.

Step 3 Remove zero pairs.

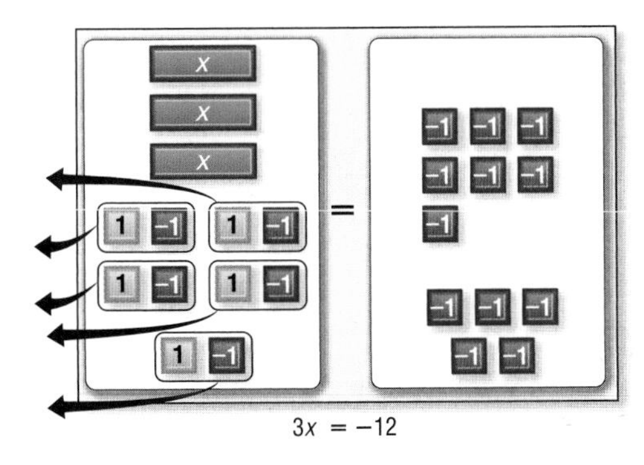

$3x = -12$

Group the tiles to form zero pairs and remove the zero pairs.

Step 4 Group the tiles.

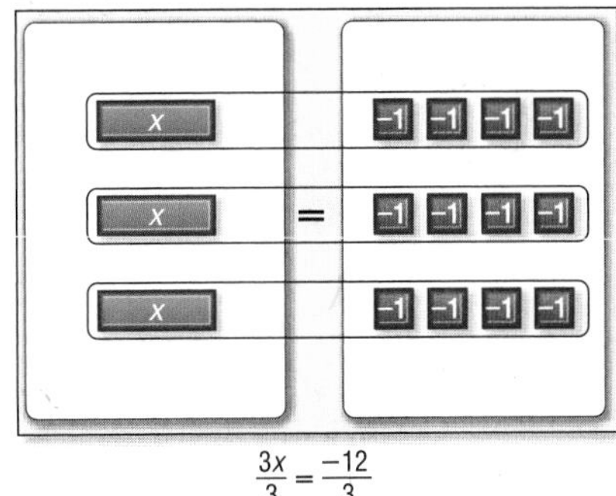

$\frac{3x}{3} = \frac{-12}{3}$

Separate the tiles into 3 equal groups to match the 3 x-tiles. Each x tile is paired with 4 negative 1 tiles. Thus, $x = -4$.

MODEL **Use algebra tiles to solve each equation.**

1. $2x - 3 = -9$

2. $3x + 5 = 14$

3. $3x - 2 = 10$

4. $-8 = 2x + 4$

5. $3 + 4x = 11$

6. $2x + 7 = 1$

7. $9 = 4x - 7$

8. $7 + 3x = -8$

9. $3x - 1 = -10$

10. MAKE A CONJECTURE What steps would you use to solve $7x - 12 = -61$?

2-4 Solving Multi-Step Equations

Main Ideas

- Solve equations involving more than one operation.
- Solve consecutive integer problems.

New Vocabulary

multi-step equations
consecutive integers
number theory

GET READY for the Lesson

An alligator hatchling 8 inches long grows about 12 inches per year. The expression $8 + 12a$ represents the length in inches of an alligator that is a years old.

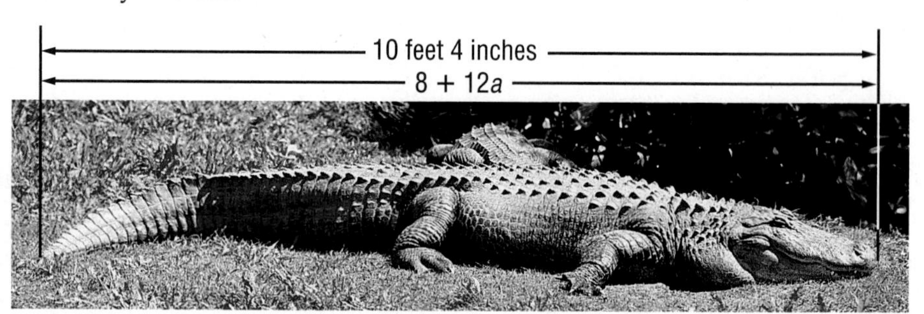

Since 10 feet 4 inches equals $10(12) + 4$ or 124 inches, the equation $8 + 12a = 124$ can be used to estimate the age of the alligator in the photograph. Notice that this equation involves more than one operation.

Solve Multi-Step Equations To solve equations with more than one operation, often called **multi-step equations**, undo operations by working backward.

EXAMPLE Solve Using Addition and Division

1 **Solve $7m - 17 = 60$. Check your solution.**

$7m - 17 = 60$ Original equation

$7m - 17 + 17 = 60 + 17$ Add 17 to each side.

$7m = 77$ Simplify.

$\frac{7m}{7} = \frac{77}{7}$ Divide each side by 7.

$m = 11$ Simplify.

CHECK $7m - 17 = 60$ Original equation

$7(11) - 17 \stackrel{?}{=} 60$ Substitute 11 for m.

$77 - 17 \stackrel{?}{=} 60$ Multiply.

$60 = 60$ 

CHECK Your Progress

Solve each equation. Check your solution.

1A. $2a - 6 = 4$ **1B.** $8 = 3r + 7$ **1C.** $\frac{t}{8} + 21 = 14$

EXAMPLE Solve Using Multiplication and Addition

2 **Solve $\frac{p-15}{9} = -6$.**

$$\frac{p-15}{9} = -6 \qquad \text{Original equation}$$

$$9\left(\frac{p-15}{9}\right) = 9(-6) \qquad \text{Multiply each side by 9.}$$

$$p - 15 = -54 \qquad \text{Simplify.}$$

$$p - 15 + 15 = -54 + 15 \qquad \text{Add 15 to each side.}$$

$$p = -39 \qquad \text{Simplify.}$$

CHECK Your Progress

Solve each equation. Check your solution.

2A. $\frac{k-12}{5} = 4$ **2B.** $\frac{n+1}{-2} = 15$

Real-World EXAMPLE Write and Solve a Multi-Step Equation

3 **SKIING Hugo is buying a pair of water skis that are on sale for $\frac{2}{3}$ of the original price. After he uses a $25 gift certificate, the total cost before taxes is $115. What was the original price of the skis? Write an equation for the problem. Then solve the equation.**

Words	Two-thirds	of	the price	minus	25	is	115.
Variable	Let p = original price of the skis.						
Equation	$\frac{2}{3}$	$\cdot$	p	$-$	25	=	115

Study Tip

Leading coefficients

Use the same steps to solve a multi-step equation if the leading coefficient is a fraction or an integer.

$$\frac{2}{3}p - 25 = 115 \qquad \text{Original equation}$$

$$\frac{2}{3}p - 25 + 25 = 115 + 25 \qquad \text{Add 25 to each side.}$$

$$\frac{2}{3}p = 140 \qquad \text{Simplify.}$$

$$\frac{3}{2}\left(\frac{2}{3}p\right) = \frac{3}{2}(140) \qquad \text{Multiply each side by } \frac{3}{2}.$$

$$p = 210 \qquad \text{Simplify.}$$

The original price of the skis was $210.

CHECK Your Progress

3. Write an equation for the following problem. Then solve the equation. *Sixteen is equal to 7 increased by the product of 3 and a number.*

Solve Consecutive Integer Problems **Consecutive integers** are integers in counting order, such as 7, 8, and 9. Beginning with an even integer and counting by two will result in *consecutive even integers*. Beginning with an odd integer and counting by two will result in *consecutive odd integers*.

Consecutive Even Integers	Consecutive Odd Integers
$-4, -2, 0, 2, 4$	$-3, -1, 1, 3, 5$

The study of numbers and the relationships between them is called **number theory**.

EXAMPLE 4 Solve a Consecutive Integer Problem

NUMBER THEORY Write an equation for the problem below. Then solve the equation and answer the problem.

Find three consecutive even integers whose sum is −42.

Let n = the least even integer.

Then $n + 2$ = the next greater even integer, and $n + 4$ = the greatest of the three even integers.

Words	The sum of three consecutive even integers	is	−42.
Equation	$n + (n + 2) + (n + 4)$	$=$	-42

$n + (n + 2) + (n + 4) = -42$	Original equation
$3n + 6 = -42$	Simplify.
$3n + 6 - 6 = -42 - 6$	Subtract 6 from each side.
$3n = -48$	Simplify.
$\frac{3n}{3} = \frac{-48}{3}$	Divide each side by 3.
$n = -16$	Simplify.

$n + 2 = -16 + 2$ or -14 $\quad$ $n + 4 = -16 + 4$ or -12

The consecutive even integers are −16, −14, and −12.

CHECK −16, −14, and −12 are consecutive even integers.

$-16 + (-14) + (-12) = -42$ ✓

Study Tip

Representing Consecutive Integers

You can use the same expressions to represent either consecutive even integers or consecutive odd integers. It is the value of n (odd or even) that differs between the two expressions.

CHECK Your Progress

4. Write an equation for the following problem. Then solve the equation and answer the problem.
Find three consecutive integers whose sum is 21.

Online Personal Tutor at algebra1.com

CHECK Your Understanding

Examples 1–2 (pp. 92–93)

Solve each equation. Check your solution.

1. $4g - 2 = -6$
2. $18 = 5p + 3$
3. $9 = 1 + \frac{m}{7}$
4. $\frac{3}{2}a - 8 = 11$
5. $20 = \frac{n - 3}{8}$
6. $\frac{b + 4}{-2} = -17$

Example 3 (p. 93)

7. **NUMBER THEORY** Twelve decreased by twice a number equals –34. Write an equation for this situation and then find the number.

8. **WORLD CULTURES** The English alphabet contains 2 more than twice as many letters as the Hawaiian alphabet. How many letters are there in the Hawaiian alphabet?

Example 4 (p. 94)

Write an equation and solve each problem.

9. Find three consecutive integers with a sum of 42.
10. Find three consecutive even integers with a sum of -12.

Exercises

HOMEWORK HELP	
For Exercises	See Examples
11–22	1–2
23, 24	3
25–28	4

Solve each equation. Check your solution.

11. $5n + 6 = -4$
12. $-11 = 7 + 3c$
13. $15 = 4a - 5$
14. $7g - 14 = -63$
15. $\frac{a}{7} - 3 = -2$
16. $\frac{c}{-3} + 5 = 7$
17. $9 + \frac{y}{5} = 6$
18. $\frac{t}{8} - 6 = -12$
19. $\frac{r + 1}{3} = 8$
20. $11 = \frac{5 + m}{-2}$
21. $14 = \frac{d - 6}{2}$
22. $\frac{17 - s}{4} = -10$

Write an equation and solve each problem.

23. Six less than two thirds of a number is negative ten. Find the number.
24. Twenty-nine is thirteen added to four times a number. What is the number?
25. Find three consecutive odd integers with a sum of 51.
26. Find three consecutive even integers with a sum of -30.
27. Find four consecutive integers with a sum of 94.
28. Find four consecutive integers with a sum of 26.

29. **ANALYZE TABLES** Adele Jones is on a business trip and plans to rent a subcompact car from Speedy Rent-A-Car. Her company has given her a budget of $60 per day for car rental. What is the maximum distance Ms. Jones can drive in one day and still stay within her budget?

Speedy Rent-A-Car Price List

Subcompact
$14.95 per day plus $0.10 per mile
Compact
$19.95 per day plus $0.12 per mile
Full Size
$22.95 per day plus $0.15 per mile

Write an equation and solve each problem.

30. NUMBER THEORY Maggie was thinking of a number. If she multiplied the number by 3, subtracted 8, added 2 times the original number, added -4, and then subtracted the original number, the result was 48. Write an equation that the number satisfies. Then solve the equation.

31. MOUNTAIN CLIMBING A general rule for those climbing more than 7000 feet above sea level is to allow a total of $\left(\frac{a - 7000}{2000} + 2\right)$ weeks of camping during the ascension. In this expression, a represents the altitude in feet. If a group of mountain climbers have allowed for 9 weeks of camping, how high can they climb without worrying about altitude sickness?

Real-World Link

Many mountain climbers experience altitude sickness caused by a decrease in oxygen. Climbers can acclimate themselves to these higher altitudes by camping for one or two weeks at various altitudes as they ascend the mountain.

Source: *Shape*

Solve each equation. Check your solution.

32. $-3d - 1.2 = 0.9$

33. $-2.5r - 32.7 = 74.1$

34. $0.2n + 3 = 8.6$

35. $-9 - \frac{p}{4} = 5$

36. $\frac{-3j - (-4)}{-6} = 12$

37. $3.5x + 5 - 1.5x = 8$

38. If $3a - 9 = 6$, what is the value of $5a + 2$?

39. If $2x + 1 = 5$, what is the value of $3x - 4$?

SHOE SIZE For Exercises 40 and 41, use the following information.
If ℓ represents the length of a person's foot in inches, the expression $2\ell - 12$ can be used to estimate his or her shoe size.

40. What is the approximate length of the foot of a person who wears size 8?

41. Measure your foot and use the expression to determine your shoe size. How does this number compare to the size of shoe you are wearing?

42. GEOMETRY A rectangle is cut from the corner of a 10-inch by 10-inch piece of paper. The area of the remaining piece of paper is $\frac{4}{5}$ of the area of the original piece of paper. If the width of the rectangle removed from the paper is 4 inches, what is the length of the rectangle?

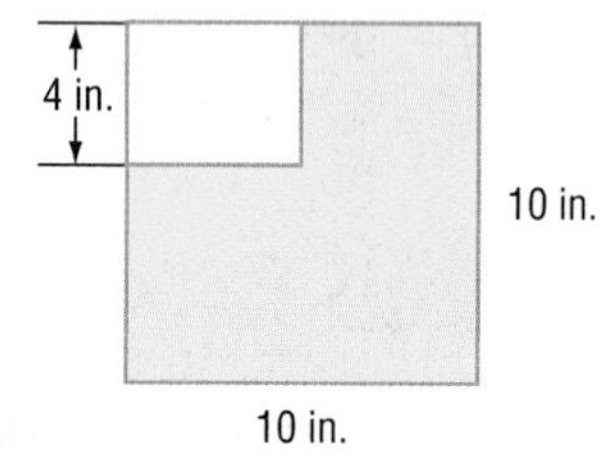

H.O.T. Problems

43. OPEN ENDED Write a problem that can be modeled by the equation $2x + 40 = 60$. Then solve the equation and explain the solution in the context of the problem.

44. REASONING Describe the steps used to solve $\frac{w + 3}{5} - 4 = 6$.

45. CHALLENGE Determine whether the following statement is *sometimes*, *always*, or *never* true. Explain your reasoning.
The sum of two consecutive even numbers equals the sum of two consecutive odd numbers.

46. *Writing in Math* Use the information about alligators on page 92 to explain how equations can be used to estimate the age of an animal. Include an explanation of how to solve the equation representing the alligator and an estimate of the age of the alligator.

STANDARDIZED TEST PRACTICE

47. REVIEW A hang glider 25 meters above the ground started to descend at a constant rate of 2 meters per second. Which equation could be used to determine h, the hang glider's height after t seconds of descent?

A $h = 25t + 2$

B $h = -25t + 2$

C $h = 2t + 25$

D $h = -2t + 25$

48. REVIEW Two rectangular walls each with a length of 12 feet and a width of 23 feet need to be painted. It costs \$0.08 per square foot for paint. How much money will it cost to paint the two walls?

F \$22.08 H \$34.50

G \$23.04 J \$44.16

49. Maddie works at Game Exchange. They are having a sale on video games and DVDs.

Item	Price	Special
Video Games	\$20	Buy 2 Get 1 Free
DVDs	\$15	Buy 1 Get 1 Free

She purchases four video games and uses her employee discount of 15%. If sales tax is 7.25%, how much does she spend on the games?

A \$54.70

B \$55.35

C \$60

D \$64.35

Spiral Review

Solve each equation. Check your solution. (Lesson 2-3)

50. $-7t = 91$ **51.** $\frac{r}{15} = -8$ **52.** $26 = \frac{2}{3}b$

53. TRANSPORTATION In 2005, there were 9 more models of sport utility vehicles than there were in 2000. Write and solve an equation to find how many models of sport utility vehicles there were in 2000. (Lesson 2-2)

Sport Utility Vehicles	
Year	**Number of Models**
2000	m
2005	56

Use the Distributive Property to rewrite each expression. Then find the product. (Lesson 1-5)

54. $17 \cdot 9$ **55.** $13(101)$ **56.** $18 \cdot 2\frac{1}{9}$

Write an algebraic expression for each verbal expression. (Lesson 1-1)

57. the product of 5 and m plus half of n

58. the sum of 3 times a and the square of b

GET READY for the Next Lesson

PREREQUISITE SKILL Simplify each expression. (Lesson 1-5)

59. $5d - 2d$ **60.** $11m - 5m$ **61.** $8t + 6t$

62. $7g - 15g$ **63.** $-9f + 6f$ **64.** $-3m + (-7m)$

2-5 Solving Equations with the Variable on Each Side

Main Ideas

- Solve equations with the variable on each side.
- Solve equations involving grouping symbols.

New Vocabulary

identity

GET READY for the Lesson

In 2003, about 46.9 million U.S. households had dial-up Internet service and about 26 million had broadband service. During the next five years, it was projected that the number of dial-up users would decrease an average of 3 million per year and the number of broadband users would increase an average of 8 million per year. The following expressions represent the number of dial-up and broadband Internet users x years after 2003.

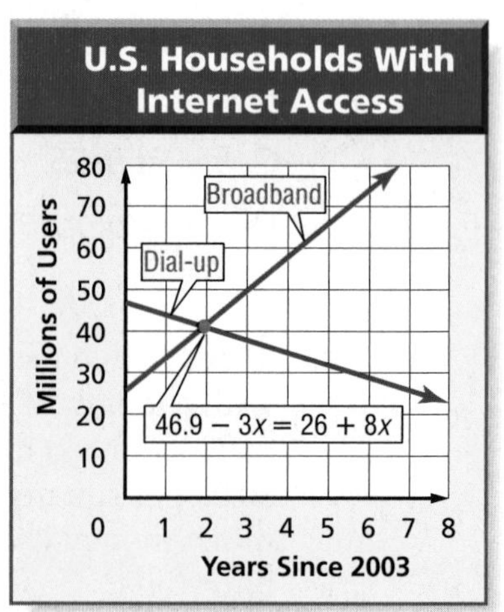

Source: Strategy Analytics

Dial-Up Internet Users: $46.9 - 3x$
Broadband Internet Users: $26 + 8x$

The equation $46.9 - 3x = 26 + 8x$ represents the time at which the number of dial-up and broadband Internet users are equal. Notice that this equation has the variable x on each side.

Variables On Each Side To solve equations with variables on each side, first use the Addition or Subtraction Property of Equality to write an equivalent equation that has all of the variables on one side.

EXAMPLE Solve an Equation with Variables on Each Side

1 Solve $-2 + 10p = 8p - 1$. Check your solution.

$-2 + 10p = 8p - 1$	Original equation
$-2 + 10p - 8p = 8p - 1 - 8p$	Subtract $8p$ from each side.
$-2 + 2p = -1$	Simplify.
$-2 + 2p + 2 = -1 + 2$	Add 2 to each side.
$2p = 1$	Simplify.
$\frac{2p}{2} = \frac{1}{2}$	Divide each side by 2.
$p = \frac{1}{2}$ or 0.5	Simplify.

The solution is $\frac{1}{2}$ or 0.5. Check by substituting into the original equation.

Study Tip

Solving Equations

The equation in Example 1 can also be solved by first subtracting $10p$ from each side.

Concepts in Motion

BrainPOP® algebra1.com

CHECK Your Progress **Solve each equation. Check your solution.**

1A. $3w + 2 = 7w$ **1B.** $\frac{s}{2} + 1 = \frac{1}{4}s - 6$

Grouping Symbols When solving equations that contain grouping symbols, first use the Distributive Property to remove the grouping symbols.

Review Vocabulary

Distributive Property: For any numbers a, b, and c, $a(b + c) = ab + ac$ and $a(b - c) = ab - ac$. (Lesson 1–5)

EXAMPLE Solve an Equation with Grouping Symbols

2 **Solve $4(2r - 8) = \frac{1}{7}(49r + 70)$. Check your solution.**

$4(2r - 8) = \frac{1}{7}(49r + 70)$	Original equation
$8r - 32 = 7r + 10$	Distributive Property
$8r - 32 - 7r = 7r + 10 - 7r$	Subtract $7r$ from each side.
$r - 32 = 10$	Simplify.
$r - 32 + 32 = 10 + 32$	Add 32 to each side.
$r = 42$	Simplify.

CHECK

$4(2r - 8) = \frac{1}{7}(49r + 70)$	Original equation
$4[2(42) - 8] \stackrel{?}{=} \frac{1}{7}[49(42) + 70]$	Substitute 42 for r.
$4(84 - 8) \stackrel{?}{=} \frac{1}{7}(2058 + 70)$	Multiply.
$4(76) \stackrel{?}{=} \frac{1}{7}(2128)$	Add and subtract.
$304 = 304$ ✓	

CHECK Your Progress **Solve each equation. Check your solution.**

2A. $8s - 10 = 3(6 - 2s)$

2B. $7(n - 1) = -2(3 + n)$

Some equations may have no solution. That is, there is no value of the variable that will result in a true equation. Other equations are true for all values of the variables. An equation like this is called an **identity**.

EXAMPLE No Solutions or Identity

3 **Solve each equation.**

a. $\mathbf{2m + 5 = 5(m - 7) - 3m}$

$2m + 5 = 5(m - 7) - 3m$	Original equation
$2m + 5 = 5m - 35 - 3m$	Distributive Property
$2m + 5 = 2m - 35$	Simplify.
$2m + 5 - 2m = 2m - 35 - 2m$	Subtract $2m$ from each side.
$5 = -35$	This statement is false.

Since $5 = -35$ is a false statement, this equation has no solution.

b. $\mathbf{3(r + 1) - 5 = 3r - 2}$

$3(r + 1) - 5 = 3r - 2$	Original equation
$3r + 3 - 5 = 3r - 2$	Distributive Property
$3r - 2 = 3r - 2$	Reflexive Property of Equality

Since the expressions on each side of the equation are the same, this equation is an identity. It is true for all values of r.

CHECK Your Progress

Solve each equation.

3A. $7x + 5(x - 1) = -5 + 12x$ **3B.** $6(y - 5) = 2(10 + 3y)$

CONCEPT SUMMARY *Steps for Solving Equations*

Step 1 Simplify the expressions on each side. Use the Distributive Property as needed.

Step 2 Use the Addition and/or Subtraction Properties of Equality to get the variables on one side and the numbers without variables on the other side. Simplify.

Step 3 Use the Multiplication or Division Property of Equality to solve.

STANDARDIZED TEST EXAMPLE

4 **Find the value of x so that the figures have the same area.**

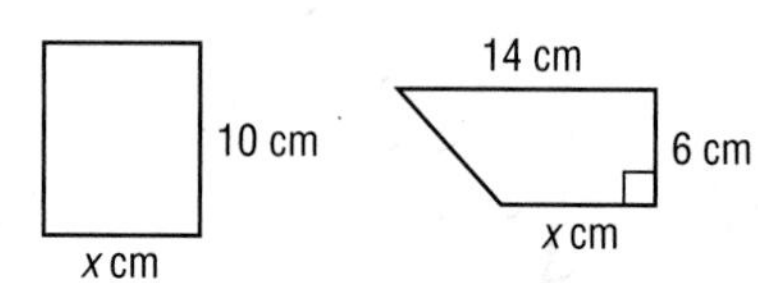

A 5
B 6
C 7
D 8

Read the Test Item

The equation $10x = \frac{1}{2}(14 + x)(6)$ represents this situation.

Solve the Test Item

You can solve the equation or substitute each value into the equation and see if it makes the equation true. We will solve by substitution.

A
$$10x = \frac{1}{2}(14 + x)(6)$$
$$10(5) \stackrel{?}{=} \frac{1}{2}(14 + 5)(6)$$
$$50 \stackrel{?}{=} \frac{1}{2}(19)(6)$$
$$50 \neq 57$$

B
$$10x = \frac{1}{2}(14 + x)(6)$$
$$10(6) \stackrel{?}{=} \frac{1}{2}(14 + 6)(6)$$
$$60 \stackrel{?}{=} \frac{1}{2}(20)(6)$$
$$60 = 60 \checkmark$$

Since the value 6 results in a true statement, you do not need to check 7 and 8. The answer is B.

Test-Taking Tip

Substitution If you are asked to solve a complicated equation, it sometimes takes less time to check each possible answer rather than to actually solve the equation.

CHECK Your Progress

4. SHOPPING A purse is on sale for one fourth off the original price, or \$12 off. What was the original price of the purse?

F \$12 **G** \$36 **H** \$48 **J** \$60

Online **Personal Tutor at** algebra1.com

✓CHECK Your Understanding

Examples 1 and 2 (pp. 98–99)

Solve each equation. Check your solution.

1. $20c + 5 = 5c + 65$

2. $\frac{3}{8} - \frac{1}{4}t = \frac{1}{2}t - \frac{3}{4}$

3. $3(a - 5) = -6$

4. $6 = 3 + 5(d - 2)$

5. **NUMBER THEORY** Four times the greater of two consecutive integers is 1 more than five times the lesser number. Find the integers.

Example 3 (pp. 99–100)

Solve each equation. Check your solution.

6. $5 + 2(n + 1) = 2n$

7. $7 - 3r = r - 4(2 + r)$

8. $14v + 6 = 2(5 + 7v) - 4$

9. $5h - 7 = 5(h - 2) + 3$

Example 4 (p. 100)

10. **STANDARDIZED TEST EXAMPLE** Find the value of x so that the figures have the same perimeter.

Triangle sides: $x + 4$, $5x + 1$, $2x + 5$. Rectangle sides: $2x$, $x + 9$.

A 4 **C** 6

B 5 **D** 7

Exercises

HOMEWORK HELP	
For Exercises	See Examples
11–18, 25–28	1
19–24, 29, 30	2
31–34	3
43, 44	4

Solve each equation. Check your solution.

11. $3k - 5 = 7k - 21$

12. $5t - 9 = -3t + 7$

13. $8s + 9 = 7s + 6$

14. $3 - 4q = 10q + 10$

15. $\frac{3}{4}n + 16 = 2 - \frac{1}{8}n$

16. $\frac{1}{4} - \frac{2}{3}y = \frac{3}{4} - \frac{1}{3}y$

17. $\frac{c + 1}{8} = \frac{c}{4}$

18. $\frac{3m - 2}{5} = \frac{7}{10}$

19. $8 = 4(3c + 5)$

20. $7(m - 3) = 7$

21. $6(r + 2) - 4 = -10$

22. $5 - \frac{1}{2}(x - 6) = 4$

23. $4(2a - 1) = -10(a - 5)$

24. $2(w - 3) + 5 = 3(w - 1)$

25. One half of a number increased by 16 is four less than two thirds of the number. Find the number.

26. The sum of one half of a number and 6 equals one third of the number. What is the number?

27. Two less than one third of a number equals 3 more than one fourth of the number. Find the number.

28. Two times a number plus 6 is three less than one fifth of the number. What is the number?

29. **NUMBER THEORY** Twice the greater of two consecutive odd integers is 13 less than three times the lesser number. Find the integers.

30. **NUMBER THEORY** Three times the greatest of three consecutive even integers exceeds twice the least by 38. What are the integers?

EXTRA PRACTICE
See pages 721, 745.

Math Online
Self-Check Quiz at algebra1.com

Solve each equation. Check your solution.

31. $4(f - 2) = 4f$

32. $\frac{3}{2}y - y = 4 + \frac{1}{2}y$

33. $3(1 + d) - 5 = 3d - 2$

34. $-3(2n - 5) = 0.5(-12n + 30)$

Real-World Link

The normal values for resting heart rates in beats per minute:

- infants: 100–160
- children 1–10 years: 70–120
- children over 10 and adults: 60–80
- well-trained athletes: 40–60

Source: MedlinePlus.com

35. **HEALTH** When exercising, a person's pulse rate should not exceed a certain limit. This maximum rate is represented by the expression $0.8(220 - a)$, where a is age in years. Find the age of a person whose maximum pulse is 152.

36. **HARDWARE** Traditionally, nails are given names such as 2-penny, 3-penny, and so on. These names describe the lengths of the nails. Use the diagram to find the name of a nail that is $2\frac{1}{2}$ inches long.

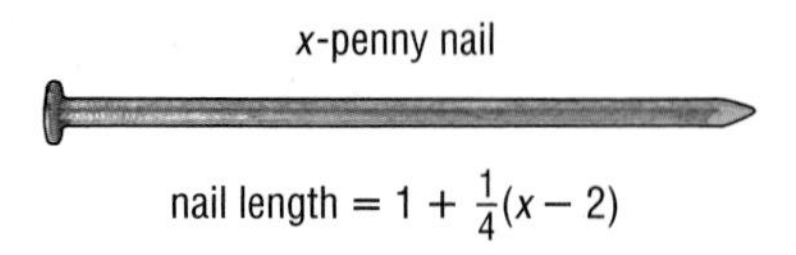

Source: *World Book Encyclopedia*

Solve each equation. Check your solution.

37. $\frac{1}{4}(7 + 3g) = -\frac{g}{8}$

38. $\frac{1}{6}(a - 4) = \frac{1}{3}(2a + 4)$

39. $1.03p - 4 = -2.15p + 8.72$

40. $18 - 3.8t = 7.36 - 1.9t$

41. $5.4w + 8.2 = 9.8w - 2.8$

42. $2[s + 3(s - 1)] = 18$

43. **ANALYZE TABLES** The table shows the households that had Brand A and Brand B of personal computers in a recent year and the average growth rates. How long will it take for the two brands to be in the same number of households?

Brand of Computer	Millions of Households with Computer	Growth Rate (million households per year)
A	4.9	0.275
B	2.5	0.7

44. **GEOMETRY** The rectangle and square shown at right have the same perimeter. Find the dimensions of each figure.

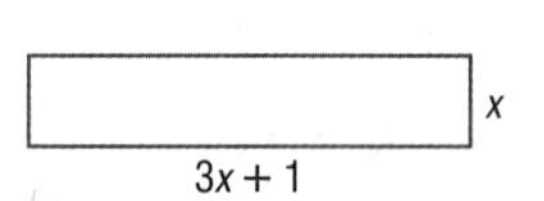

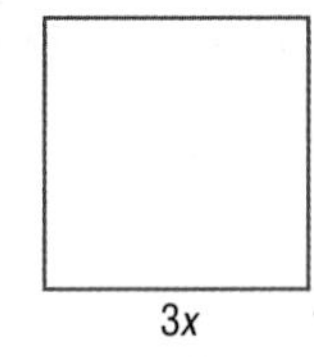

H.O.T. Problems

45. **CHALLENGE** Write an equation that has one or more grouping symbols, the variable on each side of the equals sign, and a solution of −2. Discuss the steps you used to write the equation.

46. **OPEN ENDED** Find a counterexample to the statement *All equations have a solution*. Explain your reasoning.

47. **REASONING** Determine whether each solution is correct. If the solution is not correct, describe the error and give the correct solution.

a.
$$2(g + 5) = 22$$
$$2g + 5 = 22$$
$$2g + 5 - 5 = 22 - 5$$
$$2g = 17$$
$$g = 8.5$$

b.
$$5d = 2d - 18$$
$$5d - 2d = 2d - 18 - 2d$$
$$3d = -18$$
$$d = -6$$

48. ***Writing in Math*** Use the information about Internet users on page 98 to explain how an equation can be used to determine when two populations are equal. Include the steps for solving the equation and the year when the number of dial-up Internet users will equal the number of broadband Internet users according to the model. Explain why this method can be used to predict events.

STANDARDIZED TEST PRACTICE

49. Which equation represents the second step of the solution process?

Step 1 $4(2x + 7) - 6 = 3x$
Step 2 ☐
Step 3 $5x + 28 - 6 = 0$
Step 4 $5x = -22$
Step 5 $x = -4.4$

A $4(2x - 6) + 7 = 3x$

B $4(2x + 1) = 3x$

C $8x + 7 - 6 = 3x$

D $8x + 28 - 6 = 3x$

50. REVIEW Tanya sells cosmetics door-to-door. She makes \$5 an hour and 15% commission on the total dollar value on whatever she sells. If Tanya's commission is increased to 17%, how much money will she make if she sells \$300 dollars worth of product and works 30 hours?

F \$201

G \$226

H \$255

J \$283

Spiral Review

Solve each equation. Check your solution. (Lesson 2-4)

51. $\frac{2}{9}v - 6 = 14$

52. $\frac{x - 3}{7} = -2$

53. $5 - 9w = 23$

54. HEALTH A female burns 4.5 Calories per minute pushing a lawn mower. Write an equation to represent the number of Calories *C* burned if Ebony pushes a lawn mower for *m* minutes. How long will it take Ebony to burn 150 Calories mowing the lawn? (Lesson 2-3)

55. A teacher took a survey of his students to find out how many televisions they had in their homes. The results are shown in the table. Write a set of ordered pairs representing the data in the table and draw a graph showing the relationship between students and the number of televisions in their homes. (Lesson 1-9)

Televisions	Number of Students
1	8
2	12
3	4
4	2

Write an algebraic expression for each verbal expression. Then simplify, indicating the properties used. (Lesson 1-6)

56. twice the product of *p* and *q* increased by the product of *p* and *q*

57. three times the square of *x* plus the sum of *x* squared and seven times *x*

Find the solution set for each inequality, given the replacement set. (Lesson 1-3)

58. $3x + 2 > 2; \{0, 1, 2\}$

59. $2y^2 - 1 > 0; \{1, 3, 5\}$

Evaluate each expression when $a = 5$, $b = 8$, and $c = 1$. (Lesson 1-2)

60. $5(b - a)$

61. $\frac{3a^2}{b + c}$

62. $(a + 2b) - c$

GET READY for the Next Lesson

PREREQUISITE SKILL Simplify each fraction. (Pages 694–695)

63. $\frac{28}{49}$

64. $\frac{36}{60}$

65. $\frac{8}{120}$

66. $\frac{108}{9}$

CHAPTER 2

Mid-Chapter Quiz

Lessons 2-1 through 2-5

Translate each equation into a verbal sentence. (Lesson 2-1)

1. $2 = x - 9$
2. $8 + 7t = 22$
3. $a = 1 + \frac{3}{5}b$
4. $n^2 - 6 = 5n$

GEOMETRY For Exercises 5 and 6, use the following information.

The surface area S of a sphere equals four times π times the square of the radius r. (Lesson 2-1)

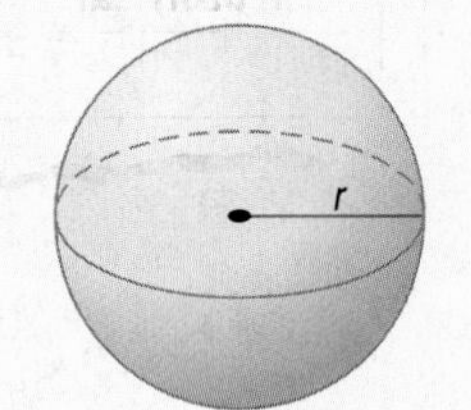

5. Write the formula for the surface area of a sphere.

6. What is the surface area of a sphere if the radius is 7 centimeters?

Solve each equation. Check your solution. (Lesson 2-2)

7. $d + 18 = -27$
8. $m - 77 = -61$
9. $-12 + a = -36$
10. $t - (-16) = 9$

PALEONTOLOGY For Exercises 11 and 12, use the following information. (Lesson 2-2)

The skeleton of a juvenile dinosaur was recently found in Illinois. If the dinosaur had been fully grown, it would have been $4\frac{1}{2}$ feet taller.

11. Write an equation to find the height x of the juvenile dinosaur if a fully grown dinosaur is 12 feet tall.

12. Solve the equation to find the height of the dinosaur.

Solve each equation. Check your solution. (Lesson 2-3)

13. $\frac{2}{3}p = 18$
14. $-17y = 391$
15. $5x = -45$
16. $-\frac{2}{5}d = -10$

17. TECHNOLOGY In a phone survey of teens who have Internet access, three fourths, or 825 of those surveyed, said they use instant messaging. How many teens were surveyed? (Lesson 2-3)

Solve each equation. Check your solution. (Lesson 2-4)

18. $-3x - 7 = 18$
19. $5 = \frac{m - 5}{4}$
20. $4h + 5 = 11$
21. $5d - 6 = 3d + 9$

22. MULTIPLE CHOICE Coach Bronson recorded the heights of 130 freshmen and 95 seniors. Which expression could be used to find the average height of these freshmen and seniors? (Lesson 2-4)

Students	Average Height
freshmen	5 feet 5 inches
seniors	5 feet 8 inches

A $\frac{65(130) + 68(95)}{225}$

B $[65(130) + 68(95)] \cdot 225$

C $65(130) + 68(95) \cdot \frac{1}{225}$

D $\frac{68(130) + 65(95)}{225}$

Solve each equation. Then check your solution. (Lesson 2-5)

23. $7 + 2(w + 1) = 2w + 9$

24. $-8(4 + 9r) = 7(-2 - 11r)$

25. NUMBER THEORY Two thirds of a number equals 3 increased by one half of the number. Find the number. (Lesson 2-5)

26. MULTIPLE CHOICE The sides of the hexagon are the same length. If the perimeter of the hexagon is $18x + 9$ square centimeters, what is the length of each side? (Lesson 2-5)

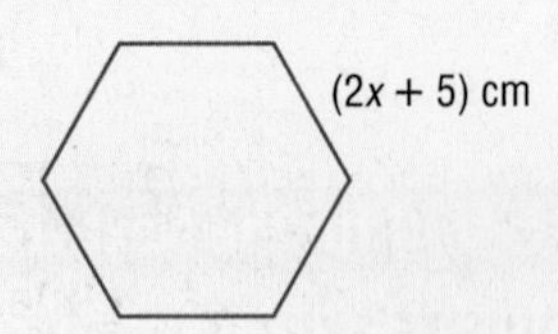

F 3.5 cm
G 7 cm
H 12 cm
J 33.5 cm

2-6 Ratios and Proportions

Main Ideas

- Determine whether two ratios form a proportion.
- Solve proportions.

New Vocabulary

ratio
proportion
extremes
means
rate
scale

GET READY for the Lesson

The ingredients in the recipe will make 4 servings of honey frozen yogurt. Keri can use ratios and equations to find the amount of each ingredient needed to make enough yogurt for her club meeting.

Honey Frozen Yogurt	
2 cups 2% milk	2 eggs, beaten
$\frac{3}{4}$ cup honey	2 cups plain low-fat
1 dash salt	yogurt
	1 tablespoon vanilla

Ratios and Proportions A **ratio** is a comparison of two numbers by division. The ratio of x to y can be expressed in the following ways.

$$x \text{ to } y \qquad x{:}y \qquad \frac{x}{y}$$

The recipe above states that for 4 servings you need 2 cups of milk. The ratio of servings to milk may be written as 4 to 2, 4:2, or $\frac{4}{2}$. In simplest form, the ratio is written as 2 to 1, 2:1, or $\frac{2}{1}$.

Suppose you wanted to double the recipe to have 8 servings. The amount of milk required would be 4 cups. The ratio of servings to milk is $\frac{8}{4}$. When this ratio is simplified, the ratio is $\frac{2}{1}$. Notice that this ratio is equal to the original ratio. An equation stating that two ratios are equal is called a **proportion**. So, we can state that $\frac{4}{2} = \frac{8}{4}$ is a proportion.

$$\frac{4}{2} \overset{\div 2}{=} \frac{2}{1} \qquad \frac{8}{4} \overset{\div 4}{=} \frac{2}{1}$$

Study Tip

Whole Numbers

A ratio that is equivalent to a whole number is written with a denominator of 1.

EXAMPLE Determine Whether Ratios Form a Proportion

1 Determine whether the ratios $\frac{4}{5}$ and $\frac{24}{30}$ form a proportion.

$$\frac{4}{5} \overset{\div 1}{=} \frac{4}{5} \qquad \frac{24}{30} \overset{\div 6}{=} \frac{4}{5}$$

The ratios are equal. Therefore, they form a proportion.

CHECK Your Progress

Determine whether each pair of ratios forms a proportion. Write *yes* or *no*.

1A. $\frac{6}{10}, \frac{2}{5}$

1B. $\frac{1}{6}, \frac{5}{30}$

There are special names for the terms in a proportion.

0.8 and 0.7 are called the **means.** They are the middle terms of the proportion.

$$0.4 : 0.8 = 0.7 : 1.4$$

0.4 and 1.4 are called the **extremes**. They are the first and last terms of the proportion.

Vocabulary Link

Extremes

Everyday Use something at one end or the other of a range, as in extremes of heat and cold

Math Use the first and last terms of a proportion

KEY CONCEPT — *Means–Extremes Property of Proportion*

Words In a proportion, the product of the extremes is equal to the product of the means.

Symbols If $\frac{a}{b} = \frac{c}{d}$, then $ad = bc$.

Examples Since $\frac{2}{4} = \frac{1}{2}$, $2(2) = 4(1)$ or $4 = 4$.

Another way to determine whether two ratios form a proportion is to use cross products. If the cross products are equal, then the ratios form a proportion.

EXAMPLE Use Cross Products

Study Tip

Cross Products

When you find cross products, you are said to be *cross multiplying*.

2 **Use cross products to determine whether each pair of ratios forms a proportion.**

a. $\frac{0.4}{0.8}, \frac{1.4}{0.7}$

$\frac{0.4}{0.8} \stackrel{?}{=} \frac{0.7}{1.4}$ Write the equation.

$0.4(1.4) \stackrel{?}{=} 0.8(0.7)$ Find the cross products.

$0.56 = 0.56$ Simplify.

The cross products are equal, so the ratios form a proportion.

b. $\frac{6}{8}, \frac{24}{28}$

$\frac{6}{8} \stackrel{?}{=} \frac{24}{28}$ Write the equation.

$6(28) \stackrel{?}{=} 8(24)$ Find the cross products.

$168 \neq 192$ Simplify.

The cross products are not equal, so the ratios do not form a proportion.

CHECK Your Progress

2A. $\frac{0.2}{1.8}, \frac{1}{0.9}$

2B. $\frac{15}{36}, \frac{35}{42}$

Solve Proportions To solve proportions that involve a variable, use cross products and the techniques used to solve other equations.

EXAMPLE Solve a Proportion

3 Solve the proportion $\frac{n}{15} = \frac{24}{16}$.

$\frac{n}{15} = \frac{24}{16}$	Original equation
$16(n) = 15(24)$	Find the cross products.
$16n = 360$	Simplify.
$\frac{16n}{16} = \frac{360}{16}$	Divide each side by 16.
$n = 22.5$	Simplify.

CHECK Your Progress

Solve each proportion. If necessary, round to the nearest hundredth.

3A. $\frac{r}{8} = \frac{25}{40}$

3B. $\frac{3.2}{4} = \frac{2.6}{n}$

Concepts in Motion
Interactive Lab algebra1.com

The ratio of two measurements having different units of measure is called a **rate**. For example, a price of $1.99 per dozen eggs, a speed of 55 miles per hour, and a salary of $30,000 per year are all rates. Proportions are often used to solve problems involving rates.

Real-World EXAMPLE

4 BICYCLING Trent goes on a 30-mile bike ride every Saturday. He rides the distance in 4 hours. At this rate, how far can he ride in 6 hours?

Explore Let m represent the number of miles Trent can ride in 6 hours.

Plan Write a proportion for the problem using rates.

miles → $\frac{30}{4} = \frac{m}{6}$ ← miles
hours → ← hours

Solve Estimate: If he rides 30 miles in 4 hours, then he would ride 60 miles in 8 hours. So, in 6 hours, he would ride between 30 and 60 miles.

$\frac{30}{4} = \frac{m}{6}$	Original proportion
$30(6) = 4(m)$	Find the cross products.
$180 = 4m$	Simplify.
$\frac{180}{4} = \frac{4m}{4}$	Divide each side by 4.
$45 = m$	Simplify.

Check Check the reasonableness of the solution. If Trent rides 30 miles in 4 hours, he rides 7.5 miles in 1 hour. So, in 6 hours, Trent can ride 6×7.5 or 45 miles. The answer is correct.

Study Tip

To ensure that the proportion is set up correctly, you should label the units. For example, $\frac{15 \text{ mi}}{2 \text{ hrs}} = \frac{25 \text{ mi}}{x \text{ hrs}}$.

4. EXERCISE It takes 7 minutes for Isabella to walk around the track twice. At this rate, how many times can she walk around the track in a half hour?

A ratio or rate called a **scale** compares the size of a model to the actual size of the object using a proportion. Maps and blueprints are two common scale drawings.

Real-World Link

Crater Lake is a volcanic crater in Oregon that was formed by an explosion 42 times the blast of Mount St. Helens.

Source: travel.excite.com

Real-World EXAMPLE

5 **CRATER LAKE The scale of a map for Crater Lake National Park is 2 inches = 9 miles. The distance between Discovery Point and Phantom Ship Overlook on the map is about $1\frac{3}{4}$ inches. What is the distance d between these two places?**

scale → $\frac{2}{9} = \frac{1\frac{3}{4}}{d}$ ← scale
actual → ← actual

$2(d) = 9\left(1\frac{3}{4}\right)$ Find the cross products.

$2d = \frac{63}{4}$ Simplify.

$2d \div 2 = \frac{63}{4} \div 2$ Divide each side by 2.

$d = \frac{63}{8}$ or $7\frac{7}{8}$ Simplify.

The actual distance is about $7\frac{7}{8}$ miles.

Concepts in Motion
BrainPOP®
algebra1.com

CHECK Your Progress

5. AIRPLANES On a model airplane, the scale is 5 centimeters = 2 meters. If the wingspan of the model is 28.5 centimeters, what is the wingspan of the actual airplane?

CHECK Your Understanding

Examples 1, 2 (pp. 105–106)

Determine whether each pair of ratios forms a proportion. Write *yes* or *no*.

1. $\frac{4}{11}, \frac{12}{33}$
2. $\frac{16}{17}, \frac{8}{9}$
3. $\frac{2.1}{3.5}, \frac{0.5}{0.7}$

Example 3 (p. 107)

Solve each proportion. If necessary, round to the nearest hundredth.

4. $\frac{3}{4} = \frac{6}{x}$
5. $\frac{a}{45} = \frac{5}{15}$
6. $\frac{0.6}{1.1} = \frac{n}{8.47}$

Example 4 (p. 107)

7. **TRAVEL** The Lehmans' minivan requires 5 gallons of gasoline to travel 120 miles. How much gasoline will they need for a 350-mile trip?

Example 5 (p. 108)

8. **BLUEPRINTS** On a blueprint for a house, 2.5 inches equals 10 feet. If the length of a wall is 12 feet, how long is the wall on the blueprint?

Exercises

HOMEWORK HELP	
For Exercises	See Examples
9–14	1, 2
15–26	3
27, 28	4
29, 30	5

Determine whether each pair of ratios forms a proportion. Write yes or no.

9. $\frac{3}{2}, \frac{21}{14}$
10. $\frac{8}{9}, \frac{12}{18}$
11. $\frac{2.3}{3.4}, \frac{3.0}{3.6}$
12. $\frac{5}{2}, \frac{4}{1.6}$
13. $\frac{21.1}{14.4}, \frac{1.1}{1.2}$
14. $\frac{4.2}{5.6}, \frac{1.68}{2.24}$

Solve each proportion. If necessary, round to the nearest hundredth.

15. $\frac{4}{x} = \frac{2}{10}$
16. $\frac{1}{y} = \frac{3}{15}$
17. $\frac{6}{5} = \frac{x}{15}$
18. $\frac{20}{28} = \frac{n}{21}$
19. $\frac{6}{8} = \frac{7}{a}$
20. $\frac{16}{7} = \frac{9}{b}$
21. $\frac{w}{2} = \frac{4.5}{6.8}$
22. $\frac{t}{0.3} = \frac{1.7}{0.9}$
23. $\frac{2}{0.21} = \frac{8}{n}$
24. $\frac{2.4}{3.6} = \frac{s}{1.8}$
25. $\frac{1}{0.19} = \frac{12}{n}$
26. $\frac{7}{1.066} = \frac{z}{9.65}$

27. **WORK** Jun earns $152 in 4 days. At that rate, how many days will it take him to earn $532?

28. **DRIVING** Lanette drove 248 miles in 4 hours. At that rate, how long will it take her to drive an additional 93 miles?

29. **MODELS** A collector's model racecar is scaled so that 1 inch on the model equals $6\frac{1}{4}$ feet on the actual car. If the model is $\frac{2}{3}$ inch high, how high is the actual car?

30. **GEOGRAPHY** On a map of Illinois, the distance between Chicago and Algonquin is 3.2 centimeters. If 2 centimeters = 40 kilometers, what is the approximate distance between the two cities?

Solve each proportion. If necessary, round to the nearest hundredth.

31. $\frac{6}{14} = \frac{7}{x - 3}$
32. $\frac{5}{3} = \frac{6}{x + 2}$
33. $\frac{3 - y}{4} = \frac{1}{9}$

34. **PETS** A research study shows that three out of every twenty pet owners bought their pets from breeders. Of the 122 animals cared for by a veterinarian, how many would you expect to have been bought from breeders?

ANALYZE TABLES For Exercises 35 and 36, use the table.

35. Write a ratio of the number of gold medals won to the total number of medals won for each country.

All-Time Summer Olympic Medal Standings, 1896–2004

Country	Gold	Silver	Bronze	Total
United States	907	697	615	2219
USSR/UT/Russia	525	436	409	1370
Germany/E. Ger/W. Ger	388	408	434	1230
Great Britain	189	242	237	668
France	199	202	230	631
Italy	189	154	168	511
Sweden	140	157	179	476

Source: infoplease.com and athens2004.com

36. Do any two of the ratios you wrote for Exercise 35 form a proportion? If so, explain the real-world meaning of the proportion.

EXTRA PRACTICE
See pages 721, 745.

Math Online
Self-Check Quiz at algebra1.com

H.O.T. Problems

37. **OPEN ENDED** Find an example of ratios used in an advertisement. Analyze the ratios and describe how they are used to sell the product.

38. **REASONING** Explain the difference between a ratio and a proportion.

39. **CHALLENGE** Consider the proportion $a:b:c = 3:1:5$. What is the value of $\frac{2a + 3b}{4b + 3c}$? (*Hint*: Choose different values of a, b, and c for which the proportion is true and evaluate the expression.)

40. ***Writing in Math*** Use the information about recipes on page 105 to explain how ratios are used in recipes. Include an explanation of how to use a proportion to determine how much honey is needed if you use 3 eggs, and a description of how to alter the recipe to get 5 servings.

STANDARDIZED TEST PRACTICE

41. In the figure, $x:y = 2:3$ and $y:z = 3:5$. If $x = 10$, find the value of z.

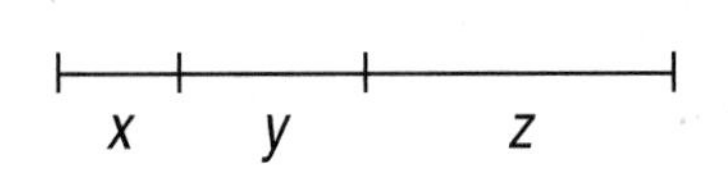

A 15

B 20

C 25

D 30

42. **REVIEW** If $\triangle LMN$ is similar to $\triangle LPO$, what is the length of side z?

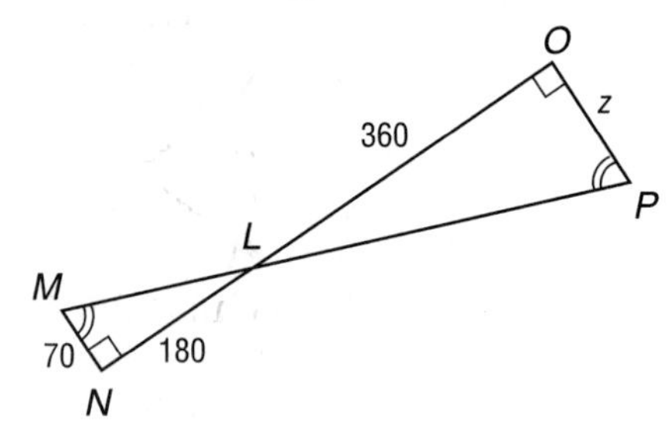

F 240 units

G 140 units

H 120 units

J 70 units

Spiral Review

Solve each equation. Then check your solution. (Lessons 2-4 and 2-5)

43. $8y - 10 = -3y + 2$

44. $17 + 2n = 21 + 2n$

45. $-6(d - 3) = -d$

46. $5 - 9w = 23$

47. $\frac{m}{-5} + 6 = 31$

48. $\frac{z - 7}{5} = -3$

49. Sketch a reasonable graph for the temperature in the following statement. *In August, Joel enters his house and turns on the air conditioner.* (Lesson 1-9)

Evaluate each expression. (Lesson 1-2)

50. $12(5) - 6(4)$

51. $7(0.2 + 0.5) - 0.6$

52. $[6^2 - 3(2 + 5)] \div 5$

GET READY for the Next Lesson

PREREQUISITE SKILL Find each percent. (Pages 702–703)

53. Eighteen is what percent of 60?

54. What percent of 14 is 4.34?

55. Six is what percent of 15?

56. What percent of 2 is 8?

2-7 Percent of Change

Main Ideas

- Find percents of increase and decrease.
- Solve problems involving percents of change.

New Vocabulary

percent of change
percent of increase
percent of decrease

GET READY for the Lesson

Phone companies began using area codes in 1947. The graph shows the number of area codes in use in different years. The growth in the number of area codes can be described by using a percent of change.

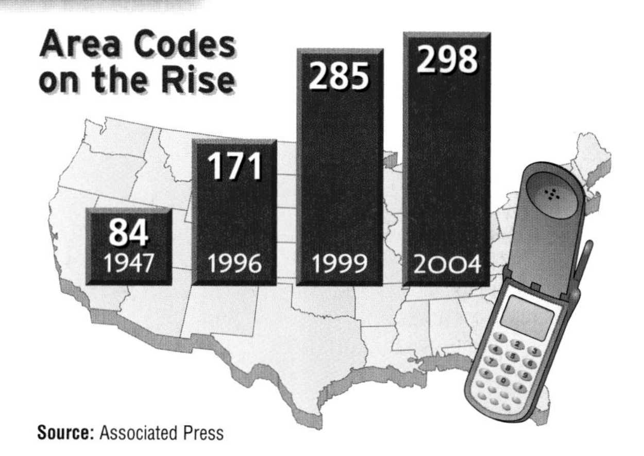

Source: Associated Press

Percent of Change When an increase or decrease is expressed as a percent, the percent is called the **percent of change**. If the new number is greater than the original number, the percent of change is a **percent of increase**. If the new number is less than the original, the percent of change is a **percent of decrease**.

EXAMPLE Find Percent of Change

State whether each percent of change is a percent of increase or a percent of decrease. Then find each percent of change.

a. original: \$25
new: \$28

Since the new amount is greater than the original, this is a percent of increase. Find the *amount* of change.

$$28 - 25 = 3$$

Use the original number, 25, as the base.

change → / original amount →
$$\frac{3}{25} = \frac{r}{100}$$
$$3(100) = 25(r)$$
$$300 = 25r$$
$$\frac{300}{25} = \frac{25r}{25}$$
$$12 = r$$

The percent of increase is 12%.

b. original: 30
new: 12

This is a percent of decrease because the new amount is less than the original. Find the amount of change.

$$30 - 12 = 18$$

Use the original number, 30, as the base.

change → / original amount →
$$\frac{18}{30} = \frac{r}{100}$$
$$18(100) = 30(r)$$
$$1800 = 30r$$
$$\frac{1800}{30} = \frac{30r}{30}$$
$$60 = r$$

The percent of decrease is 60%.

Review Vocabulary

Percent Proportion
percent proportion:

$$\frac{\text{part}}{\text{base}} = \frac{\text{percent}}{100}$$

(page 702)

Real-World Link

On November 12, 1892, the Allegheny Athletic Association paid William "Pudge" Heffelfinger \$500 to play football. This game is considered the start of professional football.

Source: *World Book Encyclopedia*

1A. original: 66; new: 30 **1B.** original: 9.8; new: 12.1

Real-World EXAMPLE Find the Missing Value

2 FOOTBALL The National Football League's (NFL) fields are 120 yards long. The Canadian Football League's (CFL) fields are 25% longer. How long is a CFL field?

Let ℓ = the length of a CFL field. Since 25% is a percent of increase, an NFL field is shorter than a CFL field. Therefore, $\ell - 120$ represents the change.

change → / original amount → $\frac{\ell - 120}{120} = \frac{25}{100}$	Percent proportion
$(\ell - 120)(100) = 120(25)$	Find the cross products.
$100\ell - 12{,}000 = 3000$	Distributive Property
$100\ell - 12{,}000 + 12{,}000 = 3000 + 12{,}000$	Add 12,000 to each side.
$100\ell = 15{,}000$	Simplify.
$\frac{100\ell}{100} = \frac{15{,}000}{100}$	Divide each side by 100.
$\ell = 150$	Simplify.

The length of the field used by the CFL is 150 yards.

2. TUITION A recent percent of increase in tuition at Northwestern University was 5.4%. If the new cost is \$29,940 per year, find the original cost per year.

Online Personal Tutor at algebra1.com

Solve Problems Sales tax is a tax that is added to the cost of the item. It is an example of a percent of increase.

Real-World EXAMPLE

3 SALES TAX A concert ticket costs \$45. If the sales tax is 6.25%, what is the total price of the ticket?

The tax is 6.25% of the price of the ticket.

6.25% of \$45 = 0.0625×45 6.25% = 0.0625

= 2.8125 Use a calculator.

Round \$2.8125 to \$2.81. Add this amount to the original price.

\$45.00 + \$2.81 = \$47.81 The total price of the ticket is \$47.81.

CHECK Your Progress

3. TAXES A new DVD costs \$24.99. If the sales tax is 7.25%, what is the total cost?

Cross-Curricular Project

Math Online A percent of increase or decrease can be used to describe trends in populations. Visit algebra1.com to continue work on your project.

Discount is the amount by which the regular price of an item is reduced. It is an example of a percent of decrease.

4 **DISCOUNT A sweater is on sale for 35% off the original price. If the original price of the sweater is $38, what is the discounted price?**

The discount is 35% of the original price.

35% of $38 = 0.35 × 38   35% = 0.35
= 13.30   Use a calculator.

Subtract $13.30 from the original price.

$38.00 − $13.30 = $24.70   The discounted price of the sweater is $24.70.

CHECK Your Progress

4. **SALES** A picture frame originally priced at $14.89 is on sale for 40% off. What is the discounted price?

CHECK Your Understanding

State whether each percent of change is a percent of increase or a percent of decrease. Then find each percent of change. Round to the nearest whole percent.

Example 1 (p. 111)

1. original: 72
new: 36
2. original: 45
new: 50
3. original: 14 books
new: 16 books
4. original: 150 T-shirts
new: 120 T-shirts

Example 2 (p. 112)

5. **GEOGRAPHY** The distance from El Monte to Fresno is 211 miles. The distance from El Monte to Oakland is about 64.5% longer. To the nearest mile, what is the distance from El Monte to Oakland?

Find the total price of each item.

Example 3 (p. 112)

6. software: $39.50
sales tax: 6.5%
7. compact disc: $15.99
sales tax: 5.75%

Example 4 (p. 113)

Find the discounted price of each item.

8. jeans: $45.00
discount: 25%
9. book: $19.95
discount: 33%

Exercises

HOMEWORK HELP	
For Exercises	See Examples
10–17	1
18, 19	2
20–25	3
26–31	4

State whether each percent of change is a percent of increase or a percent of decrease. Then find each percent of change. Round to the nearest whole percent.

10. original: 50
new: 70
11. original: 25
new: 18
12. original: 58
new: 152
13. original: 13.7
new: 40.2
14. original: 15.6 meters
new: 11.4 meters
15. original: 132 students
new: 150 students
16. original: $85
new: $90
17. original: 40 hours
new: 32.5 hours

18. **EDUCATION** According to the Census Bureau, the average income of a person with a high school diploma is \$27,351. The income of a person with a bachelor's degree is about 57% higher. What is the average income of a person with a bachelor's degree?

19. **BOATS** A 36-foot sailboat that is new costs 86% more than the same boat in good used condition. What is the cost of a new 36-foot sailboat?

Buying a Sailboat

Type of Boat	Cost (\$)
Used	70,000
New	x

Real-World Career...

Military

A military career can involve many different duties like working in a hospital, programming computers, or repairing helicopters. The military provides training and work in these fields and others.

For more information, go to algebra1.com.

Find the total price of each item.

20. umbrella: \$14.00
tax: 5.5%

21. backpack: \$35.00
tax: 7%

22. candle: \$7.50
tax: 5.75%

23. hat: \$18.50
tax: 6.25%

24. clock radio: \$39.99
tax: 6.75%

25. sandals: \$29.99
tax: 5.75%

Find the discounted price of each item.

26. shirt: \$45.00
discount: 40%

27. socks: \$6.00
discount: 20%

28. watch: \$37.55
discount: 35%

29. gloves: \$24.25
discount: 33%

30. suit: \$175.95
discount: 45%

31. coat: \$79.99
discount: 30%

Find the final price of each item.

32. lamp: \$120.00
discount: 20%
tax: 6%

33. dress: \$70.00
discount: 30%
tax: 7%

34. camera: \$58.00
discount: 25%
tax: 6.5%

35. **MILITARY** In 2000, the United States had 2.65 million active-duty military personnel. In 2004, there were 1.41 million active-duty military personnel. What was the percent of decrease? Round to the nearest whole percent.

36. **THEME PARKS** In 2003, 162.3 million people visited theme parks in the United States. In 2004, the number of visitors increased by about 4%. About how many people visited theme parks in the United States in 2004?

37. **ANALYZE TABLES** What are the projected 2050 populations for each country in the table? Which is projected to be the most populous?

Country	1997 Population (billions)	Projected Percent of Increase for 2050
China	1.24	22.6%
India	0.97	57.8%
United States	0.27	44.4%

Source: *USA TODAY*

38. **RESEARCH** Use the Internet or other reference to find the tuition for the last several years at a college of your choice. Find the percent of change for the tuition during these years. Predict the tuition for the year you plan to graduate from high school.

H.O.T. Problems

39. **CHALLENGE** Is the following equation *sometimes*, *always*, or *never* true? Explain your reasoning.

$$x\% \text{ of } y = y\% \text{ of } x$$

40. **OPEN ENDED** Give a counterexample to the statement *The percent of change must always be less than 100%.*

41. **FIND THE ERROR** Laura and Cory are writing proportions to find the percent of change if the original number is 20 and the new number is 30. Who is correct? Explain your reasoning.

Laura	Cory
Amount of change: 30 - 20 = 10	Amount of change: 30 - 20 = 10
$\frac{10}{20} = \frac{r}{100}$	$\frac{10}{30} = \frac{r}{100}$

42. ***Writing in Math*** Use the data on page 111 to find the percent of increase in the number of area codes from 1999 to 2004. Explain why knowing a percent of change can be more informative than knowing how much the quantity changed.

STANDARDIZED TEST PRACTICE

43. The number of students at Franklin High School increased from 840 to 910 over a 5-year period. What was the percent of increase?

A 8.3% **C** 18.5%

B 14.0% **D** 92.3%

44. **REVIEW** The rectangle has a perimeter of P centimeters. Which equation could be used to find the length of the rectangle?

2.4 cm

F $P = 2.4\ell$ **H** $P = 2.4 + 2\ell$

G $P = 4.8 + \ell$ **J** $P = 4.8 + 2\ell$

Spiral Review

Solve each proportion. (Lesson 2-6)

45. $\frac{a}{45} = \frac{3}{15}$

46. $\frac{2}{3} = \frac{8}{d}$

47. $\frac{5.2}{10.4} = \frac{t}{48}$

Solve each equation. Check your solution. (Lesson 2-5)

48. $6n + 3 = -3$

49. $7 + 5c = -23$

50. $18 = 4a - 2$

51. **SALES** As a salesperson, Mr. Goetz is paid a monthly salary and a commission on sales, as shown in the table. How much must Mr. Goetz sell to earn $2000 this month? (Lesson 2-4)

Mr. Goetz's Income	
Monthly salary	$500
Commission on sales	2%

Evaluate each expression. (Lesson 1-2)

52. $(3 + 6) \div 3^2$

53. $6(12 - 7.5) - 7$

54. $20 \div 4 \cdot 8 \div 10$

GET READY for the Next Lesson

PREREQUISITE SKILL Solve each equation. Check your solution. (Lesson 2-5)

55. $7y + 7 = 3y - 5$

56. $7(d - 3) - 2 = 5$

57. $-8 = 4 - 2(a - 5)$

READING MATH

Sentence Method and Proportion Method

Recall that you can solve percent problems using two different methods. With either method, it is helpful to use "clue" words such as *is* and *of*. In the sentence method, *is* means equals and *of* means multiply. With the proportion method, the "clue" words indicate where to place the numbers in the proportion.

Sentence Method

15% of 40 is what number?

$0.15 \cdot 40 = ?$

Proportion Method

15% of 40 is what number?

$\frac{\text{(is) } P}{\text{(of) } B} = \frac{R\text{(percent)}}{100} \rightarrow \frac{P}{40} = \frac{15}{100}$

You can use the proportion method to solve percent of change problems. In this case, use the proportion $\frac{\text{difference}}{\text{original}} = \frac{\%}{100}$. When reading a percent of change problem, or any other word problem, look for the important numerical information.

Example **In life skills class, Kishi heated 20 milliliters of water. She let the water boil for 10 minutes. Afterward, only 17 milliliters of water remained, due to evaporation. What is the percent of decrease in the amount of water?**

$\frac{\text{difference}}{\text{original}} = \frac{\%}{100} \rightarrow \frac{20 - 17}{20} = \frac{r}{100}$ Percent proportion

$\frac{3}{20} = \frac{r}{100}$ Simplify.

$3(100) = 20(r)$ Find the cross products.

$300 = 20r$ Simplify.

$\frac{300}{20} = \frac{20r}{20}$ Divide each side by 20.

$15 = r$ Simplify.

There was a 15% decrease in the amount of water.

Reading to Learn

Give the original number and the amount of change. Then write and solve a percent proportion.

1. Monsa needed to lose weight for wrestling. At the start of the season, he weighed 166 pounds. By the end of the season, he weighed 158 pounds. What is the percent of decrease in Monsa's weight?
2. On Carla's last Algebra test, she scored 94 points out of 100. On her first Algebra test, she scored 75 points out of 100. What is the percent of increase in her score?
3. An online bookstore tracks daily book sales. A certain book sold 12,476 copies on Monday. After the book was mentioned on a national news program, sales were 37,884 on Tuesday. What is the percent of increase from Monday to Tuesday?

2-8 Solving for a Specific Variable

Main Ideas

- Solve equations for given variables.
- Use formulas to solve real-world problems.

New Vocabulary

dimensional analysis

GET READY for the Lesson

Suppose the designer of the Magnum XL-200 decided to adjust the height of the second hill so that the coaster would have a speed of 49 feet per second when it reached the top. If we ignore friction, the equation $g(195 - h) = \frac{1}{2}v^2$ can be used to find the height of the second hill. In this equation, g represents the acceleration due to gravity (32 feet per second squared), h is the height of the second hill, and v is the velocity of the coaster when it reaches the top of the second hill.

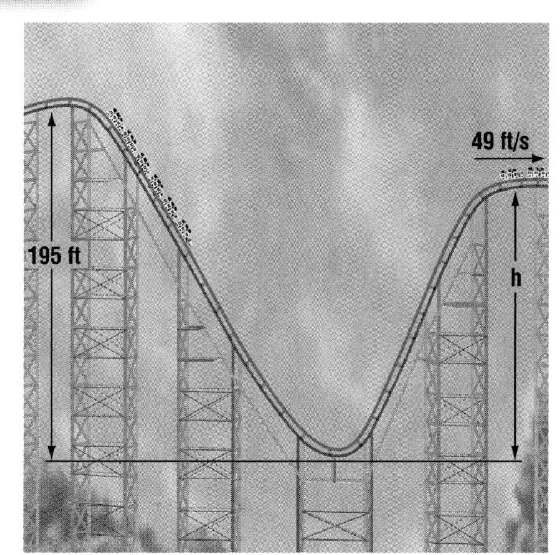

Solve for Variables Some equations such as the one above contain more than one variable. It is often useful to solve these equations for one of the variables.

EXAMPLE Solve an Equation for a Specific Variable

Solve $3x - 4y = 7$ for y.

$3x - 4y = 7$	Original equation
$3x - 4y - 3x = 7 - 3x$	Subtract $3x$ from each side.
$-4y = 7 - 3x$	Simplify.
$\frac{-4y}{-4} = \frac{7 - 3x}{-4}$	Divide each side by -4.
$y = \frac{7 - 3x}{-4}$ or $\frac{3x - 7}{4}$	Simplify.

The value of y is $\frac{3x - 7}{4}$.

CHECK Your Progress

Solve each equation for the variable indicated.

1A. $15 = 3n + 6p$, for n **1B.** $\frac{k - 2}{5} = 11j$, for k

It is sometimes helpful to use the Distributive Property to isolate the variable for which you are solving an equation or formula.

EXAMPLE Solve an Equation for a Specific Variable

2 **Solve $2m - t = sm + 5$ for m.**

$2m - t = sm + 5$	Original equation
$2m - t - sm = sm + 5 - sm$	Subtract sm from each side.
$2m - t - sm = 5$	Simplify.
$2m - t - sm + t = 5 + t$	Add t to each side.
$2m - sm = 5 + t$	Simplify.
$m(2 - s) = 5 + t$	Use the Distributive Property.
$\frac{m(2 - s)}{2 - s} = \frac{5 + t}{2 - s}$	Divide each side by $2 - s$.
$m = \frac{5 + t}{2 - s}$	Simplify.

The value of m is $\frac{5 + t}{2 - s}$. Since division by 0 is undefined, $2 - s \neq 0$ or $s \neq 2$.

CHECK Your Progress

Solve each equation for the variable indicated.

2A. $d + 5c = 3d - 1$, for d

2B. $6q - 18 = qr + s$, for q

Online Personal Tutor at algebra1.com

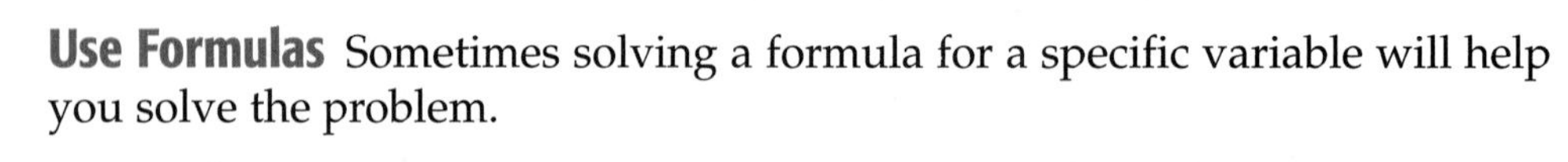
Use Formulas Sometimes solving a formula for a specific variable will help you solve the problem.

Real-World Link

The largest yo-yo in the world is 32.7 feet in circumference. It was launched by crane from a height of 189 feet.

Source: www.guinnessworldrecords.com

Real-World EXAMPLE

3 **YO-YOS Use the information about the largest yo-yo at the left. The formula for the circumference of a circle is $C = 2\pi r$, where C represents circumference and r represents radius.**

a. Solve the formula for r.

$C = 2\pi r$	Circumference formula
$\frac{C}{2\pi} = \frac{2\pi r}{2\pi}$	Divide each side by 2π.
$\frac{C}{2\pi} = r$	Simplify.

b. Find the radius of the yo-yo.

$\frac{C}{2\pi} = r$	Formula for radius
$\frac{32.7}{2\pi} = r$	$C = 32.7$
$5.2 \approx r$	The yo-yo has a radius of about 5.2 feet.

CHECK Your Progress

The formula for the volume of a rectangular prism is $V = \ell wh$, where ℓ is the length, w is the width, and h is the height.

3A. Solve the formula for w.

3B. Find the width of a rectangular prism that has a volume of 79.04 cubic centimeters, a length of 5.2 centimeters, and a height of 4 centimeters.

When using formulas, you may want to use **dimensional analysis**, which is the process of carrying units throughout a computation.

Real-World EXAMPLE

4 PHYSICAL SCIENCE The formula $s = \frac{1}{2}at^2$ represents the distance s that a free-falling object will fall near a planet or the Moon in a given time t. In the formula, a represents the acceleration due to gravity.

a. Solve the formula for a.

$s = \frac{1}{2}at^2$ Original formula

$\frac{2}{t^2}(s) = \frac{2}{t^2}\left(\frac{1}{2}at^2\right)$ Multiply each side by $\frac{2}{t^2}$.

$\frac{2s}{t^2} = a$ Simplify.

b. A free-falling object near the Moon drops 20.5 meters in 5 seconds. What is the value of a for the Moon?

$a = \frac{2s}{t^2}$ Formula for a

$= \frac{2(20.5 \text{ m})}{(5 \text{ s})^2}$ $s = 20.5$ m and $t = 5$ s.

$= \frac{1.64 \text{ m}}{\text{s}^2}$ or 1.64 m/s^2 Use a calculator.

Acceleration due to gravity on the Moon is 1.64 meters per second squared.

CHECK Your Progress

The formula $s = vt + \frac{1}{2}at^2$ represents the distance s an object travels with an initial velocity v, time t, and constant rate of acceleration a.

4A. Solve the formula for v.

4B. A sports car accelerates at a rate of 8 ft/s^2 and travels 100 feet in about 2.8 seconds. What is the initial velocity to the nearest tenth?

CHECK Your Understanding

Examples 1, 2 (pp. 117–118)

Solve each equation or formula for the variable specified.

1. $-3x + b = 6x$, for x

2. $4z + b = 2z + c$, for z

3. $\frac{y + a}{3} = c$, for y

4. $p = a(b + c)$, for a

Example 3 (p. 118)

GEOMETRY For Exercises 5 and 6, use the formula for the area of a triangle.

5. Solve the formula for h.

6. What is the height of a triangle with an area of 28 square feet and a base of 8 feet?

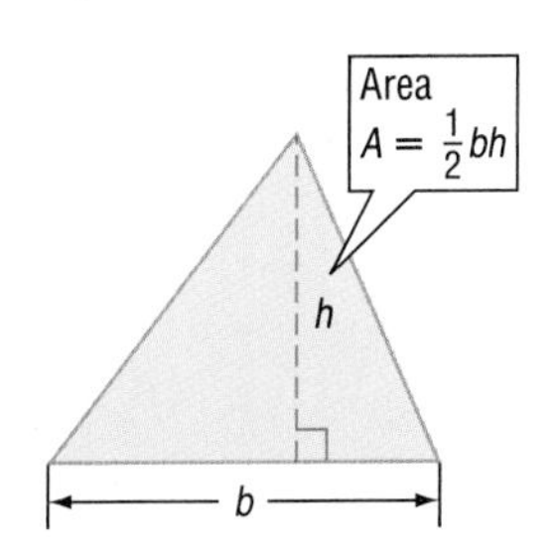

Example 4 (p. 119)

7. SWIMMING A swimmer swims about one third of a lap per minute. At this rate, how many minutes would it take to swim 8 laps? (*Hint:* Use $d = rt$.)

Exercises

HOMEWORK HELP	
For Exercises	See Examples
8–15	1
16–21	2
22–25	3, 4

Solve each equation or formula for the variable specified.

8. $y = mx + b$, for m

9. $v = r + at$, for a

10. $km + 5x = 6y$, for m

11. $4b - 5 = -t$, for b

12. $c = \frac{3}{4}y + b$, for y

13. $\frac{3}{5}m + a = b$, for m

14. $\frac{3ax - n}{5} = -4$, for x

15. $\frac{by + 2}{3} = c$, for y

16. $5g + h = g$, for g

17. $8t - r = 12t$, for t

18. $3y + z = am - 4y$, for y

19. $9a - 2b = c + 4a$, for a

20. $at + b = ar - c$, for a

21. $2g - m = 5 - gh$, for g

GEOMETRY For Exercises 22 and 23, use the formula for the area of a trapezoid.

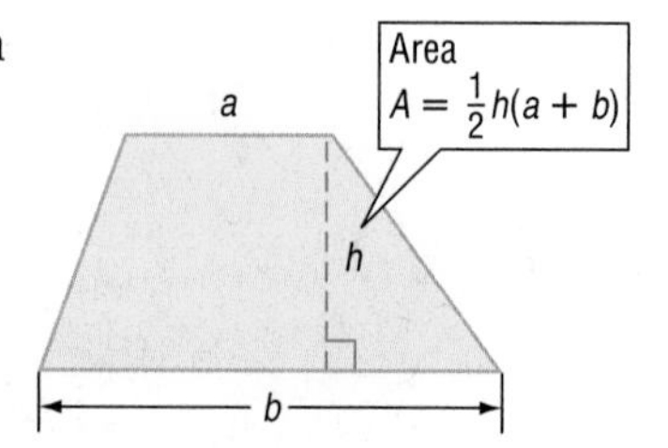

22. Solve the formula for h.

23. What is the height of a trapezoid with an area of 60 square meters and bases of 8 meters and 12 meters?

WORK For Exercises 24 and 25, use the following information.

The formula $s = \frac{w - 10e}{m}$ is often used by placement services to find keyboarding speeds. In the formula, s represents the speed in words per minute, w represents the number of words typed, e represents the number of errors, and m represents the number of minutes typed.

24. Solve the formula for e.

25. If Mateo typed 410 words in 5 minutes and received a keyboard speed of 76 words per minute, how many errors did he make?

Solve each equation or formula for the variable specified.

26. $S = \frac{n}{2}(A + t)$, for A

27. $p(t + 1) = -2$, for t

28. $\frac{5x + y}{a} = 2$, for a

Write an equation and solve for the variable specified.

29. Seven less than a number t equals another number r plus 6. Solve for t.

30. Five minus twice a number p equals 6 times another number q plus 1. Solve for p.

31. Five eighths of a number x is 3 more than one half of another number y. Solve for y.

DANCING For Exercises 32 and 33, use the following information.

The formula $P = \frac{1.2W}{H^2}$ represents the amount of pressure exerted on the floor by a dancer's heel. In this formula, P is the pressure in pounds per square inch, W is the weight of a person wearing the shoe in pounds, and H is the width of the heel of the shoe in inches.

32. Solve the formula for W.

33. Find the weight of the dancer if the heel is 3 inches wide and the pressure exerted is 30 pounds per square inch.

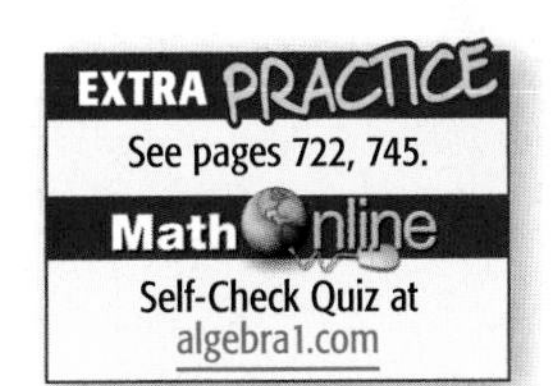

34. PACKAGING The Yummy Ice Cream Company wants to package ice cream in cylindrical containers that have a volume of 5453 cubic centimeters. The marketing department decides the diameter of the base of the containers should be 20 centimeters. How tall should the containers be? (Hint: $V = \pi r^2 h$)

H.O.T. Problems

35. CHALLENGE Write a formula for the area of the arrow. Describe how you found it.

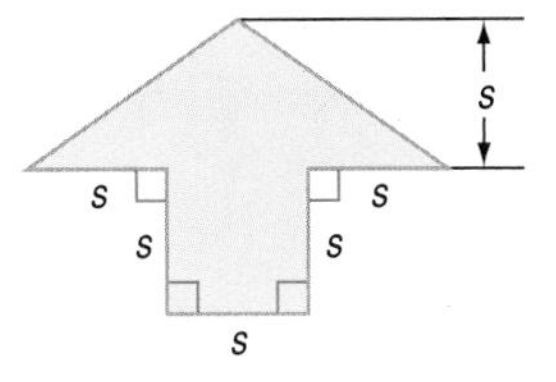

36. REASONING Describe the possible values of t if $s = \frac{r}{t-2}$. Explain your reasoning.

37. OPEN ENDED Write a formula for A, the area of a geometric figure such as a triangle or rectangle. Then solve the formula for a variable other than A.

38. *Writing in Math* Use the information on page 117 to explain how equations are used to design roller coasters. Include a list of steps you could use to solve the equation for h, and the height of the second hill of the roller coaster.

STANDARDIZED TEST PRACTICE

39. If $2x + y = 5$, what is the value of $4x$?

A $10 - y$

B $10 - 2y$

C $\frac{5 - y}{2}$

D $\frac{10 - y}{2}$

40. REVIEW What is the base of the triangle if the area is 56 meters squared?

F 4 m

G 8 m

H 16 m

J 28 m

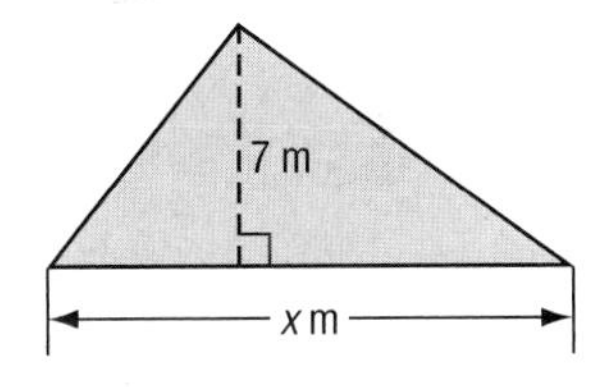

Spiral Review

41. FOOD In order for a food to be marked "reduced fat," it must have at least 25% less fat than the same full-fat food. If one ounce of reduced-fat cheese has 7.5 grams of fat, what is the least amount of fat in one ounce of regular cheese? (Lesson 2-7)

Solve each proportion. (Lesson 2-6)

42. $\frac{2}{9} = \frac{5}{a}$ **43.** $\frac{15}{32} = \frac{t}{8}$ **44.** $\frac{x+1}{8} = \frac{3}{4}$

GET READY for the Next Lesson

PREREQUISITE SKILL Use the Distributive Property to rewrite each expression without parentheses. (Lesson 1-5)

45. $6(2 - t)$ **46.** $(5 + 2m)3$ **47.** $-7(3a + b)$ **48.** $\frac{2}{3}(6h - 9)$

2-9 Weighted Averages

Main Ideas

- Solve mixture problems.
- Solve uniform motion problems.

New Vocabulary

weighted average
mixture problem
uniform motion problem

GET READY for the Lesson

In an individual figure skating competition, the score for the long program is worth twice the score for the short program. Suppose a skater scores 5.5 in the short program and 5.8 in the long program. The final score is determined using a weighted average.

$$\frac{5.5(1) + 5.8(2)}{1 + 2} = \frac{5.5 + 11.6}{3}$$

$$= \frac{17.1}{3} \text{ or } 5.7$$ The final score would be 5.7.

Mixture Problems The skater's average score is an example of a weighted average. The **weighted average** M of a set of data is the sum of the product of the number of units and the value per unit divided by the sum of the number of units. **Mixture problems**, in which two or more parts are combined into a whole, are solved using weighted averages.

Real-World EXAMPLE Prices

1 TRAIL MIX How many pounds of mixed nuts selling for \$4.75 per pound should be mixed with 10 pounds of dried fruit selling for \$5.50 per pound to obtain a trail mix that sells for \$4.95 per pound?

Let w = the number of pounds of mixed nuts. Make a table.

	Units (lb)	Price per Unit (lb)	Total Price
Dried Fruit	10	\$5.50	$5.50(10)$
Mixed Nuts	w	\$4.75	$4.75w$
Trail Mix	$10 + w$	\$4.95	$4.95(10 + w)$

Price of dried fruit	plus	price of nuts	equals	price of trail mix.
$5.50(10)$	$+$	$4.75w$	$=$	$4.95(10 + w)$

$5.50(10) + 4.75w = 4.95(10 + w)$ Original equation

$55.00 + 4.75w = 49.50 + 4.95w$ Distributive Property

$55.00 + 4.75w - 4.75w = 49.50 + 4.95w - 4.75w$ Subtract $4.75w$ from each side.

$55.00 = 49.50 + 0.20w$ Simplify.

$55.00 - 49.50 = 49.50 + 0.20w - 49.50$ Subtract 49.50 from each side.

$5.50 = 0.20w$ Simplify.

$\frac{5.50}{0.20} = \frac{0.20w}{0.20}$ Divide each side by 0.20.

$27.5 = w$ Simplify.

27.5 pounds of mixed nuts should be used for the trail mix.

1. COFFEE How many pounds of coffee beans that sell for \$9.50 per pound should be mixed with 2 pounds of coffee beans that sell for \$11.75 per pound to obtain a mix that sells for \$10 per pound?

Percent problems can also be solved using weighted averages.

Real-World EXAMPLE Percent Mixture Problems

Study Tip

Mixture Problems

When you organize the information in mixture problems, remember that the final mixture must contain the sum of the parts in the correct quantities and at the correct percents.

2 SCIENCE A chemistry experiment calls for a 30% solution of copper sulfate. Kendra has 40 milliliters of 25% solution. How many milliliters of 60% solution should she add to make a 30% solution?

Let x = the amount of 60% solution to be added. Make a table.

	Amount of Solution (mL)	Amount of Copper Sulfate
25% Solution	40	$0.25(40)$
60% Solution	x	$0.60x$
30% Solution	$40 + x$	$0.30(40 + x)$

Write and solve an equation using the information in the table.

Amount of copper sulfate in 25% solution	plus	amount of copper sulfate in 60% solution	equals	amount of copper sulfate in 30% solution.
$0.25(40)$	$+$	$0.60x$	$=$	$0.30(40 + x)$

$0.25(40) + 0.60x = 0.30(40 + x)$ Original equation

$10 + 0.60x = 12 + 0.30x$ Distributive Property

$10 + 0.60x - 0.30x = 12 + 0.30x - 0.30x$ Subtract $0.30x$ from each side.

$10 + 0.30x = 12$ Simplify.

$10 - 0.30x - 10 = 12 - 10$ Subtract 10 from each side.

$0.30x = 2$ Simplify.

$\frac{0.30x}{0.30} = \frac{2}{0.30}$ Divide each side by 0.30.

$x \approx 6.67$ Simplify.

Kendra should add 6.67 milliliters of the 60% solution to the 40 milliliters of the 25% solution. Check by substituting 6.67 for x in the original equation.

CHECK Your Progress

2. ANTIFREEZE One type of antifreeze is 40% glycol, and another type of antifreeze is 60% glycol. How much of each kind should be used to make 100 gallons of antifreeze that is 48% glycol?

Online Personal Tutor at algebra1.com

Study Tip

Uniform Motion Problems

Uniform motion problems are sometimes referred to as **rate problems**.

Uniform Motion Problems Motion problems are another application of weighted averages. **Uniform motion problems** are problems where an object moves at a certain speed, or rate. The formula $d = rt$ is used to solve these problems. In the formula, d represents distance, r represents rate, and t represents time.

Real-World EXAMPLE Speed of One Vehicle

3 TRAVEL On Alberto's drive to his aunt's house, the traffic was light and he drove the 45-mile trip in one hour. However, the return trip took him two hours. What was his average speed for the round trip?

To find the average speed for each leg of the trip, rewrite $d = rt$ as $r = \frac{d}{t}$.

Concepts in Motion
Animation
algebra1.com

Going	**Returning**
$r = \frac{d}{t}$	$r = \frac{d}{t}$
$= \frac{45 \text{ miles}}{1 \text{ hour}}$ or 45 miles per hour	$= \frac{45 \text{ miles}}{2 \text{ hours}}$ or 22.5 miles per hour

You may think that the average speed of the trip would be $\frac{45 + 22.5}{2}$ or 33.75 miles per hour. However, Alberto did not drive at these speeds for equal amounts of time. You must find the weighted average for the trip.

Round Trip Let M = the average speed.

$M = \frac{45(1) + 22.5(2)}{1 + 2}$ Definition of weighted average

$= \frac{90}{3}$ or 30 Simplify.

Alberto's average speed was 30 miles per hour.

CHECK Your Progress

3. EXERCISE Austin jogged 2.5 miles in 16 minutes and then walked 1 mile in 10 minutes. What was his average speed?

Real-World Link

Under ideal conditions, a siren can be heard from up to 440 feet. However, under normal conditions, a siren can be heard from only 125 feet.

Source: U.S. Department of Transportation

Real-World EXAMPLE Speeds of Two Vehicles

4 SAFETY Use the information about sirens at the left. A car and an emergency vehicle are heading toward each other. The car is traveling at a speed of 30 miles per hour or about 44 feet per second. The emergency vehicle is traveling at a speed of 50 miles per hour or about 74 feet per second. If the vehicles are 1000 feet apart and the conditions are ideal, in how many seconds will the driver of the car first hear the siren?

Draw a diagram. The driver can hear the siren when the total distance traveled by the two vehicles equals 1000 − 440 or 560 feet.

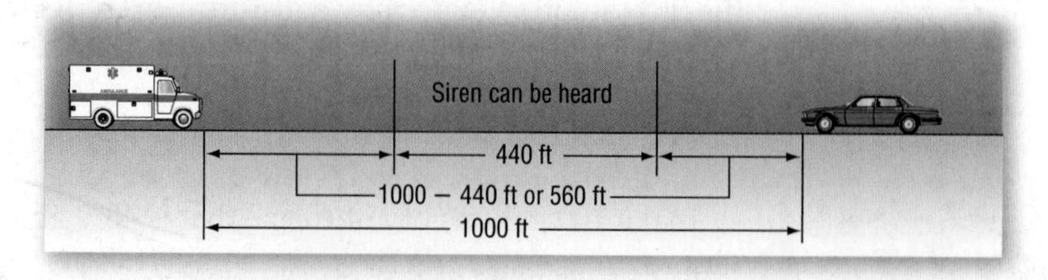

Let t = the number of seconds until the driver can hear the siren.

	r	t	$d = rt$
Car	44	t	$44t$
Emergency Vehicle	74	t	$74t$

Write and solve an equation.

Distance traveled by car	plus	distance traveled by emergency vehicle	equals	560 feet.
$44t$	$+$	$74t$	$=$	560

$44t + 74t = 560$ Original equation

$118t = 560$ Simplify.

$\frac{118t}{118} = \frac{560t}{118}$ Divide each side by 118.

$t \approx 4.75$ The driver will hear the siren in about 4.75 seconds.

CHECK Your Progress

4. **CYCLING** Two cyclists begin traveling in opposite directions on a circular bike trail that is 5 miles long. One cyclist travels 12 miles per hour, and the other travels 18 miles per hour. How long will it be before they meet?

CHECK Your Understanding

Example 1 (p. 122)

BUSINESS For Exercises 1–3, use the following information.

The Candle Supply Store sells votive wax for \$0.90 a pound and low-shrink wax for \$1.04 a pound. How many pounds of low-shrink wax should be mixed with 8 pounds of votive wax to obtain a blend that sells for \$0.98 a pound?

1. Copy and complete the table representing the problem.

	Number of Pounds	Price per Pound	Total Price
Votive Wax	8	\$0.90	0.90(8)
Low-Shrink Wax	p		
Blend	$8 + p$		

2. Write an equation to represent the problem.
3. How many pounds of low-shrink wax should be mixed with 8 pounds of votive wax?

4. **COFFEE** A specialty coffee store wants to create a special mix using two coffees, one priced at \$6.40 per pound and the other priced at \$7.28 per pound. How many pounds of the \$7.28 coffee should be mixed with 9 pounds of the \$6.40 coffee to sell the mixture for \$6.95 per pound?

Example 2 (p. 123)

FOOD For Exercises 5–7, use the following information.

How many quarts of pure pineapple juice should Theo add to a 20% pineapple drink to create 5 quarts of a 50% pineapple juice mixture?

5. Copy and complete the table representing the problem.

	Quarts	Total Amount of Juice
20% Juice	$5-n$	
100% Juice	n	
50% Juice		

6. Write an equation to represent the problem.
7. How much pure pineapple juice and 20% juice should Theo use?

8. METALS An alloy of metals is 25% copper. Another alloy is 50% copper. How much of each should be used to make 1000 grams of an alloy that is 45% copper?

Example 3 (p. 124)

9. TRAVEL A boat travels 16 miles due north in 2 hours and 24 miles due west in 2 hours. What is the average speed of the boat?

10. EXERCISE Felisa jogged 3 miles in 25 minutes and then jogged 3 more miles in 30 minutes. What was her average speed in miles per minute?

Example 4 (pp. 124–125)

11. CYCLING A cyclist begins traveling 18 miles per hour. At the same time and place, an in-line skater follows the cyclist's path and begins traveling 6 miles per hour. After how long will they be 24 miles apart?

12. RESCUE A fishing boat radioed the Coast Guard for a helicopter to pick up a sick crew member. At the time of the message, the boat is 660 kilometers from the helicopter and heading toward it. The average speed of the boat is 30 kilometers per hour, and the average speed of the helicopter is 300 kilometers per hour. How long will it take the helicopter to reach the boat?

Exercises

HOMEWORK HELP

For Exercises	See Examples
13–24	1
25, 26	2
27, 28	3
29–34	4

13. GRADES In Ms. Martinez's science class, a test is worth three times as much as a quiz. If a student has test grades of 85 and 92 and quiz grades of 82, 75, and 95, what is the student's average grade?

14. ANALYZE TABLES At Westbridge High School, a student's grade point average (GPA) is based on the student's grade and the class credit rating. Brittany's grades for this quarter are shown. Find Brittany's GPA if a grade of A equals 4 and a B equals 3.

Class	Credit Rating	Grade
Algebra 1	1	A
Science	1	A
English	1	B
Spanish	1	A
Music	$\frac{1}{2}$	B

METALS For Exercises 15–18, use the following information.

In 2005, the international price of gold was $432 per ounce, and the international price of silver was $7.35 per ounce. Suppose gold and silver were mixed to obtain 15 ounces of an alloy worth $177.21 per ounce.

15. Copy and complete the table representing the problem.

	Number of Ounces	Price per Ounce	Value
Gold	g		
Silver	$15 - g$		
Alloy			

16. Write an equation to represent the problem.

17. How much gold was used in the alloy?

18. How much silver was used in the alloy?

19. FUND-RAISING The Madison High School marching band sold solid-color gift wrap for $4 per roll and print gift wrap for $6 per roll. The total number of rolls sold was 480, and the total amount of money collected was $2340. How many rolls of each kind of gift wrap were sold?

BUSINESS For Exercises 20–23, use the following information.

Party Supplies Inc. sells metallic balloons for $2 each and helium balloons for $3.50 per dozen. Yesterday, they sold 36 more metallic balloons than dozens of helium balloons. The total sales for both types of balloons were $281.00.

20. Copy and complete the table representing the problem.

	Number	Price	Total Price
Metallic Balloons	b		
Dozens of Helium Balloons	$b - 36$		

21. Write an equation to represent the problem.

22. How many metallic balloons were sold?

23. How many dozen helium balloons were sold?

24. MONEY Lakeisha spent $4.57 on color copies and black-and-white copies for her project. She made 7 more black-and-white copies than color copies. How many color copies did she make?

Type of Copy	Cost Per Page
color	$0.44
black-and-white	$0.07

25. FOOD Refer to the graphic at the left. How much whipping cream and 2% milk should be mixed to obtain 35 gallons of milk with 4% butterfat?

26. CHEMISTRY Hector is performing a chemistry experiment that requires 140 milliliters of a 30% copper sulfate solution. He has a 25% copper sulfate solution and a 60% copper sulfate solution. How many milliliters of each solution should he mix to obtain the needed solution?

27. TRAVEL A boat travels 36 miles in 1.5 hours and then 14 miles in 0.75 hour. What is the average speed of the boat?

28. EXERCISE An inline skater skated 1.5 miles in 28 minutes and then 1.2 more miles in 10 minutes. What was the average speed in miles per minute?

TRAVEL For Exercises 29–31, use the following information.

Two trains leave Smithville at the same time, one traveling east and the other west. The eastbound train travels at 40 miles per hour, and the westbound train travels at 30 miles per hour. Let h represent the hours since departure.

29. Copy and complete the table representing the situation.

	r	t	$d = rt$
Eastbound Train			
Westbound Train			

30. Write an equation to determine when the trains will be 245 miles apart.

31. In how many hours will the trains be 245 miles apart?

ANIMALS For Exercises 32–34, use the graphic at the right.

Let t represent the number of seconds until the cheetah catches its prey.

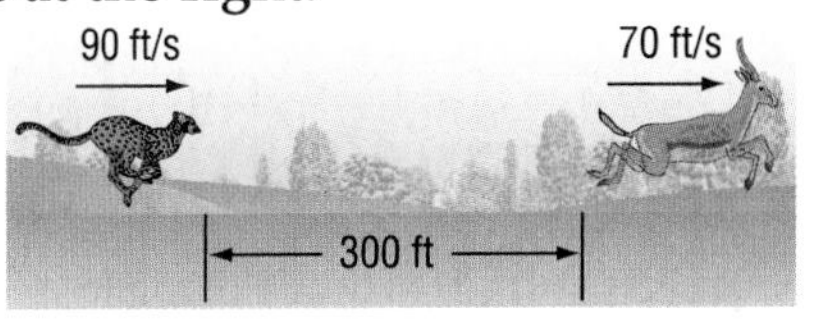

32. Copy and complete the table representing the situation.

33. Write an equation to determine when the cheetah will catch its prey.

34. When will the cheetah catch its prey?

	r	t	$d = rt$
Cheetah			
Prey			

35. **TRACK AND FIELD** A sprinter has a bad start, and his opponent is able to start 1 second before him. If the sprinter averages 8.2 meters per second and his opponent averages 8 meters per second, will he be able to catch his opponent before the end of the 200-meter race? Explain.

36. **TRAVEL** A subway travels 60 miles per hour from Glendale to Midtown. Another subway, traveling at 45 miles per hour, takes 11 minutes longer for the same trip. How far apart are Glendale and Midtown?

H.O.T. Problems

37. **OPEN ENDED** Describe a real-world example of a weighted average.

38. **CHALLENGE** Write a mixture problem for $1.00x + 0.28(40) = 0.40(x + 40)$.

39. ***Writing in Math*** Use the information on page 122 to explain how scores are calculated in a figure-skating competition. Include an explanation of how a weighted average can be used to find a skating score and a demonstration of how to find the weighted average of a skater who received a 4.9 in the short program and a 5.2 in the long program.

STANDARDIZED TEST PRACTICE

40. **REVIEW** Eula Jones is investing \$6000, part at 4.5% interest and the rest at 6% interest. If d represents the amount invested at 4.5%, which expression represents the amount of interest earned in one year by the account paying 6%?

A $0.06d$
B $0.06(d - 6000)$
C $0.06(d + 6000)$
D $0.06(6000 - d)$

41. Todd drove from Boston to Cleveland, a distance of 616 miles. His breaks, gasoline, and food stops took 2 hours. If his trip took 16 hours altogether, what was his average speed?

F 38.5 mph
G 40 mph
H 44 mph
J 47.5 mph

Spiral Review

Solve each equation for the variable specified. (Lesson 2-8)

42. $a + 6 = \frac{b - 1}{4}$, for b

43. $3t - 4 = 6t - s$, for t

State whether each percent of change is a percent of increase or a percent of decrease. Then find the percent of change. Round to the nearest whole percent. (Lesson 2-7)

44. original: \$25
new: \$14

45. original: 35
new: 42

46. original: 244
new: 300

47. **MONEY** Tyler had \$80 in his savings account. After his mother made a deposit, his new balance was \$115. Write and solve an equation to find the amount of the deposit. (Lesson 2-2)

Cross-Curricular Project

Algebra and Social Studies

You're Only as Old as You Feel! It's time to complete your project. Use the information and data you have gathered about living to be 100 to prepare a portfolio or Web page. Be sure to include graphs and/or tables in the presentation.

Cross-Curricular Project at algebra1.com

EXTEND 2-9

Spreadsheet Lab

Finding a Weighted Average

You can use a spreadsheet to calculate weighted averages. It allows you to make calculations and print almost anything you can organize in a table.

The basic unit in a spreadsheet is called a cell. A cell may contain numbers, words, or a formula. Each cell is named by the column and row that describe its location. For example, cell B4 is in column B, row 4.

EXAMPLE

Greta Norris manages a sales firm. Each of her employees earns a different rate of commission. She has entered the commission rate and the sales for each employee for October in a spreadsheet. What was the average commission rate?

Commission

◇	A	B	C	D
1	Employee	Rate	Sales	Commission
2	Mark	0.05	1588	=C2*B2
3	Andrea	0.075	1800	=C3*B3
4	Tamika	0.1	2008	=C4*B4
5	Daniel	0.1	2105	=C5*B5
6	Lasanda	0.05	1725	=C6*B6
7	Diego	0.06	1988	=C7*B7
8	Total		=SUM(C2:C7)	=SUM(D2:D7)
9	Weighted Average Commission Rate	=D8 / C8		
10				

Sheet 1 / Sheet 2 / Sheet 3

The spreadsheet shows the formula to calculate the weighted average. It multiplies the rate of commission by the amount of sales and finds the total amount of sales and amount of commission. Then it divides the amount of commission by the amount of sales. To the nearest tenth, the average rate of commission is 7.4%.

EXERCISES

For Exercises 1–3, use the table at the right.

November Coffee Sales

Product	Pounds Sold	Price Per Pound ($)
Hawaiian Cafe	56	16.95
Mocha Java	97	12.59
House Blend	124	10.75
Decaf Espresso	71	10.15
Breakfast Blend	69	11.25
Italian Roast	45	9.95

1. What is the average price of a pound of coffee?
2. How does the November weighted average change if all of the coffee prices are increased by $1.00? by 10%?
3. Find the weighted average of a pound of coffee if the shop sold 50 pounds of each type of coffee. How does the weighted average compare to the average of the per-pound coffee prices? Explain.

CHAPTER 2

Study Guide and Review

FOLDABLES™ Study Organizer

GET READY to Study

Be sure the following Key Concepts are noted in your Foldable.

Key Concepts

Writing Equations (Lesson 2-1)

- Four-Step Problem Solving Plan:
  - Step 1 Explore the problem.
  - Step 2 Plan the solution.
  - Step 3 Solve the problem.
  - Step 4 Check the solution.

Solving Equations (Lessons 2-2 to 2-5)

- Addition and Subtraction Properties of Equality: If an equation is true and the same number is added to or subtracted from each side, the resulting equation is true.
- Multiplication and Division Properties of Equality: If an equation is true and each side is multiplied by the same number or divided by the same non-zero number, the resulting equation is true.
- Steps for Solving Equations:
  - Step 1 Use the Distributive Property if necessary.
  - Step 2 Simplify expressions on each side.
  - Step 3 Use the Addition and/or Subtraction Properties of Equality to get the variables on one side and the numbers without variables on the other side.
  - Step 4 Simplify the expression on each side of the equals sign.
  - Step 5 Use the Multiplication or Division Property of Equality to solve.

Ratios and Proportions (Lesson 2-6)

- The Means-Extremes Property of Proportion states that in a proportion, the product of the extremes is equal to the product of the means.

Key Vocabulary

consecutive integers (p. 94)
defining a variable (p. 71)
dimensional analysis (p. 119)
equivalent equations (p. 79)
extremes (p. 106)
formula (p. 72)
four-step problem solving plan (p. 71)
identity (p. 99)
means (p. 106)
mixture problem (p. 122)
multi-step equations (p. 92)
number theory (p. 94)
percent of change (p. 111)
percent of decrease (p. 111)
percent of increase (p. 111)
proportion (p. 105)
rate (p. 107)
ratio (p. 105)
scale (p. 108)
solve an equation (p. 79)
uniform motion problem (p. 123)
weighted average (p. 122)

Vocabulary Check

State whether each sentence is *true* or *false*. If *false*, replace the underlined word or number to make a true sentence.

1. An example of consecutive integers is -10 and -9.
2. A comparison of two numbers by division is called a ratio.
3. In the proportion $1:4 = x:2$, 4 and x are called the means.
4. Equations with more than one operation are called equivalent equations.
5. Dimensional analysis is the process of carrying units throughout a computation.
6. Mixture problems are solved using number theory.
7. An equation that is true for every value of the variable is called an identity.

Lesson-by-Lesson Review

2–1 Writing Equations (pp. 70–76)

Translate each sentence into an equation.

8. The sum of z and one fifth of w is ninety-six.

9. The product of m and n is as much as three times the sum of m and 8.

10. The difference of p and thirteen is identical to the square of p.

Translate each equation into a verbal sentence.

11. $\frac{56}{g} = 7 - 3g$ **12.** $8 + 2k = \frac{1}{4}k$

13. DESERTS One third of Earth's land surface is desert. There are 50 million square kilometers of desert. Define a variable and then write and solve an equation to find the total number of square kilometers of land surface.

Example 1 **Translate the following sentence into an equation.**

Forty-one increased by twice a number m is the same as three times the sum of a number m and seven.

$41 + 2m = 3(m + 7)$

Example 2 **Translate $2y - 5 = \frac{1}{2}y$ into a verbal sentence.**

Twice a number y minus five equals one half y.

2–2 Solving Equations by Using Addition and Subtraction (pp. 78–84)

Solve each equation.

14. $h - 15 = -22$ **15.** $\frac{3}{5} = \frac{2}{3} + a$

16. $16 - (-q) = 83$ **17.** $-55 = x + (-7)$

Write an equation for each problem. Then solve the equation and check your solution.

18. Nine subtracted from a number equals −15. Find the number.

19. The sum of a number and −71 is 29. What is the number?

20. CANS The can opener was invented in 1858. That was 48 years after cans were first introduced. Write and solve an equation to determine in what year the can was introduced.

Example 3 **Solve $13 + p = -5$.**

$13 + p = -5$	Original equation
$13 + p - 13 = -5 - 13$	Subtract 13 from each side.
$p = -18$	Check this result.

Example 4 **Write an equation for the problem below. Then solve the equation.**

The difference of a number and 62 is −47. What is the number?

$x - 62 = -47$	Original equation
$x - 62 + 62 = -47 + 62$	Add 62 to each side.
$x = 15$	Check this result.

Study Guide and Review

2-3 Solving Equations by Using Multiplication and Division (pp. 85–90)

Solve each equation. Check your solution.

21. $\frac{a}{-7} = -29$

22. $-18 = \frac{3}{5}m$

23. $14p = 42$

24. $\frac{y}{32} = \frac{5}{-8}$

Write an equation for each problem. Then solve the equation.

25. Three eighths of a number equals 9. What is the number?

26. The quotient of a number and -4 is -72. Find the number.

27. DINING China produces 45 billion chopsticks per year from 25 million trees. Write and solve an equation to find the number of chopsticks that can be produced from 1 tree.

Example 5 **Solve $\frac{k}{-3} = -9$.**

$\frac{k}{-3} = -9$	Original equation
$\frac{k}{-3}(-3) = -9(-3)$	Multiply each side by -3.
$k = 27$	Check this result.

Example 6 **Write an equation for the problem below. Then solve the equation.**

48 is negative twelve times a number. Find the number.

$48 = -12n$	Original equation
$\frac{48}{-12} = \frac{-12n}{-12}$	Divide each side by -12.
$-4 = n$	Check this result.

2-4 Solving Multi-Step Equations (pp. 92–97)

Solve each equation. Check your solution.

28. $5 = 4t - 7$

29. $6 + \frac{y}{3} = -45$

30. $9 = \frac{d + 5}{8}$

31. $\frac{c}{-4} - 2 = -36$

Write an equation and solve each problem.

32. 22 increased by six times a number is -20. What is the number?

33. Find three consecutive odd integers with a sum of 39.

34. THEATER Each row in a theater has eight more seats than the previous row. If the first row has 14 seats, which row has 46 seats?

Example 7 **Write and solve an equation to find three consecutive even integers with a sum of 150.**

Let n = the least even integer.
Then $n + 2$ = the next greater even integer and $n + 4$ = the greatest of the three even integers.

$n + (n + 2) + (n + 4) = 150$	Original equation
$3n + 6 = 150$	Simplify.
$3n = 144$	Subtract 6 from each side.
$n = 48$	Divide each side by 3.

The integers are 48, 50, and 52.

Mixed Problem Solving
For mixed problem-solving practice, see page 745.

2-5 Solving Equations with the Variable on Each Side (pp. 98–103)

Solve each equation. Check your solution.

35. $5b + 3 = 9b - 17$

36. $\frac{2}{3}n - 3 = \frac{1}{3}(2n - 9)$

37. FUND-RAISING The Band Boosters pay $200 to rent a concession stand at a university football game. They purchase cans of soft drinks for $0.25 each and sell them at the game for $1.50 each. How many cans of soft drinks must they sell to break even?

Example 8 **Solve $7x + 56 = 5x - 11$.**

$7x + 56 = 5x - 11$	Original equation
$7x + 56 - 5x = 5x - 11 - 5x$	Subtract 5x from each side.
$2x + 56 = -11$	Simplify.
$2x + 56 - 56 = -11 - 56$	Subtract 56 from each side.
$2x = -67$	Simplify.
$\frac{2x}{2} = \frac{-67}{2}$	Divide each side by 2.
$x = -33.5$	Check this result.

2-6 Ratios and Proportions (pp. 105–110)

Solve each proportion. If necessary, round to the nearest hundredth.

38. $\frac{14}{x} = \frac{20}{8}$

39. $\frac{0.47}{6} = \frac{1.41}{m}$

40. WORLD RECORDS Dustin drank 14 ounces of tomato juice through a straw in 33 seconds. At this rate, approximately how long would it take him to drink a bottle of water that contains 16.9 ounces?

Example 9 **Solve the proportion $\frac{h}{15} = \frac{7}{21}$.**

$\frac{h}{15} = \frac{7}{21}$	Original equation
$h(21) = 15(7)$	Find the cross products.
$21h = 105$	Distributive Property
$\frac{21h}{21} = \frac{105}{21}$	Divide each side by 21.
$h = 5$	Simplify.

2-7 Percent of Change (pp. 111–115)

State whether each percent of change is a percent of increase or a percent of decrease. Then find each percent of change. Round to the nearest whole percent.

41. original: 54
new: 46

42. original: 17
new: 33

43. TIPS Felicia's meal cost $23.74. How much money should she leave for a 15% tip?

Example 10 **Find the total price of a CD that costs $18.75 with 6.5% sales tax.**

6.5% of $18.75 $= 0.065 \times 18.75$	$6.5\% = 0.065$
$= 1.21875$	Multiply.

Round $1.21875 to $1.22. Add this amount to the original price.

$18.75 + $1.22 = $19.97

The total price of the CD is $19.97.

2-8 Solving for a Specific Variable (pp. 117–121)

Solve each equation or formula for the variable specified.

44. $d = 5v + m$, for v

45. $\frac{7}{4}k - g = s$, for k

46. $\frac{ac - 3}{6} = w$, for c

47. $9h + z = pq + 2h$, for h

48. TRAVEL In the formula $d = rt$, d is the distance, r is the rate, and t is the time spent traveling. Solve this equation for r. Then find Aida's rate if she drove 219 miles in $3\frac{3}{4}$ hours.

Example 11 Solve $8q + 3b = 12$ for b.

$8q + 3b = 12$	Original equation
$8q + 3b - 8q = 12 - 8q$	Subtract $8q$ from each side.
$3b = 12 - 8q$	Simplify.
$\frac{3b}{3} = \frac{12 - 8q}{3}$	Divide each side by 3.
$b = \frac{12 - 8q}{3}$	Simplify.

The value of b is $\frac{12 - 8q}{3}$.

2-9 Weighted Averages (pp. 122–128)

49. COFFEE Jerome blends Brand A of coffee that sells for \$9.50 a pound with Brand B of coffee that sells for \$12.00 a pound. If the 45 pounds of blend sells for \$10.50 a pound, how many pounds of each type of coffee did Jerome use?

50. PUNCH Raquel is mixing lemon-lime soda and a fruit juice blend that is 45% juice. If she uses 3 quarts of soda, how many quarts of fruit juice must be added to produce punch that is 30% juice?

51. JOGGING Delmar ran 4 miles in 22 minutes, stopped and rested, and ran an additional 4 miles in 28 minutes. Find his average speed.

52. TRAVEL Connor is driving 65 miles per hour on the highway. Ed is 15 miles behind him driving at 70 miles per hour. After how many hours will Ed catch up to Connor?

Example 12 MANUFACTURING Percy mixes 750 liters of water with 250 liters of 2% bleach. What percent of bleach is in the resultant mixture?

Let x = the percent of bleach in the resultant mixture. Make a table.

	Amount of Solution (L)	Amount of Resultant Mixture
Water (0% bleach)	750	750(0)
Bleach (2% bleach)	250	250(0.02)
Resultant Mixture (x% bleach)	1000	1000(x)

Write and solve an equation using the information in the table.

$750(0) + 250(0.02) = 1000(x)$	Original equation
$5 = 1000x$	Simplify.
$0.005 = x$	Divide each side by 1000.

There is 0.5% bleach in the resultant mixture.

Practice Test

Translate each sentence into an equation.

1. The sum of x and four times y is equal to twenty.
2. Two thirds of n is negative eight fifths.

Solve each equation. Then check your solution.

3. $-15 + k = 8$
4. $k - 16 = -21$
5. $-1.2x = 7.2$
6. $5a = 125$
7. $\frac{t-7}{4} = 11$
8. $\frac{3}{4}y = 27$
9. $-12 = 7 - \frac{y}{3}$
10. $-\frac{2}{3}z = -\frac{4}{9}$

11. **MULTIPLE CHOICE** The perimeter of the larger square is 11.6 centimeters greater than the perimeter of the smaller square. What are the side lengths of the smaller square?

A 1.1 cm

B 4.2 cm

C 4.5 cm

D 16.8 cm

Solve each equation. Then check your solution.

12. $-3(x + 5) = 8x + 18$
13. $2p + 1 = 5p - 11$

14. **POSTAGE** What was the percent of increase in the price of a first-class stamp from 2001 to 2006? Round to the nearest whole percent.

First-Class Stamps	
Year	**Cost of Stamp**
2001	34¢
2006	39¢

15. **MULTIPLE CHOICE** If $\frac{4}{5}$ of $\frac{3}{4} = \frac{2}{5}$ of $\frac{x}{4}$, find the value of x.

F 12

G 6

H 3

J $\frac{3}{2}$

Solve each proportion.

16. $\frac{2}{10} = \frac{1}{a}$
17. $\frac{3}{5} = \frac{24}{x}$
18. $\frac{n}{4} = \frac{3.25}{52}$
19. $\frac{5}{12} = \frac{10}{x-1}$

State whether each percent of change is a percent of increase or a percent of decrease. Then find the percent of change. Round to the nearest whole percent.

20. original: 45
    new: 9
21. original: 12
    new: 20

Solve each equation or formula for the variable specified.

22. $h = at - 0.25vt^2$, for a
23. $a(y + 1) = b$, for y

24. **SALES** At The Central Perk coffee shop, Destiny sold 30 more cups of cappuccino than espresso, for a total of $178.50 worth of espresso and cappuccino. How many cups of each were sold?

Coffee	Cost per Cup ($)
espresso	2.00
cappuccino	2.50

25. **BOATING** *The Yankee Clipper* leaves the pier at 9:00 A.M. at 8 knots (nautical miles per hour). A half hour later, *The River Rover* leaves the same pier in the same direction traveling at 10 knots. At what time will *The River Rover* overtake *The Yankee Clipper*?

CHAPTER 2

Standardized Test Practice

Cumulative, Chapters 1–2

Read each question. Then fill in the correct answer on the answer document provided by your teacher or on a sheet of paper.

1. Martin's Car Rental rents cars for \$20 per 100 miles driven. How many miles could be driven for \$125?

A 6.25
B 100
C 625
D 6250

2. Marcus's telephone company charges \$30 per month and \$0.05 per minute. How much did it cost him in January if he talked for 120 minutes?

F \$24
G \$36
H \$60
J \$90

3. GRIDDABLE Solve the equation.

$$\frac{1}{2}x + 12 = \frac{3}{4}x - 16$$

4. Leland's lawn service charged Mr. Mackenzie \$65 to plant a new tree plus \$18.50 per hour to mow and fertilize his lawn. The total charge was \$157.50. For about how many hours did Leland's lawn service work on Mr. Mackenzie's lawn?

A 4
B 5
C 6
D 8

5. Lauren has \$35 to spend on beads for making necklaces. If she spends \$15 for string and each bead costs \$2.50, solve the inequality $2.50n + 15 \leq 35$ to determine how many beads she can buy.

F more than 14
G less than 14
H more than 8
J less than 8

6. Margot and Lainy are selling bracelets at the farmers' market. It costs \$25 to rent a booth. They sold each of their bracelets for \$15. If the profit was \$355, which equation represents the number of bracelets n they sold?

A $355 = 15n - 25$
B $355 = 25 + 15n$
C $25 = 355 + 15n$
D $355 = (25 + n)15$

TEST-TAKING TIP

Question 6 When you write an equation, check that the given values make a true statement. For example, in Question 6, substitute the values for the number of bracelets and the cost of the booth into your equation.

7. GRIDDABLE Brianne makes baby blankets for a baby store. She works on the blankets 30 hours per week. The store pays her \$9.50 per hour plus 30% of the profit. If her hourly rate is increased by \$0.75 and her commission is raised to 40%, how much will she earn in dollars if a total of \$300 in blankets is sold?

8. Maya is selling T-shirts to raise money for the student council. She makes \$6.75 for every T-shirt that she sells. If she wants to raise \$800 selling the T-shirts, what is a reasonable number of shirts she should sell?

F 100
G 120
H 500
J 5400

9. GRIDDABLE Mr. Hiskey is making a doll-sized model of his house for his daughter. The model is $\frac{1}{16}$ the size of the actual house. If the door in the model is 6 inches tall, how many inches tall is the actual door on the house?

Math Online Standardized Test Practice at algebra1.com

Preparing for Standardized Tests
For test-taking strategies and more practice, see pages 756–773.

10. Carter is riding his bicycle. He leaves his house and stops at a park that is 25 miles from his house. After the stop, he continues to ride his bicycle at a constant rate of 18 miles per hour away from his house. Which equation could be used to determine the time in hours t it will take him to reach a distance of 120 miles from his house?

A $120 = 18t + 25$

B $120 = t(18 + 25)$

C $120 = 18(t + 25)$

D $120 = 25t + 18$

11. The amount of water needed to fill a balloon best represents the balloon's _____.

F volume

G surface area

H circumference

J perimeter

12. The scale factor of two similar triangles is 5:2. The perimeter of the smaller triangle is 108 inches. What is the perimeter of the larger triangle?

A 540 in.

B 270 in.

C 216 in.

D 43.2 in.

13. Bailey estimates that his income has gone up $2000 each year from 1999 to 2005. What additional information is needed to calculate his income in 2005?

F The range of his income from 1999 to 2005.

G His income in 1999.

H How much he thinks his income will go up in the next year.

J The national average income.

14. Rectangle *WXYZ* is shown below.

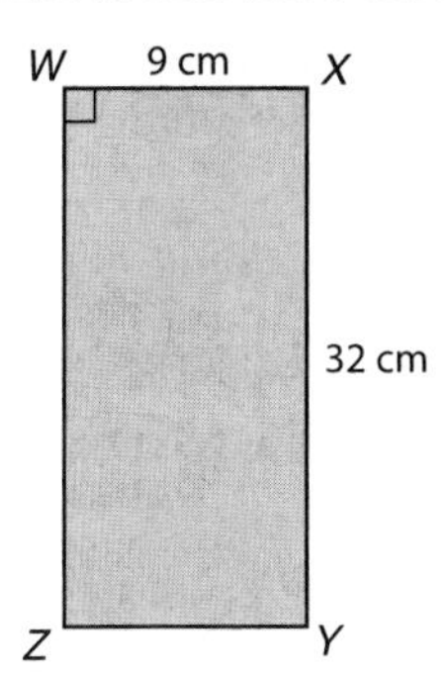

If each dimension of the rectangle is doubled, what is the perimeter of *WXYZ*?

A 82 cm

B 100 cm

C 164 cm

D 288 cm

Pre-AP

Record your answers on a sheet of paper. Show your work.

15. Latoya bought 48 one-foot-long sections of fencing. She plans to use the fencing to enclose a rectangular area for a garden.

a. Using ℓ for the length and w for the width of the garden, write an equation for its perimeter.

b. If the length ℓ in feet and width w in feet are positive integers, what is the greatest possible area of this garden?

c. If the length and width in feet are positive integers, what is the least possible area of the garden?

d. How do the shapes of the gardens with the greatest and least areas compare?

NEED EXTRA HELP?															
If You Missed Question...	1	2	3	4	5	6	7	8	9	10	11	12	13	14	15
Go to Lesson or Page...	2-6	2-4	2-5	2-4	1-5	1-3	2-4	2-3	2-6	2-1	2-1	2-6	2-7	707	2-8

Unit 2
Linear Functions

Focus
Use linear functions and inequalities to represent and model real-world situations.

CHAPTER 3
Functions and Patterns

BIG Idea Understand the concepts of a relation and a function, and determine whether a given relation defined by a graph, a set of ordered pairs, or a symbolic expression is a function.

CHAPTER 4
Analyzing Linear Equations

BIG Idea Graph a linear equation, compute an equation's *x*- and *y*-intercepts, and derive linear equations by using the point-slope formula.

CHAPTER 5
Solving Systems of Linear Equations

BIG Idea Solve a system of two linear equations in two variables algebraically and interpret the answer graphically.

CHAPTER 6
Solving Linear Inequalities

BIG Idea Solve absolute value inequalities, multistep problems involving linear inequalities in one variable, and solve a system of two linear inequalities in two variables.

Cross-Curricular Project

Algebra and Sports

The Spirit of the Games The first Olympic Games featured only one event–a foot race. In 2004, the Olympic Games featured thousands of competitors in about 300 events. The 2004 summer games were held in Athens, Greece. In this project, you will explore how linear functions can be used to represent times in Olympic events.

Math Online Log on to algebra1.com to begin.

CHAPTER 3

Functions and Patterns

BIG Ideas

- Understand the skills required to manipulate symbols to solve problems and simplify expressions.
- Understand the meaning of the slope and intercepts of the graphs of linear functions.

Key Vocabulary

arithmetic sequence (p. 165)
function (p. 149)
inverse (p. 145)
y-intercept (p. 156)

Real-World Link

Currency A function is a rule or a formula. You can use a function to describe real-world situations like converting between currencies. For example, in Japan, an item that costs 10,000 yen is equivalent to about 87 U.S. dollars.

Functions and Patterns Make this Foldable to help you organize your notes about graphing relations and functions. Begin with three sheets of notebook paper.

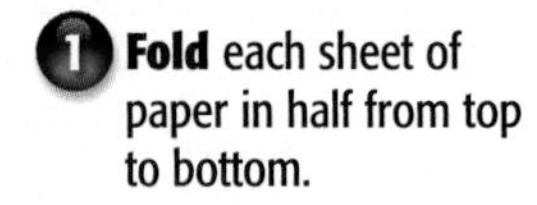

1. **Fold** each sheet of paper in half from top to bottom.

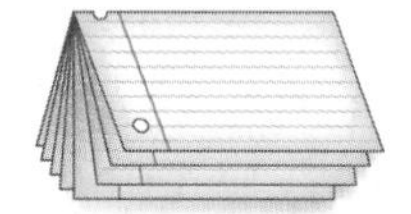

2. **Cut** along fold. Staple the six half-sheets together to form a booklet.

3. **Cut** tabs into margin. The top tab is 2 lines deep, the next tab is 6 lines deep, and so on.

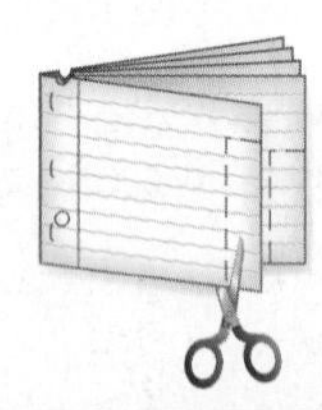

4. **Label** each of the tabs with a lesson number. Use the last page for vocabulary.

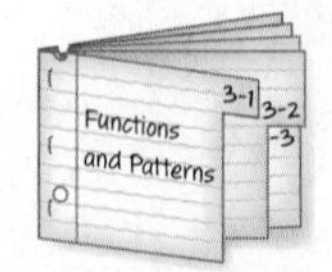

GET READY for Chapter 3

Diagnose Readiness You have two options for checking Prerequisite Skills.

Option 2

Math Online Take the Online Readiness Quiz at algebra1.com.

Option 1

Take the Quick Check below. Refer to the Quick Review for help.

QUICKCheck

Evaluate each expression if $a = -1$, $b = 4$, and $c = -3$. (Lesson 1-2)

1. $a + b - c$

2. $2c - b$

3. $3a - 6b - 2c$

4. $6a + 8b + \frac{2}{3}c$

5. FOOD Noah is buying a sandwich with 1 type of meat, 2 types of cheese, and 2 types of vegetable. Each topping costs \$1.55, \$0.65, and \$0.85 respectively. How much will Noah spend on the sandwich?

Solve each equation for y. (Lesson 2-8)

6. $2x + y = 1$

7. $x = 8 - y$

8. $6x - 3y = 12$

9. $2x + 3y = 9$

10. $9 - \frac{1}{2}y = 4x$

11. $\frac{y + 5}{3} = x + 2$

Graph each ordered pair on a coordinate grid. (Lesson 1-9)

12. $(3, 0)$

13. $(-2, 1)$

14. $(-3, 3)$

15. $(-5, 5)$

16. $(0, 6)$

17. $(2, -1)$

18. MAPS Taylor is looking at a map and needs to go 3 blocks east and 2 blocks south from where he is standing now. If he is standing at (0, 0), what will his coordinates be when he arrives at his destination?

QUICKReview

EXAMPLE 1

Evaluate $a + 2b + 3c$ if $a = -1$, $b = 4$, and $c = -3$.

$a + 2b + 3c$	Original expression
$= (-1) + 2(4) + 3(-3)$	Substitute -1 for a, 4 for b, and -3 for c.
$= -1 + 8 - 9$	Multiply.
$= -2$	Simplify.

EXAMPLE 2

Solve $x - 2 = \frac{y}{3}$ for y.

$x - 2 = \frac{y}{3}$	Original equation
$3(x - 2) = (3)\frac{y}{3}$	Multiply each side by 3.
$3x - 6 = y$	Simplify.

EXAMPLE 3

Graph $(5, -3)$ on a coordinate grid.

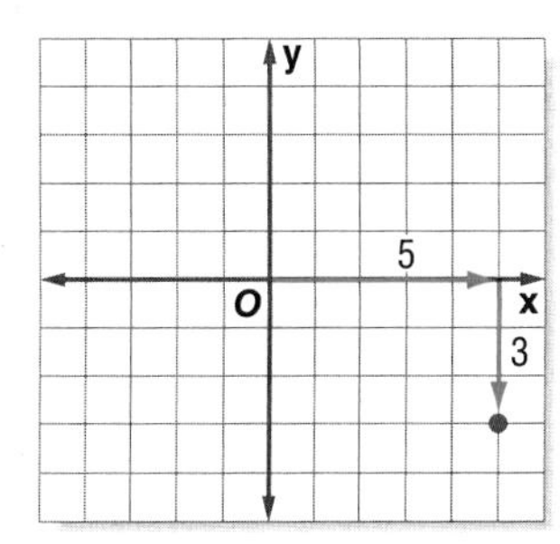

Start at the origin. Since the x-coordinate is 5, move 5 units to the right.

Since the y-coordinate is -3, move down 3 units. Draw a dot.

Algebra Lab

Modeling Relations

The observation of patterns is used in many disciplines such as science, history, economics, social studies, and mathematics. When a quantity depends on another, the pattern can be described in many ways.

ACTIVITY

Step 1 Use centimeter cubes to build a tower similar to the one shown at the right.

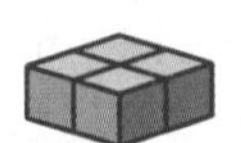

Step 2 Copy the table below. Record the number of layers in the tower and the number of cubes used to build it in the table.

Layers	Cubes
1	4
2	
3	
4	
5	
6	
7	
8	

Step 3 Add layers to the tower. Record the number of layers and the number of cubes in each tower.

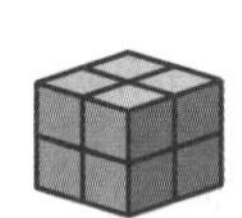

Analyze the Results

Study the data you recorded in the Activity.

1. As the number of layers in the tower increases, how does the number of cubes in the tower change?
2. If there are n layers in a tower, how many cubes are there in the tower? Explain.
3. Write the data in your table as ordered pairs (layers, cubes). Graph the ordered pairs.

Extension

4. Copy and complete the table at the right for the towers that you built. To determine the surface area, count the number of squares showing on each tower, including those on the base. (Hint: The surface area of the 1-layer tower above is 16.)
5. When a layer is added to the tower, what is the effect on the surface area of the tower? Explain.

Layers	Surface Area
1	16
2	
3	
4	
5	
6	
7	
8	

3-1 Representing Relations

Main Ideas

- Represent relations as sets of ordered pairs, tables, mappings, and graphs.
- Find the inverse of a relation.

New Vocabulary

mapping
inverse

GET READY for the Lesson

Ken Griffey, Jr.'s batting statistics for home runs and strikeouts can be represented as a set of ordered pairs. The number of home runs are the first coordinates, and the number of strikeouts are the second coordinates.

You can plot the ordered pairs on a graph to look for patterns.

Ken Griffey, Jr.		
Year	**Home Runs**	**Strikeouts**
1998	56	121
1999	48	108
2000	40	117
2001	22	72
2002	8	39
2003	13	44
2004	20	67

Source: baseball-reference.com

Study Tip

Look Back
To review **relations,** see Lesson 1-9.

Represent Relations Recall that a *relation* is a set of ordered pairs. A relation can also be represented by a table, a graph, or a mapping. A **mapping** illustrates how each element of the domain is paired with an element in the range.

Ordered Pairs

(1, 2)
(−2, 3)
(0, −3)

Table

x	y
1	2
−2	3
0	−3

Graph

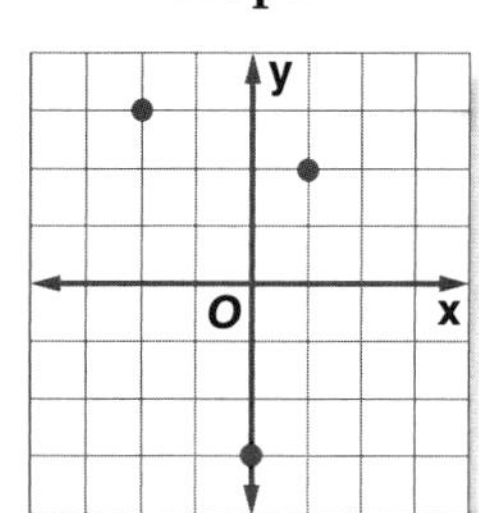

Mapping

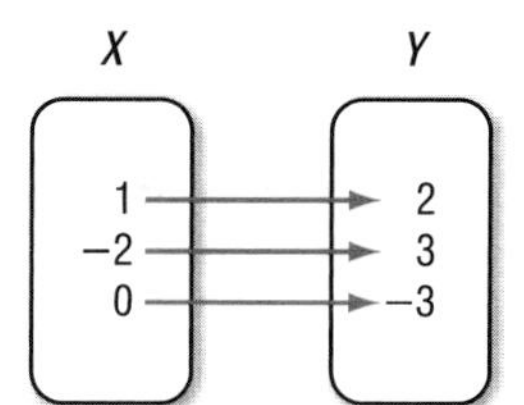

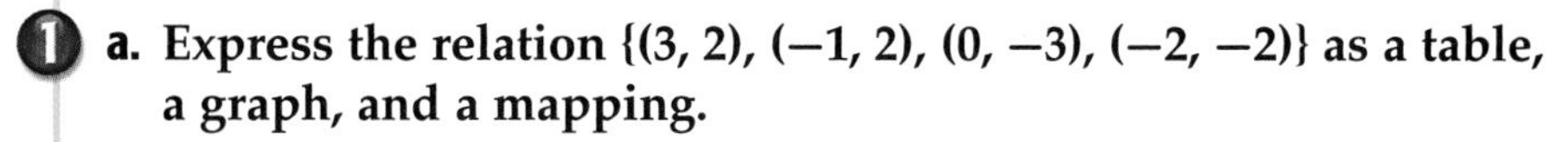

EXAMPLE Represent a Relation

1 **a. Express the relation {(3, 2), (−1, 2), (0, −3), (−2, −2)} as a table, a graph, and a mapping.**

Table

List the x-coordinates in the first column and the corresponding y-coordinates in the second column.

x	y
3	2
−1	2
0	−3
−2	−2

Graph

Graph each ordered pair on a coordinate plane.

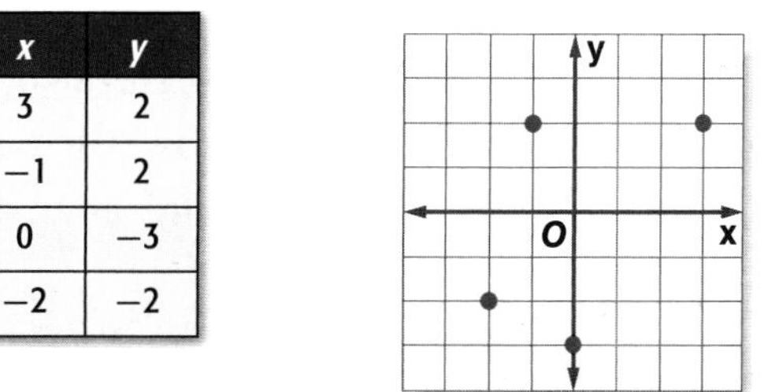

Mapping

List the x-values in set X and the y-values in set Y. Draw arrows from the x-values in X to the corresponding y-values.

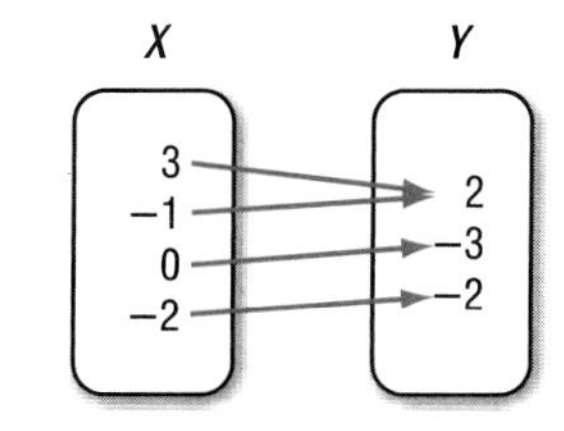

(continued on the next page)

Concepts in Motion
Animation algebra1.com

Study Tip

Domain and Range

When writing the elements of the domain and range, if a value is repeated, you need to list it only once.

b. Determine the domain and range.

The domain for this relation is {−2, −1, 0, 3}.

The range is {−3, −2, 2}.

CHECK Your Progress

1A. Express the relation {(−4, 8), (−1, 9), (−4, 7), (6, 9)} as a table, a graph, and a mapping.

1B. Determine the domain and range of the relation.

Recall that the domain of a relation is the set of values of the independent variable and the range is the set of values of the dependent variable. This is useful when using relations that represent real-life situations.

Real-World Link

An average student between the ages of 12 and 17 spends 10.3 hours per week with friends outside of school and about 7.8 hours per week talking to friends using the telephone, E-mail, instant messaging, or text messaging.

Source: www.pewinternet.org

Real-World EXAMPLE

2 **ANALYZE TABLES The table shows the results of a recent survey in which students were asked about their use of text messaging.**

Percent of Students Who Use Text Messaging						
Age of Students	12	13	14	15	16	17
Percent	17	30	34	45	46	54

Source: www.pewinternet.org

a. Determine the domain and range of the relation.

The domain is {12, 13, 14, 15, 16, 17} because age is the independent variable. It is unaffected by the percents.

The range is {17, 30, 34, 45, 46, 54} because the percent of teens who use text messaging depends on the age of the teens.

b. Graph the data.

- The values of the x-axis need to go from 12 to 17. It is not practical to begin the scale at 0. Begin at 12 and extend to 17 to include all of the data. The units can be 1 unit per grid square.
- The values on the y-axis need to go from 17 to 54. In this case, you can begin the scale at 0 and extend to 60. You can use units of 10.

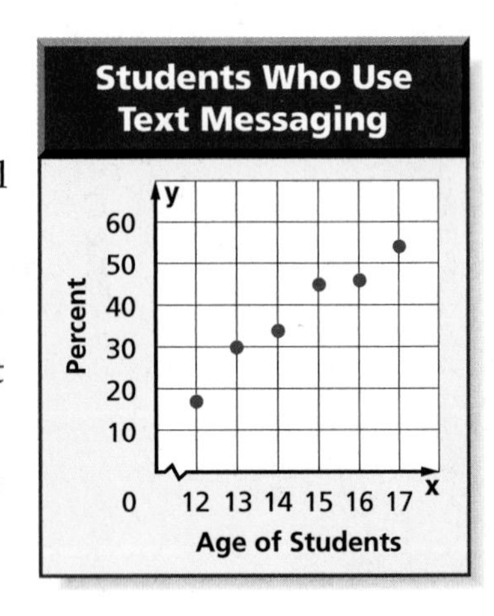

c. What conclusions might you make from the graph of the data?

There is a steady increase in the percent of students who use text messaging as the students get older.

CHECK Your Progress

MONEY Leticia earns $7 for walking 1 dog, $28 for walking 4 dogs, $42 for walking 6 dogs, and $49 for walking 7 dogs.

2A. Determine the domain and range of the relation.

2B. Graph the data.

Inverse Relations The **inverse** of any relation is obtained by switching the coordinates in each ordered pair. The domain of a relation becomes the range of the inverse and the range of a relation becomes the domain of the inverse.

KEY CONCEPT ***Inverse of a Relation***

Relation Q is the inverse of relation S if and only if for every ordered pair (a, b) in S, there is an ordered pair (b, a) in Q.

Relation	Inverse of Relation
{(0, 2), (−5, 4)}	{(2, 0), (4, −5)}

EXAMPLE Inverse Relation

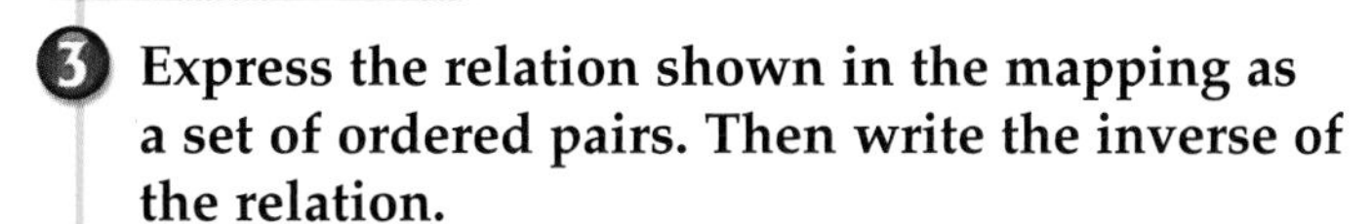

3 **Express the relation shown in the mapping as a set of ordered pairs. Then write the inverse of the relation.**

Relation Notice that both 2 and 3 in the domain are paired with −4 in the range. {(2, −4), (3, −4), (5, −7), (6, −8)}

Inverse Exchange x and y in each ordered pair to write the inverse relation. {(−4, 2), (−4, 3), (−7, 5), (−8, 6)}

The mapping of the inverse is shown at the right. Compare this to the mapping of the relation.

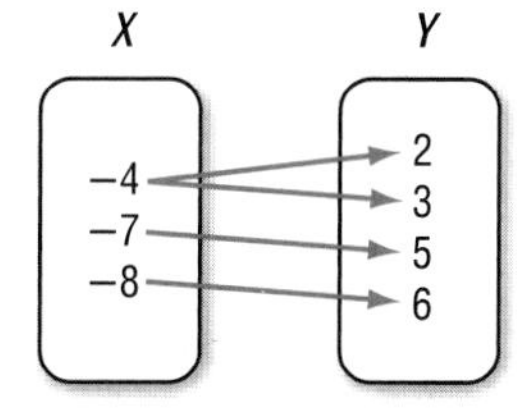

Study Tip

Functions The relation in Example 3 is a function, but the inverse of the relation is not. You will learn more about functions in Lesson 3-2.

CHECK Your Progress

3. Express the relation shown in the table as a set of ordered pairs. Then write the inverse of the relation.

x	1	3	5	7
y	2	4	6	8

ALGEBRA LAB

Relations and Inverses

- Graph the relation {(3, 4), (−2, 5), (−4, −3), (5, −6), (−1, 0), (0, 2)} on grid paper using a colored pencil. Connect the points in order.
- Use a different colored pencil to graph the inverse of the relation, connecting the points in order.
- Fold the grid paper so that the positive y-axis lies on top of the positive x-axis. Hold the paper up to a light to view the points.

ANALYZE

1. What do you notice about the location of the points?

2. Unfold the paper. Describe the transformation of each point and its inverse.

3. What do you think are the ordered pairs that represent the points on the fold line? Describe these in terms of x and y.

4. How could you graph the inverse of a function without writing ordered pairs first?

CHECK Your Understanding

Example 1 (pp. 143–144)

Express each relation as a table, a graph, and a mapping. Then determine the domain and range.

1. $\{(5, -2), (8, 3), (-7, 1)\}$

2. $\{(6, 4), (3, -3), (-1, 9), (5, -3)\}$

Example 2 (p. 144)

COOKING **For Exercises 3 and 4, use the table.** Recipes often have different cooking times for high altitudes because water boils at a lower temperature.

3. Determine the domain and range of the relation and then graph the data.

4. Use your graph to estimate the boiling point of water at an altitude of 7000 feet.

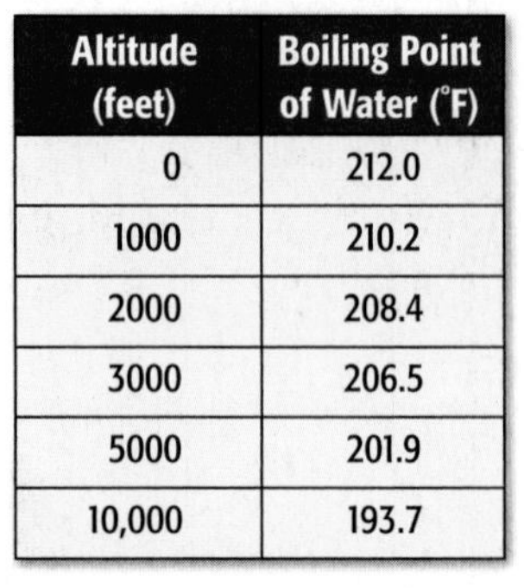

Altitude (feet)	Boiling Point of Water (°F)
0	212.0
1000	210.2
2000	208.4
3000	206.5
5000	201.9
10,000	193.7

Source: Stevens Institute of Technology

Example 3 (p. 145)

Express the relation shown in each table, mapping, or graph as a set of ordered pairs. Then write the inverse of the relation.

5.

x	y
3	−2
−6	7
4	3
−6	5

6.

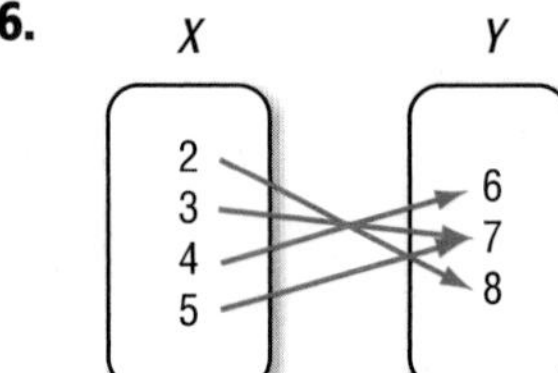

7.

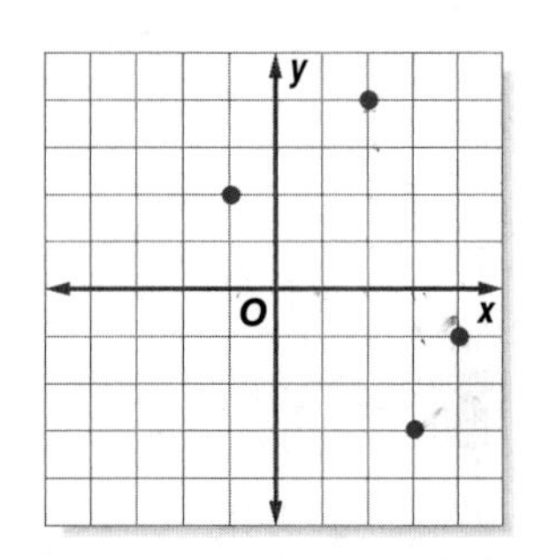

Exercises

HOMEWORK HELP	
For Exercises	See Examples
8–13	1
14–21	2
22–27	3

Express each relation as a table, a graph, and a mapping. Then determine the domain and range.

8. $\{(0, 0), (6, -1), (5, 6), (4, 2)\}$

9. $\{(3, 8), (3, 7), (2, -9), (1, -9)\}$

10. $\{(4, -2), (3, 4), (1, -2), (6, 4)\}$

11. $\{(0, 2), (-5, 1), (0, 6), (-1, 9)\}$

12. $\{(3, 4), (4, 3), (2, 2), (5, -4), (-4, 5)\}$

13. $\{(7, 6), (3, 4), (4, 5), (-2, 6), (-3, 2)\}$

ANALYZE GRAPHS **For Exercises 14–17, use the graph of the average number of students per computer in U.S. public schools.**

14. Name three ordered pairs from the graph.

15. Determine the domain of the relation.

16. What are the least and greatest range values?

17. What conclusions can you make from the graph of the data?

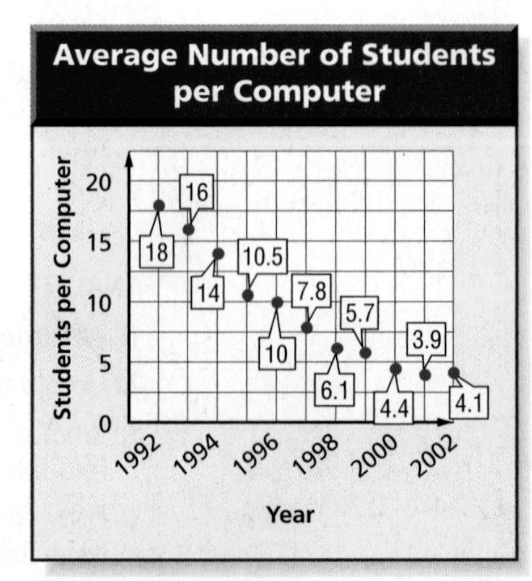

Source: Quality Education Data

FOOD For Exercises 18–21, use the graph that shows the projected annual production of apples from 2007–2014.

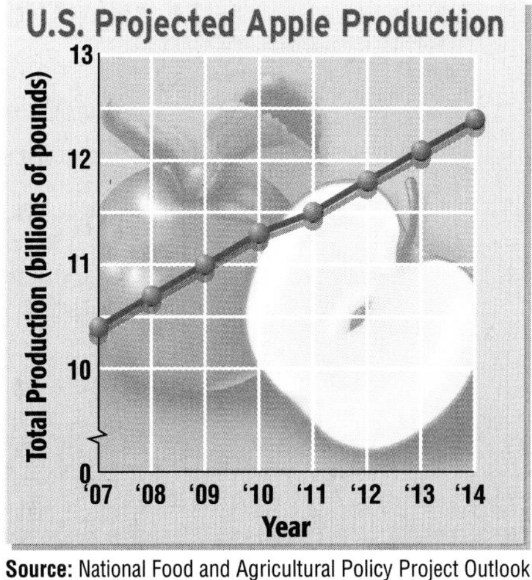

Source: National Food and Agricultural Policy Project Outlook

18. Estimate the domain and range.

19. Which year is projected to have the lowest production? the highest?

20. Describe any patterns that you see.

21. What is a reasonable range value for a domain value of 2015? Explain what this ordered pair represents.

Express the relation shown in each table, mapping, or graph as a set of ordered pairs. Then write the inverse of the relation.

22.

x	y
0	3
−5	2
4	7
−8	2

23.

x	y
0	0
4	7
8	10.5
12	18
16	14.5

24.

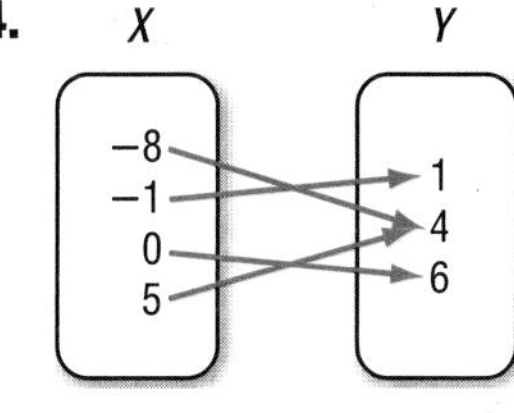

25.

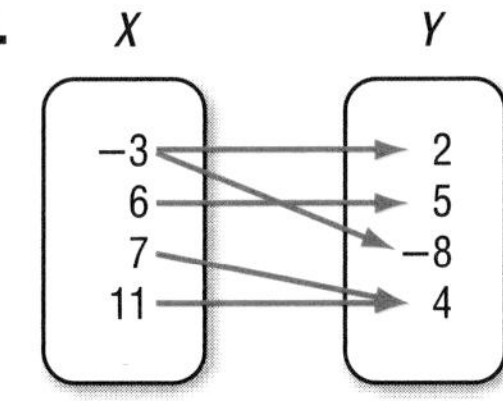

26.

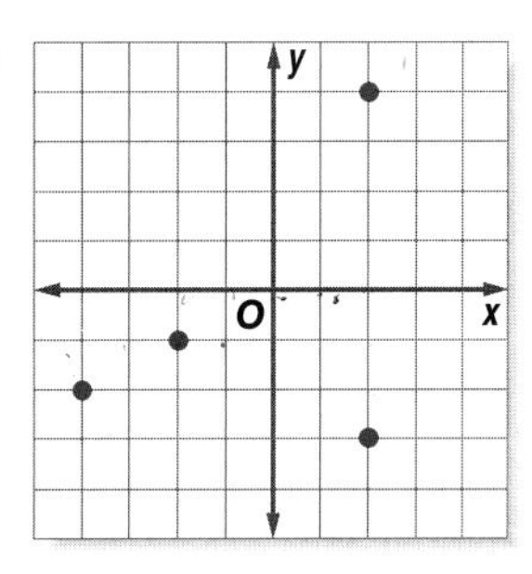

27.

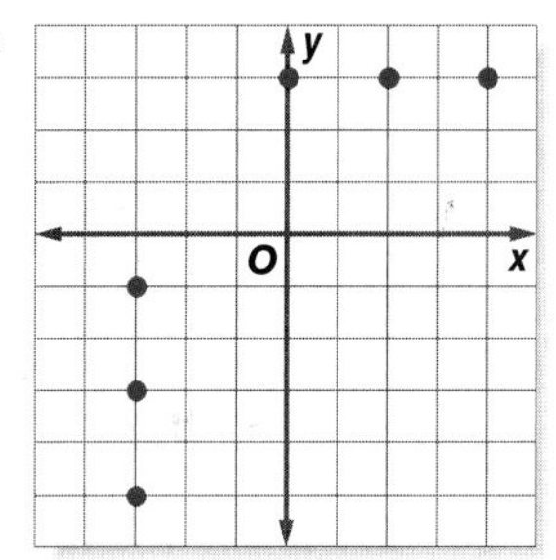

Real-World Link

Studies have shown that students who own fish score higher on both math and verbal SATs, with a combined score 200 points higher than non-pet owners.

Source: diveintofish.com/media/med_facts.html

Express each relation as a set of ordered pairs and describe the domain and range. Then write the inverse of the relation.

28.

Buying Aquarium Fish	
Number of Fish	Total Cost ($)
1	2.50
2	5.50
5	10.00
8	18.75

29.

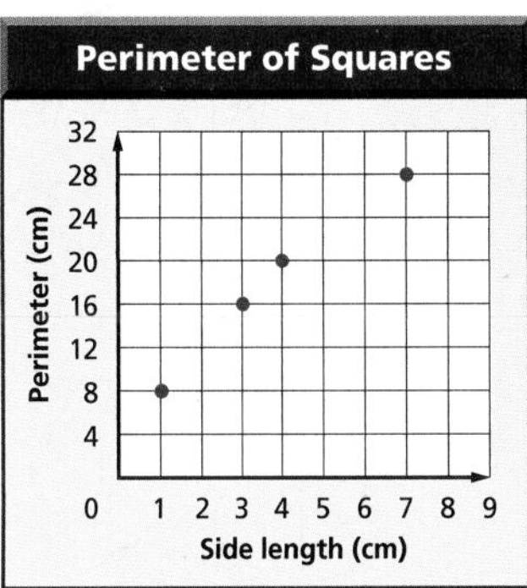

BIOLOGY For Exercises 30–33, use the fact that a person typically has about 2 pounds of muscle for each 5 pounds of body weight.

30. Make a table to show the relation between body and muscle weight for people weighing 100, 105, 110, 115, 120, 125, and 130 pounds.

31. State the domain and range and then graph the relation.

32. What are the domain and range of the inverse?

33. Graph the inverse relation.

H.O.T. Problems

34. CHALLENGE Find a counterexample to disprove the following.

The domain of relation F contains the same elements as the range of relation G. The range of relation F contains the same elements as the domain of relation G. Therefore, relation G must be the inverse of relation F.

35. OPEN ENDED Describe a real-life situation that can be represented using a relation and discuss how one of the quantities in the relation depends on the other. Then give an example of such a relation in three different ways.

36. *Writing in Math* Use the information about batting statistics on page 143 to explain how relations can be used to represent baseball statistics. Include a graph of the relation of the number of Ken Griffey, Jr.'s, home runs and his strikeouts. Describe the relationship between the quantities.

STANDARDIZED TEST PRACTICE

37. What is the domain of the function that contains the points at (0, -3), (−2, 4), (4, −3), and (−3, 1)?

A $\{-3, -2\}$

B $\{-3, 1, 4\}$

C $\{-3, -2, 0, 1\}$

D $\{-3, -2, 0, 4\}$

38. REVIEW Kara deposited \$2000 into a savings account that pays 1.5% interest compounded annually. If she does not deposit any more money into her account, how much will she earn in interest at the end of one year?

F \$30

G \$35

H \$300

J \$350

Spiral Review

39. CHEMISTRY Jamaal has 20 milliliters of a 30% solution of nitric acid. How many milliliters of a 15% solution should he add to obtain a 25% solution of nitric acid? (Lesson 2-9)

Solve each equation or formula for the variable specified. (Lesson 2-8)

40. $3x + b = 2x + 5$ for x

41. $6w - 3h = b$ for h

42. HOURLY PAY Dominique earned \$9.75 per hour before her employer increased her hourly rate to \$10.15 per hour. What was the percent of increase in her salary? (Lesson 2-7)

GET READY for the Next Lesson

PREREQUISITE SKILL Evaluate each expression. (Lesson 1-2)

43. $12 \div 4 + 15 \cdot 3$

44. $12(19 - 15) - 3 \cdot 8$

45. $(25 - 4) \div (2^2 - 1)$

3-2 Representing Functions

Main Ideas

- Determine whether a relation is a function.
- Find functional values.

New Vocabulary

function
vertical line test
function notation
function value

GET READY for the Lesson

The table shows barometric pressures and temperatures recorded by the National Climatic Data Center over a three-day period.

Pressure (millibars)	1013	1006	997	995	995	1000	1006	1011	1016	1019
Temperature (C)	3	4	10	13	8	4	1	−2	−6	−9

Notice that when the pressure is 995 and 1006 millibars, there is more than one value for the temperature.

Study Tip

Look Back

To review **relations and functions,** see Lesson 1-9.

Identify Functions Recall that relations in which each element of the domain is paired with exactly one element of the range are called **functions**.

KEY CONCEPT *Function*

A function is a relation in which each element of the domain is paired with *exactly* one element of the range.

EXAMPLE Identify Functions

Determine whether each relation is a function. Explain.

a.

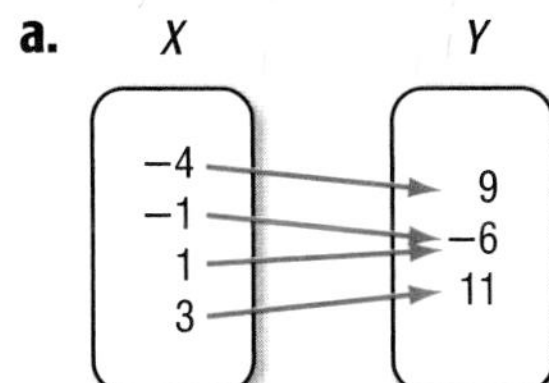

For each element of the domain, there is only one corresponding element in the range. So, this mapping represents a function. It does not matter if two elements of the domain are paired with the same element in the range.

b.

x	y
−3	6
2	5
3	1
2	4

The element 2 in the domain is paired with both 5 and 4 in the range. So, if x is 2, there are two possible values for y. The relation in this table does not represent a function.

CHECK Your Progress

1. $\{(-2, 4), (1, 5), (3, 6), (5, 8), (7, 10)\}$

You can use the **vertical line test** to see if a graph represents a function. If no vertical line can be drawn that intersects the graph more than once, then the graph is a function. If a vertical line can be drawn so that it intersects the graph at two or more points, the graph is not a function.

Concepts in Motion
Animation
algebra1.com

Function	Not a Function	Function
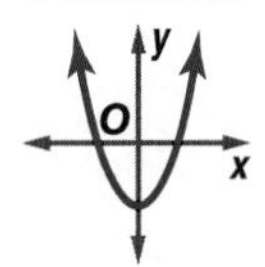	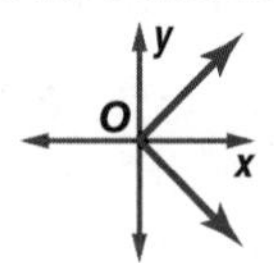	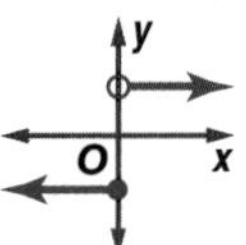

EXAMPLE Equations as Functions

2 Determine whether $2x - y = 6$ represents a function.

Make a table and plot points to graph the equation.

x	1	2	3
y	−4	−2	0

Since the equation is in the form $Ax + By = C$, the graph of the equation will be a line. Place a pencil at the left of the graph to represent a vertical line. Slowly move the pencil to the right across the graph.

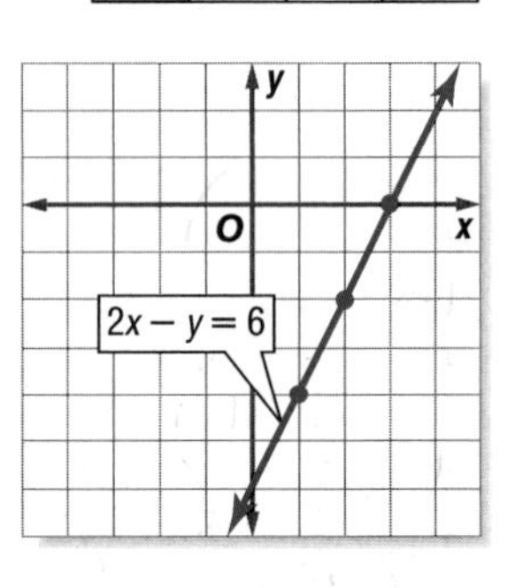

For each value of x, this vertical line passes through no more than one point on the graph. Thus, the graph represents a function.

CHECK Your Progress

2. Determine whether $x = -2$ is a function.

Review Vocabulary

Independent/ Dependent Variables In a function, the value of the dependent variable depends on the value of the independent variable. (Lesson 1-9)

Function Values Equations that are functions can be written in a form called **function notation**. For example, consider $y = 3x - 8$.

equation	**function notation**
$y = 3x - 8$	$f(x) = 3x - 8$

In a function, x represents the independent quantity, or the elements of the domain and $f(x)$ represents the dependent quantity, or the elements of the range. For example, $f(5)$ is the element in the range that corresponds to the element 5 in the domain. We say that $f(5)$ is the **function value** of f for $x = 5$.

Reading Math

Function Notation The symbol $f(x)$ is read *f of x*. The symbol $f(5)$ is read *f of 5*.

EXAMPLE Function Values

3 If $f(x) = 2x + 5$, find each value.

a. $f(-2)$

$f(-2) = 2(-2) + 5$ Replace x with −2.
$= -4 + 5$ Multiply.
$= 1$ Add.

b. $f(1) + 4$

$f(1) + 4 = [2(1) + 5] + 4$ Replace x with 1.
$= 7 + 4$ Simplify.
$= 11$ Add.

CHECK Your Progress

3A. $f(3)$

3B. $2 - f(0)$

Math Online Extra Examples at algebra1.com

Reading Math

Functions Other letters such as g and h can be used to represent functions. For example, $g(x)$ is read *g of x* and $h(t)$ is read *h of t*.

EXAMPLE Nonlinear Function Values

4 **PHYSICS The function $h(t) = -16t^2 + 68t + 2$ represents the height $h(t)$ of a football in feet t seconds after it is kicked. Find each value.**

a. $h(4)$

$h(4) = -16(4)^2 + 68(4) +$	Replace t with 4.
$= -256 + 272 + 2$	Multiply.
$= 18$	Simplify.

b. $2[h(g)]$

$2[h(g)] = 2[-16(g)^2 + 68(g) + \ ]$	Evaluate $h(g)$ by replacing t with g.
$= 2(-16g^2 + 68g + \ )$	Simplify.
$= -32g^2 + 136g + 4$	Multiply the value of $h(g)$ by 2.

CHECK Your Progress **If $f(t) = 2t^3$, find each value.**

4A. $f(4)$ **4B.** $3[f(t)]$

Test-Taking Tip

Functions
When representing functions, determine the values of the domain and range that make sense for the given situation.

STANDARDIZED TEST EXAMPLE

5 The algebraic form of a function is $s = 9h$, where s is Barbara's weekly salary and h is the number of hours that she works in a week. Which of the following represents the same function?

A For every week Barbara works, she earns $9.

C $f(h) = 9$

B

h	s
0	0
2	18
4	36

D

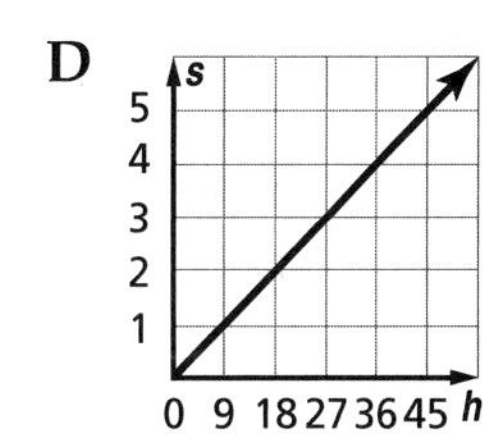

Read the Test Item

The independent variable is h and the dependent variable is s.

Solve the Test Item

Choices A and C both represent a constant weekly salary of $9. This is incorrect because the salary depends on the number of hours worked.

Choice D represents $s = \frac{1}{9}h$, which is incorrect. In choice B, the salary equals the number of hours times 9, which is correct. The answer is B.

CHECK Your Progress

5. Which statement represents the function that is described below?

For every minute that Beatriz walks, she walks 0.1 mile.

F $f(t) = 0.1t$ **G** $f(t) = t + 0.1$ **H** $f(t) = 0.1 - t$ **J** $f(t) = t - 0.1$

Online **Personal Tutor at** algebra1.com

CHECK Your Understanding

Example 1 (p. 149)

Determine whether each relation is a function.

1.

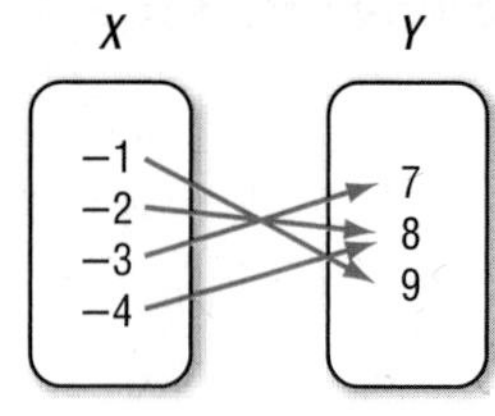

2.

x	y
−3	0
2	1
2	4
6	5

3. $\{(-4, 1), (-1, 4), (3, -2), (-4, 5)\}$

4. $y = x + 3$

Example 2 (p. 150)

5.

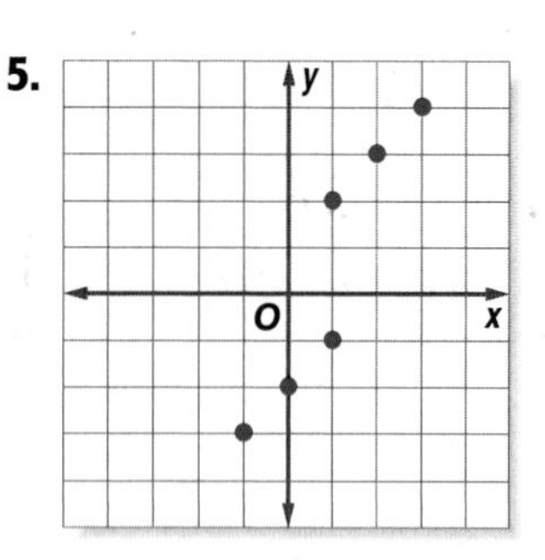

6.

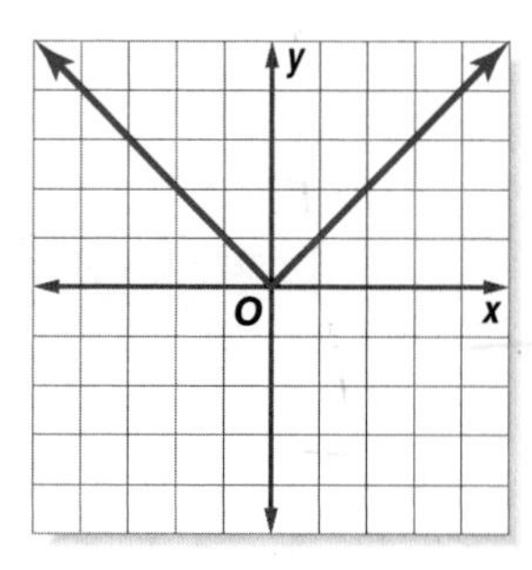

Examples 3, 4 (pp. 150–151)

If $f(x) = 4x - 5$ and $g(x) = x^2 + 1$, find each value.

7. $f(2)$

8. $f(c)$

9. $f(x + 5)$

10. $g(-1)$

11. $g(t) - 4$

12. $g(3n)$

13. **CELL PHONE PICTURES** The cost of sending cell phone pictures is given by $y = 0.25x$, where x is the number of pictures that you send. Write the equation in function notation and then find $f(5)$ and $f(12)$. What do these values represent?

Example 5 (p. 151)

14. **STANDARDIZED TEST PRACTICE** Represent the function described in Exercise 13 in two different ways.

Exercises

HOMEWORK HELP

For Exercises	See Examples
15–22	1
23–26	2
27–38	3, 4
39–44	5

Determine whether each relation is a function.

15. X: 3, 5, 6; Y: 4, 0, 2, 4

16.

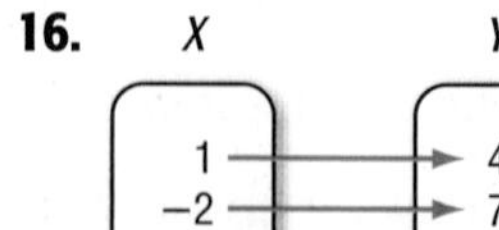

17.

x	y
2	7
4	9
5	5
8	−1

18.

x	y
−9	−5
−4	0
3	6
7	1
6	−5
3	2

19.

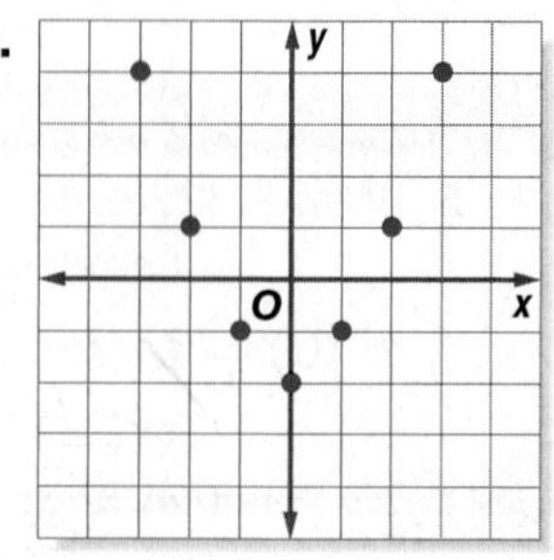

20.

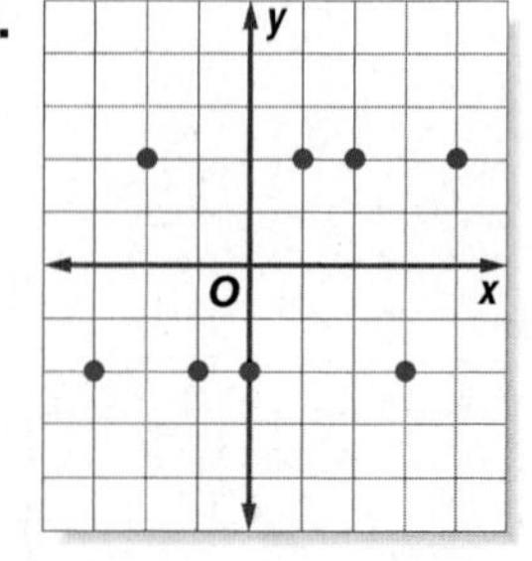

Math Online A graph of the winning Olympic swimming times will help you determine whether the winning time is a function of the year. Visit algebra1.com to continue work on your project.

Determine whether each relation is a function.

21. $\{(5, -7), (6, -7), (-8, -1), (0, -1)\}$ **22.** $\{(4, 5), (3, -2), (-2, 5), (4, 7)\}$

23. $y = -8$ **24.** $x = 15$

25. $y = 3x - 2$ **26.** $y = 3x + 2y$

If $f(x) = 3x + 7$ and $g(x) = x^2 - 2x$, find each value.

27. $f(3)$ **28.** $f(-2)$ **29.** $g(5)$

30. $g(0)$ **31.** $g(-3) + 1$ **32.** $f(8) - 5$

33. $g(2c)$ **34.** $g(4n)$ **35.** $f(k + 2)$

36. $f(a - 1)$ **37.** $3[f(r)]$ **38.** $2[g(t)]$

Meteorologist

Meteorologists study the physical characteristics of the atmosphere using instruments and tools such as weather balloons. This information is used to interpret and predict trends in the weather.

For more information, go to algebra1.com.

METEOROLOGY For Exercises 39–42, use the following information.
The temperature of the atmosphere decreases about 5°F for every 1000 feet increase in altitude. Thus, if the temperature at ground level is 77°F, the temperature at an altitude of h feet is found by using $t = 77 - 0.005h$.

39. Write the equation in function notation. Then find $f(100)$, $f(200)$, and $f(1000)$.

40. Suppose the temperature at ground level was less than 77°F. Describe how the range values in Exercise 39 would change. Explain.

41. Graph the function.

42. Use the graph of the function to estimate the temperature at 4000 feet.

EDUCATION For Exercises 43 and 44, use the following information.
The average national math test scores $f(s)$ for 17-year-olds can be represented as a function of the national science scores s by $f(s) = 0.8s + 72$.

43. Graph this function.

44. What is the science score that corresponds to a math score of 308?

Determine whether each relation is a function.

45.

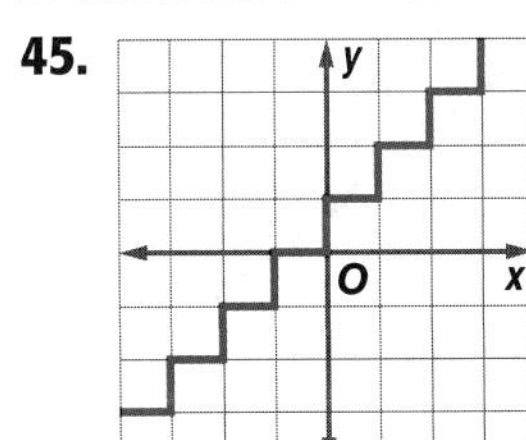

46.

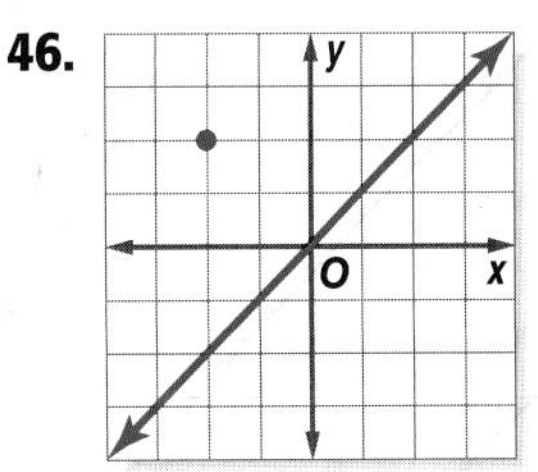

47.

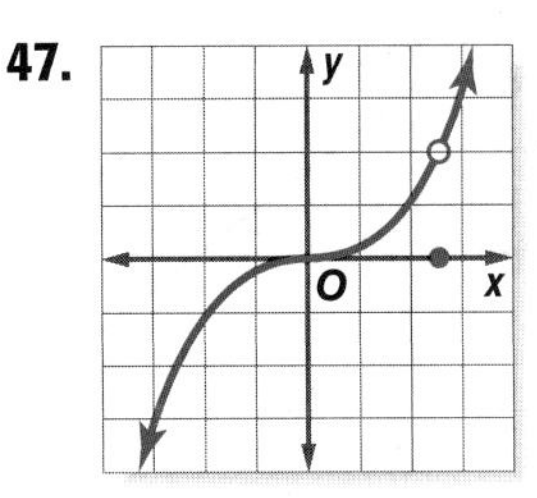

48. PARKING A parking garage charges \$2.00 for the first hour, \$2.75 for the second, \$3.50 for the third, \$4.25 for the fourth, and \$5.00 for any time over four hours. Choose the graph that best represents the information and determine whether the graph represents a function. Explain.

a.

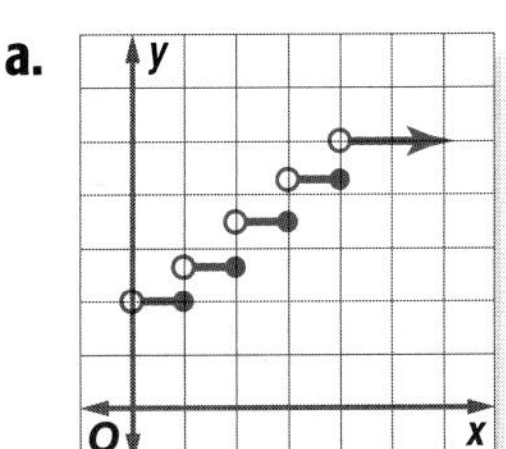

b.

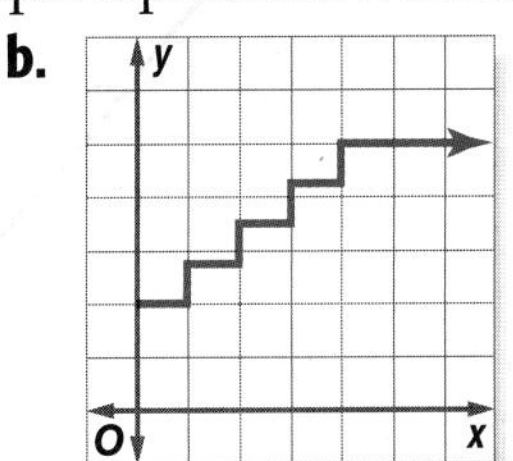

c.

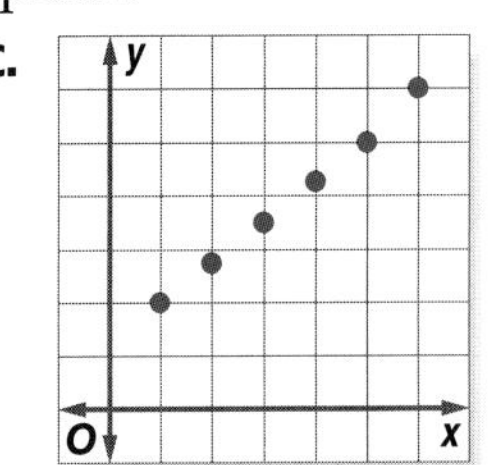

H.O.T. Problems

REASONING For Exercises 49 and 50, refer to the following information. The ordered pairs (0, 1), (3, 2), (3, –5), and (5, 4) are on the graph of a relation between x and y.

49. Determine whether y is a function of x. Explain.

50. Determine whether x is a function of y. Explain.

51. CHALLENGE State whether the following is *sometimes*, *always*, or *never* true. Explain your reasoning. *The inverse of a function is also a function.*

52. OPEN ENDED Disprove the following statement by finding a counterexample. *All linear equations are functions.*

53. *Writing in Math* Use the information on page 149 to explain how functions are used in meteorology. Describe the relationship between pressure and temperature, and investigate whether the relation is a function.

STANDARDIZED TEST PRACTICE

54. Which relation is a function?

A {(–5, 6), (4, –3), (2, –1), (4, 2)}

B {(3, –1), (3, –5), (3, 4), (3, 6)}

C {(–2, 3), (0, 3), (–2, –1), (–1, 2)}

D {(–5, 6), (4, –3), (2, –1), (4, 2)}

55. REVIEW If $a = -4$ and $b = 8$, then $3a(b + 2) + a =$

F –124

G –98

H –26

J 18

Spiral Review

56. Express the relation shown in the table as a set of ordered pairs. Then write the inverse of the relation. (Lesson 3-1)

x	y
–4	9
2	5
–2	–2
11	12

57. AIRPLANES At 1:30 P.M., an airplane leaves Tucson for Baltimore, a distance of 2240 miles. The plane flies at 280 miles per hour. A second airplane leaves Tucson at 2:15 P.M. and is scheduled to land in Baltimore 15 minutes before the first airplane. At what rate must the second airplane travel to arrive on schedule? (Lesson 2-9)

56. RUNNING Lacey can run a 10K race (about 6.2 miles) in 45 minutes. If she runs a 26-mile marathon at the same pace, how long will it take her to finish? (Lesson 2-6)

GET READY for the Next Lesson

PREREQUISITE SKILL Solve each equation. (Lesson 2-4)

59. $r - 9 = 12$

60. $-4 = 5n + 6$

61. $3 - 8w = 35$

62. $\frac{g}{4} + 2 = 5$

3-3 Linear Functions

Main Ideas

- Identify linear equations, intercepts, and zeros.
- Graph linear equations.

New Vocabulary

linear equation
standard form
x-intercept
y-intercept
zero

GET READY for the Lesson

It is recommended that no more than 30% of a person's daily caloric intake come from fat. Since each gram of fat contains 9 Calories, the most grams of fat f that you should have each day is given by $f = 0.3\left(\frac{C}{9}\right)$ or $f = \frac{C}{30}$. C is the total number of Calories C that you consume. The graph of this equation shows the maximum number of grams of fat you should consume based on the total number of Calories you consume.

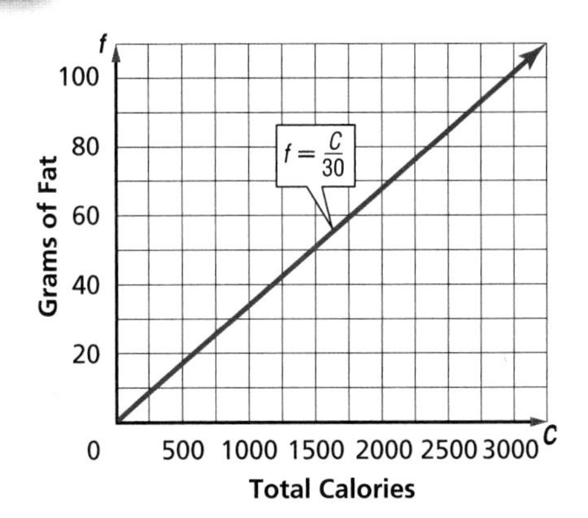

Identify Linear Equations, Intercepts, and Zeros A **linear equation** is the equation of a line. Linear equations can often be written in the form $Ax + By = C$. This is called the **standard form** of a linear equation.

KEY CONCEPT — *Standard Form of a Linear Equation*

The standard form of a linear equation is

$$Ax + By = C,$$

where $A \geq 0$, A and B are not both zero, and A, B, and C are integers with a greatest common factor of 1.

EXAMPLE Identify Linear Equations

Determine whether each equation is a linear equation. If so, write the equation in standard form.

a. $y = 5 - 2x$

Rewrite the equation so that both variables are on the same side of the equation.

$y = 5 - 2x$	Original equation
$y + 2x = 5 - 2x + 2x$	Add $2x$ to each side.
$2x + y = 5$	Simplify.

The equation is now in standard form where $A = 2$, $B = 1$, and $C = 5$. This is a linear equation.

b. $2xy - 5y = 6$

Since the term $2xy$ has two variables, the equation cannot be written in the form $Ax + By = C$. Therefore, this is not a linear equation.

CHECK Your Progress

Determine whether each equation is a linear equation. If so, write the equation in standard form.

1A. $\frac{1}{3}y = -1$

1B. $y = x^2 + 3$

> **Study Tip**
>
> **Linear Functions**
>
> The graph of a linear function has at most one x-intercept and one y-intercept, unless it is the function $f(x) = 0$, in which case every point of the graph is an x-intercept.

The x-coordinate of the point at which the graph of an equation crosses the x-axis is an **x-intercept**. The y-coordinate of the point at which the graph crosses the y-axis is called a **y-intercept**.

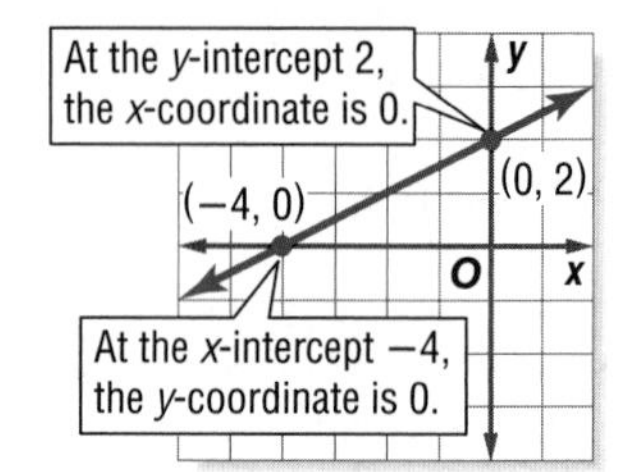

Values of x for which $f(x) = 0$ are called **zeros** of the function f. The zero of a linear function is its x-intercept.

Real-World EXAMPLE

2 **ANALYZE GRAPHS** **High school students in Palo Alto, California, can buy ticket booklets for lunch, as shown in the graph.**

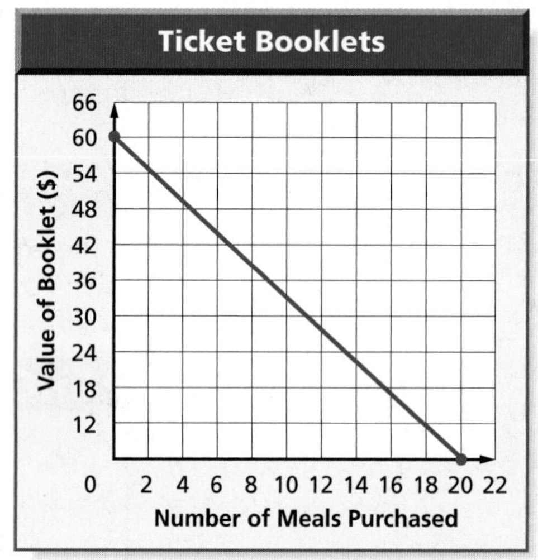

a. Determine the x-intercept, y-intercept, and zero.

The x-intercept is 20 because it is the x-coordinate of the point where the line crosses the x-axis. The zero of the function is also 20.

The y-intercept is 60 because it is the y-coordinate of the point where the line crosses the y-axis.

> **Study Tip**
>
> **Intercepts**
>
> Usually, the individual coordinates are called the x- and y-intercepts.
>
> - The x-intercept 20 is located at (20, 0).
> - The y-intercept 60 is located at (0, 60).

b. Describe what the intercepts mean.

The x-intercept 20 means that after 20 meals are purchased, the meal ticket booklet has a value of \$0.

The y-intercept 60 means before any meals are purchased, the booklet has a value of \$60.

HEALTH **Use the graph at the right that shows the cost of a gym membership.**

2A. Determine the x-intercept, y-intercept, and zero.

2B. Describe what the intercept(s) mean.

Personal Tutor at algebra1.com

Study Tip

Defining Variables

Time is the *independent* variable, and volume of water is the *dependent* variable.

Real-World EXAMPLE

3 **ANALYZE TABLES** **It is recommended that a swimming pool be drained at a maximum rate of 720 gallons per hour. The table shows the function relating the volume of water in a pool and the time in hours that the pool has been draining.**

Draining a Pool	
Time, x (h)	Volume, y (gal)
0	10,080
2	8640
6	5760
10	2880
12	1440
14	0

a. Determine the x-intercept, y-intercept, and zero of the graph of the function.

x-intercept $= 14$ — 14 is the value of x when $y = 0$.

y-intercept $= 10{,}080$ — 10,080 is the value of y when $x = 0$.

zero $= 14$ — The zero of the function is the x-intercept of the graph.

b. Describe what the intercepts mean.

The x-intercept 14 means that after 14 hours, the water in the pool has a volume of 0 gallons, or the pool is completely drained.

The y-intercept 10,080 means that the pool contained 10,080 gallons of water at time 0, or before it started to drain. This is shown in the graph.

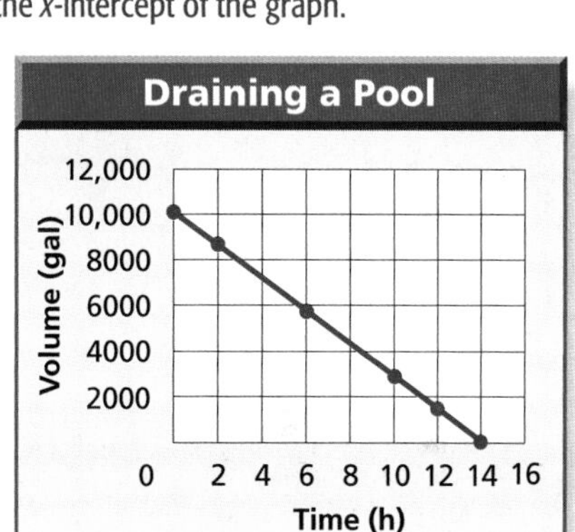

CHECK Your Progress

3. Use the table to determine the x-intercept, y-intercept, and zero of the graph of the function.

x	−3	−2	−1	0	1
y	2	0	−2	−4	−6

Concepts in Motion
Animation
algebra1.com

Graph Linear Equations The graph of an equation represents all of its solutions. So, every ordered pair that satisfies the equation represents a point on the line. An ordered pair that does not satisfy the equation represents a point *not* on the line.

EXAMPLE Graph by Making a Table

4 **Graph $y = \frac{1}{2}x - 3$.**

Select values from the domain and make a table. Then graph the ordered pairs. The domain is all real numbers, so there are infinitely many solutions. Draw a line through the points.

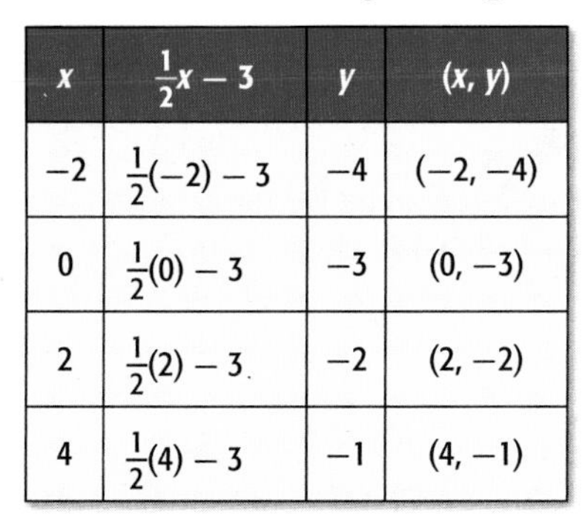

x	$\frac{1}{2}x - 3$	y	(x, y)
−2	$\frac{1}{2}(-2) - 3$	−4	(−2, −4)
0	$\frac{1}{2}(0) - 3$	−3	(0, −3)
2	$\frac{1}{2}(2) - 3$	−2	(2, −2)
4	$\frac{1}{2}(4) - 3$	−1	(4, −1)

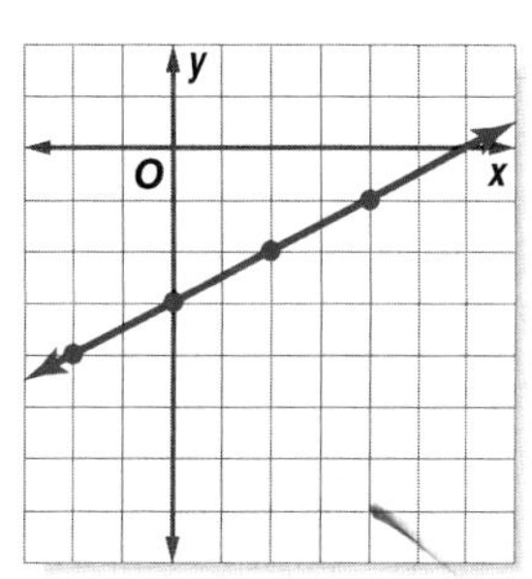

This line represents all of the solutions of $y = \frac{1}{2}x - 3$.

Study Tip

Rewriting Equations

When appropriate, first solve an equation for y in order to find values for y more easily.

$3x + y = -1 \rightarrow$
$y = -1 - 3x$

CHECK Your Progress **Graph each equation.**

4A. $3x + y = -1$ **4B.** $y = -2$

EXAMPLE 5 Graph by Using Intercepts

5 Graph $3x + 2y = 9$ using the x-intercept and y-intercept.

To find the x-intercept, let $y = 0$.

$3x + 2y = 9$	Original equation
$3x + 2(0) = 9$	Replace y with 0.
$3x = 9$	Divide each side by 3.
$x = 3$	

To find the y-intercept, let $x = 0$.

$3x + 2y = 9$	Original equation
$3(0) + 2y = 9$	Replace x with 0.
$2y = 9$	Divide each side by 2.
$y = 4.5$	

Concepts in Motion
Interactive Lab
algebra1.com

The x-intercept is 3, so the graph intersects the x-axis at (3, 0). The y-intercept is 4.5, so the graph intersects the y-axis at (0, 4.5). Plot these points. Then draw the line through them.

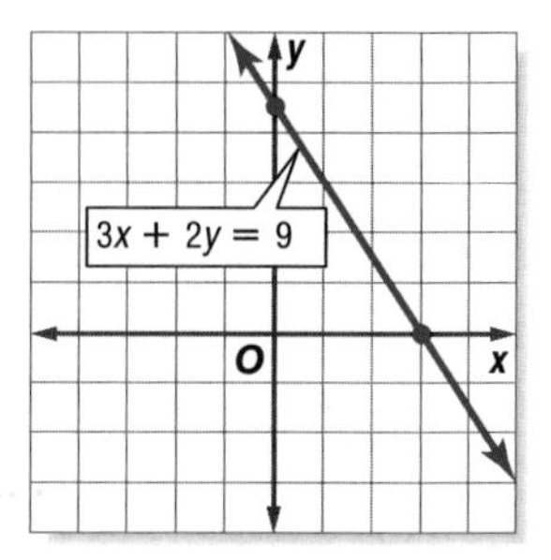

CHECK Your Progress

5. Graph $y = -x - 5$ using the x- and y-intercepts.

CHECK Your Understanding

Example 1 (pp. 155–156) **Determine whether each equation is a linear equation. If so, write the equation in standard form.**

1. $x + y^2 = 25$ **2.** $3y + 2 = 0$ **3.** $\frac{3}{5}x - \frac{2}{5}y = 5$

Examples 2, 3 (pp. 156–157) **Determine the x-intercept and y-intercept of each linear function and describe what the intercepts mean.**

4.

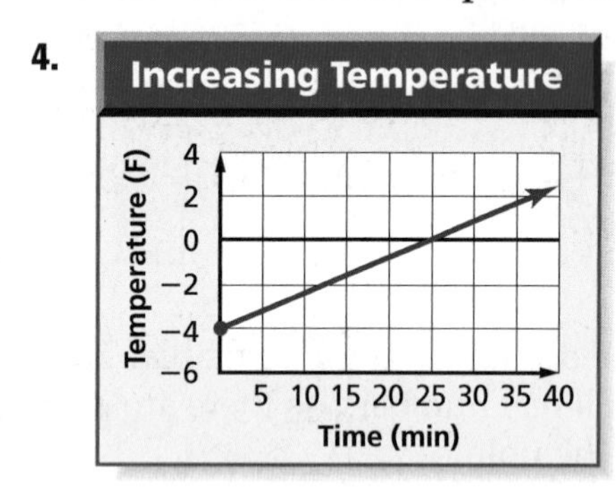

5.

Position of Scuba Diver	
Time, x (s)	**Depth, y (m)**
0	−24
3	−18
6	−12
9	−6
12	0

6. What are the zeros of the functions represented in Exercises 4 and 5?

Example 4 (p. 157) **Graph each equation by making a table.**

7. $x - y = 0$ **8.** $x = 3$

Example 5 (p. 158) **Graph each equation by using the x- and y-intercepts.**

9. $y = -3 - x$ **10.** $x + 4y = 10$

11. RODEOS Tickets for a rodeo cost \$5 for children and \$10 for adults. The equation $5x + 10y = 60$ represents the number of children x and adults y who can attend the rodeo for \$60. Use the x- and y-intercepts to graph the equation. What do these values mean?

Exercises

HOMEWORK HELP	
For Exercises	See Examples
12–17	1
18–23	2, 3
24–38	4, 5

Determine whether each equation is a linear equation. If so, write the equation in standard form.

12. $3x = 5y$
13. $6 - y = 2x$
14. $6xy + 3x = 4$
15. $y + 5 = 0$
16. $7y = 2x + 5x$
17. $y = 4x^2 - 1$

Determine the x-intercept, y-intercept, and zero of each linear function.

18.

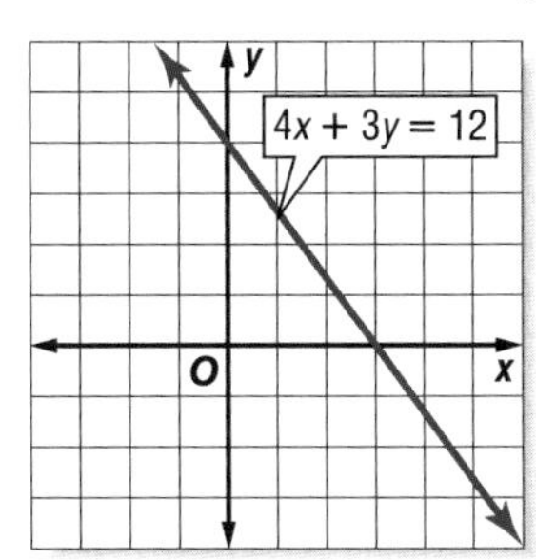

19.

x	y
−3	−1
−2	0
−1	1
0	2
1	3

Determine the x-intercept and y-intercept of each linear function and describe what the intercepts mean.

20.

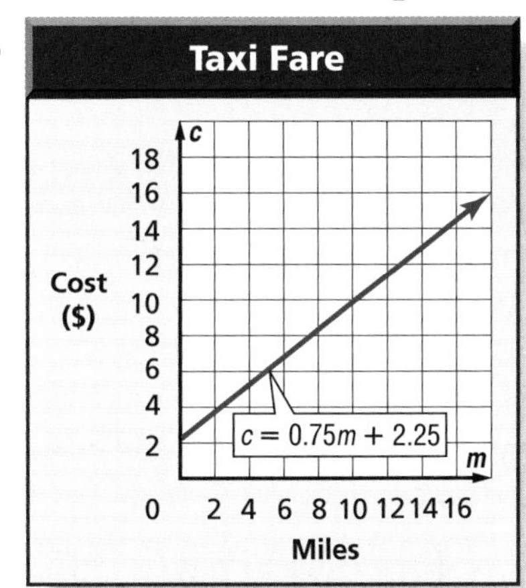

21.

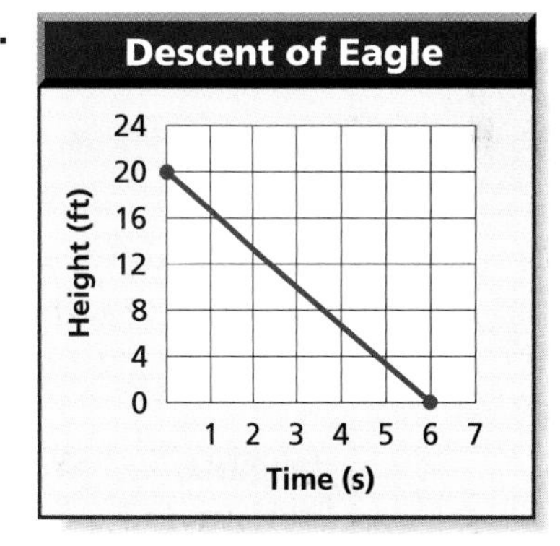

22.

Swimming to Burn Calories	
Time, x (min)	Calories Burned, y
0	0
10	106
15	159
20	212
30	318

23.

Eva's Distance from Home	
Time, x (min)	Distance, y (mi)
0	4
2	3
4	2
6	1
8	0

Graph each equation.

24. $y = -1$
25. $y = 2x$
26. $y = x - 6$
27. $y = 5 - x$
28. $y = 4 - 3x$
29. $x = 3y$
30. $x = 4y - 6$
31. $x - y = -3$
32. $4x + 6y = 8$

METEOROLOGY For Exercises 33–35, use the following information.
The distance d in miles that the sound of thunder travels in t seconds is given by the equation $d = 0.21t$.

33. Make a table of values.
34. Graph the equation.
35. Use the graph to estimate how long it will take you to hear thunder from a storm that is 3 miles away.

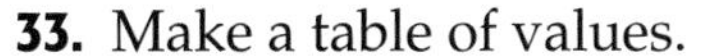

EXTRA PRACTICE
See pages 723, 746.
Math Online
Self-Check Quiz at algebra1.com

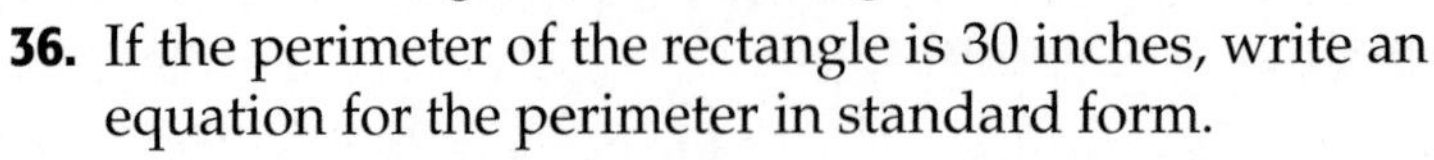

GEOMETRY For Exercises 36–38, refer to the figure at right.
The perimeter P of a rectangle is given by $2\ell + 2w = P$, where ℓ is the length of the rectangle and w is the width.

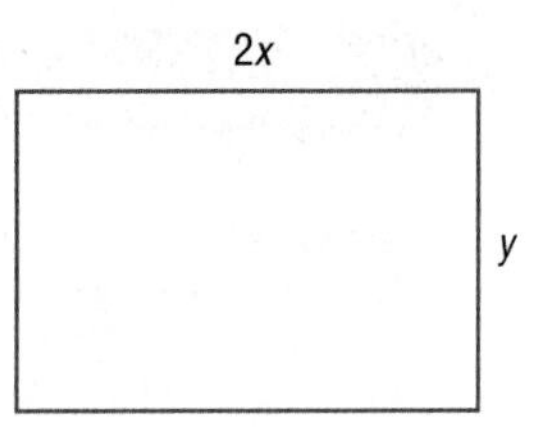

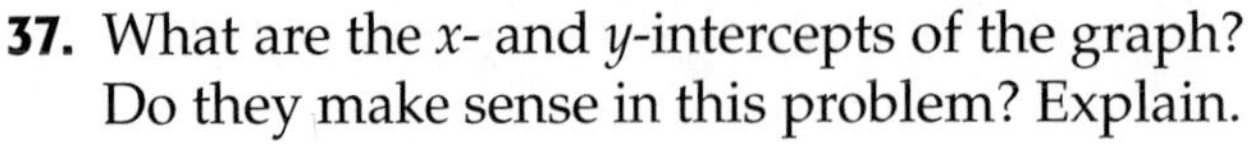

36. If the perimeter of the rectangle is 30 inches, write an equation for the perimeter in standard form.

37. What are the x- and y-intercepts of the graph? Do they make sense in this problem? Explain.

38. Graph the equation.

Real-World Link

How heavy is air? The atmospheric pressure is a measure of the weight of air. At sea level, air pressure is 14.7 pounds per square inch.

Source: www.brittanica.com

Determine whether each equation is a linear equation. If so, write the equation in standard form.

39. $x + \frac{1}{y} = 7$

40. $\frac{x}{2} = 10 + \frac{2y}{3}$

41. $7n - 8m = 4 - 2m$

42. $3a + b - 2 = b$

43. $2r - 3rs + 5s = 1$

44. $\frac{3m}{4} = \frac{2n}{3} - 5$

Graph each equation.

45. $1.5x + y = 4$

46. $75 = 2.5x + 5y$

47. $\frac{4x}{3} = \frac{3y}{4} + 1$

48. $y + \frac{1}{3} = \frac{1}{4}x - 3$

49. $\frac{1}{2}x + y = 4$

50. $1 = x - \frac{2}{3}y$

51. Find the x- and y-intercepts of the graph of $4x - 7y = 14$.

52. Graph $5x + 3y = 15$. Where does the line intersect the x-axis? Where does the line intersect the y-axis? What is the slope?

OCEANOGRAPHY For Exercises 53 and 54, use the information at left and below.
Under water, pressure increases 4.3 pounds per square inch (psi) for every 10 feet you descend. This can be expressed by the equation $p = 0.43d + 14.7$, where p is the pressure in pounds per square inch and d is the depth in feet.

53. Graph the equation and find the y-intercept.

54. Divers cannot work at depths below about 400 feet. Given this information, determine a reasonable domain and range for this situation.

H.O.T. Problems

55. REASONING Verify that the point at $(-4, 2)$ lies on the line with the equation $y = \frac{1}{2}x + 4$.

OPEN ENDED Describe a linear equation in the form $Ax + By = C$ for each condition.

56. $A = 0$

57. $B = 0$

58. $C = 0$

59. CHALLENGE Demonstrate how you can determine whether a point at (x, y) is *above, below,* or *on* the line given by $2x - y = 8$ without graphing it. Give an example of each.

60. ***Writing in Math*** Use the information about nutrition on page 155 to explain how linear equations can be used in nutrition. Explain how you could use the Nutrition Information labels on packages to limit your fat intake.

STANDARDIZED TEST PRACTICE

61. What are the x- and y-intercept points of the function graphed?

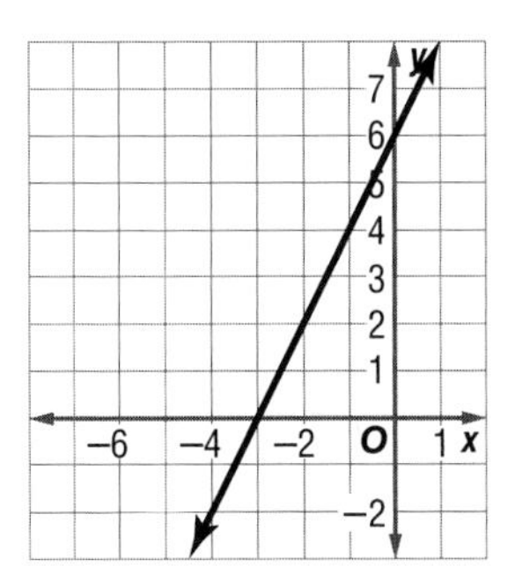

A $(-3, 0)$ and $(0, 6)$

B $(-3, 0)$ and $(6, 0)$

C $(0, -3)$ and $(0, 6)$

D $(0, -3)$ and $(6, 0)$

62. Which is the best estimate for the x-intercept of the graph of the linear function represented in the table?

x	y
0	5
1	3
2	1
3	−1
4	−3

F between 0 and 1

G between 1 and 2

H between 2 and 3

J between 3 and 4

63. REVIEW A candle is 24 centimeters high and burns 3 centimeters per hour, as shown in the graph.

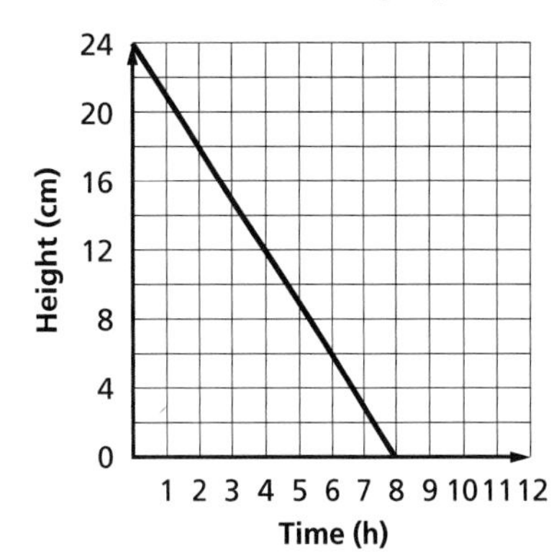

If the height of the candle is 8 centimeters, approximately how long has the candle been burning?

A 0 hours

B 24 minutes

C 64 minutes

D $5\frac{1}{2}$ hours

Spiral Review

If $f(x) = 3x - 2$ and $g(x) = x^2 - 5$, find each value. (Lesson 3-2)

64. $f(4)$

65. $g(-3)$

66. $2[f(6)]$

67. NUTRITION The cost of buying energy bars for a camping trip is given by $y = 2.25x$. Write the equation in function notation and then find $f(4)$ and $f(7)$. What do these values represent? (Lesson 3-2)

Solve each equation. Then check your solution. (Lesson 2-5)

68. $3n - 12 = 5n - 20$

69. $6(x + 3) = 3x$

70. $2(x - 2) = 3x - (4x - 5)$

71. BALLOONS Brandon slowly fills a deflated balloon with air. Without tying the balloon, he lets it go. Draw a graph to represent this situation. (Lesson 1-9)

GET READY for the Next Lesson

PREREQUISITE SKILL Find each difference.

72. $12 - 16$

73. $-5 - (-8)$

74. $16 - (-4)$

75. $-9 - 6$

EXTEND 3-3

Graphing Calculator Lab
Graphing Linear Functions

The power of a graphing calculator is the ability to graph different types of equations accurately and quickly. Often linear equations are graphed in the standard viewing window. The **standard viewing window** is [−10, 10] by [−10, 10] with a scale of 1 on each axis. To quickly choose the standard viewing window on a TI-83/84 Plus, press ZOOM 6.

ACTIVITY 1

Graph $2x - y = 3$ on a TI-83/84 Plus graphing calculator.

Step 1 Enter the equation in the Y= list.

- The Y= list shows the equation or equations that you will graph.
- Equations must be entered with the y isolated on one side of the equation. Solve the equation for y, then enter it into the calculator.

$2x - y = 3$	Original equation
$2x - y - 2x = 3 - 2x$	Subtract $2x$ from each side.
$-y = -2x + 3$	Simplify.
$y = 2x - 3$	Multiply each side by −1.

KEYSTROKES: Y= 2 X,T,θ,*n* − 3

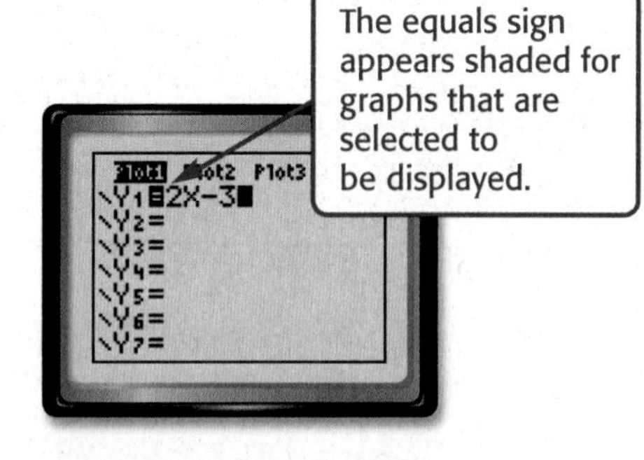

Step 2 Graph the equation in the standard viewing window.

KEYSTROKES: ZOOM 6

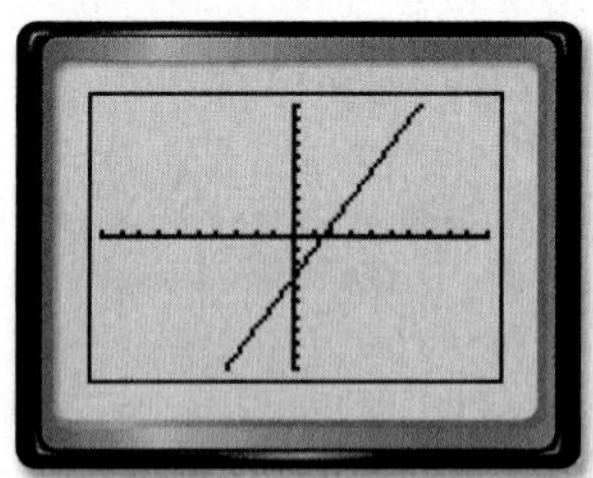

[−10, 10] scl: 1 by [−10, 10] scl: 1

Sometimes a complete graph is not displayed using the standard viewing window. A **complete graph** includes all of the important characteristics of the graph on the screen. These include the origin and the x- and y-intercepts. Notice that the graph of $2x - y = 3$ is a complete graph because all of these points are visible.

When a complete graph is not displayed using the standard viewing window, you will need to change the viewing window to accommodate these important features. You can use what you have learned about intercepts to help you choose an appropriate viewing window.

Math Online Other Calculator Keystrokes at algebra1.com

ACTIVITY 2

Graph $y = 3x - 15$ on a graphing calculator.

Step 1 Enter the equation in the Y= list and graph in the standard viewing window.

Clear the previous equation from the Y= list. Then enter the new equation and graph.

KEYSTROKES: [Y=] [CLEAR] 3 [X,T,θ,n] [−] 15 [ZOOM] 6

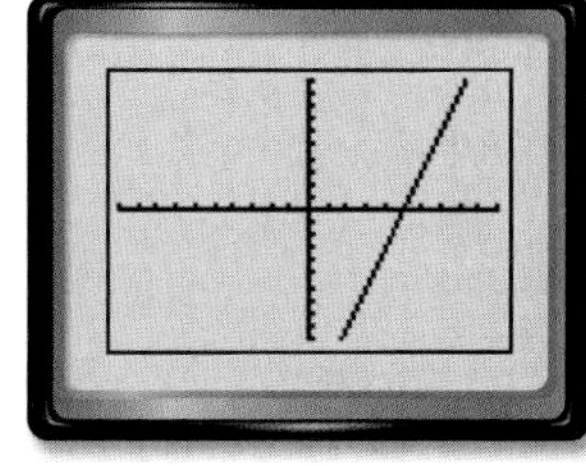

[−10, 10] scl: 1 by [−10, 10] scl: 1

Step 2 Modify the viewing window and graph again.

The origin and the x-intercept are displayed in the standard viewing window. But notice that the y-intercept is outside of the viewing window. Find the y-intercept.

$y = 3x - 15$ — Original equation
$= 3(0) - 15$ — Replace x with 0.
$= -15$ — Simplify.

Since the y-intercept is -15, choose a viewing window that includes a number less than -15. The window [−10, 10] by [−20, 5] with a scale of 1 on each axis is a good choice.

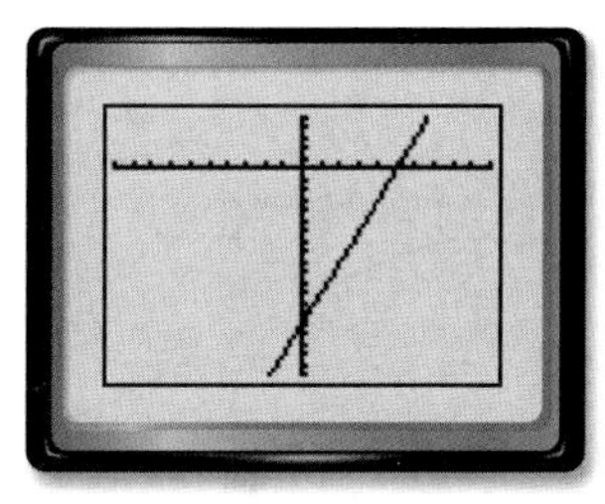

[−10, 10] scl: 1 by [−20, 5] scl: 1

KEYSTROKES: [WINDOW] −10 [ENTER] 10 [ENTER] 1 [ENTER] −20 [ENTER] 5 [ENTER] 1 [GRAPH]

EXERCISES

Graph each linear equation in the standard viewing window. Determine whether the graph is complete. If the graph is not complete, choose a viewing window that will show a complete graph and graph the equation again.

1. $y = x + 2$ **2.** $y = 4x + 5$ **3.** $y = 6 - 5x$ **4.** $2x + y = 6$

5. $x + y = -2$ **6.** $x - 4y = 8$ **7.** $y = 5x + 9$ **8.** $y = 10x - 6$

9. $y = 3x - 18$ **10.** $3x - y = 12$ **11.** $4x + 2y = 21$ **12.** $3x + 5y = -45$

For Exercises 13–15, consider the linear equation $y = 2x + b$.

13. Choose several different positive and negative values for b. Graph each equation in the standard viewing window.

14. For which values of b is the complete graph in the standard viewing window?

15. How is the value of b related to the y-intercept of the graph of $y = 2x + b$?

Chapter 3 Mid-Chapter Quiz

Lessons 3-1 through 3-3

State the domain, range, and inverse of each relation. (Lesson 3-1)

1. $\{(1, 3), (4, 6), (2, 3), (1, 5)\}$

2. $\{(-5, 8), (-1, 0), (-1, 4), (2, 7), (6, 3)\}$

3.

x	y
11	5
15	3
−8	22
11	31

4.

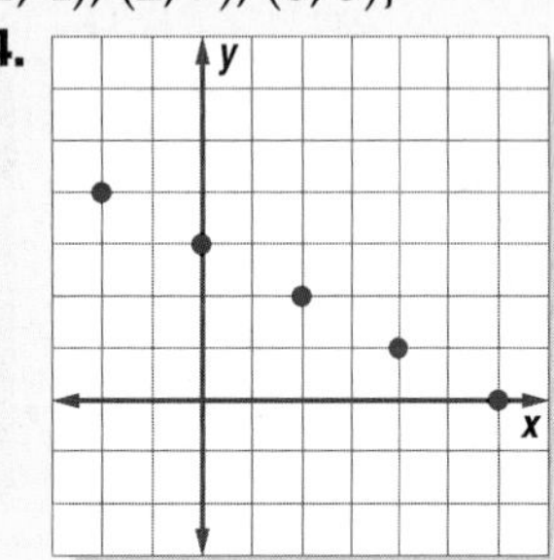

5. MULTIPLE CHOICE What are the domain and range of the relation? (Lesson 3-1)

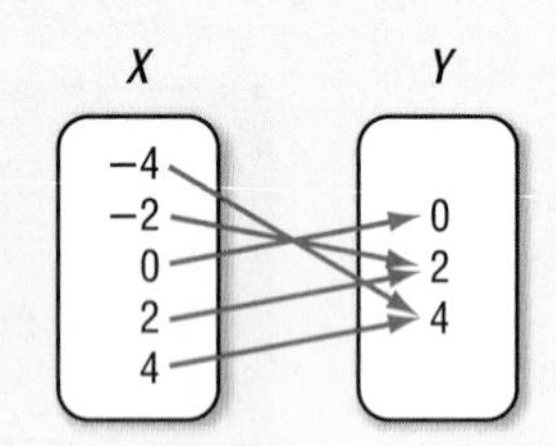

A $D = \{0, 2, 4\}$; $R = \{-4, -2, 0, 2, 4\}$

B $D = \{-4, -2, 0, 2, 4\}$; $R = \{0, 2, 4\}$

C $D = \{0, 2, 4\}$; $R = \{-4, -2, 0\}$

D $D = \{-4, -2, 0, 2, 4\}$; $R = \{-4, -2, 0, 2, 4\}$

CHEERLEADING For Exercises 6–8, use the following information. The cost of a cheerleading camp is shown in the table. (Lesson 3-1)

Number of Cheerleaders	Total Cost ($)
1	70
2	140
3	210
4	280

6. Determine the domain and range of the relation.

7. Graph the data.

8. Describe the independent and dependent quantities in this situation.

If $f(x) = 3x + 5$, find each value. (Lesson 3-2)

9. $f(-4)$ **10.** $f(2a)$ **11.** $f(x + 2)$

Determine whether each relation is a function. (Lesson 3-2)

12. $\{(3, 4), (5, 3), (-1, 4), (6, 2)\}$

13. $\{(-1, 4), (-2, 5), (7, 2), (3, 9), (-2, 1)\}$

14. MULTIPLE CHOICE Which is a true statement about the relation? (Lesson 3-2)

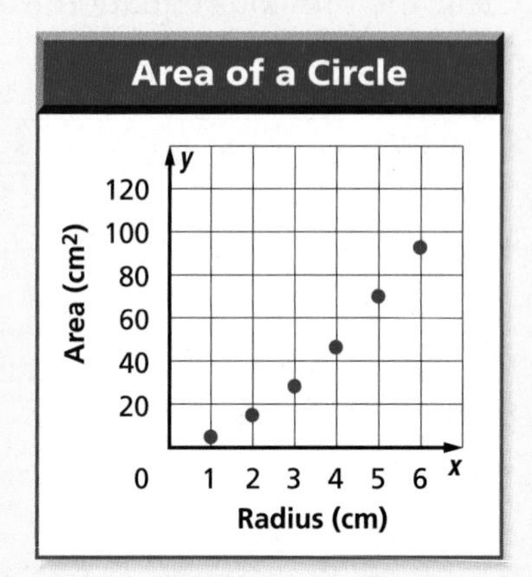

F As the radius increases, the area decreases.

G The relation is a linear function.

H The area is a function of the radius.

J The relation is not a function.

Graph each equation. (Lesson 3-3)

15. $y = x - 2$ **16.** $3x + 2y = 6$

17. MULTIPLE CHOICE If $(a, -7)$ is a solution to the equation $8a + 3b = 3$, what is a? (Lesson 3-3)

A 2 B 3 C 3.5 D −6.5

ENTERTAINMENT For Exercises 18–20, use the following information.

The equation $200x + 80y = 600$ represents the number of premium tickets x and the number of discount tickets y to a car race that can be bought with $600. (Lesson 3-3)

18. Graph the function.

19. Describe a domain and range that makes sense for this situation. Explain.

20. Describe what the x-and y-intercepts represent in the context of this situation.

3-4 Arithmetic Sequences

Main Ideas

- Recognize arithmetic sequences.
- Extend and write formulas for arithmetic sequences

New Vocabulary

sequence
terms
arithmetic sequence
common difference

GET READY for the Lesson

A probe to measure air quality is attached to a hot-air balloon. The probe has an altitude of 6.3 feet after the first second, 14.5 feet after the next second, 22.7 feet after the third second, and so on. You can make a table and look for a pattern in the data.

Time (s)	1	2	3	4	5	6	7	8
Altitude (ft)	6.3	14.5	22.7	30.9	39.1	47.3	55.5	63.7

+ 8.2 + 8.2 + 8.2 + 8.2 + 8.2 + 8.2 + 8.2

Recognize Arithmetic Sequences A **sequence** is a set of numbers, called **terms,** in a specific order. If the difference between successive terms is constant, then it is called an **arithmetic sequence**. The difference between the terms is called the **common difference**.

KEY CONCEPT *Arithmetic Sequence*

An arithmetic sequence is a numerical pattern that increases or decreases at a constant rate or value called the common difference.

Study Tip

Ellipsis
The three dots after the last number in a sequence are called an *ellipsis*. The ellipsis indicates that there are more terms in the sequence that are not listed.

EXAMPLE Identify Arithmetic Sequences

1 Determine whether each sequence is arithmetic. Explain.

a. 1, 2, 4, 8, …

1 2 4 8

+1 +2 +4

This is not an arithmetic sequence because the difference between terms is not constant.

b. $\frac{1}{2}, \frac{1}{4}, 0, -\frac{1}{4}, \ldots$

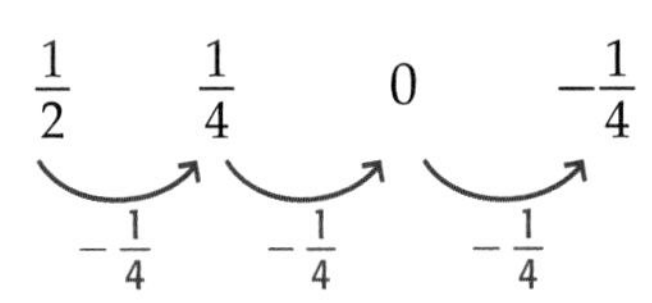

This is an arithmetic sequence because the difference between terms is constant.

CHECK Your Progress

1A. −26, −22, −18, −14, …

1B. 1, 4, 9, 25, …

Write Arithmetic Sequences You can use the common difference of an arithmetic sequence to find the next term in the sequence.

> **KEY CONCEPT** *Writing Arithmetic Sequences*
>
> **Words** Each term of an arithmetic sequence after the first term can be found by adding the common difference to the preceding term.
>
> **Symbols** An arithmetic sequence, $a_1, a_2, \ldots$, can be found as follows:
>
> $$a_1, a_2 = a_1 + d, a_3 = a_2 + d, a_4 = a_3 + d \ldots,$$
>
> where d is the common difference, a_1 is the first term, a_2 is the second term, and so on.

Real-World EXAMPLE

2 MONEY The arithmetic sequence 74, 67, 60, 53, ... represents the amount of money that Tiffany owes her mother at the end of each week. Find the next three terms.

Find the common difference by subtracting successive terms.

74 67 60 53 ? ? ?

−7 −7 −7 −7 −7 −7

The common difference is −7.

Add −7 to the last term of the sequence to get the next term in the sequence. Continue adding −7 until the next three terms are found.

53 46 39 32 The next three terms are 46, 39, 32.

−7 −7 −7

CHECK Your Progress

2. Find the next four terms of the arithmetic sequence 9.5, 11.0, 12.5, 14.0, ...

Each term in an arithmetic sequence can be expressed in terms of the first term a_1 and the common difference d.

Study Tip

Formulas The formula for the *n*th term of an arithmetic sequence is called a *recursive formula*. This means that each succeeding term is formulated from one or more of the previous terms.

Term	Symbol	In Terms of a_1 and d	Numbers
first term	a_1	a_1	8
second term	a_2	$a_1 + d$	$8 + 1(3) = 11$
third term	a_3	$a_1 + 2d$	$8 + 2(3) = 14$
fourth term	a_4	$a_1 + 3d$	$8 + 3(3) = 17$
⋮	⋮	⋮	⋮
*n*th term	a_n	$a_1 + (n - 1)d$	$8 + (n - 1)(3)$

This leads to the formula that can be used to find any term in an arithmetic sequence.

KEY CONCEPT *nth Term of an Arithmetic Sequence*

The *n*th term a_n of an arithmetic sequence with first term a_1 and common difference d is given by

$$a_n = a_1 + (n - 1)d,$$

where n is a positive integer.

Real-World EXAMPLE

3 SHIPPING The arithmetic sequence 12, 23, 34, 45, ... represents the total number of ounces that a box weighs after each additional book is added.

a. Write an equation for the *n*th term of the sequence.

In this sequence, the first term, a_1, is 12. Find the common difference.

12 23 34 45

+ 11 + 11 + 11

The common difference is 11.

Use the formula for the *n*th term to write an equation.

$a_n = a_1 + (n - 1)d$	Formula for *n*th term
$= 12 + (n - 1)11$	$a_1 = 12, d = 11$
$= 12 + 11n - 11$	Distributive Property
$= 11n + 1$	Simplify.

b. Find the 10th term in the sequence.

$a_n = 11n + 1$	Equation for the *n*th term
$a_{10} = 11(10) + 1$	Replace *n* with 10.
$a_{10} = 111$	Simplify.

c. Graph the first five terms of the sequence.

n	$11n + 1$	a_n	(n, a_n)
1	$11(1) + 1$	12	(1, 12)
2	$11(2) + 1$	23	(2, 23)
3	$11(3) + 1$	34	(3, 34)
4	$11(4) + 1$	45	(4, 45)
5	$11(5) + 1$	56	(5, 56)

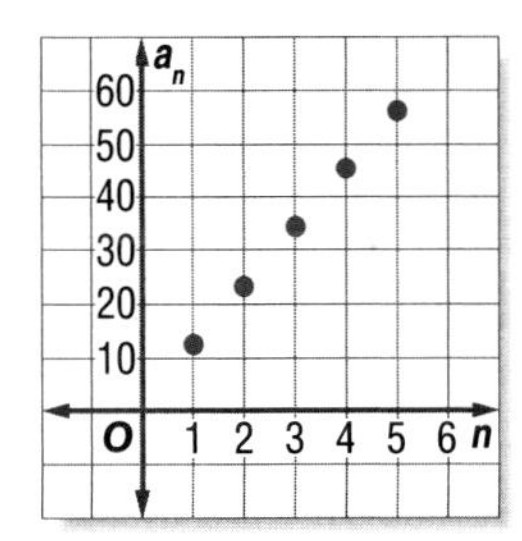

Study Tip

Graph of a Sequence
Notice that the points on the graph fall on a line. The graph of an arithmetic sequence is linear.

CHECK Your Progress

Consider the arithmetic sequence 3, −10, −23, −36, ….

3A. Write an equation for the *n*th term of the sequence.

3B. Find the 15th term in the sequence.

3C. Graph the first five terms of the sequence.

Online **Personal Tutor at** algebra1.com

CHECK Your Understanding

Example 1 (p. 165)

Determine whether each sequence is an arithmetic sequence. If it is, state the common difference.

1. 24, 16, 8, 0, ...
2. $3\frac{1}{4}, 6\frac{1}{2}, 13, 26, \ldots$

Example 2 (p. 166)

Find the next three terms of each arithmetic sequence.

3. 7, 14, 21, 28, ...
4. −34, −29, −24, −19, ...

Example 3 (p. 167)

Find the nth term of each arithmetic sequence described.

5. $a_1 = 3, d = 4, n = 8$
6. $a_1 = 10, d = -5, n = 21$
7. 23, 25, 27, 29, ... for $n = 12$
8. −27, −19, −11, −3, ... for $n = 17$

9. **FITNESS** Latisha is beginning an exercise program that calls for 20 minutes of walking each day for the first week. Each week thereafter, she has to increase her walking by 7 minutes a day. Which week of her exercise program will be the first one in which she will walk over an hour a day?

Write an equation for the nth term of each arithmetic sequence. Then graph the first five terms of the sequence.

10. 6, 12, 18, 24, ...
11. 12.1, 17.2, 22.3, 27.4, ...

Exercises

HOMEWORK HELP	
For Exercises	See Examples
12–15	1
16–19	2
20–33	3

Determine whether each sequence is an arithmetic sequence. If it is, state the common difference.

12. 7, 6, 5, 4, ...
13. 10, 12, 15, 18, ...
14. −15, −11, −7, −3, ...
15. −0.3, 0.2, 0.7, 1.2, ...

Find the next three terms of each arithmetic sequence.

16. 4, 7, 10, 13, ...
17. 18, 24, 30, 36, ...
18. −66, −70, −74, −78, ...
19. −31, −22, −13, −4, ...

GEOMETRY For Exercises 20 and 21, use the diagram below that shows the perimeter of the pattern consisting of trapezoids.

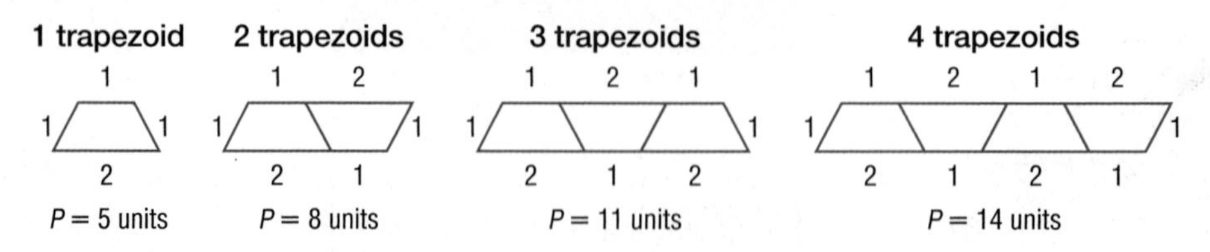

20. Write a formula that can be used to find the perimeter of a pattern containing n trapezoids.
21. What is the perimeter of the pattern containing 12 trapezoids?

Find the nth term of each arithmetic sequence described.

22. $a_1 = 8, d = 3, n = 16$
23. $a_1 = 52, d = 12, n = 102$
24. $a_1 = \frac{5}{8}, d = \frac{1}{8}, n = 22$
25. −9, −7, −5, −3, ... for $n = 18$
26. −7, −3, 1, 5, ... for $n = 35$
27. 0.5, 1, 1.5, 2, ... for $n = 50$

Real-World Link

The open-air theaters of ancient Greece held about 20,000 people. They became the models for amphitheaters, Roman coliseums, and modern sports arenas.

Source: encarta.msn.com

THEATER **For Exercises 28 and 29, use the following information.**
The Coral Gables Actors' Playhouse has 7 rows of seats in the orchestra section. The number of seats in each row forms an arithmetic sequence, as shown in the table. On opening night, 368 tickets were sold for the orchestra section.

Row	Number of Seats
7	76
6	68
5	60

28. Write a formula to find the number of seats in any given row of the orchestra section of the theater.

29. How many seats are in the first row? Was this section oversold?

Write an equation for the *n*th term of the arithmetic sequence. Then graph the first five terms in the sequence.

30. $-3, -6, -9, -12, \ldots$

31. $8, 9, 10, 11, \ldots$

32. $2, 8, 14, 20, \ldots$

33. $-18, -16, -14, -12, \ldots$

Find the next three terms of each arithmetic sequence.

34. $2\frac{1}{3}, 2\frac{2}{3}, 3, 3\frac{1}{3}, \ldots$

35. $\frac{7}{12}, 1\frac{1}{3}, 2\frac{1}{12}, 2\frac{5}{6}$

36. 200 is the __?__th term of 24, 35, 46, 57....

37. –34 is the __?__th term of 30, 22, 14, 6, ...

38. Find the value of y that makes $y + 4, 6, y, \ldots$ an arithmetic sequence.

39. Find the value of y that makes $y + 8, 4y + 6, 3y, \ldots$ an arithmetic sequence.

ANALYZE TABLES **For Exercises 40–43, use the following information.**
Taylor and Brooklyn are recording how far a ball rolls down a ramp during each second. The table shows the data they have collected.

Time (s)	1	2	3	4	5	6
Distance traveled (cm)	9	13	17	21	25	29

40. Do the distances traveled by the ball form an arithmetic sequence? Justify your answer.

41. Write an equation for the sequence. How far will the ball have traveled after 35 seconds?

42. Graph the sequence.

43. Suppose that for each second, the ball rolls twice the distance shown in the table. Is the graph representing this sequence linear? If so, describe how its rate of change is different from the rate of change shown in your original graph.

GAMES **For Exercises 44 and 45, use the following information.**
Contestants on a game show win money by answering 10 questions. The value of each question increases by $1500.

44. If the first question is worth $2500, find the value of the 10th question.

45. If the contestant answers all 10 questions correctly, how much money will he or she win?

H.O.T. Problems

46. OPEN ENDED Create an arithmetic sequence with a common difference of -10.

47. CHALLENGE Is $2x + 5, 4x + 5, 6x + 5, 8x + 5, \ldots$ an arithmetic sequence? Explain your reasoning.

48. FIND THE ERROR Marisela and Richard are finding the common difference for the arithmetic sequence -44, -32, -20, -8. Who is correct? Explain.

Marisela
$-32 - (-44) = 12$
$-20 - (-32) = 12$
$-8 - (-20) = 12$

Richard
$-44 - (-32) = -12$
$-32 - (-20) = -12$
$-20 - (-8) = -12$

49. ***Writing in Math*** Refer to the data about measuring air quality on page 165. Write a formula for the arithmetic sequence that represents the altitude of the probe after each second, and an explanation of how you could use this information to predict the altitude of the probe after 15 seconds.

STANDARDIZED TEST PRACTICE

50. REVIEW Luis deposits $25 each week into a savings account from his part-time job. If he has $350 in savings now, how much will he have in 12 weeks?

A $600
B $625
C $650
D $675

51. What is the slope of a line that contains the point at $(1, -5)$ and has the same y-intercept as $2x - y = 9$?

F -9
G -7
H 2
J 4

52. REVIEW Which is a true statement about the relation graphed?

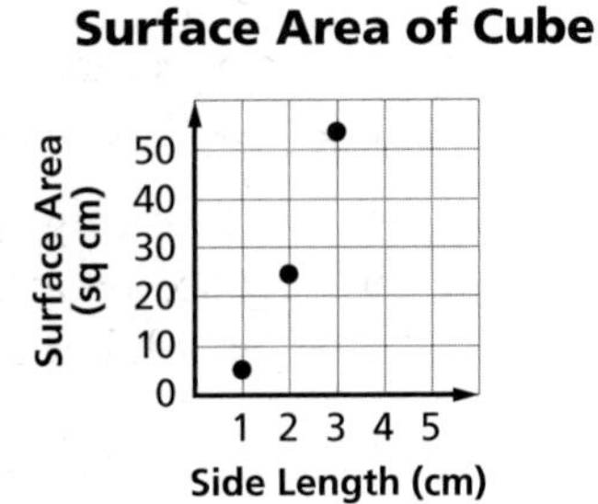

A As the side length of a cube increases, the surface area decreases.

B Surface area is the independent quantity.

C The surface area of a cube is a function of the side length.

D The relation is not a function.

Spiral Review

Determine whether each equation is a linear equation. If so, write the equation in standard form. (Lesson 3-3)

53. $x^2 + 3x - y = 8$

54. $y - 8 = 10 - x$

55. $2y = y + 2x - 3$

56. TAX The amount of sales tax in California is given by $y = 0.0725x$, where x is the cost of an item that you buy. Write the equation in function notation and then find $f(40)$. What does this value represent? (Lesson 3-2)

57. Translate the sentence *The sum of twice r and three times s is identical to thirteen* into an algebraic equation. (Lesson 2-1)

GET READY for the Next Lesson

PREREQUISITE SKILL Graph each point on the same coordinate plane. (Lesson 1-9)

58. $J(3, 0)$

59. $L(-3, -4)$

60. $M(3, 5)$

61. $N(5, -1)$

READING MATH

Inductive and Deductive Reasoning

Throughout your life, you have used reasoning skills, possibly without even knowing it. As a child, you used inductive reasoning to conclude that your hand would hurt if you touched the stove while it was hot. Now, you use inductive reasoning when you decide, after many trials, that one of the worst ways to prepare for an exam is by studying only an hour before you take it. **Inductive reasoning** is used to derive a general rule after observing many individual events.

Inductive reasoning involves:

- observing many examples
- looking for a pattern
- making a conjecture
- checking the conjecture
- discovering a likely conclusion

With **deductive reasoning**, you use a general rule to help you decide about a specific event. You come to a conclusion by accepting facts. There is no conjecturing involved. Read the two statements below.

1) If a person wants to play varsity sports, he or she must have a C average in academic classes.
2) Jolene is playing on the varsity tennis team.

If these two statements are accepted as facts, then the obvious conclusion is that Jolene has at least a C average in her academic classes. This is an example of deductive reasoning.

Reading to Learn

1. Explain the difference between inductive and deductive reasoning. Then give an example of each.

2. When Sherlock Holmes reaches a conclusion about a murderer's height because he knows the relationship between a man's height and the distance between his footprints, what kind of reasoning is he using? Explain.

3. When you examine a sequence of numbers and decide that it is an arithmetic sequence, what kind of reasoning are you using? Explain.

4. Once you have found the common difference for an arithmetic sequence, what kind of reasoning do you use to find the 100th term in the sequence?

5. a. Copy and complete the following table.

3^1	3^2	3^3	3^4	3^5	3^6	3^7	3^8	3^9
3	9	27						

b. Write the sequence of numbers representing the numbers in the ones place.

c. Find the number in the ones place for the value of 3100. Explain your reasoning. State the type of reasoning that you used.

6. A sequence contains all numbers less than 50 that are divisible by 5. You conclude that 35 is in the sequence. Is this an example of inductive or deductive reasoning? Explain.

3-5 Proportional and Nonproportional Relationships

Main Ideas

- Write an equation for a proportional relationship.
- Write an equation for a nonproportional relationship.

New Vocabulary

inductive reasoning

GET READY for the Lesson

Water is one of the few substances that expands when it freezes. The table shows volumes of water and the corresponding volumes of ice.

Volume of Water (ft³)	11	22	33	44	55
Volume of Ice (ft³)	12	24	36	48	60

The relation in the table can be represented by a graph. Let w represent the volume of water, and let c represent the volume of ice. When the ordered pairs are graphed, they form a linear pattern. This pattern can be described by an equation.

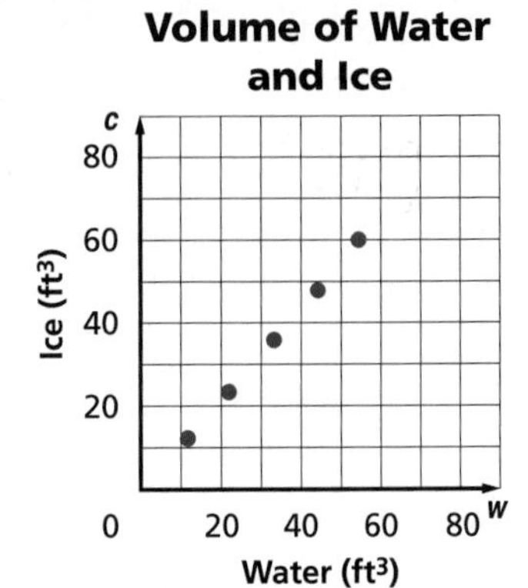

Proportional Relationships Using a pattern to find a general rule utilizes **inductive reasoning**. If the relationship between the domain and range of a relation is linear, the relationship can be described by a linear equation. If the equation is of the form $y = kx$, then the relationship is proportional. In a proportional relationship, the graph will pass through (0, 0).

Real-World EXAMPLE Proportional Relationships

1 FUEL ECONOMY The table below shows the average amount of gas Rogelio's car uses, depending on how many miles he drives.

Gallons of gasoline	1	2	3	4	5
Miles driven	28	56	84	112	140

a. Graph the data. What conclusion can you make about the relationship between the number of gallons used and the number of miles driven?

The graph shows a linear relationship between the number of gallons used g and the number of miles driven m.

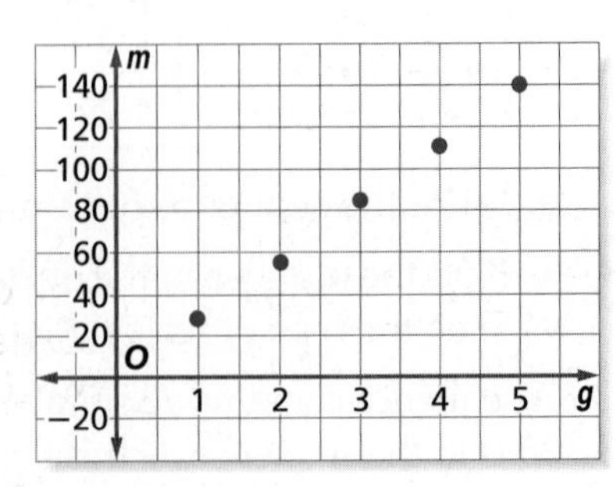

Real-World Link
Hybrid cars have a small, fuel-efficient gas engine combined with an electric motor. The electric motor is powered by batteries that recharge automatically while the car is being driven.

Source: artheasy.com/live_hybrid_cars.htm

b. Write an equation to describe this relationship.

Look at the relationship between the values in the domain and range to find a pattern that can be described by an equation.

+ 1 + 1 + 1 + 1

Gallons of gasoline	1	2	3	4	5
Miles driven	28	56	84	112	140

+ 28 + 28 + 28 + 28

The difference of the values for g is 1, and the difference of the values for m is 28. This suggests that $m = 28g$. Check to see if this equation is correct by substituting values of g into the equation.

CHECK If $g = 1$, then $m = 28(1)$ or 28. ✓
If $g = 2$, then $m = 28(2)$ or 56. ✓
If $g = 3$, then $m = 28(3)$ or 84. ✓

The equation checks. Since this relation is a function, we can write the equation as $f(g) = 28g$, where $f(g)$ represents the number of miles driven.

CHECK Your Progress

FUND-RAISING The table shows the cost of buying Spanish Club T-shirts.

Number of T-shirts	1	2	3	4
Total Cost ($)	7.50	15.00	22.50	30.00

1A. Graph the data and describe the relationship between the number of T-shirts bought and the amount spent.

1B. Write an equation to describe this relationship.

Nonproportional Relationships Some linear relationships are nonproportional. In the equation of a nonproportional situation, a constant must be added or subtracted from the variable expression.

EXAMPLE Nonproportional Relationships

2 **Write an equation in function notation for the relation graphed at the right.**

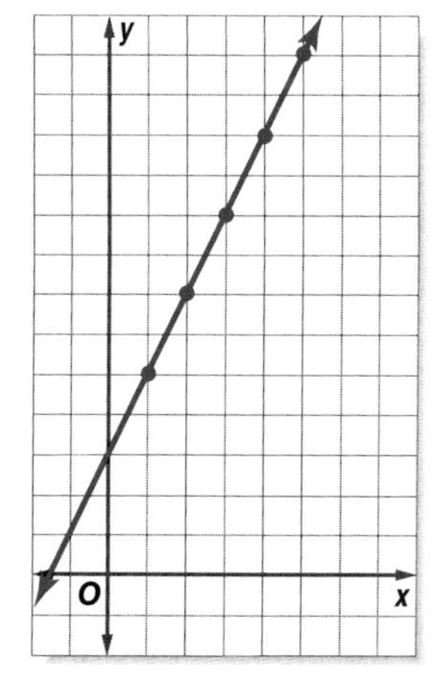

Make a table of ordered pairs.

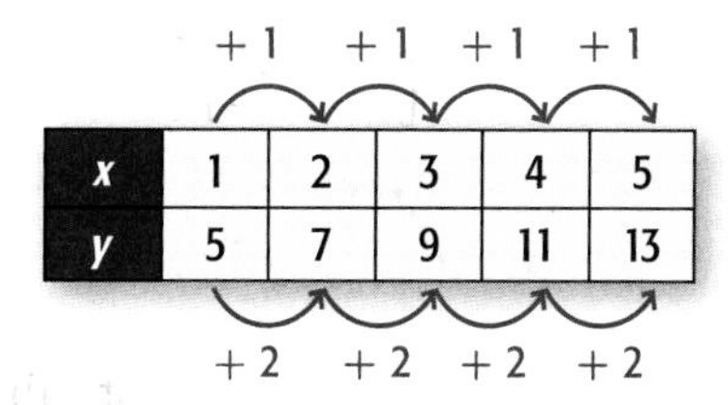

+ 1 + 1 + 1 + 1

x	1	2	3	4	5
y	5	7	9	11	13

+ 2 + 2 + 2 + 2

The difference of the x values is 1, and the difference of the y values is 2. The difference in y values is twice the difference of x values. This suggests that $y = 2x$.

Study Tip

Nonproportional Relationships
The x- and y-values in Example 2 do not have a proportional relationship because $\frac{x}{y}$ is not always the same; for example, $\frac{1}{5} \neq \frac{2}{7}$.

(continued on the next page)

CHECK Suppose the equation is $y = 2x$. If $x = 1$, then $y = 2(1)$ or 2. But the y-value for $x = 1$ is 5. This is a difference of 3. Try some other values in the domain to see if the same difference occurs.

x	1	2	3	4	5
$2x$	2	4	6	8	10
y	5	7	9	11	13

y is always 3 more than $2x$.

This pattern suggests that 3 should be added to one side of the equation in order to correctly describe the relation. Check $y = 2x + 3$.

If $x = 2$, then $y = 2(2) + 3$ or 7.
If $x = 3$, then $y = 2(3) + 3$ or 9.

Thus, $y = 2x + 3$ correctly describes this relation. Since this relation is a function, the equation in function notation is $f(x) = 2x + 3$.

CHECK Your Progress

2. Write an equation in function notation for the relation shown in the table.

x	1	2	3	4
y	3	2	1	0

Personal Tutor at algebra1.com

CHECK Your Understanding

Example 1 (pp. 172–173)

Write an equation in function notation for each relation.

1.

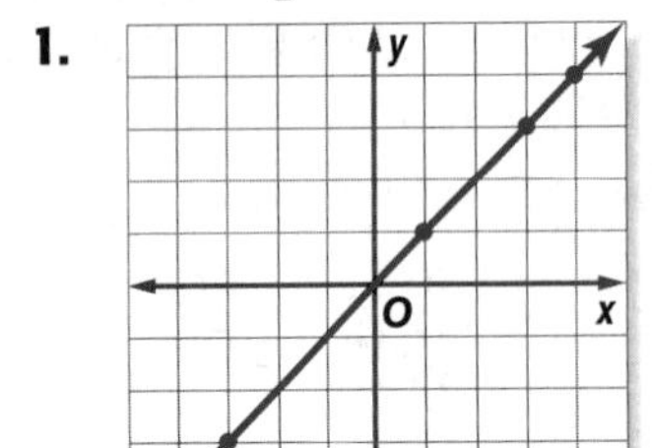

2.

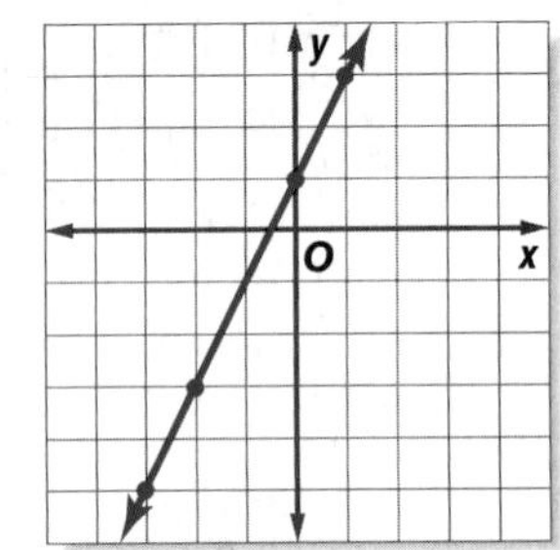

GEOMETRY For Exercises 3 and 4, use the table below that shows the perimeter of a square with sides of a given length.

Side length (in)	1	2	3	4	5
Perimeter (in)	4	8	12	16	20

3. Graph the data. What can you conclude about the relationship between side length and perimeter?

4. Write an equation to describe the relationship.

Example 2 (pp. 173–174)

ANALYZE TABLES For Exercises 5–7, use the table below that shows the underground temperature of rocks at various depths below Earth's surface.

Depth (km)	1	2	3	4	5	6
Temperature (°C)	55	90	125	160	195	230

5. Graph the data.

6. Write an equation in function notation for the relation.

7. Find the temperature of a rock that is 10 kilometers below the surface.

Exercises

HOMEWORK HELP	
For Exercises	See Examples
14, 15	1
16–22	2

Find the next three terms in each sequence.

8. 0, 2, 6, 12, 20, ...

9. 9, 7, 10, 8, 11, 9, 12, ...

10. 1, 4, 9, 16, ...

11. 0, 2, 5, 9, 14, 20, ...

12. $a + 1, a + 2, a + 3, \ldots$

13. $x + 1, 2x + 1, 3x + 1, \ldots$

Write an equation in function notation for each relation.

14.

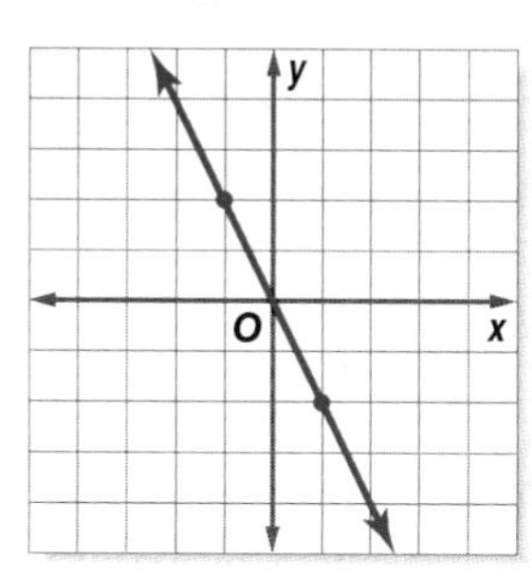

15.

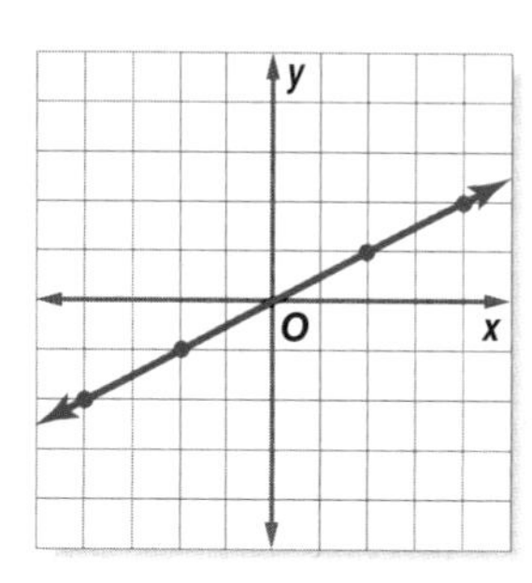

16.

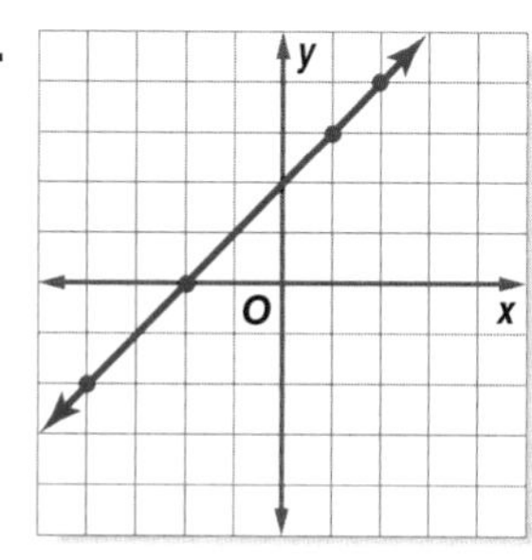

17.

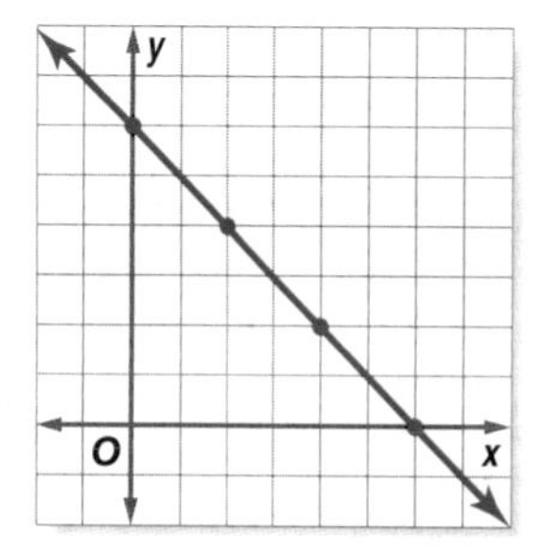

18. TRAVEL On an island cruise in Hawaii, each passenger is given a lei. A crew member hands out 3 red, 3 blue, and 3 green leis in that order. If this pattern is repeated, what color lei will the 50th person receive?

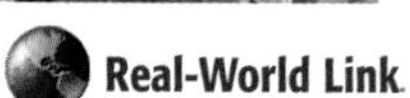
Real-World Link

Fibonacci numbers occur in many areas of nature, including pine cones, shell spirals, flower petals, branching plants, and many fruits and vegetables.

Source: mathworld.wolfram.com

NUMBER THEORY For Exercises 19 and 20, use the following information.
In 1201, Leonardo Fibonacci introduced his now famous pattern of numbers called the Fibonacci sequence.

$$1, 1, 2, 3, 5, 8, 13, \ldots$$

Notice the pattern in this sequence. After the second number, each number in the sequence is the sum of the two numbers that precede it. That is, $2 = 1 + 1$, $3 = 2 + 1$, $5 = 3 + 2$, and so on.

19. Write the first 12 terms of the Fibonacci sequence.

20. Notice that every third term is divisible by 2. What do you notice about every fourth term? every fifth term?

Write an equation in function notation for each relation.

21.

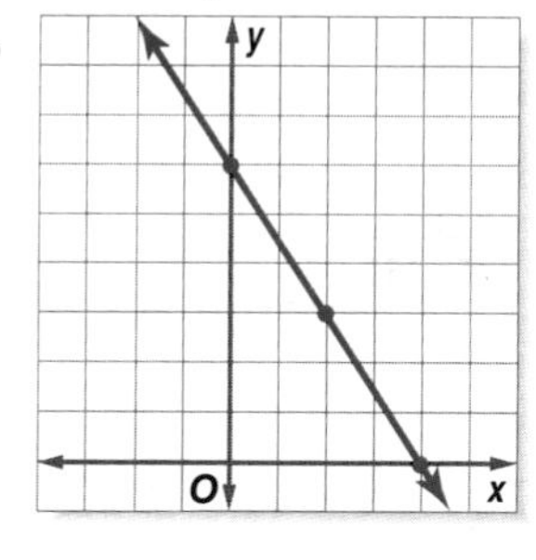

22.

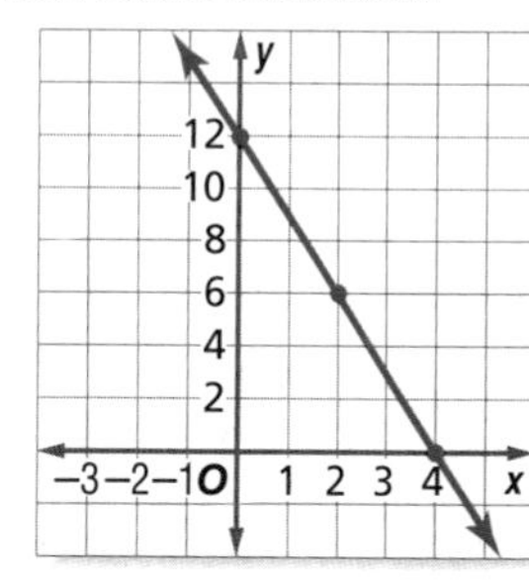

FITNESS For Exercises 23 and 24, use the table below that shows the maximum heart rate to maintain during aerobic activities such as biking.

Age (yr)	20	30	40	50	60	70
Pulse rate (beats/min)	175	166	157	148	139	130

Source: Ontario Association of Sport and Exercise Sciences

23. Write an equation in function notation for the relation.

24. What would be the maximum heart rate to maintain in aerobic training for a 10-year-old? an 80-year-old?

H.O.T. Problems

25. CHALLENGE Describe how inductive reasoning can be used to write an equation from a pattern.

26. OPEN ENDED Create a number sequence in which the first term is 4. Explain the pattern that you used.

27. *Writing in Math* Use the information about science on page 172 to explain how writing equations from patterns is important in science. Explain the relationship between the volumes of water and ice.

STANDARDIZED TEST PRACTICE

28. The table below shows the cost C of renting a pontoon boat for h hours.

Hours	1	2	3
Cost	7.25	14.50	21.75

Which equation best represents the data?

A $C = 7.25h$

B $C = h + 7.25$

C $C = 21.75 - 7.25h$

D $C = 7.25h + 21.75$

29. REVIEW Donald can ride 8 miles on his bicycle in 30 minutes. At this rate, how long would it take him to ride 30 miles?

F 8 hours

G 6 hours 32 minutes

H 2 hours

J 1 hour 53 minutes

Spiral Review

Find the next three terms of each arithmetic sequence. (Lesson 3-4)

30. 9, 5, 1, −3, …

31. −25, −19, −13, −7, …

32. 22, 34, 46, 58, …

Graph each equation. (Lesson 3-3)

33. $y = x + 3$

34. $y = 2x - 4$

35. $2x + 5y = 10$

36. IN THE MEDIA The following statement appeared on a news Web site shortly after a giant lobster named Bubba was found near Nantucket, Massachusetts. Approximately how much did Bubba weigh? (Lesson 2-3)

"At Tuesday's price of $14.98 a pound, Bubba would retail for about $350."

Source: cnn.com

CHAPTER 3 Study Guide and Review

Download Vocabulary Review from algebra1.com

FOLDABLES™ Study Organizer — GET READY to Study

Be sure the following Key Concepts are noted in your Foldable.

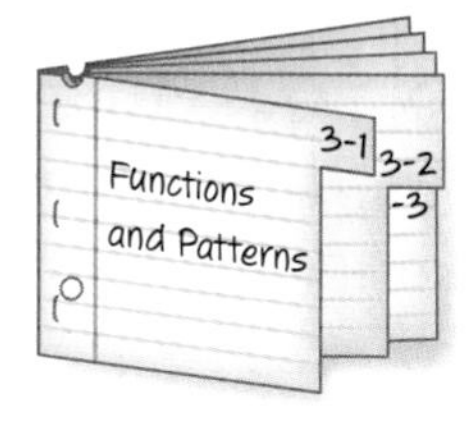

Key Concepts

Representing Relations and Functions

(Lessons 3-1 and 3-2)

- Relation Q is the inverse of relation S if, and only if, for every ordered pair (a, b) in S, there is an ordered pair (b, a) in Q.
- A function is a relation in which each element of the domain is paired with *exactly* one element of the range.

Linear Functions (Lesson 3-3)

- The standard form of a linear equation is $Ax + By = C$, where $A \geq 0$, A and B are not both zero, and A, B, and C are integers with the greatest common factor of 1.

Arithmetic Sequences (Lesson 3-4)

- An arithmetic sequence is a numerical pattern that increases or decreases at a constant rate or value called the common difference.
- The nth term a_n of an arithmetic sequence with first term a_1 and common difference d is given by $a_n = a_1 + (n - 1)d$, where n is a positive integer.

Number Patterns (Lesson 3-5)

- When you make a conclusion based on a pattern of examples, you are using inductive reasoning.

Key Vocabulary

arithmetic sequence (p. 165)
common difference (p. 165)
function (p. 149)
function notation (p. 150)
function value (p. 150)
inductive reasoning (p. 172)
inverse (p. 144)
linear equation (p. 155)
mapping (p. 143)
sequence (p. 165)
standard form (p. 155)
terms (p. 165)
vertical line test (p. 150)
x-intercept (p. 156)
y-intercept (p. 156)
zero (p. 156)

Vocabulary Check

State whether each sentence is *true* or *false*. If *false*, replace the underlined word or number to make a true sentence.

1. The mapping of a relation is obtained by switching the coordinates of each ordered pair.
2. The function value of $g(x)$ for $x = 8$ is $g(8)$.
3. To determine if a graph represents a function, you can use the vertical line test.
4. A relation is a set of ordered pairs.
5. In a function, $f(x)$ represents the elements of the domain.
6. The x-coordinate of the point at which a graph of an equation crosses the x-axis is an x-intercept.
7. A linear equation is the equation of a line.
8. The difference between the terms of an arithmetic sequence is called the inverse.
9. The regular form of a linear equation is $Ax + By = C$.
10. Values of x for which $f(x) = 0$ are called zeros of the function f.

Math Online Vocabulary Review at algebra1.com

Study Guide and Review

Lesson-by-Lesson Review

3-1 Representing Relations (pp. 143–148)

Express each relation as a table, a graph, and a mapping. Then determine the domain and range.

11. {(−2, 6), (3, −2), (3, 0), (4, 6)}

12. {(2, 5), (−3, 1), (4, −2), (2, 3)}

RIDES For Exercises 13 and 14, use the table. It shows the angles of descent and the vertical drops for five roller coasters.

Angle of Descent (°)	Vertical Drop (ft)
45	72
52	137
55	118
60	195
80	300

13. Determine the domain and range.

14. Graph the data. What conclusions might you make from the graph?

Example 1 **Express the relation {(3, 2), (5, 3), (4, 3), (5, 2)} as a table, a graph, and a mapping. Then determine the domain and range.**

Table

x	y
3	2
5	3
4	3
5	2

Graph

Mapping

X Y

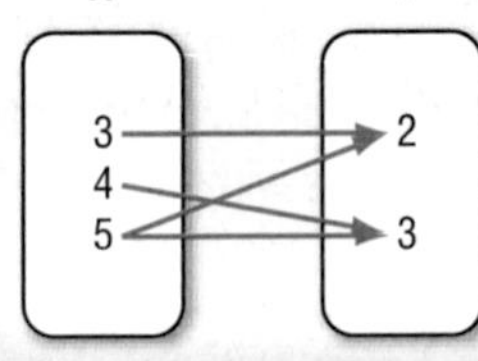

The domain is {3, 4, 5}. The range is {2, 3}.

3-2 Representing Functions (pp. 149–154)

Determine whether each relation is a function.

15. {(5, 3), (1, 4), (−6, 5), (1, 6), (−2, 7)}

16. {(2, 3), (−3, −4), (−1, 3), (6, 7)}

If $f(x) = x^2 - x + 1$, find each value.

17. $f(-1)$ **18.** $f(5) - 3$ **19.** $f(a)$

20. DOLPHINS The amount of food that an adult bottlenose dolphin eats per day can be approximated by $y = 0.05x$, where x is the dolphin's body weight in pounds. Write the equation in function notation and then find $f(460)$. What does this value represent?

Example 2 **Determine whether the relation shown is a function. Explain.**

x	y
0	−4
1	−1
2	2
6	3

Since each element of the domain is paired with exactly one element of the range, the relation is a function.

Example 3 **If $g(x) = 2x - 1$, find $g(-6)$.**

$g(-6) = 2(-6) - 1$ Replace x with −6.

$= -12 - 1$ Multiply.

$= -13$ Subtract.

Mixed Problem Solving
For mixed problem-solving practice, see page 746.

3-3 Linear Functions (pp. 155–161)

Determine the x-intercept, y-intercept, and zero of each linear function.

21.

x	y
−8	0
−4	3
0	6
4	9

22. 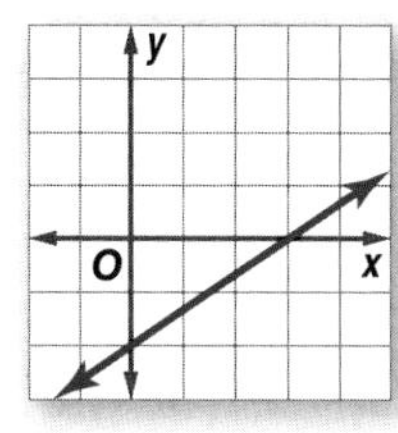

Graph each equation.

23. $y = -x + 2$

24. $x + 5y = 4$

25 $2x - 3y = 6$

26. $5x + 2y = 10$

27. SOUND The distance d in kilometers that sound waves travel through water is given by $d = 1.6t$, where t is the time in seconds. Graph the equation. Estimate how far sound can travel through water in 7 seconds.

Example 4 **Graph $3x - y = 4$ by using the x- and y-intercepts.**

Find the x-intercept.

$3x - y = 4$

$3x - 0 = 4$ Let $y = 0$.

$3x = 4$

$x = \frac{4}{3}$

Find the y-intercept.

$3x - y = 4$

$3(0) - y = 4$ Let $x = 0$.

$-y = 4$

$y = -4$

x-intercept: $\frac{4}{3}$, y-intercept: -4

The graph intersects the x-axis at $\left(\frac{4}{3}, 0\right)$ and the y-axis at $(0, -4)$. Plot these points. Then draw the line through them.

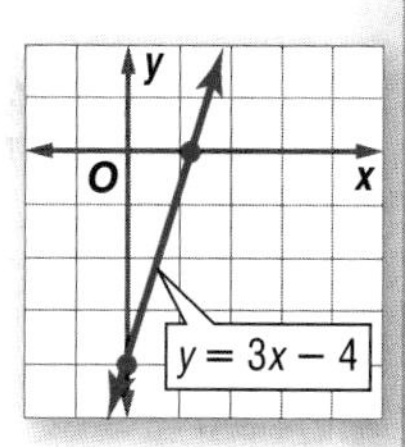

3-4 Arithmetic Sequences (pp. 165–170)

Find the next three terms of each arithmetic sequence.

28. 6, 11, 16, 21, …

29. 1.4, 1.2, 1.0, 0.8, …

30. −3, −11, −19, −27, …

Find the nth term of each arithmetic sequence described.

31. $a_1 = 6, d = 5, n = 11$

32. 28, 25, 22, 19, … for $n = 8$

33. MONEY The table represents Tiffany's income. Write an equation for this sequence and use the equation to find her income if she works 20 hours.

Hours Worked	1	2	3	4
Income ($)	20.50	29	37.50	46

Example 5 **Find the next three terms of the arithmetic sequence 10, 23, 36, 49, ….**

Find the common difference.

10 23 36 49
+ 13 + 13 + 13

So, $d = 13$.

Add 13 to the last term of the sequence. Continue adding 13 until the next three terms are found.

49 62 75 88
+ 13 + 13 + 13

The next three terms are 62, 75, and 88.

3-5 Proportional and Nonproportional Relationships (pp. 172–176)

Write an equation in function notation for each relation.

34.

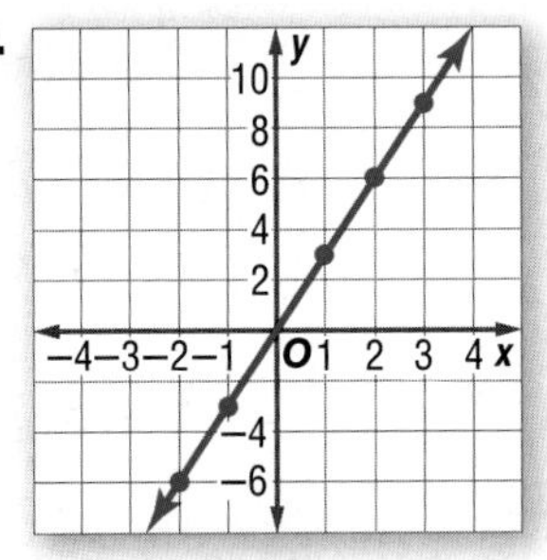

35.

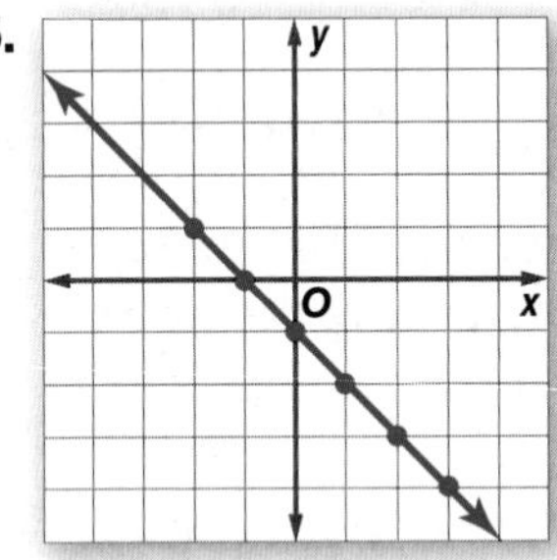

ANALYZE TABLES For Exercises 36–38, use the table below that shows the cost of picking your own strawberries at a local farm.

Number of Pounds	1	2	3	4
Total Cost ($)	1.25	2.50	3.75	5.00

36. Graph the data.

37. Write an equation in function notation to describe this relationship.

38. How much would 6 pounds of strawberries cost if you picked them yourself?

Example 6 **Write an equation in function notation for the relation graphed at the right.**

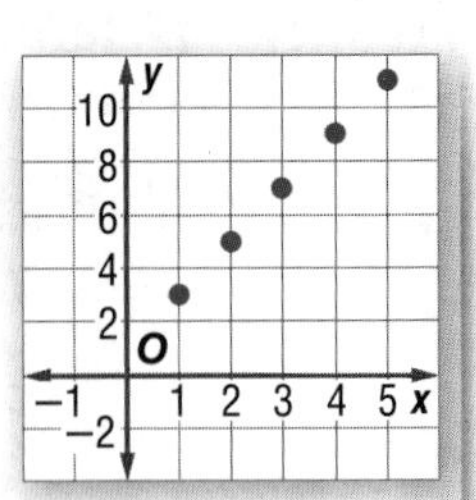

Make a table of ordered pairs for several points on the graph.

x	1	2	3	4	5
y	3	5	7	9	11

The difference in y-values is twice the difference of x-values. This suggests that $y = 2x$. However, $3 \neq 2(1)$. Compare the values of y to the values of $2x$.

x	1	2	3	4	5
$2x$	2	4	6	8	10
y	3	5	7	9	11

y is always 1 more than $2x$.

The difference between y and $2x$ is always 1. So the equation is $y = 2x + 1$. Since this relation is also a function, it can be written as $f(x) = 2x + 1$.

CHAPTER 3 Practice Test

Express the relation shown in each table, mapping, or graph as a set of ordered pairs. Then write the inverse of the relation.

1.

x	$f(x)$
0	−1
2	4
4	5
6	10

2.

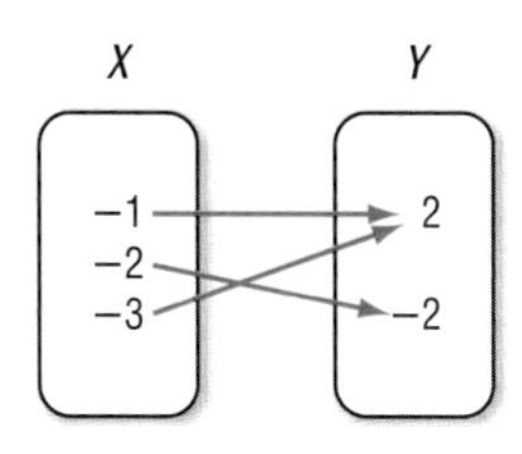

3. 

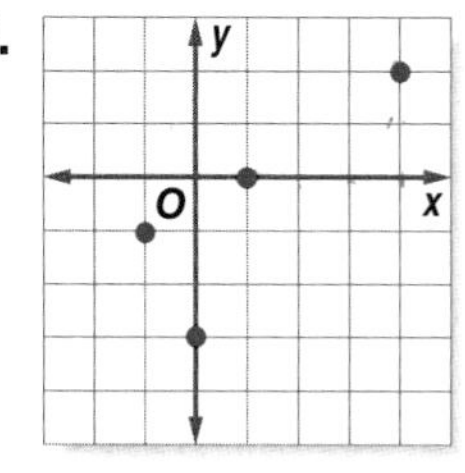

Determine whether each relation is a function.

4. {(2, 4), (3, 2), (4, 6), 5, 4)}

5. {(3, 1), (2, 5), (4, 0), (3, −2)}

6. $8y = 7 + 3x$

If $f(x) = -2x + 5$ and $g(x) = x^2 - 4x + 1$, find each value.

7. $g(-2)$

8. $f\left(\frac{1}{2}\right)$

9. $g(3a) + 1$

10. $f(x + 2)$

TEMPERATURE **The equation to convert Celsius temperature C to Kelvin temperature K is shown in the graph.**

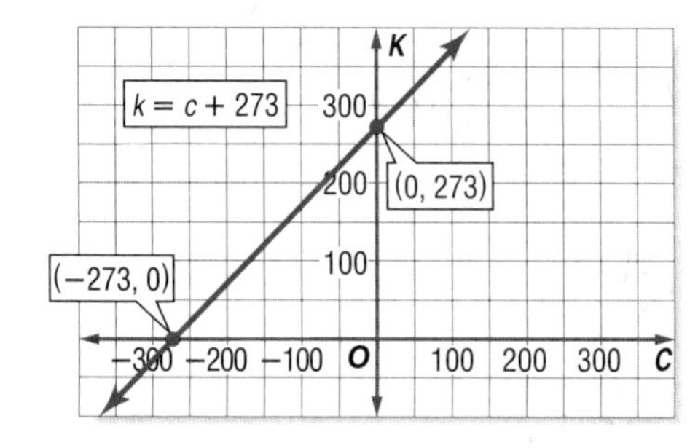

11. State the independent and dependent variables. Explain.

12. Determine the x-intercept and y-intercept and describe what the intercepts mean.

13. **MULTIPLE CHOICE** If $f(x) = 3x - 2$, find $f(8) - f(-5)$.

A 7
B 9
C 37
D 39

Graph each equation.

14. $y = x + 2$

15. $y = 4x$

16. $x + 2y = -1$

17. $-3x = 5 - y$

Find the next three terms in each sequence.

18. 5, −10, 15, −20, 25, …

19. 5, 5, 6, 8, 11, 15, …

BIOLOGY **For Exercises 20 and 21, use the following information.**

The amount of blood in the body can be predicted by the equation $y = 0.07w$, where y is the number of pints of blood and w is the weight of a person in pounds.

20. Graph the equation.

21. Predict the weight of a person whose body holds 12 pints of blood.

Determine whether each sequence is an arithmetic sequence. If it is, state the common difference.

22. −40, −32, −24, −16, …

23. 0.75, 1.5, 3, 6, 12, …

24. 5, 17, 29, 41, …

25. **MULTIPLE CHOICE** In each figure, only one side of each regular pentagon touches. Each side of each pentagon is 1 centimeter. If the pattern continues, what is the perimeter of a figure that has 6 pentagons?

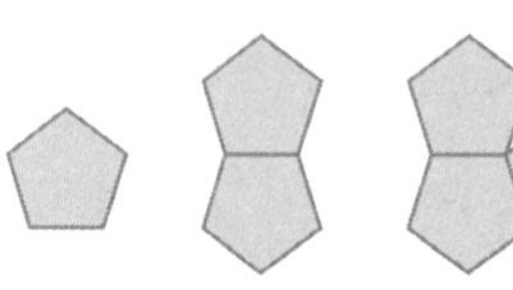

F 15 cm
G 25 cm
H 20 cm
J 30 cm

CHAPTER 3

Standardized Test Practice

Cumulative, Chapters 1–3

Read each question. Then fill in the correct answer on the answer document provided by your teacher or on a sheet of paper.

1. The function $c(x) = 0.50(x - 100) + 20$ represents the charge for renting a car at Scott's Rental Cars when a car is driven x miles. Which statement best represents the formula for this charge?

A The charge consists of a set fee of $.50 and 20 dollars for each mile over 100 miles.

B The charge consists of a set fee of $100 and $0.50 for each mile over 20 miles.

C The charge consists of a set fee of $20 and $0.50 per mile for each mile.

D The charge consists of a set fee of $20 and $0.50 per mile over 100 miles.

2. Which problem is best represented by the number sentence $20 + 2(20 - x) = 54$?

F Kayla babysat for 2 weeks. For the first week she babysat for 20 hours. For the second week she babysat for less than 20 hours. She babysat for a total of 54 hours in those two weeks. How many less hours did she babysit for in the second week?

G Jocelyn ran 20 miles the first week of a training program. The second and third week she ran less than 20 miles. In the second and third week she ran the same number of miles. In the three weeks she ran a total of 54 miles. How many miles less did she run each of the second and third weeks?

H Steven earned $20 at his job and washed 2 cars in less than 20 minutes. He earned a total of $54. How much did he earn per washed car?

J Ely earned $20 walking dogs and sold 2 magazine subscriptions for $20 each. Now he has $54. How much did he earn?

3. GRIDDABLE What is the value of the expression $(4 \cdot 1)^2 - \frac{(2 + 6)}{(4 \cdot 2)}$?

4. Which is the best representation of the function $y = |x|$?

A
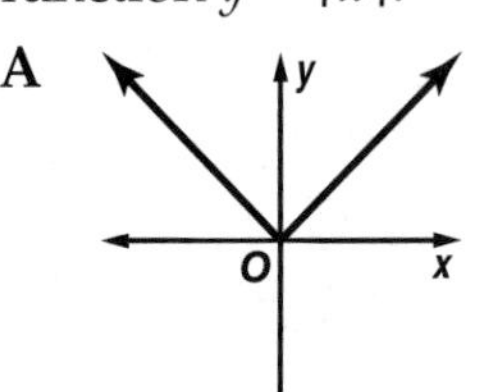

C
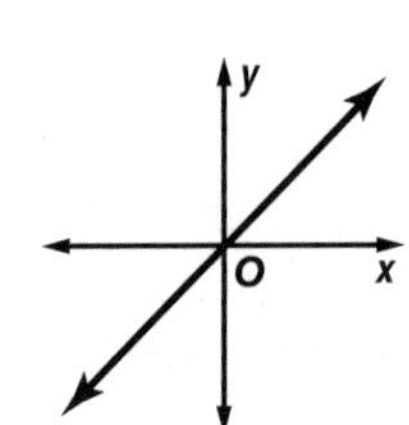

B
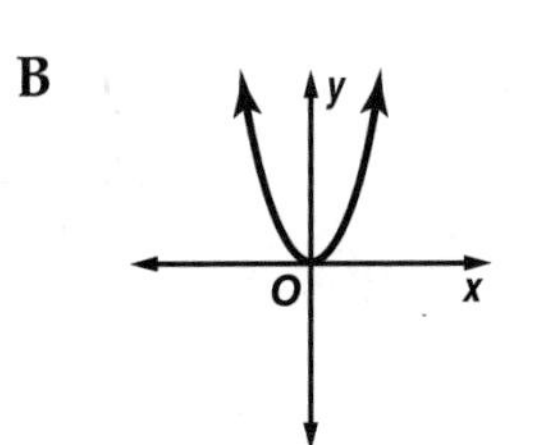

D
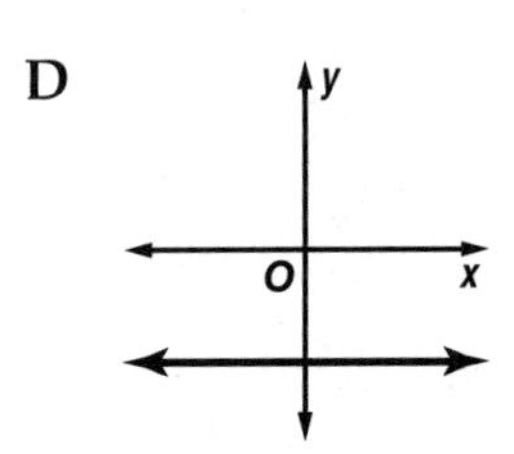

5. Which is NOT a correct representation of the function $f(x) = \{(3, 1), (6, 2), (9, 3), (12, 4)\}$?

F
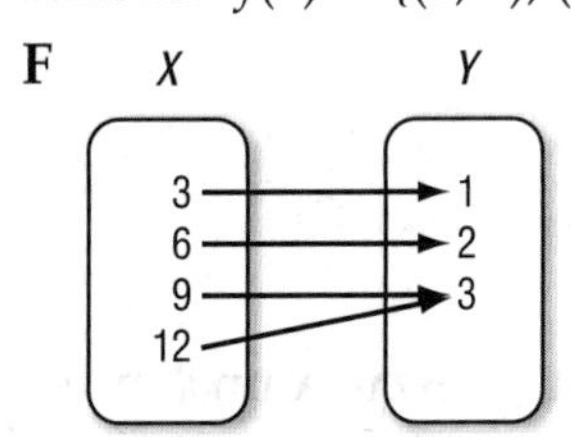

G
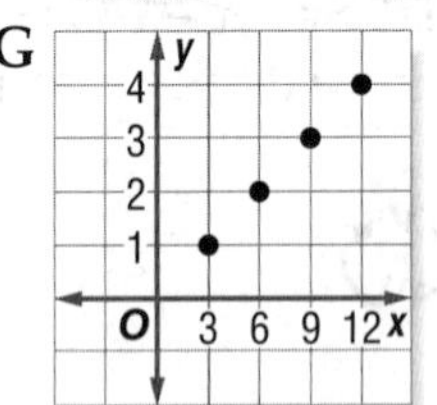

H $y = \frac{1}{3}x$ and the domain is $\{3, 6, 9, 12\}$

J y is a natural number less than or equal to 4 and x is three times y

TEST-TAKING TIP

Question 5 Always read every answer choice, particularly in questions that ask, "Which is NOT a correct representation of the function?"

For test-taking strategies and more practice, see pages 756–773.

6. Michael wants to write an expression that will always produce an odd integer. Which of the following will always produce an odd integer, n?

A $n + 1$

B $2n + 2$

C $2n + 1$

D $3n + 1$

7. GRIDDABLE Marcus and Peter are swimming laps together. Marcus gains 4 laps on Peter in 2 hours. How many laps will he gain in 45 minutes?

8. Thomas recorded data on a game at the carnival which awards points for throwing a dart at a dart board. If the dart lands on a yellow space you get x points and if the dart lands on a red space you receive y points. Amy threw 12 darts that landed in the yellow space and 9 darts that landed in the red space. Which expression gives the average point per dart throw?

F $21\left(\frac{12}{x} + \frac{9}{y}\right)$

G $\frac{9x + 12y}{21}$

H $\frac{12x + 9y}{21}$

J $\frac{x + y}{21}$

9. Carmen wrapped a ribbon around the girth of a cube-shaped present. She used 48 inches of ribbon to fit exactly around the present.

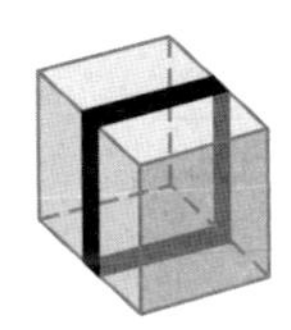

What is the volume of the present?

A 12 in^3

B 48 in^3

C 144 in^3

D 1728 in^3

10. The area of a trapezoid A is $\frac{1}{2}$ times the sum of the bases a and b times the height. Which equation best represents this relationship?

F $A = \frac{1}{2}a + bh$

G $A = \frac{1}{2}(a + bh)$

H $A = \frac{1}{2}(a + b)h$

J $A = \frac{1}{2}(a + b + h)$

11. The odometer on Jenna's car is broken. It advances 1.1 miles for every mile Jenna drives. If the odometer showed that she drove 290.4 miles since she last filled the gas tank, how many miles did she actually drive?

A 264 miles

B 289.3 miles

C 291.5 miles

D 319.4 miles

Pre-AP

Record your answers on a sheet of paper. Show your work.

12. A car company lists the stopping distances of a car at different speeds.

Speed (ft/s)	Minimum Stopping Distance (ft)
10	2
20	8
40	31
60	70
100	194

a. Does the table of values represent a function? Explain.

b. Is this a proportional relationship? Explain.

NEED EXTRA HELP?												
If You Missed Question...	1	2	3	4	5	6	7	8	9	10	11	12
Go to Lesson or Page...	2-4	2-4	1-2	3-3	3-1	2-1	2-6	2-1	708	2-1	2-6	3-2

CHAPTER 4

Analyzing Linear Equations

BIG Ideas

- Find the slope of a line.
- Write linear equations in slope-intercept and point-slope forms.
- Write equations for parallel and perpendicular lines.
- Draw a scatter plot and write an equation for a line of fit.

Key Vocabulary

point-slope form (p. 220)

rate of change (p. 187)

slope (p. 189)

slope-intercept form (p. 204)

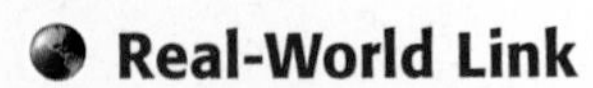

Real-World Link

Space Exploration Linear equations are used to model a variety of real-world situations, including the cost of the U.S. space program.

Analyzing Linear Equations Make this Foldable to help you organize information about writing linear equations. Begin with four sheets of grid paper.

1. **Fold** each sheet of grid paper in half along the width. Then cut along the crease.

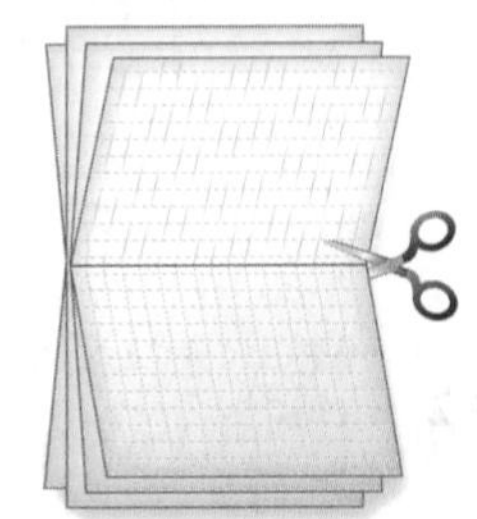

2. **Staple** the eight half-sheets together to form a booklet.

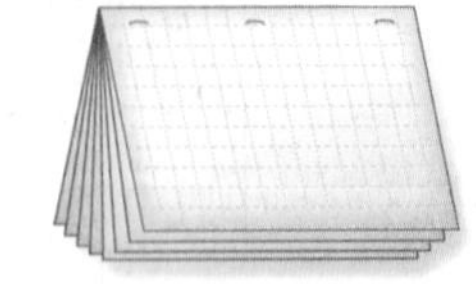

3. **Cut** seven lines from the bottom of the top sheet, six lines from the second sheet, and so on.

4. **Label** each of the tabs with a lesson number. The last tab is for the vocabulary.

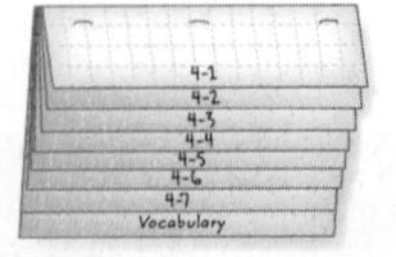

GET READY for Chapter 4

Diagnose Readiness You have two options for checking Prerequisite Skills.

Option 2

Math Online Take the Online Readiness Quiz at algebra1.com.

Option 1

Take the Quick Check below. Refer to the Quick Review for help.

QUICKCheck

Simplify. (Prerequisite Skill)

1. $\frac{2}{10}$ **2.** $\frac{8}{12}$ **3.** $\frac{2}{-8}$

4. $\frac{-4}{8}$ **5.** $\frac{-5}{-15}$ **6.** $\frac{-7}{-28}$

7. $\frac{9}{3}$ **8.** $\frac{18}{12}$ **9.** $-\frac{26}{10}$

Evaluate $\frac{a-b}{c-d}$ for the values given. (Lesson 1-2)

10. $a = 6, b = 5, c = 8, d = 4$

11. $a = -2, b = 1, c = 4, d = 0$

12. $a = -3, b = -3, c = 4, d = 7$

13. CELL PHONES The average cost per minute of using a cell phone decreased \$0.44 between 1996 and 2004. On average, how much did the cost decrease each year? (Lesson 1-2)

Write the ordered pair for each point. (Lesson 1-9)

14. *A* **15.** *B*

16. *C* **17.** *D*

18. *F* **19.** *G*

QUICKReview

EXAMPLE 1

Simplify $-\frac{4}{28}$.

$-\frac{4}{28} = \frac{-4 \div 4}{28 \div 4}$ Divide −4 and 28 by their GCF, 4.

$= \frac{-1}{7}$ or $-\frac{1}{7}$ Simplify. Since the signs are different, the quotient is negative.

EXAMPLE 2

Evaluate $\frac{a-b}{c-d}$ if $a = 2, b = 5, c = -3, d = -12$.

$\frac{a-b}{c-d}$ Original expression

$= \frac{2-5}{(-3)-(-12)}$ Substitute 2 for *a*, 5 for *b*, −3 for *c*, and −12 for *d*.

$= \frac{-3}{9}$ Simplify.

$= \frac{-3 \div 3}{9 \div 3}$ Divide −3 and 9 by their GCF, 3.

$= \frac{-1}{3}$ or $-\frac{1}{3}$ Simplify. The signs are different, so the quotient is negative.

EXAMPLE 3

Write the ordered pair for *A*.

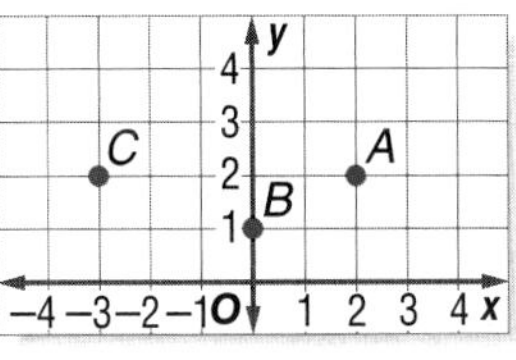

Step 1 Begin at point *A*.

Step 2 Follow along a vertical line through the point to find the *x*-coordinate on the *x*-axis. The *x*-coordinate is 2.

Step 3 Follow along a horizontal line through the point to find the *y*-coordinate on the *y*-axis. The *y*-coordinate is 2.

The ordered pair for point *A* is (2, 2).

Algebra Lab

Steepness of a Line

In mathematics, you can measure the steepness of a line using a ratio.

SET UP the Lab

- Stack three books on your desk.
- Lean a ruler on the books, creating a ramp.
- Tape the ruler to the desk.

ACTIVITY

Step 1 Measure and record the rise and the run of the ramp. Then calculate the ratio $\frac{\text{rise}}{\text{run}}$. Record the data in a table like the one at the right.

rise	run	$\frac{\text{rise}}{\text{run}}$

Step 2 Keeping the rise the same, move the books to make the ramp steeper. Measure the rise and run, and calculate the ratio $\frac{\text{rise}}{\text{run}}$. Repeat three times and record the data.

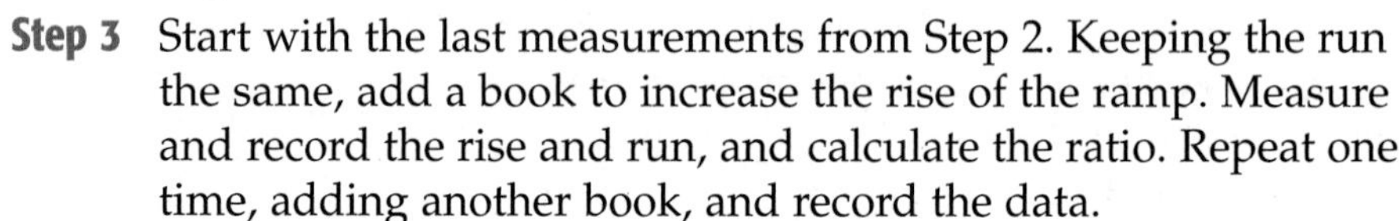

Step 3 Start with the last measurements from Step 2. Keeping the run the same, add a book to increase the rise of the ramp. Measure and record the rise and run, and calculate the ratio. Repeat one time, adding another book, and record the data.

ANALYZE THE RESULTS

1. Examine the ratios you recorded in Step 2. How do they change as the ramp becomes steeper?
2. Examine the ratios you recorded in Step 3. What happens to the ratio when the run stays the same and the rise increases?
3. **MAKE A PREDICTION** Suppose you want to construct a skateboard ramp that is not as steep as the one shown at the left. List three different sets of $\frac{\text{rise}}{\text{run}}$ measurements that will result in a less steep ramp. Verify your predictions by calculating the ratio $\frac{\text{rise}}{\text{run}}$ of each ramp.

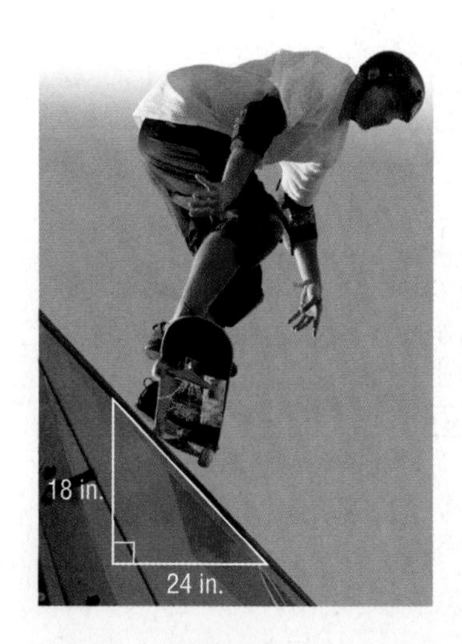

4. Copy the coordinate graph and draw a line through the origin with a $\frac{\text{rise}}{\text{run}}$ ratio greater than the original line. Then draw a line through the origin with a ratio less than the original line. Explain your reasoning using the words *rise* and *run*.

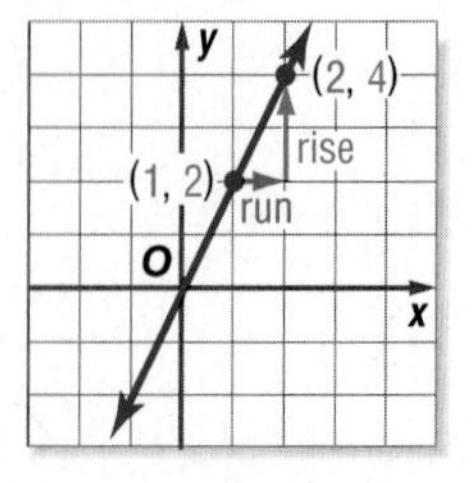

4-1 Rate of Change and Slope

Main Ideas

- Use rate of change to solve problems.
- Find the slope of a line.

New Vocabulary

rate of change
slope

GET READY for the Lesson

Houses in the north have steeper roofs so that snow does not pile up. A roof *pitch* describes how steep it is. It is the number of units the roof rises for each unit of run. In the photo, the roof rises 8 feet for each 12 feet of run.

$$\frac{\text{rise}}{\text{run}} = \frac{8}{12} \text{ or } \frac{2}{3}$$

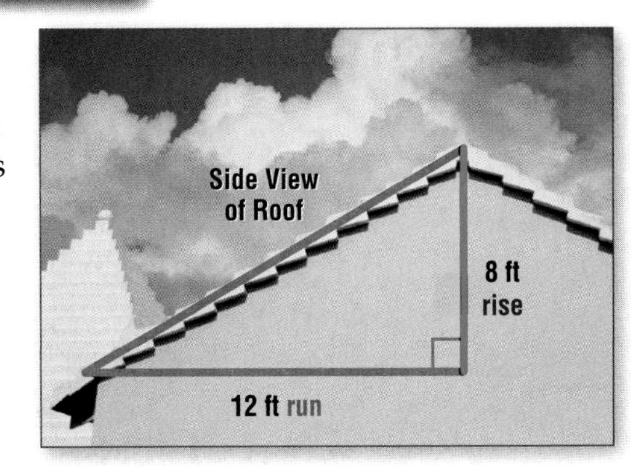

Rate of Change **Rate of change** is a ratio that describes, on average, how much one quantity changes with respect to a change in another quantity. If x is the independent variable and y is the dependent variable, then

$$\text{rate of change} = \frac{\text{change in } y}{\text{change in } x}.$$

Study Tip

Independent Quantities

Rates of change often include *time* as the independent variable.

The table at the right shows the distance a person has walked for various amounts of time.

Time Walking (s)	Distance Walked (ft)
x	y
1	4
2	8
3	12

(+1, +1 for x; +4, +4 for y)

Each time x increases by 1 second, y increases by 4 feet.

$$\text{rate of change} = \frac{\text{change in } y}{\text{change in } x}$$

$$= \frac{\text{change in distance}}{\text{change in time}}$$

$$= \frac{4}{1} \begin{array}{l} \leftarrow \text{feet} \\ \leftarrow \text{seconds} \end{array}$$

The rate of change is $\frac{4}{1}$. This means that the person walked 4 feet per second.

Real-World EXAMPLE

1 ENTERTAINMENT Use the table to find the rate of change. Explain the meaning of the rate of change.

Number of Computer Games	Total Cost ($)
x	y
2	78
4	156
6	234

Each time x increases by 2 games, y increases by $78.

(continued on the next page)

$$\text{rate of change} = \frac{\text{change in } y}{\text{change in } x}$$

$$= \frac{\text{change in cost}}{\text{change in number of games}}$$

$$= \frac{156 - 78}{4 - 2}$$

$$= \frac{78}{2} \text{ or } \frac{39}{1} \begin{array}{l} \leftarrow \text{dollars} \\ \leftarrow \text{games} \end{array}$$

The rate of change is $\frac{39}{1}$. This means that it costs $39 per game.

CHECK Your Progress

REMODELING The table shows how the area changes with the number of floor tiles.

1A. Find the rate of change.

1B. Explain the meaning of the rate of change.

Floor Tiles	Area (in²)
x	y
3	48
6	96
9	144

So far, you have seen rates of change that are *constant*. Many real-world situations involve rates of change that are not constant.

Real-World Link

Walt Disney World's Magic Kingdom is the most visited theme park in the United States. An estimated 15 million people pass through its gates each year.

Source: *Amusement Business* Magazine

Real-World EXAMPLE

2 ENTERTAINMENT The graph shows the number of people who visited U.S. theme parks in recent years.

a. Find the rates of change for 1996–1998 and 2000–2002.

Source: *tia.org*

rate of change

$$= \frac{\text{change in attendance}}{\text{change in time}} \begin{array}{l} \leftarrow \text{people} \\ \leftarrow \text{years} \end{array}$$

1996–1998:

$$\frac{\text{change in attendance}}{\text{change in time}} = \frac{81.8 - 78.8}{1998 - 1996} \quad \text{Substitute.}$$

$$= \frac{3}{2} \text{ or } 1.5 \quad \text{Simplify.}$$

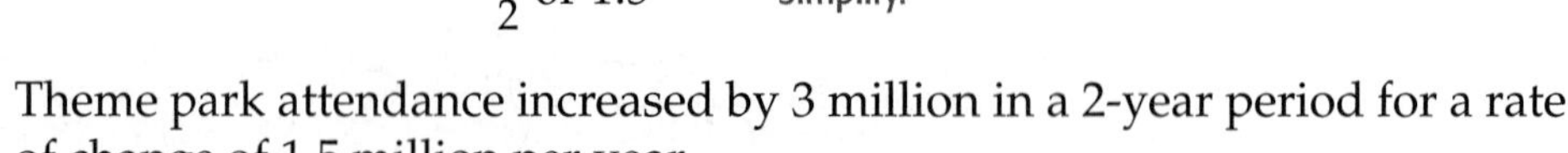

Theme park attendance increased by 3 million in a 2-year period for a rate of change of 1.5 million per year.

2000–2002:

$$\frac{\text{change in attendance}}{\text{change in time}} = \frac{92.4 - 84.6}{2002 - 2000} \quad \text{Substitute.}$$

$$= \frac{7.8}{2} \text{ or } 3.9 \quad \text{Simplify.}$$

Over this 2-year period, attendance increased by 7.8 million, for a rate of change of 3.9 million per year.

b. Explain the meaning of the rate of change in each case.

For 1996–1998, on average, 1.5 million more people went to a theme park each year than the last.

For 2000–2002, on average, 3.9 million more people attended theme parks each year than the last.

c. How are the different rates of change shown on the graph?

There is a greater vertical change for 2000–2002 than for 1996–1998. Therefore, the section of the graph for 2000–2002 is steeper.

CHECK Your Progress

2. Refer to the graph. Without calculating, find the 2-year period that has the least rate of change. Then calculate to verify your answer.

Online Personal Tutor at algebra1.com

Find Slope The **slope** of a line is the ratio of the change in the y-coordinates (rise) to the change in the x-coordinates (run) as you move in the positive direction.

Vocabulary Link
Slope
Everyday Use A hill used for snow skiing is often called a slope.
Math Use Slope is used to describe steepness.

Slope can be used to describe a rate of change. This number describes how steep the line is. The greater the absolute value of the slope, the steeper the line.

The graph shows a line that passes through (1, 3) and (4, 5).

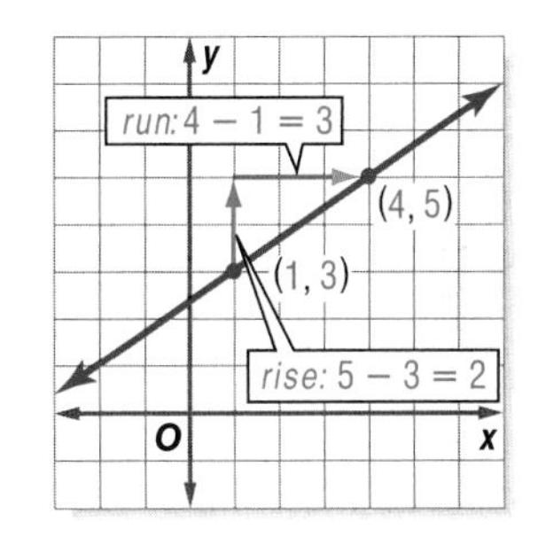

$$\text{slope} = \frac{\text{rise}}{\text{run}}$$

$$= \frac{\text{change in } y\text{-coordinates}}{\text{change in } x\text{-coordinates}}$$

$$= \frac{5-3}{4-1} \text{ or } \frac{2}{3}$$

So, the slope of the line is $\frac{2}{3}$.

Any two points on a line can be used to determine the slope.

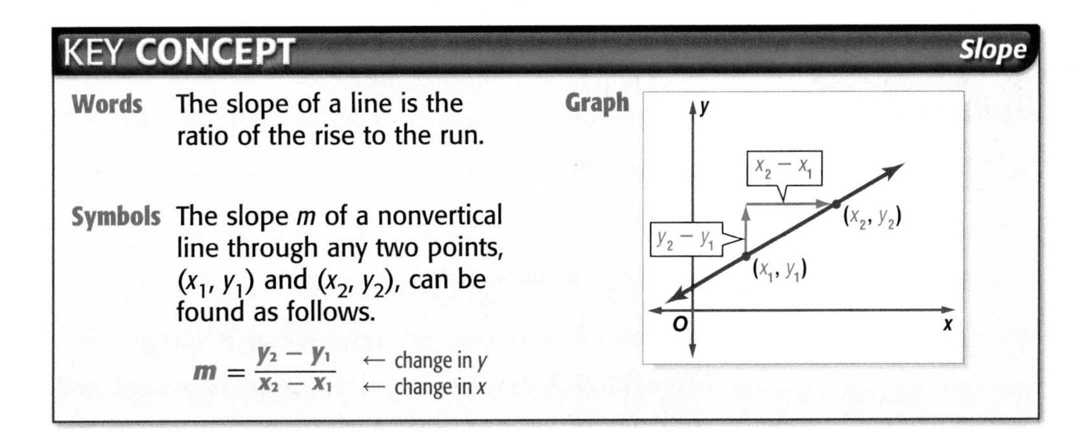

KEY CONCEPT *Slope*

Words The slope of a line is the ratio of the rise to the run.

Symbols The slope m of a nonvertical line through any two points, (x_1, y_1) and (x_2, y_2), can be found as follows.

$$m = \frac{y_2 - y_1}{x_2 - x_1} \quad \begin{array}{l} \leftarrow \text{change in } y \\ \leftarrow \text{change in } x \end{array}$$

Graph

The slope of a line can be positive, negative, zero, or undefined.

EXAMPLE **Positive Slope**

> **Study Tip**
>
> **Common Misconception**
> It may make your calculations easier to choose the point on the left as (x_1, y_1). However, either point may be chosen as (x_1, y_1).

3 **Find the slope of the line that passes through (−1, 2) and (3, 4).**

Let $(-1, 2) = (x_1, y_1)$ and $(3, 4) = (x_2, y_2)$.

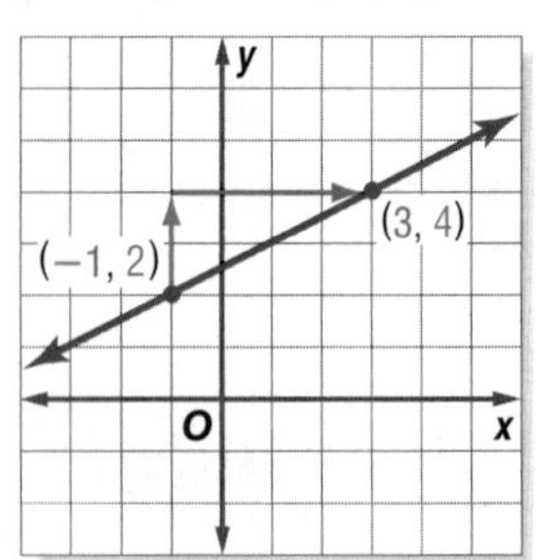

$m = \frac{y_2 - y_1}{x_2 - x_1}$ — rise/run

$= \frac{4 - 2}{3 - (-1)}$ — Substitute.

$= \frac{2}{4}$ or $\frac{1}{2}$ — Simplify.

CHECK Your Progress

Find the slope of the line that passes through each set of points.

3A. (3, 6), (4, 8) **3B.** (−4, 2), (2, 10)

EXAMPLE **Negative Slope**

4 **Find the slope of the line that passes through (−1, −2)and (−4, 1).**

Let $(-1, -2) = (x_1, y_1)$ and $(-4, 1) = (x_2, y_2)$.

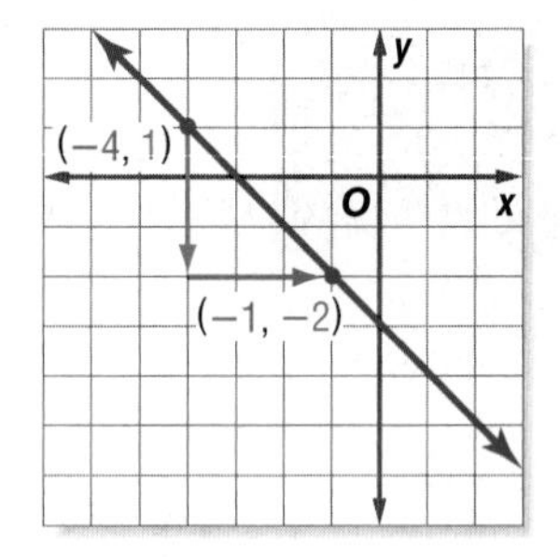

$m = \frac{y_2 - y_1}{x_2 - x_1}$ — rise/run

$= \frac{1 - (-2)}{-4 - (-1)}$ — Substitute.

$= \frac{3}{-3}$ or -1 — Simplify.

CHECK Your Progress

Find the slope of the line that passes though each set of points.

4A. (−2, 2), (−6, 4) **4B.** (4, 3), (−1, 11)

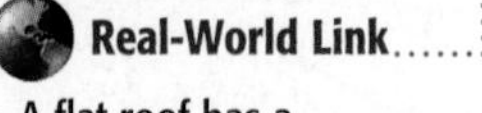
Real-World Link

A flat roof has a slope of zero.

EXAMPLE **Zero Slope**

5 **Find the slope of the line that passes through (1, 2) and (−1, 2).**

Let $(1, 2) = (x_1, y_1)$ and $(-1, 2) = (x_2, y_2)$.

y
(−1, 2) (1, 2)
O x

$m = \frac{y_2 - y_1}{x_2 - x_1}$ — rise/run

$= \frac{2 - 2}{-1 - 1}$ — Substitute.

$= \frac{0}{-2}$ or 0 — Simplify.

CHECK Your Progress

Find the slope of the line that passes through each set of points.

5A. (6, 7), (−2, 7) **5B.** (−4, −2), (0, −2)

EXAMPLE Undefined Slope

6 Find the slope of the line that passes through (1, −2) and (1, 3).

Let $(1, -2) = (x_1, y_1)$ and $(1, 3) = (x_2, y_2)$.

$$m = \frac{y_2 - y_1}{x_2 - x_1} \quad \frac{\text{rise}}{\text{run}}$$

$$= \frac{3 - (-2)}{1 - 1} \text{ or } \frac{5}{0}$$

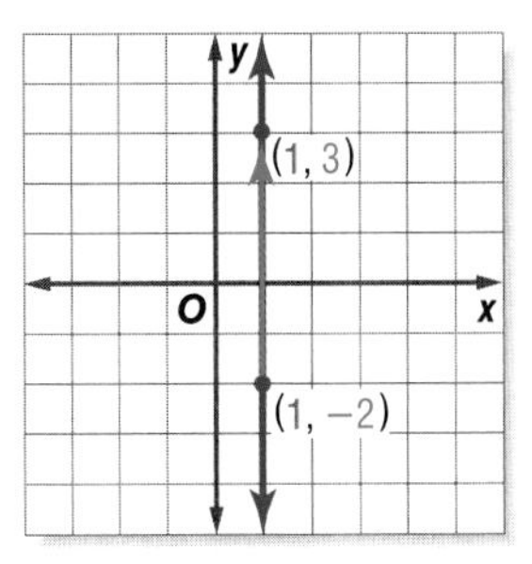

Since division by zero is undefined, the slope is undefined.

CHECK Your Progress

Find the slope of the line that passes through each set of points.

6A. (3, 2), (3, −1) **6B.** (−2, −1), (−2, 5)

KEY CONCEPT Slope

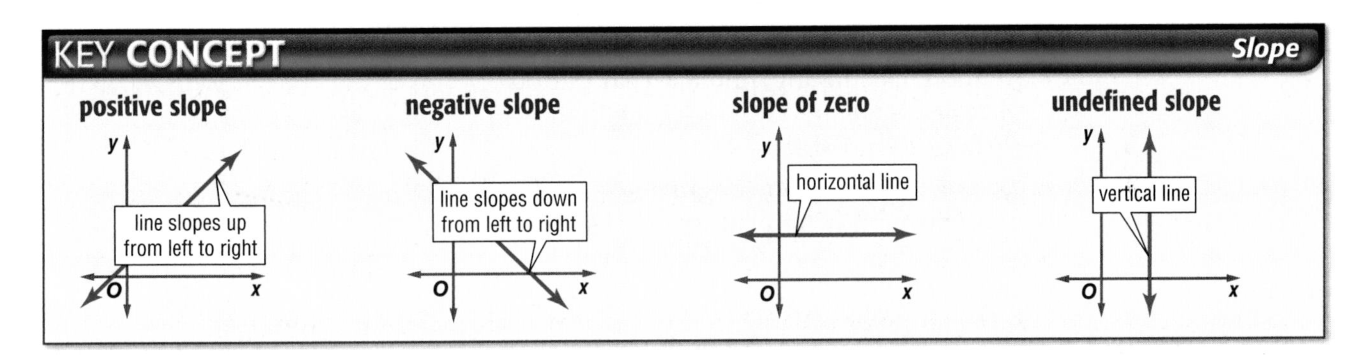

Given the slope of a line and one point on the line, you can find other points on the line.

EXAMPLE Find Coordinates Given Slope

7 Find the value of r so that the line through $(r, 6)$ and $(10, -3)$ has a slope of $-\frac{3}{2}$.

$m = \frac{y_2 - y_1}{x_2 - x_1}$	Slope Formula
$-\frac{3}{2} = \frac{-3 - 6}{10 - r}$	Let $(r, 6) = (x_1, y_1)$ and $(10, -3) = (x_2, y_2)$.
$-\frac{3}{2} = \frac{-9}{10 - r}$	Subtract.
$-3(10 - r) = 2(-9)$	Find the cross products.
$-30 + 3r = -18$	Simplify.
$3r = 12$	Add 30 to each side and simplify.
$r = 4$	Divide each side by 3 and simplify.

So, the line goes through (4, 6).

CHECK Your Progress

Find the value of r so the line that passes through each pair of points has the given slope.

7A. $(1, 4), (-1, r); m = 2$ **7B.** $(r, -6), (5, -8); m = -8$

CHECK Your Understanding

Example 1 (pp. 187–188)

Find the rate of change represented in each table or graph.

1.

x	y
3	−6
5	2
7	10
9	18
11	26

2.

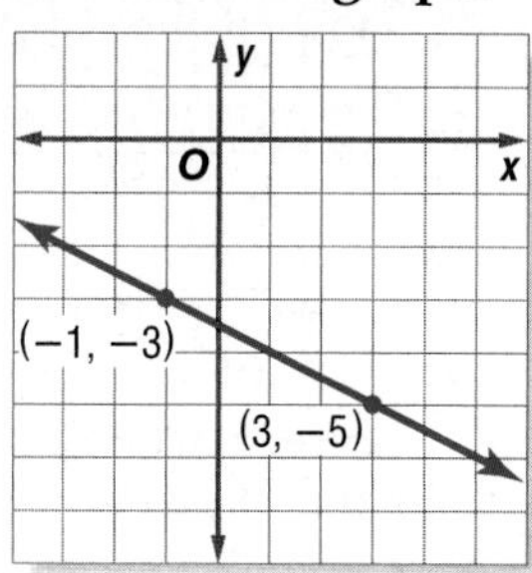

Example 2 (pp. 188–189)

SPORTS For Exercises 3–5, use the graph at the right.

3. Find the rate of change for prices from 2002 to 2004. Explain the meaning of the rate of change.

4. Without calculating, find a 2-year period that had a greater rate of change than 2002 to 2004. Explain.

5. Between which years might a new stadium have been built? Explain your reasoning.

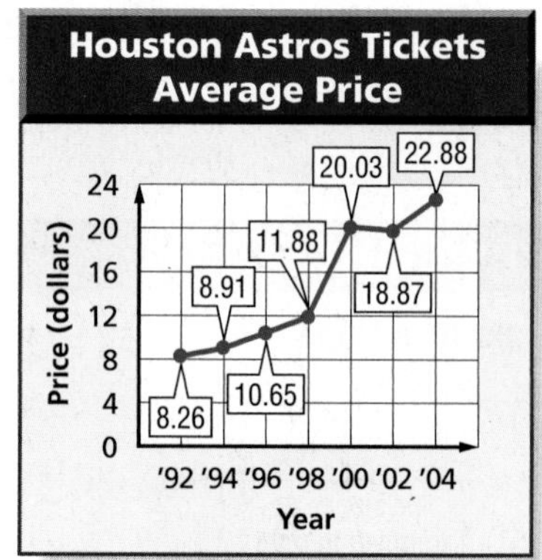

Source: *Team Marketing Report*

Examples 3–6 (pp. 190–191)

Find the slope of the line that passes through each pair of points.

6. $(1, 1), (3, 4)$ **7.** $(0, 0), (5, 4)$ **8.** $(-2, 2), (-1, -2)$

9. $(9, -4), (7, -1)$ **10.** $(3, 5), (-2, 5)$ **11.** $(-1, 3), (-1, 0)$

Example 7 (p. 191)

Find the value of r so the line that passes through each pair of points has the given slope.

12. $(6, -2), (r, -6), m = 4$ **13.** $(9, r), (6, 3), m = -\frac{1}{3}$

Exercises

HOMEWORK HELP

For Exercises	See Examples
14–17	1
18–19	2
20–31	3–6
32–35	7

Find the rate of change represented in each table or graph.

14.

x	y
5	2
10	3
15	4
20	5

15.

x	y
1	15
2	9
3	3
4	−3

16.

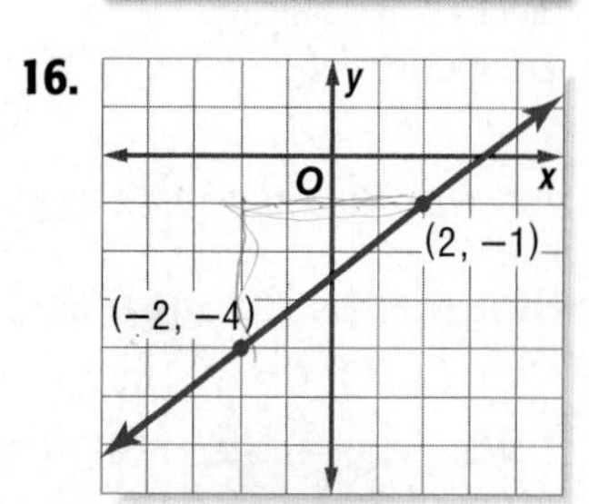

17.

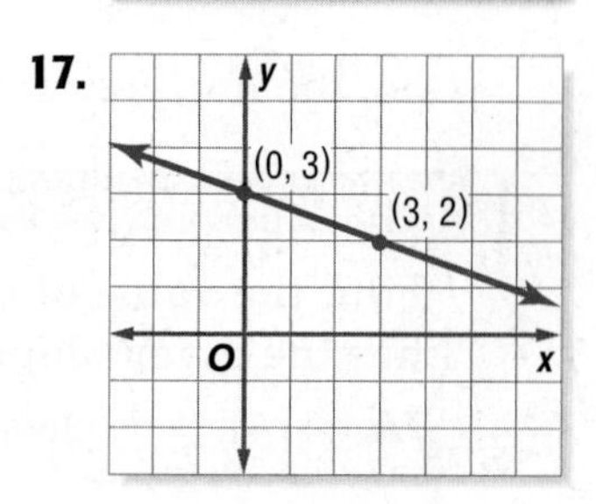

18. SPORTS What was the annual rate of change from 1995 to 2003? Explain the meaning of the rate of change.

Women Competing in Triathlons	
Year	**Number of Women**
1995	4600
2003	19,100

19. CELL PHONES In 2000, 5% of 13- to 17-year-olds had cell phones. By 2004, 56% of teens had cell phones. Find the annual rate of change in the percent of teens with cell phones from 2000 to 2004. Describe what the rate of change means.

Find the slope of the line that passes through each pair of points.

20. $(2, 3), (9, 7)$ **21.** $(-3, 6), (2, 4)$ **22.** $(2, 6), (-1, 3)$

23. $(-3, 3), (1, 3)$ **24.** $(-2, 1), (-2, 3)$ **25.** $(-3, 9), (-7, 6)$

26. $(5, 7), (-2, -3)$ **27.** $(2, -1), (5, -3)$ **28.** $(-4, -1), (-3, -3)$

29. $(-3, -4), (5, -1)$ **30.** $(-2, 3), (8, 3)$ **31.** $(-5, 4), (-5, -1)$

Find the value of r so the line that passes through each pair of points has the given slope.

32. $(6, 2), (9, r), m = -1$ **33.** $(r, -5), (3, 13), m = 8$

34. $(5, r), (2, -3), m = \frac{4}{3}$ **35.** $(-2, 8), (r, 4), m = -\frac{1}{2}$

Find the slope of the line that passes through each pair of points.

36.

x	y
4.5	−1
5.3	2

37.

x	y
0.75	1
0.75	−1

38.

x	y
$2\frac{1}{2}$	$-1\frac{1}{2}$
$-\frac{1}{2}$	$\frac{1}{2}$

39. DRIVING When driving up a certain hill, you rise 15 feet for every 1000 feet you drive forward. What is the slope of the road?

CONSTRUCTION Use a ruler to estimate the slope of each object.

40.

41.

Real-World Link

Tony Hawk made skateboarding history by landing a 900 at the 1999 X Games. That's two and a half rotations in mid-air!

Source: espn.go.com

42. Find the slope of the line that passes through the origin and (r, s).

43. What is the slope of the line that passes through (a, b) and $(a, -b)$?

Find the value of r so the line that passes through each pair of points has the given slope.

44. $\left(\frac{1}{2}, -\frac{1}{4}\right), \left(r, -\frac{5}{4}\right), m = 4$ **45.** $\left(\frac{2}{3}, r\right), \left(1, \frac{1}{2}\right), m = \frac{1}{2}$

46. $(4, r), (r, 2), m = -\frac{5}{4}$ **47.** $(r, 5), (-2, r), m = -\frac{2}{9}$

ANALYZE TABLES **For Exercises 48–50, use the table that shows Karen's height at various ages.**

Age (years)	12	14	16	18	20
Height (inches)	60	64	66	67	67

48. Make a broken-line graph of the data.

49. Use the graph to determine in which two-year period Karen grew the fastest. Explain your reasoning.

50. Discuss the rate of change associated with the horizontal section of the graph.

ANALYZE GRAPHS **For Exercises 51–53, use the graph that shows public school enrollment.**

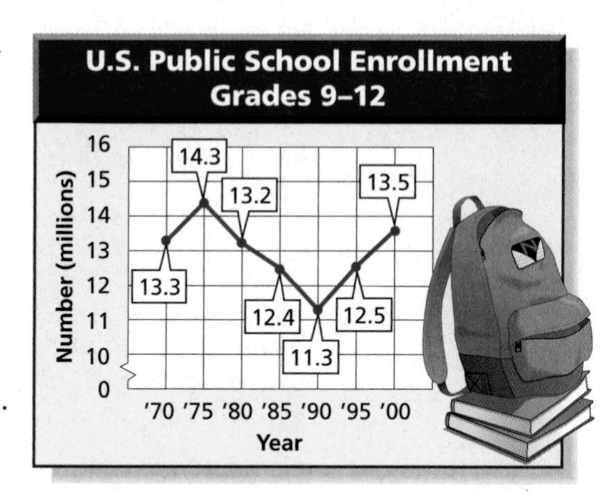

51. For which 5-year period was the rate of change the greatest? the least?

52. Find the rate of change from 1985 to 1990.

53. Explain the meaning of the part of the graph with a negative slope.

GROWTH RATE **For Exercises 54–56, use the following information.**
After her last haircut, May's hair was 8 inches long. In three months, it grew another inch. Assume that the hair growth continues at the same rate.

54. Make a table that shows May's hair length for each of the three months and for the next three months.

55. Draw a graph showing the relationship between May's hair length and time in months.

56. What is the slope of the graph? What does it represent?

57. **CONSTRUCTION** The slope of a stairway determines how easy it is to climb the stairs. Suppose the vertical distance between two floors is 8 feet 9 inches. Find the total run of the ideal stairway in feet and inches. (*Hint:* Do not include any part of the top or bottom floor in the run.)

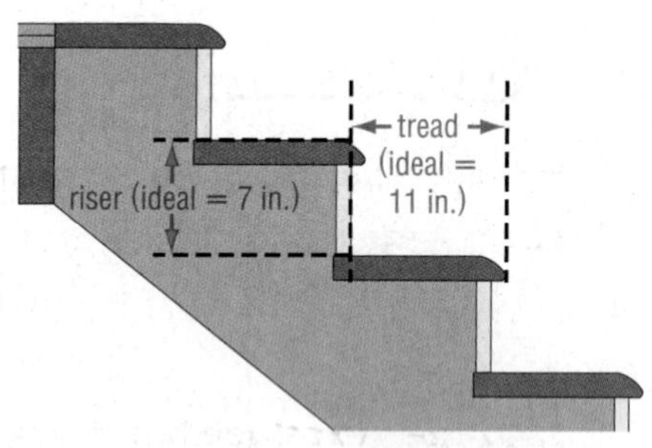

58. **RESEARCH** Use the Internet or another reference to find the population of your city or town in 1930, 1940, ..., 2000. Between which two decades was the rate of change the greatest? Explain.

H.O.T. Problems

59. **CHALLENGE** Develop a strategy for determining whether the slope of the line through $(-4, -5)$ and $(4, 5)$ is positive or negative without calculating.

60. **OPEN ENDED** Integrate what you know about rate of change to describe the function at the right.

Time (wk)	Height of Plant (in.)
4	9.0
6	13.5
8	18.0

61. **CHALLENGE** Determine whether $Q(2, 3)$, $R(-1, -1)$, and $S(-4, -2)$ lie on the same line that passes through $(-2, -2)$ and $(4, 0)$. Explain your reasoning.

62. **FIND THE ERROR** Carlos and Allison are finding the slope of the line that passes through (2, 6) and (5, 3). Who is correct? Explain your reasoning.

Carlos
$\frac{3-6}{5-2} = \frac{-3}{3}$ or -1

Allison
$\frac{6-3}{5-2} = \frac{3}{3}$ or 1

63. ***Writing in Math*** Discuss how to find the slope of a roof and compare the appearance of roofs with different slopes.

STANDARDIZED TEST PRACTICE

64. A music store has x CDs in stock. If 350 are sold and $3y$ are added to stock, which expression represents the number of CDs in stock?

A $350 + 3y - x$

B $x - 350 + 3y$

C $x + 350 + 3y$

D $3y - 350 - x$

65. **REVIEW** A recipe for fruit punch calls for 2 ounces of orange juice for every 8 ounces of lemonade. If Jennifer uses 64 ounces of lemonade, which proportion can she use to find x, the number of ounces of orange juice she should add to make the fruit punch?

F $\frac{2}{x} = \frac{64}{6}$

G $\frac{8}{x} = \frac{64}{2}$

H $\frac{2}{8} = \frac{x}{64}$

J $\frac{6}{2} = \frac{x}{64}$

Spiral Review

Write an equation in function notation for each relation. (Lesson 3-5)

66.

Number of Lunches	1	2	3	4	5
Total Cost ($)	5	10	15	20	25

67.

Time (s)	7	9	11	14	16
Altitude (ft)	4	2	0	−3	−5

Find the next three terms of each arithmetic sequence. (Lesson 3-4)

68. 8, 20, 32, 44, ...

69. −9, −6, −3, 0, ...

70. 35, 31, 27, 23, ...

71. −56, −47, −38, −29, ...

72. **FOOD** Garrett is making $\frac{1}{3}$-pound hamburgers. One pound of hamburger costs $3.19. How much will it cost to make 18 hamburgers? (Lesson 2-3)

GET READY for the Next Lesson

PREREQUISITE SKILL Find each quotient. (Pages 690–691)

73. $6 \div \frac{2}{3}$

74. $\frac{3}{4} \div \frac{1}{6}$

75. $\frac{3}{4} \div 6$

76. $18 \div \frac{7}{8}$

4-2 Slope and Direct Variation

Main Ideas

- Write and graph direct variation equations.
- Solve problems involving direct variation.

New Vocabulary

direct variation
constant of variation
family of graphs
parent graph

GET READY for the Lesson

It costs $2.25 per ringtone that you download for your cell phone. If you graph the ordered pairs, the slope of the line is 2.25.

Number of Ringtones	Total Cost ($)
x	y
0	0
1	2.25
2	4.50
3	6.75
4	9.00

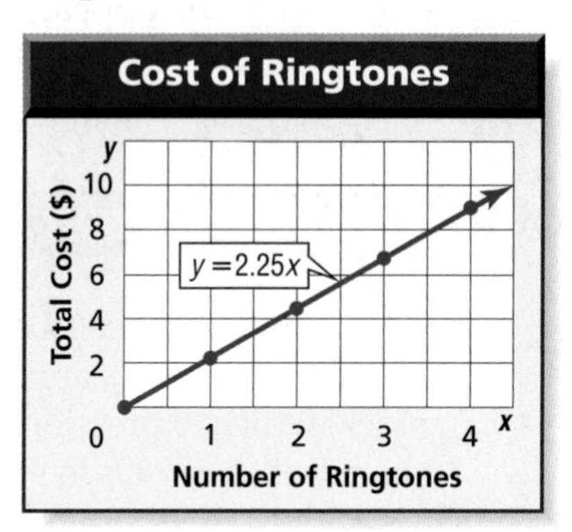

The total cost y depends *directly* on the number of ringtones that you download x. The rate of change is constant.

Direct Variation A **direct variation** is described by an equation of the form $y = kx$, where $k \neq 0$. The equation $y = kx$ represents a constant rate of change and k is the **constant of variation.**

EXAMPLE Slope and Constant of Variation

1 Name the constant of variation for each equation. Then find the slope of the line that passes through each pair of points.

a.

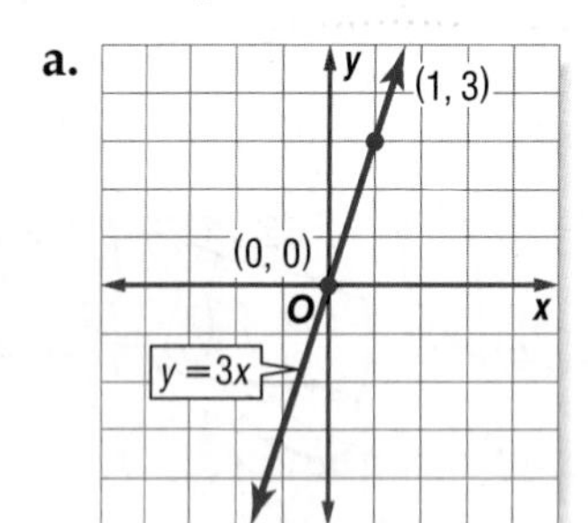

The constant of variation is 3.

$m = \frac{y_2 - y_1}{x_2 - x_1}$ Slope formula

$= \frac{3 - 0}{1 - 0}$ $(x_1, y_1) = (0, 0)$ $(x_2, y_2) = (1, 3)$

$= 3$ The slope is 3.

b.

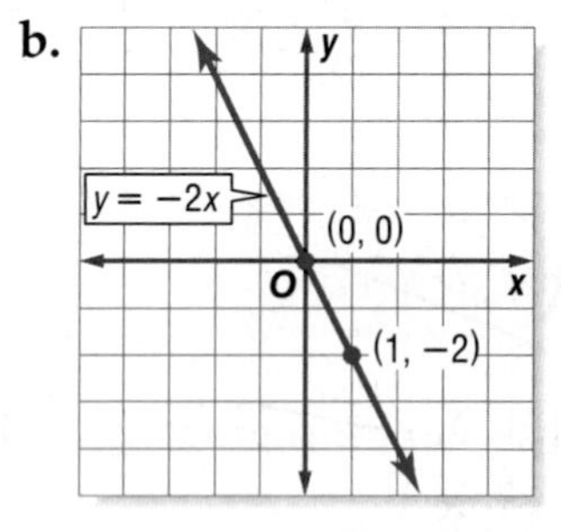

The constant of variation is −2.

$m = \frac{y_2 - y_1}{x_2 - x_1}$ Slope formula

$= \frac{-2 - 0}{1 - 0}$ $(x_1, y_1) = (0, 0)$ $(x_2, y_2) = (1, -2)$

$= -2$ The slope is −2.

Study Tip

Constant of Variation

Compare the constants of variation with the slopes of the graphs. What is the slope of the graph of $y = kx$?

CHECK Your Progress

1. Name the constant of variation for $y = \frac{1}{4}x$. Then find the slope of the line that passes through (0, 0) and (4, 1).

Since (0, 0) is a solution of $y = kx$, the graph of $y = kx$ always passes through the origin.

EXAMPLE Graph a Direct Variation

2 **Graph each equation.**

a. $y = 4x$

Step 1 Write the slope as a ratio.

$$4 = \frac{4}{1} \quad \frac{\text{rise}}{\text{run}}$$

Step 2 Graph (0, 0).

Step 3 From the point (0, 0), move up 4 units and right 1 unit. Draw a dot.

Step 4 Draw a line containing the points.

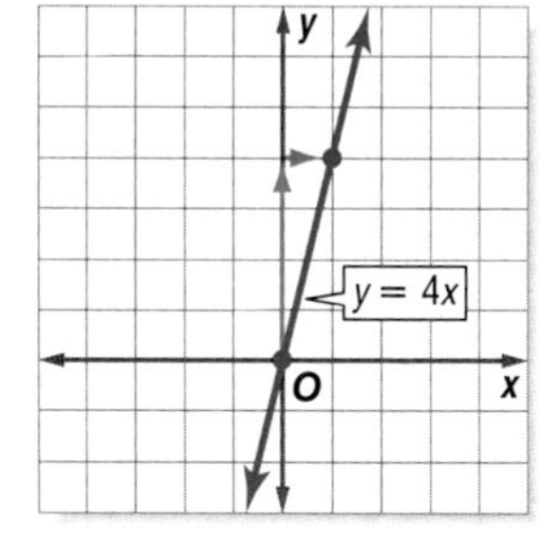

b. $y = -\frac{1}{3}x$

Step 1 Write the slope as a ratio.

$$-\frac{1}{3} = \frac{-1}{3} \quad \frac{\text{rise}}{\text{run}}$$

Step 2 Graph (0, 0).

Step 3 From the point (0, 0), move down 1 unit and right 3 units. Draw a dot.

Step 4 Draw a line containing the points.

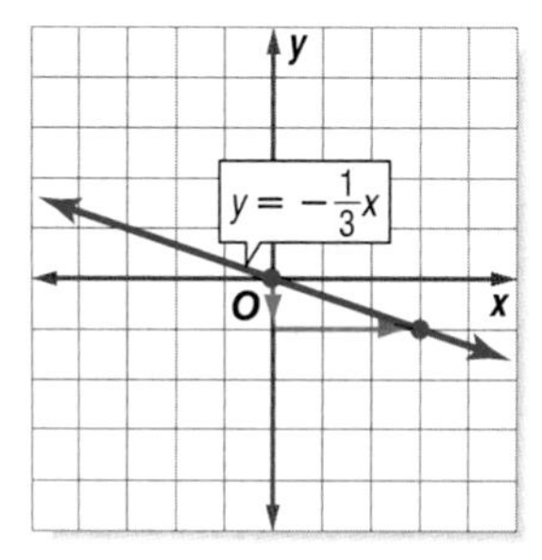

Study Tip

Direct Proportions

In direct variation equations, the variables are directly proportional to each other. That is, the ratio $x : y$ is a constant.

CHECK Your Progress

Graph each equation.

2A. $y = 6x$ **2B.** $y = \frac{2}{3}x$ **2C.** $y = -5x$ **2D.** $y = -\frac{3}{4}x$

A **family of graphs** includes graphs and equations of graphs that have at least one characteristic in common. The **parent graph** is the simplest graph in a family.

GRAPHING CALCULATOR LAB

Graphs of $y = mx$

The calculator screen shows the graphs of $y = x$, $y = 2x$, and $y = 4x$.

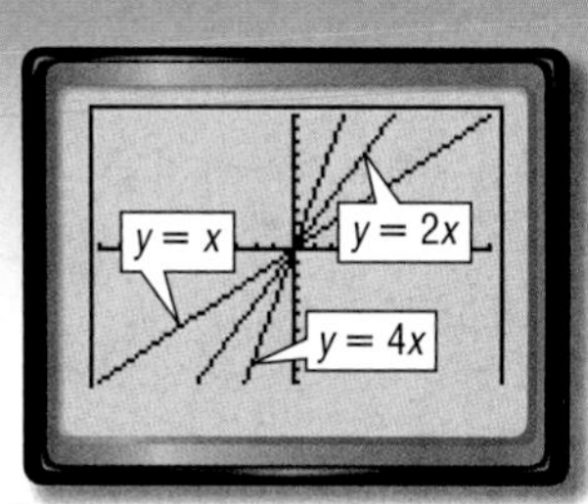

[−10, 10] scl: 1 by [−10, 10] scl: 1

THINK AND DISCUSS

1. Describe any similarities and differences among the graphs.
2. Write an equation with a graph that has a steeper slope than $y = 4x$. Check your answer by graphing $y = 4x$ and your equation.
3. Write an equation with a graph that lies between the graphs of $y = x$ and $y = 2x$. Check your answer by graphing the equations.
4. What characteristics do the graphs have in common? How are they different?
5. These equations are all of the form $y = mx$. How does the graph change as the absolute value of m increases?

The results of the Graphing Calculator Lab lead to some general observations about the graphs of direct variation equations.

CONCEPT SUMMARY — ***Direct Variation Graphs***

- Direct variation equations are of the form $y = kx$, where $k \neq 0$.
- The graph of $y = kx$ always passes through the origin.
- The slope is positive if $k > 0$.
- The slope is negative if $k < 0$.

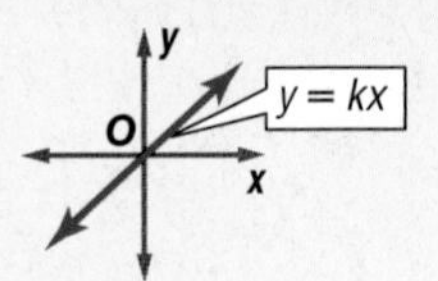

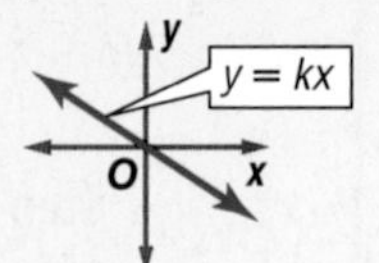

If you know that y varies directly as x, you can write a direct variation equation that relates the two quantities.

EXAMPLE 3 Write and Solve a Direct Variation Equation

Suppose y varies directly as x, and $y = 28$ when $x = 7$.

a. Write a direct variation equation that relates x and y.

Find the value of k.

$y = kx$	Direct variation formula
$28 = k(7)$	Replace y with 28 and x with 7.
$\frac{28}{7} = \frac{k(7)}{7}$	Divide each side by 7.
$4 = k$	Simplify.

Therefore, the direct variation equation is $y = 4x$.

b. Use the direct variation equation to find x when $y = 52$.

$y = 4x$	Direct variation equation
$52 = 4x$	Replace y with 52.
$\frac{52}{4} = \frac{4x}{4}$	Divide each side by 4.
$13 = x$	Simplify.

Therefore, $x = 13$ when $y = 52$.

> **Study Tip**
>
> **Direct Variation**
> Once you have the value of k, you can use it to find the value of x or y when given the value of the other variable.

CHECK Your Progress

Suppose y varies directly as x, and $y = 6$ when $x = -18$.

3A. Write a direct variation equation that relates x and y.

3B. Find y when $x = -2$.

Solve Problems One of the most common applications of direct variation is the formula $d = rt$. Distance d varies directly as time t, and the rate r is the constant of variation.

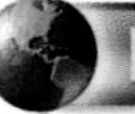

Real-World EXAMPLE

4 BIOLOGY The migration of snow geese varies directly as the number of hours. A flock of snow geese migrated 375 miles in 7.5 hours.

a. Write a direct variation equation for the distance *d* flown in time *t*.

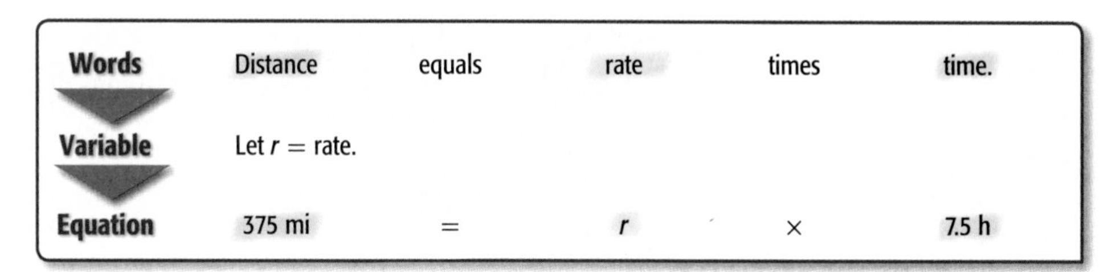

Words	Distance	equals	rate	times	time.
Variable	Let r = rate.				
Equation	375 mi	=	r	×	7.5 h

Real-World Link

Snow geese migrate more than 3000 miles from their winter home in the southwest United States to their summer home in the Canadian arctic.

Souce: Audubon Society

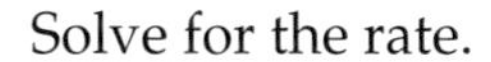

Solve for the rate.

$375 = r(7.5)$ Original equation

$\frac{375}{7.5} = \frac{r(7.5)}{7.5}$ Divide each side by 7.5.

$50 = r$ Simplify.

Therefore, the direct variation equation is $d = 50t$. What does the 50 represent?

b. Graph the equation.

The graph of $d = 50t$ passes through the origin with slope 50.

$m = \frac{50}{1}$ $\frac{\text{rise}}{\text{run}}$

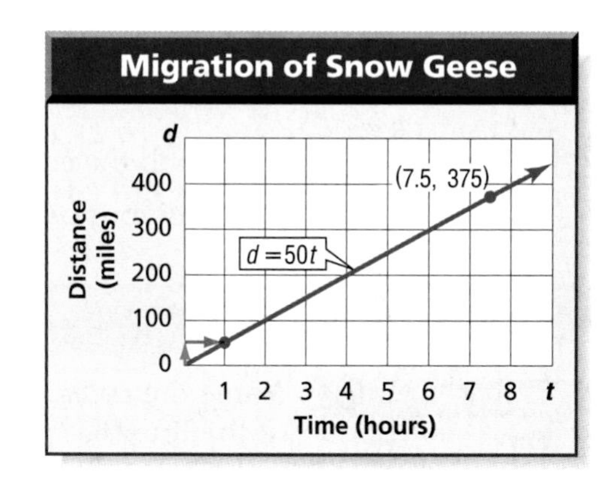

c. Estimate how many hours of flying time it would take the geese to migrate 3000 miles.

$d = 50t$ Original equation

$3000 = 50t$ Replace d with 3000.

$\frac{3000}{50} = \frac{50t}{50}$ Divide each side by 50.

$t = 60$ Simplify.

At this rate, it would take 60 hours of flying time to migrate 3000 miles.

CHECK Your Progress

HOT AIR BALLOONS A hot air balloon's ascent varies directly as the number of minutes. A hot air balloon ascended 350 feet in 5 minutes.

4A. Write a direct variation for the distance d ascended in time t.

4B. Graph the equation.

4C. Estimate how many minutes it would take the hot air balloon to ascend 2100 feet.

Online Personal Tutor at algebra1.com

CHECK Your Understanding

Example 1 (p. 196) **Name the constant of variation for each equation. Then find the slope of the line that passes through each pair of points.**

1.

2.

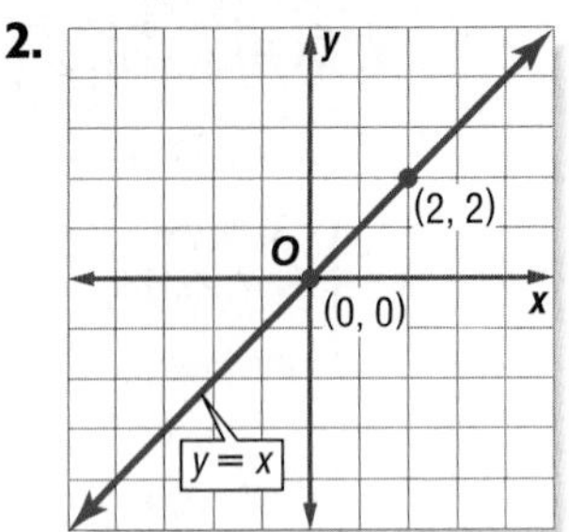

Example 2 (p. 197) **Graph each equation.**

3. $y = 2x$ **4.** $y = \frac{1}{2}x$ **5.** $y = -3x$ **6.** $y = -\frac{5}{3}x$

Example 3 (p. 198) **Suppose y varies directly as x. Write a direct variation equation that relates x and y. Then solve.**

7. If $y = 27$ when $x = 6$, find x when $y = 45$.

8. If $y = -7$ when $x = 14$, find y when $x = -16$.

Example 4 (pp. 198–199) **JOBS For Exercises 9–11, use the following information.**
Suppose your pay varies directly as the number of hours you work. Your pay for 7.5 hours is \$45.

9. Write a direct variation equation relating your pay to the hours worked.

10. Graph the equation.

11. Find your pay if you work 30 hours.

Exercises

HOMEWORK HELP

For Exercises	See Examples
12–17	1
18–25	2
26–29	3
30, 31	4

Name the constant of variation for each equation. Then find the slope of the line that passes through each pair of points.

12.

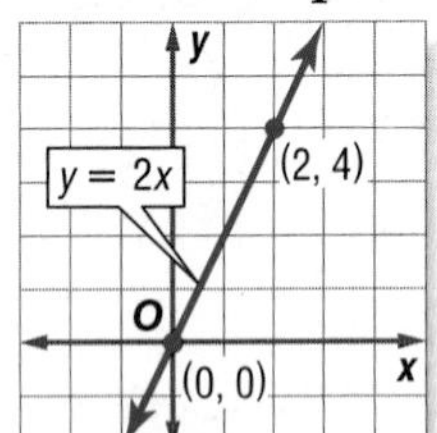

13.

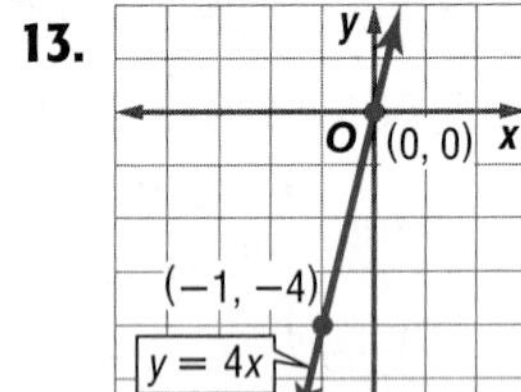

14.

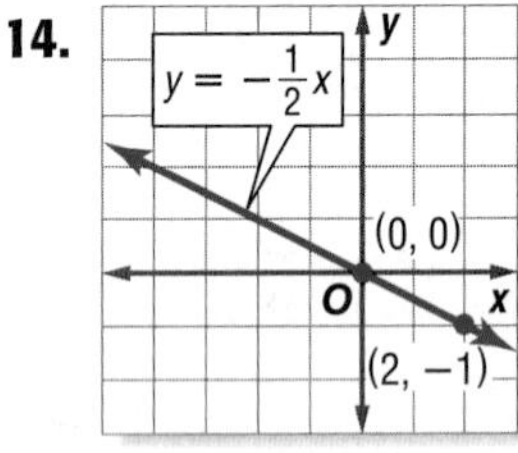

15.

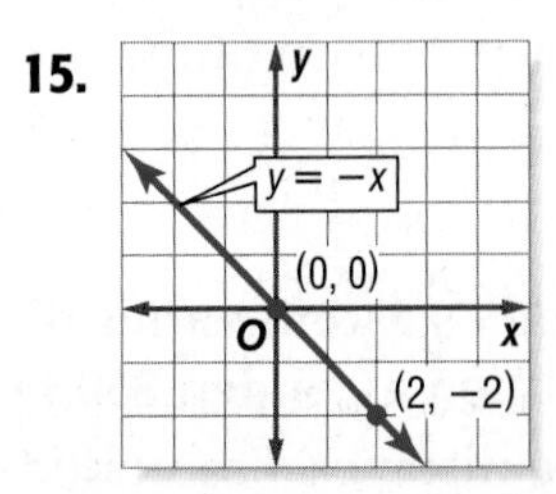

16.

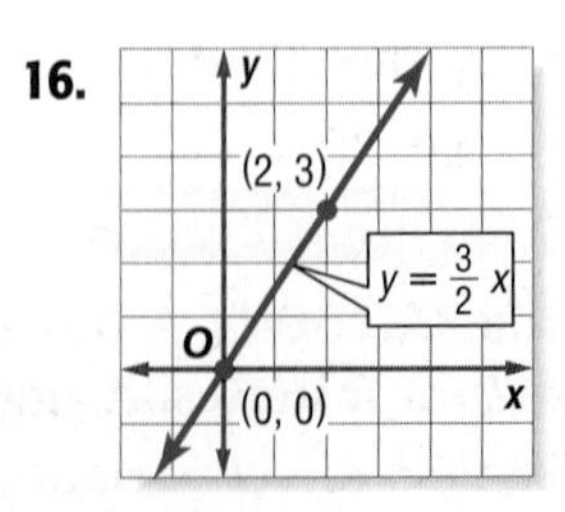

17.

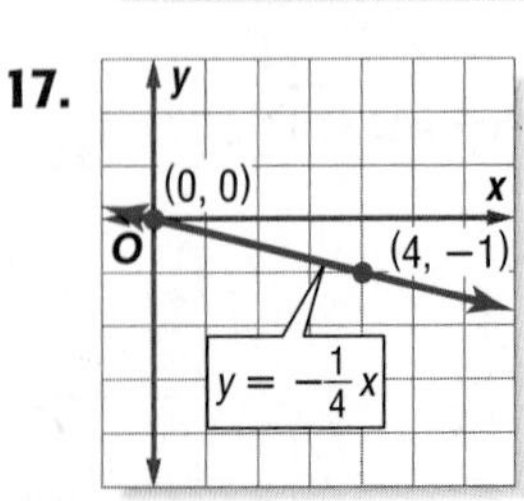

Graph each equation.

18. $y = 3x$ **19.** $y = -x$ **20.** $y = -4x$ **21.** $y = \frac{5}{2}x$

22. $y = \frac{1}{5}x$ **23.** $y = -\frac{2}{3}x$ **24.** $y = -\frac{4}{3}x$ **25.** $y = -\frac{9}{2}x$

Suppose y varies directly as x. Write a direct variation equation that relates x and y. Then solve.

26. If $y = 8$ when $x = 4$, find y when $x = 5$.

27. If $y = -16$ when $x = 4$, find x when $y = 20$.

28. If $y = 4$ when $x = 12$, find y when $x = -24$.

29. If $y = 12$ when $x = 15$, find x when $y = 21$.

SPORTS For Exercises 30 and 31, use the following information.
The distance a golf ball travels at an altitude of 7000 feet varies directly with the distance the ball travels at sea level, as shown in the table.

Hitting a Golf Ball		
Altitude (ft)	0 (sea level)	7000
Distance (yd)	200	210

30. Write and graph an equation that relates the distance a golf ball travels at an altitude of 7000 feet y with the distance at sea level x.

31. What would be a person's average driving distance at 7000 feet if his average driving distance at sea level is 180 yards?

Real-World Career...
Veterinarian
A veterinarian uses math to compare the age of an animal to the age of a human on the basis of bone and tooth growth and to determine the amount of medicine to prescribe based on the weight of the animal.

Math Online
For more information, go to algebra1.com.

ANALYZE TABLES For Exercises 32 and 33, use the following information.
Most animals age more rapidly than humans do. The chart shows equivalent ages for horses and humans.

Horse age (x)	0	1	2	3	4	5
Human age (y)	0	3	6	9	12	15

32. Write an equation that relates human age to horse age.

33. Find the equivalent horse age for a human who is 16 years old.

Suppose y varies directly as x. Write a direct variation equation that relates x and y. Then solve.

34. If $y = 2.5$ when $x = 0.5$, find y when $x = 20$.

35. If $y = -6.6$ when $x = 9.9$, find y when $x = 6.6$.

36. If $y = 2\frac{2}{3}$ when $x = \frac{1}{4}$, find y when $x = 1\frac{1}{8}$.

37. If $y = 6$ when $x = \frac{2}{3}$, find x when $y = 12$.

ANALYZE GRAPHS Which line in the graph represents the sprinting speed of each animal?

38. elephant, 25 mph

39. reindeer, 32 mph

40. lion, 50 mph

41. grizzly bear, 30 mph

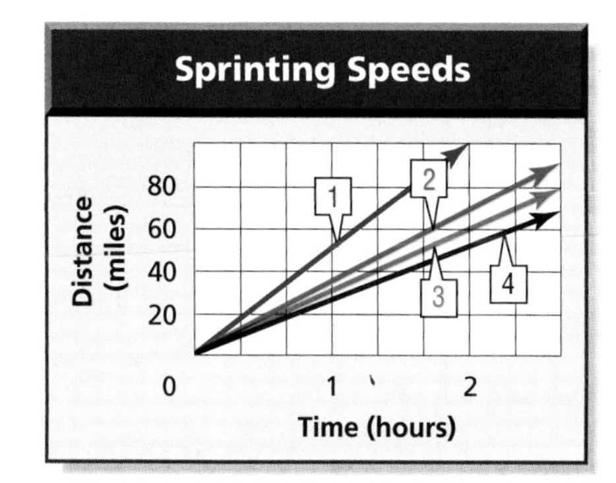

Write a direct variation equation that relates the variables. Then graph the equation.

42. GEOMETRY The circumference C of a circle is about 3.14 times the diameter d.

43. GEOMETRY The perimeter P of a square is 4 times the length of a side s.

44. RETAIL The total cost is C for n yards of ribbon priced at \$0.99 per yard.

45. RETAIL Kona coffee beans are \$14.49 per pound. The cost of p pounds is C.

GRAPHING CALCULATOR **Consider the graphs of $y = -1x$, $y = -2x$, and $y = -4x$.**

46. Graph these three equations on the same screen.

47. Describe the similarities and differences between these graphs and the graphs in the Graphing Calculator Lab on page 195.

48. Write an equation whose graph is steeper than the graph of $y = -4x$.

49. Make a conjecture about how you can determine without graphing which of two direct variation equations has the graph with a steeper slope.

EXTRA PRACTICE See pages 725, 747.

Math Online Self-Check Quiz at algebra1.com

H.O.T. Problems

50. Which One Doesn't Belong? Identify the equation that does not belong with the other three. Explain your reasoning.

$9 = rs$ | $9a = 0$ | $z = \frac{1}{9}x$ | $s = \frac{9}{t}$

51. OPEN ENDED Model a real-world situation using a direct variation equation. Graph the equation and describe the rate of change.

52. CHALLENGE Suppose y varies directly as x. If the value of x is doubled, what can you conclude about the value of y? Explain your reasoning.

53. *Writing in Math* Write an equation that relates the total cost y to the number of ringtones x for ringtones that cost \$1.50 each. Compare the steepness of the graph of this equation to the graph at the top of page 196.

STANDARDIZED TEST PRACTICE

54. Which equation *best* represents the graph?

A $y = 2x$

B $y = -2x$

C $y = \frac{1}{2}x$

D $y = -\frac{1}{2}x$

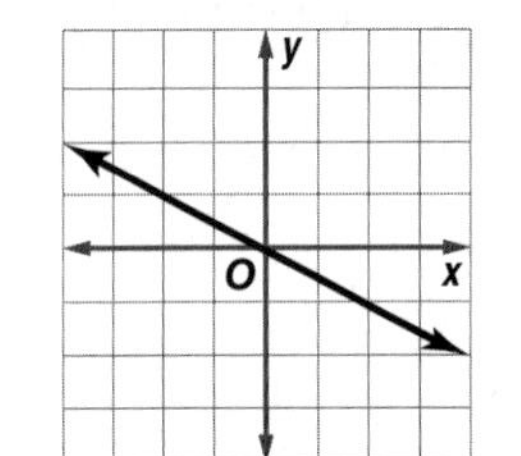

55. REVIEW Roberto receives an employee discount of 12%. If he spent \$355 at the store, how much was his discount to the nearest dollar?

F \$3

G \$4

H \$30

J \$43

Spiral Review

56. Find the value of r so that the line that passes through (1, 7) and (r, 3) has a slope of 2. (Lesson 4-1)

Write an equation in function notation for the relation shown in each table. (Lesson 3-5)

57.

x	0	1	2	3	4	5
y	1	5	9	13	17	21

58.

x	2	4	6	8	10	12
y	8	6	4	2	0	−2

59. BASKETBALL A school purchased five new basketballs for \$149.95. At that rate, how much more money will it cost the school to have 12 new basketballs in all? (Lesson 2-6)

GET READY for the Next Lesson

PREREQUISITE SKILL **Solve each equation for y.** (Lesson 2-8)

60. $4x = y + 3$

61. $2y = 4x + 10$

62. $9x + 3y = 12$

EXPLORE 4-3

Graphing Calculator Lab

Investigating Slope-Intercept Form

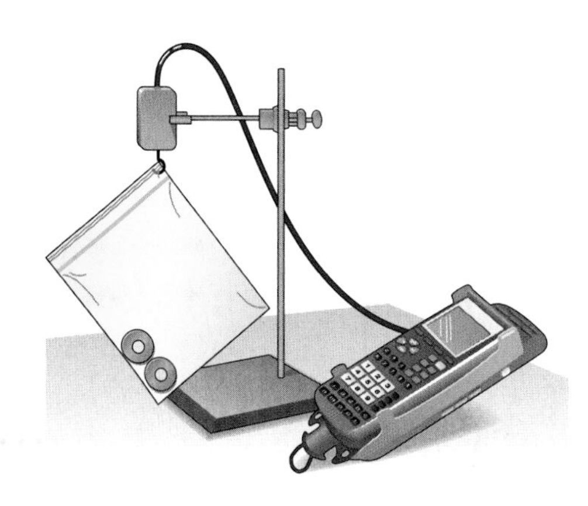

SET UP the Lab

- Cut a small hole in a top corner of a plastic sandwich bag. Hang the bag from the end of the force sensor.
- Connect the force sensor to your data collection device.

ACTIVITY

Step 1 Use the sensor to collect the weight with 0 washers in the bag. Record the data pair in the calculator.

Step 2 Place one washer in the plastic bag. Wait for the bag to stop swinging, then measure and record the weight.

Step 3 Repeat the experiment, adding different numbers of washers to the bag. Each time, record the data.

ANALYZE THE RESULTS

1. The domain contains values represented by the independent variable, washers. The range contains values represented by the dependent variable, weight. Use the graphing calculator to create a scatterplot using the ordered pairs (washers, weight).
2. Write a sentence that describes the points on the graph.
3. Describe the position of the point on the graph that represents the trial with no washers in the bag.
4. The rate of change can be found by using the formula for slope.
   $\frac{\text{rise}}{\text{run}} = \frac{\text{change in weight}}{\text{change in number of washers}}$
   Find the rate of change in the weight as more washers are added.
5. Explain how the rate of change is shown on the graph.

The graph shows sample data from a washer experiment. Describe the graph for each situation.

[0, 20] scl: 2 by [0, 1] scl: 0.25

6. A bag that hangs weighs 0.8 N when empty and increases in weight at the rate of the sample.
7. A bag that has the same weight when empty as the sample and increases in weight at a faster rate.
8. A bag that has the same weight when empty as the sample and increases in weight at a slower rate.

4-3 Graphing Equations in Slope-Intercept Form

Main Ideas

- Write and graph linear equations in slope-intercept form.
- Model real-world data with an equation in slope-intercept form.

New Vocabulary

slope-intercept form

GET READY for the Lesson

An online store charges \$3 per order plus \$0.99 per book for shipping.

Number of Books	Shipping Cost
1	3.99
2	4.98
3	5.97
4	6.96
5	7.95
6	8.94
7	9.93

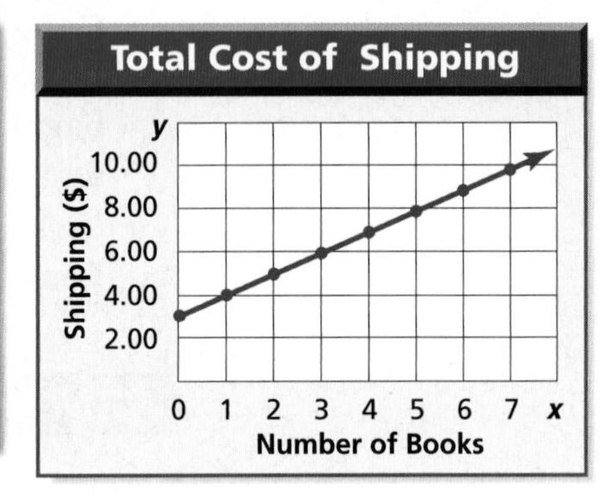

The slope of the line is 0.99. It crosses the y-axis at (0, 3).

The equation of the line is $y = 0.99x + 3$.

↑ charge per book, \$0.99 ↑ flat fee, \$3.00

Slope-Intercept Form An equation of the form $y = mx + b$, where m is the slope and b is the y-intercept, is in **slope-intercept form.**

KEY CONCEPT *Slope-Intercept Form*

Words The linear equation $y = mx + b$ is written in slope-intercept form, where m is the slope and b is the y-intercept.

Symbols $y = mx + b$

slope ↑ ↑ y-intercept

Graph

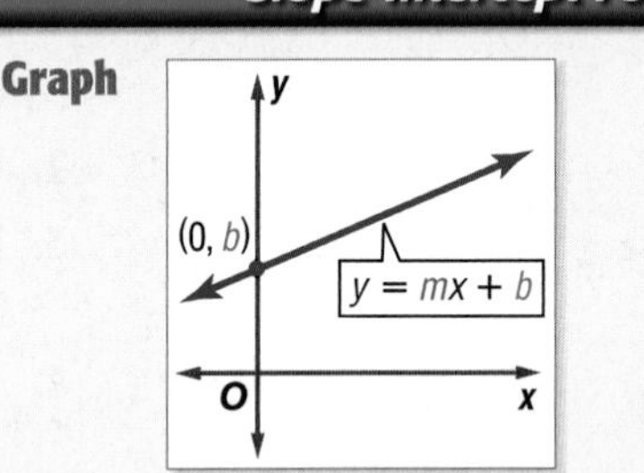

Concepts in Motion
BrainPOP®
algebra1.com

EXAMPLE Write an Equation Given Slope and y-Intercept

1 **Write an equation in slope-intercept form of the line with a slope of 3 and a y-intercept of −5.**

$y = mx + b$ Slope-intercept form

$y = 3x + (-5)$ Replace m with 3 and b with −5.

$y = 3x - 5$ Rewrite.

CHECK Your Progress

1. Write an equation of the line with a slope of $-\frac{1}{2}$ and a y-intercept of 3.

EXAMPLE Write an Equation From a Graph

2 **Write an equation in slope-intercept form of the line shown in the graph.**

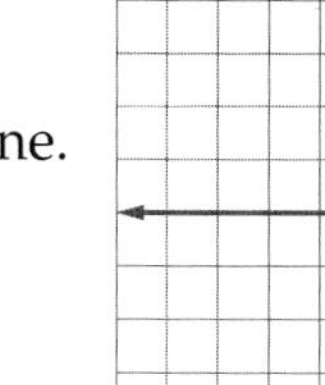

Step 1 Find the slope using two points on the line. Let $(x_1, y_1) = (0, 3)$ and $(x_2, y_2) = (2, -1)$.

$$m = \frac{y_2 - y_1}{x_2 - x_1} \qquad \frac{\text{rise}}{\text{run}}$$

$$= \frac{-1 - 3}{2 - 0} \qquad x_1 = 0, x_2 = 2; \; y_1 = 3, y_2 = -1$$

$$= \frac{-4}{2} \text{ or } -2 \qquad \text{Simplify.}$$

The slope is -2.

Step 2 The line crosses the y-axis at $(0, 3)$. So, the y-intercept is 3.

Step 3 Finally, write the equation.

$y = mx + b$ — Slope-intercept form

$y = -2x + 3$ — Replace m with -2 and b with 3.

The equation of the line is $y = -2x + 3$.

Study Tip

Vertical Lines

The equation of a vertical line *cannot* be written in slope-intercept form. *Why?*

y, $x = a$, O, $(a, 0)$, x

Horizontal Lines

The equation of a horizontal line *can* be written in slope-intercept form as $y = 0x + b$ or $y = b$.

y, $y = b$, $(0, b)$, O, x

CHECK Your Progress

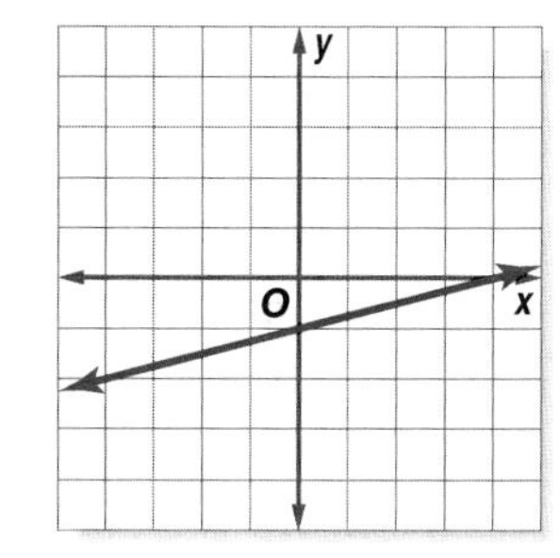

2. Write an equation in slope-intercept form of the line shown at the right.

Online Personal Tutor at algebra1.com

EXAMPLE Graph Equations

3 **Graph each equation.**

a. $y = -\frac{2}{3}x + 1$

Step 1 The y-intercept is 1. So, graph $(0, 1)$.

Step 2 The slope is $-\frac{2}{3}$ or $\frac{-2}{3}$. $\frac{\text{rise}}{\text{run}}$

From $(0, 1)$, move down 2 units and right 3 units. Draw a dot.

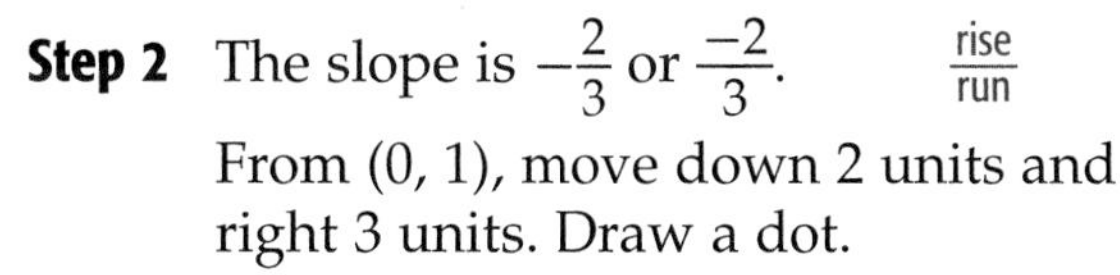

Step 3 Draw a line through the points.

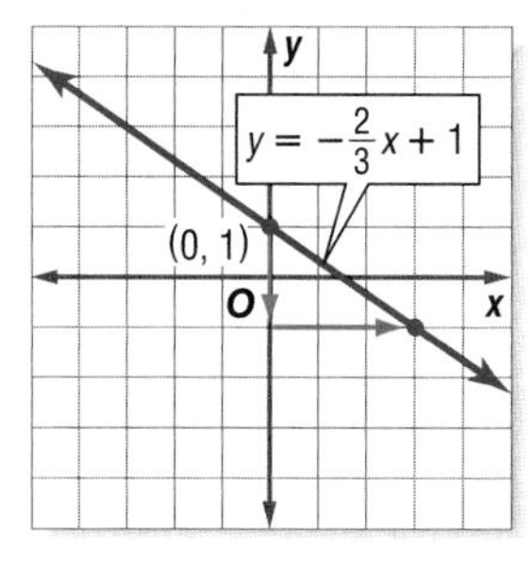

b. $5x - 3y = 6$

Step 1 Solve for y to write the equation in slope-intercept form.

$5x - 3y = 6$	Original equation
$5x - 3y - 5x = 6 - 5x$	Subtract $5x$ from each side.
$-3y = 6 - 5x$	Simplify.
$-3y = -5x + 6$	$6 - 5x = 6 + (-5x)$ or $-5x + 6$
$\frac{-3y}{-3} = \frac{-5x + 6}{-3}$	Divide each side by -3.
$\frac{-3y}{-3} = \frac{-5x}{-3} + \frac{6}{-3}$	Divide each term in the numerator by -3.
$y = \frac{5}{3}x - 2$	Simplify.

(continued on the next page)

Step 2 The y-intercept of $y = \frac{5}{3}x - 2$ is -2. So, graph $(0, -2)$.

Step 3 The slope is $\frac{5}{3}$. From $(0, -2)$, move up 5 units and right 3 units. Draw a dot.

Step 4 Draw a line containing the points.

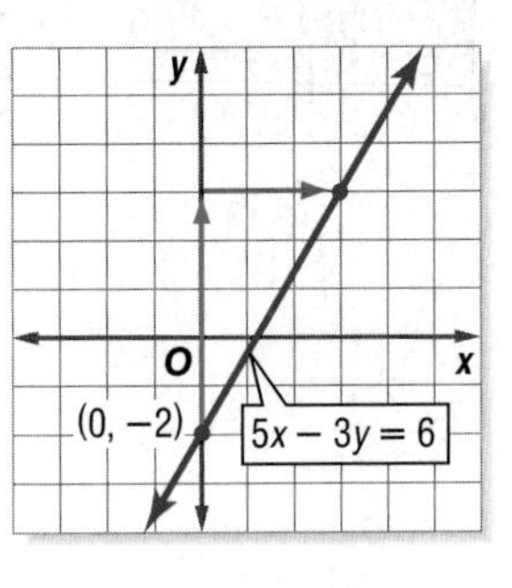

CHECK Your Progress

Graph each equation.

3A. $y = 2x - 3$ **3B.** $y = \frac{1}{4}x + 5$ **3C.** $4x + 3y = -12$ **3D.** $2x - 3y = 6$

Model Real-World Data If a quantity changes at a constant rate over time, it can be modeled by a linear equation. The y-intercept represents a starting point, and the slope represents the rate of change.

Real-World EXAMPLE

Real-World Link

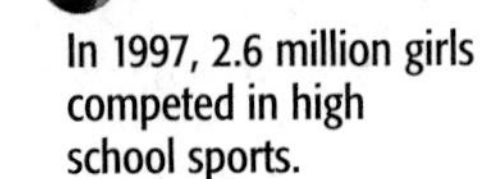

In 1997, 2.6 million girls competed in high school sports.

Source: www.nfhs.org

4 **SPORTS** **Use the information at the left about high school sports.**

a. The number of girls competing in high school sports has increased by an average of 0.06 million per year since 1997. Write a linear equation to find the number of girls in high school sports in any year after 1997.

Words	Number of girls competing	equals	rate of change	times	number of years after 1997	plus	amount at start.
Variables	Let G = number of girls competing.				Let n = number of years after 1997.		
Equation	G	$=$	0.06	$\cdot$	n	$+$	2.6

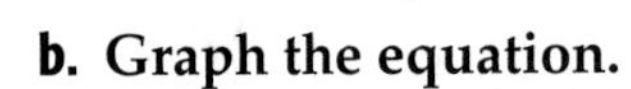

b. Graph the equation.

The graph passes through (0, 2.6) with slope 0.06.

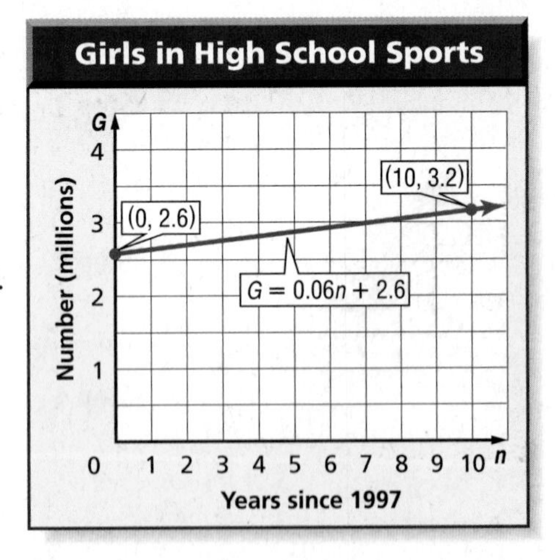

c. Find the number of girls competing in 2007.

The year 2007 is 10 years after 1997.

$G = 0.06n + 2.6$ Write the equation.

$= 0.06(\mathbf{10}) + 2.6$ Replace n with 10.

$= 3.2$ Simplify.

So, 3.2 million girls competed in high school sports in 2007.

CHECK Your Progress

FUND-RAISERS **The band boosters are selling submarine sandwiches for \$5 each. The cost of the ingredients to make the sandwiches was \$1160.**

4A. Write an equation for the profit P made on s sandwiches.

4B. Graph the equation.

4C. Find the total profit if 1400 sandwiches are sold.

CHECK Your Understanding

Example 1 (p. 204) **Write an equation in slope-intercept form of the line with the given slope and y-intercept.**

1. slope: -3, y-intercept: 1

2. slope: 4, y-intercept: -2

Example 2 (p. 205) **Write an equation in slope-intercept form of the line shown in each graph.**

3.

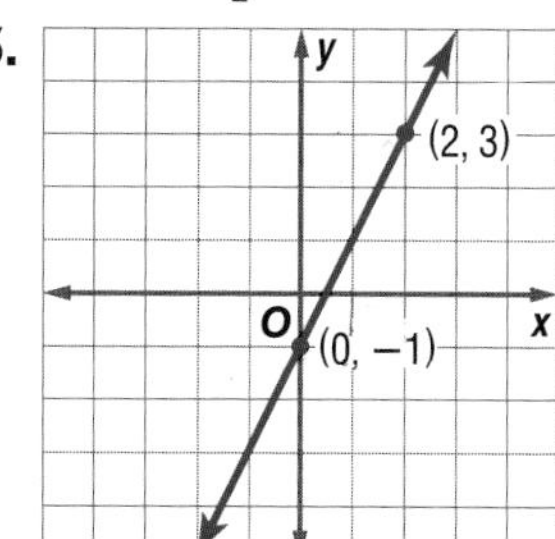

4.

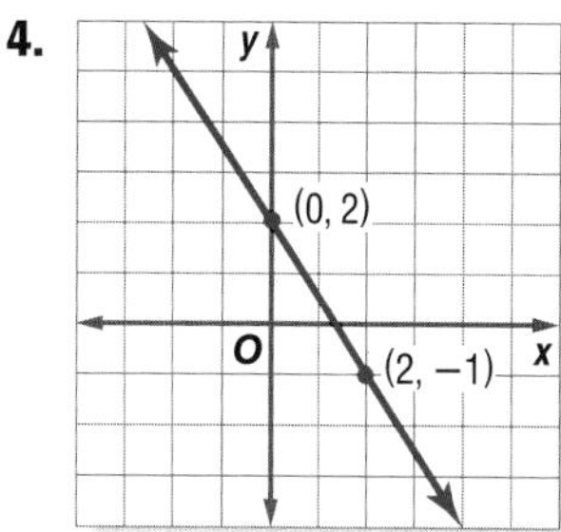

Example 3 (p. 205) **Graph each equation.**

5. $y = 2x - 3$

6. $y = -3x + 1$

7. $2x + y = 5$

8. $3x - 2y = 2$

Example 4 (p. 206) **MONEY For Exercises 9–11, use the following information.**
Suppose you have already saved \$50 toward the cost of a new television. You plan to save \$5 more each week for the next several weeks.

9. Write an equation for the total amount T that you will have w weeks from now.

10. Graph the equation.

11. Find the total amount saved after 7 weeks.

Exercises

HOMEWORK HELP	
For Exercises	See Examples
12–17	1
18–23	2
24–32	3
33–38	4

Write an equation in slope-intercept form of the line with the given slope and y-intercept.

12. slope: -2, y-intercept: 6

13. slope: 3, y-intercept: -5

14. slope: $\frac{1}{2}$, y-intercept: 3

15. slope: $-\frac{3}{5}$, y-intercept: 12

16. slope: 0, y-intercept: 3

17. slope: -1, y-intercept: 0

Write an equation in slope-intercept form of the line shown in each graph.

18.

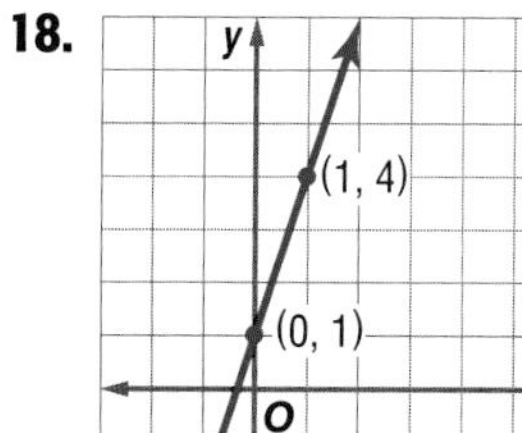

19.

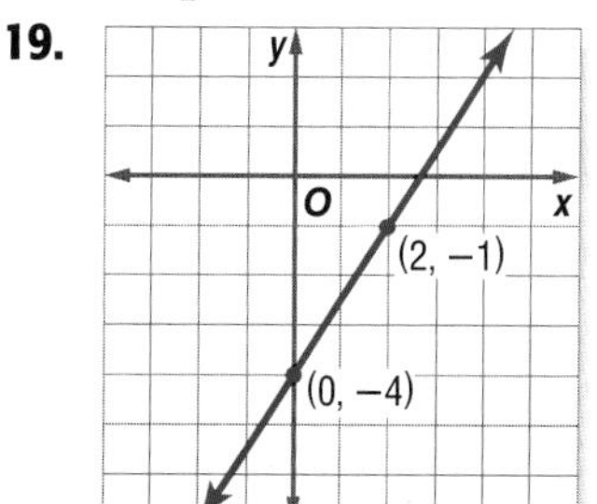

20.

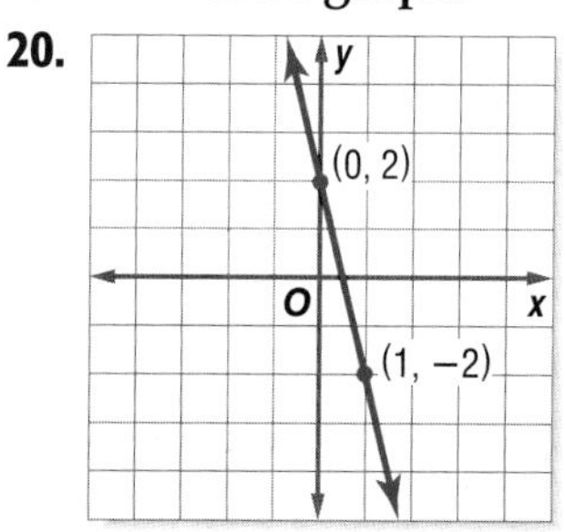

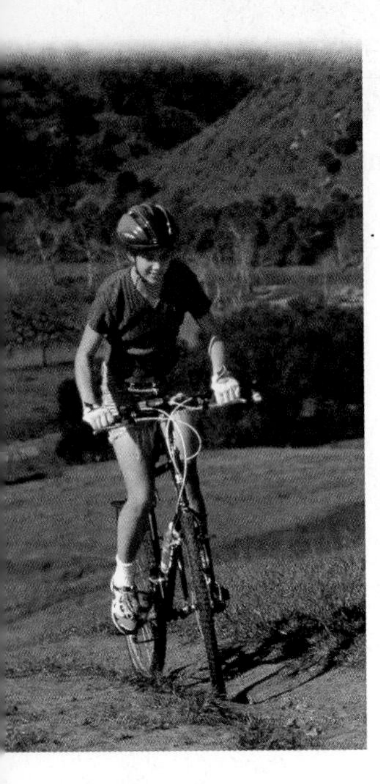

Write an equation in slope-intercept form of the line shown in each graph.

21.

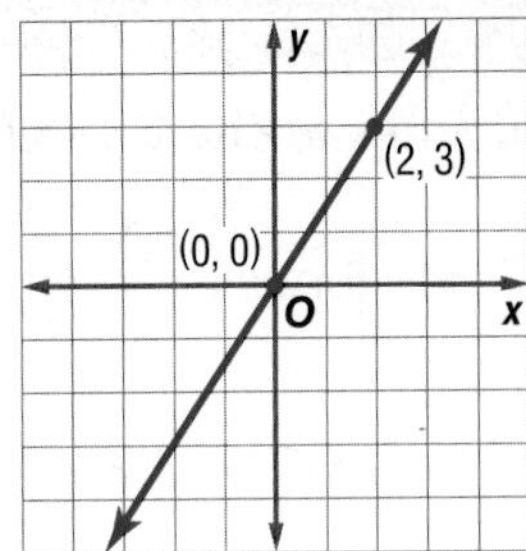

22.

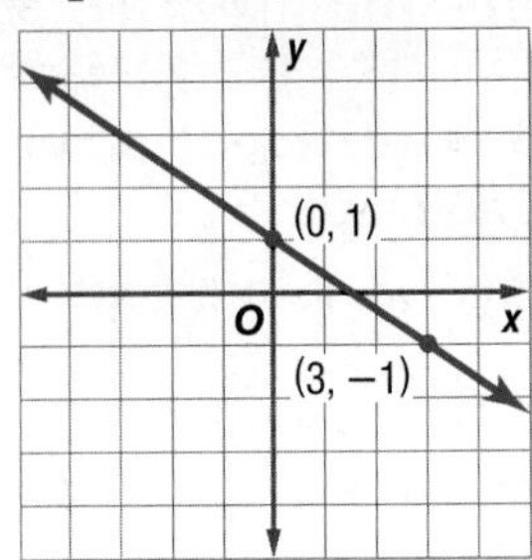

23.

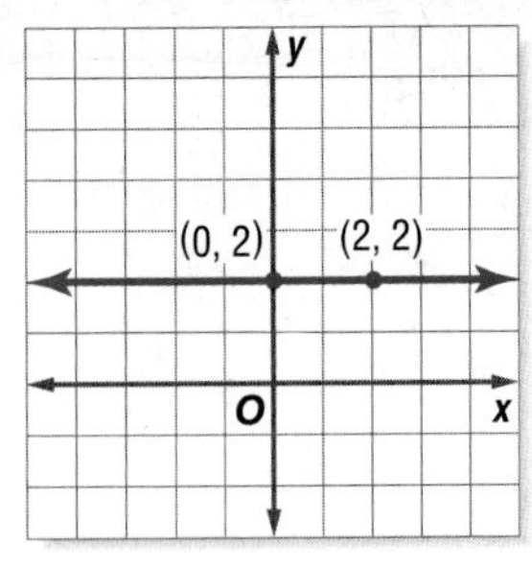

Graph each equation.

24. $y = x - 2$

25. $y = 3x + 1$

26. $y = -4x + 1$

27. $y = \frac{1}{2}x + 4$

28. $y = -\frac{1}{3}x - 3$

29. $3x + y = -2$

30. $2x - y = -3$

31. $3y = 2x + 3$

32. $2x + 3y = 6$

Real-World Link

More than 3 million teens participate in mountain bicycling each year.

Source: *Statistical Abstract of the United States*

BICYCLES For Exercises 33 and 34, use the following information.
A rental company on Padre Island charges \$8 per hour for a mountain bicycle plus a \$5 fee for a helmet.

33. Write a linear equation in slope-intercept form for the total rental cost for a helmet and bicycle for t hours. Then graph the equation.

34. Find the cost of a 2-hour rental.

ANALYZE GRAPHS For Exercises 35 and 36, use the following information.
In 2003, book sales in the United States totaled \$23.4 billion. Suppose sales continue to increase by about \$1.2 billion each year.

35. Write an equation in slope-intercept form to find the total sales S for the number of years t since 2003.

36. If the trend continues, what will sales be in 2007?

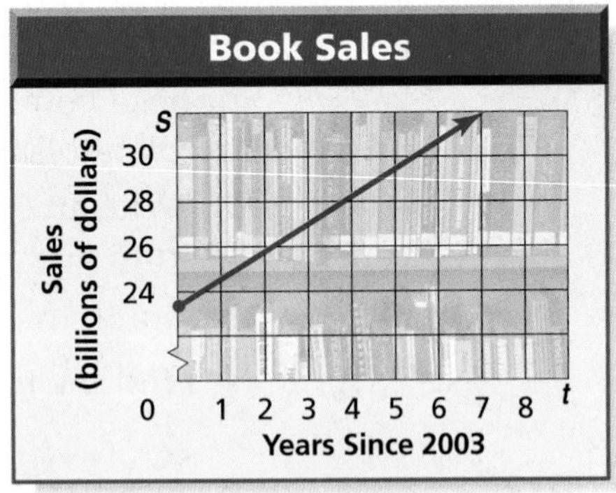

Source: Association of American Publishers

COLLEGE For Exercises 37 and 38, use the following information
For Kentucky residents, the average tuition per year at the University of Kentucky is \$258.15 per credit hour for part-time students. Housing costs \$2125 per year.

37. Write an equation in slope-intercept form for the tuition T for c credit hours.

38. Find the cost of tuition in a year for a student taking 32 credit hours.

EXTRA PRACTICE
See pages 715, 737.

Math Online
Self-Check Quiz at algebra1.com

Write an equation of the line with the given slope and y-intercept.

39. slope: -1, y-intercept: 0

40. slope: 0.5; y-intercept: 7.5

41. slope: 0, y-intercept: 7

42. slope: -1.5, y-intercept: -0.25

43. Write an equation of a horizontal line that crosses the y-axis at $(0, -5)$.

44. Write an equation of a line that passes through the origin with slope 3.

H.O.T. Problems

45. CHALLENGE Summarize the characteristic that the graphs of $y = 2x + 3$, $y = 4x + 3$, $y = -x + 3$, and $y = -10x + 3$ have in common.

46. OPEN ENDED Draw a graph representing a real-world linear function and write an equation for the graph. Describe verbally what the graph represents, including the slope and y-intercept.

47. ***Writing in Math*** Use the data about online shipping costs on page 204 to explain how y-intercepts can be used to describe real-world costs. Write a description of a situation in which the y-intercept of its graph is \$25.

STANDARDIZED TEST PRACTICE

48. Which statement is *most* strongly supported by the graph?

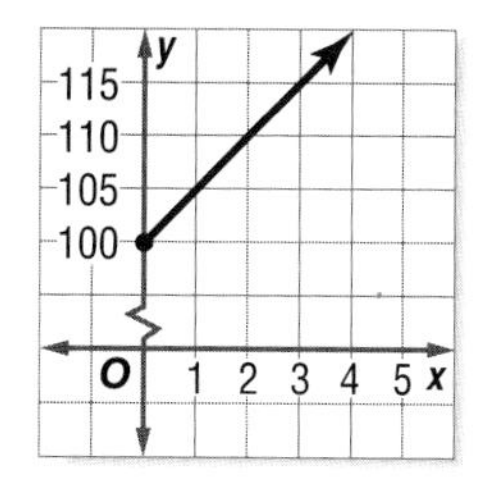

A You have \$100 and plan to spend \$5 each week.

B You have \$100 and plan to save \$5 each week.

C You need \$100 for a new CD player and plan to save \$5 each week.

D You need \$100 for a new CD player and plan to spend \$5 each week.

49. REVIEW Sam is going to put a border around a poster he is making for a class project. x represents the poster's width, and y represents the poster's length. Which equation represents how much border Sam will use if he doubles both the length and the width of the poster?

F $4xy$

G $4(x + y)$

H $(x + y)^4$

J $16(x + y)$

Spiral Review

Suppose y varies directly as x. Write a direct variation equation that relates x and y. Then solve. (Lesson 4-2)

50. If $y = -54$ when $x = 9$, find x when $y = -42$.

51. If $y = 45$ when $x = 60$, find x when $y = 8$.

Find the rate of change represented in each table or graph. (Lesson 4-1)

52.

x	y
1	4
2	7
3	10
4	13
5	16

53.

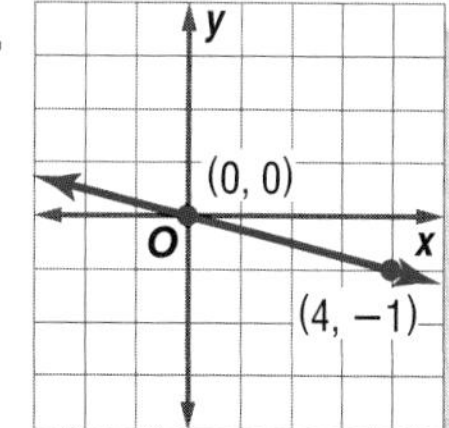

54.

x	y
8	50
13	40
18	30
23	20
28	10

55. LIFE SCIENCE A *Laysan albatross* tracked by biologists flew more than 24,843 miles in just 90 days in flights across the North Pacific to find food for its chick. At this rate, how far could the bird fly in a week? (Lesson 2-6)

GET READY for the Next Lesson

PREREQUISITE SKILL Find the slope of the line that passes through each pair of points. (Lesson 4-1)

56. $(-1, 2), (1, -2)$

57. $(-3, -1), (2, 3)$

58. $(5, 8), (-2, 8)$

EXTEND 4-3

Graphing Calculator Lab

The Family of Linear Graphs

A family of people is a group related by birth, marriage, or adoption. Recall that a *family of graphs* includes graphs with at least one characteristic in common.

You can use a graphing calculator to investigate how changing the parameters m and b in $y = mx + b$ affects the graphs in the family of linear functions.

Concepts in MOtion Animation algebra1.com

ACTIVITY 1 **Changing b in $y = mx + b$**

Graph $y = x$, $y = x + 4$, and $y = x - 2$ in the standard viewing window.

Enter the equations in the Y= list as Y1, Y2, and Y3. Then graph the equations.

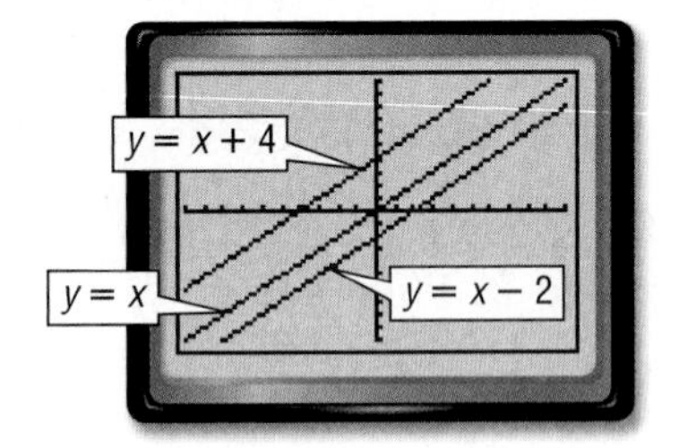

KEYSTROKES: *Review graphing on pages 162 and 163.*

1A. How do the slopes of the graphs compare?

1B. Compare the graph of $y = x + 4$ and the graph of $y = x$. How would you obtain the graph of $y = x + 4$ from the graph of $y = x$?

1C. How would you obtain the graph of $y = x - 2$ from the graph of $y = x$?

Changing m in $y = mx + b$ affects the graphs in a different way than changing b. First, investigate positive values of m.

ACTIVITY 2 **Changing m in $y = mx + b$, Positive Values**

Graph $y = x$, $y = 2x$, and $y = \frac{1}{3}x$ in the standard viewing window.

Enter the equations in the Y= list and graph.

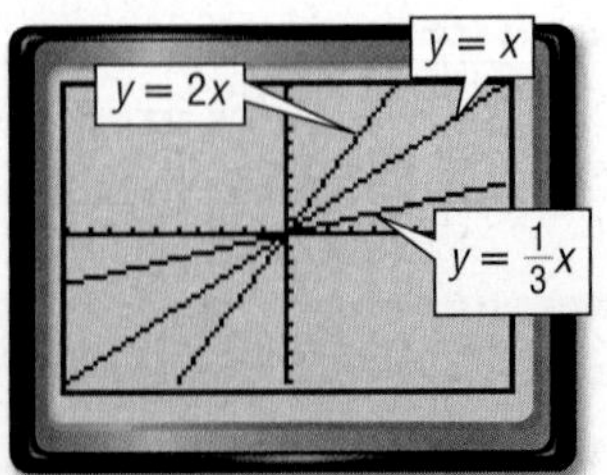

2A. How do the y-intercepts of the graphs compare?

2B. Compare the graphs of $y = 2x$ and $y = x$.

2C. Which is steeper, the graph of $y = \frac{1}{3}x$ or the graph of $y = x$?

Does changing m to a negative value affect the graph differently than changing it to a positive value?

Math Online Other Calculator Keystrokes at algebra1.com

ACTIVITY 3 Changing m in $y = mx + b$, Negative Values

Graph $y = x$, $y = -x$, $y = -3x$, and $y = -\frac{1}{2}x$ in the standard viewing window.

Enter the equations in the Y= list and graph.

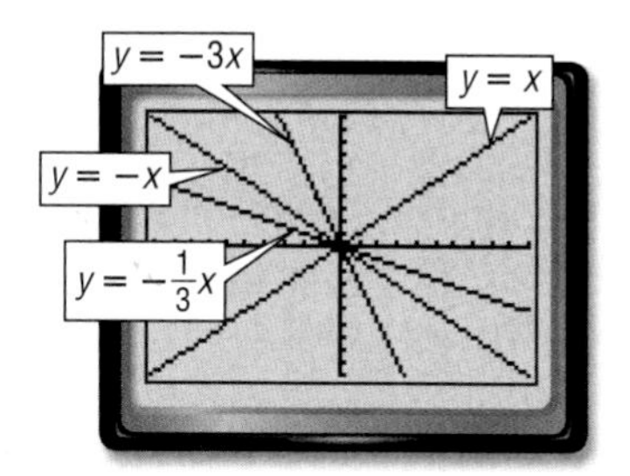

3A. How are the graphs with negative values of m different than graphs with a positive m?

3B. Compare the graphs of $y = -x$, $y = -3x$, and $y = -\frac{1}{2}x$. Which is steepest?

ANALYZE THE RESULTS

Graph each set of equations on the same screen. Describe the similarities or differences among the graphs.

1. $y = 2x$
$y = 2x + 3$
$y = 2x - 7$

2. $y = x + 1$
$y = 2x + 1$
$y = \frac{1}{4}x + 1$

3. $y = x + 4$
$y = 2x + 4$
$y = 0.75x + 4$

4. $y = \frac{1}{2}x + 2$
$y = \frac{1}{2}x - 5$
$y = \frac{1}{2}x + 4$

5. $y = -2x - 2$
$y = -4x - 2$
$y = -\frac{1}{3}x - 2$

6. $y = 3x$
$y = 3x + 6$
$y = 3x - 7$

7. Families of graphs have common characteristics. What do the graphs of all equations of the form $y = mx + b$ have in common?

8. How does the value of b affect the graph of $y = mx + b$?

9. What is the result of changing the value of m on the graph of $y = mx + b$ if m is positive? if m is negative?

10. How can you determine which graph is steepest by examining the following equations?

$y = 3x$, $y = -4x - 7$, $y = \frac{1}{2}x + 4$

11. Explain how knowing about the effects of m and b can help you sketch the graph of an equation.

Nonlinear functions can also be defined in terms of a family of graphs. Graph each set of equations on the same screen. Describe the similarities or differences among the graphs.

12. $y = x^2$
$y = -3x^2$
$y = (-3x)^2$

13. $y = x^2$
$y = x^2 + 3$
$y = (x - 2)^2$

14. $y = x^2$
$y = 2x^2 + 4$
$y = (3x)^2 - 5$

15. Describe the similarities and differences in the classes of functions $f(x) = x^2 + c$ and $f(x) = (x + c)^2$, where c is any real number.

CHAPTER 4

Mid-Chapter Quiz

Lessons 4-1 through 4-3

Find the slope of the line that passes through each pair of points. (Lesson 4-1)

1. $(-4, 6), (-3, 8)$ **2.** $(8, 3), (-11, 3)$

POPULATION GROWTH For Exercises 3–5, use the following information.

The graph shows the population growth in the USA for 1960 through 2005. (Lesson 4-1)

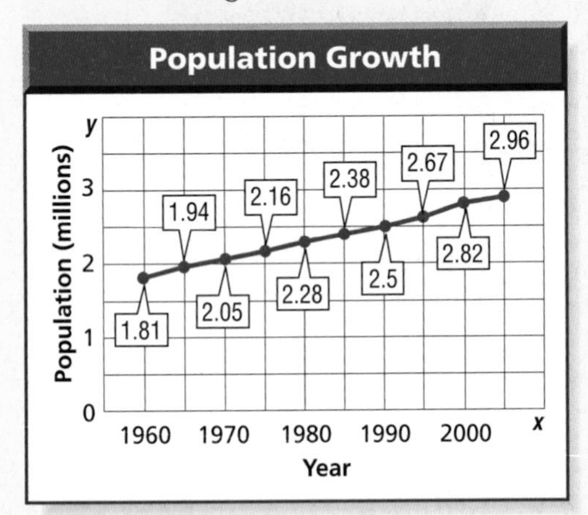

Source: U.S. Census Bureau

3. For which 5-year time period was the rate of change the greatest? the least?

4. Find the rate of change from 1980 to 1990.

5. Explain the meaning of the slope from 1960 to 2005.

6. What is the slope of the *line* containing the points shown in the table? (Lesson 4-1)

x	y
-4	-3
2	6
6	12

7. Find the value of r so the line that passes through $(5, -3)$ and $(r, -5)$ has slope 2. (Lesson 4-1)

8. Suppose that y varies directly as x, and $y = 24$ when $x = 8$. Write a direct variation equation that relates x and y. Use the equation to find y when $x = -3$. (Lesson 4-2)

Graph each equation. (Lessons 4-2 and 4-3)

9. $y = -7x$ **10.** $y = \frac{3}{4}x + 2$ **11.** $x - y = 5$

12. MULTIPLE CHOICE Megan works at a sporting goods store, and her salary is shown in the graph. Which is a valid conclusion that can be made from the graph? (Lesson 4-2)

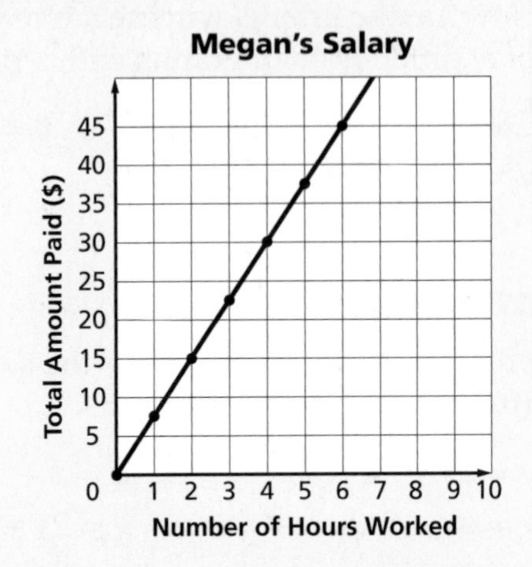

A Megan earns about \$7 per hour.

B Megan earns about \$30 for every 2 hours that she works.

C Megan earns about \$52 per week.

D Megan earns about \$60 for each shift that she works.

For Exercises 13–15, use the following information.

Suppose you have already saved \$75 toward the cost of a new television. You plan to save \$5 more each week for the next several weeks. (Lesson 4-3)

13. Write an equation for the total amount T you will have w weeks from now.

14. Graph the equation.

15. Find the total amount saved after 10 weeks.

16. MULTIPLE CHOICE Which equation describes a line that has a y-intercept of 3 and a slope of 2? (Lesson 4-3)

F $y = 3 + 2x$

G $y = (3 + x)2$

H $y = 3x + 2$

J $y = (3x + 1)2$

Writing Equations in Slope-Intercept Form

Main Ideas

- Write an equation of a line given the slope and one point on a line.
- Write an equation of a line given two points on the line.

New Vocabulary

linear extrapolation

GET READY for the Lesson

In 2006, the population of a city was about 263 thousand. At that time, the population was growing at a rate of about 7 thousand per year.

Year	Population (thousands)
2005	256
2006	263
2007	270

If you could write an equation based on the slope, 7 (thousand), and the point (2006, 263), you could predict the population for another year.

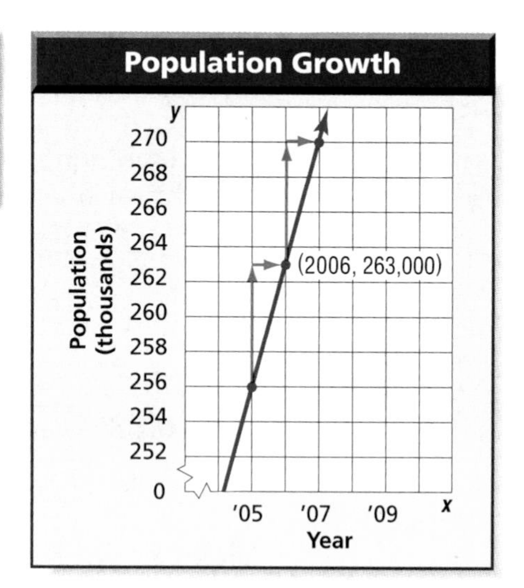

Write an Equation Given the Slope and One Point You have learned how to write an equation of a line when you know the slope and a specific point, the y-intercept. The following example shows how to write an equation when you know the slope and any point on the line.

EXAMPLE Write an Equation Given Slope and One Point

1 Write an equation of a line that passes through (1, 5) with slope 2.

Step 1 Find the y-intercept by replacing m with 2 and (x, y) with (1, 5) in the slope-intercept form and solving for b.

$y = mx + b$ Slope-intercept form

$5 = 2(1) + b$ Replace m with 2, y with 5, and x with 1.

$5 = 2 + b$ Multiply.

$5 - 2 = 2 + b - 2$ Subtract 2 from each side.

$3 = b$ Simplify.

Step 2 Write the slope-intercept form using $m = 2$ and $b = 3$.

$y = mx + b$ Slope-intercept form

$y = 2x + 3$ Replace m with 2 and b with 3.

Therefore, an equation of the line is $y = 2x + 3$.

(continued on the next page)

CHECK You can check your result by graphing $y = 2x + 3$ on a graphing calculator.

Use the CALC menu to verify.

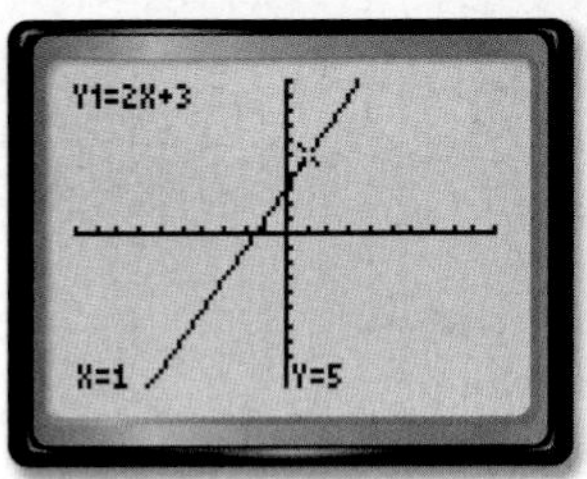

[−10, 10] scl: 1 by [−10, 10] scl: 1

CHECK Your Progress

1. Write an equation of the line that passes through $(-4, 7)$ with slope -1.

Write an Equation Given Two Points If you know two points on a line, first find the slope. Then follow the steps in Example 1.

STANDARDIZED TEST EXAMPLE — Write an Equation Given Two Points

2 The table shows the coordinates of two points on the graph of a linear function. Which equation describes the function?

x	y
−3	−1
6	−4

A $y = -\frac{1}{3}x - 2$

B $y = 3x - 2$

C $y = -\frac{1}{3}x + 2$

D $y = \frac{1}{3}x - 2$

Read the Test Item

The table represents the ordered pairs $(-3, -1)$ and $(6, -4)$.

Solve the Test Item

Step 1 Find the slope of the line containing the points.

$m = \frac{y_2 - y_1}{x_2 - x_1}$ Slope formula

$= \frac{-4 - (-1)}{6 - (-3)}$ $(x_1, y_1) = (-3, -1)$ and $(x_2, y_2) = (6, -4)$

$= \frac{-3}{9}$ or $-\frac{1}{3}$ Simplify.

Step 2 Use the slope and one of the two points to find the y-intercept.

$y = mx + b$ Slope-intercept form

$-4 = -\frac{1}{3}(6) + b$ Replace m with $-\frac{1}{3}$, x with 6, and y with -4.

$-4 = -2 + b$ Multiply.

$-2 = b$ Add 2 to each side.

Step 3 Write the slope-intercept form using $m = -\frac{1}{3}$ and $b = -2$.

$y = mx + b$ Slope-intercept form

$y = -\frac{1}{3}x - 2$ Replace m with $-\frac{1}{3}$, and b with -2. The answer is A.

Test-Taking Tip

Check Results You can check your result by graphing. The line should pass through $(-3, -1)$ and $(6, -4)$.

CHECK Your Progress

2. The graph of an equation contains the points at $(-1, 12)$ and $(4, -8)$. Which equation describes this function?

F $y = -\frac{1}{4}x - 8$

G $y = 4x + 8$

H $y = \frac{1}{4}x - 8$

J $y = -4x + 8$

Online Personal Tutor at algebra1.com

Real-World EXAMPLE

3 BASEBALL After 22 games in 2005, J. D. Drew of the Los Angeles Dodgers had 10 runs batted in. After 36 games, he had 15 runs batted in. Write a linear equation to estimate the number of runs batted in for any number of games that season.

Real-World Link

In 2005, J. D. Drew played a total of 72 games.

Source: MLB.com

Explore You know the number of runs batted in after 22 and 36 games.

Plan Let x represent the number of games. Let y represent the number of runs batted in. Write an equation of the line that passes through (22, 10) and (36, 15).

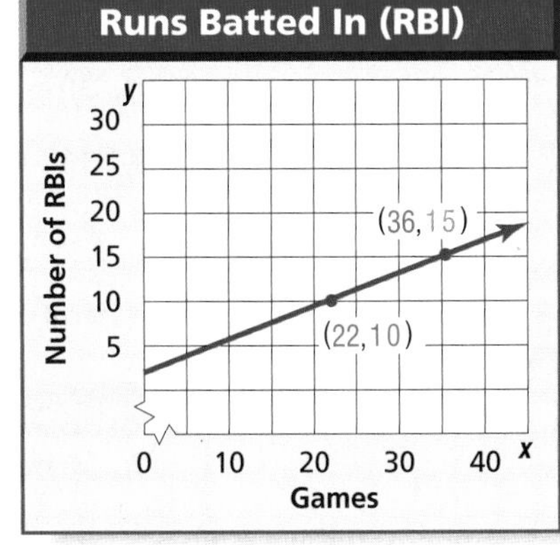

Solve Find the slope.

$m = \frac{y_2 - y_1}{x_2 - x_1}$ Slope formula

$= \frac{15 - 10}{36 - 22}$ Let $(x_1, y_1) = (22, 10)$ and $(x_2, y_2) = (36, 15)$.

$= \frac{5}{14}$ or about 0.357 Simplify.

Choose (36, 15) and find the y-intercept of the line.

$y = mx + b$ Slope-intercept form

$15 = 0.357(36) + b$ Replace m with 0.357, x with 36, and y with 15.

$15 = 12.852 + b$ Multiply.

$2.148 = b$ Subtract 12.852 from each side and simplify.

Write the slope-intercept form using $m = 0.357$, and $b = 2.148$.

$y = mx + b$ Slope-intercept form

$y = 0.357x + 2.148$ Replace m with 0.357 and b with 2.148.

Therefore, the equation is $y = 0.357x + 2.148$.

Check Check your result by substituting the coordinates of the point not chosen, (22, 10), into the equation.

$y = 0.357x + 2.148$ Original equation

$10 \stackrel{?}{=} 0.357(22) + 2.148$ Replace y with 10 and x with 22.

$10 \approx 10.002$ Simplify.

The slope was rounded, so the answers vary slightly. The answer checks.

CHECK Your Progress

3. MONEY As a part-time job, Ethan makes deliveries for a caterer. In addition to his weekly salary, he is also paid \$16 per delivery. Last week, he made 5 deliveries and his total salary was \$215. Write a linear equation to find Ethan's total weekly salary S if he makes d deliveries.

When you use a linear equation to predict values that are beyond the range of the data, you are using **linear extrapolation.**

Vocabulary Link
Extra
Everyday Use beyond or outside, as in extracurricular activities
Math Use beyond the range of data

Real-World EXAMPLE

4 SPORTS Use the equation in Example 3 and the information in the margin to estimate Drew's runs batted in during the 2005 season.

$y = 0.357x + 2.148$ Original equation

$= 0.357(72) + 2.148$ Replace x with 72.

≈ 28 Simplify.

Using the equation, an estimate for the number of RBIs is 28.

CHECK Your Progress

4. **MONEY** Use the equation that you wrote in Check Your Progress 3 to predict how much money Ethan will earn in a week if he makes 8 deliveries.

Be cautious when making a prediction or an estimate using just two given points. The model may be *approximately* correct, but still give inaccurate predictions. For example, in 2005, J. D. Drew had 36 runs batted in, which was 8 more than the estimate.

CHECK Your Understanding

Example 1 (p. 213)

Write an equation of the line that passes through each point with the given slope.

1. $(4, -2), m = 2$
2. $(3, 7), m = -3$
3. $(-3, 5), m = -1$

Example 2 (p. 214)

Write an equation of the line that passes through each pair of points.

4. $(5, 1), (8, -2)$
5. $(6, 0), (0, 4)$
6. $(5, 2), (-7, -4)$

7. **STANDARDIZED TEST PRACTICE** The table of ordered pairs shows the coordinates of the two points on the graph of a line. Which equation describes the line?

x	y
−5	2
0	7

A $y = x + 7$
B $y = x - 7$
C $y = -5x + 2$
D $y = 5x + 2$

Examples 3 and 4 (pp. 215–216)

CANOE RENTAL For Exercises 8 and 9, use the information at the right and below.
Ilia and her friends rented a canoe for 3 hours and paid a total of $45.

8. Write a linear equation to find the total cost C of renting the canoe for h hours.
9. How much would it cost to rent the canoe for 8 hours?

Exercises

HOMEWORK HELP

For Exercises	See Examples
10–17	1
18–25	2
26–29	3, 4

Write an equation of the line that passes through each point with the given slope.

10.

11.

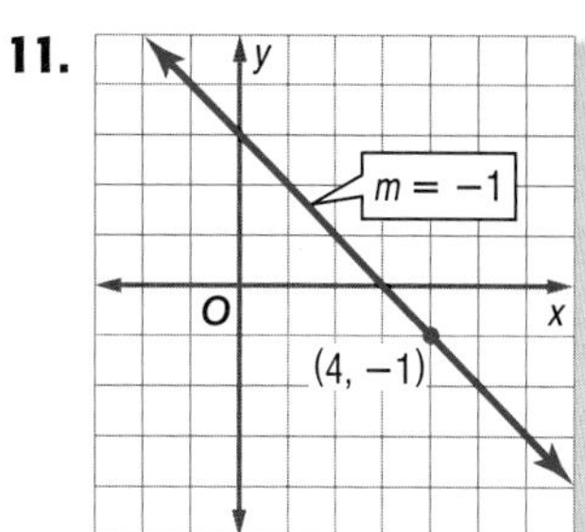

12. $(5, -2), m = 3$ **13.** $(5, 4), m = -5$ **14.** $(3, 0), m = -2$

15. $(5, 3), m = \frac{1}{2}$ **16.** $(-3, -1), m = -\frac{2}{3}$ **17.** $(-3, -5), m = -\frac{5}{3}$

Write an equation of the line that passes through each pair of points.

18. $(4, 2), (-2, -4)$ **19.** $(3, -2), (6, 4)$

20. $(-1, 3), (2, -3)$ **21.** $(2, -2), (3, 2)$

22. $(7, -2), (-4, -2)$ **23.** $(0, 5), (-3, 5)$

24. $(1, 1), (7, 4)$ **25.** $(5, 7), (0, 6)$

POPULATION For Exercises 26 and 27, use the data at the top of page 213.

26. Write a linear equation to find the city's population P for any year t.

27. Predict what the city's population will be in 2010.

DOGS For Exercises 28 and 29, refer to the information below.
In 2001, there were about 62.5 thousand golden retrievers registered in the United States. In 2002, the number was 56.1 thousand.

28. Write a linear equation to predict the number of golden retrievers G that will be registered in year t.

29. Predict the number of golden retrievers that will be registered in 2007.

Write an equation of the line that passes through each pair of points.

30. $(5, -2), (7, 1)$ **31.** $\left(-\frac{5}{4}, 1\right), \left(-\frac{1}{4}, \frac{3}{4}\right)$ **32.** $\left(\frac{5}{12}, -1\right), \left(-\frac{3}{4}, \frac{1}{6}\right)$

33. Write an equation of the line that has an x-intercept -3 and a y-intercept 5.

EXTRA PRACTICE
See pages 725, 747.

Math Online
Self-Check Quiz at algebra1.com

For Exercises 34 and 35, consider line ℓ that passes through (14, 2) and (28, 6).

34. Write an equation for line ℓ and describe the slope.

35. Where does line ℓ intersect the x-axis? the y-axis?

H.O.T. Problems

36. CHALLENGE The x-intercept of a line is p, and the y-intercept is q. Use symbols to describe an equation of the line.

37. OPEN ENDED Create a real-world situation that fits the graph at the right. Then draw and label the graph to represent this situation. Define the two quantities and describe the functional relationship between them. Write an equation to represent this relationship and describe what the slope and y-intercept mean.

H.O.T. Problems

38. REASONING Tell whether the statement is *sometimes, always,* or *never* true. Explain your reasoning.

You can write the equation of a line given its x- and y-intercepts.

39. *Writing in Math* Use the information about population on page 213 to explain how the slope-intercept form can be used to make predictions. Discuss how slope-intercept form is used in linear extrapolation.

STANDARDIZED TEST PRACTICE

40. Which equation *best* describes the relationship between the values of x and y shown in the table?

A $y = x - 5$

B $y = 2x - 5$

C $y = 3x - 7$

D $y = x^2 - 7$

x	y
−1	−7
0	−5
2	−1
4	3

41. REVIEW Mrs. Aguilar's bedroom is shaped like a rectangle that measures 13 feet by 11 feet. She wants to purchase carpet for the bedroom that costs $2.95 per square foot, including tax. How much will it cost to carpet her bedroom?

F $70.80

G $141.60

H $145.95

J $421.85

Spiral Review

Graph each equation. (Lesson 4-3)

42. $y = 3x - 2$

43. $x + y = 6$

44. $x + 2y = 8$

45. HEALTH Each time your heart beats, it pumps 2.5 ounces of blood. Write a direct variation equation that relates the total volume of blood V with the number of times your heart beats b. (Lesson 4-2)

Determine the x-intercept and y-intercept of each linear function and describe what the intercepts mean. (Lesson 3-3)

46.

Kwame's Bike Ride	
Time, x (min)	Distance, y (mi)
0	0
5	1
10	2
15	3

47.

Tara's Walk Home	
Time, x (min)	Distance, y (mi)
0	3
15	2
30	1
45	0

Determine the domain and range of each relation. (Lesson 3-1)

48. $\{(0, 8), (9, -2), (4, 2)\}$

49. $\{(-2, 1), (5, 1), (-2, 7), (0, -3)\}$

GET READY for the Next Lesson

PREREQUISITE SKILL **Find each difference.** (Pages 688–689)

50. $4 - 7$

51. $5 - 12$

52. $2 - (-3)$

53. $-1 - 4$

EXTEND 4-4

Graphing Calculator Lab
Step Functions

The graph of a step function is a series of line segments. One step function is called the *greatest integer function.* The greatest integer function is written as $f(x) = [\![x]\!]$, where $f(x)$ is the greatest integer less than or equal to x.

ACTIVITY 1

Graph $f(x) = [\![x]\!]$ in the standard viewing window.

The calculator may need to be changed to dot mode for the function to graph correctly. Press MODE then use the arrow and ENTER keys to select DOT.

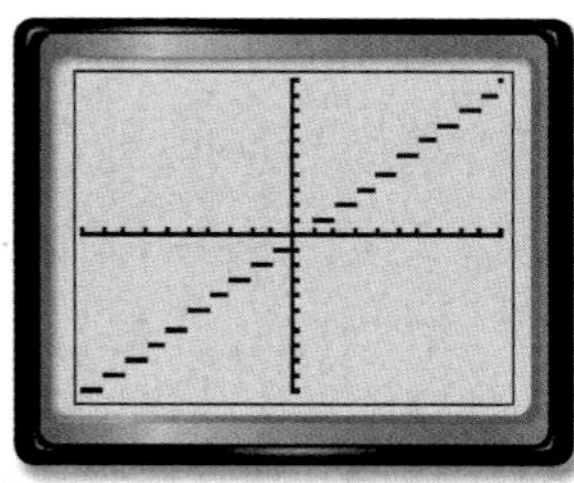
[−10, 10] scl: 1 by [−10, 10] scl: 1

Enter the equation in the Y= list. Then graph the equation.

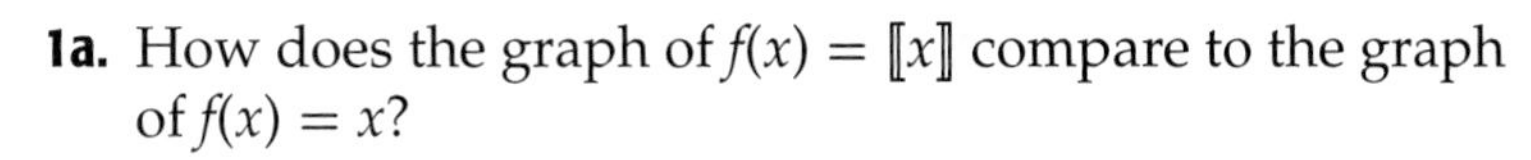
KEYSTROKES: Y= MATH ▶ 5 X,T,θ,n) Zoom 6

1a. How does the graph of $f(x) = [\![x]\!]$ compare to the graph of $f(x) = x$?

1b. What are the domain and range of the function $f(x) = [\![x]\!]$? Explain.

Step functions are often used in real-world situations involving time or money.

ACTIVITY 2

TAXES A city collects $0.02 in income tax for every $1 of income. Write and graph a function for the taxes y of an income of x dollars.

Because the city does not collect taxes for fractions of a dollar, the situation can be described by the step function $y = 0.02[\![x]\!]$.

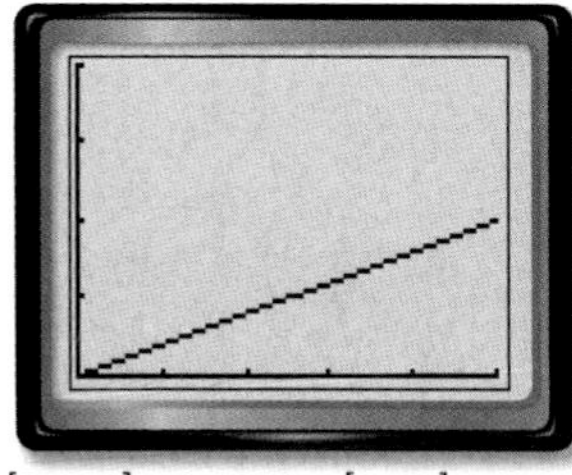
[0, 500] scl: 100 by [0, 20] scl: 5

KEYSTROKES: Y= 0.02 MATH ▶ 5 X,T,θ,n) GRAPH

2a. Explain why the situation is correctly modeled by a step function instead of the linear function $y = 0.02x$.

2b. Use the CALC function to find the income tax on $458.67.

ANALYZE THE RESULTS

1. A parking garage charges $4 for every hour or fraction of an hour. Is this situation modeled by a linear function or a step function? Explain your reasoning.

2. **MAKE A CONJECTURE** Explain why the greatest integer function is sometimes called the *floor function.*

4-5 Writing Equations in Point-Slope Form

Main Ideas

- Write the equation of a line in point-slope form.
- Write linear equations in different forms.

New Vocabulary

point-slope form

GET READY for the Lesson

The graph shows a line with slope 2. Another point on the line is (x, y).

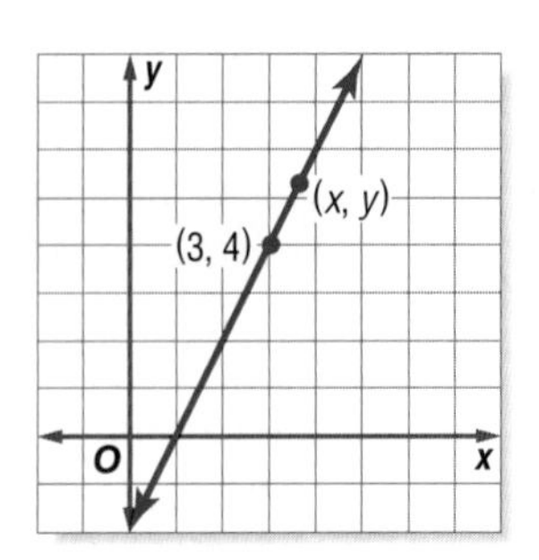

$m = \frac{y_2 - y_1}{x_2 - x_1}$	Slope formula
$2 = \frac{y - 4}{x - 3}$	$(x_2, y_2) = (x, y)$ $(x_1, y_1) = (3, 4)$
$2(x - 3) = \frac{y - 4}{x - 3}(x - 3)$	Multiply each side by $(x - 3)$.
$2(x - 3) = y - 4$	Simplify.
$y - 4 = 2(x - 3)$	Symmetric Property of Equality

In $y - 4 = 2(x - 3)$: 4 is the *y*-coordinate, 2 is the slope, 3 is the *x*-coordinate.

Point-Slope Form The equation above was generated using the coordinates of a known point and the slope of the line. It is written in **point-slope form.**

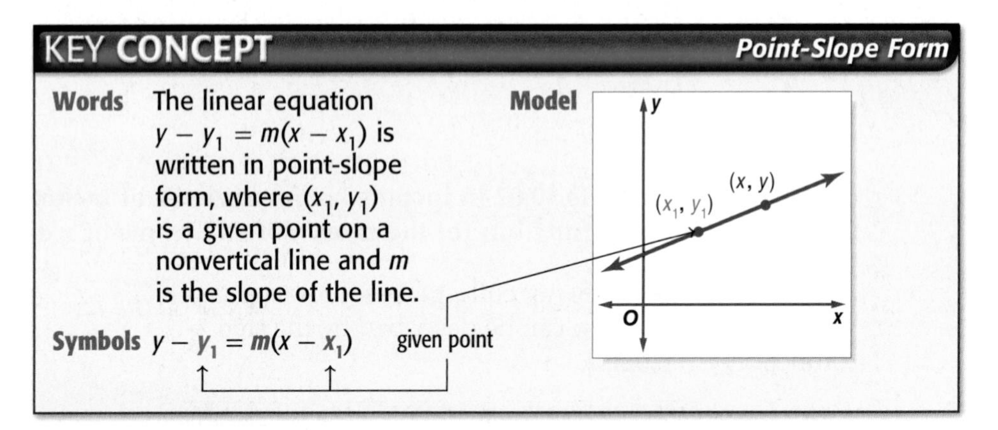

KEY CONCEPT — *Point-Slope Form*

Words The linear equation $y - y_1 = m(x - x_1)$ is written in point-slope form, where (x_1, y_1) is a given point on a nonvertical line and m is the slope of the line.

Symbols $y - y_1 = m(x - x_1)$ given point

Model

EXAMPLE Write an Equation Given Slope and a Point

1 **Write the point-slope form of an equation for the line that passes through (−1, 5) with slope −3.**

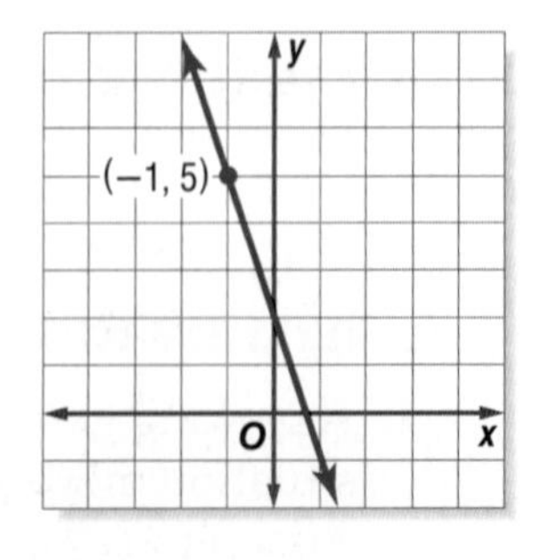

$y - y_1 = m(x - x_1)$	Point-slope form
$y - 5 = -3[x - (-1)]$	$(x_1, y_1) = (-1, 5)$
$y - 5 = -3(x + 1)$	Simplify.

CHECK Your Progress

1. Write the point-slope form of an equation for the line that passes through (1, −4) with slope $-\frac{8}{3}$.

EXAMPLE Write an Equation of a Horizontal Line

Study Tip

Vertical and Horizontal Lines

Vertical lines cannot be written in point-slope form because the slope is undefined. Horizontal lines can be written in point-slope form because their slope is 0.

2 **Write the point-slope form of an equation for the horizontal line that passes through (6, –2).**

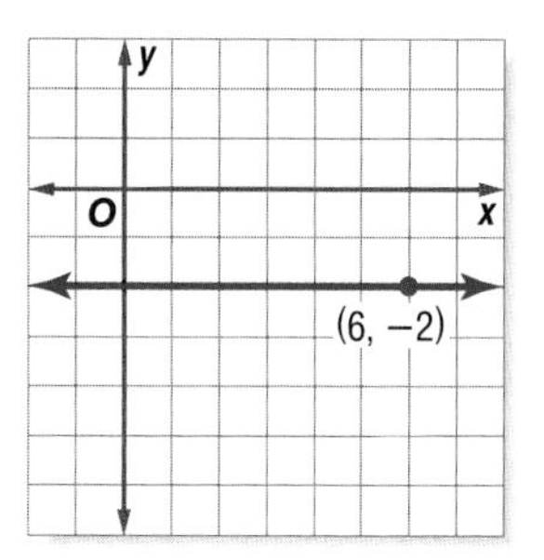

$y - y_1 = m(x - x_1)$	Point-slope form
$y - (-2) = 0(x - 6)$	$(x_1, y_1) = (6, -2)$
$y + 2 = 0$	Simplify.

CHECK Your Progress

2. Write the point-slope form of an equation for the horizontal line that passes through (–4, 4).

Forms of Linear Equations You have learned how to write linear equations given the slope and one point or two points.

CONCEPT SUMMARY — *Writing Equations*

Given the Slope and One Point

Step 1 Substitute the values of m, x, and y into the slope-intercept form and solve for b. Or use the point-slope form. Substitute the value of m and let x and y be (x_1, y_1).

Step 2 Write the slope-intercept form using the values of m and b.

Given Two Points

Step 1 Find the slope.

Step 2 Choose one of the two points to use.

Step 3 Follow the steps for writing an equation given the slope and one point.

Study Tip

Look Back

To review **standard form**, see Lesson 3-5.

Linear equations in point-slope form can be written in slope-intercept or standard form.

EXAMPLE Write an Equation in Standard Form

3 **Write $y + 5 = -\frac{5}{4}(x - 2)$ in standard form.**

In standard form, the variables are on the left side of the equation. A, B, and C are all integers.

$y + 5 = -\frac{5}{4}(x - 2)$	Original equation
$4(y + 5) = 4\left(-\frac{5}{4}\right)(x - 2)$	Multiply each side by 4 to eliminate the fraction.
$4y + 20 = -5(x - 2)$	Distributive Property
$4y + 20 = -5x + 10$	Distributive Property
$4y + 20 - 20 = -5x + 10 - 20$	Subtract 20 from each side.
$4y = -5x - 10$	Simplify.
$4y + 5x = -5x - 10 + 5x$	Add $5x$ to each side.
$5x + 4y = -10$	Simplify.

The standard form of the equation is $5x + 4y = -10$.

CHECK Your Progress

3. Write $y - 1 = 7(x + 5)$ in standard form.

EXAMPLE Write an Equation in Slope-Intercept Form

4 **Write $y - 2 = \frac{1}{2}(x + 5)$ in slope-intercept form.**

$y - 2 = \frac{1}{2}(x + 5)$	Original equation
$y - 2 = \frac{1}{2}x + \frac{5}{2}$	Distributive Property
$y - 2 + 2 = \frac{1}{2}x + \frac{5}{2} + 2$	Add 2 to each side.
$y = \frac{1}{2}x + \frac{9}{2}$	$2 = \frac{4}{2}$ and $\frac{4}{2} + \frac{5}{2} = \frac{9}{2}$

The slope-intercept form of the equation is $y = \frac{1}{2}x + \frac{9}{2}$.

CHECK Your Progress

4. Write $y + 6 = -3(x - 4)$ in slope-intercept form.

EXAMPLE Write an Equation in Point-Slope Form

Study Tip

Hypotenuse

The *hypotenuse* is the side of a right triangle opposite the right angle.

5 **GEOMETRY** **The figure shows right triangle ABC.**

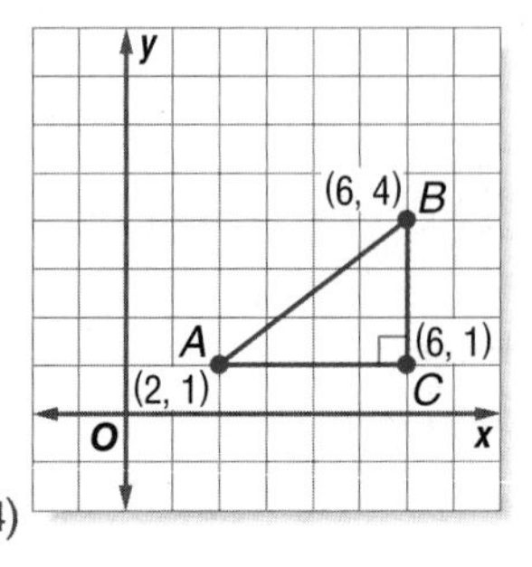

a. Write the point-slope form of the line containing hypotenuse $\overline{AB}$.

Step 1 First, find the slope of $\overline{AB}$.

$m = \frac{y_2 - y_1}{x_2 - x_1}$ Slope formula

$= \frac{4 - 1}{6 - 2}$ or $\frac{3}{4}$ $(x_1, y_1) = (2, 1)$ and $(x_2, y_2) = (6, 4)$

Step 2 You can use either point for (x_1, y_1) in the point-slope form.

Method 1 Use (6, 4).	**Method 2** Use (2, 1).
$y - y_1 = m(x - x_1)$	$y - y_1 = m(x - x_1)$
$y - 4 = \frac{3}{4}(x - 6)$	$y - 1 = \frac{3}{4}(x - 2)$

Study Tip

Standard Form

Regardless of which point is used to find the point-slope form, the standard form results in the same equation.

b. Write each equation in standard form.

$y - 4 = \frac{3}{4}(x - 6)$	Original equation	$y - 1 = \frac{3}{4}(x - 2)$
$4(y - 4) = 4\left(\frac{3}{4}\right)(x - 6)$	Multiply each side by 4.	$4(y - 1) = 4\left(\frac{3}{4}\right)(x - 2)$
$4y - 16 = 3(x - 6)$	Multiply.	$4y - 4 = 3(x - 2)$
$4y - 16 = 3x - 18$	Distributive Property	$4y - 4 = 3x - 6$
$4y = 3x - 2$	Add to each side.	$4y = 3x - 2$
$-3x + 4y = -2$	Subtract 3x from each side.	$-3x + 4y = -2$
$3x - 4y = 2$	Multiply each side by −1.	$3x - 4y = 2$

CHECK Your Progress

GEOMETRY **Triangle JKL has vertices $J(1, 2,)$ $K(0, -3)$, and $L(-4, 1)$.**

5A. Write the point-slope form of the line containing side $\overline{JK}$.

5B. Write the standard form of the line containing $\overline{JK}$.

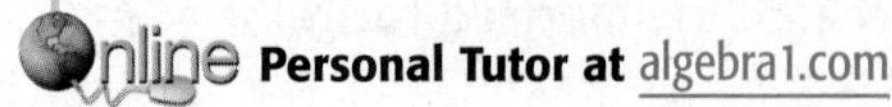

CHECK Your Understanding

Examples 1, 2 (pp. 220–221)

Write the point-slope form of an equation for the line that passes through each point with the given slope.

1.

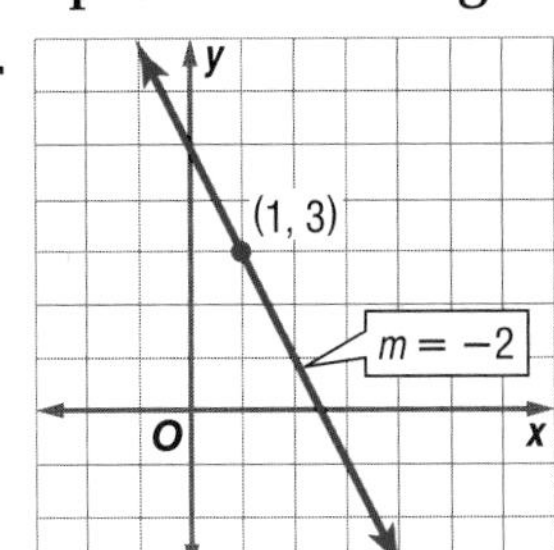

2.

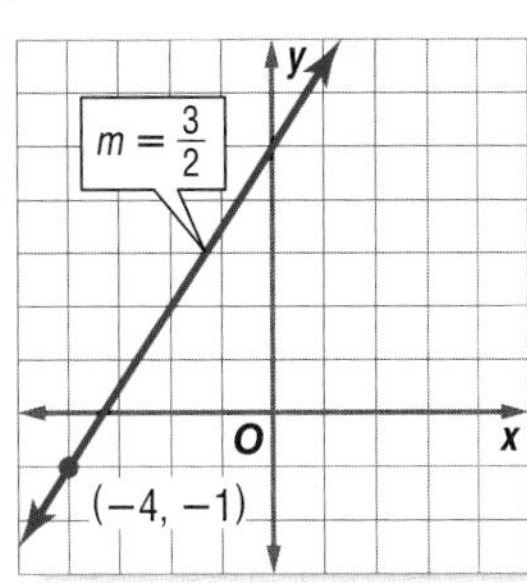

3.

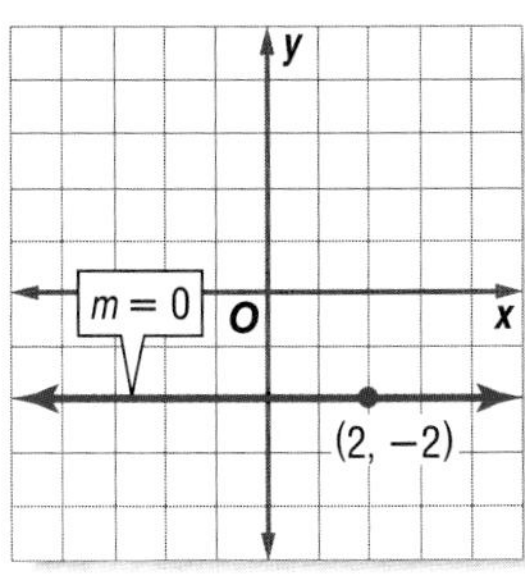

Example 3 (p. 221)

Write each equation in standard form.

4. $y - 5 = 4(x + 2)$

5. $y + 3 = -\frac{3}{4}(x - 1)$

6. $y + 2 = \frac{5}{3}(x + 6)$

Example 4 (p. 222)

Write each equation in slope-intercept form.

7. $y + 6 = 2(x - 2)$

8. $y + 3 = -\frac{2}{3}(x - 6)$

9. $y - 9 = x + 4$

Example 5 (p. 222)

GEOMETRY For Exercises 10 and 11, use parallelogram *ABCD*.

10. Write the point-slope form of the line containing $\overline{AD}$.

11. Write the standard form of the line containing $\overline{AD}$.

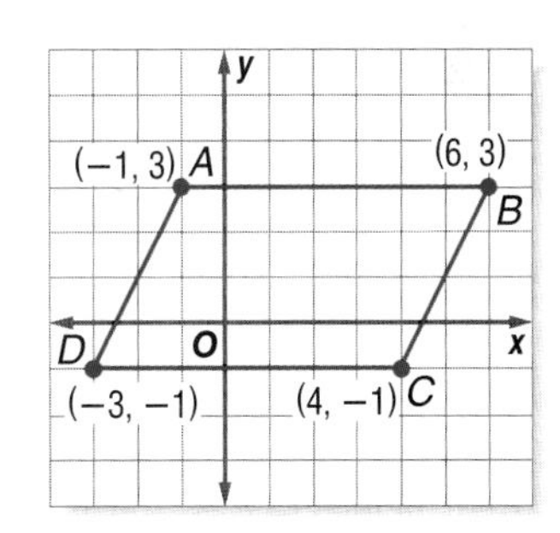

Exercises

HOMEWORK HELP	
For Exercises	See Examples
12–19	1, 2
20–27	3
28–35	4
36–41	5

Write the point-slope form of an equation for the line that passes through each point with the given slope.

12. $(6, 1), m = -4$

13. $(-4, -3), m = 1$

14. $(9, -5), m = 0$

15. $(-7, 6), m = 0$

16. $(-4, 8), m = \frac{7}{2}$

17. $(1, -3), m = -\frac{5}{8}$

18. Write the point-slope form of an equation for the horizontal line that passes through $(5, -9)$.

19. A horizontal line passes through $(0, 7)$. Write the point-slope form of its equation.

Write each equation in standard form.

20. $y - 13 = 4(x - 2)$

21. $y - 5 = -2(x + 6)$

22. $y + 3 = -5(x + 1)$

23. $y + 7 = \frac{1}{2}(x + 2)$

24. $y - 1 = \frac{5}{6}(x - 4)$

25. $y - 2 = -\frac{2}{5}(x - 8)$

26. $2y + 3 = -\frac{1}{3}(x - 2)$

27. $4y - 5x = 3(4x - 2y + 1)$

Write each equation in slope-intercept form.

28. $y - 2 = 3(x - 1)$

29. $y - 5 = 6(x + 1)$

30. $y + 2 = -2(x - 5)$

31. $y + 3 = \frac{1}{2}(x + 4)$

32. $y - 1 = \frac{2}{3}(x + 9)$

33. $y + 3 = -\frac{1}{4}(x + 2)$

34. $y + 3 = -\frac{1}{3}(2x + 6)$

35. $y + 4 = 3(3x + 3)$

BUSINESS For Exercises 36–38, use the following information.
A home security company provides security systems for \$5 per week, plus an installation fee. The total cost for installation and 12 weeks of service is \$210.

36. Write the point-slope form of an equation to find the total fee y for any number of weeks x. (*Hint:* The point (12, 210) is a solution to the equation.)

37. Write the equation in slope-intercept form.

38. What is the flat fee for installation?

Real-World Link

In 1907, movie theaters were called nickelodeons. There were about 5000 movie screens, and the average movie ticket cost 5 cents.

Source: National Association of Theatre Owners

MOVIES For Exercises 39–41, use the following information.
Between 2001 and 2003, the number of movie screens in the United States increased an average of 410 each year. In 2001, there were about 35,170 movie screens.

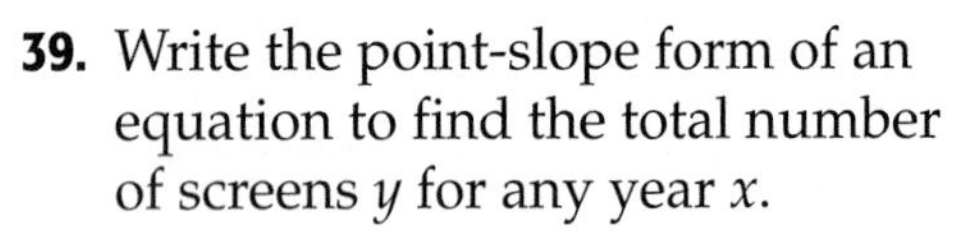

39. Write the point-slope form of an equation to find the total number of screens y for any year x.

40. Write the equation in slope-intercept form.

41. Predict the number of movie screens in the United States in 2007.

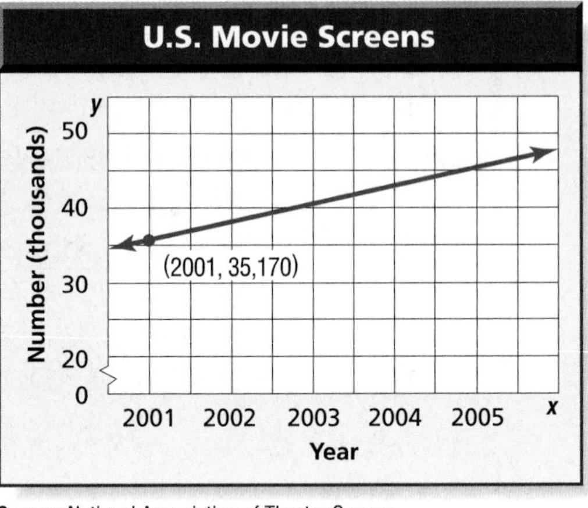

Source: National Association of Theatre Owners

Write each equation in standard form.

42. $y + 4 = -\frac{1}{3}(x - 12)$

43. $y - 3 = 2.5(x + 1)$

44. $y - 6 = 1.7(x + 7)$

Write each equation in slope-intercept form.

45. $y + \frac{1}{2} = x - \frac{1}{2}$

46. $y - \frac{7}{2} = \frac{1}{2}(x - 4)$

47. $y + \frac{1}{4} = -3\left(x + \frac{1}{2}\right)$

48. Write the point-slope form, slope-intercept form, and standard form of an equation for a line that passes through (5, −3) with slope 10.

49. Line ℓ passes through (1, −6) with slope $\frac{3}{2}$. Write the point-slope form, slope-intercept form, and standard form of an equation for line ℓ.

H.O.T. Problems

50. FIND THE ERROR Tanya and Akira wrote the point-slope form of an equation for a line that passes through (−2, −6) and (1, 6). Tanya says that Akira's equation is wrong. Tanya says they are both correct. Who is correct? Explain.

Tanya	Akira
$y + 6 = 4(x + 2)$	$y - 6 = 4(x - 1)$

51. **OPEN ENDED** Compose a real-life scenario that has a constant rate of change and whose value at a particular time is (x, y). Represent this situation using an equation in slope-intercept form and an equation in point-slope form.

52. **REASONING** Find an equation for the line that passes through $(-4, 8)$ and $(3, -7)$. What is the slope? Where does the line intersect the x-axis? the y-axis?

53. **REASONING** Barometric pressure is a linear function of altitude. At an altitude of 2 kilometers, the barometric pressure is 600 mmHg, At 7 kilometers, the barometric pressure is 300 mmHg. Find a formula for the barometric pressure as a function of altitude.

54. **CHALLENGE** A line contains the points $(9, 1)$ and $(5, 5)$. Make a convincing argument that the same line intersects the x-axis at $(10, 0)$.

55. ***Writing in Math*** Demonstrate how you can use the slope formula to write the point-slope form of an equation of a line.

STANDARDIZED TEST PRACTICE

56. What is the equation of the line that passes through $(0, 1)$, and that has a slope of 3?

A $y = 3x - 1$

B $y = 3x - 2$

C $y = 3x + 4$

D $y = 3x + 1$

57. **REVIEW** What is the slope of the line that passes through $(1, 3)$ and $(-3, 1)$?

F -2

G $-\frac{1}{2}$

H $\frac{1}{2}$

J 2

Spiral Review

Write the slope-intercept form of an equation of the line that satisfies each condition. (Lessons 4-3 and 4-4)

58. passes through $(2, -4)$ and $(0, 6)$

59. a horizontal line through $(1, -1)$

60. slope -2 and y-intercept -5

61. passes through $(-2, 4)$ with slope 3

62. **WATER** The table shows the number of gallons of water that a standard showerhead uses. Write an equation in function notation to describe the relation. (Lesson 3-5)

Number of Minutes	1	2	3	4
Number of Gallons	6	12	18	24

Solve each equation. (Lesson 2-3)

63. $4a - 5 = 15$

64. $7 + 3c = -11$

65. $\frac{2}{9}v - 6 = 14$

66. Evaluate $(25 - 4) \div (2^2 - 1^3)$. (Lesson 1-3)

GET READY for the Next Lesson

PREREQUISITE SKILL Write the slope-intercept form of an equation for the line that passes through each pair of points. (Lesson 4-3)

67. $(5, -1), (-3, 3)$

68. $(0, 2), (8, 0)$

69. $(2, 1), (3, -4)$

READING MATH

Understanding the Questions

Describe what canyon hiking is.

- hiking in a deep narrow valley that has steep sides

Explain what canyon hiking involves.

- Depending on the canyon, the hike may require a rope and training in basic rope work, or advanced training in rope work, rappelling, setting up anchors, and so on to descend canyon walls safely.

Notice that both responses above give information about canyon hiking. However, the second response provides more in-depth information than the first. Often in mathematics you are asked to *describe, explain, compare and contrast,* or *justify* statements. As in the situation above, these terms require different levels of response.

Question	What Your Answer Should Show
Describe	**KNOWLEDGE:** recalling information
Explain	**COMPREHENSION:** understanding information
Compare and Contrast	**ANALYSIS:** taking apart information
Justify	**EVALUATION:** making choices based on information

Reading to Learn

1. Describe the information that is needed to write an equation of a line. Explain the steps that you can take to write an equation of a line.

2. Compare and contrast equations that are written in slope-intercept form, point-slope form, and standard form.

3. The graph shows the number of members in U.S. Lacrosse.

a. Describe the trend.

b. Explain possible reasons for the trend.

c. Use the graph to justify a city's decision to have a lacrosse field included in a new sports complex.

Source: U.S. Lacrosse Participation Survey, 2004

4. *Distinguish, summarize, define, predict,* and *demonstrate* are other terms used in mathematics. Write a brief definition of each term as it applies to mathematics and determine whether it requires an answer that shows knowledge, comprehension, analysis, or evaluation.

4-6 Statistics: Scatter Plots and Lines of Fit

Main Ideas

- Interpret points on a scatter plot.
- Use lines of fit to make and evaluate predictions.

New Vocabulary

scatter plot
line of fit
best-fit line
linear interpolation

GET READY for the Lesson

The points of a set of real-world data do not always lie on one line. But, you may be able to draw a line that seems to be close to all the points. The line in the graph shows a linear relationship between the year x and the number of whooping cranes sighted in January of each year, y. Generally, as the years increase, the number of whooping cranes also increases.

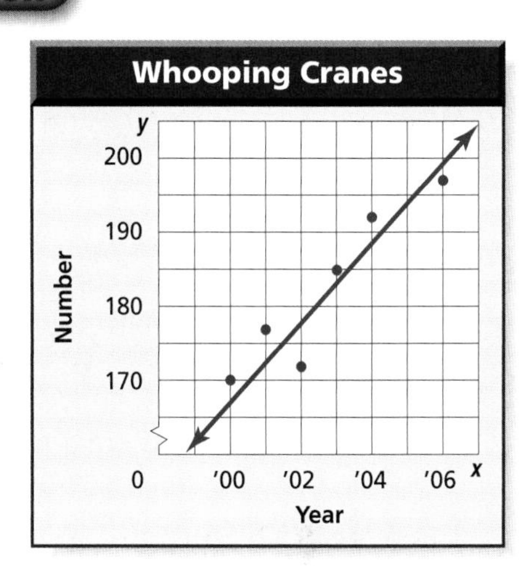

Interpret Points On a Scatter Plot A **scatter plot** is a graph in which two sets of data are plotted as ordered pairs in a coordinate plane. Scatter plots are used to investigate a relationship between two quantities. If the pattern in a scatter plot is linear, you can draw a line to summarize the data. This can help identify trends in the data and the type of correlation.

KEY CONCEPT — Scatter Plots

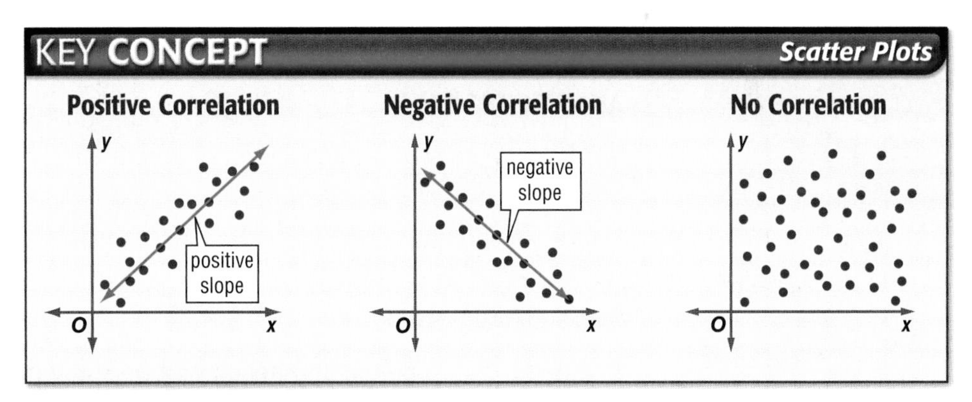

Real-World EXAMPLE

1 NUTRITION Determine whether the graph shows a *positive correlation*, a *negative correlation*, or *no correlation*. If there is a positive or negative correlation, describe its meaning in the situation.

The graph shows a positive correlation. As the number of fat grams increases, the number of Calories increases.

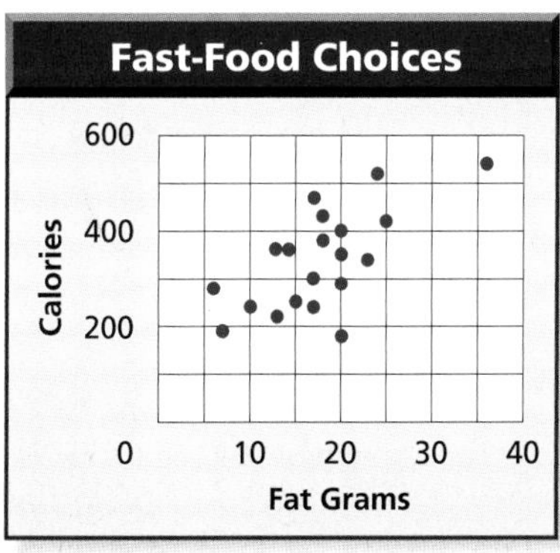

Source: Olen Publishing Co.

Study Tip

Correlations

positive: as x increases, y increases

negative: as x increases, y decreases

no correlation: no relationship between x and y

CHECK Your Progress

1. **CARS** The graph shows the weight and the highway gas mileage of selected cars. Determine whether the graph shows a *positive correlation*, a *negative correlation*, or *no correlation*. If there is a positive or negative correlation, describe its meaning in the situation.

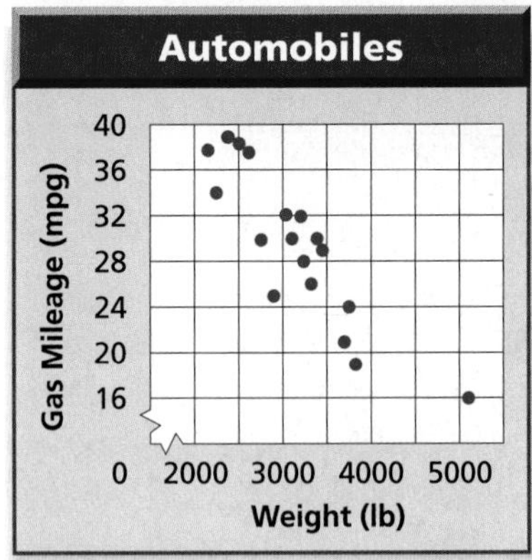

Source: Yahoo!

Is there a relationship between the length of a person's foot and his or her height? Make a scatter plot and then look for a pattern.

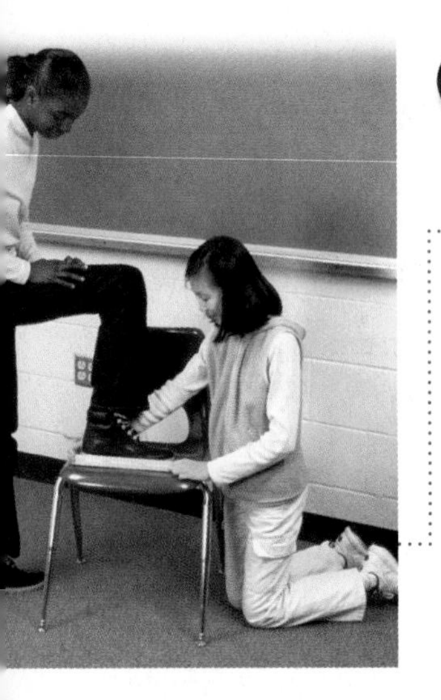

ALGEBRA LAB

Making Predictions

COLLECT AND ORGANIZE THE DATA

- Measure your partner's foot and height in centimeters. Then trade places.
- Add the points (foot length, height) to a class scatter plot.

ANALYZE THE DATA

1. Is there a correlation between foot length and height for the members of your class? If so, describe it.
2. Draw a line that summarizes the data and shows how the height changes as the foot length changes.

MAKE A CONJECTURE

3. Use the line to predict the height of a person whose foot length is 25 centimeters. Explain your method.

Make and Evaluate Predictions If the data points do not all lie on a line, but are close to a line, you can draw a **line of fit.** This line describes the trend of the data. Once you have a line of fit, you can find an equation of the line.

In this lesson, you will use a graphical method to find a line of fit. In Extend Lesson 4-6, you will use a graphing calculator to find a line of fit. The calculator uses a statistical method to find the line that most closely approximates the data. This line is called the **best-fit line.**

Math Online **Extra Examples at algebra1.com**

 **Real-World Link**

The difference in height between the top of the hill and the bottom is the vertical drop.

Source: ultimaterollercoaster.com

Real-World EXAMPLE

2 ROLLER COASTERS **The table shows the largest vertical drops of nine roller coasters in the United States and the number of years after 1988 that they were opened.**

Years since 1988	1	3	5	8	12	12	12	13	15
Vertical Drop (ft)	151	155	225	230	306	300	255	255	400

Source: ultimaterollercoaster.com

a. Draw a scatter plot and determine what relationship exists, if any, in the data.

As the number of years increases, the vertical drop of roller coasters increases. There is a positive correlation between the variables.

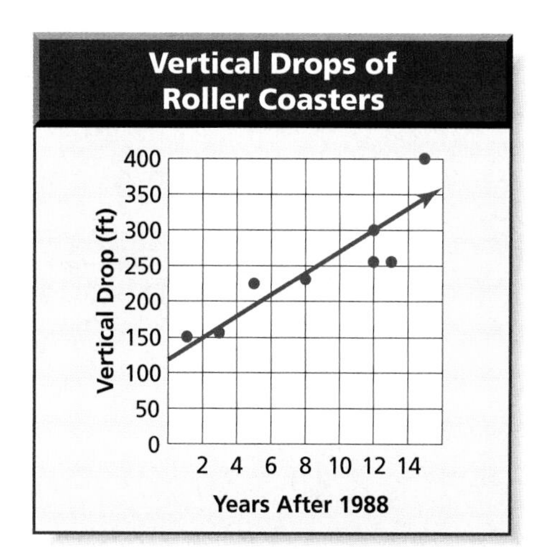

b. Draw a line of fit.

No one line will pass through all of the data points. Draw a line that passes close to the points. One line of fit is shown in the scatter plot.

c. Write the slope-intercept form of an equation for the line of fit.

The line of fit shown above passes through (2, 150) and (12, 300).

Step 1 Find the slope.

$m = \frac{y_2 - y_1}{x_2 - x_1}$ Slope formula

$= \frac{300 - 150}{12 - 2}$ $(x_1, y_1) = (2, 150)$, $(x_2, y_2) = (12, 300)$

$= \frac{150}{10}$ or 15 Simplify.

Step 2 Use $m = 15$ and either the point-slope form or the slope-intercept form to write the equation of the line of fit.

$y - y_1 = m(x - x_1)$

$y - 150 = 15(x - 2)$

$y - 150 = 15x - 30$

$y = 15x + 120$

A slope of 15 means that the vertical drops increased an average of 15 feet per year. A y-intercept of 120 means that a roller coaster that opened in 1988 has a vertical drop of approximately 120 feet.

Study Tip

Lines of Fit

When you use the graphical method, the line of fit is an approximation. So, you may draw another line of fit using other points that is equally valid. Some valid lines of fit may not contain any of the data points.

CHECK Your Progress

EAGLES **The table shows an estimate for the number of bald eagle pairs in the United States for certain years since 1985.**

Year since 1985	3	5	7	9	11	14	15
Bald Eagle Pairs	2500	3000	3700	4500	5000	5800	6500

Source: U.S. Fish and Wildlife Service

2A. Draw a scatter plot and determine what relationship exists, if any, in the data.

2B. Draw a line of fit for the scatter plot.

2C. Write the slope-intercept form of an equation for the line of fit.

Concepts in Motion
Interactive Lab algebra1.com

Online **Personal Tutor at** algebra1.com

Linear extrapolation is used to predict values that are *outside* the range of the data. You can also use a linear equation to predict values that are *inside* the range of the data. This is called **linear interpolation.**

Real-World EXAMPLE

> **Study Tip**
>
> **Limits of the Model**
> Notice that the equation cannot be used in extreme cases. For example, it is not reasonable that a roller coaster be 1000 feet tall.

3 ROLLER COASTERS Use the equation for the line of fit in Example 2. Estimate the largest vertical drop of a roller coaster that is opened in 2007.

$y = 15x + 120$ Original equation

$= 15(19) + 120$ Replace x with 2007 − 1988 or 19.

$= 405$ Simplify.

In 2007, the largest vertical drop is estimated to be 405 feet.

CHECK Your Progress

3. EAGLES Use the equation for the line of fit in Check Your Progress 2B on page 229 to estimate the number of bald eagle pairs in 2008.

CHECK Your Understanding

Example 1 (p. 228)

Determine whether each graph shows a *positive correlation,* a *negative correlation,* or *no correlation.* If there is a positive or negative correlation, describe its meaning in the situation.

1.

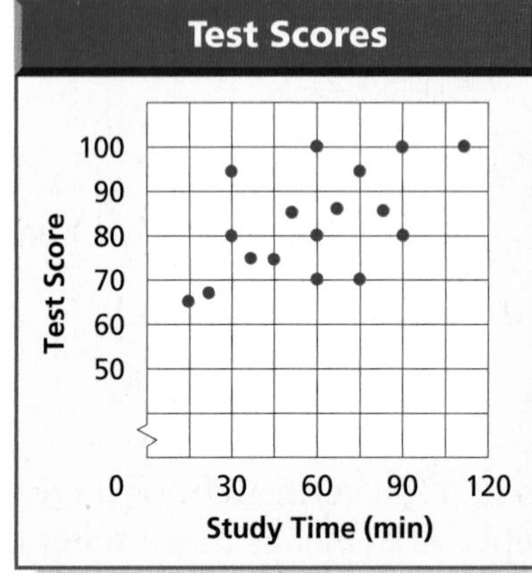

2.

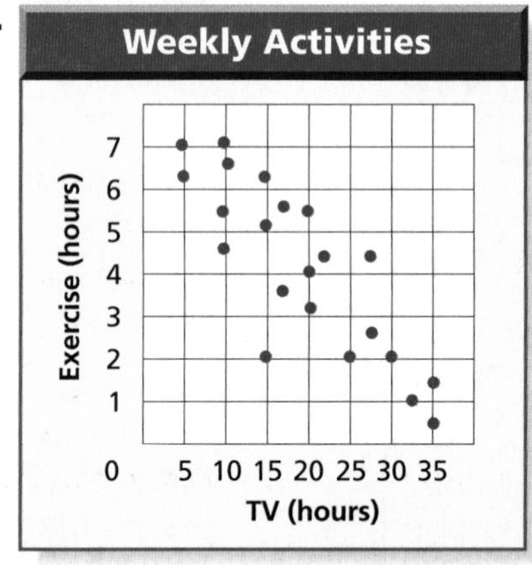

BIOLOGY For Exercises 3–7, use the table that shows the average body temperature in degrees Celsius of nine insects at a given air temperature.

Temperature (°C)									
Air	25.7	30.4	28.7	31.2	31.5	26.2	30.1	31.5	18.2
Body	27.0	31.5	28.9	31.0	31.5	25.6	28.4	31.7	18.7

Example 2 (p. 229)

3. Draw a scatter plot and determine what relationship exists, if any, in the data.

4. Draw a line of fit for the scatter plot.

5. Write the slope-intercept form of an equation for the line of fit.

Example 3 (p. 230)

6. Predict the body temperature of an insect if the air temperature is 40.2°C.

7. Suppose the air temperature is −50°C. According to your judgment, do you think the equation can give a reasonable estimate for the body temperature of an insect? Explain.

Exercises

HOMEWORK HELP	
For Exercises	See Examples
8–11	1
12–27	2, 3

Determine whether each graph shows a *positive correlation*, a *negative correlation*, or *no correlation*. If there is a positive or negative correlation, describe its meaning in the situation.

8.

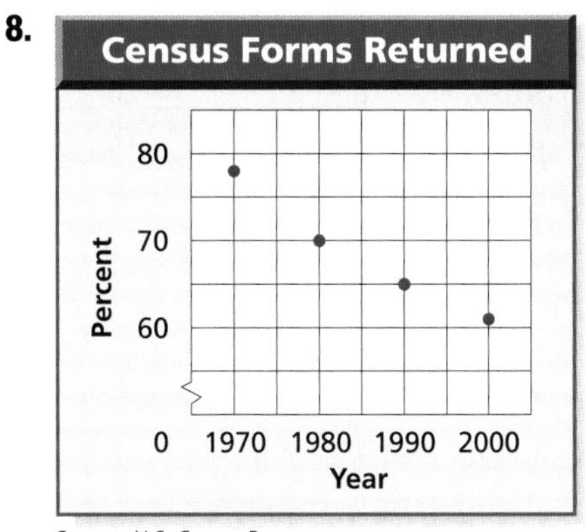

Source: U.S. Census Bureau

9.

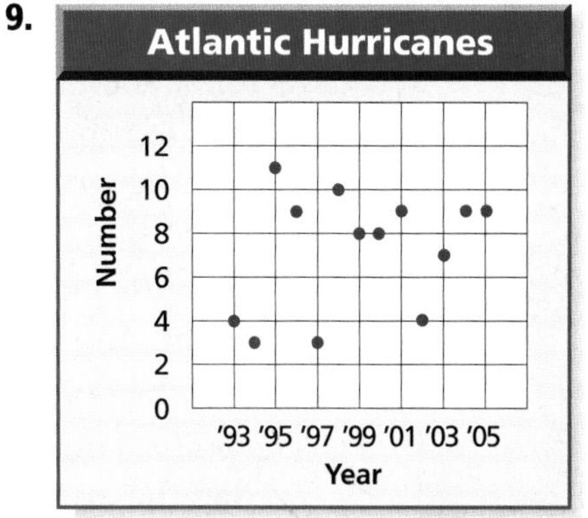

Source: *USA TODAY*

10.

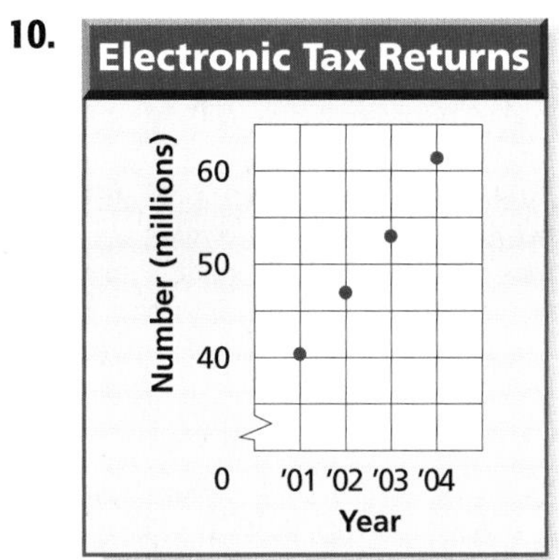

Source: IRS

11.

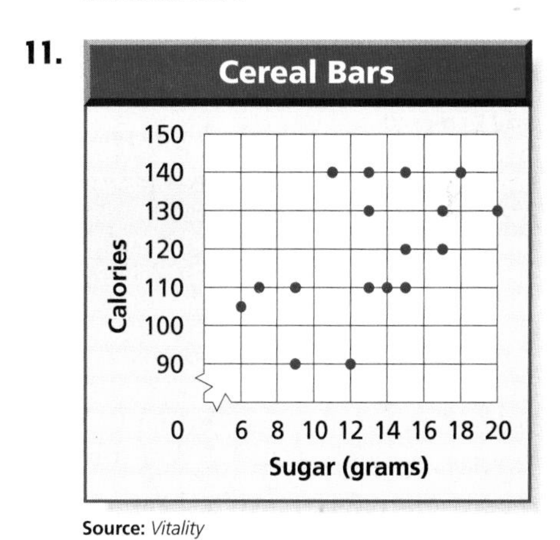

Source: *Vitality*

BIRDS **For Exercises 12–14, refer to the graph at the top of page 227 about whooping cranes.**

12. Use the points (2000, 170) and (2003, 185) to write an equation for a line of fit.

13. Predict the number of whooping cranes in 2008.

14. Is it reasonable to use the equation to estimate the number of whooping cranes in any year, such as in 1900? Explain.

USED CARS **For Exercises 15–17, use the scatter plot that shows the ages and prices of used cars from classified ads.**

15. Use the points (2, 9600) and (5, 6000) to write the slope-intercept form of an equation for the line of fit shown in the scatter plot.

16. Predict the price of a car that is 7 years old.

17. Can you use the equation to make a decision about buying a used car that is 50 years old? Explain.

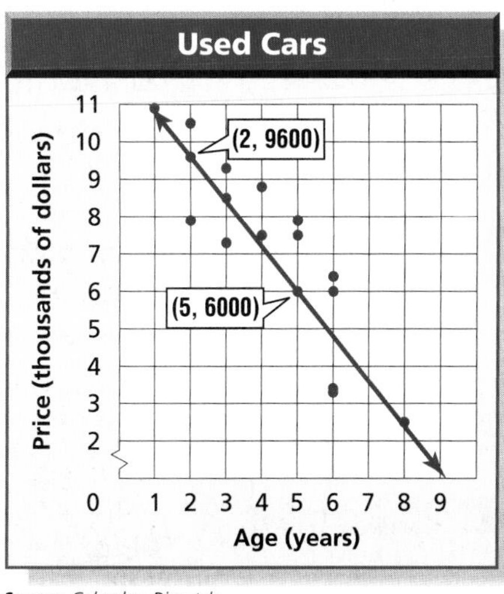

Source: *Columbus Dispatch*

Cross-Curricular Project

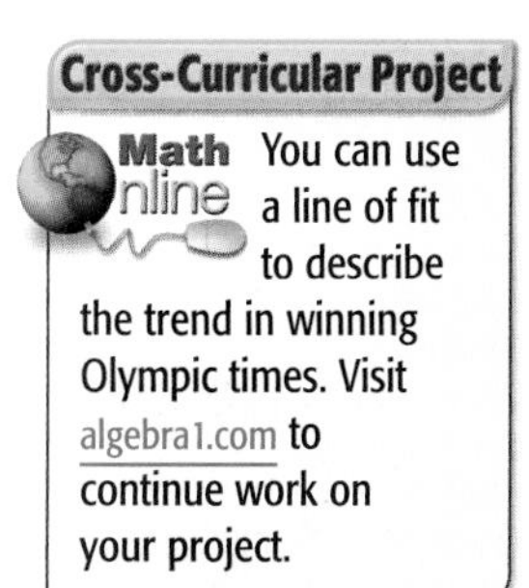

You can use a line of fit to describe the trend in winning Olympic times. Visit algebra1.com to continue work on your project.

PHYSICAL SCIENCE **For Exercises 18–22, use the following information.**
Hydrocarbons are composed of only carbon and hydrogen atoms. The table gives the number of carbon atoms and the boiling points for several hydrocarbons.

Hydrocarbons			
Name	Formula	Number of Carbon Atoms	Boiling Point (°C)
Ethane	C_2H_6	2	−89
Propane	C_3H_8	3	−42
Butane	C_4H_{10}	4	−1
Hexane	C_6H_{12}	6	69
Octane	C_8H_{18}	8	126

18. Draw a scatter plot comparing the numbers of carbon atoms to the boiling points.

19. Draw a line of fit for the data.

20. Write the slope-intercept form of an equation for the line of fit.

21. Predict the boiling point for pentane (C_5H_{12}), which has 5 carbon atoms.

22. The boiling point of heptane is 98.4°C. Use the equation of the line of fit to predict the number of carbon atoms in heptane.

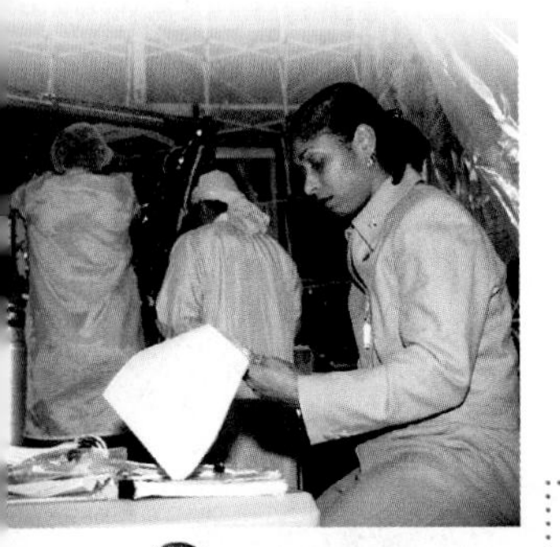

Real-World Career
Aereospace Engineers
Aerospace engineers specialize in a type of aircraft such as commercial airplanes, military aircraft, and spacecraft.

For more information, go to algebra1.com.

SPACE **For Exercises 23–27, use the table that shows the amount the United States government has spent on space and other technologies in selected years.**

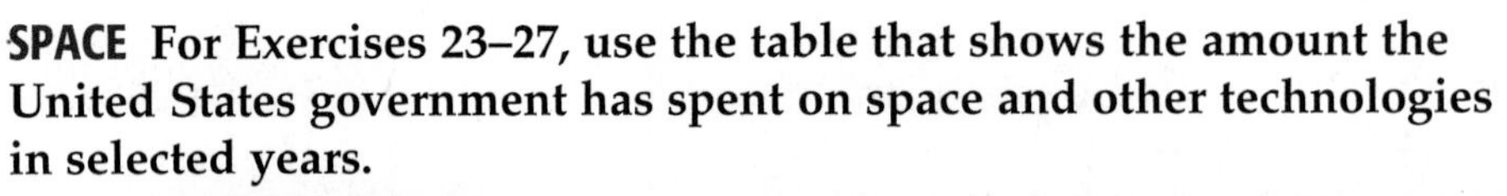

Federal Spending on Space and Other Technologies									
Year	1980	1985	1990	1995	1996	1997	1998	1999	2004
Spending (billions of dollars)	4.5	6.6	11.6	12.6	12.7	13.1	12.9	12.4	15.4

Source: U.S. Office of Management and Budget

23. Draw a scatter plot and determine what relationship, if any, exists in the data.

24. Draw a line of fit for the scatter plot.

25. Let x represent the number of years since 1980. Let y represent spending in billions of dollars. Write the slope-intercept form of the equation for the line of fit.

26. Predict the amount that will be spent on space and other technologies in 2007.

27. Make a critical judgment about the amount that will be spent on space and other technologies in the next century. Would the equation that you wrote be a useful model?

GEOGRAPHY **For Exercises 28–31, use the following information.**
The *latitude* of a place on Earth is the measure of its distance from the equator.

28. MAKE A CONJECTURE What do you think is the relationship between a city's latitude and its January temperature?

29. RESEARCH Use the Internet or other reference to find the latitude of 15 cities in the northern hemisphere and the corresponding January mean temperatures.

30. Make a scatter plot and draw a line of fit for the data.

31. Write an equation for the line of fit.

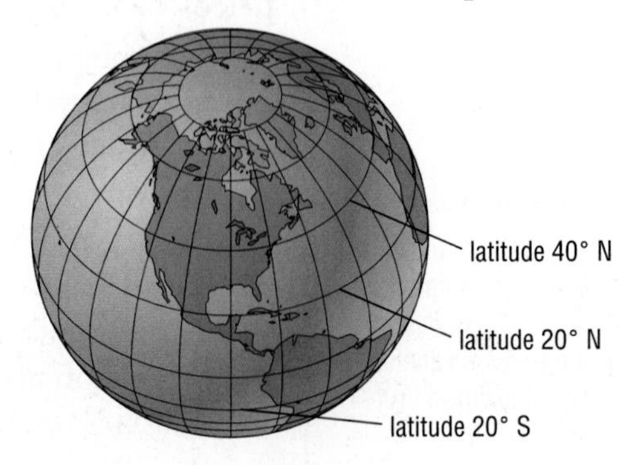

H.O.T. Problems

32. OPEN ENDED Sketch scatter plots that have each type of correlation: positive, negative, and none. Associate each graph with a real-life situation.

33. REASONING Compare and contrast interpolation and extrapolation.

34. CHALLENGE A test contains 20 true-false questions. Draw a scatter plot that shows the relationship between the number of correct answers x and the number of incorrect answers y.

35. ***Writing in Math*** Draw a scatter plot that shows a person's height and his or her age, with a description of any trends. Explain how you could use the scatter plot to predict a person's age given his or her height. How can the information from a scatter plot be used to identify trends and make decisions?

STANDARDIZED TEST PRACTICE

36. REVIEW Mr. Hernandez collected data on the heights and average stride lengths of a random sample of students in grades 8, 9, and 10. He then graphed the data on a scatter plot. What correlation did he most likely see?

A positive　　**C** constant

B negative　　**D** no correlation

37. Which equation *best* fits the data in the table?

x	y
1	5
2	7
3	7
4	11

F $y = x + 4$

G $y = 2x + 3$

H $y = 7$

J $y = 4x - 5$

Spiral Review

Write the point-slope form of an equation for the line that passes through each point with the given slope. (Lesson 4-5)

38. $(1, -2)$; $m = 3$　　**39.** $(-2, 3)$; $m = -2$　　**40.** $(-3, -3)$; $m = 1$

41. COMMUNICATION A calling plan charges a rate per minute plus a flat fee. A 10-minute call to the Czech Republic costs \$3.19. A 15-minute call costs \$4.29. Write a linear equation in slope-intercept form to represent the total cost C of an m-minute call. Then find the cost of a 12-minute call. (Lesson 4-4)

42. EXERCISE The statement below was found in *Healthy Fun* magazine.

A typical 100-pound kid can burn more than 350 calories per hour riding a bike.

At this rate, about how many Calories would be burned riding a bike 25 minutes? (Lesson 2-6)

GET READY for the Next Lesson

PREREQUISITE SKILL Write the multiplicative inverse of each number. (Pages 698–699)

43. 10　　**44.** -1　　**45.** $\frac{2}{3}$　　**46.** $-\frac{1}{9}$　　**47.** $\frac{3}{4}$

EXTEND 4-6

Graphing Calculator Lab

Regression and Median-Fit Lines

One type of equation of best-fit you can find is a linear **regression equation.**

ACTIVITY 1

MUSIC The table shows the percent of music sales that were made on the Internet in the United States for the period 1997–2004.

Year	1997	1998	1999	2000	2001	2002	2003	2004
Sales	0.3	1.1	2.4	3.2	2.9	3.4	5.0	5.9

Source: Recording Industry Association of America

Find and graph a linear regression equation. Then predict the percent of music sales that will be made on the Internet in 2010.

Step 1 **Find a regression equation.**

Enter the years in L1 and the earnings in L2. Find the regression equation.

KEYSTROKES: [STAT] [ENTER] 1997 [ENTER] … [▶] 0.3 [ENTER] … [STAT] [▶] 4 [ENTER]

The equation is about $y = 0.73x - 1459.25$.

r is the **linear correlation coefficient.** The closer the absolute value of r is to 1, the better the equation models the data.

Step 2 **Graph the regression equation.**

Use STAT PLOT to graph the scatter plot.

KEYSTROKES: [2nd] [STAT] [ENTER] [ENTER]

Copy the equation to the Y= list and graph.

KEYSTROKES: [Y=] [VARS] 5 [▶] [▶] 1 [GRAPH]

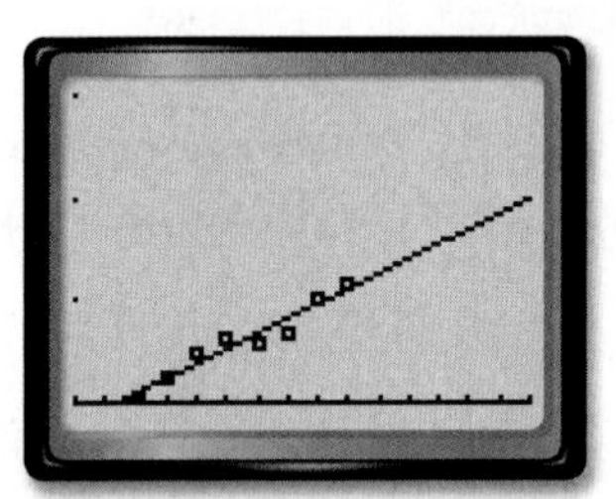

[1995, 2010] scl: 1 by [0, 15] scl: 5

Step 3 **Predict using the regression equation.**

Find y when $x = 2010$.

KEYSTROKES: [2nd] [CALC] 1 2010 [ENTER]

According to the regression equation, in 2010 about 9.97% of music sales will be made on the Internet.

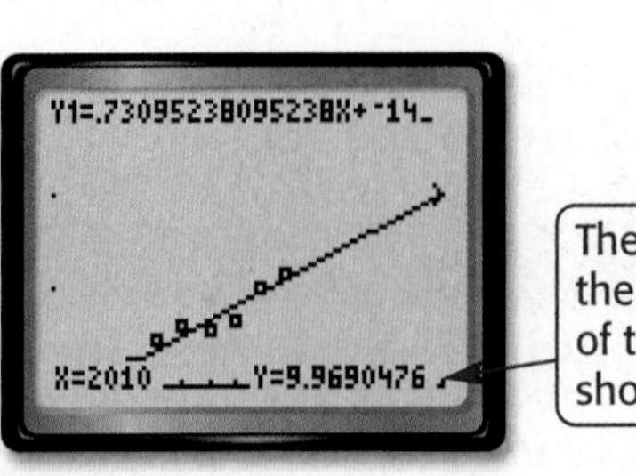

Math Online Other Calculator Keystrokes at algebra1.com

A second type of best-fit line that can be found using a graphing calculator is a **median-fit line.** The equation of a median-fit line is calculated using the medians of the coordinates of the data points.

ACTIVITY 2

Find and graph a median-fit equation for the data on music sales. Then predict the percent of sales that will be made on the Internet in 2010. Compare this prediction to the one made using the regression equation.

Step 1 Find a median-fit equation.

The data are already in Lists 1 and 2. Find the median-fit equation by using *Med-Med* on the STAT CALC menu.

KEYSTROKES: [STAT] [▶] 3 [ENTER]

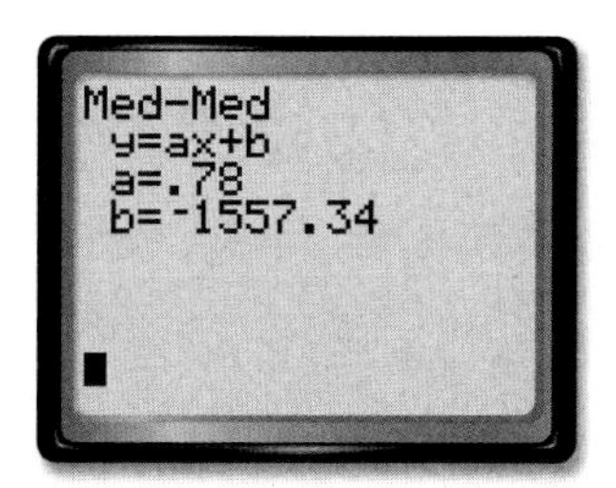

The median-fit equation is $y = 0.78x - 1557.34$.

Step 2 Graph the median-fit equation.

Copy the equation to the Y= list and graph.

KEYSTROKES: [Y=] [CLEAR] [VARS] 5 [▶] [▶] 1 [GRAPH]

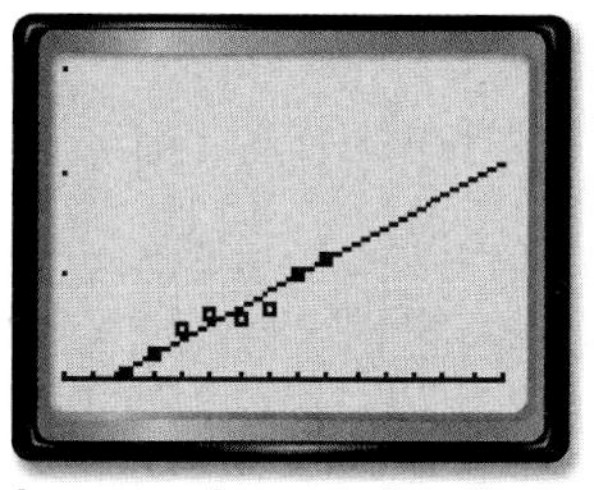

[1995, 2010] scl: 1 by [0, 15] scl: 5

Step 3 Predict using the median-fit equation.

KEYSTROKES: [2nd] [CALC] 1 2010 [ENTER]

According to the median-fit equation, about 10.46% of music sales will be made on the Internet in 2010. This is slightly more than the predicted value found using the regression equation.

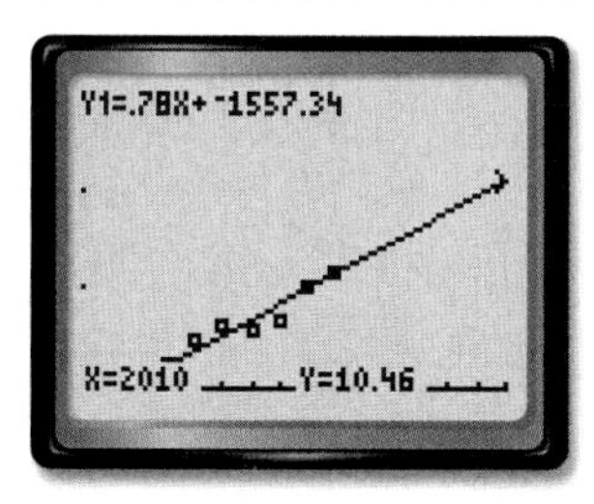

ANALYZE THE RESULTS

Refer to the data on roller coasters in Example 2 on page 229.

1. Find regression and median-fit equations for the data.

2. What is the correlation coefficient of the regression equation? What does it tell you about the data?

3. Use the regression and median-fit equations to predict the largest vertical drop for a roller coaster in 2007. Compare these to the number found in Example 3 on page 230.

4-7 Geometry: Parallel and Perpendicular Lines

Main Ideas

- Write an equation of the line that passes through a given point, parallel to a given line.
- Write an equation of the line that passes through a given point, perpendicular to a given line.

New Vocabulary

parallel lines
perpendicular lines

GET READY for the Lesson

The graph shows a family of linear graphs whose slope is 1. Note that the lines do not appear to intersect.

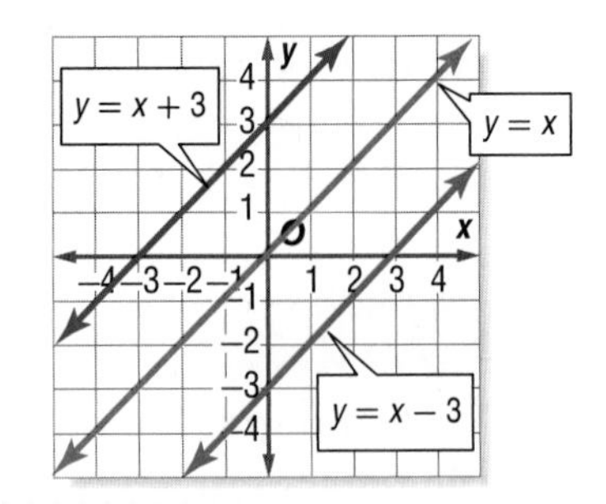

Parallel Lines Lines in the same plane that do not intersect are called **parallel lines.** Parallel lines have the same slope.

KEY CONCEPT *Parallel Lines in a Coordinate Plane*

Words If two nonvertical lines have the same slope, then they are parallel. All vertical lines are parallel.

Model

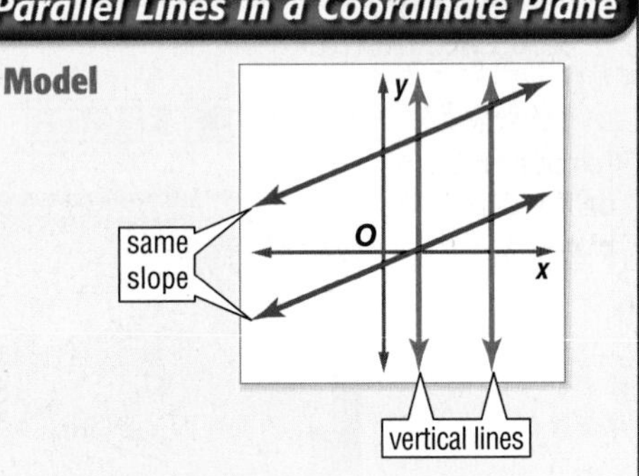

You can write the equation of a line parallel to a given line if you know a point on the line and an equation of the given line.

EXAMPLE Parallel Line Through a Given Point

1 **Write the slope-intercept form of an equation for the line that passes through (−1, −2) and is parallel to the graph of $y = -3x - 2$.**

The line parallel to $y = -3x - 2$ has the same slope, −3. Replace m with −3, and (x_1, y_1) with (−1, −2) in the point-slope form.

$y - y_1 = m(x - x_1)$	Point-slope form
$y - (-2) = -3[x - (-1)]$	Replace m with −3, y_1 with −2, and x_1 with −1.
$y + 2 = -3(x + 1)$	Simplify.
$y + 2 = -3x - 3$	Distributive Property
$y + 2 - 2 = -3x - 3 - 2$	Subtract 2 from each side.
$y = -3x - 5$	Write the equation in slope-intercept form.

CHECK Your Progress

1. Write the point-slope form of an equation for the line that passes through (4, −1) and is parallel to the graph of $y = \frac{1}{4}x + 7$.

Perpendicular Lines Lines that intersect at right angles are called **perpendicular lines.** There is a relationship between the slopes of perpendicular lines.

ALGEBRA LAB

Perpendicular Lines

- A scalene triangle is one in which no two sides are equal in length. Cut out a scalene right triangle ABC so that $\angle C$ is a right angle. Label the vertices and the sides as shown.

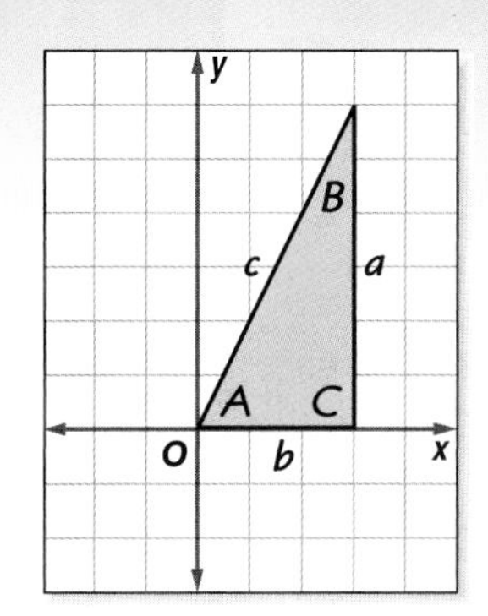

- Draw a coordinate plane on grid paper. Place $\triangle ABC$ on the coordinate plane so that A is at the origin and side b lies along the positive x-axis.

ANALYZE THE RESULTS

1. Name the coordinates of B.
2. What is the slope of side c?
3. Rotate the triangle 90° counterclockwise so that A is still at the origin and side b is along the positive y-axis. Name the coordinates of B.
4. What is the slope of side c?
5. Repeat the activity for two other scalene right triangles.
6. For each triangle and its rotation, what is the relationship between the first position of side c and the second?
7. For each triangle and its rotation, describe the relationship between the coordinates of B in the first and second positions.
8. Describe the relationship between the slopes of c in each position.

MAKE A CONJECTURE

9. Describe the relationship between the slopes of any two perpendicular lines.

Study Tip

Look Back

To review **rotations on the coordinate plane,** see Lesson 3-2.

The results of the Algebra Lab suggests an important property of perpendicular lines.

Review Vocabulary

Reciprocals $\frac{1}{4}$ and 4 are reciprocals because their product is 1. (Lesson 1-4)

KEY CONCEPT *Perpendicular Lines in a Coordinate Plane*

Words If the product of the slopes of two nonvertical lines is –1, then the lines are perpendicular. In this case, the slopes are *opposite reciprocals* of each other. Vertical lines and horizontal lines are also perpendicular.

Model

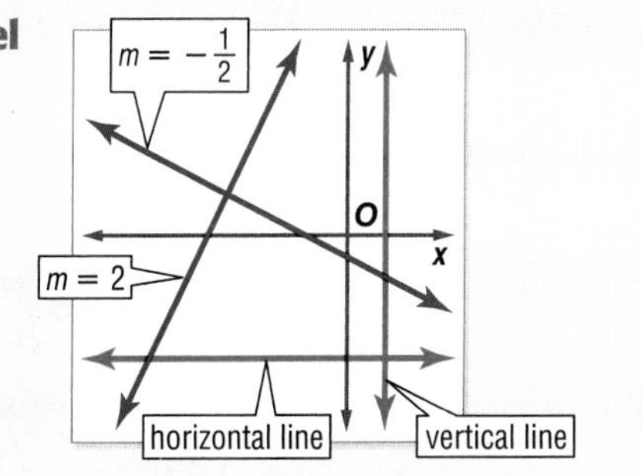

Real-World EXAMPLE — Determine Whether Lines are Perpendicular

2 **DESIGN The outline of a new company logo is shown on a coordinate plane. Is $\angle DFE$ a right angle?**

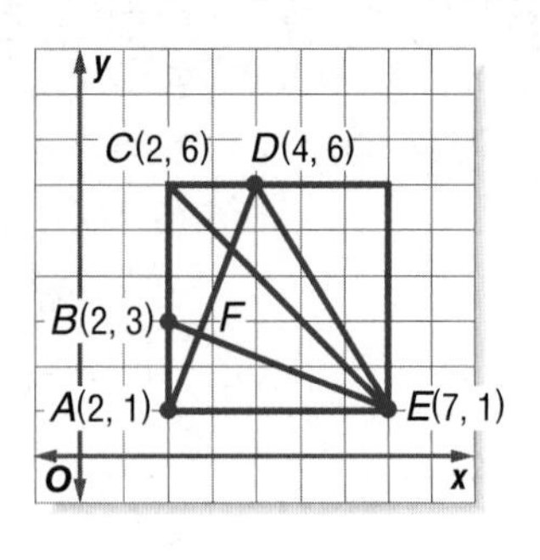

If $\overline{BE}$ and $\overline{AD}$ are perpendicular, then $\angle DFE$ is a right angle. Find the slopes of $\overline{BE}$ and $\overline{AD}$.

slope of $\overline{BE}$: $m = \frac{1-3}{7-2}$ or $-\frac{2}{5}$

slope of $\overline{AD}$: $m = \frac{6-1}{4-2}$ or $\frac{5}{2}$

The line segments are perpendicular because $-\frac{2}{5} \cdot \frac{5}{2} = -1$. Therefore, $\angle DFE$ is a right angle.

CHECK Your Progress

2. **CONSTRUCTION** On the plans for a tree house, a beam represented by $\overline{QR}$ has endpoints $Q(-6, 2)$ and $R(-1, 8)$. A beam represented by $\overline{ST}$ has endpoints $S(-3, 6)$ and $T(-8, 5)$. Are the beams perpendicular? Explain.

You can write the equation of a line perpendicular to a given line if you know a point on the line and the equation of the given line.

EXAMPLE — Perpendicular Line Through a Given Point

3 **Write the slope-intercept form of an equation for the line that passes through $(-3, -2)$ and is perpendicular to the graph of $x + 4y = 12$.**

Step 1 Find the slope of the given line.

$x + 4y = 12$	Original equation
$x + 4y - x = 12 - x$	Subtract $1x$ from each side.
$4y = -1x + 12$	Simplify.
$\frac{4y}{4} = \frac{-1x + 12}{4}$	Divide each side by 4.
$y = -\frac{1}{4}x + 3$	Simplify.

Step 2 The slope of the given line is $-\frac{1}{4}$. So, the slope of the line perpendicular to this line is the opposite reciprocal of $-\frac{1}{4}$, or 4.

Step 3 Use the point-slope form to find the equation.

$y - y_1 = m(x - x_1)$	Point-slope form
$y - (-2) = 4[x - (-3)]$	$(x_1, y_1) = (-3, -2)$ and $m = 4$
$y + 2 = 4(x + 3)$	Simplify.
$y + 2 = 4x + 12$	Distributive Property
$y + 2 - 2 = 4x + 12 - 2$	Subtract 2 from each side.
$y = 4x + 10$	Simplify.

Study Tip

Graphing Calculator

If you graph perpendicular lines on a graphing calculator, the lines will not appear to be perpendicular if the scales on the axes are not set correctly. After graphing, press ZOOM 5 to set the axes for a correct representation.

CHECK Your Progress

3. Write the slope-intercept form of an equation for the line that passes through $(4, 7)$ and is perpendicular to the graph of $y = \frac{2}{3}x - 1$.

Online Personal Tutor at algebra1.com

EXAMPLE Perpendicular Line Through a Given Point

4 **Write an equation in slope-intercept form for a line perpendicular to the graph of $y = -\frac{1}{3}x + 2$ that passes through the x-intercept of that line.**

Step 1 Find the slope of the perpendicular line. The slope of the given line is $-\frac{1}{3}$, therefore a perpendicular line has slope 3.

Step 2 Find the x-intercept of the given line.

$y = -\frac{1}{3}x + 2$	Original equation
$0 = -\frac{1}{3}x + 2$	Replace y with 0.
$-2 = -\frac{1}{3}x$	Subtract 2 from each side.
$6 = x$	Multiply each side by -3.

The x-intercept is at (6, 0).

Step 3 Substitute the slope and the given point into the point-slope form of a linear equation. Then write in slope-intercept form.

$y - y_1 = m(x - x_1)$	Point-slope form
$y - 0 = 3(x - 6)$	Replace x_1 with 6, y_1 with 0, and m with 3.
$y = 3x - 18$	Distributive Property

CHECK Your Progress

4. Write an equation in slope-intercept form for a line perpendicular to the graph of $3x + 2y = 8$ that passes through the y-intercept of that line.

CHECK Your Understanding

Example 1 (pp. 236–237) **Write the slope-intercept form of an equation for the line that passes through the given point and is parallel to the graph of each equation.**

1.

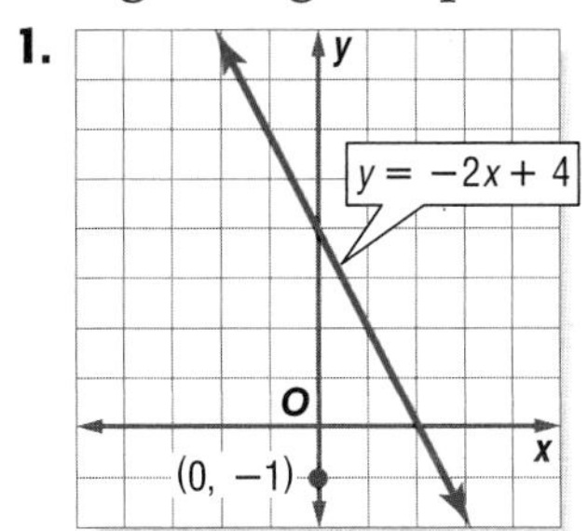

2.

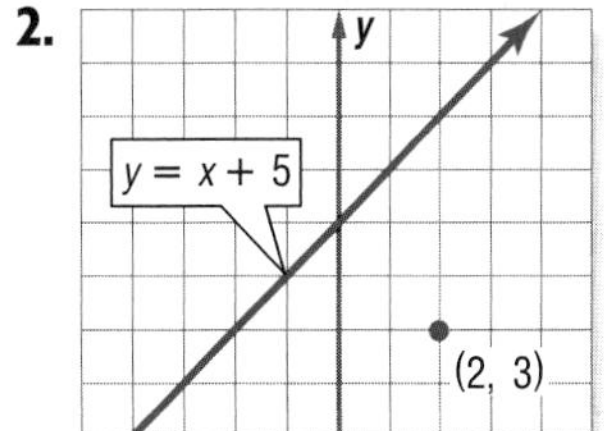

3. $(1, -3), y = 2x - 1$

4. $(-2, 2), -3x + y = 4$

Example 2 (p. 238)

5. GARDENS A garden is in the shape of a quadrilateral with vertices $A(-2, 1)$, $B(3, -3)$, $C(5, 7)$, and $D(-3, 4)$. Two paths represented by $\overline{AC}$ and $\overline{BD}$ cut across the garden. Are the paths perpendicular? Explain.

Example 3 (p. 238) **Write the slope-intercept form of an equation for the line that passes through the given point and is perpendicular to the graph of the equation.**

6. $(-3, 1), y = \frac{1}{3}x + 2$

7. $(6, -2), y = \frac{3}{5}x - 4$

8. $(2, -2), 2x + y = 5$

Example 4 (p. 239)

9. Write the slope-intercept form for an equation of a line that is perpendicular to the graph of $y = 6x - 6$ and passes through the x-intercept of that line.

Exercises

HOMEWORK HELP	
For Exercises	See Examples
10–16	1
18, 19	2
17, 20–25	3
26, 27	4

Write the slope-intercept form of an equation for the line that passes through the given point and is parallel to the graph of each equation.

10. $(-3, 2), y = x - 6$

11. $(2, -1), y = 2x + 2$

12. $(-5, -4), y = \frac{1}{2}x + 1$

13. $(3, 3), y = \frac{2}{3}x - 1$

14. $(-4, -3), y = -\frac{1}{3}x + 3$

15. $(-1, 2), y = -\frac{1}{2}x - 4$

16. GEOMETRY A *parallelogram* is a quadrilateral in which opposite sides are parallel. Determine whether $ABCD$ is a parallelogram. Explain your reasoning.

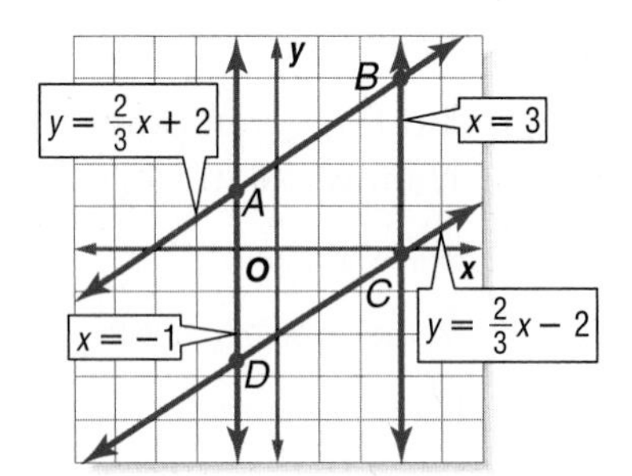

17. GEOMETRY The line with equation $y = 3x - 4$ contains side $\overline{AC}$ of right triangle ABC. If the vertex of the right angle C is at $(3, 5)$, what is an equation of the line that contains side $\overline{BC}$?

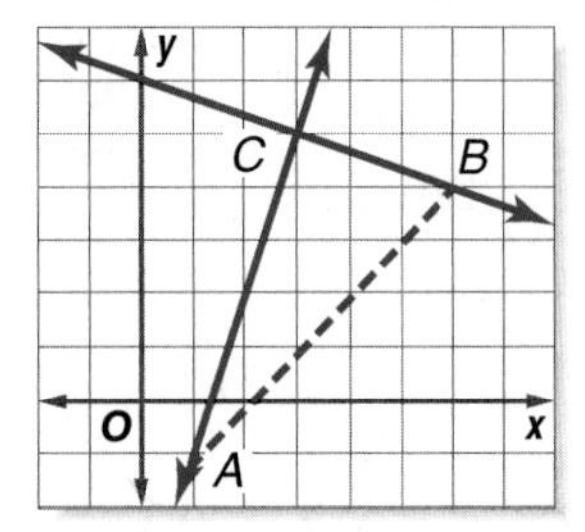

18. Determine whether $y = -6x + 4$ and $y = \frac{1}{6}x$ are perpendicular. Explain.

19. MAPS On a map, Elmwood Drive passes through $R(4, -11)$ and $S(0, -9)$ and Taylor Road passes through $J(6, -2)$ and $K(4, -5)$. If they are straight lines, are the two streets perpendicular? Explain.

Write the slope-intercept form of an equation for the line that passes through the given point and is perpendicular to the graph of the equation.

20. $(-2, 0), y = x - 6$

21. $(1, 1), y = 4x + 6$

22. $(-3, 1), y = -3x + 7$

23. $(1, -3), y = \frac{1}{2}x + 4$

24. $(-2, 7), 2x - 5y = 3$

25. $(4, 7), 3y - 2x = -3$

26. Find an equation for the line that has a y-intercept of -2 and is perpendicular to the graph of $3x + 6y = 2$.

27. Write an equation of the line that is perpendicular to the line through $(9, 10)$ and $(3, -2)$ and passes through the x-intercept of that line.

Determine whether the graphs of each pair of equations are *parallel*, *perpendicular*, or *neither*.

28. $y = -2x + 11$
$y + 2x = 23$

29. $3y = 2x + 14$
$2x - 3y = 2$

30. $y = -5x$
$y = 5x - 18$

EXTRA PRACTICE
See pages 726, 747.

Math Online
Self-Check Quiz at algebra1.com

31. GEOMETRY Determine the relationship between the diagonals $\overline{AC}$ and $\overline{BD}$ of square $ABCD$ with $A(1, 3)$, $B(3, -1)$, $C(-1, -3)$ and $D(-3, 1)$.

32. Write an equation of the line that is parallel to the graph of $y = 7x - 3$ and passes through the origin.

H.O.T. Problems

33. CHALLENGE The line that passes through the points $(3a, 4)$ and $(-1, 2)$ is parallel to the graph of $-4x + 2y = 6$. Find the value of a.

34. OPEN ENDED Draw two segments on the coordinate plane that appear to be perpendicular. Describe how you could check your accuracy without measuring.

35. ***Writing in Math*** Illustrate how you can determine whether two lines are parallel or perpendicular. Write an equation for a graph parallel to the line graphed at the right, and an equation with a graph perpendicular to the line graphed. Explain.

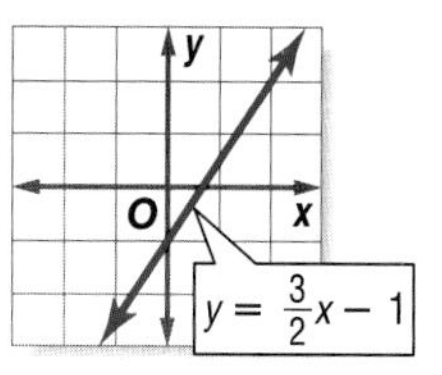

STANDARDIZED TEST PRACTICE

36. Which equation represents a line that is perpendicular to the graph and passes through the point at (2, 0)?

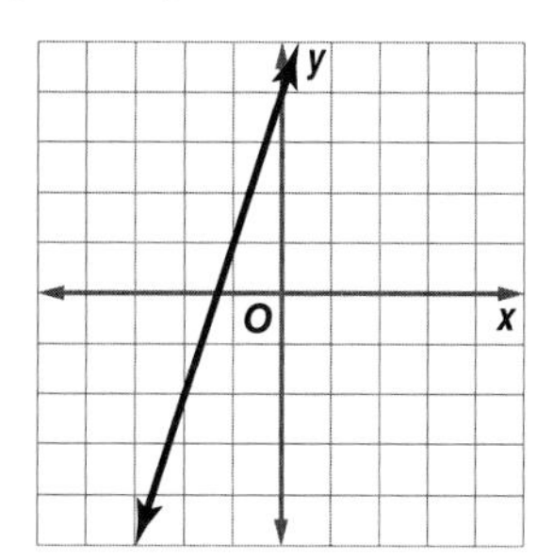

A $y = 3x - 6$

B $y = -3x + 6$

C $y = -\frac{1}{3}x + \frac{2}{3}$

D $y = \frac{1}{3}x - \frac{2}{3}$

37. REVIEW If $\triangle JKL$ is similar to $\triangle JNM$ what is the length of side a?

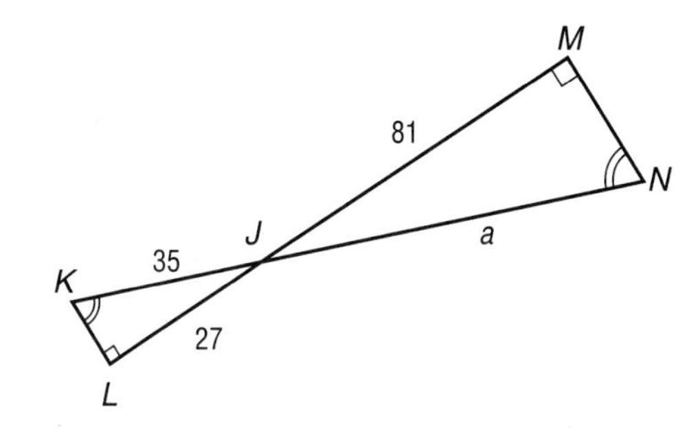

F 62.5 units

G 105 units

H 125 units

J 155.5 units

Spiral Review

38. TECHNOLOGY Would a scatter plot showing the relationship between the year and the amount of memory available on a personal computer manufactured that year show a *positive*, *negative*, or *no* correlation? (Lesson 4-6)

Write the point-slope form of an equation for the line that passes through each point with the given slope. (Lesson 4-5)

39. $(3, 5), m = -2$

40. $(-4, 7), m = 5$

41. $(-1, -3), m = -\frac{1}{2}$

42. ARCHITECTURE An architect is building a scale model of a sports complex. If the tallest building of the complex is 160 feet and the scale is 1 inch = 8 feet, how tall is the highest point of the scale model? (Lesson 2-6)

43. Solve $\frac{6c - t}{7} = b$ for c. (Lesson 2-8)

CHAPTER 4 Study Guide and Review

FOLDABLES Study Organizer — GET READY to Study

Be sure the following Key Concepts are noted in your Foldable.

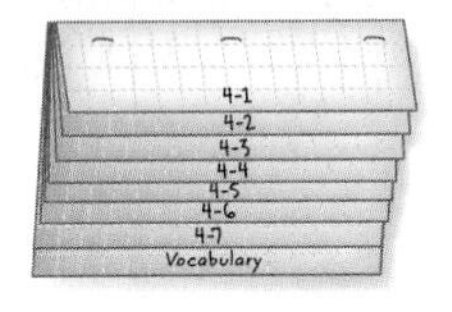

Key Concepts

Rate of Change and Slope (Lesson 4-1)

- If x is the independent variable and y is the dependent variable, then rate of change equals $\frac{\text{change in } y}{\text{change in } x}$.
- The slope of a line is the ratio of the rise to the run; $m = \frac{y_2 - y_1}{x_2 - x_1}$.

Slope and Direct Variation (Lesson 4-2)

- A direct variation is described by an equation of the form $y = kx$, where $k \neq 0$.

Linear Equations in Slope-Intercept and Point-Slope Form (Lessons 4-3, 4-4, 4-5)

- The linear equation $y = mx + b$ is in slope-intercept form, where m = slope and b = y-intercept.
- The linear equation $y - y_1 = m(x - x_1)$ is in point-slope form, where (x_1, y_1) is a point on a nonvertical line and m is the slope.

Scatter Plots and Lines of Fit (Lesson 4-6)

- A line of fit describes the trend of the data, and its equation can be used to make predictions.
- The correlation between x and y is positive if as x increases, y increases, and negative if as x increases, y decreases. There is no correlation between x and y if no relationship exists between x and y.

Parallel and Perpendicular Lines (Lesson 4-7)

- Two nonvertical lines are parallel if they have the same slope. Two nonvertical lines are perpendicular if the product of their slopes is -1.

Key Vocabulary

- best-fit line (p. 228)
- constant of variation (p. 196)
- direct variation (p. 196)
- family of graphs (p. 197)
- linear extrapolation (p. 216)
- linear interpolation (p. 230)
- line of fit (p. 228)
- parallel lines (p. 236)
- parent graph (p. 197)
- perpendicular lines (p. 237)
- point-slope form (p. 220)
- rate of change (p. 187)
- scatter plot (p. 227)
- slope (p. 189)
- slope-intercept form (p. 204)

Vocabulary Check

State whether each sentence is *true* or *false*. If *false*, replace the underlined word or number to make a true sentence.

1. Any two points on a line can be used to determine the slope.
2. The equation $y - 2 = -3(x - 1)$ is written in point-slope form.
3. The equation of a vertical line can be written in slope-intercept form.
4. When you use a linear equation to predict values that are beyond the range of the data, you are using linear interpolation.
5. The lines with equations $y = -5x + 7$ and $y = -5x - 6$ are perpendicular.
6. The lines with the equations $4x - y = 8$ and $y = -\frac{1}{4}x$ are parallel.
7. The slope of the line $y = 5$ is 5.
8. The line that most closely approximates a set of data is called a best-fit line.
9. An equation of the form $y = kx$, where $k \neq 0$, describes a linear extrapolation.
10. The y-intercept of the equation $3x - 2y = 24$ is $\frac{3}{2}$.

Lesson-by-Lesson Review

4–1 Rate of Change and Slope (pp. 187–195)

Find the rate of change represented in each table or graph.

11.

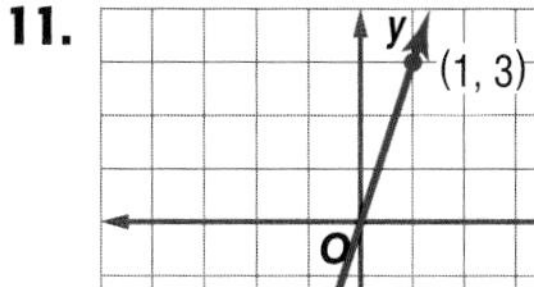
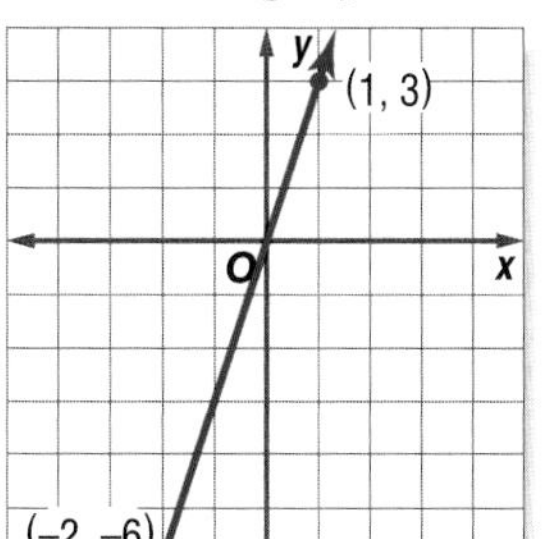

12.

x	y
−2	−3
4	−3

Find the slope of the line that passes through each pair of points.

13. (0, 5), (6, 2) **14.** (−6, 4), (−6, −2)

15. DIGITAL CAMERAS The average cost of using an online photo finisher decreased from \$0.50 per print to \$0.27 per print between 2002 and 2005. Find the average rate of change in the cost. Explain what the rate of change means.

Example 1 Find the slope of the line that passes through (0, −4) and (3, 2).

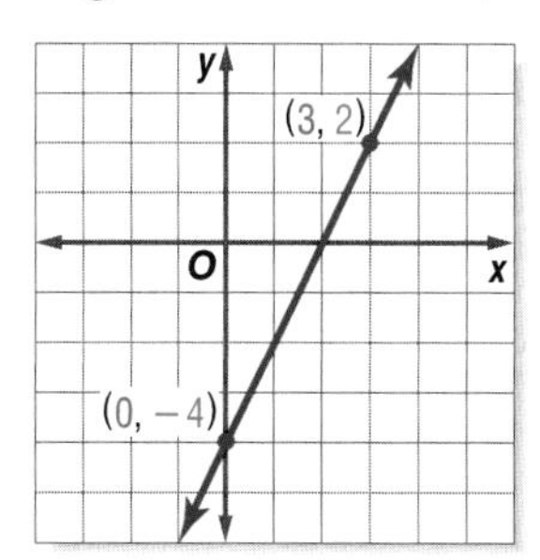

Let (0, −4) = (x_1, y_1) and (3, 2) = (x_2, y_2).

$m = \frac{y_2 - y_1}{x_2 - x_1}$ Slope formula

$= \frac{2 - (-4)}{3 - 0}$ $x_1 = 0, x_2 = 3, y_1 = -4, y_2 = 2$

$= \frac{6}{3}$ or 2 Simplify.

4–2 Slope and Direct Variation (pp. 196–202)

Graph each equation.

16. $y = x$ **17.** $y = \frac{4}{3}x$ **18.** $y = -2x$

Suppose y varies directly as x. Write a direct variation equation that relates x and y. Then solve.

19. If $y = 15$ when $x = 2$, find y when $x = 8$.

20. $y = -6$ when $x = 9$, find x when $y = -3$.

21. $y = 4$ when $x = -4$, find y when $x = 7$.

22. JOBS Suppose you earn \$127 for working 20 hours. Write a direct variation equation relating your earnings to the number of hours worked.

Example 2 Suppose y varies directly as x, and $y = -24$ when $x = 8$. Write a direct variation equation that relates x and y.

Find the constant of variation.

$y = kx$ Direct variation equation

$-24 = k(8)$ Replace y with −24 and x with 8.

$\frac{-24}{8} = \frac{k(8)}{8}$ Divide each side by 8.

$-3 = k$ Simplify.

So, the direct variation equation is $y = -3x$.

Study Guide and Review

4-3 Slope-Intercept Form (pp. 204–209)

Write an equation in slope-intercept form of the line with the given slope and y-intercept.

23. slope: 3, y-intercept: 2

24. slope: 1, y-intercept: -3

25. slope: 0, y-intercept: 4

26. slope: $\frac{1}{3}$, y-intercept: 2

Graph each equation.

27. $y = \frac{2}{3}x + 1$

28. $6x + 2y = -8$

29. $y = -x - 5$

30. $5x - 3y = -3$

31. WIRELESS PHONES A wireless phone-service provider charges a \$0.35 daily fee plus \$0.10 per minute. Write a linear equation to find the daily cost y for any number of minutes x.

Example 3 Graph $-3x + y = -1$.

Write in slope-intercept form.

$-3x + y = -1$	Original equation
$-3x + y + 3x = -1 + 3x$	Add $3x$ to each side.
$y = 3x - 1$	Simplify.

Step 1 The y-intercept is -1. So, graph $(0, -1)$.

Step 2 The slope is 3 or $\frac{3}{1}$. From $(0, -1)$, move up 3 units and right 1 unit. Then draw a line.

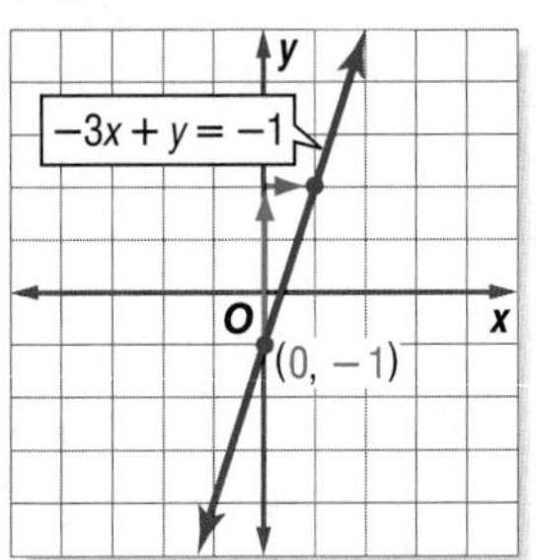

4-4 Writing Equations in Slope-Intercept Form (pp. 213–218)

Write an equation of the line that passes through each point with the given slope.

32. $(-3, 3)$, $m = 1$

33. $(4, -3)$, $m = -\frac{3}{5}$

34. $(8, -1)$, $m = 0$

35. $(0, 6)$, $m = -2$

Write an equation of the line that passes through each pair of points.

36. $(-4, 2)$, $(1, 12)$

37. $(5, 0)$, $(4, 5)$

38. MUSIC The table shows the average time Americans spent annually listening to recorded music. Write an equation to predict the number of hours h for any year y.

Year	Amount of Time (h)
1999	290
2006	195

Example 4 Write an equation of the line that passes through $(-2, -3)$ with slope $\frac{1}{2}$.

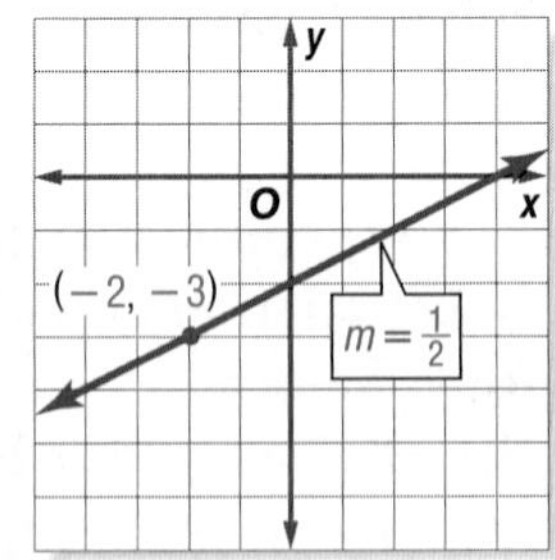

$y = mx + b$	Slope-intercept form
$-3 = \frac{1}{2}(-2) + b$	Replace m with $\frac{1}{2}$, y with -3, and x with -2.
$-3 = -1 + b$	Multiply.
$-3 + 1 = -1 + b + 1$	Add 1 to each side.
$-2 = b$	Simplify.

Therefore, the equation is $y = \frac{1}{2}x - 2$.

Mixed Problem Solving
For mixed problem-solving practice, see page 747.

4–5 Writing Equations in Point-Slope Form (pp. 220–225)

Write the point-slope form of an equation for the line that passes through each point with the given slope.

39.

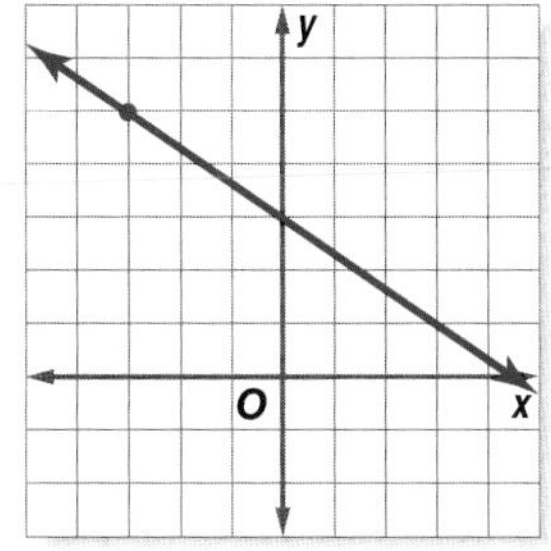

40. $(4, 6), m = 5$

41. $(5, -3), m = \frac{1}{2}$

42. $\left(\frac{1}{4}, -2\right), m = 0$

Write each equation in standard form.

43. $y + 4 = 1.5(x - 4)$

44. $y - 6 = \frac{2}{3}(x + 9)$

Write each equation in slope-intercept form.

45. $y - 1 = 2(x + 1)$

46. $y + 3 = \frac{1}{2}(x - 5)$

47. LAWN CARE A lawn care company charges \$25 per month for lawn maintenance, plus an initial service fee. The total cost for service fee and 8 months of maintenance is \$165. Write the point-slope form of an equation to find the total cost y for any number of months x. (*Hint:* (8, 165) is a solution of the equation.)

Example 5 **Write the point-slope form of an equation for a line that passes through (−2, 5) with slope 3.**

$y - y_1 = m(x - x_1)$	Point-slope form
$y - 5 = 3[x - (-2)]$	$(x_1, y_1) = (-2, 5)$
$y - 5 = 3(x + 2)$	Subtract.

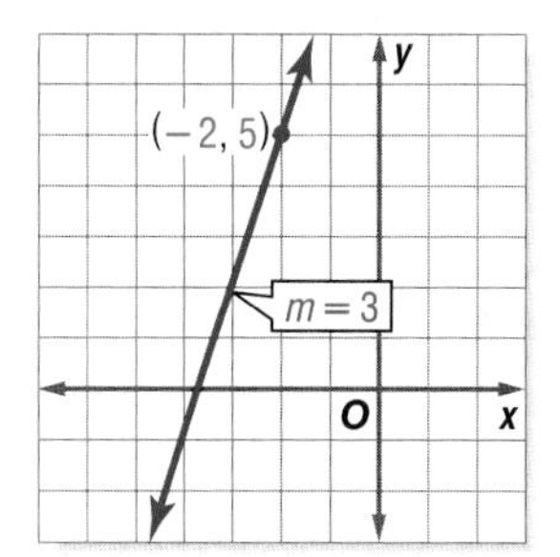

Example 6 **Write $y + 4 = \frac{1}{2}(x - 6)$ in slope-intercept form and in standard form.**

$y + 4 = \frac{1}{2}(x - 6)$	Original equation
$2(y + 4) = 2\left(\frac{1}{2}\right)(x - 6)$	Multiply each side by 2 to eliminate the fraction.
$2y + 8 = x - 6$	Distributive Property
$2y = x - 14$	Subtract 8 from each side.
$\frac{2y}{2} = \frac{x - 14}{2}$	Divide each side by 2.
$y = \frac{1}{2}x - 7$	Simplify.

The slope-intercept form is $y = \frac{1}{2}x - 7$.

$2y = x - 14$	Return to equation.
$2y - x = x - 14 - x$	Subtract x from each side.
$-x + 2y = -14$	Simplify.
$\frac{-x + 2y}{-1} = \frac{-14}{-1}$	Divide each side by −1.
$x - 2y = 14$	Simplify.

The standard form is $x - 2y = 14$.

Study Guide and Review

4-6 Statistics: Scatter Plots and Lines of Fit (pp. 227–233)

USE TABLES For Exercises 48–52, use the table that shows the length and weight of several humpback whales.

Length (ft)	40	42	45	46	50	52	55
Weight (long tons)	25	29	34	35	43	45	51

48. Draw a scatter plot with length on the x-axis and weight on the y-axis.

49. Draw a line of fit for the data.

50. Write the slope-intercept form of an equation for the line of fit.

51. Predict the weight of a 48-foot humpback whale.

52. Most newborn humpback whales are about 12 feet in length. Use the equation of the line of fit to predict the weight of a newborn humpback whale. Do you think your prediction is accurate? Explain.

Example 7 **Use the table shown to draw a scatter plot and predict the future stock price.**

Month	1	5	10	15	20	48
Price	$7	$17	$23	$35	$47	?

Step 1 Draw the scatter plot and find the line of fit.

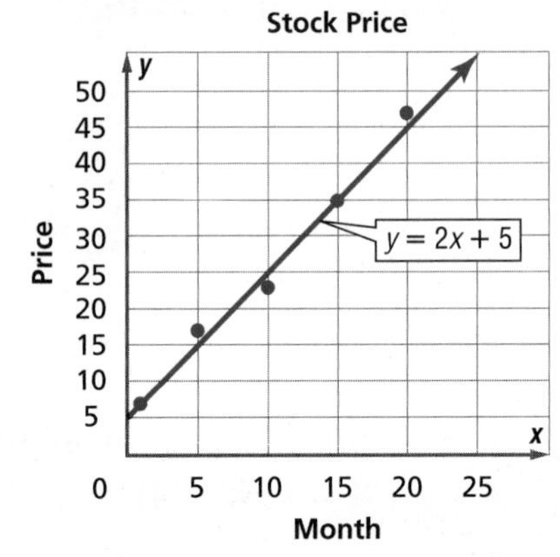

Step 2 Use line of fit to make predictions.

$y = 2x + 5$ Line of fit equation

$= 2(48) + 5$ Substitute 48 for x.

$= 96 + 5$ or 101 Simplify.

If the trend continues, the price will be $101.

4-7 Geometry: Parallel and Perpendicular Lines (pp. 236–241)

Write the slope-intercept form of an equation for the line that passes through the given point and satisfies each condition.

53. $(4, 6)$; parallel to $y = 3x - 2$

54. $(3, 0)$; parallel to $3x + 9y = 1$

55. $(2, -5)$; perpendicular to $5y = -x + 1$

56. $(0, -3)$; perpendicular to $y = -2x - 7$

57. GEOMETRY Determine if triangle ABC with vertices $A(-2, 0)$, $B(3, 3)$, and $C(-5, 5)$ is a right triangle. Explain.

Example 8 **Write the slope-intercept form of an equation for a line that passes through (5, −2) and is parallel to $y = 2x + 7$.**

The line parallel to $y = 2x + 7$ has the same slope, 2.

$y - y_1 = m(x - x_1)$ Point-slope form

$y - (-2) = 2(x - 5)$ Replace m with 2, y_1 with −2, and x_1 with 5.

$y + 2 = 2x - 10$ Simplify.

$y = 2x - 12$ Subtract 2 from each side.

CHAPTER 4 Practice Test

Find the slope of the line that passes through each pair of points.

1. $(5, 8), (-3, 7)$
2. $(5, -2), (3, -2)$
3. $(-4, 7), (8, -1)$
4. $(6, -3), (6, 4)$

5. **MULTIPLE CHOICE** Which is the slope of the linear function shown in the graph?

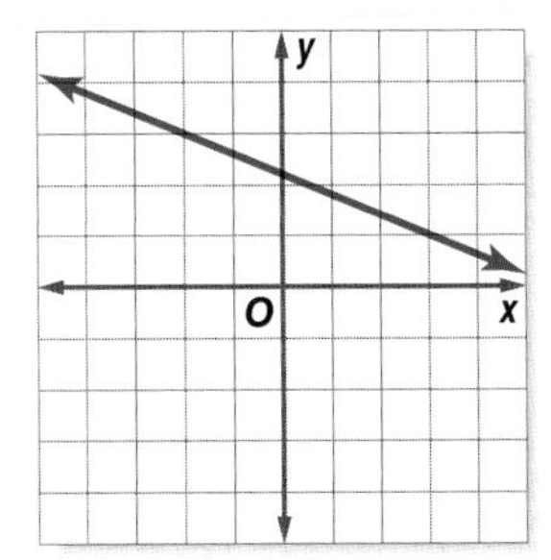

A $-\frac{5}{2}$

B $-\frac{2}{5}$

C $\frac{5}{2}$

D $\frac{2}{5}$

6. **BUSINESS** A Web design company advertises that it will design and maintain a Website for your business for $9.95 per month. Write a direct variation equation to find the total cost C for any number of months m.

Graph each equation.

7. $y = 3x - 1$
8. $y = 2x + 3$
9. $2x + 3y = 9$
10. $4y - 2x = 12$

Suppose y varies directly as x. Write a direct variation equation that relates x and y.

11. $y = 6$ when $x = 9$
12. $y = -8$ when $x = 8$
13. $y = -5$ when $x = -2$
14. $y = 2$ when $x = -12$

15. Write the point-slope form of an equation for a line that passes through $(-4, 3)$ with slope -2.

16. **MULTIPLE CHOICE** The temperature is 80°F at noon and is expected to rise 4° each hour during the afternoon. Which equation could be used to determine h, the number of hours it will take to reach a temperature of 96°?

F $96 = 4 + 80h$

G $96 = 4(h + 80)$

H $96 = 80 + 4h$

J $96 = (4 + 80)h$

Write the slope-intercept form of an equation of the line that satisfies each condition.

17. has slope -4 and y-intercept 3
18. passes through $(-2, -5)$ and $(8, -3)$
19. parallel to $3x + 7y = 4$ and passes through $(5, -2)$
20. perpendicular to the graph of $5x - 3y = 9$ and passes through the origin

ANALYZE TABLES For Exercises 21–25, use the table that shows the relationship between dog years and human years.

Dog Years	1	2	3	4	5	6	7
Human Years	15	24	28	32	37	42	47

21. Draw a scatter plot and determine what relationship, if any, exists in the data.
22. Draw a line of fit for the scatter plot.
23. Write the slope-intercept form of an equation for the line of fit.
24. Determine how many human years are comparable to 13 dog years.
25. Is it reasonable to use the equation for the line of fit to estimate the age in human years of a dog 20 years old? Explain.

Standardized Test Practice

Cumulative, Chapters 1-4

Read each question. Then fill in the correct answer on the answer document provided by your teacher or on a sheet of paper.

1. Which of the equations below represents the third step of the solution process?

Step 1 $-2(3a - 1) + 2 = 16$

Step 2 $-6a + 2 + 2 = 16$

Step 3 ☐

Step 4 $-6a = 12$

Step 5 $a = -2$

A $-2(3a - 1) = 16$
C $-2(3a + 1) = 16$
B $-6a + 4 = 16$
D $-6a - 1 + 4 = 16$

2. Corey collected data on the number of hours that his class spent studying per day and the number of hours they spent watching TV. If he posts the data on a scatterplot, what correlation will he mostly likely see between the number of hours spent studying and the number of hours spent watching TV?

F Negative
H Positive
G No correlation
J Constant

3. Petra is starting a jewelry business. She has \$200 to make jewelry for her business. If each bead costs \$12, which table best describes a, the amount of money remaining after she buys n beads?

A

n	a
0	\$200.00
1	\$188.00
2	\$176.00
3	\$164.00
4	\$152.00

C

n	a
0	\$200.00
1	\$176.00
4	\$140.00
6	\$116.00
8	\$92.00

B

n	a
0	\$200.00
2	\$188.00
4	\$176.00
6	\$164.00
8	\$152.00

D

n	a
0	\$188.00
1	\$176.00
2	\$164.00
3	\$140.00
4	\$128.00

4. Which is always a correct conclusion about the quantities in the function $y = 2x$?

F The value of y will always be positive.

G The value of y will always be greater than the value of x.

H The variable y is always twice x.

J As the value of x increases the value of y decreases.

TEST-TAKING TIP

Read each question carefully. Be sure you understand what the question asks. Look for words such as not, estimate, always and approximately.

5. **GRIDDABLE** The figure shows a pattern of squares made of dots.

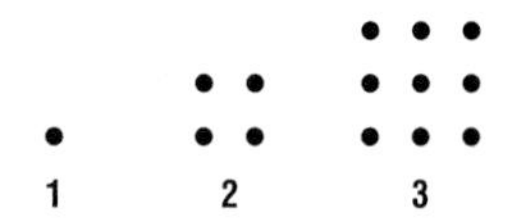

How many dots will be in the sixth square?

6. Vera is drawing a mural on her wall. She placed a grid over the mural.

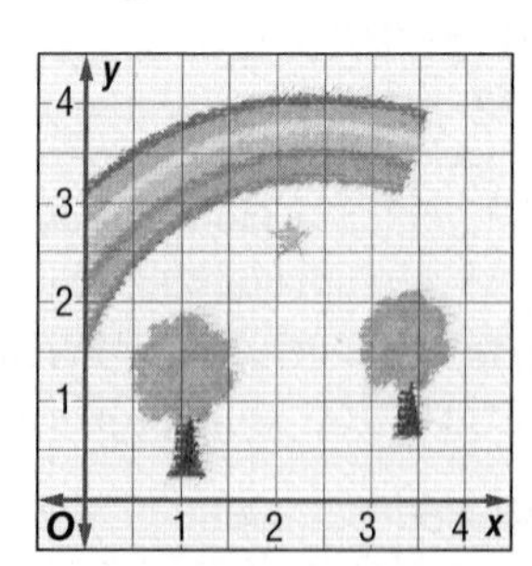

Which coordinate point best describes the star on the mural?

A (4, 5)
B (5, 4)
C (2, 2.5)
D (2.5, 2)

7. What is the equation of the line shown below?

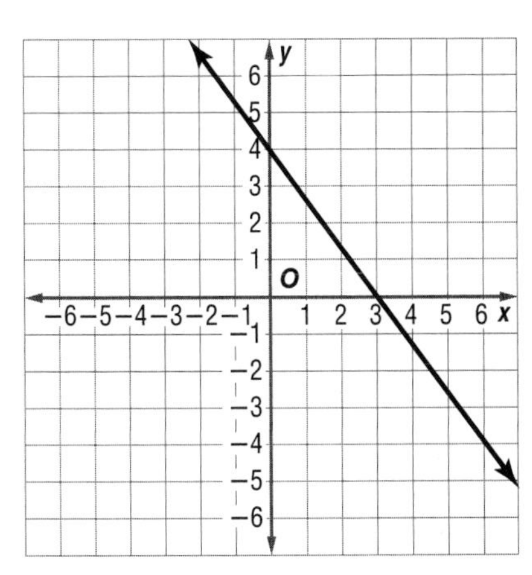

F $y = -\frac{4}{3}x + 4$
H $y = -\frac{4}{3}x + 3$
G $y = -\frac{3}{4}x + 4$
J $y = -\frac{3}{4}x + 3$

8. **GRIDDABLE** Meagan makes and sells jewelry at the local farmer's market. At last year's farmer's market, Meagan sold 75% of her jewelry. If she sold the same percentage this year and had 1,000 pieces of jewelry, how many pieces of jewelry did she sell?

9. The data in the table show the cost of renting a boat by the hour, including a deposit.

Renting a Boat

Hours (h)	1	3	5
Cost in dollars (c)	30	60	90

If hours h were graphed on the horizontal axis and cost c were graphed on the vertical axis, what would be the equation of the line that fits the data?

A $c = 15h$
C $c = 15h - 15$
B $c = \frac{1}{15}h + 15$
D $c = 15h + 15$

10. The graph shows the levels of the water in a pool from the time it begins to drain to the time it is empty. Which of the following best describes the slope of the line segment?

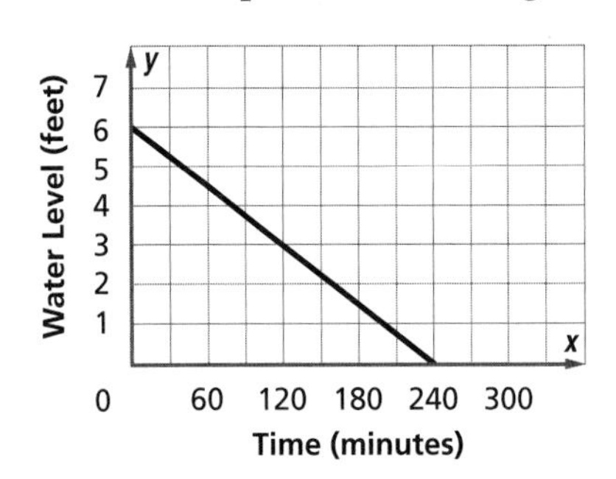

F The water drains about 1 foot per 40 minutes.
G The water drains about 40 feet per minute.
H The water drains about 1 foot per 2 minutes.
J The water drains about 2 feet per minute.

Pre-AP

Record your answers on a sheet of paper. Show your work.

11. A friend wants to enroll for cellular phone service. Three different plans are available.

Plan 1 charges $0.59 per minute.

Plan 2 charges a monthly fee of $10, plus $0.39 per minute.

Plan 3 charges a monthly fee of $59.95.

a. For each plan, write an equation that represents the monthly cost C for m number of minutes per month.

b. Graph each of the three equations.

c. Your friend expects to use 100 minutes per month. In which plan do you think that your friend should enroll? Explain.

NEED EXTRA HELP?											
If You Missed Question...	1	2	3	4	5	6	7	8	9	10	11
Go to Lesson or Page...	2-4	4-7	3-1	4-1	3-5	1-9	4-3	694	4-6	4-3	4-3

CHAPTER 5

Solving Systems of Linear Equations

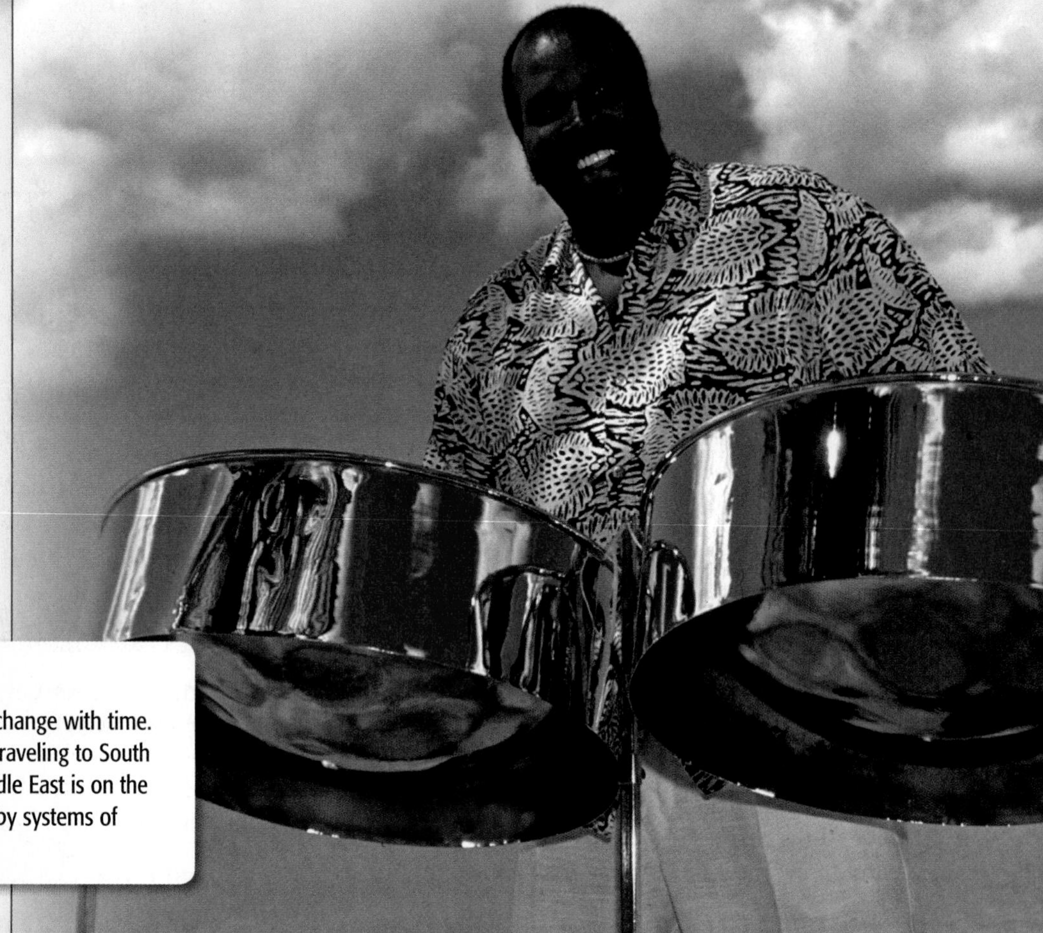

BIG Ideas

- Solve systems of linear equations by graphing.
- Solve systems of linear equations algebraically.
- Apply systems of linear equations.

Key Vocabulary

elimination (p. 266)
substitution (p. 260)
system of equations (p. 253)

Real-World Link

Travel Trends in the travel industry change with time. For example, the number of tourists traveling to South America, the Caribbean, and the Middle East is on the rise. These changes can be modeled by systems of linear equations.

FOLDABLES™ Study Organizer

Solving Systems of Linear Equations Make this Foldable to record information about solving systems of equations and inequalities. Begin with five sheets of grid paper.

1. **Fold** each sheet in half along the width.

2. **Unfold** and cut four rows from left side of each sheet, from the top to the crease.

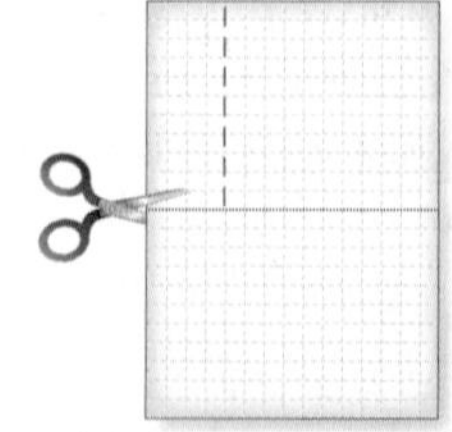

3. **Stack** the sheets and staple to form a booklet.

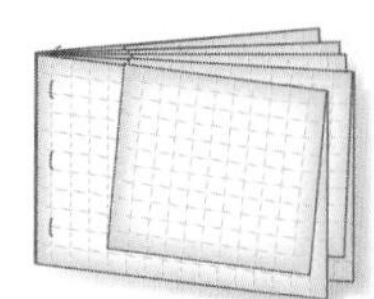

4. **Label** each page with a lesson number and title.

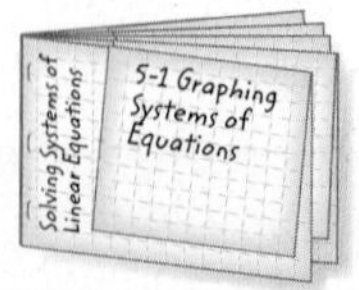

GET READY for Chapter 5

Diagnose Readiness You have two options for checking Prerequisite Skills.

Option 2

Math Online Take the Online Readiness Quiz at algebra1.com.

Option 1

Take the Quick Check below. Refer to the Quick Review for help.

QUICKCheck

Graph each equation. (Lesson 3-3)

1. $y = 1$

2. $y = -2x$

3. $y = 4 - x$

4. $y = 2x + 3$

5. $y = 5 - 2x$

6. $y = \frac{1}{2}x + 2$

7. HOUSES The number on Craig's house is 7. The numbers of the houses on his block increase by 2. Graph the equation that models the house numbers on Craig's block.

Solve each equation or formula for the variable specified. (Lesson 2-8)

8. $4x + a = 6x$, for x

9. $8a + y = 16$, for a

10. $\frac{7bc - d}{10} = 12$, for b

11. $\frac{7m + n}{q} = 2m$, for q

Simplify each expression. If not possible, write *simplified*. (Lesson 1-6)

12. $(3x + y) - (2x + y)$

13. $(7x - 2y) - (7x + 4y)$

14. MOWING Jake and his brother charge x dollars to cut and y dollars to weed an average lawn. Simplify the expression that gives the total amount that their business earns in a weekend if Jake cuts and weeds 7 lawns and his brother cuts and weeds 10 lawns.

QUICKReview

EXAMPLE 1

Graph $y = \frac{3}{4}x - 3$.

Step 1 The y-intercept is -3. So, graph $(0, -3)$.

Step 2 The slope is $\frac{3}{4}$. From $(0, -3)$, move up 3 units and right 4 units. Draw a dot.

Step 3 Draw the line.

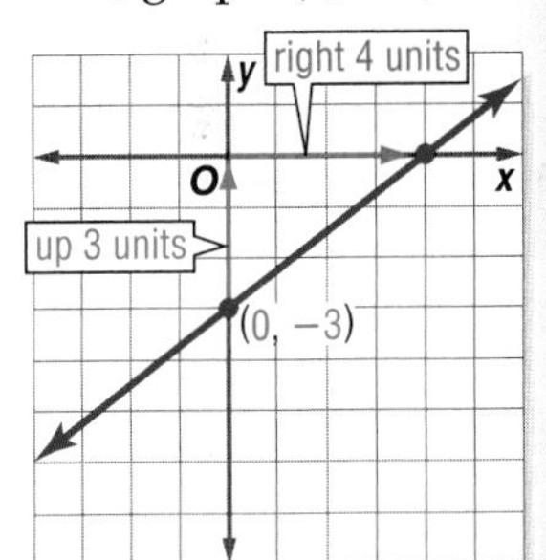

EXAMPLE 2

Solve $\frac{2y}{3s} = \frac{2y}{13x}$ for x.

$\frac{2y}{3s} = \frac{2y}{13x}$	Original equation
$2y \cdot 13x = 3s \cdot 2y$	Find the cross products.
$26yx = 6sy$	Simplify.
$x = \frac{6sy}{26y}$	Divide each side by $26y$.
$x = \frac{3s}{13}$	Simplify.

EXAMPLE 3

Simplify $3(x - y) - (x - y)$. If not possible, write *simplified*.

$3(x - y) - (x - y)$	Original expression
$= 3x - 3y - x + y$	Distributive Property
$= 2x - 2y$	Combine like terms.
$= 2(x - y)$	Factor out a 2.

Spreadsheet Lab

Systems of Equations

You can use a spreadsheet to investigate when two quantities will be equal. Enter each formula into the spreadsheet and look for the row in which both formulas have the same result.

Concepts in Motion
Interactive Lab
algebra1.com

EXAMPLE

Bill Winters is considering two job offers in telemarketing departments. The salary at the first job is $400 per week plus 10% commission on Mr. Winters' sales. At the second job, the salary is $375 per week plus 15% commission. For what amount of sales would the weekly salary be the same at either job?

Enter different amounts for Mr. Winters' weekly sales in column A. Then enter the formula for the salary at the first job in each cell in column B. In each cell of column C, enter the formula for the salary at the second job.

Job Salaries.xls

◇	A	B	C
1	Sales	Salary 1	Salary 2
2	0	400	375
3	100	410	390
4	200	420	405
5	300	430	420
6	400	440	435
7	500	450	450
8	600	460	465
9	700	470	480
10	800	480	495
11	900	490	510
12	1000	500	525

Sheet 1 / Sheet 2 / Sheet 3

The spreadsheet shows that for sales of $500 the total weekly salary for each job is $450.

Exercises

For Exercises 1–4, use the spreadsheet of weekly salaries above.

1. If x is the amount of Mr. Winters' weekly sales and y is his total weekly salary, write a linear equation for the salary at the first job.
2. Write a linear equation for the salary at the second job.
3. Which ordered pair is a solution for both of the equations you wrote for Exercises 1 and 2?

   a. (100, 410) **b.** (300, 420) **c.** (500, 450) **d.** (900, 510)
4. Use the graphing capability of the spreadsheet program to graph the salary data using a line graph. At what point do the two lines intersect? What is the significance of that point in the real-world situation?
5. How could you find the sales for which Mr. Winters' salary will be equal without using a spreadsheet?

5-1 Graphing Systems of Equations

Main Ideas

- Determine whether a system of linear equations has no, one, or infinitely many solutions.
- Solve systems of equations by graphing.

New Vocabulary

system of equations
consistent
independent
dependent
inconsistent

Concepts in Motion
Animation
algebra1.com

GET READY for the Lesson

If x is the number of years since 2000 and y is units sold in millions, the following equations represent the sales of CD singles and music videos.

CD singles: $y = 34.2 - 14.9x$
music videos: $y = 3.3 + 4.7x$

The point at which the graphs of the two equations intersect represents the time when the CD units sold equaled the music videos sold. The ordered pair of this point is a solution of both equations.

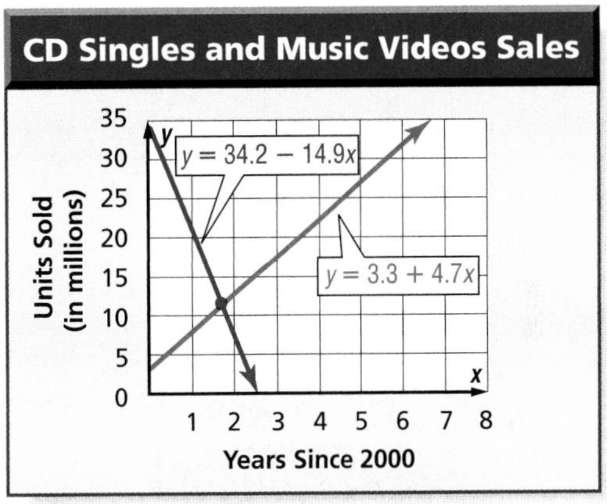

Source: The Recording Industry Association of America

Number of Solutions Two equations, such as $y = 34.2 - 14.9x$ and $y = 3.3 + 4.7x$, together are called a **system of equations**. A solution of a system is an ordered pair that satisfies both equations. A system of two linear equations can have no, one, or an infinite number of solutions.

- If the graphs intersect or coincide, the system of equations is **consistent**. That is, it has at least one ordered pair that satisfies both equations.
- If a consistent system has exactly one solution, it is **independent**. If it has infinite solutions, it is **dependent**.
- If the graphs are parallel, the system of equations is said to be **inconsistent**. There are *no* ordered pairs that satisfy both equations.

KEY CONCEPT — *Graphing Systems of Equations*

Graph of a System	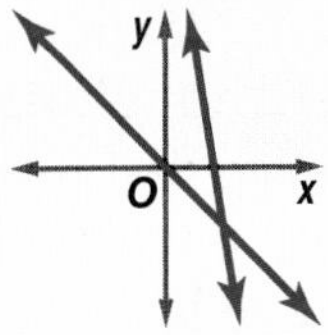	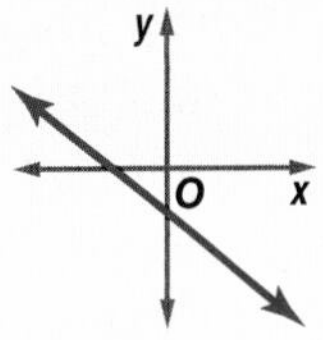	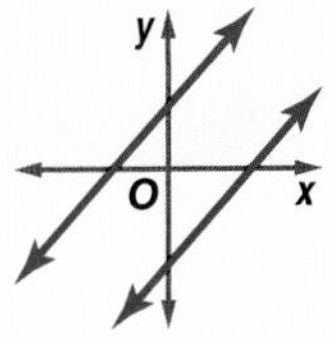
Number of Solutions	exactly one solution	infinitely many	no solutions
Terminology	consistent and independent	consistent and dependent	inconsistent

EXAMPLE 1 Number of Solutions

Use the graph at the right to determine whether each system has *no* solution, *one* solution, or *infinitely many* solutions.

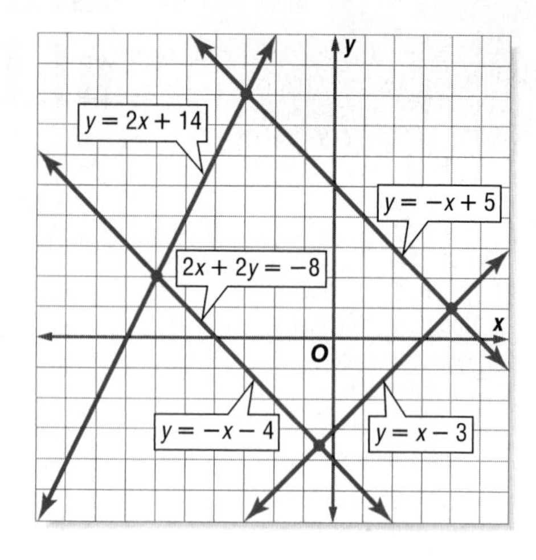

a. $y = -x + 5$

$y = x - 3$

Since the graphs are intersecting lines, there is one solution.

b. $y = -x + 5$

$2x + 2y = -8$

Since the graphs are parallel, there are no solutions.

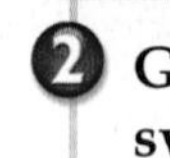

CHECK Your Progress

1A. $2x + 2y = -8$

$y = -x - 4$

1B. $y = 2x + 14$

$y = -x + 5$

Solve By Graphing One method of solving systems of equations is to carefully graph the equations on the same coordinate plane.

EXAMPLE 2 Solve a System of Equations

Graph each system of equations. Then determine whether the system has *no* solution, *one* solution, or *infinitely many* solutions. If the system has one solution, name it.

a. $y = -x + 8$

$y = 4x - 7$

The graphs appear to intersect at (3, 5). Check by replacing x with 3 and y with 5.

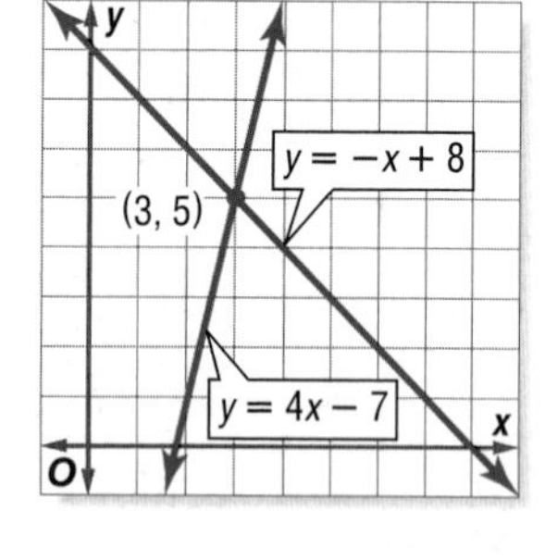

CHECK

$y = -x + 8$	$y = 4x - 7$
$5 \stackrel{?}{=} -3 + 8$	$5 \stackrel{?}{=} 4(3) - 7$
$5 = 5$ ✓	$5 = 5$ ✓

The solution is (3, 5).

Study Tip

Look Back

To review **graphing linear equations,** see Lesson 3-3.

b. $x + 2y = 5$

$2x + 4y = 2$

The graphs are parallel lines. Since they do not intersect, there are no solutions to this system of equations. Notice that the lines have the same slope but different y-intercepts. *Recall that a system of equations that has no solution is said to be inconsistent.*

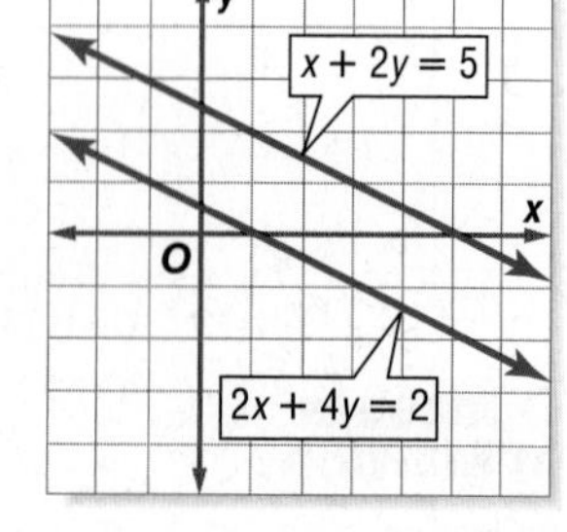

CHECK Your Progress

2A. $x - y = 2$

$3y + 2x = 9$

2B. $y = -2x - 3$

$2x + y = -3$

Online **Personal Tutor at** algebra1.com

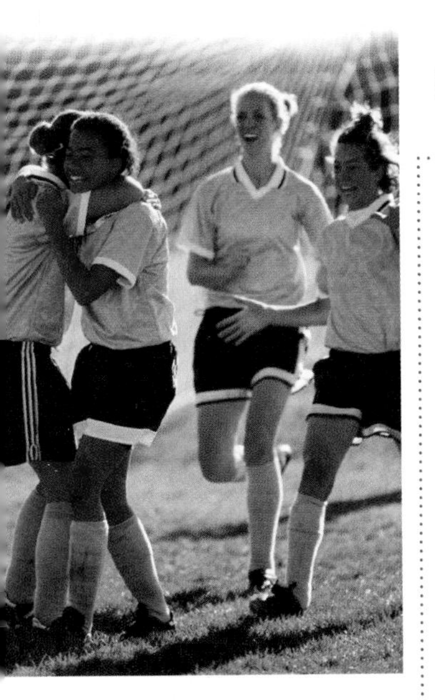

Real-World EXAMPLE — Write and Solve a System of Equations

3 **SPORTS** **The number of girls participating in high school soccer and track and field has steadily increased during the past few years. Use the information in the table to predict the year in which the number of girls participating in these two sports will be the same.**

High School Sport	Number of Girls Participating in 2004 (thousands)	Average Rate of Increase (thousands per year)
soccer	309	8
track and field	418	3

Source: National Federation of State High School Associations

Words	Number of girls participating	equals	rate of increase	times	number of years after 2004	plus	number participating in 2004.
Variables	Let y = number of girls competing.				Let x = number of years after 2004.		
Equations	soccer: y	$=$	8	$\times$	x	$+$	309
	track and field: y	$=$	3	$\times$	x	$+$	418

Graph the equations $y = 8x + 309$ and $y = 3x + 418$. The graphs appear to intersect at (22, 485). Check by replacing x with 22 and y with 485 in each equation.

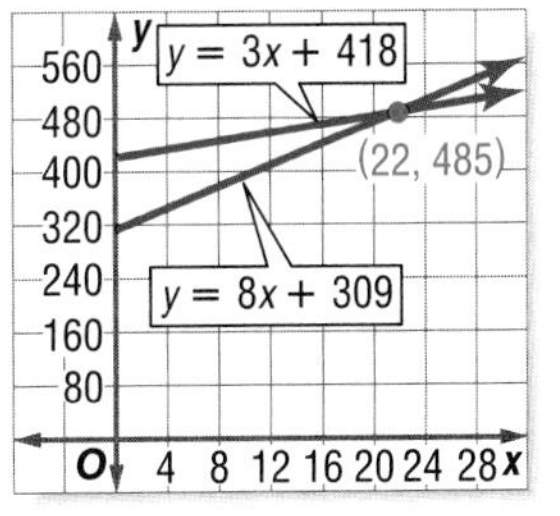

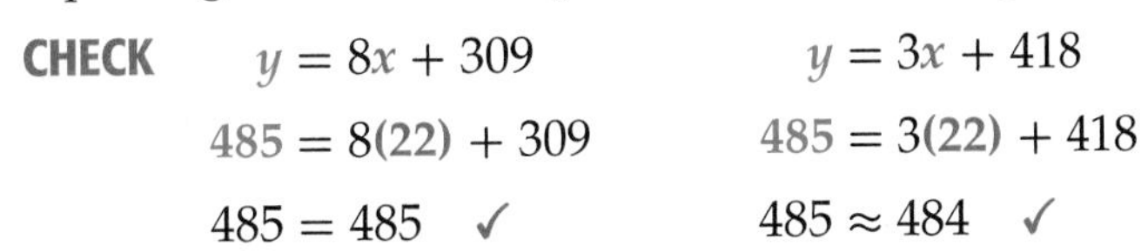

CHECK

$y = 8x + 309$ | $y = 3x + 418$

$485 = 8(22) + 309$ | $485 = 3(22) + 418$

$485 = 485$ ✓ | $485 \approx 484$ ✓

The solution means that approximately 22 years after 2004, or in 2026, the number of girls participating in high school soccer and track and field will be the same, about 485,000.

Real-World Link

In 2004, 2.9 million girls participated in high school sports. This was an all-time high for female participation.

Source: National Federation of State High School Associations

CHECK Your Progress

3. GARDENS A rectangular garden has a border around it consisting of 60 bricks. The width of the border has $\frac{2}{3}$ the number of bricks as the length. How many bricks are along one length of the garden?

CHECK Your Understanding

Example 1 (p. 254)

Use the graph to determine whether each system has *no* solution, *one* solution, or *infinitely many* solutions.

1. $y = x - 4$
$y = \frac{1}{3}x - 2$

2. $y = \frac{1}{3}x + 2$
$y = \frac{1}{3}x - 2$

3. $x - y = 4$
$y = x - 4$

4. $x - y = 4$
$y = -\frac{1}{3}x + 4$

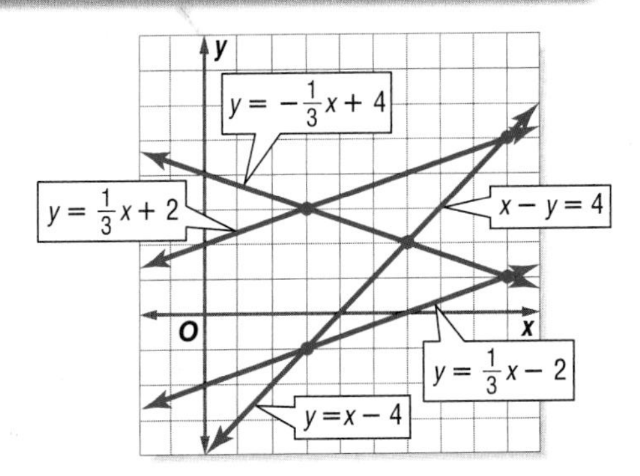

Example 2 (p. 254) **Graph each system of equations. Then determine whether the system has *no* solution, *one* solution, or *infinitely many* solutions. If the system has one solution, name it.**

5. $y = 3x - 4$
   $y = -3x - 4$

6. $x + y = 2$
   $y = 4x + 7$

7. $x + y = 4$
   $x + y = 1$

8. $2x + 4y = 2$
   $3x + 6y = 3$

Example 3 (p. 255)

9. **GEOMETRY** The length of the rectangle is 1 meter less than twice its width. What are the dimensions of the rectangle?

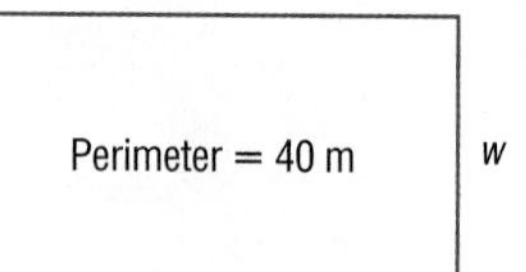

Exercises

HOMEWORK	
For Exercises	See Examples
10–15	1
16–27	2
28–31	3

Use the graph to determine whether each system has *no* solution, *one* solution, or *infinitely many* solutions.

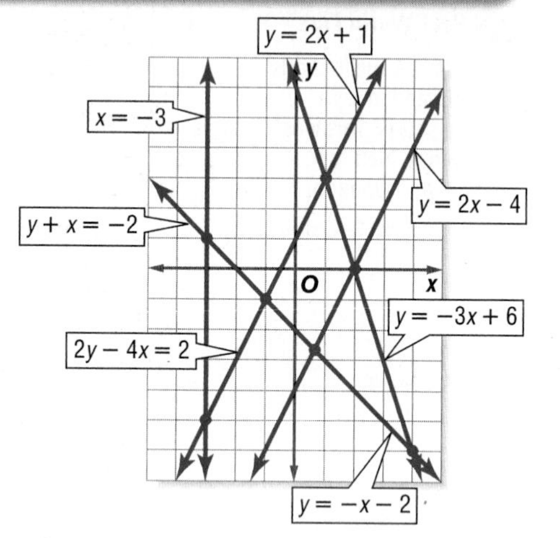

10. $x = -3$
    $y = 2x + 1$

11. $y = -x - 2$
    $y = 2x - 4$

12. $y = 2x + 1$
    $2y - 4x = 2$

13. $y = 2x + 1$
    $y = 2x - 4$

14. $y + x = -2$
    $y = -x - 2$

15. $2y - 4x = 2$
    $y = 2x - 4$

Graph each system of equations. Then determine whether the system has *no* solution, *one* solution, or *infinitely many* solutions. If the system has one solution, name it.

16. $y = -6$
    $4x + y = 2$

17. $x = 2$
    $3x - y = 8$

18. $y = \frac{1}{2}x$
    $2x + y = 10$

19. $y = -x$
    $y = 2x - 6$

20. $y = 2x + 6$
    $y = -x - 3$

21. $x - 2y = 2$
    $3x + y = 6$

22. $x + y = 2$
    $2y - x = 10$

23. $3x + 2y = 12$
    $3x + 2y = 6$

24. $2x + 3y = 4$
    $-4x - 6y = -8$

25. $2x + y = -4$
    $5x + 3y = -6$

26. $4x + 3y = 24$
    $5x - 8y = -17$

27. $3x + y = 3$
    $2y = -6x + 6$

SAVINGS For Exercises 28 and 29, use the following information.
Monica and Max Gordon each want to buy a scooter. Monica has already saved \$25 and plans to save \$5 per week until she can buy the scooter. Max has \$16 and plans to save \$8 per week.

28. In how many weeks will Monica and Max have saved the same amount of money?

29. How much will each person have saved at that time?

Cross-Curricular Project

Math Online You can graph a system of equations to predict when men's and women's Olympic times will be the same. Visit algebra1.com to continue work on your project.

BALLOONING For Exercises 30 and 31, use the information in the graphic at the right.

30. In how many minutes will the balloons be at the same height?

31. How high will the balloons be at that time? Is your answer reasonable? Explain.

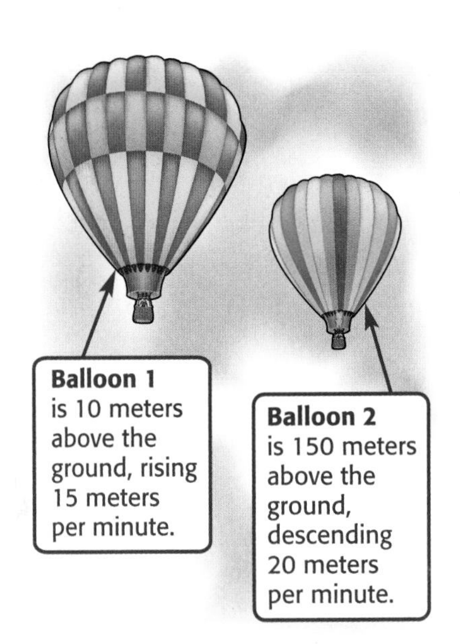

Graph each system of equations. Then determine whether the system has *no* solution, *one* solution, or *infinitely many* solutions. If the system has one solution, name it.

32. $y = 0.6x - 5$
$2y = 1.2x$

33. $6 - \frac{3}{8}y = x$
$\frac{2}{3}x + \frac{1}{4}y = 4$

ANALYZE GRAPHS For Exercises 34–36, use the graph at the right.

34. Which company had the greater profit during the ten years?

35. Which company had a greater rate of growth?

36. If the profit patterns continue, will the profits of the two companies ever be equal? Explain.

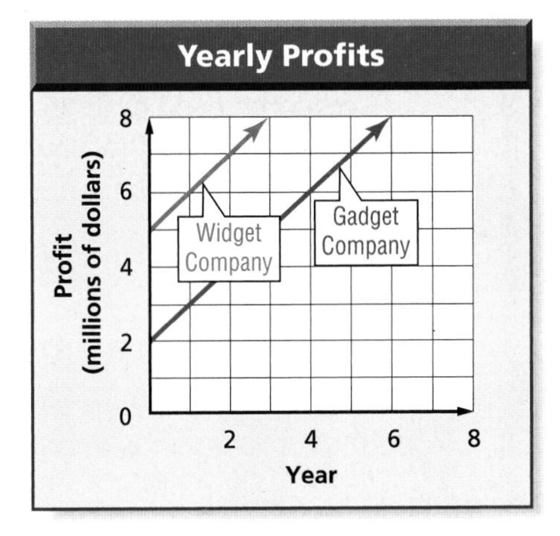

POPULATION For Exercises 37–39, use the following information.

The U.S. Census Bureau divides the country into four sections. They are the Northeast, the Midwest, the South, and the West. The populations and rates of growth for the Midwest and the West are shown in the table.

Section	2000 Population (millions)	Average Rate of Increase (millions per year)
Midwest	64.4	0.3
West	63.2	1.0

Source: U.S. Census Bureau

37. Write an equation to represent the population of the Midwest for the years since 2000.

38. Write an equation to represent the population of the West for the years since 2000.

39. Graph the population equations. Assume that the rate of growth of each of these areas remained the same. Estimate the solution and interpret what it means.

H.O.T. Problems

40. CHALLENGE The solution of the system of equations $Ax + y = 5$ and $Ax + By = 20$ is $(2, -3)$. What are the values of A and B? Justify your reasoning.

41. OPEN ENDED Write three equations such that they form a system of equations with $y = 5x - 3$. The systems should have *no, one,* and *infinitely many* solutions, respectively.

42. **REASONING** Determine whether a system of two linear equations with (0, 0) and (2, 2) as solutions *sometimes*, *always*, or *never* has other solutions. Explain.

43. ***Writing in Math*** Use the information on page 253 to explain how graphs can be used to compare the sales of two products. Include an estimate of the year in which the CD units sold equaled the music videos sold. Then determine the reasonableness of your solution in the context of the problem.

STANDARDIZED TEST PRACTICE

44. A buffet restaurant has one price for adults and another price for children. The Taylor family has two adults and three children, and their bill was \$40.50. The Wong family has three adults and one child. Their bill was \$38. Which system of equations could be used to determine the buffet price for an adult and the price for a child?

A $x + y = 40.50$
$x + y = 38$

B $2x + 3y = 40.50$
$x + 3y = 38$

C $2x + 3y = 40.50$
$3x + y = 38$

D $2x + 2y = 40.50$
$3x + y = 38$

45. **REVIEW** Francisco has 3 dollars more than $\frac{1}{4}$ the number of dollars that Kayla has. Which expression represents how much money Francisco has?

F $3\left(\frac{1}{4}k\right)$

G $3 - \frac{1}{4}k$

H $\frac{1}{4}k + 3$

J $\frac{1}{4} + 3k$

Spiral Review

Write the slope-intercept form of an equation for the line that passes through the given point and is parallel to the graph of each equation. (Lesson 4-7)

46.

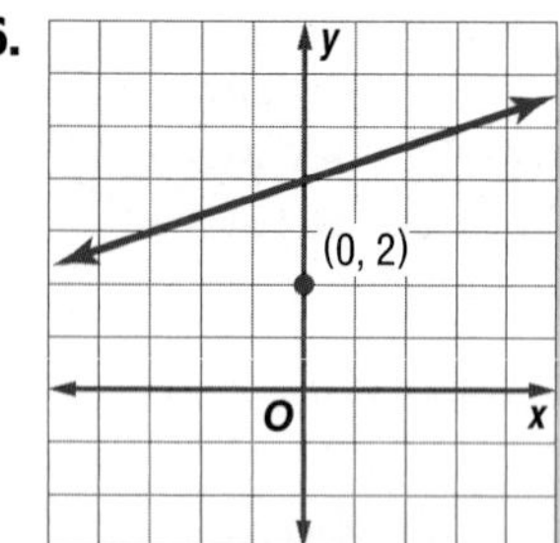

47.

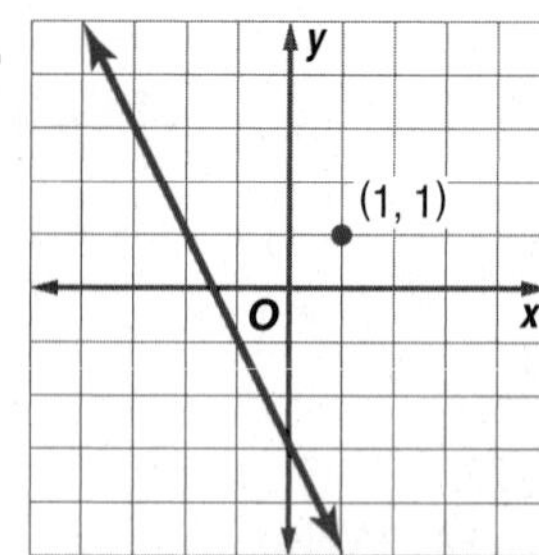

48. **BIOLOGY** The table shows the date of the month that 10 students were born, and their heights. Draw a scatter plot and determine what relationship exists, if any, in the data. Explain. (Lesson 4-6)

Date of Birth	12	28	24	15	3	11	20	5	3	9
Height (in.)	60	58	62	60	59	64	66	65	67	62

GET READY for the Next Lesson

PREREQUISITE SKILL Solve each equation for the variable specified. (Lesson 2-8)

49. $12x - y = 10x$, for y

50. $6a + b = 2a$, for a

51. $\frac{7m - n}{q} = 10$, for q

EXTEND 5-1

Graphing Calculator Lab
Systems of Equations

You can use a TI-83/84 Plus graphing calculator to solve a system of equations.

EXAMPLE

Solve the system of equations. State the decimal solution to the nearest hundredth.

$2.93x + y = 6.08$

$8.32x - y = 4.11$

Step 1 Solve each equation for y. Enter them into the calculator.

$2.93x + y = 6.08$	First equation
$2.93x + y - 2.93x = 6.08 - 2.93x$	Subtract $2.93x$ from each side.
$y = 6.08 - 2.93x$	Simplify.
$8.32x - y = 4.11$	Second equation
$8.32x - y - 8.32x = 4.11 - 8.32x$	Subtract $8.32x$ from each side.
$-y = 4.11 - 8.32x$	Simplify.
$(-1)(-y) = (-1)(4.11 - 8.32x)$	Multiply each side by -1.
$y = -4.11 + 8.32x$	Simplify.

Step 2 Enter these equations in the Y= list and graph.

KEYSTROKES: *Review on pages 162–163.*

Step 3 Use the CALC menu to find the point of intersection.

KEYSTROKES: 2nd [CALC] 5 ENTER ENTER ENTER

The solution is approximately (0.91, 3.43).

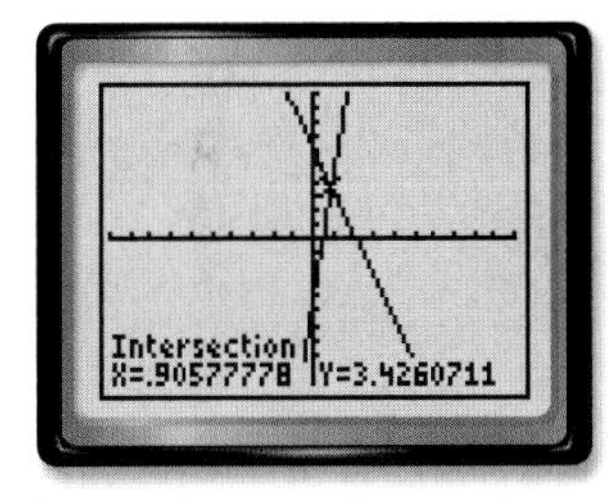

[−10, 10] scl: 1 by [−10, 10] scl: 1

Exercises

Use a graphing calculator to solve each system of equations. Write decimal solutions to the nearest hundredth.

1. $y = 3x - 4$
$y = -0.5x + 6$

2. $y = 2x + 5$
$y = -0.2x - 4$

3. $x + y = 5.35$
$3x - y = 3.75$

4. $0.35x - y = 1.12$
$2.25x + y = -4.05$

5. $1.5x + y = 6.7$
$5.2x - y = 4.1$

6. $5.4x - y = 1.8$
$6.2x + y = -3.8$

7. $5x - 4y = 26$
$4x + 2y = 53.3$

8. $2x + 3y = 11$
$4x + y = -6$

9. $0.22x + 0.15y = 0.30$
$-0.33x + y = 6.22$

Online **Other Calculator Keystrokes at** algebra1.com

5-2 Substitution

Main Ideas

- Solve systems of equations algebraically by using substitution.
- Solve real-world problems involving systems of equations.

New Vocabulary

substitution

GET READY for the Lesson

Americans spend more time online than they spend reading daily newspapers. If x represents the number of years since 2000 and y represents the average number of hours per person per year, the following system represents the situation.

reading daily newspapers: $y = -2.8x + 150.4$
online: $y = 14.4x + 102.8$

The solution of the system represents the year that the number of hours spent on each activity will be the same. To solve this system, you could graph the equations and find the point of intersection. However, the exact coordinates of the point would be very difficult to determine from the graph. You could find a more accurate solution by using algebraic methods.

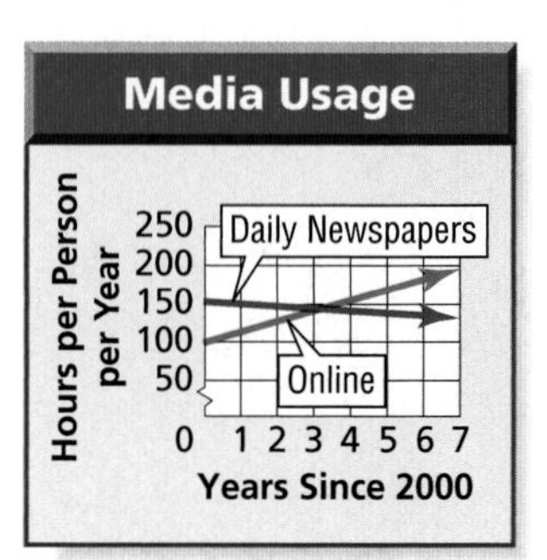

Substitution The exact solution of a system of equations can be found by using algebraic methods. One such method is called **substitution**.

ALGEBRA LAB

Using Substitution

Use algebra tiles and an equation mat to solve the system of equations.
$3x + y = 8$ and $y = x - 4$

MODEL AND ANALYZE

Since $y = x - 4$, use 1 positive x-tile and 4 negative 1-tiles to represent y. Use algebra tiles to represent $3x + y = 8$.

1. Use what you know about equation mats to solve for x. What is the value of x?
2. Use $y = x - 4$ to solve for y.
3. What is the solution of the system of equations?
4. **MAKE A CONJECTURE** Explain how to solve the following system of equations using algebra tiles. $4x + 3y = 10$ and $y = x + 1$
5. Why do you think this method is called substitution?

EXAMPLE Solve Using Substitution

1 **Use substitution to solve each system of equations.**

a. $y = 3x$

$x + 2y = -21$

Since $y = 3x$, substitute $3x$ for y in the second equation.

$x + 2y = -21$	Second equation
$x + 2(3x) = -21$	$y = 3x$
$x + 6x = -21$	Simplify.
$7x = -21$	Combine like terms.
$x = -3$	Divide each side by 7 and simplify.

Use $y = 3x$ to find the value of y.

$y = 3x$	First equation
$y = 3(-3)$ or -9	$x = -3$

The solution is $(-3, -9)$. Check the solution by graphing.

b. $x + 5y = -3$

$3x - 2y = 8$

Solve the first equation for x since the coefficient of x is 1.

$x + 5y = -3$	First equation
$x + 5y - 5y = -3 - 5y$	Subtract $5y$ from each side.
$x = -3 - 5y$	Simplify.

Find the value of y by substituting $-3 - 5y$ for x in the second equation.

$3x - 2y = 8$	Second equation
$3(-3 - 5y) - 2y = 8$	$x = -3 - 5y$
$-9 - 15y - 2y = 8$	Distributive Property
$-9 - 17y = 8$	Combine like terms.
$-9 - 17y + 9 = 8 + 9$	Add 9 to each side.
$-17y = 17$	Simplify.
$\frac{-17y}{-17} = \frac{17}{-17}$	Divide each side by -17.
$y = -1$	Simplify.

Substitute -1 for y in either equation to find the value of x.

$x + 5y = -3$	First equation
$x + 5(-1) = -3$	$y = -1$
$x - 5 = -3$	Simplify.
$x = 2$	Add 5 to each side.

The solution is $(2, -1)$. The graph verifies the solution.

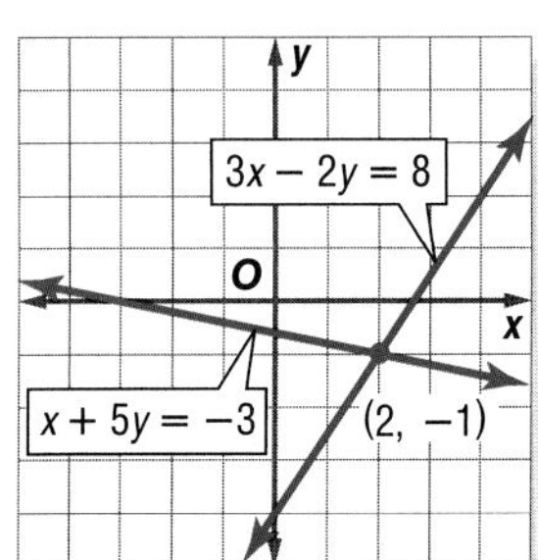

Study Tip

Looking Back

To review **solving linear equations,** see Lesson 2-5.

Study Tip

Substituting

When substituting to find the value of the second variable, choose the equation that is easier to solve.

CHECK Your Progress

1A. $4x + 5y = 11$

$y = 3x - 13$

1B. $x - 3y = -9$

$5x - 2y = 7$

In general, if you solve a system of linear equations and the result is a true statement (an identity such as $-4 = -4$), the system has an infinite number of solutions. If the result is a false statement (such as $-4 = 5$), there is no solution.

EXAMPLE **Infinitely Many or No Solutions**

Use substitution to solve the system of equations.

$6x - 2y = -4$

$y = 3x + 2$

Study Tip

Dependent Systems There are infinitely many solutions of the system in Example 2, because the slope-intercept form of both equations is $y = 3x + 2$. That is, the equations are equivalent, and they have the same graph.

Since $y = 3x + 2$, substitute $3x + 2$ for y in the first equation.

$6x - 2y = -4$ First equation

$6x - 2(3x + 2) = -4$ $y = 3x + 2$

$6x - 6x - 4 = -4$ Distributive Property

$-4 = -4$ Simplify.

The statement $-4 = -4$ is true. So, there are infinitely many solutions.

CHECK Your Progress

2A. $2x - y = 8$
$y = 2x - 3$

2B. $4x - 3y = 1$
$6y - 8x = -2$

Real-World Problems Sometimes it is helpful to organize data tables, charts, graphs, or diagrams before solving a problem.

EXAMPLE **Write and Solve a System of Equations**

3 **METAL ALLOYS A metal alloy is 25% copper. Another metal alloy is 50% copper. How much of each alloy should be used to make 1000 grams of a metal alloy that is 45% copper?**

Study Tip

Alternative Method Using a system of equations is an alternative method for solving the weighted average problems that you studied in Lesson 2-9.

Let a = the number of grams of the 25% copper alloy and b = the number of grams of the 50% copper alloy. Use a table to organize the information.

	25% Copper	50% Copper	45% Copper	
Total Grams	a	b	1000	$\rightarrow a + b = 1000$
Grams of Copper	$0.25a$	$0.50b$	$0.45(1000)$	$\rightarrow 0.25a + 0.5b = 0.45(1000)$

$a + b = 1000$ First equation

$a + b - b = 1000 - b$ Subtract b from each side.

$a = 1000 - b$ Simplify.

$0.25a + 0.50b = 0.45(1000)$ Second equation

$0.25(1000 - b) + 0.50b = 0.45(1000)$ $a = 1000 - b$

$250 - 0.25b + 0.50b = 450$ Distributive Property

$250 + 0.25b = 450$ Combine like terms.

$250 + 0.25b - 250 = 450 - 250$ Subtract 250 from each side.

$0.25b = 200$ Simplify.

$\frac{0.25b}{0.25} = \frac{200}{0.25}$ Divide each side by 0.25.

$b = 800$ Simplify.

$$a + b = 1000 \quad \text{First equation}$$
$$a + 800 = 1000 \quad b = 800$$
$$a = 200 \quad \text{Subtract 800 from each side and simplify.}$$

200 grams of the 25% alloy and 800 grams of the 50% alloy should be used.

CHECK Your Progress

3. **BASEBALL** The New York Yankees and the Cincinnati Reds together have won a total of 31 World Series. The Yankees have won 5.2 times as many as the Reds. How many World Series did each team win?

Online Personal Tutor at algebra1.com

CHECK Your Understanding

Examples 1–2 (pp. 261–262)

Use substitution to solve each system of equations. If the system does *not* have exactly one solution, state whether it has *no* solution or *infinitely many* solutions.

1. $2x + 7y = 3$
   $x = 1 - 4y$
2. $6x - 2y = -4$
   $y = 3x + 2$
3. $y = \frac{3}{5}x$
   $3x - 5y = 15$
4. $x + 3y = 12$
   $x - y = 8$
5. $a + b = 1$
   $5a + 3b = -1$
6. $\frac{1}{3}x - y = 2$
   $x - 3y = 6$

Example 3 (p. 262)

7. **TRANSPORTATION** The Thrust SSC is the world's fastest land vehicle. Suppose the driver of a car with a top speed of 200 miles per hour requests a race against the SSC. The car gets a head start of one-half hour. If there is unlimited space to race, at what distance will the SSC pass the car?

Exercises

HOMEWORK HELP	
For Exercises	See Examples
8–19	1–2
20, 21	3

Use substitution to solve each system of equations. If the system does *not* have exactly one solution, state whether it has *no* solution or *infinitely many* solutions.

8. $y = 5x$
   $2x + 3y = 34$
9. $x = 4y$
   $2x + 3y = 44$
10. $x = 4y + 5$
    $x = 3y - 2$
11. $y = 2x + 3$
    $y = 4x - 1$
12. $4c = 3d + 3$
    $c = d - 1$
13. $x = \frac{1}{2}y + 3$
    $2x - y = 6$
14. $8x + 2y = 13$
    $4x + y = 11$
15. $2x - y = -4$
    $-3x + y = -9$
16. $3x - 5y = 11$
    $x - 3y = 1$
17. $2x + 3y = 1$
    $-x + \frac{1}{3}y = 5$
18. $c - 5d = 2$
    $2c + d = 4$
19. $5r - s = 5$
    $-4r + 5s = 17$

Review Vocabulary

Supplementary Angles two angles with measures that have the sum of 180 degrees

20. **GEOMETRY** Angles X and Y are supplementary, and the measure of angle X is 24 degrees greater than the measure of angle Y. Find the angle measures.

21. CHEMISTRY MX Labs needs to make 500 gallons of a 34% acid solution. The only solutions available are a 25% acid solution and a 50% acid solution. How many gallons of each solution should be mixed to make the 34% solution?

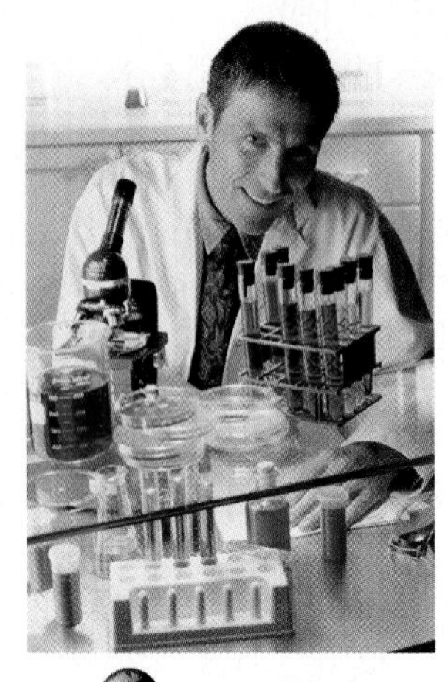

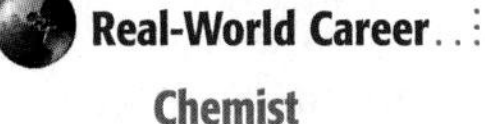

Real-World Career

Chemist

Many chemists study the properties and composition of substances. This research can be applied to the develpment of products including cosmetics, pharmaceuticals, and cleaning products.

For more information, go to algebra1.com.

Use substitution to solve each system of equations. If the system does *not* have exactly one solution, state whether it has *no* solutions or *infinitely many* solutions.

22. $x - 3y = 0$
$3x + y = 7$

23. $-0.3x + y = 0.5$
$0.5x - 0.3y = 1.9$

24. $0.5x - 2y = 17$
$2x + y = 104$

25. $y = \frac{1}{2}x + 3$
$y = 2x - 1$

JOBS For Exercises 26 and 27, use the following information.

Shantel Jones has a job offer in which she will receive $600 per month plus a commission of 2% of the total price of the cars she sells. At her current job, she receives $1000 per month plus a commission of 1.5% of her total sales.

26. What is the total price of the cars that Ms. Jones must sell each month to make the same income from either dealership?

27. How much must Ms. Jones sell to make the new job a better deal?

28. LANDSCAPING A blue spruce grows an average of 6 inches per year. A hemlock grows an average of 4 inches per year. If a blue spruce is 4 feet tall and a hemlock is 6 feet tall, write a system of equations to represent their growth. Find and interpret the solution in the context of the situation.

29. ANALYZE TABLES The table shows the approximate number of tourists in two areas during a recent year and the average rates of change in tourism. If the trends continue, in how many years would you expect the number of tourists to the regions to be equal?

Destination	Number of Tourists	Average Rates of Change in Tourists (millions per year)
South America and the Caribbean	40.3 million	increase of 0.8
Middle East	17.0 million	increase of 1.8

30. RESEARCH Use the Internet or other resource to find the pricing plans for various cell phones. Determine the number of minutes you would need to use the phone for two plans to cost the same amount of money.

H.O.T. Problems

31. FIND THE ERROR In the system $a + b = 7$ and $1.29a + 0.49b = 6.63$, a represents the pounds of apples bought and b represents pounds of bananas. Josh and Lydia are finding and interpreting the solution. Who is correct? Explain.

Josh

$1.29a + 0.49b = 6.63$
$1.29a + 0.49(7 - a) = 6.63$
$1.29a + 3.43 - 0.49a = 6.63$
$0.8a = 3.2$
$a = 4$

$a + b = 7$, so $b = 3$. The solution $(4, 3)$ means that 4 apples and 3 bananas were bought.

Lydia

$1.29a + 0.49b = 6.63$
$1.29(7 - b) + 0.49b = 6.63$
$9.03 - 1.29b + 0.49b = 6.63$
$-0.8b = -2.4$
$b = 3$

The solution $b = 3$ means that 3 apples and 3 bananas were bought.

32. **CHALLENGE** Solve the system of equations below. Write the solution as an ordered triple of the form (x, y, z). Describe the steps that you used.

$$2x + 3y - z = 17 \qquad y = -3z - 7 \qquad 2x = z + 2$$

33. **OPEN ENDED** Create a system of equations that has one solution. Illustrate how the system could represent a real-world situation and describe the significance of the solution in the context of the situation.

34. ***Writing in Math*** Use the data about the time Americans spend online and reading newspapers on page 260 to explain how substitution can be used to analyze problems. Then solve the system and interpret its meaning in the situation.

A STANDARDIZED TEST PRACTICE

35. The debate team plans to make and sell trail mix for a fundraiser.

Item	Cost Per Pound
sunflower seeds	\$4.00
raisins	\$1.50

The number of pounds of raisins in the mix is to be 3 times the pounds of sunflower seeds. If they can spend \$34, which system can be used to find r, the pounds of raisins, and s, pounds of sunflower seeds, they should buy?

A $3s = r$
$4s + 1.5r = 34$

B $3r = s$
$4s + 1.5r = 34$

C $3s = r$
$4r + 1.5s = 34$

D $3r = s$
$4r + 1.5s = 34$

36. What is the solution to this system of equations?

$$\begin{cases} x + 4y = 1 \\ 2x - 3y = -9 \end{cases}$$

F $(2, -8)$

G $(-3, 1)$

H no solution

J infinitely many solutions

37. **REVIEW** What is the value of x if $4x - 3 = -2x$?

A -2

B $-\frac{1}{2}$

C $\frac{1}{2}$

D 2

Spiral Review

Graph each system of equations. Then determine whether the system has *no* solution, *one* solution, or *infinitely many* solutions. If the system has one solution, name it. (Lesson 5-1)

38. $x + y = 3$
$x + y = 4$

39. $x + 2y = 1$
$2x + y = 5$

40. $2x + y = 3$
$4x + 2y = 6$

41. Draw a scatter plot that shows a positive correlation. (Lesson 4-7)

42. **RECYCLING** When a pair of blue jeans is made, the leftover denim scraps can be recycled. One pound of denim is left after making every fifth pair of jeans. How many pounds of denim would be left from 250 pairs of jeans? (Lesson 2-6)

GET READY for the Next Lesson

PREREQUISITE SKILL Simplify each expression. (Lesson 1-5)

43. $6a - 9a$

44. $8t + 4t$

45. $-7g - 8g$

46. $7d - (2d + b)$

47. $(2x + 5y) + (x - 2y)$

48. $(3m + 2n) - (5m + 7n)$

5-3 Elimination Using Addition and Subtraction

Main Ideas

- Solve systems of equations algebraically by using elimination with addition.
- Solve systems of equations algebraically by using elimination with subtraction.

New Vocabulary

elimination

GET READY for the Lesson

The winter solstice marks the shortest day and longest night of the year in the Northern Hemisphere. On that day in Seward, Alaska, the difference between the number of hours of darkness n and the number of hours of daylight d is 12. The following system of equations represents the situation.

$$n + d = 24$$
$$n - d = 12$$

Notice that if you add these equations, the variable d is eliminated.

$$\begin{array}{rl} n + d &= 24 \\ (+)\ n - d &= 12 \\ \hline 2n \quad\quad &= 36 \end{array}$$

Elimination Using Addition Sometimes adding two equations together will eliminate one variable. Using this step to solve a system of equations is called **elimination**.

EXAMPLE Elimination Using Addition

Use elimination to solve the system of equations.

$\mathbf{3x - 5y = -16}$

$\mathbf{2x + 5y = 31}$

Since the coefficients of the y-terms, -5 and 5, are additive inverses, you can eliminate these terms by adding the equations.

$$\begin{array}{rcl} 3x - 5y &=& -16 \\ (+)\ 2x + 5y &=& 31 \\ \hline 5x \quad\quad &=& 15 \end{array}$$

Write the equations in column form and add.

The y variable is eliminated.

$\frac{5x}{5} = \frac{15}{5}$ Divide each side by 5.

$x = 3$ Simplify.

Now substitute 3 for x in either equation to find the value of y.

$3x - 5y = -16$ First equation

$3(3) - 5y = -16$ Replace x with 3.

$9 - 5y = -16$ Simplify.

$9 - 5y - 9 = -16 - 9$ Subtract 9 from each side.

$-5y = -25$ Simplify.

$\frac{-5y}{-5} = \frac{-25}{-5}$ Divide each side by −5.

$y = 5$ Simplify.

The solution is (3, 5).

CHECK Your Progress

Use elimination to solve each system of equations.

1A. $-4x + 3y = -3$
$4x - 5y = 5$

1B. $4y + 3x = 22$
$3x - 4y = 14$

EXAMPLE Write and Solve a System of Equations

Study Tip

Eliminating Variables

When solving systems by elimination, be sure to add like terms.

2 **Twice one number added to another number is 18. Four times the first number minus the other number is 12. Find the numbers.**

Let x represent the first number and y represent the second number.

Twice one number	added to	another number	is	18.
$2x$	$+$	y	$=$	18

Four times the first number	minus	the other number	is	12.
$4x$	$-$	y	$=$	12

Use elimination to solve the system.

$$\begin{aligned} 2x + y &= 18 \\ (+)\ 4x - y &= 12 \\ \hline 6x &= 30 \end{aligned}$$

Write the equations in column form and add.

The variable y is eliminated.

$\frac{6x}{6} = \frac{30}{6}$ Divide each side by 6.

$x = 5$ Simplify.

Now substitute 5 for x in either equation to find the value of y.

$4x - y = 12$ Second equation

$4(5) - y = 12$ Replace x with 5.

$20 - y = 12$ Simplify.

$-y = -8$ Simplify.

$\frac{-y}{-1} = \frac{-8}{-1}$ Divide each side by −1.

$y = 8$

The numbers are 5 and 8.

CHECK Your Progress

2. The sum of two numbers is −10. Negative three times the first number minus the second number equals 2. Find the numbers.

Online **Personal Tutor at** algebra1.com

Elimination Using Subtraction Sometimes subtracting one equation from another will eliminate one variable.

EXAMPLE Elimination Using Subtraction

3 **Use elimination to solve the system of equations.**

$5s + 2t = 6$

$9s + 2t = 22$

Since the coefficients of the t-terms, 2 and 2, are the same, you can eliminate the t-terms by subtracting the equations.

$$\begin{array}{rl} 5s + 2t = 6 & \text{Write the equations in column form and subtract.} \\ (-)\ 9s + 2t = 22 & \\ \hline -4s = -16 & \text{The variable } t \text{ is eliminated.} \\ \frac{-4s}{-4} = \frac{-16}{-4} & \text{Divide each side by } -4. \\ s = 4 & \text{Simplify.} \end{array}$$

Now substitute 4 for s in either equation to find the value of t.

$$\begin{array}{rl} 5s + 2t = 6 & \text{First equation} \\ 5(4) + 2t = 6 & s = 4 \\ 20 + 2t = 6 & \text{Simplify.} \\ 2t = -14 & \text{Subtract 20 from each side and simplify.} \\ \frac{2t}{2} = -\frac{14}{2} & \text{Divide each side by 2.} \\ t = -7 & \text{The solution is } (4, -7). \end{array}$$

CHECK Your Progress

Use elimination to solve each system of equations.

3A. $8b + 3c = 11$
$8b + 7c = 7$

3B. $12n - m = -14$
$6n - m = -8$

CHECK Your Understanding

Examples 1, 3 (pp. 266–268)

Use elimination to solve each system of equations.

1. $2a - 3b = -11$
$a + 3b = 8$

2. $4x + y = -9$
$4x + 2y = -10$

3. $6x + 2y = -10$
$2x + 2y = -10$

4. $-4m + 2n = 6$
$-4m + n = 8$

Example 2 (p. 267)

5. The sum of two numbers is 24. Five times the first number minus the second number is 12. What are the two numbers?

6. FOOTBALL In 2003, Rich Gannon, the Oakland Raiders quarterback, earned about $4 million more than Charles Woodson, the Raiders cornerback. Together they cost the Raiders approximately $9 million. How much did each make?

Math Online **Extra Examples at** algebra1.com

Exercises

HOMEWORK HELP	
For Exercises	See Examples
7–18	1, 3
19–24	2

Use elimination to solve each system of equations.

7. $x + y = -3$
$x - y = 1$

8. $s - t = 4$
$s + t = 2$

9. $3m - 2n = 13$
$m + 2n = 7$

10. $-4x + 2y = 8$
$4x - 3y = -10$

11. $3a + b = 5$
$2a + b = 10$

12. $2m - 5n = -6$
$2m - 7n = -14$

13. $3r - 5s = -35$
$2r - 5s = -30$

14. $13a + 5b = -11$
$13a + 11b = 7$

15. $3x - 5y = 16$
$-3x + 2y = -10$

16. $6s + 5t = 1$
$6s - 5t = 11$

17. $4x - 3y = 12$
$4x + 3y = 24$

18. $a - 2b = 5$
$3a - 2b = 9$

19. The sum of two numbers is 48, and their difference is 24. What are the numbers?

20. Find the two numbers whose sum is 51 and whose difference is 13.

21. Three times one number added to another number is 18. Twice the first number minus the other number is 12. Find the numbers.

22. One number added to twice another number is 13. Four times the first number added to twice the other number is -2. What are the numbers?

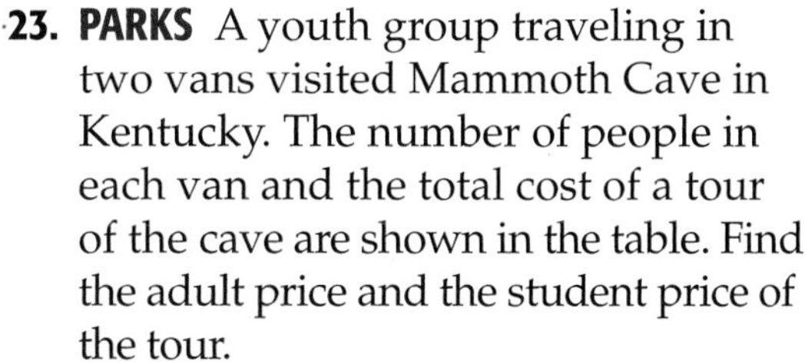

23. PARKS A youth group traveling in two vans visited Mammoth Cave in Kentucky. The number of people in each van and the total cost of a tour of the cave are shown in the table. Find the adult price and the student price of the tour.

Van	Number of Adults	Number of Students	Total Cost
A	2	5	\$77
B	2	7	\$95

24. ARCHITECTURE The total height of an office building b and the granite statue that stands on top of it g is 326.6 feet. The difference in heights between the building and the statue is 295.4 feet. How tall is the statue?

Real-World Link

On average, 2 million people visit Mammoth Cave National Park each year. On a busy summer day, about 5000 to 7000 people come to the park.

Source: Mammoth Cave National Park

ANALYZE GRAPHS For Exercises 25–27, use the information in the graph.

25. Let x represent the number of years since 2000 and y represent population in billions. Write an equation to represent the population of China.

26. Write an equation to represent the population of India.

27. Use elimination to find the solution to the system of equations. Interpret the solution.

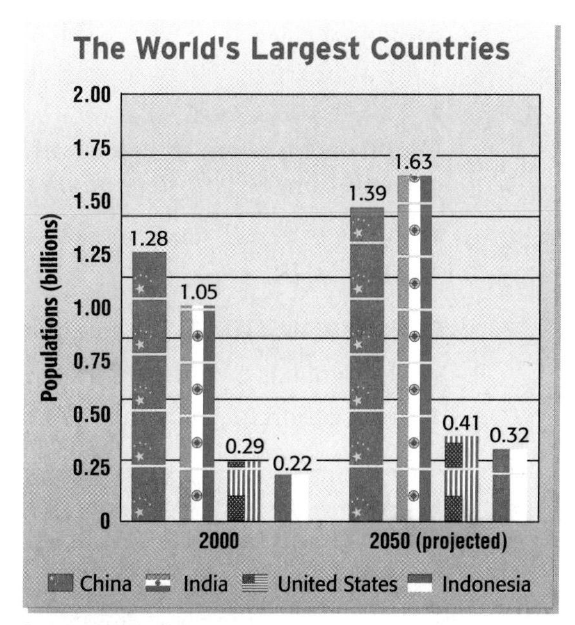

Source: Population Reference Bureau

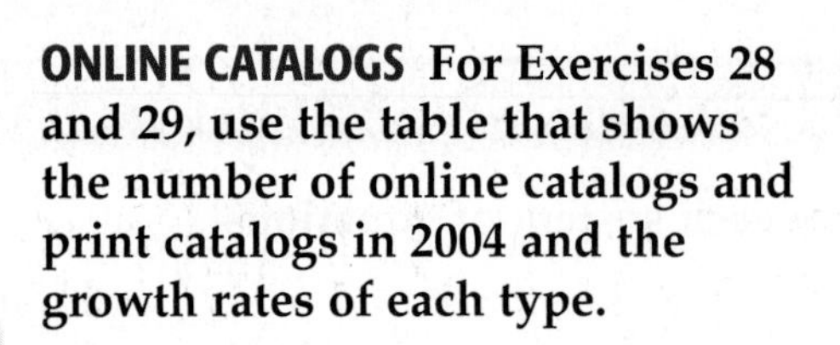

ONLINE CATALOGS For Exercises 28 and 29, use the table that shows the number of online catalogs and print catalogs in 2004 and the growth rates of each type.

Catalogs	Number in 2004	Growth Rate (number per year)
Online	7440	1293
Print	3805	−1364

Source: MediaPost Publications

28. Let x represent the number of years since 2004 and y represent the number of catalogs. Write a system of equations to represent this situation.

29. Solve the system of equations. Analyze the solution in terms of the situation. Determine the reasonableness of the solution.

H.O.T. Problems

30. OPEN ENDED Create a system of equations that can be solved by adding to eliminate a variable. Formulate a general rule for creating such systems.

31. CHALLENGE The graphs of $Ax + By = 15$ and $Ax - By = 9$ intersect at (2, 1). Find A and B and describe the steps that you used to find the values.

32. ***Writing in Math*** Use the information on page 266 to explain how to use elimination to solve a system of equations. Include a step-by-step solution of the Seward daylight problem.

STANDARDIZED TEST PRACTICE

33. What is the solution to this system of equations?

$2x - 3y = -9$
$-x + 3y = 6$

A (3, 3) **C** (−3, 3)

B (−3, 1) **D** (1, −3)

34. REVIEW Rhiannon is paid \$52 for working 4 hours. At this rate, how many hours of work will it take her to earn \$845?

F 13 hours **H** 3380 hours

G 65 hours **J** 10,985 hours

Spiral Review

Use substitution to solve each system of equations. If the system does *not* have exactly one solution, state whether it has *no* solution or *infinitely many* solutions. (Lesson 5-2)

35. $y = 5x$
$x + 2y = 22$

36. $x = 2y + 3$
$3x + 4y = -1$

37. $2y - x = -5$
$4y - 3x = -1$

Graph each system of equations. Then determine whether the system has *no* solution, *one* solution, or *infinitely many* solutions. If the system has one solution, name it. (Lesson 5-1)

38. $x - y = 3$
$3x + y = 1$

39. $2x - 3y = 7$
$3y = 7 + 2x$

40. $4x + y = 12$
$x = 3 - \frac{1}{4}y$

41. PHYSICAL SCIENCE If x cubic centimeters of water is frozen, the ice that is formed has a volume of $\left(x + \frac{1}{11}x\right)$ cubic centimeters. Simplify the expression for the volume of the ice. (Lesson 1-5)

GET READY for the Next Lesson

PREREQUISITE SKILL Rewrite each expression without parentheses. (Lesson 1-5)

42. $2(3x + 4y)$ **43.** $6(2a - 5b)$ **44.** $-3(-2m + 3n)$ **45.** $-5(4t - 2s)$

CHAPTER 5

Mid-Chapter Quiz

Lessons 5-1 through 5-3

Graph each system of equations. Then determine whether the system has *no* solution, *one* solution, or *infinitely many* solutions. If the system has one solution, name it. (Lesson 5-1)

1. $y = -x - 1$
$y = x + 5$

2. $x + y = 3$
$x - y = 1$

3. $3x - 2y = -6$
$3x - 2y = 6$

4. $3x + 2y = 4$
$6x + 4y = 8$

WORLD RECORDS For Exercises 5 and 6, use the following information.

A swimmer broke a world record by crossing the Atlantic Ocean on a raft. He traveled about 44 miles each day by swimming s hours and floating on a raft f hours. The rates that he traveled are shown in the table. (Lesson 5-1)

Activity	Rate
swimming	3 mph
floating	1 mph

5. Write a system of equations to represent this situation.

6. Graph the system. Describe what the solution means in the context of the problem.

7. MULTIPLE CHOICE The graphs of the linear equations $y = -\frac{1}{2}x + 4$ and $y = x - 2$ are shown below. If $-\frac{1}{2}x + 4 = x - 2$, what is the value of x? (Lesson 5-1)

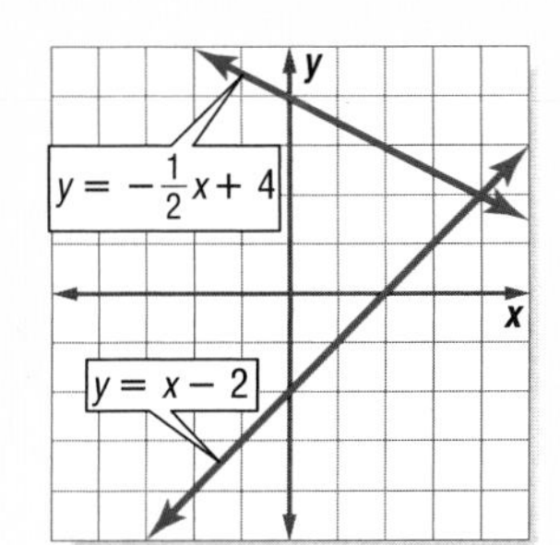

A 2

B 4

C 6

D 8

Use substitution to solve each system of equations. If the system does *not* have exactly one solution, state whether it has *no* solution or *infinitely many* solutions. (Lesson 5-2)

8. $y = 6x$
$4x + y = 10$

9. $c + d = 0$
$3c + d = -8$

10. $x - 2y = 5$
$3x - 5y = 8$

11. $x + y = 2$
$y = 2 - x$

12. MULTIPLE CHOICE Sydney has \$115 and she is earning \$50 per week at her summer job. Felipe has \$130 and is earning \$50 per week at his summer job. Which is a true statement about the system of equations that represents this situation? (Lesson 5-2)

F The system has 1 solution, which represents their hourly rates.

G The system has 1 solution, which represents the number of weeks in which they will have earned the same amount of money.

H The system has infinitely many solutions.

J The system has no solution.

Use elimination to solve each system of equations. (Lesson 5-3)

13. $a + b = 9$
$a - b = 7$

14. $3x + y = 1$
$-6x + y = 10$

15. $5x + 4y = 2$
$3x - 4y = 14$

16. $2s - 3t = 13$
$2s + 2t = -2$

WATER PARKS For Exercises 17 and 18, use the following information.

The cost of two groups going to a water park is shown in the table. (Lesson 5-3)

Group	Total Cost (\$)
2 adults, 1 child	68.97
2 adults, 4 children	125.94

17. Define variables and write a system of equations that you can use to find the admission cost of an adult and a child.

18. Solve the system of equations and explain what the solution means.

5-4 Elimination Using Multiplication

Main Ideas

- Solve systems of equations algebraically by using elimination with multiplication.
- Solve real-world problems involving systems of equations.

GET READY for the Lesson

The Finneytown Bakery is making peanut butter cookies and loaves of quick bread. The preparation and baking times for each are shown.

For these two items, the management has allotted 800 minutes of employee time and 900 minutes of oven time. The following system of equations can be used to determine how many of each to bake.

$20c + 10b = 800$
$10c + 30b = 900$

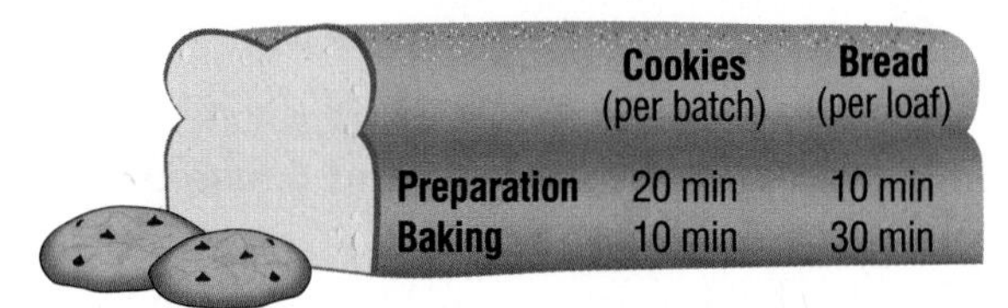

	Cookies (per batch)	Bread (per loaf)
Preparation	20 min	10 min
Baking	10 min	30 min

Elimination Using Multiplication Neither variable in the system above can be eliminated by simply adding or subtracting the equations. However, you can use the Multiplication Property of Equality so that adding or subtracting eliminates one of the variables.

EXAMPLE Multiply One Equation to Eliminate

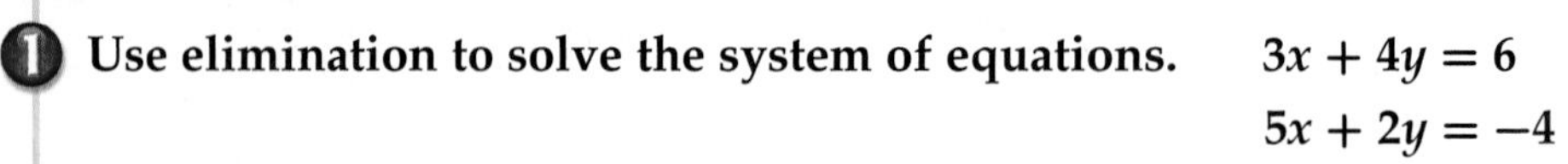

1 **Use elimination to solve the system of equations.** $3x + 4y = 6$
$5x + 2y = -4$

Multiply the second equation by -2 so the coefficients of the y-terms are additive inverses. Then add the equations.

$$\begin{array}{llrl} 3x + 4y = 6 & & 3x + 4y = 6 & \\ 5x + 2y = -4 & \text{Multiply by } -2. & (+)\ -10x - 4y = 8 & \\ \hline & & -7x \quad = 14 & \text{Add the equations.} \\ & & x = -2 & \text{Divide each side by } -7. \end{array}$$

Now substitute -2 for x in either equation to find the value of y.

$$\begin{aligned} 3x + 4y &= 6 && \text{First equation} \\ 3(-2) + 4y &= 6 && x = -2 \\ -6 + 4y &= 6 && \text{Simplify.} \\ 4y &= 12 && \text{Add 6 to each side and simplify.} \\ \frac{4y}{4} &= \frac{12}{4} && \text{Divide each side by 4 and simplify.} \\ y &= 3 && \text{The solution is } (-2, 3). \end{aligned}$$

CHECK Your Progress

Use elimination to solve each system of equations.

1A. $6x - 2y = 10$
$3x - 7y = -19$

1B. $9p + q = 13$
$3p + 2q = -4$

Sometimes you have to multiply each equation by a different number in order to solve the system.

EXAMPLE Multiply Both Equations to Eliminate

Use elimination to solve the system of equations. $3x + 4y = -25$
$2x - 3y = 6$

Method 1 Eliminate x.

$3x + 4y = -25$ Multiply by 2. $6x + 8y = -50$
$2x - 3y = 6$ Multiply by -3. $(+)\ -6x + 9y = -18$

$$
\begin{aligned}
17y &= -68 && \text{Add the equations.}\\
\frac{17y}{17} &= \frac{-68}{17} && \text{Divide each side by 17.}\\
y &= -4 && \text{Simplify.}
\end{aligned}
$$

Now substitute -4 for y in either equation to find the value of x.

$$
\begin{aligned}
2x - 3y &= 6 && \text{Second equation}\\
2x - 3(-4) &= 6 && y = -4\\
2x + 12 &= 6 && \text{Simplify.}\\
2x + 12 - 12 &= 6 - 12 && \text{Subtract 12 from each side.}\\
2x &= -6 && \text{Simplify.}\\
\frac{2x}{2} &= \frac{-6}{2} && \text{Divide each side by 2 and simplify.}\\
x &= -3 && \text{The solution is } (-3, -4).
\end{aligned}
$$

Method 2 Eliminate y.

$3x + 4y = -25$ Multiply by 3. $9x + 12y = -75$
$2x - 3y = 6$ Multiply by 4. $(+)\ 8x - 12y = 24$

$$
\begin{aligned}
17x &= -51 && \text{Add the equations.}\\
\frac{17x}{17} &= \frac{-51}{17} && \text{Divide each side by 17.}\\
x &= -3 && \text{Simplify.}
\end{aligned}
$$

Now substitute -3 for x in either equation to find the value of y.

$$
\begin{aligned}
2x - 3y &= 6 && \text{Second equation}\\
2(-3) - 3y &= 6 && x = -3\\
-6 - 3y &= 6 && \text{Simplify.}\\
-6 - 3y + 6 &= 6 + 6 && \text{Add 6 to each side.}\\
-3y &= 12 && \text{Simplify.}\\
\frac{-3y}{-3} &= \frac{12}{-3} && \text{Divide each side by } -3.\\
y &= -4 && \text{Simplify.}
\end{aligned}
$$

The solution is $(-3, -4)$, which matches the result obtained with Method 1.

Study Tip

Using Multiplication

There are many other combinations of multipliers that could be used to solve the system in Example 2. For instance, the first equation could be multiplied by -2 and the second by 3.

CHECK Your Progress

Use elimination to solve each system of equations.

2A. $5x - 3y = 6$
$2x - 5y = 10$

2B. $6a + 2b = 2$
$4a + 3b = 8$

STANDARDIZED TEST EXAMPLE Writing Systems of Equations

3 **Anita has a total of 28 e-mail addresses of family and friends stored on her PDA. Twice the number of family addresses minus the number of friends' addresses is 2. Which system of equations can be used to find how many e-mail addresses of family *a* and friends *b* that Anita has stored?**

A $a + b = 28$, $2a - b = 2$ **B** $a + b = 28$, $2a + b = 2$ **C** $a - b = 28$, $2a - b = 2$ **D** $a - b = 28$, $a - 2b = 2$

Read the Test Item

You are asked to find a system of equations to represent this situation using *a*, the number of family addresses, and *b*, the number of friends' addresses.

Solve the Test Item

Represent the situation algebraically by writing two equations.

The total of family's and friends' e-mail addresses is 28.

$$a + b = 28$$

One equation is $a + b = 28$.

Twice the number of family addresses minus the number of friends' addresses is 2.

$$2a - b = 2$$

The second equation is $2a - b = 2$.

The system of equations that represents this situation is $a + b = 28$ and $2a - b = 2$. The answer is A.

Test-Taking Tip

Checking the Complete Answer Since choices A and B both have $a + b = 28$ as one of the equations, be careful to choose the answer with the correct second equation.

CHECK Your Progress

3. The cost of 4 notebooks and 5 pens is $20. The cost of 6 notebooks and 2 pens is $19. Which system of equations can be used to find the cost of a notebook *n* and the cost of a pen *p*?

F $4n + 5p = 20$, $2n + 6p = 19$ **G** $5n + 4p = 20$, $6n + 2p = 19$ **H** $4n + 5p = 20$, $6n + 2p = 19$ **J** $5n + 4p = 20$, $2n + 6p = 19$

Real-World Problems When formulating systems of linear equations to solve real-world problems, be sure to determine the reasonableness of the solutions. Recall that only positive values will make sense as solutions for most real-world problems.

Real-World Link

About 203 million tons of freight are transported on the Ohio River each year, making it the second most used commercial river in the United States.

Source: *World Book Encyclopedia*

EXAMPLE Write and Solve a System of Equations

4 **TRANSPORTATION A coal barge on the Ohio River travels 24 miles upstream in 3 hours. The return trip takes the barge only 2 hours. Find the rate of the barge in still water.**

ESTIMATE The average speed upstream is $\frac{24}{3}$, or 8 mph. The average speed downstream is $\frac{24}{2}$ or 12 mph. So, the rate in still water should be between 8 and 12 miles per hour.

WORDS You know the distance traveled each way and the times spent traveling.

VARIABLES Let b = the rate of the barge in still water and c = the rate of the current. Use the formula rate × time = distance, or $rt = d$.

EQUATIONS

	r	t	d	$rt = d$	
Downstream	$b + c$	2	24	$(b + c)2 = 24$	→ $2b + 2c = 24$
Upstream	$b - c$	3	24	$(b - c)3 = 24$	→ $3b - 3c = 24$

Use elimination with multiplication to solve this system. Since the problem asks for b, eliminate c.

$2b + 2c = 24$ Multiply by 3. $\quad 6b + 6c = 72$

$3b - 3c = 24$ Multiply by 2. $\quad (+)\ 6b - 6c = 48$

$12b = 120$ Add the equations.

$\frac{12b}{12} = \frac{120}{12}$ Divide each side by 12.

$b = 10$ Simplify.

The rate of the barge in still water is 10 miles per hour. This solution is close to the estimate since it is between 8 and 12 miles per hour. So, the solution is reasonable.

CHECK Your Progress

4. **CANOEING** A canoe travels 4 miles upstream in 1 hour. The return trip takes the canoe 1.5 hours. Find the rate of the boat in still water.

CHECK Your Understanding

Use elimination to solve each system of equations.

Examples 1 (p. 272)

1. $2x - y = 6$
   $3x + 4y = -2$

2. $2x + 7y = 1$
   $x + 5y = 2$

Example 2 (p. 273)

3. $4x + 7y = 6$
   $6x + 5y = 20$

4. $9a - 2b = -8$
   $-7a + 3b = 12$

Example 3 (p. 274)

5. **MULTIPLE CHOICE** At a restaurant, the cost for 2 burritos and 1 tortilla salad is \$20.57. The cost for 3 burritos and 3 tortilla salads is \$36.24. Which pair of equations can be used to determine b, the cost of a burrito, and t, the cost of a tortilla salad?

   A $b + t = 20.57$
   $3b + 3t = 36.24$

   B $2b + t = 20.57$
   $3b + 3t = 36.24$

   C $2b + t = 20.57$
   $b + t = 36.24$

   D $b + 2t = 20.57$
   $b + t = 36.24$

Example 4 (pp. 274–275)

6. **BUSINESS** The owners of the River View Restaurant have hired enough servers to handle 17 tables of customers, and the fire marshal has approved the restaurant for a limit of 56 customers. How many two-seat tables and how many four-seat tables should the owners purchase?

Exercises

HOMEWORK	
For Exercises	See Examples
7–14	1, 2
15, 16	4
17, 18	3

Use elimination to solve each system of equations.

7. $x + y = 3$
$2x - 3y = 16$

8. $-5x + 3y = 6$
$x - y = 4$

9. $2x + y = 5$
$3x - 2y = 4$

10. $4x - 3y = 12$
$x + 2y = 14$

11. $8x - 3y = -11$
$2x - 5y = 27$

12. $5x - 2y = -15$
$3x + 3y = 12$

13. $4x - 7y = 10$
$3x + 2y = -7$

14. $2x - 3y = 2$
$5x + 4y = 28$

15. $12x - 3y = -3$
$6x + y = 1$

16. $-4x + 2y = 0$
$10x + 3y = 8$

17. NUMBER THEORY Seven times a number plus three times another number equals negative one. The sum of the two numbers is negative three. What are the numbers?

18. BASKETBALL In basketball, a free throw is 1 point and a field goal is either 2 or 3 points. In a recent season, Tim Duncan of the San Antonio Spurs scored a total of 1342 points. The total number of 2-point field goals and 3-point field goals was 517, and he made 305 of the 455 free throws that he attempted. Find the number of 2-point field goals and 3-point field goals Duncan made that season.

Use elimination to solve each system of equations.

19. $1.8x - 0.3y = 14.4$
$x - 0.6y = 2.8$

20. $0.4x + 0.5y = 2.5$
$1.2x - 3.5y = 2.5$

21. $3x - \frac{1}{2}y = 10$
$5x + \frac{1}{4}y = 8$

22. $2x + \frac{2}{3}y = 4$
$x - \frac{1}{2}y = 7$

23. NUMBER THEORY The sum of the digits of a two-digit number is 14. If the digits are reversed, the new number is 18 less than the original number. Find the original number.

ANALYZE TABLES For Exercises 24 and 25, use the information below.
At an entertainment center, two groups of people bought batting tokens and miniature golf games, as shown in the table.

Group	Number of Batting Tokens	Number of Miniature Golf Games	Total Cost
A	16	3	$30
B	22	5	$43

24. Define variables and formulate a system of linear equations from this situation.

25. Solve the system of equations and explain what the solution represents in terms of the situation.

26. **TRANSPORTATION** Traveling against the wind, a plane flies 1080 miles from Omaha, Nebraska, to San Diego, California, in the time shown at right. On the return trip, the plane is traveling with a wind that is twice as fast. Find the rate of the plane in still air.

Trip	Time
traveling against the wind	2 h 30 min
traveling with the wind	2 h

27. **GEOMETRY** The graphs of $x + 2y = 6$ and $2x + y = 9$ contain two of the sides of a triangle. A vertex of the triangle is at the intersection of the graphs. What are the coordinates of the vertex?

28. **TESTS** Mrs. Henderson discovered that she had accidentally reversed the digits of a test score and shorted a student 36 points. Mrs. Henderson told the student that the sum of the digits was 14 and agreed to give the student his correct score plus extra credit if he could determine his actual score without looking at his test. What was his actual score on the test?

H.O.T. Problems

29. **REASONING** Explain why multiplication is sometimes needed to solve a system of equations by elimination.

30. **FIND THE ERROR** David and Yoomee are solving a system of equations. Who is correct? Explain your reasoning.

David

$2r + 7s = 11 \Rightarrow$
$r - 9s = -7 \Rightarrow$

$2r + 7s = 11$
$(-)\ 2r - 18s = -14$
$25s = 25$
$s = 1$

$2r + 7s = 11$
$2r + 7(1) = 11$
$2r + 7 = 11$
$2r = 4$
$\frac{2r}{2} = \frac{4}{2}$
$r = 2$

The solution is (2, 1).

Yoomee

$2r + 7s = 11$
$(-)\ r - 9s = 7$
$r \quad = 18$

$2r + 7s = 11$
$2(18) + 7s = 11$
$36 + 7s = 11$
$7s = -25$
$\frac{7s}{7} = -\frac{25}{7}$
$s = -3.6$

The solution is (18, -3.6).

31. **OPEN ENDED** Formulate a system of equations that could be solved by multiplying one equation by 5 and then adding the two equations together to eliminate one of the variables.

32. **CHALLENGE** The solution of the system $4x + 5y = 2$ and $6x - 2y = b$ is $(3, a)$. Find the values of a and b. Discuss the steps that you used.

33. ***Writing in Math*** Use the information about the bakery on page 272 to explain how a manager can use a system of equations to plan employee time. Include a demonstration of how to solve the system of equations given in the problem and an explanation of how a restaurant manager would schedule oven and employee time.

STANDARDIZED TEST PRACTICE

34. If $5x + 3y = 12$ and $4x - 5y = 17$, what is the value of y?

A -1 **C** $(-1, 3)$

B 3 **D** $(3, -1)$

35. What is the solution to the system of equations?

$x + 2y = -1$

$2x + 4y = -2$

F $(-1, -1)$

G $(2, 1)$

H no solution

J infinitely many solutions

36. REVIEW What is the surface area of the rectangular solid shown below?

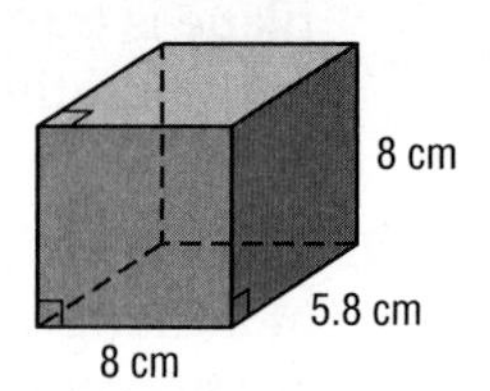

A 249.6 cm^2

B 278.4 cm^2

C 313.6 cm^2

D 371.2 cm^2

Spiral Review

Use elimination to solve each system of equations. (Lesson 5-3)

37. $x + y = 8$
$x - y = 4$

38. $2r + s = 5$
$r - s = 1$

39. $x + y = 18$
$x + 2y = 25$

Use substitution to solve each system of equations. If the system does *not* have exactly one solution, state whether it has *no* solution or *infinitely many* solutions. (Lesson 5-2)

40. $2x + 3y = 3$
$x = -3y$

41. $x + y = 0$
$3x + y = -8$

42. $x - 2y = 7$
$-3x + 6y = -21$

43. PAINTING A ladder reaches a height of 16 feet on a wall. If the bottom of the ladder is placed 4 feet away from the wall, what is the slope of the ladder as a positive number? Explain the meaning of the slope. (Lesson 4-1)

Determine the *x*-intercept, *y*-intercept, and zero of each linear function. (Lesson 3-3)

44.

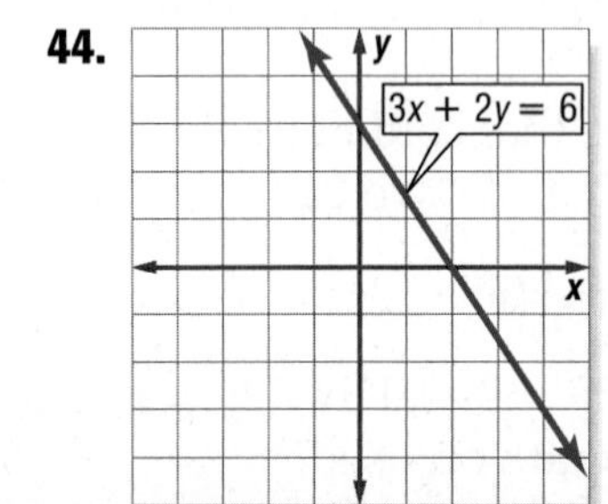

45.

x	y
−8	0
−4	1
0	2
4	3

GET READY for the Next Lesson

PREREQUISITE SKILL **Solve each equation.** (Lesson 2-4)

46. $3(x + 5) - x = 1$

47. $14 = 5(n - 1) + 9$

48. $y - 2(y + 8) = 6$

READING MATH

Making Concept Maps

After completing a chapter, it is wise to review each lesson's main topics and vocabulary. In Lesson 5-1, the new vocabulary words were *system of equations, consistent, inconsistent, independent,* and *dependent*. They are all related in that they explain how many and what kind of solutions a system of equations has.

A graphic organizer called a *concept map* is a convenient way to show these relationships. A concept map is shown below for the vocabulary words for Lesson 5-1. The main ideas are placed in boxes. Any information that describes how to move from one box to the next is placed along the arrows.

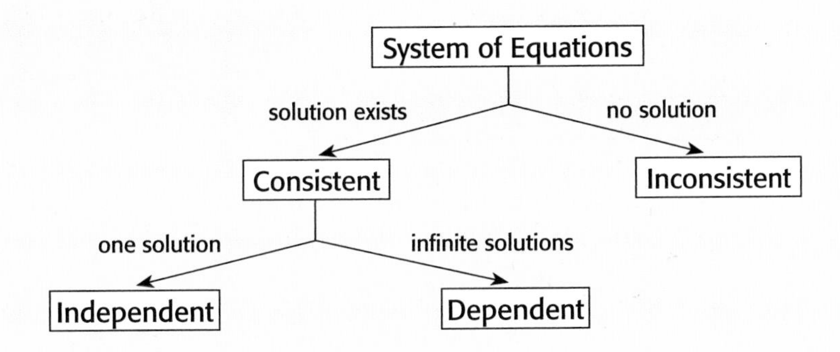

Concept maps are used to organize information. They clearly show how ideas are related to one another. They also show the flow of mental processes needed to solve problems.

Reading to Learn

Review Lessons 5-2, 5-3, and 5-4.

1. Write a couple of sentences describing the information in the concept map above.
2. How do you decide whether to use substitution or elimination? Give an example of a system that you would solve using each method.
3. How do you decide whether to multiply an equation by a factor?
4. How do you decide whether to add or subtract two equations?
5. Copy and complete the concept map at the right for solving systems of equations by using either substitution or elimination.

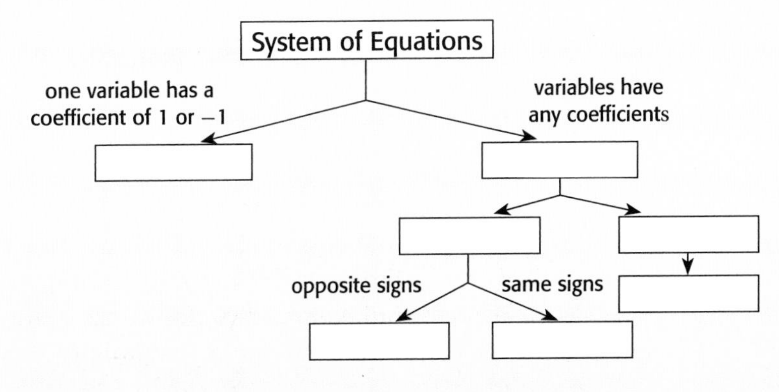

Applying Systems of Linear Equations

Main Ideas

- Determine the best method for solving systems of equations.
- Apply systems of linear equations.

GET READY for the Lesson

Northern California is home to several caves and caverns. Mercer Caverns is in Calaveras County near Murphys. Walter J. Mercer discovered the cavern in 1885. He descended to the base of the first chamber. Tours are currently offered at the caverns that descend approximately 4 times the depth that Mr. Mercer descended initially.

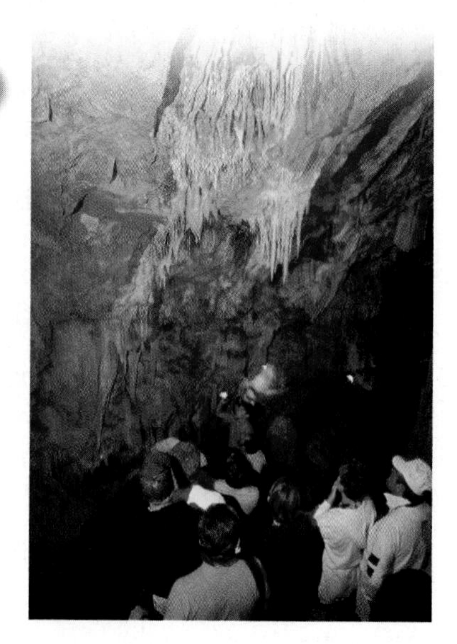

Tour	Depth (ft)
Mr. Mercer's tour	x
Current tour	y

The total depth of both tours is approximately 160 feet. How could you use a system of equations to determine the depth of each tour?

Determine the Best Method You have learned five methods for solving systems of linear equations. The table summarizes the methods and the systems for which each method works best.

KEY CONCEPT *Solving Systems of Equations*

Method	The Best Time to Use
Graphing	to estimate the solution, since graphing usually does not give an exact solution
Substitution	if one of the variables in either equation has a coefficient of 1 or -1
Elimination Using Addition	if one of the variables has opposite coefficients in the two equations
Elimination Using Subtraction	if one of the variables has the same coefficient in the two equations
Elimination Using Multiplication	if none of the coefficents are 1 or -1 and neither of the variables can be eliminated by simply adding or subtracting the equations

For an exact solution, an algebraic method is best. It is also usually a quicker method for solving linear equations than graphing. Graphing, with or without technology, is a good way to estimate solutions.

Real-World EXAMPLE — Determine the Best Method

1 SHOPPING **At a sale, Sarah bought 4 T-shirts and 3 pairs of jeans for \$181. Jenna bought 1 T-shirt and 2 pairs of jeans for \$94. The T-shirts were all the same price and the jeans were all the same price, so the following system of equations represents this situation. Determine the best method to solve the system of equations. Then solve the system.**

$4x + 3y = 181$

$x + 2y = 94$

- Since neither the coefficients of x nor the coefficients of y are the same or additive inverses, you cannot add or subtract to eliminate.
- Since the coefficient of x in the second equation is 1, you can use the substitution method. You could also use elimination using multiplication.

The following solution uses substitution. Which method would you prefer?

$x + 2y = 94$	Second equation
$x + 2y - 2y = 94 - 2y$	Subtract $2y$ from each side.
$x = 94 - 2y$	Simplify.
$4x + 3y = 181$	First equation
$4(94 - 2y) + 3y = 181$	$x = 94 - 2y$
$376 - 8y + 3y = 181$	Distributive Property
$376 - 5y = 181$	Combine like terms.
$376 - 5y - 376 = 181 - 376$	Subtract 376 from each side.
$-5y = -195$	Simplify.
$y = 39$	Divide each side by -5 and simplify.
$x + 2y = 94$	Second equation
$x + 2(39) = 94$	$y = 39$
$x + 78 = 94$	Simplify.
$x + 78 - 78 = 94 - 78$	Subtract 78 from each side.
$x = 16$	Simplify.

The solution is (16, 39). This means that the cost of a T-shirt was \$16 and the cost of a pair of jeans was \$39.

Study Tip

Alternative Method

This system could also be solved easily by multiplying the second equation by 4 and then subtracting the equations.

CHECK Your Progress

Determine the best method to solve each system of equations. Then solve the system.

1A. $5x + 7y = 2$
$-2x + 7y = 9$

1B. $3x - 4y = -10$
$5x + 8y = -2$

Apply Systems of Linear Equations When applying systems of linear equations to problem situations, it is important to analyze each solution in the context of the situation.

Real-World Link

Digital photography can trace its roots back almost 40 years, when NASA needed a technology for spacecraft to send images back to Earth.

Source: technology.com

Real-World EXAMPLE

2 PHOTOGRAPHY **Since 2000, the number of film cameras sold has decreased at an average rate of 2.5 million per year. At the same time, the number of digital cameras sold has increased at an average rate of 2.6 million per year. Use the table to estimate the year in which the sales of digital cameras equaled the sales of film cameras.**

Cameras Sold in 2000	
Type of Camera	Number Sold (millions)
film	20.0
digital	4.7

Source: Mediamark Research, Inc.

Explore You know the number of each type of camera sold in 2000 and the rates of change in numbers sold.

Plan Write an equation to represent the number of cameras sold for each type of camera. Then solve.

Solve Let x = the number of years after 2000 and let y = the total number of cameras sold.

	number sold		number sold in 2000		rate of change times number of years after 2000
film cameras	y	$=$	20.0	$+$	$-2.5x$
digital cameras	y	$=$	4.7	$+$	$2.6x$

You can use elimination by subtraction to solve this system.

$$y = 20.0 + -2.5x$$ Write the equations in column form and subtract.

$$(-)\ y = 4.7 + 2.6x$$

$$0 = 15.3 - 5.1x$$ The variable y is eliminated.

$$5.1x = 15.3$$ Add $5.1x$ to each side and simplify.

$$x = 3$$ Divide each side by 5.1 and simplify.

This means that 3 years after 2000, or in 2003, the sales of digital cameras equaled the sales of film cameras.

Check Does this solution make sense in the context of the problem? After 1 year, the number of film cameras would be about $20 - 3$ or 17. The number of digital cameras would be about $4.7 + 3$ or 7.7. Continue estimating.

Check by sketching a graph of the equations. The graphs appear to intersect at (3, 12.5), which verifies the solution of $x = 3$.

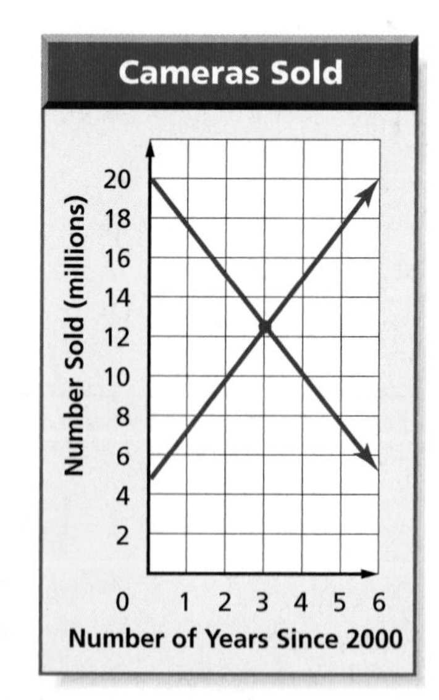

In 2003, approximately 12.5 million film and digital cameras were sold.

Study Tip

Elimination

Equations do not have to be written in standard form to use elimination by addition or subtraction.

CHECK Your Progress

2. VOLUNTEERING Jared has volunteered 50 hours and plans to continue volunteering 3 hours each week. Clementine just started volunteering 5 hours each week. Find the number of weeks in which Jared and Clementine will have both volunteered the same number of hours.

CHECK Your Understanding

Example 1 (p. 281)

Determine the best method to solve each system of equations. Then solve the system.

1. $4x + 3y = 19$
   $3x - 4y = 8$

2. $3x - 7y = 6$
   $2x + 7y = 4$

3. $y = 4x + 11$
   $3x - 2y = -7$

4. $5x - 2y = 12$
   $3x - 2y = -2$

Example 2 (p. 282)

5. **FUND-RAISING** For a Future Teachers of America fund-raiser, Denzell sold subs and pizzas as shown in the table. He sold 11 more subs than pizzas and earned a total of $233. Write and solve a system of equations to represent this situation. Then describe what the solution means.

Item	Selling Price
pizza	$5.00
sub	$3.00

Exercises

HOMEWORK HELP	
For Exercises	See Examples
6–11	1
12–13	2

Determine the best method to solve each system of equations. Then solve the system.

6. $9x - 8y = 42$
   $4x + 8y = -16$

7. $y = 3x$
   $3x + 4y = 30$

8. $x = 4y + 8$
   $2x - 8y = -3$

9. $x - y = 2$
   $5x + 3y = 18$

10. $y = 2x + 9$
    $2x - y = -9$

11. $6x - y = 9$
    $6x - y = 11$

12. **ENTERTAINMENT** Miranda has a total of 40 DVDs of movies and television shows. The number of movies is 4 less than 3 times the number of television shows. Write and solve a system of equations to find how many movies and television shows that she has on DVD.

13. **YEARBOOKS** The *break-even point* is the point at which income equals expenses. McGuffey High School is paying $13,200 for the writing and research of their yearbook, plus a printing fee of $25 per book. If they sell the books for $40 each, how many will they have to sell to break even? Explain.

Determine the best method to solve each system of equations. Then solve the system.

14. $2.3x - 1.9y = -2.5$
    $x - 0.4y = 3.6$

15. $1.6x - 0.7y = -11$
    $3.2x + 2.1y = -15$

16. $\frac{1}{2}x - \frac{1}{4}y = 6$
    $\frac{5}{8}x - \frac{1}{4}y = 8$

For Exercises 17 and 18, use the table and the information at the right.

Mara and Ling each recycled aluminum cans and newspaper, as shown in the table. Mara earned $3.77 and Ling earned $4.65.

Material	Pounds Recycled	
	Mara	Ling
aluminum cans	9	9
newspaper	26	114

17. Define variables and formulate a system of linear equations from this situation.

18. What was the price per pound of aluminum? Determine the reasonableness of your solution.

H.O.T. Problems

19. **OPEN ENDED** Formulate a system of equations that represents a situation in your school. Describe the method that you would use to solve the system. Then solve the system and explain what the solution means.

20. **Which One Doesn't Belong?** Identify the system of equations that is not the same as the other three. Explain your reasoning.

$x - y = 3$ $x + \frac{1}{2}y = 1$	$-x + y = 0$ $5x = 2y$	$y = x - 4$ $y = \frac{2}{x}$	$y = x + 1$ $y = 3x$

21. ***Writing in Math*** Suppose that in a system of equations, x represents the time spent riding a bike, y represents the distance traveled, and you determine the solution to be $(1, -7)$. Use this problem to discuss the importance of analyzing solutions in the context of real-world problems.

STANDARDIZED TEST PRACTICE

22. Marcus descends at a rate of 2 feet per second from the surface of the ocean. Toshiko is 45 feet below sea level and she is rising to the surface at a rate of 3 feet per second. Which graph represents when the two divers will be at the same depth?

A

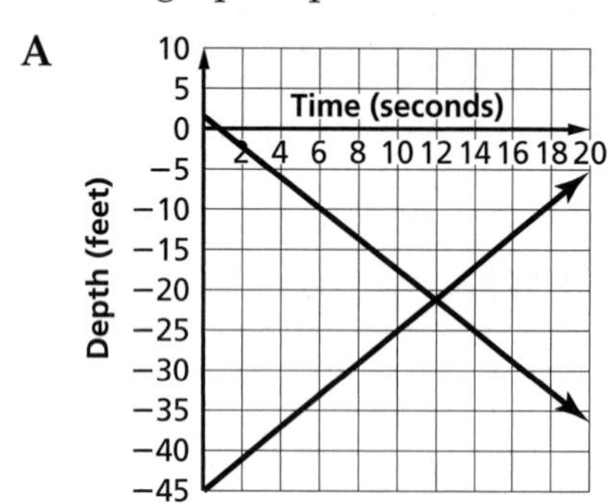

C

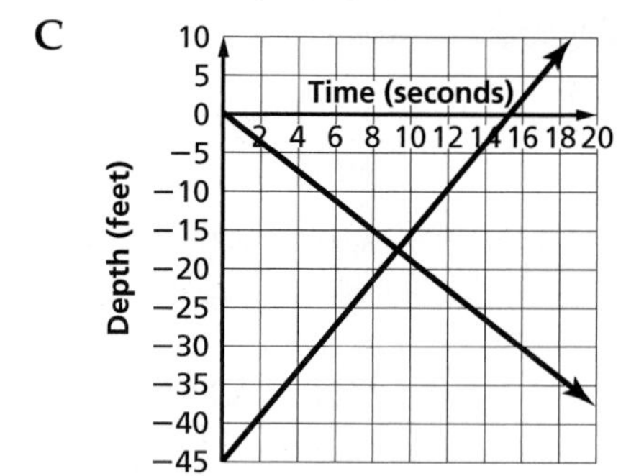

B

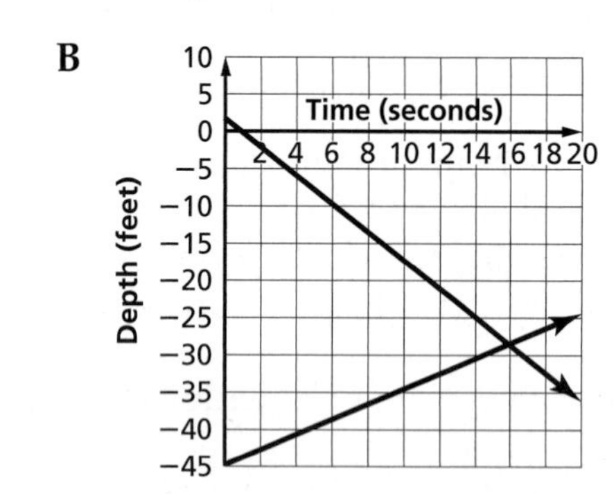

D

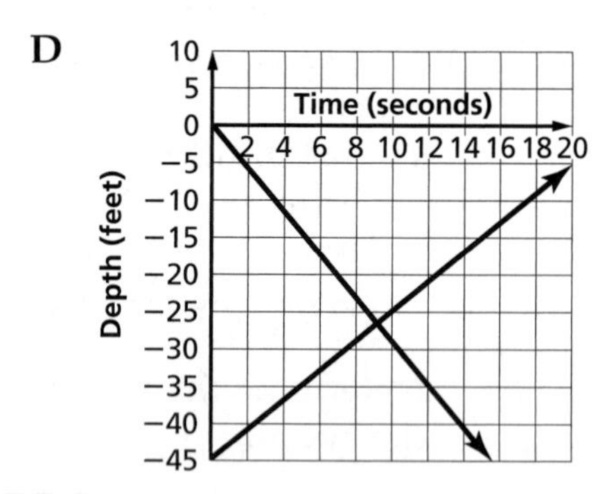

Spiral Review

Use elimination to solve each system of equations. (Lesson 5-4)

23. $x + y = -3$
$3x + 2y = -6$

24. $4x - 5y = 22$
$3x - 10y = 4$

25. $2x - 7y = -3$
$5x + 2y = -27$

26. **BUSINESS** In 2003, the United States produced about 2 million more motor vehicles than Japan. Together, the two countries produced about 22 million motor vehicles. How many vehicles were produced in each country? (Lesson 5-3)

CHAPTER 5 Study Guide and Review

Download Vocabulary Review from algebra1.com

GET READY to Study

Be sure the following Key Concepts are noted in your Foldable.

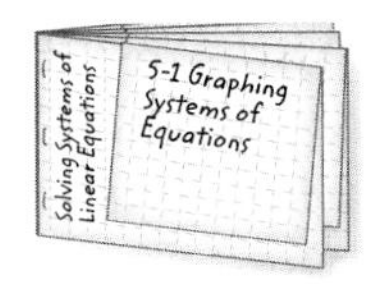

Key Concepts

Graphing Systems of Equations (Lesson 5-1)

- A system of intersecting lines has exactly one solution and is consistent and independent.
- A system whose graphs coincide has infinitely many solutions and is consistent and dependent.
- A system of parallel lines has no solution and is inconsistent.
- Graphing is best used to estimate the solution of a system of equations, since graphing usually does not give an exact solution.

Solving Systems of Equations Using Algebra (Lessons 5-2, 5-3, 5-4, and 5-5)

- Substitution is best used if one of the variables in either equation has a coefficient of 1 or −1.
- Elimination using addition is best used if one of the variables has opposite coefficients in the two equations.
- Elimination using subtraction is best used if one of the variables has the same coefficient in the two equations.
- Elimination using multiplication is best used if none of the coefficients are 1 or −1 and neither of the variables can be eliminated by simply adding or subtracting the equations.

Key Vocabulary

consistent (p. 253)
dependent (p. 253)
elimination (p. 266)
inconsistent (p. 253)
independent (p. 253)
substitution (p. 260)
system of equations (p. 253)

Vocabulary Check

State whether each sentence is *true* or *false*. If *false*, replace the underlined word or phrase to make a true sentence.

1. Two or more equations together are called a system of equations.
2. The best method for solving the system $3x - y = 9$ and $6x + y = 12$ is to use elimination using subtraction.
3. The system $2x + y = 5$ and $4x + 2y = 10$ is dependent.
4. If the graphs of the equations in a system have the same slope and different y-intercepts, the graph of the system is a pair of intersecting lines.
5. If a system has infinitely many solutions, it is inconsistent and independent.
6. The best method for solving the system $x = 4y$ and $2x + 3y = 6$ is to use substitution.
7. The system $y = 3x - 1$ and $y = 3x + 4$ is consistent.
8. Adding or subtracting two equations to solve a system of equations is known as substitution.
9. A system of equations whose solution is $(3, -5)$ is said to be independent.
10. If the graphs of the equations in a system have the same slope and y-intercept(s), the system has exactly one solution.

CHAPTER 5

Study Guide and Review

Lesson-by-Lesson Review

5-1 Graphing Systems of Equations (pp. 253–258)

Graph each system of equations. Then determine whether the system has *no* solution, *one* solution, or *infinitely many* solutions. If the system has one solution, name it.

11. $x - y = 3$
$x + y = 5$

12. $9x + 2 = 3y$
$y - 3x = 8$

13. $2x - 3y = 4$
$6y = 4x - 8$

14. $3x - y = 8$
$3x = 4 - y$

15. RACE In a race, Pablo is 3 miles behind Marc. Pablo increases his speed to 5 miles per hour, while Marc continues to run at 4 miles per hour. At this rate, how many miles will Pablo have to run to catch up to Marc? How long will this take?

Example 1 **Graph the system of equations. Then determine whether the system has *no* solution, *one* solution, or *infinitely many* solutions. If the system has one solution, name it.**

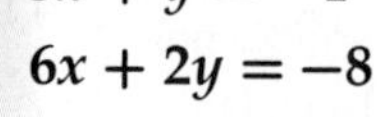

$3x + y = -4$
$6x + 2y = -8$

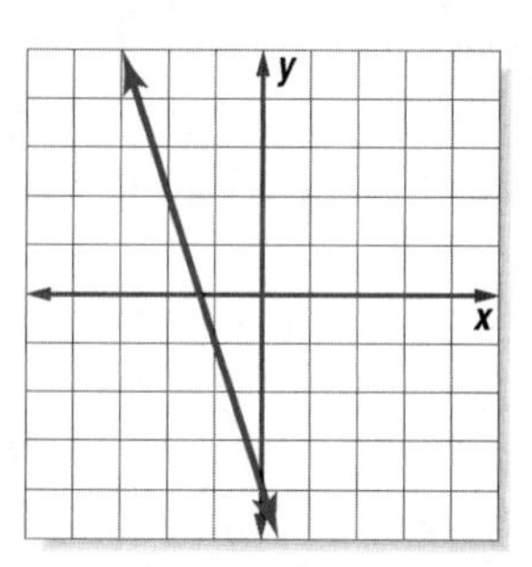

When the lines are graphed, they coincide. There are infinitely many solutions.

5-2 Substitution (pp. 260–265)

Use substitution to solve each system of equations. If the system does *not* have exactly one solution, state whether it has *no* solution or *infinitely many* solutions.

16. $2m + n = 1$
$m - n = 8$

17. $x = 3 - 2y$
$2x + 4y = 6$

18. $3x - y = 1$
$-12x + 4y = 3$

19. $6m - 2n = 24$
$n = 12 - 3m$

20. PHONES The table shows the long-distance plans of Companies A and B. For how many minutes is the cost the same for the two long-distance companies?

Company	Flat Fee	Rate
A	$0	$0.06/minute
B	$5.80	$0.04/minute

Example 2 **Use substitution to solve the system of equations.**

$y = x - 1$
$4x - y = 19$

Since $y = x - 1$, substitute $x - 1$ for y in the second equation.

$4x - y = 19$	Second equation
$4x - (x - 1) = 19$	$y = x - 1$
$4x - x + 1 = 19$	Distributive Property
$3x + 1 = 19$	Combine like terms.
$3x = 18$	Subtract 1 from each side.
$x = 6$	Divide each side by 3.

You can find the value of y by replacing x with 6 in the first equation.

$y = x - 1$	First equation
$= 6 - 1$	$x = 6$
$= 5$	The solution is (6, 5).

Mixed Problem Solving
For mixed problem-solving practice, see page 748.

5–3 Elimination Using Addition and Subtraction (pp. 266–270)

Use elimination to solve each system of equations.

21. $x + 2y = 6$
$x - 3y = -4$

22. $2m - n = 5$
$2m + n = 3$

23. $3x - y = 11$
$x + y = 5$

24. $3x + 1 = -7y$
$6x + 7y = -16$

25. AIRPORTS The Detroit Wayne County Airport and the Denver International Airport appeared in the top 20 rankings of busiest airports by number of passengers. The sum of their rankings was 29, and the difference was 9. If Denver was busier than Detroit, what were their rankings?

Example 3 **Use elimination to solve the system of equations.**

$2m - n = 4$
$m + n = 2$

Eliminate the n-terms by adding the equations.

$$\begin{array}{rl} 2m - n = 4 & \text{Write the equations in column} \\ (+)\ m + n = 2 & \text{form and add.} \\ \hline 3m = 6 & \text{Notice the variable } n \text{ is eliminated.} \\ m = 2 & \text{Divide each side by 3.} \end{array}$$

Substitute 2 for m in either equation to find n.

$$\begin{array}{rl} m + n = 2 & \text{Second equation} \\ 2 + n = 2 & \text{Replace } m \text{ with 2.} \\ n = 0 & \text{Subtract 2 from each side.} \end{array}$$

The solution is (2, 0).

5–4 Elimination Using Multiplication (pp. 272–278)

Use elimination to solve each system of equations.

26. $x - 5y = 0$
$2x - 3y = 7$

27. $x - 2y = 5$
$3x - 5y = 8$

28. $2x + 3y = 8$
$x - y = 2$

29. $-5x + 8y = 21$
$10x + 3y = 15$

30. ENTERTAINMENT The cost for tickets to see a play are shown in the table. A group of 11 adults and students bought tickets for the play. If the total cost was $156, how many of each type of ticket did they buy?

Ticket	Cost
adult	$15
student	$12

Example 4 **Use elimination to solve the system of equations.**

$x + 2y = 7$
$3x + y = 1$

Multiply the second equation by -2 so the coefficients of the y-terms are additive inverses. Then add the equations.

$$\begin{array}{lll} x + 2y = 7 & \text{Multiply} & x + 2y = 7 \\ 3x + y = 1 & \text{by } -2. & (+)\ -6x - 2y = -2 \\ \hline & & -5x = 5 \\ & & \frac{-5x}{-5} = \frac{5}{-5} \\ & & x = -1 \end{array}$$

$$\begin{array}{rl} x + 2y = 7 & \text{First equation} \\ -1 + 2y = 7 & x = -1 \\ 2y = 8 & \text{Add 1 to each side.} \\ \frac{2y}{2} = \frac{8}{2} & \text{Divide each side by 2.} \\ y = 4 & \text{Simplify.} \end{array}$$

The solution is $(-1, 4)$.

5-5 Applying Systems of Linear Equations (pp. 280–284)

Determine the best method to solve each system of equations. Then solve the system.

31. $y = 2x$
$x + 2y = 8$

32. $9x + 8y = 5$
$18x + 15y = 6$

33. $3x + 5y = 2x$
$x + 3y = y$

34. $2x + y = 3x - 15$
$x + 5 = 4y + 2x$

35. SAVINGS Raul invests \$1500 into two savings accounts, one earning 4% annual interest and the other earning 6% annual interest. At the end of one year, Raul has earned \$72 in interest. How much did he invest at each rate?

Example 5 **Determine the best method to solve the system. Then solve the system.**

$4x - 3y = 7$
$3x + 1 = y$

Since the coefficient of y in the second equation is 1, you can use the substitution method.

$4x - 3y = 7$	First equation
$4x - 3(3x + 1) = 7$	$y = 3x + 1$
$4x - 9x - 3 = 7$	Distributive Property
$-5x - 3 = 7$	Combine like terms.
$-5x - 3 + 3 = 7 + 3$	Add 3 to each side.
$-5x = 10$	Simplify.
$\frac{-5x}{-5} = \frac{10}{-5}$	Divide each side by -5.
$x = -2$	Simplify.
$3x + 1 = y$	Second equation
$3(-2) + 1 = y$	$x = -2$
$-5 = y$	Simplify.

The solution is $(-2, -5)$.

CHAPTER 5 Practice Test

Graph each system of equations. Then determine whether the system has *no* solution, *one* solution, or *infinitely many* solutions. If the system has one solution, name it.

1. $y = x + 2$
   $y = 2x + 7$

2. $x + 2y = 11$
   $x = 14 - 2y$

3. $3x + y = 5$
   $2y - 10 = -6x$

4. **MULTIPLE CHOICE** Which graph represents a system of equations with no solution?

A

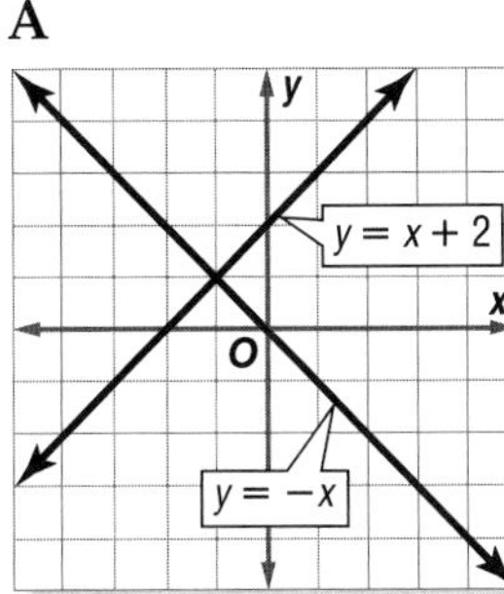

C

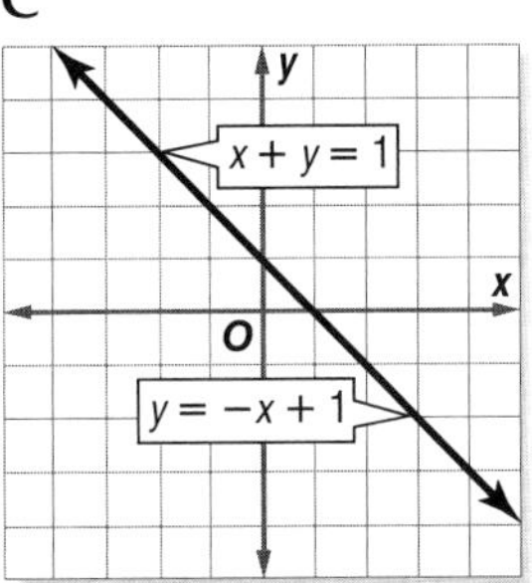

B

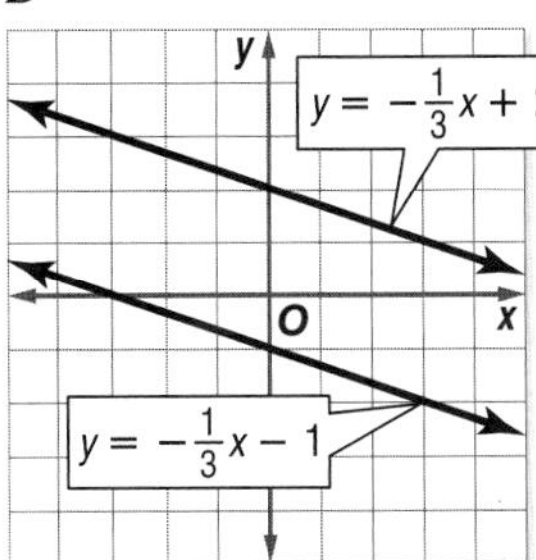

D

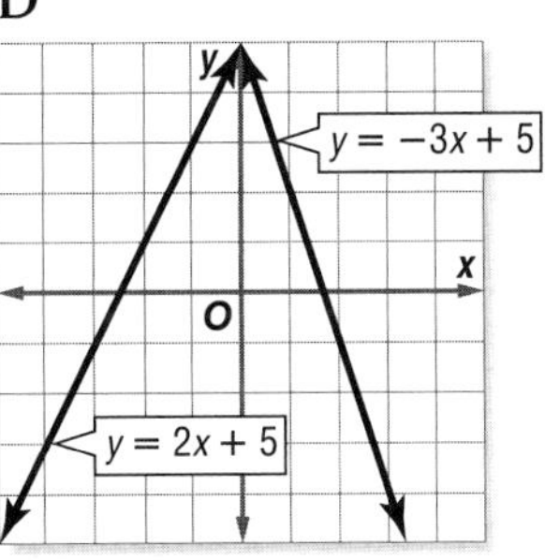

Use substitution or elimination to solve each system of equations.

5. $2x + 5y = 16$
   $5x - 2y = 11$

6. $y + 2x = -1$
   $y - 4 = -2x$

7. $2x + y = -4$
   $5x + 3y = -6$

8. $y = 7 - x$
   $x - y = -3$

9. $x = 2y - 7$
   $y - 3x = -9$

10. $x + y = -10$
    $x - y = -2$

11. $3x + y = 10$
    $3x - 2y = 16$

12. $5x - 3y = 12$
    $-2x + 3y = -3$

13. **MULTIPLE CHOICE** The units digit of a two-digit number exceeds twice the tens digit by 1. The sum of its digits is 10. Find the number.

    F 7   H 37

    G 19   J 39

14. **GEOMETRY** The difference between the length and width of a rectangle is 7 centimeters. Find the dimensions of the rectangle if its perimeter is 50 centimeters.

15. **FINANCE** Last year, Evelina invested \$10,000, part at 6% annual interest and the rest at 8% annual interest. If she received \$760 in interest at the end of the year, how much did she invest at each rate?

GEOMETRY For Exercises 16 and 17, use the graphs of $y = 2x + 6$, $3x + 2y = 19$, and $y = 2$, which contain the sides of a triangle.

16. Find the coordinates of the vertices of the triangle.

17. Find the area of the triangle.

18. **MULTIPLE CHOICE** At a movie theater, the costs for various amounts of popcorn and hot pretzels are shown below.

Boxes of Popcorn	Hot Pretzels	Total Cost
1	1	\$6.25
2	4	\$18.00

Which pair of equations can be used to find p, the cost of a box of popcorn, and z, the cost of a hot pretzel?

A $p + z = 6.25$
  $2p + 2z = 18$

B $p + z = 6.25$
  $4p + 4z = 18$

C $p + z = 6.25$
  $2p + 4z = 18$

D $p + z = 6.25$
  $4p + 2z = 18$

CHAPTER 5

Standardized Test Practice

Cumulative, Chapters 1–5

Read each question. Then fill in the correct answer on the answer document provided by your teacher or on a sheet of paper.

1. In the distance formula $d = rt$, r represents the rate of change, or slope. Which ray on the graph best represents a slope of 25 mph?

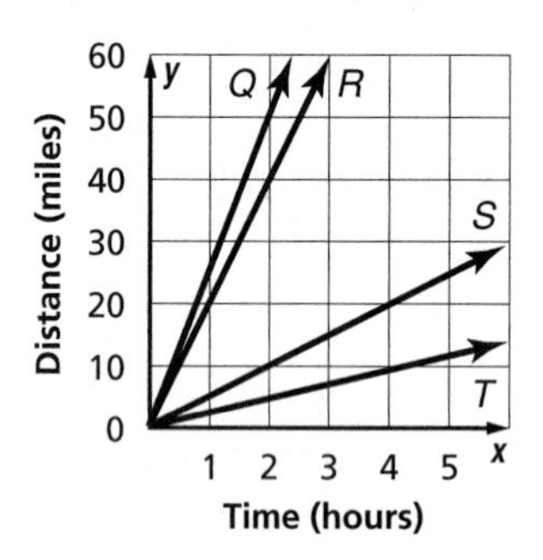

A Q **C** S

B R **D** T

2. Marla has a part-time job at a grocery store. The table shows the number of hours that Marla works in one week and the pay she receives.

Hours Worked	Weekly Pay (Dollars)
5	60
15	140
20	180

Which equation best describes the relationship between Marla's total pay, p and the number of hours she works, h?

F $p = 9h$

G $p = 8h + 20$

H $p = 10h - 20$

J $p = 15 + 8.25h$

3. GRIDDABLE Penny's Pizza Place estimates that 40% of their sales go toward labor costs. If the pizza place makes $2056.58 on Monday, approximately how much, in dollars, went to labor costs?

4. The graphs of the linear equations $y = 3x + 1$ and $y = 5x - 3$ are shown below.

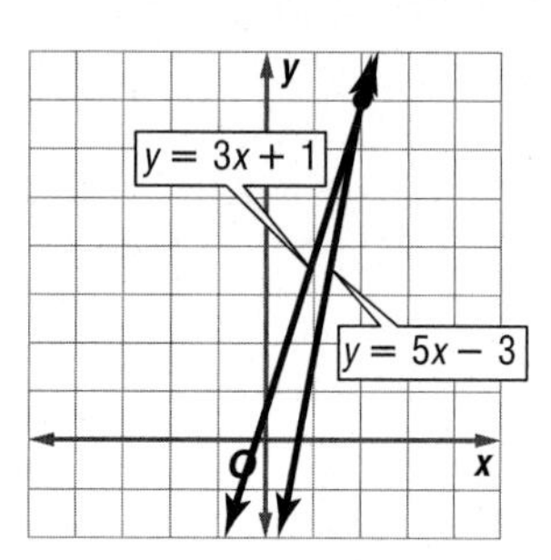

If $3x + 1 = 5x - 3$, what is the value of x?

A -3

B 1

C 2

D 7

TEST-TAKING TIP

Question 4 When solving an equation, check to make sure that your answer works. For example, in Question 4, once you get a solution for x, substitute it back into the equation to make sure you get a true statement.

5. Katelyn is training for a marathon. This week she ran 6 less than 2 times the number of miles that she ran last week. In these two weeks she ran a total of 42 miles. Which system of equations can be used to find x, the number of miles she ran this week and y, the number of miles she ran last week?

F $x + y = 42$
$x = 2y - 6$

G $x + y = 42$
$y = 2x - 6$

H $x + y = 42$
$x = 2y + 6$

J $x + y = 42$
$y = 2x + 6$

Preparing for Standardized Tests
For test-taking strategies and more practice, see pages 756–773.

6. Crazy Rides Amusement Park sells adult tickets for \$25 and children's tickets for \$15. On opening day the park sold 15 more children's tickets than adult tickets and made \$4625. Which system of equations could be used to find the number of adult tickets a, and children's tickets c, that the park sold that day?

A $25a + 15c = 4625$
$a = c + 15$

B $25a + 15c = 15$
$c = a + 4625$

C $25a + 15c = 4625$
$15 = a + c$

D $25a + 15c = 4625$
$c = a + 15$

7. What are the x- and y-intercepts of the function graphed below?

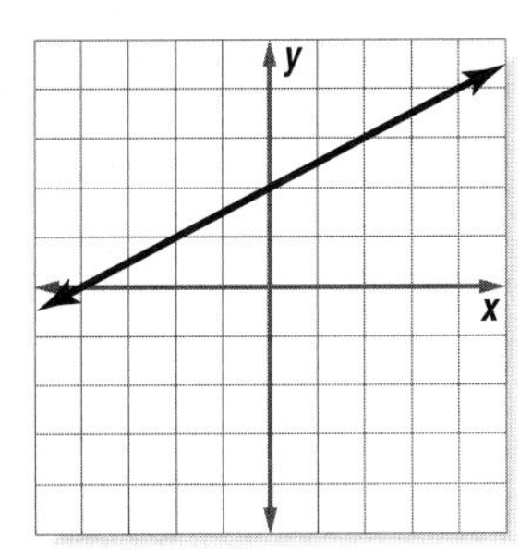

F (2, 0) and (−4, 0)
G (2, 0) and (0, −4)
H (0, 2) and (−4, 0)
J (0, 2) and (0, −4)

8. GRIDDABLE A rectangular prism has a volume of 400 cubic units. If the prism is dilated by a scale factor of 2, what is the volume of the resulting prism in cubic units?

9. Which circle has a center located at coordinates (−2, −1)?

A

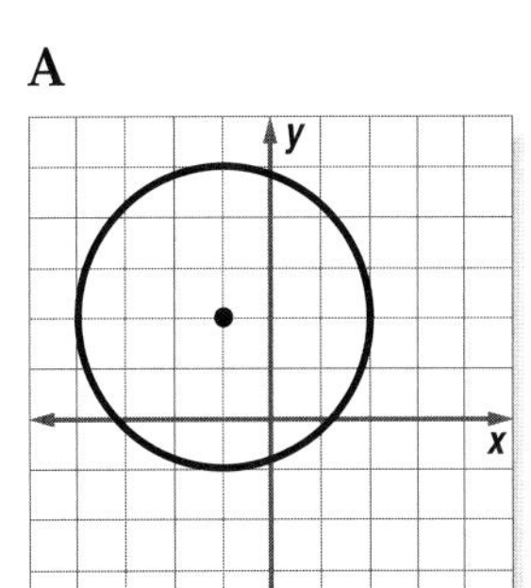

C

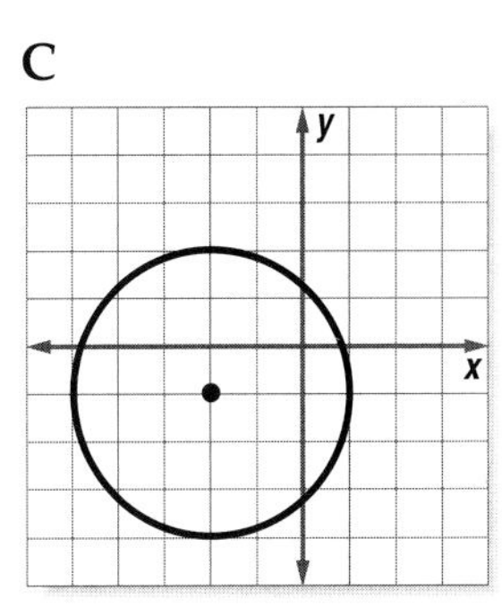

B

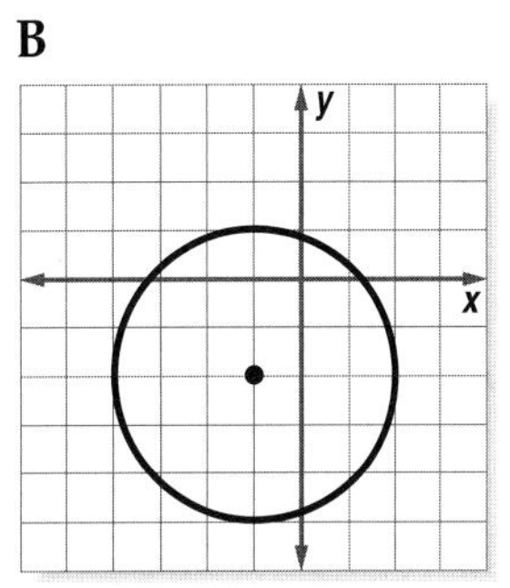

D

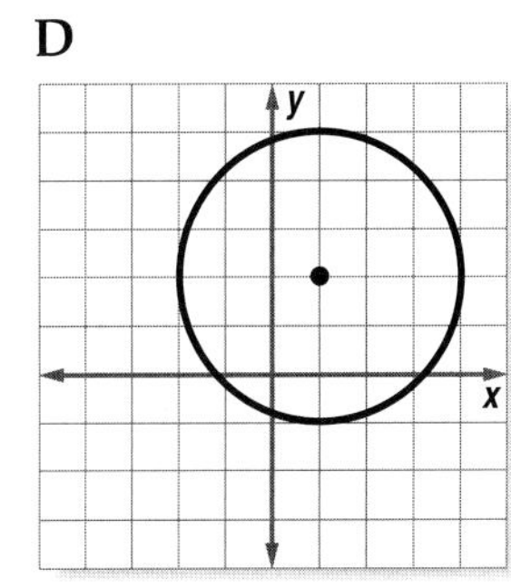

Pre-AP

Record your answers on a sheet of paper. Show your work.

10. The manager of a movie theater found that Saturday's ticket sales were \$3675. He knew that a total of 650 tickets were sold. Adult tickets cost \$7.50, and children's tickets cost \$4.50.

a. Write equations to represent the number of tickets sold and the amount of money collected. Let a represent the number of adult tickets, and let c represent the number of children's tickets.

b. How many of each kind of ticket were sold? Show your work. Include all of the steps.

Need extra help?										
If You Missed Question...	1	2	3	4	5	6	7	8	9	10
Go to Lesson or Page...	4-1	4-4	702	5-1	5-1	5-1	3-3	708	1-9	5-2

CHAPTER 6

Solving Linear Inequalities

BIG Ideas

- Solve linear inequalities and graph in the coordinate plane.
- Solve absolute value equations and inequalities.
- Solve systems of linear inequalities by graphing.

Key Vocabulary

compound inequality (p. 315)

intersection (p. 315)

set-builder notation (p. 295)

union (p. 316)

Real-World Link

Roller Coasters Inequalities are used to represent various real-world situations in which a quantity must fall within a range of possible values. For example, a roller coaster must gain enough speed on the first hill to propel it through the entire ride.

Solving Linear Inequalities Make this Foldable to record information about solving linear inequalities. Begin with two sheets of notebook paper.

1. **Fold** one sheet in half along the width. Cut along the fold from the edges to margin.

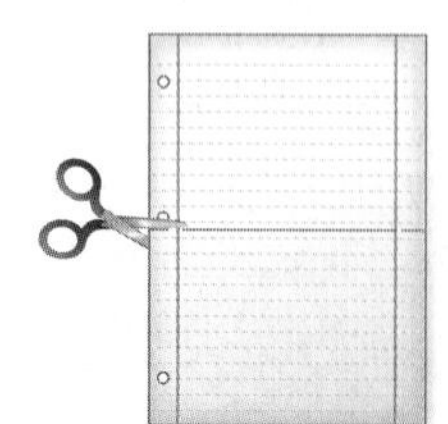

2. **Fold** the second sheet in half along the width. Cut along the fold between the margins.

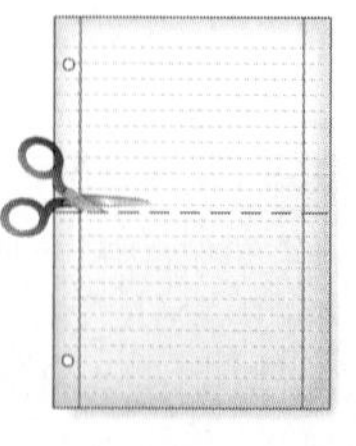

3. **Insert** the first sheet through the second sheet and align the folds.

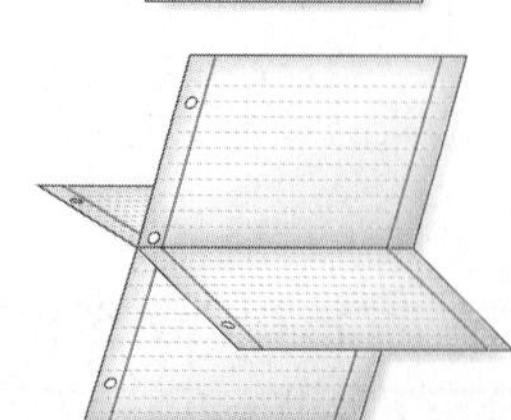

4. **Label** each page with a lesson number and title.

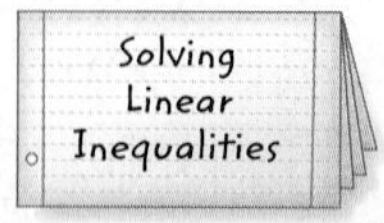

GET READY for Chapter 6

Diagnose Readiness You have two options for checking Prerequisite Skills.

Option 2

Math Online Take the Online Readiness Quiz at algebra1.com.

Option 1

Take the Quick Check below. Refer to the Quick Review for help.

QUICKCheck

Solve each equation. (Lessons 2-4 and 2-5)

1. $18 = 27 + f$

2. $d - \frac{2}{3} = \frac{1}{2}$

3. $5m + 7 = 4m - 12$

4. $3y + 4 = 16$

5. $\frac{1}{2}k - 4 = 7$

6. $4.3b + 1.8 = 8.25$

7. $6s - 12 = 2(s + 2)$

8. $n - 3 = \frac{n+1}{2}$

9. **NUMBER THEORY** Three times the lesser of two consecutive integers is 4 more than 2 times the greater number. Find the integers.

Find each value. (Prerequisite Skill)

10. $|20|$

11. $|-1.5|$

12. $|14 - 7|$

13. $|1 - 16|$

14. $|2 - 3|$

15. $|7 - 10|$

Graph each equation. (Lesson 3-3)

16. $2x + 2y = 6$

17. $x - 3y = -3$

18. $y = 2x - 3$

19. $y = -4$

20. $x = -\frac{1}{2}y$

21. $3x - 6 = 2y$

22. $15 = 3(x + y)$

23. $2 - x = 2y$

24. **CRAFTS** Rosa and Taylor are making scarves for the upcoming craft fair. Rosa can make x scarves per hour and Taylor can make y scarves per hour. Rosa can only work 8 hours and Taylor can only work 10. In total, they need to complete 25 scarves. Express Taylor's rate in terms of Rosa's rate.

QUICKReview

EXAMPLE 1

Solve $\frac{1}{2} = \frac{1}{8}t + 1$.

$\frac{1}{2} = \frac{1}{8}t + 1$	Original equation
$-\frac{1}{2} = \frac{1}{8}t$	Subtract 1 from each side.
$8\left(-\frac{1}{2}\right) = 8\left(\frac{1}{8}t\right)$	Multiply each side by 8.
$-4 = t$	Simplify.

EXAMPLE 2

Find the value of $|15 - 20|$.

$|15 - 20| = |-5|$	Evaluate inside the absolute value.
$= 5$	-5 is five units from zero in the negative direction.

EXAMPLE 3

Graph $y - x = 1$.

Step 1 Put the equation in slope intercept form.

$$y - x = 1 \rightarrow y = x + 1$$

Step 2 The y-intercept is 1. So, graph the point (0, 1).

Step 3 The slope is 1 or $\frac{1}{1}$. $\frac{\text{rise}}{\text{run}}$ From (0, 1), move up 1 unit and right 1 unit. Draw a dot.

Step 4 Draw a line connecting the points.

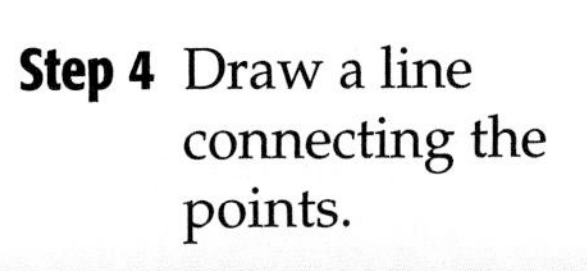

Solving Inequalities by Addition and Subtraction

Main Ideas

- Solve linear inequalities by using addition.
- Solve linear inequalities by using subtraction.

New Vocabulary

set-builder notation

GET READY for the Lesson

The data in the graph show that more high schools offer girls' track and field than girls' volleyball.

$$15{,}151 > 14{,}083$$

If 20 schools added girls' track and field and 20 schools added girls' volleyball, there would still be more schools offering girls' track and field than schools offering girls' volleyball.

$$15{,}151 + 20 \ \underline{\ \ ?\ \ }\ 14{,}083 + 20$$

$$15{,}171 > 14{,}103$$

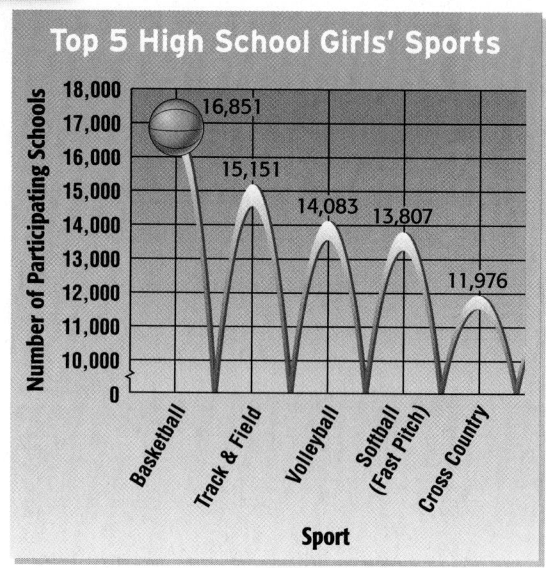

Source: National Federation of State High School Associations

Review Vocabulary

Inequality

An open sentence that contains the symbol $<$, $\leq$, $>$, or $\geq$ is called an inequality.
Example: $x \geq 4$
(Lesson 1-3)

Solve Inequalities by Addition The sports application illustrates the **Addition Property of Inequalities.**

KEY CONCEPT — Addition Property of Inequalities

Words If any number is added to each side of a true inequality, the resulting inequality is also true.

Symbols For all numbers a, b, and c, the following are true.
1. If $a > b$, then $a + c > b + c$.
2. If $a < b$, then $a + c < b + c$.

This property is also true when $>$ and $<$ are replaced with $\geq$ and $\leq$.

EXAMPLE Solve by Adding

1 Solve $t - 45 \leq 13$. Check your solution.

$t - 45 \leq 13$	*Original inequality*
$t - 45 + 45 \leq 13 + 45$	Add 45 to each side.
$t \leq 58$	Simplify.

The solution is the set {all numbers less than or equal to 58}.

CHECK To check, substitute 58, a number less than 58, and a number greater than 58.

CHECK Your Progress **Solve each inequality.**

1A. $22 > m - 8$ **1B.** $d - 14 \geq -19$

Reading Math

Set-Builder Notation
$\{t \mid t \leq 58\}$ is read *the set of all numbers t such that t is less than or equal to 58.*

The solution in Example 1 was expressed as a set. A more concise way of writing a solution set is to use **set-builder notation**. The solution in set-builder notation is $\{t \mid t \leq 58\}$.

The solution can also be graphed on a number line.

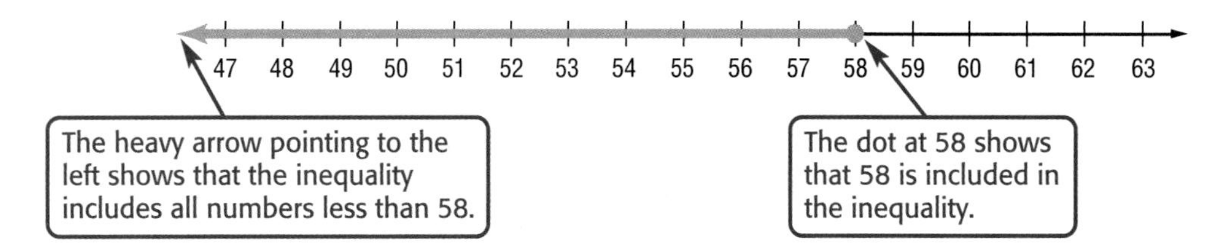

Solve Inequalities by Subtraction

Subtraction can also be used to solve inequalities.

KEY CONCEPT — *Subtraction Property of Inequalities*

Words If any number is subtracted from each side of a true inequality, the resulting inequality is also true.

Symbols For all numbers a, b, and c, the following are true.
1. If $a > b$, then $a - c > b - c$.
2. If $a < b$, then $a - c < b - c$.

This property is also true when $>$ and $<$ are replaced with $\geq$ and $\leq$.

Real-World EXAMPLE Solve by Subtracting

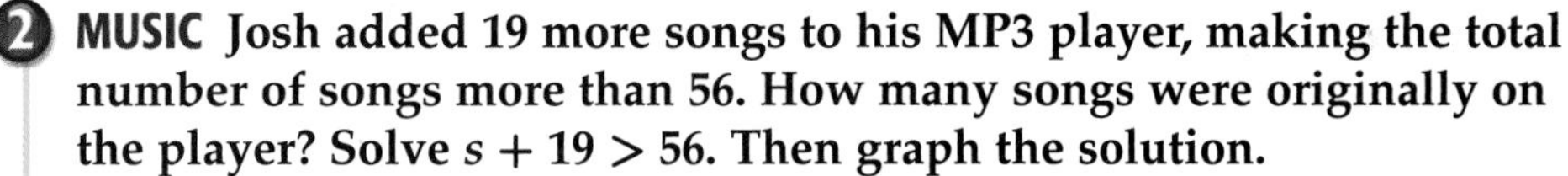

2 **MUSIC Josh added 19 more songs to his MP3 player, making the total number of songs more than 56. How many songs were originally on the player? Solve $s + 19 > 56$. Then graph the solution.**

$s + 19 > 56$	Original inequality
$s + 19 - 19 > 56 - 19$	Subtract 19 from each side.
$s > 37$	Simplify.

The solution set is $\{s \mid s > 37\}$. So, Josh had more than 37 songs originally on the music player.

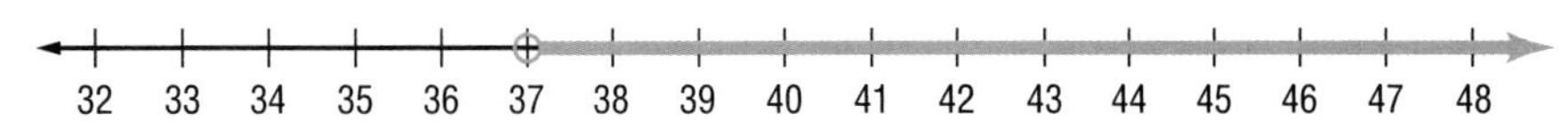

CHECK Your Progress

2. **TEMPERATURE** The temperature t in a swimming pool increased 4°F since this morning. The temperature is now less than 81°F. What was the temperature this morning? Solve $t + 4 < 81$. Then graph the solution.

Terms with variables can also be subtracted from each side to solve inequalities.

EXAMPLE Variables on Each Side

3 **Solve $5p + 7 > 6p$. Then graph the solution.**

$5p + 7 > 6p$ Original inequality

$5p + 7 - 5p > 6p - 5p$ Subtract $5p$ from each side.

$7 > p$ Simplify.

Since $7 > p$ is the same as $p < 7$, the solution set is $\{p \mid p < 7\}$.

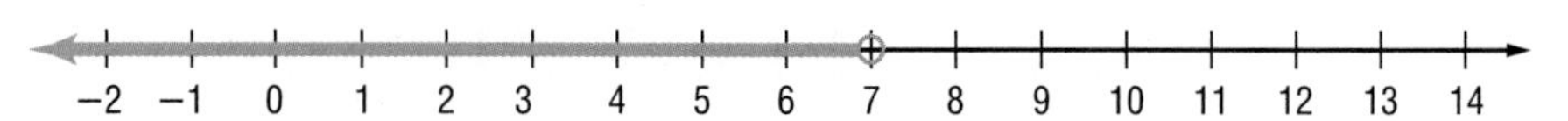

CHECK Your Progress

Solve each inequality. Graph the solution on a number line.

3A. $9n - 1 < 10n$

3B. $5h \leq 12 + 4h$

Here are some phrases that indicate inequalities in verbal problems.

Inequalities			
$<$	$>$	$\leq$	$\geq$
less than	greater than	at most	at least
fewer than	more than	no more than	no less than
		less than or equal to	greater than or equal to

Real-World EXAMPLE Write an Inequality to Solve a Problem

4 **OLYMPICS** **Irina Tchachina scored a total of 107.325 points in the four events of rhythmic gymnastics. Alina Kabaera scored a total of 81.300 in the clubs, hoop, and ball events. How many points did she need to score in the ribbon event to get ahead of Tchachina and win the gold medal?**

Words	Kabaera's total	is greater than	Tchachina's total.
Variable	Let r = Kabaera's score in the ribbon event.		
Inequality	$81.300 + r$	$>$	107.325

Solve the inequality. **ESTIMATE:** $110 - 80 = 30$

$81.300 + r > 107.325$ Original inequality

$81.300 + r - 81.300 > 107.325 - 81.300$ Subtract 81.300 from each side.

$r > 26.025$ Simplify.

Kabaera needed to score more than 26.025 points to win the gold medal. The solution is close to the estimate, so the answer is reasonable.

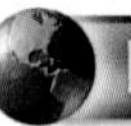

Real-World Link

Alina Kabaera of the Russian Federation won the gold medal in rhythmic gymnastics at the 2004 Summer Olympics in Athens, and her teammate Irina Tchachina won the silver medal.

Source: www.athens2004.com

CHECK Your Progress

4. SHOPPING Terrell has \$65 to spend. He bought a T-shirt for \$18 and a belt for \$14. If Terrell still wants jeans, how much can he spend on the jeans?

online Personal Tutor at algebra1.com

CHECK Your Understanding

Examples 1–3 (pp. 294–296)

Solve each inequality. Check your solution, and then graph it on a number line.

1. $t - 5 \geq 7$
2. $-7 \geq -2 + x$
3. $a + 4 < 2$
4. $9 \leq b + 4$
5. $10 > n - 1$
6. $k + 24 > -5$
7. $8r + 6 < 9r$
8. $7p \leq 6p - 2$

Example 4 (p. 296)

Define a variable, write an inequality, and solve each problem. Check your solution.

9. A number decreased by 8 is at most 14.
10. Twice a number is greater than −5 plus the number.
11. **BIOLOGY** Adult Nile crocodiles weigh up to 2200 pounds. If a young Nile crocodile weighs 157 pounds, how many pounds might it be expected to gain in its lifetime?

Exercises

HOMEWORK HELP

For Exercises	See Examples
12–29	1–3
30–35	4

Solve each inequality. Check your solution, and then graph it on a number line.

12. $t + 14 \geq 18$
13. $d + 5 \leq 7$
14. $n - 7 < -3$
15. $-5 + s > -1$
16. $5 < 3 + g$
17. $13 > 18 + r$
18. $2 \leq -1 + m$
19. $-23 \geq q - 30$
20. $11 + m \geq 15$
21. $h - 26 < 4$
22. $8 \leq r - 14$
23. $-7 > 20 + c$
24. $2y > -8 + y$
25. $3f < -3 + 2f$
26. $3b \leq 2b - 5$
27. $4w \geq 3w + 1$
28. $6x + 5 \geq 7x$
29. $-9 + 2a < 3a$

Define a variable, write an inequality, and solve each problem. Check your solution.

30. The sum of a number and 13 is at least 27.
31. A number decreased by 5 is less than 33.
32. Twice a number is more than the sum of that number and 14.
33. The sum of two numbers is at most 18, and one of the numbers is −7.

For Exercises 34–36, define a variable, write an inequality, and solve each problem. Then interpret your solution.

Real-World Link

One common species of bees is the honeybee. A honeybee colony may have 60,000 to 80,000 bees.

Source: Penn State, Cooperative Extension Service

34. **BIOLOGY** There are 3500 species of bees and more than 600,000 species of insects. How many species of insects are not bees?
35. **TECHNOLOGY** A recent survey found that more than 21 million people between the ages of 12 and 17 use the Internet. Of those online teens, about 16 million said they use the Internet at school. How many teens who are online do not use the Internet at school?
36. **ANALYZE TABLES** Chapa is limiting her fat intake to no more than 60 grams per day. Today, she has had two breakfast bars and a slice of pizza. How many more grams of fat can she have today?

Food	Grams of Fat
breakfast bar	3
slice of pizza	21

Solve each inequality. Check your solution, and then graph it on a number line.

37. $y - (-2.5) > 8.1$

38. $5.2r + 6.7 \geq 6.2r$

39. $a + \frac{1}{4} > \frac{1}{8}$

40. $\frac{3}{2}p - \frac{2}{3} \leq \frac{4}{9} + \frac{1}{2}p$

Define a variable, write an inequality, and solve each problem. Check your solution.

41. Thirty is no greater than the sum of a number and -8.

42. Four times a number is less than or equal to the sum of 3 times the number and -2.

For Exercises 43 and 44, define a variable, write an inequality, and solve each problem. Then interpret your solution.

43. MONEY City Bank requires a minimum balance of \$1500 to maintain free checking services. If Mr. Hayashi is going to write checks for the amounts listed in the table, how much money should he have in his account before writing the checks in order to have free checking?

Check	Amount
750	\$1300
751	\$947

44. SOCCER The Centerville High School soccer team has a goal of winning at least 60% of their 18 games this season. In the first three weeks, the team has won 4 games. How many more games must the team win to meet their goal?

45. GEOMETRY The length of the base of the triangle is less than the height of the triangle. What are the possible values of x? Formulate a linear inequality to solve the problem. Determine whether your answers are reasonable.

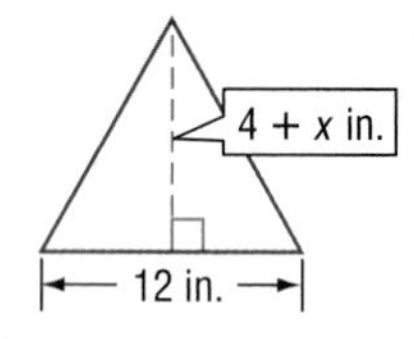

46. If $d + 5 \geq 17$, then complete each inequality.

a. $d \geq$ _?_

b. $d +$ _?_ ≥ 20

c. $d - 5 \geq$ _?_

H.O.T. Problems

47. REASONING Compare and contrast the graphs of $a < 4$ and $a \leq 4$.

48. CHALLENGE Determine whether each statement is *always, sometimes,* or *never* true. Explain.

a. If $a < b$ and $c < d$, then $a + c < b + d$.

b. If $a < b$ and $c < d$, then $a + c \geq b + d$.

c. If $a < b$ and $c < d$, then $a - c < b - d$.

49. OPEN ENDED Formulate three linear inequalities that are equivalent to $y < -3$.

50. *Writing in Math* Use the information about sports on page 294 to explain how inequalities can be used to describe school sports. Include an inequality describing the number of schools needed to add girls' track and field so that the number is greater than the number of schools currently participating in girls' basketball.

STANDARDIZED TEST PRACTICE

51. Based on the graph below, which statement is true?

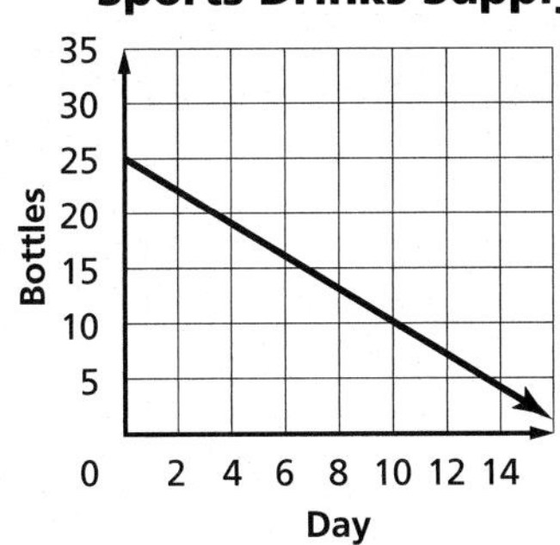

A Maria started with 30 bottles of sports drinks.

B On day 10, Maria will have drunk 10 bottles of sports drinks.

C Maria will be out of sports drinks on day 14.

D Maria drank 5 bottles in the first 2 days.

52. What is the solution set to the inequality $7 + x < 5$?

F $x < 2$ **H** $x > 2$

G $x < -2$ **J** $x > -2$

53. REVIEW Miss Miller wants to calculate the cost of buying tile to cover her rectangular kitchen floor. She knows the cost per square foot of tile, and she knows the dimensions of the kitchen. Which formula should she use to find the area of the floor?

A $A = \ell w$ **C** $P = 2\ell + 2w$

B $V = Bh$ **D** $c^2 = a^2 + b^2$

Spiral Review

Determine the best method to solve each system of equations. Then solve the system. (Lesson 5-5)

54. $4x - 3y = -1$
$3x + 3y = 15$

55. $x = 8y$
$2x + 3y = 38$

56. $2a + 5b = -3$
$3a + 4b = -8$

57. NUTRITION The costs for various items at a mall food court are shown in the table. What is the cost of an iced tea and a vegetable wrap? (Lesson 5-4)

Number of Iced Teas	Number of Vegetable Wraps	Total Cost
2	2	\$10.50
3	4	\$19.50

VOLUNTEERING For Exercises 58 and 59, use the graph. It shows the total hours that Estella spent volunteering. (Lesson 3-5)

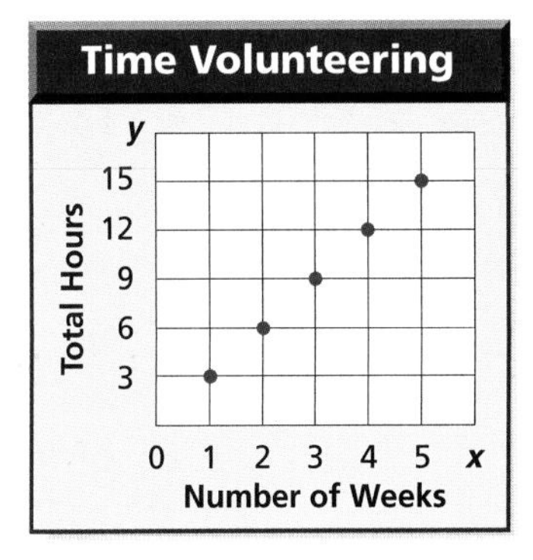

58. Write an equation in function notation for the relation.

59. What would be the total hours that Estella spent volunteering after 12 weeks?

GET READY for the Next Lesson

PREREQUISITE SKILL Solve each equation. (Lesson 2-3)

60. $6g = 42$

61. $3m = 435$

62. $\frac{t}{9} = 14$

63. $\frac{2}{3}y = -14$

EXPLORE 6-2

Algebra Lab
Solving Inequalities

You can use algebra tiles to solve inequalities.

Concepts in Motion
Animation
algebra1.com

ACTIVITY Solve $-2x \geq 6$.

Step 1 Model the inequality.

Use a self-adhesive note to cover the equals sign on the equation mat. Then write a ≥ symbol on the note. Model the inequality.

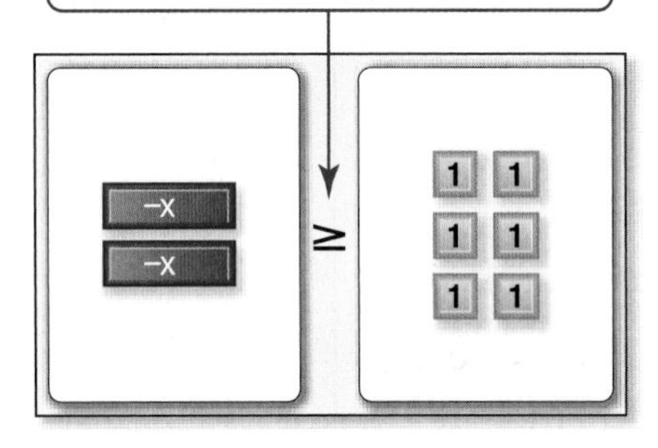

$-2x \geq 6$

Step 2 Remove the zero pairs.

Since you do not want to solve for a negative *x*-tile, eliminate the negative *x*-tiles by adding 2 positive *x*-tiles to each side. Remove the zero pairs.

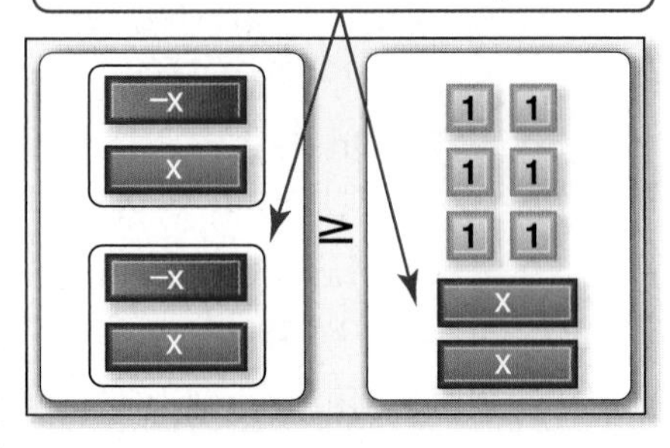

$-2x + 2x \geq 6 + 2x$

Step 3 Remove the zero pairs.

Add 6 negative 1-tiles to each side to isolate the *x*-tiles. Remove the zero pairs.

$-6 \geq 2x$

Step 4 Group the tiles.

Separate the tiles into 2 groups.

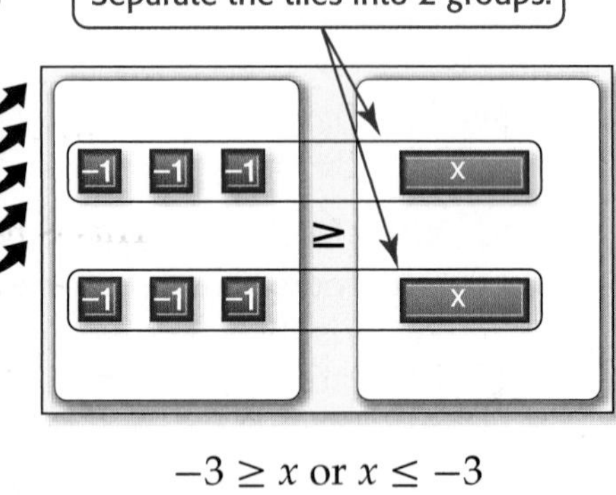

$-3 \geq x$ or $x \leq -3$

Model and Analyze the Results

Use algebra tiles to solve each inequality.

1. $-4x < 12$ **2.** $-2x > 8$ **3.** $-3x \geq -6$ **4.** $-5x \leq -5$

5. In Exercises 1–4, is the coefficient of x in each inequality *positive* or *negative*?

6. Compare the inequality symbols and locations of the variable in Exercises 1–4 with those in their solutions. What do you find?

7. Model the solution for $2x \geq 6$. What do you find? How is this different than solving $-2x \geq 6$?

6-2 Solving Inequalities by Multiplication and Division

Main Ideas

- Solve linear inequalities by using multiplication.
- Solve linear inequalities by using division.

GET READY for the Lesson

Isabel Franco is stacking cases of drinks to sell at a basketball game. A case of bottled water is 8 inches high. A case of sports drinks is 10 inches high. Notice that 8 inches is less than 10 inches, or 8 in. < 10 in. How would the height of stacks of three cases of each compare?

8 in.

10 in.

8 in. < 10 in.	The height of the water is less than the height of the sports drinks.
8 in. × 3 < 10 in. × 3	Multiply to find the height of 3 cases of each.
24 in. < 30 in.	The height of 3 cases of water is less than the height of 3 cases of sports drinks.

Study Tip

Look Back

To review solving linear equations by multiplication, see Lesson 2-3.

Solve Inequalities by Multiplication If each side of an inequality is multiplied by a positive number, the inequality remains true.

$8 > 5$	Original inequality	$5 < 9$	Original inequality
$8(2)$ _?_ $5(2)$	Multiply each side by 2.	$5(4)$ _?_ $9(4)$	Multiply each side by 4.
$16 > 10$	Simplify.	$20 < 36$	Simplify.

This is *not* true when multiplying by negative numbers.

$5 > 3$	Original inequality	$-6 < 8$	Original inequality
$5(-2)$ _?_ $3(-2)$	Multiply each side by −2.	$-6(-5)$ _?_ $8(-5)$	Multiply each side by −5.
$-10 < -6$	Simplify.	$30 > -40$	Simplify.

If each side of an inequality is multiplied by a negative number, the direction of the inequality symbol changes. These examples illustrate the **Multiplication Property of Inequalities**.

KEY CONCEPT — *Multiplying by a Positive Number*

Words If each side of a true inequality is multiplied by the same positive number, the resulting inequality is also true.

Symbols If a and b are any numbers and c is a positive number, the following are true.
If $a > b$, then $ac > bc$, and if $a < b$, then $ac < bc$.

This property also holds for inequalities involving ≥ and ≤.

Real-World EXAMPLE Write and Solve an Inequality

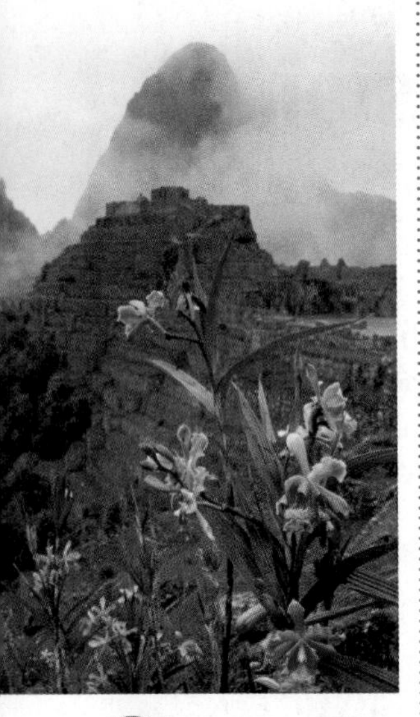

Real-World Link

More than 30,000 different orchid species flower in the wild on every continent except Antarctica.

Source: www.alohaorchid.com

1 **BIOLOGY** **Mt. Kinabalue in Malaysia has the greatest concentration of wild orchids on Earth. It contains more than 750 species, which is approximately one fourth of all orchid species in Malaysia. How many orchid species are there in Malaysia?**

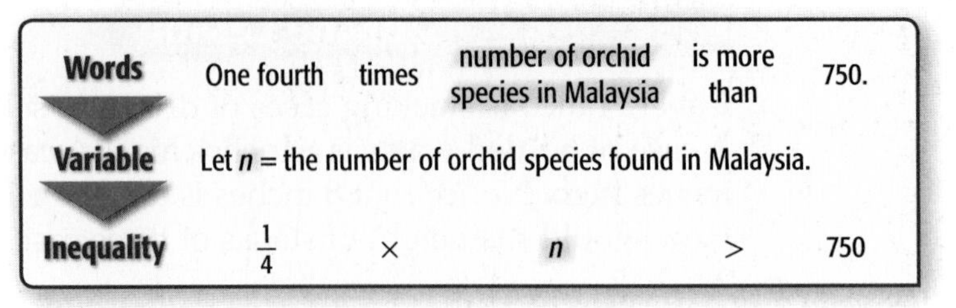

Words	One fourth	times	number of orchid species in Malaysia	is more than	750.
Variable	Let n = the number of orchid species found in Malaysia.				
Inequality	$\frac{1}{4}$	$\times$	n	$>$	750

$\frac{1}{4}n > 750$ Original inequality

$(4)\frac{1}{4}n > (4)750$ Multiply each side by 4 and do not change the inequality's direction.

$n > 3000$ Simplify.

The solution set is $\{n \mid n > 3000\}$. This means that there are more than 3000 orchid species in Malaysia.

CHECK Your Progress

1A. **SURVEYS** Of the students surveyed at Madison High School, fewer than eighty-four said they have never purchased an item online. This is about three eighths of those surveyed. How many students were surveyed?

1B. **CANDY** Fewer than 42 employees at the factory stated that they preferred chocolate candy over fruit candy. This is about two thirds of the employees. How many employees are there?

Graph 3 and 5 on a number line.

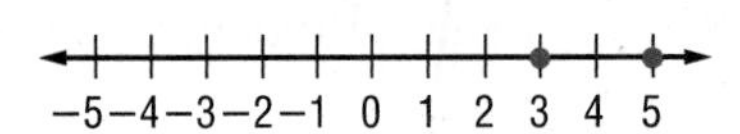

Since 3 is to the left of 5, $3 < 5$.

Multiply each number by -1.

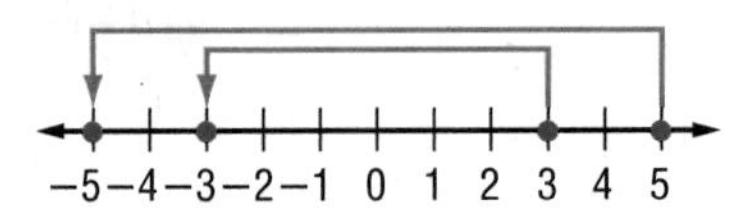

Since -3 is to the right of -5, $-3 > -5$

Notice that the numbers being compared switched positions as a result of being multiplied by a negative number. In other words, their order reversed. This suggests the following property.

KEY CONCEPT — *Multiplying by a Negative Number*

Words If each side of a true inequality is multiplied by the same negative number, the direction of the inequality symbol must be *reversed* so that the resulting inequality is also true.

Symbols If a and b are any numbers and c is a negative number, the following are true.
If $a > b$, then $ac < bc$, and if $a < b$, then $ac > bc$.

This property also holds for inequalities involving $\geq$ and $\leq$.

EXAMPLE Multiply by a Negative Number

2 **Solve $-\frac{2}{5}p < -14$.**

$-\frac{2}{5}p < -14$ Original inequality

$\left(-\frac{5}{2}\right)\left(-\frac{2}{5}p\right) > \left(-\frac{5}{2}\right)(-14)$ Multiply each side by $-\frac{5}{2}$ and change $<$ to $>$.

$p > 35$ Simplify.

The solution set is $\{p \mid p > 35\}$.

Study Tip

Common Misconception

You may be tempted to change the direction of the inequality when solving if there is a negative number anywhere in the inequality. Remember that you only change the direction of the inequality when multiplying or dividing by a negative.

CHECK Your Progress **Solve each inequality.**

2A. $-\frac{n}{6} \leq 8$ **2B.** $-\frac{4}{3}p > -10$

Solve Inequalities By Division You can also solve an inequality by dividing each side by the same number. Consider the inequality $6 < 15$.

Divide each side by 3.

$6 < 15$

$6 \div 3 \ \underline{\ ?\ }\ 15 \div 3$

$2 < 5$

The direction of the inequality symbol remains the same.

Divide each side by -3.

$6 < 15$

$6 \div (-3) \ \underline{\ ?\ }\ 15 \div (-3)$

$-2 > -5$

The direction of the inequality symbol is reversed.

These examples illustrate the **Division Property of Inequalities**.

KEY CONCEPT *Dividing by a Positive Number*

Words If each side of a true inequality is divided by the same positive number, the resulting inequality is also true.

Symbols If a and b are any numbers and c is a positive number, the following are true.

If $a > b$, then $\frac{a}{c} > \frac{b}{c}$, and if $a < b$, then $\frac{a}{c} < \frac{b}{c}$.

This property also holds for inequalities involving $\geq$ and $\leq$.

KEY CONCEPT *Dividing by a Negative Number*

Words If each side of a true inequality is divided by the same negative number, the direction of the inequality symbol must be *reversed* so that the resulting inequality is also true.

Symbols If a and b are any numbers and c is a negative number, the following are true.

If $a > b$, then $\frac{a}{c} < \frac{b}{c}$, and if $a < b$, then $\frac{a}{c} > \frac{b}{c}$.

This property also holds for inequalities involving $\geq$ and $\leq$.

EXAMPLE 3 Divide to Solve an Inequality

3 **Solve each inequality.**

a. $14h > 91$

$14h > 91$	Original inequality
$\frac{14h}{14} > \frac{91}{14}$	Divide each side by 14 and do not change the direction of the inequality sign.
$h > 6.5$	Simplify.

The solution set is $\{h \mid h > 6.5\}$.

b. $-5t \geq 275$

$-5t \geq 275$	Original inequality
$\frac{-5t}{-5} \leq \frac{275}{-5}$	Divide each side by -5 and change $\geq$ to $\leq$.
$t \leq -55$	Simplify.

The solution set is $\{t \mid t \leq -55\}$.

Study Tip

Alternate Method

Since dividing is the same as multiplying by the reciprocal, you could also solve $-5t \geq 275$ by multiplying each side by $-\frac{1}{5}$.

CHECK Your Progress

3A. $9r < 27$

3B. $-15 \geq 3t$

3C. $32 < -8k$

3D. $-5g \geq 40$

STANDARDIZED TEST EXAMPLE 4 Write an Inequality

4 Homemade bracelets are for sale for $4.75 each. Which inequality can be used to find how many bracelets Caitlin can buy for herself and her friends if she wants to spend no more than $22?

A $4.75b \geq 22$

B $4.75b \leq 22$

C $4.75 \leq 22b$

D $4.75 \geq 22b$

Read the Test Item

You want to find the inequality that represents the number of bracelets that can be bought for $22 or less.

Solve the Test Item

If b represents the number of bracelets, then $4.75b$ represents the total cost of b bracelets. So, you can eliminate Choices C and D.

No more than $22 indicates less than or equal to 22, so the inequality that represents this situation is $4.75b \leq 22$. The answer is B.

Test-Taking Tip

Clue Words

Always look for clue words like *no more than* in problems involving inequalities. They will indicate which symbol should be used.

CHECK Your Progress

4. Write an inequality for the sentence below.
Eighteen is greater than or equal to −9 times a number.

F $-9n \geq 18$

G $-9 + n \geq 18$

H $18 \geq -9n$

J $18 \geq n - 9$

CHECK Your Understanding

Examples 1, 2 (pp. 302–303)

Solve each inequality. Check your solution.

1. $\frac{t}{9} < -12$
2. $30 > \frac{1}{2}n$
3. $-\frac{3}{4}r \leq -6$
4. $-\frac{c}{8} \geq 7$

Define a variable, write an inequality, and solve each problem. Then check your solution.

5. The opposite of four times a number is more than 12.
6. Half of a number is at least 26.

7. **FUND-RAISING** The Jefferson High School Band Boosters raised more than \$5500 for their Music Scholarship Fund. This money came from sales of their Marching Band Performances DVD, which sold for \$15. How many DVDs did they sell? Define a variable and write an inequality to solve the problem. Interpret your solution.

Example 3 (pp. 303–304)

Solve each inequality. Check your solution.

8. $7m \geq 42$
9. $12x > -60$
10. $75 < -15g$
11. $-21 \leq -3s$

Example 4 (p. 304)

12. **STANDARDIZED TEST PRACTICE** The area of the rectangle is less than 85 square feet. What is the width of the rectangle?

20 ft

w ft

A $w > 4\frac{1}{4}$ ft B $w \geq 4\frac{1}{4}$ ft C $w < 4\frac{1}{4}$ ft D $w \leq 4\frac{1}{4}$ ft

Exercises

HOMEWORK HELP	
For Exercises	See Examples
13–20	1, 2
21–24	1
25–32	3
33, 34	4

Solve each inequality. Check your solution.

13. $\frac{1}{4}m < -17$
14. $\frac{b}{10} \leq 5$
15. $-7 > -\frac{r}{7}$
16. $-\frac{a}{11} > 9$
17. $\frac{5}{8}y \geq -15$
18. $6 > \frac{2}{3}v$
19. $-10 \geq \frac{x}{-2}$
20. $-\frac{3}{4}q \leq -33$

Define a variable, write an inequality, and solve each problem. Then check your solution.

21. Seven times a number is greater than 28.
22. Negative seven times a number is at least 14.
23. Twenty-four is at most a third of a number.
24. Two thirds of a number is less than −15.

Solve each inequality. Check your solution.

25. $6g \leq 144$
26. $84 < 7t$
27. $-14d \geq 84$
28. $14n \leq -98$
29. $32 > -2y$
30. $-64 \geq -16z$
31. $-26 < 26s$
32. $-6x > -72$

For Exercises 33 and 34, define a variable, write an inequality, and solve each problem. Then interpret your solution.

33. **FUND-RAISING** The Middletown High School girls basketball team wants to make at least \$2000 on their annual mulch sale. The team makes \$2.50 on each bag of mulch sold. How many bags of mulch should the team sell?

34. **EVENT PLANNING** Shaniqua is planning the prom. The hall does not charge a rental fee as long as at least \$4000 is spent on food. If she has chosen a buffet that costs \$28.95 per person, how many people must attend the prom to avoid a rental fee for the hall?

Solve each inequality. Check your solution.

35. $-\frac{2}{3}b \leq -9$

36. $25f \geq 9$

37. $-2.5w < 6.8$

38. $-0.8s > 6.4$

39. $\frac{15c}{-7} > \frac{3}{14}$

40. $\frac{4m}{5} < \frac{-3}{15}$

41. Solve $-\frac{m}{9} \leq -\frac{1}{3}$. Then graph the solution.

42. Solve $\frac{x}{4} > \frac{3}{16}$. Then graph the solution.

43. If $2a \geq 7$, then complete each inequality.

a. $a \geq$ _?_ **b.** $-4a \leq$ _?_ **c.** _?_ $a \leq -21$

Define a variable, write an inequality, and solve each problem. Check your solution.

44. Twenty-five percent of a number is greater than or equal to 90.

45. Forty percent of a number is less than or equal to 45.

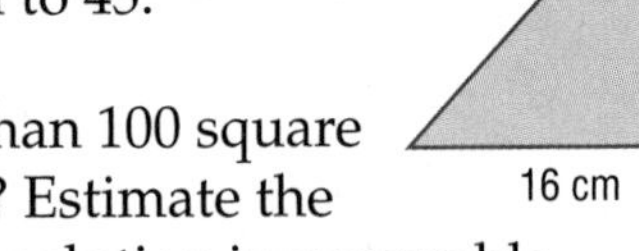

46. GEOMETRY The area of the triangle is greater than 100 square centimeters. What is the height of the triangle? Estimate the height first, and then determine whether your solution is reasonable.

For Exercises 47–50, define a variable, write an inequality, and solve each problem. Then interpret your solution.

47. LANDSCAPING Morris is planning a circular flower garden with a low fence around the border. If he can use up to 38 feet of fence, what radius can he use for the garden? (*Hint:* $C = 2\pi r$)

48. DRIVING Average speed is calculated by dividing distance by time. If the speed limit on the interstate is 65 miles per hour, how far can a person travel legally in $1\frac{1}{2}$ hours?

Real-World Link

Dr. Harry Wegeforth founded the San Diego Zoo in 1916 with just 50 animals. Today, the zoo has over 4000 animals.

Source: www.sandiegozoo.org

49. ANALYZE TABLES The annual membership to the San Diego Zoo for 2 adults and 2 children is \$128. The regular admission to the zoo is shown in the table. How many times should such a family plan to visit the zoo in a year to make a membership less expensive than paying regular admission?

Regular Admission

Visitor	Price
adult	\$19.50
child	\$11.75

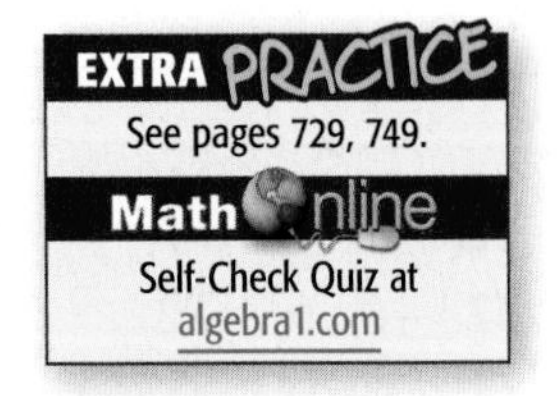

50. CITY PLANNING A city parking lot can have no more than 20% of the parking spaces limited to compact cars. If a certain parking lot has 35 spaces for compact cars, how many spaces must the lot have to conform to the code?

H.O.T. Problems

51. REASONING Explain why you can use either the Multiplication Property of Inequalities or the Division Property of Inequalities to solve $-7r \leq 28$.

52. OPEN ENDED Describe a real-life situation that can be represented by the inequality $\frac{3}{4}a > 9$.

53. CHALLENGE Give a counterexample to show that each statement is not always true.

a. If $a > b$, then $a^2 > b^2$.

b. If $a < b$ and $c < d$, then $ac < bd$.

54. FIND THE ERROR Ilonia and Zachary are solving $-9b \leq 18$. Who is correct? Explain your reasoning.

Ilonia	Zachary
$-9b \leq 18$	$-9b \leq 18$
$\frac{-9b}{-9} \geq \frac{18}{-9}$	$\frac{-9b}{-9} \leq \frac{18}{-9}$
$b \geq -2$	$b \leq -2$

55. ***Writing in Math*** Use the information about the cases of beverages on page 301 to explain how inequalities can be used in storage. Include an inequality representing a stack of cases of water or sports drinks that can be no higher than 3 feet and an explanation of how to solve the inequality.

STANDARDIZED TEST PRACTICE

56. Juan's long-distance phone company charges 9¢ for each minute. Which inequality can be used to find how long can he talk to his friend if he does not want to spend more than \$2.50 on the call?

A $0.09 \geq 2.50m$

B $0.09 \leq 2.50m$

C $0.09m \geq 2.50$

D $0.09m \leq 2.50$

57. REVIEW The table shows the results of a number cube being rolled. What is the experimental probability of rolling a 3?

Outcome	Frequency
1	4
2	8
3	2
4	0
5	5
6	1

F $\frac{2}{3}$ G $\frac{1}{3}$ H 0.2 J 0.1

Spiral Review

Solve each inequality. Check your solution, and graph it on a number line. (Lesson 6-1)

58. $s - 7 < 12$ **59.** $g + 3 \leq -4$ **60.** $7 > n + 2$

61. GYMS To join a gym, Cristina paid an initial fee of \$120, plus \$10 per month. Jackson pays \$20 per month without an initial fee. After how many months will they have paid the same amount? How much will it be? (Lesson 5-5)

Write an equation in standard form for a line that passes through each pair of points. (Lesson 4-4)

62. $(-1, 3), (2, 4)$ **63.** $(5, -2), (-1, -2)$ **64.** $(3, 3), (-1, 2)$

GET READY for the Next Lesson

PREREQUISITE SKILL Solve each equation. (Lessons 2-4 and 2-5)

65. $5x - 3 = 32$ **66.** $14g + 5 = 54$ **67.** $6y - 1 = 4y + 23$ **68.** $2(p - 4) = 7(p + 3)$

6-3 Solving Multi-Step Inequalities

Main Ideas

- Solve linear inequalities involving more than one operation.
- Solve linear inequalities involving the Distributive Property.

GET READY for the Lesson

The normal body temperature of a camel is 97.7°F in the morning. If it has had no water by noon, its body temperature can be greater than 104°F. If F represents temperature in degrees Fahrenheit, the inequality $F > 104$ represents the temperature of a camel at noon.

If C represents degrees Celsius, then $F = \frac{9}{5}C + 32$. You can solve $\frac{9}{5}C + 32 > 104$ to find the temperature in degrees Celsius of a camel at noon.

Solve Multi-Step Inequalities The inequality $\frac{9}{5}C + 32 > 104$ involves more than one operation. It can be solved by undoing the operations in the same way you would solve a multi-step equation.

Real-World EXAMPLE **Multi-Step Inequality**

1 **SCIENCE** **Find the body temperature in degrees Celsius of a camel that has had no water by noon.**

$\frac{9}{5}C + 32 > 104$	Original inequality
$\frac{9}{5}C + 32 - 32 > 104 - 32$	Subtract 32 from each side.
$\frac{9}{5}C > 72$	Simplify.
$\left(\frac{5}{9}\right)\frac{9}{5}C > \left(\frac{5}{9}\right)(72)$	Multiply each side by $\frac{5}{9}$.
$C > 40$	Simplify.

The body temperature of a camel that has had no water by noon is greater than 40°C.

CHECK Your Progress

1. **MONEY** ABC Cellphones advertises a plan with 400 minutes per month for less than the competition. The price includes the $3.50 local tax. If the competition charges $43.50, what does ABC Cellphones charge for each minute?

EXAMPLE Inequality Involving a Negative Coefficient

2 **Solve $-7b + 19 < -16$.**

$-7b + 19 < -16$	Original inequality
$-7b + 19 - 19 < -16 - 19$	Subtract 19 from each side.
$-7b < -35$	Simplify.
$\frac{-7b}{-7} > \frac{-35}{-7}$	Divide each side by -7 and change $<$ to $>$.
$b > 5$	The solution set is $\{b \mid b > 5\}$.

Concepts in Motion
BrainPOP®
algebra1.com

CHECK Your Progress Solve each inequality.

2A. $23 \geq 10 - 2w$

2B. $43 > -4y + 11$

EXAMPLE Write and Solve an Inequality

3 **Define a variable, write an inequality, and solve the problem below. Then check your solution.** ***Three times a number minus eighteen is at least five times the number plus twenty-one.***

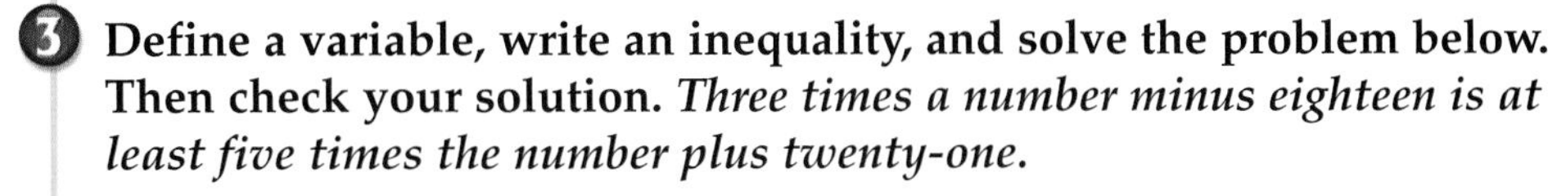

Three times a number	minus	eighteen	is at least	five times the number	plus	twenty-one.
$3n$	$-$	18	$\geq$	$5n$	$+$	21

$-2n - 18 \geq 21$	Subtract $5n$ from each side.
$-2n \geq 39$	Add 18 to each side.
$n \leq -19.5$	Divide each side by -2 and change $\geq$ to $\leq$.

The solution set is $\{n \mid n \leq -19.5\}$. Check your solution.

CHECK Your Progress

3. Write an inequality for the sentence below. Then solve the inequality. *Two more than half of a number is greater than twenty-seven.*

Personal Tutor at algebra1.com

You can solve an inequality in one variable by using a graphing calculator.

GRAPHING CALCULATOR LAB

Solving Inequalities

On a TI-83/84 Plus, clear the Y= list. Enter $6x + 9 < -4x + 29$ as Y1. (The symbol $<$ is item 5 on the TEST menu.) Press GRAPH.

THINK AND DISCUSS

1. Describe what is shown on the screen.

2. Use the TRACE function to scan the values along the graph. What do you notice about the values of y on the graph?

3. Solve the inequality algebraically. How does your solution compare to the pattern you noticed in Exercise 2?

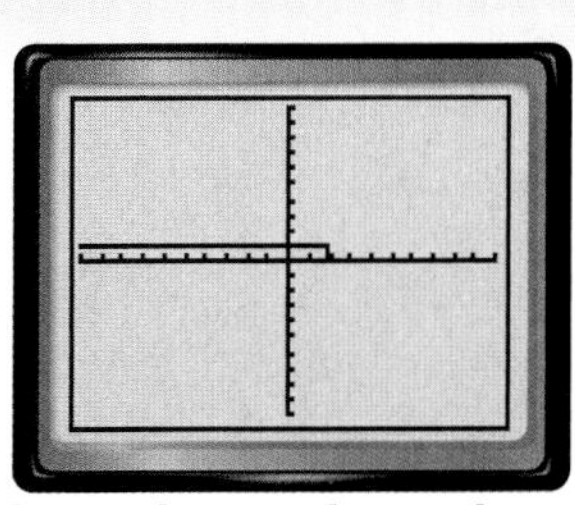

[−10, 10] scl: 1 by [−10, 10] scl: 1

Solve Inequalities Involving the Distributive Property When solving equations that contain grouping symbols, first use the Distributive Property to remove the grouping symbols.

EXAMPLE **Distributive Property**

4 **Solve $3d - 2(8d - 9) > -2d - 4$.**

$3d - 2(8d - 9) > -2d - 4$	Original inequality
$3d - 16d + 18 > -2d - 4$	Distributive Property
$-13d + 18 > -2d - 4$	Combine like terms.
$-13d + 18 + 13d > -2d - 4 + 13d$	Add $13d$ to each side.
$18 > 11d - 4$	Simplify.
$18 + 4 > 11d - 4 + 4$	Add 4 to each side.
$22 > 11d$	Simplify.
$\frac{22}{11} > \frac{11d}{11}$	Divide each side by 11.
$2 > d$	Simplify.

Since $2 > d$ is the same as $d < 2$, the solution set is $\{d \mid d < 2\}$.

Concepts in Motion
Interactive Lab
algebra1.com

CHECK Your Progress **Solve each inequality.**

4A. $6(5z - 3) \leq 36z$ **4B.** $2(h + 6) > -3(8 - h)$

If solving an inequality results in a statement that is always true, the solution set is the set of all real numbers. This is written as $\{x \mid x \text{ is a real number}\}$. If solving an inequality results in a statement that is never true, the solution set is the empty set, written as the symbol $\varnothing$. The empty set has no members.

EXAMPLE **Empty Set and All Reals**

5 **Solve $8(t + 2) - 3(t - 4) < 5(t - 7) + 8$.**

$8(t + 2) - 3(t - 4) < 5(t - 7) + 8$	Original inequality
$8t + 16 - 3t + 12 < 5t - 35 + 8$	Distributive Property
$5t + 28 < 5t - 27$	Combine like terms.
$5t + 28 - 5t < 5t - 27 - 5t$	Subtract $5t$ from each side.
$28 < -27$	Simplify.

Since the inequality results in a false statement, the solution is the empty set $\varnothing$.

CHECK Your Progress **Solve each inequality.**

5A. $18 - 3(8c + 4) \geq -6(4c - 1)$ **5B.** $46 \leq 8m - 4(2m + 5)$

CHECK Your Understanding

Examples 1, 2 (pp. 308–309)

Solve each inequality. Check your solution.

1. $6h - 10 \geq 32$

2. $-3 \leq \frac{2}{3}r + 9$

3. $-4y - 23 < 19$

4. $7b + 11 > 9b - 13$

5. CANOEING A certain canoe was advertised as having an "800 pound capacity," meaning that it can hold at most 800 pounds. If four people plan to use the canoe and take 60 pounds of supplies, write and solve an inequality to find the average weight per person.

Example 3 (p. 309)

6. Define a variable, write an inequality, and solve the problem below. Then check your solution.
Seven minus two times a number is less than three times the number plus thirty-two.

Examples 4, 5 (p. 310)

Solve each inequality. Check your solution.

7. $-6 \leq 3(5v - 2)$

8. $-5(g + 4) > 3(g - 4)$

9. $3 - 8x \geq 9 + 2(1 - 4x)$

Exercises

HOMEWORK HELP	
For Exercises	See Examples
10–19	1, 2
20–25	3
26–33	4, 5

Solve each inequality. Check your solution.

10. $5b - 1 \geq -11$

11. $21 > 15 + 2a$

12. $-9 \geq \frac{2}{5}m + 7$

13. $\frac{w}{8} - 13 > -6$

14. $-3t + 6 \leq -3$

15. $59 > -5 - 8f$

16. $-2 - \frac{d}{5} < 23$

17. $-\frac{3}{2}a + 4 > 10$

18. $9r + 15 \leq 24 + 10r$

19. $13k - 11 > 7k + 37$

Define a variable, write an inequality, and solve each problem. Then check your solution.

20. One eighth of a number decreased by five is at least thirty.

21. Two thirds of a number plus eight is greater than twelve.

22. Negative four times a number plus nine is no more than the number minus twenty-one.

23. Ten is no more than 4 times the sum of twice a number and three.

For Exercises 24 and 25, define a variable, write an inequality, and solve each problem. Then interpret your solution.

24. SALES A salesperson is paid \$22,000 a year plus 5% of the amount of sales made. What is the amount of sales needed to have an annual income greater than \$35,000?

25. ANIMALS Keith's dog weighs 90 pounds. The veterinarian told him that a healthy weight for his dog would be less than 75 pounds. If Keith's dog can lose an average of 1.25 pounds per week on a certain diet, how long will it take the dog to reach a healthy weight?

Solve each inequality. Check your solution.

26. $5(2h - 6) > 4h$

27. $21 \geq 3(a - 7) + 9$

28. $2y + 4 > 2(3 + y)$

29. $3(2 - b) < 10 - 3(b - 6)$

30. $7 + t \leq 2(t + 3) + 2$

31. $8a + 2(1 - 5a) \leq 20$

32. Solve $4(t - 7) \leq 2(t + 9)$. Show each step and justify your work.

33. Solve $-5(k + 4) > 3(k - 4)$. Show each step and justify your work.

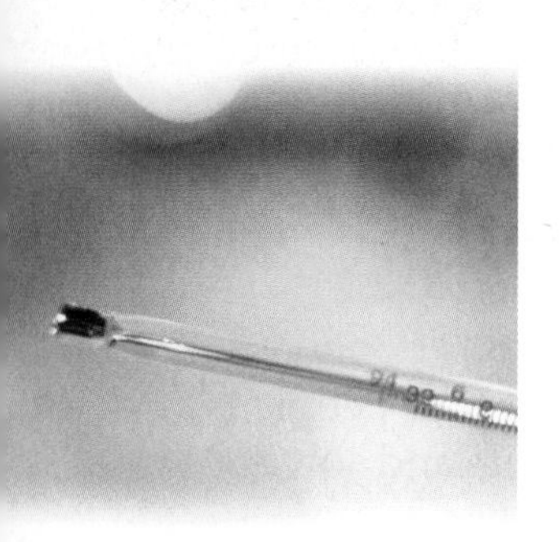

SCHOOL For Exercises 34–36, use the following information.

Carmen's scores on three math tests are shown in the table. The fourth and final test of the grading period is tomorrow. She needs an average (mean) of at least 92 to receive an A for the grading period.

Test	Score
1	91
2	95
3	88

34. If s is her score on the fourth test, write an inequality to represent the situation.

35. If Carmen wants an A in math, what must she score on the test?

36. Is 150 a solution to the inequality that you wrote in Exercise 35? Is this a reasonable solution to the problem? Explain your reasoning.

Real-World Link

Mercury is a metal that is a liquid at room temperature. In fact, its melting point is −38°C. Mercury is used in thermometers because it expands evenly as it is heated.

Source: *World Book Encyclopedia*

37. **MONEY** Nicholas has \$13 to order a pizza. The pizza costs \$7.50 plus \$1.25 per topping. He plans to tip 15% of the total cost of the pizza. Write and solve an inequality to find how many toppings he can order.

38. **PHYSICAL SCIENCE** The melting point for an element is the temperature where the element changes from a solid to a liquid. If C represents degrees Celsius and F represents degrees Fahrenheit, then $C = \frac{5(F - 32)}{9}$. Refer to the information at the left to write and solve an inequality that can be used to find the temperatures in degrees Fahrenheit for which mercury is a solid.

39. **Solve for x in each case.**

a. $3x - 6 > 2 + 4(x - 2)$
b. $2(x - 4) \leq 2 + 3(x - 6)$
c. $3x - 6 = 3(2 + 2(3 + x) + 4) - 1$
d. $\frac{3}{3x - 6} = \frac{4}{2x + 4}$

40. **NUMBER THEORY** Find all sets of two consecutive positive odd integers with a sum no greater than 18.

Solve each inequality. Check your solution.

41. $\frac{5b + 8}{3} < 3b$
42. $3.1v - 1.4 \geq 1.3v + 6.7$
43. $0.3(d - 2) - 0.8d > 4.4$

EXTRA PRACTICE See pages 729, 749.

Math Online Self-Check Quiz at algebra1.com

44. Define a variable, write an inequality, and solve the problem below. Then check your solution.
Three times the sum of a number and seven is greater than five times the number less thirteen.

Graphing Calculator

Use a graphing calculator to solve each inequality.

45. $3x + 7 > 4x + 9$
46. $13x - 11 \leq 7x + 37$
47. $2(x - 3) < 3(2x + 2)$

H.O.T. Problems

48. **REASONING** Explain how you could solve $-3p + 7 \geq 2$ without multiplying or dividing each side by a negative number.

49. **CHALLENGE** Create a multi-step inequality that has no solution and one that has infinitely many solutions. Investigate the best method for solving each one.

50. **OPEN ENDED** Create a multi-step inequality with the solution graphed below. Explain how you know.

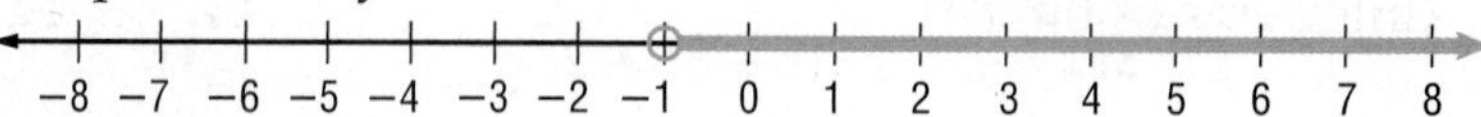

51. Which One Doesn't Belong? Identify the inequality that does not belong with the other three. Explain.

$4y + 9 > -3$ | $3y - 4 > 5$ | $2y + 1 > 7$ | $-5y + 2 < -13$

52. ***Writing in Math*** Use the information about camels on page 308 to explain how linear inequalities can be used in science.

STANDARDIZED TEST PRACTICE

53. What is the solution set of the inequality $4t + 2 < 8t - (6t - 10)$?

A $t < -6.5$

B $t > -6.5$

C $t < 4$

D $t > 4$

54. REVIEW What is the volume of the triangular prism?

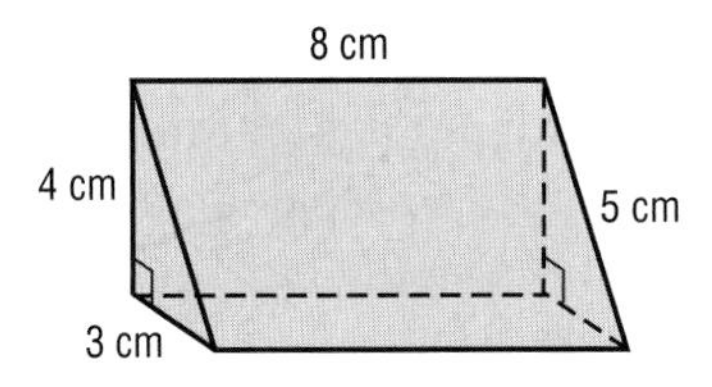

F 120 cm^3 H 48 cm^3

G 96 cm^3 J 30 cm^3

Spiral Review

55. BUSINESS The charge per mile for a compact rental car is \$0.12. Mrs. Ludlow must rent a car for a business trip. She has a budget of \$50 for mileage charges. How many miles can she travel without going over her budget? (Lesson 6-2)

Solve each inequality. Check your solution, and then graph it on a number line. (Lesson 6-1)

56. $d + 13 \geq 22$ **57.** $t - 5 < 3$ **58.** $4 > y + 7$

Write the standard form of an equation of the line that passes through the given point and has the given slope. (Lesson 4-5)

59. $(1, -3), m = 2$ **60.** $(-2, -1), m = -\frac{2}{3}$ **61.** $(3, 6), m = 0$

CABLE TV For Exercises 62 and 63, use the graph at the right. (Lesson 4-1)

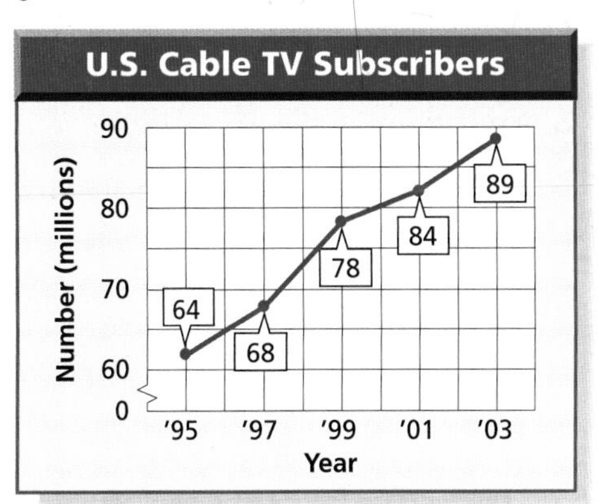

62. Find the rate of change for 2001–2003. Explain the meaning of the rate of change.

63. Without calculating, find a 2-year period that had a greater rate of change than 2001–2003. Explain.

GET READY for the Next Lesson

PREREQUISITE SKILL Graph each set of numbers on a number line. (Lesson 1-8)

64. $\{-1, 0, 3, 4\}$ **65.** $\{-5, -4, -1, 1\}$

66. {integers less than 5} **67.** {integers between 1 and 6}

READING MATH

Compound Statements

Two simple statements connected by the words *and* or *or* form a compound statement. Before you can determine whether a compound statement is true or false, you must understand what the words *and* and *or* mean. Consider the statement below.

A triangle has three sides, and a hexagon has five sides.

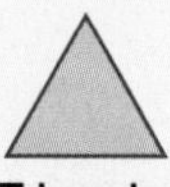
Triangle

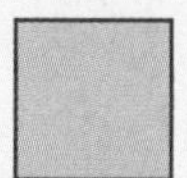
Square

Pentagon

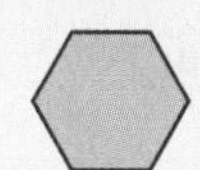
Hexagon

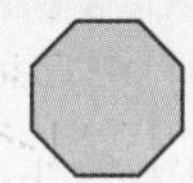
Octagon

For a compound statement connected by the word *and* to be true, both simple statements must be true. In this case, it is true that a triangle has three sides. However, it is false that a hexagon has five sides; it has six. Thus, the compound statement is false.

A compound statement connected by the word *or* may be *exclusive* or *inclusive*. For example, the statement "With your dinner, you may have soup *or* salad," is exclusive. In everyday language, *or* means one or the other, but not both. However, in mathematics, *or* is inclusive. It means one or the other or both. Consider the statement below.

A triangle has three sides, or a hexagon has five sides.

For a compound statement connected by the word *or* to be true, at least one of the simple statements must be true. Since it is true that a triangle has three sides, the compound statement is true.

Reading to Learn

Determine whether each compound statement is *true* or *false*. Explain your answer.

1. A hexagon has six sides, *or* an octagon has seven sides.
2. An octagon has eight sides, *and* a pentagon has six sides.
3. A pentagon has five sides, *and* a hexagon has six sides.
4. A triangle has four sides, *or* an octagon does *not* have seven sides.
5. A pentagon has three sides, *or* an octagon has ten sides.
6. A square has four sides, *or* a hexagon has six sides.
7. $5 < 4$ or $8 < 6$
8. $-1 > 0$ and $1 < 5$
9. $4 > 0$ and $-4 < 0$
10. $0 = 0$ or $-2 > -3$
11. $5 \neq 5$ or $-1 > -4$
12. $0 > 3$ and $2 > -2$

6-4 Solving Compound Inequalities

Main Ideas

- Solve compound inequalities containing the word *and* and graph their solution sets.
- Solve compound inequalities containing the word *or* and graph their solution sets.

New Vocabulary

compound inequality
intersection
union

GET READY for the Lesson

The Mind Eraser Roller Coaster at Six Flags in Baltimore, Maryland, is an inverted steel track roller coaster. The trains are suspended underneath the tracks. To ride this coaster, you must be at least 52 inches tall and no more than 72 inches.

Let h represent the height of a rider. You can write two inequalities to represent the height restrictions.

at least 52 inches	no more than 72 inches
$h \geq 52$	$h \leq 72$

The inequalities $h \geq 52$ and $h \leq 72$ can be combined and written without using *and*.

$$52 \leq h \leq 72$$

Inequalities Containing *and* When considered together, two inequalities such as $w \geq 40$ and $w \leq 250$ form a **compound inequality**. A compound inequality containing *and* is true only if both inequalities are true. Its graph is the **intersection** of the graphs of the two inequalities. In other words, the solution must be a solution of *both* inequalities.

The intersection can be found by graphing each inequality and then determining where the graphs overlap.

EXAMPLE Graph an Intersection

1 Graph the solution set of $x < 3$ and $x \geq -2$.

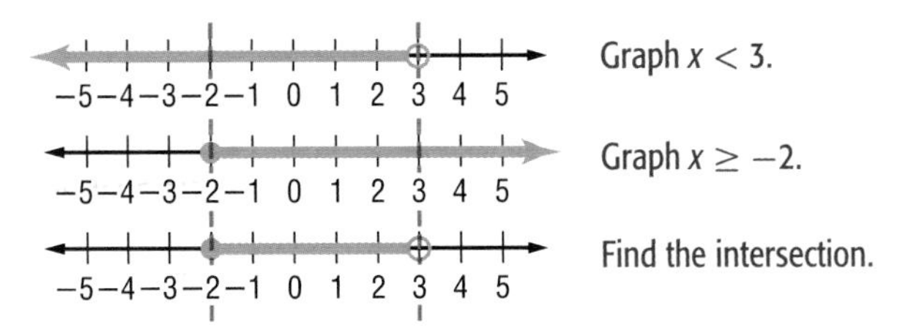

The solution set is $\{x \mid -2 \leq x < 3\}$. Note that the graph of $x \geq -2$ includes the point -2. The graph of $x < 3$ does *not* include 3.

Reading Math

Compound Inequalities

The statement $-2 \leq x < 3$ can be read *negative 2 is less than or equal to x, which is less than 3.*

CHECK Your Progress

Graph the solution set of each compound inequality.

1A. $a > -5$ and $a < 0$

1B. $p \leq 6$ and $p > 2$

Reading Math

Keywords
When solving problems involving inequalities,

- within is meant to be inclusive. Use $\leq$ or $\geq$.
- between is meant to be exclusive. Use < or >.

EXAMPLE Solve and Graph an Intersection

2 **Solve $-5 < x - 4 < 2$. Then graph the solution set.**

First express $-5 < x - 4 < 2$ using *and*. Then solve each inequality.

$-5 < x - 4$	and	$x - 4 < 2$	Write the inequalities.
$-5 + 4 < x - 4 + 4$		$x - 4 + 4 < 2 + 4$	Add 4 to each side.
$-1 < x$		$x < 6$	Simplify.

The solution set is $\{x \mid -1 < x < 6\}$. Now graph the solution set.

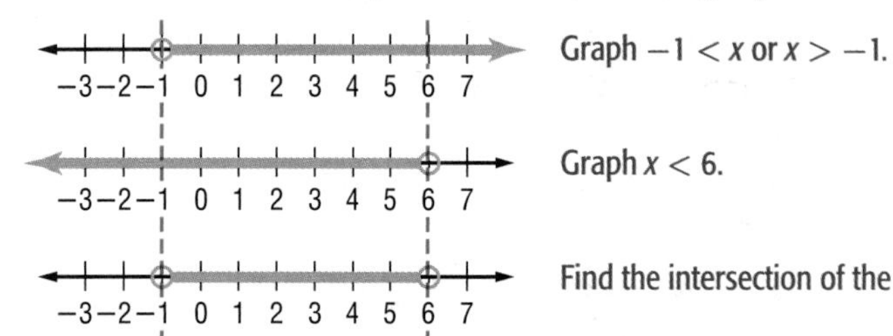

CHECK Your Progress

Solve each compound inequality. Then graph the solution set.

2A. $y - 3 \geq -11$ and $y - 3 \leq -8$ **2B.** $6 \leq r + 7 < 10$

Online **Personal Tutor at** algebra1.com

Inequalities Containing *or* Another type of compound inequality contains the word *or*. A compound inequality containing *or* is true if one or more of the inequalities is true. Its graph is the **union** of the graphs of the two inequalities. In other words, its solution is a solution of *either* inequality, not necessarily both. The union can be found by graphing each inequality.

Real-World Career... Pilot
A pilot uses math to calculate the altitude that will provide the smoothest flight.

Math Online
For more information, go to algebra1.com.

Real-World EXAMPLE Write and Graph a Compound Inequality

3 **AVIATION A pilot flying at 30,000 feet is told by the control tower that he should increase his altitude to at least 33,000 feet or decrease his altitude to no more than 26,000 feet to avoid turbulence. Write and graph a compound inequality that describes the optimum altitude.**

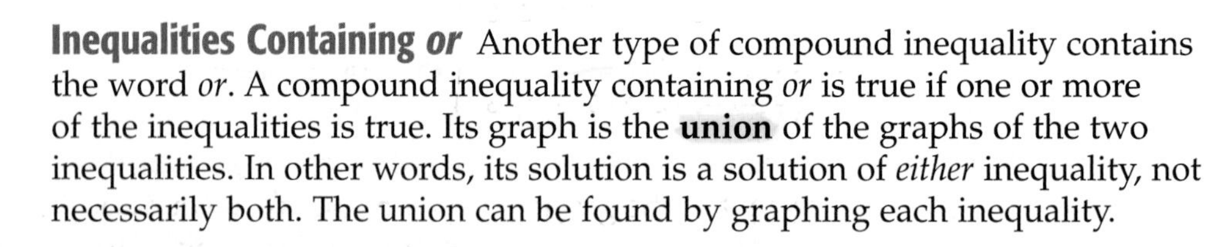

Words	The plane's altitude	is at least	33,000 feet	or	the altitude	is no more than	26,000 feet.
Variable	Let a be the plane's altitude.						
Inequality	a	$\geq$	33,000	or	a	$\leq$	26,000

Now, graph the solution set.

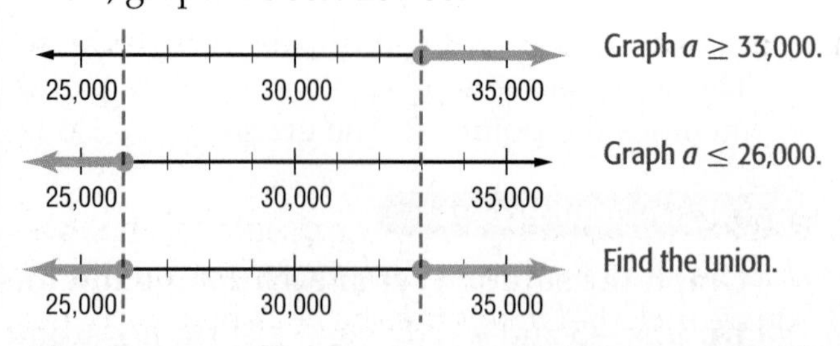

The compound inequality $a \geq 33{,}000$ or $a \leq 26{,}000$ is graphed above.

CHECK Your Progress

3. **SHOPPING** A store is offering a \$30 mail-in rebate on all color printers. Luisana is looking at different color printers that range in price from \$175 to \$260. How much can she expect to spend after the mail-in rebate?

Online **Personal Tutor at** algebra1.com

EXAMPLE Solve and Graph a Union

4 Solve $-3h + 4 < 19$ or $7h - 3 > 18$. Then graph the solution set.

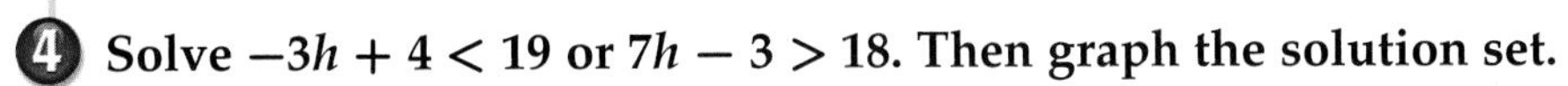

$-3h + 4 < 19$	or	$7h - 3 > 18$	Write the inequalities.
$-3h + 4 - 4 < 19 - 4$		$7h - 3 + 3 > 18 + 3$	Add or subtract.
$-3h < 15$		$7h > 21$	Simplify.
$\frac{-3h}{-3} > \frac{15}{-3}$		$\frac{7h}{7} > \frac{21}{7}$	Divide.
$h > -5$		$h > 3$	Simplify.

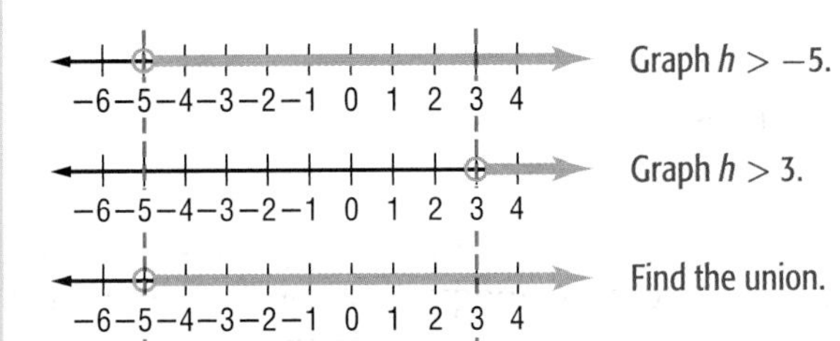

Graph $h > -5$.

Graph $h > 3$.

Find the union.

Concepts in Motion
Animation
algebra1.com

Notice that the graph of $h > -5$ contains every point in the graph of $h > 3$. So, the union is the graph of $h > -5$. The solution set is $\{h \mid h > -5\}$.

CHECK Your Progress

Solve each compound inequality. Then graph the solution set.

4A. $a + 1 < 4$ or $a - 1 \geq 3$

4B. $x \leq 9$ or $2 + 4x < 10$

CHECK Your Understanding

Example 1 (p. 315)

Graph the solution set of each compound inequality.

1. $a \leq 6$ and $a \geq -2$
2. $y < 12$ and $y > 9$

Examples 2, 4 (pp. 316–317)

Solve each compound inequality. Then graph the solution set.

3. $6 < w + 3$ and $w + 3 < 11$
4. $n - 7 \leq -5$ or $n - 7 \geq 1$
5. $3z + 1 < 13$ or $z \leq 1$
6. $-8 < x - 4 \leq -3$

Example 3 (pp. 316–317)

7. **BIKES** The recommended air pressure for the tires of a mountain bike is at least 35 pounds per square inch (psi), but no more than 80 pounds per square inch. If a bike's tires have 24 pounds per square inch, what increase in air pressure is needed so the tires are in the recommended range?

Exercises

HOMEWORK HELP	
For Exercises	See Examples
8–13	1
14–23	2, 4
24–25	3

Graph the solution set of each compound inequality.

8. $x > 5$ and $x \leq 9$

9. $s < -7$ and $s \leq 0$

10. $r < 6$ or $r > 6$

11. $m \geq -4$ or $m > 6$

12. $7 < d < 11$

13. $-1 \leq g < 3$

Solve each compound inequality. Then graph the solution set.

14. $k + 2 > 12$ and $k + 2 \leq 18$

15. $f + 8 \leq 3$ and $f + 9 \geq -4$

16. $d - 4 > 3$ or $d - 4 \leq 1$

17. $h - 10 < -21$ or $h + 3 < 2$

18. $3 < 2x - 3 < 15$

19. $4 < 2y - 2 < 10$

20. $3t - 7 \geq 5$ and $2t + 6 \leq 12$

21. $8 > 5 - 3q$ and $5 - 3q > -13$

22. $-1 + x \leq 3$ or $-x \leq -4$

23. $3n + 11 \leq 13$ or $-3n \geq -12$

24. ANALYZE TABLES The Fujita Scale (F-scale) is the official classification system for tornado damage. One factor used to classify a tornado is wind speed. Use the information in the table to write an inequality for the range of wind speeds of an F3 tornado.

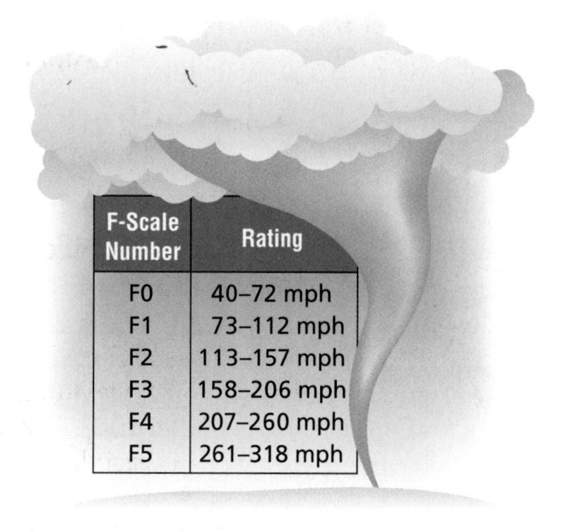

F-Scale Number	Rating
F0	40–72 mph
F1	73–112 mph
F2	113–157 mph
F3	158–206 mph
F4	207–260 mph
F5	261–318 mph

25. BIOLOGY Each type of fish thrives in a specific range of temperatures. The optimum temperatures for sharks range from 18°C to 22°C, inclusive. Write an inequality to represent temperatures where sharks will *not* thrive. (*Hint:* The word *inclusive* means that 18°C and 22°C are included in the optimum temperature range.)

Real-World Link The average life span of a shark is 25 years, but some sharks can live to be 100.

Source: about.com

Write a compound inequality for each graph.

26. −5 −4 −3 −2 −1 0 1 2 3 4 5

27.

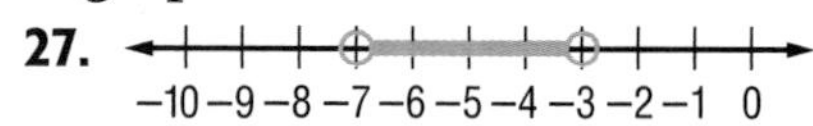

28.

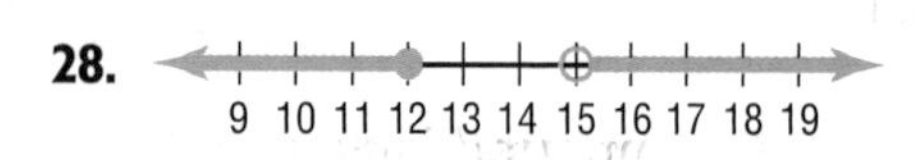

29.

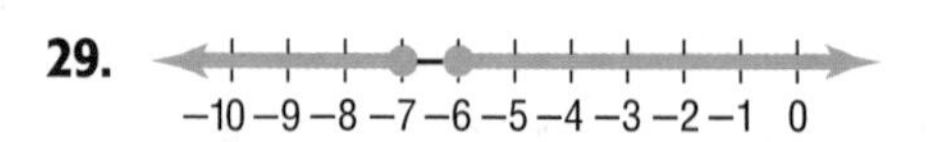

30. −9 −8 −7 −6 −5 −4 −3 −2 −1 0 1

31. −1 0 1 2 3 4 5 6 7 8 9

Solve each compound inequality. Then graph the solution set.

32. $2p - 2 \leq 4p - 8 \leq 3p - 3$

33. $3g + 12 \leq 6 - g \leq 3g - 18$

34. $4c < 2c - 10$ or $-3c < -12$

35. $0.5b > -6$ or $3b + 16 < -8 + b$

36. HEALTH About 20% of the time you sleep is spent in rapid eye movement (REM) sleep, which is associated with dreaming. If an adult sleeps 7 to 8 hours, how much time is spent in REM sleep?

37. FUND-RAISING Rashid is selling potted flowers for his school's fund-raiser. He can earn prizes depending on how much he sells. So far, he has sold \$70 worth of flowers. How much more does he need to sell to earn a prize in category D?

Sales (\$)	Prize
0–25	A
26–60	B
61–120	C
121–180	D
180+	E

Define a variable, write an inequality, and solve each problem. Then check your solution.

38. Eight less than a number is no more than 14 and no less than 5.

39. The sum of 3 times a number and 4 is between −8 and 10.

40. The product of −5 and a number is greater than 35 or less than 10.

41. One half a number is greater than 0 and less than or equal to 1.

A dog's sense of smell is about 1000 times better than a human's, but its eyesight is not as good.

Source: about.com

HEARING For Exercises 42–44, use the table.

What Humans and Dogs Can Hear	
Species	**Sound Waves (hertz)**
humans	20–20,000
dogs	15–50,000

42. Write a compound inequality for the hearing range of humans and one for the hearing range of dogs.

43. What is the union of the two graphs? the intersection?

44. Write an inequality or inequalities for the range of sounds that dogs can hear, but humans cannot.

45. RESEARCH Use the Internet or other resource to find the altitudes in miles of the layers of Earth's atmosphere, troposphere, stratosphere, mesosphere, thermosphere, and exosphere. Write inequalities for the range of altitudes for each layer.

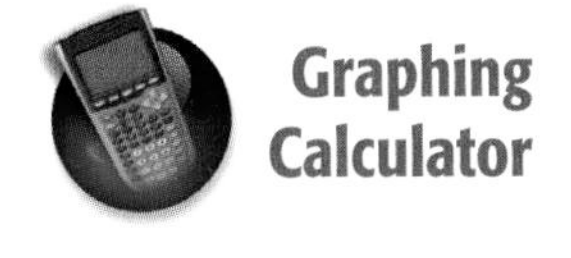

46. In Lesson 6-3, you learned how to use a graphing calculator to find the values of x that make a given inequality true. You can also use this method to test compound inequalities. The words *and* and *or* can be found in the LOGIC submenu of the TEST menu of a TI-83/84 Plus. Use this method to solve each of the following compound inequalities using your graphing calculator.

a. $x + 4 < -2$ or $x + 4 > 3$ **b.** $x - 3 \leq 5$ and $x + 6 \geq 4$

H.O.T. Problems

47. OPEN ENDED Create an example of a compound inequality containing *and* that has no solution.

48. REASONING Formulate a compound inequality to represent *\$7 is less than t, which is less than \$12.* Interpret what the solution means.

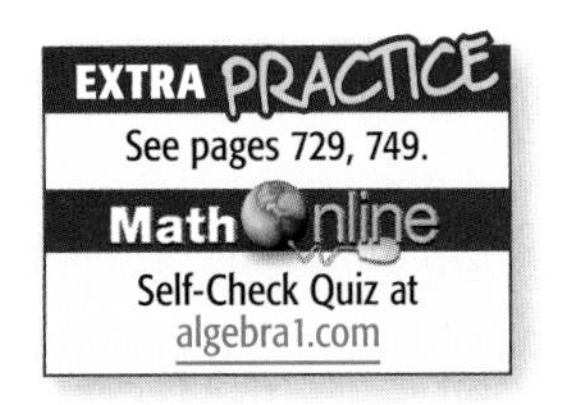

49. CHALLENGE Select compound inequalities that represent the values of x which make the following expressions *false.*

a. $x < 5$ or $x > 8$ **b.** $x \leq 6$ and $x \geq 1$

50. *Writing in Math* Use the information about the roller coaster on page 315 to explain how compound inequalities can be used to describe weight restrictions at amusement parks. Include a compound inequality describing a possible height restriction for riders of the roller coaster. Describe what this represents.

STANDARDIZED TEST PRACTICE

51. REVIEW Ten pounds of fresh tomatoes make about 15 cups of cooked tomatoes. How many cups does one pound of tomatoes make?

A $1\frac{1}{2}$ cups

B 5 cups

C 3 cups

D 4 cups

52. What is the solution set of the inequality $-7 < x + 2 < 4$?

F $-5 < x < 6$

G $-9 < x < 2$

H $-5 < x < 2$

J $-9 < x < 6$

53. REVIEW The scatterplot below shows the number of hay bales used by the Crosley farm during the last year.

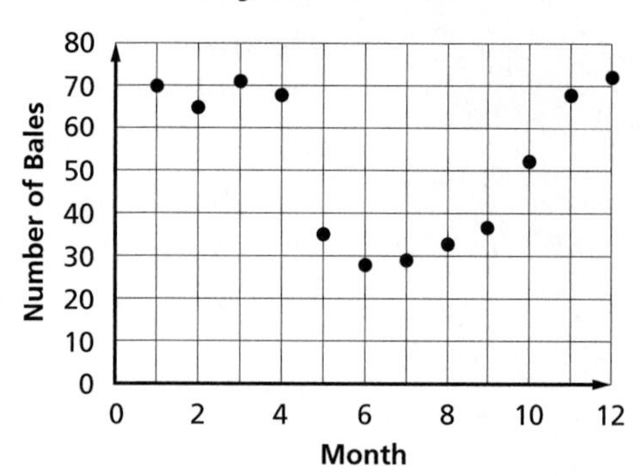

Which is an invalid conclusion?

A The Crosleys used less hay in the summer than they did in the winter.

B The Crosleys used a total of 629 bales of hay.

C The Crosleys used about 52 bales each month.

D The Crosleys used the most hay in February.

Spiral Review

54. MONEY In the summer, Richard earns \$200 per month at his part-time job at a restaurant, plus an average of \$18 for each lawn that he mows. If his goal is to earn at least \$280 this month, how many lawns will he have to mow? (Lesson 6-3)

Solve each inequality. Check your solution. (Lesson 6-2)

55. $18d \geq 90$ **56.** $-7v < 91$ **57.** $\frac{t}{13} < 13$ **58.** $-\frac{3}{8}b > 9$

Solve. Assume that y varies directly as x. (Lesson 5-2)

59. If $y = -8$ when $x = -3$, find x when $y = 6$.

60. If $y = 2.5$ when $x = 0.5$, find y when $x = 20$.

GET READY for the Next Lesson

PREREQUISITE SKILL **Solve each equation.** (Lesson 2-4)

61. $3n - 14 = 1$ **62.** $2t + 5 = 7$ **63.** $8w - 13 = 3$ **64.** $5d + 9 = -6$

65. $12 = 3n + 15$ **66.** $17 = 4p - 3$ **67.** $-3 = 4x + 5$ **68.** $-14 = 4 + 3w$

CHAPTER 6 Mid-Chapter Quiz

Lessons 6-1 through 6-4

1. **MULTIPLE CHOICE** Which graph represents all the values of m such that $m + 3 > 7$? (Lesson 6-1)

A 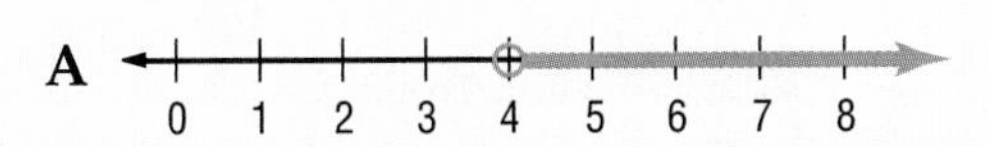

B

0 1 2 3 4 5 6 7 8

C

0 1 2 3 4 5 6 7 8

D

0 1 2 3 4 5 6 7 8

Solve each inequality. Check your solution, then graph it on a number line. (Lesson 6-1)

2. $8 + x < 9$
3. $h - 16 > -13$
4. $r + 3 \leq -1$
5. $4 \geq p + 9$
6. $-3 < a - 5$
7. $7g \leq 6g - 1$

8. **MULTIPLE CHOICE** Which inequality does NOT have the same solution as $-14w < 14$? (Lesson 6-2)

F $8w > -8$

H $\frac{w}{5} > -\frac{1}{5}$

G $11 > -11w$

J $-\frac{3}{2} < \frac{2}{3}w$

Solve each inequality. Check your solution. (Lesson 6-2)

9. $12 \geq 3a$
10. $15z \geq 105$
11. $\frac{v}{5} < 4$
12. $-\frac{3}{7}q > 15$
13. $-156 < 12r$
14. $-\frac{2}{5}w \leq -\frac{1}{2}$

15. **MONEY** Javier is saving \$175 each month to buy a used all-terrain vehicle. How long will it take him to save at least \$2900? Define a variable and write an inequality to solve the problem. Interpret your solution. (Lesson 6-2)

Solve each inequality. Check your solution. (Lesson 6-3)

16. $5 - 4b > -23$
17. $\frac{1}{2}n + 3 \geq -5$
18. $3(t + 6) < 9$
19. $9x + 2 > 20$
20. $2m + 5 \leq 4m - 1$
21. $a < \frac{2a - 15}{3}$

22. **MULTIPLE CHOICE** What is the first step in solving $\frac{y - 5}{9} \geq 13$? (Lesson 6-3)

A Add 5 to each side.

B Subtract 5 from each side.

C Divide each side by 9.

D Multiply each side by 9.

23. **PHYSICAL SCIENCE** Chlorine is a gas for all temperatures greater than $-31°$F. If F represents temperature in degrees Fahrenheit, the inequality $F > -31$ represents the temperatures for which chlorine is a gas. Solve $\frac{9}{5}C + 32 > -31$ to find the temperatures in degrees Celsius for which chlorine is a gas. (Lesson 6-3)

Write a compound inequality for each graph. (Lesson 6-4)

24. −5 −4 −3 −2 −1 0 1 2 3 4 5

25. −3 −2 −1 0 1 2 3 4 5 6 7

26. **MULTIPLE CHOICE** Some parts of a state get less than 9 inches of annual rainfall. Other parts of the state get more than 57 inches. Which inequality represents this situation? (Lesson 6-4)

F $9 < r < 57$

G $9 > r > 57$

H $r < 9$ or $r > 57$

J $r < 9$ and $r > 57$

Solve each compound inequality. Then graph the solution set. (Lesson 6-4)

27. $x - 2 < 7$ and $x + 2 > 5$
28. $2b + 5 \leq -1$ or $b - 4 \geq -4$
29. $4m - 5 > 7$ or $4m - 5 < -9$
30. $a - 4 < 1$ and $a + 2 > 1$

Solving Open Sentences Involving Absolute Value

Main Ideas

- Solve absolute value equations.
- Graph absolute value functions.

New Vocabulary

absolute value
absolute value function
piecewise function

GET READY for the Lesson

In an international survey of students from 25 high schools in 9 different countries, 52% of those surveyed chose cell phones as the technology that is most important to them.

Suppose the survey had a 3-point margin of error. This means that the result may be 3 percentage points higher or lower. So, the number of students favoring cell phones over other technology may be as high as 55% or as low as 49%.

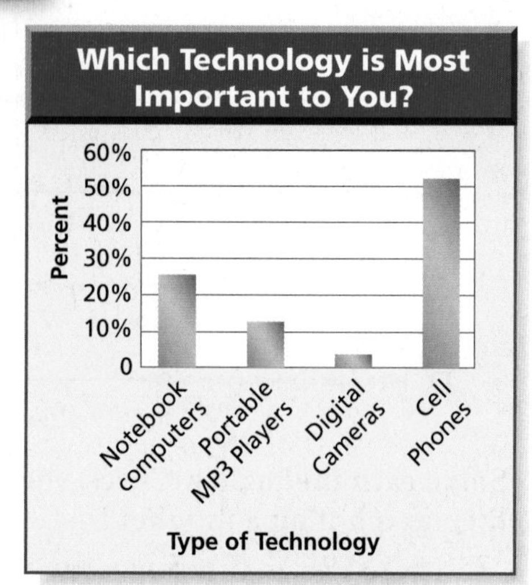

Source: www.ucdsb.on.ca/athens/surveys.index.htm

Absolute Value Equations The margin of error of the data in the bar graph is an example of absolute value. The distance between 52 and 55 on a number line is the same as the distance between 49 and 52.

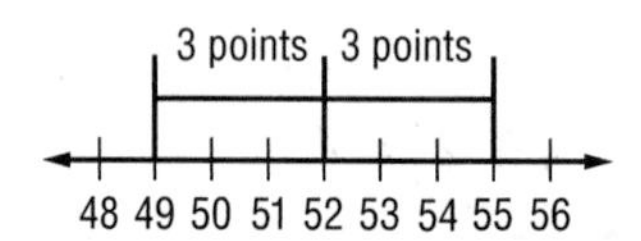

Study Tip

Absolute Value
Since distance cannot be less than zero, absolute values are always greater than or equal to zero.

The **absolute value** of any number n is its distance from zero on a number line. The absolute value of n is written as $|n|$. There are three types of open sentences involving absolute value. They are $|x| = n$, $|x| < n$, and $|x| > n$. Consider the first type. $|x| = 5$ means the distance between 0 and x is 5 units.

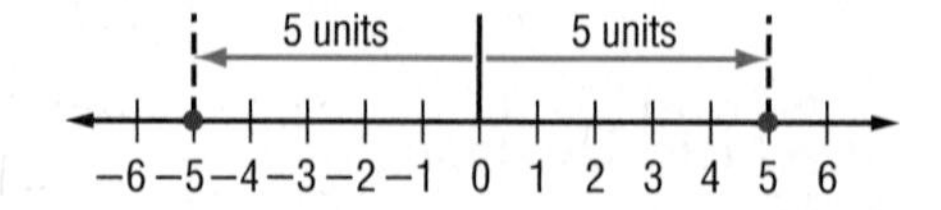

If $|x| = 5$, then $x = -5$ or $x = 5$. The solution set is $\{-5, 5\}$.

KEY CONCEPT *Solving Absolute Value Equations*

When solving equations that involve absolute value, there are two cases to consider.

Case 1 The expression inside the absolute value symbols is positive.

Case 2 The expression inside the absolute value symbols is negative.

Real-World EXAMPLE Solve an Absolute Value Equation

1 **a. SNAKES The temperature of an enclosure for a pet snake should be about 80°F, give or take 5°. Solve $|a - 80| = 5$ to find the maximum and minimum of the temperatures.**

Method 1 Graphing

$|a - 80| = 5$ means that the distance between a and 80 is 5 units. To find a on the number line, start at 80 and move 5 units in either direction.

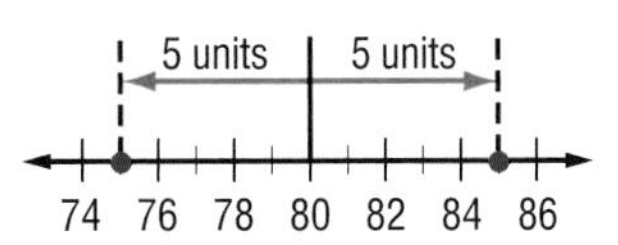

The distance from 80 to 75 is 5 units.
The distance from 80 to 85 is 5 units.
The solution set is {75, 85}.

Method 2 Compound Sentence

Write $|a - 80| = 5$ as $a - 80 = 5$ or $a - 80 = -5$.

Case 1		Case 2
$a - 80 = 5$		$a - 80 = -5$
$a - 80 + 80 = 5 + 80$	Add 80 to each side.	$a - 80 + 80 = -5 + 80$
$a = 85$	Simplify.	$a = 75$

The solution set is {75, 85}. The maximum and minimum temperatures are 85°F and 75°F.

b. Solve $|b - 1| = -3$.

$|b - 1| = -3$ means that the distance between b and 1 is -3. Since distance cannot be negative, the solution is the empty set ∅.

Study Tip

Absolute Value

Since $|a| = 3$ means $a = 3$ or $-a = 3$, the second equation can be written as $a = -3$. So, $|a - 4| = 3$ means $a - 4 = 3$ or $-(a - 4) = 3$. These can be written as $a - 4 = 3$ or $a - 4 = -3$.

Solve each open sentence. Then graph the solution set.

1A. $|y + 2| = 4$

1B. $|3n - 4| = -1$

EXAMPLE Write an Absolute Value Equation

2 **Write an open sentence involving absolute value for the graph.**

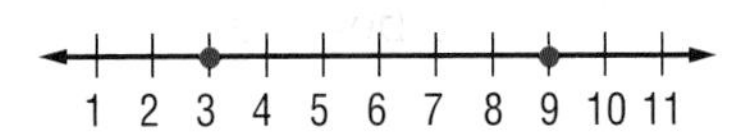

Find the point that is the same distance from 3 and from 9. This is the midpoint between 3 and 9, which is 6.

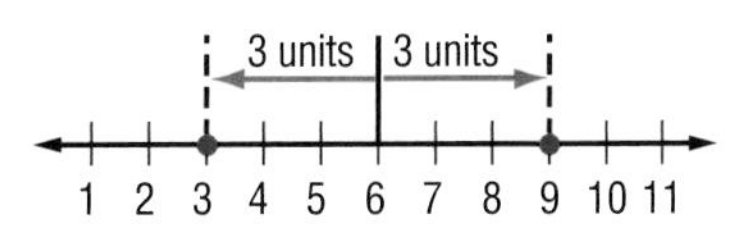

The distance from 6 to 3 is 3 units.
The distance from 6 to 9 is 3 units.

So, an equation is $|x - 6| = 3$.

CHECK Your Progress

2. Write an open sentence involving absolute value for the graph.

−1 0 1 2 3 4 5 6

Graphing Absolute Value Functions An **absolute value function** is a function written as $f(x) = |x|$, when $f(x) \geq 0$ for all values of x. To graph $f(x) = |x|$, make a table of values and plot the ordered pairs on a corrdinate plane. Notice that for negative values of x, the slope of the line is -1. When the x-values are positive, the slope of the line is 1. An absolute value function can be written using two or more expressions. This is an example of a **piecewise function.**

$f(x) = \|x\|$	
x	$f(x)$
−3	3
−2	2
−1	1
−0	0
−1	1
−2	2
−3	3

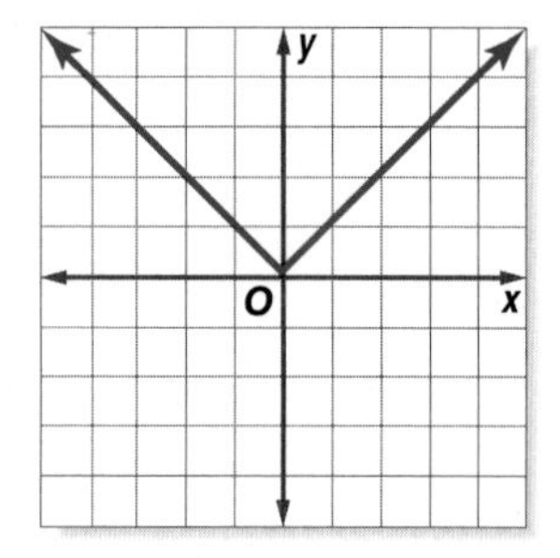

The absolute value function $f(x) = |x|$ can be written as $f(x) = \begin{cases} -x \text{ if } x < 0 \\ x \text{ if } x \geq 0 \end{cases}$.

EXAMPLE Graphing Absolute Value Functions

Study Tip

Absolute Value

When plotting points on a graph, choose values of x that are less than and that are greater than the minimum point of the graph.

3 **Graph $f(x) = |x - 4|$.**

First, find the minimum point on the graph. Since $f(x)$ cannot be negative, the minimum point of the graph is where $f(x) = 0$.

$f(x) = |x - 4|$ Original function

$0 = x - 4$ Set $f(x) = 0$.

$4 = x$ Add 4 to each side.

The minimum point of the graph is at (4, 0).

Next fill out a table of values. Include values for $x > 4$ and $x < 4$.

$f(x) = \|x - 4\|$	
x	$f(x)$
−2	6
0	4
2	2
4	0
5	1
6	2
7	3
8	4

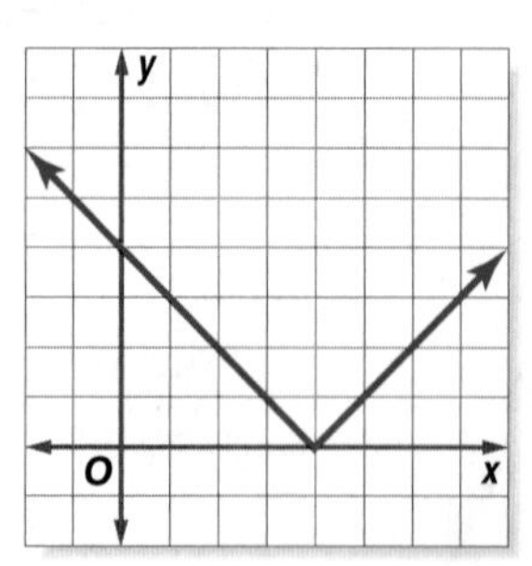

CHECK Your Progress

3. Graph $f(x) = |2x + 1|$.

CHECK Your Understanding

Example 1 (p. 323)

Solve each open sentence. Then graph the solution set.

1. $|r + 3| = 10$ **2.** $|2x - 8| = 6$ **3.** $|4n - 1| = -6$

Example 2 (p. 323)

4. Write an open sentence involving absolute value for the graph.

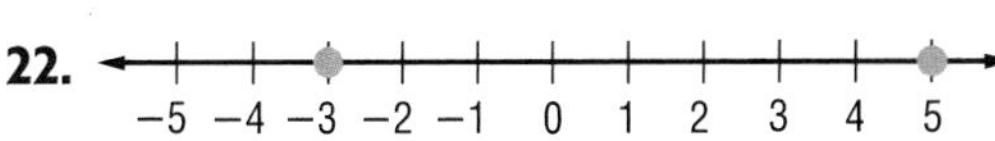

Examples 3 (p. 324)

Graph each function.

5. $f(x) = |x - 3|$ **6.** $g(x) = |2x + 4|$

Exercises

HOMEWORK HELP	
For Exercises	See Examples
7–18	1
19–22	2
23–30	3

Solve each open sentence. Then graph the solution set.

7. $|x - 5| = 8$ **8.** $|b + 9| = 2$ **9.** $|v - 2| = -5$

10. $|2p - 3| = 17$ **11.** $|5c - 8| = 12$ **12.** $|6y - 7| = -1$

13. $\left|\frac{1}{2}x + 5\right| = -3$ **14.** $|-2x + 6| = 6$ **15.** $\left|\frac{3}{4}x - 3\right| = 9$

16. $\left|-\frac{1}{2}x - 2\right| = 10$ **17.** $|-4x + 6| = 12$ **18.** $|5x - 3| = 12$

Write an open sentence involving absolute value for each graph.

19. −5 −4 −3 −2 −1 0 1 2 3 4 5

20. −5 −4 −3 −2 −1 0 1 2 3 4 5

21. −2 −1 0 1 2 3 4 5 6 7 8

22. −5 −4 −3 −2 −1 0 1 2 3 4 5

Graph each function.

23. $f(x) = |2x - 1|$ **24.** $f(x) = |x + 5|$ **25.** $g(x) = |-3x - 5|$

26. $g(x) = |-x - 3|$ **27.** $f(x) = \left|\frac{1}{2}x - 2\right|$ **28.** $f(x) = \left|\frac{1}{3}x + 2\right|$

29. $g(x) = |x + 2| + 3$ **30.** $g(x) = |2x - 3| + 1$

Solve for x.

31. $2|x| - 3 = 8$ **32.** $4 - 3|x| = 10$

Determine the domain and range for each absolute value function.

33.

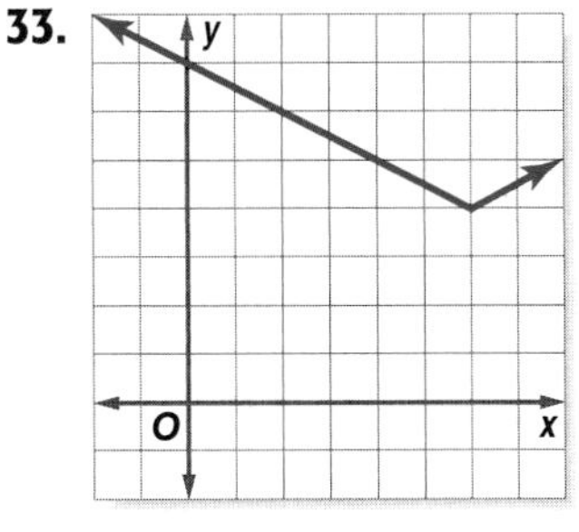

34.

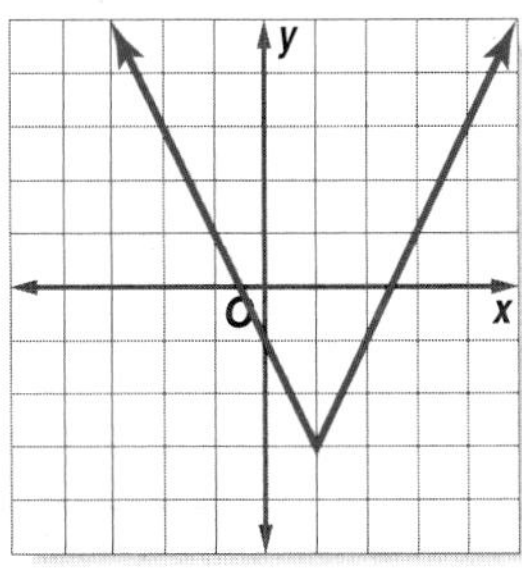

35.

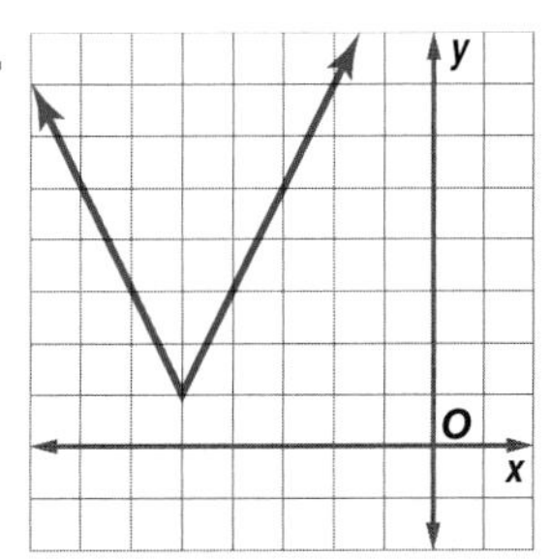

36.

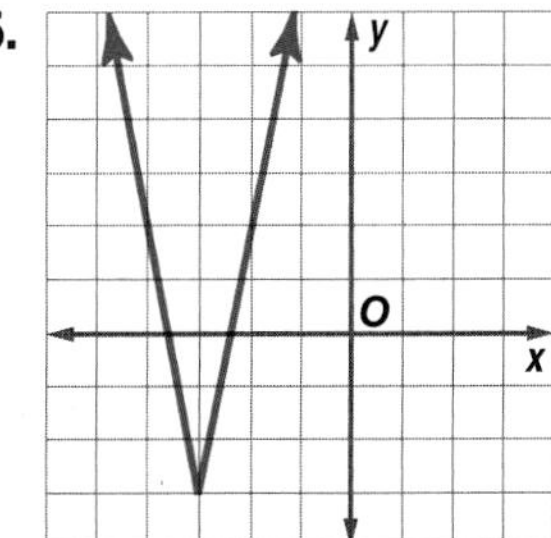

37. **ANALYZE GRAPHS** The circle graph at the right shows the results of a survey that asked teens "When do you think humans will be able to live on the moon?" If the margin of error is ± 3 percentage points, what is the range of the percent of teens who say humans will live on the moon in 2100?

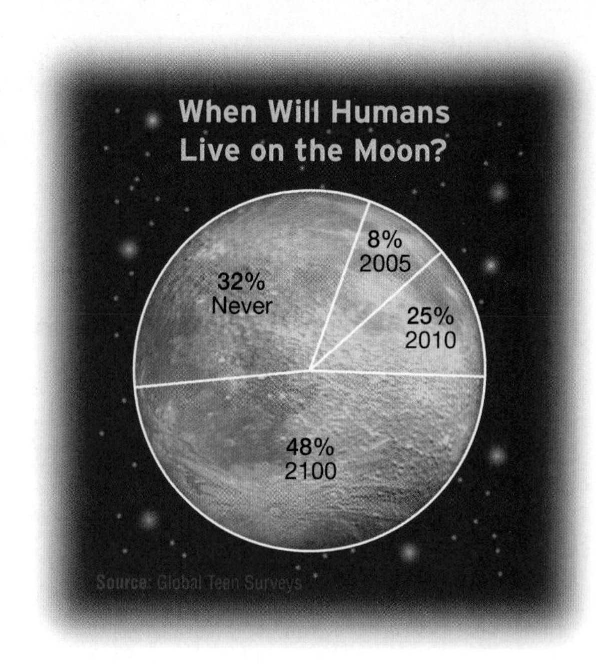

38. **PHYSICS** As part of a physics lab, Tiffany and Curtis determined that an object was traveling at 25 miles per hour. If the margin of error is 6%, determine the slowest and the fastest rate of the object.

PING PONG For Exercises 39 and 40, use the following information.
Esmerelda dropped a ping pong ball from a height of 4 feet. She tracked the height of the ball and the elapsed time. The ball hit the floor at 2 seconds and then bounced. At 4 seconds, the height was 4 feet.

39. Draw a graph of the path of the ping pong ball, assuming that the ball traveled at a constant rate. Let the x-axis represent the time and the y-axis represent the height.

40. Write a piecewise function to describe the path of the ping pong ball.

H.O.T. Problems

41. **OPEN ENDED** Describe a real-world situation that could be represented by the absolute value equation $|x - 4| = 10$.

REASONING Determine whether the following statements are *sometimes*, *always*, or *never* true, where c is a whole number. Explain your reasoning.

42. The value of $|x + 1|$ is greater than zero.
43. The solution of $|x + c| = 0$ is greater than 0.
44. The inequality $|x| + c < 0$ has no solution.
45. The value of $|x + c| + c$ is greater than zero.
46. **CHALLENGE** Use the sentence $x = 3 \pm 1.2$.
    a. Describe the values of x.
    b. Translate the sentence into an expression involving absolute value.

47. **FIND THE ERROR** Leslie and Holly are solving $|x + 3| = 2$. Who is correct? Explain your reasoning.

Leslie	
$x + 3 = 2$ or	$x + 3 = -2$
$x + 3 - 3 = 2 - 3$	$x + 3 - 3 = -2 - 3$
$x = -2$	$x = -5$

Holly	
$x + 3 = 2$ or	$x - 3 = 2$
$x + 3 - 3 = 2 - 3$	$x + 3 + 3 = 2 + 3$
$x = -1$	$x = 5$

48. ***Writing in Math*** Use the data about the technology survey on page 322 to explain how absolute value is used in surveys.

STANDARDIZED TEST PRACTICE

49. Assume n is an integer and solve for n.

$$|2n - 3| = 5$$

A $\{-4, -1\}$ C $\{-1, 4\}$

B $\{1, 1\}$ D $\{4, 4\}$

50. If p is an integer, which of the following is the solution set for $2|p| = 16$?

F $\{0, 8\}$

G $\{-8, 8\}$

H $\{-8, 0\}$

J $\{-8, 0, 8\}$

51. REVIEW What is the measure of $\angle 1$ in the figure shown below?

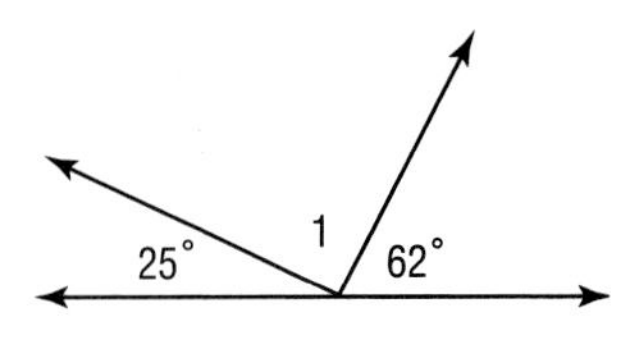

A 83°

B 87°

C 90°

D 93°

Spiral Review

52. FITNESS To achieve the maximum benefits from aerobic activity, your heart rate should be in your target zone. Your target zone is the range between 60% and 80% of your maximum heart rate. If Rafael's maximum heart rate is 190 beats per minute, what is his target zone? (Lesson 6-4)

Solve each inequality. Check your solution. (Lesson 6-3)

53. $2m + 7 > 17$ **54.** $-2 - 3x \geq 2$ **55.** $\frac{2}{3}w - 3 \leq 7$

Find the slope and y-intercept of each equation. (Lesson 4-4)

56. $2x + y = 4$ **57.** $2y - 3x = 4$ **58.** $\frac{1}{2}x + \frac{3}{4}y = 0$

Express the relation shown in each mapping as a set of ordered pairs. Then state the domain, range, and inverse. (Lesson 4-3)

59.

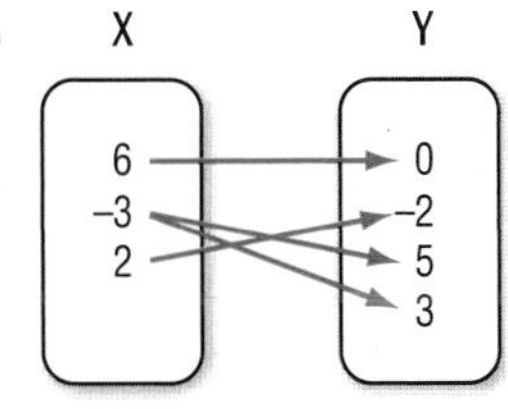

60.

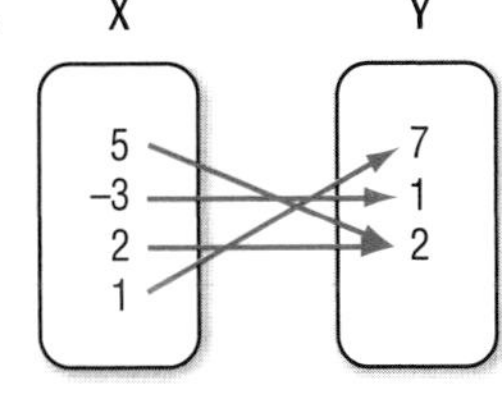

61.

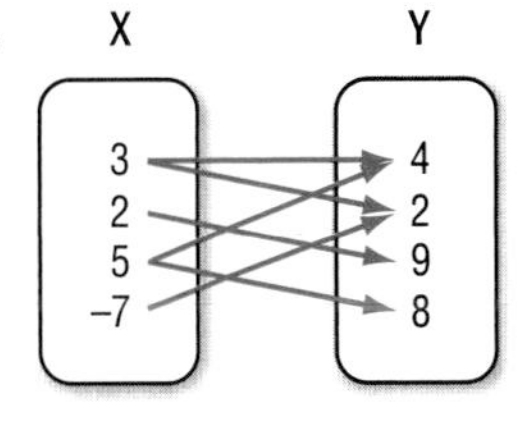

Solve each equation for the variable specified. (Lesson 2-8)

62. $I = prt$, for r **63.** $ex - 2y = 3z$, for x **64.** $\frac{a+5}{3} = 7x$, for x

GET READY for the Next Lesson

PREREQUISITE SKILL **Solve each inequality. Check your solution.** (Lesson 6-1)

65. $m + 4 \geq 6$ **66.** $p - 3 < 2$ **67.** $z + 5 \leq 11$

68. $8 + w < -12$ **69.** $4 - r \geq -3$ **70.** $6 - v \geq -2$

Graphing Calculator Lab

Graphing Absolute Value Functions

The absolute value function $y = |x|$ is the parent function of the family of absolute value functions. The graphs of absolute value functions are similar to the graphs of linear functions.

ACTIVITY 1

Graph $y = |x|$ in the standard viewing window.

Enter the equation in the Y= list. Then graph the equation.

KEYSTROKES: [Y=] [MATH] [▶] 1 [X,T,θ,n] [)] [ZOOM] 6

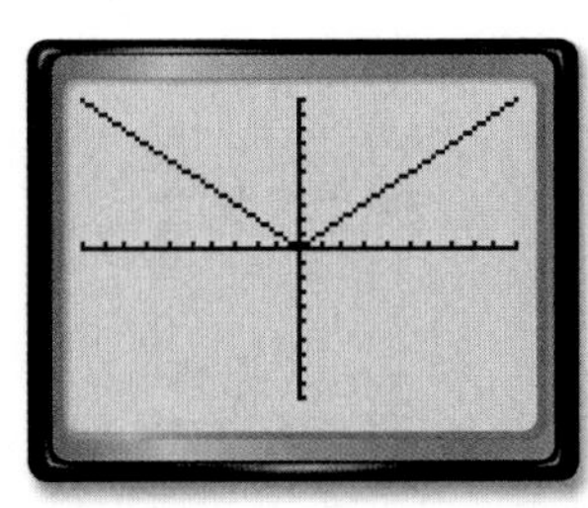

[−10, 10] scl: 1 by [−10, 10] scl: 1

1A. How does the graph of $y = |x|$ compare to the graph of $y = x$?

1B. What are the domain and range of the function $y = |x|$? Explain.

The graphs of absolute value functions are affected by changes in parameters in a way similar to the way changes in parameters affect the graphs of linear functions.

ACTIVITY 2

Graph $y = |x| - 3$ and $y = |x| + 1$ in the standard viewing window.

Enter the equations in the Y= list. Then graph.

KEYSTROKES: [Y=] [MATH] [▶] 1 [X,T,θ,n] [)] [−] 3 [ENTER] [MATH] [▶] 1 [X,T,θ,n] [)] [+] 1 [ZOOM] 6

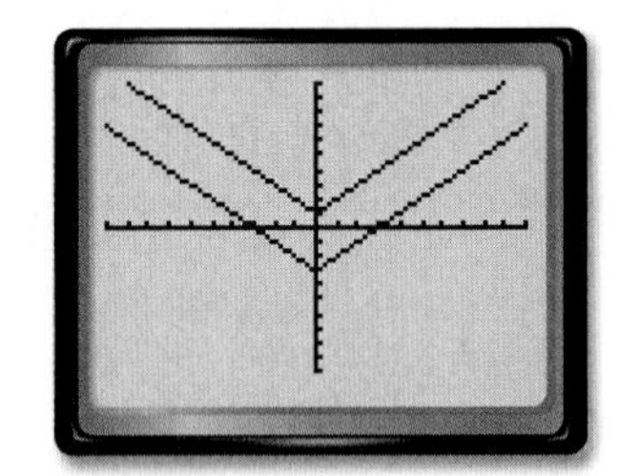

[−10, 10] scl: 1 by [−10, 10] scl: 1

2A. Compare and contrast the graphs to the graph of $y = |x|$.

2B. How does the value of c affect the graph of $y = |x| + c$?

ANALYZE THE RESULTS

1. Write the function shown in the graph.

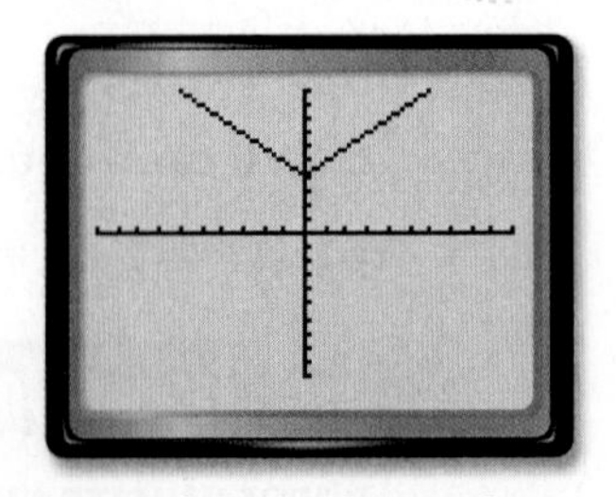

[−10, 10] scl: 1 by [−10, 10] scl: 1

2. Graph $y = -|x|$ in the standard viewing window. How is this graph related to the graph of $y = |x|$?
3. **MAKE A CONJECTURE** Describe the transformation of the parent graph $y = |x + c|$. Use a graphing calculator with different values of c to test your conjecture.
4. Determine whether the following statement is *always, sometimes,* or *never* true. Justify your answer. The y-values of the function $y = -|x - 1| - 1$ are negative.

Other Calculator Keystrokes at algebra1.com

6-6 Solving Inequalities Involving Absolute Value

Main Ideas

- Solve absolute value inequalities.
- Apply absolute value inequalities in real-world problems.

GET READY for the Lesson

To make baby carrots for snacks, long carrots are sliced into 2-inch sections and then peeled. If the machine that slices the carrots is accurate to within $\frac{1}{8}$ of an inch, the length of a baby carrot ranges from $1\frac{7}{8}$ inch to $2\frac{1}{8}$ inch.

Absolute Value Inequalities Consider the inequality $|x| < n$. $|x| < 5$ means that the distance from 0 to x is less than 5 units.

Therefore, $x > -5$ and $x < 5$. The solution set is $\{x \mid -5 < x < 5\}$. When solving inequalities of the form $|x| < n$, consider the two cases.

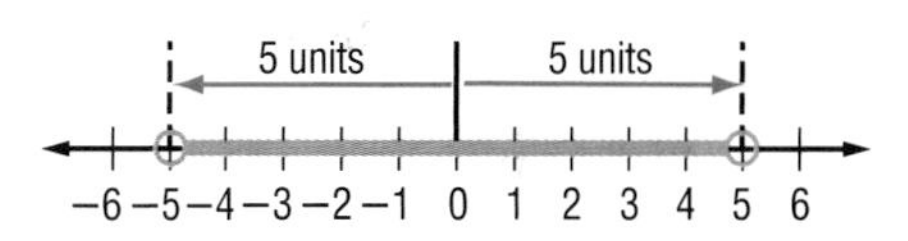

Case 1 The expression inside the absolute value symbols is positive.

Case 2 The expression inside the absolute value symbols is negative.

To solve, find the *intersection* of the solutions of these two cases.

EXAMPLE Solve an Absolute Value Inequality (<)

1 **Solve each open sentence. Then graph the solution set.**

a. $|t + 5| < 9$

Write $|t + 5| < 9$ as $t + 5 < 9$ and $-(t + 5) < 9$.

Case 1 $t + 5$ is positive.

$$t + 5 < 9$$
$$t + 5 - 5 < 9 - 5$$
$$t < 4$$

Case 2 $t + 5$ is negative.

$$-(t + 5) < 9$$
$$t + 5 > -9$$
$$t + 5 - 5 > -9 - 5$$
$$t > -14$$

Therefore, $t < 4$ and $t > -14$.

The solution set is $\{t \mid -14 < t < 4\}$.

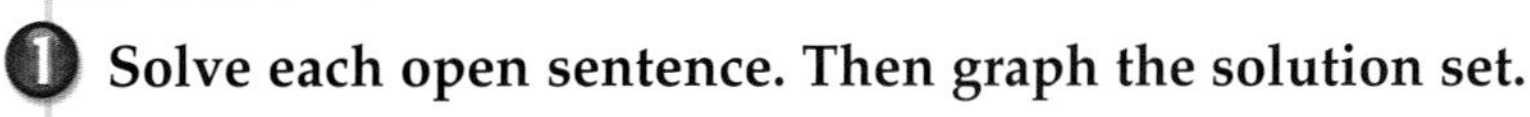

b. $|x + 2| < -1$

Since $|x + 2|$ cannot be negative, $|x + 2|$ cannot be less than -1. So, the solution set is the empty set $\varnothing$.

CHECK Your Progress

1A. $|n - 8| \leq 2$

1B. $|2c - 5| < -3$

Study Tip

Less Than

When an absolute value is on the left and the inequality symbol is $<$ or $\leq$, the compound sentence uses *and*.

Math Online Extra Examples at algebra1.com

Consider the inequality $|x| > n$. $|x| > 5$ means that the distance from 0 to x is greater than 5 units.

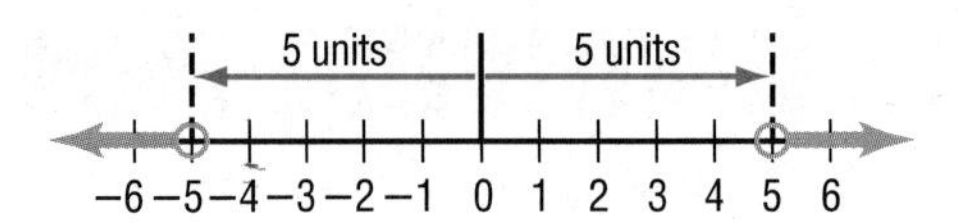

Therefore, $x < -5$ or $x > 5$. The solution set is $\{x \mid x < -5 \text{ or } x > 5\}$.

When solving inequalities of the form $|x| > n$, consider the two cases.

Case 1 The expression inside the absolute value symbols is positive.

Case 2 The expression inside the absolute value symbols is negative.

To solve, find the *union* of the solutions of these two cases.

EXAMPLE Solve an Absolute Value Inequality (>)

2 Solve each open sentence. Then graph the solution set.

a. $|2x + 8| \geq 6$

Case 1 $2x + 8$ is positive.

$2x + 8 \geq 6$	Definition of absolute value
$2x + 8 - 8 \geq 6 - 8$	Subtract 8 from each side.
$2x \geq -2$	Simplify.
$\frac{2x}{2} \geq \frac{-2}{2}$	Divide each side by 2.
$x \geq -1$	Simplify.

Case 2 $2x + 8$ is negative.

$-(2x + 8) \geq 6$	Definition of absolute value
$2x + 8 \leq -6$	Divide each side by −1 and reverse the symbol.
$2x + 8 - 8 \leq -6 - 8$	Subtract 8 from each side.
$2x \leq -14$	Simplify.
$\frac{2x}{2} \leq \frac{-14}{2}$	Divide each side by 2.
$x \leq -7$	Simplify.

The solution set is $\{x \mid x \leq -7 \text{ or } x \geq -1\}$.

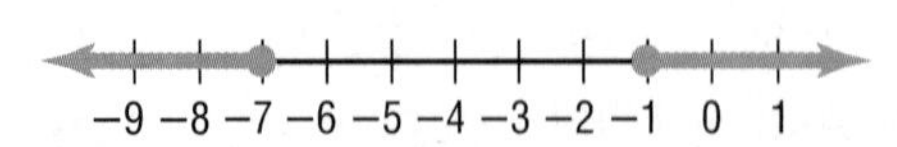

b. $|2y - 1| \geq -4$

Since $|2y - 1|$ is always greater than or equal to 0, the solution set is $\{y \mid y \text{ is a real number}\}$. Its graph is the entire number line.

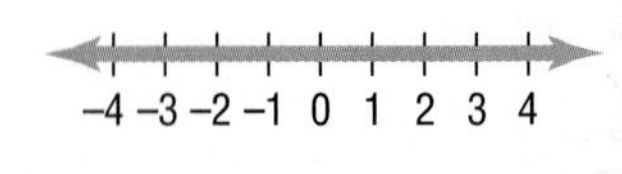

> **Study Tip**
>
> **Greater Than**
>
> When the absolute value is on the left and the inequality symbol is $>$ or $\geq$, the compound sentence uses *or*.

CHECK Your Progress

2A. $|2k + 1| > 7$

2B. $|r - 6| \geq -5$

Personal Tutor at algebra1.com

In general, there are three rules to remember when solving equations and inequalities involving absolute value.

CONCEPT SUMMARY — ***Absolute Value Equations and Inequalities***

If $|x| = n$, then $x = -n$ or $x = n$.

If $|x| < n$, then $x < n$ and $x > -n$.

If $|x| > n$, then $x > n$ or $x < -n$.

These properties are also true when $>$ or $<$ is replaced with $\geq$ or $\leq$.

Applying Absolute Value Inequalities Many situations can be represented using an absolute value inequality.

Real-World EXAMPLE

3 BIOLOGY The pH is a measure of the acidity of a solution. The pH of a healthy human stomach is about 2.5 and is within 0.5 pH of this value. Find the range of pH levels of a healthy stomach.

The difference between the actual pH of a stomach and the ideal pH of a stomach is less than or equal to 0.5. Let x be the actual pH of a stomach. Then $|x - 2.5| \leq 0.5$.

Solve each case of the inequality.

Case 1

$$x - 2.5 \leq 0.5$$
$$x - 2.5 + 2.5 \leq 0.5 + 2.5$$
$$x \leq 3.0$$

Case 2

$$-(x - 2.5) \leq 0.5$$
$$x - 2.5 \geq -0.5$$
$$x - 2.5 + 2.5 \geq -0.5 + 2.5$$
$$x \geq 2.0$$

The range of pH levels of a healthy stomach is $\{x \mid 2.0 \leq x \leq 3.0\}$.

3. **CHEMISTRY** The melting point of ice is 0° Celsius. During a chemistry experiment, Jill observed ice melting within 2 degrees. Write the range of temperatures that Jill observed ice melting.

CHECK Your Understanding

Examples 1, 2 (p. 330)

Solve each open sentence. Then graph the solution set.

1. $|c - 2| < 6$
2. $|x + 5| \leq 3$
3. $|m - 4| \leq -3$
4. $|10 - w| > 15$
5. $|2g + 5| \geq 7$
6. $|3p + 2| \geq -8$

Example 3 (p. 331)

7. **MANUFACTURING** A manufacturer produces bolts which must have a diameter within 0.001 centimeter of 1.5 centimeters. What are the acceptable measurements for the diameter of the bolts?

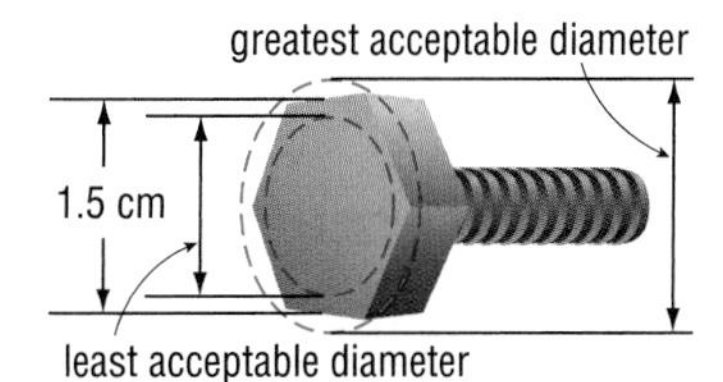

Exercises

HOMEWORK HELP	
For Exercises	See Examples
8–6	1, 2
17, 18, 31–34	3

Solve each open sentence. Then graph the solution set.

8. $|z - 2| \leq 5$ **9.** $|t + 8| < 2$ **10.** $|6 - d| \leq -4$

11. $|v + 3| > 1$ **12.** $|w - 6| \geq 3$ **13.** $|3a - 9| > -2$

14. $|3k + 4| \geq 8$ **15.** $|2n + 1| < 9$ **16.** $|4q + 7| \leq -13$

17. SCUBA DIVING The pressure of a typical scuba tank should be within 500 pounds per square inch (psi) of 2500 psi. Write the range of optimum pressures for scuba tanks.

18. ANIMALS A sheep's normal body temperature is 39°C. However, a healthy sheep may have body temperatures 1°C above or below this temperature. What is the range of body temperatures for a sheep?

Write an open sentence involving absolute value for each graph.

19. −5 −4 −3 −2 −1 0 1 2 3 4 5

20. −8 −7 −6 −5 −4 −3 −2 −1 0 1 2

21. −5 −4 −3 −2 −1 0 1 2 3 4 5

22. 3 4 5 6 7 8 9 10 11 12 13

Solve each open sentence. Then graph the solution set.

23. $\left|\frac{5h + 2}{6}\right| = 7$ **24.** $\left|\frac{2 - 3x}{5}\right| \geq 2$

25. $|3s + 2| > -7$ **26.** $|6r + 8| < -4$

Real-World Link

Always inflate your tires to the pressure that is recommended by the manufacturer. The pressure stamped on the tire is the *maximum* pressure and should only be used under certain circumstances.

Source: www.etires.com

Express each statement using an inequality involving absolute value. Do *not* solve.

27. The pH of a swimming pool must be within 0.3 of a pH of 7.5.

28. The temperature inside a refrigerator should be within 1.5 degrees of 38°F.

29. Ramona's bowling score was within 6 points of her average score of 98.

30. The cruise control of a car set at 55 miles per hour should keep the speed within 3 miles per hour of 55.

31. DRIVING Tires should be kept within 2 pounds per square inch (psi) of the manufacturer's recommended tire pressure. If the recommendation for a tire is 30 psi, what is the range of acceptable pressures?

32. PHYSICAL SCIENCE Li-Cheng must add 3.0 milliliters of sodium chloride to a solution. The sodium chloride must be within 0.5 milliliter of the required amount. How much sodium chloride can she add and obtain the correct results?

33. MINIATURE GOLF Ginger played miniature golf. Her score was within 5 strokes of her average score of 52. Determine the range of scores for Ginger's game.

34. MUSIC DOWNLOADS Carlos is allowed to download $10 worth of music each month. This month he has spent within $3 of his allowance. What is the range of money he has spent on music downloads this month?

EXTRA PRACTICE

See pages 730, 749.

Math Online

Self-Check Quiz at algebra1.com

H.O.T. Problems

35. **REASONING** Compare and contrast the solution of $|x - 2| > 6$ and the solution of $|x - 2| < 6$.

36. **OPEN ENDED** Formulate an absolute value inequality to represent a real-world situation and graph its solution set. Interpret the solution.

37. **CHALLENGE** Translate the inequality $x < 2 \pm 0.3$ into an absolute value inequality.

38. ***Writing in Math*** Refer to the information on page 329. Describe how the definition of absolute value can be applied to manufacturing baby carrots. Write an absolute value inequality to represent the range of lengths for a baby carrot.

STANDARDIZED TEST PRACTICE

39. What is the solution to the inequality $-6 < |x| < 6$?

A $-x \geq 0$

B $x \leq 0$

C $-x < 6$

D $-x > 6$

40. Which inequality *best* represents the statement below?

A jar contains 832 gumballs. Amanda's guess was within 46 pieces.

F $|g - 832| \leq 46$

G $|g + 832| \leq 46$

H $|g - 832| \geq 46$

J $|g + 832| \geq 46$

41. **REVIEW** An 84-centimeter piece of wire is cut into equal segments and then attached at the ends to form the edges of a cube. What is the volume of the cube?

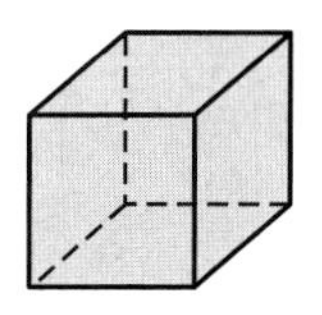

A 294 cm^3

B 343 cm^3

C 1158 cm^3

D 2744 cm^3

Spiral Review

Solve each open sentence. Then graph the solution set. (Lesson 6-5)

42. $|x + 3| = 5$

43. $|2x + 3| = -4$

44. $|3x - 2| = 4$

45. **SHOPPING** A catalog company varies the costs to ship merchandise based on the amount of the order. The cost for shipping is shown in the table. Write a compound inequality for each shipping cost. (Lesson 6-4)

Shipping Costs	
Merchandise	**Shipping**
\$0–\$25	\$5
\$25.01–\$50	\$8

Write an equation in slope-intercept form of the line with the given slope and y-intercept. (Lesson 4-3)

46. slope: -3, y-intercept: 4

47. slope: $\frac{1}{2}$, y-intercept: $\frac{3}{4}$

GET READY for the Next Lesson

PREREQUISITE SKILL Graph each equation. (Lesson 3-3)

48. $y = 3x + 4$

49. $x + y = 3$

50. $y - 2x = -1$

51. $2y - x = -6$

6-7 Graphing Inequalities in Two Variables

Main Ideas

- Graph inequalities on the coordinate plane.
- Solve real-world problems involving linear inequalities.

New Vocabulary

half-plane
boundary

GET READY for the Lesson

Hannah budgets \$30 a month for lunch. On most days, she brings her lunch. She can also buy lunch at the cafeteria or at a fast-food restaurant. She spends an average of \$3 for lunch at the cafeteria and an average of \$4 for lunch at a restaurant. How many times a month can Hannah buy her lunch and remain within her budget?

My Monthly Budget

Lunch (school days)	\$30
Entertainment	\$55
Clothes	\$50
Fuel	\$60

Words	The cost of eating in the cafeteria	plus	the cost of eating in a restaurant	is less than or equal to	\$30.
Variables	Let x = the number of days she buys lunch at the cafeteria.		Let y = the number of days she buys lunch at a restaurant.		
Inequality	$3x$	$+$	$4y$	$\leq$	30

There are many solutions for this inequality. Each solution represents a different combination of lunches bought in the cafeteria and in a restaurant.

Graph Linear Inequalities The solution set for an inequality in two variables contains many ordered pairs when the domain and range are the set of real numbers. The graphs of all of these ordered pairs fill a region on the coordinate plane called a **half-plane**. An equation defines the **boundary** or edge for each half-plane.

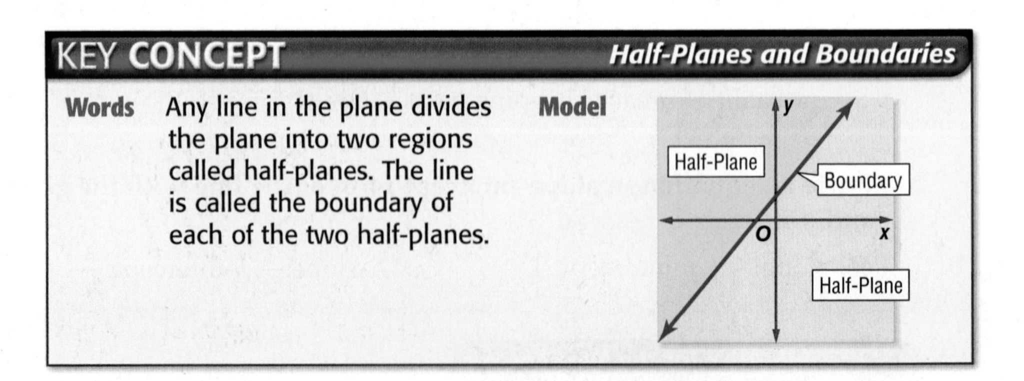

The boundary may or may not be included in the graph of an inequality. Graphing the boundary is the first step in graphing a linear inequality.

Study Tip

Dashed Line
Like a circle on a number line, a dashed line on a coordinate plane indicates that the boundary is not part of the solution set.

Solid Line
Like a dot on a number line, a solid line on a coordinate plane indicates that the boundary is included.

Consider the graph of $y > 4$. First determine the boundary by graphing $y = 4$, the equation you obtain by replacing the inequality sign with an equals sign. Since the inequality involves y-values greater than 4, but not equal to 4, the line should be dashed. The boundary divides the coordinate plane into two half-planes.

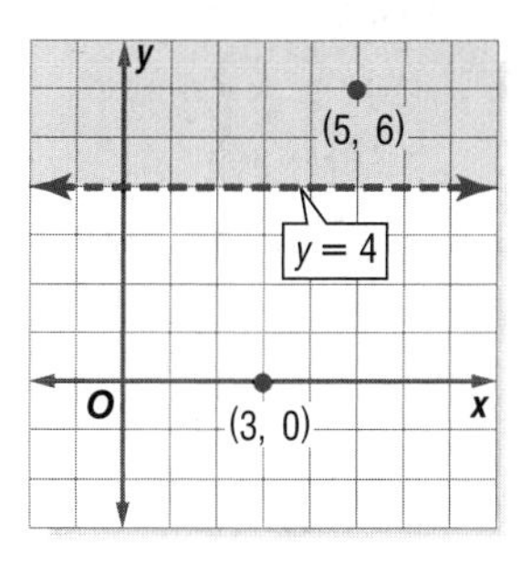

To determine which half-plane contains the solution, choose a point from each half-plane and test it in the inequality.

Try (3, 0).		Try (5, 6).	
$y > 4$	$y = 0$	$y > 4$	$y = 6$
$0 > 4$	false	$6 > 4$	true

The half-plane that contains (5, 6) contains the solution. Shade that half-plane.

EXAMPLE Graph an Inequality

1 Graph $y - 2x \le -4$.

Step 1 Solve for y in terms of x.

$$y - 2x \le -4 \quad \text{Original inequality}$$
$$y - 2x + 2x \le -4 + 2x \quad \text{Add } 2x \text{ to each side.}$$
$$y \le 2x - 4 \quad \text{Simplify.}$$

Step 2 Graph $y = 2x - 4$. Since $y \le 2x - 4$ means $y < 2x - 4$ or $y = 2x - 4$, the boundary is included in the solution set. The boundary should be drawn as a solid line.

Study Tip

Origin as the Test Point
Use the origin as a standard test point because the values often simplify calculations.

Step 3 Select a point in one of the half-planes and test it. Let's use (0, 0).

$$y - 2x \le -4 \quad \text{Original inequality}$$
$$0 - 2(0) \le -4 \quad x = 0, y = 0$$
$$0 \le -4 \quad \text{false}$$

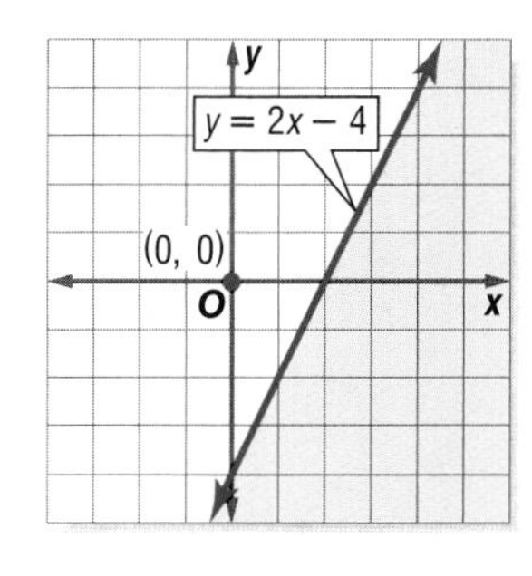

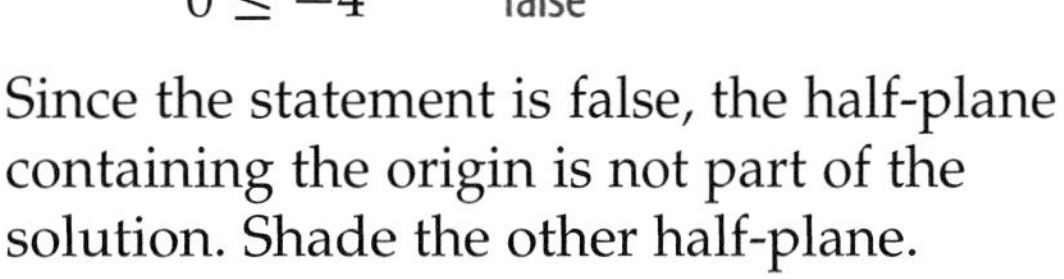

Since the statement is false, the half-plane containing the origin is not part of the solution. Shade the other half-plane.

CHECK Test a point in the other half plane, for example, (3, −3).

$$y - 2x \le -4 \quad \text{Original inequality}$$
$$-3 - 2(3) \le -4 \quad x = 3, y = -3$$
$$-9 \le -4 \checkmark \quad \text{Simplify.}$$

Since the statement is true, the half-plane containing (3, −3) should be shaded. The graph of the solution is correct.

CHECK Your Progress **Graph each inequality.**

1A. $x \le -1$

1B. $y > \frac{1}{2}x + 3$

Solve Real-World Problems When solving real-world inequalities, the domain and range of the inequality are often restricted to nonnegative numbers or whole numbers.

Real-World EXAMPLE Write and Solve an Inequality

2 **ADVERTISING Rosa Padilla sells radio advertising in 30-second and 60-second time slots. During every hour, there are up to 15 minutes available for commercials. How many commercial slots can she sell for one hour of broadcasting?**

Real-World Link

A typical one-hour program on television contains 40 minutes of the program and 20 minutes of commercials. During peak periods, a 30-second commercial can cost an average of $2.3 million.

Source: www.superbowl-ads.com

Explore You know the length of the time slots in seconds and the number of minutes each hour available for commercials.

Plan Let x = the number of 30-second commercials. Let y = the number of 60-second or 1-minute commercials. Write an open sentence representing this situation.

$\frac{1}{2}$ min	times	the number of 30-s commercials	plus	1 min	times	the number of 1-min commercials	is up to	15 min.
$\frac{1}{2}$	$\cdot$	x	$+$	1	$\cdot$	y	$\leq$	15

Solve Solve for y in terms of x.

$$\frac{1}{2}x + y \leq 15 \quad \text{Original inequality}$$

$$\frac{1}{2}x + y - \frac{1}{2}x \leq 15 - \frac{1}{2}x \quad \text{Subtract } \frac{1}{2}x \text{ from each side.}$$

$$y \leq 15 - \frac{1}{2}x \quad \text{Simplify.}$$

Since the open sentence includes the equation, graph $y = 15 - \frac{1}{2}x$ as a solid line. Test a point in one of the half-planes, for example (0, 0). Shade the half-plane containing (0, 0) since $0 \leq 15 - \frac{1}{2}(0)$ is true.

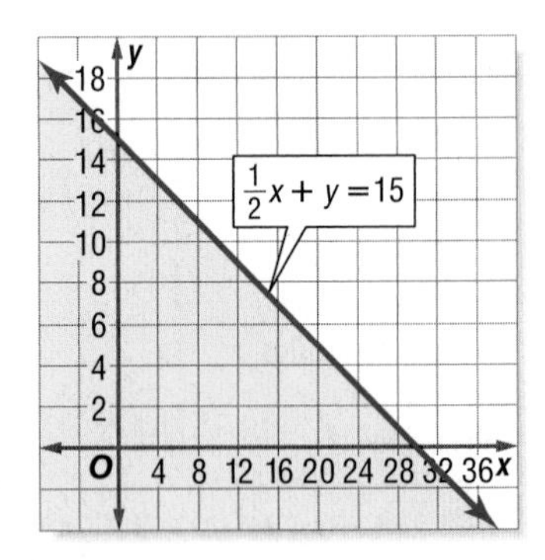

Check Examine the solution.

- Rosa cannot sell a negative number of commercials. Therefore, the domain and range contain only nonnegative numbers.
- She also cannot sell half of a commercial. Thus, only points in the shaded half-plane with x- and y-coordinates that are whole numbers are possible solutions.

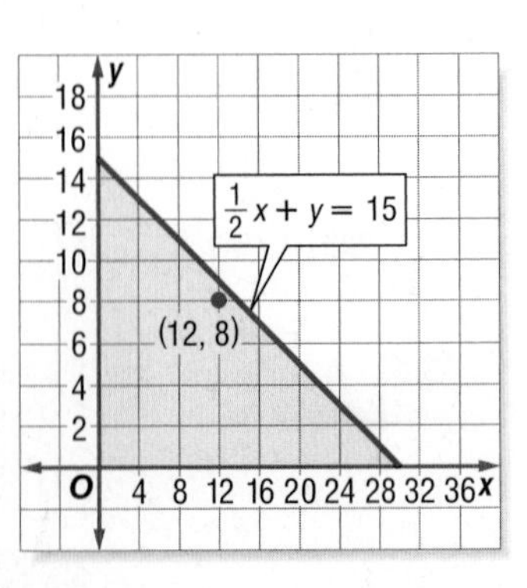

One solution is (12, 8). This represents twelve 30-second commercials and eight 60-second commercials in a one hour period.

2. MARATHONS Neil wants to run a marathon at a pace of at least 6 miles per hour. Write an inequality for the miles y he will run in x hours and graph the solution set.

Online Personal Tutor at algebra1.com

CHECK Your Understanding

Example 1 (p. 335)

Graph each inequality.

1. $y \geq 4$
2. $y \leq 2x - 3$
3. $y > x + 3$
4. $4 - 2x < -2$
5. $1 - y > x$
6. $x + 2y \leq 5$

Example 2 (pp. 336–337)

7. **ENTERTAINMENT** Coach Washington wants to take her softball team out for pizza and soft drinks after the last game of the season. She doesn't want to spend more than \$60. Write an inequality that represents this situation and graph the solution set.

Exercises

HOMEWORK HELP	
For Exercises	See Examples
8–17	1
18–21	2

Graph each inequality.

8. $y < -3$
9. $x \geq 2$
10. $5x + 10y > 0$
11. $y < x$
12. $2y - x \leq 6$
13. $6x + 3y > 9$
14. $3y - 4x \geq 12$
15. $y \leq -2x - 4$
16. $8x - 6y < 10$
17. $3x - 1 \geq y$

POSTAGE For Exercises 18 and 19, use the following information.
The U.S. Postal Service limits the size of packages. The length of the longest side plus the distance around the thickest part must be less than or equal to 108 inches.

18. Write an inequality that represents this situation.
19. Are there any restrictions on the domain or range? Explain.

Cross-Curricular Project
Math Online A linear inequality can be used to represent trends in Olympic times. Visit algebra1.com to continue work on your project.

ANALYZE TABLES For Exercises 20–22, use the table.
A delivery truck with a 4000-pound weight limit is transporting televisions and microwaves.

Item	Weight (lb)
television	77
microwave	55

20. Define variables and write an inequality for this situation.
21. Will the truck be able to deliver 35 televisions and 25 microwaves at once?
22. Write two possible solutions to the inequality. Are there solutions that make the inequality mathematically true, but are not reasonable in the context of the problem? Explain.

Determine which ordered pairs are part of the solution set for each inequality.

23. $y \leq 3 - 2x$, $\{(0, 4), (-1, 3), (6, -8), (-4, 5)\}$

24. $y < 3x$, $\{(-3, 1), (-3, 2), (1, 1), (1, 2)\}$

25. $x + y < 11$, $\{(5, 7), (-13, 10), (4, 4), (-6, -2)\}$

26. $2x - 3y > 6$, $\{(3, 2), (-2, -4), (6, 2), (5, 1)\}$

Match each inequality with its graph.

27. $2y + x \leq 6$

28. $\frac{1}{2}x - y > 4$

29. $y > 3 + \frac{1}{2}x$

30. $4y + 2x \geq 16$

a.

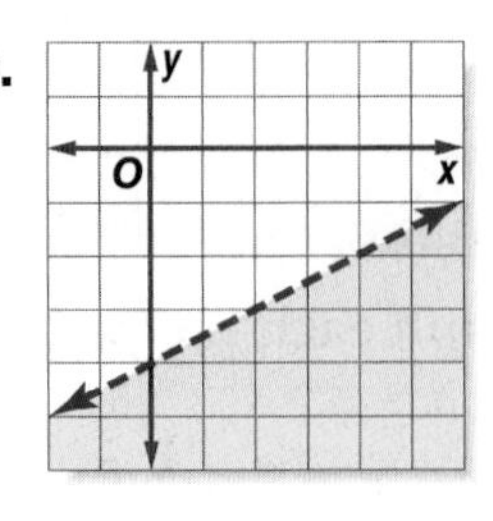

b.

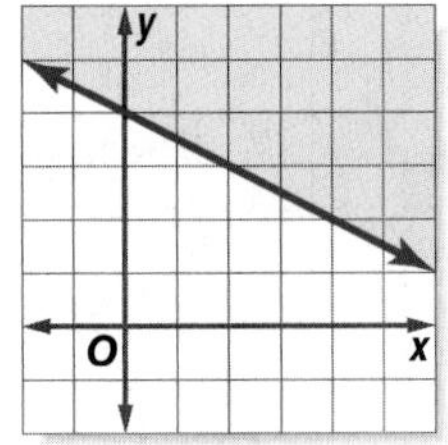

c.

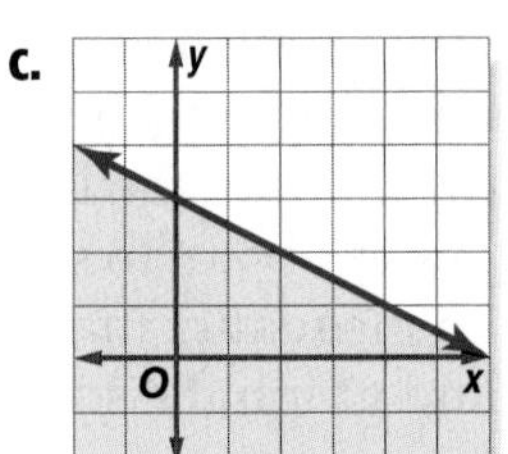

d.

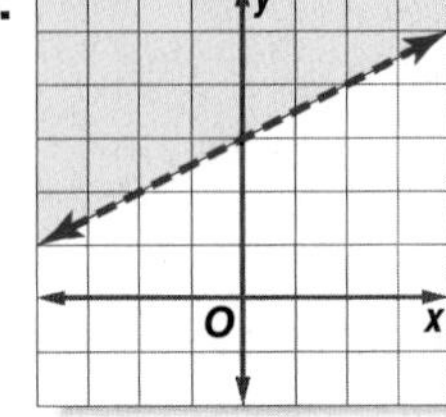

Determine which ordered pairs are part of the solution set for each inequality.

31. $|x - 3| \geq y$, $\{(6, 4), (-1, 8), (-3, 2), (5, 7)\}$

32. $|y + 2| < x$, $\{(2, -4), (-1, -5), (6, -7), (0, 0)\}$

Graph each inequality.

33. $3(x + 2y) > -18$

34. $\frac{1}{2}(2x + y) < 2$

MUSEUMS For Exercises 35 and 36, use the table at the right.
A Cub Scout troop plans to visit a flight museum. The troop leaders can spend up to $96 on admission.

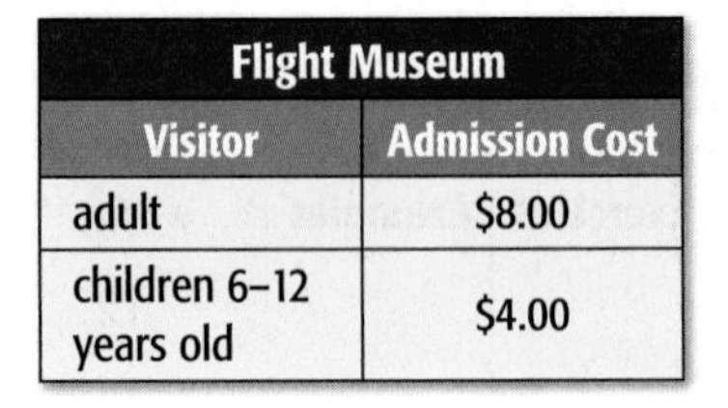

Flight Museum	
Visitor	**Admission Cost**
adult	$8.00
children 6–12 years old	$4.00

35. Write an inequality for this situation.

36. Will the entire troop be able to go to the museum if there are 3 adults and 16 Cub Scouts who are all under 12 years old? Explain.

H.O.T. Problems

37. REASONING Compare and contrast the graph of $y = x + 2$ and the graph of $y < x + 2$.

38. OPEN ENDED Create a linear inequality in two variables and graph it.

39. REASONING Explain why it is usually only necessary to test one point when graphing an inequality.

40. CHALLENGE Graph the intersection of the graphs of $y \leq x - 1$ and $y \geq -x$.

41. ***Writing in Math*** Use the information about budgets on page 334 to explain how inequalities are used in finances. Include an explanation of the restrictions placed on the domain and range of the inequality that describes the number of times Hannah can buy lunch. Describe three possible solutions of the inequality.

STANDARDIZED TEST PRACTICE

42. Which inequality is shown on the graph at the right?

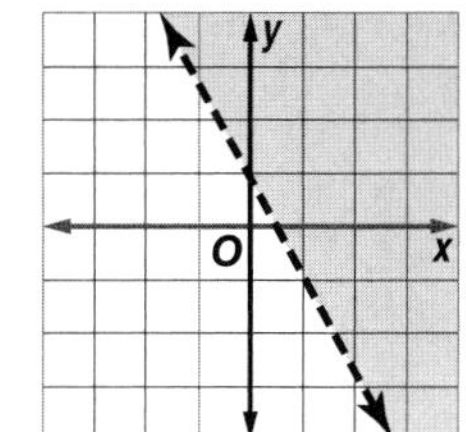

A $2x + y < 1$

B $2x + y > 1$

C $2x + y \leq 1$

D $2x + y \geq 1$

43. REVIEW The perimeters of two similar polygons are 250 centimeters and 300 centimeters, respectively. What is the scale factor of the two polygons?

F $\frac{5}{6}$

G $\frac{3}{4}$

H $\frac{1}{2}$

J $\frac{1}{4}$

Spiral Review

BIOLOGY For Exercises 44 and 45, use the following information.
The *average* length of a human pregnancy is 280 days. However, a healthy, full-term pregnancy can be 14 days longer or shorter. (Lesson 6-6)

44. If d is the length in days, write an absolute value inequality for the length of a full-term pregnancy.

45. Solve the inequality for the length of a full-term pregnancy.

Solve each open sentence. Then graph the solution. (Lesson 6-4)

46. $|y + 0.5| = 6.5$

47. $|m - 0.5| = 2.5$

Write an equation in slope-intercept form of the line that passes through the given point and is parallel to the graph of each equation. (Lesson 4-7)

48. $(1, -3); y = 3x - 2$

49. $(0, 4); x + y = -3$

50. $(-1, 2); 2x - y = 1$

Find the next two terms in each sequence. (Lesson 3-4)

51. 7, 13, 19, 25, …

52. 243, 81, 27, 9, …

53. 3, 6, 12, 24, …

State whether each percent of change is a percent of *increase* or *decrease*. Then find the percent of change. Round to the nearest whole percent. (Lesson 2-7)

54. original: 200
new: 172

55. original: 100
new: 142

56. original: 53
new: 75

GET READY for the Next Lesson

PREREQUISITE SKILL **Graph each equation.** (Lesson 4-3)

57. $y = 3x + 1$

58. $x - y = -4$

59. $5x + 2y = 6$

EXTEND 6-7

Graphing Calculator Lab
Graphing Inequalities

To graph inequalities, graphing calculators shade between two functions. Enter a lower boundary as well as an upper boundary for each inequality.

ACTIVITY 1 **Graph two different inequalities on your graphing calculator.**

Step 1 Graph $y \leq 3x + 1$.

- Clear all functions from the Y= list.

KEYSTROKES: [Y=] [CLEAR]

- Graph $y \leq 3x + 1$ in the standard window.

KEYSTROKES: [2nd] [DRAW] 7 −10 [,] 3 [X,T,θ,n] [+] 1 [)] [ENTER]

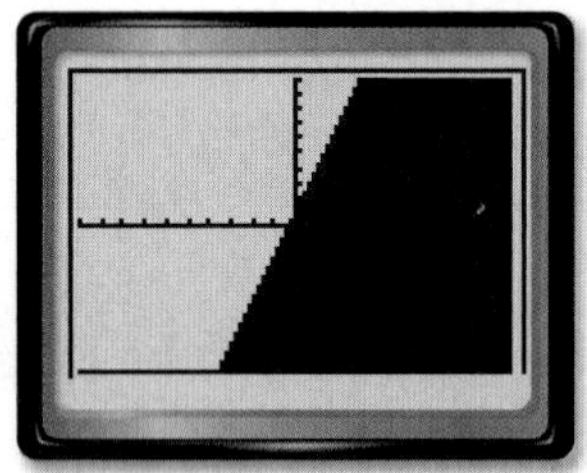

[−10, 10] scl: 1 by [−10, 10] scl: 1

The lower boundary is Ymin or −10. The upper boundary is $y = 3x + 1$. All ordered pairs for which y is *less than or equal to* $3x + 1$ lie *below or on* the line and are solutions.

Step 2 Graph $y - 3x \geq 1$.

- Clear the drawing that is displayed.

KEYSTROKES: [2nd] [DRAW] 1

- Rewrite $y - 3x \geq 1$ as $y \geq 3x + 1$ and graph it.

KEYSTROKES: [2nd] [DRAW] 7 3 [X,T,θ,n] [+] 1 [,] 10 [)] [ENTER]

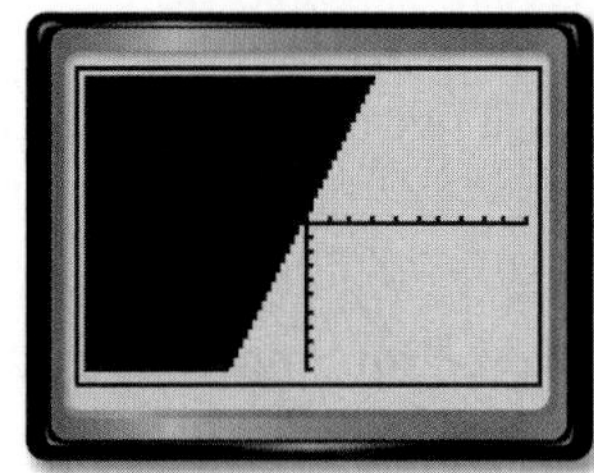

[−10, 10] scl: 1 by [−10, 10] scl: 1

The lower boundary is $y = 3x + 1$. The upper boundary is Ymax or 10. All ordered pairs for which y is *greater than or equal to* $3x + 1$ lie *above or on* the line and are solutions.

Exercises

1. Compare and contrast the two graphs shown above.

2. Graph the inequality $y \geq -2x + 4$ in the standard viewing window.

  a. What functions do you enter as the lower and upper boundaries?

  b. Using your graph, name four solutions of the inequality.

3. Suppose student movie tickets cost \$4 and adult movie tickets cost \$8. You would like to buy at least 10 tickets, but spend no more than \$80.

  a. Let x = number of student tickets and y = number of adult tickets. Write two inequalities, one representing the total number of tickets and the other representing the total cost of the tickets.

  b. Which inequalities would you use as the lower and upper boundaries?

  c. Graph the inequalities. Use the viewing window [0, 20] scl: 1 by [0, 20] scl: 1.

  d. Name four possible combinations of student and adult tickets.

Math Online Other Calculator Keystrokes at algebra1.com

6-8 Graphing Systems of Inequalities

Main Ideas

- Solve systems of inequalities by graphing.
- Solve real-world problems involving systems of inequalities.

New Vocabulary

system of inequalities

GET READY for the Lesson

Joshua's doctor recommends the following.

- Get between 60 and 80 grams of protein per day.
- Keep daily fat intake between 60 and 75 grams.

The green section of the graph indicates the appropriate amounts of protein and fat for Joshua.

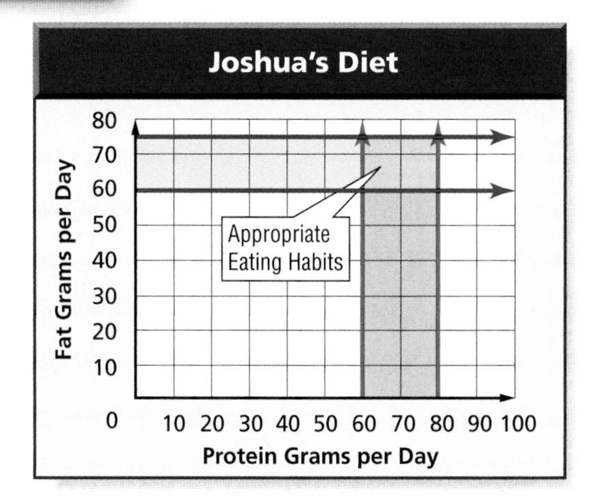

Systems of Inequalities A **system of inequalities** is a set of two or more inequalities with the same variables. To solve a system of inequalities like the one above, find the ordered pairs that satisfy all the inequalities. The solution set is represented by the intersection, or overlap, of the graphs.

EXAMPLE Solve by Graphing

Solve each system of inequalities by graphing.

a. $y < -x + 1$
$y \leq 2x + 3$

The solution includes the ordered pairs in the intersection of the graphs of $y < -x + 1$ and $y \leq 2x + 3$. This region is shaded in green at the right. The graph of $y = -x + 1$ is dashed and is *not* included in the graph of $y < -x + 1$. The graph of $y = 2x + 3$ is solid and *is* included in the graph of $y \leq 2x + 3$.

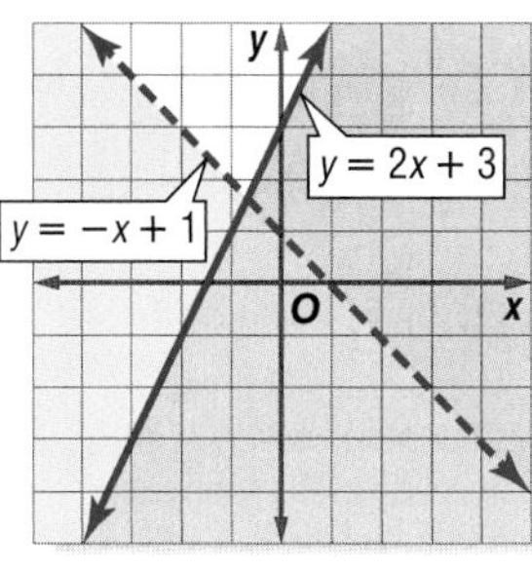

b. $x - y < -1$
$x - y > 3$

The graphs of $x - y = -1$ and $x - y = 3$ are parallel lines. Because the two regions have no points in common, the system of inequalities has no solution.

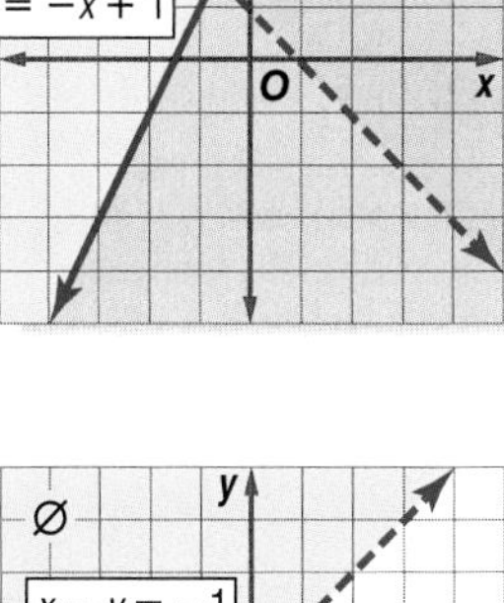

Concepts in Motion
Animation algebra1.com

CHECK Your Progress

1A. $y \leq 3$
$x + y \geq 1$

1B. $2x + y \geq 2$
$2x + y < 4$

 **Personal Tutor at** algebra1.com

You can use a TI-83/84 Plus to solve systems of inequalities.

GRAPHING CALCULATOR LAB

Graphing Systems of Inequalities

To graph the system $y \geq 4x - 3$ and $y \leq -2x + 9$ on a TI-83/4 Plus, select the SHADE feature in the DRAW menu. Enter the function that is the lower boundary of the region to be shaded, followed by the upper boundary. (Note that inequalities that have $>$ or $\geq$ are lower boundaries and inequalities that have $<$ or $\leq$ are upper boundaries.)

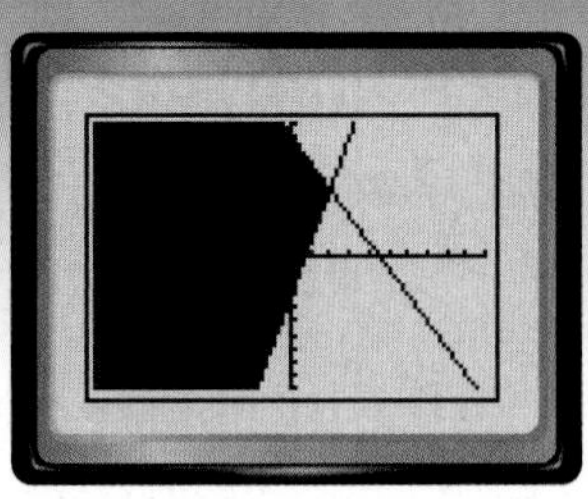

[−10, 10] scl: 1 by [−10, 10] scl: 1

THINK AND DISCUSS

1. To graph the system $y \leq 3x + 1$ and $y \geq -2x - 5$ on a graphing calculator, which function should you enter first?
2. Use a graphing calculator to graph the system $y \leq 3x + 1$ and $y \geq -2x - 5$.
3. Explain how you could use a graphing calculator to graph the system $2x + y \geq 7$ and $x - 2y \geq 5$. Then use a graphing calculator to graph the system.

Real-World Problems In real-world problems involving systems of inequalities, sometimes only whole-number solutions make sense.

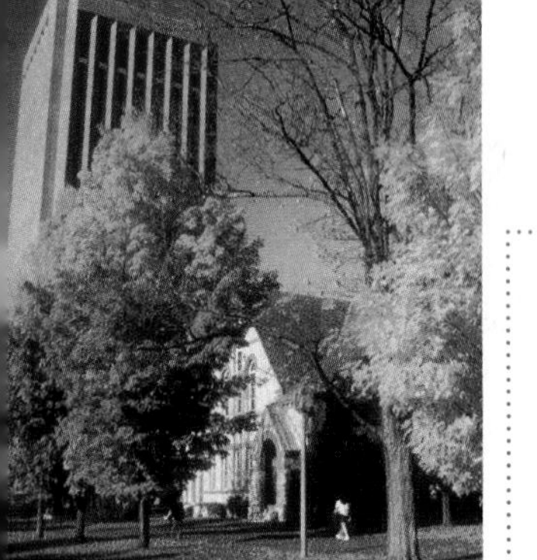

Real-World Link

The average first-year student at University of Massachusetts Amherst had a high school GPA of 3.38.

Source: umass.edu/oir

Real-World EXAMPLE Use a System of Inequalities

2 COLLEGE The middle 50% of first-year students attending the University of Massachusetts at Amherst scored between 520 and 630, inclusive, on the math portion of the SAT. They scored between 510 and 620, inclusive, on the critical reading portion of the test. Graph the scores that a student would need to be in the middle 50% of first year students.

Words The math score is between 520 and 630, inclusive. The critical reading score is between 510 and 620, inclusive.

Variables Let m = the math score and let c = the critical reading score.

Inequalities $520 \leq m \leq 630$

$510 \leq c \leq 620$

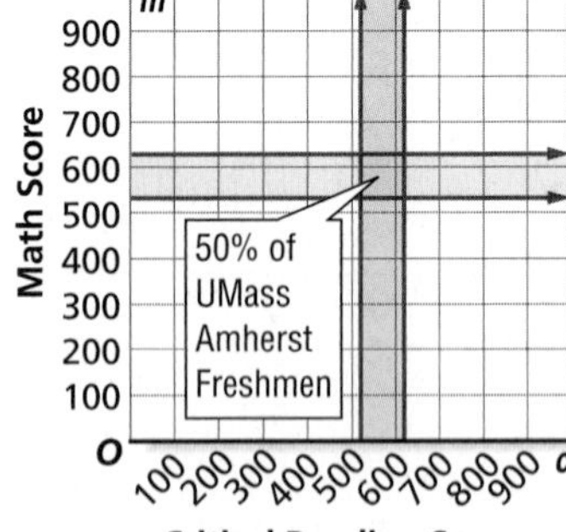

The solution is the set of all ordered pairs that are in the intersection of the graphs of these inequalities. However, since SAT scores are whole numbers, only whole-number solutions make sense in this problem.

CHECK Your Progress

2. **HEALTH** The LDL or "bad" cholesterol of a teenager should be less than 110. The HDL or "good" cholesterol of a teenager should be between 35 and 59. Make a graph showing appropriate levels of cholesterol for a teenager.

CHECK Your Understanding

Example 1 (p. 341)

Solve each system of inequalities by graphing.

1. $x > 5$
   $y \leq 4$
2. $y > 3$
   $y > -x + 4$
3. $y \leq -x + 3$
   $y \leq x + 3$
4. $2x + y \geq 4$
   $y \leq -2x - 1$

Example 2 (p. 342)

HEALTH For Exercises 5 and 6, use the following information.
Natasha exercises every day by walking and jogging at least 3 miles. Natasha walks at a rate of 4 miles per hour and jogs at a rate of 8 miles per hour. Suppose she only has a half-hour to exercise today.

5. Draw a graph showing the possible amount of time she can spend walking and jogging.
6. List three possible solutions.

Exercises

HOMEWORK HELP	
For Exercises	See Examples
7–18	1
19–22	2

Solve each system of inequalities by graphing.

7. $y < 0$
   $x \geq 0$
8. $x > -4$
   $y \leq -1$
9. $y \geq -2$
   $y - x < 1$
10. $x \geq 2$
    $y + x \leq 5$
11. $x \leq 3$
    $x + y > 2$
12. $y \geq 2x + 1$
    $y \leq -x + 1$
13. $y < 2x + 1$
    $y \geq -x + 3$
14. $y - x < 1$
    $y - x > 3$
15. $y - x < 3$
    $y - x \geq 2$
16. $2x + y \leq 4$
    $3x - y \geq 6$
17. $3x - 4y < 1$
    $x + 2y \leq 7$
18. $x + y > 4$
    $-2x + 3y < -12$

MANUFACTURING For Exercises 19 and 20, use the following information.
The Natural Wood Company has machines that sand and varnish desks and tables. The table below gives the time requirements of the machines.

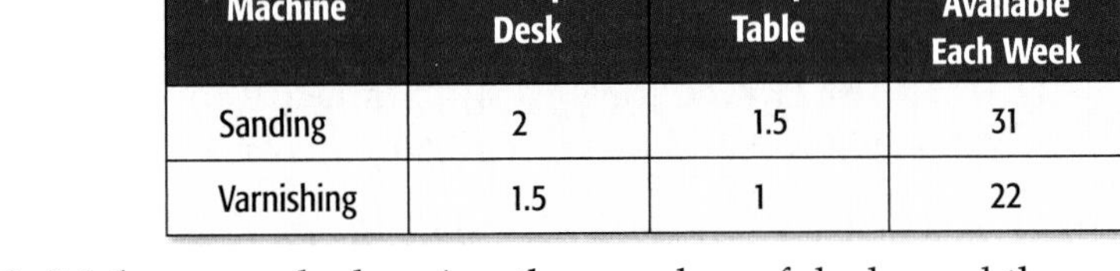

Machine	Hours per Desk	Hours per Table	Total Hours Available Each Week
Sanding	2	1.5	31
Varnishing	1.5	1	22

19. Make a graph showing the number of desks and the number of tables that can be made in a week.
20. List three possible solutions.

Real-World Career
Visual Artist
A visual artist uses math to create art to communicate ideas. The work of fine artists is made for display. Illustrators and graphic designers produce art for clients.

For more information, go to algebra1.com.

ART For Exercises 21 and 22, use the following information.
A painter has 32 units of yellow dye and 54 units of blue dye for mixing to make two shades of green. The units needed to make a gallon of light green and a gallon of dark are shown in the table.

Color	Units of Yellow Dye	Units of Blue Dye
light green	4	1
dark green	1	6

21. Make a graph showing the numbers of gallons of the two greens she can make.
22. List three possible solutions.

Solve each system of inequalities by graphing.

23. $x \geq 0$
$x - 2y \leq 2$
$3x + 4y \leq 12$

24. $y \leq x + 3$
$2x - 7y \leq 4$
$3x + 2y \leq 6$

25. $x < 2$
$4y > x$
$2x - y < -9$
$x + 3y < 9$

Write a system of inequalities for each graph.

26.

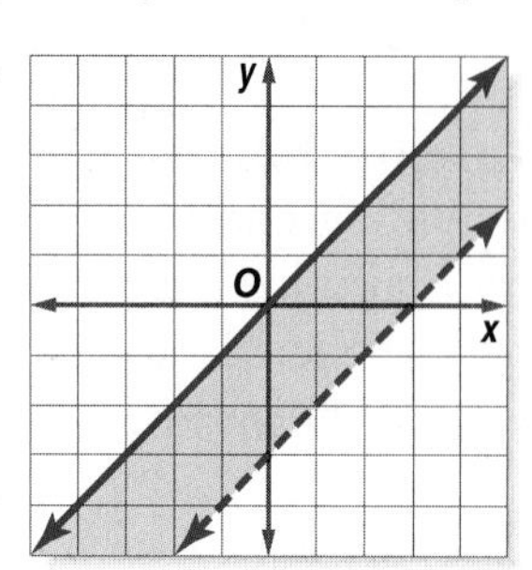

27.

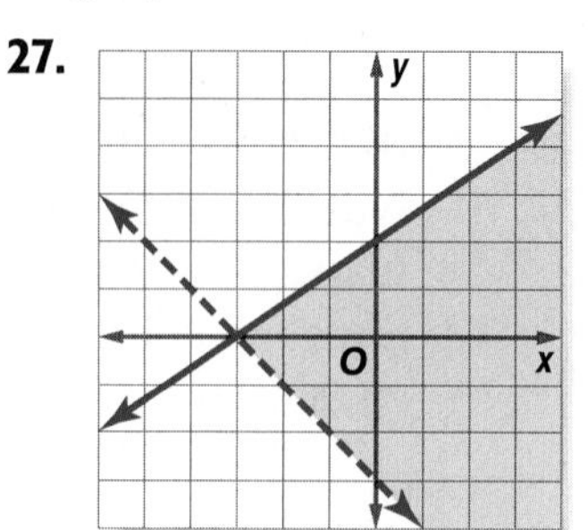

EXTRA PRACTICE
See pages 730, 749.
Math Online
Self-Check Quiz at
algebra1.com

AGRICULTURE For Exercises 28 and 29, use the following information.
To ensure a growing season of sufficient length, Mr. Hobson has at most 16 days left to plant his corn and soybean crops. He can plant corn at a rate of 250 acres per day and soybeans at a rate of 200 acres per day.

28. If he has at most 3500 acres available, make a graph showing how many acres of each type of crop he can plant.

29. Name one solution and explain what it means.

Graphing Calculator

Use a graphing calculator to solve each system of inequalities.

30. $y \leq x + 9$
$y \geq -x - 4$

31. $y \leq 2x + 10$
$y \geq 7x + 15$

32. $3x - y \leq 6$
$x - y \geq -1$

Sketch the region in the plane that satisfies both inequalities.

33. $3y - x \geq 6$
$y < -2x - 1$

34. $3y - x \leq 9$
$4y + x \leq 12$

H.O.T. Problems

35. OPEN ENDED Draw the graph of a system of inequalities that has no solution.

36. CHALLENGE Create a system of inequalities equivalent to $|x| \leq 4$.

37. FIND THE ERROR Jocelyn and Sonia are solving the system of inequalities $x + 2y \geq -2$ and $x - y > 1$. Who is correct? Explain your reasoning.

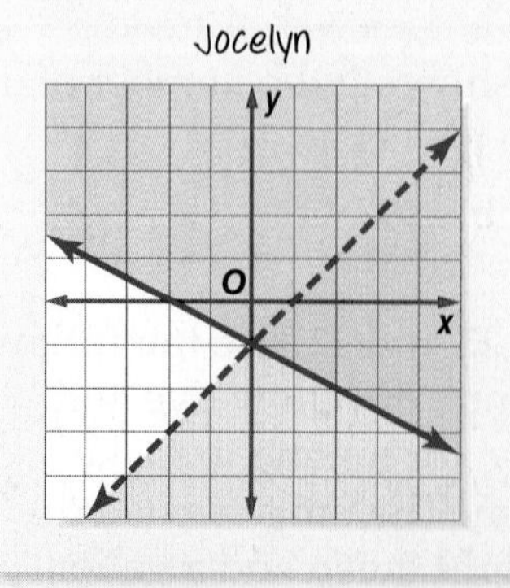

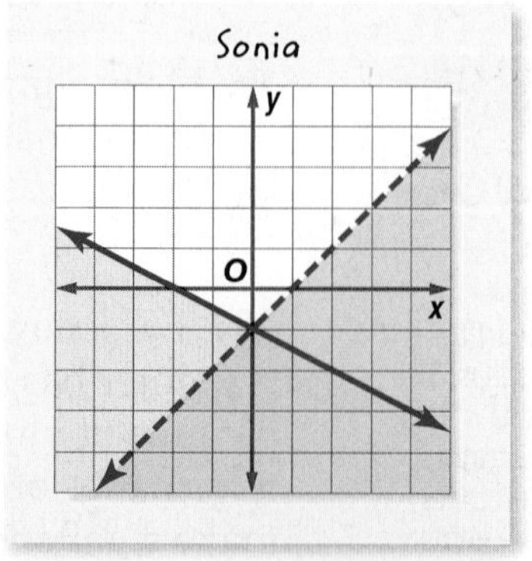

38. ***Writing in Math*** Use the information about nutrition on page 341 to explain how you can use a system of inequalities to plan a sensible diet. Include two appropriate Calorie and fat intakes for a day and the system of inequalities that is represented by the graph.

STANDARDIZED TEST PRACTICE

39. Which system of inequalities is *best* represented by the graph?

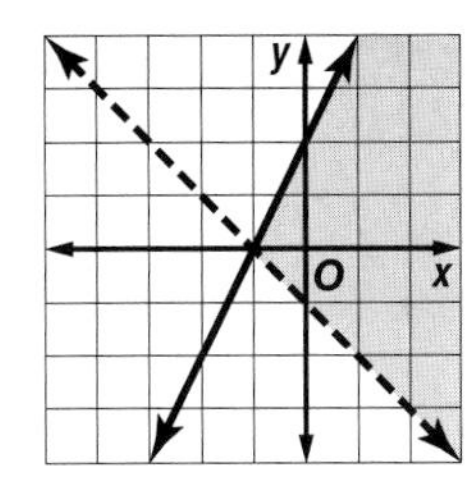

A $y \leq 2x + 2$
$y > -x - 1$

B $y \geq 2x + 2$
$y < -x - 1$

C $y < 2x + 2$
$y \leq -x - 1$

D $y > 2x + 2$
$y \leq -x - 1$

40. REVIEW The table shows the cost of organic wheat flour, depending on the amount purchased. Which conclusion can be made based on the information in the table?

Number of Pounds	Cost ($)
2	3.40
10	16.00
25	37.50
50	75.00

F The cost of 4 pounds of flour would be more than $7.

G The cost of 100 pounds of flour would be less than $150.

H The cost of flour is always more than $1.60 per pound.

J The cost of flour is always less than $1.50 per pound.

Spiral Review

Graph each inequality. (Lesson 6-6)

41. $y < \frac{1}{2}x - 4$

42. $2y + x \leq 6$

43. $x + 2y \geq 8$

44. FISH The temperature of a freshwater tropical fish tank should be within 1.5 degrees of 76.5°F. Express this using an inequality involving absolute value. (Lesson 6-6)

Use elimination to solve each system of equations. (Lesson 5-4)

45. $2x + 3y = 1$
$4x - 5y = 13$

46. $5x - 2y = -3$
$3x + 6y = -9$

47. $-3x + 2y = 12$
$2x - 3y = -13$

48. $6x - 2y = 4$
$5x - 3y = -2$

Determine whether each relation is a function. (Lesson 3-2)

49. $y = -15$

50. $x = 5$

51. $\{(1, 0), (1, 4), (-1, 1)\}$

52. $\{(6, 3), (5, -2), (2, 3)\}$

Cross-Curricular Project

Math and Science

The Spirit of the Games It's time to complete your project. Use the information and data you have gathered about the Olympics to prepare a portfolio or Web page. Be sure to include graphs and/or tables in the presentation.

Math Online Cross-Curricular Project at algebra1.com

CHAPTER 6 Study Guide and Review

FOLDABLES Study Organizer — GET READY to Study

Be sure the following Key Concepts are noted in your Foldable.

Solving Linear Inequalities

Key Concepts

Solving Inequalities by Adding, Subtracting, Multiplying, or Dividing (Lessons 6-1 and 6-2)

- If any number is added to or subtracted from each side of a true inequality, the resulting inequality is also true.
- If each side of a true inequality is multiplied or divided by the same positive number, the resulting inequality is also true.

Multi-Step and Compound Inequalities (Lessons 6-3 and 6-4)

- If each side of a true inequality is multiplied or divided by the same negative number, the direction of the inequality symbol must be *reversed* so that the resulting inequality is also true.

Absolute Value Equations and Inequalities (Lessons 6-5 and 6-6)

- The absolute value of any number n is its distance from zero on a number line and is written as $|n|$.
- If $|x| = n$, then $x = -n$ or $x = n$.
  If $|x| < n$, then $x < n$ or $x > -n$.
  If $|x| > n$, then $x > n$ or $x < -n$.

Inequalities in Two Variables (Lesson 6-7)

- Any line in the plane divides the plane into two regions called *half-planes*. The line is called the boundary of each of the two half-planes.

Systems of Inequalities (Lesson 6-8)

- A system of inequalities is a set of two or more inequalities with the same variables.

Key Vocabulary

boundary (p. 334)
compound inequality (p. 315)
half-plane (p. 334)
intersection (p. 315)
piecewise function (p. 324)
set-builder notation (p. 295)
system of inequalities (p. 341)
union (p. 316)

Vocabulary Check

State whether each sentence is *true* or *false*. If *false*, replace the underlined word or phrase to make a true sentence.

1. The edge of a half-plane is called a boundary.
2. The symbol ∅ means intersection.
3. The phrase at least is represented by the same symbol as the phrase greater than.
4. To solve a system of inequalities, find the ordered pairs that satisfy all the inequalities involved.
5. The union can be found by graphing each inequality and then determining where the graphs overlap.
6. The solution $\{x \mid x < 5\}$ is written in set-builder notation.
7. A compound inequality containing and is true if one or more of the inequalities is true.
8. When solving $4x > -12$, the direction of the inequality symbol should be reversed.
9. The graph of $y > 3x - 6$ has a dashed line boundary.
10. A union is formed when a line in the plane divides the plane into two regions.

Vocabulary Review at algebra1.com

Lesson-by-Lesson Review

6–1 Solving Inequalities by Addition and Subtraction (pp. 294–299)

Solve each inequality. Check your solution, and then graph it on a number line.

11. $x - 9 < 16$ **12.** $-11 \geq -5 + p$

13. $12w + 4 \leq 13w$ **14.** $8g > 7g - 1$

For Exercises 15 and 16, define a variable, write an inequality, and solve each problem. Check your solution.

15. Sixteen is less than the sum of a number and 31.

16. TOMATOES There are more than 10,000 varieties of tomatoes. One seed company produces seed packages for 200 varieties of tomatoes. For how many varieties do they not provide seeds?

Example 1 **Solve $-2 \leq h + 17$. Check your solution, and then graph it on a number line.**

$-2 \leq h + 17$ Original inequality

$-2 - 17 \leq h + 17 - 17$ Subtract 17 from each side.

$-19 \leq h$ Simplify.

Since $-19 \leq h$ is the same as $h \geq -19$, the solution set is $\{h \mid h \geq -19\}$.

−21 −20 −19 −18 −17 −16 −15 −14 −13 −12

6–2 Solving Inequalities by Multiplication and Division (pp. 301–307)

Solve each inequality. Check your solution.

17. $15v > 60$ **18.** $3 \leq -\frac{d}{13}$

19. $-9m < 99$ **20.** $-15 \geq \frac{3}{5}k$

For Exercises 21 and 22, define a variable, write an inequality, and solve the problem. Check your solution.

21. Eighty percent of a number is greater than or equal to 24.

22. FISHING About 41.6 million tons of fish were caught in China in a recent year. If this is over 35% of the world's catch, how many fish were caught in the world that year?

Example 2 **Solve $-14g \geq 126$.**

$-14g \geq 126$ Original inequality

$\frac{-14g}{-14} \leq \frac{126}{-14}$ Divide each side by −14 and change ≥ to ≤.

$g \leq -9$ Simplify.

The solution set is $\{g \mid g \leq -9\}$.

Example 3 **Solve $\frac{3}{4}w < 15$.**

$\frac{3}{4}w < 15$ Original inequality

$\left(\frac{4}{3}\right)\frac{3}{4}w < \left(\frac{4}{3}\right)15$ Multiply each side by $\frac{4}{3}$.

$w < 20$ Simplify.

The solution set is $\{w \mid w < 20\}$.

CHAPTER 6

Study Guide and Review

6–3 Solving Multi-Step Inequalities (pp. 308–313)

Solve each inequality. Check your solution.

23. $5 - 6y > -19$

24. $\frac{1 - 7n}{5} \geq 10$

25. $-5x + 3 \leq 3x + 19$

26. $7(g + 8) < 3(g + 2) + 4g$

For Exercises 27 and 28, define a variable, write an inequality, and solve the problem. Check your solution.

27. Two thirds of a number decreased by 27 is at least 9.

28. CATS Dexter has \$20 to spend at the pet store. He plans to buy a toy for his cat that costs \$3.75 and several bags of cat food. If each bag of cat food costs \$2.99, what is the greatest number of bags that he can buy?

Example 4 **Solve $4(n - 1) < 7n + 8$.**

$4(n - 1) < 7n + 8$	Original Inequality
$4n - 4 < 7n + 8$	Distributive Property
$4n - 4 - 7n < 7n + 8 - 7n$	Subtract $7n$ from each side.
$-3n - 4 < 8$	Simplify.
$-3n - 4 + 4 < 8 + 4$	Add 4 to each side.
$-3n < 12$	Simplify.
$\frac{-3n}{-3} > \frac{12}{-3}$	Divide and change $<$ to $>$.
$n > -4$	Simplify.

The solution set is $\{n \mid n > -4\}$.

6–4 Solving Compound Inequalities (pp. 315–320)

Graph the solution set of each compound inequality.

29. $10 - 2y > 12$ and $7y < 4y + 9$

30. $a - 3 \leq 8$ or $a + 5 \geq 21$

31. $3w + 8 \leq 2$ or $w + 12 \geq 2 - w$

32. $-1 < p + 3 < 5$

33. FAIRS A vendor at the state fair is trying to guess Martin's age within 2 years. The vendor guesses that Martin is 35 years old. If m represents Martin's age, write a compound inequality that represents the possible range of m if the vendor is correct. Then graph the solution set.

Example 5 **Graph the solution set of $x \geq -1$ and $x > 3$.**

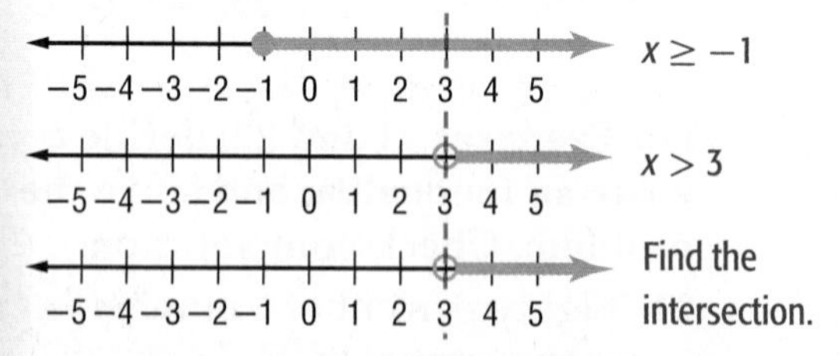

Example 6 **Graph the solution set of $x \leq -2$ or $x > 4$.**

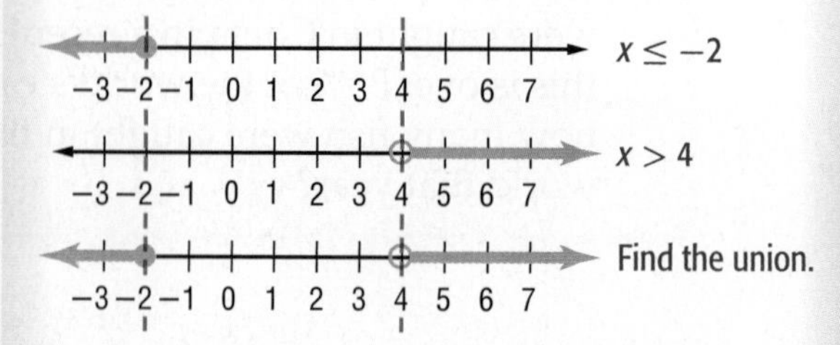

Mixed Problem Solving
For mixed problem-solving practice, see page 749.

6–5 Solving Open Sentences Involving Absolute Value (pp. 322–327)

Solve each open sentence. Then graph the solution set.

34. $|x + 4| = 3$ **35.** $|2x - 3| = 5$

36. TESTS Kent has an A in math class. If his score on the next test is 98%, plus or minus 2 percentage points, he will maintain an A average. Write an open sentence to find the highest and lowest scores he can earn on the next test.

37. CARS The stated capacity of a fuel tank in a passenger car is accurate within 3%. Write an open sentence to find the greatest and least capacity for a fuel tank if the stated capacity is 13.6 gallons.

Example 7 **Solve $|x + 6| = 15$. Then graph the solution set.**

$|x + 6| = 15$ is $x + 6 = 15$ or $x + 6 = -15$.

$x + 6 = 15$ $\quad$ $x + 6 = -15$

$x + 6 - 6 = 15 - 6$ $\quad$ $x + 6 - 6 = -15 - 6$

$x = 9$ $\quad$ $x = -21$

The solution set is $\{-21, 9\}$.

−24 −20 −16 −12 −8 −4 0 4 8 12 16

6–6 Solving Inequalities Involving Absolute Value (pp. 329–333)

Solve each open sentence. Then graph the solution set.

38. $|3d + 8| < 23$ **39.** $|g + 2| \geq -9$

40. $|m - 1| > -6$ **41.** $|2x - 5| \geq 7$

42. $|4h - 3| < 13$ **43.** $|w + 8| \leq 11$

44. AIRPLANES For the average commercial airplane to take off from the runway, its speed should be within 10 miles per hour of 170 miles per hour. Define a variable, write an open sentence, and find this range of takeoff speeds.

Example 8 **Solve $|2x - 3| < 5$. Then graph the solution set.**

$|2x - 3| < 5$ is $2x - 3 < 5$ and $2x - 3 > -5$.

$2x - 3 < 5$ $\quad$ $2x - 3 > -5$

$2x - 3 + 3 < 5 + 3$ $\quad$ $2x - 3 + 3 > -5 + 3$

$2x < 8$ $\quad$ $2x > -2$

$\frac{2x}{2} < \frac{8}{2}$ $\quad$ $\frac{2x}{2} > \frac{-2}{2}$

$x < 4$ $\quad$ $x > -1$

$x < 4$ and $x > -1$

The solution set is $\{x \mid -1 < x < 4\}$.

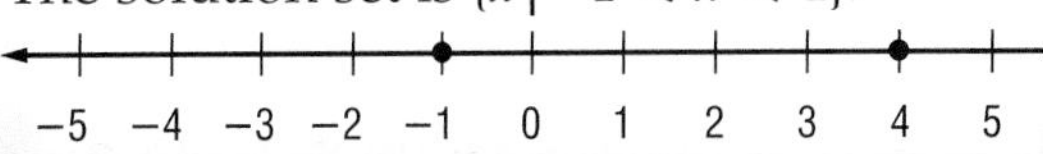

6–7 Graphing Inequalities in Two Variables (pp. 334–339)

Graph each inequality.

45. $y - 2x < -3$ **46.** $x + 2y \geq 4$

47. $y \leq 5x + 1$ **48.** $2x - 3y > 6$

49. MOVING A moving company charges $95 an hour and $0.08 a mile to move items from Brianna's old apartment to her new house. If Brianna has only $500 for moving expenses, write an inequality for this situation. Can she afford to hire this moving company if she knows it will take 5 hours and the distance between houses is 75 miles?

Example 9 **Graph $y \geq x - 2$.**

Since the boundary is included in the solution, draw a solid line.

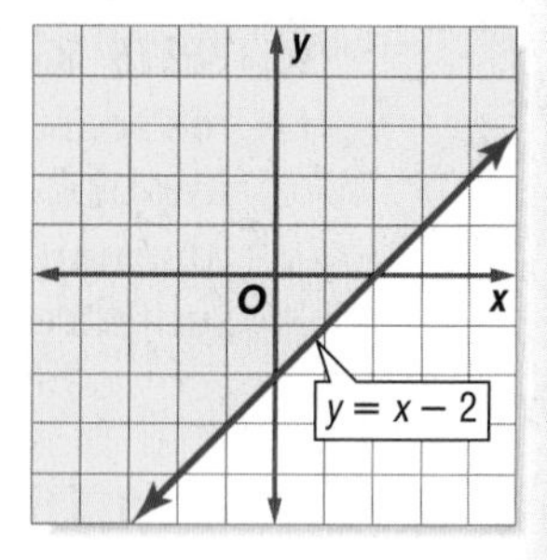

Test the point (0, 0).

$y \geq x - 2$ Original inequality

$0 \geq 0 - 2$ $x = 0, y = 0$

$0 \geq -2$ true

The half-plane that contains (0, 0) should be shaded.

6–8 Graphing Systems of Inequalities (pp. 341–345)

Solve each system of inequalities by graphing.

50. $y < 3x$
$x + 2y \geq -21$

51. $y > -x - 1$
$y \leq 2x + 1$

52. $2x + y < 9$
$x + 11y < -6$

53. $y \geq 1$
$y + x \leq 3$

54. TREES Justin wants to plant peach and apple trees in his backyard. He can fit at most 12 trees. Each peach tree costs $60, and each apple tree costs $75. If he only has $800 to spend, make a graph showing the number of each kind of tree that he can buy. Then list three possible solutions.

Example 10 **Solve the system of inequalities by graphing.**

$x \geq -3$

$y \leq x + 2$

The solution includes the ordered pairs in the intersection of the graphs $x \geq -3$ and $y \leq x + 2$. This region is shaded in green. The graphs of $x \geq -3$ and $y \leq x + 2$ are boundaries of this region.

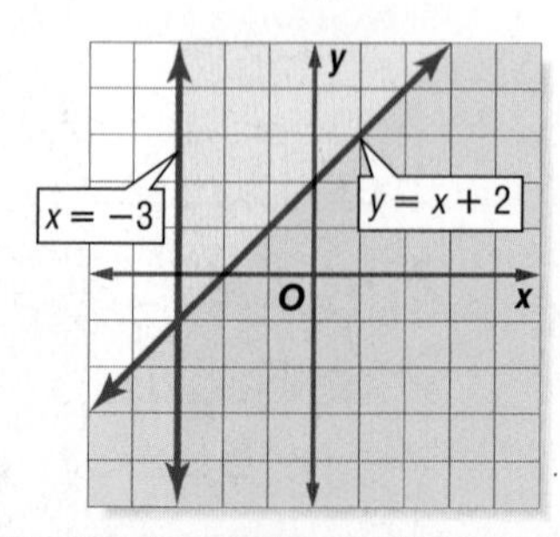

Chapter 6 Practice Test

Solve each inequality. Check your solution.

1. $-23 \geq g - 6$
2. $9p < 8p - 18$
3. $4m - 11 \geq 8m + 7$
4. $3(k - 2) > 12$
5. **REAL ESTATE** A homeowner is selling her house. She must pay 7% of the selling price to her real estate agent after the house is sold. Define a variable and write and solve an inequality to find what the selling price of her house must be to have at least \$140,000 after the agent is paid. Round to the nearest dollar.
6. Solve $6 + |r| = 3$.
7. Solve $|d| > -2$.

Solve each compound inequality. Then graph the solution set.

8. $r + 3 > 2$ and $4r < 12$
9. $3n + 2 \geq 17$ or $3n + 2 \leq -1$

Solve each open sentence. Then graph the solution set.

10. $|4x + 3| = 9$
11. $|6 - 4m| = 8$
12. $|2a - 5| < 7$
13. $|7 - 3s| \geq 2$

For Exercises 14–17, define a variable, write an inequality, and solve each problem. Check your solution.

14. One fourth of a number is no less than −3.
15. Three times a number subtracted from 14 is less than two.
16. Five less than twice a number is between 13 and 21.
17. **TRAVEL** Mary's car gets the gas mileage shown in the table. If her car's tank holds 15 gallons, what is the range of distance that Mary can drive her car on one tank of gasoline?

Gas Mileage	Miles Per Gallon
minimum	18
maximum	21

Graph each inequality.

18. $y \geq 3x - 2$
19. $2x + 3y < 6$
20. $x - 2y > 4$
21. **MULTIPLE CHOICE** Ricardo purchased x bottles of paint and y paint brushes. He spent less than \$20, not including tax. If $3x + 2y < 20$ represents this situation, which point represents a reasonable number of bottles of paint and paint brushes that Ricardo could have purchased?

Product	Cost
bottle of paint	\$3.00
paint brush	\$2.00

A (2, 7)

B (5, 4)

C (2, 8)

D (5, 2)

Solve each system of inequalities by graphing.

22. $y > -4$
$y < -1$
23. $y \leq 3$
$y > -x + 2$
24. **MULTIPLE CHOICE** Which graph represents $y > 2x + 1$ and $y < -x - 2$?

F

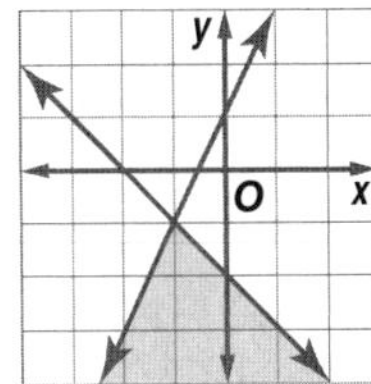

H

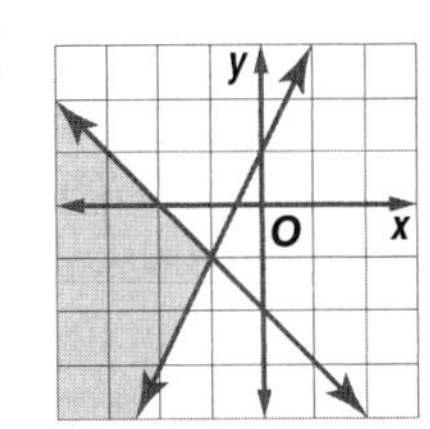

G

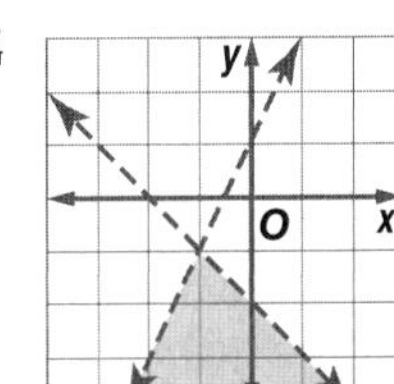

J

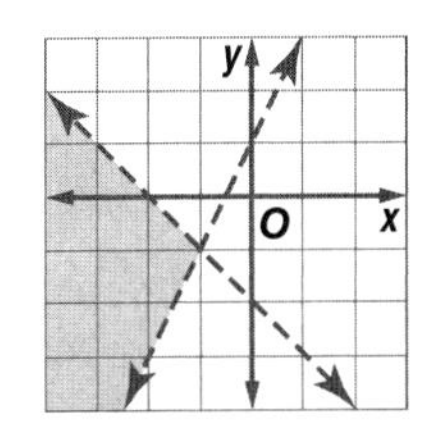

Math Online Chapter Test at algebra1.com

CHAPTER 6

Standardized Test Practice

Cumulative, Chapters 1–6

Read each question. Then fill in the correct answer on the answer document provided by your teacher or on a sheet of paper.

1. Perry measured the distance that a stick floated down a stream and the time that it took to float that distance. He recorded this in the table below.

Time, x (minutes)	Distance, y (meters)
0	0
2	6
3	9
5	15
10	30

Which equation best represents the relationship between distance floated, y, and the time, x?

A $y = 3x$ **C** $y = 3x^2$

B $y = \frac{3}{x}$ **D** $y = -3x$

TEST-TAKING TIP

QUESTION 1 When you write an equation, check that the given values make a true statement. For example, in Question 1, substitute the values of the coordinates of one of the points into your equation to check.

2. Mr. Carter's history class earned \$500 to visit the local museum. The class needed to purchase tickets and rent a bus. The tickets cost \$8 each and the bus costs \$50 to rent. Which inequality represents the number of students x, that can go on the trip?

F $50 - 8x \leq 500$ **H** $50x + 8 \geq 500$

G $50x - 8 \geq 500$ **J** $50 + 8x \leq 500$

3. GRIDDABLE The drama club sold 180 tickets to their first performance. They charged \$5 for each adult ticket and \$4 for each student ticket. They made a total of \$798. How many student tickets did they sell?

4. If the variables a and b are related so that $a + b > a - b$, which statement must be true about a and b?

A The variable a is greater than the variable b.

B The variable a is a negative number.

C The variable b is a negative number.

D The variable b is a positive number.

5. Which graph best represents the temperature of a glass of water after an ice cube is placed in it?

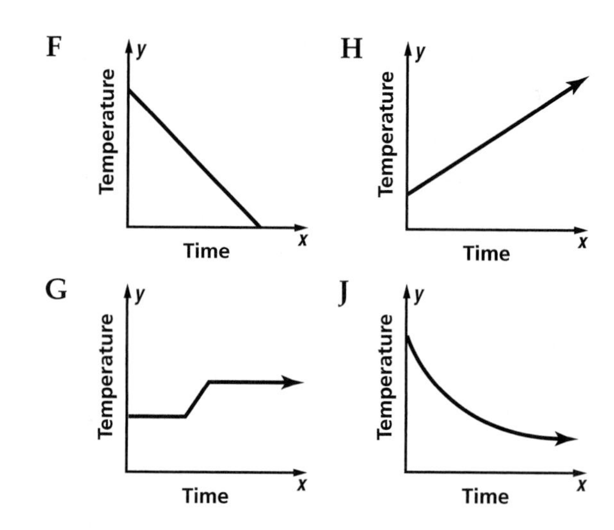

6. Jonah has 80 sports trading cards. The number of baseball trading cards is 16 less than twice the number of basketball trading cards. Which system of equations can be used to find how many baseball trading cards, x, and basketball cards, y, Jonah has?

A $x + y = 80$
$y = 2x + 16$

B $x + y = 80$
$x = 2y - 16$

C $y - x = 80$
$x = 2y - 16$

D $y - 80 = x$
$x = 2y$

Math Online Standardized Test Practice at algebra1.com

7. Find the volume of the figure shown. Round to the nearest tenth.

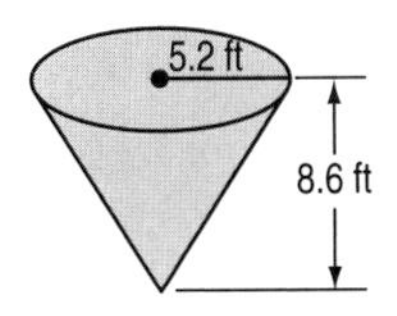

F 243.5 ft^3

G 421.5 ft^3

H 487.0 ft^3

J 730.6 ft^3

8. Airplane A is descending from an altitude of 13,000 feet at a rate of 1300 feet per minute. Airplane B is ascending from the ground at a rate of 1000 feet per minute. Which graph below accurately represents the point when the airplanes will reach the same altitude?

A
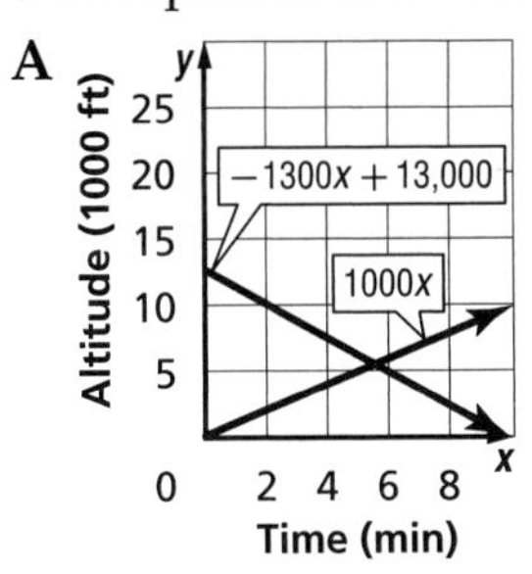

C
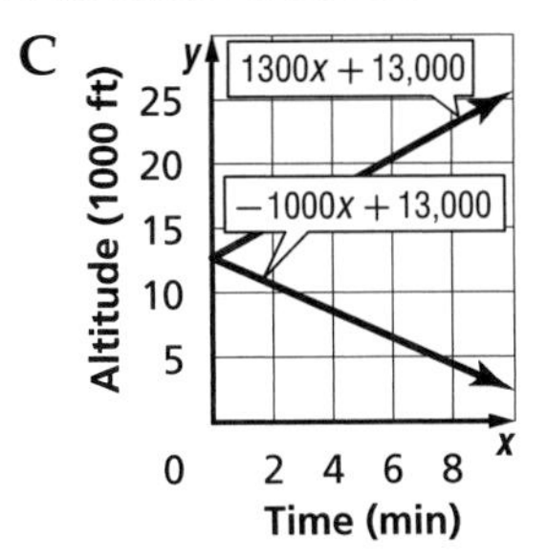

B
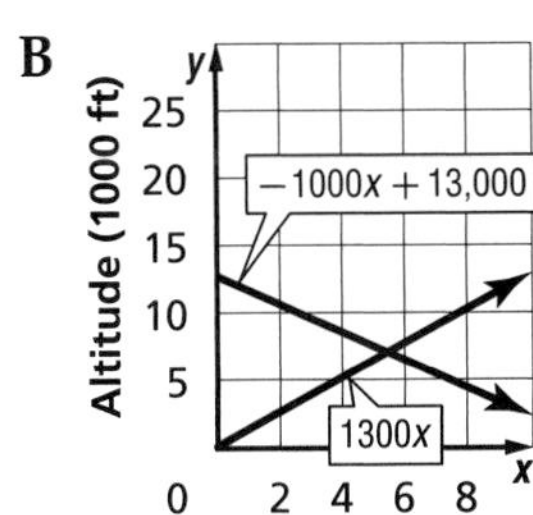

D
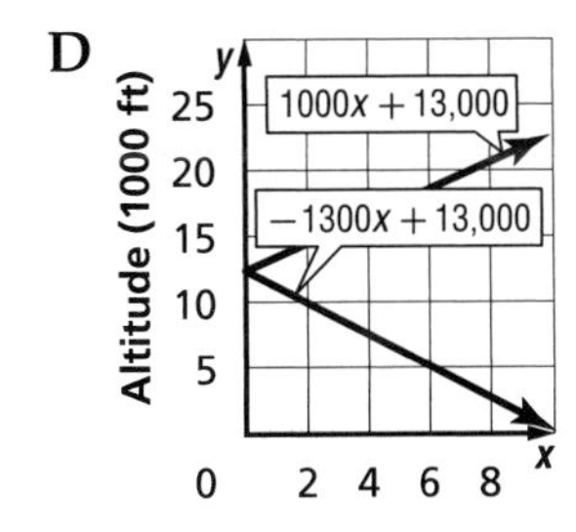

9. Which graph best represents the solution to this system of inequalities?

$$3x \geq y + 2$$
$$-2x \geq -4y + 8$$

A
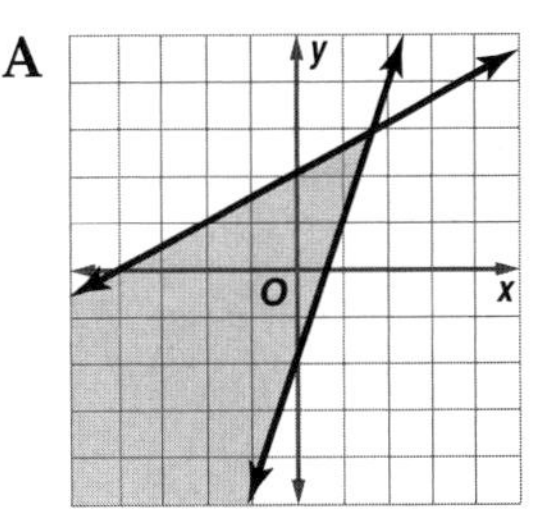

C
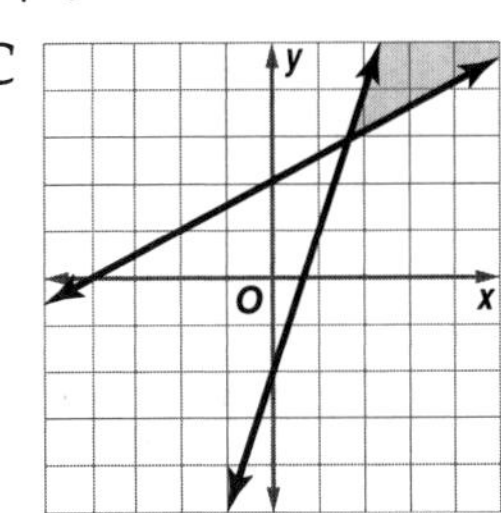

B
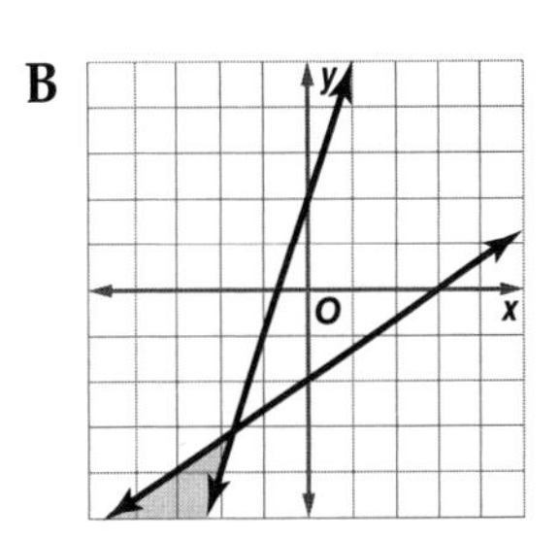

D
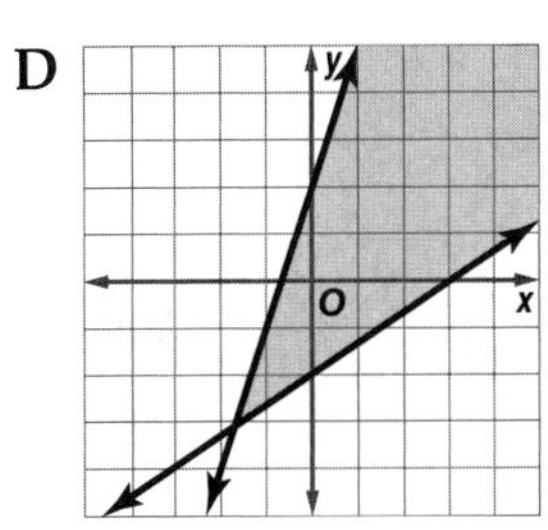

Pre-AP

Record your answers on a sheet of paper. Show your work.

10. The Carlson family is building a house on a lot that is 91 feet long and 158 feet wide.

a. Town law states that the sides of a house cannot be closer than 10 feet to the edges of a lot. Write an inequality for the possible lengths of the Carlson family's house, and solve the inequality.

b. The Carlson family wants their house to be at least 2800 square feet and no more than 3200 square feet. They also want their house to have the maximum possible length. Write an inequality for the possible widths of their house, and solve the inequality. Round your answer to the nearest whole number of feet.

NEED EXTRA HELP?										
If You Missed Question...	1	2	3	4	5	6	7	8	9	10
Go to Lesson or Page...	4-4	6-3	5-4	6-1	1-9	5-1	708	5-5	6-8	6-1, 6-2, 6-4

UNIT 3

Polynomials and Nonlinear Functions

Focus

Use quadratic and other nonlinear functions to represent and model problem situations and to analyze and interpret relationships.

CHAPTER 7 Polynomials

BIG Idea Understand there are situations modeled by functions that are not linear, and model the situations.

CHAPTER 8 Factoring

BIG Idea Use algebraic skills to simplify algebraic expressions, and solve equations and inequalities in problem situations.

CHAPTER 9

Quadratic and Exponential Functions

BIG Idea Understand there is more than one way to solve a quadratic equation and solve them using appropriate methods.

BIG Idea Understand there are situations modeled by functions that are neither linear nor quadratic, and model the situations.

Cross-Curricular Project

Algebra and Physical Science

Out of this World You can probably name the planets in the solar system, but can you name planets outside of our system? In recent years, planets in other systems have been discovered. In August, 2004, a team of astronomers discovered a small planet orbiting a star known as 55 Cancri. Star 55 Cancri has three other planets, making it the first known four-planet system outside our system. In this project, you will examine how exponents, factors, and graphs are useful in presenting information about planets.

Math Online **Log on to algebra1.com to begin.**

CHAPTER 7

Polynomials

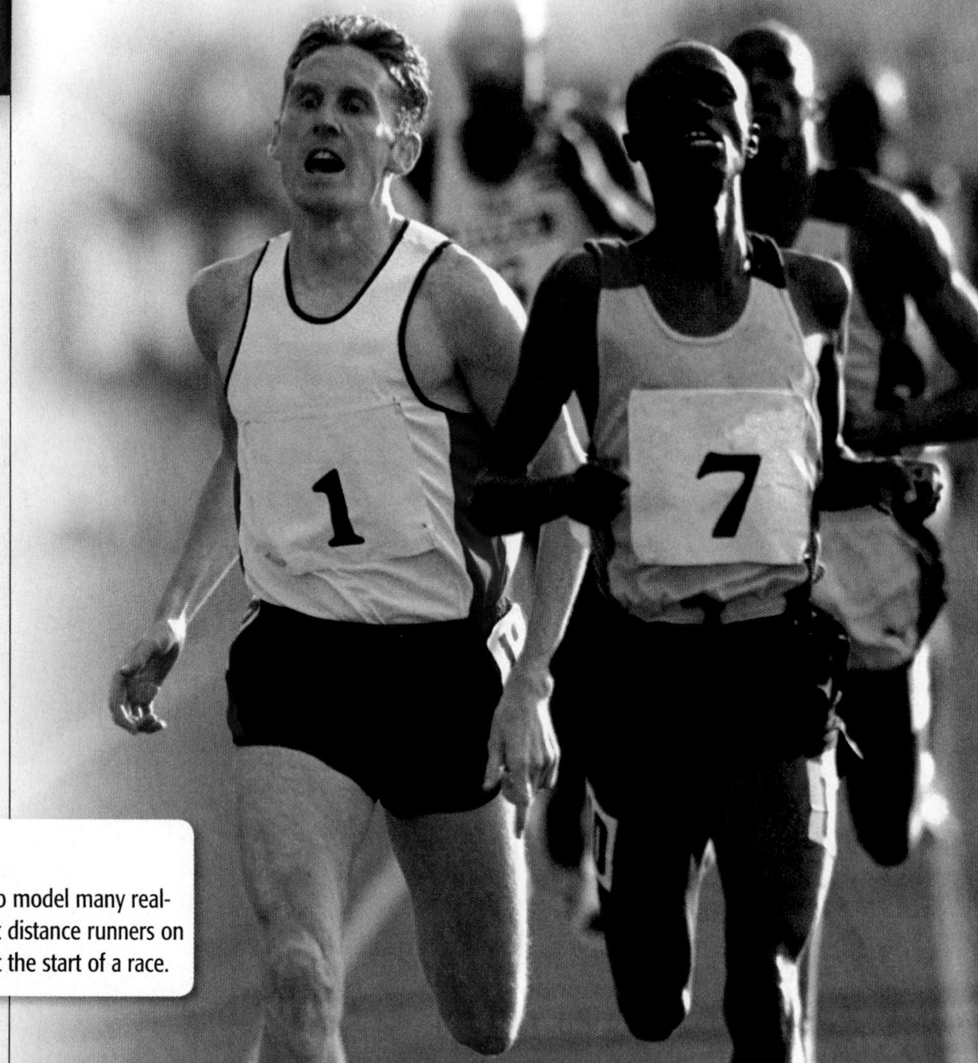

BIG Ideas

- Find products and quotients of monomials.
- Find the degree of a polynomial and arrange the terms in order.
- Add, subtract, and multiply polynomial expressions.
- Find special products of binomials.

Key Vocabulary

binomial (p. 376)
FOIL method (p. 399)
monomial (p. 358)
polynomial (p. 376)

Real-World Link

Running Polynomials can be used to model many real-world situations, such as the way that distance runners on a curved track should be staggered at the start of a race.

FOLDABLES™ Study Organizer

Polynomials Make this Foldable to help you organize information about polynomials. Begin with a sheet of 11″ by 17″ paper.

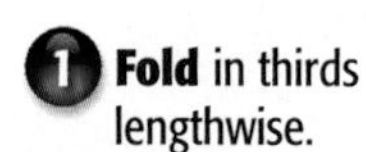

1 Fold in thirds lengthwise.

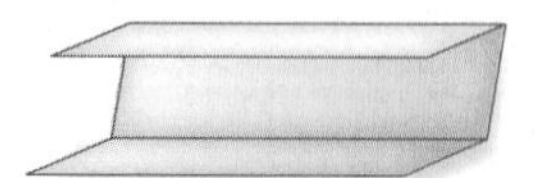

2 Open and fold a 2″ tab along the width. Then fold the rest in fourths.

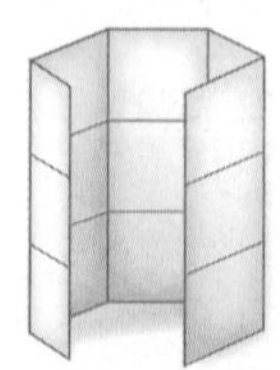

3 Draw lines along folds and label as shown.

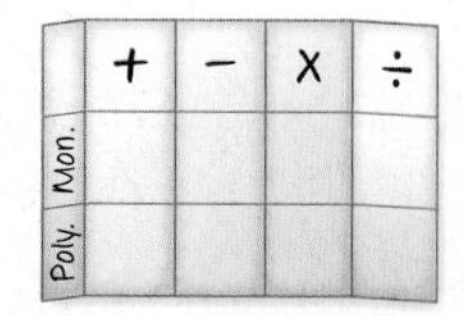

GET READY for Chapter 7

Diagnose Readiness You have two options for checking Prerequisite Skills.

Option 2

Math Online Take the Online Readiness Quiz at algebra1.com.

Option 1

Take the Quick Check below. Refer to the Quick Review for help.

QUICKCheck

Write each expression using exponents. (Lesson 1-1)

1. $2 \cdot 2 \cdot 2 \cdot 2 \cdot 2$
2. $3 \cdot 3 \cdot 3 \cdot 3$
3. $5 \cdot 5$
4. $x \cdot x \cdot x$
5. $a \cdot a \cdot a \cdot a \cdot a \cdot a$
6. $x \cdot x \cdot y \cdot y \cdot y$
7. $\frac{1}{2} \cdot \frac{1}{2} \cdot \frac{1}{2} \cdot \frac{1}{2} \cdot \frac{1}{2}$
8. $\frac{a}{b} \cdot \frac{a}{b} \cdot \frac{c}{d} \cdot \frac{c}{d} \cdot \frac{c}{d} \cdot \frac{c}{d}$

Evaluate each expression. (Lesson 1-1)

9. 3^2
10. 4^3
11. $(-6)^2$
12. $(-3)^3$
13. $\left(\frac{2}{3}\right)^4$
14. $\left(-\frac{7}{8}\right)^2$
15. **PROBABILITY** The probability of correctly guessing the outcome of a flipped penny six times is $\left(\frac{1}{2}\right)^6$. Express this probability as a fraction without exponents.

Find the area or volume of each figure. (Prerequisite Skill)

16. 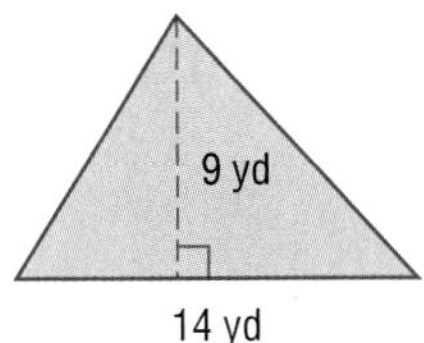

17. 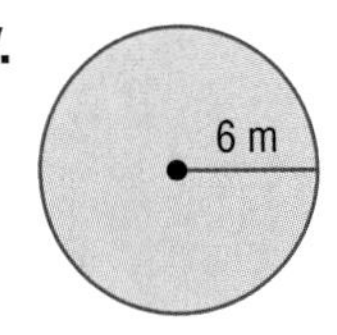

18. / 19. 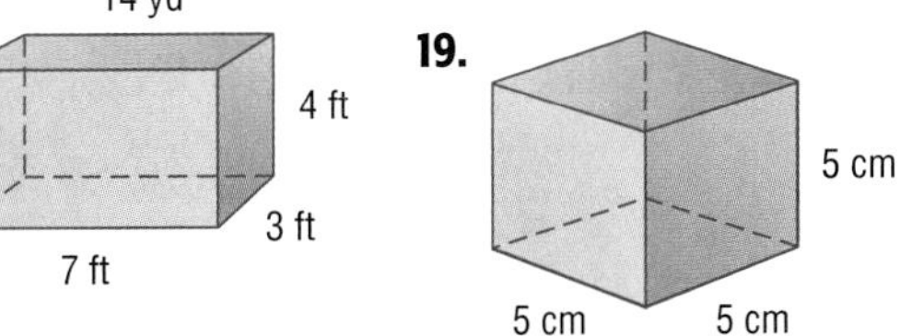

QUICKReview

EXAMPLE 1

Express $6 \cdot 6 \cdot 6 \cdot x \cdot x + y \cdot y \cdot y \cdot y \cdot z$ using exponents.

3 factors of six is 6^3. 4 factors of y is y^4.

2 factors of x is x^2. 1 factor of z is z^1 or z.

So, $6 \cdot 6 \cdot 6 \cdot x \cdot x + y \cdot y \cdot y \cdot y \cdot z = 6^3x^2 + y^4z$.

EXAMPLE 2

Evaluate $\left(\frac{8}{11}\right)^2$.

$\left(\frac{8}{11}\right)^2$ Original expression

$= \frac{8^2}{11^2}$ Power of a Quotient Rule

$= \frac{64}{121}$ Simplify.

EXAMPLE 3

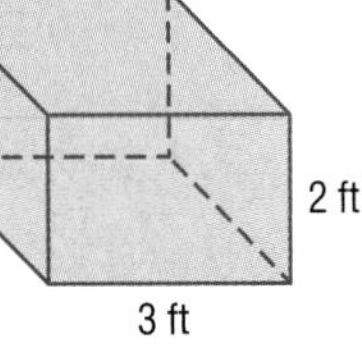

Find the volume of the figure.

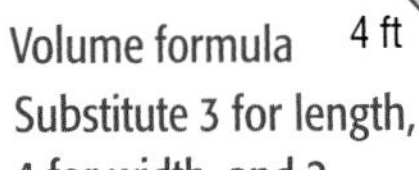

$V = \ell wh$ Volume formula

$= 3 \cdot 4 \cdot 2$ Substitute 3 for length, 4 for width, and 2 for height.

$= 24 \text{ ft}^3$ Evaluate volume.

The volume of the box is 24 cubic feet.

7-1 Multiplying Monomials

Main Ideas

- Multiply monomials.
- Simplify expressions involving powers of monomials.

New Vocabulary

monomial

constant

GET READY for the Lesson

The table shows the braking distance for a vehicle at certain speeds. If s represents the speed in miles per hour, then the approximate number of feet that the driver must apply the brakes is $\frac{1}{20}s^2$. Notice that when speed is doubled, the braking distance is quadrupled.

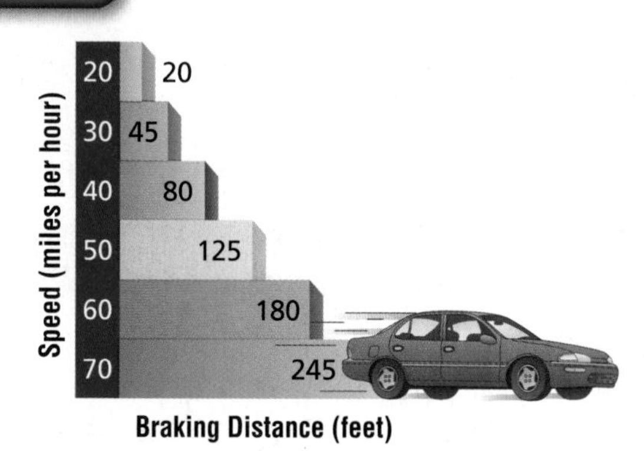

Source: *British Highway Code*

Multiply Monomials A **monomial** is a number, a variable, or a product of a number and one or more variables like $\frac{1}{20}s^2$. An expression like $\frac{x}{2y}$, which involves the division of variables is not a monomial. Monomials that are real numbers are called **constants.**

EXAMPLE Identify Monomials

1 **Determine whether each expression is a monomial. Explain your reasoning.**

	Expression	Monomial?	Reason
a.	-5	yes	-5 is a real number and an example of a constant.
b.	$p + q$	no	The expression involves the addition, not the product, of two variables.
c.	x	yes	Single variables are monomials.

CHECK Your Progress

1A. $-x + 5$ **1B.** $23abcd^2$ **1C.** $\frac{xyz^3}{2}$ **1D.** $\frac{ab}{c}$

 **Personal Tutor at** algebra1.com

Recall that an expression of the form x^n is called a *power* and represents the product you obtain when x is used as a factor n times. The word *power* is also used to refer to the exponent itself. The number x is the *base*, and the number n is the *exponent*.

$$2^5 = \overbrace{2 \cdot 2 \cdot 2 \cdot 2 \cdot 2}^{\text{5 factors}} \text{ or } 32$$

(exponent: 5; base: 2)

In the following examples, the definition of a power is used to find the products of powers. Look for a pattern in the exponents.

$$2^3 \cdot 2^5 = \overbrace{2 \cdot 2 \cdot 2}^{3 \text{ factors}} \cdot \overbrace{2 \cdot 2 \cdot 2 \cdot 2 \cdot 2}^{5 \text{ factors}} \text{ or } 2^8 \quad (3 + 5 \text{ or } 8 \text{ factors})$$

$$3^2 \cdot 3^4 = \overbrace{3 \cdot 3}^{2 \text{ factors}} \cdot \overbrace{3 \cdot 3 \cdot 3 \cdot 3}^{4 \text{ factors}} \text{ or } 3^6 \quad (2 + 4 \text{ or } 6 \text{ factors})$$

These examples suggest the property for multiplying powers.

KEY CONCEPT — *Product of Powers*

Words To multiply two powers that have the same base, add their exponents.

Symbols For any number a and all integers m and n, $a^m \cdot a^n = a^{m+n}$.

Example $a^4 \cdot a^{12} = a^{4+12}$ or a^{16}

EXAMPLE Product of Powers

2 Simplify each expression.

a. $(5x^7)(x^6)$

$(5x^7)(x^6) = (5)(1)(x^7)(x^6)$ Group the coefficients and the variables.

$= (5 \cdot 1)(x^{7+6})$ Product of Powers

$= 5x^{13}$ Simplify.

Study Tip

Power of 1

A variable with no exponent indicated can be written as a power of 1, for example, $x = x^1$ and $ab = a^1b^1$.

b. $(4ab^6)(-7a^2b^3)$

$(4ab^6)(-7a^2b^3) = (4)(-7)(a \cdot a^2)(b^6 \cdot b^3)$ Group the coefficients and the variables.

$= -28(a^{1+2})(b^{6+3})$ Product of Powers

$= -28a^3b^9$ Simplify.

CHECK Your Progress

2A. $(3y^4)(7y^5)$ **2B.** $(-4rs^2t^3)(-6r^5s^2t^3)$

Powers of Monomials You can also look for a pattern to discover the property for finding the power of a power.

$$(4^2)^5 = \overbrace{(4^2)(4^2)(4^2)(4^2)(4^2)}^{5 \text{ factors}}$$
$$= 4^{2+2+2+2+2}$$
$$= 4^{10}$$

Apply rule for Product of Powers.

$$(z^8)^3 = \overbrace{(z^8)(z^8)(z^8)}^{3 \text{ factors}}$$
$$= z^{8+8+8}$$
$$= z^{24}$$

These examples suggest the property for finding the power of a power.

KEY CONCEPT — *Power of a Power*

Words To find the power of a power, multiply the exponents.

Symbols For any number a and all integers m and n, $(a^m)^n = a^{m \cdot n}$.

Example $(k^5)^9 = k^{5 \cdot 9}$ or k^{45}

EXAMPLE Power of a Power

3 **Simplify $[(3^2)^3]^2$.**

$[(3^2)^3]^2 = (3^{2 \cdot 3})^2$ Power of a Power

$= (3^6)^2$ Simplify.

$= 3^{6 \cdot 2}$ Power of a Power

$= 3^{12}$ or 531,441 Simplify.

CHECK Your Progress

3. Simplify $[(2^2)^2]^4$.

Look for a pattern in these examples.

$(xy)^4 = (xy)(xy)(xy)(xy)$
$= (x \cdot x \cdot x \cdot x)(y \cdot y \cdot y \cdot y)$
$= x^4y^4$

$(6ab)^3 = (6ab)(6ab)(6ab)$
$= (6 \cdot 6 \cdot 6)(a \cdot a \cdot a)(b \cdot b \cdot b)$
$= 6^3a^3b^3$ or $216a^3b^3$

These examples suggest the following property.

Study Tip

Powers of Monomials

Sometimes the rules for the Power of a Power and the Power of a Product are combined into one rule.
$(a^mb^n)^p = a^{mp}b^{np}$

KEY CONCEPT *Power of a Product*

Words To find the power of a product, find the power of each factor and multiply.

Symbols For all numbers a and b and any integer m, $(ab)^m = a^mb^m$.

Example $(-2xy)^3 = (-2)^3x^3y^3$ or $-8x^3y^3$

EXAMPLE Power of a Product

4 **GEOMETRY Express the area of the square as a monomial.**

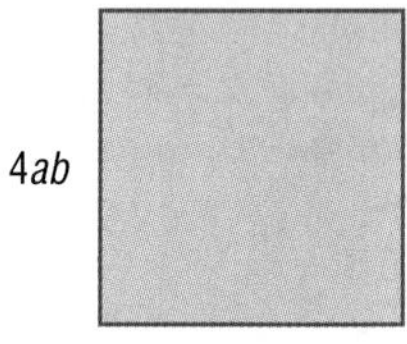

Area $= s^2$ Formula for the area of a square

$= (4ab)^2$ Replace s with $4ab$.

$= 4^2a^2b^2$ Power of a Product

$= 16a^2b^2$ Simplify.

The area of the square is $16a^2b^2$ square units.

CHECK Your Progress

4. Express the area of a square with sides of length $2xy^2$ as a monomial.

CONCEPT SUMMARY *Simplifying Expressions*

To *simplify* an expression involving monomials, write an equivalent expression in which:

- each base appears exactly once,
- there are no powers of powers, and
- all fractions are in simplest form.

EXAMPLE Simplify Expressions

5 **Simplify $(3xy^4)^2[(-2y)^2]^3$.**

$(3xy^4)^2[(-2y)^2]^3 = (3xy^4)^2(-2y)^6$	Power of a Power
$= (3)^2x^2(y^4)^2(-2)^6y^6$	Power of a Product
$= 9x^2y^8(64)y^6$	Power of a Power
$= 9(64)x^2 \cdot y^8 \cdot y^6$	Commutative Property
$= 576x^2y^{14}$	Product of Powers

CHECK Your Progress

5. Simplify $\left(\frac{1}{2}a^2b^2\right)^3[(-4b)^2]^2$.

CHECK Your Understanding

Example 1 (p. 358)

Determine whether each expression is a monomial. Write *yes* or *no*. Explain.

1. $5 - 7d$ **2.** $\frac{4a}{3b}$ **3.** n

Examples 2, 3 (pp. 359–360)

Simplify.

4. $x(x^4)(x^6)$ **5.** $(4a^4b)(9a^2b^3)$ **6.** $[(2^3)^2]^3$

7. $[(3^2)^2]^2$ **8.** $(3y^5z)^2$ **9.** $(-2f^2g)^3$

Example 4 (p. 360)

GEOMETRY **Express the area of each triangle as a monomial.**

10.

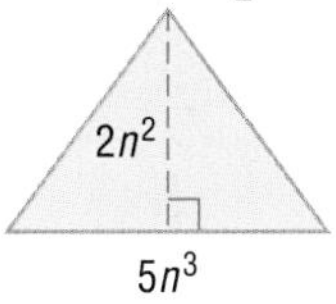

11.

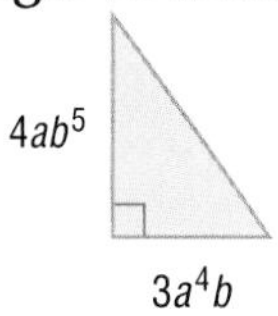

Example 5 (p. 361)

Simplify.

12. $(-2v^3w^4)^3(-3vw^3)^2$ **13.** $(5x^2y)^2(2xy^3z)^3(4xyz)$

Exercises

HOMEWORK HELP

For Exercises	See Examples
14–19	1
20–23	2
24–28	3
29–30	4
31–34	5

Determine whether each expression is a monomial. Write *yes* or *no*. Explain.

14. 12 **15.** $4x^3$ **16.** $a - 2b$

17. $4n + 5m$ **18.** $\frac{x}{y^2}$ **19.** $\frac{1}{5}abc^{14}$

Simplify.

20. $(ab^4)(ab^2)$ **21.** $(p^5q^4)(p^2q)$ **22.** $(-7c^3d^4)(4cd^3)$

23. $(-3j^7k^5)(-8jk^8)$ **24.** $(9pq^7)^2$ **25.** $(7b^3c^6)^3$

26. $[(3^2)^4]^2$ **27.** $[(4^2)^3]^2$ **28.** $[(-2xy^2)^3]^2$

GEOMETRY **Express the area of each figure as a monomial.**

29.

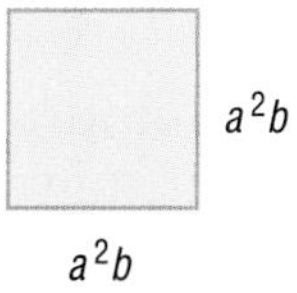

30.

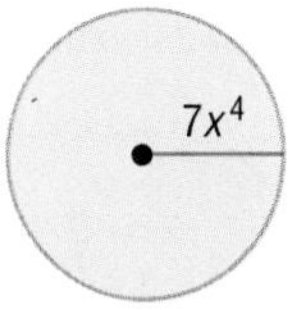

Simplify.

31. $(4cd)^2(-3d^2)^3$

32. $(-2x^5)^3(-5xy^6)^2$

33. $(2ag^2)^4(3a^2g^3)^2$

34. $(2m^2n^3)^3(3m^3n)^4$

35. Simplify the expression $(-2b^3)^4 - 3(-2b^4)^3$.

36. Simplify the expression $2(-5y^3)^2 + (-3y^3)^3$.

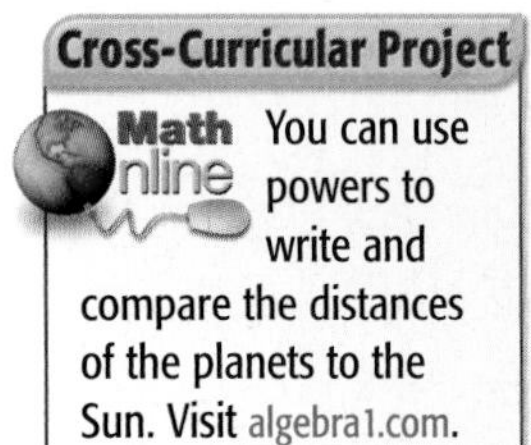

Cross-Curricular Project

Math Online You can use powers to write and compare the distances of the planets to the Sun. Visit algebra1.com.

37. CHEMISTRY Lemon juice is 10^2 times as acidic as tomato juice. Tomato juice is 10^3 times as acidic as egg whites. How many times as acidic is lemon juice as egg whites? Write as a monomial.

38. GEOLOGY The seismic waves of a magnitude 6 earthquake are 10^2 times as great as a magnitude 4 earthquake. The seismic waves of a magnitude 4 earthquake are 10 times as great as a magnitude 3 earthquake. How many times as great are the seismic waves of a magnitude 6 earthquake as those of a magnitude 3 earthquake? Write as a monomial.

Simplify.

39. $(5a^2b^3c^4)(6a^3b^4c^2)$

40. $(10xy^5z^3)(3x^4y^6z^3)$

41. $(0.5x^3)^2$

42. $(0.4h^5)^3$

43. $\left(-\frac{3}{4}c\right)^3$

44. $\left(\frac{4}{5}a^2\right)^2$

45. $(8y^3)(-3x^2y^2)\left(\frac{3}{8}xy^4\right)$

46. $\left(\frac{4}{7}m\right)^2(49m)(17p)\left(\frac{1}{34}p^5\right)$

GEOMETRY **Express the volume of each solid as a monomial.**

47.

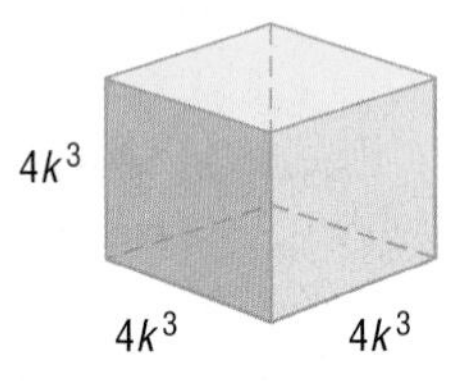

48.

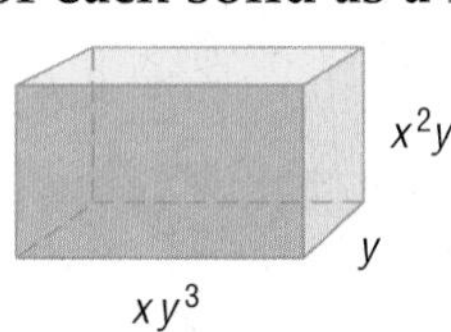

49.

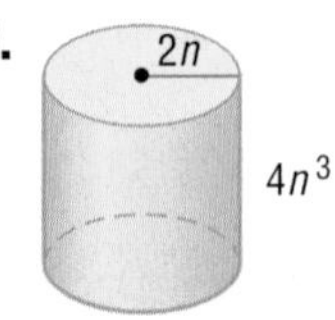

TELEPHONES **For Exercises 50 and 51, use the following information.**
The first transatlantic telephone cable has 51 amplifiers along its length. Each amplifier strengthens the signal on the cable 10^6 times.

50. After it passes through the second amplifier, the signal has been boosted $10^6 \cdot 10^6$ times. Simplify this expression.

51. Represent the number of times the signal has been boosted after it has passed through the first four amplifiers as a power of 10^6. Then simplify the expression.

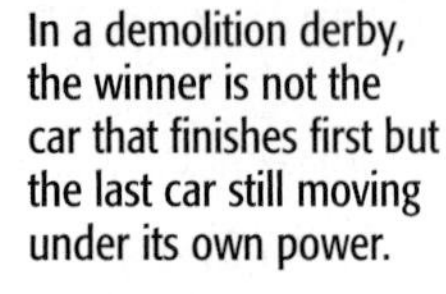

Real-World Link

In a demolition derby, the winner is not the car that finishes first but the last car still moving under its own power.

Source: *Smithsonian Magazine*

DEMOLITION DERBY **For Exercises 52 and 53, use the following information.**
When a car hits an object, the damage is measured by the collision impact. For a certain car, the collision impact I is given by $I = 2s^2$, where s represents the speed in kilometers per minute.

52. What is the collision impact if the speed of the car is 1 kilometer per minute? 2 kilometers per minute? 4 kilometers per minute?

53. As the speed doubles, explain what happens to the collision impact.

TESTING For Exercises 54 and 55, use the following information.

A history test covers two chapters. There are 2^{12} ways to answer the 12 true-false questions on the first chapter and 2^{10} ways to answer the 10 true-false questions on the second chapter.

54. How many ways are there to answer all 22 questions on the test?

55. If a student guesses on each question, what is the probability of answering all questions correctly?

H.O.T. Problems

56. OPEN ENDED Write three different expressions that are equivalent to x^6.

CHALLENGE Determine whether each statement is *true* or *false*. If true, explain your reasoning. If false, give a counterexample.

57. For any real number a, $(-a)^2 = -a^2$.

58. For all real numbers a and b, and all integers m, n, and p, $(a^m b^n)^p = a^{mp} b^{np}$.

59. For all real numbers a, b, and all integers n, $(a + b)^n = a^n + b^n$.

60. FIND THE ERROR Nathan and Poloma are simplifying $(5^2)(5^9)$. Who is correct? Explain your reasoning.

Nathan

$(5^2)(5^9) = (5 \cdot 5)^{2+9}$

$= 25^{11}$

Poloma

$(5^2)(5^9) = 5^{2+9}$

$= 5^{11}$

61. REASONING Compare each pair of monomials. Explain why each pair is or is not equivalent.

a. $5m^2$ and $(5m)^2$

b. $(yz)^4$ and y^4z^4

c. $-3a^2$ and $(-3a)^2$

d. $2(c^7)^3$ and $8c^{21}$

62. ***Writing in Math*** Use the data about braking distances on page 358 to explain why doubling speed quadruples braking distance.

STANDARDIZED TEST PRACTICE

63. The length of a rectangle is three times the width of the rectangle. If the width of the rectangle is y units, what is the area of the rectangle?

A $3y$ units^2

B $3y^2$ units^2

C $y + 3$ units^2

D $3y(y + 3)$ units^2

64. REVIEW The vertices of $\triangle ABC$ have coordinates $A(4, 5)$, $B(1, 3)$, and $C(4, 0)$. What will the coordinates of A' be if the triangle is translated 3 units down and 2 units to the left?

F (1, 3)

G (2, 2)

H (7, 7)

J (8, 0)

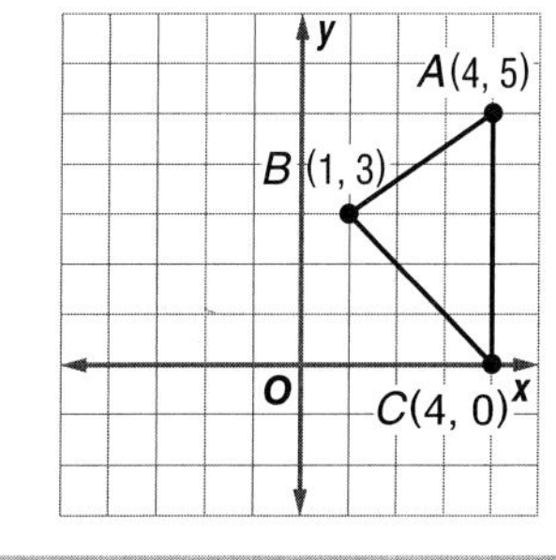

Spiral Review

Solve each system of inequalities by graphing. (Lesson 6-8)

65. $y \leq 2x + 2$
$y \geq -x - 1$

66. $y \geq x - 2$
$y < 2x - 1$

67. $x > -2$
$y < x + 3$

Determine which ordered pairs are part of the solution set for each inequality. (Lesson 6-7)

68. $y \leq 2x$, $\{(1, 4), (-1, 5), (5, -6), (-7, 0)\}$

69. $y < 8 - 3x$, $\{(-4, 2), (-3, 0), (1, 4), (1, 8)\}$

Solve each compound inequality. Then graph the solution set. (Lesson 6-4)

70. $4 + h \leq -3$ or $4 + h \geq 5$

71. $4 < 4a + 12 < 24$

72. $14 < 3h + 2$ and $3h + 2 < 2$

73. $2m - 3 > 7$ or $2m + 7 > 9$

Use elimination to solve each system of equations. (Lesson 5-4)

74. $-4x + 5y = 2$
$x + 2y = 6$

75. $3x + 4y = -25$
$2x - 3y = 6$

76. $x + y = 20$
$4 = 0.4x + 0.15y$

Write an equation in function notation for each relation. (Lesson 3-5)

77.

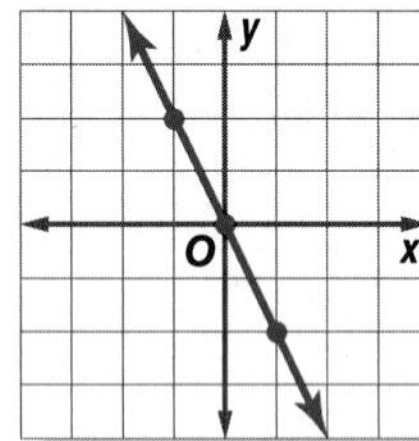

78.

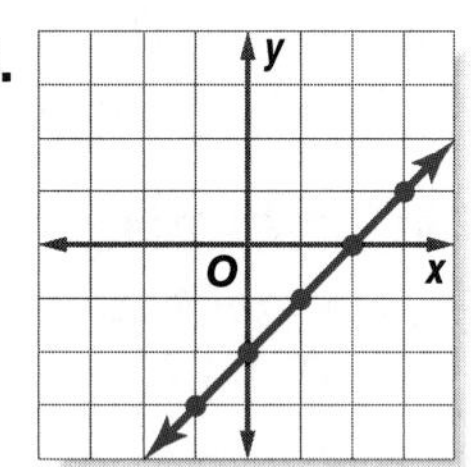

Express the relation shown in each table, mapping, or graph as a set of ordered pairs. Then write the inverse of the relation. (Lesson 3-1)

79.

x	y
−5	2
−2	3
0	5
4	9

80.

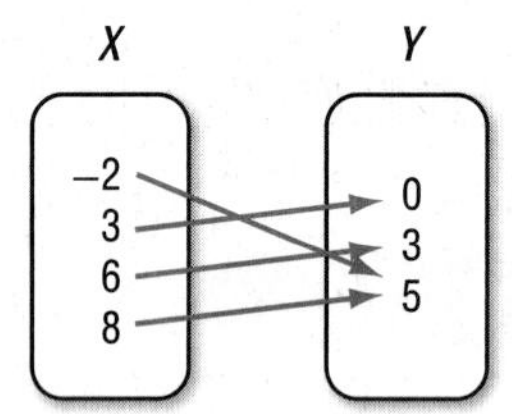

81.

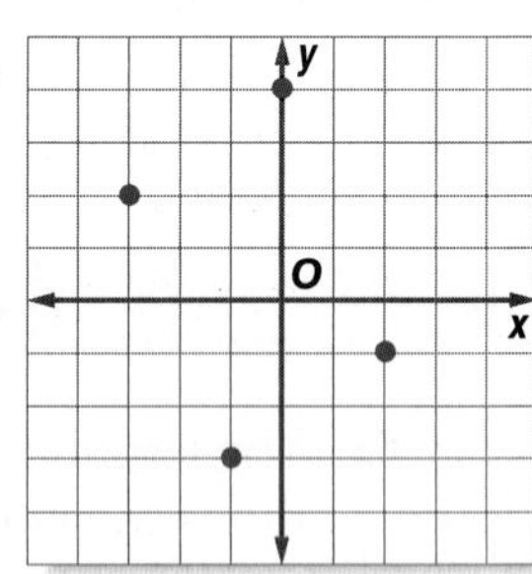

82. TRANSPORTATION Two trains leave York at the same time, one traveling north, the other south. The northbound train travels at 40 miles per hour and the southbound at 30 miles per hour. In how many hours will the trains be 245 miles apart? (Lesson 2-9)

GET READY for the Next Lesson

PREREQUISITE SKILL Simplify. (Pages 694–695)

83. $\frac{2}{6}$

84. $\frac{3}{15}$

85. $\frac{10}{5}$

86. $\frac{27}{9}$

87. $\frac{14}{36}$

88. $\frac{9}{48}$

89. $\frac{44}{32}$

90. $\frac{45}{18}$

Algebra Lab

Investigating Surface Area and Volume

ACTIVITY

- Cut out the pattern shown from a sheet of centimeter grid paper. Fold along the dashed lines and tape the edges together to form a rectangular prism.
- Find the surface area SA of the prism by counting the squares on all the faces of the prism or by using the formula $SA = 2w\ell + 2wh + 2\ell h$, where w is the width, ℓ is the length, and h is the height of the prism.
- Find the volume V of the prism by using the formula $V = \ell wh$.
- Now construct another prism with dimensions that are 2 times each of the dimensions of the first prism, or 4 centimeters by 10 centimeters by 6 centimeters.
- Finally, construct a third prism with dimensions that are 3 times each of the dimensions of the first prism.

Analyze the Results

1. Copy and complete the table using the prisms you made.

Prism	Dimensions	Surface Area (cm^2)	Volume (cm^3)	Surface Area Ratio $\left(\frac{SA \text{ of New}}{SA \text{ of Original}}\right)$	Volume Ratio $\left(\frac{V \text{ of New}}{V \text{ of Original}}\right)$
Original	2 by 5 by 3	62	30	___	___
A	4 by 10 by 6				
B	6 by 15 by 9				

2. **MAKE A CONJECTURE** Suppose you multiply each dimension of a prism by 2. What is the ratio of the surface area of the new prism to the surface area of the original prism? What is the ratio of the volumes?

3. If you multiply each dimension of a prism by 3, what is the ratio of the surface area of the new prism to the surface area of the original? What is the ratio of the volumes?

4. Suppose you multiply each dimension of a prism by a. Make a conjecture about the ratios of surface areas and volumes.

5. Repeat the activity using cylinders. To start, make a cylinder with radius 4 centimeters and height 5 centimeters. To compute surface area SA and volume V, use the formulas $SA = 2\pi r^2 + 2\pi rh$ and $V = \pi r^2 h$, where r is the radius and h is the height of the cylinder. Do the conjectures you made in Exercise 4 hold true for cylinders? Explain.

7-2 Dividing Monomials

Main Ideas

- Simplify expressions involving the quotient of monomials.
- Simplify expressions containing negative exponents.

New Vocabulary

zero exponent

negative exponent

GET READY for the Lesson

To test whether a solution is a base or an acid, chemists use a pH test. This test measures the concentration c of hydrogen ions (in moles per liter) in the solution.

$$c = \left(\frac{1}{10}\right)^{pH}$$

The table gives examples of solutions with various pH levels. You can find the quotient of powers and use negative exponents to compare measures on the pH scale.

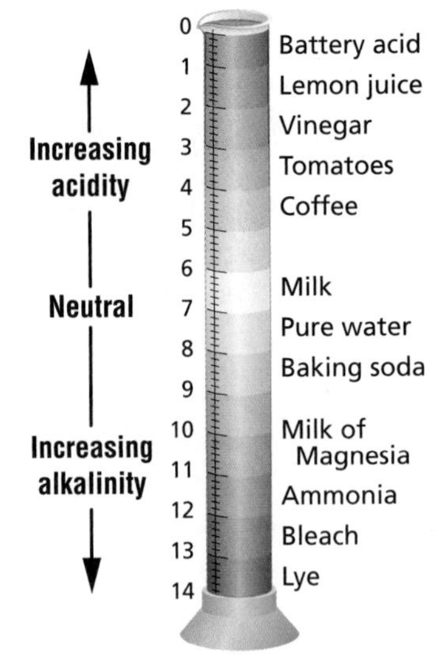

Source: U.S. Geological Survey

Quotients of Monomials Look for a pattern in the examples below.

$$\frac{4^5}{4^3} = \frac{\overbrace{\cancel{4} \cdot \cancel{4} \cdot \cancel{4} \cdot 4 \cdot 4}^{5\text{ factors}}}{\underbrace{\cancel{4} \cdot \cancel{4} \cdot \cancel{4}}_{3\text{ factors}}} = \underbrace{4 \cdot 4}_{5-3\text{ or }2\text{ factors}} \text{ or } 4^2$$

$$\frac{3^6}{3^2} = \frac{\overbrace{\cancel{3} \cdot \cancel{3} \cdot 3 \cdot 3 \cdot 3 \cdot 3}^{6\text{ factors}}}{\underbrace{\cancel{3} \cdot \cancel{3}}_{2\text{ factors}}} = \underbrace{3 \cdot 3 \cdot 3 \cdot 3}_{6-2\text{ or }4\text{ factors}} \text{ or } 3^4$$

KEY CONCEPT — Quotient of Powers

Words To divide two powers with the same base, subtract the exponents.

Symbols For all integers m and n and any nonzero number a, $\frac{a^m}{a^n} = a^{m-n}$.

Example $\frac{b^{15}}{b^7} = b^{15-7}$ or b^8

EXAMPLE Quotient of Powers

1 Simplify $\frac{a^5b^8}{ab^3}$. Assume that no denominator is equal to zero.

$\frac{a^5b^8}{ab^3} = \left(\frac{a^5}{a}\right)\left(\frac{b^8}{b^3}\right)$ Group powers that have the same base.

$= (a^{5-1}), (b^{8-3})$ or a^4b^5 Quotient of Powers

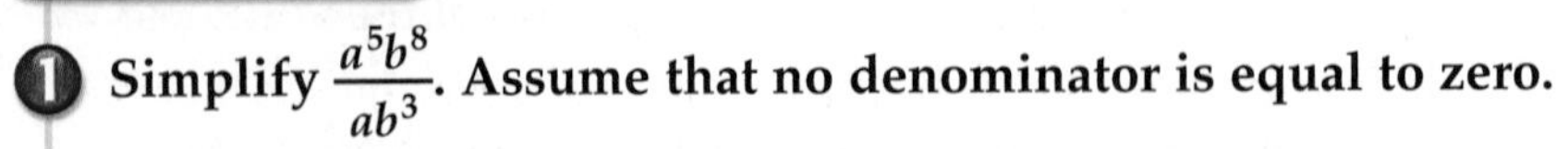

CHECK Your Progress

1. Simplify $\frac{x^3y^4}{x^2y}$. Assume that no denominator is equal to zero.

Concepts in Motion
BrainPop®
algebra1.com

Look for a pattern in the example below.

$$\left(\frac{2}{5}\right)^3 = \underbrace{\left(\frac{2}{5}\right)\left(\frac{2}{5}\right)\left(\frac{2}{5}\right)}_{3\text{ factors}} = \frac{\overbrace{2 \cdot 2 \cdot 2}^{3\text{ factors}}}{\underbrace{5 \cdot 5 \cdot 5}_{3\text{ factors}}} \text{ or } \frac{2^3}{5^3}$$

KEY CONCEPT *Power of a Quotient*

Words To find the power of a quotient, find the power of the numerator and the power of the denominator.

Symbols For any integer m and any real numbers a and b, $b \neq 0$, $\left(\frac{a}{b}\right)^m = \frac{a^m}{b^m}$.

EXAMPLE Power of a Quotient

2 **Simplify $\left(\frac{2p^2}{3}\right)^4$.**

$\left(\frac{2p^2}{3}\right)^4 = \frac{(2p^2)^4}{3^4}$ Power of a Quotient

$= \frac{2^4(p^2)^4}{3^4}$ Power of a Product

$= \frac{16p^8}{81}$ Power of a Power

CHECK Your Progress **Simplify each expression.**

2A. $\left(\frac{3x^4}{4}\right)^3$

2B. $\left(\frac{5x^5y}{6}\right)^2$

Negative Exponents A graphing calculator can be used to investigate expressions with 0 as an exponent and negative exponents.

Study Tip

Graphing Calculator

To express a value as a fraction, press MATH ENTER ENTER.

GRAPHING CALCULATOR LAB

Zero Exponent and Negative Exponents

Use the ^ key to evaluate expressions with exponents.

1. Copy and complete the table.

Power	2^4	2^3	2^2	2^1	2^0	2^{-1}	2^{-2}	2^{-3}	2^{-4}
Value									

2. Describe the relationship between each pair of values.

a. 2^4 and 2^{-4} **b.** 2^3 and 2^{-3} **c.** 2^2 and 2^{-2} **d.** 2^1 and 2^{-1}

3. **Make a conjecture** as to the fractional value of 5^{-1}. Verify your conjecture using a calculator.

4. What is the value of 5^0?

5. What happens when you evaluate 0^0?

Study Tip

Alternative Method

Another way to look at the problem of simplifying $\frac{2^4}{2^4}$ is to recall that any nonzero number divided by itself is 1: $\frac{2^4}{2^4} = \frac{16}{16}$ or 1.

To understand why a calculator gives a value of 1 for 2^0, study the two methods used to simplify $\frac{2^4}{2^4}$.

Method 1

$\frac{2^4}{2^4} = 2^{4-4}$ Quotient of Powers

$= 2^0$ Subtract.

Method 2

$\frac{2^4}{2^4} = \frac{\overset{1}{\cancel{2}} \cdot \overset{1}{\cancel{2}} \cdot \overset{1}{\cancel{2}} \cdot \overset{1}{\cancel{2}}}{\underset{1}{\cancel{2}} \cdot \underset{1}{\cancel{2}} \cdot \underset{1}{\cancel{2}} \cdot \underset{1}{\cancel{2}}}$ Definition of powers

$= 1$ Simplify.

Since $\frac{2^4}{2^4}$ cannot have two different values, we can conclude that $2^0 = 1$.

KEY CONCEPT *Zero Exponent*

Words Any nonzero number raised to the zero power is 1.

Symbols For any nonzero number a, $a^0 = 1$.

Example $(-0.25)^0 = 1$

EXAMPLE Zero Exponent

3 **Simplify each expression. Assume that no denominator is equal to zero.**

a. $\left(-\frac{3x^5y}{8xy^7}\right)^0$

$\left(-\frac{3x^5y}{8xy^7}\right)^0 = 1$ $a^0 = 1$

b. $\frac{t^3s^0}{t}$

$\frac{t^3s^0}{t} = \frac{t^3(1)}{t}$ $a^0 = 1$

$= \frac{t^3}{t}$ Simplify.

$= t^2$ Quotient of Powers

CHECK Your Progress

3A. $\frac{x^0y^4}{y^2}$

3B. $\left(\frac{2x^3y^2z^5}{10xy^3z^4}\right)^0$

To investigate the meaning of a negative exponent, we can simplify expressions like $\frac{8^2}{8^5}$ in two ways.

Method 1

$\frac{8^2}{8^5} = 8^{2-5}$ Quotient of Powers

$= 8^{-3}$ Subtract.

Method 2

$\frac{8^2}{8^5} = \frac{\overset{1}{\cancel{8}} \cdot \overset{1}{\cancel{8}}}{\underset{1}{\cancel{8}} \cdot \underset{1}{\cancel{8}} \cdot 8 \cdot 8 \cdot 8}$ Definition of powers

$= \frac{1}{8^3}$ Simplify.

Since $\frac{8^2}{8^5}$ cannot have two different values, we can conclude that $8^{-3} = \frac{1}{8^3}$.

KEY CONCEPT — *Negative Exponent*

Words For any nonzero number a and any integer n, a^{-n} is the reciprocal of a^n. In addition, the reciprocal of a^{-n} is a^n.

Symbols For any nonzero number a and any integer n, $a^{-n} = \frac{1}{a^n}$ and $\frac{1}{a^{-n}} = a^n$.

Examples $5^{-2} = \frac{1}{5^2}$ or $\frac{1}{25}$ $\qquad \frac{1}{m^{-3}} = m^3$

Study Tip

Common Misconception

Do not confuse a negative number with a number raised to a negative power.

$3^{-1} = \frac{1}{3} \quad -3 \neq \frac{1}{3}$

An expression is simplified when it contains only positive exponents.

EXAMPLE Negative Exponents

4 **Simplify each expression. Assume that no denominator is equal to zero.**

a. $\dfrac{b^{-3}c^2}{d^{-5}}$

$$\frac{b^{-3}c^{c}}{d^{-5}} = \left(\frac{b^{-3}}{1}\right)\left(\frac{c^2}{1}\right)\left(\frac{1}{d^{-5}}\right) \quad \text{Write as a product of fractions.}$$

$$= \left(\frac{1}{b^3}\right)\left(\frac{c^2}{1}\right)\left(\frac{d^5}{1}\right) \quad a^{-n} = \frac{1}{a^n}$$

$$= \frac{c^2d^5}{b^3} \quad \text{Multiply fractions.}$$

b. $\dfrac{-3a^{-4}b^7}{21a^2b^7c^{-5}}$

$$\frac{-3a^{-4}b^7}{21a^2b^7c^{-5}} = \left(\frac{-3}{21}\right)\left(\frac{a^{-4}}{a^2}\right)\left(\frac{b^7}{b^7}\right)\left(\frac{1}{c^{-5}}\right) \quad \text{Group powers with the same base.}$$

$$= \frac{-1}{7}\left(a^{-4-2}\right)\left(b^{7-7}\right)\left(c^5\right) \quad \text{Quotient of Powers and Negative Exponent Properties}$$

$$= \frac{-1}{7}a^{-6}b^0c^5 \quad \text{Simplify.}$$

$$= \frac{-1}{7}\left(\frac{1}{a^6}\right)(1)c^5 \quad \text{Negative Exponent and Zero Exponent Properties}$$

$$= -\frac{c^5}{7a^6} \quad \text{Multiply fractions.}$$

c. $\dfrac{-3q^{-2}rs^4}{-12qr^{-3}s^{-5}}$

$$\frac{-3q^{-2}rs^4}{-12qr^{-3}s^{-5}} = \left(\frac{-3}{-12}\right)\left(\frac{q^{-2}}{q}\right)\left(\frac{r}{r^{-3}}\right)\left(\frac{s^4}{s^{-5}}\right) \quad \text{Group powers with the same base.}$$

$$= \frac{1}{4}q^{-3}r^4s^9 \quad \text{Simplify.}$$

$$= \frac{r^4s^9}{4q^3} \quad \text{Negative Exponent Property}$$

CHECK Your Progress

4A. $\dfrac{r^{-5}s^4}{t^{-3}}$

4B. $\dfrac{24x^{-2}y^4}{-6x^{-3}y^{-2}z^{-1}}$

Test-Taking Tip

Some problems can be solved using estimation. The area of the circle is less than the area of the square. Therefore, the ratio of the two areas must be less than 1. Use 3 as an approximate value for π to determine which of the choices is less than 1.

A STANDARDIZED TEST EXAMPLE Apply Properties of Exponents

5 **Write the ratio of the area of the circle to the area of the square in simplest form.**

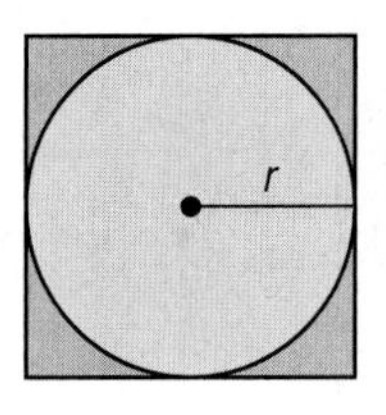

A $\frac{\pi}{2}$ C $\frac{2\pi}{1}$

B $\frac{\pi}{4}$ D $\frac{\pi}{3}$

Read the Test Item

A ratio is a comparison of two quantities. It can be written in fraction form.

Solve the Test Item

- area of circle: πr^2
  length of square: diameter of circle or $2r$
  area of square: $(2r)^2$

- $\frac{\text{area of circle}}{\text{area of square}} = \frac{\pi r^2}{(2r)^2}$ Substitute.

  $= \frac{\pi}{4} r^{2-2}$ Quotient of Powers

  $= \frac{\pi}{4} r^0$ or $\frac{\pi}{4}$ $r^0 = 1$

The answer is B.

CHECK Your Progress

5. Write the ratio of the area of the circle to the area of the square in simplest form.

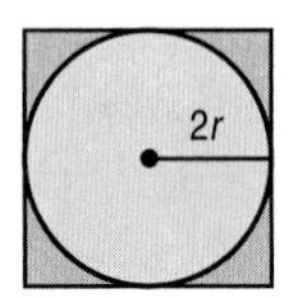

F $\frac{\pi}{3}$ G $\frac{\pi}{2}$ H $\frac{\pi}{4}$ J $\frac{3\pi}{2}$

Online Personal Tutor at algebra1.com

CHECK Your Understanding

Simplify. Assume that no denominator is equal to zero.

Example 1 (p. 367)

1. $\frac{7^8}{7^2}$ **2.** $\frac{x^8y^{12}}{x^2y^7}$ **3.** $\frac{5pq^7}{10p^6q^3}$

Example 2 (p. 367)

4. $\left(\frac{2c^3d}{7z^2}\right)^3$ **5.** $\left(\frac{4a^2b}{2c^3}\right)^2$ **6.** $\left(\frac{3mn^3}{6n^2}\right)^2$

Example 3 (p. 369)

7. $y^0(y^5)(y^{-9})$ **8.** $\frac{(4m^{-3}n^5)^0}{mn}$ **9.** $\frac{(3x^2y^5)^0}{(21x^5y^2)^0}$

Example 4 (p. 370)

10. 13^{-2} **11.** $\frac{c^{-5}}{d^3g^{-8}}$ **12.** $\frac{(cd^{-2})^3}{(c^4d^9)^{-2}}$

Example 5 (p. 370)

13. STANDARDIZED TEST PRACTICE Find the ratio of the volume of the cylinder to the volume of the sphere.

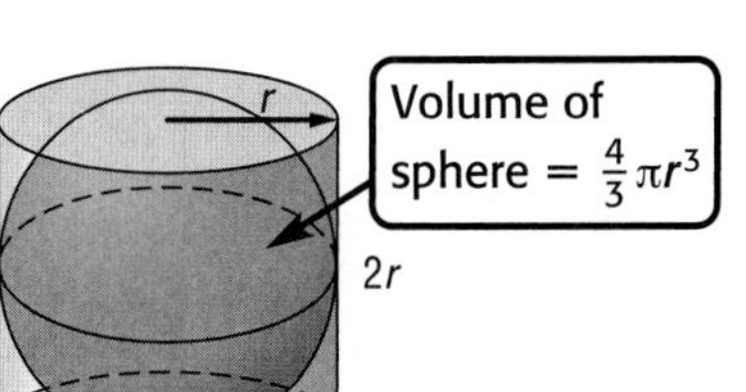

Volume of cylinder = $\pi r^2 h$

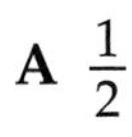

A $\frac{1}{2}$ C $\frac{4}{3}$

B $\frac{3}{4}$ D $\frac{3}{2}$

Exercises

HOMEWORK HELP	
For Exercises	See Examples
14–17	1
18–19	2
20–21	3
22–28	4
29–30	5

Simplify. Assume that no denominator is equal to zero.

14. $\frac{4^{12}}{4^2}$ **15.** $\frac{3^{13}}{3^7}$ **16.** $\frac{p^7n^3}{p^4n^2}$

17. $\frac{y^3z^9}{yz^2}$ **18.** $\left(\frac{5b^4n}{2a^6}\right)^2$ **19.** $\left(\frac{3m^7}{4x^5y^3}\right)^4$

20. $\left(\frac{r^{-2}t^5}{t^{-1}}\right)^0$ **21.** $\left(\frac{4c^{-2}d}{b^{-2}c^3d^{-1}}\right)^0$ **22.** 6^{-2}

23. 5^{-3} **24.** $\left(\frac{4}{5}\right)^{-2}$ **25.** $\left(\frac{3}{2}\right)^{-3}$

26. $n^2(p^{-4})(n^{-5})$ **27.** $\frac{28a^7c^{-4}}{7a^3b^0c^{-8}}$ **28.** $x^3y^0x^{-7}$

29. The area of the rectangle is $24x^5y^3$ square units. Find the length of the rectangle.

30. The area of the triangle is $100a^3b$ square units. Find the height of the triangle.

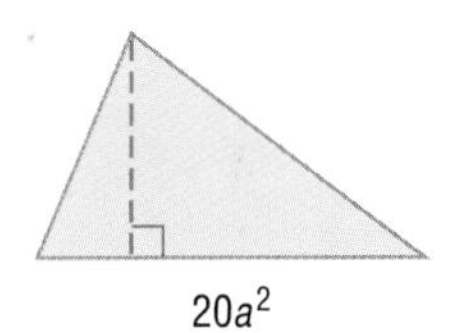

Simplify. Assume that no denominator is equal to zero.

31. $\frac{-2a^3}{10a^8}$ **32.** $\frac{15b}{45b^5}$ **33.** $\frac{30h^{-2}k^{14}}{5hk^{-3}}$

34. $\frac{18x^3y^4z^7}{-2x^2yz}$ **35.** $\frac{-19y^0z^4}{-3z^{16}}$ **36.** $\frac{(5r^{-2})^{-2}}{(2r^3)^2}$

37. $\frac{p^{-4}q^{-3}}{(p^5q^2)^{-1}}$ **38.** $\left(\frac{r^{-2}t^5}{t^{-1}}\right)^0$ **39.** $\left(\frac{5b^{-2}n^4}{n^2z^{-3}}\right)^{-1}$

PROBABILITY For Exercises 40 and 41, use the following information.

If you toss a coin, the probability of getting heads is $\frac{1}{2}$. If you toss a coin 2 times, the probability of getting heads each time is $\frac{1}{2} \cdot \frac{1}{2}$ or $\left(\frac{1}{2}\right)^2$.

40. Write an expression to represent the probability of tossing a coin n times and getting n heads.

41. Express your answer to Exercise 40 as a power of 2.

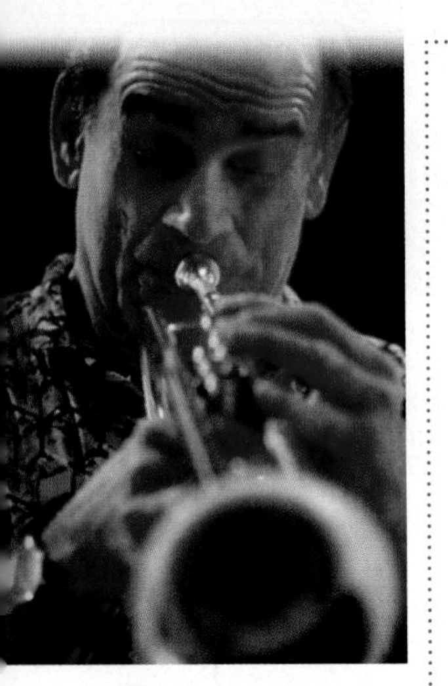

Real-World Link

Timbre is the quality of the sound produced by a musical instrument. Sound quality is what distinguishes the sound of a note played on a flute from the sound of the same note played on a trumpet with the same frequency and intensity.

Source: www.school.discovery.com

SOUND **For Exercises 42–44, use the following information.**
The intensity of sound can be measured in watts per square meter. The table gives the watts per square meter for some common sounds.

Watts per Square Meter	Common Sounds
10^2	jet plane (30 m away)
10^1	pain level
10^0	amplified music (2 m away)
10^{-2}	noisy kitchen
10^{-3}	heavy traffic
10^{-6}	normal conversation
10^{-7}	average home
10^{-9}	soft whisper
10^{-12}	barely audible

42. How many times more intense is the sound from heavy traffic than the sound from normal conversation?

43. What sound is 10,000 times as loud as a noisy kitchen?

44. How does the intensity of a whisper compare to that of normal conversation?

LIGHT **For Exercises 45 and 46, use the table at the right.**

45. Express the range of the wavelengths of visible light using positive exponents. Then evaluate each expression.

46. Express the range of the wavelengths of X rays using positive exponents. Then evaluate each expression.

Spectrum of Electromagnetic Radiation

Region	Wavelength (om)
Radio	greater than 10
Microwave	10^1 to 10^{-2}
Infrared	10^{-2} to 10^{-5}
Visible	10^{-5} to 10^{-4}
Ultraviolet	10^{-4} to 10^{-7}
X rays	10^{-7} to 10^{-9}
Gamma Rays	less than 10^{-9}

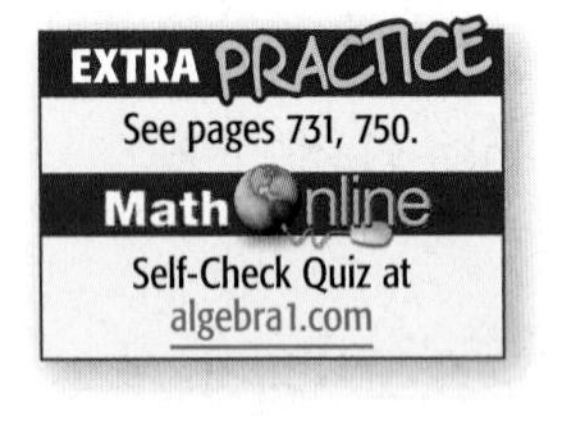

47. COMPUTERS In 1993, the processing speed of a desktop computer was about 10^8 instructions per second. By 2004, it had increased to 10^{10} instructions per second. How many times faster is the newer computer?

H.O.T. Problems

48. OPEN ENDED Name two monomials whose product is $54x^2y^3$.

49. ALTERNATIVE METHODS Describe a method of simplifying $\frac{a^3b^5}{ab^2}$ using negative exponents instead of the Quotient of Powers Property.

CHALLENGE **Simplify. Assume that no denominator equals zero.**

50. $a^n(a^3)$ **51.** $(5^{4x-3})(5^{2x+1})$ **52.** $\frac{c^{x+7}}{c^{x-4}}$

53. REASONING Write a convincing argument to show why $3^0 = 1$ using the following pattern: $3^5 = 243, 3^4 = 81, 3^3 = 27, 3^2 = 9$.

54. FIND THE ERROR Jamal and Angelina are simplifying $\frac{-4x^3}{x^5}$. Who is correct? Explain your reasoning.

Jamal

$\frac{-4x^3}{x^5} = -4x^{3-5}$

$= -4x^{-2}$

$= \frac{-4}{x^2}$

Angelina

$\frac{-4x^3}{x^5} = \frac{x^{3-5}}{4}$

$= \frac{x^{-2}}{4}$

$= \frac{1}{4x^2}$

55. *Writing in Math* Use the information about pH levels on page 366 to explain how you can use the properties of exponents to compare measures on the pH scale. Demonstrate an example comparing two pH levels using the properties of exponents.

STANDARDIZED TEST PRACTICE

56. How many times greater is the volume of the larger cube than the volume of the smaller cube?

A 2

B 4

C 8

D 16

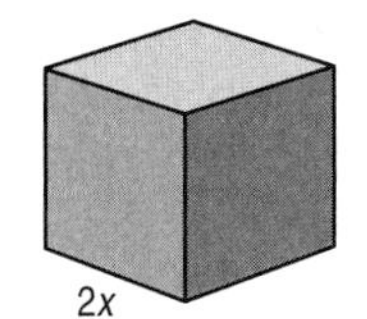

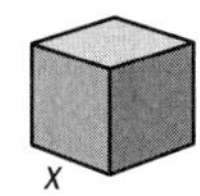

57. REVIEW $\triangle QRS$ is similar to $\triangle TUV$. What is the length of $\overline{UV}$?

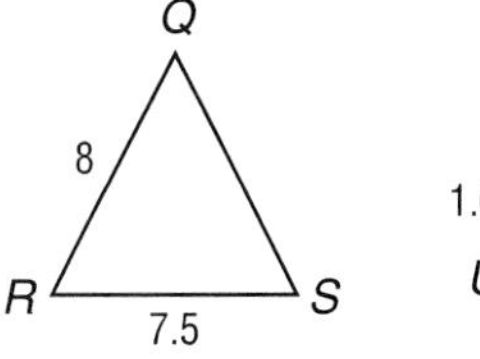

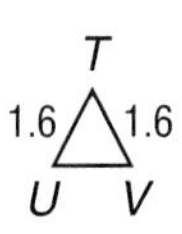

F 11.0 **G** 2.3 **H** 1.7 **J** 1.5

Spiral Review

Simplify. (Lesson 7-1)

58. $(m^3n)(mn^2)$

59. $(3x^4y^3)(4x^4y)$

60. $(a^3x^2)^4$

61. $(3cd^5)^2$

62. $[(2^3)^2]^2$

63. $(-3ab)^3(2b^3)^2$

NUTRITION For Exercises 64 and 65, use the following information. Between the ages of 11 and 18, you should get at least 1200 milligrams of calcium each day. One ounce of mozzarella cheese has 147 milligrams of calcium, and one ounce of Swiss cheese has 219 milligrams. Suppose you want to eat no more than 8 ounces of cheese. (Lesson 6-8)

64. Draw a graph showing the possible amounts of each type of cheese you can eat and still get your daily requirement of calcium. Let x be the amount of mozzarella cheese and y be the amount of Swiss cheese.

65. List three possible solutions.

GET READY for the Next Lesson

PREREQUISITE SKILL Evaluate each expression when $a = 5$, $b = -2$, and $c = 3$. (Lesson 1-2)

66. $5b^2$

67. $b^3 + 3ac$

68. $-2b^4 - 5b^3 - b$

READING MATH

Mathematical Prefixes and Everyday Prefixes

You may have noticed that many prefixes used in mathematics are also used in everyday language. You can use the everyday meaning of these prefixes to better understand their mathematical meaning. The table shows four mathematical prefixes along with their meaning and an example of an everyday word using that prefix.

Prefix	Everyday Meaning	Example
mono-	**1.** one; single; alone	**monologue** A continuous series of jokes or comic stories delivered by one comedian.
bi-	**1.** two **2.** both **3.** both sides, parts, or directions	**bicycle** A vehicle with two wheels behind one another.
tri-	**1.** three **2.** occurring at intervals of three **3.** occurring three times during	**trilogy** A group of three dramatic or literary works related in subject or theme.
poly-	**1.** more than one; many; much	**polygon** A closed plane figure bounded by three or more line segments.

Source: *The American Heritage Dictionary of the English Language*

You can use your everyday understanding of prefixes to help you understand mathematical terms that use those prefixes.

Reading to Learn

1. Give an example of a geometry term that uses one of these prefixes. Then define that term.

2. **MAKE A CONJECTURE** Given your knowledge of the meaning of the word monomial, make a conjecture as to the meaning of each of the following mathematical terms.

a. binomial **b.** trinomial **c.** polynomial

3. Research the following prefixes and their meanings.

a. semi- **b.** hexa- **c.** octa-

d. penta- **e.** tri- **f.** quad-

Algebra Lab
Polynomials

Algebra tiles can be used to model polynomials. A polynomial is a monomial or the sum of monomials. The diagram at the right shows the models.

Polynomial Models	
Polynomials are modeled using three types of tiles.	1 x x^2
Each tile has an opposite.	−1 $-x$ $-x^2$

ACTIVITY

Use algebra tiles to model each polynomial.

- $4x$

  To model this polynomial, you will need 4 green x-tiles.

- $2x^2 - 3$

  To model this polynomial, you will need 2 blue x^2-tiles and 3 red −1-tiles.

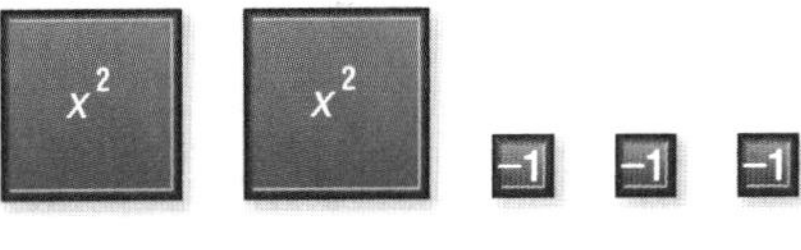

- $-x^2 + 3x + 2$

  To model this polynomial, you will need 1 red $-x^2$-tile, 3 green x-tiles, and 2 yellow 1-tiles.

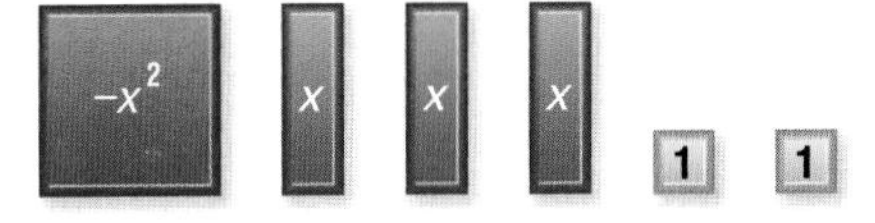

Model and Analyze

Use algebra tiles to model each polynomial. Then draw a diagram of your model.

1. $-2x^2$

2. $5x - 4$

3. $3x^2 - x$

4. $x^2 + 4x + 3$

Write an algebraic expression for each model.

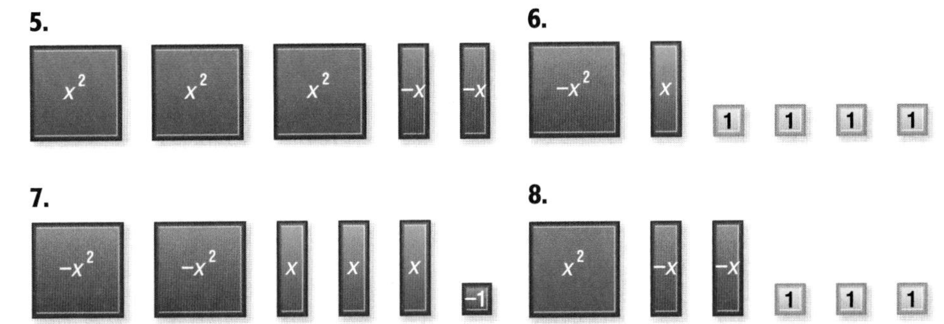

9. MAKE A CONJECTURE Write a sentence or two explaining why algebra tiles are sometimes called *area tiles.*

7-3 Polynomials

Main Ideas

- Find the degree of a polynomial.
- Arrange the terms of a polynomial in ascending or descending order.

New Vocabulary

polynomial
binomial
trinomial
degree of a monomial
degree of a polynomial

GET READY for the Lesson

The number of hours H spent per person per year playing video games from 2000 through 2005 is shown in the table. These data can be modeled by the equation

$$H = \frac{1}{4}(t^4 - 9t^3 + 24t^2 + 19t + 280),$$

where t is the number of years since 2000. The expression $t^4 - 9t^3 + 24t^2 + 19t + 280$ is an example of a polynomial.

Video Game Usage

Year	Hours per Year
2000	70
2001	79
2002	90
2003	97
2004	103
2005	115

Source: U.S. Census Bureau

Degree of a Polynomial A **polynomial** is a monomial or a sum of monomials. Some polynomials have special names. A **binomial** is the sum of *two* monomials, and a **trinomial** is the sum of *three* monomials.

Monomial	Binomial	Trinomial
7	$3 + 4y$	$x + y + z$
$4ab^3c^2$	$7pqr + pq^2$	$3v^2 - 2w + ab^3$

EXAMPLE Identify Polynomials

1 State whether each expression is a polynomial. If it is a polynomial, identify it as a *monomial, binomial,* or *trinomial*.

	Expression	Polynomial?	Monomial, Binomial, or Trinomial?
a.	$2x - 3yz$	Yes, $2x - 3yz = 2x + (-3yz)$, the sum of two monomials.	binomial
b.	$8n^3 + 5n^{-2}$	No. $5n^{-2} = \frac{5}{n^2}$, which is not a monomial.	none of these
c.	-8	Yes, -8 is a real number.	monomial
d.	$4a^2 + 5a + a + 9$	Yes, the expression simplifies to $4a^2 + 6a + 9$, so it is the sum of three monomials.	trinomial

1A. x

1B. $-3y^2 - 2y + 4y - 1$

1C. $5rs + 7tuv$

1D. $10x^{-4} - 8x^3$

EXAMPLE Write a Polynomial

2 **GEOMETRY** **Write a polynomial to represent the area of the shaded region.**

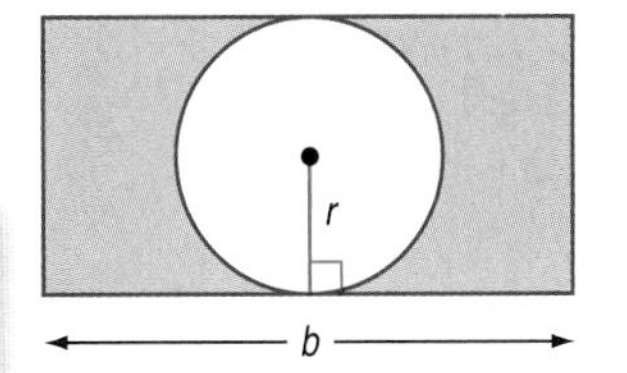

Words	The area of the shaded region is the area of the rectangle minus the area of the circle.
Variables	area of shaded region = A width of rectangle = $2r$ rectangle area = $b(2r)$ circle area = πr^2
	Area of shaded region = rectangle area − circle area.
Equation	$A = b(2r) - \pi r^2$ $A = 2br - \pi r^2$

The polynomial representing the area of the shaded region is $2br - \pi r^2$.

CHECK Your Progress

2. Write a polynomial to represent the area of the shaded region.

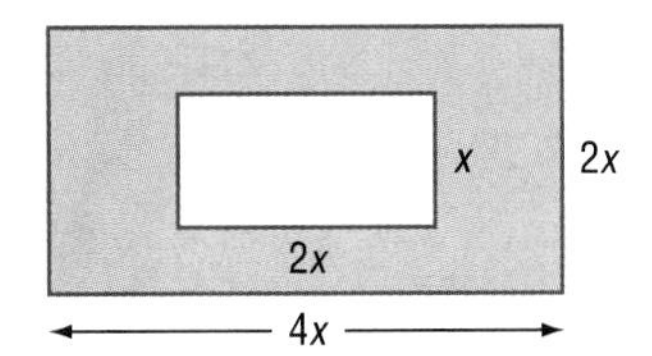

 **Personal Tutor at** algebra1.com

Concepts in Motion
Interactive Lab
algebra1.com

The **degree of a monomial** is the sum of the exponents of all its variables.

The **degree of a polynomial** is the greatest degree of any term in the polynomial. To find the degree of a polynomial, you must find the degree of each term.

Monomial	Degree
$8y^4$	4
$3a$	1
$-2xy^2z^3$	1 + 2 + 3 or 6
7	0

EXAMPLE Degree of a Polynomial

Reading Math

Degrees of 1 and 0

- Since $a = a^1$, the monomial $3a$ can be rewritten as $3a^1$. Thus $3a$ has degree 1.
- Since $x^0 = 1$, the monomial 7 can be rewritten as $7x^0$. Thus 7 has degree 0.

3 **Find the degree of each polynomial.**

	Polynomial	Terms	Degree of Each Term	Degree of Polynomial
a.	$5mn^2$	$5mn^2$	3	3
b.	$-4x^2y^2 + 3x^2 + 5$	$-4x^2y^2, 3x^2, 5$	4, 2, 0	4
c.	$3a + 7ab - 2a^2b + 16$	$3a, 7ab, -2a^2b, 16$	1, 2, 3, 0	3

CHECK Your Progress

3A. $7xy^5z$ **3B.** $12m^3n^2 - 8mn^2 + 3$ **3C.** $2rs - 3rs^2 - 7r^2s^2 - 13$

Write Polynomials in Order The terms of a polynomial are usually arranged so that the powers of one variable are in *ascending* (increasing) order or *descending* (decreasing) order.

EXAMPLE Arrange Polynomials in Ascending Order

Concepts in Motion
BrainPOP®
algebra1.com

4 **Arrange the terms of each polynomial so that the powers of x are in ascending order.**

a. $7x^2 + 2x^4 - 11$

$7x^2 + 2x^4 - 11 = 7x^2 + 2x^4 - 11x^0$ $\quad x^0 = 1$

$= -11 + 7x^2 + 2x^4$ $\quad$ Compare powers of x: $0 < 2 < 4$.

b. $2xy^3 + y^2 + 5x^3 - 3x^2y$

$2xy^3 + y^2 + 5x^3 - 3x^2y$

$= 2x^1y^3 + y^2 + 5x^3 - 3x^2y^1$ $\quad x = x^1$

$= y^2 + 2xy^3 - 3x^2y + 5x^3$ $\quad$ Compare powers of x: $0 < 1 < 2 < 3$.

CHECK Your Progress

4A. $3x^2y^4 + 2x^4y^2 - 4x^3y + x^5 - y^2$ **4B.** $7x^3 - 4xy^4 + 3x^2y^3 - 11x^6y$

EXAMPLE Arrange Polynomials in Descending Order

5 **Arrange the terms of each polynomial so that the powers of x are in descending order.**

a. $6x^2 + 5 - 8x - 2x^3$

$6x^2 + 5 - 8x - 2x^3 = 6x^2 + 5x^0 - 8x^1 - 2x^3$ $\quad x^0 = 1$ and $x = x^1$

$= -2x^3 + 6x^2 - 8x + 5$ $\quad 3 > 2 > 1 > 0$

b. $3a^3x^2 - a^4 + 4ax^5 + 9a^2x$

$3a^3x^2 - a^4 + 4ax^5 + 9a^2x$

$= 3a^3x^2 - a^4x^0 + 4a^1x^5 + 9a^2x^1$ $\quad a = a^1, x^0 = 1$, and $x = x^1$

$= 4ax^5 + 3a^3x^2 + 9a^2x - a^4$ $\quad 5 > 2 > 1 > 0.$

CHECK Your Progress

5A. $4x^2 + 2x^3y + 5 - x$ **5B.** $x + 2x^7y - 5x^4y^8 - x^2y^2 + 3$

CHECK Your Understanding

Example 1 (p. 376)

State whether each expression is a polynomial. If the expression is a polynomial, identify it as a *monomial*, a *binomial*, or a *trinomial*.

1. $5x - 3xy + 2x$ **2.** $\frac{2z}{5}$ **3.** $9a^2 + 7a - 5$

Example 2 (p. 377)

4. GEOMETRY Write a polynomial to represent the area of the shaded region.

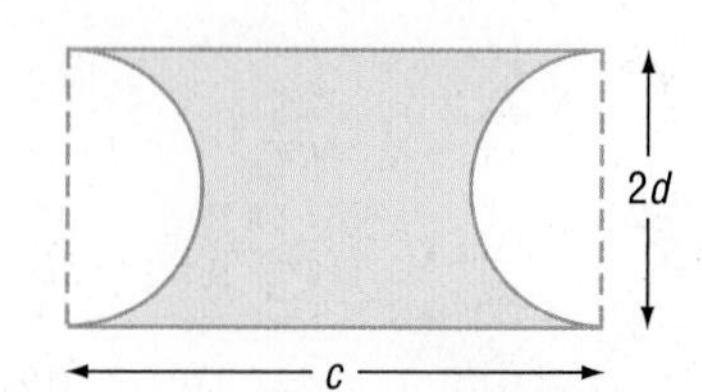

Example 3 (p. 377)

Find the degree of each polynomial.

5. 1 **6.** $3x + 2$ **7.** $2x^2y^3 + 6x^4$

Example 4 (p. 378)

Arrange the terms of each polynomial so that the powers of x are in ascending order.

8. $6x^3 - 12 + 5x$ **9.** $-7a^2x^3 + 4x^2 - 2ax^5 + 2a$

Example 5 (p. 378)

Arrange the terms of each polynomial so that the powers of x are in descending order.

10. $2c^5 + 9cx^2 + 3x$ **11.** $y^3 + x^3 + 3x^2y + 3xy^2$

Exercises

HOMEWORK HELP

For Exercises	See Examples
12–17	1
18–21	2
22–33	3
34–41	4
42–49	5

State whether each expression is a polynomial. If the expression is a polynomial, identify it as a *monomial*, a *binomial*, or a *trinomial*.

12. 14 **13.** $\frac{6m^2}{p} + p^3$ **14.** $7b - 3.2c + 8b$

15. $\frac{1}{3}x^2 + x - 2$ **16.** $6gh^2 - 4g^2h + g$ **17.** $-4 + 2a + \frac{5}{a^2}$

GEOMETRY **Write a polynomial to represent the area of each shaded region.**

18.

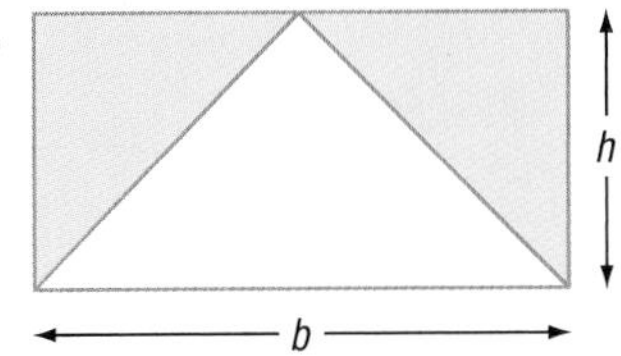

19.

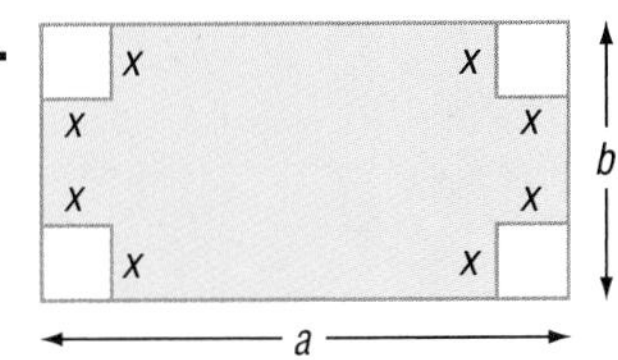

20.

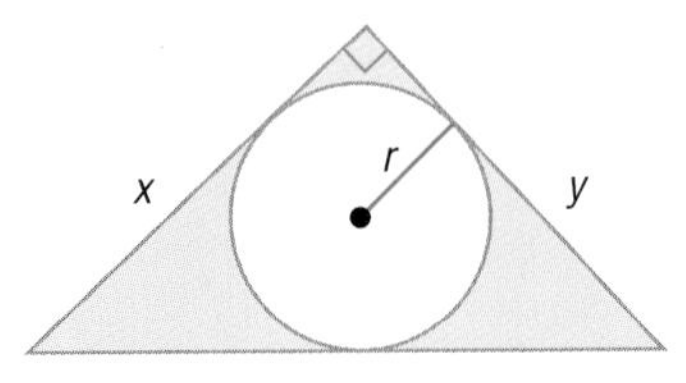

21.

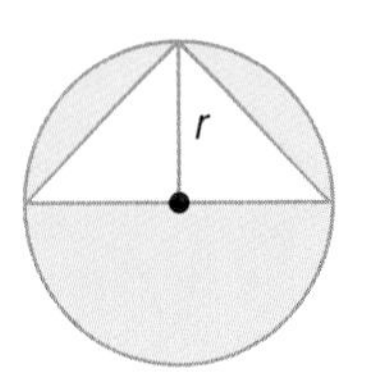

Find the degree of each polynomial.

22. $5x^3$ **23.** $9y$ **24.** $4ab$

25. -13 **26.** $c^4 + 7c^2$ **27.** $6n^3 - n^2p^2$

28. $15 - 8ag$ **29.** $3a^2b^3c^4 - 18a^5c$ **30.** $2x^3 - 4y + 7xy$

31. $3z^5 - 2x^2y^3z - 4x^2z$ **32.** $7 + d^5 - b^2c^2d^3 + b^6$ **33.** $11r^2t^4 - 2s^4t^5 + 24$

Arrange the terms of each polynomial so that the powers of x are in ascending order.

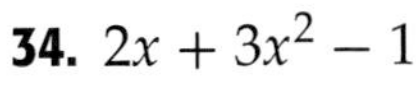

34. $2x + 3x^2 - 1$ **35.** $9x^3 + 7 - 3x^5$

36. $c^2x^3 - c^3x^2 + 8c$ **37.** $x^3 + 4a + 5a^2x^6$

38. $4 + 3ax^5 + 2ax^2 - 5a^7$ **39.** $10x^3y^2 - 3x^9y + 5y^4 + 2x^2$

40. $3xy^2 - 4x^3 + x^2y + 6y$ **41.** $-8a^5x + 2ax^4 - 5 - a^2x^2$

EXTRA PRACTICE
See pages 731, 750.

Math Online
Self-Check Quiz at algebra1.com

Arrange the terms of each polynomial so that the powers of x are in descending order.

42. $5 + x^5 + 3x^3$

43. $2x - 1 + 6x^2$

44. $4a^3x^2 - 5a + 2a^2x^3$

45. $b^2 + x^2 - 2xb$

46. $c^2 + cx^3 - 5c^3x^2 + 11x$

47. $9x^2 + 3 + 4ax^3 - 2a^2x$

48. $8x - 9x^2y + 7y^2 - 2x^4$

49. $4x^3y + 3xy^4 - x^2y^3 + y^4$

50. MONEY Write a polynomial to represent the value of q quarters, d dimes and n nickels.

Real-World Link

From 1980 to 1999, the number of triplet and higher births rose approximately 532% (from 1377 to 7321 births). This steep climb in multiple births coincides with the increased use of fertility drugs.

Source: National Center for Health and Statistics

51. MULTIPLE BIRTHS The rate of quadruplet births Q in the United States in recent years can be modeled by $Q = -0.5t^3 + 11.7t^2 - 21.5t + 218.6$, where t represents the number of years since 1992. For what values of t does this model no longer give realistic data? Explain your reasoning.

PACKAGING For Exercises 52 and 53, use the following information.

A convenience store sells milkshakes in cups with semispherical lids. The volume of a cylinder is the product of π, the square of the radius r, and the height h. The volume of a sphere is the product of $\frac{4}{3}$, π, and the cube of the radius.

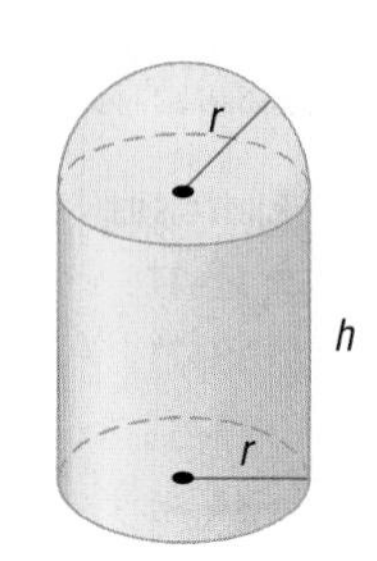

52. Write a polynomial that represents the volume of the container.

53. If the height of the container is 6 inches and the radius is 2 inches, find the volume of the container.

54. Write two polynomials that represent the perimeter and area of the rectangle shown at right.

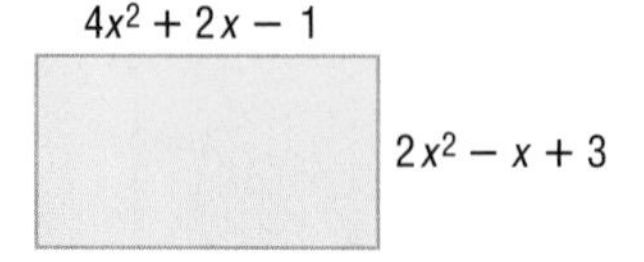

H.O.T. Problems

55. OPEN ENDED Give an example of a monomial of degree zero.

56. REASONING Explain why a polynomial cannot contain a variable term with a negative power.

57. CHALLENGE Tell whether the following statement is *true* or *false*. Explain your reasoning.

The degree of a binomial can never be zero.

58. REASONING Determine whether each statement is *true* or *false*. If false, give a counterexample.

a. All binomials are polynomials.

b. All polynomials are monomials.

c. All monomials are polynomials.

59. ***Writing in Math*** Use the information about video game usage on page 376 to explain how polynomials can be useful in modeling data. Include a discussion of the accuracy of the equation by evaluating the polynomial for $t = \{0, 1, 2, 3, 4, 5\}$ and an example of how and why someone might use this equation.

STANDARDIZED TEST PRACTICE

60. Which expression could be used to represent the area of the shaded region of the rectangle, reduced to simplest terms?

A $3x^2 + 7x$
C $3x^2 - 7x$
B $4x^2 + 8x$
D $4x^2 + 7x$

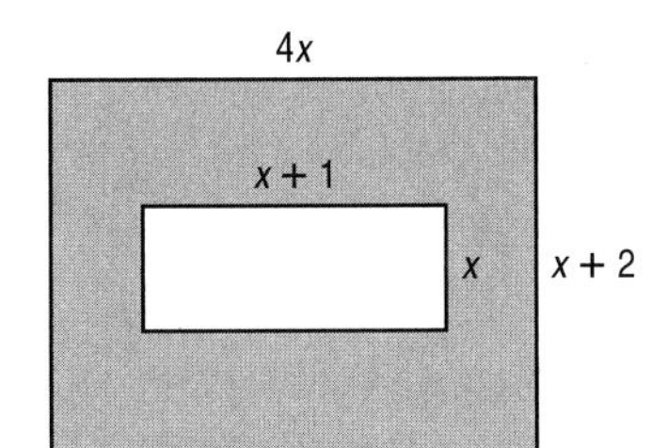

61. REVIEW Lawanda rolled a six-sided game cube 30 times and recorded her results in the table below. Each side of the cube is a different color. Which color has the same experimental probability and theoretical probability?

F Purple
G Red
H White
J Orange

Color	Rolls
Red	4
Blue	8
White	4
Orange	3
Green	6
Purple	5

Spiral Review

Simplify. Assume that no denominator is equal to zero. (Lesson 7-2)

62. $a^0b^{-2}c^{-1}$
63. $\frac{-5n^5}{n^8}$
64. $\left(\frac{4x^3y^2}{3z}\right)^2$
65. $\frac{(-y)^5m^8}{y^3m^{-7}}$

Determine whether each expression is a monomial. Write *yes* or *no*. (Lesson 7-1)

66. $3a + 4b$
67. $\frac{6}{n}$
68. $\frac{v^2}{3}$

Determine whether each relation is a function. (Lesson 3-6)

69.

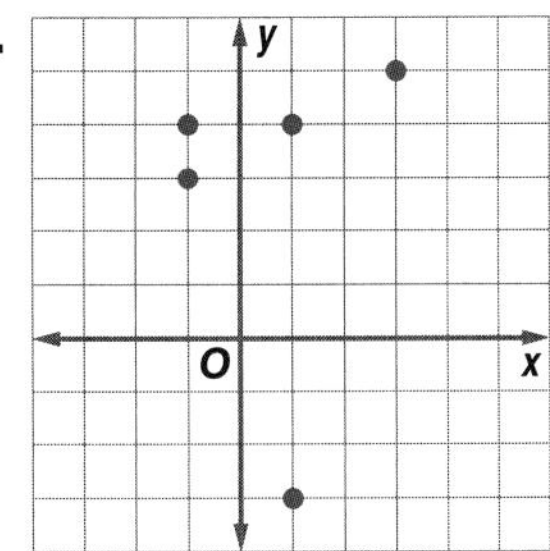

70.

x	y
−2	−2
0	1
3	4
5	−2

71. MAPS The scale of a road map is 1.5 inches = 100 miles. The distance between New Hartford, Connecticut, and Westerly, Rhode Island, by highway on the map is about 1.0 inch. What is the distance between these two cities? (Lesson 2-6)

Find each square root. Round to the nearest hundredth if necessary. (Lesson 1-8)

72. $\pm\sqrt{121}$
73. $\sqrt{3.24}$
74. $-\sqrt{52}$

GET READY for the Next Lesson

PREREQUISITE SKILL Simplify each expression, if possible. If not possible, write *in simplest form*. (Lesson 1-5)

75. $3n + 5n$
76. $9a^2 + 3a - 2a^2$
77. $-3a + 5b + 4a - 7b$
78. $4x + 3y - 6 + 7x + 8 - 10y$

EXPLORE 7-4

Algebra Lab

Adding and Subtracting Polynomials

Monomials such as $5x$ and $-3x$ are called *like terms* because they have the same variable to the same power. When you use algebra tiles, you can recognize like terms because these tiles have the same size and shape as each other.

Polynomial Models	
Like terms are represented by tiles that have the same shape and size.	x x −x (like terms)
A *zero pair* may be formed by pairing one tile with its opposite. You can remove or add zero pairs without changing the value of the polynomial.	x −x → 0

ACTIVITY 1 **Use algebra tiles to find $(3x^2 - 2x + 1) + (x^2 + 4x - 3)$.**

Step 1 Model each polynomial.

$3x^2 - 2x + 1 \longrightarrow$ $3x^2$, $-2x$, 1

$x^2 + 4x - 3 \longrightarrow$ x^2, $4x$, -3

Step 2 Combine like terms and remove zero pairs.

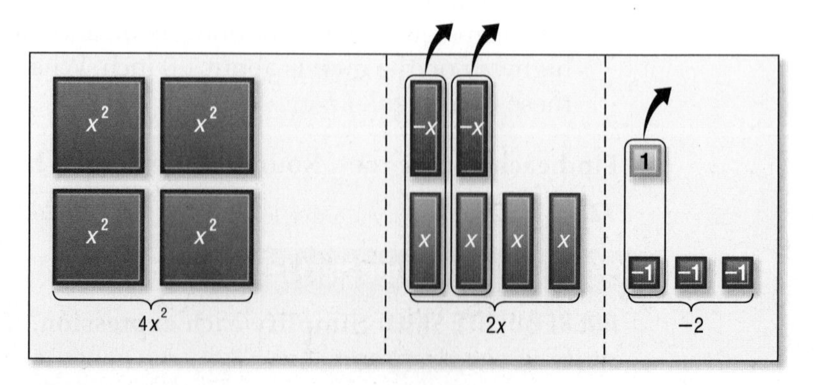

Step 3 Write the polynomial for the tiles that remain.

$(3x^2 - 2x + 1) + (x^2 + 4x - 3) = 4x^2 + 2x - 2$

ACTIVITY 2 **Use algebra tiles to find $(5x + 4) - (-2x + 3)$.**

Step 1 Model the polynomial $5x + 4$.

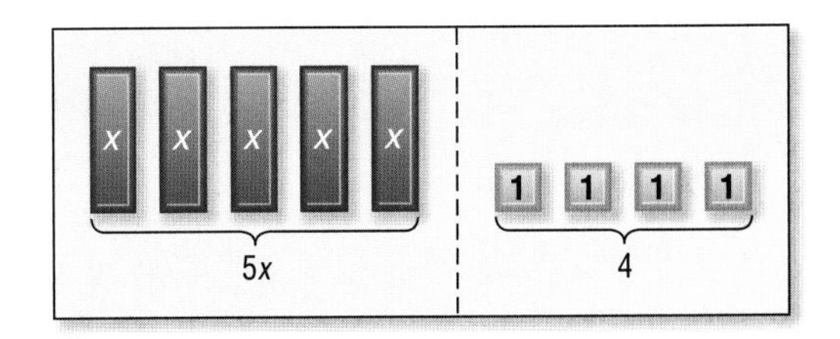

Step 2 To subtract $-2x + 3$, you must remove 2 $-x$-tiles and 3 1-tiles. You can remove the 1-tiles, but there are no $-x$-tiles. Add 2 zero pairs of x-tiles. Then remove the 2 $-x$-tiles.

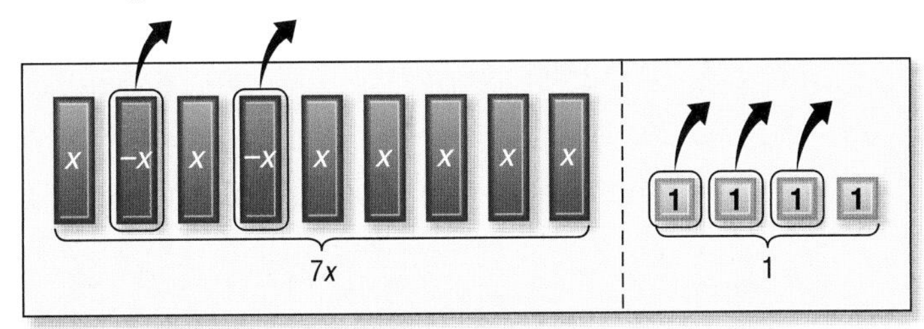

Step 3 The remaining tiles model $7x + 1$.

Recall that you can subtract a number by adding its additive inverse or opposite. Similarly, you can subtract a polynomial by adding its opposite.

ACTIVITY 3 **Use algebra tiles and the additive inverse, or opposite, to find $(5x + 4) - (-2x + 3)$.**

Step 1 To find the difference of $5x + 4$ and $-2x + 3$, add $5x + 4$ and the opposite of $-2x + 3$. The opposite of $-2x + 3$ is $2x - 3$.

Step 2 Write the polynomial for the tiles that remain.
$(5x + 4) - (-2x + 3) = 7x + 1$
Notice that this is the same answer as in Activity 2.

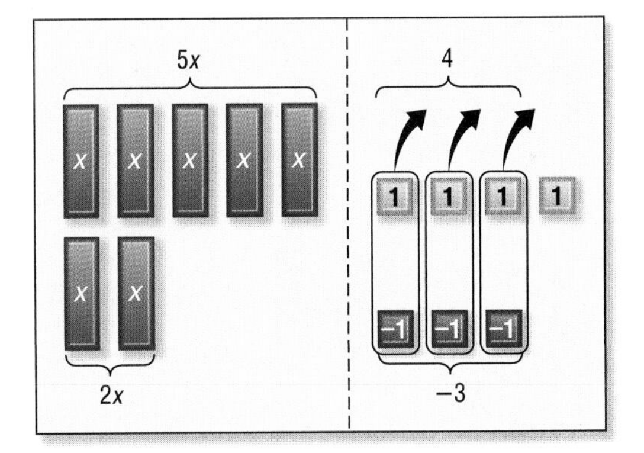

Concepts in Motion
Animation algebra1.com

MODEL AND ANALYZE

Use algebra tiles to find each sum or difference.

1. $(5x^2 + 3x - 4) + (2x^2 - 4x + 1)$

2. $(2x^2 + 5) + (3x^2 + 2x + 6)$

3. $(-4x^2 + x) + (5x - 2)$

4. $(3x^2 + 4x + 2) - (x^2 - 5x - 5)$

5. $(-x^2 + 7x) - (-x^2 + 3x)$

6. $(8x + 4) - (6x^2 + x - 3)$

7. Find $(2x^2 - 3x + 1) - (2x + 3)$ using each method from Activities 2 and 3. Illustrate and explain how zero pairs are used in each case.

7-4 Adding and Subtracting Polynomials

Main Ideas

- Add polynomials.
- Subtract polynomials.

GET READY for the Lesson

From 2000 to 2003, the amount of sales (in millions of dollars) of rap/hip-hop music R and country music C in the United States can be modeled by the following equations, where t is the number of years since 2000.

$R = -132.32t^3 + 624.74t^2 - 773.61t + 1847.67$

$C = -3.42t^3 + 8.6t^2 - 94.95t + 1532.56$

The total music sales T of rap/hip-hop and country music is $R + C$.

Add Polynomials To add polynomials, you can group like terms horizontally or write them in column form, aligning like terms. Adding polynomials involves adding like terms.

Study Tip

Adding Columns

When adding like terms in column form, remember that you are adding integers. Rewrite each monomial to eliminate subtractions. For example, you could rewrite $3x^2 - 4x + 8$ as $3x^2 + (-4x) + 8$.

EXAMPLE Add Polynomials

1 **Find $(3x^2 - 4x + 8) + (2x - 7x^2 - 5)$.**

Method 1 Horizontal

$(3x^2 - 4x + 8) + (2x - 7x^2 - 5)$

$= [3x^2 + (-7x^2)] + (-4x + 2x) + [8 + (-5)]$ Group like terms.

$= -4x^2 - 2x + 3$ Add like terms.

Method 2 Vertical

$$\begin{array}{r} 3x^2 - 4x + 8 \\ (+)\ -7x^2 + 2x - 5 \\ \hline -4x^2 - 2x + 3 \end{array}$$

Notice that terms are in descending order with like terms aligned.

CHECK Your Progress

1. Find $(5x^2 - 3x + 4) + (6x - 3x^2 - 3)$.

Subtract Polynomials Recall that you can subtract a real number by adding its opposite or additive inverse. Similarly, you can subtract a polynomial by adding its additive inverse.

To find the additive inverse of a polynomial, replace each term with its additive inverse.

Polynomial	Additive Inverse
$-5m + 3n$	$5m - 3n$
$2y^2 - 6y + 11$	$-2y^2 + 6y - 11$
$7a + 9b - 4$	$-7a - 9b + 4$

EXAMPLE Subtract Polynomials

2 **Find $(3n^2 + 13n^3 + 5n) - (7n + 4n^3)$.**

Study Tip

Inverse of a Polynomial

When finding the additive inverse of a polynomial, remember to find the additive inverse of *every* term.

Method 1 Horizontal

Subtract $7n + 4n^3$ by adding its additive inverse.

$(3n^2 + 13n^3 + 5n) - (7n + 4n^3)$

$= (3n^2 + 13n^3 + 5n) + (-7n - 4n^3)$ — The additive inverse of $7n + 4n^3$ is $-7n - 4n^3$.

$= 3n^2 + [13n^3 + (-4n^3)] + [5n + (-7n)]$ — Group like terms.

$= 3n^2 + 9n^3 - 2n$ — Combine like terms.

Method 2 Vertical

Align like terms in columns and subtract by adding the additive inverse.

$$\begin{array}{lr} & 3n^2 + 13n^3 + 5n \\ (-) & 4n^3 + 7n \\ \hline \end{array} \quad \text{Add the opposite.} \quad \begin{array}{lr} & 3n^2 + 13n^3 + 5n \\ (+) & -4n^3 - 7n \\ \hline & 3n^2 + 9n^3 - 2n \end{array}$$

Thus, $(3n^2 + 13n^3 + 5n) - (7n + 4n^3) = 3n^2 + 9n^3 - 2n$ or, arranged in descending order, $9n^3 + 3n^2 - 2n$.

CHECK Your Progress

2. Find $(4x^3 - 3x^2 + 6x - 4) - (-2x^3 + x^2 - 2)$.

When polynomials are used to model real-world data, their sums and differences can have real-world meaning, too.

Real-World Career... Teacher

The educational requirements for a teaching license vary by state. In the 2004–2005 school year, the average public K–12 teacher salary was \$47,808.

For more information, go to algebra1.com.

Real-World EXAMPLE

3 **EDUCATION The total number of public school teachers T consists of two groups, elementary E and secondary S. From 1992 through 2003, the number (in thousands) of secondary teachers and total teachers could be modeled by the following equations, where n is the number of years since 1992.**

$S = 29n + 949 \quad T = 58n + 2401$

a. Find an equation that models the number of elementary teachers E for this time period.

Subtract the polynomial for S from the polynomial for T.

Total	$58n + 2401$		$58n + 2401$
− Secondary	$(-)\ 29n + 949$	Add the opposite.	$(+)\ -29n - 949$
Elementary			$29n + 1452$

An equation is $E = 29n + 1452$.

b. Use the equation to predict the number of elementary teachers in 2015.

The year 2015 is 2015 − 1992 or 23 years after the year 1992.

If this trend continues, the number of elementary teachers in 2015 would be $29(23) + 1452$ or about 2,119,000.

3. WIRELESS DEVICES An electronics store sells cell phones and pagers. The equations below represent the monthly sales m of cell phones C and pagers P. Write an equation that represents the total monthly sales T of wireless devices. Use the equation to predict the number of wireless devices sold in 10 months.

$C = 7m + 137 \quad P = 4m + 78$

CHECK Your Understanding

Examples 1, 2 (pp. 384, 385)

Find each sum or difference.

1. $(4p^2 + 5p) + (-2p^2 + p)$
2. $(5y^2 - 3y + 8) + (4y^2 - 9)$
3. $(8cd - 3d + 4c) + (-6 + 2cd)$
4. $(-8xy + 3x^2 - 5y) + (4x^2 - 2y + 6xy)$
5. $(6a^2 + 7a - 9) - (-5a^2 + a - 10)$
6. $(g^3 - 2g^2 + 5g + 6) - (g^2 + 2g)$
7. $(3ax^2 - 5x - 3a) - (6a - 8a^2x + 4x)$
8. $(4rst - 8r^2s + s^2) - (6rs^2 + 5rst - 2s^2)$

Example 3 (p. 385)

POPULATION For Exercises 9 and 10, use the following information.
From 1980 through 2003, the female population F and the male population M of the United States (in thousands) are modeled by the following equations, where n is the number of years since 1980.

$F = 1{,}379n + 115{,}513 \qquad M = 1{,}450n + 108{,}882$

9. Find an equation that models the total population T of the United States in thousands for this time period.
10. If this trend continues, what will the population of the U. S. be in 2010?

Exercises

HOMEWORK HELP	
For Exercises	See Examples
11–16	1
17–22	2
23–24	3

Find each sum or difference.

11. $(6n^2 - 4) + (-2n^2 + 9)$
12. $(9z - 3z^2) + (4z - 7z^2)$
13. $(3 + a^2 + 2a) + (a^2 - 8a + 5)$
14. $(-3n^2 - 8 + 2n) + (5n + 13 + n^2)$
15. $(x + 5) + (2y + 4x - 2)$
16. $(2b^3 - 4b + b^2) + (-9b^2 + 3b^3)$
17. $(11 + 4d^2) - (3 - 6d^2)$
18. $(4g^3 - 5g) - (2g^3 + 4g)$
19. $(-4y^3 - y + 10) - (4y^3 + 3y^2 - 7)$
20. $(4x + 5xy + 3y) - (3y + 6x + 8xy)$
21. $(3x^2 + 8x + 4) - (5x^2 - 4)$
22. $(5ab^2 + 3ab) - (2ab^2 + 4 - 8ab)$

GEOMETRY The measures of two sides of a triangle are given. If P is the perimeter, find the measure of the third side.

23. $P = 7x + 3y$

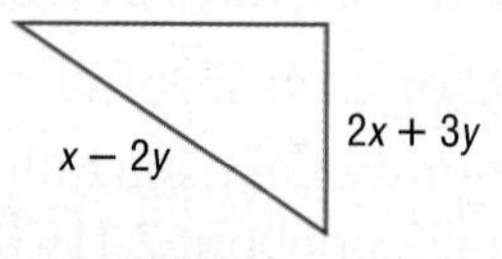

24. $P = 10x^2 - 5x + 16$

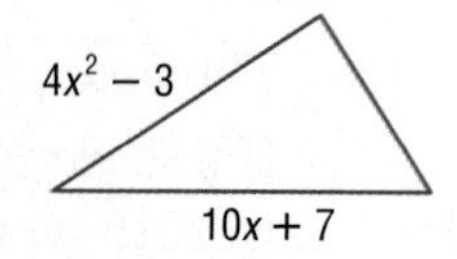

Find each sum or difference.

25. $(3a + 2b - 7c) + (6b - 4a + 9c) + (-7c - 3a - 2b)$

26. $(5x^2 - 3) + (x^2 - x + 11) + (2x^2 - 5x + 7)$

27. $(3y^2 - 8) + (5y + 9) - (y^2 + 6y - 4)$

28. $(9x^3 + 3x - 13) - (6x^2 - 5x) + (2x^3 - x^2 - 8x + 4)$

In 2002, attendance at movie theaters was at its highest point in 40 years with 1.63 billion tickets sold for a record \$9.52 billion in gross income.

Source: The National Association of Theatre Owners

MOVIES For Exercises 29 and 30, use the following information.
From 1995 to 2004, the number of indoor movie screens I and total movie screens T in the U.S. could be modeled by the following equations, where n is the number of years since 1995.

$$I = -194.8n^2 + 2{,}658n + 26{,}933 \qquad T = -193n^2 + 2{,}616n + 27{,}793$$

29. Find an equation that models the number of outdoor movie screens D.

30. If this trend continues, how many outdoor screens will there be in 2010?

POSTAL SERVICE For Exercises 31–33, use the following information.
The U.S. Postal Service restricts the sizes of boxes shipped by parcel post. The sum of the length and the girth of the box must not exceed 108 inches.

Suppose you want to make an open box using a 60-by-40-inch piece of cardboard by cutting squares out of each corner and folding up the flaps. The lid will be made from another piece of cardboard. You do not know how big the squares should be, so for now call the length of the side of each square x.

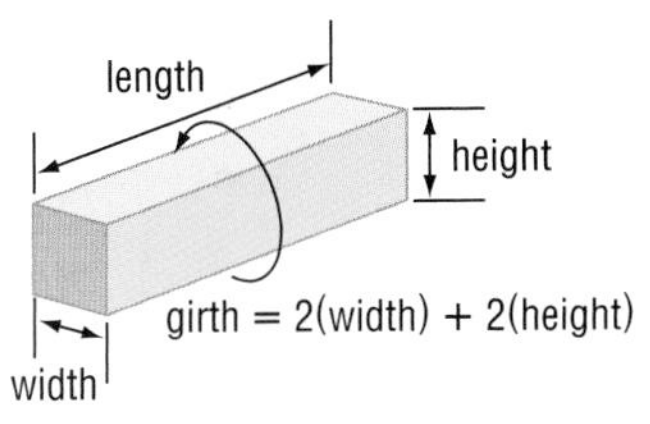

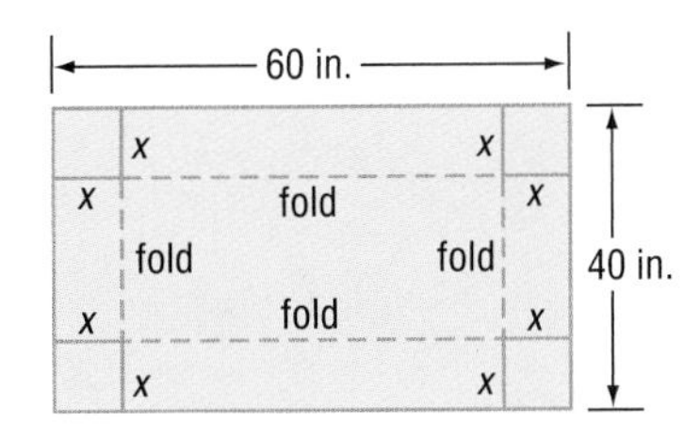

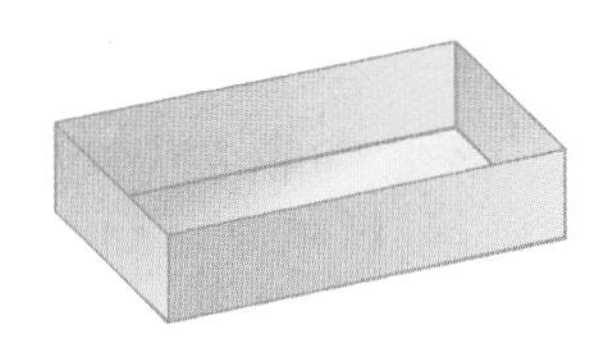

31. Write polynomials to represent the length, width, and girth of the box formed.

32. Write and solve an inequality to find the least possible value of x you could use in designing this box so it meets postal regulations.

33. What is the greatest integral value of x you could use to design this box if it does not have to meet regulations?

H.O.T. Problems

34. REASONING Explain why $5xy^2$ and $3x^2y$ are *not* like terms.

35. OPEN ENDED Write two polynomials with a difference of $2x^2 + x + 3$.

36. FIND THE ERROR Esteban and Kendra are finding $(5a - 6b) - (2a + 5b)$. Who is correct? Explain your reasoning.

Esteban

$(5a - 6b) - (2a + 5b)$
$= (-5a + 6b) + (-2a - 5b)$
$= -7a + b$

Kendra

$(5a - 6b) - (2a + 5b)$
$= (5a - 6b) + (-2a - 5b)$
$= 3a - 11b$

CHALLENGE **For Exercises 37–39, suppose x is an integer.**

37. Write an expression for the next integer greater than x.

38. Show that the sum of two consecutive integers, x and the next integer after x, is always odd. (*Hint:* A number is considered even if it is divisible by 2.)

39. What is the least number of consecutive integers that must be added together to always arrive at an even integer?

40. ***Writing in Math*** Use the information about music sales on page 384 to explain how you can use polynomials to model sales. Include an equation that models total music sales, and an example of how and why someone might use this equation in your answer.

STANDARDIZED TEST PRACTICE

41. The perimeter of the rectangle shown below is $16a + 2b$. Which expression represents the length of the rectangle?

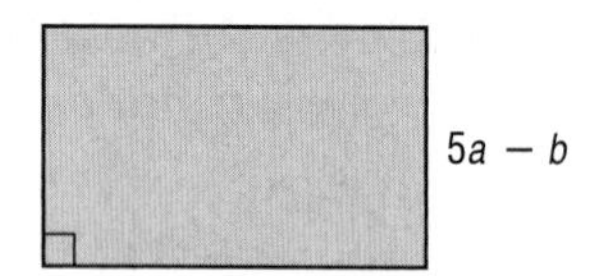

A $3a + 2b$

B $10a + 2b$

C $2a - 3b$

D $6a + 4b$

42. REVIEW The scale factor of two similar polygons is 4:5. The perimeter of the larger polygon is 200 inches. What is the perimeter of the smaller polygon?

F 250 inches

G 160 inches

H 80 inches

J 40 inches

Spiral Review

Find the degree of each polynomial. (Lesson 7-3)

43. $15t^3y^2$

44. 24

45. $m^2 + n^3$

46. $4x^2y^3z - 5x^3z$

Simplify. Assume no denominator is equal to zero. (Lesson 7-2)

47. $\frac{49a^4b^7c^2}{7ab^4c^3}$

48. $\frac{-4n^3p^{-5}}{n^{-2}}$

49. $\frac{(8n^7)^2}{(3n^2)^{-3}}$

KEYBOARDING **For Exercises 50–53, use the table that shows keyboarding speeds of 12 students in words per minute (wpm) and weeks of experience.** (Lesson 4-7)

Experience (weeks)	4	7	8	1	6	3	5	2	9	6	7	10
Keyboarding Speed (wpm)	33	45	46	20	40	30	38	22	52	44	42	55

50. Make a scatter plot of these data. Then draw a line of fit.

51. Find the equation of the line.

52. Use the equation to predict the speed of a student after a 12-week course.

53. Why is this equation not used to predict the speed for any number of weeks of experience?

GET READY for the Next Lesson

PREREQUISITE SKILL. **Simplify.** (Lesson 1-5)

54. $6(3x - 8)$

55. $-2(b + 9)$

56. $-7(-5p + 4q)$

57. $9(3a + 5b - c)$

58. $8(x^2 + 3x - 4)$

59. $-3(2a^2 - 5a + 7)$

CHAPTER 7

Mid-Chapter Quiz

Lessons 7-1 through 7-4

Simplify. (Lesson 7-1)

1. $n^3(n^4)(n)$

2. $4ad(3a^3d)$

3. $(-2w^3z^4)^3(-4wz^3)^2$

4. MULTIPLE CHOICE Ruby says that $(xy)^2 = x^2 + 2xy + y^2$ for every value of x and y, but Ebony disagrees. What does $(xy)^2$ really equal? (Lesson 7-1)

A $2x^2y$

B $2xy$

C xy^2

D x^2y^2

5. MULTIPLE CHOICE Which expression represents the volume of the cube? (Lesson 7-1)

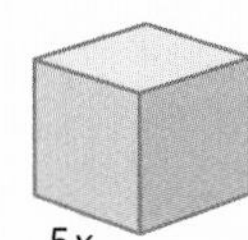

F $15x^3$

G $25x^2$

H $25x^3$

J $125x^3$

Simplify. Assume that no denominator is equal to zero. (Lesson 7-2)

6. $\dfrac{25p^{10}}{15p^3}$

7. $\left(\dfrac{6k^3}{7np^4}\right)^2$

8. $\dfrac{4x^0y^2}{(3y^{-3}z^5)^{-2}}$

9. $\dfrac{\left(m^2np^3\right)^{-3}}{\left(m^5n^3p^6\right)^{-4}}$

10. GEOMETRY The area of the rectangle is $66a^3b^5c^7$ square inches. Find the length of the rectangle. (Lesson 7-2)

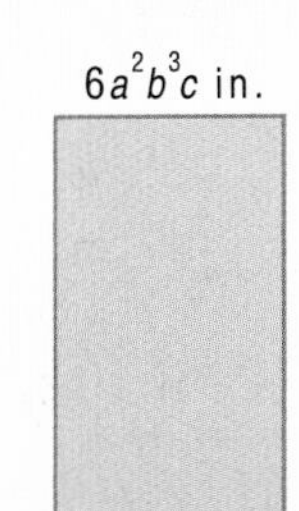

11. MULTIPLE CHOICE The wavelength of a microwave is 10^{-2} centimeters, and the wavelength of an X ray is 10^{-8} centimeters. How many times greater is the length of a microwave than an X ray? (Lesson 7-3)

A 10^{10}

B 10^6

C 10^{-6}

D 10^{-10}

Find the degree of each polynomial. (Lesson 7-3)

12. $5x^4$

13. $-9n^3p^4$

14. $7a^2 - 2ab^2$

15. $-6 - 8x^2y^2 + 5y^3$

GEOMETRY For Exercises 16 and 17, use the figure below. (Lesson 7-3)

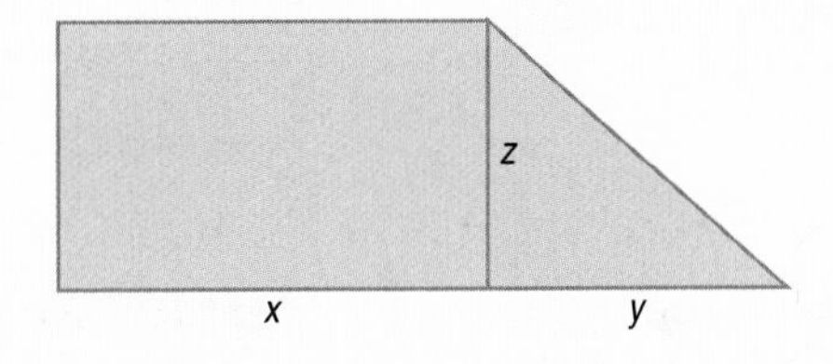

16. Write a polynomial that represents the area of the figure.

17. If $x = 7$ feet, $y = 3$ feet, and $z = 2$ feet, find the area of the figure.

Arrange the terms of each polynomial so that the powers of x are in ascending order. (Lesson 7-4)

18. $4x^2 + 9x - 12 + 5x^3$

19. $2xy^4 + x^3y^5 + 5x^5y - 13x^2$

20. MULTIPLE CHOICE If three consecutive integers are x, $x + 1$, and $x + 2$, what is the sum of these three integers? (Lesson 7-4)

F $2x + 3$

G $3x + 3$

H $x(x + 1)(x + 2)$

J $x^3 + 3x^2 + 2x$

7-5 Multiplying a Polynomial by a Monomial

Main Ideas

- Find the product of a monomial and a polynomial.
- Solve equations involving polynomials.

GET READY for the Lesson

The algebra tiles shown are grouped together to form a rectangle with a width of $2x$ and a length of $x + 3$. Notice that the rectangle consists of 2 blue x^2-tiles and 6 green x-tiles. The area of the rectangle is the sum of these algebra tiles or $2x^2 + 6x$.

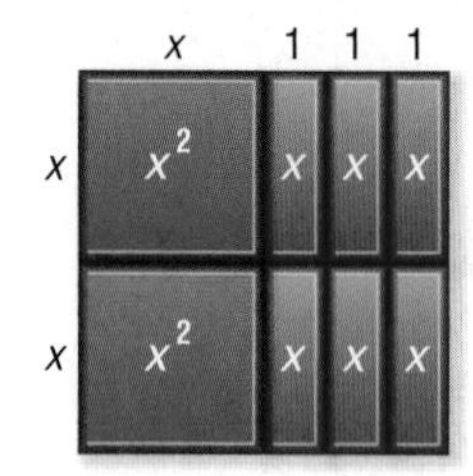

Product of Monomial and Polynomial The Distributive Property can be used to multiply a polynomial by a monomial.

EXAMPLE Multiply a Polynomial by a Monomial

1 **Find $-2x^2(3x^2 - 7x + 10)$.**

Method 1 Horizontal

$-2x^2(3x^2 - 7x + 10)$

$= -2x^2(3x^2) - (-2x^2)(7x) + (-2x^2)(10)$ Distributive Property

$= -6x^4 - (-14x^3) + (-20x^2)$ Multiply.

$= -6x^4 + 14x^3 - 20x^2$ Simplify.

Method 2 Vertical

$$\begin{array}{r} 3x^2 - 7x + 10 \\ (\times) \qquad\quad -2x^2 \\ \hline -6x^4 + 14x^3 - 20x^2 \end{array}$$

Distributive Property

Multiply.

Review Vocabulary

Distributive Property: For any numbers a, b, and c, $a(b + c) = ab + ac$ and $a(b - c) = ab - ac$. (Lesson 1–5)

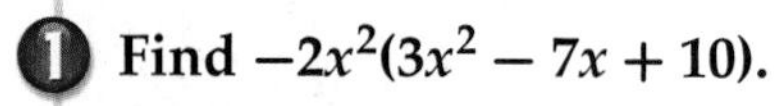

CHECK Your Progress

1. Find $5a^2(-4a^2 + 2a - 7)$.

EXAMPLE Simplify Expressions

2 **Simplify $4(3d^2 + 5d) - d(d^2 - 7d + 12)$.**

$4(3d^2 + 5d) - d(d^2 - 7d + 12)$

$= 4(3d^2) + 4(5d) + (-d)(d^2) - (-d)(7d) + (-d)(12)$ Distributive Property

$= 12d^2 + 20d + (-d^3) - (-7d^2) + (-12d)$ Product of Powers

$= 12d^2 + 20d - d^3 + 7d^2 - 12d$ Simplify.

$= -d^3 + (12d^2 + 7d^2) + (20d - 12d)$ Commutative and Associative Properties

$= -d^3 + 19d^2 + 8d$ Combine like terms.

CHECK Your Progress

2. Simplify $3(5x^2 + 2x - 4) - x(7x^2 + 2x - 3)$.

Real-World Link

About 98% of long-distance companies service their calls using the network of one of three companies. Since the quality of phone service is basically the same, a company's rates are the primary factor in choosing a long-distance provider.

Source: Chamberland Enterprises

Real-World EXAMPLE

3 **PHONE SERVICE Greg pays a fee of $20 a month for local calls. Long-distance rates are 6¢ per minute for in-state calls and 5¢ per minute for out-of-state calls. Suppose Greg makes 300 minutes of long-distance phone calls in January and m of those minutes are for in-state calls.**

a. Find an expression for Greg's phone bill for January.

Words	Bill	=	service fee	+	in-state minutes	·	6¢ per minute	+	out-of-state minutes	· 5¢ per minute.
Variables	If m = number of minutes of in-state calls, then $300 - m$ = number of minutes of out-of-state calls. Let B = phone bill for the month of January.									
Equation	B	=	20	+	m	·	0.06	+	$(300 - m)$	· 0.05

$$
\begin{aligned}
B &= 20 + m \cdot 0.06 + (300 - m) \cdot 0.05 && \text{Write the equation.}\\
&= 20 + 0.06m + 300(0.05) - m(0.05) && \text{Distributive Property}\\
&= 20 + 0.06m + 15 - 0.05m && \text{Simplify.}\\
&= 35 + 0.01m && \text{Simplify.}
\end{aligned}
$$

Greg's bill for January is $35 + 0.01m$, for m minutes of in-state calls.

b. Evaluate the expression to find the cost if Greg had 37 minutes of in-state calls in January.

$$
\begin{aligned}
35 + 0.01m &= 35 + 0.01(37) && m = 37\\
&= 35 + 0.37 && \text{Multiply.}\\
&= 35.37 && \text{Add.}
\end{aligned}
$$

Greg's bill was $35.37.

CHECK Your Progress

3. A parking garage charges $30 per month plus $0.50 per daytime hour and $0.25 per hour during nights and weekends. Suppose Juana parks in the garage for 47 hours in January and h of those are night and weekend hours. Find an expression for her January bill. Then find the cost if Juana had 12 hours of night and weekend hours.

Online **Personal Tutor at** algebra1.com

Solve Equations with Polynomial Expressions Many equations contain polynomials that must be added, subtracted, or multiplied.

EXAMPLE Polynomials on Both Sides

4 **Solve $y(y - 12) + y(y + 2) + 25 = 2y(y + 5) - 15$.**

$$
\begin{aligned}
y(y - 12) + y(y + 2) + 25 &= 2y(y + 5) - 15 && \text{Original equation}\\
y^2 - 12y + y^2 + 2y + 25 &= 2y^2 + 10y - 15 && \text{Distributive Property}\\
2y^2 - 10y + 25 &= 2y^2 + 10y - 15 && \text{Combine like terms.}\\
-10y + 25 &= 10y - 15 && \text{Subtract } 2y^2 \text{ from each side.}\\
-20y + 25 &= -15 && \text{Subtract } 10y \text{ from each side.}\\
-20y &= -40 && \text{Subtract 25 from each side.}\\
y &= 2 && \text{Divide each side by } -20.
\end{aligned}
$$

CHECK $y(y - 12) + y(y + 2) + 25 = 2y(y + 5) - 15$ — Original equation

$2(2 - 12) + 2(2 + 2) + 25 \stackrel{?}{=} 2(2)(2 + 5) - 15$ — $y = 2$

$2(-10) + 2(4) + 25 \stackrel{?}{=} 4(7) - 15$ — Simplify.

$-20 + 8 + 25 \stackrel{?}{=} 28 - 15$ — Multiply.

$13 = 13$ ✓ — Add and subtract.

4. Solve $2x(x + 4) + 7 = (x + 8) + 2x(x + 1) + 12$.

CHECK Your Understanding

Example 1 (p. 390)

Find each product.

1. $-3y(5y + 2)$

2. $9b^2(2b^3 - 3b^2 + b - 8)$

3. $2x(4a^4 - 3ax + 6x^2)$

4. $-4xy(5x^2 - 12xy + 7y^2)$

Example 2 (p. 390)

Simplify.

5. $t(5t - 9) - 2t$

6. $x(3x + 4) + 2(7x - 3)$

7. $5n(4n^3 + 6n^2 - 2n + 3) - 4(n^2 + 7n)$

8. $4y^2(y^2 - 2y + 5) + 3y(2y^2 - 2)$

Example 3 (p. 391)

SAVINGS For Exercises 9–11, use the following information.
Matthew's grandmother left him \$10,000 for college. Matthew puts some of the money into a savings account earning 3% interest per year. With the rest, he buys a certificate of deposit (CD) earning 5% per year.

9. If Matthew puts x dollars into the savings account, write an expression to represent the amount of the CD.

10. Write an equation for the total amount of money T Matthew will have saved for college after one year.

11. If Matthew puts \$3000 in savings, how much money will he have in one year?

Example 4 (p. 391)

Solve each equation.

12. $-2(w + 1) + w = 7 - 4w$

13. $3(y - 2) + 2y = 4y + 14$

14. $a(a + 3) + a(a - 6) + 35 = a(a - 5) + a(a + 7)$

15. $n(n - 4) + n(n + 8) = n(n - 13) + n(n + 1) + 16$

Exercises

HOMEWORK HELP

For Exercises	See Examples
16–25	1
26–31	2
32–35	3
36–43	4

Find each product.

16. $r(5r + r^2)$

17. $w(2w^3 - 9w^2)$

18. $-4x(8 + 3x)$

19. $5y(-2y^2 - 7y)$

20. $7ag(g^3 + 2ag)$

21. $-3np(n^2 - 2p)$

22. $-2b^2(3b^2 - 4b + 9)$

23. $6x^3(5 + 3x - 11x^2)$

24. $8x^2y(5x + 2y^2 - 3)$

25. $-cd^2(3d + 2c^2d - 4c)$

Simplify.

26. $d(-2d + 4) + 15d$

27. $-x(4x^2 - 2x) - 5x^3$

28. $3w(6w - 4) + 2(w^2 - 3w + 5)$

29. $5n(2n^3 + n^2 + 8) + n(4 - n)$

30. $10(4m^3 - 3m + 2) - 2m(-3m^2 - 7m + 1)$

31. $4y(y^2 - 8y + 6) - 3(2y^3 - 5y^2 + 2)$

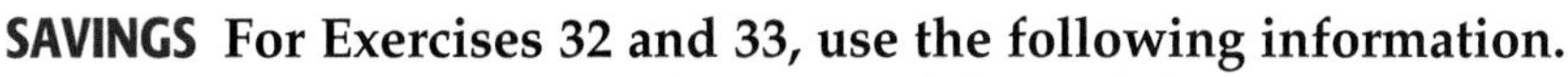

SAVINGS For Exercises 32 and 33, use the following information.
Marta has \$6000 to invest. She puts x dollars of this money into a savings account that earns 2% interest per year. With the rest, she buys a certificate of deposit that earns 4% per year.

32. Write an equation for the amount of money T Marta will have in one year.

33. Suppose at the end of one year, Marta has a total of \$6210. How much money did Marta invest in each account?

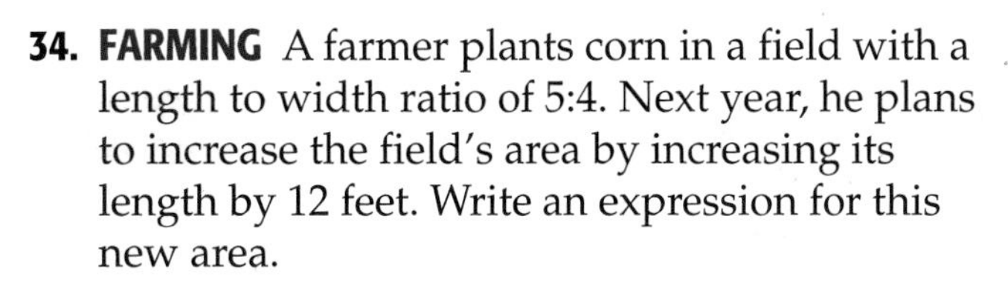

34. FARMING A farmer plants corn in a field with a length to width ratio of 5:4. Next year, he plans to increase the field's area by increasing its length by 12 feet. Write an expression for this new area.

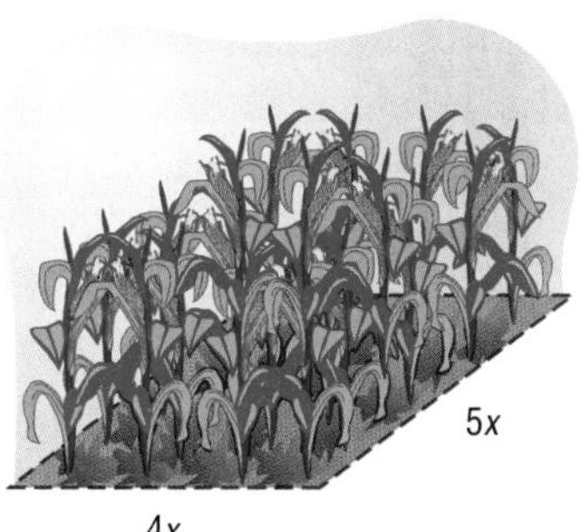

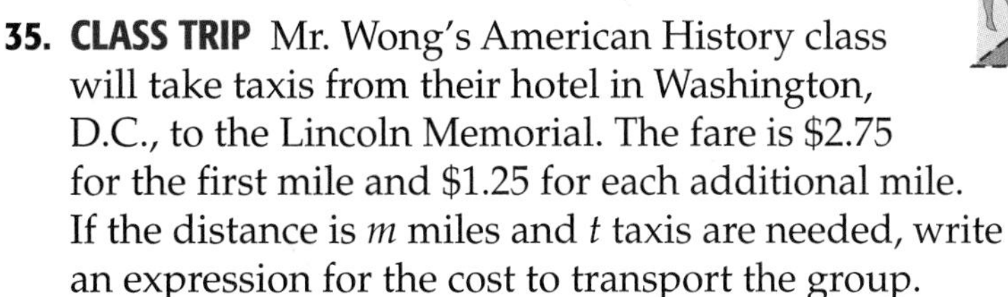

35. CLASS TRIP Mr. Wong's American History class will take taxis from their hotel in Washington, D.C., to the Lincoln Memorial. The fare is \$2.75 for the first mile and \$1.25 for each additional mile. If the distance is m miles and t taxis are needed, write an expression for the cost to transport the group.

Real-World Link

Inside the Lincoln Memorial is a 19-foot marble statue of the United States' 16th president. The statue is flanked on either side by the inscriptions of Lincoln's Second Inaugural Address and Gettysburg Address.

Source: washington.org

Solve each equation.

36. $2(4x - 7) = 5(-2x - 9) - 5$

37. $4(3p + 9) - 5 = -3(12p - 5)$

38. $d(d - 1) + 4d = d(d - 8)$

39. $c(c + 3) - c(c - 4) = 9c - 16$

40. $a(3a - 2) + 2a(a + 4) = a(a + 2) + 4a(a - 3) + 48$

41. $3(4w - 2) + 6(w + 4) - 3 = 4w - 7(w + 2) + 5(3w + 7)$

Expand and simplify.

42. $4(x + 2) - 6$

43. $3x - 2(x + 1)$

Find each product.

44. $-\frac{3}{4}hk^2(20k^2 + 5h - 8)$

45. $\frac{2}{3}a^2b(6a^3 - 4ab + 9b^2)$

46. $-5a^3b(2b + 5ab - b^2 + a^3)$

47. $4p^2q^2(2p^2 - q^2 + 9p^3 + 3q)$

Simplify.

48. $-3c^2(2c + 7) + 4c(3c^2 - c + 5) + 2(c^2 - 4)$

49. $4x^2(x + 2) + 3x(5x^2 + 2x - 6) - 5(3x^2 - 4x)$

Solve each equation.

50. $2n(n + 4) + 18 = n(n + 5) + n(n - 2) - 7$

51. $3g(g - 4) - 2g(g - 7) = g(g + 6) - 28$

GEOMETRY Find the area of each shaded region in simplest form.

52.

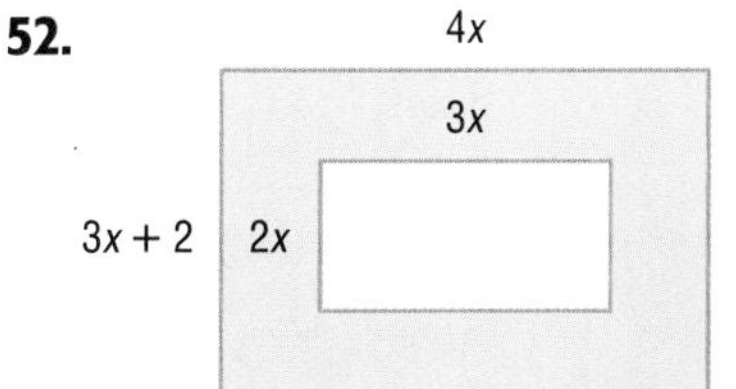

53.

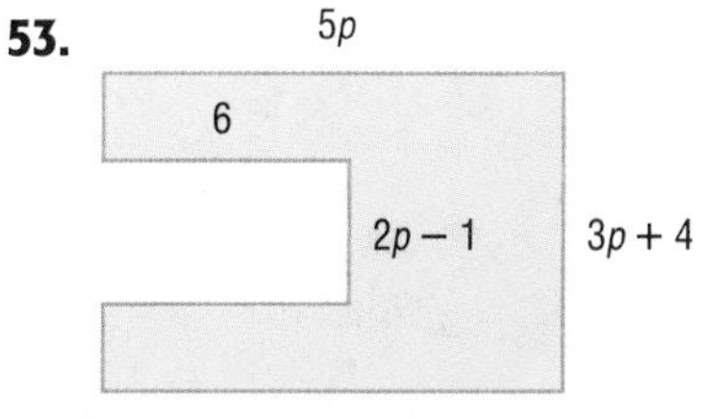

VOLUNTEERING **For Exercises 54 and 55, use the following information.**
Loretta is making baskets of apples and oranges for homeless shelters. She wants to place a total of 10 pieces of fruit in each basket. Apples cost 25¢ each, and oranges cost 20¢ each.

54. If a represents the number of apples Loretta uses, write a polynomial model in simplest form for the total amount of money T Loretta will spend.

55. If Loretta uses 4 apples in each basket, find the total cost for fruit.

Real-World Link

Approximately one third of young people in grades 7–12 suggested that "working for the good of my community and country" and "helping others or volunteering" were important future goals.

Source: Primeday/Roper National Youth Opinion Survey

SALES **For Exercises 56 and 57, use the following information.**
A store advertises that all sports equipment is 30% off the retail price. In addition, the store asks customers to select and pop a balloon to receive a coupon for an additional n percent off one of their purchases.

56. Write an expression for the cost of a pair of inline skates with retail price p.

57. Use this expression to calculate the cost, not including sales tax, of a $200 pair of inline skates for an additional 10% off.

58. SPORTS You may have noticed that when runners race around a curved track, their starting points are staggered. This is so each contestant runs the same distance to the finish line.

If the radius of the inside lane is x and each lane is 2.5 feet wide, how far apart should the officials start the runners in the inside lane and the outside (6th) lane? (*Hint*: Circumference = $2\pi r$, where r is the radius of the circle)

NUMBER THEORY **For Exercises 59 and 60, let x be an odd integer.**

59. Write an expression for the next odd integer.

60. Find the product of x and the next odd integer.

H.O.T. Problems

61. OPEN ENDED Write a monomial and a trinomial involving one variable. Then find their product.

CHALLENGE **For Exercises 62–64, use the following information.**
An even number can be represented by $2x$, where x is any integer.

62. Show that the product of two even integers is always even.

63. Write a representation for an odd integer.

64. Show that the product of an even and an odd integer is always even.

65. ***Writing in Math*** Use the information about the area of a rectangle on page 390 to explain how the product of a monomial and a polynomial relate to finding the area of a rectangle. Include the product of $2x$ and $x + 3$ derived algebraically in your answer.

STANDARDIZED TEST PRACTICE

66. A plumber charges $70 for the first thirty minutes of each house call plus $4 for each additional minute that she works. The plumber charges Ke-Min $122 for her time. What amount of time, in minutes, did the plumber work?

A 43

B 48

C 58

D 64

67. REVIEW What is the slope of this line?

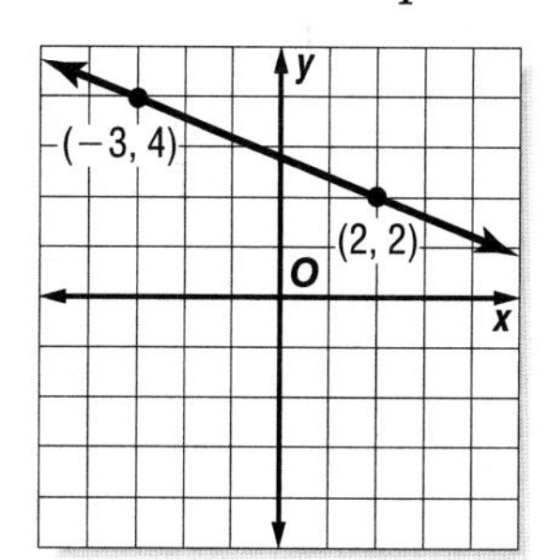

F $-\frac{5}{2}$ **G** -2 **H** $-\frac{1}{2}$ **J** $-\frac{2}{5}$

Spiral Review

Find each sum or difference. (Lesson 7-4)

68. $(4x^2 + 5x) + (-7x^2 + x)$

69. $(3y^2 + 5y - 6) - (7y^2 - 9)$

70. $(5b - 7ab + 8a) - (5ab - 4a)$

71. $(6p^3 + 3p^2 - 7) + (p^3 - 6p^2 - 2p)$

State whether each expression is a polynomial. If the expression is a polynomial, identify it as a *monomial,* a *binomial,* or a *trinomial.* (Lesson 7-3)

72. $4x^2 - 10ab + 6$

73. $4c + ab - c$

74. $\frac{7}{y} + y^2$

75. $\frac{n^2}{3}$

Define a variable, write an inequality, and solve each problem. Then check your solution. (Lesson 6-3)

76. Six increased by ten times a number is less than nine times the number.

77. Nine times a number increased by four is no less than seven decreased by thirteen times the number.

Write an equation of the line that passes through each pair of points. (Lesson 4-4)

78. $(-3, -8), (1, 4)$

79. $(-4, 5), (2, -7)$

80. $(3, -1), (-3, 2)$

Solve each equation. (Lesson 2-5)

81. $2(x + 3) + 3 = 4x - 5$

82. $3(y - 3) - 6 = 9y - 15$

83. $2(3a + 6) - 3 = 6a + 12$

84. BASKETBALL Tremaine scored 54 three-point field goals, 84 two-point field goals, and 106 free throws in 23 games. How many points did he score on average per game? (Lesson 2-4)

GET READY for the Next Lesson

PREREQUISITE SKILL Simplify. (Lesson 7-1)

85. $(a)(a)$

86. $2x(3x^2)$

87. $-3y^2(8y^2)$

88. $4y(3y) - 4y(6)$

89. $-5n(2n^2) - (-5n)(8n) + (-5n)(4)$

90. $3p^2(6p^2) - 3p^2(8p) + 3p^2(12)$

EXPLORE 7-6

Algebra Lab

Multiplying Polynomials

ACTIVITY 1 **Use algebra tiles to find $(x + 2)(x + 5)$.**

The rectangle will have a width of $x + 2$ and a length of $x + 5$. Use algebra tiles to mark off the dimensions on a product mat. Then complete the rectangle with algebra tiles.

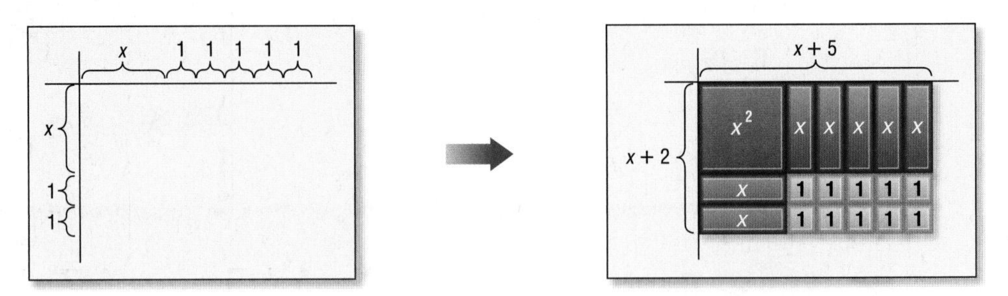

The rectangle consists of 1 blue x^2-tile, 7 green x-tiles, and 10 yellow 1-tiles. The area of the rectangle is $x^2 + 7x + 10$. Therefore, $(x + 2)(x + 5) = x^2 + 7x + 10$.

ACTIVITY 2 **Use algebra tiles to find $(x - 1)(x - 4)$.**

Step 1 The rectangle will have a width of $x - 1$ and a length of $x - 4$. Use algebra tiles to mark off the dimensions on a product mat. Then begin to make the rectangle with algebra tiles.

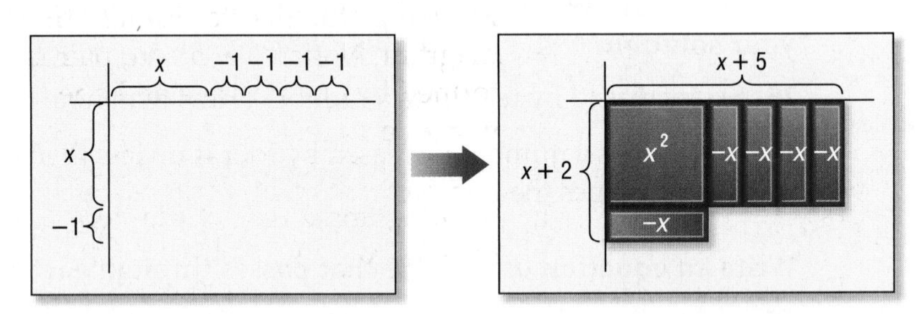

Step 2 Determine whether to use 4 yellow 1-tiles or 4 red −1-tiles to complete the rectangle. Remember that the numbers at the top and side give the dimensions of the tile needed. The area of each tile is the product of −1 and −1 or 1. This is represented by a yellow 1-tile. Fill in the space with 4 yellow 1-tiles to complete the rectangle.

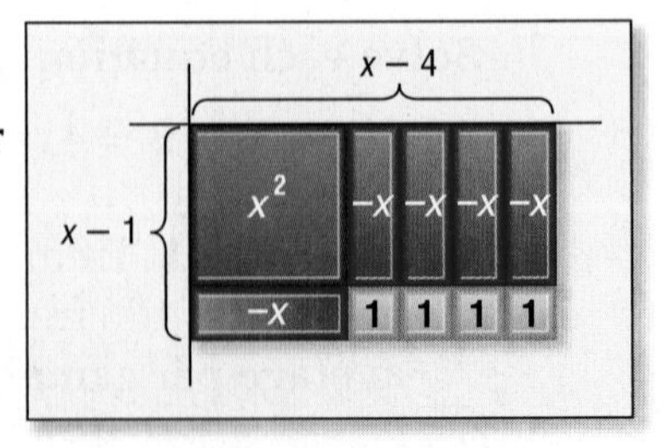

The rectangle consists of 1 blue x^2-tile, 5 red $-x$-tiles, and 4 yellow 1-tiles. The area of the rectangle is $x^2 - 5x + 4$. Therefore, $(x - 1)(x - 4) = x^2 - 5x + 4$.

ACTIVITY 3 **Use algebra tiles to find $(x - 3)(2x + 1)$.**

Step 1 The rectangle will have a width of $x - 3$ and a length of $2x + 1$. Mark off the dimensions on a product mat. Then make the rectangle.

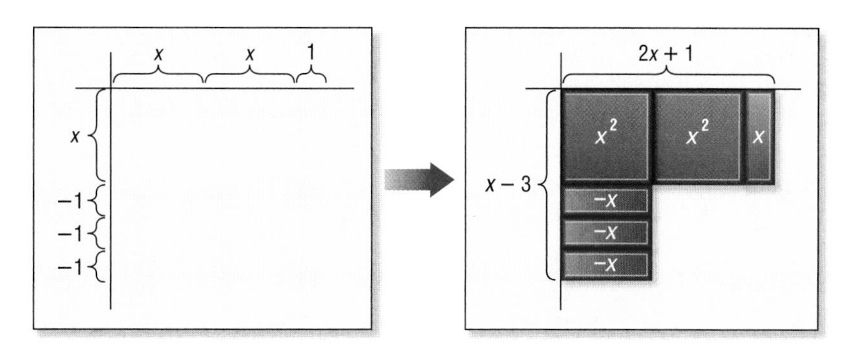

Step 2 Determine what color x-tiles and what color 1-tiles to use to complete the rectangle. The area of each $-x$-tile is the product of x and -1. This is represented by a red $-x$-tile. The area of the -1-tile is represented by the product of 1 and -1 or -1. This is represented by a red -1-tile. Complete the rectangle with 3 red $-x$-tiles and 3 red -1-tiles.

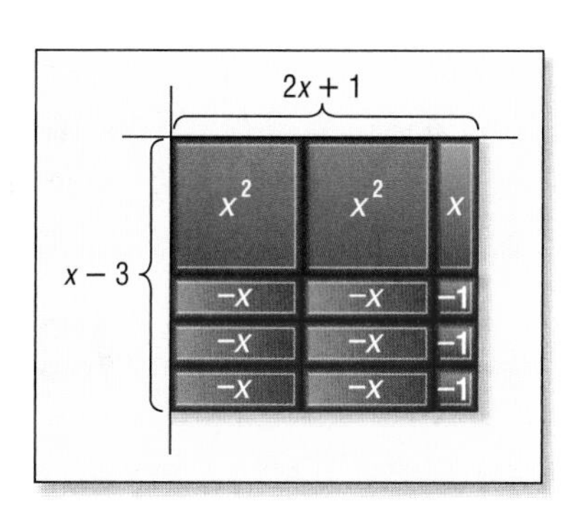

Step 3 Rearrange the tiles to simplify the diagram. Notice that a zero pair is formed by one positive and one negative x-tile.

There are 2 blue x^2-tiles, 5 red $-x$-tiles, and 3 red -1-tiles left. $(x - 3)(2x + 1) = 2x^2 - 5x - 3$.

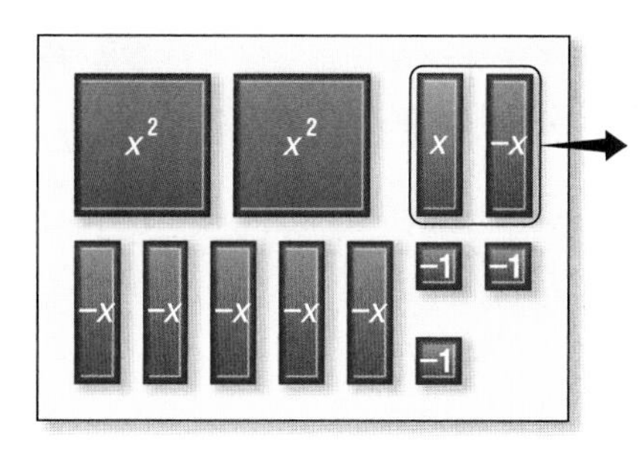

Concepts in Motion
Animation
algebra1.com

Model

Use algebra tiles to find each product.

1. $(x + 2)(x + 3)$ **2.** $(x - 1)(x - 3)$ **3.** $(x + 1)(x - 2)$

4. $(x + 1)(2x + 1)$ **5.** $(x - 2)(2x - 3)$ **6.** $(x + 3)(2x - 4)$

Analyze the Results

7. You can also use the Distributive Property to find the product of two binomials. The figure at the right shows the model for $(x + 3)(x + 4)$ separated into four parts. Write a sentence or two explaining how this model shows the use of the Distributive Property.

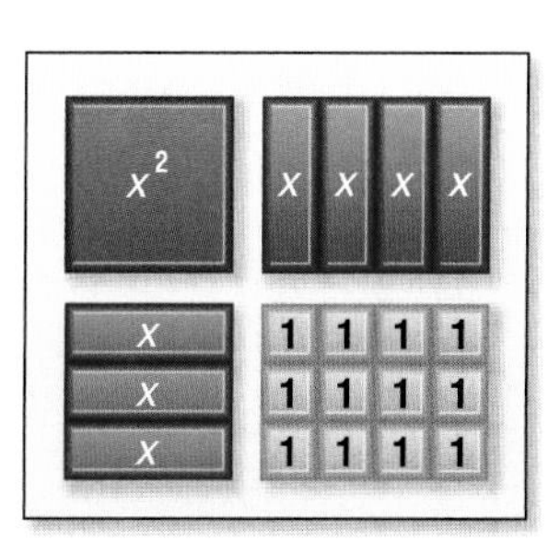

7-6 Multiplying Polynomials

Main Ideas

- Multiply two binomials by using the FOIL method.
- Multiply two polynomials by using the Distributive Property.

New Vocabulary

FOIL method

GET READY for the Lesson

To compute 24 × 36, we multiply each digit in 24 by each digit in 36, paying close attention to the place value of each digit.

Step 1 Multiply by the ones.	**Step 2** Multiply by the tens.	**Step 3** Add like place values.
$\begin{array}{r} 24 \\ \times\ 36 \\ \hline 144 \end{array}$	$\begin{array}{r} 24 \\ \times\ 36 \\ \hline 144 \\ 720 \\ \hline \end{array}$	$\begin{array}{r} 24 \\ \times\ 36 \\ \hline 144 \\ +\ 720 \\ \hline 864 \end{array}$
$6 \times 24 = 6(20 + 4)$ $= 120 + 24$ or 144	$30 \times 24 = 30(20 + 4)$ $= 600 + 120$ or 720	

You can multiply two binomials in a similar way.

Multiply Binomials To multiply two binomials, apply the Distributive Property twice as you do when multiplying two-digit numbers.

EXAMPLE The Distributive Property

1 Find $(x + 3)(x + 2)$.

Method 1 Vertical

Multiply by 2.	Multiply by x.	Combine like terms.
$\begin{array}{r} x + 3 \\ (\times)\ x + 2 \\ \hline 2x + 6 \end{array}$	$\begin{array}{r} x + 3 \\ (\times)\ x + 2 \\ \hline 2x + 6 \\ x^2 + 3x \quad \\ \hline \end{array}$	$\begin{array}{r} x + 3 \\ (\times)\ x + 2 \\ \hline 2x + 6 \\ x^2 + 3x \quad \\ \hline x^2 + 5x + 6 \end{array}$
$2(x + 3) = 2x + 6$	$x(x + 3) = x^2 + 3x$	

Method 2 Horizontal

$$\begin{aligned} (x + 3)(x + 2) &= x(x + 2) + 3(x + 2) && \text{Distributive Property} \\ &= x(x) + x(2) + 3(x) + 3(2) && \text{Distributive Property} \\ &= x^2 + 2x + 3x + 6 && \text{Multiply.} \\ &= x^2 + 5x + 6 && \text{Combine like terms.} \end{aligned}$$

CHECK Your Progress

Find each product.

1A. $(m + 4)(m + 5)$

1B. $(y - 2)(y + 8)$

There is a shortcut version of the Distributive Property called the **FOIL method.** You can use the FOIL method to multiply two binomials.

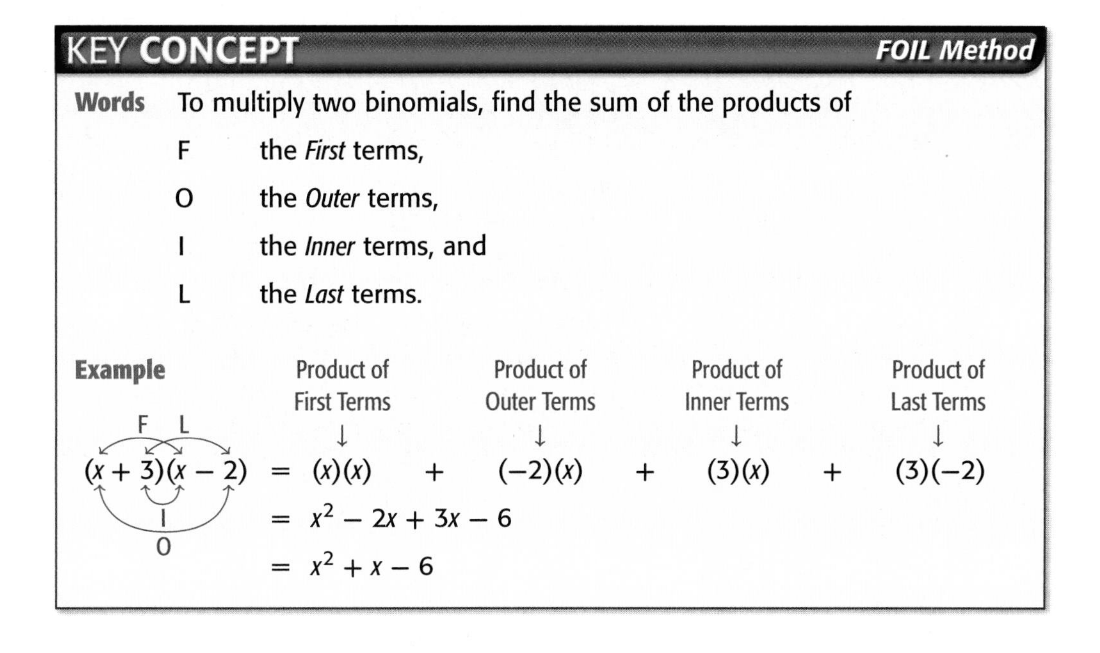

KEY CONCEPT *FOIL Method*

Words To multiply two binomials, find the sum of the products of

F the *First* terms,

O the *Outer* terms,

I the *Inner* terms, and

L the *Last* terms.

Example

	Product of First Terms		Product of Outer Terms		Product of Inner Terms		Product of Last Terms
$(x + 3)(x - 2) =$	$(x)(x)$	$+$	$(-2)(x)$	$+$	$(3)(x)$	$+$	$(3)(-2)$

$= x^2 - 2x + 3x - 6$

$= x^2 + x - 6$

Concepts in Motion
Animation
algebra1.com

EXAMPLE FOIL Method

2 **Find each product.**

a. $(x - 5)(x + 7)$

$(x - 5)(x + 7) = (x)(x) + (x)(7) + (-5)(x) + (-5)(7)$ FOIL method

$= x^2 + 7x - 5x - 35$ Multiply.

$= x^2 + 2x - 35$ Combine like terms.

b. $(2y + 3)(6y - 7)$

$(2y + 3)(6y - 7)$

$= (2y)(6y) + (2y)(-7) + (3)(6y) + (3)(-7)$ FOIL method

$= 12y^2 - 14y + 18y - 21$ Multiply.

$= 12y^2 + 4y - 21$ Combine like terms.

Study Tip

Checking Your Work

You can check your products in Examples 2a and 2b by reworking each problem using the Distributive Property.

CHECK Your Progress

2A. $(x + 3)(x - 4)$

2B. $(4a - 5)(3a + 2)$

The FOIL method can be used to find an expression that represents the area of geometric shapes when the lengths of the sides are given as binomials.

EXAMPLE 3 FOIL Method

3 GEOMETRY The area A of a trapezoid is one half the height h times the sum of the bases, b_1 and b_2. Write an expression for the area of the trapezoid.

$3x - 7$

$x + 2$

$2x + 1$

Explore Identify the height and bases.

$$h = x + 2$$
$$b_1 = 3x - 7$$
$$b_2 = 2x + 1$$

Plan Now write and apply the formula.

Area	equals	one-half	height	times	sum of bases.
A	$=$	$\frac{1}{2}$	h	$\cdot$	$(b_1 + b_2)$

Solve

$A = \frac{1}{2}h(b_1 + b_2)$	Original formula
$= \frac{1}{2}(x + 2)[(3x - 7) + (2x + 1)]$	Substitution
$= \frac{1}{2}(x + 2)(5x - 6)$	Add polynomials in the brackets.
$= \frac{1}{2}[x(5x) + x(-6) + 2(5x) + 2(-6)]$	FOIL Method
$= \frac{1}{2}(5x^2 - 6x + 10x - 12)$	Multiply.
$= \frac{1}{2}(5x^2 + 4x - 12)$	Combine like terms.
$= \frac{5}{2}x^2 + 2x - 6$	Distributive Property

Check The area of the trapezoid is $\frac{5}{2}x^2 + 2x - 6$ square units.

CHECK Your Progress

3. Write an expression for the area of a triangle with a base of $2x + 3$ and a height of $3x - 1$.

Online **Personal Tutor at** algebra1.com

Study Tip

Common Misconception

A common mistake when multiplying polynomials horizontally is to combine terms that are not alike. For this reason, you may prefer to multiply polynomials in column form, aligning like terms.

Multiply Polynomials The Distributive Property can be used to multiply any two polynomials.

EXAMPLE 4 The Distributive Property

4 Find each product.

a. $(4x + 9)(2x^2 - 5x + 3)$

$(4x + 9)(2x^2 - 5x + 3)$

$= 4x(2x^2 - 5x + 3) + 9(2x^2 - 5x + 3)$	Distributive Property
$= 8x^3 - 20x^2 + 12x + 18x^2 - 45x + 27$	Distributive Property
$= 8x^3 - 2x^2 - 33x + 27$	Combine like terms.

b. $(y^2 - 2y + 5)(6y^2 - 3y + 1)$

$$(y^2 - 2y + 5)(6y^2 - 3y + 1)$$
$$= y^2(6y^2 - 3y + 1) - 2y(6y^2 - 3y + 1) + 5(6y^2 - 3y + 1)$$
$$= 6y^4 - 3y^3 + y^2 - 12y^3 + 6y^2 - 2y + 30y^2 - 15y + 5$$
$$= 6y^4 - 15y^3 + 37y^2 - 17y + 5$$

CHECK Your Progress

4A. $(3x - 5)(2x^2 + 7x - 8)$ **4B.** $(m^2 + 2m - 3)(4m^2 - 7m + 5)$

CHECK Your Understanding

Examples 1–2 (pp. 398–399)

Find each product.

1. $(y + 4)(y + 3)$
2. $(x - 2)(x + 6)$
3. $(a - 8)(a + 5)$
4. $(4h + 5)(h + 7)$
5. $(9p - 1)(3p - 2)$
6. $(2g + 7)(5g - 8)$

Example 3 (p. 400)

7. **GEOMETRY** The area A of a triangle is half the product of the base b times the height h. Write a polynomial expression that represents the area of the triangle at the right.

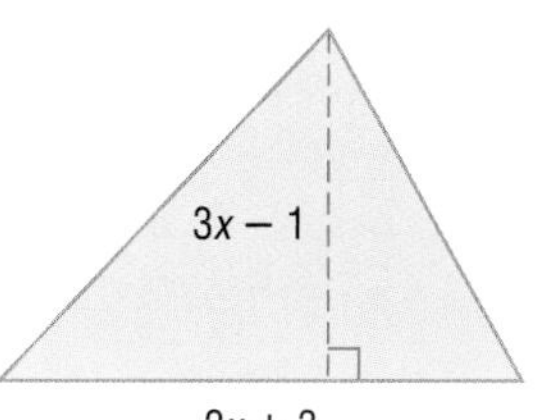

Example 4 (p. 400)

Find each product.

8. $(3k - 5)(2k^2 + 4k - 3)$
9. $(4x^2 - 2)(2x^2 + 6x + 1)$
10. $(y^2 - 5y + 3)(4y^2 + 2y - 6)$
11. $(3m^2 + 2m - 7)(5m^2 + m + 9)$

Exercises

HOMEWORK HELP

For Exercises	See Examples
12–29	1, 2
30–33	3
34–41	4

Find each product.

12. $(b + 8)(b + 2)$
13. $(n + 6)(n + 7)$
14. $(x - 4)(x - 9)$
15. $(a - 3)(a - 5)$
16. $(y + 4)(y - 8)$
17. $(p + 2)(p - 10)$
18. $(2w - 5)(w + 7)$
19. $(k + 12)(3k - 2)$
20. $(8d + 3)(5d + 2)$
21. $(4g + 3)(9g + 6)$
22. $(7x - 4)(5x - 1)$
23. $(6a - 5)(3a - 8)$
24. $(2n + 3)(2n + 3)$
25. $(5m - 6)(5m - 6)$
26. $(10r - 4)(10r + 4)$
27. $(7t + 5)(7t - 5)$
28. $(8x + 2y)(5x - 4y)$
29. $(11a - 6b)(2a + 3b)$

GEOMETRY Write an expression to represent the area of each figure.

30.

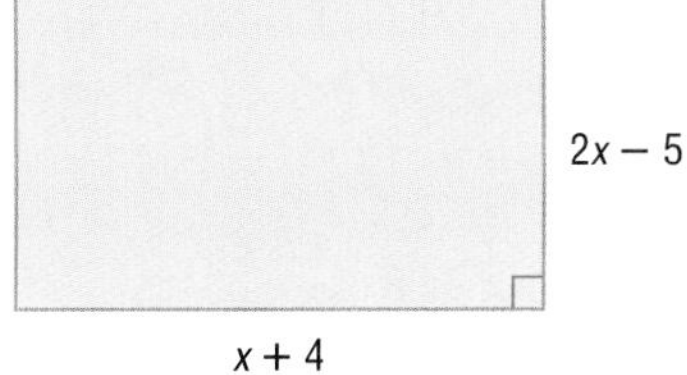

31.

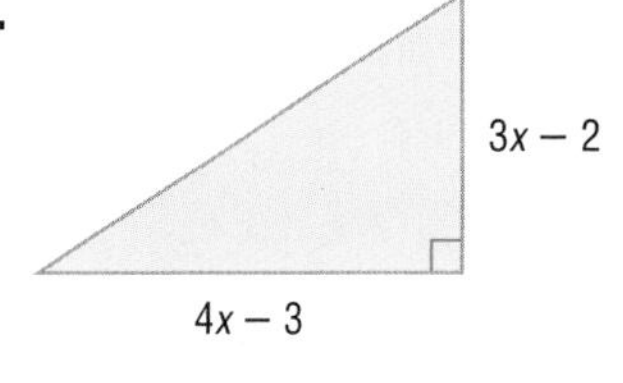

32.

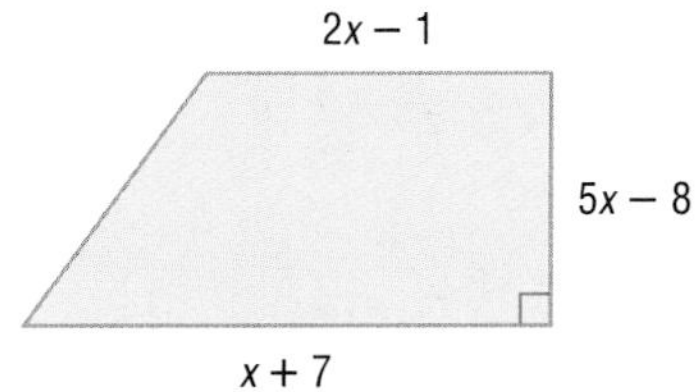

33.

3x + 4

Find each product.

34. $(p + 4)(p^2 + 2p - 7)$

35. $(a - 3)(a^2 - 8a + 5)$

36. $(2x - 5)(3x^2 - 4x + 1)$

37. $(3k + 4)(7k^2 + 2k - 9)$

38. $(n^2 - 3n + 2)(n^2 + 5n - 4)$

39. $(y^2 + 7y - 1)(y^2 - 6y + 5)$

Simplify.

40. $(m + 2)[(m^2 + 3m - 6) + (m^2 - 2m + 4)]$

41. $[(t^2 + 3t - 8) - (t^2 - 2t + 6)](t - 4)$

GEOMETRY The volume V of a prism equals the area of the base B times the height h. Write an expression to represent the volume of each prism.

42.

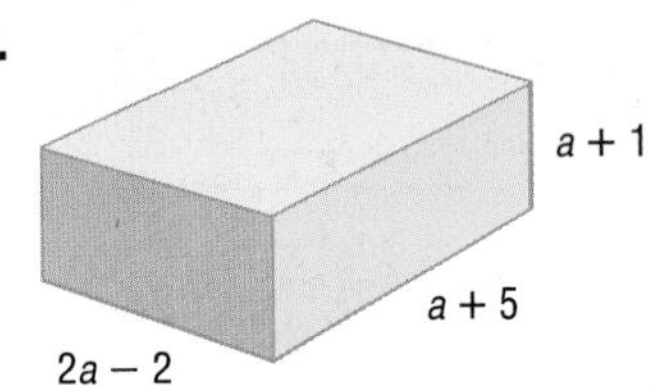

43.

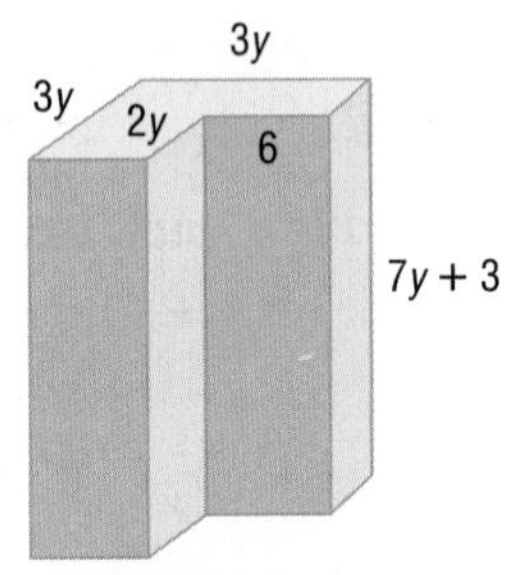

44. BASKETBALL The dimensions of a professional basketball court are represented by a width of $5y - 6$ feet and a length of $2y + 10$ feet. Find an expression for the area of the court.

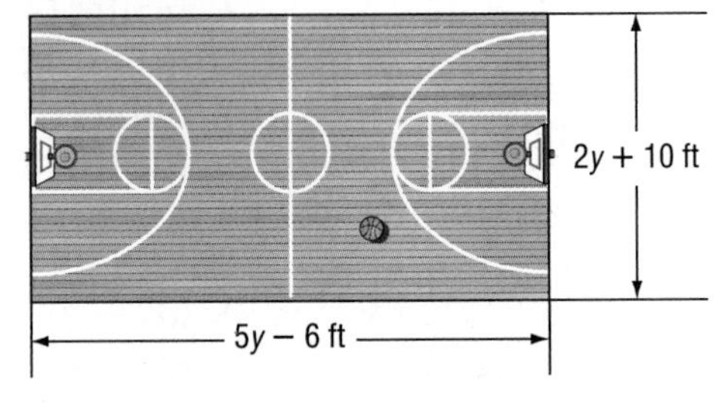

Real-World Link

More than 200 million people a year pay to see basketball games. That is more admissions than for any other American sport.

Source: *Compton's Encyclopedia*

OFFICE SPACE For Exercises 45–47, use the following information.

LaTanya's modular office is square. Her office in the company's new building will be 2 feet shorter in one direction and 4 feet longer in the other.

45. Write expressions for the dimensions of LaTanya's new office.

46. Write a polynomial expression for the area of her new office.

47. Suppose her office is presently 9 feet by 9 feet. Will her new office be bigger or smaller than her old office and by how much? Explain.

48. POOL CONSTRUCTION A homeowner is installing a swimming pool in his backyard. He wants its length to be 4 feet longer than its width. Then he wants to surround it with a concrete walkway 3 feet wide. If he can only afford 300 square feet of concrete for the walkway, what should the dimensions of the pool be?

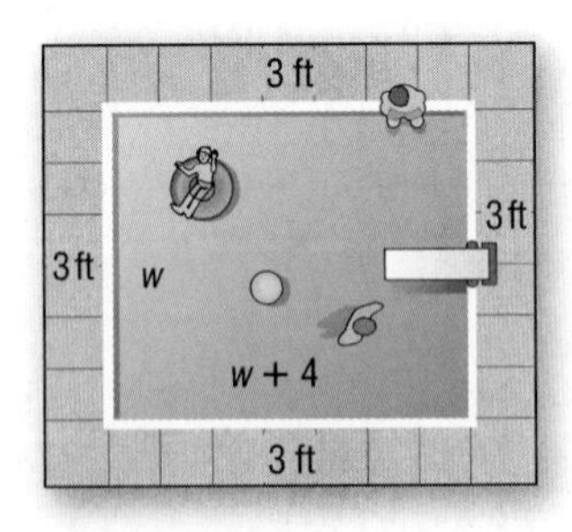

EXTRA PRACTICE
See pages 732, 750.

Math Online
Self-Check Quiz at algebra1.com

49. REASONING Compare and contrast the procedure used to multiply a trinomial by a binomial using the vertical method with the procedure used to multiply a three-digit number by a two-digit number.

50. ALGEBRA TILES Draw a diagram to show how you would use algebra tiles to find the product of $2x - 1$ and $x + 3$.

51. CHALLENGE Determine whether the following statement is *sometimes*, *always*, or *never* true. Explain your reasoning.

The product of a binomial and a trinomial is a polynomial with four terms.

52. OPEN ENDED Write a binomial and a trinomial involving a single variable. Then find their product.

53. ***Writing in Math*** Using the information about multiplying binomials on page 398 explain how multiplying binomials is similar to multiplying two-digit numbers. Include a demonstration of a horizontal method for multiplying 24×36 in your answer.

STANDARDIZED TEST PRACTICE

54. A rectangle's width is represented by x and its length by y. Which expression best represents the area of the rectangle if the length and width are doubled?

A $2xy$

B $2(xy)^2$

C $4xy$

D $(xy)^2$

55. REVIEW Tania's age is 4 years less than twice her little brother Billy's age. If Tania is 12, which equation can be used to determine Billy's age?

F $x = 12$

G $12 = 4 - 2x$

H $12(2) - 4 = x$

J $12 = 2x - 4$

Spiral Review

Simplify. (Lesson 7-5)

56. $3x(2x - 4) + 6(5x^2 + 2x - 7)$

57. $4a(5a^2 + 2a - 7) - 3(2a^2 - 6a - 9)$

58. GEOMETRY The sum of the degree measures of the angles of a triangle is 180. (Lesson 7-4)

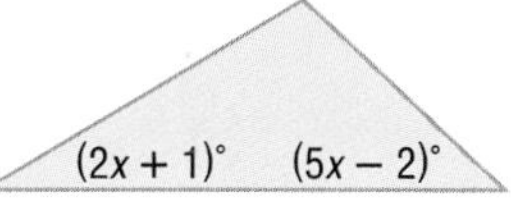

a. Write an expression to represent the measure of the third angle of the triangle.

b. If $x = 15$, find the measures of the three angles of the triangle.

If $f(x) = 2x - 5$ and $g(x) = x^2 + 3x$, find each value. (Lesson 3-6)

59. $f(-4)$

60. $g(-2) + 7$

61. $f(a + 3)$

Solve each equation or formula for the variable specified. (Lesson 2-8)

62. $a = \frac{v}{t}$ for t

63. $ax - by = 2cz$ for y

64. $4x + 3y = 7$ for y

Solve each equation. (Lesson 2-4)

65. $\frac{d - 2}{3} = 7$

66. $3n + 6 = -15$

67. $35 + 20h = 100$

GET READY for the Next Lesson

PREREQUISITE SKILL Simplify. (Lesson 7–1)

68. $(6a)^2$

69. $(7x)^2$

70. $(9b)^2$

71. $(4y^2)^2$

72. $(2v^3)^2$

73. $(3g^4)^2$

7-7 Special Products

Main Ideas

- Find squares of sums and differences.
- Find the product of a sum and a difference.

New Vocabulary

difference of two squares

GET READY for the Lesson

In the previous lesson, you learned how to multiply two binomials using the FOIL method. You may have noticed that the *Outer* and *Inner* terms often combine to produce a trinomial product. This is not always the case, however. Notice that the product of $x + 3$ and $x - 3$ is a binomial.

$(x + 5)(x - 3)$	$(x + 3)(x - 3)$
F O I L	F O I L
$= x^2 - 3x + 5x - 15$	$= x^2 - 3x + 3x - 9$
$= x^2 + 2x - 15$	$= x^2 + 0x - \mathbf{9}$
	$= x^2 - 9$

Squares of Sums and Differences While you can always use the FOIL method to find the product of two binomials, some pairs of binomials have products that follow a specific pattern. One such pattern is the *square of a sum*, $(a + b)^2$ or $(a + b)(a + b)$.

$$(a + b)^2 = a^2 + ab + ab + b^2$$
$$= a^2 + 2ab + b^2$$

KEY CONCEPT	Square of a Sum
Words	The square of $a + b$ is the square of a plus twice the product of a and b plus the square of b.
Symbols	$(a + b)^2 = (a + b)(a + b) = a^2 + 2ab + b^2$
Example	$(x + 7)^2 = x^2 + 2(x)(7) + 7^2 = x^2 + 14x + 49$

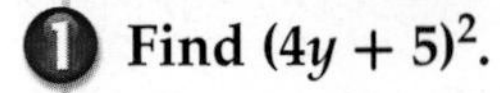

EXAMPLE Square of a Sum

1 Find $(4y + 5)^2$.

$(a + b)^2 = a^2 + 2ab + b^2$

$(4y + 5)^2 = (4y)^2 + 2(4y)(5) + 5^2$ $\quad a = 4y$ and $b = 5$

$= 16y^2 + 40y + 25$ $\quad$ Check by using FOIL.

Concepts in Motion
Animation
algebra1.com

Math Online Extra Examples at algebra1.com

CHECK Your Progress Find each product.

1A. $(8c + 3d)^2$ **1B.** $(3x + 4y)^2$

Study Tip

$(a + b)^2$
In the pattern for $(a + b)^2$, a and b can be numbers, variables, or expressions with numbers and variables.

To find the pattern for the *square of a difference*, $(a - b)^2$, write $a - b$ as $a + (-b)$ and square it using the square of a sum pattern.

$$(a - b)^2 = [a + (-b)]^2$$

$$= a^2 + 2(a)(-b) + (-b)^2 \quad \text{Square of a Sum}$$

$$= a^2 - 2ab + b^2 \quad \text{Simplify. Note that } (-b)^2 = (-b)(-b) \text{ or } b^2.$$

KEY CONCEPT *Square of a Difference*

Words The square of $a - b$ is the square of a minus twice the product of a and b plus the square of b.

Symbols $(a - b)^2 = (a - b)(a - b) = a^2 - 2ab + b^2$

Example $(x - 4)^2 = x^2 - 2(x)(4) + 4^2 = x^2 - 8x + 16$

EXAMPLE Square of a Difference

2 **Find $(5m^3 - 2n)^2$.**

$$(a - b)^2 = a^2 - 2ab + b^2$$

$$(5m^3 - 2n)^2 = (5m^3)^2 - 2(5m^3)(2n) + (2n)^2 \quad a = 5m^3 \text{ and } b = 2n$$

$$= 25m^6 - 20m^3n + 4n^2 \quad \text{Simplify.}$$

CHECK Your Progress Find each product.

2A. $(6p - 1)^2$ **2B.** $(a - 2b)^2$

Real-World Career
Geneticist
Laboratory geneticists work in medicine to find cures for disease, in agriculture to breed new crops and livestock, and in police work to identify criminals.

For more information, go to algebra1.com

Real-World EXAMPLE

3 **GENETICS The Punnett square shows the possible gene combinations between two hamsters. Each hamster passes on one *dominant* gene *G* for golden coloring and one *recessive* gene *g* for cinnamon coloring.**

Show how combinations can be modeled by the square of a binomial. Then determine what percent of the offspring will be pure golden, hybrid golden, and pure cinnamon.

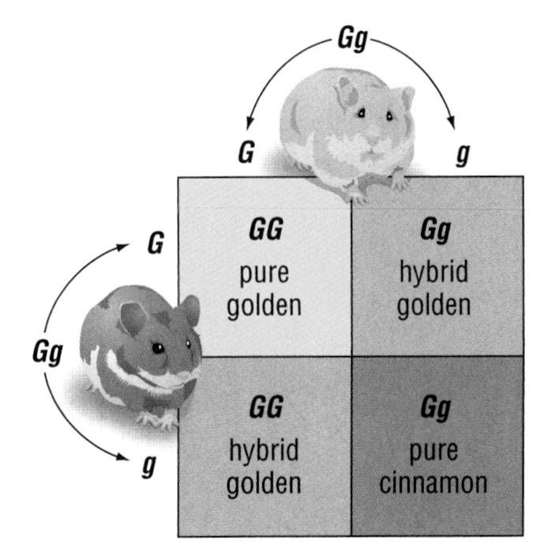

Each parent has half the genes necessary for golden coloring and half the genes necessary for cinnamon coloring. The makeup of each parent can be modeled by $0.5G + 0.5g$. Their offspring can be modeled by the product of $0.5G + 0.5g$ and $0.5G + 0.5g$ or $(0.5G + 0.5g)^2$.

Use this product to determine possible colors of the offspring.

$$(a + b)^2 = a^2 + 2ab + b^2 \quad \text{Square of a Sum}$$

$$(0.5G + 0.5g)^2 = (0.5G)^2 + 2(0.5G)(0.5g) + (0.5g)^2 \quad a = 0.5G \text{ and } b = 0.5g$$

$$= 0.25G^2 + 0.5Gg + 0.25g^2 \quad \text{Simplify.}$$

$$= 0.25GG + 0.5Gg + 0.25gg \quad G^2 = GG \text{ and } g^2 = gg$$

Thus, 25% of the offspring are *GG* or pure golden, 50% are *Gg* or hybrid golden, and 25% are *gg* or pure cinnamon.

CHECK Your Progress

3. Andrew has a garden that is x feet long and x feet wide. He decides that he wants to add 3 feet to the length and the width in order to grow more vegetables. Show how the new area of the garden can be modeled by the square of a binomial.

Online **Personal Tutor at** algebra1.com

Product of a Sum and a Difference You can use the diagram below to find the pattern for the product of the sum and difference of the *same two terms*, $(a + b)(a - b)$. Recall that $a - b$ can be rewritten as $a + (-b)$.

Concepts in Motion
Animation
algebra1.com

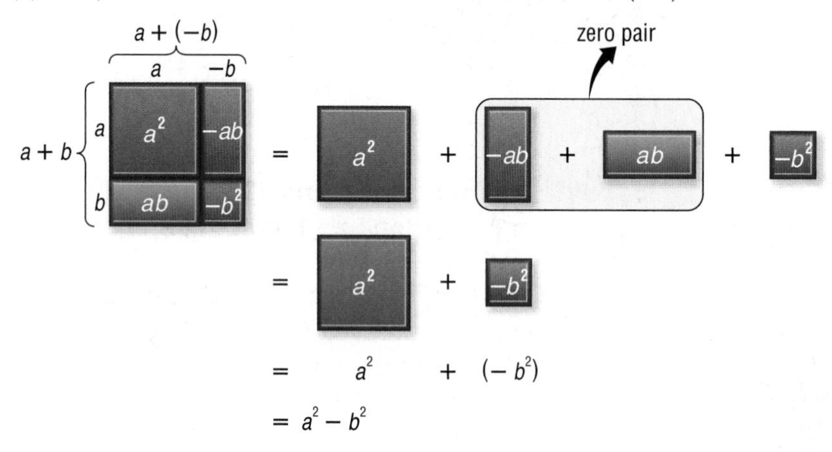

The resulting product, $a^2 - b^2$, is called the **difference of two squares.**

KEY CONCEPT — *Product of a Sum and a Difference*

Words The product of $a + b$ and $a - b$ is the square of a minus the square of b.

Symbols $(a + b)(a - b) = (a - b)(a + b) = a^2 - b^2$

Example $(x + 9)(x - 9) = x^2 - 9^2 = x^2 - 81$

EXAMPLE Product of a Sum and a Difference

4 **Find $(11v - 8w^2)(11v + 8w^2)$**

$$(a - b)(a + b) = a^2 - b^2$$

$$(11v - 8w^2)(11v + 8w^2) = (11v)^2 - (8w^2)^2 \quad a = 11v \text{ and } b = 8w^2$$

$$= 121v^2 - 64w^4 \quad \text{Simplify.}$$

CHECK Your Progress **Find each product.**

4A. $(3n + 2)(3n - 2)$ **4B.** $(4c - 7d)(4c + 7d)$

CHECK Your Understanding

Examples 1–2 (pp. 404–405)

Find each product.

1. $(a + 6)^2$ **2.** $(2a + 7b)^2$ **3.** $(3x + 9y)^2$

4. $(4n - 3)(4n - 3)$ **5.** $(x^2 - 6y)^2$ **6.** $(9 - p)^2$

Example 3 (pp. 405–406)

GENETICS For Exercises 7 and 8, use the following information.
Dalila has brown eyes and Bob has blue eyes. Brown genes *B* are dominant over blue genes *b*. A person with genes *BB* or *Bb* has brown eyes. Someone with genes *bb* has blue eyes. Suppose Dalila's genes for eye color are *Bb*.

7. Write an expression for the possible eye coloring of Dalila and Bob's children.

8. What is the probability that a child of Dalila and Bob would have blue eyes?

Example 4 (p. 406)

Find each product.

9. $(8x - 5)(8x + 5)$ **10.** $(3a + 7b)(3a - 7b)$ **11.** $(4y^2 + 3z)(4y^2 - 3z)$

Exercises

HOMEWORK HELP	
For Exercises	See Examples
12–14	1
15–17	2
18–19	3
20–22	4

Find each product.

12. $(k + 8)(k + 8)$ **13.** $(y + 4)^2$ **14.** $(2g + 5)^2$

15. $(a - 5)(a - 5)$ **16.** $(n - 12)^2$ **17.** $(7 - 4y)^2$

GENETICS For Exercises 18 and 19, use the following information and the Punnett square.
Cystic fibrosis is inherited from parents only if both parents have the abnormal *CF* gene. Children of two parents with the *CF* gene will either be affected with the disease, a carrier but not affected, or not have the gene.

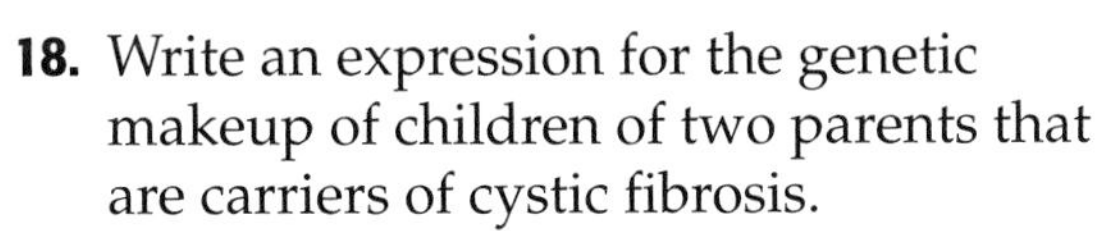

18. Write an expression for the genetic makeup of children of two parents that are carriers of cystic fibrosis.

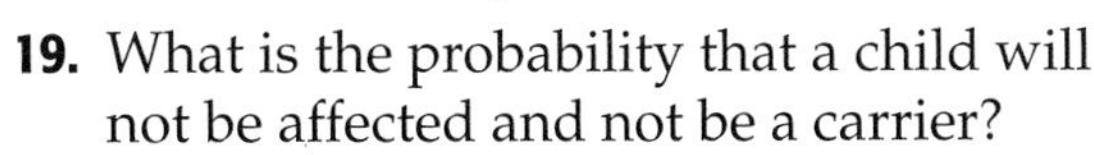

19. What is the probability that a child will not be affected and not be a carrier?

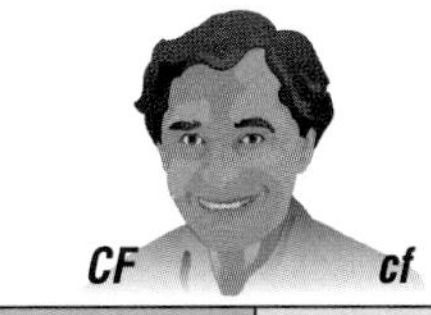

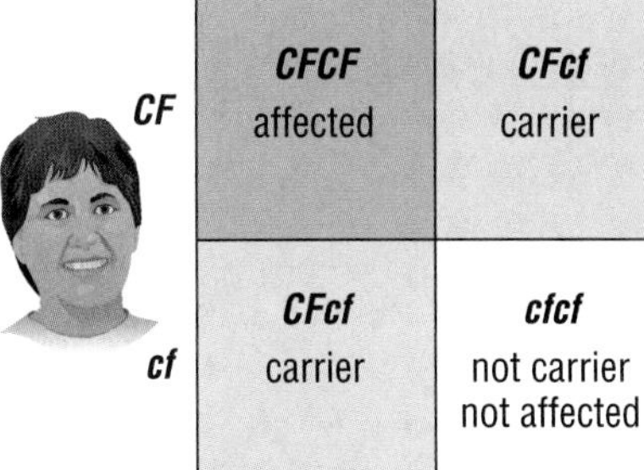

Find each product.

20. $(b + 7)(b - 7)$ **21.** $(c - 2)(c + 2)$ **22.** $(11r + 8)(11r - 8)$

Find each product.

23. $(9x + 3)^2$ **24.** $(4 - 6h)^2$ **25.** $(12p - 3)(12p + 3)$

26. $(a + 5b)^2$ **27.** $(m + 7n)^2$ **28.** $(2x - 9y)^2$

29. $(3n - 10p)^2$ **30.** $(5w + 14)(5w - 14)$ **31.** $(4d - 13)(4d + 13)$

32. $(x^3 + 4y)^2$ **33.** $(3a^2 - b^2)^2$ **34.** $(8a^2 - 9b^3)(8a^2 + 9b^3)$

35. $(5x^4 - y)(5x^4 + y)$ **36.** $\left(\frac{2}{3}x - 6\right)^2$ **37.** $\left(\frac{4}{5}x + 10\right)^2$

38. $(2n + 1)(2n - 1)(n + 5)$ **39.** $(p + 3)(p - 4)(p - 3)(p + 4)$

EXTRA PRACTICE
See pages 732, 750.
Math Online
Self-Check Quiz at algebra1.com

MAGIC TRICK **For Exercises 40–43, use the following information.**
Madison says that she can perform a magic trick with numbers. She asks you to pick an integer, any integer. Square that integer. Then, add twice your original number. Next add 1. Take the square root of the result. Finally, subtract your original number. Then Madison exclaims with authority, "Your answer is 1!"

40. Pick an integer and follow Madison's directions. Is your result 1?

41. Let a represent the integer you chose. Then, find a polynomial representation for the first three steps of Madison's directions.

42. The polynomial you wrote in Exercise 41 is the square of what binomial sum?

43. Take the square root of the perfect square you wrote in Exercise 42, then subtract a, your original integer. What is the result?

Real-World Link
Cael Sanderson of Iowa State University is the only wrestler in NCAA Division 1 history to be undefeated for four years. He compiled a 159–0 record from 1999–2002.

Source: www.teamsanderson.cc

WRESTLING **For Exercises 44–46, use the following information.**
A high school wrestling mat must be a square with 38-foot sides and contain two circles as shown. Suppose the inner circle has a radius of s feet, and the outer circle's radius is nine feet longer than the inner circle.

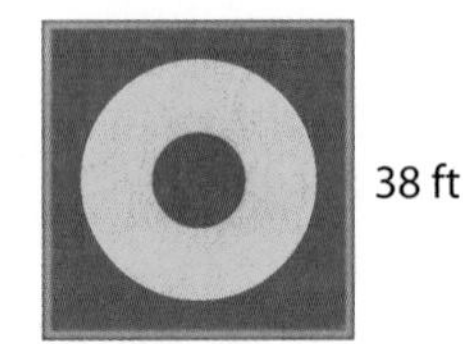

44. Write an expression for the area of the larger circle.

45. Write an expression for the area of the square outside the circle.

46. Use the expression to find the area if $s = 1$.

47. GEOMETRY The area of the shaded region models the difference of two squares, $a^2 - b^2$. Show that the area of the shaded region is also equal to $(a - b)(a + b)$. (*Hint*: Divide the shaded region into two trapezoids as shown.)

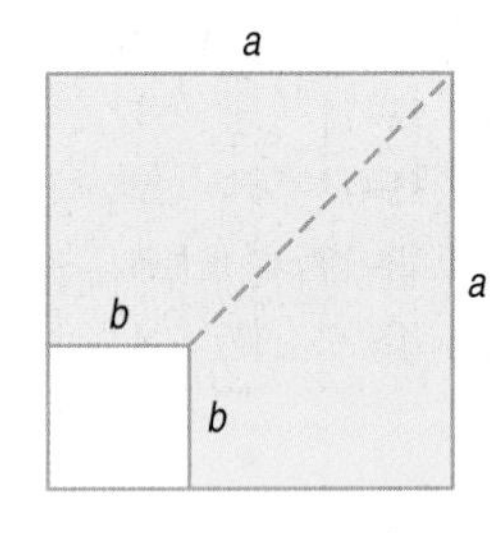

H.O.T. Problems

48. REASONING Compare and contrast the pattern for the square of a sum with the pattern for the square of a difference.

49. ALGEBRA TILES Draw a diagram to show how you would use algebra tiles to model the product of $x - 3$ and $x - 3$, or $(x - 3)^2$.

50. OPEN ENDED Write two binomials whose product is a difference of squares. Then multiply to verify your answer.

51. CHALLENGE Does a pattern exist for the cube of a sum, $(a + b)^3$?

a. Investigate this question by finding the product of $(a + b)(a + b)(a + b)$.

b. Use the pattern you discovered in part a to find $(x + 2)^3$.

c. Draw a diagram of a geometric model for the cube of a sum.

52. ***Writing in Math*** Using the information about the product of two binomials on page 404 distinguish when the product of two binomials is also a binomial. Include an example of two binomials whose product is a binomial and an example of two binomials whose product is not a binomial in your answer.

STANDARDIZED TEST PRACTICE

53. The base of a triangle is represented by $x - 4$, and the height is represented by $x + 4$. Which of the following represents the area of the triangle?

A $x^2 - 16$

B $\frac{1}{2}x^2 + 4x - 8$

C $x^2 + 8x - 16$

D $\frac{1}{2}x^2 - 8$

54. REVIEW The sum of a number and 8 is −19. Which equation shows this relationship?

F $8n = -19$

G $n + 8 = -19$

H $n - 8 = 19$

J $n - 19 = 8$

Spiral Review

Find each product. (Lesson 7-6)

55. $(x + 2)(x + 7)$

56. $(c - 9)(c + 3)$

57. $(4y - 1)(5y - 6)$

58. $(3n - 5)(8n + 5)$

59. $(x - 2)(3x^2 - 5x + 4)$

60. $(2k + 5)(2k^2 - 8k + 7)$

Solve. (Lesson 7-5)

61. $6(x + 2) + 4 = 5(3x - 4)$

62. $-3(3a - 8) + 2a = 4(2a + 1)$

63. $p(p + 2) + 3p = p(p - 3)$

64. $y(y - 4) + 2y = y(y + 12) - 7$

Use elimination to solve each system of equations. (Lesson 5-3, 5-4)

65. $\frac{3}{4}x + \frac{1}{5}y = 5$
$\frac{3}{4}x - \frac{1}{5}y = -5$

66. $2x - y = 10$
$5x + 3y = 3$

67. $2x = 4 - 3y$
$3y - x = -11$

Write the slope-intercept form of an equation that passes through the given point and is perpendicular to the graph of each equation. (Lesson 4-6)

68. $5x + 5y = 35, (-3, 2)$

69. $2x - 5y = 3, (-2, 7)$

70. $5x + y = 2, (0, 6)$

Find the *n*th term of each arithmetic sequence described. (Lesson 3-7)

71. $a_1 = 3, d = 4, n = 18$

72. $-5, 1, 7, 13, \ldots$ for $n = 12$

73. PHYSICAL FITNESS Mitchell likes to exercise regularly. He likes to warm up by walking two miles. Then he runs five miles. Finally, he cools down by walking another mile. Identify the graph that best represents Mitchell's heart rate as a function of time. (Lesson 1-9)

a.

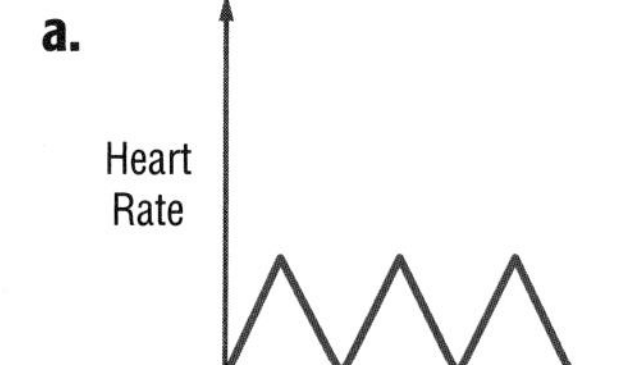

b.

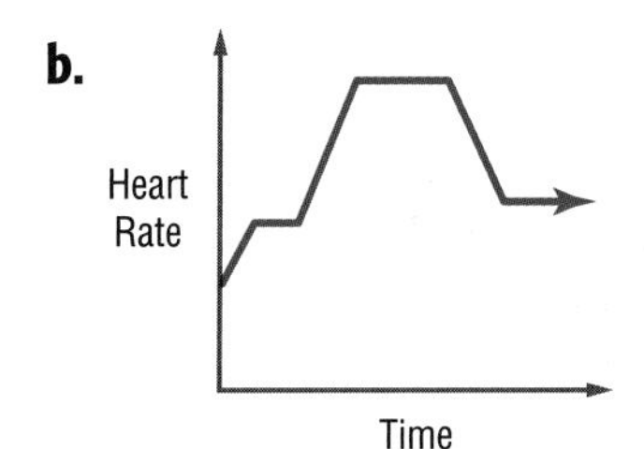

c.

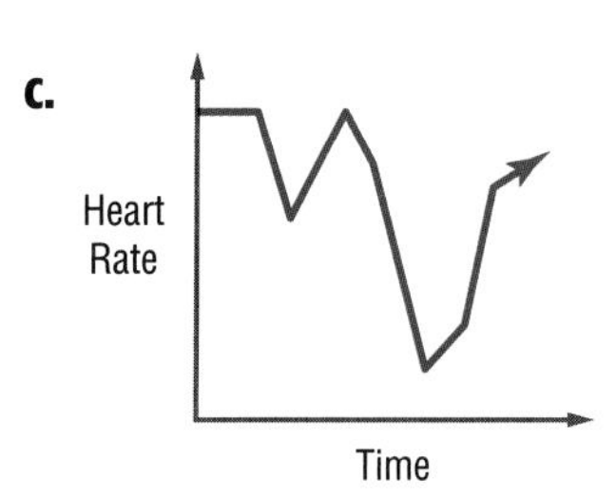

CHAPTER 7

Study Guide and Review

FOLDABLES Study Organizer — GET READY to Study

Be sure the following Key Concepts are noted in your Foldable.

	+	−	x	÷
Mon.				
Poly.				

Key Concepts

Multiplying Monomials (Lesson 7-1)

- To multiply two powers that have the same base, add exponents.
- To find the power of a power, multiply exponents.
- The power of a product is the product of the powers.

Dividing Monomials (Lesson 7-2)

- To divide two powers that have the same base, subtract the exponents.
- To find the power of a quotient, find the power of the numerator and the power of the denominator.
- Any nonzero number raised to the zero power is 1.
- For any nonzero number a and any integer n,

$$a^{-n} = \frac{1}{a^n} \text{ and } \frac{1}{a^{-n}} = a^n.$$

Polynomials (Lesson 7-3)

- The degree of a monomial is the sum of the exponents of all its variables.
- The degree of a polynomial is the greatest degree of any term. To find the degree of a polynomial, you must find the degree of each term.

Operations with Polynomials

(Lessons 7-4 to 7-7)

- The Distributive Property can be used to multiply a polynomial by a monomial.
- Square of a Sum: $(a + b)^2 = a^2 + 2ab + b^2$
- Square of a Difference: $(a - b)^2 = a^2 - 2ab + b^2$
- Product of a Sum and a Difference: $(a + b)(a - b) = (a - b)(a + b) = a^2 - b^2$

Key Vocabulary

binomial (p. 376)
constant (p. 358)
degree of a monomial (p. 377)
degree of a polynomial (p. 377)
difference of two squares (p. 406)
FOIL method (p. 399)
monomial (p. 358)
negative exponent (p. 369)
polynomial (p. 376)
Power of a Power (p. 359)
Power of a Product (p. 360)
Power of a Quotient (p. 367)
Product of Powers (p. 359)
Quotient of Powers (p. 366)
trinomial (p. 376)
zero exponent (p. 368)

Vocabulary Check

Choose a term from the vocabulary list that best matches each example.

1. $4^{-3} = \frac{1}{4^3}$
2. $(n^3)^5 = n^{15}$
3. $\frac{4x^2y}{8xy^3} = \frac{x}{2y^2}$
4. $4x^2$
5. $x^2 - 3x + 1$
6. $2^0 = 1$
7. $x^4 - 3x^3 + 3x^2 - 1$
8. $x^2 + 2$
9. $(a^3b)(2ab^2) = 2a^4b^3$
10. $(x + 3)(x - 4) = x^2 - 4x + 3x - 12$

Math Online Vocabulary Review at algebra1.com

Lesson-by-Lesson Review

7–1 Multiplying Monomials (pp. 358–364)

Simplify.

11. $y^3 \cdot y^3 \cdot y$

12. $(3ab)(-4a^2b^3)$

13. $(-4a^2x)(-5a^3x^4)$

14. $(4a^2b)^3$

15. $(-3xy)^2(4x)^3$

16. $(-2c^2d)^4(-3c^2)^3$

17. $-\frac{1}{2}(m^2n^4)^2$

18. $(5a^2)^3 + 7(a^6)$

19. GEOMETRY A cone has a radius of $4x^3$ and a height of $3b^2$. Use the formula $V = \frac{1}{3}(\pi r^2 \ell)$ to find the volume of the cone.

Example 1 **Simplify $(2ab^2)(3a^2b^3)$.**

$(2ab^2)(3a^2b^3)$

$= (2 \cdot 3)(a \cdot a^2)(b^2 \cdot b^3)$ Commutative Property

$= 6a^3b^5$ Product of Powers

Example 2 **Simplify $(2x^2y^3)^3$.**

$(2x^2y^3)^3 = 2^3(x^2)^3(y^3)^3$ Power of a Product

$= 8x^6y^9$ Power of a Power

7–2 Dividing Monomials (pp. 366–373)

Simplify. Assume that no denominator is equal to zero.

20. $\frac{(3y)^0}{6a}$

21. $\left(\frac{3bc^2}{4d}\right)^3$

22. $x^{-2}y^0z^3$

23. $\frac{27b^{-2}}{14b^{-3}}$

24. $\frac{(3a^3bc^2)^2}{18a^2b^3c^4}$

25. $\frac{-16a^3b^2x^4y}{-48a^4bxy^3}$

26. $\frac{(-a)^5b^8}{a^5b^2}$

27. $\frac{(4a^{-1})^{-2}}{(2a^4)^2}$

28. $\left(\frac{5xy^{-2}}{35x^{-2}y^6}\right)^0$

29. $\frac{12}{3}\left(\frac{m}{n^3}\right)\left(\frac{n^4}{m^3}\right)$

30. GEOMETRY The area of a triangle is $50a^2b$ square feet. The base of the triangle is $5a$ feet. What is the height of the triangle?

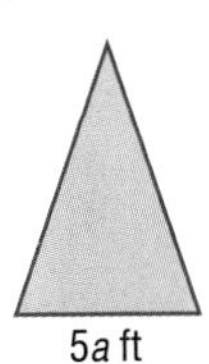

Example 3 **Simplify $\frac{2x^6y}{8x^2y^2}$. Assume that no denominator is equal to zero.**

$\frac{2x^6y}{8x^2y^2} = \left(\frac{2}{8}\right)\left(\frac{x^6}{x^2}\right)\left(\frac{y}{y^2}\right)$ Group the powers with the same base.

$= \left(\frac{1}{4}\right)(x^{6-2})(y^{1-2})$ Quotient of Powers

$= \frac{x^4}{4y}$ Simplify.

Example 4 **Simplify $\frac{m^{-4}n^3p^0}{mn^{-2}}$. Assume that no denominator is zero.**

$\frac{m^{-4}n^3p^0}{mn^{-2}} = \left(\frac{m^{-4}}{m}\right)\left(\frac{n^3}{n^{-2}}\right)(p^0)$ Group the powers with the same base.

$= (m^{-4-1})(n^{3+2})$ Quotient of Powers and Zero Exponent

$= \frac{n^5}{m^5}$ Simplify.

Study Guide and Review

7-3 Polynomials (pp. 376–381)

Find the degree of each polynomial.

31. $n - 2p^2$

32. $29n^2 + 17n^2t^2$

33. $4xy + 9x^3z^2 + 17rs^3$

34. $-6x^5y - 2y^4 + 4 - 8y^2$

Arrange the terms of each polynomial so that the powers of x are in descending order.

35. $3x^4 - x + x^2 - 5$

36. $-2x^2y^3 - 27 - 4x^4 + xy + 5x^3y^2$

37. CONSTRUCTION Ben is building a brick patio with pavers using the drawing below. Write a polynomial to represent the area of the patio.

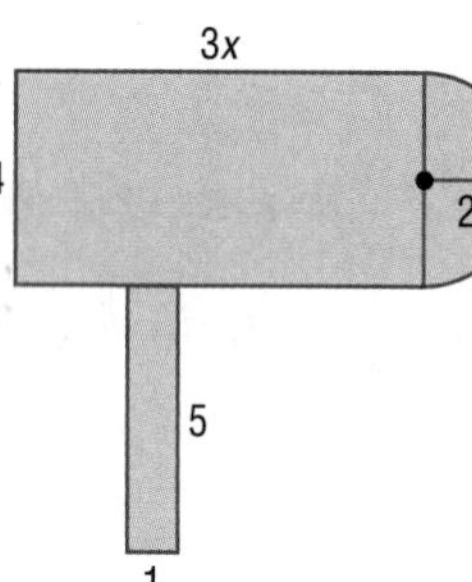

Example 5 Find the degree of $2xy^3 + x^2y$.

Polynomial:	$2xy^3 + x^2y$
Terms:	$2xy^3, x^2y$
Degree of Each Term:	4, 3
Degree of Polynomial:	4

Example 6 Arrange the terms of $4x^2 + 9x^3 - 2 - x$ so that the powers of x are in descending order.

$4x^2 + 9x^3 - 2 - x$

$= 4x^2 + 9x^3 - 2x^0 - x^1$ $x^0 = 1$ and $x = x^1$

$= 9x^3 + 4x^2 - x - 2$ $3 > 2 > 1 > 0$

7-4 Adding and Subtracting Polynomials (pp. 384–388)

Find each sum or difference.

38. $(2x^2 - 5x + 7) - (3x^3 + x^2 + 2)$

39. $(x^2 - 6xy + 7y^2) + (3x^2 + xy - y^2)$

40. $(7z^2 + 4) - (3z^2 + 2z - 6)$

41. $(13m^4 - 7m - 10) + (8m^4 - 3m + 9)$

42. $(11m^2n^2 + 4mn - 6) + (5m^2n^2 + 6mn + 17)$

43. $(-5p^2 + 3p + 49) - (2p^2 + 5p + 24)$

44. GARDENING Kyle is planting flowers around the perimeter of his rectangular garden. If the perimeter of his garden is $110x$ and one side measures $25x$, find the length of the other side.

Example 7 Find $(7r^2 + 9r) - (12r^2 - 4)$.

$(7r^2 + 9r) - (12r^2 - 4)$

$= 7r^2 + 9r + (-12r^2 + 4)$ The additive inverse of $12r^2 - 4$ is $-12r^2 + 4$.

$= (7r^2 - 12r^2) + 9r + 4$ Group like terms.

$= -5r^2 + 9r + 4$ Add like terms.

Mixed Problem Solving
For mixed problem-solving practice, see page 750.

7-5 Multiplying a Polynomial by a Monomial (pp. 390–395)

Simplify.

45. $b(4b - 1) + 10b$

46. $x(3x - 5) + 7(x^2 - 2x + 9)$

47. $8y(11y^2 - 2y + 13) - 9(3y^3 - 7y + 2)$

48. $2x(x - y^2 + 5) - 5y^2(3x - 2)$

Solve each equation.

49. $m(2m - 5) + m = 2m(m - 6) + 16$

50. $2(3w + w^2) - 6 = 2w(w - 4) + 10$

51. SHOPPING Nicole bought x shirts for \$15.00 each, y pants for \$25.72 each, and z belts for \$12.53 each. Sales tax on these items was 7%. Write an expression to find the total cost of Nicole's purchases.

Example 8 Simplify $x^2(x + 2) + 3(x^3 + 4x^2)$.

$x^2(x + 2) + 3(x^3 + 4x^2)$
$= x^2(x) + x^2(2) + 3(x^3) + 3(4x^2)$
$= x^3 + 2x^2 + 3x^3 + 12x^2$ Multiply.
$= 4x^3 + 14x^2$ Combine like terms.

Example 9 Solve $x(x - 10) + x(x + 2) + 3 = 2x(x + 1) - 7$.

$x(x - 10) + x(x + 2) + 3 = 2x(x + 1) - 7$
$x^2 - 10x + x^2 + 2x + 3 = 2x^2 + 2x - 7$
$2x^2 - 8x + 3 = 2x^2 + 2x - 7$
$-8x + 3 = 2x - 7$
$-10x = -10$
$x = 1$

7-6 Multiplying Polynomials (pp. 398–403)

Find each product.

52. $(r - 3)(r + 7)$

53. $(4a - 3)(a + 4)$

54. $(5r - 7s)(4r + 3s)$

55. $(3x + 0.25)(6x - 0.5)$

56. $(2k + 1)(k^2 + 7k - 9)$

57. $(4p - 3)(3p^2 - p + 2)$

58. MANUFACTURING A company is designing a box in the shape of a rectangular prism for dry pasta. The length is 2 inches more than twice the width and the height is 3 inches more than the length. Write an expression for the volume of the box.

Example 10 Find $(3x + 2)(x - 2)$.

F L
$(3x + 2)(x - 2)$
I
O

F O I L
$= (3x)(x) + (3x)(-2) + (2)(x) + (2)(-2)$
$= 3x^2 - 6x + 2x - 4$ Multiply
$= 3x^2 - 4x - 4$ Combine like terms.

Example 11 Find $(2y - 5)(4y^2 + 3y - 7)$.

$(2y - 5)(4y^2 + 3y - 7)$
$= 2y(4y^2 + 3y - 7) - 5(4y^2 + 3y - 7)$
$= 8y^3 + 6y^2 - 14y - 20y^2 - 15y + 35$
$= 8y^3 - 14y^2 - 29y + 35$

7-7 Special Products (pp. 404–409)

Find each product.

59. $(x - 6)(x + 6)$

60. $(4x + 7)^2$

61. $(8x - 5)^2$

62. $(5x - 3y)(5x + 3y)$

63. $(6a - 5b)^2$

64. $(3m + 4n)^2$

65. GENETICS Emily and Santos are both able to roll their tongues. Tongue rolling genes R are dominant over nonrolling genes r. A person with genes RR or Rr is able to roll their tongue. Someone with genes rr cannot roll their tongue. Suppose Emily's and Santos's genes for tongue rolling are Rr. Write an expression for the possible tongue-rolling abilities of Santos's and Emily's children. What is the probability that a child of Emily and Santos could not roll their tongue?

Example 12 Find $(r - 5)^2$.

$(a - b)^2 = a^2 - 2ab + b^2$ Square of a Difference

$(r - 5)^2 = r^2 - 2(r)(5) + 5^2$ $a = r$ and $b = 5$

$= r^2 - 10r + 25$ Simplify.

Example 13 Find $(2c + 9)(2c - 9)$.

$(a + b)(a - b) = a^2 - b^2$

$(2c + 9)(2c - 9) = (2c)^2 - 9^2$ $a = 2c$ and $b = 9$

$= 4c^2 - 81$ Simplify.

Practice Test

Simplify. Assume that no denominator is equal to zero.

1. $(a^2b^4)(a^3b^5)$
2. $(-12abc)(4a^2b^4)$
3. $\left(\frac{3}{5}m\right)^2$
4. $(-3a)^4(a^5b)^2$
5. $(-5a^2)(-6b^3)^2$
6. $\frac{mn^4}{m^3n^2}$
7. $\frac{9a^2bc^2}{63a^4bc}$
8. $\frac{48a^2bc^5}{(3ab^3c^2)^2}$

Find the degree of each polynomial. Then arrange the terms so that the powers of y are in descending order.

9. $2y^2 + 8y^4 + 9y$
10. $5xy - 7 + 2y^4 - x^2y^3$

Find each sum or difference.

11. $(5a + 3a^2 - 7a^3) + (2a - 8a^2 + 4)$
12. $(x^3 - 3x^2y + 4xy^2 + y^3) - (7x^3 + x^2y - 9xy^2 + y^3)$

13. **GEOMETRY** The measures of two sides of a triangle are given. If the perimeter is represented by $11x^2 - 29x + 10$, find the measure of the third side.

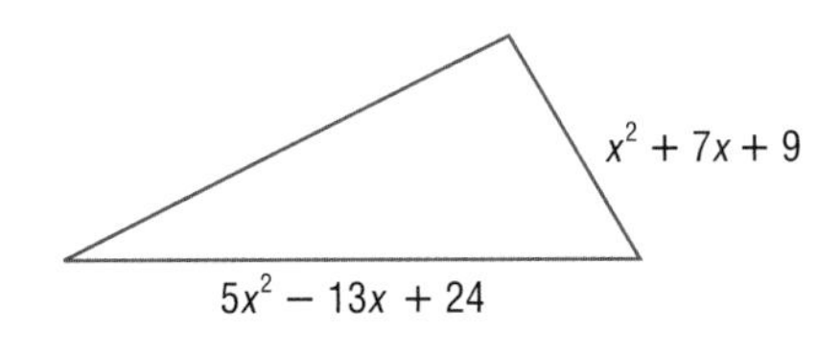

14. **MULTIPLE CHOICE** What is the area of the square with sides that measure $x - 6$?

A $4x - 24$

B $x^2 - 12x + 36$

C $x^2 + 12x + 36$

D $x^2 - 36$

Simplify.

15. $(h - 5)^2$
16. $(4x - y)(4x + y)$
17. $3x^2y^3(2x - xy^2)$
18. $(2a^2b + b^2)^2$
19. $(4m + 3n)(2m - 5n)$
20. $(2c + 5)(3c^2 - 4c + 2)$

Solve each equation.

21. $2x(x - 3) = 2(x^2 - 7) + 2$
22. $3a(a^2 + 5) - 11 = a(3a^2 + 4)$

23. **MULTIPLE CHOICE** If $x^2 + 2xy + y^2 = 8$, find $3(x + y)^2$.

F 2

G 4

H 12

J 24

GENETICS The Punnett square shows the possible gene combinations of a cross between two pea plants. Each plant passes on one dominant gene T for tallness and one recessive gene t for shortness.

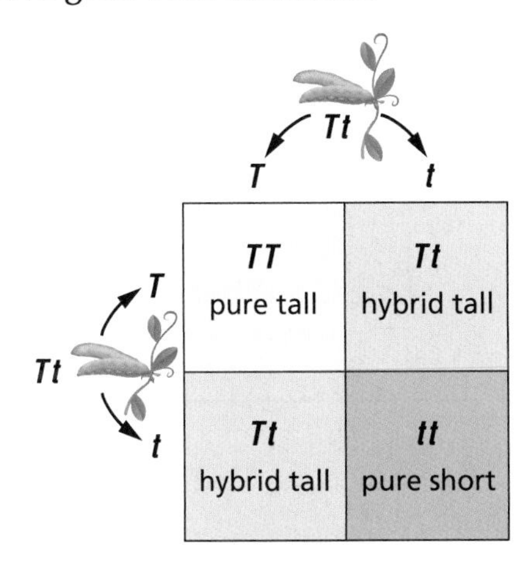

24. Show how the possible combinations can be modeled by the square of a binomial.
25. What is the probability that the offspring will be pure tall (TT), hybrid tall (Tt), and pure short (tt)?

CHAPTER 7

Standardized Test Practice

Cumulative, Chapters 1–7

Read each question. Then fill in the correct answer on the answer document provided by your teacher or on a sheet of paper.

1. Which expression describes the area in square units of a rectangle that has a width of $2a^2b$ and a length of $4a^3b^4$?

A $8a^6b^4$ C $6a^6b^4$

B $8a^5b^5$ D $6a^5b^5$

TEST-TAKING TIP

QUESTION 1 On multiple-choice questions, try to compute the answer first. Then compare your answer to the given answer choices. If you don't find your answer among the choices, check your calculations.

2. GRIDDABLE What is the value of r in the equation below?

$$r = \frac{5(2) - 3}{2(7 - 6)}$$

3. Which statement is true for the graph below?

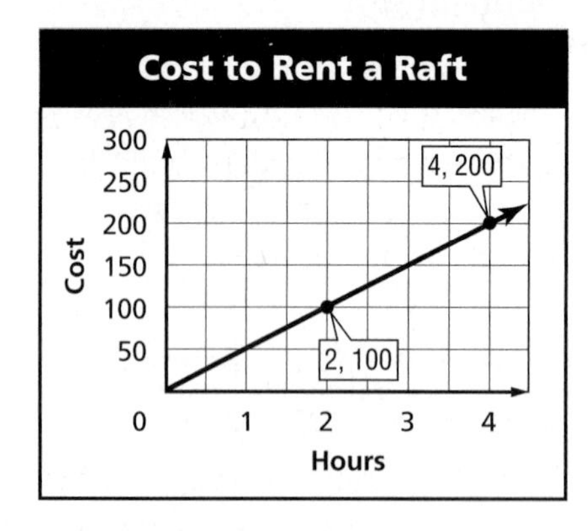

F It will cost Chelsea \$200 to rent a raft for 4 hours.

G It will cost Maria \$150 to rent a raft for 2 hours.

H It will cost John \$100 to rent a raft for 1 hour.

J It will cost Marcus \$100 to rent a raft for 3 hours.

4. The graph of the linear equation $y = \frac{1}{2}x + 3$ is shown below.

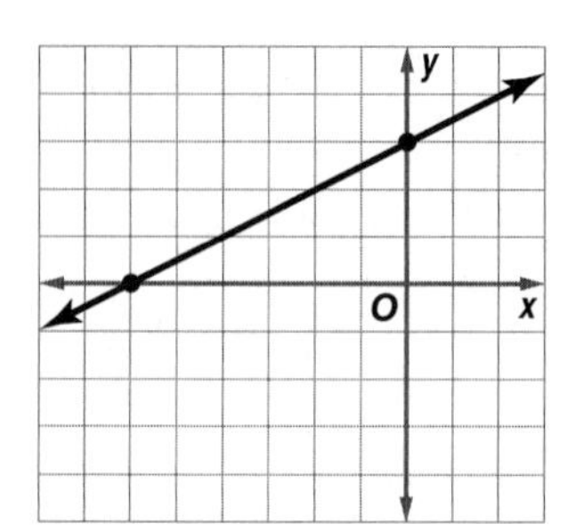

Which point is not in the solution set of $y > \frac{1}{2}x + 3$?

A (1, 5) C (−3, 4)

B (−6, 1) D (−2, 1)

5. Which point on the grid below satisfies the conditions $x < 5$ and $y > -6$?

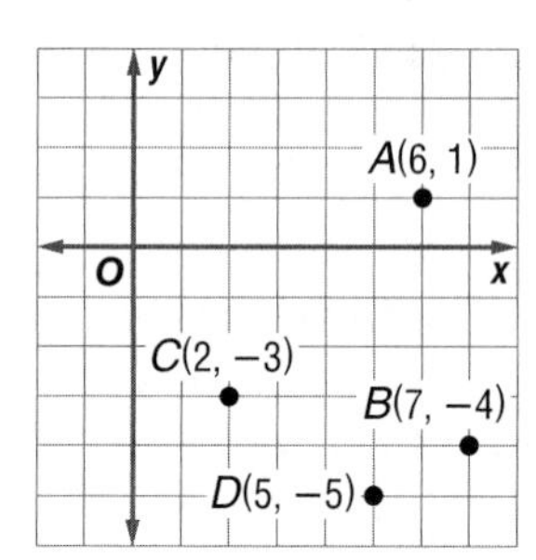

F Point A

G Point B

H Point C

J Point D

6. Maya used 18 square feet of material to make a blanket. The blanket is in the shape of a square. She would like to put a binding around the outside of the blanket. How can she find the length of a side?

A Find the square root of the area.

B Multiply the area by 4.

C Divide the area by 4.

D Divide the are by 2.

Preparing for Standardized Tests
For test-taking strategies and more practice, see pages 756–773.

7. Describe the effect on the area of a circle when the radius is tripled.

F The area is reduced by $\frac{1}{3}$.

G The area remains constant.

H The area is tripled.

J The area is increased nine times.

8. GRIDDABLE Bradley needs to stain his deck. The deck measures 14 by 16 feet. If the stain costs $1.25 per square foot, including tax, how much will it cost to stain his deck?

9. The table shows the sum of the interior angle measures in certain convex polygons.

Convex Polygon	Number of Sides	Sum of Angle Measures
triangle	3	180°
quadrilateral	4	360°
pentagon	5	540°
hexagon	6	720°

Based on the table, what is the sum of the angle measures of an octagon?

A 720°

B 900°

C 1000°

D 1080°

10. Carlos builds circular parachutes. The area of one parachute is 100 square feet. If he triples the radius of the parachute to build a new parachute, what will the area of the new parachute be?

F 100 ft^2

G 300 ft^2

H 600 ft^2

J 900 ft^2

11. The net of a square pyramid is shown below.

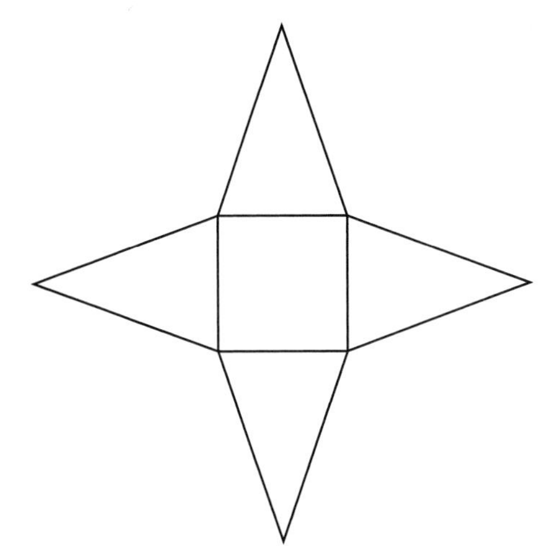

Measure the dimensions of the pyramid to the nearest $\frac{1}{8}$ inch. Find the surface area of the pyramid to the nearest square inch.

A 3 in^2 C 6 in^2

B 4 in^2 D 8 in^2

Pre-AP

Record your answers on a sheet of paper. Show your work.

12. Two cars leave at the same time and both drive to Nashville. The cars' distance from Knoxville, in miles, can be represented by the two equations below, where t represents time in hours.

Car A: $A = 65t + 10$; Car B: $B = 55t + 20$

a. Which car is faster? Explain.

b. How far did Car B travel after 2 hours?

c. Find an expression that models the distance between the two cars.

d. How far apart are the cars after $3\frac{1}{2}$ hours?

NEED EXTRA HELP?												
If You Missed Question...	1	2	3	4	5	6	7	8	9	10	11	12
Go to Lesson or Page...	7-1	1-3	3-3	6-6	6-7	7-2	7-2	7-2	7-1	7-2	7-1	7-4

CHAPTER 8

Factoring

BIG Ideas

- Find the prime factorization of integers and monomials.
- Factor polynomials.
- Use the Zero Product Property to solve equations.

Key Vocabulary

factored form (p. 421)

perfect square trinomials (p. 454)

prime polynomial (p. 443)

Real-World Link

Dolphins Factoring is used to solve problems involving vertical motion. For example, factoring can be used to determine how long a dolphin that jumps out of the water is in the air.

FOLDABLES™ Study Organizer

Factoring Make this Foldable to help you organize your notes on factoring. Begin with a sheet of plain $8\frac{1}{2}''$ by 11″ paper.

1. **Fold** in thirds and then in half along the width.

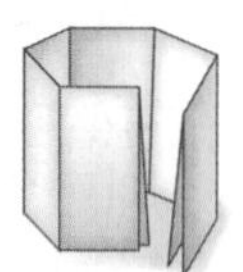

2. **Open.** Fold lengthwise, leaving a $\frac{1}{2}''$ tab on the right.

3. **Open.** Cut short side along folds to make tabs.

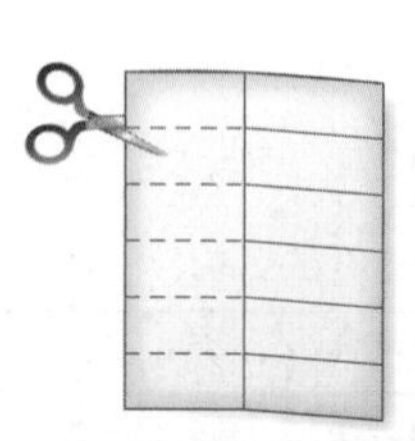

4. **Label** each tab as shown.

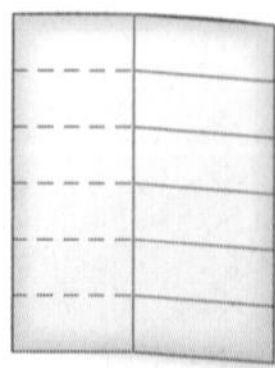

GET READY for Chapter 8

Diagnose Readiness You have two options for checking Prerequisite Skills.

Option 2

Math Online Take the Online Readiness Quiz at algebra1.com.

Option 1

Take the Quick Check below. Refer to the Quick Review for help.

QUICKCheck

Rewrite each expression using the Distributive Property. Then simplify. (Lesson 1-5)

1. $3(4 - x)$
2. $a(a + 5)$
3. $-7(n^2 - 3n + 1)$
4. $6y(-3y - 5y - 5y^2 + y^3)$
5. **JOBS** In a typical week, Mr. Jackson averages 4 hours using e-mail, 10 hours of meeting in person, and 20 hours on the telephone. Write an expression that could be used to determine how many hours he will spend on these activities over the next month.

Find each product. (Lesson 7-6)

6. $(x + 4)(x + 7)$
7. $(3n - 4)(n + 5)$
8. $(6a - 2b)(9a + b)$
9. $(-x - 8y)(2x - 12y)$
10. **TABLE TENNIS** The dimensions of a homemade table tennis table are represented by a width of $2x + 3$ and a length of $x + 1$. Find an expression for the area of the table tennis table.

Find each product. (Lesson 7-7)

11. $(y + 9)^2$
12. $(3a - 2)^2$
13. $(3m + 5n)^2$
14. $(6r - 7s)^2$

QUICKReview

EXAMPLE 1

Rewrite $n\left(n - 3n^2 + 2 + \frac{4}{n}\right)$ using the Distributive Property. Then simplify.

$n\left(n - 3n^2 + 2 + \frac{4}{n}\right)$ Original expression

$= (n)(n) + (n)(-3n^2) + (n)(2) + (n)\left(\frac{4}{n}\right)$ Distribute n to each term inside the parentheses.

$= n^2 - 3n^3 + 2n + 4$ Multiply.

$= -3n^3 + n^2 + 2n + 4$ Rewrite in descending order with respect to the exponents.

EXAMPLE 2

Find $(x + 2)(3x - 1)$.

$(x + 2)(3x - 1)$ Original expression

$= (x)(3x) + (x)(-1) + (2)(3x) + (2)(-1)$ FOIL Method

$= 3x^2 - x + 6x - 2$ Multiply.

$= 3x^2 + 5x - 2$ Combine like terms.

EXAMPLE 3

Find $(3 - g)^2$.

$(3 - g)^2 = (3 - g)(3 - g)$ Laws of Exponents

$= 3^2 - 3g - 3g + g^2$ Multiply.

$= 3^2 - 6g + g^2$ Combine like terms.

$= 9 - 6g + g^2$ Simplify.

8-1 Monomials and Factoring

Main Ideas

- Find prime factorizations of monomials.
- Find the greatest common factors of monomials.

New Vocabulary

prime number
composite number
prime factorization
factored form
greatest common factor (GCF)

GET READY for the Lesson

In the search for extraterrestrial life, scientists listen to radio signals coming from faraway galaxies. How can they be sure that a particular radio signal was deliberately sent by intelligent beings instead of coming from some natural phenomenon? What if that signal began with a series of beeps in a pattern composed of the first 30 prime numbers ("beep-beep," "beep-beep-beep," and so on)?

Prime Factorization Numbers that are multiplied are *factors* of the resulting product. Numbers that have whole number factors can be represented geometrically. Consider all of the possible rectangles with whole number dimensions that have areas of 18 square units.

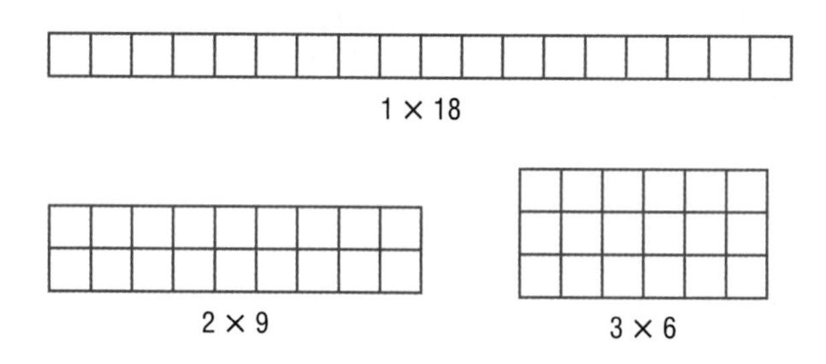

The number 18 has six factors: 1, 2, 3, 6, 9, and 18.

Study Tip

Prime Numbers
Before deciding that a number is prime, try dividing it by all of the prime numbers that are less than the square root of that number.

KEY CONCEPT — *Prime and Composite Numbers*

Words	Examples
A whole number, greater than 1, for which the only factors are 1 and itself, is called a **prime number**.	2, 3, 5, 7, 11, 13, 17, 19
A whole number, greater than 1, that has more than two factors is called a **composite number**.	4, 6, 8, 9, 10, 12, 14, 15

0 and 1 are neither prime nor composite.

A whole number expressed as the product of prime factors is called the **prime factorization** of the number. Two methods of factoring 90 are shown.

Method 1 Find the least prime factors.

$90 = 2 \cdot 45$ — The least prime factor of 90 is 2.
$\quad = 2 \cdot 3 \cdot 15$ — The least prime factor of 45 is 3.
$\quad = 2 \cdot 3 \cdot 3 \cdot 5$ — The least prime factor of 15 is 3.

Method 2 Use a factor tree.

Concepts in Motion
Animation
algebra1.com

$90 = 9 \cdot 10$

$9 = 3 \cdot 3$, $10 = 2 \cdot 5$

$3 \cdot 3 \cdot 2 \cdot 5$

All of the factors in the last step are prime. Thus, the prime factorization of 90 is $2 \cdot 3 \cdot 3 \cdot 5$ or $2 \cdot 3^2 \cdot 5$.

Usually the factors are ordered from the least prime factor to the greatest.

Factoring a monomial is similar to factoring a whole number. A monomial is in **factored form** when it is expressed as the product of prime numbers and variables, and no variable has an exponent greater than 1.

EXAMPLE Prime Factorization of a Monomial

Cross-Curricular Project

Math Online Finding the GCF of distances will help you make a scale model of the solar system. Visit algebra1.com to continue work on your project.

1 **Factor $-12a^2b^3$ completely.**

$-12a^2b^3 = -1 \cdot 12a^2b^3$ — Express -12 as $-1 \cdot 12$

$= -1 \cdot 2 \cdot 6 \cdot a \cdot a \cdot b \cdot b \cdot b$ — $12 = 2 \cdot 6$, $a^2 = a \cdot a$, and $b^3 = b \cdot b \cdot b$

$= -1 \cdot 2 \cdot 2 \cdot 3 \cdot a \cdot a \cdot b \cdot b \cdot b$ — $6 = 2 \cdot 3$

Thus, $-12a^2b^3$ in factored form is $-1 \cdot 2 \cdot 2 \cdot 3 \cdot a \cdot a \cdot b \cdot b \cdot b$.

CHECK Your Progress

Factor each monomial completely.

1A. $38rs^2t$ **1B.** $-66pq^2$

Greatest Common Factor Two or more numbers may have some common prime factors. Consider the prime factorization of 48 and 60.

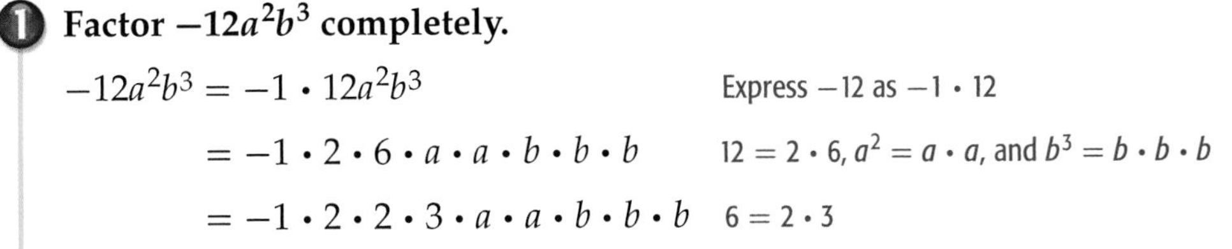

The common prime factors of 48 and 60 are 2, 2, and 3.

The product of the common prime factors, $2 \cdot 2 \cdot 3$ or 12, is called the greatest common factor of 48 and 60. The **greatest common factor (GCF)** is the greatest number that is a factor of both original numbers. The GCF of two or more monomials can be found in a similar way.

KEY CONCEPT — *Greatest Common Factor (GCF)*

- The GCF of two or more monomials is the product of their common factors when each monomial is written in factored form.
- If two or more integers or monomials have a GCF of 1, then the integers or monomials are said to be *relatively prime*.

Study Tip

Alternative Method

You can also find the greatest common factor by listing the factors of each number and finding which of the common factors is the greatest. Consider Example 2.

15: (1), 3, 5, 15
16: (1), 2, 4, 8, 16

The only common factor, and therefore the greatest common factor, is 1.

EXAMPLE Finding GCF

2 **GEOMETRY The areas of two rectangles are 15 square inches and 16 square inches, respectively. The length and width of both figures are whole numbers. If the rectangles have the same width, what is the greatest possible value for their widths?**

Find the GCF of 15 and 16.

$15 = 3 \cdot 5$ Factor each number.

$16 = 2 \cdot 2 \cdot 2 \cdot 2$ There are no common prime factors.

The GCF of 15 and 16 is 1, so 15 and 16 are relatively prime. The width of the rectangles is 1 inch.

CHECK Your Progress

2. What is the greatest possible value for the widths if the rectangles described above have areas of 84 square inches and 70 square inches, respectively?

EXAMPLE GCF of a Set of Monomials

3 **Find the GCF of $36x^2y$ and $54xy^2z$.**

$36x^2y = (2) \cdot 2 \cdot (3) \cdot (3) \cdot (x) \cdot x \cdot (y)$ Factor each number.

$54xy^2z = (2) \cdot (3) \cdot (3) \cdot 3 \cdot (x) \cdot (y) \cdot y \cdot z$ Circle the common prime factors.

The GCF of $36x^2y$ and $54xy^2z$ is $2 \cdot 3 \cdot 3 \cdot x \cdot y$ or $18xy$.

CHECK Your Progress

Find the GCF of each set of monomials.

3A. $17d^3, 5d^2$

3B. $22p^2q, 32pr^2t$

Personal Tutor at algebra1.com

CHECK Your Understanding

Example 1 (p. 421)

Factor each monomial completely.

1. $4p^2$

2. $39b^3c^2$

3. $-100x^3yz^2$

4. **GARDENING** Corey is planting 120 jalapeno pepper plants in a rectangular arrangement in his garden. In what ways can he arrange them so that he has the same number of plants in each row, at least 4 rows of plants, and at least 6 plants in each row?

Examples 2, 3 (p. 422)

Find the GCF of each set of monomials.

5. 10, 15

6. 54, 63

7. $18xy, 36y^2$

8. $25n, 21m$

9. $12qr, 8r^2, 16rs$

10. $42a^2b, 6a^2, 18a^3$

Exercises

HOMEWORK HELP	
For Exercises	See Examples
11–18	1
19–27	2, 3

Factor each monomial completely.

11. $66d^4$ **12.** $85x^2y^2$ **13.** $-49a^3b^2$

14. $50gh$ **15.** $160pq^2$ **16.** $-243n^3m$

17. GEOMETRY A rectangle has an area of 96 square millimeters and its length and width are both whole numbers. What are the minimum and maximum values for the perimeter of the rectangle? Explain your reasoning.

18. MARCHING BANDS The number of members in two high school marching bands is shown in the table. During the halftime show, the bands plan to march into the stadium from opposite ends using formations with the same number of rows. If the bands match up in the center of the field, what is the maximum number of rows, and how many band members will be in each row?

High School	Number of Band Members
Logan	75
Northeast	90

Find the GCF of each set of monomials.

19. 27, 72 **20.** 32, 48 **21.** 18, 35

22. $15a$, $28b^2$ **23.** $24d^2$, $30c^2d$ **24.** $20gh$, $36g^2h^2$

25. $15r^2s$, $35s^2$, $70rs$ **26.** $28a^2b^2$, $63a^3b^2$, $91b^3$ **27.** $14m^2n^2$, $18mn$, $2m^2n^3$

28. NUMBER THEORY *Twin primes* are two consecutive odd numbers that are prime. The first pair of twin primes is 3 and 5. List the next five pairs of twin primes.

29. GEOMETRY The area of a triangle is 20 square centimeters. What are possible whole-number dimensions for the base and height of the triangle?

30. RESEARCH Use the Internet or another source to investigate *Mersenne primes*. Describe what they are, and then list three Mersenne primes.

H.O.T. Problems

31. REASONING Determine whether the following statement is *true* or *false*. If false, provide a counterexample. *All prime numbers are odd.*

32. CHALLENGE Suppose 6 is a factor of ab, where a and b are natural numbers. Make a valid argument to explain why each assertion is *true* or provide a counterexample to show that an assertion is *false.*

a. 6 must be a factor of *a or* of *b*.

b. 3 must be a factor of *a or* of *b*.

c. 3 must be a factor of *a and* of *b*.

33. OPEN ENDED Name two monomials whose GCF is $5x^2$. Justify your choices.

34. *Writing in Math* Use the information about signals on page 420 to explain how prime numbers are related to the search for extraterrestrial life. Include a list of the first 30 prime numbers and an explanation of how you found them.

STANDARDIZED TEST PRACTICE

35. If a line passes through A and B, approximately where will the line cross the x-axis?

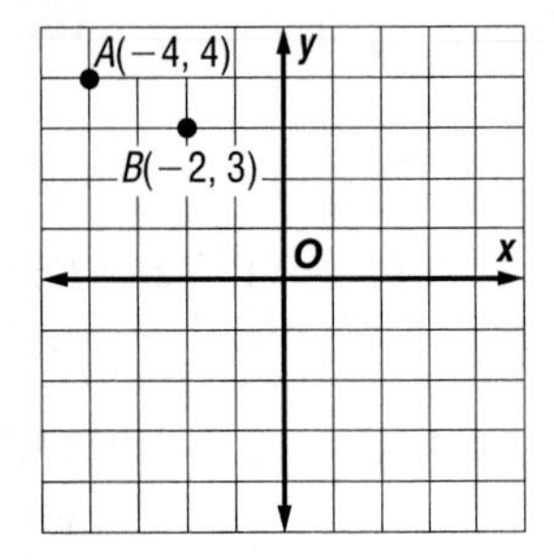

A between −1 and 0

B between 1 and 2

C between 2.5 and 3.5

D between 3.5 and 4.5

36. REVIEW A shoe store organizes its sale shoes by size. The chart below shows how many pairs of shoes in different styles are on each size rack.

Style	Number of Pairs of Shoes
athletic shoes	15
loafers	8
sandals	22
boots	5

If Bethany chooses a pair without looking, what is the probability that she will choose a pair of boots?

F $\frac{5}{8}$

G $\frac{3}{5}$

H $\frac{4}{25}$

J $\frac{1}{10}$

Spiral Review

Find each product. (Lessons 7-6 and 7-7)

37. $(2x - 1)^2$

38. $(3a + 5)(3a - 5)$

39. $(7p^2 + 4)(7p^2 + 4)$

40. $(6r + 7)(2r - 5)$

41. $(10h + k)(2h + 5k)$

42. $(b + 4)(b^2 + 3b - 18)$

43. VIDEOS Professional closed-captioning services cost \$10 per video minute plus a fee of \$50. A company budgeted \$500 for closed-captioning for an instructional video. Define a variable. Then write and solve an inequality to find the number of video minutes for which they can have closed-captioning and stay within their budget. (Lesson 6-3)

Find the value of r so that the line that passes through the given points has the given slope. (Lesson 4-1)

44. $(1, 2), (-2, r), m = 3$

45. $(-5, 9), (r, 6), m = -\frac{3}{5}$

46. RETAIL SALES A department store buys clothing at wholesale prices and then marks the clothing up 25% to sell at retail price to customers. If the retail price of a jacket is \$79, what was the wholesale price? (Lesson 2-7)

GET READY for the Next Lesson

PREREQUISITE SKILL Use the Distributive Property to rewrite each expression. (Lesson 1-5)

47. $5(2x + 8)$

48. $a(3a + 1)$

49. $2g(3g - 4)$

50. $-4y(3y - 6)$

51. $7b + 7c$

52. $2x + 3x$

Algebra Lab
Factoring Using the Distributive Property

Sometimes you know the product of binomials and are asked to find the factors. This is called factoring. You can use algebra tiles to factor binomials.

Use algebra tiles to factor $3x + 6$.

Step 1 Model the polynomial $3x + 6$.

Step 2 Arrange the tiles into a rectangle. The total area of the rectangle represents the product, and its length and width represent the factors.

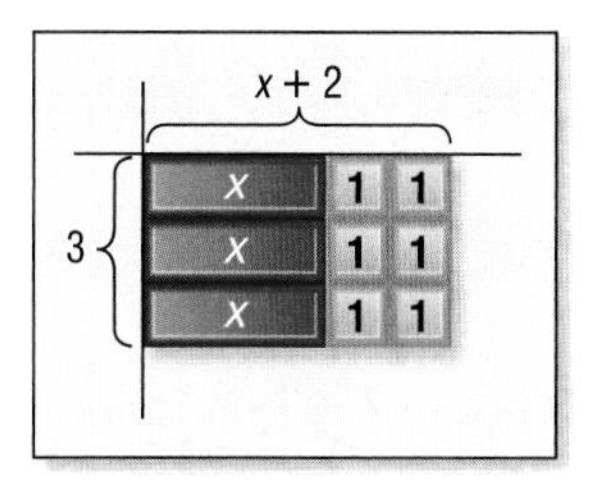

The rectangle has a width of 3 and a length of $x + 2$. So, $3x + 6 = 3(x + 2)$.

ACTIVITY 2 **Use algebra tiles to factor $x^2 - 4x$.**

Step 1 Model the polynomial $x^2 - 4x$.

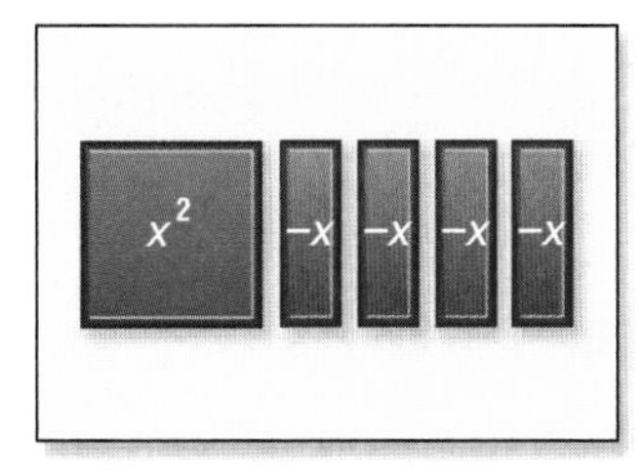

Step 2 Arrange the tiles into a rectangle.

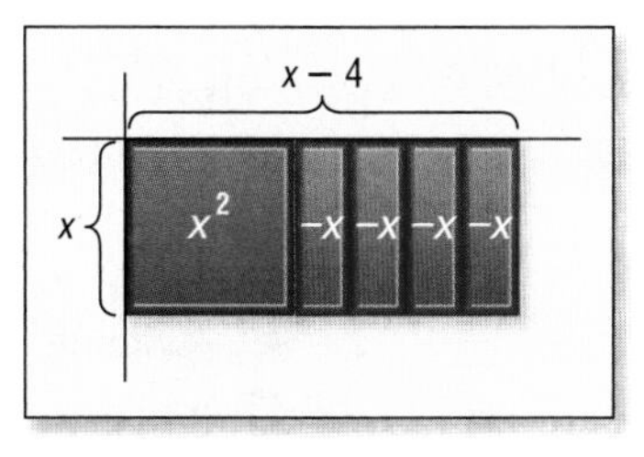

The rectangle has a width of x and a length of $x - 4$. So, $x^2 - 4x = x(x - 4)$.

Analyze the Results

Use algebra tiles to factor each binomial.

1. $2x + 10$ **2.** $6x - 8$ **3.** $5x^2 + 2x$ **4.** $9 - 3x$

Tell whether each binomial can be factored. Justify your answer with a drawing.

5. $4x - 10$ **6.** $3x - 7$ **7.** $x^2 + 2x$ **8.** $x^2 + 4$

9. **MAKE A CONJECTURE** Explain how you can use algebra tiles to determine whether a binomial can be factored. Include an example of one binomial that can be factored and one that cannot.

8-2 Factoring Using the Distributive Property

Main Ideas

- Factor polynomials by using the Distributive Property.
- Solve quadratic equations of the form $ax^2 + bx = 0$.

New Vocabulary

factoring
factoring by grouping
Zero Products Property
roots

GET READY for the Lesson

Roger Clemens, pitcher for the Houston Astros, has had fastballs clocked at 98 miles per hour or about 151 feet per second. If he threw a ball directly upward with the same velocity, the height h of the ball in feet above the point at which he released it could be modeled by the formula $h = 151t - 16t^2$, where t is the time in seconds. You can use factoring and the Zero Product Property to determine how long the ball would remain in the air before returning to his glove.

Factor by Using the Distributive Property In Chapter 7, you used the Distributive Property to multiply a polynomial by a monomial.

$$\begin{aligned} 2a(6a + 8) &= 2a(6a) + 2a(8) \\ &= 12a^2 + 16a \end{aligned}$$

You can reverse this process to express a polynomial as the product of a monomial factor and a polynomial factor.

$$\begin{aligned} 12a^2 + 16a &= 2a(6a) + 2a(8) \\ &= 2a(6a + 8) \end{aligned}$$

Thus, a *factored form* of $12a^2 + 16a$ is $2a(6a + 8)$. **Factoring** a polynomial means to find its *completely* factored form.

EXAMPLE Use the Distributive Property

1 Use the Distributive Property to factor each polynomial.

a. $\mathbf{12a^2 + 16a}$

First, find the GCF of $12a^2$ and $16a$.

$12a^2 = (2) \cdot (2) \cdot 3 \cdot (a) \cdot a$ Factor each monomial.

$16a = (2) \cdot (2) \cdot 2 \cdot 2 \cdot (a)$ Circle the common prime factors.

GCF: $2 \cdot 2 \cdot a$ or $4a$

Write each term as the product of the GCF and its remaining factors. Then use the Distributive Property to factor out the GCF.

$12a^2 + 16a = 4a(3 \cdot a) + 4a(2 \cdot 2)$ Rewrite each term using the GCF.

$= 4a(3a) + 4a(4)$ Simplify remaining factors.

$= 4a(3a + 4)$ Distributive Property

Thus, the completely factored form of $12a^2 + 16a$ *is* $4a(3a + 4)$.

b. $\mathbf{18cd^2 + 12c^2d + 9cd}$

$18cd^2 = 2 \cdot 3 \cdot 3 \cdot c \cdot d \cdot d$ Factor each monomial.

$12c^2d = 2 \cdot 2 \cdot 3 \cdot c \cdot c \cdot d$ Circle the common prime factors.

$9cd = 3 \cdot 3 \cdot c \cdot d$

GCF: $3 \cdot c \cdot d$ or $3cd$

$18cd^2 + 12c^2d + 9cd = 3cd(6d) + 3cd(4c) + 3cd(3)$ Rewrite each term using the GCF.

$= 3cd(6d + 4c + 3)$ Distributive Property

CHECK Your Progress

1A. $16a + 4b$

1B. $3p^2q - 9pq^2 + 36pq$

Using the Distributive Property to factor polynomials having four or more terms is called **factoring by grouping** because pairs of terms are grouped together and factored. The Distributive Property is then applied a second time to factor a common binomial factor.

Study Tip

Factoring by Grouping

Sometimes you can group terms in more than one way when factoring a polynomial. For example, the polynomial in Example 2 could have been factored in the following way.

$4ab + 8b + 3a + 6$

$= (4ab + 3a) + (8b + 6)$

$= a(4b + 3) + 2(4b + 3)$

$= (4b + 3)(a + 2)$

Notice that this result is the same as in Example 2.

EXAMPLE Use Grouping

2 **Factor** $\mathbf{4ab + 8b + 3a + 6}$**.**

$4ab + 8b + 3a + 6$

$= (4ab + 8b) + (3a + 6)$ Group terms with common factors.

$= 4b(a + 2) + 3(a + 2)$ Factor the GCF from each grouping.

$= (a + 2)(4b + 3)$ Distributive Property

CHECK Your Progress **Factor each polynomial.**

2A. $6x^2 - 15x - 8x + 20$

2B. $rs + 5s - r - 5$

Recognizing binomials that are additive inverses is often helpful when factoring by grouping. For example, $7 - y$ and $y - 7$ are additive inverses. By rewriting $7 - y$ as $-1(y - 7)$, factoring by grouping is possible in the following example.

EXAMPLE Use the Additive Inverse Property

3 **Factor** $\mathbf{35x - 5xy + 3y - 21}$**.**

$35x - 5xy + 3y - 21 = (35x - 5xy) + (3y - 21)$ Group terms with common factors.

$= 5x(7 - y) + 3(y - 7)$ Factor the GCF from each grouping.

$= 5x(-1)(y - 7) + 3(y - 7)$ $7 - y = -1(y - 7)$

$= -5x(y - 7) + 3(y - 7)$ $5x(-1) = -5x$

$= (y - 7)(-5x + 3)$ Distributive Property

CHECK Your Progress **Factor each polynomial.**

3A. $c - 2cd + 8d - 4$

3B. $3p - 2p^2 - 18p + 27$

CONCEPT SUMMARY — *Factoring by Grouping*

Words A polynomial can be factored by grouping if *all* of the following situations exist.

- There are four or more terms.
- Terms with common factors can be grouped together.
- The two common binomial factors are identical or are additive inverses of each other.

Symbols $ax + bx + ay + by = x(a + b) + y(a + b)$

$$= (a + b)(x + y)$$

Solve Equations by Factoring Some equations can be solved by factoring. Consider the following products.

$6(0) = 0 \qquad 0(-3) = 0 \qquad (5 - 5)(0) = 0 \qquad -2(-3 + 3) = 0$

Notice that in each case, *at least one* of the factors is zero. These examples illustrate the **Zero Product Property**.

Study Tip

Zero Product Property

If the product of two factors is equal to a *nonzero* value, then you *cannot* use the Zero Product Property. You must first multiply all the factors, and then put all the terms on one side of the equation, with zero on the other. Then you must factor the new expression and use the Zero Product Property.

KEY CONCEPT — *Zero Product Property*

Word If the product of two factors is 0, then at least one of the factors must be 0.

Symbols For any real numbers a and b, if $ab = 0$, then either $a = 0$, $b = 0$, or both a and b equal zero.

The solutions of an equation are called the **roots** of the equation.

EXAMPLE Solve an Equation

4 Solve each equation. Check the solutions.

a. $(d - 5)(3d + 4) = 0$

If $(d - 5)(3d + 4) = 0$, then according to the Zero Product Property either $d - 5 = 0$ or $3d + 4 = 0$.

$(d - 5)(3d + 4) = 0$	Original equation
$d - 5 = 0 \quad \text{or} \quad 3d + 4 = 0$	Set each factor equal to zero.
$d = 5 \qquad\qquad 3d = -4$	Solve each equation.
$d = -\frac{4}{3}$	

The roots are 5 and $-\frac{4}{3}$.

CHECK Substitute 5 and $-\frac{4}{3}$ for d in the original equation.

$$(d - 5)(3d + 4) = 0$$
$$(5 - 5)[3(5) + 4] \stackrel{?}{=} 0$$
$$(0)(19) \stackrel{?}{=} 0$$
$$0 = 0 \ \checkmark$$

$$(d - 5)(3d + 4) = 0$$
$$\left[\left(-\frac{4}{3}\right) - 5\right]\left[3\left(-\frac{4}{3}\right) + 4\right] \stackrel{?}{=} 0$$
$$\left(-\frac{19}{3}\right)(0) \stackrel{?}{=} 0$$
$$0 = 0 \ \checkmark$$

Study Tip

Common Misconception

You may be tempted to try to solve the equation in Example 4b by dividing each side of the equation by x. Remember, however, that x is an *unknown* quantity. If you divide by x, you may actually be dividing by zero, which is undefined.

b. $x^2 = 7x$

Write the equation so that it is of the form $ab = 0$.

$x^2 = 7x$	Original equation
$x^2 - 7x = 0$	Subtract $7x$ from each side.
$x(x - 7) = 0$	Factor using the GCF of x^2 and $-7x$, which is x.
$x = 0$ or $x - 7 = 0$	Zero Product Property
$x = 7$	Solve each equation.
The roots are 0 and 7.	Check by substituting 0 and 7 for x in the original equation.

CHECK Your Progress

4A. $3n(n + 2) = 0$ **4B.** $7d^2 - 35d = 0$ **4C.** $x^2 = -10x$

Online Personal Tutor at algebra1.com

CHECK Your Understanding

Examples 1–3 (pp. 426–427)

Factor each polynomial.

1. $9x^2 + 36x$

2. $4r^2 + 8rs + 28r$

3. $5y^2 - 15y + 4y - 12$

4. $5c - 10c^2 + 2d - 4cd$

Example 4 (pp. 428–429)

Solve each equation. Check the solutions.

5. $h(h + 5) = 0$

6. $(n - 4)(n + 2) = 0$

7. $5m = 3m^2$

8. PHYSICAL SCIENCE A flare is launched from a life raft. The height h of the flare in feet above the sea is modeled by the formula $h = 100t - 16t^2$, where t is the time in seconds after the flare is launched. Let $h = 0$ and solve $0 = 100t - 16t^2$ for t. How many seconds will it take for the flare to return to the sea? Explain your reasoning.

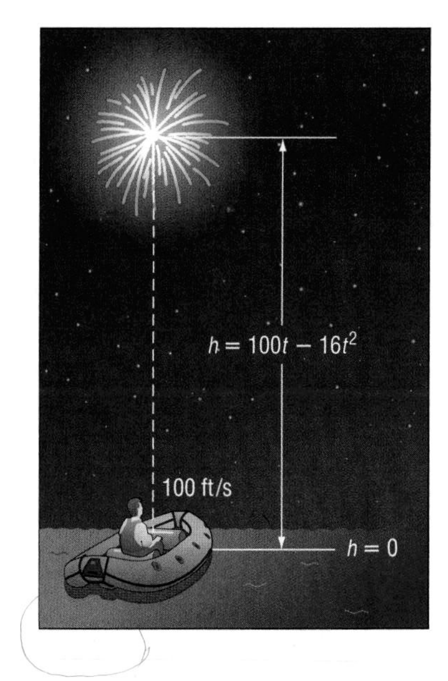

Exercises

HOMEWORK HELP	
For Exercises	See Examples
9–14	1
15–20	2, 3
21–30	4

Factor each polynomial.

9. $5x + 30y$

10. $a^5b - a$

11. $14gh - 18h$

12. $8bc^2 + 24bc$

13. $15x^2y^2 + 25xy + x$

14. $12ax^3 + 20bx^2 + 32cx$

15. $x^2 + 2x + 3x + 6$

16. $12y^2 + 9y + 8y + 6$

17. $18x^2 - 30x - 3x + 5$

18. $2my + 7x + 7m + 2xy$

19. $8ax - 6x - 12a + 9$

20. $10x^2 - 14xy - 15x + 21y$

Solve each equation. Check the solutions.

21. $x(x - 24) = 0$

22. $a(a + 16) = 0$

23. $(q + 4)(3q - 15) = 0$

24. $(3y + 9)(y - 7) = 0$

25. $(2b - 3)(3b - 8) = 0$

26. $(4n + 5)(3n - 7) = 0$

27. $3z^2 + 12z = 0$

28. $2x^2 = 5x$

29. BASEBALL Malik popped a ball straight up with an initial upward velocity of 45 feet per second. The height h, in feet, of the ball above the ground is modeled by the equation $h = 2 + 48t - 16t^2$. How long was the ball in the air if the catcher catches the ball when it is 2 feet above the ground? Is your answer reasonable in the context of this situation?

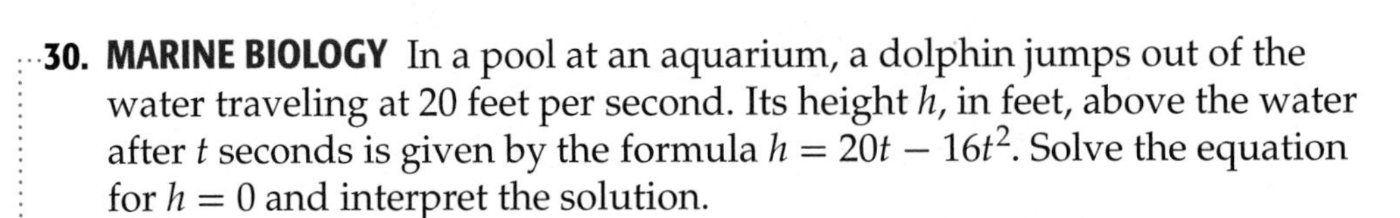

30. MARINE BIOLOGY In a pool at an aquarium, a dolphin jumps out of the water traveling at 20 feet per second. Its height h, in feet, above the water after t seconds is given by the formula $h = 20t - 16t^2$. Solve the equation for $h = 0$ and interpret the solution.

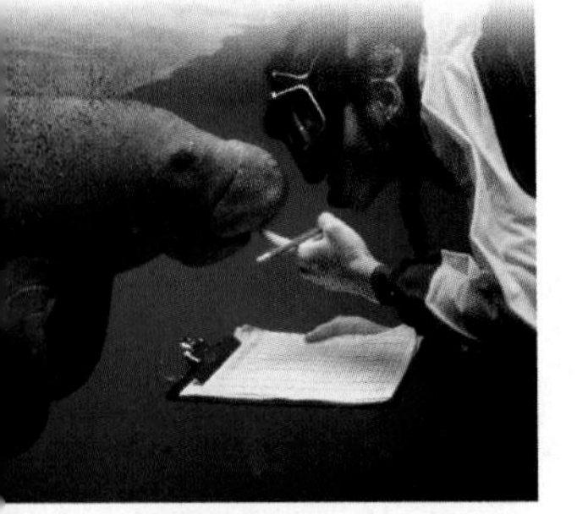

Real-World Career

Marine Biologist
A marine biologist uses math to study and analyze factors that affect organisms living in and near the ocean.

Math Online
For more information, go to algebra1.com.

Factor each polynomial.

31. $12x^2y^2z + 40xy^3z^2$

32. $18a^2bc^2 - 48abc^3$

GEOMETRY Find an expression for the area of a square with the given perimeter.

33. $P = (12x + 20y)$ in.

34. $P = (36a - 16b)$ cm

35. GEOMETRY The expression $\frac{1}{2}n^2 - \frac{3}{2}n$ can be used to find the number of diagonals in a polygon that has n sides. Write the expression in factored form and find the number of diagonals in a decagon (10-sided polygon).

SOFTBALL For Exercises 36 and 37, use the following information.
Alisha is scheduling the games for a softball league. To find the number of games she needs to schedule, she uses the equation $g = \frac{1}{2}n^2 - \frac{1}{2}n$, where g represents the number of games needed for each team to play each other exactly once and n represents the number of teams.

36. Write this equation in factored form.

37. How many games are needed for 7 teams to play each other exactly 3 times?

GEOMETRY Write an expression in factored form for the area of each shaded region.

38.

39.

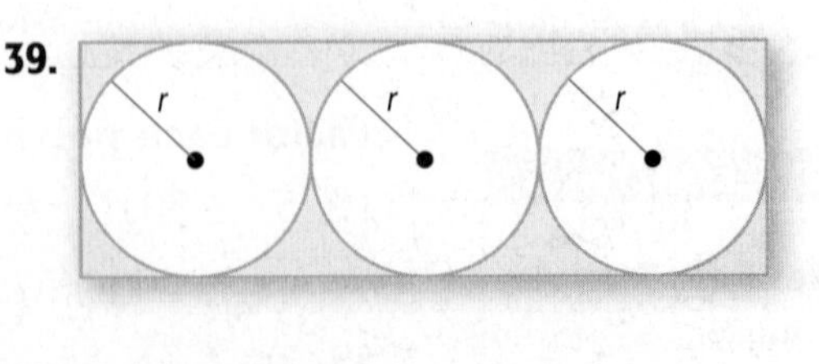

H.O.T. Problems

40. REASONING Represent $4x^2 + 12x$ as a product of factors in three different ways. Then decide which of the three is the completely factored form. Explain your reasoning.

41. **OPEN ENDED** Write an equation that can be solved by using the Zero Product Property. Describe how to solve the equation and then find the roots.

42. **REASONING** Explain why $(x - 2)(x + 4) = 0$ cannot be solved by dividing each side by $x - 2$.

43. **CHALLENGE** Factor $a^{x+y} + a^x b^y - a^y b^x - b^{x+y}$. Describe your steps.

44. ***Writing in Math*** Use the information about Roger Clemens on page 426 to explain how you can determine how long a baseball will remain in the air. Explain how to use factoring and the Zero Product Property to solve the problem. Then interpret each solution in the context of the problem.

STANDARDIZED TEST PRACTICE

45. Which of the following shows $16x^2 - 4x$ factored completely?

A $4x(x)$

B $4x(4x - 1)$

C $x(4x - 1)$

D $x(x - 4)$

46. **REVIEW** The frequency table shows the results of a survey in which students were asked to name the colors of their bicycles. Which measure of data describes the most popular color for a bicycle?

Color	Frequency
black	|||| I
blue	|||| |||| |||| ||||
red	|||| III
silver	|||| |||| |||| I
Total	50

F mean

G median

H mode

J range

Spiral Review

Find the GCF of each set of monomials. (Lesson 8-1)

47. $9a, 8ab$

48. $16h, 28hk^2$

49. $3x^2y^2, 9xy, 15x^3y$

Find each product. (Lesson 7-7)

50. $(4s^3 + 3)^2$

51. $(2p + 5q)(2p - 5q)$

52. $(3k + 8)(3k + 8)$

53. **FINANCE** Michael uses at most 60% of his annual Flynn Company stock dividend to purchase more shares of Flynn Company stock. If his dividend last year was \$885 and Flynn Company stock is selling for \$14 per share, what is the greatest number of shares that he can purchase? (Lesson 6-2)

GET READY for the Next Lesson

PREREQUISITE SKILL Find each product. (Lesson 7-6)

54. $(n + 8)(n + 3)$

55. $(x - 4)(x - 5)$

56. $(b - 10)(b + 7)$

57. $(3a + 1)(6a - 4)$

58. $(5p - 2)(9p - 3)$

59. $(2y - 5)(4y + 3)$

Algebra Lab
Factoring Trinomials

You can use algebra tiles to factor trinomials. If a polynomial represents the area of a rectangle formed by algebra tiles, then the rectangle's length and width are *factors* of the polynomial.

ACTIVITY 1 **Use algebra tiles to factor $x^2 + 6x + 5$.**

Step 1 Model $x^2 + 6x + 5$.

Step 2 Place the x^2-tile at the corner of the product mat. Arrange the 1-tiles into a rectangular array. Because 5 is prime, the 5 tiles can be arranged in a rectangle in one way, a 1-by-5 rectangle.

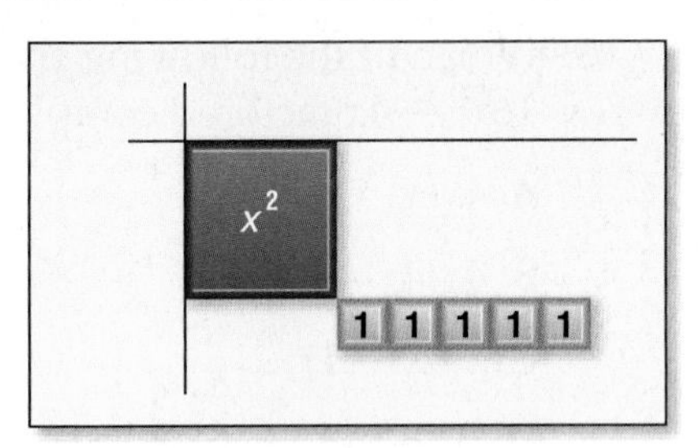

Step 3 Complete the rectangle with the x-tiles.

The rectangle has a width of $x + 1$ and a length of $x + 5$. Therefore, $x^2 + 6x + 5 = (x + 1)(x + 5)$.

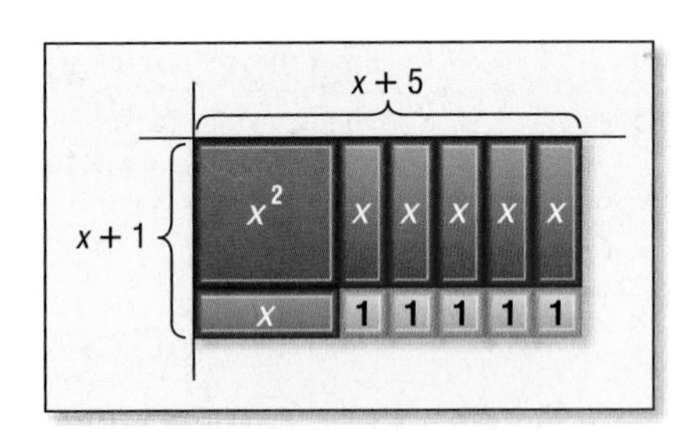

ACTIVITY 2 **Use algebra tiles to factor $x^2 + 7x + 6$.**

Step 1 Model $x^2 + 7x + 6$.

Step 2 Place the x^2-tile at the corner of the product mat. Arrange the 1-tiles into a rectangular array. Since $6 = 2 \times 3$, try a 2-by-3 rectangle. Try to complete the rectangle. Notice that there are two extra x-tiles.

Concepts in Motion
Animation algebra1.com

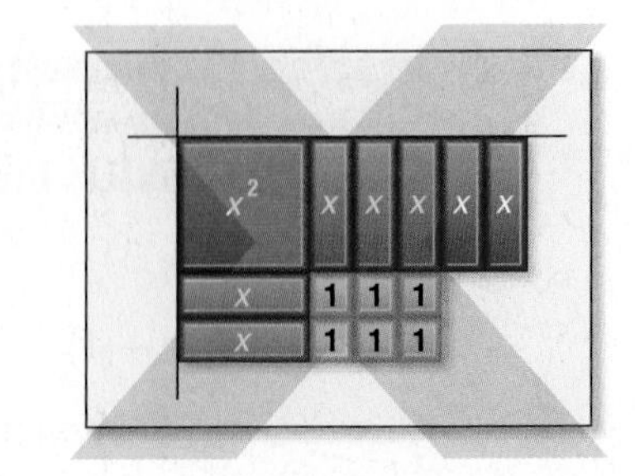

Step 3 Arrange the 1-tiles into a 1-by-6 rectangular array. This time you can complete the rectangle with the x-tiles.

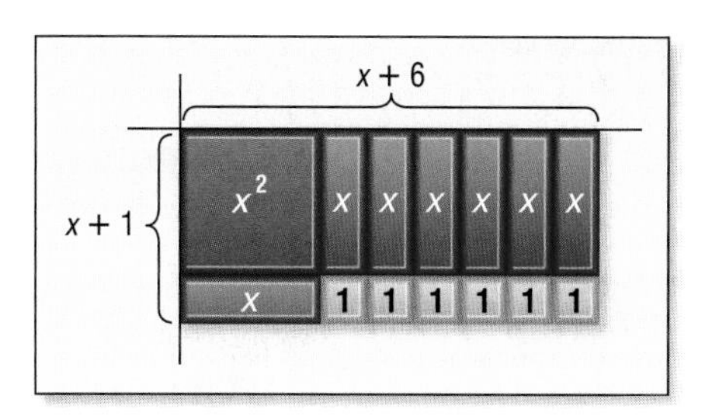

The rectangle has a width of $x + 1$ and a length of $x + 6$. Therefore, $x^2 + 7x + 6 = (x + 1)(x + 6)$.

ACTIVITY 3 **Use algebra tiles to factor $x^2 - 2x - 3$.**

Step 1 Model the polynomial $x^2 - 2x - 3$.

Step 2 Place the x^2-tile at the corner of the product mat. Arrange the 1-tiles into a 1-by-3 rectangular array as shown.

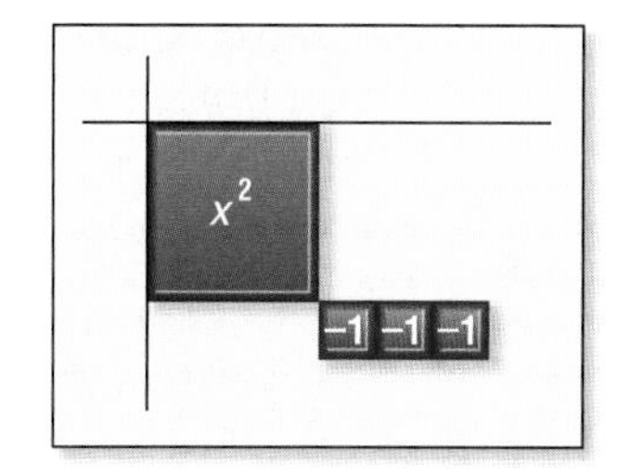

Step 3 Place the x-tile as shown. Recall that you can add zero-pairs without changing the value of the polynomial. In this case, add a zero pair of x-tiles.

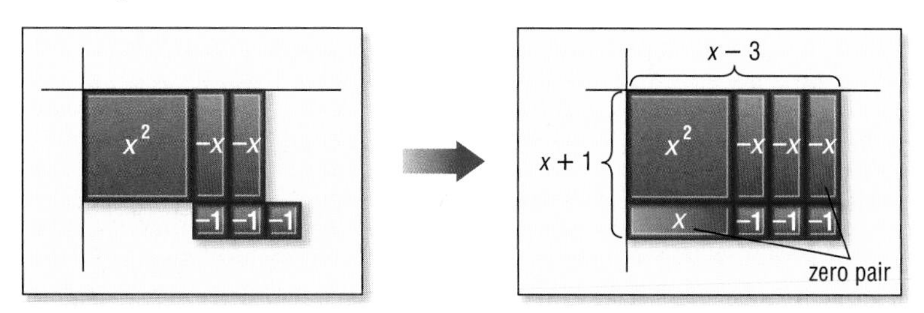

The rectangle has a width of $x + 1$ and a length of $x - 3$. Therefore, $x^2 - 2x - 3 = (x + 1)(x - 3)$.

ANALYZE THE RESULTS

Use algebra tiles to factor each trinomial.

1. $x^2 + 4x + 3$ **2.** $x^2 + 5x + 4$ **3.** $x^2 - x - 6$ **4.** $x^2 - 3x + 2$

5. $x^2 + 7x + 12$ **6.** $x^2 - 4x + 4$ **7.** $x^2 - x - 2$ **8.** $x^2 - 6x + 8$

9. Examine the dimensions of the rectangles in each factored model. How does the sum of the dimensions compare to the coefficient of the x-term? Explain how you could use this observation to factor trinomials.

8-3 Factoring Trinomials: $x^2 + bx + c$

Main Ideas

- Factor trinomials of the form $x^2 + bx + c$.
- Solve equations of the form $x^2 + bx + c = 0$.

GET READY for the Lesson

Tamika has enough bricks to make a 30-foot border around the rectangular vegetable garden she is planting. The nursery says that the plants will need a space of 54 square feet to grow. What should the dimensions of her garden be?

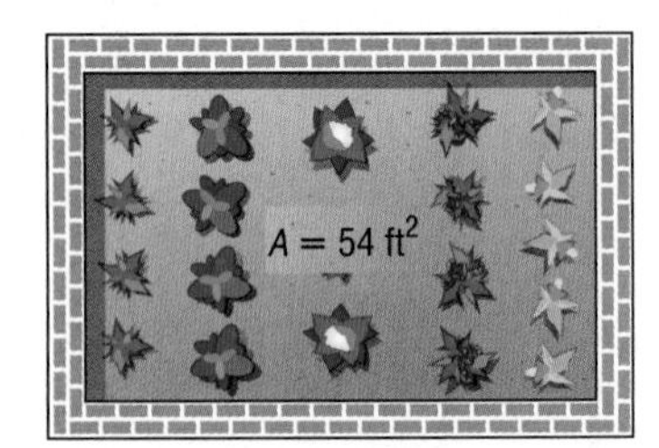

$P = 30$ ft

To solve this problem, you need to find two numbers with a product of 54 and a sum of 15, half the perimeter of the garden.

Factor $x^2 + bx + c$ When two numbers are multiplied, each number is a factor of the product. Similarly, when two binomials are multiplied, each binomial is a factor of the product. To factor certain types of trinomials, you will use the pattern for multiplying two binomials. Study the following example.

$$\begin{aligned} (x+2)(x+3) &= \overset{F}{(x \cdot x)} + \overset{O}{(x \cdot 3)} + \overset{I}{(x \cdot 2)} + \overset{L}{(2 \cdot 3)} && \text{Use the FOIL method.} \\ &= x^2 + 3x + 2x + 6 && \text{Simplify.} \\ &= x^2 + (3+2)x + 6 && \text{Distributive Property} \\ &= x^2 + 5x + 6 && \text{Simplify.} \end{aligned}$$

Observe the following pattern in this multiplication.

$$\begin{aligned} (x+2)(x+3) &= x^2 + (3+2)x + (2 \cdot 3) \\ (x+m)(x+n) &= x^2 + (n+m)x + mn && \text{Let } 2 = m \text{ and } 3 = n. \\ &= x^2 + \underbrace{(m+n)}x + \underbrace{mn} && \text{Commutative } (+) \\ x^2 &+ \quad bx \quad + \quad c && b = m + n \text{ and } c = mn \end{aligned}$$

Notice that the coefficient of the middle term is the sum of m and n and the last term is the product of m and n. This pattern can be used to factor trinomials of the form $x^2 + bx + c$.

KEY CONCEPT — *Factoring $x^2 + bx + c$*

Words To factor trinomials of the form $x^2 + bx + c$, find two integers, m and n, with a sum equal to b and a product equal to c. Then write $x^2 + bx + c$ as $(x + m)(x + n)$.

Symbols $x^2 + bx + c = (x + m)(x + n)$ when $m + n = b$ and $mn = c$.

Example $x^2 + 5x + 6 = (x + 2)(x + 3)$, since $2 + 3 = 5$ and $2 \cdot 3 = 6$

EXAMPLE *b* and *c* are Positive

1 Factor $x^2 + 6x + 8$.

In this trinomial, $b = 6$ and $c = 8$. You need to find two numbers with a sum of 6 and a product of 8. Make an organized list of the factors of 8, and look for the pair of factors with a sum of 6.

Factors of 8	Sum of Factors
1, 8	9
2, 4	6

The correct factors are 2 and 4.

$x^2 + 6x + 8 = (x + m)(x + n)$ Write the pattern.

$= (x + 2)(x + 4)$ $m = 2$ and $n = 4$

CHECK You can check this result by multiplying the two factors.

$(x + 2)(x + 4) = \overset{F}{x^2} + \overset{O}{4x} + \overset{I}{2x} + \overset{L}{8}$ FOIL method

$= x^2 + 6x + 8$ ✓ Simplify.

CHECK Your Progress **Factor each trinomial.**

1A. $a^2 + 8a + 15$ **1B.** $9 + 10t + t^2$

When factoring a trinomial where b is negative and c is positive, use what you know about the product of binomials to narrow the list of possible factors.

EXAMPLE *b* is Negative and *c* is Positive

2 Factor $x^2 - 10x + 16$.

In this trinomial, $b = -10$ and $c = 16$. This means that $m + n$ is negative and mn is positive. So m and n must both be negative. Make a list of the negative factors of 16, and look for the pair with the sum of -10.

Study Tip

Testing Factors
Once you find the correct factors, there is no need to test any other factors. Therefore, it is not necessary to test -4 and -4 in Example 2.

Factors of 16	Sum of Factors
−1, −16	−17
−2, −8	−10
−4, −4	−8

The correct factors are −2 and −8.

$x^2 - 10x + 16 = (x + m)(x + n)$ Write the pattern.

$= (x - 2)(x - 8)$ $m = -2$ and $n = -8$

CHECK You can check this result by using a graphing calculator. Graph $y = x^2 - 10x + 16$ and $y = (x - 2)(x - 8)$ on the same screen. Since only one graph appears, the two graphs must coincide. Therefore, the trinomial has been factored correctly. ✓

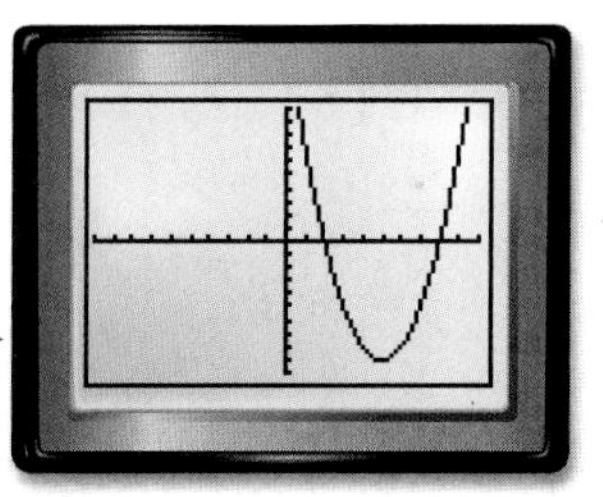

[−10, 10] scl: 1 by [−10, 10] scl: 1

CHECK Your Progress **Factor each trinomial.**

2A. $21 - 22m + m^2$ **2B.** $s^2 - 11s + 28$

Study Tip

Alternate Method

You can use the opposite of FOIL to factor trinomials. For instance, consider Example 3.

$x^2 + 2x - 15$

$(x + \blacksquare)(x + \blacksquare)$

Try factor pairs of -15 until the sum of the products of the Inner and Outer terms is $2x$.

EXAMPLE c is Negative

3 **Factor each trinomal.**

a. $x^2 + 2x - 15$

Since $b = 2$ and $c = -15$, $m + n$ is positive and mn is negative. So either m or n is negative, but not both. List the factors of -15, where one factor of each pair is negative. Look for the pair of factors with a sum of 2.

Factors of −15	Sum of Factors
1, −15	−14
−1, 15	14
3, −5	−2
−3, 5	2

The correct factors are −3 and 5.

$x^2 + 2x - 15 = (x + m)(x + n)$ Write the pattern.

$= (x - 3)(x + 5)$ $m = -3$ and $n = 5$

b. $x^2 - 7x - 18$

Since $b = -7$ and $c = -18$, $m + n$ is negative and mn is negative. So either m or n is negative, but not both.

Factors of −18	Sum of Factors
1, −18	−17
−1, 18	17
2, −9	−7

The correct factors are 2 and −9.

$x^2 - 7x - 18 = (x + m)(x + n)$ Write the pattern.

$= (x + 2)(x - 9)$ $m = 2$ and $n = -9$

CHECK Your Progress **Factor each trinomial.**

3A. $h^2 + 3h - 40$

3B. $r^2 - 2r - 24$

Solve Equations by Factoring Some equations of the form $x^2 + bx + c = 0$ can be solved by factoring and then using the Zero Product Property.

EXAMPLE Solve an Equation by Factoring

4 **Solve $x^2 + 5x - 6 = 0$. Check the solutions.**

$x^2 + 5x - 6 = 0$ Original equation

$(x - 1)(x + 6) = 0$ Factor.

$x - 1 = 0$ or $x + 6 = 0$ Zero Product Property

$x = 1$ $\quad x = -6$ Solve each equation.

The roots are 1 and −6. Check by substituting 1 and −6 for x in the original equation.

CHECK Your Progress

Solve each equation. Check the solutions.

4A. $x^2 + 16x = -28$

4B. $g^2 + 6g = 27$

Real-World EXAMPLE Solve a Real-World Problem by Factoring

5 **YEARBOOK DESIGN** **A sponsor for the school yearbook has asked that the length and width of a photo in their ad be increased by the same amount in order to double the area of the photo. If the original photo is 12 centimeters wide by 8 centimeters long, what should be the new dimensions of the enlarged photo?**

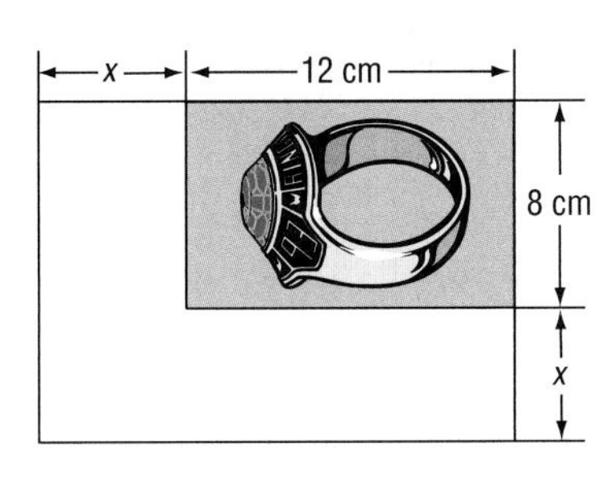

Explore Begin by making a diagram like the one shown above, labeling the appropriate dimensions.

Plan Let x = the amount added to each dimension of the photo.

The new length	times	the new width	equals	twice the old area.
$x + 12$	$\cdot$	$x + 8$	$=$	$2(8)(12)$

Solve

$(x + 12)(x + 8) = 2(8)(12)$	Write the equation.
$x^2 + 20x + 96 = 192$	Multiply.
$x^2 + 20x - 96 = 0$	Rewrite the equation so that one side equals 0.
$(x + 24)(x - 4) = 0$	Factor.
$x + 24 = 0$ or $x - 4 = 0$	Zero Product Property
$x = -24$ $\quad x = 4$	Solve each equation.

Check The solution set is $\{-24, 4\}$. In the context of the situation, only 4 is a valid solution because dimensions cannot be negative. Thus, the new dimensions of the photo should be $4 + 12$ or 16 centimeters, and $4 + 8$ or 12 centimeters.

CHECK Your Progress

5. GEOMETRY The height of a parallelogram is 18 centimeters less than its base. If the parallelogram has an area of 175 square centimeters, what is its height?

Online Personal Tutor at algebra1.com

CHECK Your Understanding

Examples 1–3 (pp. 435–436)

Factor each trinomial.

1. $x^2 + 11x + 24$

2. $n^2 - 3n + 2$

3. $w^2 + 13w - 48$

4. $p^2 - 2p - 35$

5. $y^2 + y - 20$

6. $72 + 27a + a^2$

Example 4 (p. 436)

Solve each equation. Check the solutions.

7. $n^2 + 7n + 6 = 0$

8. $a^2 + 5a - 36 = 0$

9. $y^2 + 9 = 10y$

10. $d^2 - 3d = 70$

Example 5 (p. 437)

11. NUMBER THEORY Find two consecutive integers x and $x + 1$ with a product of 156.

Exercises

HOMEWORK HELP	
For Exercises	See Examples
12–23	1–3
24–31	4
32, 33	5

Factor each trinomial.

12. $x^2 + 12x + 27$
13. $c^2 + 12c + 35$
14. $y^2 + 13y + 30$
15. $d^2 - 7d + 10$
16. $p^2 - 17p + 72$
17. $g^2 - 19g + 60$
18. $x^2 + 6x - 7$
19. $n^2 + 3n - 54$
20. $y^2 - y - 42$
21. $z^2 - 18z - 40$
22. $-72 + 6w + w^2$
23. $-30 + 13x + x^2$

Solve each equation. Check the solutions.

24. $b^2 + 20b + 36 = 0$
25. $y^2 + 4y - 12 = 0$
26. $d^2 + 2d - 8 = 0$
27. $m^2 - 19m + 48 = 0$
28. $z^2 = 18 - 7z$
29. $h^2 + 15 = -16h$
30. $24 + k^2 = 10k$
31. $c^2 - 50 = -23c$

32. GEOMETRY The triangle has an area of 40 square centimeters. Find the height h of the triangle.

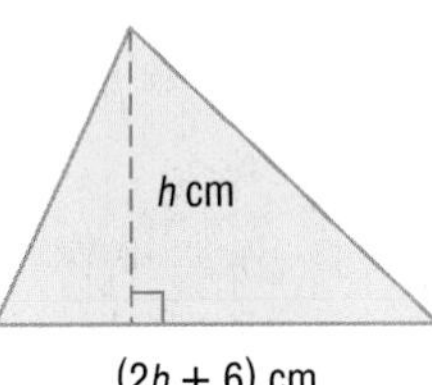

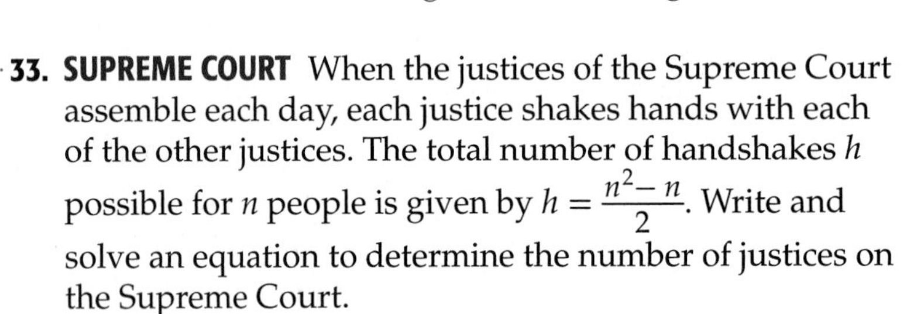

33. SUPREME COURT When the justices of the Supreme Court assemble each day, each justice shakes hands with each of the other justices. The total number of handshakes h possible for n people is given by $h = \frac{n^2 - n}{2}$. Write and solve an equation to determine the number of justices on the Supreme Court.

Real-World Link

The "Conference handshake" has been a tradition since the late 19th century. Each day, there is a total of 36 handshakes by the justices.

Source: supremecourtus.gov

RUGBY For Exercises 34 and 35, use the following information.
The length of a Rugby League field is 52 meters longer than its width w.

34. Write an expression for the area of the field.

35. The area of a Rugby League field is 8160 square meters. Find the dimensions of the field.

GEOMETRY Find an expression for the perimeter of a rectangle with the given area.

36. area $= x^2 + 24x - 81$
37. area $= x^2 + 13x - 90$

EXTRA PRACTICE
See pages 733, 751.
Math Online
Self-Check Quiz at algebra1.com

SWIMMING For Exercises 38–40, use the following information.
The length of a rectangular swimming pool is 20 feet greater than its width. The area of the pool is 525 square feet.

38. Define a variable and write an equation for the area of the pool.

39. Solve the equation.

40. Interpret the solutions. Do they both make sense in the context of the problem? Explain.

H.O.T. Problems

41. REASONING Explain why, when factoring $x^2 + 6x + 9$, it is not necessary to check the sum of the factor pairs -1 and -9 or -3 and -3.

42. OPEN ENDED Give an example of an equation that can be solved using the factoring techniques presented in this lesson. Then solve your equation.

43. FIND THE ERROR Peter and Aleta are solving $x^2 + 2x = 15$. Who is correct? Explain your reasoning.

Peter	Aleta
$x^2 + 2x = 15$	$x^2 + 2x = 15$
$x(x + 2) = 15$	$x^2 + 2x - 15 = 0$
$x = 15$ or $x + 2 = 15$	$(x - 3)(x + 5) = 0$
$x = 13$	$x - 3 = 0$ or $x + 5 = 0$
	$x = 3$ $x = -5$

CHALLENGE Find all values of k so that each trinomial can be factored using integers.

44. $x^2 + kx - 19$

45. $x^2 + kx + 14$

46. $x^2 - 8x + k, k > 0$

47. $x^2 - 5x + k, k > 0$

48. ***Writing in Math*** Use the information about Tamika's garden on page 434 to explain how factoring can be used to find the dimensions of a garden. Explain how your method is related to the process used to factor trinomials of the form $x^2 + bx + c$.

STANDARDIZED TEST PRACTICE

49. Which is a factor of $x^2 + 9x + 18$?

A $x + 2$

B $x - 2$

C $x + 3$

D $x - 3$

50. REVIEW An 8-foot by 5-foot section of wall is to be covered by square tiles that measure 4 inches on each side. If the tiles are not cut, how many of them will be needed to cover the wall?

F 30

G 240

H 360

J 1440

Spiral Review

Solve each equation. Check the solutions. (Lesson 8-2)

51. $(x + 3)(2x - 5) = 0$

52. $7b(b - 4) = 0$

53. $5y^2 = -9y$

Find the GCF of each set of monomials. (Lesson 8-1)

54. 24, 72

55. $9pq^5, 21p^3q^3$

56. $30x^2, 75x^3y^4, 20x^4z$

57. MUSIC Albertina practices the guitar 20 minutes each day. She wants to add 5 minutes to her practice time each day until she is practicing at least 45 minutes daily. How many days will it take her to reach her goal? (Lesson 6-3)

GET READY for the Next Lesson

PREREQUISITE SKILL Factor each polynomial. (Lesson 8-2)

58. $3y^2 + 2y + 9y + 6$

59. $3a^2 + 2a + 12a + 8$

60. $4x^2 + 3x + 8x + 6$

61. $2p^2 - 6p + 7p - 21$

62. $3b^2 + 7b - 12b - 28$

63. $4g^2 - 2g - 6g + 3$

CHAPTER 8

Mid-Chapter Quiz

Lessons 8-1 through 8-3

Factor each monomial completely. (Lesson 8-1)

1. $35mn$ **2.** $27r^2$

3. $20xy^3$ **4.** $78a^2bc^3$

5. THEATER Drama students have 140 chairs to place in front of an outdoor stage. In what ways can they arrange the chairs so that there is the same number in each row, at least 6 rows of chairs, and at least 6 chairs in each row? (Lesson 8-1)

Find the GCF of each set of monomials. (Lesson 8-1)

6. $24ab^2, 21a^3$ **7.** $18n, 25p^2$

8. $15q^2r^2, 5r^2s$ **9.** $42x^2y, 30xy^2$

Factor each polynomial. (Lesson 8-2)

10. $3m + 18n$

11. $4xy^2 - xy$

12. $32a^2b + 40b^3 - 8a^2b^2$

13. $6pq + 16p - 15q - 40$

14. PHOTOS Olinda is placing matting x inches wide around a photo that is 5 inches long and 3 inches wide. Write an expression in factored form for the area of the matting. (Lesson 8-2)

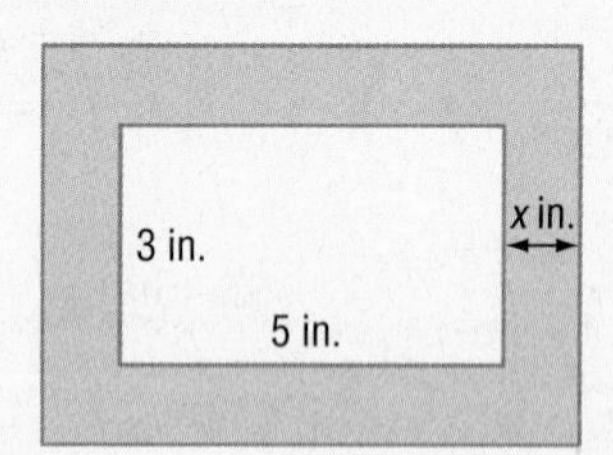

15. FOOTBALL In a football game, Darryl punts the ball downfield. The height h of the football above the ground after t seconds can be modeled by $h = 76.8t - 16t^2$. How long was the football in the air? (Lesson 8-2)

16. MULTIPLE CHOICE What are the roots of $d^2 - 12d = 0$? (Lesson 8-2)

A 0 and −12 C −12 and 12

B 0 and 12 D 12 and 12

17. GEOMETRY Write an expression in factored form for the area of the shaded region. (Lesson 8-2)

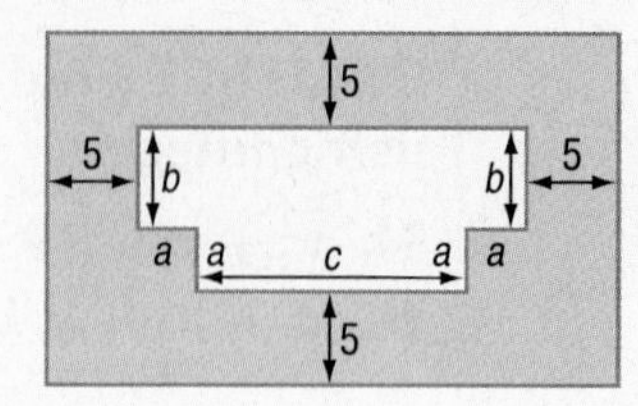

Solve each equation. Check the solutions. (Lesson 8-2)

18. $(8n + 5)(n - 4) = 0$

19. $9x^2 - 27x = 0$

20. $10x^2 = -3x$

Factor each trinomial. (Lesson 8-3)

21. $n^2 - 2n - 48$

22. $x^2 - 4xy + 3y^2$

23. $a^2 + 5ab + 4b^2$

24. $s^2 - 13st + 36t^2$

Solve each equation. Check the solutions. (Lesson 8-3)

25. $a^2 + 7a + 10 = 0$

26. $n^2 + 4n - 21 = 0$

27. $x^2 - 2x - 6 = 74$

28. $x^2 - x + 56 = 17x$

29. GEOMETRY The rectangle has an area of 180 square feet. Find the width w of the rectangle. (Lesson 8-3)

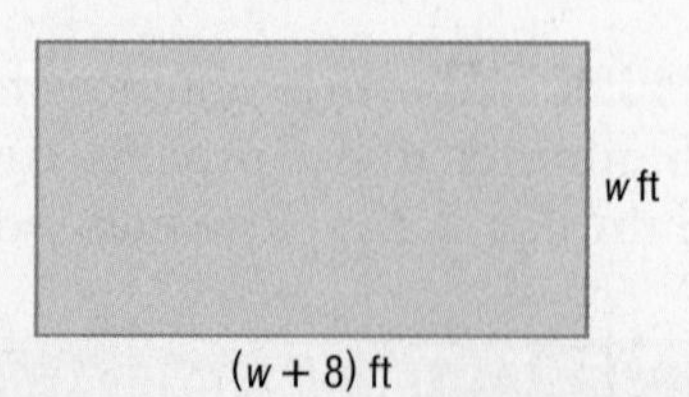

30. MULTIPLE CHOICE Which represents one of the roots of $0 = x^2 + 3x - 18$? (Lesson 8-3)

F −6 H 6

G −3 J $\frac{1}{3}$

8-4 Factoring Trinomials: $ax^2 + bx + c$

Main Ideas

- Factor trinomials of the form $ax^2 + bx + c$.
- Solve equations of the form $ax^2 + bx + c = 0$.

New Vocabulary

prime polynomial

GET READY for the Lesson

The factors of $2x^2 + 7x + 6$ are the dimensions of the rectangle formed by the algebra tiles shown below.

The process you use to form the rectangle is the same mental process you can use to factor this trinomial algebraically.

Factor $ax^2 + bx + c$ For trinomials of the form $x^2 + bx + c$, the coefficient of x^2 is 1. To factor trinomials of this form, you find the factors of c with a sum of b. We can modify this approach to factor trinomials for which the leading coefficient is not 1.

ALGEBRA LAB

1. Complete the following table.

Product of Two Binomials	Use FOIL. $ax^2 + mx + nx + c$	$ax^2 + bx + c$	$m \cdot n$	$a \cdot c$
$(2x + 3)(x + 4)$	$2x^2 + 8x + 3x + 12$	$2x^2 + 11x + 12$	24	24
$(x + 1)(3x + 5)$				
$(2x - 1)(4x + 1)$				
$(3x + 5)(4x - 2)$				

2. How are m and n related to a and c?

3. How are m and n related to b?

You can use the pattern in the Algebra Lab and the method of factoring by grouping to factor trinomials. Consider $6x^2 + 17x + 5$. Find two numbers, m and n, with the product of $6 \cdot 5$ or 30 and the sum of 17. The correct factors are 2 and 15.

$6x^2 + 17x + 5 = 6x^2 + mx + nx + 5$ — Write the pattern.

$= 6x^2 + 2x + 15x + 5$ — $m = 2$ and $n = 15$

$= (6x^2 + 2x) + (15x + 5)$ — Group terms with common factors.

$= 2x(3x + 1) + 5(3x + 1)$ — Factor the GCF from each grouping.

$= (3x + 1)(2x + 5)$ — $3x + 1$ is the common factor.

Therefore, $6x^2 + 17x + 5 = (3x + 1)(2x + 5)$.

EXAMPLE Factor $ax^2 + bx + c$

1 **Factor each trinomial.**

a. $7x^2 + 29x + 4$

In this trinomial, $a = 7$, $b = 29$, and $c = 4$. You need to find two numbers with a sum of 29 and a product of $7 \cdot 4$ or 28. Make an organized list of the factors of 28 and look for the pair of factors with the sum of 29.

Factors of 28	Sum of Factors
1, 28	29

The correct factors are 1 and 28.

$$\begin{aligned} 7x^2 + 29x + 4 &= 7x^2 + mx + nx + 4 && \text{Write the pattern.} \\ &= 7x^2 + 1x + 28x + 4 && m = 1 \text{ and } n = 28 \\ &= (7x^2 + 1x) + (28x + 4) && \text{Group terms with common factors.} \\ &= x(7x + 1) + 4(7x + 1) && \text{Factor the GCF from each grouping.} \\ &= (7x + 1)(x + 4) && \text{Distributive Property} \end{aligned}$$

b. $10x^2 - 43x + 28$

In this trinomial, $a = 10$, $b = -43$ and $c = 28$. Since b is negative, $m + n$ is negative. Since c is positive, mn is positive. So, both m and n are negative.

Study Tip

Finding Factors
Put pairs in an organized list so you do not miss any possible pairs of factors.

List the negative factors of $10 \cdot 28$ or 280. →

Factors of 280	Sum of Factors
−1, −280	−281
−2, −140	−142
−4, −70	−74
−5, −56	−61
−7, −40	−47
−8, −35	−43

← Look for the pairs of factors with a sum of −43.

The correct factors are −8 and −35.

$$\begin{aligned} &10x^2 - 43x + 28 \\ &= 10x^2 + mx + nx + 28 && \text{Write the pattern.} \\ &= 10x^2 + (-8)x + (-35)x + 28 && m = -8 \text{ and } n = -35 \\ &= (10x^2 - 8x) + (-35x + 28) && \text{Group terms with common factors.} \\ &= 2x(5x - 4) + 7(-5x + 4) && \text{Factor the GCF from each grouping.} \\ &= 2x(5x - 4) + 7(-1)(5x - 4) && -5x + 4 = (-1)(5x - 4) \\ &= 2x(5x - 4) + (-7)(5x - 4) && 7(-1) = -7 \\ &= (5x - 4)(2x - 7) && \text{Distributive Property} \end{aligned}$$

Study Tip

Factoring Completely
Always check for a GCF first before trying to factor a trinomial.

c. $3x^2 + 24x + 45$

The GCF of the terms $3x^2$, $24x$, and 45 is 3. Factor this out first.

$3x^2 + 24x + 45 = 3(x^2 + 8x + 15)$ Distributive Property

Now factor $x^2 + 8x + 15$. Since the leading coefficient is 1, find two factors of 15 with a sum of 8. The correct factors are 3 and 5.

So, $x^2 + 8x + 15 = (x + 3)(x + 5)$. Thus, the complete factorization of $3x^2 + 24x + 45$ is $3(x + 3)(x + 5)$.

CHECK Your Progress

1A. $5x^2 + 13x + 6$ **1B.** $6x^2 + 22x - 8$ **1C.** $10y^2 - 35y + 30$

A polynomial that cannot be written as a product of two polynomials with integral coefficients is called a **prime polynomial**.

EXAMPLE Determine Whether a Polynomial Is Prime

2 **Factor $2x^2 + 5x - 2$.**

In this trinomial, $a = 2$, $b = 5$, and $c = -2$. Since b is positive, $m + n$ is positive. Since c is negative, mn is negative. So either m or n is negative, but not both. Therefore, make a list of the factors of $2(-2)$ or -4, where one factor in each pair is negative. Look for a pair of factors with a sum of 5.

Factors of −4	Sum of Factors
1, −4	−3
−1, 4	3
−2, 2	0

There are no factors with a sum of 5. Therefore, $2x^2 + 5x - 2$ cannot be factored using integers. Thus, $2x^2 + 5x - 2$ is a prime polynomial.

CHECK Your Progress

2A. Is $4r^2 - r + 7$ prime?

2B. Is $2x^2 + 3x - 5$ prime?

Solve Equations by Factoring Some equations of the form $ax^2 + bx + c = 0$ can be solved by factoring and then using the Zero Product Property.

EXAMPLE Solve Equations by Factoring

3 **Solve $8a^2 - 9a - 5 = 4 - 3a$. Check the solutions.**

$8a^2 - 9a - 5 = 4 - 3a$ — Write the equation.

$8a^2 - 6a - 9 = 0$ — Rewrite so that one side equals 0.

$(4a + 3)(2a - 3) = 0$ — Factor the left side.

$4a + 3 = 0$ or $2a - 3 = 0$ — Zero Product Property

$4a = -3$ $\quad$ $2a = 3$ — Solve each equation.

$a = -\frac{3}{4}$ $\quad$ $a = \frac{3}{2}$

The roots are $-\frac{3}{4}$ and $\frac{3}{2}$.

CHECK Check each solution in the original equation.

$$8a^2 - 9a - 5 = 4 - 3a \qquad 8a^2 - 9a - 5 = 4 - 3a$$

$$8\left(-\frac{3}{4}\right)^2 - 9\left(-\frac{3}{4}\right) - 5 \stackrel{?}{=} 4 - 3\left(-\frac{3}{4}\right) \qquad 8\left(\frac{3}{2}\right)^2 - 9\left(\frac{3}{2}\right) - 5 \stackrel{?}{=} 4 - 3\left(\frac{3}{2}\right)$$

$$\frac{9}{2} + \frac{27}{4} - 5 \stackrel{?}{=} 4 + \frac{9}{4} \qquad 18 - \frac{27}{2} - 5 \stackrel{?}{=} 4 - \frac{9}{2}$$

$$\frac{25}{4} = \frac{25}{4} \checkmark \qquad -\frac{1}{2} = -\frac{1}{2} \checkmark$$

CHECK Your Progress

3A. $3x^2 - 5x = 12$

3B. $2x^2 - 30x + 88 = 0$

A model for the vertical motion of a projected object is given by the equation $h = -16t^2 + vt + s$, where h is the height in feet, t is the time in seconds, v is the initial upward velocity in feet per second, and s is the initial height of the object in feet.

Real-World EXAMPLE

Study Tip

Factoring When *a* Is Negative

When factoring a trinomial of the form $ax^2 + bx + c$ where a is negative, it is helpful to factor out a negative monomial.

4 **PEP RALLY At a pep rally, small foam footballs are launched by cheerleaders using a sling-shot. How long is a football in the air if a student catches it on its way down 26 feet above the gym floor?**

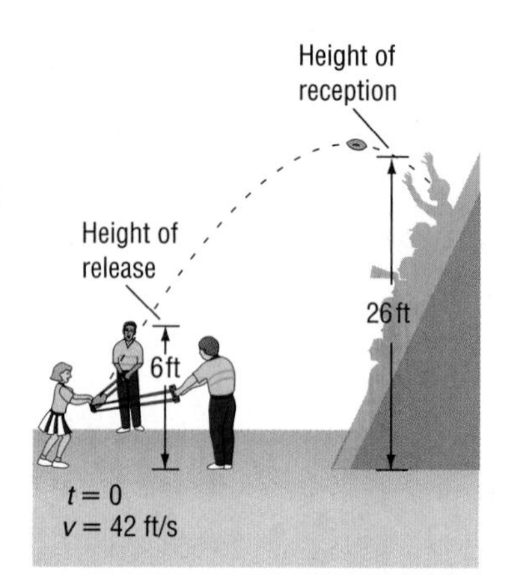

$h = -16t^2 + vt + s$	Vertical motion model
$26 = -16t^2 + 42t + 6$	$h = 26$, $v = 42$, $s = 6$
$0 = -16t^2 + 42t - 20$	Subtract 26 from each side.
$0 = -2(8t^2 - 21t + 10)$	Factor out -2.
$0 = 8t^2 - 21t + 10$	Divide each side by -2.
$0 = (8t - 5)(t - 2)$	Factor $8t^2 - 21t + 10$.
$8t - 5 = 0$ or $t - 2 = 0$	Zero Product Property
$8t = 5$ $t = 2$	Solve each equation.
$t = \frac{5}{8}$	

The solutions are $\frac{5}{8}$ second and 2 seconds.

The first time represents how long it takes the football to reach a height of 26 feet on its way up. The later time represents how long it takes the ball to reach a height of 26 feet again on its way down. Thus, the football will be in the air for 2 seconds before the student catches it.

CHECK Your Progress

4. Six times the square of a number plus 11 times the number equals 2. What are possible values of x?

Online Personal Tutor at algebra1.com

CHECK Your Understanding

Examples 1–2 (pp. 442–443)

Factor each trinomial, if possible. If the trinomial cannot be factored using integers, write *prime*.

1. $3a^2 + 8a + 4$ **2.** $2t^2 - 11t + 7$ **3.** $2p^2 + 14p + 24$

4. $2x^2 + 13x + 20$ **5.** $6x^2 + 15x - 9$ **6.** $4n^2 - 4n - 35$

Example 3 (p. 443)

Solve each equation. Check the solutions.

7. $3x^2 + 11x + 6 = 0$ **8.** $10p^2 - 19p + 7 = 0$ **9.** $6n^2 + 7n = 20$

Example 4 (p. 444)

10. CLIFF DIVING Suppose a diver leaps from the edge of a cliff 80 feet above the ocean with an initial upward velocity of 8 feet per second. How long will it take the diver to enter the water below?

Exercises

HOMEWORK HELP	
For Exercises	See Examples
11–22	1–2
23–30	3
31–32	4

Factor each trinomial, if possible. If the trinomial cannot be factored using integers, write *prime*.

11. $2x^2 + 7x + 5$
12. $6p^2 + 5p - 6$
13. $5d^2 + 6d - 8$
14. $8k^2 - 19k + 9$
15. $9g^2 - 12g + 4$
16. $2a^2 - 9a - 18$
17. $2x^2 - 3x - 20$
18. $5c^2 - 17c + 14$
19. $3p^2 - 25p + 16$
20. $10n^2 - 11n - 6$
21. $6r^2 - 14r - 12$
22. $30x^2 - 25x - 30$

Solve each equation. Check the solutions.

23. $5x^2 + 27x + 10 = 0$
24. $24x^2 - 14x - 3 = 0$
25. $12a^2 - 13a = 35$
26. $6x^2 - 14x = 12$
27. $21x^2 - 6 = 15x$
28. $24x^2 - 46x = 18$
29. $17x^2 - 11x + 2 = 2x^2$
30. $24x^2 - 30x + 8 = -2x$

31. ROCK CLIMBING Damaris is rock climbing at Joshua Tree National Park in the Mojave Desert. She launches a grappling hook from a height of 6 feet with an initial upward velocity of 56 feet per second. The hook just misses the stone ledge that she wants to scale. As it falls, the hook anchors on a ledge 30 feet above the ground. How long was the hook in the air?

32. GYMNASTICS The feet of a gymnast making a vault leave the horse at a height of 8 feet with an initial upward velocity of 8 feet per second. Use the model for vertical motion to find the time t in seconds it takes for the gymnast's feet to reach the mat. (*Hint*: Let $h = 0$, the height of the mat.)

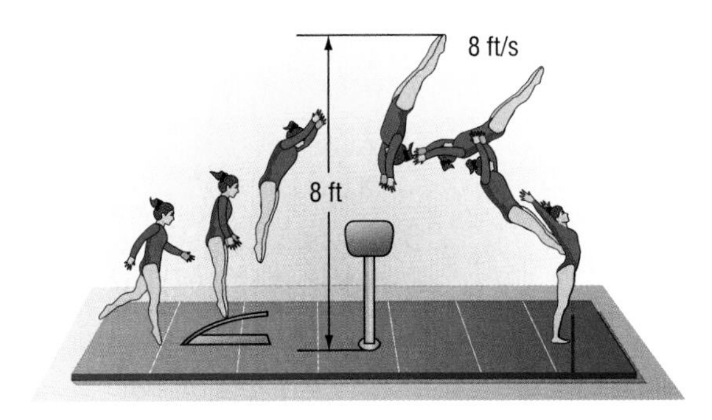

33. GEOMETRY A square has an area of $9x^2 + 30xy + 25y^2$ square inches. What is the perimeter of the square? Explain.

Solve each equation. Check the solutions.

34. $\frac{x^2}{12} - \frac{2x}{3} - 4 = 0$
35. $t^2 - \frac{t}{6} = \frac{35}{6}$
36. $(3y + 2)(y + 3) = y + 14$
37. $(4a - 1)(a - 2) = 7a - 5$

GEOMETRY For Exercises 38 and 39, use the following information.

A rectangle 35 square inches in area is formed by cutting off strips of equal width from a rectangular piece of paper.

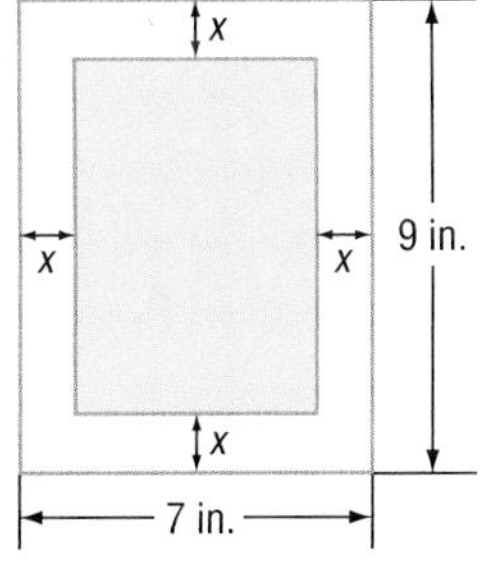

38. Find the width of each strip.
39. Find the dimensions of the new rectangle.

H.O.T. Problems

40. OPEN ENDED Create a trinomial that can be factored using a pair of numbers with a sum of 9 and a product of 14.

41. CHALLENGE Find all values of k so that $2x^2 + kx + 12$ can be factored as two binomials using integers.

42. FIND THE ERROR Dasan and Luther are factoring $2x^2 + 11x + 18$. Who is correct? Explain your reasoning.

Dasan	Luther
$2x^2 + 11x + 18 = 2(x^2 + 11x + 18)$ $= 2(x + 9)(x + 2)$	$2x^2 + 11x + 18$ is prime.

43. ***Writing in Math*** Explain how to determine which values should be chosen for m and n when factoring a polynomial of the form $ax^2 + bx + c$.

STANDARDIZED TEST PRACTICE

44. Which of the following shows $6x^2 + 24x + 18$ factored completely?

A $(3x + 6)^2$

B $(3x + 3)(2x + 6)$

C $(3x + 2)(2x + 9)$

D $(6x + 3)(x + 6)$

45. REVIEW An oak tree grew 18 inches per year from 1985 to 2006. If the tree was 25 feet tall in 1985, what was the height of the tree in 2001?

F 31.0 ft

G 49.0 ft

H 56.5 ft

J 80.5 ft

Spiral Review

Factor each trinomial, if possible. If the trinomial cannot be factored using integers, write *prime*. (Lesson 8-3)

46. $a^2 - 4a - 21$ **47.** $t^2 + 2t + 2$ **48.** $d^2 + 15d + 44$

Solve each equation. Check the solutions. (Lesson 8-2)

49. $(y - 4)(5y + 7) = 0$ **50.** $(2k + 9)(3k + 2) = 0$ **51.** $12u = u^2$

CAMERAS For Exercises 52 and 53, use the graph at the right. (Lessons 2-7 and 7-3)

52. Find the percent of increase in the number of digital cameras sold from 1999 to 2003.

53. Use the answer from Exercise 52 to verify the statement that digital camera sales increased more than 9 times from 1999 to 2003 is correct.

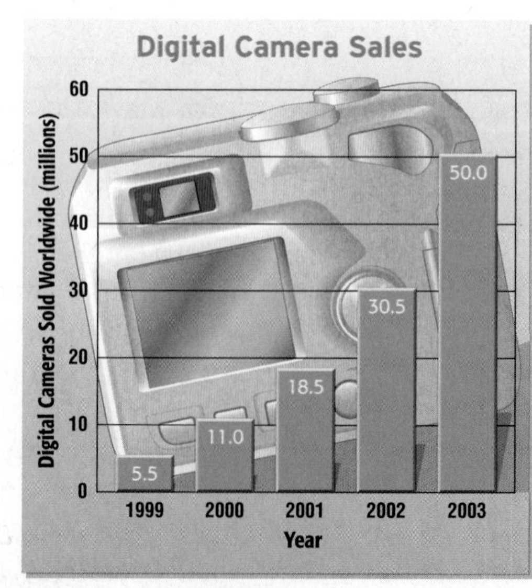

Source: Digital Photography Review

GET READY for the Next Lesson

PREREQUISITE SKILL Find the principal square root of each number. (Lesson 1-8)

54. 49 **55.** 36 **56.** 100

57. 121 **58.** 169 **59.** 225

8-5 Factoring Differences of Squares

Main Ideas

- Factor binomials that are the differences of squares.
- Solve equations involving the differences of squares.

GET READY for the Lesson

A basketball player's *hang time* is the length of time he or she is in the air after jumping. Given the maximum height h a player can jump, you can determine his or her hang time t in seconds by solving $4t^2 - h = 0$. If h is a perfect square, this equation can be solved by factoring, using the pattern for the difference of squares.

Factor $a^2 - b^2$ A geometric model can be used to factor the difference of squares.

ALGEBRA LAB

Difference of Squares

Step 1 Use a straightedge to draw two squares similar to those shown below. Choose any measures for a and b.

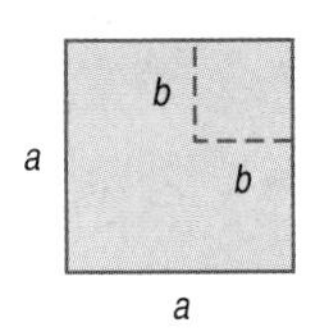

Notice that the area of the large square is a^2, and the area of the small square is b^2.

Step 2 Cut the small square from the large square.

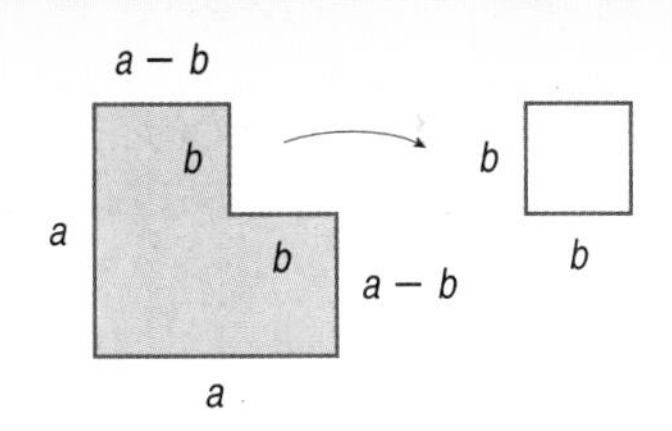

The area of the remaining irregular region is $a^2 - b^2$.

Step 3 Cut the irregular region into two congruent pieces as shown below.

a − b
Region 1
b
a
b
a − b
Region 2
a

Step 4 Rearrange the two pieces to form a rectangle with length $a + b$ and width $a - b$.

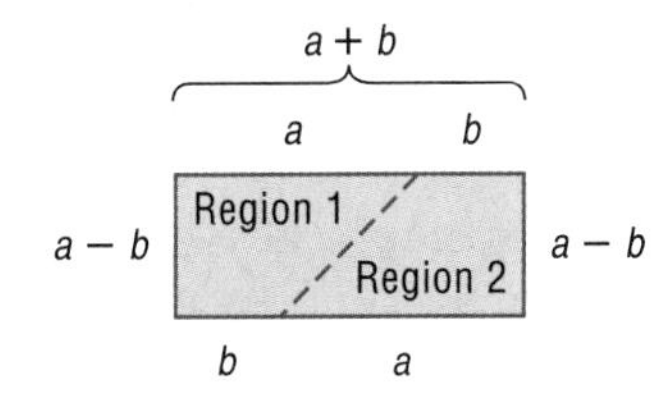

ANALYZE THE RESULTS

1. Write an expression representing the area of the rectangle.

2. Explain why $a^2 - b^2 = (a + b)(a - b)$.

Study Tip

Look Back

To review the **product of a sum and a difference**, see Lesson 7-7.

Concepts in Motion

Animation
algebra1.com

The Algebra Lab leads to the following rule for finding the difference of two squares.

> **KEY CONCEPT** — *Difference of Squares*
>
> **Symbols** $a^2 - b^2 = (a + b)(a - b)$ or $(a - b)(a + b)$
>
> **Examples** $x^2 - 9 = (x + 3)(x - 3)$ or $(x - 3)(x + 3)$

EXAMPLE Factor the Difference of Squares

 **Factor each binomial.**

a. $n^2 - 25$

$n^2 - 25 = n^2 - 5^2$ — Write in the form $a^2 - b^2$.

$= (n + 5)(n - 5)$ — Factor the difference of squares.

b. $36x^2 - 49y^2$

$36x^2 - 49y^2 = (6x)^2 - (7y)^2$ — $36x^2 = 6x \cdot 6x$ and $49y^2 = 7y \cdot 7y$

$= (6x + 7y)(6x - 7y)$ — Factor the difference of squares.

c. $48a^3 - 12a$

If the terms of a binomial have a common factor, the GCF should be factored out first before trying to apply any other factoring technique.

$48a^3 - 12a = 12a(4a^2 - 1)$ — The GCF of $48a^3$ and $-12a$ is $12a$.

$= 12a[(2a) - 1^2]$ — $4a^2 = 2a \cdot 2a$ and $1 = 1 \cdot 1$

$= 12a(2a + 1)(2a - 1)$ — Factor the difference of squares.

CHECK Your Progress

1A. $81 - t^2$ **1B.** $64g^2 - h^2$

1C. $9x^3 - 4x$ **1D.** $-4y^3 + 9y$

Occasionally, the difference of squares pattern needs to be applied more than once to factor a polynomial completely.

EXAMPLE Apply a Factoring Technique More Than Once

Study Tip

Common Misconception
Remember that the sum of two squares, $x^2 + a^2$, is not factorable. $x^2 + a^2$ is a prime polynomial.

2 **Factor $x^4 - 81$.**

$x^4 - 81 = [(x^2)^2 - 9^2]$ — $x^4 = x^2 \cdot x^2$ and $81 = 9 \cdot 9$

$= (x^2 + 9)(x^2 - 9)$ — Factor the difference of squares.

$= (x^2 + 9)(x^2 - 3^2)$ — $x^2 = x \cdot x$ and $9 = 3 \cdot 3$

$= (x^2 + 9)(x + 3)(x - 3)$ — Factor the difference of squares.

CHECK Your Progress

Factor each binomial.

2A. $y^4 - 1$ **2B.** $4a^4 - 4b^4$

Math Online **Extra Examples at** algebra1.com

Study Tip

Look Back

To review **factoring by grouping**, see Lesson 8-2.

EXAMPLE Apply Several Different Factoring Techniques

3 **Factor $5x^3 + 15x^2 - 5x - 15$.**

$5x^3 + 15x^2 - 5x - 15$	Original polynomial
$= 5(x^3 + 3x^2 - x - 3)$	Factor out the GCF.
$= 5[(x^3 - x) + (3x^2 - 3)]$	Group terms with common factors.
$= 5[x(x^2 - 1) + 3(x^2 - 1)]$	Factor each grouping.
$= 5(x^2 - 1)(x + 3)$	$x^2 - 1$ is the common factor.
$= 5(x + 1)(x - 1)(x + 3)$	Factor the difference of squares, $x^2 - 1$, into $(x + 1)(x - 1)$.

CHECK Your Progress **Factor each polynomial.**

3A. $2x^3 + x^2 - 50x - 25$ **3B.** $r^3 + 6r^2 + 11r + 6$

Solve Equations by Factoring You can apply the Zero Product Property to an equation that is written as the product of factors set equal to 0.

STANDARDIZED TEST EXAMPLE Solve Equations by Factoring

4 In the equation $y = x^2 - \frac{9}{16}$, which is a value of x when $y = 0$?

A $-\frac{9}{4}$ **B** 0 **C** $\frac{3}{4}$ **D** $\frac{9}{4}$

Read the Test Item

Factor $x^2 - \frac{9}{16}$ as the difference of squares. Then find the values of x.

Solve the Test Item

$y = x^2 - \frac{9}{16}$	Original equation
$0 = x^2 - \frac{9}{16}$	Replace y with 0.
$0 = x^2 - \left(\frac{3}{4}\right)^2$	$x^2 = x \cdot x$ and $\frac{9}{16} = \frac{3}{4} \cdot \frac{3}{4}$
$0 = \left(x + \frac{3}{4}\right)\left(x - \frac{3}{4}\right)$	Factor the difference of squares.
$0 = x + \frac{3}{4}$ or $0 = x - \frac{3}{4}$	Zero Product Property
$-\frac{3}{4} = x$ $\quad$ $\frac{3}{4} = x$	Solve each equation.

The solutions are $-\frac{3}{4}$ and $\frac{3}{4}$. The correct answer is C.

Test-Taking Tip

When working with a difference of two squares, the solutions will be a number and its opposite. Therefore, choices A and D can be eliminated because if one of them is a solution then the other is also a solution.

CHECK Your Progress

4. Which are the solutions of $18x^3 = 50x$?

F $0, \frac{5}{3}$ **G** $-\frac{5}{3}, \frac{5}{3}$ **H** $-\frac{5}{3}, 0, \frac{5}{3}$ **J** $-\frac{5}{3}, 1, \frac{5}{3}$

Online **Personal Tutor at** algebra1.com

EXAMPLE 5 Use Differences of Two Squares

GEOMETRY The area of the shaded part of the square is 72 square inches. Find the dimensions of the square.

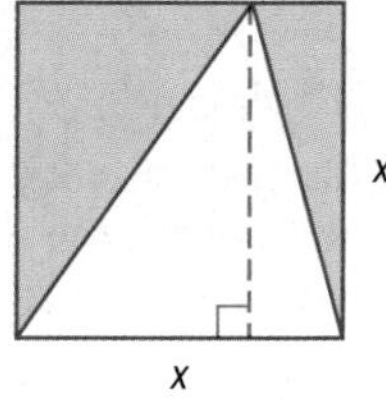

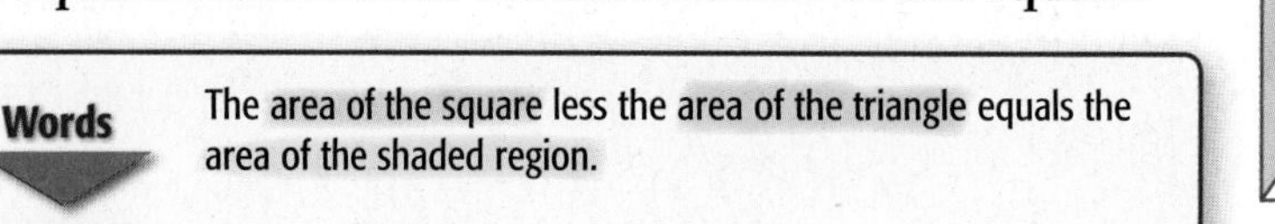

Words The area of the square less the area of the triangle equals the area of the shaded region.

Variable Let $x =$ the side length of the square.

Equation $x^2 - \frac{1}{2}x^2 = 72$

$x^2 - \frac{1}{2}x^2 = 72$	Original equation
$\frac{1}{2}x^2 = 72$	Combine like terms.
$\frac{1}{2}x^2 - 72 = 0$	Subtract 72 from each side.
$x^2 - 144 = 0$	Multiply each side by 2 to remove the fraction.
$(x - 12)(x + 12) = 0$	Factor the difference of squares.

$x - 12 = 0$ or $x + 12 = 0$ Zero Product Property

$x = 12$ $x = -12$ Solve each equation.

Since length cannot be negative, the only reasonable solution is 12. The dimensions of the square are 12 inches by 12 inches. Is this solution reasonable in the context of the original problem?

CHECK Your Progress

5. DRIVING The formula $\frac{1}{24}s^2 = d$ approximates a vehicle's speed s in miles per hour given the length d in feet of skid marks on dry concrete. If skid marks on dry concrete are 54 feet long, how fast was the car traveling when the brakes were applied?

CHECK Your Understanding

Examples 1–3 (pp. 448–449)

Factor each polynomial, if possible. If the polynomial cannot be factored, write *prime*.

1. $n^2 - 81$

2. $4 - 9a^2$

3. $2x^5 - 98x^3$

4. $32x^4 - 2y^4$

5. $4t^2 - 27$

6. $x^3 - 3x^2 - 9x + 27$

Example 4 (p. 449)

Solve each equation by factoring. Check the solutions.

7. $4y^2 = 25$

8. $x^2 - \frac{1}{36} = 0$

9. $121a = 49a^3$

Example 5 (p. 450)

10. GEOMETRY A corner is cut off a 2-inch by 2-inch square piece of paper as shown. What value of x will result in an area that is $\frac{7}{9}$ the area of the original square?

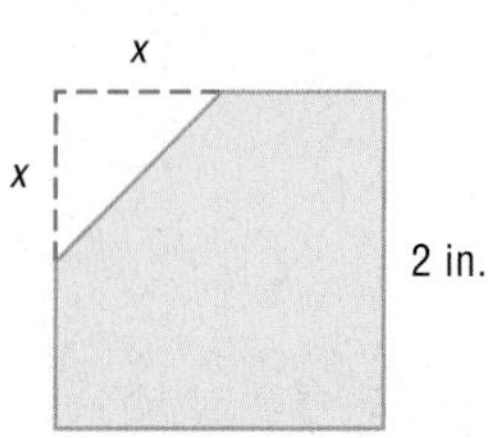

Exercises

HOMEWORK HELP	
For Exercises	See Examples
11–22	1–3
23–30	4
31, 32	5

Factor each polynomial, if possible. If the polynomial cannot be factored, write *prime*.

11. $x^2 - 49$ **12.** $n^2 - 36$ **13.** $81 + 16k^2$

14. $-16 + 49h^2$ **15.** $75 - 12p^2$ **16.** $-18r^3 + 242r$

17. $144a^2 - 49b^2$ **18.** $9x^2 - 10y^2$ **19.** $n^3 + 5n^2 - 4n - 20$

20. $3x^3 + x^2 - 75x - 25$ **21.** $z^4 - 16$ **22.** $256g^4 - 1$

Solve each equation by factoring. Check the solutions.

23. $25x^2 = 36$ **24.** $9y^2 = 64$ **25.** $12 - 27n^2 = 0$ **26.** $50 - 8a^2 = 0$

27. $w^2 - \frac{4}{49} = 0$ **28.** $\frac{81}{100} - p^2 = 0$ **29.** $36 - \frac{1}{9}r^2 = 0$ **30.** $\frac{1}{4}x^2 - 25 = 0$

31. BOATING The basic breaking strength b in pounds for a natural fiber line is determined by the formula $900c^2 = b$, where c is the circumference of the line in inches. What circumference of natural line would have 3600 pounds of breaking strength?

32. GEOMETRY Find the dimensions of a rectangle with the same area as the shaded region in the drawing. Assume that the dimensions of the rectangle must be represented by binomials with integral coefficients.

(4n + 1) cm
(4n + 1) cm
5 cm
5 cm

33. AERODYNAMICS The pressure difference P above and below a wing is described by the formula $P = \frac{1}{2}dv_1^{\,2} - \frac{1}{2}dv_2^{\,2}$, where d is the density of the air, v_1 is the velocity of the air passing above, and v_2 is the velocity of the air passing below. Write this formula in factored form.

34. PACKAGING The width of a box is 9 inches more than its length. The height of the box is 1 inch less than its length. If the box has a volume of 72 cubic inches, what are the dimensions of the box?

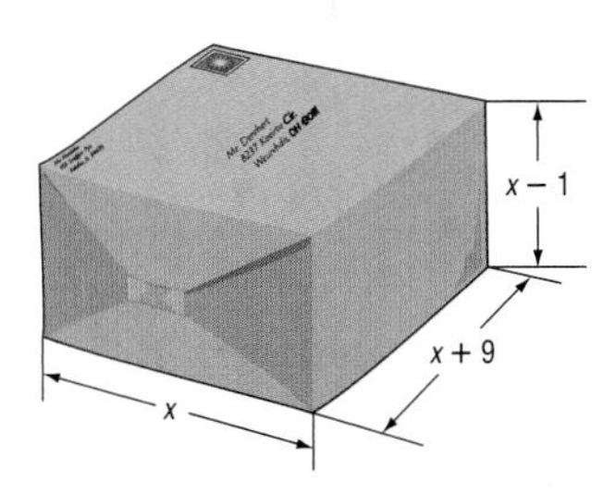

H.O.T. Problems

35. OPEN ENDED Create a binomial that is the difference of two squares. Then factor your binomial.

36. CHALLENGE Show that $a^2 - b^2 = (a + b)(a - b)$ algebraically. (*Hint:* Rewrite $a^2 - b^2$ as $a^2 - ab + ab - b^2$.)

37. FIND THE ERROR Manuel and Jessica are factoring $64x^2 + 16y^2$. Who is correct? Explain your reasoning.

Manuel
$64x^2 + 16y^2$
$= 16(4x^2 + y^2)$

Jessica
$64x^2 + 16y^2$
$= 16(4x^2 + y^2)$
$= 16(2x + y)(2x - y)$

38. REASONING The following statements appear to prove that 2 is equal to 1. Find the flaw in this "proof."

Suppose a and b are real numbers such that $a = b, a \neq 0, b \neq 0$.

(1)	$a = b$	Given.
(2)	$a^2 = ab$	Multiply each side by a.
(3)	$a^2 - b^2 = ab - b^2$	Subtract b^2 from each side.
(4)	$(a - b)(a + b) = b(a - b)$	Factor.
(5)	$a + b = b$	Divide each side by $a - b$.
(6)	$a + a = a$	Substitution Property; $a = b$
(7)	$2a = a$	Combine like terms.
(8)	$2 = 1$	Divide each side by a.

39. ***Writing in Math*** Use the information about basketball on page 447 to explain how to determine a basketball player's hang time. Include a maximum height that is a perfect square and that would be considered a reasonable distance for a student athlete to jump. Describe how to find the hang time for this height.

STANDARDIZED TEST PRACTICE

40. What are the solutions to the quadratic equation $25b^2 - 1 = 0$?

A $0, \frac{1}{5}$

B $-\frac{1}{5}, 0$

C $\frac{1}{5}, 1$

D $-\frac{1}{5}, \frac{1}{5}$

41. REVIEW Carla's Candle Shop sells 3 small candles for a total of \$5.94. Which expression can be used to find the total cost c of x candles?

F $\frac{5.94}{x}$

G $5.94x$

H $\frac{x}{1.98}$

J $1.98x$

Spiral Review

Factor each trinomial, if possible. If the trinomial cannot be factored using integers, write *prime*. (Lesson 8-4)

42. $2n^2 + 5n + 7$ **43.** $6x^2 - 11x + 4$ **44.** $21p^2 + 29p - 10$

Solve each equation. Check the solutions. (Lesson 8-3)

45. $y^2 + 18y + 32 = 0$ **46.** $k^2 - 8k = -15$ **47.** $b^2 - 8 = 2b$

48. STATISTICS Amy's scores on the first three of four 100-point biology tests were 88, 90, and 91. To get a B+ in the class, her average must be between 88 and 92, inclusive, on all tests. What score must she receive on the fourth test to get a B+ in biology? (Lesson 5-4)

GET READY for the Next Lesson

PREREQUISITE SKILL Find each product. (Lesson 7-7)

49. $(x + 1)(x + 1)$ **50.** $(x + 8)^2$ **51.** $(3x - 4)(3x - 4)$ **52.** $(5x - 2)^2$

READING MATH

Proofs

When you solve an equation by factoring, you are using a deductive argument. Each step can be justified by an algebraic property.

Solve $4x^2 - 324 = 0$.

$4x^2 - 324 = 0$	Original equation
$(2x)^2 - 18^2 = 0$	$4x^2 = (2x)^2$ and $324 = 18^2$
$(2x + 18)(2x - 18) = 0$	Factor the difference of squares.
$2x + 18 = 0$ or $2x - 18 = 0$	Zero Product Property
$x = -9$ $\quad$ $x = 9$	Solve each equation.

Notice that the column on the left is a step-by-step process that leads to a solution. The column on the right contains the reasons for each statement. A *two-column proof* is a deductive argument that contains statements and reasons.

Two-Column Proof

Given: a, x, and y are real numbers such that $a \neq 0$, $x \neq 0$, and $y \neq 0$.
Prove: $ax^4 - ay^4 = a(x^2 + y^2)(x + y)(x - y)$

There is a reason for each statement.

The first statement contains the given information.

The last statement is what you want to prove.

Statements	Reasons
1. a, x, and y are real numbers such that $a \neq 0$, $x \neq 0$, and $y \neq 0$.	**1.** Given
2. $ax^4 - ay^4 = a(x^4 - y^4)$	**2.** The GCF of ax^4 and ay^4 is a.
3. $ax^4 - ay^4 = a[(x^2)^2 - (y^2)^2]$	**3.** $x^4 = (x^2)^2$ and $y^4 = (y^2)^2$
4. $ax^4 - ay^4 = a(x^2 + y^2)(x^2 - y^2)$	**4.** Factor the difference of squares.
5. $ax^4 - ay^4 = a(x^2 + y^2)(x + y)(x - y)$	**5.** Factor the difference of squares.

Reading to Learn

1. Solve $\frac{1}{16}t^2 - 100 = 0$ by using a two-column proof.
2. Write a two-column proof using the following information. (*Hint:* Group terms with common factors.)

   Given: c and d are real numbers such that $c \neq 0$ and $d \neq 0$.

   Prove: $c^3 - cd^2 - c^2d + d^3 = (c + d)(c - d)(c - d)$
3. Explain how the process used to write two-column proofs can be useful in solving Find the Error exercises, such as Exercise 37 on page 451.

8-6 Perfect Squares and Factoring

Main Ideas

- Factor perfect square trinomials.
- Solve equations involving perfect squares.

New Vocabulary

perfect square trinomials

GET READY for the Lesson

The senior class has decided to build an outdoor pavilion. It will have an 8-foot by 8-foot portrayal of the school's mascot in the center. The class is selling bricks with students' names on them to finance the project. If they sell enough bricks to cover 80 square feet and want to arrange the bricks around the art, how wide should the border of bricks be?

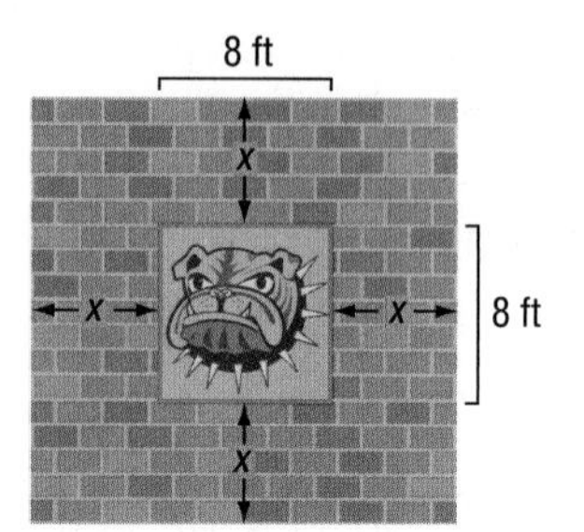

To solve this problem, you need to solve the equation $(8 + 2x)^2 = 144$.

Factor Perfect Square Trinomials Numbers like 16, 49, and 144 are perfect squares, since each can be expressed as the square of an integer.

$$16 = 4 \cdot 4 \text{ or } 4^2 \qquad 49 = 7 \cdot 7 \text{ or } 7^2 \qquad 144 = 12 \cdot 12 \text{ or } 12^2$$

Products of the form $(a + b)^2$ and $(a - b)^2$, such as $(8 + 2x)^2$, are also perfect squares. Recall that these are special products that follow specific patterns.

$$(a + b)^2 = (a + b)(a + b) \qquad (a - b)^2 = (a - b)(a - b)$$
$$= a^2 + ab + ab + b^2 \qquad = a^2 - ab - ab + b^2$$
$$= a^2 + 2ab + b^2 \qquad = a^2 - 2ab + b^2$$

These patterns can help you factor **perfect square trinomials**, which are trinomials that are the squares of binomials.

Squaring a Binomial	Factoring a Perfect Square
$(x + 7)^2 = x^2 + 2(x)(7) + 7^2$ $= x^2 + 14x + 49$	$x^2 + 14x + 49 = x^2 + 2(x)(7) + 7^2$ $= (x + 7)^2$
$(3x - 4)^2 = (3x)^2 - 2(3x)(4) + 4^2$ $= 9x^2 - 24x + 16$	$9x^2 - 24x + 16 = (3x)^2 - 2(3x)(4) + 4^2$ $= (3x - 4)^2$

For a trinomial to be factorable as a perfect square, three conditions must be satisfied as illustrated in the example below.

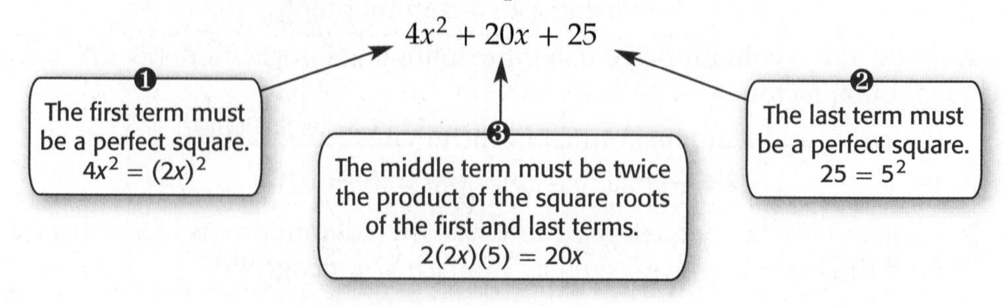

KEY CONCEPT — *Factoring Perfect Square Trinomials*

Words If a trinomial can be written in the form $a^2 + 2ab + b^2$ or $a^2 - 2ab + b^2$, then it can be factored as $(a + b)^2$ or as $(a - b)^2$, respectively.

Symbols $a^2 + 2ab + b^2 = (a + b)^2$ and $a^2 - 2ab + b^2 = (a - b)^2$

EXAMPLE Factor Perfect Square Trinomials

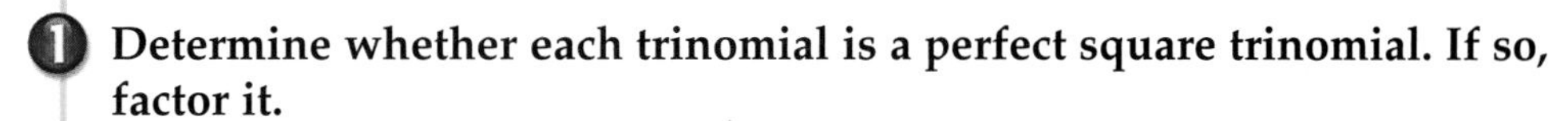

1 Determine whether each trinomial is a perfect square trinomial. If so, factor it.

a. $16x^2 + 32x + 64$

1. Is the first term a perfect square? Yes, $16x^2 = (4x)^2$.
2. Is the last term a perfect square? Yes, $64 = 8^2$.
3. Is the middle term equal to $2(4x)(8)$? No, $32x \neq 2(4x)(8)$.

$16x^2 + 32x + 64$ is not a perfect square trinomial.

b. $9y^2 - 12y + 4$

1. Is the first term a perfect square? Yes, $9y^2 = (3y)^2$.
2. Is the last term a perfect square? Yes, $4 = 2^2$.
3. Is the middle term equal to $2(3y)(2)$? Yes, $12y = 2(3y)(2)$.

$9y^2 - 12y + 4$ is a perfect square trinomial.

$9y^2 - 12y + 4 = (3y)^2 - 2(3y)(2) + 2^2$ Write as $a^2 - 2ab + b^2$.

$= (3y - 2)^2$ Factor using the pattern.

CHECK Your Progress

1A. $n^2 - 24n + 144$ **1B.** $x^2 + 9x + 81$

You have learned various techniques for factoring polynomials. The Concept Summary can help you decide when to use a specific technique.

CONCEPT SUMMARY — *Factoring Polynomials*

Number of Terms	Factoring Technique		Example
2 or more	greatest common factor		$3x^2 + 6x^2 - 15x = 3x(x^2 + 2x - 5)$
2	difference of squares	$a^2 - b^2 = (a + b)(a - b)$	$4x^2 - 25 = (2x + 5)(2x - 5)$
3	perfect square trinomial	$a^2 + 2ab + b^2 = (a + b)^2$ $a^2 - 2ab + b^2 = (a - b)^2$	$x^2 + 6x + 9 = (x + 3)^2$ $4x^2 - 4x + 1 = (2x - 1)^2$
	$x^2 + bx + c$	$x^2 + bx + c = (x + m)(x + n)$ when $m + n = b$ and $mn = c$.	$x^2 - 9x + 20 = (x - 5)(x - 4)$
	$ax^2 + bx + c$	$ax^2 + bx + c = ax^2 + mx + nx + c$ when $m + n = b$ and $mn = ac$. Then use factoring by grouping.	$6x^2 - x - 2 = 6x^2 + 3x - 4x - 2$ $= 3x(2x + 1) - 2(2x + 1)$ $= (2x + 1)(3x - 2)$
4 or more	factoring by grouping	$ax + bx + ay + by$ $= x(a + b) + y(a + b)$ $= (a + b)(x + y)$	$3xy - 6y + 5x - 10$ $= (3xy - 6y) + (5x - 10)$ $= 3y(x - 2) + 5(x - 2)$ $= (x - 2)(3y + 5)$

EXAMPLE Factor Completely

Study Tip

Factoring Methods

When there is a GCF other than 1, it is usually easier to factor it out first. Then, check the appropriate factoring methods in the order shown in the table.

2 Factor each polynomial.

a. $4x^2 - 36$

First check for a GCF. Then, since the polynomial has two terms, check for the difference of squares.

$4x^2 - 36 = 4(x^2 - 9)$	4 is the GCF.
$= 4(x^2 - 3^2)$	$x^2 = x \cdot x$ and $9 = 3 \cdot 3$
$= 4(x + 3)(x - 3)$	Factor the difference of squares.

b. $25x^2 + 5x - 6$

This is not a perfect square trinomial. It is of the form $ax^2 + bx + c$. Are there two numbers m and n with a product of $25(-6)$ or -150 and a sum of 5? Yes, the product of 15 and -10 is -150 and the sum is 5.

$25x^2 + 5x - 6 = 25x^2 + mx + nx - 6$	Write the pattern.
$= 25x^2 + 15x - 10x - 6$	$m = 15$ and $n = -10$
$= (25x^2 + 15x) + (-10x - 6)$	Group terms with common factors.
$= 5x(5x + 3) - 2(5x + 3)$	Factor out the GCF from each grouping.
$= (5x + 3)(5x - 2)$	$5x + 3$ is the common factor.

CHECK Your Progress

2A. $2x^2 - 32$

2B. $9t^2 - 3t - 20$

Solve Equations with Perfect Squares When solving equations involving repeated factors, it is only necessary to set one of the repeated factors equal to zero.

EXAMPLE Solve Equations with Repeated Factors

3 Solve $x^2 - x + \frac{1}{4} = 0$.

$x^2 - x + \frac{1}{4} = 0$	Original equation
$x^2 - 2(x)\left(\frac{1}{2}\right) + \left(\frac{1}{2}\right)^2 = 0$	Recognize $x^2 - x + \frac{1}{4}$ as a perfect square trinomial.
$\left(x - \frac{1}{2}\right)^2 = 0$	Factor the perfect square trinomial.
$x - \frac{1}{2} = 0$	Set repeated factor equal to zero.
$x = \frac{1}{2}$	Solve for x.

CHECK Your Progress

Solve each equation. Check the solutions.

3A. $a^2 + 12a + 36 = 0$

3B. $y^2 - \frac{4}{3}y + \frac{4}{9} = 0$

Reading Math

Square Root Solutions $\pm\sqrt{36}$ is read as *plus or minus the square root of 36.*

You have solved equations like $x^2 - 36 = 0$ by factoring. You can also use the definition of a square root to solve this equation.

$x^2 - 36 = 0$	Original equation
$x^2 = 36$	Add 36 to each side.
$x = \pm\sqrt{36}$	Take the square root of each side.

Remember that there are two square roots of 36, namely 6 and -6. Therefore, the solution set is $\{-6, 6\}$. You can express this as $\{\pm 6\}$.

KEY CONCEPT *Square Root Property*

Symbols For any number $n > 0$, if $x^2 = n$, then $x = \pm\sqrt{n}$.

Example $x^2 = 9$

$x = \pm\sqrt{9}$ or ± 3

EXAMPLE

4 **PHYSICAL SCIENCE During an experiment, a ball is dropped from a height of 205 feet. The formula $h = -16t^2 + h_0$ can be used to approximate the number of seconds t it takes for the ball to reach height h from an initial height h_0 in feet. Find the time it takes the ball to reach the ground.**

$h = -16t^2 + h_0$	Original formula
$0 = -16t^2 + 205$	Replace h with 0 and h_0 with 205.
$-205 = -16t^2$	Subtract 205 from each side.
$12.8125 = t^2$	Divide each side by -16.
$\pm 3.6 \approx t$	Take the square root of each side.

Since a negative number does not make sense in this situation, the solution is 3.6. This means that it takes about 3.6 seconds for the ball to reach the ground.

CHECK Your Progress

4. Find the time it takes a ball to reach the ground if it is dropped from a bridge that is half as high as the one described above.

EXAMPLE Use the Square Root Property to Solve Equations

5 **Solve each equation. Check the solutions.**

a. $(a + 4)^2 = 49$

$(a + 4)^2 = 49$	Original equation
$a + 4 = \pm\sqrt{49}$	Square Root Property
$a + 4 = \pm 7$	$49 = 7 \cdot 7$
$a = -4 \pm 7$	Subtract 4 from each side.
$a = -4 + 7$ or $a = -4 - 7$	Separate into two equations.
$= 3$ $\quad$ $= -11$	Simplify.
The roots are -11 and 3.	Check in the original equation.

(continued on the next page)

b. $(x - 3)^2 = 5$

$(x - 3)^2 = 5$ Original equation

$x - 3 = \pm\sqrt{5}$ Square Root Property

$x = 3 \pm \sqrt{5}$ Add 3 to each side.

Since 5 is not a perfect square, the roots are $3 \pm \sqrt{5}$. Using a calculator, the roots are $3 + \sqrt{5}$ or about 5.24 and $3 - \sqrt{5}$ or about 0.76.

Concepts in Motion
Interactive Lab
algebra1.com

CHECK Your Progress

5A. $z^2 + 2z + 1 = 16$ **5B.** $(y - 8)^2 = 7$

CHECK Your Understanding

Example 1 (p. 455)

Determine whether each trinomial is a perfect square trinomial. If so, factor it.

1. $y^2 + 8y + 16$ **2.** $9x^2 - 30x + 10$

Example 2 (p. 456)

Factor each polynomial, if possible. If the polynomial cannot be factored, write *prime*.

3. $2x^2 + 18$ **4.** $c^2 - 5c + 6$

5. $8x^2 - 18x - 35$ **6.** $9g^2 + 12g - 4$

Examples 3, 5 (pp. 456–458)

Solve each equation. Check the solutions.

7. $4y^2 + 24y + 36 = 0$ **8.** $3n^2 = 48$

9. $a^2 - 6a + 9 = 16$ **10.** $(m - 5)^2 = 13$

Example 4 (p. 457)

11. HISTORY Galileo showed that objects of different weights fall at the same velocity by dropping two objects of different weights from the top of the Leaning Tower of Pisa. A model for the height h in feet of an object dropped from an initial height h_0 feet is $h = -16t^2 + h_0$, where t is the time in seconds after the object is dropped. Use this model to determine approximately how long it took for objects to hit the ground if Galileo dropped them from a height of 180 feet.

Exercises

HOMEWORK HELP

For Exercises	See Examples
12–15	1
16–23	2
24–33	3, 5
34–37	4

Determine whether each trinomial is a perfect square trinomial. If so, factor it.

12. $4y^2 - 44y + 121$ **13.** $2c^2 + 10c + 25$

14. $9n^2 + 49 + 42n$ **15.** $25a^2 - 120ab + 144b^2$

Factor each polynomial, if possible. If the polynomial cannot be factored, write *prime*.

16. $4k^2 - 100$ **17.** $4a^2 - 36b^2$

18. $x^2 + 6x - 9$ **19.** $50g^2 + 40g + 8$

20. $9t^3 + 66t^2 - 48t$ **21.** $20n^2 + 34n + 6$

22. $5y^2 - 90$ **23.** $18y^2 - 48y + 32$

Solve each equation. Check the solutions.

24. $3x^2 + 24x + 48 = 0$

25. $7r^2 = 70r - 175$

26. $49a^2 + 16 = 56a$

27. $18y^2 + 24y + 8 = 0$

28. $y^2 - \frac{2}{3}y + \frac{1}{9} = 0$

29. $a^2 + \frac{4}{5}a + \frac{4}{25} = 0$

30. $x^2 + 10x + 25 = 81$

31. $(w + 3)^2 = 2$

32. $p^2 + 2p + 1 = 6$

33. $x^2 - 12x + 36 = 11$

34. FORESTRY The number of board feet B that a log will yield can be estimated by using the formula $B = \frac{L}{16}(D^2 - 8D + 16)$, where D is the diameter in inches and L is the log length in feet. For logs that are 16 feet long, what diameter will yield approximately 256 board feet?

Real-World Link

Some amusement park free-fall rides can seat 4 passengers across per coach and reach speeds of up to 62 miles per hour.

Source: pgathrills.com

FREE-FALL RIDE For Exercises 35 and 36, use the following information.
The height h in feet of a car above the exit ramp of an amusement park's free-fall ride can be modeled by $h = -16t^2 + s$, where t is the time in seconds after the car drops and s is the starting height of the car in feet.

35. How high above the car's exit ramp should the ride's designer start the drop in order for riders to experience free fall for at least 3 seconds?

36. Approximately how long will riders be in free fall if their starting height is 160 feet above the exit ramp?

37. HUMAN CANNONBALL A circus acrobat is shot out of a cannon with an initial upward velocity of 64 feet per second. If the acrobat leaves the cannon 6 feet above the ground, will he reach a height of 70 feet? If so, how long will it take him to reach that height? Use the model for vertical motion.

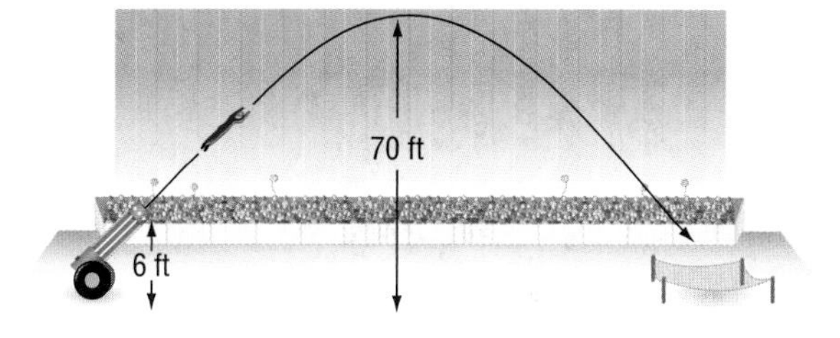

Factor each polynomial, if possible. If the polynomial cannot be factored, write *prime*.

38. $4a^3 + 3a^2b^2 + 8a + 6b^2$

39. $5a^2 + 7a + 6b^2 - 4b$

40. $x^2y^2 - y^2 - z^2 + x^2z^2$

41. $4m^4n + 6m^3n - 16m^2n^2 - 24mn^2$

42. GEOMETRY The volume of a rectangular prism is $x^3y - 63y^2 + 7x^2 - 9xy^3$ cubic meters. Find the dimensions of the prism if they can be represented by binomials with integral coefficients.

43. GEOMETRY If the area of the square shown is $16x^2 - 56x + 49$ square inches, what is the area of the rectangle in terms of x?

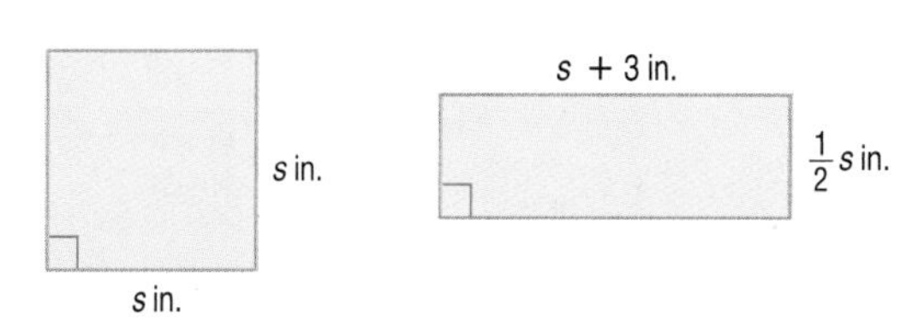

H.O.T. Problems

44. REASONING Determine whether the following statement is *sometimes*, *always*, or *never* true. Explain your reasoning.

$$a^2 - 2ab - b^2 = (a - b)^2, b \neq 0$$

45. **OPEN ENDED** Create a polynomial that requires at least two different factoring techniques to factor it completely. Then factor the polynomial completely, describing the techniques that were used.

46. **Which One Doesn't Belong?** Identify the trinomial that does not belong with the other three. Explain your reasoning.

$4x^2 - 36x + 81$	$25x^2 + 10x + 1$	$4x^2 + 10x + 4$	$9x^2 - 24x + 16$

CHALLENGE Determine all values of k that make each of the following a perfect square trinomial.

47. $4x^2 + kx + 1$ 48. $x^2 - 18x + k$ 49. $x^2 + 20x + k$

50. ***Writing in Math*** Use the information about the project on page 454 to explain how factoring can be used to design a pavilion. Explain how the equation $(8 + 2x)^2 = 144$ models the given situation, solve this equation, and interpret its solutions.

STANDARDIZED TEST PRACTICE

51. What are the solutions for the equation $3(5x - 1)^2 = 27$?

A $-\frac{9}{5}$ and 2

B -2 and $\frac{9}{5}$

C $-\frac{2}{5}$ and $\frac{4}{5}$

D $-\frac{1}{5}$ and $\frac{3}{5}$

52. **REVIEW** Marta has a bag of 8 marbles. There are 3 red marbles, 2 blue marbles, 2 white marbles, and 1 black marble. If she picks one marble without looking, what is the probability that it is either black or white?

F $\frac{1}{8}$ H $\frac{3}{8}$

G $\frac{1}{4}$ J $\frac{5}{8}$

Spiral Review

Solve each equation. Check the solutions. (Lessons 8-4 and 8-5)

53. $9x^2 - 16 = 0$ 54. $49m^2 = 81$ 55. $8k^2 + 22k - 6 = 0$ 56. $12w^2 + 23x = -5$

Solve each inequality. Check the solution. (Lesson 6-2)

57. $\frac{r}{5} > -11$ 58. $8 > \frac{2}{3}n$ 59. $76 < 4t$ 60. $-14c \leq 84$

61. **BUSINESS** Jake's Garage charges $180 for a two-hour repair job and $375 for a five-hour repair job. Write a linear equation that Jake can use to bill customers for repair jobs of any length of time. (Lesson 4-3)

62. **MODEL TRAINS** One of the most popular sizes of model trains is called the HO. Every dimension of the HO model measures $\frac{1}{87}$ times that of a real engine. The HO model of a modern diesel locomotive is about 8 inches long. About how many feet long is the real locomotive? (Lesson 3-6)

CHAPTER 8 Study Guide and Review

FOLDABLES Study Organizer — GET READY to Study

Be sure the following Key Concepts are noted in your Foldable.

Key Concepts

Monomials and Factoring (Lesson 8-1)

- The greatest common factor (GCF) of two or more monomials is the product of their common prime factors.

Factoring Using the Distributive Property (Lesson 8-2)

- Using the Distributive Property to factor polynomials with four or more terms is called factoring by grouping.

$$ax + bx + ay + by = x(a + b) + y(a + b)$$
$$= (a + b)(x + y)$$

- Factoring can be used to solve some equations. According to the Zero Product Property, for any real numbers a and b, if $ab = 0$, then either $a = 0$, $b = 0$, or both a and b equal zero.

Factoring Trinomials and Differences of Squares (Lessons 8-3, 8-4, and 8-5)

- To factor $x^2 + bx + c$, find m and n with a sum of b and a product of c. Then write $x^2 + bx + c$ as $(x + m)(x + n)$.
- To factor $ax^2 + bx + c$, find m and n with a product of ac and a sum of b. Then write as $ax^2 + mx + nx + c$ and factor by grouping.
  $a^2 - b^2 = (a + b)(a - b)$ or $(a - b)(a + b)$

Perfect Squares and Factoring (Lesson 8-6)

- $a^2 + 2ab + b^2 = (a + b)^2$ and $a^2 - 2ab + b^2 = (a - b)^2$
- For a trinomial to be a perfect square, the first and last terms must be perfect squares, and the middle term must be twice the product of the square roots of the first and last terms.
- For any number $n > 0$, if $x^2 = n$, then $x = \pm\sqrt{n}$.

Key Vocabulary

composite number (p. 420)
factored form (p. 421)
factoring (p. 426)
factoring by grouping (p. 427)
greatest common factor (p. 422)
perfect square trinomials (p. 454)
prime factorization (p. 421)
prime number (p. 420)
prime polynomial (p. 443)
roots (p. 428)

Vocabulary Check

State whether each sentence is *true* or *false*. If *false*, replace the underlined word, phrase, expression, or number to make a true sentence.

1. The number 27 is an example of a prime number.
2. $2x$ is the greatest common factor of $12x^2$ and $14xy$.
3. 66 is an example of a perfect square.
4. 61 is a factor of 183.
5. The prime factorization of 48 is $3 \cdot 4^2$.
6. $x^2 - 25$ is an example of a perfect square trinomial.
7. The number 35 is an example of a composite number.
8. $x^2 - 3x - 70$ is an example of a prime polynomial.
9. Expressions with four or more unlike terms can sometimes be factored by grouping.
10. $(b - 7)(b + 7)$ is the factorization of a difference of squares.

Lesson-by-Lesson Review

8-1 Monomials and Factoring (pp. 420–424)

Factor each monomial completely.

11. $28n^3$ **12.** $-33a^2b$

13. $150st$ **14.** $-378pq^2r^2$

Find the GCF of each set of monomials.

15. 35, 30 **16.** 12, 18, 40

17. $12ab, 4a^2b^2$ **18.** $16mrt, 30m^2r$

19. $20n^2, 25np^5$

20. $60x^2y^2, 15xyz, 35xz^3$

21. HOME IMPROVEMENT A landscape architect is designing a stone path to cover an area 36 inches by 120 inches. What is the maximum size square stone that can be used so that none of the stones have to be cut?

Example 1 **Factor $68cd^2$ completely.**

$68cd^2 = 4 \cdot 17 \cdot c \cdot d \cdot d$ $\quad 68 = 4 \cdot 17, d^2 = d \cdot d$

$= 2 \cdot 2 \cdot 17 \cdot c \cdot d \cdot d$ $\quad 4 = 2 \cdot 2$

Thus, $68cd^2$ in factored form is $2 \cdot 2 \cdot 17 \cdot c \cdot d \cdot d$.

Example 2 **Find the GCF of $15x^2y$ and $45xy^2$.**

$15x^2y = 3 \cdot 5 \cdot x \cdot x \cdot y$ $\quad$ Factor each number.

$45xy^2 = 3 \cdot 3 \cdot 5 \cdot x \cdot y \cdot y$ $\quad$ Circle the common prime factors.

The GCF is $3 \cdot 5 \cdot x \cdot y$ or $15xy$.

8-2 Factoring Using the Distributive Property (pp. 426–431)

Factor each polynomial.

22. $13x + 26y$ **23.** $a^2 - 4ac + ab - 4bc$

24. $24a^2b^2 - 18ab$ **25.** $26ab + 18ac + 32a^2$

26. $4rs + 12ps + 2mr + 6mp$

27. $24am - 9an + 40bm - 15bn$

Solve each equation. Check the solutions.

28. $x(2x - 5) = 0$

29. $4x^2 = -7x$

30. $(3n + 8)(2n - 6) = 0$

31. EXERCISE A gymnast jumps on a trampoline traveling at 12 feet per second. Her height h in feet above the trampoline after t seconds is given by the formula $h = 12t - 16t^2$. How long is the gymnast in the air before returning to the trampoline?

Example 3 **Factor $2x^2 - 3xz - 2xy + 3yz$.**

$2x^2 - 3xz - 2xy + 3yz$

$= (2x^2 - 3xz) + (-2xy + 3yz)$

$= x(2x - 3z) - y(2x - 3z)$

$= (x - y)(2x - 3z)$

Example 4 **Solve $x^2 = 5x$. Check the solutions.**

Write the equation so that it is of the form $ab = 0$.

$x^2 = 5x$ $\quad$ Original equation

$x^2 - 5x = 0$ $\quad$ Subtract $5x$ from each side.

$x(x - 5) = 0$ $\quad$ Factor using the GCF, x.

$x = 0$ or $x - 5 = 0$ $\quad$ Zero Product Property

$x = 5$ $\quad$ Solve the equation.

The roots are 0 and 5. Check by substituting 0 and 5 for x in the original equation.

Mixed Problem Solving
For mixed problem-solving practice, see page 751.

8–3 Factoring Trinomials: $x^2 + bx + c$ (pp. 434–439)

Factor each trinomial.

32. $y^2 + 7y + 12$ **33.** $x^2 - 9x - 36$

34. $b^2 + 5b - 6$ **35.** $18 - 9r + r^2$

Solve each equation. Check the solutions.

36. $y^2 + 13y + 40 = 0$

37. $x^2 - 5x - 66 = 0$

38. SOCCER In order for a town to host an international soccer game, its field's length must be 110–120 yards, and its width must be 70–80 yards. Green Meadows soccer field is 30 yards longer than it is wide. Write an expression for the area of the rectangular field. If the area of the field is 8800 square yards, will Green Meadows be able to host an international game? Explain.

Example 5 Factor $x^2 - 9x + 20$.

$b = -9$ and $c = 20$, so $m + n$ is negative and mn is positive. Therefore, m and n must both be negative. List the negative factors of 20, and look for the pair of factors with a sum of -9.

Factors of 20	Sum of Factors
−1, −20	−21
−2, −10	−12
−4, −5	−9

The correct factors are -4 and -5.

$x^2 - 9x + 20 = (x + m)(x + n)$ Write the pattern.

$= (x - 4)(x - 5)$ $m = -4$ and $n = -5$

8–4 Factoring Trinomials: $ax^2 + bx + c$ (pp. 441–446)

Factor each trinomial, if possible. If the trinomial cannot be factored using integers, write *prime*.

39. $2a^2 - 9a + 3$ **40.** $2m^2 + 13m - 24$

41. $12b^2 + 17b + 6$ **42.** $3n^2 - 6n - 45$

Solve each equation. Check the solutions.

43. $2r^2 - 3r - 20 = 0$ **44.** $40x^2 + 2x = 24$

45. BASEBALL Victor hit a baseball into the air that modeled the equation $h = -16t^2 + 36t + 1$, where h is the height in feet and t is the time in seconds. How long was the ball in the air if Casey caught the ball 9 feet above the ground on its way down?

Example 6 Factor $12x^2 + 22x - 14$.

$12x^2 + 22x - 14 = 2(6x^2 + 11x - 7)$ Factor.

So, $a = 6$, $b = 11$, and $c = -7$. Since b is positive, $m + n$ is positive. Since c is negative, mn is negative. So either m or n is negative. List the factors of $6(-7)$ or -42, where one factor in each pair is negative. The correct factors are -3 and 14.

$6x^2 + 11x - 7 = 6x^2 + mx + nx - 7$

$= 6x^2 - 3x + 14x - 7$

$= (6x^2 - 3x) + (14x - 7)$

$= 3x(2x - 1) + 7(2x - 1)$

$= (2x - 1)(3x + 7)$

Thus, the complete factorization of $12x^2 + 22x - 14$ is $2(2x - 1)(3x + 7)$.

Study Guide and Review

8–5 Factoring Differences of Squares (pp. 447–452)

Factor each polynomial, if possible. If the polynomial cannot be factored, write *prime*.

46. $64 - 4s^2$

47. $2y^3 - 128y$

48. $9b^2 - 20$

49. $\frac{1}{4}n^2 - \frac{9}{16}r^2$

Solve each equation by factoring. Check the solutions.

50. $b^2 - 16 = 0$

51. $25 - 9y^2 = 0$

52. $16a^2 = 81$

53. $\frac{25}{49} - r^2 = 0$

54. EROSION A boulder breaks loose from the face of a mountain and falls toward the water 576 feet below. The distance d that the boulder falls in t seconds is given by the equation $d = 16t^2$. How long does it take the boulder to hit the water?

Example 7 **Solve $y^2 + 9 = 90$ by factoring.**

$y^2 + 9 = 90$	Original equation
$y^2 - 81 = 0$	Subtract 90 from each side.
$y^2 - (9)^2 = 0$	$y^2 = y \cdot y$ and $81 = 9 \cdot 9$
$(y + 9)(y - 9) = 0$	Factor the difference of squares.
$y + 9 = 0$ or $y - 9 = 0$	Zero Product Property
$y = 9 \qquad y = -9$	Solve each equation.

The roots are -9 and 9.

8–6 Perfect Squares and Factoring (pp. 454–460)

Factor each polynomial, if possible. If the polynomial cannot be factored, write *prime*.

55. $a^2 + 18a + 81$

56. $9k^2 - 12k + 4$

57. $4 - 28r + 49r^2$

58. $32n^2 - 80n + 50$

Solve each equation. Check the solutions.

59. $6b^3 - 24b^2 + 24b = 0$

60. $144b^2 = 36$

61. $49m^2 - 126m + 81 = 0$

62. $(c - 9)^2 = 144$

63. PICTURE FRAMING A picture that measures 7 inches by 7 inches is being framed. The area of the frame is 32 square inches. What is the width of the frame?

Example 8 **Solve $(x - 4)^2 = 121$.**

$(x - 4)^2 = 121$	Original equation
$x - 4 = \pm\sqrt{121}$	Square Root Property
$x - 4 = \pm 11$	$121 = 11 \cdot 11$
$x = 4 \pm 11$	Add 4 to each side.
$x = 4 + 11$ or $x = 4 - 11$	Separate into two equations.
$= 15 \qquad = -7$	

The roots are -7 and 15.

CHAPTER 8 Practice Test

Factor each monomial completely.

1. $9g^2h$
2. $-40ab^3c$

Find the GCF of each set of monomials.

3. $16c^2, 4cd^2$
4. $12r, 35st$
5. $10xyz, 15x^2y$
6. $18a^2b^2, 28a^3b^2$

Factor each polynomial, if possible. If the polynomial cannot be factored using integers, write *prime*.

7. $x^2 + 14x + 24$
8. $28m^2 + 18m$
9. $a^2 - 11ab + 18b^2$
10. $2h^2 - 3h - 18$
11. $6x^3 + 15x^2 - 9x$
12. $15a^2b + 5a^2 - 10a$

13. **MULTIPLE CHOICE** What are the roots of $x^2 - 3x - 4 = 0$?

    A -4 and -1

    B -4 and 1

    C 4 and -1

    D 4 and 1

14. **GEOMETRY** When the length and width of the rectangle are increased by the same amount, the area is increased by 26 square inches. What are the dimensions of the new rectangle?

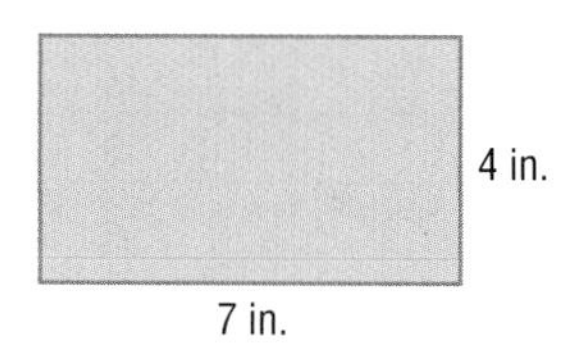

Factor each polynomial, if possible. If the polynomial cannot be factored using integers, write *prime*.

15. $a^2 - 4$
16. $t^2 - 16t + 64$
17. $64p^2 - 63p + 16$
18. $36m^2 + 60mn + 25n^2$
19. $x^3 - 4x^2 - 9x + 36$
20. $4my - 20m + 3py - 15p$

21. **ART** An artist is designing square tiles like the one shown at the right. The area of the shaded part of each tile is 98 square centimeters. Find the dimensions of the tile.

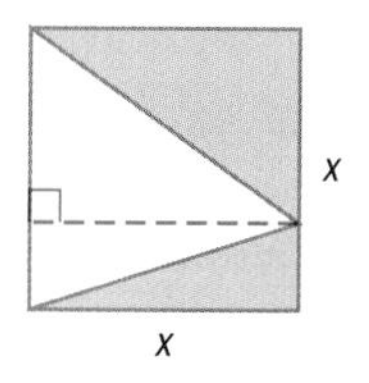

22. **CONSTRUCTION** A sidewalk will be built along the inside edges of all four sides of the rectangular lawn described in the table. The remaining lawn will have an area of 425 square feet. How wide will the walk be?

Dimensions of Lawn	
length	32 ft
width	24 ft

Solve each equation. Check the solutions.

23. $(4x - 3)(3x + 2) = 0$
24. $4x^2 = 36$
25. $18s^2 + 72s = 0$
26. $t^2 + 25 = 10t$
27. $a^2 - 9a - 52 = 0$
28. $x^3 - 5x^2 - 66x = 0$
29. $2x^2 = 9x + 5$
30. $3b^2 + 6 = 11b$

31. **GEOMETRY** The parallelogram has an area of 52 square centimeters. Find the height h of the parallelogram.

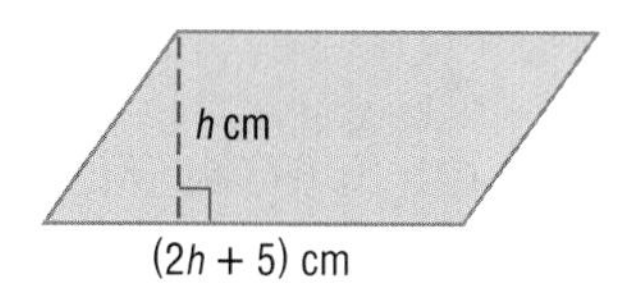

32. **MULTIPLE CHOICE** Which represents one of the roots of $0 = 2x^2 + 9x - 5$?

    F -5

    G $-\frac{1}{2}$

    H $\frac{5}{2}$

    J 5

Chapter Test at algebra1.com

CHAPTER 8

Standardized Test Practice

Cumulative, Chapters 1–8

Read each question. Then fill in the correct answer on the answer document provided by your teacher or on a sheet of paper.

1. Marlo bought 6 notebooks, 12 pencils, 8 pens, 1 backpack, 2 binders and 1 calendar. According to the chart below, which equation best represents the total amount she spent?

Item	Cost
Notebooks	2 for \$4.50
Pencils	4 for \$1.25
Pens	2 for \$1.00
Backpacks	2 for \$35.00
Binders	1 for \$2.50
Calendars	3 for \$21.00

A Cost $= 6(4.50) + 12(1.25) + 8(1.00) + 1(35.00) + 2(2.50) + 1(21.00)$

B Cost $= 2(4.50) + 4(1.25) + 2(1.00) + 2(35.00) + 1(2.50) + 3(21.00)$

C Cost $= 3(4.50) + 3(1.25) + 4(1.00) + \frac{1}{2}(35.00) + 2(2.50) + \frac{1}{3}(21.00)$

D Cost $= \frac{1}{3}(4.50) + \frac{1}{3}(1.25) + \frac{1}{4}(1.00) + 2(35.00) + \frac{1}{2}(2.50) + 3(21.00)$

2. The area of a rectangle is $24a^6b^{13}$ square units. If the width of the rectangle is $8a^5b^7$ units, how many units long is the rectangle? ($a \neq 0$ and $b \neq 0$)

F $3a^{11}b^{20}$ H $3ab^6$
G $16a^{11}b^{20}$ J $32ab^6$

TEST-TAKING TIP

Question 2 When answering a multiple-choice question, first find an answer on your own. Then, compare your answer to the choices given in the item. If your answer does not match any of the answer choices, check your calculations.

3. GRIDDABLE Kayla is making a 120-inch by 144-inch quilt with quilt squares that measure 6 inches on a side. If the squares are not cut, how many of them will be needed to make the quilt?

4. The area of a rectangle is $2x^2 - 5x - 3$, and the width is $2x + 1$. Which expression best describes the rectangle's length?

A $x + 3$ B $2x - 3$ C $x - 3$ D $2x - 1$

5. Aliya used algebra tiles to model the trinomial $x^2 - 3x - 4$ as shown below.

x^2	$-x$	$-x$	$-x$	$-x$
x	-1	-1	-1	-1

What are the factors of this trinomial?

F $(x - 4)(x + 1)$ H $(x - 2)(x + 2)$
G $(x + 4)(x - 1)$ J $(x - 2)(x - 2)$

6. A music store surveyed 100 of its customers about their preferred styles of music. The results of the survey are shown in the table.

Favorite Style of Music	
Style	**Frequency**
Country	25
Rock	38
Jazz	18
Classical	12
Other	7

What conclusion can be drawn if the store only uses this data to order new CDs?

A More than half of each order should be country and rock CDs.

B More than half of each order should be rock CDs.

C Only country, rock, and jazz CDs should be ordered.

D About a fourth of each order should be classical music CDs.

Math Online Standardized Test Practice at algebra1.com

Preparing for Standardized Tests
For test-taking strategies and more practice, see pages 756–773.

7. At Haulalani's Sandwich Shop, Haulalani made a chart of the percentage of each type of sandwich sold. Below is her chart.

Sandwich Type	Percent of Sales
Turkey	34
Ham	28
Roast beef	16
Veggie	9
Other	13

Which circle graph represents this situation?

F
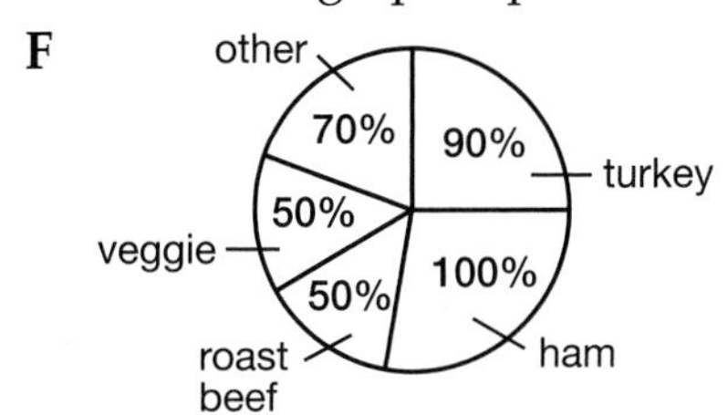

G
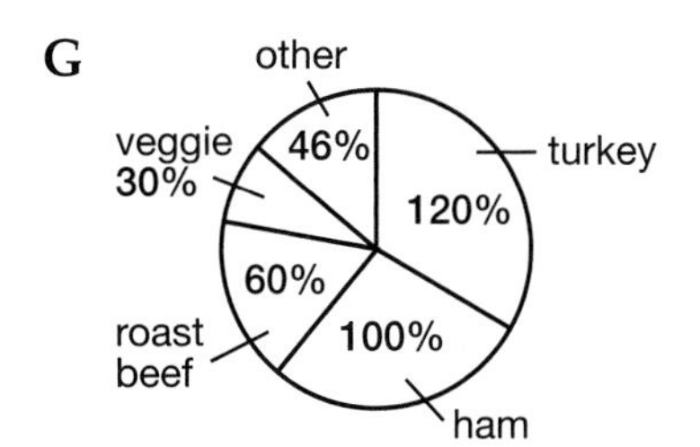

H
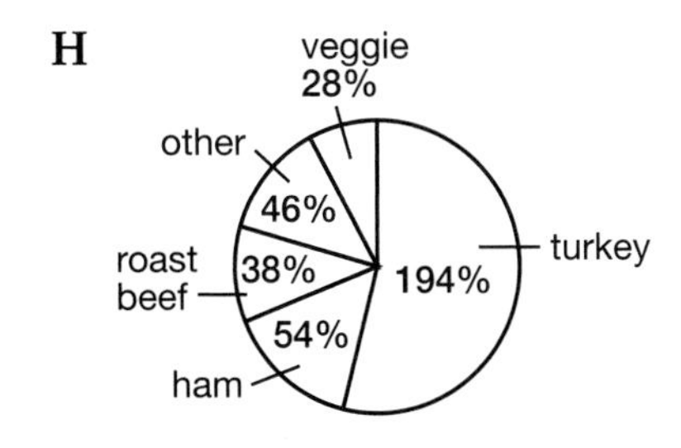

J
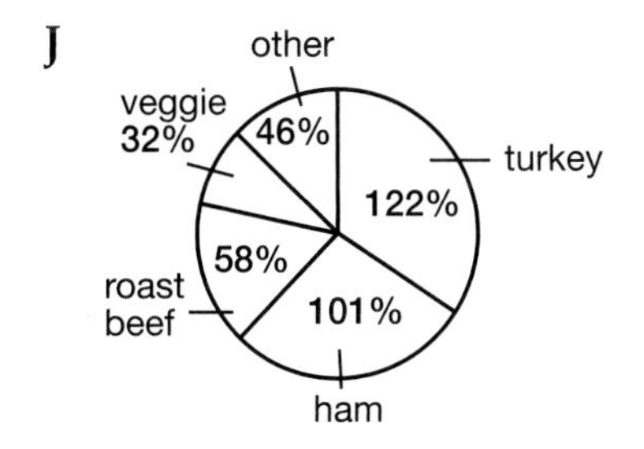

8. Which of the following shows $16x^2 + 24x + 9$ factored completely?

A $(4x + 3)^2$

B $(4x + 9)(4x + 1)$

C $(16x + 9)(x + 1)$

D $16x^2 + 24x + 9$

9. The slope of the line below is $\frac{3}{4}$.

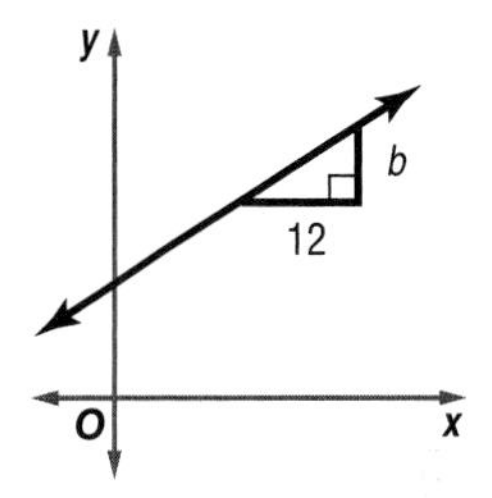

What is the value of b?

F 3 G 4 H 9 J 16

10. GRIDDABLE A cylindrical grain silo has a radius of 5.5 feet and a height of 20 feet. If grain is poured in at 5 cubic feet per minute, about how long, in minutes, will it take to fill the empty silo? Round to the nearest tenth.

Pre-AP

Record your answers on a sheet of paper. Show your work.

11. Madison is building a fenced, rectangular dog pen. The width of the pen will be 3 yards less than the length. The total area enclosed is 28 square yards.

a. Using L to represent the length of the pen, write an equation showing the area of the pen in terms of its length.

b. What is the length of the pen?

c. How many yards of fencing will Madison need to enclose the pen completely?

NEED EXTRA HELP?											
If You Missed Question...	1	2	3	4	5	6	7	8	9	10	11
Go to Lesson or Page...	2-1	7-2	7-2	8-4	8-3	714	714	8-6	4-1	708	8-4

CHAPTER 9

Quadratic and Exponential Functions

BIG Ideas

- Graph quadratic functions.
- Solve quadratic equations.
- Graph exponential functions.
- Solve problems involving growth and decay.

Key Vocabulary

completing the square (p. 487)
exponential function (p. 502)
parabola (p. 471)
Quadratic Formula (p. 493)

Real-World Link

Dinosaurs Exponential decay is one type of exponential function. Carbon dating uses exponential decay to determine the age of fossils and dinosaurs.

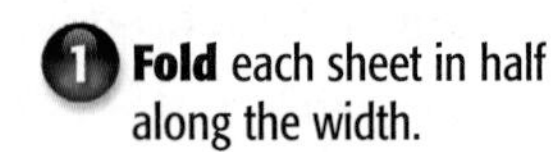

Quadratic and Exponential Functions Make this Foldable to help you organize your notes. Begin with three sheets of grid paper.

1. **Fold** each sheet in half along the width.

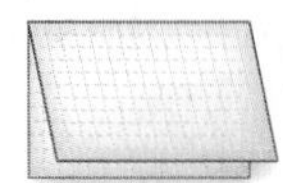

2. **Unfold** each sheet and tape to form one long piece.

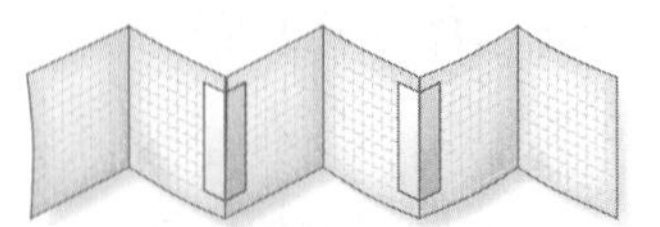

3. **Label** each page with the lesson number as shown. Refold to form a booklet.

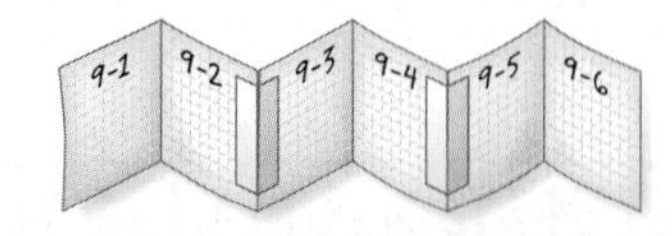

GET READY for Chapter 9

Diagnose Readiness You have two options for checking Prerequisite Skills.

Option 2

Math Online Take the Online Readiness Quiz at algebra1.com.

Option 1

Take the Quick Check below. Refer to the Quick Review for help.

QUICKCheck

Use a table of values to graph each equation. (Lesson 3-3)

1. $y = x + 5$
2. $y = 2x - 3$
3. $y = 0.5x + 1$
4. $y = -3x - 2$
5. $2x - 3y = 12$
6. $5y = 10 + 2x$
7. **SAVINGS** Suppose you have already saved \$200 toward the cost of a car. You plan to save \$35 each month for the next several months. Graph the equation for the total amount T you will have in m months.

Determine whether each trinomial is a perfect square trinomial. If so factor it. (Lesson 8-6)

8. $t^2 + 12t + 36$
9. $a^2 - 14a + 49$
10. $m^2 - 18m + 81$
11. $y^2 + 8y + 12$
12. $9b^2 - 6b + 1$
13. $6x^2 + 4x + 1$
14. $4p^2 + 12p + 9$
15. $16s^2 - 24s + 9$

Find the next three terms of each arithmetic sequence. (Lesson 3-4)

16. 5, 9, 13, 17, ...
17. 12, 5, −2, −9, ...
18. −4, −1, 2, 5, ...
19. 24, 32, 40, 48, ...
20. **GEOMETRY** Write a formula that can be used to find the number of sides of a pattern containing n triangles.

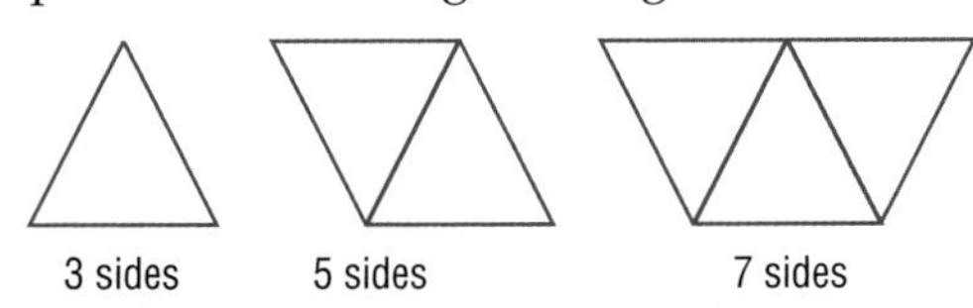

QUICKReview

EXAMPLE 1

Use a table of values to graph $y = 2x - 2$.

x	$y = 2x - 2$	y
−1	2(−1) − 2	−4
0	2(0) − 2	−2
1	2(1) − 2	0
2	2(2) − 2	2

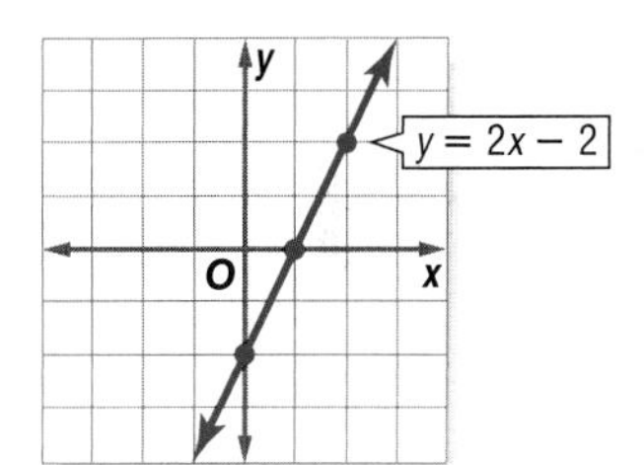

EXAMPLE 2

Determine whether $x^2 - 22x + 121$ is a perfect square trinomial. If so, factor it.

1. Is the first term a perfect square? yes
2. Is the last term a perfect square? yes
3. Is the middle term equal to $2(1x)(11)$? yes

$x^2 - 22x + 121$ is a perfect square trinomial.

$x^2 - 22x + 121 = (x - 11)^2$

EXAMPLE 3

Find the next three terms of the arithmetic sequence −104, −4, 96, 196, ...

Find the common difference by subtracting successive terms.

$-4 - (-104) = 100$

The common difference is 100.

Add to find the next three terms.

$196 + 100 = 296$, $296 + 100 = 396$,
$396 + 100 = 496$

The next three terms are 296, 396, 496.

Graphing Calculator Lab

Exploring Graphs of Quadratic Functions

Not all functions are linear. The graphs of nonlinear functions have different shapes. One type of nonlinear function is a *quadratic function*. The graph of a quadratic function is a *parabola*. You use a data collection device to conduct an experiment and investigate quadratic functions.

SET UP the Lab

- Set up the data collection device to collect data every 0.2 second for 4 seconds.
- Connect the motion sensor to your data collection device. Position the motion detector on the floor pointed upward.

ACTIVITY

Step 1 Have one group member hold a ball about 3 feet above the motion detector. Another group member will operate the data collection device.

Step 2 When the person operating the data collection device says "go," he or she should press the start button to begin data collection. At the same time, the ball should be tossed straight upward.

Step 3 Try to catch the ball at about the same height at which it was tossed. Stop collecting data when the ball is caught.

Analyze the Results

1. The domain contains values represented by the independent variable, time. The range contains values represented by the dependent variable, distance. Use the graphing calculator to graph the data.
2. Write a sentence that describes the shape of the graph. Is the graph linear? Explain.
3. Describe the position of the point on the graph that represents the starting position of the ball.
4. Use the TRACE feature of the calculator to find the maximum height of the ball. At what time was the maximum height achieved?
5. Repeat the experiment and toss the ball higher. Compare and contrast the new graph and the first graph.
6. Conduct an experiment in which the motion detector is held at a height of 4 feet and pointed downward at a dropped ball. How does the graph for this experiment compare to the other graphs?

9-1 Graphing Quadratic Functions

Main Ideas

- Graph quadratic functions.
- Find the equation of the axis of symmetry and the coordinates of the vertex of a parabola.

New Vocabulary

quadratic function
parabola
minimum
maximum
vertex
symmetry
axis of symmetry

GET READY for the Lesson

Boston's Fourth of July celebration includes a fireworks display set to music. If a rocket (firework) is launched with an initial velocity of 39.2 meters per second at a height of 1.6 meters above the ground, the equation $h = -4.9t^2 + 39.2t + 1.6$ represents the rocket's height h in meters after t seconds. The rocket will explode at approximately the highest point.

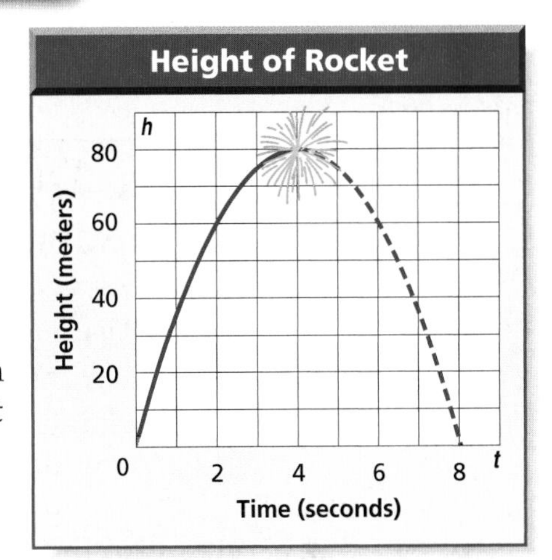

Graph Quadratic Functions The function describing the height of the rocket is an example of a quadratic function. A **quadratic function** can be written in the form $y = ax^2 + bx + c$, where $a \neq 0$. This form of equation is called *standard form.* The graph of a quadratic function is called a **parabola.**

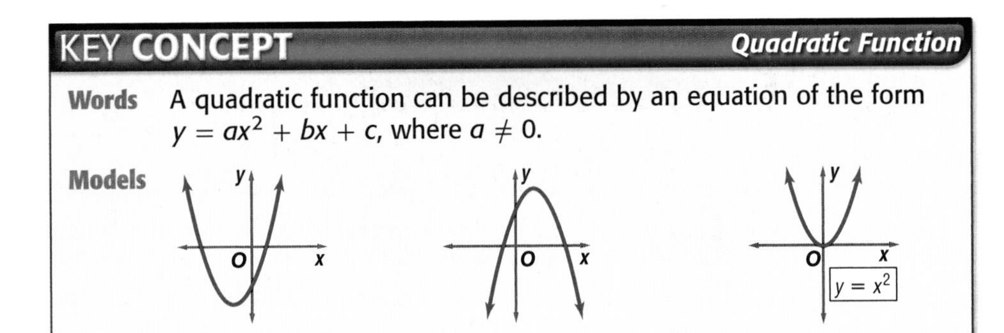

KEY CONCEPT *Quadratic Function*

Words A quadratic function can be described by an equation of the form $y = ax^2 + bx + c$, where $a \neq 0$.

Models

Study Tip

Parent Graph
The parent graph of the family of quadratic functions is $y = x^2$.

EXAMPLE Graph Opens Upward

1 Use a table of values to graph $y = 2x^2 - 4x - 5$. What are the domain and range of this function?

Graph these ordered pairs and connect them with a smooth curve. Because the parabola extends infinitely, the domain is all real numbers. The range is all real numbers greater than or equal to −7.

x	y
−2	11
−1	1
0	−5
1	−7
2	−5
3	1
4	11

1. Use a table of values to graph $y = x^2 + 3$. What are the domain and range of this function?

Consider the standard form $y = ax^2 + bx + c$. Notice that the value of a in Example 1 is positive and the curve opens upward. The graph of any quadratic function in which a is positive opens upward. The lowest point, or **minimum**, of this graph is located at $(1, -7)$.

Real-World EXAMPLE — Graph Opens Downward

2 FLYING DISKS The equation $y = -x^2 + 4x + 3$ represents the height y of a flying disk x seconds after it is tossed.

a. Use a table of values to graph $y = -x^2 + 4x + 3$.

Graph these ordered pairs and connect them with a smooth curve.

x	y
−1	−2
0	3
1	6
2	7
3	6
4	3
5	−2

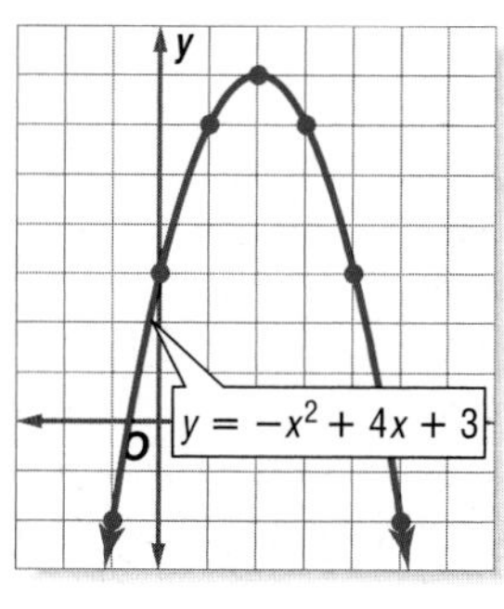

b. What are the domain and range of this function?

D: $\{x|x \text{ is a real number.}\}$

R: $\{y|y \leq 7\}$

c. Describe reasonable domain and range values for this situation.

The flying disk is in the air for about 4.6 seconds, so a reasonable domain is $\{x|0 \leq x \leq 4.6\}$. The height of the flying disk ranges from 0 to 7 feet, so a reasonable range is $\{y|0 \leq y \leq 7\}$.

CHECK Your Progress

2. Use a table of values to graph $y = -2x^2 + x + 1$. What are the domain and range of this function?

> **Reading Math**
>
> **Vertex** The plural of vertex is *vertices*. In math, vertex has several meanings. For example, there are the vertex of an angle, the vertices of a polygon, and the vertex of a parabola.

Notice that the value of a in Example 2 is negative and the curve opens downward. The graph of any quadratic function in which a is negative opens downward. The highest point, or **maximum**, of the graph is located at $(2, 3)$. The maximum or minimum point of a parabola is called the **vertex**.

Symmetry and Vertices Parabolas possess a geometric property called **symmetry**. Symmetrical figures are those in which each half of the figure matches the other exactly.

The line that divides a parabola into two halves is called the **axis of symmetry**. Each point on the parabola that is on one side of the axis of symmetry has a corresponding point on the parabola on the other side of the axis. The vertex is the only point on the parabola that is on the axis of symmetry. Notice the relationship between the values a and b and the equation of the axis of symmetry.

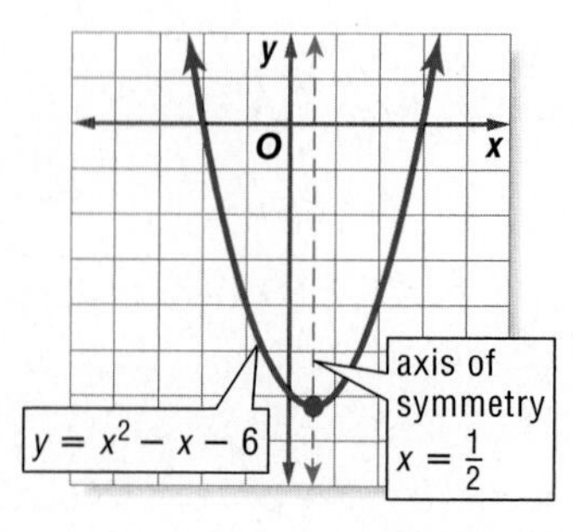

KEY CONCEPT — *Axis of Symmetry of a Parabola*

Words The equation of the axis of symmetry for the graph of $y = ax^2 + bx + c$, where $a \neq 0$, is $x = -\frac{b}{2a}$.

Model

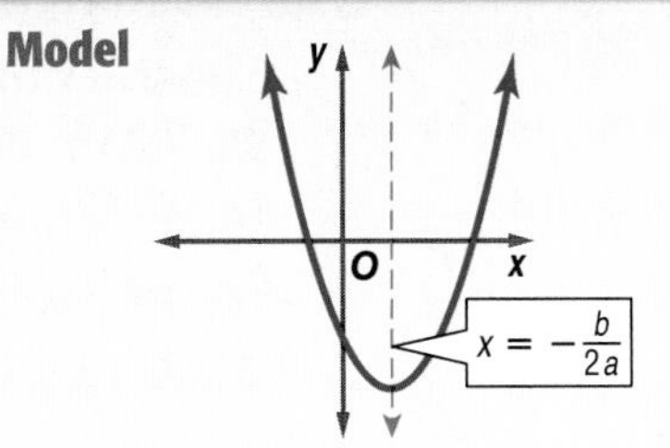

EXAMPLE Vertex and Axis of Symmetry

3 **Consider the graph of $y = -3x^2 - 6x + 4$.**

a. Write the equation of the axis of symmetry.

In $y = -3x^2 - 6x + 4$, $a = -3$ and $b = -6$.

$x = -\frac{b}{2a}$ Equation for the axis of symmetry of a parabola

$x = -\frac{-6}{2(-3)}$ or -1 $a = -3$ and $b = -6$

The equation of the axis of symmetry is $x = -1$.

b. Find the coordinates of the vertex.

Since the equation of the axis of symmetry is $x = -1$ and the vertex lies on the axis, the x-coordinate for the vertex is -1.

$y = -3x^2 - 6x + 4$ Original equation

$y = -3(-1)^2 - 6(-1) + 4$ $x = -1$

$y = -3 + 6 + 4$ Simplify.

$y = 7$ Add.

The vertex is at $(-1, 7)$.

Study Tip

Coordinates of Vertex

Notice that you can find the x-coordinate by knowing the axis of symmetry. However, to find the y-coordinate, you must substitute the value of x into the quadratic equation.

c. Identify the vertex as a maximum or minimum.

Since the coefficient of the x^2 term is negative, the parabola opens downward and the vertex is a maximum point.

d. Graph the function.

You can use the symmetry of the parabola to help you draw its graph. On a coordinate plane, graph the vertex and the axis of symmetry. Choose a value for x other than -1. For example, choose 1 and find the y-coordinate that satisfies the equation.

$y = -3x^2 - 6x + 4$ Original equation

$y = -3(1)^2 - 6(1) + 4$ Let $x = 1$.

$y = -5$ Simplify.

Graph $(1, -5)$. Since the graph is symmetrical about its axis of symmetry $x = -1$, you can find another point on the other side of the axis of symmetry. The point at $(1, -5)$ is 2 units to the right of the axis. Go 2 units to the left of the axis and plot the point $(-3, -5)$. Repeat this for several other points. Then sketch the parabola.

CHECK Your Progress

Consider the graph of $y = x^2 + 2x + 18$.

3A. Write the equation of the axis of symmetry.

3B. Find the coordinates of the vertex.

3C. Identify the vertex as a maximum or minimum.

3D. Graph the function.

STANDARDIZED TEST PRACTICE

Match Equations and Graphs

4 Which is the graph of $y + 1 = (x + 1)^2$?

A

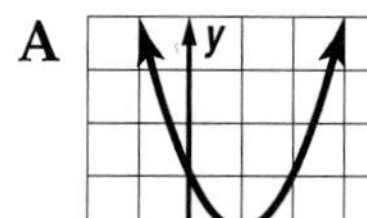

B

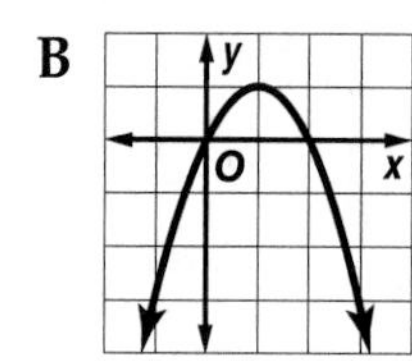

C

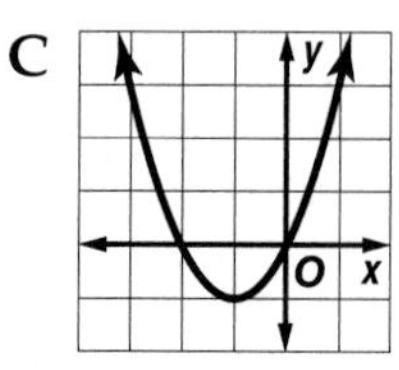

D

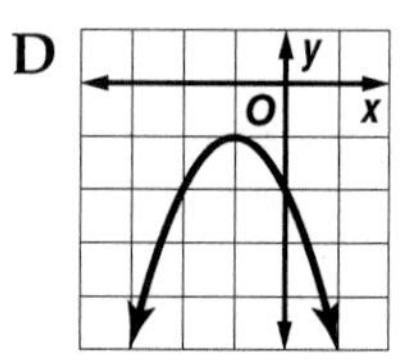

Test-Taking Tip

Substituting Values The ordered pair (0, 0) satisfies the equation $y + 1 = (x + 1)^2$. Since the point at (0, 0) is on the graph, choices A and D can be eliminated.

Read the Test Item

You are given a quadratic function, and you are asked to choose its graph.

Solve the Test Item

Step 1 First write the equation in standard form.

$y + 1 = (x + 1)^2$	Original equation
$y + 1 = x^2 + 2x + 1$	$(x + 1)^2 = x^2 + 2x + 1$
$y + 1 - 1 = x^2 + 2x + 1 - 1$	Subtract 1 from each side.
$y = x^2 + 2x$	Simplify.

Step 2 Then find the axis of symmetry of the graph of $y = x^2 + 2x$.

$x = -\frac{b}{2a}$	Equation for the axis of symmetry
$x = -\frac{2}{2(1)}$ or -1	$a = 1$ and $b = 2$

The axis of symmetry is $x = -1$. Look at the graphs. Since only choices C and D have $x = -1$ as their axis of symmetry, you can eliminate choices A and B. Since the coefficient of the x^2 term is positive, the graph opens upward. Eliminate choice D. The answer is C.

CHECK Your Progress

4. Which is the equation of the graph?

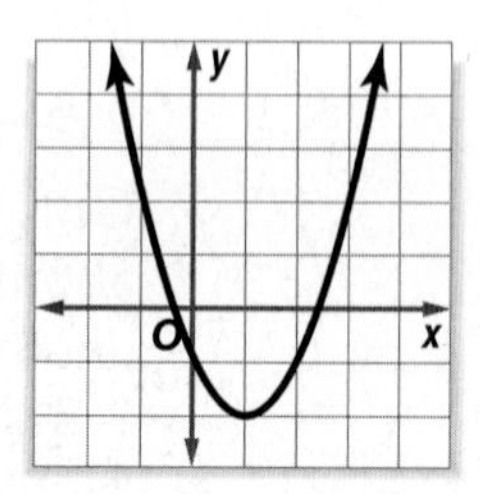

F $y - 1 = (x + 2)^2$

G $y - 1 = (x - 2)^2$

H $y + 2 = (x - 1)^2$

J $y - 2 = (x + 1)^2$

Online Personal Tutor at algebra1.com

CHECK Your Understanding

Examples 1, 2 (pp. 471–472)

Use a table of values to graph each function.

1. $y = x^2 - 5$

2. $y = x^2 + 2$

3. $y = -x^2 + 4x + 5$

4. $y = x^2 + x - 1$

Example 3 (pp. 473–474)

Write the equation of the axis of symmetry, and find the coordinates of the vertex of the graph of each function. Identify the vertex as a maximum or minimum. Then graph the function.

5. $y = x^2 + 4x - 9$

6. $y = -x^2 + 5x + 6$

7. $y = -(x - 2)^2 + 1$

8. $y = (x + 3)^2 - 4$

Example 4 (p. 474)

9. STANDARDIZED TEST PRACTICE Which is the graph of $y = -\frac{1}{2}x^2 + 1$?

A

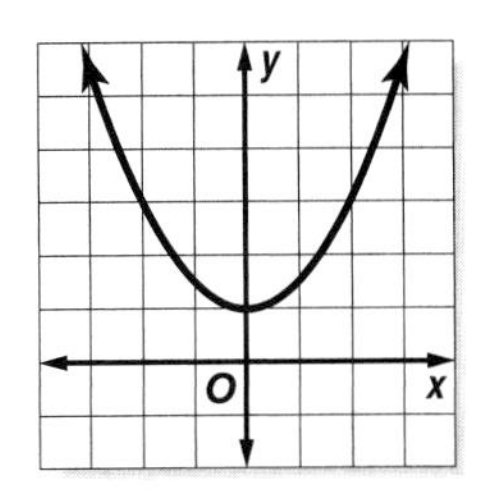

C

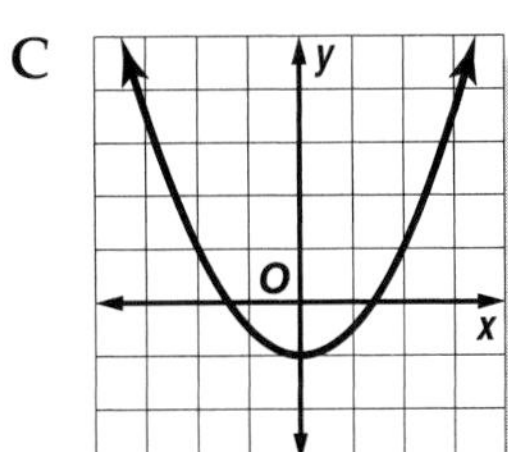

B

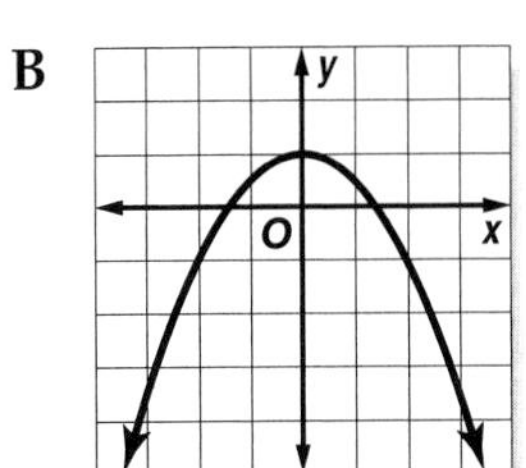

D

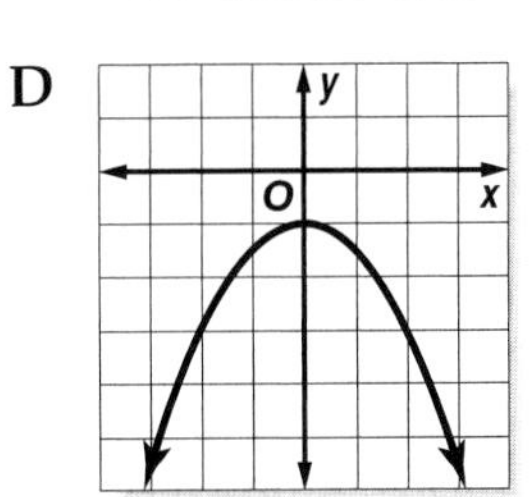

Exercises

HOMEWORK HELP

For Exercises	See Examples
10–15	1, 2
16–29	3

Use a table of values to graph each function.

10. $y = x^2 - 3$

11. $y = -x^2 + 7$

12. $y = x^2 - 2x - 8$

13. $y = x^2 - 4x + 3$

14. $y = -3x^2 - 6x + 4$

15. $y = -3x^2 + 6x + 1$

Write the equation of the axis of symmetry, and find the coordinates of the vertex of the graph of each function. Identify the vertex as a maximum or minimum. Then graph the function.

16. $y = 4x^2$

17. $y = -2x^2$

18. $y = x^2 + 2$

19. $y = -x^2 + 5$

20. $y = -x^2 + 2x + 3$

21. $y = -x^2 - 6x + 15$

22. $y = 3x^2 - 6x + 4$

23. $y = 9 - 8x + 2x^2$

24. What is the equation of the axis of symmetry of the graph of $y = -3x^2 + 2x - 5$?

25. Find the equation of the axis of symmetry of the graph of $y = 4x^2 - 5x + 16$.

ENTERTAINMENT **For Exercises 26 and 27, use the following information.**

A carnival game involves striking a lever that forces a weight up a tube. If the weight reaches 20 feet to ring the bell, the contestant wins a prize. The equation $h = -16t^2 + 32t + 3$ gives the height of the weight if the initial velocity is 32 feet per second.

26. Find the maximum height of the weight.

27. Will a prize be won? Explain.

PETS **For Exercises 28 and 29, use the following information.**

Miriam has 40 meters of fencing to build a pen for her dog.

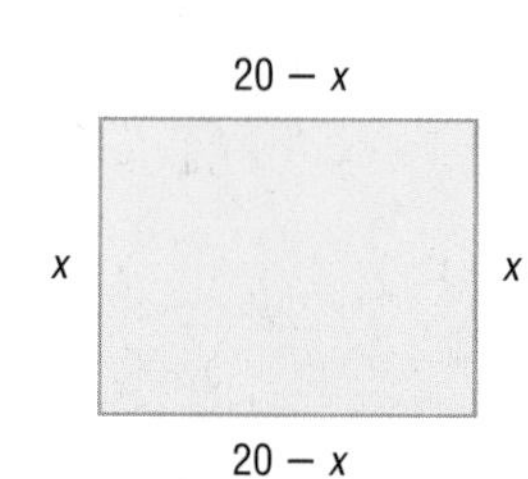

28. Use the diagram to write an equation for the area A of the pen. Describe a reasonable domain and range for this situation.

29. What value of x will result in the greatest area? What is the greatest possible area of the pen?

Real-World Link

The Gateway Arch is part of a tribute to Thomas Jefferson, the Louisiana Purchase, and the pioneers who settled the West. Each year about 2.5 million people visit the arch.

Source: *World Book Encyclopedia*

Write the equation of the axis of symmetry, and find the coordinates of the vertex of the graph of each function. Identify the vertex as a maximum or minimum. Then graph the function.

30. $y = -2(x - 4)^2 - 3$

31. $y + 2 = x^2 - 10x + 25$

32. $y - 5 = \frac{1}{3}(x + 2)^2$

33. $y + 1 = \frac{1}{3}(x + 1)^2$

34. The vertex of a parabola is at $(-4, -3)$. If one x-intercept is -11, what is the other x-intercept?

35. What is the equation of the axis of symmetry of a parabola if its x-intercepts are -6 and 4?

ARCHITECTURE **For Exercises 36–38, use the following information.**

The shape of the Gateway Arch in St. Louis, Missouri, is a *catenary* curve. It resembles a parabola with the equation $h = -0.00635x^2 + 4.0005x - 0.07875$, where h is the height in feet and x is the distance from one base in feet.

36. What is the equation of the axis of symmetry?

37. What is the distance from one end of the arch to the other?

38. What is the maximum height of the arch?

FOOTBALL **For Exercises 39–41, use the following information.**

A football is kicked from ground level at an initial velocity of 90 feet per second. The equation $h = -16t^2 + 90t$ gives the height h of the football after t seconds.

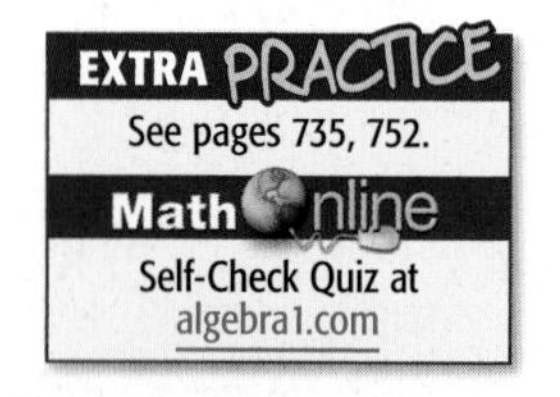

39. What is the height of the ball after one second?

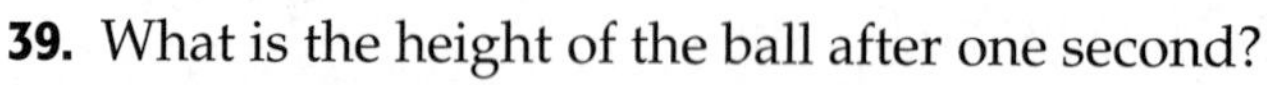

40. When is the ball 126 feet high?

41. When is the height of the ball zero feet? Describe the events these represent.

H.O.T. Problems

42. OPEN ENDED Sketch a parabola that models a real-life situation and describe what the vertex represents. Determine reasonable domain and range values for this type of situation.

43. REASONING Sketch the parent graph of the function $y = 3x^2 - 5x - 2$.

REASONING Let $f(x) = x^2 - 9$.

44. What is the domain of $f(x)$?

45. What is the range of $f(x)$?

46. For what values of x is $f(x)$ negative?

47. When x is a real number, what are the domain and range of $f(x) = \sqrt{x^2 - 9}$?

48. REASONING Determine the range of $f(x) = (x - 5)^2 - 6$.

49. CHALLENGE Write and graph a quadratic equation whose graph has the axis of symmetry $x = -\frac{3}{8}$. Summarize the steps that you took to determine the equation.

50. *Writing in Math* Use the information about a rocket's path on page 471 to explain how a fireworks display can be coordinated with recorded music. Include an explanation of how to determine when the rocket will explode and how to find the height of the rocket when it explodes.

STANDARDIZED TEST PRACTICE

51. In the graph of the function $y = x^2 - 3$, which describes the shift in the vertex of the parabola if, in the function, -3 is changed to 1?

A 2 units up

B 4 units up

C 2 units down

D 4 units down

52. REVIEW The costs of two packs of Brand A gum and two packs of Brand B gum are shown in the table. What percent of the cost of Brand B gum does James save by buying two packs of Brand A gum?

Gum	Cost of Two Packs
Brand A	\$1.98
Brand B	\$2.50

F 11.6%

G 20.8%

H 26.3%

J 79.2%

Spiral Review

Factor each polynomial, if possible. (Lessons 8-5 and 8-6)

53. $x^2 + 6x - 9$

54. $a^2 + 22a + 121$

55. $4m^2 - 4m + 1$

56. $4q^2 - 9$

57. $2a^2 - 25$

58. $1 - 16g^2$

Find each sum or difference. (Lesson 7-5)

59. $(13x + 9y) + 11y$

60. $(8 - 2c^2) + (1 + c^2)$

61. $(7p^2 - p - 7) - (p^2 + 11)$

62. RECREATION At a recreation facility, 3 members and 3 nonmembers pay a total of \$180 to take an aerobics class. A group of 5 members and 3 nonmembers pay \$210 to take the same class. How much does it cost each to take an aerobics class? (Lesson 5-3)

GET READY for the Next Lesson

PREREQUISITE SKILL Find the x-intercept of the graph of each equation. (Lesson 3-3)

63. $3x + 4y = 24$

64. $2x - 5y = 14$

65. $-2x - 4y = 7$

EXTEND 9-1

Graphing Calculator Lab

The Family of Quadratic Functions

The parent function of the family of quadratic functions is $y = x^2$.

Concepts in MOtion
Animation algebra1.com

ACTIVITY 1

Graph each group of equations on the same screen. Use the standard viewing window. Compare and contrast the graphs.

KEYSTROKES: *Review graphing equations on pages* 162 *and* 163.

a. $y = x^2, y = 2x^2, y = 4x^2$

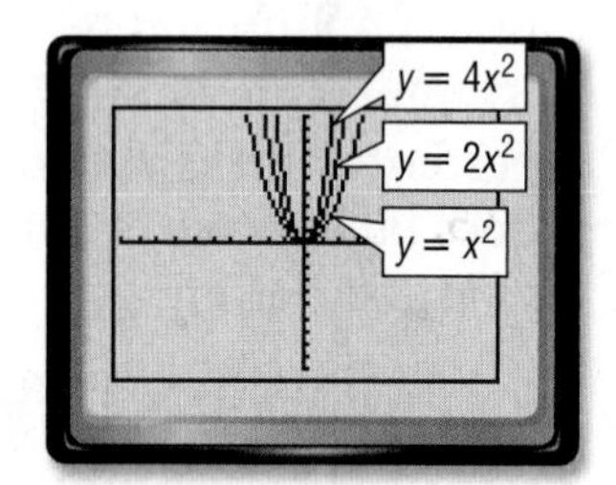

Each graph opens upward and has its vertex at the origin. The graphs of $y = 2x^2$ and $y = 4x^2$ are narrower than the graph of $y = x^2$.

b. $y = x^2, y = 0.5x^2, y = 0.2x^2$

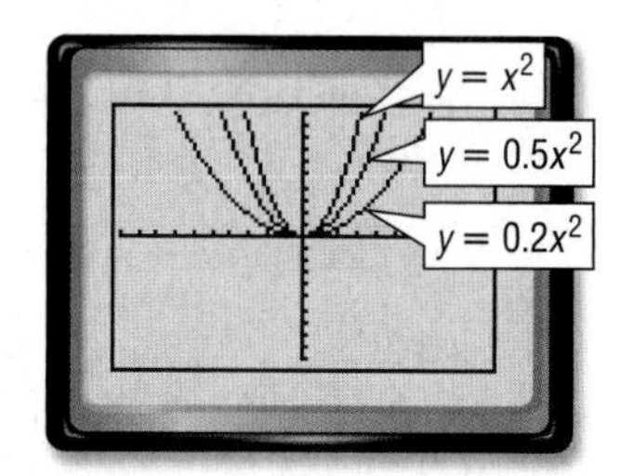

Each graph opens upward and has its vertex at the origin. The graphs of $y = 0.5x^2$ and $y = 0.2x^2$ are wider than the graph of $y = x^2$.

1A. How does the value of a in $y = ax^2$ affect the shape of the graph?

c. $y = x^2, y = x^2 + 3, y = x^2 - 2,$
$y = x^2 - 4$

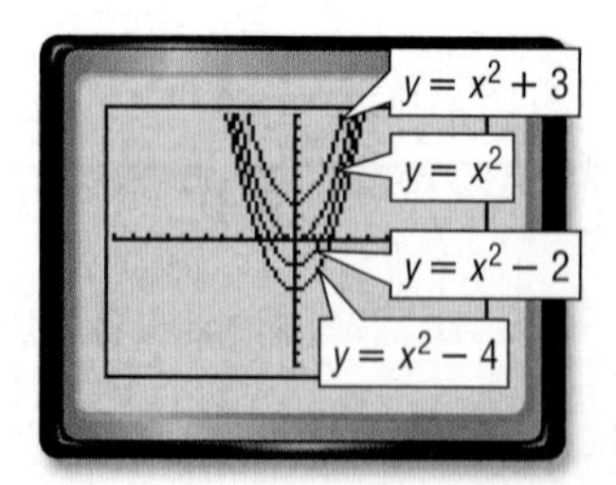

Each graph opens upward and has the same shape as $y = x^2$. However, each parabola has a different vertex, located along the y-axis.

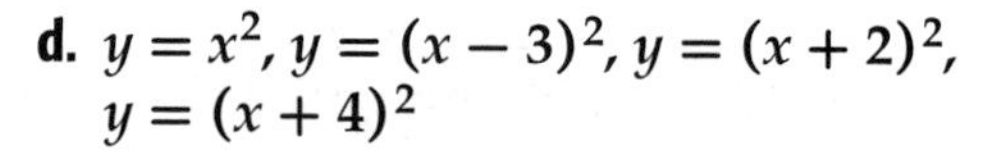

d. $y = x^2, y = (x - 3)^2, y = (x + 2)^2,$
$y = (x + 4)^2$

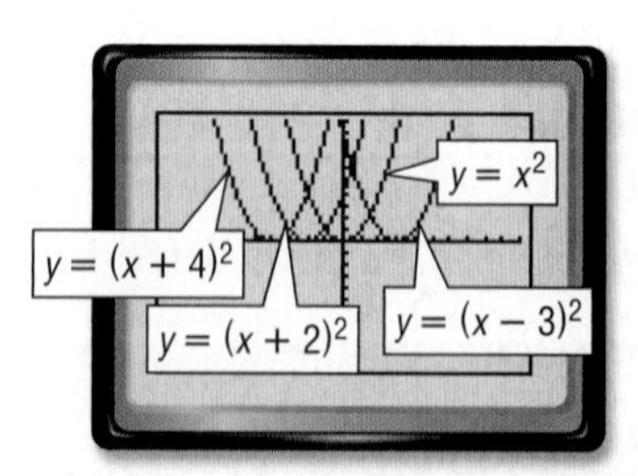

Each graph opens upward and has the same shape as $y = x^2$. However, each parabola has a different vertex located along the x-axis.

1B. How does the value of the constant affect the position of the graph?

1C. How is the location of the vertex related to the equation of the graph?

Math Online **Other Calculator Keystrokes at** algebra1.com

Suppose you graph the same equation using different windows. How will the appearance of the graph change?

ACTIVITY 2

Graph $y = x^2 - 7$ in each viewing window. What conclusions can you draw about the appearance of a graph in the window used?

a. **standard viewing window**

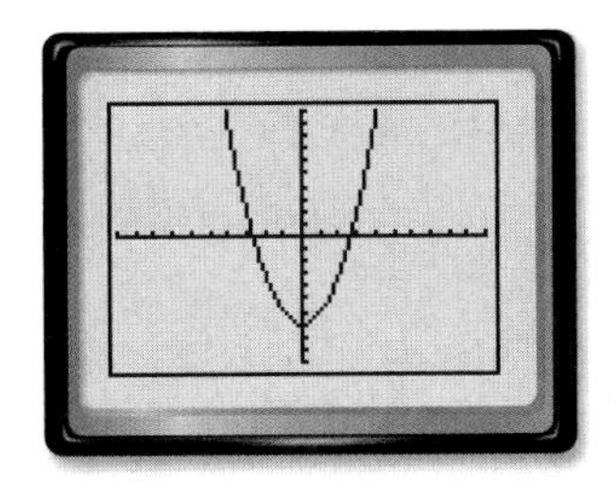

b. **[−10, 10] scl: 1 by [−200, 200] scl: 50**

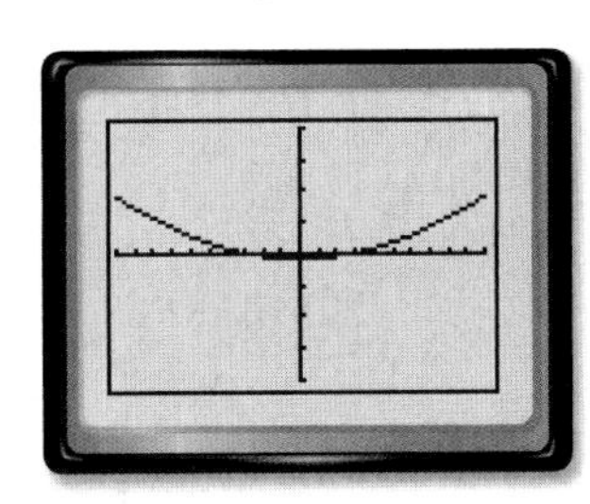

c. **[−50, 50] scl: 5 by [−10, 10] scl: 1**

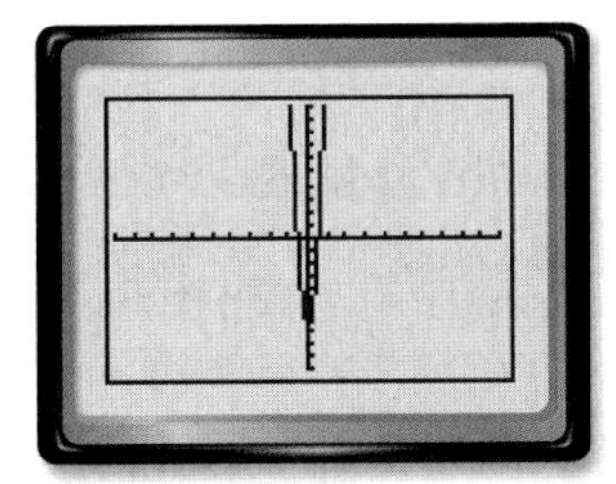

d. **[−0.5, 0.5] scl: 0.1 by [−10, 10] scl: 1**

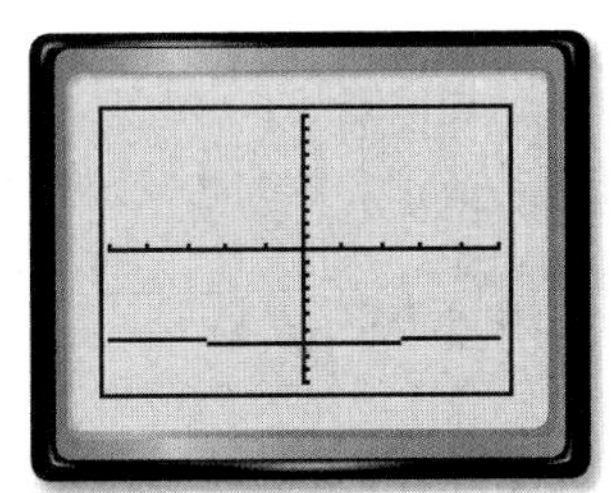

Without knowing the window, graph **b** might be of the family $y = ax^2$, where $0 < a < 1$. Graph **c** looks like a member of $y = ax^2 - 7$, where $a > 1$. Graph **d** looks more like a line. However, all are graphs of the same equation.

EXERCISES

Graph each family of equations on the same screen. Compare and contrast the graphs.

1. $y = -x^2$
$y = -3x^2$
$y = -6x^2$

2. $y = -x^2$
$y = -0.6x^2$
$y = -0.4x^2$

3. $y = -x^2$
$y = -(x + 5)^2$
$y = -(x - 4)^2$

4. $y = -x^2$
$y = -x^2 + 7$
$y = -x^2 - 5$

Use the graphs on page 478 and Exercises 1–4 above to predict the appearance of the graph of each equation. Then draw the graph.

5. $y = -0.1x^2$ **6.** $y = (x + 1)^2$ **7.** $y = 4x^2$ **8.** $y = x^2 - 6$

Describe how each change in $y = x^2$ would affect the graph of $y = x^2$. Be sure to consider all values of a, h, and k.

9. $y = ax^2$ **10.** $y = (x + h)^2$ **11.** $y = x^2 + k$ **12.** $y = (x + h)^2 + k$

9-2 Solving Quadratic Equations by Graphing

Main Idea

- Solve quadratic equations by graphing.
- Estimate solutions of quadratic equations by graphing.

New Vocabulary

quadratic equation
roots
zeros
double root

GET READY for the Lesson

A golf ball follows a path much like a parabola. Because of this property, quadratic functions can be used to simulate parts of a computer golf game. One of the x-intercepts of the quadratic function represents the location where the ball will hit the ground.

Solve by Graphing A **quadratic equation** is an equation that can be written in the form $ax^2 + bx + c = 0$, where $a \neq 0$. The value of the related quadratic function is 0.

Quadratic Equation	Related Quadratic Function
$x^2 - 2x - 3 = 0$	$f(x) = x^2 - 2x - 3$

The solutions of a quadratic equation are called the **roots** of the equation. The roots of a quadratic equation can be found by finding the x-intercepts or **zeros** of the related quadratic function.

EXAMPLE Two Roots

1 Solve $x^2 + 6x - 7 = 0$ by graphing.

Graph the related function $f(x) = x^2 + 6x - 7$. The equation of the axis of symmetry is $x = -\frac{6}{2(1)}$ or $x = -3$. When x equals -3, $f(x)$ equals $(-3)^2 + 6(-3) - 7$ or -16. So, the coordinates of the vertex are $(-3, -16)$. Make a table of values to find other points to sketch the graph.

Concepts in Motion
Animation algebra1.com

x	$f(x)$
−8	9
−6	−7
−4	−15
−3	−16
−2	−15
0	−7
2	9

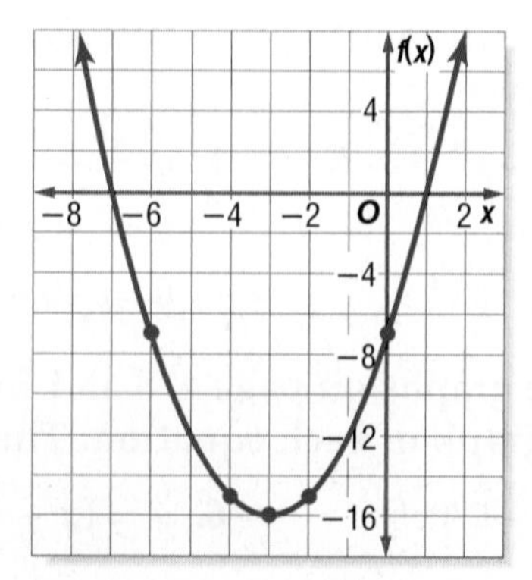

To solve $x^2 + 6x - 7 = 0$, you need to know where the value of $f(x)$ is 0. This occurs at the x-intercepts. The x-intercepts of the parabola appear to be -7 and 1.

Study Tip

***x*-intercepts**
The x-intercepts of a graph are also called the *horizontal intercepts*.

Study Tip

Common Misconception

Although solutions found by graphing may appear to be exact, you cannot be sure that they are exact. Solutions need to be verified by substituting into the equation and checking, or by using the algebraic methods that you will learn in this chapter.

CHECK Solve by factoring.

$x^2 + 6x - 7 = 0$ Original equation

$(x + 7)(x - 1) = 0$ Factor.

$x + 7 = 0$ or $x - 1 = 0$ Zero Product Property

$x = -7$ ✔ $x = 1$ ✔ The solutions are -7 and 1.

CHECK Your Progress

1. Solve $-c^2 + 5c - 4 = 0$ by graphing.

Quadratic equations always have two roots. However, these roots are not always two distinct numbers. Sometimes the two roots are the same number, called a **double root.** In other cases the roots are not real numbers.

EXAMPLE A Double Root

2 **Solve $b^2 + 4b = -4$ by graphing.**

First rewrite the equation so one side is equal to zero.

$b^2 + 4b = -4$ Original equation

$b^2 + 4b + 4 = 0$ Add 4 to each side.

Graph the related function $f(b) = b^2 + 4b + 4$.

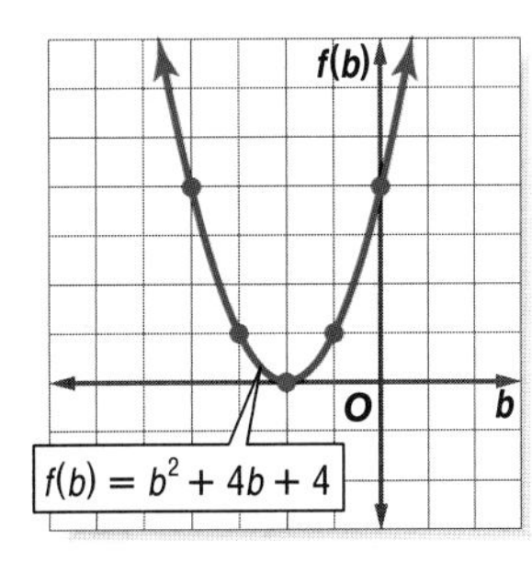

Notice that the vertex of the parabola is the b-intercept. Thus, one solution is -2. What is the other solution?

Try solving the equation by factoring.

$b^2 + 4b + 4 = 0$ Original equation

$(b + 2)(b + 2) = 0$ Factor.

$b + 2 = 0$ or $b + 2 = 0$ Zero Product Property

$b = -2$ $b = -2$ The solution is -2.

CHECK Your Progress

2. Solve $0 = x^2 - 6x + 9$ by graphing.

EXAMPLE No Real Roots

Study Tip

Empty Set

The symbol ∅, indicating an empty set, is often used to represent no real solutions.

3 **Solve $x^2 - x + 4 = 0$ by graphing.**

Graph the related function $f(x) = x^2 - x + 4$.

x	$f(x)$
−1	6
0	4
1	4
2	6

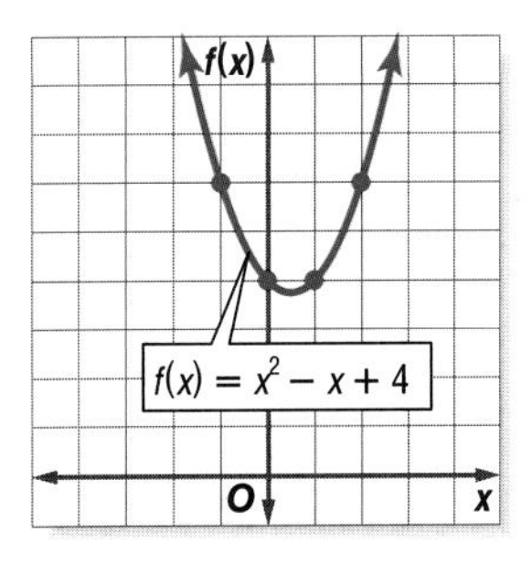

The graph has no x-intercept. Thus, there are no real number solutions for this equation.

CHECK Your Progress

3. Solve $-t^2 - 3t = 5$ by graphing.

Factoring can be used to determine whether the graph of a quadratic function intersects the x-axis in zero, one, or two points.

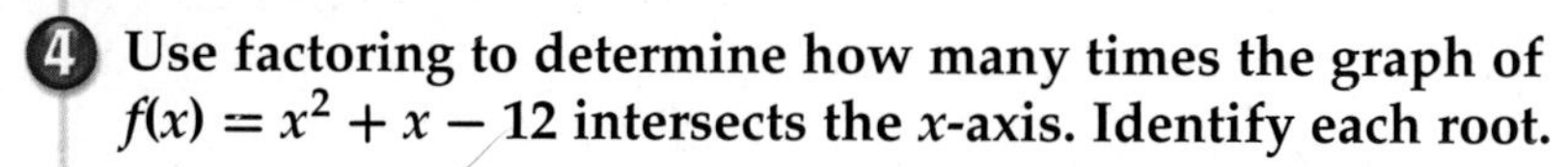

EXAMPLE Factoring

4 **Use factoring to determine how many times the graph of $f(x) = x^2 + x - 12$ intersects the x-axis. Identify each root.**

The graph intersects the x-axis when $f(x) = 0$.

$x^2 + x - 12 = 0$ Original equation

$(x - 3)(x + 4) = 0$ Factor.

Since the trinomial factors into two distinct factors, the graph of the function intersects the x-axis 2 times. The roots are $x = 3$ and $x = -4$.

CHECK Your Progress

4. Use factoring to determine how many times the graph of $f(x) = x^2 - 10x + 25$ intersects the x-axis. Identify each root.

Estimate Solutions In Examples 1 and 2, the roots of the equation were integers. Usually the roots of a quadratic equation are not integers. In these cases, use estimation to approximate the roots of the equation.

EXAMPLE Rational Roots

5 **Solve $n^2 + 6n + 7 = 0$ by graphing. If integral roots cannot be found, estimate the roots by stating the consecutive integers between which the roots lie.**

Study Tip

Location of Roots

Since quadratic functions are continuous, there must be a zero between x-values when their function values have opposite signs.

Graph the related function $f(n) = n^2 + 6n + 7$.

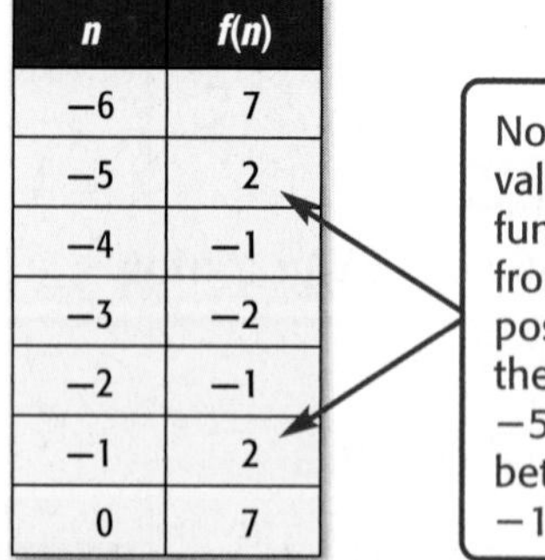

n	$f(n)$
−6	7
−5	2
−4	−1
−3	−2
−2	−1
−1	2
0	7

Notice that the value of the function changes from negative to positive between the n values of −5 and −4 and between −2 and −1.

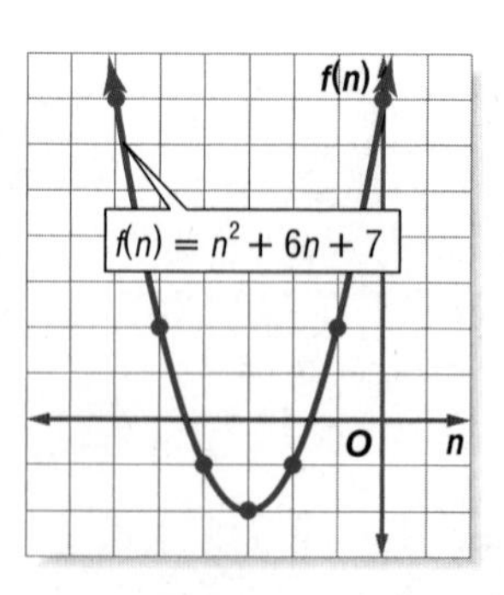

The n-intercepts are between −5 and −4 and between −2 and −1. So, one root is between −5 and −4, and the other root is between −2 and −1.

CHECK Your Progress

5. Solve $2a^2 + 6a - 3 = 0$ by graphing. If integral roots cannot be found, estimate the roots by stating the consecutive integers between which the roots lie.

Online **Personal Tutor at** algebra1.com

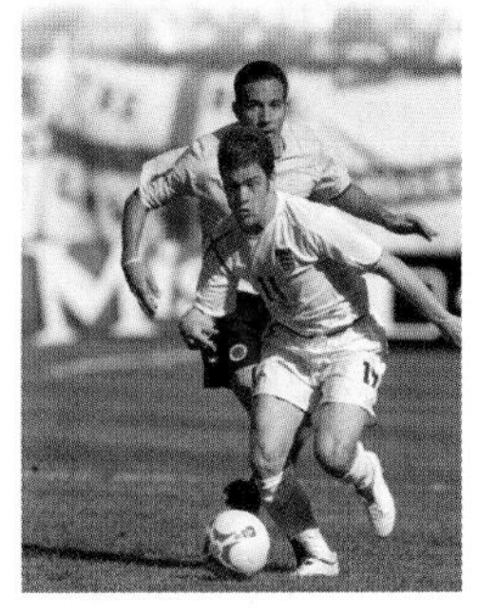

Real-World Link

The game of soccer, called "football" in countries other than North America, began in 1857 in Britain. It is played on every continent of the world.

Source: worldsoccer.about.com

Real-World EXAMPLE

6 **SOCCER If a goalie kicks a soccer ball with an upward velocity of 65 feet per second and his foot meets the ball 3 feet off the ground, the function $y = -16t^2 + 65t + 3$ represents the height of the ball y in feet after t seconds. Approximately how long is the ball in the air?**

You need to find the solution of the equation $0 = -16t^2 + 65t + 3$. Use a graphing calculator to graph the related function $y = -16t^2 + 65t + 3$. The x-intercept is about 4. Therefore, the ball is in the air about 4 seconds.

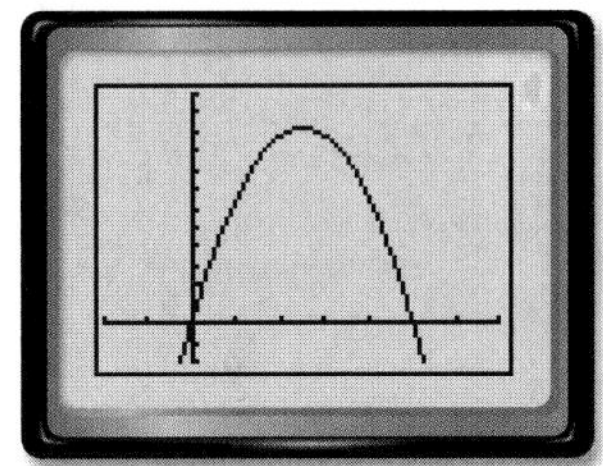

[−2, 7] scl: 1 by [−20, 80] scl: 10

CHECK Your Progress

6. **NUMBER THEORY** Use a quadratic equation to find two numbers whose sum is 5 and whose product is −24.

CHECK Your Understanding

Examples 1–3 (pp. 480–481)

Solve each equation by graphing.

1. $x^2 - 7x + 6 = 0$
2. $-a^2 - 10a = 25$
3. $c^2 + 3 = 0$

Example 4 (p. 482)

Use factoring to determine how many times the graph of each function intersects the x-axis. Identify each root.

4. $f(x) = x^2 + 2x - 24$
5. $f(x) = x^2 + 14x + 49$

Example 5 (p. 482)

Solve each equation by graphing. If integral roots cannot be found, estimate the roots by stating the consecutive integers between which the roots lie.

6. $-t^2 - 5t + 1 = 0$
7. $0 = x^2 - 16$
8. $w^2 - 3w = 5$

Example 6 (p. 483)

9. **NUMBER THEORY** Two numbers have a sum of 4 and a product of −12. Use a quadratic equation to determine the two numbers.

Exercises

HOMEWORK HELP

For Exercises	See Examples
10–17	1–3
18–21	4
22–30	5
31, 32	6

Solve each equation by graphing.

10. $c^2 - 5c - 24 = 0$
11. $5n^2 + 2n + 6 = 0$
12. $0 = x^2 + 6x + 9$
13. $-b^2 + 4b = 4$
14. $x^2 + 2x + 5 = 0$
15. $-2r^2 - 6r = 0$
16. The roots of a quadratic equation are −2 and −6. The minimum point of the graph of its related function is at (−4, −2). Sketch the graph of the function and compare the graph to the graph of the parent function $y = x^2$.
17. The roots of a quadratic equation are −6 and 0. The maximum point of the graph of its related function is at (−3, 4). Sketch the graph of the function and compare the graph to the graph of the parent function $y = x^2$.

Use factoring to determine how many times the graph of each function intersects the x-axis. Identify each root.

18. $g(x) = x^2 - 8x + 16$
19. $h(x) = x^2 + 12x + 32$
20. $f(x) = x^2 + 3x + 4$
21. $g(x) = x^2 + 3x + 4$

Solve each equation by graphing. If integral roots cannot be found, estimate the roots by stating the consecutive integers between which the roots lie.

22. $a^2 - 12 = 0$ **23.** $-n^2 + 7 = 0$ **24.** $2c^2 + 20c + 32 = 0$

25. $3s^2 + 9s - 12 = 0$ **26.** $0 = x^2 + 6x + 6$ **27.** $0 = -y^2 + 4y - 1$

28. $-a^2 + 8a = -4$ **29.** $x^2 + 6x = -7$ **30.** $m^2 - 10m = -21$

31. NUMBER THEORY Use a quadratic equation to find two numbers whose sum is 9 and whose product is 20.

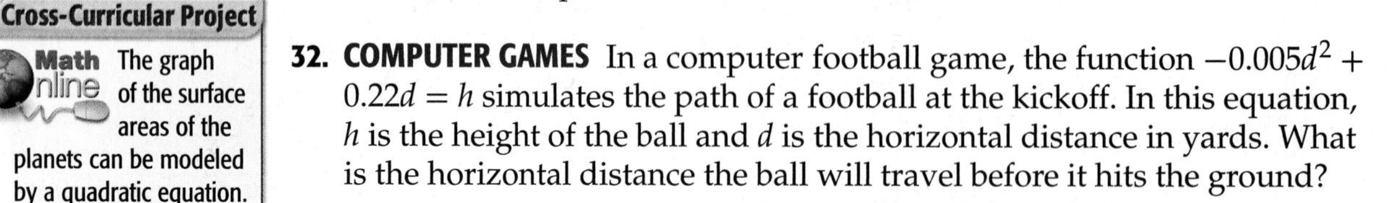

Cross-Curricular Project

Math Online The graph of the surface areas of the planets can be modeled by a quadratic equation. Visit algebra1.com to continue work on your project.

32. COMPUTER GAMES In a computer football game, the function $-0.005d^2 + 0.22d = h$ simulates the path of a football at the kickoff. In this equation, h is the height of the ball and d is the horizontal distance in yards. What is the horizontal distance the ball will travel before it hits the ground?

33. HIKING While hiking in the mountains, Monya and Kishi stop for lunch on a ledge 1000 feet above a valley. Kishi decides to climb to another ledge 20 feet above Monya. Monya throws an apple up to Kishi, but Kishi misses it. The equation $h = -16t^2 + 30t + 1000$ represents the height in feet above the valley of the apple t seconds after it was thrown. How long did it take for the apple to reach the ground?

THEATER For Exercises 34–37, use the following information.

The drama club is building a backdrop using arches whose shape can be represented by the function $f(x) = -x^2 + 2x + 8$, where x is the length in feet. The area under each arch is to be covered with fabric.

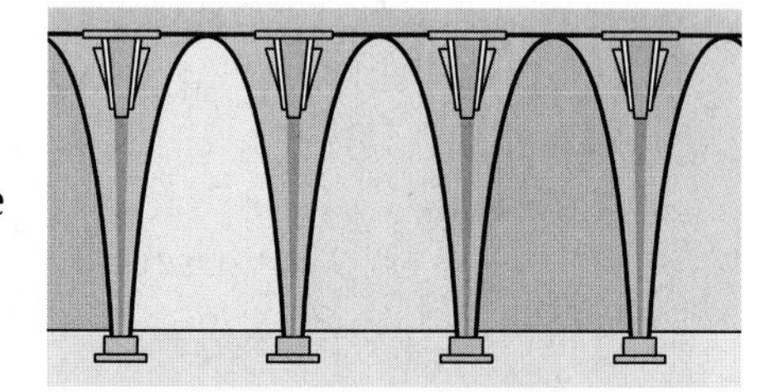

34. Graph the quadratic function and determine its x-intercepts.

35. What is the length of the segment along the floor of each arch?

36. What is the height of the arch?

37. The formula $A = \frac{2}{3}bh$ can be used to estimate the area A under a parabola. In this formula, b represents the length of the base, and h represents the height. If there are five arches, calculate the total amount of fabric that is needed.

WORK For Exercises 38–40, use the following information.

Kirk and Montega mow the soccer playing fields. They must mow an area 500 feet long and 400 feet wide. They agree that each will mow half the area. Kirk will mow around the edge in a path of equal width until half the area is left.

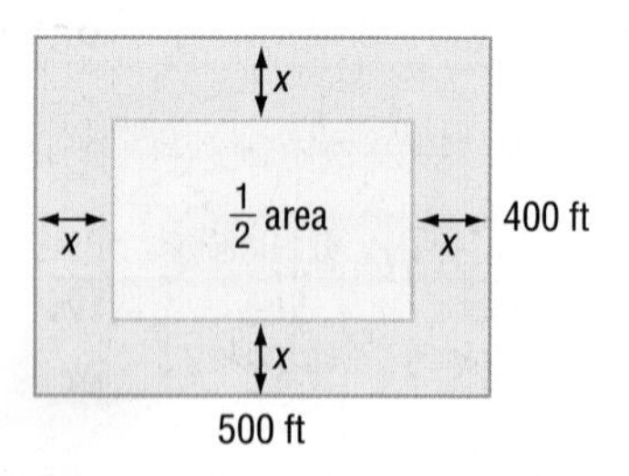

38. What is the area each person will mow?

39. Write a quadratic equation that could be used to find the width x that Kirk should mow. What width should Kirk mow?

40. The mower can mow a path 5 feet wide. To the nearest whole number, how many times should Kirk go around the field?

EXTRA PRACTICE
See pages 735, 752.
Math Online
Self-Check Quiz at algebra1.com

H.O.T. Problems

41. OPEN ENDED Draw a graph to show a counterexample to the following statement. Explain. *All quadratic equations have two different solutions.*

42. CHALLENGE Describe the zeros of $f(x) = \frac{x^3 + 2x^2 - 3x}{x + 5}$. Explain your reasoning.

Study Tip

Look Back
To review **linear inequalities,** see Lesson 6–6.

43. CHALLENGE The graph shown is a *quadratic inequality.* Similar to a linear inequality, the quadratic equation is a boundary between two half-planes. Analyze the graph and determine whether the inequality is *always, sometimes,* or *never* greater than 2. Explain.

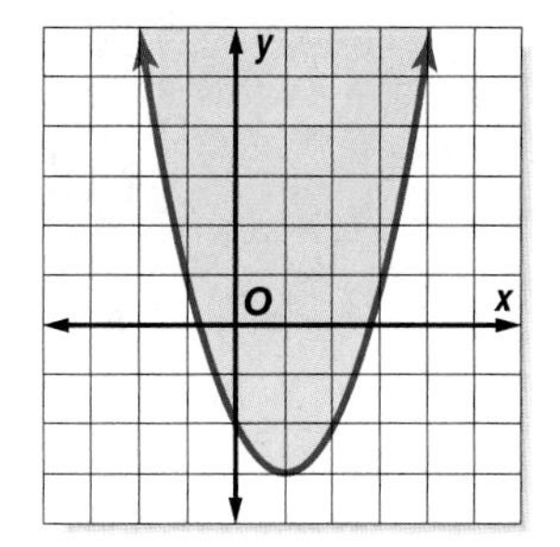

44. *Writing in Math* Use the information about computer games on page 480 to explain how quadratic equations can be used in computer simulations. Describe what the roots of a simulation equation for a computer golf game represent.

STANDARDIZED TEST PRACTICE

45. The graph of the equation $y = x^2 + 10x + 21$ is shown. For what value or values of x is $y = 0$?

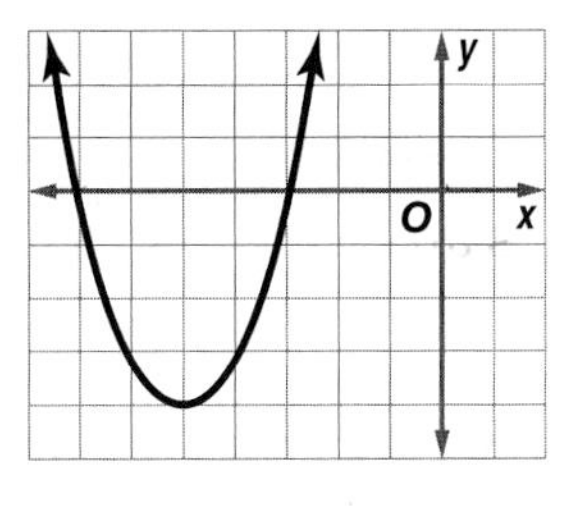

A $x = -4$

B $x = -5$

C $x = 7$ and $x = 3$

D $x = -7$ and $x = -3$

46. REVIEW Q-Mart has 1200 blue towels in stock. If they sell half of their towels every three months and do not receive any more shipments of towels, how many towels will they have left after a year?

F 60

G 75

H 150

J 300

Spiral Review

Write the equation of the axis of symmetry, and find the coordinates of the vertex of the graph of each equation. Identify the vertex as a maximum or minimum. Then graph the function. (Lesson 9-1)

47. $y = x^2 + 6x + 9$

48. $y = -x^2 + 4x - 3$

49. $y = 0.5x^2 - 6x + 5$

Solve each equation. Check your solutions. (Lesson 8-6)

50. $m^2 - 24m = -144$

51. $7r^2 = 70r - 175$

52. $4d^2 + 9 = -12d$

Simplify. Assume that no denominator is equal to zero. (Lesson 7-2)

53. $\frac{10m^4}{30m}$

54. $\frac{22a^2b^5c^7}{-11abc^2}$

55. $\frac{-9m^3n^5}{27m^{-2}n^5y^{-4}}$

56. SHIPPING An empty book crate weighs 30 pounds. The weight of a book is 1.5 pounds. For shipping, the crate must weigh at least 55 pounds and no more than 60 pounds. What is the acceptable number of books that can be packed in the crate? (Lesson 6-4)

GET READY for the Next Lesson

PREREQUISITE SKILL Determine whether each trinomial is a perfect square trinomial. If so, factor it. (Lesson 8-6)

57. $a^2 + 14 + 49$

58. $m^2 - 10m + 25$

59. $t^2 + 16t - 64$

60. $4y^2 + 12y + 9$

9-3 Solving Quadratic Equations by Completing the Square

Main Ideas

- Solve quadratic equations by finding the square root.
- Solve quadratic equations by completing the square.

New Vocabulary

completing the square

GET READY for the Lesson

Al-Khwarizmi, born in Baghdad in 780, is considered to be one of the foremost mathematicians of all time. He wrote algebra in sentences instead of using equations, and he explained the work with geometric sketches. Al-Khwarizmi would have described $x^2 + 8x = 35$ as "A square and 8 roots are equal to 35 units." He would solve the problem using the following sketch.

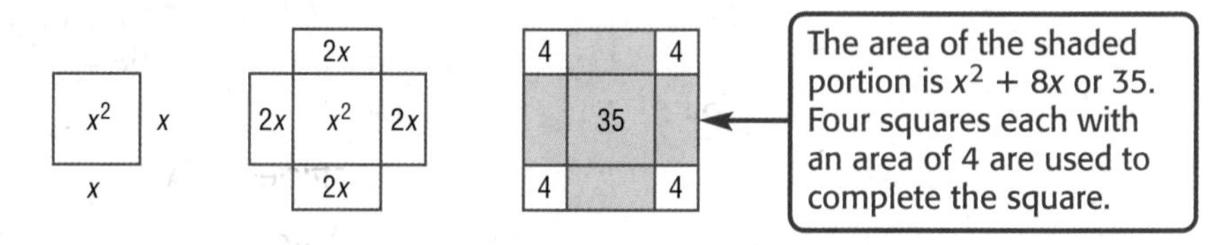

To solve problems this way today, you might use algebra tiles or a method called *completing the square*.

Find the Square Root Some equations can be solved by taking the square root of each side.

EXAMPLE Irrational Roots

1 **Solve $x^2 - 10x + 25 = 7$ by taking the square root of each side. Round to the nearest tenth if necessary.**

$x^2 - 10x + 25 = 7$	Original equation
$(x - 5)^2 = 7$	$x^2 - 10x + 25$ is a perfect square trinomial.
$\sqrt{(x - 5)^2} = \sqrt{7}$	Take the square root of each side.
$\lvert x - 5 \rvert = \sqrt{7}$	Simplify.
$x - 5 = \pm\sqrt{7}$	Definition of absolute value
$x - 5 + 5 = \pm\sqrt{7} + 5$	Add 5 to each side.
$x = 5 \pm \sqrt{7}$	Simplify.

Use a calculator to evaluate each value of x.

$x = 5 + \sqrt{7}$	or	$x = 5 - \sqrt{7}$	Write each solution.
≈ 7.6		≈ 2.4	Simplify.

The solution set is $\{2.4, 7.6\}$.

CHECK Your Progress

1. Solve $m^2 + 18m + 81 = 90$ by taking the square root of each side. Round to the nearest tenth if necessary.

Review Vocabulary

Perfect Square Trinomial a trinomial that is the square of a binomial; *Example:* $x^2 + 12x + 36$ is a perfect square trinomial because it is the square of $(x + 6)$. (Lesson 8-6)

Complete the Square In Example 1, the quadratic expression on one side of the equation was a perfect square. However, few quadratic expressions are perfect squares. To make any quadratic expression a perfect square, a method called **completing the square** may be used.

Consider the pattern for squaring a binomial such as $x + 6$.

$$(x + 6)^2 = x^2 + 2(6)(x) + 6^2$$
$$= x^2 + 12x + 36$$
$$\downarrow \quad \uparrow$$
$$\left(\frac{12}{2}\right)^2 \rightarrow 6^2$$

Notice that one half of 12 is 6 and 6^2 is 36.

KEY CONCEPT — *Completing the Square*

To complete the square for a quadratic expression of the form $x^2 + bx$, you can follow the steps below.

Step 1 Find $\frac{1}{2}$ of b, the coefficient of x.

Step 2 Square the result of Step 1.

Step 3 Add the result of Step 2 to $x^2 + bx$, the original expression.

EXAMPLE Complete the Square

2 **Find the value of c that makes $x^2 + 6x + c$ a perfect square.**

Method 1 Use algebra tiles.

Concepts in Motion
Animation
algebra1.com

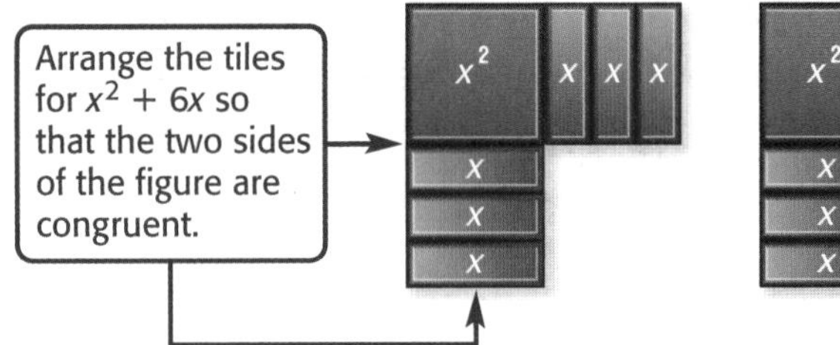

$x^2 + 6x + 9$ is a perfect square.

Method 2 Complete the square.

Step 1 Find $\frac{1}{2}$ of 6. $\qquad \frac{6}{2} = 3$

Step 2 Square the result of Step 1. $\qquad 3^2 = 9$

Step 3 Add the result of Step 2 to $x^2 + 6x$. $\qquad x^2 + 6x + 9$

Thus, $c = 9$. Notice that $x^2 + 6x + 9 = (x + 3)^2$.

CHECK Your Progress

2. Find the value of c that makes $r^2 + 8r + c$ a perfect square.

You can use the technique of completing the square to solve quadratic equations.

EXAMPLE Solve an Equation by Completing the Square

3 **Solve $a^2 - 14a + 3 = -10$ by completing the square.**

Isolate the a^2 and a terms. Then complete the square and solve.

$a^2 - 14a + 3 = -10$	Original equation
$a^2 - 14a + 3 - 3 = -10 - 3$	Subtract 3 from each side.
$a^2 - 14a = -13$	Simplify.
$a^2 - 14a + 49 = -13 + 49$	Since $\left(\frac{-14}{2}\right)^2 = 49$, add 49 to each side.
$(a - 7)^2 = 36$	Factor $a^2 - 14a + 49$.
$a - 7 = \pm 6$	Take the square root of each side.
$a = 7 \pm 6$	Add 7 to each side.
$a = 7 + 6$ or $a = 7 - 6$	Separate the solutions.
$= 13$ $\quad = 1$	Simplify.

The solution set is $\{1, 13\}$.

CHECK Your Progress

3. Solve $x^2 - 8x = 4$ by completing the square. Round to the nearest tenth if necessary.

To solve a quadratic equation in which the leading coefficient is not 1, first divide each term by the coefficient. Then complete the square.

Real-World EXAMPLE Solve a Quadratic Equation in Which $a \neq 1$

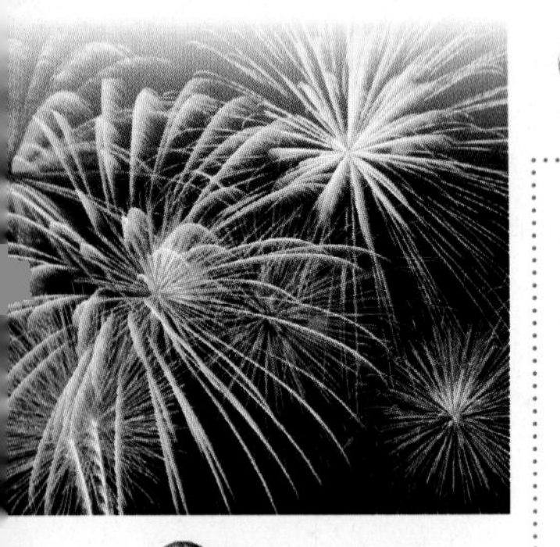

Real-World Link

One of the exploded fireworks for the Lake Toya Festival in Japan on July 15, 1988, broke a world record. The diameter of the burst was 3937 feet.

Source: *The Guinness Book of Records*

4 **ENTERTAINMENT The path of debris from fireworks when the wind is about 15 miles per hour can be modeled by the quadratic function $h = -0.04x^2 + 2x + 8$, where h is the height and x is the horizontal distance in feet. How far away from the launch site will the debris land?**

Explore You know the function that relates the horizontal and vertical distances. You want to know how far away the debris will land.

Plan The debris will hit the ground when $h = 0$. Complete the square to solve $-0.04x^2 + 2x + 8 = 0$.

Solve

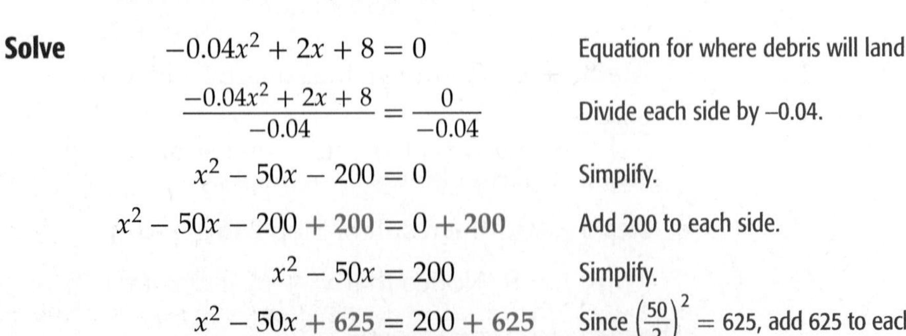

$-0.04x^2 + 2x + 8 = 0$	Equation for where debris will land
$\frac{-0.04x^2 + 2x + 8}{-0.04} = \frac{0}{-0.04}$	Divide each side by –0.04.
$x^2 - 50x - 200 = 0$	Simplify.
$x^2 - 50x - 200 + 200 = 0 + 200$	Add 200 to each side.
$x^2 - 50x = 200$	Simplify.
$x^2 - 50x + 625 = 200 + 625$	Since $\left(\frac{50}{2}\right)^2 = 625$, add 625 to each side.
$x^2 - 50x + 625 = 825$	Simplify.
$(x - 25)^2 = 825$	Factor $x^2 - 50x + 625$.
$x - 25 = \pm\sqrt{825}$	Take the square root of each side.
$x = 25 \pm \sqrt{825}$	Add 25 to each side.

Use a calculator to evaluate each value of x.

$x = 25 + \sqrt{825}$ or $x = 25 - \sqrt{825}$ Separate the solutions.

≈ 53.7 ≈ -3.7 Evaluate.

Check Since you are looking for a distance, the negative number is not reasonable. The debris will land about 53.7 feet from the launch site.

CHECK Your Progress

4. Solve $3n^2 - 18n = 30$ by completing the square. Round to the nearest tenth if necessary.

CHECK Your Understanding

Example 1 (p. 486)

Solve each equation by taking the square root of each side. Round to the nearest tenth if necessary.

1. $b^2 - 6b + 9 = 25$

2. $m^2 + 14m + 49 = 20$

Example 2 (p. 487)

Find the value of c that makes each trinomial a perfect square.

3. $a^2 - 12a + c$

4. $t^2 + 5t + c$

Example 3 (p. 488)

Solve each equation by completing the square. Round to the nearest tenth if necessary.

5. $c^2 - 6c = 7$

6. $x^2 + 7x = -12$

7. $v^2 + 14v - 9 = 6$

8. $r^2 - 4r = 2$

9. $4a^2 + 9a - 1 = 0$

10. $7 = 2p^2 - 5p + 8$

Example 4 (pp. 488–489)

11. GEOMETRY The area of a square can be doubled by increasing the length by 6 inches and the width by 4 inches. What is the length of the side of the square?

Exercises

HOMEWORK HELP	
For Exercises	See Examples
12–15	1
16–19	2
20–27	3
28–33	4

Solve each equation by taking the square root of each side. Round to the nearest tenth if necessary.

12. $b^2 - 4b + 4 = 16$

13. $t^2 + 2t + 1 = 25$

14. $g^2 - 8g + 16 = 2$

15. $w^2 + 16w + 64 = 18$

Find the value of c that makes each trinomial a perfect square.

16. $s^2 - 16s + c$

17. $y^2 - 10y + c$

18. $p^2 - 7p + c$

19. $c + 11k + k^2$

Solve each equation by completing the square. Round to the nearest tenth if necessary.

20. $s^2 - 4s - 12 = 0$

21. $d^2 + 3d - 10 = 0$

22. $y^2 - 19y + 4 = 70$

23. $d^2 + 20d + 11 = 200$

24. $a^2 - 5a = -4$

25. $p^2 - 4p = 21$

26. $x^2 + 4x + 3 = 0$

27. $d^2 - 8d + 7 = 0$

28. $5s^2 - 10s = 23$

29. $9r^2 + 49 = 42r$

30. $4h^2 + 25 = 20h$

31. $9w^2 - 12w - 1 = 0$

32. PARK PLANNING A rectangular garden of wild flowers is 9 meters long by 6 meters wide. A pathway of constant width goes around the garden. If the area of the path equals the area of the garden, what is the width of the path?

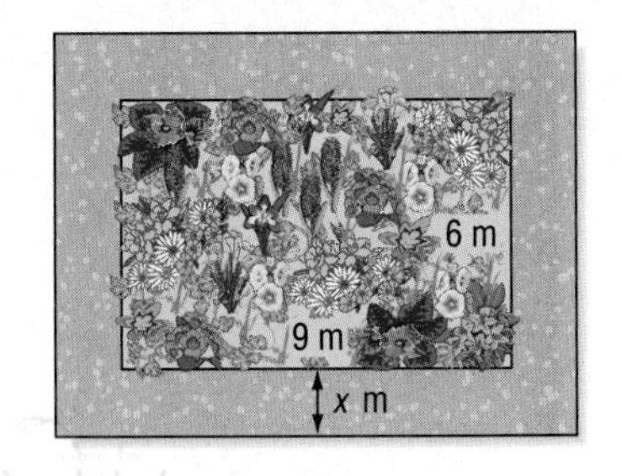

33. NUTRITION The consumption of bread and cereal in the United States is increasing and can be modeled by the function $y = 0.059x^2 - 7.423x + 362.1$, where y represents the consumption of bread and cereal in pounds and x represents the number of years since 1900. If this trend continues, in what future year will the average American consume 300 pounds of bread and cereal?

Solve each equation by completing the square. Round to the nearest tenth if necessary.

34. $0.3t^2 + 0.1t = 0.2$

35. $0.4v^2 + 2.5 = 2v$

36. $\frac{1}{2}d^2 - \frac{5}{4}d - 3 = 0$

37. $\frac{1}{3}f^2 - \frac{7}{6}f + \frac{1}{2} = 0$

Real-World Career

Photographer

Photographers must consider lighting, lens setting, and composition to create the best photograph.

Math Online
For more information, go to algebra1.com.

38. Find all values of c that make $x^2 + cx + 81$ a perfect square.

39. Find all values of c that make $x^2 + cx + 144$ a perfect square.

Solve each equation for x in terms of c by completing the square.

40. $x^2 + 4x + c = 0$

41. $x^2 - 6x + c = 0$

42. PHOTOGRAPHY Emilio is placing a photograph behind a 12-inch-by-12-inch piece of matting. The photograph is to be positioned so that the matting is twice as wide at the top and bottom as it is at the sides. If the area of the photograph is to be 54 square inches, what are the dimensions?

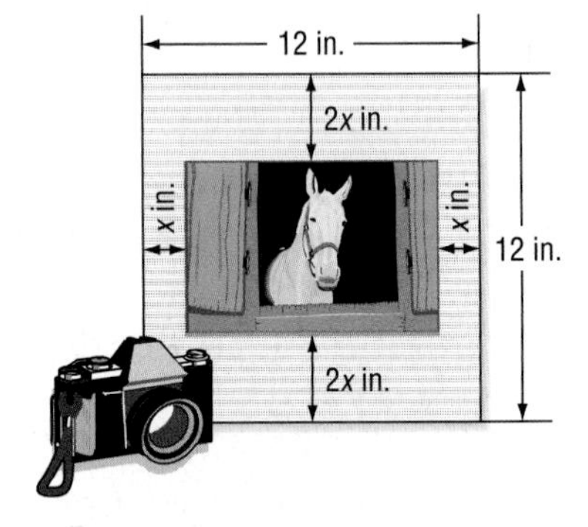

EXTRA PRACTICE
See pages 735, 752.

Math Online
Self-Check Quiz at algebra1.com

H.O.T. Problems

43. OPEN ENDED Make a square using one or more of each of the following types of tiles.

- x^2-tile
- x-tile
- 1-tile

Describe the area of your square using an algebraic expression.

44. REASONING Compare and contrast the following strategies for solving $x^2 - 5x - 7 = 0$: completing the square, graphing the related function, and factoring.

45. CHALLENGE Without graphing, describe the solution of $x^2 + 4x + 12 = 0$. Explain your reasoning. Then describe the graph of the related function.

46. Which One Doesn't Belong? Identify the expression that does not belong with the other three. Explain your reasoning.

$n^2 - n + \frac{1}{4}$	$n^2 + n + \frac{1}{4}$	$n^2 - \frac{2}{3}n + \frac{1}{9}$	$n^2 + \frac{1}{3}n + \frac{1}{9}$

47. ***Writing in Math*** Use the information about Al-Khwarizmi on page 486 to explain how ancient mathematicians used squares to solve algebraic equations. Include an explanation of Al-Khwarizmi's drawings for $x^2 + 8x = 35$ and a step-by-step algebraic solution with justification for each step of the equation.

STANDARDIZED TEST PRACTICE

48. What are the solutions to the quadratic equation $p^2 - 14p = 32$?

A 16 **C** $-2, 16$

B $-3, 14$ **D** $-4, 7$

49. **REVIEW** If $a = -5$ and $b = 6$, then $3a - 2ab =$

F -75 **H** 30

G -55 **J** 45

Spiral Review

Solve each equation by graphing. (Lesson 9-2)

50. $x^2 + 7x + 12 = 0$

51. $x^2 - 16 = 0$

52. $x^2 - 2x + 6 = 0$

PARKS For Exercises 53 and 54, use the following information. (Lesson 9-1)

A city is building a dog park that is rectangular in shape and measures 280 feet around three of the four sides as shown in the diagram.

53. If the width of the park in feet is x, write an equation that models the area A of the park.

54. Analyze the graph of the related function by finding the coordinates of the vertex and describing what this point represents.

Find the GCF for each set of monomials. (Lesson 8-1)

55. $14a^2b^3, 20a^3b^2c, 35ab^3c^2$

56. $32m^2n^3, 8m^2n, 56m^3n^2$

Write an inequality for each graph. (Lesson 6-4)

57. −6 −5 −4 −3 −2 −1 0 1 2 3 4 5 6

58. 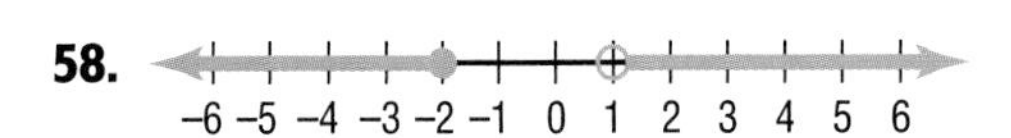

Use substitution to solve each system of equations. If the system does not have exactly one solution, state whether it has no solution or infinitely many solutions. (Lesson 5-2)

59. $y = 2x$
$x + y = 9$

60. $x = y + 3$
$2x - 3y = 5$

61. $x - 2y = 3$
$3x + y = 23$

GET READY for the Next Lesson

PREREQUISITE SKILL Evaluate $\sqrt{b^2 - 4ac}$ for each set of values. Round to the nearest tenth if necessary. (Lesson 1-2)

62. $a = 1, b = -2, c = -15$

63. $a = 2, b = 7, c = 3$

64. $a = 1, b = 5, c = -2$

65. $a = -2, b = 7, c = 5$

Mid-Chapter Quiz

Lessons 9-1 through 9-3

Write the equation of the axis of symmetry, and find the coordinates of the vertex of the graph of each function. Identify the vertex as a maximum or minimum. Then graph the function. (Lesson 9-1)

1. $y = x^2 - x - 6$
2. $y = 2x^2 + 3$
3. $y = -3x^2 - 6x + 5$
4. **MULTIPLE CHOICE** Which graph shows a function $y = x^2 + b$ when $b > 1$? (Lesson 9-1)

A

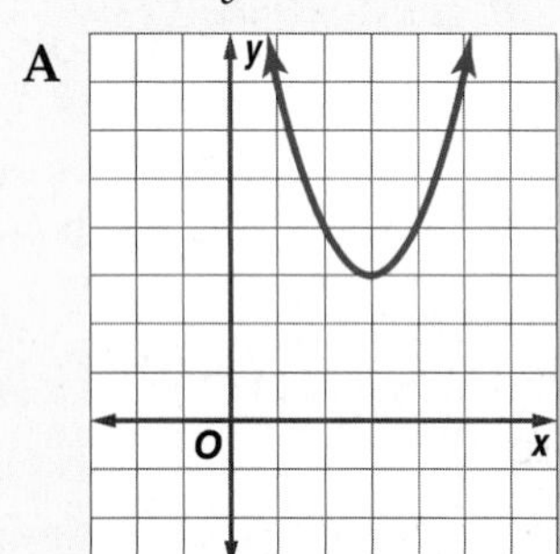

B

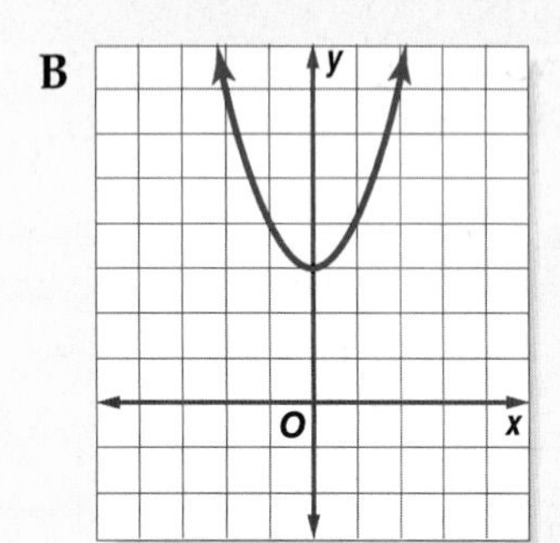

C

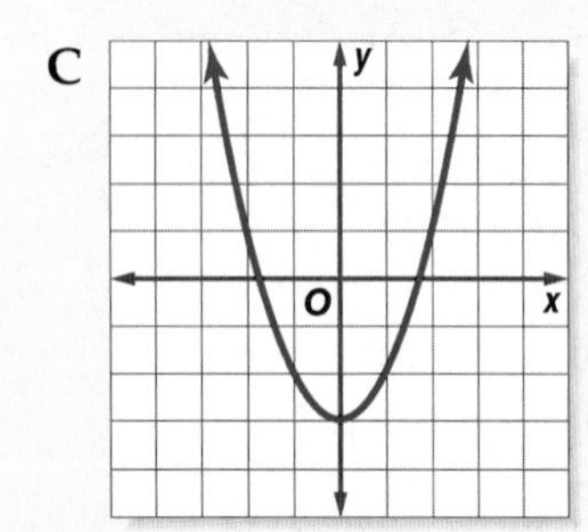

D

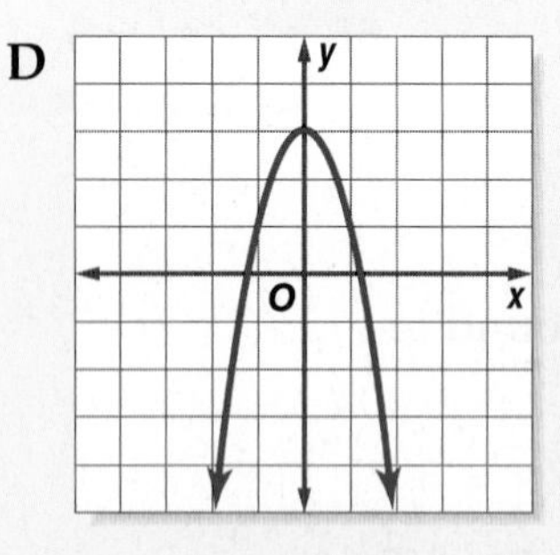

Solve each equation by graphing. If integral roots cannot be found, estimate the roots by stating the consecutive integers between which the roots lie. (Lesson 9-2)

5. $x^2 + 6x + 10 = 0$
6. $x^2 - 2x - 1 = 0$
7. $x^2 - 5x - 6 = 0$

8. **SOFTBALL** In a softball game, Lola hit the ball straight up with an initial upward velocity of 47 feet per second. The height h of the softball in feet above ground after t seconds can be modeled by the equation $h = -16t^2 + 47t + 3$. How long was the softball in the air before it hit the ground? (Lesson 9-2)

Solve each equation by completing the square. Round to the nearest tenth if necessary. (Lesson 9-3)

9. $s^2 + 8s = -15$
10. $a^2 - 10a = -24$
11. $y^2 - 14y + 49 = 5$
12. $2b^2 - b - 7 = 14$

13. **ROCKETS** A model rocket is launched from the ground with an initial upward velocity of 475 feet per second. About how many seconds will it take to reach the ground? Use the formula $h = -16t^2 + 175t$, where h is the height of the rocket and t is the time in seconds. Round to the nearest tenth if necessary. (Lesson 9-3)

14. **GEOMETRY** The length and width of the rectangle are increased by the same amount so that the new area is 154 square centimeters. Find the dimensions of the new rectangle. (Lesson 9-3)

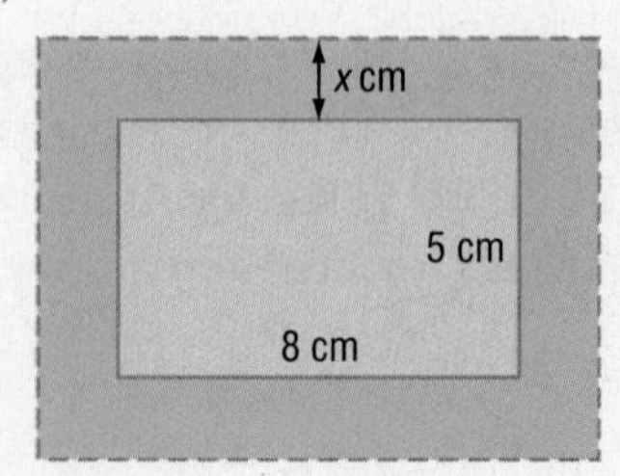

9-4 Solving Quadratic Equations by Using the Quadratic Formula

Main Ideas

- Solve quadratic equations by using the Quadratic Formula.
- Use the discriminant to determine the number of solutions for a quadratic equation.

New Vocabulary

Quadratic Formula
discriminant

GET READY for the Lesson

In the past few decades, there has been a dramatic increase in the percent of people living in the United States who were born in other countries. This trend can be modeled by the quadratic function $P = 0.006t^2 - 0.080t + 5.281$, where P is the percent born outside the United States and t is the number of years since 1960.

Percent Born Outside the U.S.

Percent: 4, 8, 12, 16, 20 (P)
Years Since 1960: 0, 10, 20, 30, 40, 50 (t)
$P = 0.006t^2 - 0.080t + 5.281$

To predict when 15% of the population will be people who were born outside of the U.S., you can solve the equation $15 = 0.006t^2 - 0.080t + 5.281$. This equation would be impossible or difficult to solve using factoring, graphing, or completing the square.

Quadratic Formula You can solve the standard form of the quadratic equation $ax^2 + bx + c = 0$ for x. The result is the **Quadratic Formula.**

KEY CONCEPT — *The Quadratic Formula*

The solutions of a quadratic equation in the form $ax^2 + bx + c = 0$, where $a \neq 0$, are given by the Quadratic Formula.

$$x = \frac{-b \pm \sqrt{b^2 - 4ac}}{2a}$$

You can solve quadratic equations by factoring, graphing, completing the square, or using the Quadratic Formula.

EXAMPLE Solve Quadratic Equations

Solve each equation. Round to the nearest tenth if necessary.

a. $x^2 - 2x - 24 = 0$

Method 1 Factoring

$x^2 - 2x - 24 = 0$ Original equation

$(x + 4)(x - 6) = 0$ Factor $x^2 - 2x - 24$.

$x + 4 = 0$ or $x - 6 = 0$ Zero Product Property

$x = -4$ $\quad x = 6$ Solve for x.

Study Tip

Quadratic Formula
The Quadratic Formula is proved in Lesson 10-1.

(continued on the next page)

Method 2 Quadratic Formula

For this equation, $a = 1$, $b = -2$, and $c = -24$.

$x = \frac{-b \pm \sqrt{b^2 - 4ac}}{2a}$	Quadratic Formula
$= \frac{-(-2) \pm \sqrt{(-2)^2 - 4(1)(-24)}}{2(1)}$	$a = 1$, $b = -2$, and $c = -24$
$= \frac{2 \pm \sqrt{4 + 96}}{2}$	Multiply.
$= \frac{2 \pm \sqrt{100}}{2}$ or $\frac{2 \pm 10}{2}$	Add and simplify.
$x = \frac{2 - 10}{2}$ or $x = \frac{2 + 10}{2}$	Separate the solutions.
$= -4$ $= 6$	Simplify.

The solution set is $\{-4, 6\}$.

b. $24x^2 - 14x = 6$

Step 1 Rewrite the equation in standard form.

$24x^2 - 14x = 6$	Original equation
$24x^2 - 14x - 6 = 0$	Subtract 6 from each side.

Step 2 Apply the Quadratic Formula.

$x = \frac{-b \pm \sqrt{b^2 - 4ac}}{2a}$	Quadratic Formula
$= \frac{-(-14) \pm \sqrt{(-14)^2 - 4(24)(-6)}}{2(24)}$	$a = 24$, $b = -14$, and $c = -6$
$= \frac{14 \pm \sqrt{196 + 576}}{48}$	Multiply.
$= \frac{14 \pm \sqrt{772}}{48}$	Add.
$x = \frac{14 - \sqrt{772}}{48}$ or $x = \frac{14 + \sqrt{772}}{48}$	Separate the solutions.
≈ -0.3 ≈ 0.9	Simplify.

Check the solutions by using the CALC menu on a graphing calculator to determine the zeros of the related quadratic function.

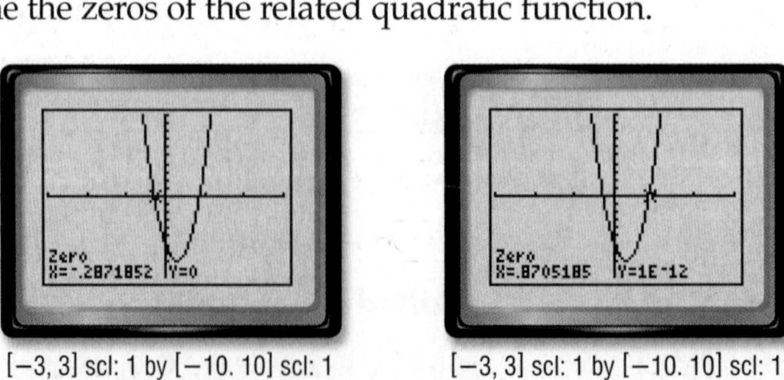

[−3, 3] scl: 1 by [−10. 10] scl: 1   [−3, 3] scl: 1 by [−10. 10] scl: 1

To the nearest tenth, the solution set is $\{-0.3, 0.9\}$.

Study Tip

The Quadratic Formula

You may want to simplify this equation by dividing each side by 2 before applying the Quadratic Formula. However, the Quadratic Formula can help you find the solution of *any* quadratic equation.

CHECK Your Progress

1A. $x^2 + 3x - 18 = 0$ **1B.** $4x^2 + 2x = 17$

Online Personal Tutor at algebra1.com

The table summarizes the five methods for solving quadratic equations.

CONCEPT SUMMARY — *Solving Quadratic Equations*

Method	Can Be Used	Comments	Lesson(s)
factoring	sometimes	Use if constant term is 0 or factors are easily determined.	8-2 to 8-6
using a table	sometimes	Not always exact; use only when an approximate solution is sufficient.	9-2
graphing	always	Not always exact; use only when an approximate solution is sufficient.	9-2
completing the square	always	Useful for equations of the form $x^2 + bx + c = 0$, where b is an even number.	9-3
Quadratic Formula	always	Other methods may be easier to use in some cases, but this method always gives accurate solutions.	9-4

Real-World EXAMPLE — Use the Quadratic Formula to Solve a Problem

Real-World Link

Astronauts have found walking on the Moon to be very different from walking on Earth because the gravitational pull of the Moon is only 1.6 meters per second squared. The gravitational pull on Earth is 9.8 meters per second squared.

Source: *World Book Encyclopedia*

2 **SPACE TRAVEL** **The height H of an object t seconds after it is propelled upward with an initial velocity v is represented by $H = -\frac{1}{2}gt^2 + vt + h$, where g is the gravitational pull and h is the initial height. Suppose an astronaut on the Moon throws a baseball upward with an initial velocity of 10 meters per second, letting go of the ball 2 meters above the ground. Use the information at the left to find how much longer the ball will stay in the air than a similarly thrown baseball on Earth.**

In order to find when the ball hits the ground, you must find when $H = 0$. Write two equations to represent the situation on the Moon and on Earth.

Baseball Thrown on the Moon

$$H = -\frac{1}{2}gt^2 + vt + h$$
$$0 = -\frac{1}{2}(1.6)t^2 + 10t + 2$$
$$0 = -0.8t^2 + 10t + 2$$

Baseball Thrown on Earth

$$H = -\frac{1}{2}gt^2 + vt + h$$
$$0 = -\frac{1}{2}(9.8)t^2 + 10t + 2$$
$$0 = -4.9t^2 + 10t + 2$$

To find accurate solutions, use the Quadratic Formula.

Moon:

$$t = \frac{-b \pm \sqrt{b^2 - 4ac}}{2a}$$
$$= \frac{-10 \pm \sqrt{10^2 - 4(-0.8)(2)}}{2(-0.8)}$$
$$= \frac{-10 \pm \sqrt{106.4}}{-1.6}$$
$t \approx 12.7$ or $t \approx -0.2$

Earth:

$$t = \frac{-b \pm \sqrt{b^2 - 4ac}}{2a}$$
$$= \frac{-10 \pm \sqrt{10^2 - 4(-4.9)(2)}}{2(-4.9)}$$
$$= \frac{-10 \pm \sqrt{139.2}}{-9.8}$$
$t \approx 2.2$ or $t \approx -0.2$

Since a negative time is not reasonable, use the positive solutions. The ball will stay in the air about 12.7 – 2.2 or 10.5 seconds longer on the Moon.

CHECK Your Progress

2. **GEOMETRY** The perimeter of a rectangle is 60 inches. Find the dimensions of the rectangle if its area is 221 square inches.

The Discriminant In the Quadratic Formula, the expression under the radical sign, $b^2 - 4ac$, is called the **discriminant**. The value of the discriminant can be used to determine the number of real roots for a quadratic equation.

KEY CONCEPT *Using the Discriminant*

Discriminant	negative	zero	positive
Example	$2x^2 + x + 3 = 0$ $x = \frac{-1 \pm \sqrt{1^2 - 4(2)(3)}}{2(2)}$ $= \frac{-1 \pm \sqrt{-23}}{4}$ There are no real roots since no real number can be the square root of a negative number.	$x^2 + 6x + 9 = 0$ $x = \frac{-6 \pm \sqrt{6^2 - 4(1)(9)}}{2(1)}$ $= \frac{-6 \pm \sqrt{0}}{2}$ $= \frac{-6}{2}$ or -3 There is a double root, -3.	$x^2 - 5x + 2 = 0$ $x = \frac{-(-5) \pm \sqrt{(-5)^2 - 4(1)(2)}}{2(1)}$ $= \frac{5 \pm \sqrt{17}}{2}$ There are two roots, $\frac{5 + \sqrt{17}}{2}$ and $\frac{5 - \sqrt{17}}{2}$.
Graph of Related Function	$f(x) = 2x^2 + x + 3$ The graph does not cross the x-axis.	$f(x) = x^2 + 6x + 9$ The graph touches the x-axis in one place.	$f(x) = x^2 - 5x + 2$ The graph crosses the x-axis twice.
Number of Real Roots	0	1	2

EXAMPLE **Use the Discriminant**

3 **State the value of the discriminant for each equation. Then determine the number of real roots of the equation.**

a. $2x^2 + 10x + 11 = 0$

$b^2 - 4ac = 10^2 - 4(2)(11)$ $a = 2, b = 10$, and $c = 11$

$= 12$ Simplify.

Since the discriminant is positive, the equation has two real roots.

b. $3m^2 + 4m = -2$

Step 1 Rewrite the equation in standard form.

$3m^2 + 4m = -2$ Original equation

$3m^2 + 4m + 2 = -2 + 2$ Add 2 to each side.

$3m^2 + 4m + 2 = 0$ Simplify.

Step 2 Find the discriminant.

$b^2 - 4ac = 4^2 - 4(3)(2)$ $a = 3, b = 4$, and $c = 2$

$= -8$ Simplify.

Since the discriminant is negative, the equation has no real roots.

CHECK Your Progress

3A. $4n^2 - 20n + 25 = 0$ **3B.** $5x^2 - 3x + 8 = 0$ **3C.** $2x^2 + 11x + 15 = 0$

CHECK Your Understanding

Example 1 (pp. 493–494)

Solve each equation by using the Quadratic Formula. Round to the nearest tenth if necessary.

1. $x^2 + 7x + 6 = 0$

2. $t^2 + 11t = 12$

3. $r^2 + 10r + 12 = 0$

4. $3v^2 + 5v + 11 = 0$

Example 2 (p. 495)

5. MANUFACTURING A pan is to be formed by cutting 2-centimeter-by-2-centimeter squares from each corner of a square piece of sheet metal and then folding the sides. If the volume of the pan is to be 441 square centimeters, what should the dimensions of the original piece of sheet metal be?

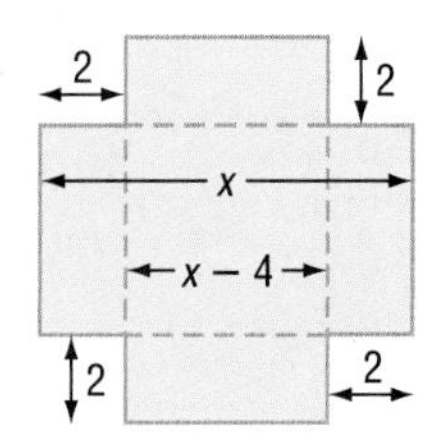

Example 3 (p. 496)

State the value of the discriminant for each equation. Then determine the number of real roots of the equation.

6. $m^2 + 5m - 6 = 0$

7. $s^2 + 8s + 16 = 0$

8. $2z^2 + z = -50$

Exercises

HOMEWORK HELP	
For Exercises	See Examples
9–20	1, 2
21, 22	3
23–28	4

Solve each equation by using the Quadratic Formula. Round to the nearest tenth if necessary.

9. $v^2 + 12v + 20 = 0$

10. $3t^2 - 7t - 20 = 0$

11. $5y^2 - y - 4 = 0$

12. $x^2 - 25 = 0$

13. $r^2 + 25 = 0$

14. $2x^2 + 98 = 28x$

15. $4s^2 + 100 = 40s$

16. $2r^2 + r - 14 = 0$

17. $2n^2 - 7n - 3 = 0$

18. $5v^2 - 7v = 1$

19. $11z^2 - z = 3$

20. $2w^2 = -(7w + 3)$

21. GEOMETRY What are the dimensions of rectangle *ABCD*?

Rectangle *ABCD*	
perimeter	42 cm
area	80 cm^2

22. PHYSICAL SCIENCE A projectile is shot vertically up in the air from ground level. Its distance s, in feet, after t seconds is given by $s = 96t - 16t^2$. Find the values of t when s is 96 feet.

State the value of the discriminant for each equation. Then determine the number of real roots of the equation.

23. $x^2 + 3x - 4 = 0$

24. $y^2 + 3y + 1 = 0$

25. $4p^2 + 10p = -6.25$

26. $1.5m^2 + m = -3.5$

27. $2r^2 = \frac{1}{2}r - \frac{2}{3}$

28. $\frac{4}{3}n^2 + 4n = -3$

Solve each equation by using the Quadratic Formula. Round to the nearest tenth if necessary.

29. $1.34d^2 - 1.1d = -1.02$

30. $-2x^2 + 0.7x = -0.3$

31. $2y^2 - \frac{5}{4}y = \frac{1}{2}$

32. $w^2 + \frac{2}{25} = \frac{3}{5}w$

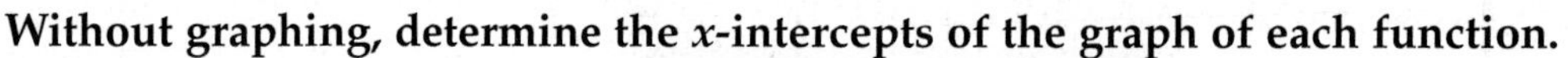

Without graphing, determine the x-intercepts of the graph of each function.

33. $f(x) = 4x^2 - 9x + 4$

34. $f(x) = 13x^2 - 16x - 4$

Without graphing, determine the number of x-intercepts of the graph of each function.

35. $f(x) = 7x^2 - 3x - 1$

36. $f(x) = x^2 + 4x + 7$

RECREATION **For Exercises 37 and 38, use the following information.**

As Darius is skiing down a ski slope, Jorge is on the chairlift on the same slope. The chair lift has stopped. Darius stops directly below Jorge and attempts to toss a disposable camera up to him. If the camera is thrown with an initial velocity of 35 feet per second, the equation for the height of the camera is $h = -16t^2 + 35t + 5$, where h represents the height in feet and t represents the time in seconds.

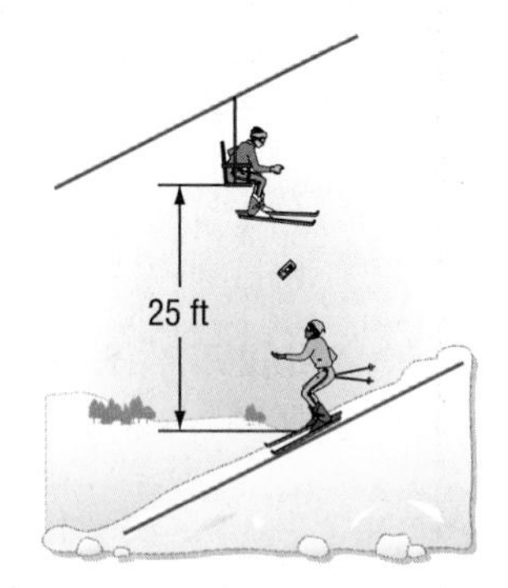

Real-World Link

Downhill skiing is the most popular type of snow skiing. Skilled skiers can obtain speeds of about 60 miles per hour as they race down mountain slopes.

Source: *World Book Encyclopedia*

37. If the chairlift is 25 feet above the ground, will Jorge have 0, 1, or 2 chances to catch the camera?

38. If Jorge is unable to catch the camera, when will it hit the ground?

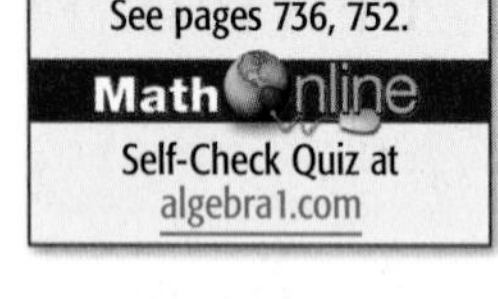

EXTRA PRACTICE
See pages 736, 752.

Math Online
Self-Check Quiz at algebra1.com

39. AMUSEMENT PARKS The Demon Drop ride at Cedar Point takes riders to the top of a tower and drops them 60 feet at speeds reaching 80 feet per second. A function that models this ride is $h = -16t^2 + 64t - 60$, where h is the height in feet and t is the time in seconds. About how many seconds does it take for riders to drop from 60 feet to 0 feet?

H.O.T. Problems

40. REASONING Use the Quadratic Formula to show that $f(x) = 3x^2 - 2x - 4$ has two real roots.

41. FIND THE ERROR Lakeisha and Juanita are determining the number of solutions of $5y^2 - 3y = 2$. Who is correct? Explain your reasoning.

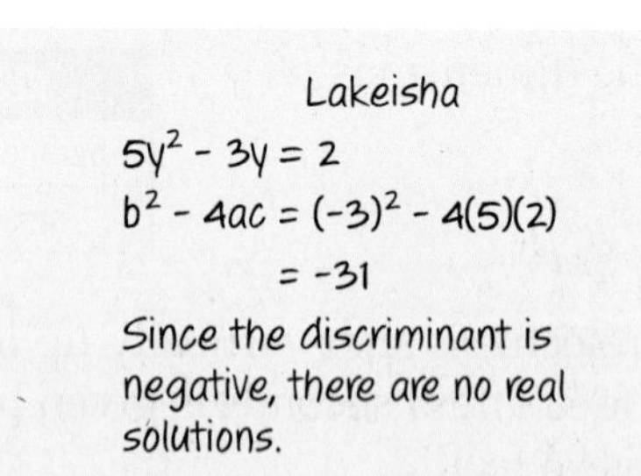

Lakeisha

$5y^2 - 3y = 2$

$b^2 - 4ac = (-3)^2 - 4(5)(2)$

$= -31$

Since the discriminant is negative, there are no real solutions.

Juanita

$5y^2 - 3y = 2$

$5y^2 - 3y - 2 = 0$

$b^2 - 4ac = (-3)^2 - 4(5)(-2)$

$= 49$

Since the discriminant is positive, there are two real roots.

42. OPEN ENDED Write a quadratic equation with no real solutions. Explain how you know there are no solutions.

43. REASONING Use factoring techniques to determine the number of real roots of the function $f(x) = x^2 - 8x + 16$. Compare this method to using the discriminant.

44. ***Writing in Math*** Describe three different ways to solve $x^2 - 2x - 15 = 0$. Which method do you prefer and why?

STANDARDIZED TEST PRACTICE

45. Which statement *best* describes why there is no real solution to the quadratic equation $y = x^2 - 6x + 13$?

A The value of $(-6)^2 - 4 \cdot 1 \cdot 13$ is a perfect square.

B The value of $(-6)^2 - 4 \cdot 1 \cdot 13$ is equal to zero.

C The value of $(-6)^2 - 4 \cdot 1 \cdot 13$ is negative.

D The value of $(-6)^2 - 4 \cdot 1 \cdot 13$ is positive.

46. REVIEW In the system of equations $6x - 3y = 12$ and $2x + 5y = 9$, which expression can be correctly substituted for y in the equation $2x + 5y = 9$?

F $12 + 2x$

G $12 - 2x$

H $-4 + 2x$

J $4 - 2x$

Spiral Review

Solve each equation by completing the square. Round to the nearest tenth if necessary. (Lesson 9-3)

47. $x^2 - 8x = -7$ **48.** $a^2 + 2a + 5 = 20$ **49.** $n^2 - 12n = 5$

Solve each equation by graphing. If integral roots cannot be found, estimate the roots by stating the consecutive integers between which the roots lie. (Lesson 9-2)

50. $x^2 - x = 6$ **51.** $2x^2 + x = 2$ **52.** $-x^2 + 3x + 6 = 0$

53. GEOMETRY The triangle has an area of 96 square centimeters. Find the base b of the triangle. (Lesson 8-3)

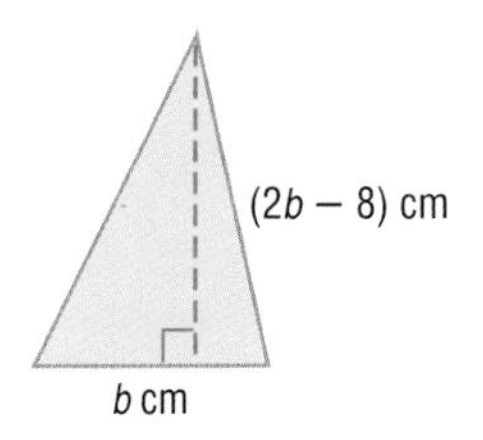

Factor each polynomial. (Lesson 8-2)

54. $24r + 6s$ **55.** $15xy^3 + y^4$ **56.** $2ax + 6xc + ba + 3bc$

Solve each inequality. Then check your solution. (Lesson 6-3)

57. $2m + 7 > 17$ **58.** $-2 - 3x \geq 2$ **59.** $-20 \geq 8 + 7k$

Write an equation of the line that passes through each point with the given slope. (Lesson 4-4)

60. $(2, 13), m = 4$ **61.** $(-2, -7), m = 0$ **62.** $(-4, 6), m = \frac{3}{2}$

GET READY for the Next Lesson

PREREQUISITE SKILL Evaluate $c(a^x)$ for each of the given values. (Lesson 1-1)

63. $a = 2, c = 1, x = 4$ **64.** $a = 7, c = 3, x = 2$ **65.** $a = 5, c = 2, x = 3$

EXTEND 9-4

Algebra Lab
Applying Quadratic Equations

Many of the real-world problems you solved in Chapters 8 and 9 were physical problems involving the path of an object that is influenced by gravity. These paths, called **trajectories,** can be modeled by a quadratic function. The formula relating the height of the object $H(t)$ and time t is shown below.

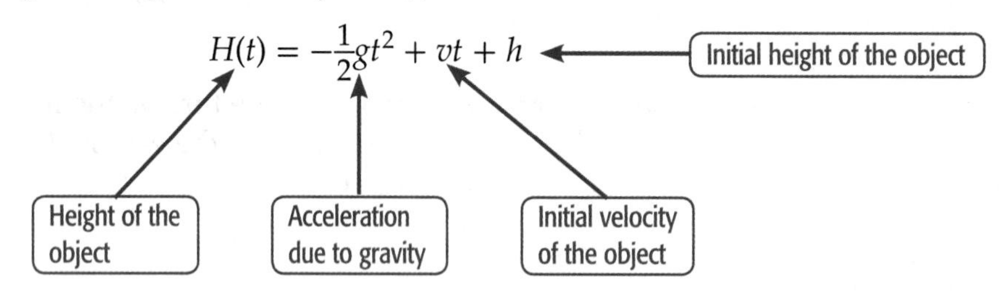

The acceleration due to gravity is 9.8 meters per second, per second. We express this by saying *9.8 meters per second squared.* Similarly, it is 32 feet per second squared.

EXAMPLE 1

Juan kicks a football at a velocity of 25 meters per second. If the ball makes contact with his foot 0.5 meter off the ground, how long will the ball stay in the air?

We want to find the time t when $H(t)$ is 0. First substitute the known values into the motion formula. Since the known measures are written in terms of meters and meters per second, use 9.8 meters per second squared for the acceleration due to gravity.

$H(t) = -\frac{1}{2}gt^2 + vt + h$ Motion Formula

$0 = -\frac{1}{2}(9.8)t^2 + 25t + 0.5$ $H(t) = 0, g = 9.8, v = 25, h = 0.5$

$0 = -4.9t^2 + 25t + 0.5$ Simplify.

Use the Quadratic Formula to solve for t.

$t = \frac{-b \pm \sqrt{b^2 - 4ac}}{2a}$ Quadratic Formula

$= \frac{-25 \pm \sqrt{25^2 - 4(-4.9)(0.5)}}{2(-4.9)}$ $a = -4.9, b = 25, c = 0.5$

$= \frac{-25 \pm \sqrt{634.8}}{-9.8}$ Simplify.

$t \approx -0.02$ or $t \approx 5.12$ Use a calculator.

Since time cannot be a negative value, discard the negative solution. The football will be in the air about 5 seconds.

If an object were projected downward, the initial velocity of the object is negative.

EXAMPLE 2

Katharine is on a bridge 12 feet above a pond. She throws a handful of fish food straight down with a velocity of 8 feet per second. In how many seconds will it reach the surface of the water?

Since the units given are in feet, use $g = 32$ ft/s^2. Katharine throws the food down, so the initial velocity is negative. When the food hits the water, $H(t)$ will be 0 feet.

$H(t) = -\frac{1}{2}gt^2 + vt + h$ — Motion Formula

$0 = -\frac{1}{2}(32)t^2 - 8t + 12$ — $H(t) = 0, g = 32, v = -8, h = 12$

$0 = -16t^2 - 8t + 12$ — Simplify.

$0 = -4t^2 - 2t + 3$ — Divide each side by 4.

Use the Quadratic Formula to solve for t.

$t = \frac{-b \pm \sqrt{b^2 - 4ac}}{2a}$ — Quadratic Formula

$= \frac{2 \pm \sqrt{(-2)^2 - 4(-4)(3)}}{2(-4)}$ — $a = -4, b = -2, c = 3$

$= \frac{2 \pm \sqrt{52}}{-8}$ — Simplify.

$t \approx -1.15$ or $t \approx 0.65$ — Use a calculator.

Discard the negative solution. The fish food will hit the water in 0.65 second.

EXERCISES

1. Darren swings at a golf ball on the ground with a velocity of 10 feet per second. How long was the ball in the air?
2. Amalia hits a volleyball at a velocity of 15 meters per second. If the ball was hit from a height of 1.8 meters, determine the time it takes for the ball to land on the floor. Assume that the ball is not hit by another player.
3. Michael is repairing the roof on a shed. He accidentally dropped a box of nails from a height of 14 feet. How long did it take for the box to land on the ground? Since the box was dropped and not thrown, $v = 0$.
4. Carmen threw a penny into a fountain. She threw it from a height of 1.2 meters and at a velocity of 6 meters per second. How long did it take for the penny to hit the surface of the water?

9-5 Exponential Functions

Main Ideas

- Graph exponential functions.
- Identify data that displays exponential behavior.

New Vocabulary

exponential function

GET READY for the Lesson

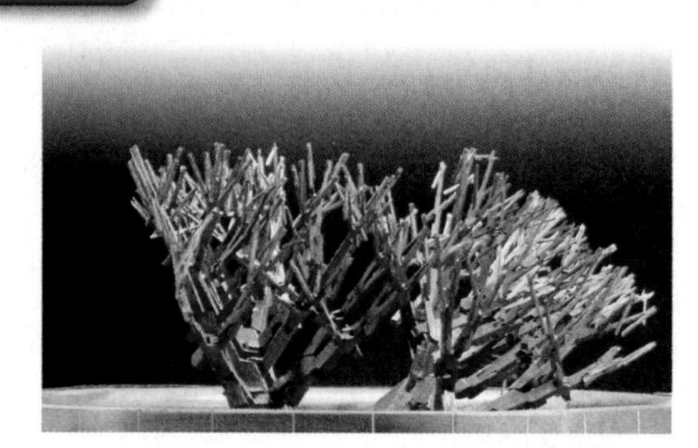

Earnest "Mooney" Warther was a whittler and a carver. For one of his most unusual carvings, Mooney carved a large pair of pliers in a tree.

From this original carving, he carved another pair of pliers in each handle of the original. Then he carved another pair of pliers in each of those handles. He continued this pattern to create the original pliers and 8 more layers of pliers. Even more amazing is the fact that all of the pliers work.

Graph Exponential Functions The number of pliers on each level is given in the table below.

Level	Number of Pliers	Power of 2
Original	1	2^0
First	1(2) = 2	2^1
Second	2(2) = 4	2^2
Third	2(2)(2) = 8	2^3
Fourth	2(2)(2)(2) = 16	2^4
Fifth	2(2)(2)(2)(2) = 32	2^5
Sixth	2(2)(2)(2)(2)(2) = 64	2^6
Seventh	2(2)(2)(2)(2)(2)(2) = 128	2^7
Eighth	2(2)(2)(2)(2)(2)(2)(2) = 256	2^8

Study the last column above. Notice that the exponent matches the level. So we can write an equation to describe y, the number of pliers for any given level x as $y = 2^x$. This function is neither linear nor quadratic. It is in the class of functions called **exponential functions** in which the variable is the exponent.

KEY CONCEPT *Exponential Function*

An exponential function is a function that can be described by an equation of the form $y = a^x$, where $a > 0$ and $a \neq 1$.

As with other functions, you can use ordered pairs to graph an exponential function.

EXAMPLE Graph an Exponential Function with $a > 1$

1 a. Graph $y = 4^x$. State the y-intercept.

x	4^x	y
-2	4^{-2}	$\frac{1}{16}$
-1	4^{-1}	$\frac{1}{4}$
0	4^0	1
1	4^1	4
2	4^2	16
3	4^3	64

Graph the ordered pairs and connect the points with a smooth curve. The y-intercept is 1.

b. Use the graph to determine the approximate value of $4^{1.8}$.

The graph represents all real values of x and their corresponding values of y for $y = 4^x$. So, the value of y is about 12 when $x = 1.8$. Use a calculator to confirm this value. $4^{1.8} \approx 12.12573253$

Study Tip

Exponential Graphs

Notice that the y-values change little for small values of x, but they increase quickly as the values of x become greater.

CHECK Your Progress

1A. Graph $y = 7^x$. State the y-intercept.

1B. Use the graph to determine the approximate value of $y = 7^{0.1}$ to the nearest tenth. Use a calculator to confirm the value.

The graphs of functions of the form $y = a^x$, where $a > 1$, all have the same shape as the graph in Example 1, rising faster and faster as you move from left to right.

EXAMPLE Graph Exponential Functions with $0 < a < 1$

2 a. Graph $y = \left(\frac{1}{2}\right)^x$. State the y-intercept.

x	$\left(\frac{1}{2}\right)^x$	y
-3	$\left(\frac{1}{2}\right)^{-3}$	8
-2	$\left(\frac{1}{2}\right)^{-2}$	4
-1	$\left(\frac{1}{2}\right)^{-1}$	2
0	$\left(\frac{1}{2}\right)^{0}$	1
1	$\left(\frac{1}{2}\right)^{1}$	$\frac{1}{2}$
2	$\left(\frac{1}{2}\right)^{2}$	$\frac{1}{4}$

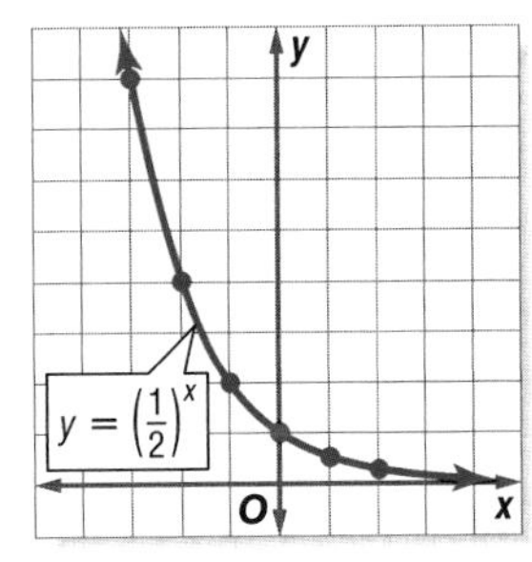

The y-intercept is 1. *Notice that the y-values decrease less rapidly as x increases.*

b. Use the graph to determine the approximate value of $\left(\frac{1}{2}\right)^{-2.5}$.

The value of y is about $5\frac{1}{2}$ when $x = -2.5$. Use a calculator to confirm this value. $\left(\frac{1}{2}\right)^{-2.5} \approx 5.656854249$

CHECK Your Progress

2A. Graph $y = \left(\frac{1}{3}\right)^x + 2$. State the y-intercept.

2B. Use the graph to determine the approximate value of $y = \left(\frac{1}{3}\right)^{-1.5} + 2$ to the nearest tenth. Use a calculator to confirm the value.

GRAPHING CALCULATOR LAB

Transformations of Exponential Functions

The graphs of $y = 2^x$, $y = 3 \cdot 2^x$, and $y = 0.5 \cdot 2^x$ are shown at the right. Notice that the y-intercept of $y = 2^x$ is 1, the y-intercept of $y = 3 \cdot 2^x$ is 3, and the y-intercept of $y = 0.5 \cdot 2^x$ is 0.5. The graph of $y = 3 \cdot 2^x$ is steeper than the graph of $y = 2^x$. The graph of $y = 0.5 \cdot 2^x$ is not as steep as the graph of $y = 2^x$.

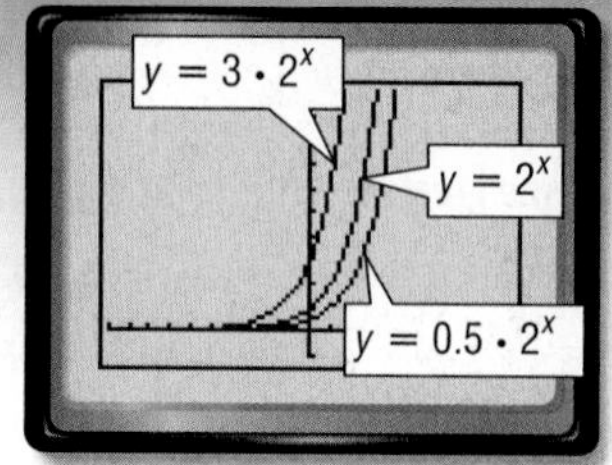

[−10, 10] scl: 1 by [−1, 10] scl: 1

THINK AND DISCUSS

Graph each set of equations on the same screen. Compare and contrast the graphs.

1. $y = 2^x$
$y = 2^x + 3$
$y = 2^x - 4$

2. $y = 2^x$
$y = 2^{x+5}$
$y = 2^{x-4}$

3. $y = 2^x$
$y = 3^x$
$y = 5^x$

4. $y = 3 \cdot 2^x$
$y = 3(2^x - 1)$
$y = 3(2^x + 1)$

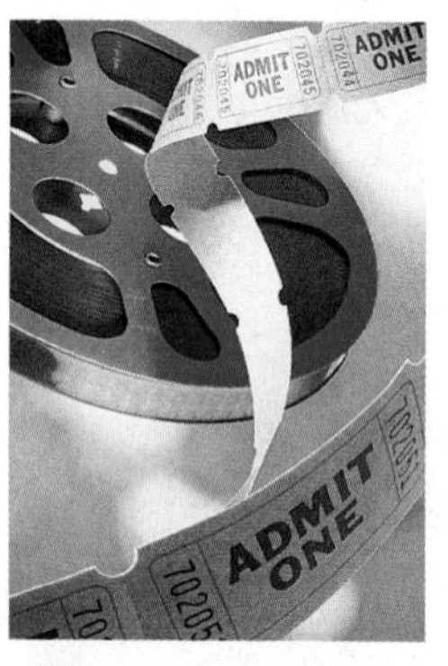

Real-World Link

The first successful photographs of motion were made in 1877. Today, the motion picture industry is big business, with the highest-grossing movie making $1,835,300,000.

Source: imdb.com

Real-World EXAMPLE — Use Exponential Functions to Solve Problems

3 **MOTION PICTURES** **Movie ticket sales decrease each weekend after an opening. The function $E = 49.9 \cdot 0.692^w$ models the earnings of a popular movie. In this equation, E represents earnings in millions of dollars and w represents the weekend number.**

a. Graph the function. What values of E and w are meaningful in the context of the problem?

Use a graphing calculator to graph the function. Only values where $E \leq 49.9$ and $w > 0$ are meaningful in the context of the problem.

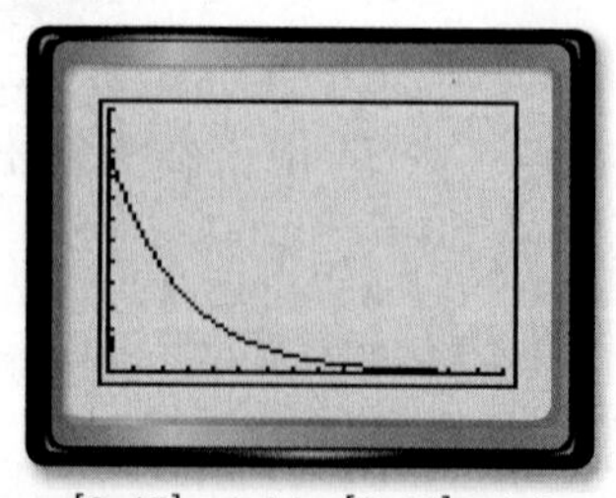

[0, 15] scl: 1 by [0, 60] scl: 5

b. How much did the movie make on the first weekend?

$E = 49.9 \cdot 0.692^w$ Original equation

$= 49.9 \cdot 0.692^1$ $w = 1$

$= 34.5308$ Use a calculator.

On the first weekend, the movie grossed about $34.53 million.

c. How much did it make on the fifth weekend?

$E = 49.9 \cdot 0.692^w$ Original equation

$= 49.9 \cdot 0.692^5$ $w = 5$

≈ 7.918282973 Use a calculator.

On the fifth weekend, the movie grossed about $7.92 million.

CHECK Your Progress

3. **BIOLOGY** A certain bacteria doubles every 20 minutes. How many will there be after 2 hours?

Study Tip

Checking Answers

The graph of an exponential function may resemble part of the graph of a quadratic function. So be sure to check for a pattern as well as looking at a graph.

Identify Exponential Behavior How do you know if a set of data is exponential? One method is to observe the shape of the graph. Another way is to use the problem-solving strategy *look for a pattern.*

EXAMPLE Identify Exponential Behavior

4 Determine whether the set of data at the right displays exponential behavior. Explain why or why not.

x	0	10	20	30	40	50
y	80	40	20	10	5	2.5

Method 1 Look for a Pattern

The domain values are at regular intervals of 10. Look for a common factor among the range values.

80 40 20 10 5 2.5

$\times\frac{1}{2}$ $\times\frac{1}{2}$ $\times\frac{1}{2}$ $\times\frac{1}{2}$ $\times\frac{1}{2}$

Since the domain values are at regular intervals and the range values differ by a common factor, the data are probably exponential. Its equation may involve $\left(\frac{1}{2}\right)^x$.

Method 2 Graph the Data

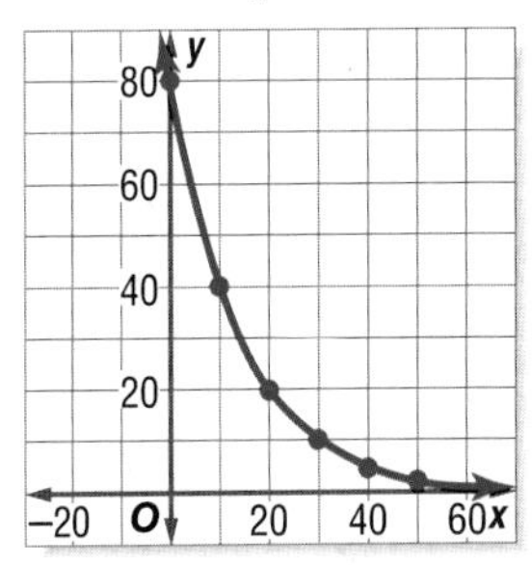

The graph shows a rapidly decreasing value of y as x increases. This is a characteristic of exponential behavior.

CHECK Your Progress

4. Determine whether the set of data displays exponential behavior. Explain why or why not.

x	0	10	20	30	40	50
y	15	21	27	33	39	45

CHECK Your Understanding

Examples 1, 2 (pp. 503–504)

Graph each function. State the y-intercept. Then use the graph to determine the approximate value of the given expression. Use a calculator to confirm the value.

1. $y = 3^x; 3^{1.2}$
2. $y = \left(\frac{1}{4}\right)^x; \left(\frac{1}{4}\right)^{1.7}$
3. $y = 9^x; 9^{0.8}$

Graph each function. State the y-intercept.

4. $y = 2 \cdot 3^x$
5. $y = 4(5^x - 10)$

Example 3 (pp. 504–505)

6. BIOLOGY The function $f(t) = 100 \cdot 1.05^t$ models the growth of a fruit fly population, where $f(t)$ is the number of flies and t is time in days.

a. What values for the domain and range are reasonable in the context of this situation? Explain.

b. After two weeks, approximately how many flies are in this population?

Example 4 (p. 505)

Determine whether the data in each table display exponential behavior. Explain why or why not.

7.

x	0	1	2	3	4	5
y	1	6	36	216	1296	7776

8.

x	4	6	8	10	12	14
y	5	9	13	17	21	25

Exercises

HOMEWORK HELP	
For Exercises	See Examples
9–18	1, 2
19–22	3
23–26	4

Graph each function. State the y-intercept. Then use the graph to determine the approximate value of the given expression. Use a calculator to confirm the value.

9. $y = 5^x; 5^{1.1}$

10. $y = 10^x; 10^{0.3}$

11. $y = \left(\frac{1}{10}\right)^x; \left(\frac{1}{10}\right)^{-1.3}$

12. $y = \left(\frac{1}{5}\right)^x; \left(\frac{1}{5}\right)^{0.5}$

13. $y = 6^x; 6^{0.3}$

14. $y = 8^x; 8^{0.8}$

Graph each function. State the y-intercept.

15. $y = 5(2^x)$

16. $y = 3(5^x)$

17. $y = 3^x - 7$

18. $y = 2^x + 4$

BIOLOGY For Exercises 19 and 20, use the following information.
A population of bacteria in a culture increases according to the model $p = 300 \cdot 2.7^{0.02t}$, where t is the number of hours and $t = 0$ corresponds to 9:00 A.M.

19. Use this model to estimate the number of bacteria at 11 A.M.

20. Graph the function and name the y-intercept. Describe what the y-intercept represents and describe a reasonable domain and range for this situation.

BUSINESS For Exercises 21 and 22, use the following information.
The amount of money spent at West Outlet Mall in Midtown continues to increase. The total $T(x)$ in millions of dollars can be estimated by the function $T(x) = 12(1.12)^x$, where x is the number of years after it opened in 1995.

21. According to the function, find the amount of sales for the mall in the years 2005, 2006, and 2007.

22. Graph the function and name the y-intercept. What does the y-intercept represent in this problem?

Determine whether the data in each table display exponential behavior. Explain why or why not.

23.

x	−2	−1	0	1
y	−5	−2	1	4

24.

x	0	1	2	3
y	1	0.5	0.25	0.125

25.

x	10	20	30	40
y	16	12	9	6.75

26.

x	−1	0	1	2
y	−0.5	1.0	−2.0	4.0

Graph each function. State the y-intercept.

27. $y = 2(3^x) - 1$

28. $y = 2(3^x + 1)$

29. $y = 3(2^x - 5)$

Real-World Link

The first Boston Marathon was held in 1886. The distance of this race was based on the Greek legend that Pheidippides ran 24.8 miles from Marathon to Athens to bring the news of victory over the Persian army.

Source: www.bostonmarathon.org

Identify each function as *linear, quadratic,* or *exponential.*

30. $y = 4^x + 3$ **31.** $y = 2x(x - 1)$ **32.** $5x + y = 8$

33.

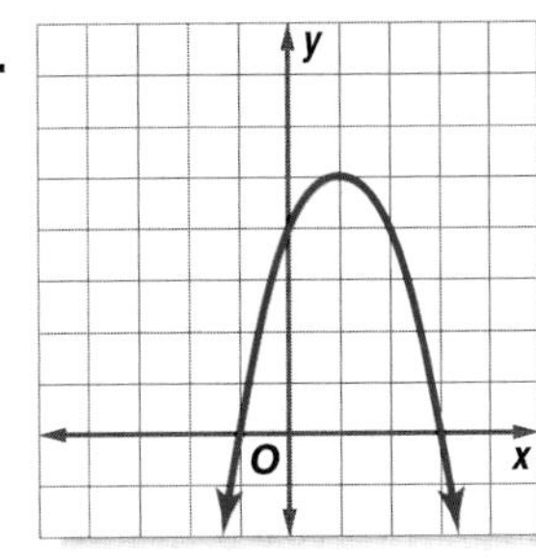

34.

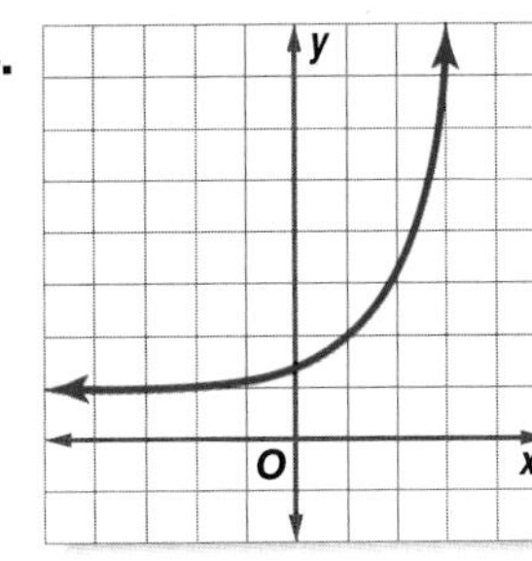

35.

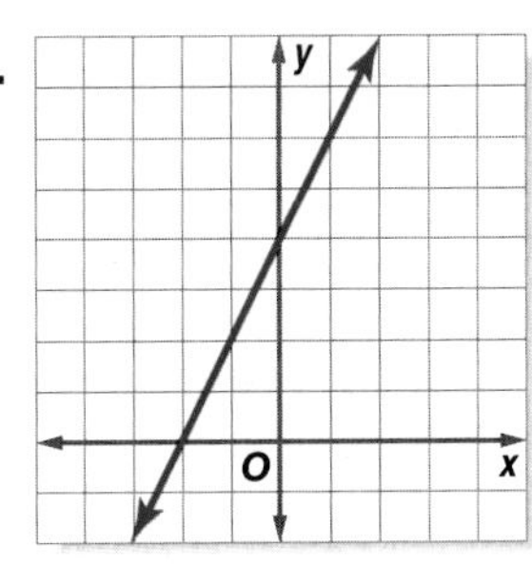

TOURNAMENTS For Exercises 36–38, use the following information.
In a quiz bowl competition, three schools compete, and the winner advances to the next round. Therefore, after each round, only $\frac{1}{3}$ of the schools remain in the competition for the next round. Suppose 729 schools start the competition.

36. Write an exponential function to describe the number of schools remaining after x rounds.

37. How many schools are left after 3 rounds?

38. How many rounds will it take to declare a champion?

ANALYZE TABLES For Exercises 39 and 40, use the following information.
A runner is training for a marathon, running a total of 20 miles per week on a regular basis. She plans to increase the distance $D(x)$ in miles according to the function $D(x) = 20(1.1)^x$, where x represents the number of weeks of training.

EXTRA PRACTICE
See pages 736, 752.
Math Online
Self-Check Quiz at
algebra1.com

39. Copy and complete the table showing the number of miles she plans to run.

40. The runner's goal is to work up to 50 miles per week. What is the first week that the total will be 50 miles or more?

Week	Distance (miles)
1	
2	
3	
4	

H.O.T. Problems

41. REASONING Determine whether the graph of $y = a^x$, where $a > 0$ and $a \neq 1$, *sometimes, always,* or *never* has an x-intercept. Explain your reasoning.

42. OPEN ENDED Choose an exponential function that represents a real-world situation and graph the function. Analyze the graph.

43. FIND THE ERROR Amalia and Hannah are graphing $y = \left(\frac{1}{3}\right)^x$. Who is correct? Explain your reasoning.

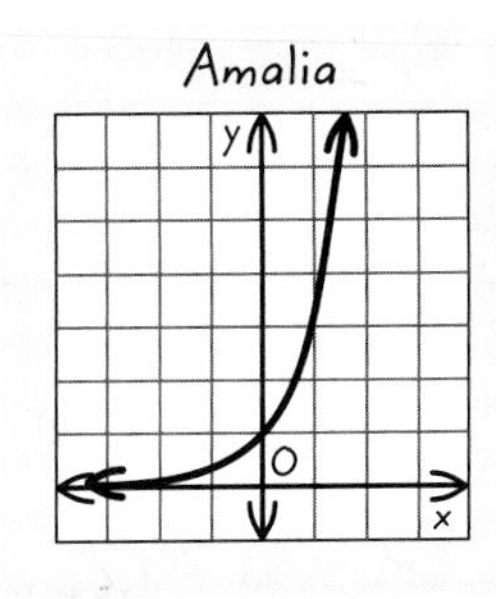

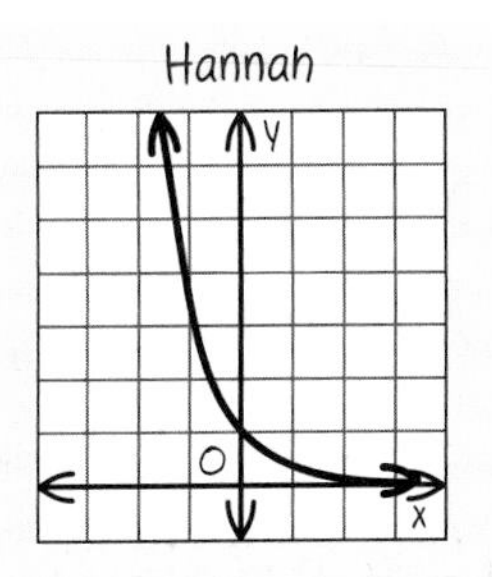

CHALLENGE Describe the graph of each equation as a transformation of the graph of $y = 5^x$.

44. $y = \left(\frac{1}{5}\right)^x$ **45.** $y = 5^x + 2$ **46.** $y = 5^x - 4$

47. ***Writing in Math*** Use the information about the carving on page 502 to explain how exponential functions can be used in art. Include the exponential function representing the pliers, an explanation of which x and y values are meaningful, and the graph of this function.

STANDARDIZED TEST PRACTICE

48. Compare the graphs of $y = 2^x$ and $y = 6^x$.

A The graph of $y = 6^x$ increases at a faster rate than the graph of $y = 2^x$.

B The graph of $y = 2^x$ increases at a faster rate than the graph of $y = 6^x$.

C The graph of $y = 6^x$ is the graph of $y = 2^x$ translated 4 units up.

D The graph of $y = 6^x$ is the graph of $y = 2^x$ translated 3 units up.

49. REVIEW $\triangle KLM$ is similar to $\triangle HIJ$. Which scale factor is used to transform $\triangle KLM$ to $\triangle HIJ$?

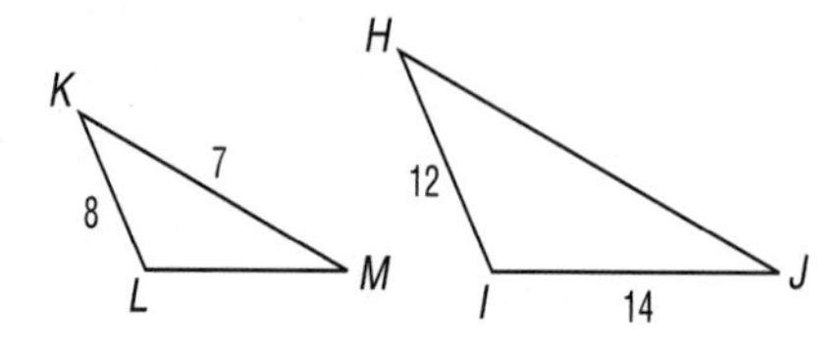

F $\frac{1}{2}$

G 1

H $1\frac{1}{2}$

J 2

Spiral Review

Solve each equation by using the Quadratic Formula. Round to the nearest tenth if necessary. (Lesson 9-4)

50. $x^2 - 9x - 36 = 0$

51. $2t^2 + 3t - 1 = 0$

52. $5y^2 + 3 = y$

Solve each equation by completing the square. Round to the nearest tenth if necessary. (Lesson 9-3)

53. $x^2 - 7x = -10$

54. $a^2 - 12a = 3$

55. $t^2 + 6t + 3 = 0$

Factor each trinomial, if possible. If the trinomial cannot be factored using integers, write *prime*. (Lesson 8-3)

56. $m^2 - 14m + 40$

57. $t^2 - 2t + 35$

58. $z^2 - 5z - 24$

Solve each inequality. (Lesson 6-1)

59. $x + 7 > 2$

60. $10 \geq x + 8$

61. $y - 7 < -12$

62. NUMBER THEORY Three times one number equals twice a second number. Twice the first number is 3 more than the second number. Find the numbers. (Lesson 5-4)

GET READY for the Next Lesson

PREREQUISITE SKILL **Evaluate $p(1 + r)^t$ for each of the given values.** (Lesson 1-1)

63. $p = 5, r = \frac{1}{2}, t = 2$

64. $p = 300, r = \frac{1}{4}, t = 3$

65. $p = 100, r = 0.2, t = 2$

66. $p = 6, r = 0.5, t = 3$

Algebra Lab

Investigating Exponential Functions

ACTIVITY

Step 1 Cut a sheet of notebook paper in half.

Step 2 Stack the two halves, one on top of the other.

Step 3 Make a table like the one at the right and record the number of sheets of paper you have in the stack after one cut.

Number of Cuts	Number of Sheets
0	1
1	
2	

Step 4 Cut the two stacked sheets in half, placing the resulting pieces in a single stack. Record the number of sheets of paper in the new stack after 2 cuts.

Step 5 Continue cutting the stack in half, each time putting the resulting piles in a single stack and recording the number of sheets in the stack. Stop when the resulting stack is too thick to cut.

ANALYZE THE RESULTS

1. Write a list of ordered pairs (x, y), where x is the number of cuts and y is the number of sheets in the stack. Notice that the list starts with the ordered pair $(0, 1)$, which represents the single sheet of paper before any cuts were made.
2. Continue the list beyond the point where you stopped cutting, until you reach the ordered pair for 7 cuts. Explain how you calculated the last y values for your list after you had stopped cutting.
3. Plot the ordered pairs in your list on a coordinate grid. Be sure to choose a scale for the y-axis so that you can plot all of the points.
4. Write a function that expresses y as a function of x.
5. Evaluate the function you wrote in Exercise 4 for $x = 8$ and $x = 9$. Does it give the correct number of sheets in the stack after 8 and 9 cuts?
6. Notebook paper usually stacks about 500 sheets to the inch. How thick would your stack of paper be if you had been able to make 9 cuts?
7. Suppose each cut takes about 5 seconds. If you had been able to keep cutting, you would have made 36 cuts in three minutes. At 500 sheets to the inch, make a conjecture as to how thick you think the stack would be after 36 cuts.
8. Calculate the thickness of your stack after 36 cuts. Write your answer in miles.
9. Use the results of the Activity to complete a table like the one at the right for 0 to 7 cuts. Then write a function to describe the area y after x cuts.

Number of Cuts	Area of Sheet
0	1
1	0.5
2	
...	

9-6 Growth and Decay

Main Ideas

- Solve problems involving exponential growth.
- Solve problems involving exponential decay.

New Vocabulary

exponential growth
compound interest
exponential decay

GET READY for the Lesson

The number of Weblogs or "blogs" increased at a monthly rate of about 13.7% between November 2003 and July 2005. Let y represent the total number of blogs in millions, and let t represent the number of months since November 2003. Then the average number per month can be modeled by $y = 1.1(1 + 0.137)^t$ or $y = 1.1(1.137)^t$.

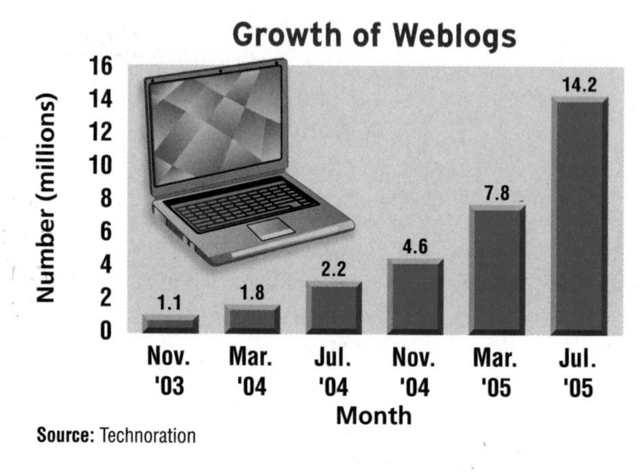

Source: Technoration

Exponential Growth The equation for the number of blogs is in the form $y = C(1 + r)^t$. This is the general equation for **exponential growth** in which the initial amount C increases by the same percent over a given period of time.

> **KEY CONCEPT** *General Equation for Exponential Growth*
>
> The general equation for exponential growth is $y = C(1 + r)^t$ where y represents the final amount, C represents the initial amount, r represents the rate of change expressed as a decimal, and t represents time.

Real-World EXAMPLE Exponential Growth

1 SPORTS In 1971, there were 294,105 females in high school sports. Since then, the number has increased an average of 8.5% per year.

a. Write an equation to represent the number of females participating in high school sports since 1971.

$y = C(1 + r)^t$	General equation for exponential growth
$= 294{,}105(1 + 0.085)^t$	$C = 294{,}105$ and $r = 8.5\%$ or 0.085
$= 294{,}105(1.085)^t$	Simplify.

An equation to represent the number of females participating in high school sports is $y = 294{,}105(1.085)^t$, where y is the number of female athletes and t is the number of years since 1971.

b. According to the equation, how many females participated in high school sports in 2005?

$y = 294{,}105(1.085)^t$	Equation for females participating in sports
$= 294{,}105(1.085)^{34}$	$t = 2005 - 1971$ or 34
$\approx 4{,}711{,}004$	Use a calculator.

In 2005, about 4,711,004 females participated.

CHECK Your Progress

TECHNOLOGY Computer use has risen 19% annually since 1980.

1A. If 18.9 million computers were in use in 1980, write an equation for the number of computers in use for t years after 1980.

1B. Predict the number of computers in 2015.

Online **Personal Tutor at** algebra1.com

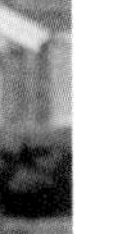

One special application of exponential growth is **compound interest.** The equation for compound interest is $A = P\left(1 + \frac{r}{n}\right)^{nt}$, where A is the current amount of the investment, P is the principal (initial amount of the investment), r represents the annual rate of interest expressed as a decimal, n represents the number of times the interest is compounded each year, and t represents the number of years that the money is invested.

Real-World EXAMPLE Compound Interest

2 COLLEGE Maria's parents invested $14,000 at 6% per year compounded monthly. How much money will there be in 10 years?

$A = P\left(1 + \frac{r}{n}\right)^{nt}$	Compound interest equation
$= 14{,}000\left(1 + \frac{0.06}{12}\right)^{12(10)}$	$P = 14{,}000$, $r = 6\%$ or 0.06, $n = 12$, and $t = 10$
$= 14{,}000(1.005)^{120}$	Simplify.
$\approx 25{,}471.55$	Use a calculator.

There will be about $25,471.55.

Real-World Link

According to the College Board, the 2004–2005 average costs for college were $14,640 for students attending 4-year public colleges and $30,295 for students at 4-year private colleges.

Source: *World Book Encyclopedia*

CHECK Your Progress

2. MONEY Determine the amount of an investment if $300 is invested at an interest rate of 3.5% compounded monthly for 22 years.

Exponential Decay A variation of the growth equation can be used as the general equation for exponential decay. In **exponential decay,** the original amount decreases by the same percent over a period of time.

KEY CONCEPT ***General Equation for Exponential Decay***

The general equation for exponential decay is $y = C(1 - r)^t$, where y represents the final amount, C represents the initial amount, r represents the rate of decay expressed as a decimal, and t represents time.

Real-World EXAMPLE Exponential Decay

3 **SWIMMING A fully inflated raft loses 6.6% of its air every day. The raft originally contains 4500 cubic inches of air.**

a. Write an equation to represent the loss of air.

$y = C(1 - r)^t$ General equation for exponential decay

$= 4500(1 - 0.066)^t$ $C = 4500$ and $r = 6.6\%$ or 0.066

$= 4500(0.934)^t$ Simplify.

An equation to represent the loss of air is $y = 4500(0.934)^t$, where y represents the amount of air in the raft in cubic inches and t represents the number of days.

b. Estimate the amount of air that will be lost after 7 days.

$y = 4500(0.934)^t$ Equation for air loss

$= 4500(0.934)^7$ $t = 7$

≈ 2790 Use a calculator.

The amount of air lost after 7 days will be about 2790 cubic inches.

CHECK Your Progress

POPULATION During the past several years, the population of Campbell County, Kentucky, has been decreasing at an average rate of about 0.3% per year. In 2000, its population was 88,647.

3A. Write an equation to represent the population since 2000.

3B. If the trend continues, predict the population in 2010.

CHECK Your Understanding

Example 1 (pp. 510–511)

ANALYZE GRAPHS For Exercises 1 and 2, use the graph at the right and the following information.
The median house price in the United States increased an average of 8.9% each year between 2002 and 2004. Assume this pattern continues.

1. Write an equation for the median house price for t years after 2004.

2. Predict the median house price in 2009.

Median House Price	
2002	$187,600
2003	$195,000
2004	$221,000

Source: RealEstateJournal.com

Example 2 (p. 511)

3. INVESTMENTS Determine the amount of an investment if $400 is invested at an interest rate of 7.25% compounded quarterly for 7 years.

Example 3 (p. 512)

4. POPULATION In 1995, the population of West Virginia reached 1,821,000, its highest in the 20th century. During the rest of the 20th century, its population decreased 0.2% each year. If this trend continues, predict the population of West Virginia in 2010.

Exercises

HOMEWORK HELP

For Exercises	See Examples
5–8	1
9, 10	2
11, 12	3

WEIGHT TRAINING For Exercises 5 and 6, use the following information.
In 1997, there were 43.2 million people who used free weights.

5. Assuming the use of free weights increases 6% annually, write an equation for the number of people using free weights t years from 1997.

6. Predict the number of people using free weights in 2007.

7. POPULATION The population of Mexico has been increasing 1.7% annually. If the population was 100,350,000 in 2000, predict the population in 2012.

8. ANALYZE GRAPHS The increase in the number of visitors to the Grand Canyon National Park is similar to an exponential function. If the average visitation has increased 5.63% annually since 1920, predict the number of visitors to the park in 2020.

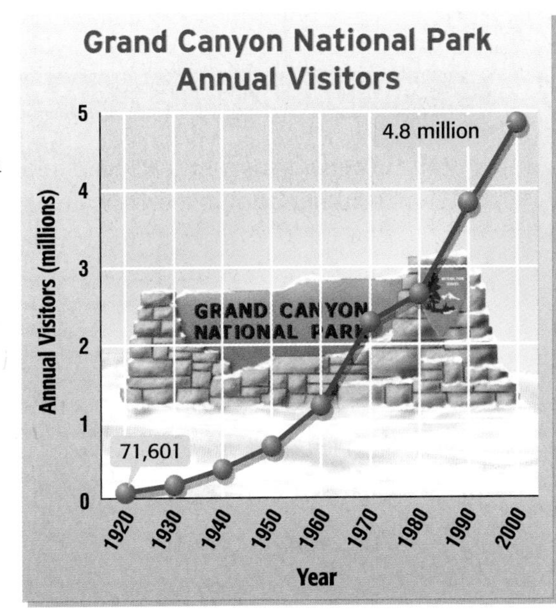

Source: Grand Canyon National Park

9. INVESTMENTS Determine the amount of an investment if \$500 is invested at an interest rate of 5.75% compounded monthly for 25 years.

10. INVESTMENTS Determine the amount of an investment if \$250 is invested at an interest rate of 7.3% compounded quarterly for 40 years.

11. POPULATION The country of Latvia has been experiencing a 1.1% annual decrease in population. In 2005, its population was 2,290,237. If the trend continues, predict Latvia's population in 2015.

12. MUSIC In 1994, the sales of music cassettes reached its peak at \$2,976,400,000. Since then, cassette sales have been declining. If the annual percent of decrease in sales is 18.6%, predict the sales of cassettes in the year 2009.

ARCHAEOLOGY For Exercises 13–15, use the following information.
The *half-life* of a radioactive element is defined as the time that it takes for one-half a quantity of the element to decay. Radioactive Carbon-14 is found in all living organisms and has a half-life of 5730 years. Archaeologists use this information to estimate the age of fossils. Consider a living organism with an original Carbon-14 content of 256 grams. The number of grams remaining in the fossil of the organism after t years would be $256(0.5)^{\frac{t}{5730}}$.

13. If the organism died 5730 years ago, what is the amount of Carbon-14 today?

14. If an organism died 10,000 years ago, what is the amount of Carbon-14 today?

15. If the fossil has 32 grams of Carbon-14 remaining, how long ago did it live? (*Hint:* Make a table.)

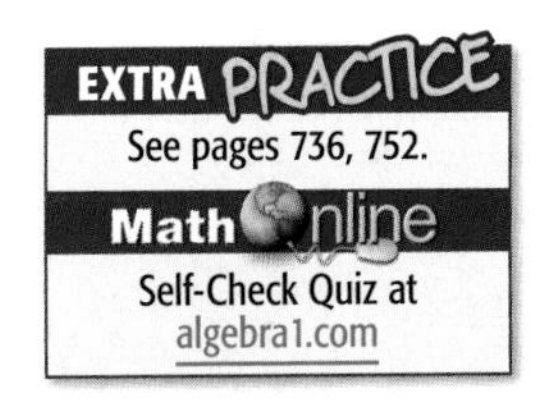

16. **RESEARCH** Find the enrollment of your school district each year for the last decade. Find the rate of change from one year to the next. Then, determine the average annual rate of change for those years. Use this information to estimate the enrollment for your school district in ten years.

H.O.T. Problems

17. **OPEN ENDED** Create a compound interest problem that could be solved by the equation $A = 500\left(1 + \frac{0.07}{4}\right)^{4(6)}$.

18. ***Writing in Math*** Use the information about Weblogs on page 510 to explain how exponential growth can be used to predict future blogs. Include an explanation of the equation $y = 1.1(1 + 0.137)^t$ and an estimate of the number of blogs in January 2010.

STANDARDIZED TEST PRACTICE

19. Lorena is investing a $5000 inheritance from her aunt in a certificate of deposit that matures in 4 years. The interest rate is 6.25% compounded quarterly. What is the balance of the account after 4 years?

A $5078.13

B $5319.90

C $5321.82

D $6407.73

20. **REVIEW** Diego is building a 10-foot ramp for loading heavy equipment into the back of a semi-truck. If the floor of the truck is 3.5 feet off the ground, about how far from the truck should the ramp be?

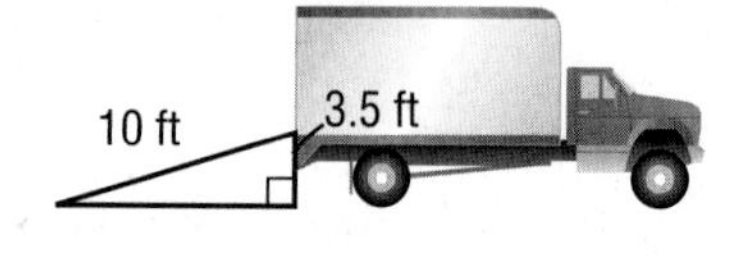

F 9 ft

G 10 ft

H 10.6 ft

J 11 ft

Spiral Review

Graph each function. State the y-intercept. (Lesson 9-5)

21. $y = \left(\frac{1}{8}\right)^x$

22. $y = 2^x - 5$

23. $y = 4(3^x - 6)$

Solve each equation by using the Quadratic Formula. Round to the nearest tenth if necessary. (Lesson 9-4)

24. $m^2 - 9m - 10 = 0$

25. $2t^2 - 4t = 3$

26. $7x^2 + 3x + 1 = 0$

27. **SKIING** A course for cross-country skiing is regulated so that the slope of any hill cannot be greater than 0.33. A hill rises 60 meters over a horizontal distance of 250 meters. Does the hill meet the requirements? Explain. (Lesson 4-1)

Cross-Curricular Project

Algebra and Science

Out of This World It is time to complete your project. Use the infomation and data you have gathered about the solar system to prepare a brochure, poster, or Web page. Be sure to include the three graphs, tables, diagrams, or calculations in the presentation.

Math Online Cross-Curricular Project at algebra1.com

EXTEND 9-6

Graphing Calculator Lab
Curve Fitting

Concepts in Motion
Interactive Lab algebra1.com

If there is a constant increase or decrease in data values, there is a linear trend. If the values are increasing or decreasing more and more rapidly, there may be a quadratic or exponential trend.

Linear Trend

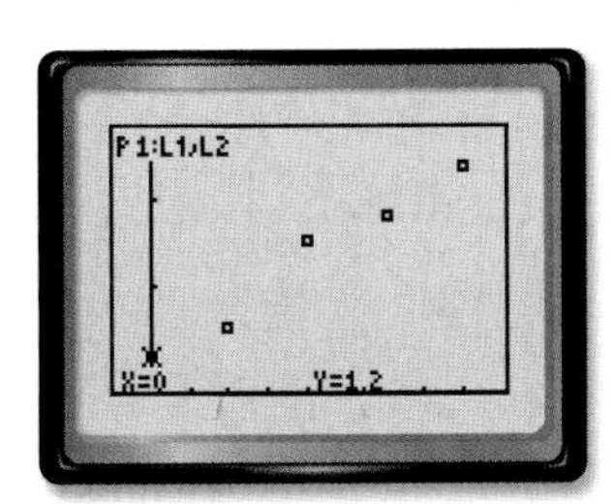

Quadratic Trend

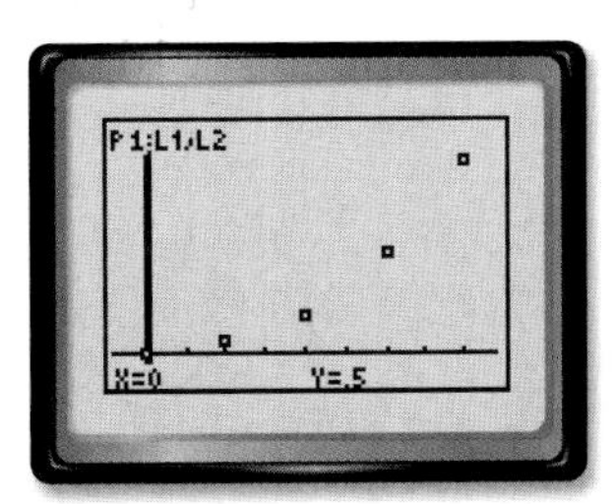

Exponential Trend

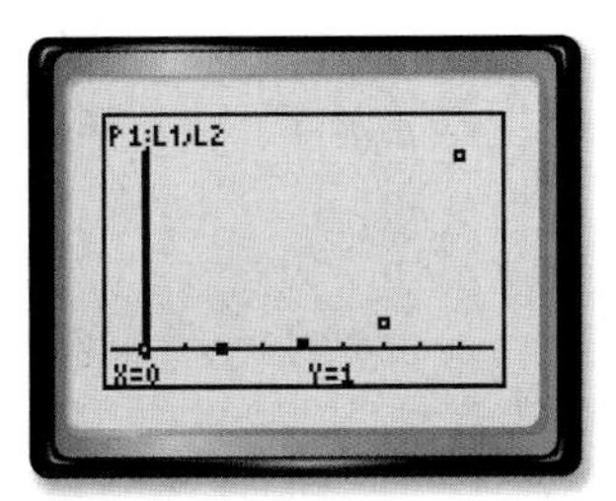

With a TI-83/84 Plus, you can find the appropriate regression equation.

ACTIVITY 1

FARMING A study is conducted in which groups of 25 corn plants are given a different amount of fertilizer and the gain in height after a certain time is recorded. The table below shows the results.

Fertilizer (mg)	0	20	40	60	80
Gain in Height (in.)	6.48	7.35	8.73	9.00	8.13

Step 1 Make a scatter plot.

- Enter the fertilizer in L1 and the height in L2.

KEYSTROKES: *Review entering a list on page 234.*

- Use STAT PLOT to graph the scatter plot.

KEYSTROKES: *Review statistical plots on page 234. Use* ZOOM *9 to graph.*

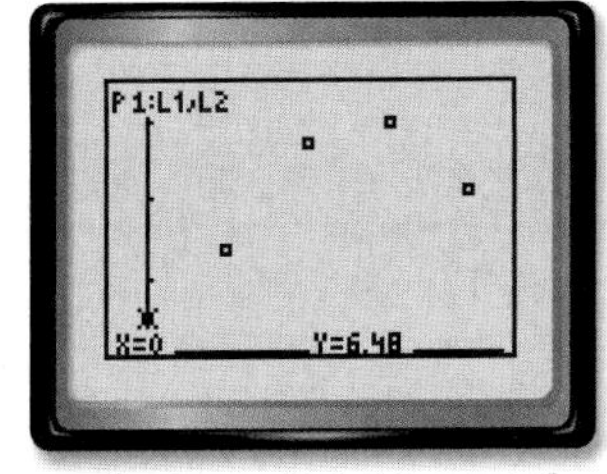

[−8, 88] scl: 5 by [6.0516, 9.4284] scl: 1

The graph appears to be a quadratic regression.

Step 2 Find the regression equation.

- Select DiagnosticOn from the CATALOG.
- Select QuadReg on the STAT CALC menu.

KEYSTROKES: STAT ▶ 5 ENTER

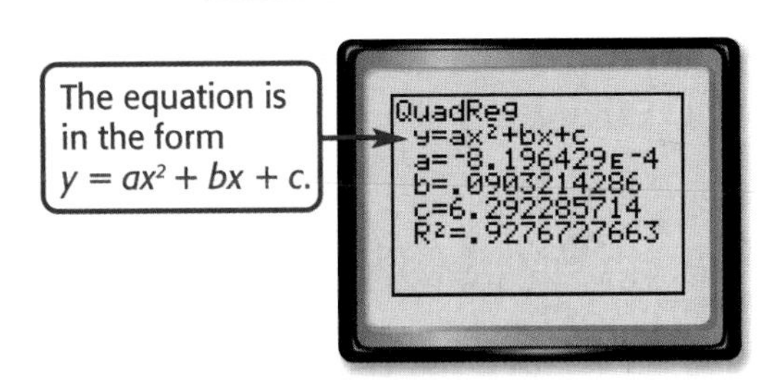

The equation is about $y = -0.0008x^2 + 0.1x + 6.3$.

R^2 is the **coefficient of determination**. The closer R^2 is to 1, the better the model. To choose a quadratic or exponential model, fit both and use the one with the R^2 value closer to 1.

Step 3 Graph the regression equation.

- Copy the equation to the Y= list and graph.

KEYSTROKES: Y= VARS 5 ▶ ▶ 1 ZOOM 9

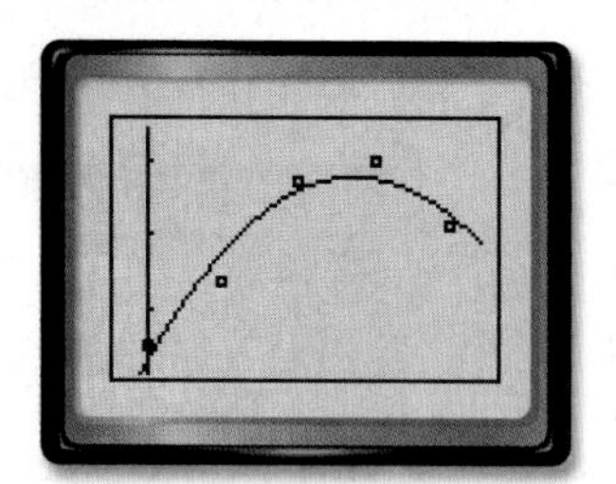

[−8, 88] scl: 5 by [6.0516, 9.4284] scl: 1

Step 4 Predict using the equation.

- Find the amount of fertilizer that produces the maximum gain in height.

KEYSTROKES: 2nd [CALC] 4

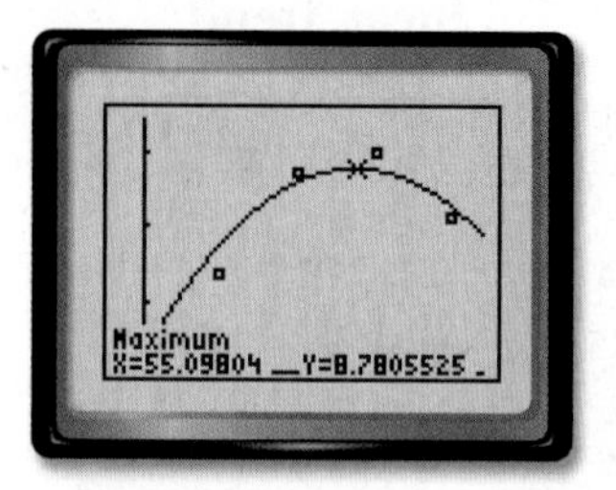

[−8, 88] scl: 5 by [6.0516, 9.4284] scl: 1

According to the graph, on average about 55 milligrams of the fertilizer produces the maximum gain.

Exercises

Plot each set of data points. Determine whether to use a *linear, quadratic,* or *exponential* regression equation. State the coefficient of determination.

1.

x	y
0.0	2.98
0.2	1.46
0.4	0.90
0.6	0.51
0.8	0.25
1.0	0.13

2.

x	y
1	25.9
2	22.2
3	20.0
4	19.3
5	18.2
6	15.9

3.

x	y
10	35
20	50
30	70
40	88
50	101
60	120

4.

x	y
1	3.67
3	5.33
5	6.33
7	5.67
9	4.33
11	2.67

TECHNOLOGY DVD players were introduced in 1997. For Exercises 5–8, use the table at the right.

5. Make a scatter plot of the data.

6. Find an appropriate regression equation, and state the coefficient of determination.

7. Use the regression equation to predict the number of DVD players that will sell in 2008.

8. Do you believe your equation would be accurate for a year beyond the range of the data, such as 2020? Explain.

Year	DVD Players Sold (millions)
1997	0.32
1998	1.09
1999	4.02
2000	8.50
2001	12.71
2002	17.09
2003	21.99

Source: Consumer Electronics Association

CHAPTER 9 Study Guide and Review

FOLDABLES™ Study Organizer — GET READY to Study

Be sure the following Key Concepts are noted in your Foldable.

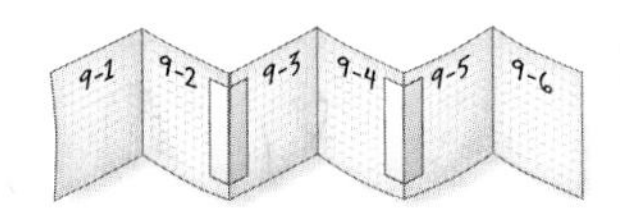

Key Concepts

Graphing Quadratic Functions (Lesson 9-1)

- A quadratic function can be described by an equation of the form $y = ax^2 + bx + c$, where $a \neq 0$.
- The axis of symmetry for the graph of $y = ax^2 + bx + c$, where $a \neq 0$, is $x = -\frac{b}{2a}$.

Solving Quadratic Equations (Lessons 9-2, 9-3, and 9-4)

- The solutions of a quadratic equation are called the roots of the equation. They are the x-intercepts or zeros of the related quadratic function.
- Quadratic equations can be solved by completing the square. To complete the square for $x^2 + bx$, find $\frac{1}{2}$ of b, square this result, and then add the final result to $x^2 + bx$.
- The solutions of a quadratic equation can be found by using the Quadratic Formula $x = \frac{-b \pm \sqrt{b^2 - 4ac}}{2a}$.

Exponential Functions (Lessons 9-5 and 9-6)

- An exponential function can be described by an equation of the form $y = a^x$, where $a > 0$ and $a \neq 1$.
- The general equation for exponential growth is $y = C(1 + r)^t$ and the general equation for exponential decay is $y = C(1 - r)^t$, where y = the final amount, C = the initial amount, r = the rate of change, and t = the time.

Key Vocabulary

axis of symmetry (p. 472)
completing the square (p. 487)
compound interest (p. 511)
discriminant (p. 496)
exponential function (p. 502)
general equation for exponential decay (p. 511)
general equation for exponential growth (p. 510)
maximum (p. 472)
minimum (p. 472)
parabola (p. 471)
quadratic equation (p. 480)
Quadratic Formula (p. 493)
quadratic function (p. 471)
roots (p. 480)
symmetry (p. 472)
vertex (p. 472)
zeros (p. 480)

Vocabulary Check

State whether each sentence is *true* or *false*. If *false*, replace the underlined word or phrase to make a true sentence.

1. The graph of a quadratic function is a <u>parabola</u>.
2. The solutions of a quadratic equation are called <u>roots</u>.
3. The <u>zeros</u> of a quadratic function can be found by using the equation $x = -\frac{b}{2a}$.
4. The <u>vertex</u> is the maximum or minimum point of a parabola.
5. The <u>exponential decay</u> equation is $y = C(1 + r)^t$.
6. An example of a <u>quadratic function</u> is $y = 8^x$.
7. <u>Symmetry</u> is a geometric property possessed by parabolas.
8. The graph of a quadratic function has a <u>minimum</u> if the coefficient of the x^2 term is negative.
9. The expression $b^2 - 4ac$ is called the <u>discriminant</u>.
10. A quadratic equation whose graph has two x-intercepts has <u>no</u> real roots.

Math Online Vocabulary Review at algebra1.com

Study Guide and Review

Lesson-by-Lesson Review

9–1 Graphing Quadratic Functions (pp. 471–477)

Write the equation of the axis of symmetry, and find the coordinates of the vertex of the graph of each function. Identify the vertex as a maximum or minimum. Then graph the function.

11. $y = x^2 + 2x$ **12.** $y = -3x^2 + 4$

13. $y = x^2 - 3x - 4$ **14.** $y = 3x^2 + 6x - 17$

15. $y = -2x^2 + 1$ **16.** $y = -x^2 - 3x$

PHYSICAL SCIENCE **For Exercises 17–20, use the following information.**

A model rocket is launched with a velocity of 64 feet per second. The equation $h = -16t^2 + 64t$ gives the height of the rocket t seconds after it is launched.

17. Write the equation of the axis of symmetry and find the coordinates of the vertex.

18. Graph the function.

19. What is the maximum height that the rocket reaches?

20. How many seconds is the rocket in the air?

Example 1 **Consider the graph of $y = x^2 - 8x + 12$.**

a. Write the equation of the axis of symmetry.

$x = -\frac{b}{2a}$

$x = -\frac{-8}{2(1)}$

$x = 4$

The equation of the axis of symmetry is $x = 4$.

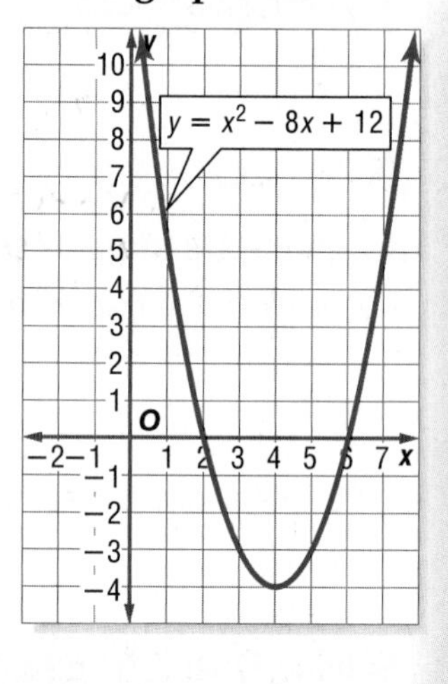

b. Find the coordinates of the vertex.

The x-coordinate for vertex is 4.

$y = x^2 - 8x + 12$	Original equation
$= (4)^2 - 8(4) + 12$	$x = 4$
$= 16 - 32 + 12$	Simplify.
$= -4$	Simplify.

The coordinates of the vertex are $(4, -4)$.

9–2 Solving Quadratic Equations by Graphing (pp. 480–485)

Solve each equation by graphing. If integral roots cannot be found, estimate the roots by stating the consecutive integers between which the roots lie.

21. $x^2 - x - 12 = 0$ **22.** $-x^2 + 6x - 9 = 0$

23. $x^2 + 4x - 3 = 0$ **24.** $2x^2 - 5x + 4 = 0$

25. $x^2 - 10x = -21$ **26.** $6x^2 - 13x = 15$

27. NUMBER THEORY Use a quadratic equation to find two numbers whose sum is 5 and whose product is -24.

Example 2 **Solve $x^2 - 3x - 4 = 0$ by graphing.**

Graph the related function $f(x) = x^2 - 3x - 4$.

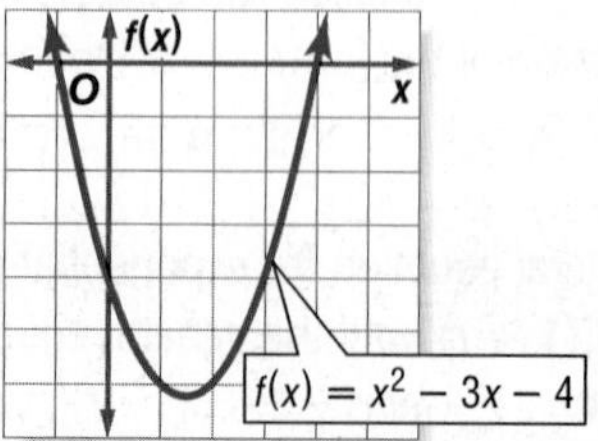

The x-intercepts are -1 and 4. Therefore, the solutions are -1 and 4.

Mixed Problem Solving
For mixed problem-solving practice, see page 752.

9-3 Solving Quadratic Equations by Completing the Square (pp. 486–491)

Solve each equation by taking the square root of each side. Round to the nearest tenth if necessary.

28. $a^2 + 6a + 9 = 4$

29. $n^2 - 2n + 1 = 25$

Solve each equation by completing the square. Round to the nearest tenth if necessary.

30. $-3x^2 + 4 = 0$

31. $x^2 - 16x + 32 = 0$

32. $m^2 - 7m = 5$

33. GEOMETRY The area of a square can be tripled by increasing the length by 6 centimeters and the width by 3 centimeters. What is the length of the side of the square?

Example 3 **Solve $y^2 + 6y + 2 = 0$ by completing the square. Round to the nearest tenth if necessary.**

Step 1 Isolate the y^2 and y terms.

$y^2 + 6y + 2 = 0$ Original equation

$y^2 + 6y = -2$ Subtract 2 from each side.

Step 2 Complete the square and solve.

$y^2 + 6y + 9 = -2 + 9$ $\left(\frac{6}{2}\right)^2 = 9$; add 9 to each side.

$(y + 3)^2 = 7$ Factor $y^2 + 6y + 9$.

$y + 3 = \pm\sqrt{7}$ Take the square root of each side.

$y = -3 \pm \sqrt{7}$ Subtract 3 from each side.

The solutions are about -5.6 and -0.4.

9-4 Solving Quadratic Equations by Using the Quadratic Formula (pp. 493–499)

Solve each equation by using the Quadratic Formula. Round to the nearest tenth if necessary.

34. $x^2 - 8x = 20$

35. $r^2 + 10r + 9 = 0$

36. $4p^2 + 4p = 15$

37. $2y^2 + 3 = -8y$

38. $2d^2 + 8d + 3 = 3$

39. $21a^2 + 5a - 7 = 0$

40. ENTERTAINMENT A stunt person attached to a safety harness drops from a height of 210 feet. A function that models the drop is $h = -16t^2 + 210$, where h is the height in feet and t is the time in seconds. About how many seconds does it take to drop from 210 feet to 30 feet?

Example 4 **Solve $2x^2 + 7x - 15 = 0$ by using the Quadratic Formula.**

For this equation, $a = 2$, $b = 7$, and $c = -15$.

$x = \frac{-b \pm \sqrt{b^2 - 4ac}}{2a}$ Quadratic Formula

$= \frac{-7 \pm \sqrt{7^2 - 4(2)(-15)}}{2(2)}$ $a = 2$, $b = 7$, and $c = -15$

$= \frac{-7 \pm \sqrt{169}}{4}$ Simplify.

$= \frac{-7 + 13}{4}$ or $\frac{-7 - 13}{4}$ Separate the solutions.

$x = 1.5$ or $x = -5$ Simplify.

The solutions are -5 and 1.5.

9-5 Exponential Functions (pp. 502–508)

Graph each function. State the y-intercept. Then use the graph to determine the approximate value of the given expression. Use a calculator to confirm the value.

41. $y = 5^x; 5^{0.7}$

42. $y = \left(\frac{1}{6}\right)^x; \left(\frac{1}{6}\right)^{0.2}$

Graph each function. State the y-intercept.

43. $y = 3^x + 6$

44. $y = 3^{x+2}$

45. BIOLOGY The population of bacteria in a Petri dish increases according to the model $p = 550 \cdot 2.7^{0.008t}$, where t is the number of hours and $t = 0$ corresponds to 1:00 P.M. Use this model to estimate the number of bacteria in the dish at 5 P.M.

Example 5 **Graph $y = 2^x - 3$. State the y-intercept.**

x	$2^x - 3$	y
−3	$2^{-3} - 3$	−2.875
−1	$2^{-1} - 3$	−2.5
0	$2^0 - 3$	−2
1	$2^1 - 3$	−1
2	$2^2 - 3$	1
3	$2^3 - 3$	5

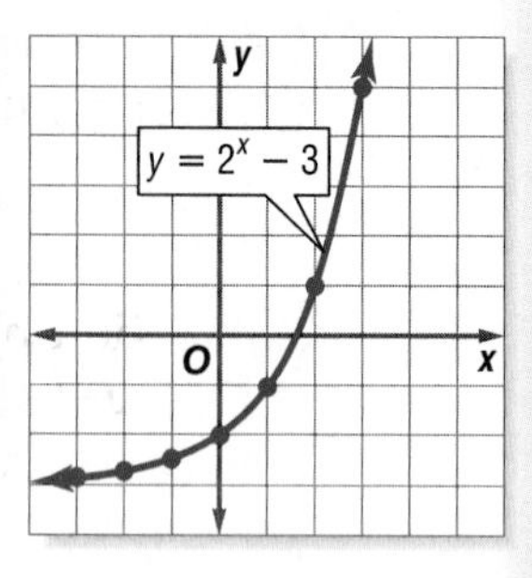

Graph the ordered pairs and connect the points with a smooth curve. The y-intercept is −2.

9-6 Growth and Decay (pp. 510–514)

Determine the final amount for each investment.

	Principal	Annual Interest Rate	Time (yr)	Type of Compounding
46.	\$2000	3%	8	quarterly
47.	\$5500	2.25%	15	monthly
48.	\$15,000	2.5%	25	monthly
49.	\$500	1.75%	40	daily

RESTAURANTS **For Exercises 50 and 51, use the following information.**

The total restaurant sales in the United States increased at an annual rate of about 5.2% between 1996 and 2004. In 1996, the total sales were \$310 billion.

50. Write an equation for the average sales per year for t years after 1996.

51. Predict the total restaurant sales in 2008.

Example 6 **Find the final amount of an investment if \$1500 is invested at an interest rate of 2.5% compounded quarterly for 10 years.**

$A = p\left(1 + \frac{r}{n}\right)^{nt}$ Compound interest equation

$= 1500\left(1 + \frac{0.025}{4}\right)^{4(10)}$ $P = 1500, r = 2.5\%$ or $0.025, n = 4$, and $t = 10$

≈ 1924.54 Simplify.

The final amount in the account is about \$1924.54.

CHAPTER 9 Practice Test

Write the equation of the axis of symmetry, and find the coordinates of the vertex of the graph of each function. Identify the vertex as a maximum or minimum. Then graph the function.

1. $y = x^2 - 4x + 13$

2. $y = -3x^2 - 6x + 4$

3. $y = 2x^2 + 3$

4. $y = -1(x - 2)^2 + 1$

Solve each equation by graphing. If integral roots cannot be found, estimate the roots by stating the consecutive integers between which the roots lie.

5. $x^2 - 2x + 2 = 0$

6. $x^2 + 6x = -7$

7. $x^2 + 24x + 144 = 0$

8. $2x^2 - 8x = 42$

9. MULTIPLE CHOICE Which function is graphed below?

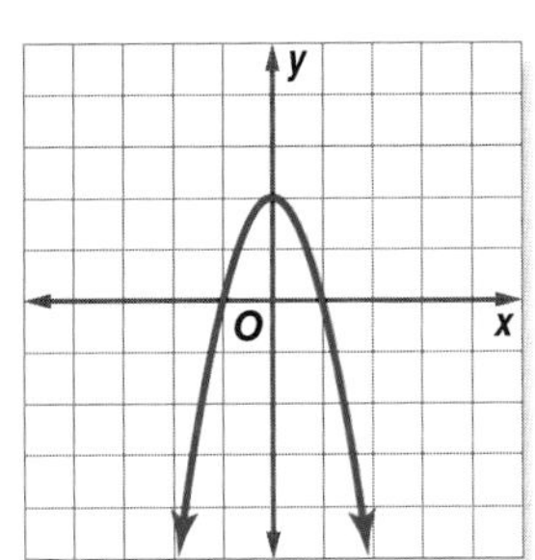

A $y = 2x^2 - 2$

B $y = 2x^2 + 2$

C $y = -2x^2 - 2$

D $y = -2x^2 + 2$

Solve each equation. Round to the nearest tenth if necessary.

10. $x^2 + 7x + 6 = 0$

11. $2x^2 - 5x - 12 = 0$

12. $6n^2 + 7n = 20$

13. $3k^2 + 2k = 5$

14. $y^2 - \frac{3}{5}y + \frac{2}{25} = 0$

15. $-3x^2 + 5 = 14x$

16. $z^2 - 13z = 32$

17. $7m^2 = m + 5$

18. MULTIPLE CHOICE Which equation best represents the parabola graphed below if it is shifted 3 units to the right?

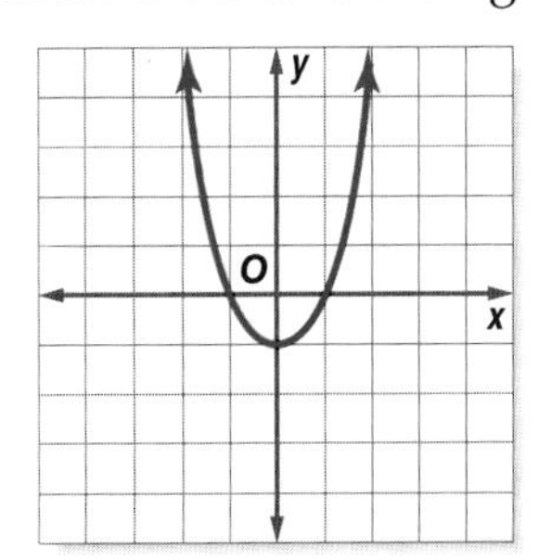

F $y = x^2 - 1$

G $y = x^2 + 2$

H $y = x^2 - 6x + 8$

J $y = x^2 + 6x + 8$

Graph each function. State the y-intercept.

19. $y = \left(\frac{1}{2}\right)^x$

20. $y = 4 \cdot 2^x$

21. $y = 0.5(4^x)$

22. $y = 5^x - 4$

23. Graph $y = \left(\frac{1}{3}\right)^x - 3$ and state the y-intercept. Then use the graph to determine the approximate value of $y = \left(\frac{1}{3}\right)^{3.5} - 3$. Use a calculator to confirm the value.

24. CARS Ley needs to replace her car. If she leases a car, she will pay $410 a month for 2 years and then has the option to buy the car for $14,458. The current price of the car is $17,369. If the car depreciates at 16% per year, how will the depreciated price compare with the buy-out price of the lease?

25. FINANCE Find the total amount of the investment shown in the table if interest is compounded quarterly.

Principal	$1500
Length of Investment	10 yr
Annual Interest Rate	6%

CHAPTER 9 Standardized Test Practice

Cumulative, Chapters 1–9

Read each question. Then fill in the correct answer on the answer document provided by your teacher or on a sheet of paper.

1. In the graph of the function $f(x) = x^2 - 3$, which describes the shift in the vertex of the parabola if, in the function, 3 is changed to 5?

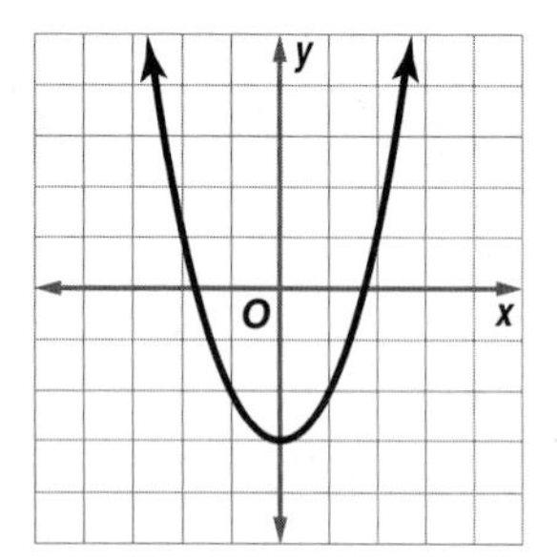

A 2 units up

B 8 units up

C 2 units down

D 8 units down

2. The area of a rectangle is given by the equation $3\ell^2 + 10\ell = 25$, in which ℓ is the rectangle's length. What is the length of the rectangle?

$3\ell^2 + 10\ell = 25$

ℓ

F $\frac{3}{2}$

G $\frac{5}{3}$

H $\frac{10}{3}$

J 5

3. Solve the equation $x(x + 2) = 224$ which represents two consecutive even integers whose product is 224. What is the smaller number?

A 2   C 16

B 14   D 24

4. Ben calculated the cost $f(x)$ to make and print x T-shirts in one month according to the function

$$f(x) = 9x - 3.00(x - 100),$$

where $x > 100$ T-shirts. The best interpretation for this function is that it costs Ben—

F \$9 for any number of T-shirts.

G \$6 for all shirts up to 100.

H \$3 less per T-shirt for all shirts over 100.

J \$3 for all shirts over 100.

5. What is the solution set of $3y^2 = 12$?

A $\{3\}$

B $\{-2, 2\}$

C $\{-2, 2, 3\}$

D $\{-2\}$

TEST-TAKING TIP

Question 5 Always write down your calculations on scrap paper or in the test booklet, even if you think you can do the calculations in your head. Writing down your calculations will help you avoid making simple mistakes.

6. Jennie hit a baseball straight up in the air. The height h, in feet, of the ball above the ground is modeled by the equation $h = -16t^2 + 64t$. How long is the ball above ground?

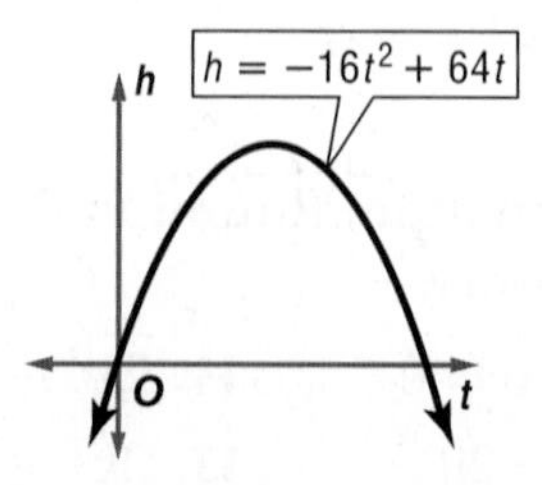

F 1 second

G 2 seconds

H 4 seconds

J 16 seconds

7. Lisa has a savings account. She withdraws half of the contents every year. After 4 years she has \$2000 left. How much did she have in the savings account originally?

A \$32,000

B \$16,000

C \$8,000

D \$2,000

8. When is this statement true?

The multiplicative inverse of a number is less than the original number.

F This statement is never true.

G This statement is always true.

H This statement is true for numbers greater than 1.

J This statement is true for numbers less than −1.

9. GRIDDABLE What is the slope of the line graphed below?

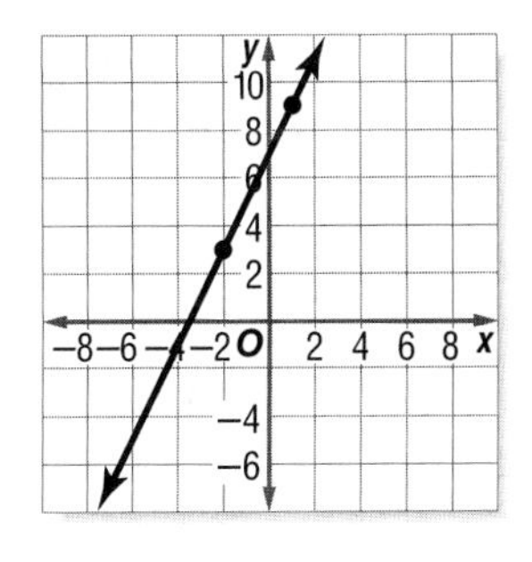

10. Which inequality is shown on the graph below?

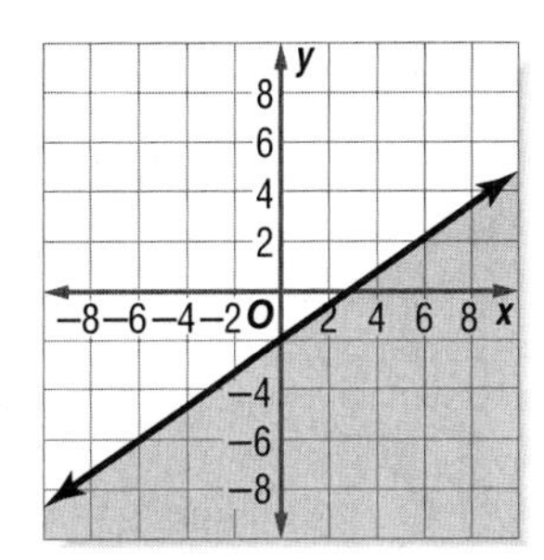

F $y \leq \frac{2}{3}x - 2$

G $y < \frac{2}{3}x - 2$

H $y \geq \frac{2}{3}x - 2$

J $y > \frac{2}{3}x - 2$

11. Mr. Collins made a deck that was 12 feet long. He is making a new deck that is 25% longer. What will be the length of the new deck?

A 9 feet

B 12 feet

C 15 feet

D 18 feet

Pre-AP

Record your answers on a sheet of paper. Show your work.

12. The path of Annika Sorenstam's golf ball through the air is modeled by the equation $h = -16t^2 + 80t + 1$, where h is the height in feet of the golf ball and t is the time in seconds.

a. Approximately how long was the ball in the air?

b. How high was the ball at its highest point?

NEED EXTRA HELP?												
If You Missed Question...	1	2	3	4	5	6	7	8	9	10	11	12
Go to Lesson or Page...	9-1	9-4	8-3	3-2	8-5	8-2	9-6	4-1	1-4	6-6	3-7	9-2

UNIT 4

Advanced Expressions and Data Analysis

Focus

Use a variety of representations, tools, and technology to model mathematical situations to solve meaningful problems.

CHAPTER 10
Radical Expressions and Triangles

BIG Idea Use algebraic skills to simplify radical expressions and solve equations in problem situations.

CHAPTER 11 Rational Expressions and Equations

BIG Idea Use algebraic skills to simplify rational expressions and solve equations.

CHAPTER 12
Statistics and Probability

BIG Idea Use graphical and numerical techniques to study patterns and analyze data.

BIG Idea Use probability models to describe everyday situations involving chance.

Cross-Curricular Project

Algebra and Physical Science

Building the Best Roller Coaster Each year, amusement park owners compete to earn part of the billions of dollars Americans spend at amusement parks. Often, the parks draw customers with new, taller, and faster roller coasters. In this project, you will examine how radical and rational functions are related to buying and building a new roller coaster and analyze data involving U.S. amusement parks.

Math Online Log on to algebra1.com to begin.

CHAPTER 10

Radical Expressions and Triangles

BIG Ideas

- Simplify and perform operations with radical expressions.
- Solve radical equations.
- Use the Pythagorean Theorem and Distance Formula.
- Use similar triangles and trigonometric ratios.

Key Vocabulary

Distance Formula (p. 555)
Pythagorean triple (p. 550)
radical equation (p. 541)
radical expression (p. 528)

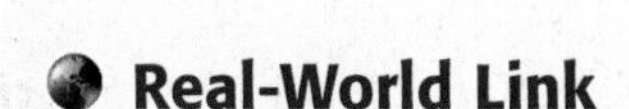

Real-World Link

Skydiving Physics problems are among the many applications of radical equations. Formulas that contain the value for the acceleration due to gravity, such as free-fall times, can be written as radical equations.

Radical Expressions and Triangles Make this Foldable to help you organize your notes. Begin with one $8\frac{1}{2}''$ by $11''$ paper.

1. **Fold** in half matching the short sides.

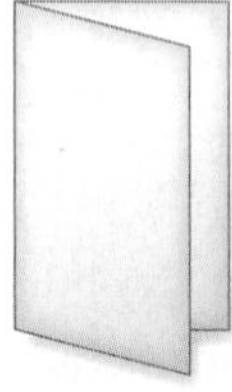

2. **Unfold** and fold the long side up 2 inches to form a pocket.

3. **Staple** or glue the outer edges to complete the pocket.

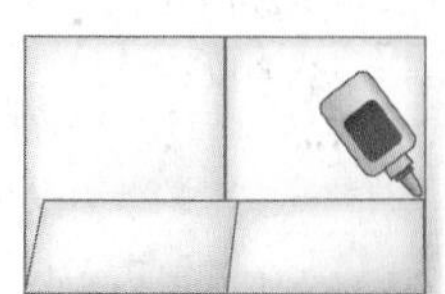

4. **Label** each side as shown. Use index cards to record examples.

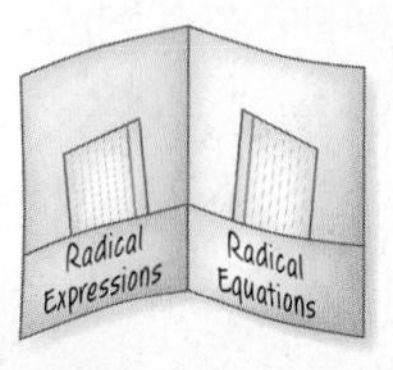

GET READY for Chapter 10

Diagnose Readiness You have two options for checking Prerequisite Skills.

Option 2

Math Online Take the Online Readiness Quiz at algebra1.com.

Option 1

Take the Quick Check below. Refer to the Quick Review for help.

QUICKCheck

Find each square root. If necessary, round to the nearest hundredth. (Lesson 1-8)

1. $\sqrt{25}$ **2.** $\sqrt{80}$ **3.** $\sqrt{56}$

4. PAINTING Todd is painting a square mural with an area of 81 square feet. What is the length of the side of the mural?

Simplify each expression. (Lesson 1-6)

5. $(10c - 5d) + (6c + 5d)$

6. $3a + 7b - 2a$

7. $(21m + 15n) - (9n - 4m)$

8. $14x - 6y + 2y$

Solve each equation. (Lesson 8-3)

9. $x^2 + 10x + 24 = 0$ **10.** $2x^2 + x + 1 = 2$

11. GEOMETRY The triangle has an area of 120 square centimeters. Find h.

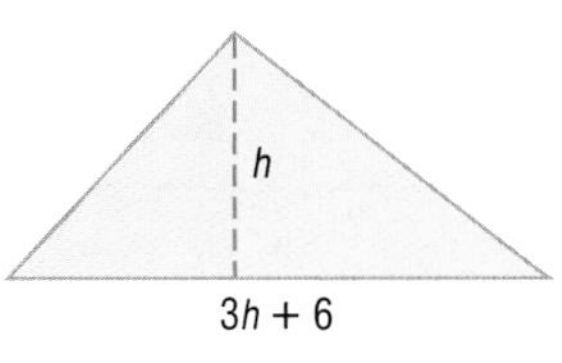

Use cross products to determine whether each pair of ratios forms a proportion. Write *yes* or *no*. (Lesson 2-6)

12. $\frac{2}{3}, \frac{8}{12}$ **13.** $\frac{4}{5}, \frac{16}{25}$ **14.** $\frac{8}{10}, \frac{12}{16}$

15. MODELS A collector's model train is scaled so that 1 inch on the model equals 3.5 feet on the actual train. If the model is 3.25 inches tall, how tall is the actual train?

QUICKReview

EXAMPLE 1

Find the square root of $\sqrt{82}$. If necessary, round to the nearest hundredth.

$\sqrt{82} = 9.05538513814\ldots$ Use a calculator.

To the nearest hundredth, $\sqrt{82} \approx 9.06$.

EXAMPLE 2

Simplify $6m + 3 + 2k - 17b - 100 - 16k + 8b + m$.

$6m + 3 + 2k - 17b - 100 - 16k + 8b + m$

$= (6m + m) + (2k - 16k) + (8b - 17b) + (3 - 100)$

$= 7m - 14k - 9b - 97$ Simplify.

EXAMPLE 3

Solve $3x^2 + 4x - 4 = 0$.

$3x^2 + 4x - 4 = 0$ Original equation

$(3x - 2)(x + 2) = 0$ Factor.

$(3x - 2) = 0$ or $(x + 2) = 0$ Zero Product Property

$x = \frac{2}{3}$ or $x = -2$ Solve each equation.

EXAMPLE 4

Use cross products to determine whether $\frac{5}{8}$ and $\frac{60}{96}$ form a proportion. Write *yes* or *no*.

$\frac{5}{8} \stackrel{?}{=} \frac{60}{96}$ Write the equation.

$5(96) \stackrel{?}{=} 8(60)$ Find the cross products.

$480 = 480$ Simplify. They form a proportion.

10-1

Simplifying Radical Expressions

Main Ideas

- Simplify radical expressions using the Product Property of Square Roots.
- Simplify radical expressions using the Quotient Property of Square Roots.

New Vocabulary

radical expression
radicand
rationalizing the denominator
conjugate

GET READY for the Lesson

A spacecraft leaving Earth must have a velocity of at least 11.2 kilometers per second (25,000 miles per hour) to enter into orbit. This velocity is called the *escape velocity*. The escape velocity of an object is given by the radical expression $\sqrt{\frac{2GM}{R}}$, where G is the gravitational constant, M is the mass of the planet or star, and R is the radius of the planet or star. Once values are substituted for the variables, the formula can be simplified.

Product Property of Square Roots A **radical expression** is an expression that contains a square root, such as $\sqrt{\frac{2GM}{R}}$. A **radicand**, the expression under the radical sign, is in simplest form if it contains no perfect square factors other than 1. The following property can be used to simplify square roots.

KEY CONCEPT — Product Property of Square Roots

Words For any numbers a and b, where $a \geq 0$ and $b \geq 0$, the square root of the product ab is equal to the product of each square root.

Symbols $\sqrt{ab} = \sqrt{a} \cdot \sqrt{b}$ **Example** $\sqrt{4 \cdot 25} = \sqrt{4} \cdot \sqrt{25}$

Reading Math

Radical Expressions $2\sqrt{3}$ is read *two times the square root of 3* or *two radical three*.

EXAMPLE 1 Simplify Square Roots

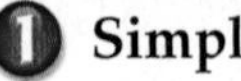

1 **Simplify $\sqrt{90}$.**

$\sqrt{90} = \sqrt{2 \cdot 3 \cdot 3 \cdot 5}$ Prime factorization of 90

$= \sqrt{3^2} \cdot \sqrt{2 \cdot 5}$ Product Property of Square Roots

$= 3\sqrt{10}$ Simplify.

CHECK Your Progress

Simplify.

1A. $\sqrt{27}$ **1B.** $\sqrt{150}$

Study Tip

Alternative Method

To find $\sqrt{3} \cdot \sqrt{15}$, you could multiply first and then use the prime factorization.

$\sqrt{3} \cdot \sqrt{15} = \sqrt{45}$

$= \sqrt{3^2} \cdot \sqrt{5}$

$= 3\sqrt{5}$

EXAMPLE Multiply Square Roots

2 **Simplify $\sqrt{3} \cdot \sqrt{15}$.**

$\sqrt{3} \cdot \sqrt{15} = \sqrt{3} \cdot \sqrt{3} \cdot \sqrt{5}$ Product Property of Square Roots

$= \sqrt{3^2} \cdot \sqrt{5}$ or $3\sqrt{5}$ Product Property; then simplify.

CHECK Your Progress **Simplify.**

2A. $\sqrt{5} \cdot \sqrt{10}$

2B. $\sqrt{6} \cdot \sqrt{8}$

When finding the principal square root of an expression containing variables, be sure that the result is not negative. Consider the expression $\sqrt{x^2}$. It may seem that $\sqrt{x^2} = x$. Let's look at $x = -2$.

$\sqrt{x^2} \stackrel{?}{=} x$

$\sqrt{(-2)^2} \stackrel{?}{=} -2$ Replace x with -2.

$\sqrt{4} \stackrel{?}{=} -2$ $(-2)^2 = 4$

$2 \neq -2$ $\sqrt{4} = 2$

For radical expressions where the exponent of the variable inside the radical is even and the resulting simplified exponent is odd, you must use absolute value to ensure nonnegative results.

$$\sqrt{x^2} = |x| \qquad \sqrt{x^4} = x^2 \qquad \sqrt{x^6} = |x^3|$$

EXAMPLE Simplify a Square Root with Variables

3 **Simplify $\sqrt{40x^4y^5z^3}$.**

$\sqrt{40x^4y^5z^3} = \sqrt{2^3 \cdot 5 \cdot x^4 \cdot y^5 \cdot z^3}$ Prime factorization

$= \sqrt{2^2} \cdot \sqrt{2} \cdot \sqrt{5} \cdot \sqrt{x^4} \cdot \sqrt{y^4} \cdot \sqrt{y} \cdot \sqrt{z^2} \cdot \sqrt{z}$ Product Property

$= 2 \cdot \sqrt{2} \cdot \sqrt{5} \cdot x^2 \cdot y^2 \cdot \sqrt{y} \cdot |z| \cdot \sqrt{z}$ Simplify.

$= 2x^2y^2|z|\sqrt{10yz}$ The absolute value of z ensures a nonnegative result.

CHECK Your Progress **Simplify.**

3A. $\sqrt{32r^2s^4t^5}, t \geq 0$

3B. $\sqrt{56xy^{10}z^5}, x \geq 0, z \geq 0$

Quotient Property of Square Roots You can divide square roots and simplify radical expressions by using the Quotient Property of Square Roots.

KEY CONCEPT *Quotient Property of Square Roots*

Words For any numbers a and b, where $a \geq 0$ and $b > 0$, the square root of the quotient $\frac{a}{b}$ is equal to the quotient of each square root.

Symbols $\sqrt{\frac{a}{b}} = \frac{\sqrt{a}}{\sqrt{b}}$

Example $\sqrt{\frac{49}{4}} = \frac{\sqrt{49}}{\sqrt{4}}$

Study Tip

Look Back
To review the **Quadratic Formula**, see Lesson 9-4.

You can use the Quotient Property of Square Roots to derive the Quadratic Formula by solving the quadratic equation $ax^2 + bx + c = 0$.

$ax^2 + bx + c = 0$	Original equation.
$x^2 + \frac{b}{a}x + \frac{c}{a} = 0$	Divide each side by a, $a \neq 0$.
$x^2 + \frac{b}{a}x = -\frac{c}{a}$	Subtract $\frac{c}{a}$ from each side.
$x^2 + \frac{b}{a}x + \frac{b^2}{4a^2} = -\frac{c}{a} + \frac{b^2}{4a^2}$	Complete the square; $\left(\frac{b}{2a}\right)^2 = \frac{b^2}{4a^2}$.
$\left(x + \frac{b}{2a}\right)^2 = \frac{-4ac + b^2}{4a^2}$	Factor $x^2 + \frac{b}{a}x + \frac{b^2}{4a^2}$.
$\left\lvert x + \frac{b}{2a}\right\rvert = \sqrt{\frac{b^2 - 4ac}{4a^2}}$	Take the square root of each side.
$x + \frac{b}{2a} = \pm\sqrt{\frac{b^2 - 4ac}{4a^2}}$	Remove the absolute value symbols and insert $\pm$.
$x + \frac{b}{2a} = \pm\frac{\sqrt{b^2 - 4ac}}{\sqrt{4a^2}}$	Quotient Property of Square Roots
$x + \frac{b}{2a} = \pm\frac{\sqrt{b^2 - 4ac}}{2a}$	$\sqrt{4a^2} = 2a$
$x = \frac{-b \pm \sqrt{b^2 - 4ac}}{2a}$	Subtract $\frac{b}{2a}$ from each side.

Study Tip

Plus or Minus Symbols
The $\pm$ symbol is used with the radical expression since both square roots lead to solutions.

If no prime factors with an exponent greater than 1 appear under the radical sign and if no radicals are left in the denominator, then a fraction containing radicals is in simplest form. **Rationalizing the denominator** of a radical expression is a method used to eliminate radicals from a denominator.

EXAMPLE Rationalizing the Denominator

4 **Simplify.**

a. $\sqrt{\frac{10}{3}}$

$\sqrt{\frac{10}{3}} = \frac{\sqrt{10}}{\sqrt{3}}$	Quotient Property of Square Roots
$= \frac{\sqrt{10}}{\sqrt{3}} \cdot \frac{\sqrt{3}}{\sqrt{3}}$	Multiply by $\frac{\sqrt{3}}{\sqrt{3}}$.
$= \frac{\sqrt{30}}{3}$	Product Property of Square Roots

b. $\frac{\sqrt{2n}}{\sqrt{6}}$

$\frac{\sqrt{2n}}{\sqrt{6}} = \frac{\sqrt{2n}}{\sqrt{6}} \cdot \frac{\sqrt{6}}{\sqrt{6}}$	Multiply by $\frac{\sqrt{6}}{\sqrt{6}}$.
$= \frac{\sqrt{12n}}{6}$	Product Property of Square Roots
$= \frac{\sqrt{2 \cdot 2 \cdot 3 \cdot n}}{6}$	Prime factorization
$= \frac{2\sqrt{3n}}{6}$	$\sqrt{2^2} = 2$
$= \frac{\sqrt{3n}}{3}$	Divide numerator and denominator by 2.

Study Tip

Calculator
Approximations for radical expressions are often used when solving real-world problems. The expression $\frac{\sqrt{30}}{3}$ can be approximated by pressing [2nd] [√] 30 [)] [÷] 3 [ENTER].

CHECK Your Progress

4A. $\frac{\sqrt{14}}{\sqrt{5}}$

4B. $\frac{\sqrt{6y}}{\sqrt{12}}$

Binomials of the form $p\sqrt{q} + r\sqrt{s}$ and $p\sqrt{q} - r\sqrt{s}$ are called **conjugates**. For example, $3 + \sqrt{2}$ and $3 - \sqrt{2}$ are conjugates. Conjugates are useful when simplifying radical expressions because if p, q, r, and s are rational numbers, the product of the two conjugates is a rational number. Use the pattern for the difference of squares $(a - b)(a + b) = a^2 - b^2$ to find their product.

$$(3 + \sqrt{2})(3 - \sqrt{2}) = 3^2 - (\sqrt{2})^2 \quad a = 3, b = \sqrt{2}$$
$$= 9 - 2 \text{ or } 7 \quad (\sqrt{2})^2 = \sqrt{2} \cdot \sqrt{2} \text{ or } 2$$

EXAMPLE Use Conjugates to Rationalize a Denominator

5 **Simplify $\frac{2}{6 - \sqrt{3}}$.**

$$\frac{2}{6 - \sqrt{3}} = \frac{2}{6 - \sqrt{3}} \cdot \frac{6 + \sqrt{3}}{6 + \sqrt{3}} \quad \frac{6 + \sqrt{3}}{6 + \sqrt{3}} = 1\text{; The conjugate of } 6 - \sqrt{3} \text{ is } 6 + \sqrt{3}.$$

$$= \frac{2(6 + \sqrt{3})}{6^2 - (\sqrt{3})^2} \quad (a - b)(a + b) = a^2 - b^2$$

$$= \frac{12 + 2\sqrt{3}}{36 - 3} \quad (\sqrt{3})^2 = 3$$

$$= \frac{12 + 2\sqrt{3}}{33} \quad \text{Simplify.}$$

Cross-Curricular Project

Math Online The speed of a roller coaster can be determined by evaluating a radical expression. Visit algebra1.com to continue work on your project.

CHECK Your Progress **Simplify.**

5A. $\frac{3}{2 + \sqrt{2}}$

5B. $\frac{7}{3 - \sqrt{7}}$

Online **Personal Tutor at** algebra1.com

When simplifying radical expressions, check the following conditions to determine if the expression is in simplest form.

CONCEPT SUMMARY *Simplest Radical Form*

A radical expression is in simplest form when the following three conditions have been met.

1. No radicands have perfect square factors other than 1.

2. No radicands contain fractions.

3. No radicals appear in the denominator of a fraction.

CHECK Your Understanding

Examples 1, 2 (pp. 528–529)

Simplify.

1. $\sqrt{20}$

2. $\sqrt{52}$

3. $2\sqrt{32}$

4. $\sqrt{2} \cdot \sqrt{8}$

5. $\sqrt{3} \cdot \sqrt{18}$

6. $3\sqrt{10} \cdot 4\sqrt{10}$

7. GEOMETRY A square has sides measuring $2\sqrt{7}$ feet each. Determine the area of the square.

Examples 3, 4
(pp. 529–530)

Simplify.

8. $\sqrt{54a^2b^2}$ **9.** $\sqrt{60x^5y^6}$ **10.** $\sqrt{88m^3n^2p^5}$

11. $\dfrac{4}{\sqrt{6}}$ **12.** $\sqrt{\dfrac{3}{10}}$ **13.** $\sqrt{\dfrac{7}{2}} \cdot \sqrt{\dfrac{5}{3}}$

14. PHYSICS The period of a pendulum is the time required for it to make one complete swing back and forth. The formula of the period P of a pendulum is $P = 2\pi\sqrt{\dfrac{\ell}{32}}$, where ℓ is the length of the pendulum in feet. If a pendulum in a clock tower is 8 feet long, find the period. Use 3.14 for π.

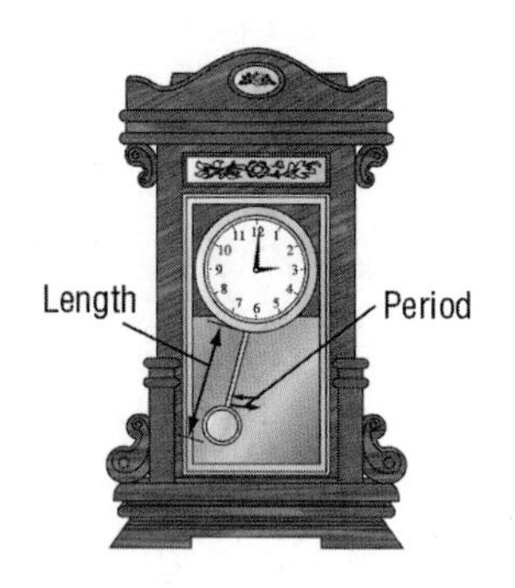

Example 5
(p. 531)

Simplify.

15. $\dfrac{8}{3-\sqrt{2}}$ **16.** $\dfrac{2\sqrt{5}}{-4+\sqrt{8}}$

Exercises

HOMEWORK HELP

For Exercises	See Examples
17–20	1
21–24	2
25–28	3
29–34	4
35–40	5

Simplify.

17. $\sqrt{18}$ **18.** $\sqrt{24}$ **19.** $\sqrt{80}$

20. $\sqrt{75}$ **21.** $\sqrt{5} \cdot \sqrt{6}$ **22.** $\sqrt{3} \cdot \sqrt{8}$

23. $7\sqrt{30} \cdot 2\sqrt{6}$ **24.** $2\sqrt{3} \cdot 5\sqrt{27}$ **25.** $\sqrt{40a^4}$

26. $\sqrt{50m^3n^5}$ **27.** $\sqrt{147x^6y^7}$ **28.** $\sqrt{72x^3y^4z^5}$

29. $\sqrt{\dfrac{2}{7}} \cdot \sqrt{\dfrac{7}{3}}$ **30.** $\sqrt{\dfrac{3}{5}} \cdot \sqrt{\dfrac{6}{4}}$ **31.** $\sqrt{\dfrac{t}{8}}$

32. $\sqrt{\dfrac{27}{p^2}}$ **33.** $\sqrt{\dfrac{5c^5}{4d^5}}$ **34.** $\dfrac{\sqrt{9x^5y}}{\sqrt{12x^2y^6}}$

35. $\dfrac{18}{6-\sqrt{2}}$ **36.** $\dfrac{3\sqrt{3}}{-2+\sqrt{6}}$ **37.** $\dfrac{10}{\sqrt{7}+\sqrt{2}}$

38. $\dfrac{2}{\sqrt{3}+\sqrt{6}}$ **39.** $\dfrac{4}{4-3\sqrt{3}}$ **40.** $\dfrac{3\sqrt{7}}{5\sqrt{3}+3\sqrt{5}}$

INVESTIGATION For Exercises 41–43, use the following information.

Police officers can use the formula $s = \sqrt{30fd}$ to determine the speed s that a car was traveling in miles per hour by measuring the distance d in feet of its skid marks. In this formula, f is the coefficient of friction for the type and condition of the road.

41. Write a simplified formula for the speed if $f = 0.6$ for a wet asphalt road.

42. What is a simplified formula for the speed if $f = 0.8$ for a dry asphalt road?

43. An officer measures skid marks that are 110 feet long. Determine the speed of the car for both wet road conditions and for dry road conditions. Write in both simplified radical form and as a decimal approximation.

44. GEOMETRY A rectangle has a width of $3\sqrt{5}$ centimeters and a length of $4\sqrt{10}$ centimeters. Find the area of the rectangle. Write as a simplified radical expression.

Real-World Career

Insurance Investigator
Insurance investigators decide whether claims are covered by the customer's policy, and investigate the circumstances of a claim.

For more information, go to algebra1.com.

45. GEOMETRY The formula for the area A of a square with side length s is $A = s^2$. Solve this equation for s, and find the side length of a square having an area of 72 square inches. Write as a simplified radical expression.

PHYSICS For Exercises 46 and 47, use the following information.
The formula for the kinetic energy of a moving object is $E = \frac{1}{2}mv^2$, where E is the kinetic energy in joules, m is the mass in kilograms, and v is the velocity in meters per second.

46. Solve the equation for v.

47. Find the velocity of an object whose mass is 0.6 kilogram and whose kinetic energy is 54 joules. Write as a simplified radical expression.

48. GEOMETRY A rectangle has a length of $\sqrt{\frac{a}{8}}$ meters and a width of $\sqrt{\frac{a}{2}}$ meters. What is the area of the rectangle?

49. SPACE EXPLORATION Refer to the application at the beginning of the lesson. Find the escape velocity for the Moon in kilometers per second if $G = \frac{6.7 \times 10^{-20}\text{ km}}{s^2\text{ kg}}$, $M = 7.4 \times 10^{22}$ kg, and $R = 1.7 \times 10^3$ km. Use a calculator and write your answer as a decimal approximation. How does this compare to the escape velocity for Earth?

50. GEOMETRY Hero's Formula can be used to calculate the area A of a triangle given the three side lengths a, b, and c. Determine the area of a triangle if the side lengths of a triangle are 13, 10, and 7 feet.

$$A = \sqrt{s(s - a)(s - b)(s - c)},\text{ where } s = \frac{1}{2}(a + b + c)$$

51. QUADRATIC FORMULA Determine the next step in the derivation of the Quadratic Formula.

Step 1 $x^2 + \frac{b}{a}x + \frac{c}{a} = 0$

Step 2 $x^2 + \frac{b}{a}x = -\frac{c}{a}$

52. QUADRATIC FORMULA Four steps in the derivation of the Quadratic Formula are shown below. Determine the correct order of the steps.

1	2
$x + \frac{b}{2a} = \pm\sqrt{\frac{b^2 - 4ac}{4a^2}}$	$x + \frac{b}{2a} = \pm\frac{\sqrt{b^2 - 4ac}}{2a}$

3	4
$\left\lvert x + \frac{b}{2a} \right\rvert = \sqrt{\frac{b^2 - 4ac}{4a^2}}$	$x^2 + \frac{b}{a}x = -\frac{c}{a}$

EXTRA PRACTICE
See pages 737, 753.
Math Online
Self-Check Quiz at algebra1.com

H.O.T. Problems

53. REASONING Kary takes any number, subtracts 4, multiplies by 4, takes the square root, and takes the reciprocal to get $\frac{1}{2}$. What number did she start with? Write a formula to describe the process.

54. OPEN ENDED Give an example of a binomial in the form $a\sqrt{b} + c\sqrt{d}$ and its conjugate. Then find their product.

55. **FIND THE ERROR** Ben is solving $(3x - 2)^2 = (2x + 6)^2$. He found that $x = -4$. Is this solution correct? Explain.

56. **CHALLENGE** Solve the equation $\left|y^3\right| = \frac{1}{3\sqrt{3}}$ for y.

57. ***Writing in Math*** Use the information about space exploration on page 580 to explain how radical expressions can be used in space exploration. Include an explanation of how you could determine the escape velocity of a planet and why you would need this information before you landed on it.

STANDARDIZED TEST PRACTICE

58. **REVIEW** If the cube has a surface area of $96a^2$, what is its volume?

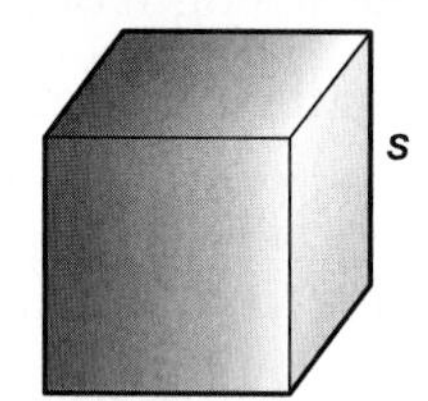

Formula for surface area of a cube = $6s^2$

A $32a^3$

B $48a^3$

C $64a^3$

D $96a^3$

59. The perimeter P of a square can be found using the formula $\frac{1}{4}P = \sqrt{A}$, where A is the area of the square. What is the perimeter of a square with an area of 81 square centimeters?

F 108 cm

G 72 cm

H 54 cm

J 36 cm

Spiral Review

Find the next three terms in each geometric sequence. (Lesson 9-7)

60. 2, 6, 18, 54

61. 1, −2, 4, −8

62. 384, 192, 96, 48

63. $\frac{1}{9}, \frac{2}{3}, 4, 24$

64. $3, \frac{3}{4}, \frac{3}{16}, \frac{3}{64}$

65. 50, 10, 2, 0.4

66. **BIOLOGY** A certain type of bacteria, if left alone, doubles its number every 2 hours. If there are 1000 bacteria at a certain point in time, how many bacteria will there be 24 hours later? (Lesson 9-6)

67. **PHYSICS** According to Newton's Law of Cooling, the difference between the temperature of an object and its surroundings decreases exponentially in time. Suppose a cup of coffee is 95°C and it is in a room that is 20°C. The cooling of the coffee can be modeled by the equation $y = 75(0.875)^t$, where y is the temperature difference and t is the time in minutes. Find the temperature of the coffee after 15 minutes. (Lesson 9-6)

Factor each trinomial, if possible. If the trinomial cannot be factored using integers, write *prime*. (Lesson 8-4)

68. $6x^2 + 7x - 5$

69. $35x^2 - 43x + 12$

70. $5x^2 + 3x + 31$

GET READY for the Next Lesson

PREREQUISITE SKILL Find each product. (Lesson 7-6)

71. $(x - 3)(x + 2)$

72. $(a + 2)(a + 5)$

73. $(2t + 1)(t - 6)$

74. $(4x - 3)(x + 1)$

75. $(5x + 3y)(3x - y)$

76. $(3a - 2b)(4a + 7b)$

EXTEND 10-1

Graphing Calculator Lab
Fractional Exponents

You have studied the properties of exponents that are whole numbers. You can use a calculator to explore the meaning of fractional exponents.

ACTIVITY

Step 1 Evaluate $9^{\frac{1}{2}}$ and $\sqrt{9}$.

KEYSTROKES: 9 [^] [(] 1 [÷] 2 [)] [ENTER] 3

KEYSTROKES: [2nd] [$\sqrt{\ }$] 9 [ENTER] 3

Record the results in a table like the one at the right.

Expression	Value	Expression	Value
$9^{\frac{1}{2}}$	3	$\sqrt{9}$	3
$16^{\frac{1}{2}}$		$\sqrt{16}$	
$8^{\frac{1}{3}}$		$\sqrt[3]{8}$	
$27^{\frac{1}{3}}$		$\sqrt[3]{27}$	
$8^{\frac{2}{3}}$		$\sqrt[3]{8^2}$	
$16^{\frac{3}{4}}$		$\sqrt[4]{8^3}$	

Step 2 Use calculator to evaluate each expression. Record each result in your table. To find a root other than a square root, choose the $\sqrt[x]{\ }$ function from the [MATH] menu.

1A. Study the table. What do you observe about the value of an expression of the form $a^{\frac{1}{n}}$?

1B. What do you observe about the value of an expression of the form $a^{\frac{m}{n}}$?

Analyze the Results

1. Recall the Power of a Power Property. For any number a and all integers m and n, $(a^m)^n = a^{m \cdot n}$. Assume that fractional exponents behave as whole number exponents and find the value of $\left(b^{\frac{1}{2}}\right)^2$.

$$\left(b^{\frac{1}{2}}\right)^2 = b^{\frac{1}{2} \cdot 2} \quad \text{Power of a Power Property}$$

$$= b^1 \text{ or } b \quad \text{Simplify.}$$

Thus, $b^{\frac{1}{2}}$ is a number whose square equals b. So it makes sense to define $b^{\frac{1}{2}} = \sqrt{b}$. Use a similar process to define $b^{\frac{1}{n}}$.

2. Define $b^{\frac{m}{n}}$. Justify your answer.

Write each expression as a power of x.

3. $\dfrac{\sqrt{x}}{(\sqrt[4]{x})(x)}$

4. $\dfrac{(x)(\sqrt[3]{x})}{(\sqrt{x})(\sqrt[5]{x})}$

Write each root as an expression using a fractional exponent. Then evaluate the expression.

5. $\sqrt{49}$

6. $\sqrt[4]{81}$

7. $\sqrt{4^3}$

8. $\sqrt[3]{125^2}$

10-2 Operations with Radical Expressions

Main Ideas

- Add and subtract radical expressions.
- Multiply radical expressions.

GET READY for the Lesson

The formula $d = \sqrt{\frac{3h}{2}}$ represents the distance d in miles that a person h feet high can see. To determine how much farther a person can see from atop the Sears Tower than from atop the Empire State Building, we can substitute the heights of both buildings into the equation.

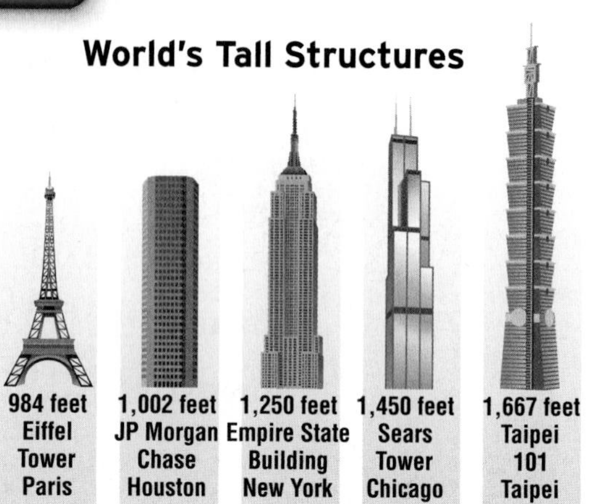

Add and Subtract Radical Expressions Radical expressions in which the radicands are alike can be added or subtracted in the same way that like monomials are added or subtracted.

Monomials	Radical Expressions
$2x + 7x = (2 + 7)x$	$2\sqrt{11} + 7\sqrt{11} = (2 + 7)\sqrt{11}$
$= 9x$	$= 9\sqrt{11}$
$15y - 3y = (15 - 3)y$	$15\sqrt{2} - 3\sqrt{2} = (15 - 3)\sqrt{2}$
$= 12y$	$= 12\sqrt{2}$

EXAMPLE Expressions with Like Radicands

1 Simplify each expression.

a. $4\sqrt{3} + 6\sqrt{3} - 5\sqrt{3}$

$4\sqrt{3} + 6\sqrt{3} - 5\sqrt{3} = (4 + 6 - 5)\sqrt{3}$ Distributive Property

$= 5\sqrt{3}$ Simplify.

b. $12\sqrt{5} + 3\sqrt{7} + 6\sqrt{7} - 8\sqrt{5}$

$12\sqrt{5} + 3\sqrt{7} + 6\sqrt{7} - 8\sqrt{5} = 12\sqrt{5} - 8\sqrt{5} + 3\sqrt{7} + 6\sqrt{7}$

$= (12 - 8)\sqrt{5} + (3 + 6)\sqrt{7}$ Distributive Property

$= 4\sqrt{5} + 9\sqrt{7}$ Simplify.

CHECK Your Progress

1A. $3\sqrt{2} - 5\sqrt{2} + \sqrt{2}$

1B. $15\sqrt{11} - 14\sqrt{13} + 6\sqrt{13} - 11\sqrt{11}$

Math Online Extra Examples at algebra1.com

In Example 1b, $4\sqrt{5} + 9\sqrt{7}$ cannot be simplified further because the radicands are different. There are no common factors, and each radicand is in simplest form. If the radicals in an expression are not in simplest form, simplify them first.

EXAMPLE **Expressions with Unlike Radicands**

2 **Simplify $2\sqrt{20} + 3\sqrt{45} + \sqrt{180}$.**

$$\begin{aligned}
2\sqrt{20} + 3\sqrt{45} + \sqrt{180} &= 2\sqrt{2^2 \cdot 5} + 3\sqrt{3^2 \cdot 5} + \sqrt{6^2 \cdot 5} \\
&= 2\left(\sqrt{2^2} \cdot \sqrt{5}\right) + 3\left(\sqrt{3^2} \cdot \sqrt{5}\right) + \sqrt{6^2} \cdot \sqrt{5} \\
&= 2\left(2\sqrt{5}\right) + 3\left(3\sqrt{5}\right) + 6\sqrt{5} \\
&= 4\sqrt{5} + 9\sqrt{5} + 6\sqrt{5} \\
&= 19\sqrt{5}
\end{aligned}$$

CHECK Your Progress

Simplify.

2A. $4\sqrt{54} + 2\sqrt{24} - \sqrt{150}$

2B. $4\sqrt{12} - 6\sqrt{48} + 5\sqrt{24}$

Multiply Radical Expressions Multiplying two radical expressions with different radicands is similar to multiplying binomials.

EXAMPLE **Multiply Radical Expressions**

Study Tip

Look Back

To review the **FOIL method**, see Lesson 7-6.

3 **GEOMETRY** **Find the area of the rectangle in simplest form.**

$4\sqrt{5} - 2\sqrt{3}$

$3\sqrt{6} - \sqrt{10}$

To find the area of the rectangle, multiply the measures of the length and width.

$$\left(4\sqrt{5} - 2\sqrt{3}\right)\left(3\sqrt{6} - \sqrt{10}\right)$$

$$\begin{aligned}
&= \underbrace{\left(4\sqrt{5}\right)\left(3\sqrt{6}\right)}_{\text{First terms}} + \underbrace{\left(4\sqrt{5}\right)\left(-\sqrt{10}\right)}_{\text{Outer terms}} + \underbrace{\left(-2\sqrt{3}\right)\left(3\sqrt{6}\right)}_{\text{Inner terms}} + \underbrace{\left(-2\sqrt{3}\right)\left(-\sqrt{10}\right)}_{\text{Last terms}} \\
&= 12\sqrt{30} - 4\sqrt{50} - 6\sqrt{18} + 2\sqrt{30} && \text{Multiply.} \\
&= 12\sqrt{30} - 4\sqrt{5^2 \cdot 2} - 6\sqrt{3^2 \cdot 2} + 2\sqrt{30} && \text{Prime factorization} \\
&= 12\sqrt{30} - 20\sqrt{2} - 18\sqrt{2} + 2\sqrt{30} && \text{Simplify.} \\
&= 14\sqrt{30} - 38\sqrt{2} && \text{Combine like terms.}
\end{aligned}$$

CHECK Your Progress

Find each product.

3A. $\left(5\sqrt{5} - 4\sqrt{3}\right)\left(6\sqrt{10} - 2\sqrt{6}\right)$

3B. $\left(6\sqrt{7} + 3\sqrt{2}\right)\left(4\sqrt{10} - 5\sqrt{6}\right)$

Online **Personal Tutor at** algebra1.com

You can use a calculator to verify that a simplified radical expression is equivalent to the original expression. Consider Example 3. First, find a decimal approximation for the original expression.

KEYSTROKES: (4 2nd [√] 5) − 2 2nd [√] 3)) (3 2nd [√] 6) − 2nd [√] 10)) ENTER

22.94104268

Next, find a decimal approximation for the simplified expression.

KEYSTROKES: 14 2nd [√] 30) − 38 2nd [√] 2) ENTER

22.94104268

Since the approximations are equal, the expressions are equivalent.

CHECK Your Understanding

Examples 1, 2 (pp. 536–537)

Simplify.

1. $4\sqrt{3} + 7\sqrt{3}$
2. $2\sqrt{6} - 7\sqrt{6}$
3. $5\sqrt{5} - 3\sqrt{20}$
4. $2\sqrt{3} + \sqrt{12}$
5. $3\sqrt{5} + 5\sqrt{6} + 3\sqrt{20}$
6. $8\sqrt{3} + \sqrt{3} + \sqrt{9}$

Example 3 (p. 537)

Find each product.

7. $\sqrt{2}\left(\sqrt{8} + 4\sqrt{3}\right)$
8. $\left(4 + \sqrt{5}\right)\left(3 + \sqrt{5}\right)$

9. **GEOMETRY** Find the perimeter and the area of the square.

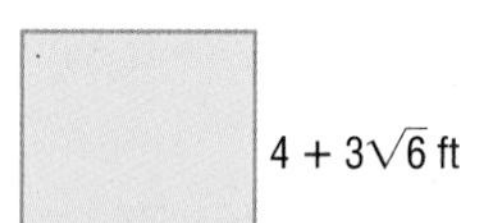

10. **ELECTRICITY** The voltage V required for a circuit is given by $V = \sqrt{PR}$, where P is the power in watts and R is the resistance in ohms. How many more volts are needed to light a 100-watt bulb than a 75-watt bulb if the resistance for both is 110 ohms?

Exercises

HOMEWORK HELP	
For Exercises	See Examples
11–18	1
19–22	2
23–33	3

Simplify.

11. $8\sqrt{5} + 3\sqrt{5}$
12. $3\sqrt{6} + 10\sqrt{6}$
13. $2\sqrt{15} - 6\sqrt{15} - 3\sqrt{15}$
14. $5\sqrt{19} + 6\sqrt{19} - 11\sqrt{19}$
15. $16\sqrt{x} + 2\sqrt{x}$
16. $3\sqrt{5b} - 4\sqrt{5b} + 11\sqrt{5b}$
17. $8\sqrt{3} - 2\sqrt{2} + 3\sqrt{2} + 5\sqrt{3}$
18. $4\sqrt{6} + \sqrt{17} - 6\sqrt{2} + 4\sqrt{17}$
19. $\sqrt{18} + \sqrt{12} + \sqrt{8}$
20. $\sqrt{6} + 2\sqrt{3} + \sqrt{12}$
21. $3\sqrt{7} - 2\sqrt{28}$
22. $2\sqrt{50} - 3\sqrt{32}$

Find each product.

23. $\sqrt{6}\left(\sqrt{3} + 5\sqrt{2}\right)$
24. $\sqrt{5}\left(2\sqrt{10} + 3\sqrt{2}\right)$
25. $\left(3 + \sqrt{5}\right)\left(3 - \sqrt{5}\right)$
26. $\left(7 - \sqrt{10}\right)^2$
27. $\left(\sqrt{6} + \sqrt{8}\right)\left(\sqrt{24} + \sqrt{2}\right)$
28. $\left(\sqrt{5} - \sqrt{2}\right)\left(\sqrt{14} + \sqrt{35}\right)$
29. $\left(2\sqrt{10} + 3\sqrt{15}\right)\left(3\sqrt{3} - 2\sqrt{2}\right)$
30. $\left(5\sqrt{2} + 3\sqrt{5}\right)\left(2\sqrt{10} - 3\right)$

31. **GEOMETRY** Find the perimeter and area of a rectangle with a length of $8\sqrt{7} + 4\sqrt{5}$ inches and a width of $2\sqrt{7} - 3\sqrt{5}$ inches.

32. **GEOMETRY** The perimeter of a rectangle is $2\sqrt{3} + 4\sqrt{11} + 6$ centimeters, and its length is $2\sqrt{11} + 1$ centimeters. Find the width.

33. **GEOMETRY** The area A of a rhombus can be found using the formula $A = \frac{1}{2}d_1d_2$, where d_1 and d_2 are the lengths of the diagonals of the rhombus. What is the area of the rhombus at the right?

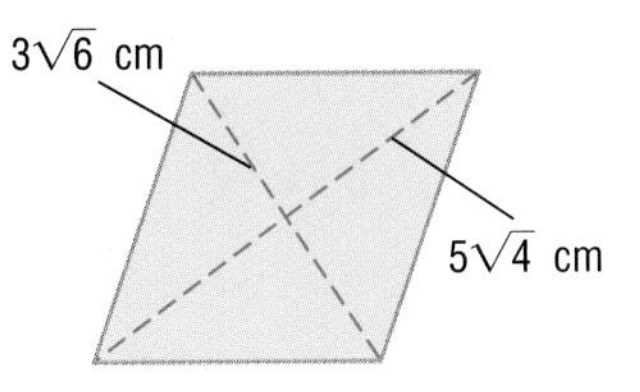

Simplify.

34. $\sqrt{2} + \sqrt{\frac{1}{2}}$

35. $\sqrt{10} - \sqrt{\frac{2}{5}}$

36. $3\sqrt{3} - \sqrt{45} + 3\sqrt{\frac{1}{3}}$

37. $6\sqrt{\frac{7}{4}} + 3\sqrt{28} - 10\sqrt{\frac{1}{7}}$

DISTANCE For Exercises 38 and 39, refer to the application at the beginning of the lesson.

38. How much farther can a person see from atop the Sears Tower than from atop the Empire State Building? Write as a simplified radical expression and as a decimal approximation.

39. A person atop the Empire State Building can see approximately 4.57 miles farther than a person atop the Texas Commerce Tower in Houston. Explain how you could find the height of the Texas Commerce Tower.

Real-World Link

The Sears Tower was the tallest building in the world from 1974 to 1996. You can see four states from its roof: Michigan, Indiana, Illinois, and Wisconsin.

Source: the-skydeck.com

ENGINEERING For Exercises 40 and 41, use the following information.

The equation $r = \sqrt{\frac{F}{5\pi}}$ relates the radius r of a drainpipe, in inches, to the flow rate F of water passing through it, in gallons per minute.

40. Find the radius of a pipe that can carry 500 gallons of water per minute. Write as a simplified radical expression, and use a calculator to find the decimal approximation. Round to the nearest whole number.

41. An engineer determines that a drainpipe must be able to carry 1000 gallons of water per minute and instructs the builder to use an 8-inch radius pipe. Can the builder use two 4-inch radius pipes instead? Justify your answer.

MOTION For Exercises 42 and 43, use the following information.

The velocity of an object dropped from a certain height can be found using the formula $v = \sqrt{2gd}$, where v is the velocity in feet per second, g is the acceleration due to gravity, and d is the distance the object drops, in feet.

42. Find the speed of an object that has fallen 25 feet and the speed of an object that has fallen 100 feet. Use 32 feet per second squared for g. Write as a simplified radical expression.

43. When you increased the distance by 4 times, what happened to the velocity? Explain.

H.O.T. Problems

44. **CHALLENGE** Determine whether the following statement is *true* or *false*. Provide an example or counterexample to support your answer.

$$x + y > \sqrt{x^2 + y^2} \text{ when } x > 0 \text{ and } y > 0$$

45. **OPEN ENDED** Choose values for x and y. Then find $\left(\sqrt{x} + \sqrt{y}\right)^2$.

46. **Which One Doesn't Belong?** Three of these expressions are equivalent. Which one is not?

$6\sqrt{6} - 24\sqrt{2} + 6\sqrt{6} - 5\sqrt{2}$ | $12\sqrt{6} - 29\sqrt{2}$ | $29\sqrt{2} - 12\sqrt{6}$ | $3\sqrt{24} - 6\sqrt{32} + 2\sqrt{54} - \sqrt{50}$

47. **CHALLENGE** Under what conditions is $\left(\sqrt{a+b}\right)^2 = \left(\sqrt{a}\right)^2 + \left(\sqrt{b}\right)^2$ true?

48. ***Writing in Math*** Use the information about the world's tall structures on page 536 to explain how you can use radicals to determine how far a person can see. Include an explanation of how this information could help determine how far apart lifeguard towers should be on a beach.

STANDARDIZED TEST PRACTICE

49. $\sqrt{3}(4 + \sqrt{12})^2 =$

A $4\sqrt{3} + 6$

B $28\sqrt{3}$

C $28 + 16\sqrt{3}$

D $48 + 28\sqrt{3}$

50. **REVIEW** Which expression is equivalent to $3^8 \cdot 3^2 \cdot 3^4$?

F 3^{14}

G 3^{64}

H 27^{14}

J 27^{64}

Spiral Review

Simplify. (Lesson 10-1)

51. $\sqrt{40}$

52. $\sqrt{128}$

53. $-\sqrt{196x^2y^3}$

54. $\dfrac{\sqrt{50}}{\sqrt{8}}$

55. $\sqrt{\dfrac{225c^4d}{18c^2}}$

56. $\sqrt{\dfrac{63a}{128a^3b^2}}$

Find the *n*th term of each geometric sequence. (Lesson 9-7)

57. $a_1 = 4, n = 6, r = 4$

58. $a_1 = -7, n = 4, r = 9$

59. $a_1 = 2, n = 8, r = -0.8$

Solve each equation by factoring. Check your solutions. (Lesson 8-5)

60. $81 = 49y^2$

61. $q^2 - \dfrac{36}{121} = 0$

62. $48n^3 - 75n = 0$

63. $5x^3 - 80x = 240 - 15x^2$

64. **RUNNING** Tyler runs 17 miles each Saturday. It takes him about 2 hours to run this distance. At this rate, how far could he run in 3 hours and 30 minutes? (Lesson 2-6)

GET READY for the Next Lesson

PREREQUISITE SKILL Find each product. (Lesson 7-7)

65. $(x - 2)^2$

66. $(x + 5)^2$

67. $(x + 6)^2$

68. $(3x - 1)^2$

69. $(2x - 3)^2$

70. $(4x + 7)^2$

10-3 Radical Equations

Main Ideas

- Solve radical equations.
- Solve radical equations with extraneous solutions.

New Vocabulary

radical equation
extraneous solution

GET READY for the Lesson

Skydivers fall 1050 to 1480 feet every 5 seconds, reaching speeds of 120 to 150 miles per hour at *terminal velocity.* It is the highest speed they can reach and occurs when the air resistance equals the force of gravity. With no air resistance, the time t in seconds that it takes an object to fall h feet can be determined by the equation $t = \frac{\sqrt{h}}{4}$. How would you find the value of h if you are given the value of t?

Radical Equations Equations like $t = \frac{\sqrt{h}}{4}$ that contain radicals with variables in the radicand are called **radical equations**. To solve these equations, first isolate the radical on one side of the equation. Then square each side of the equation to eliminate the radical.

Real-World EXAMPLE Variable in Radical

1 FREE-FALL HEIGHT Two objects are dropped simultaneously. The first object reaches the ground in 2.5 seconds, and the second object reaches the ground 1.5 seconds later. From what heights were the two objects dropped?

First Object

$t = \frac{\sqrt{h}}{4}$	Original equation
$2.5 = \frac{\sqrt{h}}{4}$	Replace t with 2.5.
$10 = \sqrt{h}$	Multiply each side by 4.
$10^2 = \left(\sqrt{h}\right)^2$	Square each side.
$100 = h$	Simplify.

Second Object

$t = \frac{\sqrt{h}}{4}$	Original equation
$4 = \frac{\sqrt{h}}{4}$	Replace t with 4.
$16 = \sqrt{h}$	Multiply each side by 4.
$16^2 = \left(\sqrt{h}\right)^2$	Square each side.
$256 = h$	Simplify.

Check the results by substituting 100 and 256 for h in the original equation.

(continued on the next page)

CHECK $t = \frac{\sqrt{h}}{4}$ Original equation

$\stackrel{?}{=} \frac{\sqrt{100}}{4}$ $h = 100$

$\stackrel{?}{=} \frac{10}{4}$ $\sqrt{100} = 10$

$= 2.5$ ✓ Simplify.

The first object was dropped from 100 feet.

CHECK $t = \frac{\sqrt{h}}{4}$ Original equation

$\stackrel{?}{=} \frac{\sqrt{256}}{4}$ $h = 256$

$\stackrel{?}{=} \frac{16}{4}$ $\sqrt{256} = 16$

$= 4$ ✓ Simplify.

The second object was dropped from 256 feet.

CHECK Your Progress

1. **DIVING** At the swim meet, Brandon dived off two platforms at different heights. On the first dive, it took him 0.78 second to reach the water. On the next dive, it took Brandon 1.43 seconds to reach the water. How much higher is the second platform than the first?

EXAMPLE Radical Equation with an Expression

2 Solve $\sqrt{x+1} + 7 = 10$.

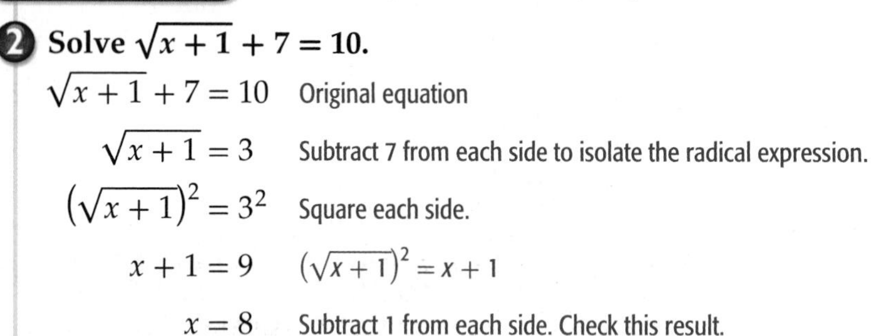

$\sqrt{x+1} + 7 = 10$	Original equation
$\sqrt{x+1} = 3$	Subtract 7 from each side to isolate the radical expression.
$(\sqrt{x+1})^2 = 3^2$	Square each side.
$x + 1 = 9$	$(\sqrt{x+1})^2 = x + 1$
$x = 8$	Subtract 1 from each side. Check this result.

CHECK Your Progress

Solve each equation. Check your solution.

2A. $\sqrt{x-3} - 2 = 4$

2B. $4 + \sqrt{x+1} = 14$

Extraneous Solutions Squaring each side of an equation sometimes produces extraneous solutions. An **extraneous solution** is a solution derived from an equation that is not a solution of the original equation. Therefore, you must check all solutions in the original equation when you solve radical equations.

EXAMPLE Variable on Each Side

3 Solve $\sqrt{x+2} = x - 4$.

$\sqrt{x+2} = x - 4$	Original equation
$(\sqrt{x+2})^2 = (x-4)^2$	Square each side.
$x + 2 = x^2 - 8x + 16$	Simplify.
$0 = x^2 - 9x + 14$	Subtract x and 2 from each side.
$0 = (x-7)(x-2)$	Factor.
$x - 7 = 0$ or $x - 2 = 0$	Zero Product Property
$x = 7$ $\quad$ $x = 2$	Solve.

Review Vocabulary

Zero Product Property
For all numbers a and b, if $ab = 0$, then $a = 0$, $b = 0$, or both a and b equal 0. (Lesson 9-2)

CHECK

$$\sqrt{x+2} = x - 4 \quad \text{Original equation}$$
$$\sqrt{7+2} \stackrel{?}{=} 7 - 4 \quad x = 7$$
$$\sqrt{9} \stackrel{?}{=} 3 \quad \text{Simplify.}$$
$$3 = 3 \ \checkmark \quad \text{True}$$
$$\sqrt{x+2} = x - 4 \quad \text{Original equation}$$
$$\sqrt{2+2} \stackrel{?}{=} 2 - 4 \quad x = 2$$
$$\sqrt{4} \stackrel{?}{=} -2 \quad \text{Simplify.}$$
$$2 \neq -2 \ \times \quad \text{False}$$

Since 2 does not satisfy the original equation, 7 is the only solution.

Solve each equation. Check your solution.

3A. $\sqrt{x+5} = x + 3$ **3B.** $x - 3 = \sqrt{x-1}$

GRAPHING CALCULATOR LAB

Solving Radical Equations

You can use a graphing calculator to solve radical equations such as $\sqrt{3x-5} = x - 5$. Clear the Y= list. Enter the left side of the equation as Y1 = $\sqrt{3x-5}$. Enter the right side of the equation as Y2 = $x - 5$. Press [Graph].

THINK AND DISCUSS

1. Sketch what is shown on the screen.
2. Use the intersect feature on the CALC menu, to find the point of intersection.
3. Solve the radical equation algebraically. How does your solution compare to the solution from the graph?

CHECK Your Understanding

Example 1 (pp. 541–542)

Solve each equation. Check your solution.

1. $\sqrt{x} = 5$ **2.** $\sqrt{2b} = -8$ **3.** $\sqrt{7x} = 7$

4. GEOMETRY The surface area of a basketball is x square inches. What is the radius of the basketball if the formula for the surface area of a sphere is $SA = 4\pi r^2$?

Examples 2, 3 (pp. 542–543)

Solve each equation. Check your solution.

5. $\sqrt{8s} + 1 = 5$ **6.** $\sqrt{7x + 18} = 9$ **7.** $\sqrt{5x+1} + 2 = 6$

8. $\sqrt{3x-5} = x - 5$ **9.** $4 + \sqrt{x-2} = x$ **10.** $\sqrt{2x+3} = x$

Exercises

HOMEWORK HELP	
For Exercises	See Examples
11–15, 29–30	1
16–22	2
23–28	3

Solve each equation. Check your solution.

11. $\sqrt{-3a} = 6$
12. $\sqrt{a} = 10$
13. $\sqrt{-k} = 4$
14. $5\sqrt{2} = \sqrt{x}$
15. $3\sqrt{7} = \sqrt{-y}$
16. $3\sqrt{4a} - 2 = 10$
17. $3 + 5\sqrt{n} = 18$
18. $\sqrt{x + 3} = -5$
19. $\sqrt{x - 5} = 2\sqrt{6}$
20. $\sqrt{3x + 12} = 3\sqrt{3}$
21. $\sqrt{2c - 4} = 8$
22. $\sqrt{4b + 1} - 3 = 0$
23. $x = \sqrt{6 - x}$
24. $x = \sqrt{x + 20}$
25. $\sqrt{5x - 6} = x$
26. $\sqrt{28 - 3x} = x$
27. $\sqrt{x + 1} = x - 1$
28. $\sqrt{1 - 2b} = 1 + b$

AVIATION For Exercises 29 and 30, use the following information.
The formula $L = \sqrt{kP}$ represents the relationship between a plane's length L and the pounds P its wings can lift, where k is a constant of proportionality calculated for a plane.

29. The length of the Douglas D-558-II, called the Skyrocket, was approximately 42 feet, and its constant of proportionality was $k = 0.1669$. Calculate the maximum takeoff weight of the Skyrocket.

30. A Boeing 747 is 232 feet long and has a takeoff weight of 870,000 pounds. Determine the value of k for this plane.

31. The square root of the sum of a number and 7 is 8. Find the number.

32. The square root of the quotient of a number and 6 is 9. Find the number.

Real-World Link

Piloted by A. Scott Crossfield on November 20, 1953, the Douglas D-558-2 Skyrocket became the first aircraft to fly faster than Mach 2, twice the speed of sound.

Source: National Air and Space Museum

Solve each equation. Check your solution.

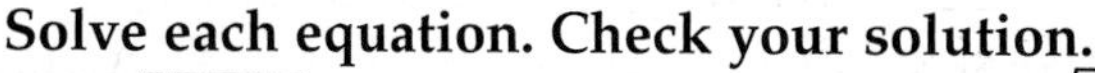

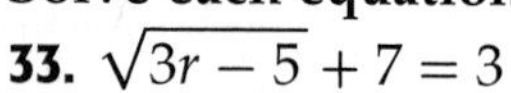

33. $\sqrt{3r - 5} + 7 = 3$
34. $\sqrt{x^2 + 9x + 14} = x + 4$
35. $5\sqrt{\frac{4t}{3}} - 2 = 0$
36. $\sqrt{\frac{4x}{5}} - 9 = 3$
37. $4 + \sqrt{m - 2} = m$
38. $\sqrt{3d - 8} = d - 2$
39. $x + \sqrt{6 - x} = 4$
40. $\sqrt{6 - 3x} = x + 16$
41. $\sqrt{2r^2 - 121} = r$
42. $\sqrt{5p^2 - 7} = 2p$

GEOMETRY For Exercises 43–46, use the figure.
The area A of a circle is equal to πr^2, where r is the radius of the circle.

43. Write an equation for r in terms of A.

44. The area of the larger circle is 96π square meters. Find the radius.

45. The area of the smaller circle is 48π square meters. Find the radius.

46. If the area of a circle is doubled, what is the change in the radius?

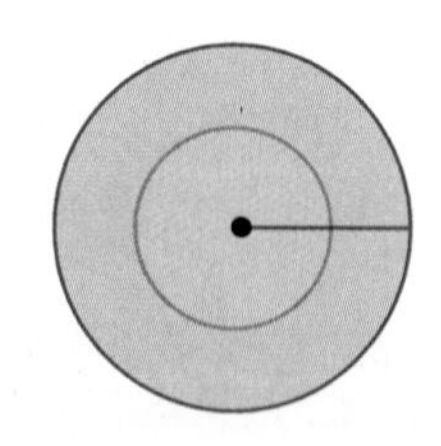

OCEANS For Exercises 47–49, use the following information.
Tsunamis, or large waves, are generated by undersea earthquakes. The speed of the tsunami in meters per second is $s = 3.1\sqrt{d}$, where d is the depth of the ocean in meters.

47. Find the speed of the tsunami if the depth of the water is 10 meters.

48. Find the depth of the water if a tsunami's speed is 240 meters per second.

49. A tsunami may begin as a 2-foot high wave traveling 500 miles per hour. It can approach a coastline as a 50-foot wave. How much speed does the wave lose if it travels from a depth of 10,000 meters to a depth of 20 meters?

EXTRA PRACTICE
See pages 737, 753.

Math Online
Self-Check Quiz at algebra1.com

50. State whether the following equation is *sometimes*, *always*, or *never* true.

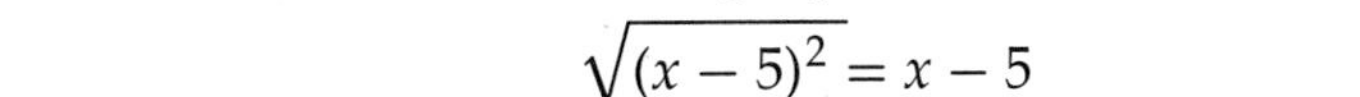

$$\sqrt{(x-5)^2} = x - 5$$

Real-World Link

The Foucault Pendulum appears to change its path during the day, but it moves in a straight line. The path under the pendulum changes because Earth is rotating beneath it.

Source: California Academy of Sciences

PHYSICAL SCIENCE For Exercises 51–53, use the following information.

The formula $P = 2\pi\sqrt{\frac{\ell}{32}}$ gives the period of a pendulum of length ℓ feet. The period P is the number of seconds it takes for the pendulum to swing back and forth once.

51. Suppose we want a pendulum to complete three periods in 2 seconds. How long should the pendulum be?

52. Two clocks have pendulums of different lengths. The first clock requires 1 second for its pendulum to complete one period. The second clock requires 2 seconds for its pendulum to complete one period. How much longer is one pendulum than the other?

53. Repeat Exercise 52 if the pendulum periods are t and $2t$ seconds.

BROADCASTING For Exercises 54–56, use the following information.

Sports broadcasts often include sound collection from the field of play. The temperature affects the speed of sound near Earth's surface. The speed V when the surface temperature t degrees Celsius can be found using the equation $V = 20\sqrt{t + 273}$.

54. Find the temperature at a baseball game if the speed of sound is 346 meters per second.

55. The speed of sound at Earth's surface is often given as 340 meters per second, but that is only accurate at a certain temperature. On what temperature is this figure based?

56. For what speeds is the surface temperature below 0°C?

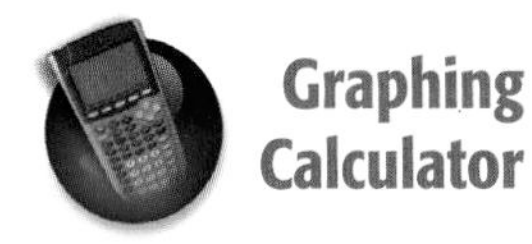

Use a graphing calculator to solve each radical equation. Round to the nearest hundredth.

57. $3 + \sqrt{2x} = 7$

58. $\sqrt{3x - 8} = 5$

59. $\sqrt{x + 6} - 4 = x$

60. $\sqrt{4x + 5} = x - 7$

61. $x + \sqrt{7 - x} = 4$

62. $\sqrt{3x - 9} = 2x + 6$

H.O.T. Problems

63. REASONING Explain why it is necessary to check for extraneous solutions in radical equations.

64. OPEN ENDED Give an example of a radical equation. Then solve the equation for the variable.

65. FIND THE ERROR Alex and Victor are solving $-\sqrt{x - 5} = -2$. Who is correct? Explain your reasoning.

Alex	Victor
$-\sqrt{x-5} = -2$	$-\sqrt{x-5} = -2$
$(-\sqrt{x-5})^2 = (-2)^2$	$-(\sqrt{x-5})^2 = (-2)^2$
$x - 5 = 4$	$-(x-5) = 4$
$x = 9$	$-x + 5 = 4$
	$x = 1$

66. CHALLENGE Solve $\sqrt{h + 9} - \sqrt{h} = \sqrt{3}$.

67. ***Writing in Math*** Use the information about skydiving on page 541 to explain how radical equations can be used to find free-fall times. Include the time it would take a skydiver to fall 10,000 feet if he falls 1200 feet every 5 seconds and also the time it would take using the equation $t = \frac{\sqrt{h}}{4}$, with an explanation of why the two methods find different times.

STANDARDIZED TEST PRACTICE

68. What is the solution for this equation?

$$\sqrt{x+3} - 2 = 7$$

A 22

B 78

C 36

D 15

69. **REVIEW** Mr. and Mrs. Hataro are putting fresh sod onto their yard. The yard is 30 feet wide and 24 feet long, and the sod comes in pieces that are 12 inches wide and 24 inches long. If they decide to cover the entire yard in sod, about how many pieces of sod will they need?

F 2.5

G 30

H 360

J 720

Spiral Review

Simplify. (Lessons 10-2 and 10-1)

70. $5\sqrt{6} + 12\sqrt{6}$

71. $\sqrt{12} + 6\sqrt{27}$

72. $\sqrt{18} + 5\sqrt{2} - 3\sqrt{32}$

73. $\sqrt{192}$

74. $\sqrt{6} \cdot \sqrt{10}$

75. $\frac{21}{\sqrt{10} + \sqrt{3}}$

Find each product. (Lesson 7-6)

76. $(r + 3)(r - 4)$

77. $(3z + 7)(2z + 10)$

78. $(2p + 5)(3p^2 - 4p + 9)$

79. PHYSICAL SCIENCE A European-made hot tub is advertised to have a temperature of 35°C to 40°C, inclusive. What is the temperature range for the hot tub in degrees Fahrenheit? Use $F = \frac{9}{5}C + 32$. (Lesson 6-4)

Write each equation in standard form. (Lesson 4-5)

80. $y = 2x + \frac{3}{7}$

81. $y - 3 = -2(x - 6)$

82. $y + 2 = 7.5(x - 3)$

83. MUSIC The table shows the number of country music radio stations in the United States. What was the percent of change in the number of stations from 2002 to 2004? (Lesson 2-7)

Year	Number of Stations
2002	2131
2004	2047

Source: M Street Corporation

GET READY for the Next Lesson

PREREQUISITE SKILL Evaluate $\sqrt{a^2 + b^2}$ for each value of a and b. (Lesson 1-2)

84. $a = 3, b = 4$

85. $a = 24, b = 7$

86. $a = 5, b = 12$

87. $a = 6, b = 8$

88. $a = 1, b = 1$

89. $a = 8, b = 12$

EXTEND 10-3

Graphing Calculator Lab

Graphs of Radical Equations

In order for a square root to be a real number, the radicand cannot be negative. When graphing a radical equation, determine when the radicand would be negative and exclude those values from the domain.

ACTIVITY 1

Concepts in Motion
Interactive Lab algebra1.com

Graph $y = \sqrt{x}$.

Enter the equation in the Y= list.

KEYSTROKES: [Y=] [2nd] [√‾] [X,T,θ,n] [Graph]

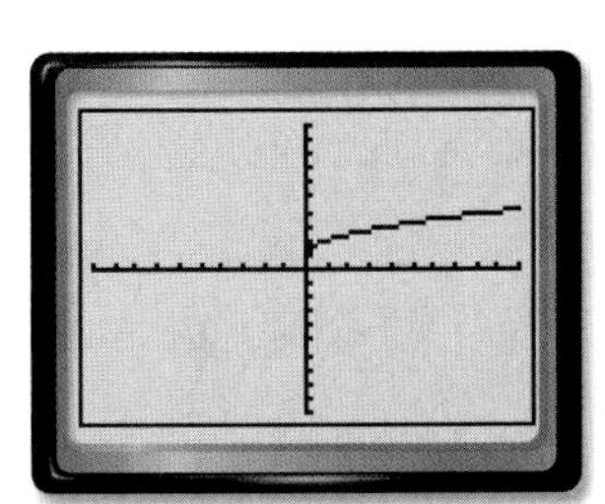

[−10, 10] scl: 1 by [−10, 10] scl: 1

1A. Examine the graph. What is the domain of the function $y = \sqrt{x}$?

1B. What is the range of $y = \sqrt{x}$?

ACTIVITY 2

Graph $y = \sqrt{x + 4}$.

Enter the equation in the Y= list.

KEYSTROKES: [Y=] [2nd] [√‾] [X,T,θ,n] [+] 4 [)] [Graph]

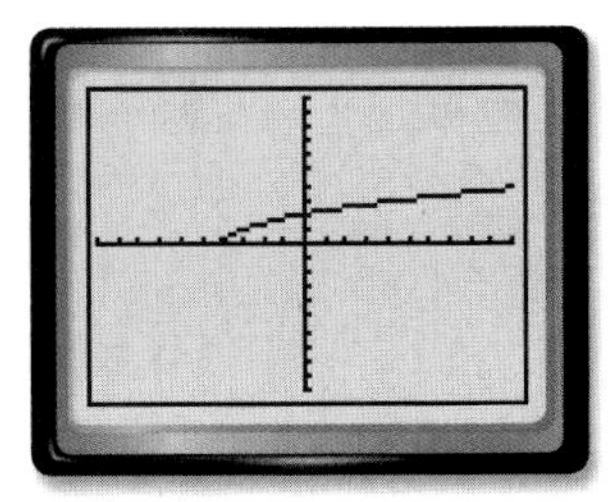

[−10, 10] scl: 1 by [−10, 10] scl: 1

2A. What are the domain and range of $y = \sqrt{x + 4}$?

2B. How does the graph of $y = \sqrt{x + 4}$ compare to the graph of the parent function $y = \sqrt{x}$?

Analyze the Results

Graph each equation and sketch the graph on your paper. State the domain of the graph. Then describe how the graph differs from the parent function $y = \sqrt{x}$.

1. $y = \sqrt{x} + 1$

2. $y = \sqrt{x} - 3$

3. $y = \sqrt{x + 2}$

4. $y = \sqrt{x - 5}$

5. $y = \sqrt{-x}$

6. $y = \sqrt{3x}$

7. $y = -\sqrt{x}$

8. $y = \sqrt{1 - x} + 6$

9. $y = \sqrt{2x + 5} - 4$

10. $y = \sqrt{|x| + 2}$

11. $y = \sqrt{|x| - 3}$

12. Is the graph of $x = y^2$ a function? Explain your reasoning.

13. Does the equation $x^2 + y^2 = 1$ determine y as a function of x? Explain.

14. Write a function whose graph is the graph of $y = \sqrt{x}$ shifted 3 units up.

Math Online **Other Calculator Keystrokes at** algebra1.com

CHAPTER 10

Mid-Chapter Quiz

Lessons 10–1 through 10–3

Simplify. (Lesson 10-1)

1. $\sqrt{48}$
2. $\sqrt{3} \cdot \sqrt{6}$
3. $\frac{3}{2 + \sqrt{10}}$

4. **MULTIPLE CHOICE** If $x = 81b^2$ and $b > 0$, then $\sqrt{x} =$ (Lesson 10-1)

   A $-9b$.

   B $9b$.

   C $3b\sqrt{27}$.

   D $27b\sqrt{3}$.

FENCE **For Exercises 5–7, use the following information.**

Hailey wants to put up a fence. She has a square backyard with an area of 160 square feet. The formula for the area A of a square with side length s is $A = s^2$. (Lesson 10-1)

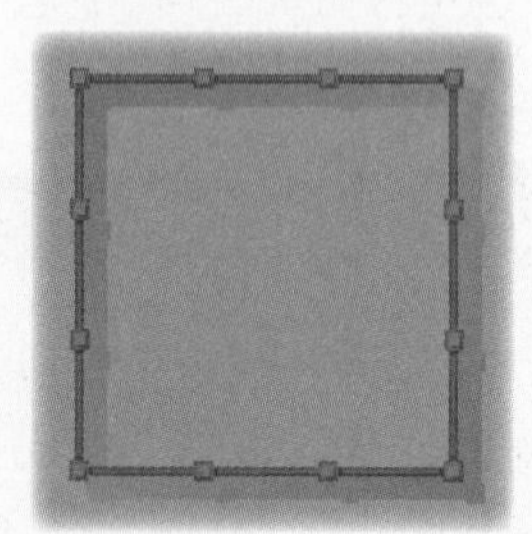

5. Solve the equation for s.
6. What is the side length of Hailey's backyard?
7. What is the perimeter of Hailey's backyard?

8. **GEOMETRY** A rectangle has a length of $\sqrt{\frac{a}{8}}$ meters and a width of $\sqrt{\frac{a}{2}}$ meters. What is the area of the rectangle?

Simplify. (Lesson 10-2)

9. $6\sqrt{5} + 3\sqrt{11} + 5\sqrt{5}$
10. $2\sqrt{3} + 9\sqrt{12}$
11. $(3 - \sqrt{6})^2$

12. **GEOMETRY** Find the area of a square with a side measure of $2 + \sqrt{7}$ centimeters. (Lesson 10-2)

SOUND **For Exercises 13 and 14, use the following information.**

The speed of sound V in meters per second near Earth's surface is given by $V = 20\sqrt{t + 273}$, where t is the surface temperature in degrees Celsius. (Lesson 10-2)

13. What is the speed of sound near Earth's surface at $-1°C$ and at $6°C$ in simplest form?
14. How much faster is the speed of sound at $6°C$ than at $-1°C$?

Solve each equation. Check your solution. (Lesson 10-3)

15. $\sqrt{15 - x} = 4$
16. $\sqrt{3x^2 - 32} = x$
17. $\sqrt{2x - 1} = 2x - 7$

18. **MULTIPLE CHOICE** The surface area S of a cone can be found by using $S = \pi r\sqrt{r^2 + h^2}$, where r is the radius of the base and h is the height of the cone. Find the height of the cone. (Lesson 10-3)

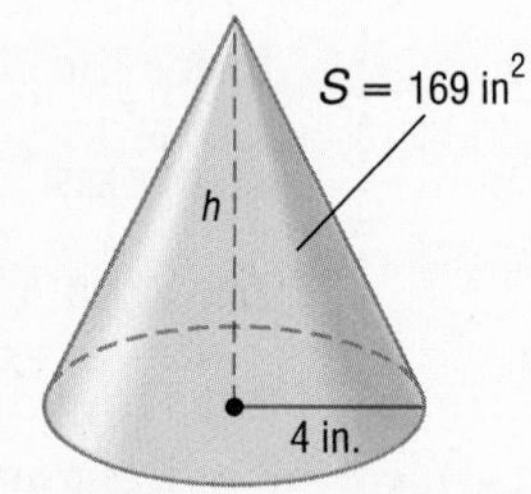

   F 2.70 in.

   G 11.03 in.

   H 12.84 in.

   J 13.30 in.

19. **PHYSICS** When an object is dropped from the top of a 250-foot tall building, the object will be h feet above the ground after t seconds, where $\frac{\sqrt{250 - h}}{4} = t$. How far above the ground will the object be after 1 second? (Lesson 10-3)

20. **SKYDIVING** The approximate time t in seconds that it takes an object to fall a distance of d feet is given by $t = \sqrt{\frac{d}{16}}$. Suppose a parachutist falls 13 seconds before the parachute opens. How far does the parachutist fall during this time period? (Lesson 10-3)

10-4 The Pythagorean Theorem

Main Ideas

- Solve problems by using the Pythagorean Theorem.
- Determine whether a triangle is a right triangle.

New Vocabulary

hypotenuse
legs
Pythagorean triple
converse

GET READY for the Lesson

The *Cyclone* roller coaster at New York's Coney Island is one of the most copied roller coasters in the world. Since it was built in 1927, seven copies of the coaster have been built. The first drop is 85 feet tall, and it descends at a 60° angle. The top speed is 60 miles per hour. You can use the Pythagorean Theorem to estimate the length of the first hill.

The Pythagorean Theorem In a right triangle, the side opposite the right angle is the **hypotenuse**. This side is always the longest side of a right triangle. The other two sides are the **legs**. To find the length of any side of a right triangle when the lengths of the other two are known, use a formula named for the Greek mathematician Pythagoras.

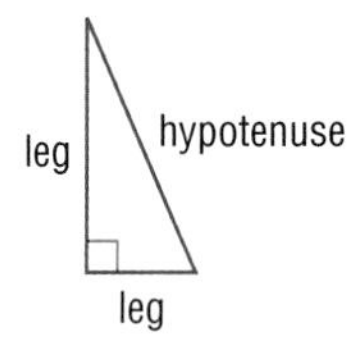

KEY CONCEPT — *The Pythagorean Theorem*

Words If a and b are the lengths of the legs of a right triangle and c is the length of the hypotenuse, then the square of the length of the hypotenuse is equal to the sum of the squares of the lengths of the legs.

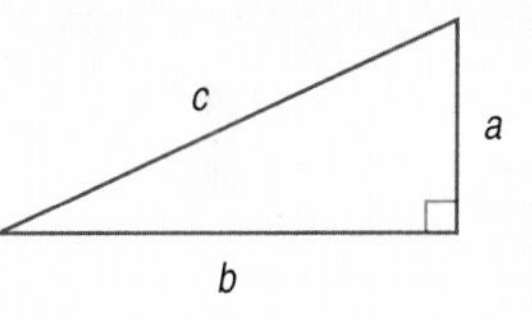

Symbols $c^2 = a^2 + b^2$

Study Tip

Triangles
Sides of a triangle are represented by lowercase letters a, b, and c.

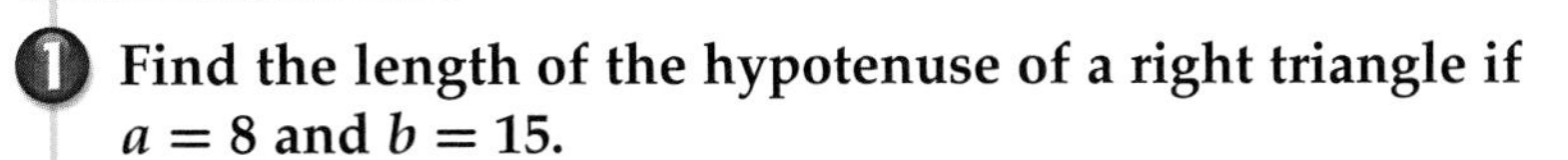

EXAMPLE Find the Length of the Hypotenuse

1 **Find the length of the hypotenuse of a right triangle if $a = 8$ and $b = 15$.**

$c^2 = a^2 + b^2$	Pythagorean Theorem
$c^2 = 8^2 + 15^2$	$a = 8$ and $b = 15$
$c^2 = 289$	Simplify.
$c = \pm\sqrt{289}$	Take the square root of each side.
$c = \pm 17$	Disregard −17. Why?

The length of the hypotenuse is 17 units.

CHECK Your Progress

Find the length of the hypotenuse of each right triangle. If necessary, round to the nearest hundredth.

1A. $a = 7, b = 13, c = ?$ **1B.** $a = 10, b = 6, c = ?$

EXAMPLE Find the Length of a Side

2 **Find the length of the missing side. If necessary, round to the nearest hundredth.**

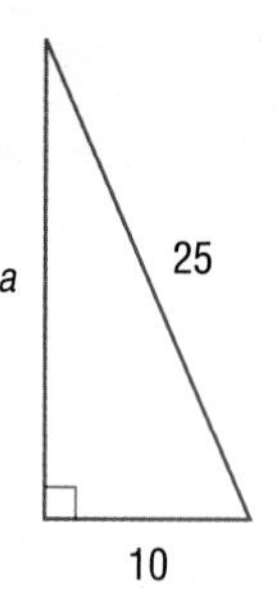

$c^2 = a^2 + b^2$	Pythagorean Theorem
$25^2 = a^2 + 10^2$	$b = 10$ and $c = 25$
$625 = a^2 + 100$	Evaluate squares.
$525 = a^2$	Subtract 100 from each side.
$\pm\sqrt{525} = a$	Use a calculator to evaluate $\sqrt{525}$.
$22.91 \approx a$	Use the positive value.

CHECK Your Progress

2A. $a = 6, c = 14, b = ?$ **2B.** $b = 11, c = 21, a = ?$

A group of three whole numbers that satisfy the Pythagorean Theorem is called a **Pythagorean triple**. Examples include (3, 4, 5) and (5, 12, 13). Multiples of Pythagorean triples also satisfy the Pythagorean Theorem, so (6, 8, 10) and (10, 24, 26) are also Pythagorean triples.

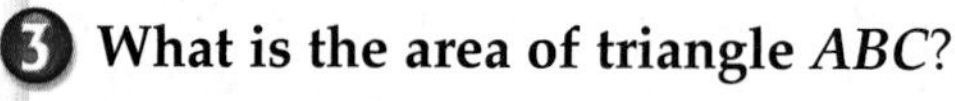

STANDARDIZED TEST EXAMPLE Pythagorean Triples

3 **What is the area of triangle *ABC*?**

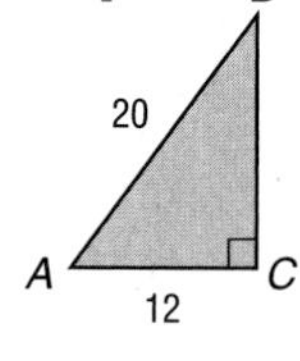

A 96 units^2 **C** 120 units^2

B 160 units^2 **D** 196 units^2

Read the Test Item

The area of a triangle is $A = \frac{1}{2}bh$. In a right triangle, the legs are the base and height. Use the given measures to find the height.

Solve the Test Item

Step 1 Check to see if the measurements are a multiple of a common Pythagorean triple. The hypotenuse is $4 \cdot 5$ units, and the leg is $4 \cdot 3$ units. This triangle is a multiple of a (3, 4, 5) triangle.

$4 \cdot 3 = 12$ $4 \cdot 4 = 16$ $4 \cdot 5 = 20$ The height is 16 units.

Step 2 Find the area of the triangle.

$A = \frac{1}{2}bh$	Area of a triangle
$A = \frac{1}{2} \cdot 12 \cdot 16$	$b = 12$ and $h = 16$
$A = 96$	Choice A is correct.

Test-Taking Tip

Memorize common Pythagorean triples and check for multiples such as (6, 8, 10). This will save you time on the test. Some other common triples are (8, 15, 17) and (7, 24, 25).

CHECK Your Progress

3. A square shopping center has a diagonal walkway from one corner to another. If the walkway is about 70 meters long, what is the approximate length of each side of the center?

F 8 meters **G** 35 meters **H** 50 meters **J** 100 meters

Online Personal Tutor at algebra1.com

Right Triangles If you exchange the hypothesis and conclusion of an if-then statement, the result is the **converse** of the statement. The following theorem, the converse of the Pythagorean Theorem, can be used to determine whether a triangle is a right triangle.

Study Tip

Converses

Converses of theorems are not necessarily true. But the converse of the Pythagorean Theorem is true.

KEY CONCEPT — *Converse of the Pythagorean Theorem*

If a and b are measures of the shorter sides of a triangle, c is the measure of the longest side, and $c^2 = a^2 + b^2$, then the triangle is a right triangle. If $c^2 \neq a^2 + b^2$, then the triangle is not a right triangle.

EXAMPLE Check for Right Triangles

4 **Determine whether the following side measures form right triangles.**

a. 20, 21, 29

Since the measure of the longest side is 29, let $c = 29$, $a = 20$, and $b = 21$. Then determine whether $c^2 = a^2 + b^2$.

$c^2 = a^2 + b^2$	Pythagorean Theorem
$29^2 \stackrel{?}{=} 20^2 + 21^2$	$a = 20, b = 21$, and $c = 29$
$841 \stackrel{?}{=} 400 + 441$	Multiply.
$841 = 841$	Add.

Since $c^2 = a^2 + b^2$, the triangle is a right triangle.

b. 8, 10, 12

Since the measure of the longest side is 12, let $c = 12$, $a = 8$, and $b = 10$. Then determine whether $c^2 = a^2 + b^2$.

$c^2 = a^2 + b^2$	Pythagorean Theorem
$12^2 \stackrel{?}{=} 8^2 + 10^2$	$a = 8, b = 10$, and $c = 12$
$144 \stackrel{?}{=} 64 + 100$	Multiply.
$144 \neq 164$	Add.

Since $c^2 \neq a^2 + b^2$, the triangle is not a right triangle.

CHECK Your Progress

4A. 9, 12, 16 **4B.** 18, 24, 30

CHECK Your Understanding

Examples 1, 2 (pp. 549–550)

If c is the measure of the hypotenuse of a right triangle, find each missing measure. If necessary, round to the nearest hundredth.

1. $a = 10, b = 24, c = ?$

2. $a = 11, c = 61, b = ?$

3. $b = 13, c = \sqrt{233}, a = ?$

4. $a = 7, b = 4, c = ?$

Find the length of each missing side. If necessary, round to the nearest hundredth.

5.

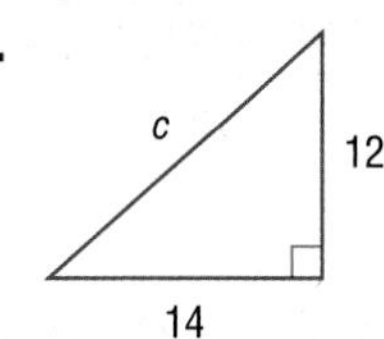

6.

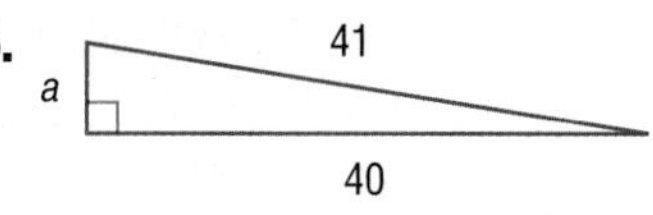

Example 3 (p. 550)

7. STANDARDIZED TEST PRACTICE In right triangle XYZ, the length of $\overline{YZ}$ is 6, and the length of the hypotenuse is 8. Find the area of the triangle.

A $6\sqrt{7}$ units^2 **B** 30 units^2 **C** 40 units^2 **D** 48 units^2

Example 4 (p. 551)

Determine whether the following side measures form right triangles. Justify your answer.

8. 4, 6, 9

9. 10, 24, 26

Exercises

HOMEWORK HELP	
For Exercises	See Examples
10–27	1, 2
28–29	3
30–35	4

If c is the measure of the hypotenuse of a right triangle, find each missing measure. If necessary, round to the nearest hundredth.

10. $a = 16, b = 63, c = ?$

11. $a = 16, c = 34, b = ?$

12. $b = 3, a = \sqrt{112}, c = ?$

13. $a = \sqrt{15}, b = \sqrt{10}, c = ?$

14. $c = 14, a = 9, b = ?$

15. $a = 6, b = 3, c = ?$

16. $b = \sqrt{77}, c = 12, a = ?$

17. $a = 4, b = \sqrt{11}, c = ?$

18. $a = \sqrt{225}, b = \sqrt{28}, c = ?$

19. $a = \sqrt{31}, c = \sqrt{155}, b = ?$

20. $a = 8x, b = 15x, c = ?$

21. $b = 3x, c = 7x, a = ?$

Find the length of each missing side. If necessary, round to the nearest hundredth.

22.

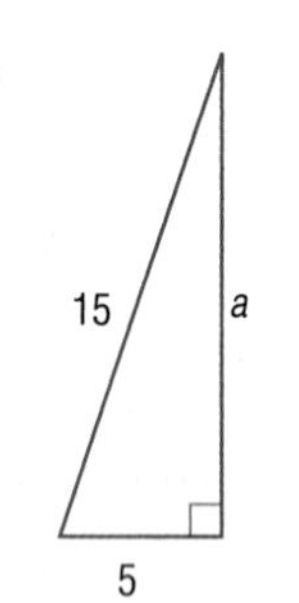

23.

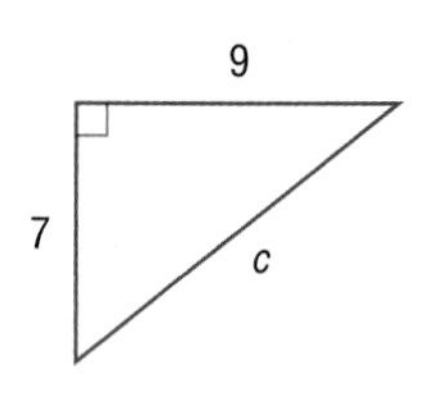

24.

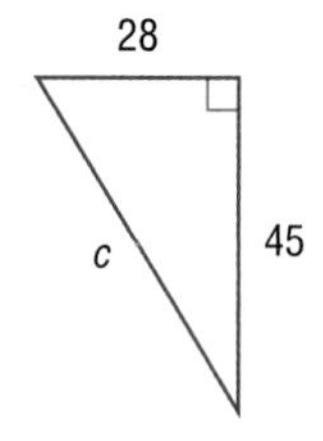

25.

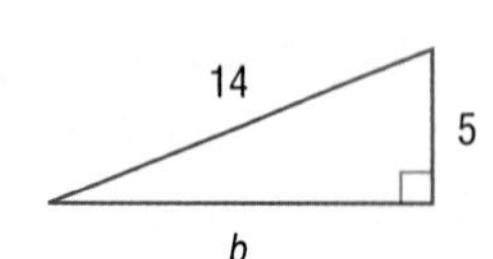

26.

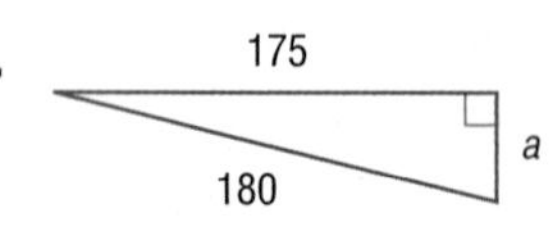

27.

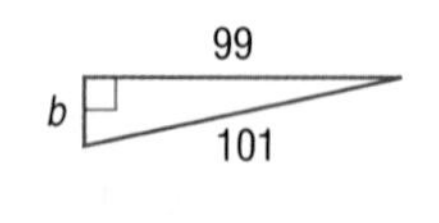

GEOMETRY For Exercises 28 and 29, refer to the triangle.

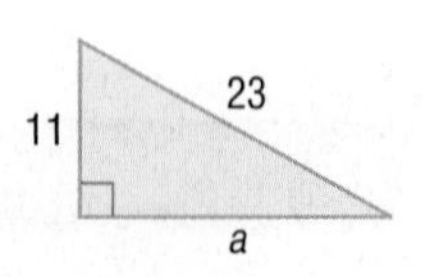

28. What is the length of side a?

29. Find the area of the triangle.

Determine whether the following side measures form right triangles. Justify your answer.

30. 30, 40, 50

31. 6, 12, 18

32. 24, 30, 36

33. 45, 60, 75

34. $15, \sqrt{31}, 16$

35. $4, 7, \sqrt{65}$

Find the length of the hypotenuse. Round to the nearest hundredth.

36.

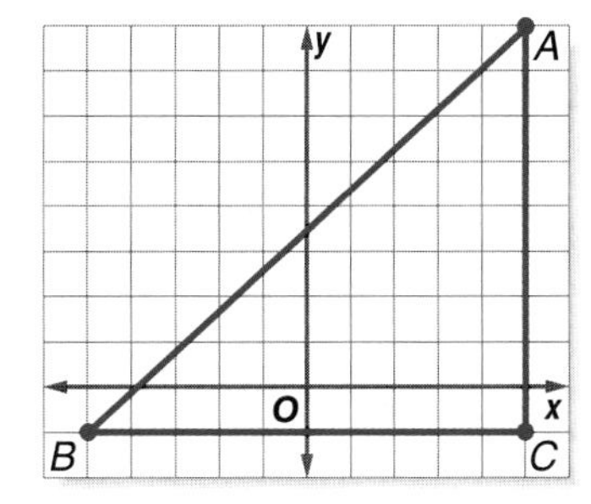

37.

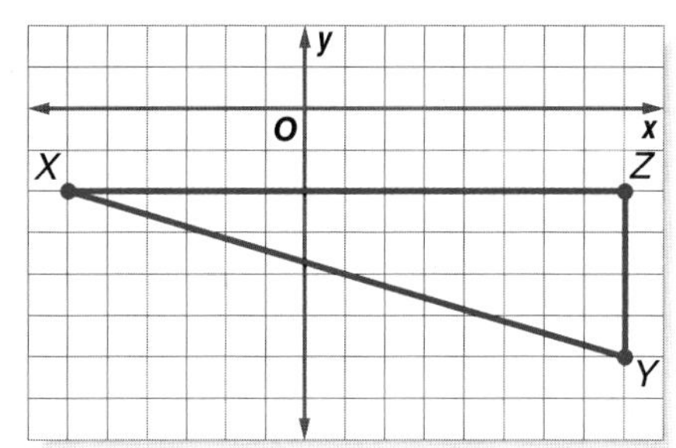

Real-World Link

Fastest, Tallest, Largest Drop: *Kingda Ka*, Jackson, New Jersey; 128 mph, 456 ft, and 418 ft, respectively

Longest: *Millennium Force*, Sandusky, Ohio; 6,595 ft

Source: ultimaterollercoaster.com

ROLLER COASTERS For Exercises 38–40, use the following information.

Suppose a roller coaster climbs 208 feet higher than its starting point making a horizontal advance of 360 feet. When it comes down, it makes a horizontal advance of 44 feet.

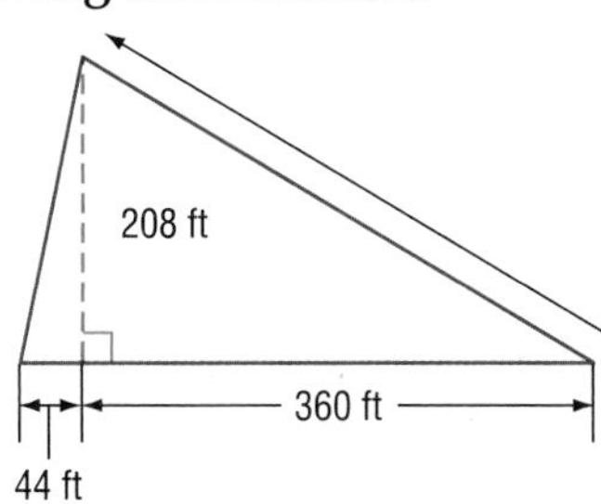

38. How far will it travel to get to the top of the ride?

39. How far will it travel on the downhill track?

40. Compare the total horizontal advance, vertical height, and total track length.

41. RESEARCH Use the Internet or other reference to find the measurements of your favorite roller coaster or a roller coaster that is in an amusement park close to you. Draw a model of the first drop. Include the height of the hill, length of the vertical drop, and steepness of the hill.

42. SAILING A sailboat's mast and boom form a right angle. The sail itself, called a *mainsail*, is in the shape of a right triangle. If the edge of the mainsail that is attached to the mast is 100 feet long and the edge of the mainsail that is attached to the boom is 60 feet long, what is the length of the longest edge of the mainsail?

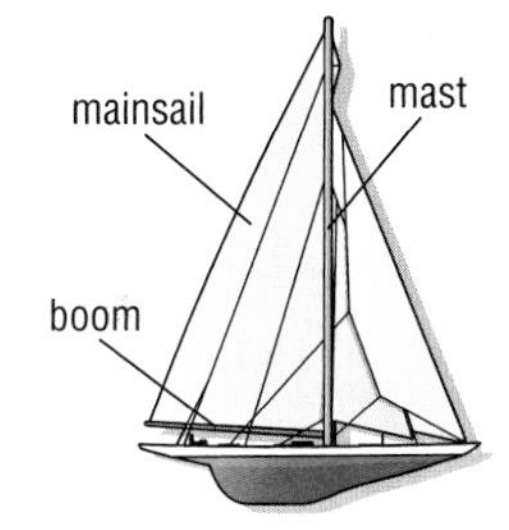

Solve each problem. If necessary, round to the nearest hundredth.

43. Find the length of a diagonal of a square if its area is 162 square feet.

44. A right triangle has one leg that is 5 centimeters longer than the other leg. The hypotenuse is 25 centimeters long. Find the length of each leg of the triangle.

45. Find the length of the diagonal of a cube if each side of the cube is 4 inches long.

46. The ratio of the length of the hypotenuse to the length of the *shorter* leg in a right triangle is 8:5. The hypotenuse measures 144 meters. Find the length of the *longer* leg.

See pages 738, 753.

Self-Check Quiz at algebra1.com

47. ROOFING A garage roof is 30 feet long and hangs an additional 2 feet over the walls. How many square feet of shingles are needed for the entire roof?

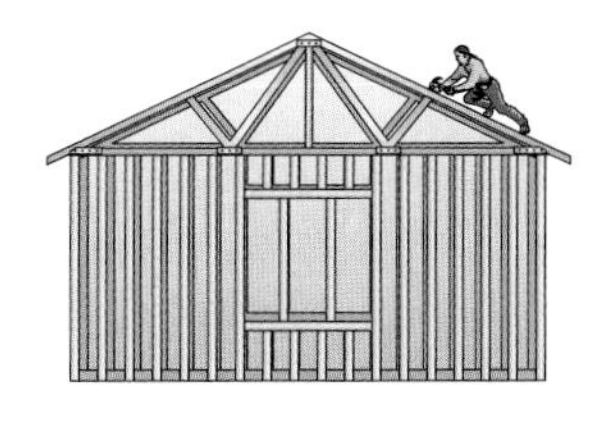

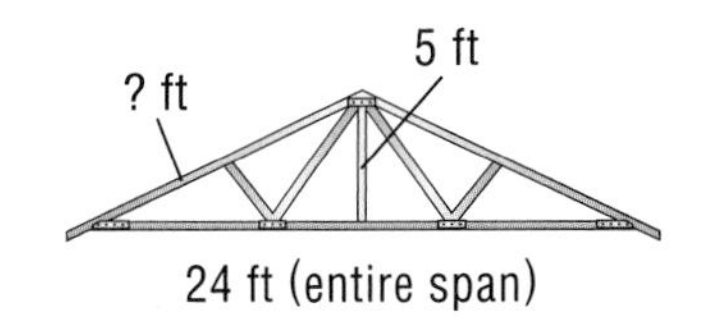

H.O.T. Problems

48. **OPEN ENDED** Draw and label a right triangle with legs and hypotenuse with rational lengths. Draw a second triangle with legs of irrational lengths and a hypotenuse of rational length.

49. **CHALLENGE** Compare the area of the largest semicircle to the areas of the two smaller semicircles at the right.

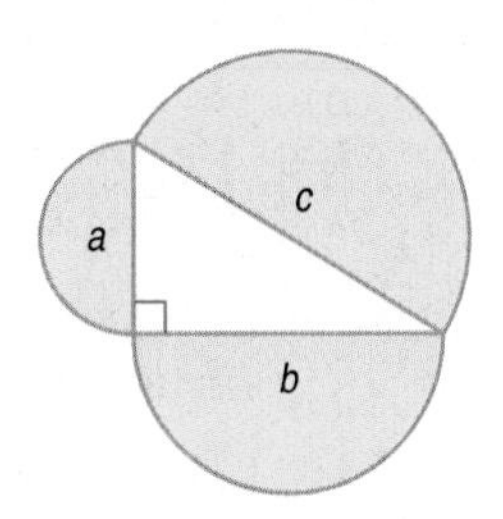

50. ***Writing in Math*** Use the information on page 549 to explain how the Pythagorean Theorem can be used in designing roller coasters. How are the height, speed, and steepness of a roller coaster related?

STANDARDIZED TEST PRACTICE

51. If the perimeter of square 1 is 160 units and the perimeter of square 2 is 120 units, what is the perimeter of square 3?

A 100 units

B 200 units

C 250 units

D 450 units

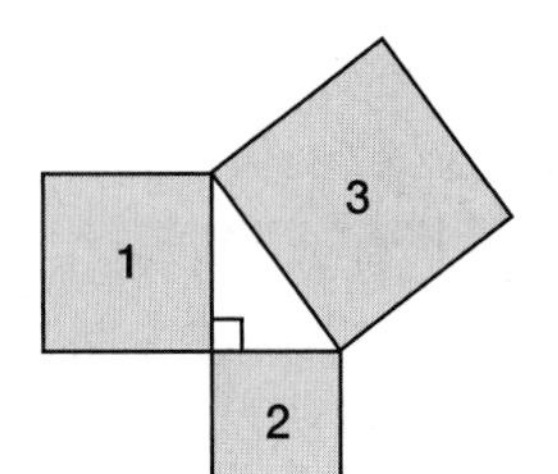

52. **REVIEW** Sara says that if x is a real number then any value of $n\sqrt[n]{x^n} = x$, but Jamal says that this is not always true. Which values could Sara use to prove that she is right?

F $x = 4$ and $n = 2$

G $x = 5$ and $n = -3$

H $x = 7$ and $n = 1$

J $x = 3$ and $n = 2$

Spiral Review

Solve each equation. Check your solution. (Lesson 10-3)

53. $\sqrt{y} = 12$

54. $3\sqrt{s} = 126$

55. $4\sqrt{2v + 1} - 3 = 17$

Simplify. (Lesson 10-2)

56. $\sqrt{72}$

57. $7\sqrt{z} - 10\sqrt{z}$

58. $\sqrt{\frac{3}{7}} + \sqrt{21}$

59. **AVIATION** Flying with the wind, a plane travels 300 miles in 40 minutes. Flying against the wind, it travels 300 miles in 45 minutes. Find its air speed. (Lesson 5-4)

Write an equation in function notation for each relation. (Lesson 3-5)

60.

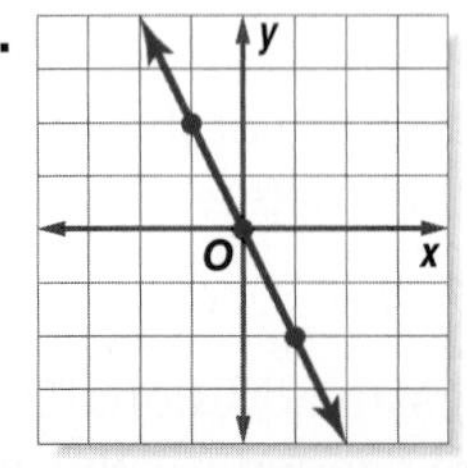

61.

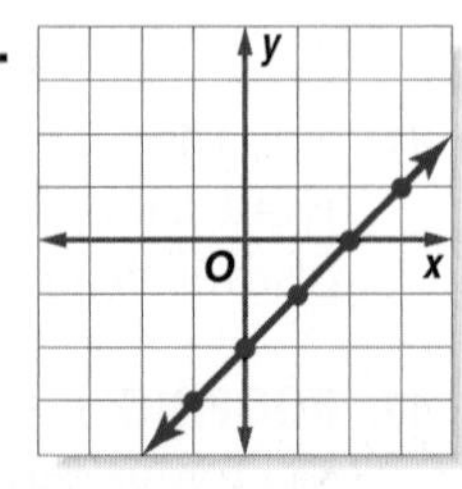

GET READY for the Next Lesson

PREREQUISITE SKILL Simplify each expression. (Lesson 10-1)

62. $\sqrt{(6 - 3)^2 + (8 - 4)^2}$

63. $\sqrt{(10 - 4)^2 + (13 - 5)^2}$

64. $\sqrt{(5 - 3)^2 + (2 - 9)^2}$

10-5 The Distance Formula

Main Ideas

- Find the distance between two points on the coordinate plane.
- Find a point that is a given distance from a second point on a plane.

New Vocabulary

Distance Formula

GET READY for the Lesson

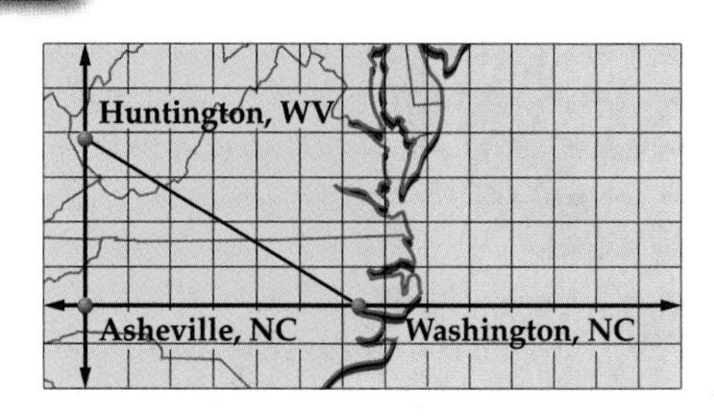

A certain helicopter can fly 450 miles before it needs to refuel. Suppose a person needs to be flown from a hospital in Washington, North Carolina, to one in Huntington, West Virginia. Each side of a square is 50 miles. If Asheville, North Carolina, is at the origin, Huntington is at (0, 196), and Washington is at (310, 0), can the helicopter make the trip one way without refueling?

The Distance Formula You can find the distance between any two points in the coordinate plane using the **Distance Formula**, which is based on the Pythagorean Theorem.

KEY CONCEPT — The Distance Formula

Words The distance d between any two points with coordinates (x_1, y_1) and (x_2, y_2) is given by $d = \sqrt{(x_2 - x_1)^2 + (y_2 - y_1)^2}$.

Model

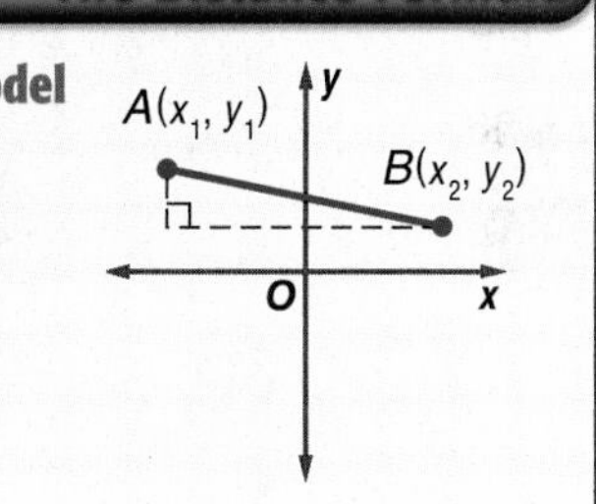

EXAMPLE Distance Between Two Points

1 Find the distance between the points at (2, 3) and (–4, 6).

$d = \sqrt{(x_2 - x_1)^2 + (y_2 - y_1)^2}$ Distance Formula

$= \sqrt{(-4 - 2)^2 + (6 - 3)^2}$ $(x_1, y_1) = (2, 3)$ and $(x_2, y_2) = (-4, 6)$

$= \sqrt{(-6)^2 + 3^2}$ Simplify.

$= \sqrt{45}$ Evaluate squares and simplify.

$= 3\sqrt{5}$ or about 6.71 units

CHECK Your Progress

1. Find the distance between the points at (4, –1) and (–2, –5).

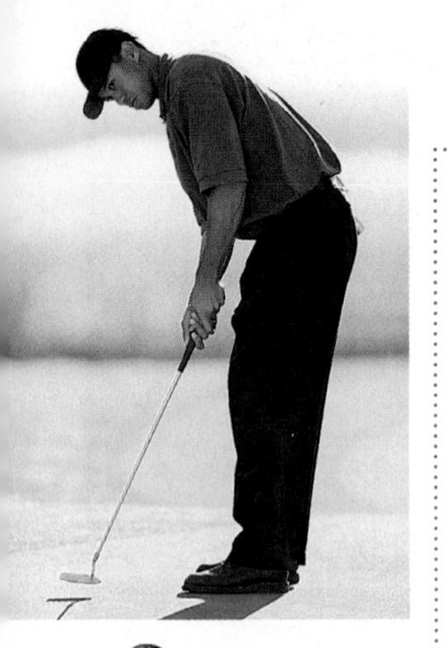

Real-World Link

There are four major tournaments that make up the "grand slam" of golf: Masters, U.S. Open, British Open, and PGA Championship. At age 24, Tiger Woods became the youngest player to win the four major events (called a career grand slam).

Source: PGA

Real-World EXAMPLE

2 **GOLF Tracy's golf ball is 20 feet short and 8 feet to the right of the cup. On her first putt, the ball lands 2 feet to the left and 3 feet beyond the cup. If the ball went in a straight line, how far did it go?**

Model the situation. If the cup is at (0, 0), then the location of the ball is (8, –20). The location after the first putt is (–2, 3).

$d = \sqrt{(x_2 - x_1)^2 + (y_2 - y_1)^2}$ — Distance Formula

$= \sqrt{(-2 - 8)^2 + [3 - (-20)]^2}$ — $(x_1, y_1) = (8, -20)$, $(x_2, y_2) = (-2, 3)$

$= \sqrt{(-10)^2 + 23^2}$ — Simplify.

$= \sqrt{629}$ or about 25

The ball traveled about 25 feet on her first putt.

CHECK Your Progress

2. Shelly hit the golf ball 12 feet past the hole and 3 feet to the left. Her first putt traveled to 2 feet beyond the cup and 1 foot to the right. How far did the ball travel on her first putt?

Find Coordinates Suppose you know the coordinates of a point, one coordinate of another point, and the distance between the two points. You can use the Distance Formula to find the missing coordinate.

EXAMPLE Find a Missing Coordinate

3 **Find the possible values of *a* if the distance between the points at (7, 5) and (*a*, –3) is 10 units.**

$d = \sqrt{(x_2 - x_1)^2 + (y_2 - y_1)^2}$ — Distance Formula

$10 = \sqrt{(a - 7)^2 + (-3 - 5)^2}$ — Let $x_2 = a$, $x_1 = 7$, $y_2 = -3$, $y_1 = 5$, and $d = 10$.

$10 = \sqrt{(a - 7)^2 + (-8)^2}$ — Simplify.

$10 = \sqrt{a^2 - 14a + 49 + 64}$ — Evaluate squares.

$10 = \sqrt{a^2 - 14a + 113}$ — Simplify.

$100 = a^2 - 14a + 113$ — Square each side.

$0 = a^2 - 14a + 13$ — Subtract 100 from each side.

$0 = (a - 1)(a - 13)$ — Factor.

$a - 1 = 0$ or $a - 13 = 0$ — Zero Product Property

$a = 1$ $\quad a = 13$ — The value of a is 1 or 13.

CHECK Your Progress

3. Find the value of a if the distance between the points at $(3, a)$ and $(-4, 5)$ is $\sqrt{58}$ units.

Online Personal Tutor at algebra1.com

CHECK Your Understanding

Example 1 (p. 555)

Find the distance between each pair of points with the given coordinates. Express in simplest radical form and as decimal approximations rounded to the nearest hundredth, if necessary.

1. $(5, -1), (11, 7)$
2. $(3, 7), (-2, -5)$
3. $(2, 2), (5, -1)$
4. $(-3, -5), (-6, -4)$

Example 2 (p. 556)

5. GEOMETRY An isosceles triangle has two sides of equal length. Determine whether $\triangle ABC$ with vertices $A(-3, 4)$, $B(5, 2)$, and $C(-1, -5)$ is isosceles.

FOOTBALL For Exercises 6 and 7, use the information at the right.

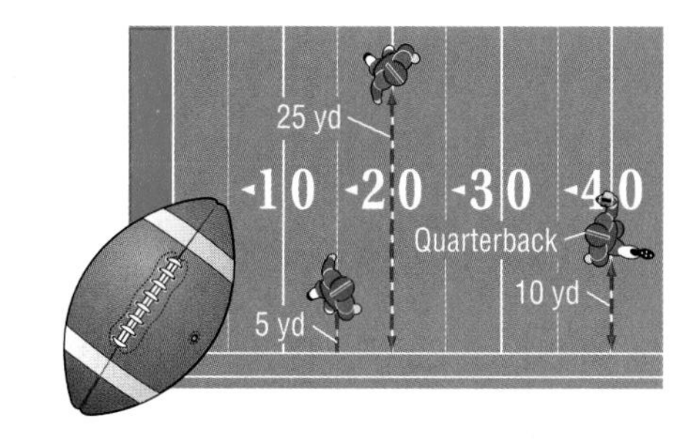

6. A quarterback can throw the football to one of the two receivers. Find the distance from the quarterback to each receiver.

7. What is the distance between the two receivers?

Example 3 (p. 556)

Find the possible values of a if the points with the given coordinates are the indicated distance apart.

8. $(3, -1), (a, 7); d = 10$
9. $(10, a), (1, -6); d = \sqrt{145}$

Exercises

HOMEWORK HELP	
For Exercises	See Examples
10–17	1
18–23	2
24–29	3

Find the distance between each pair of points whose coordinates are given. Express in simplest radical form and as decimal approximations rounded to the nearest hundredth, if necessary.

10. $(12, 3), (-8, 3)$
11. $(0, 0), (5, 12)$
12. $(6, 8), (3, 4)$
13. $(-4, 2), (4, 17)$
14. $(-3, 8), (5, 4)$
15. $(9, -2), (3, -6)$
16. $(-8, -4), (-3, -8)$
17. $(2, 7), (10, -4)$

18. FREQUENT FLYERS To determine the mileage between cities for their frequent flyer programs, some airlines superimpose a coordinate grid over the United States. The units of this grid are approximately equal to 0.316 mile. So, a distance of 3 units on the grid equals an actual distance of 3(0.316) or 0.948 mile. Suppose the locations of two airports are at (132, 428) and (254, 105). Find the actual distance between these airports to the nearest mile.

COLLEGE For Exercises 19 and 20, use the map of a college campus.

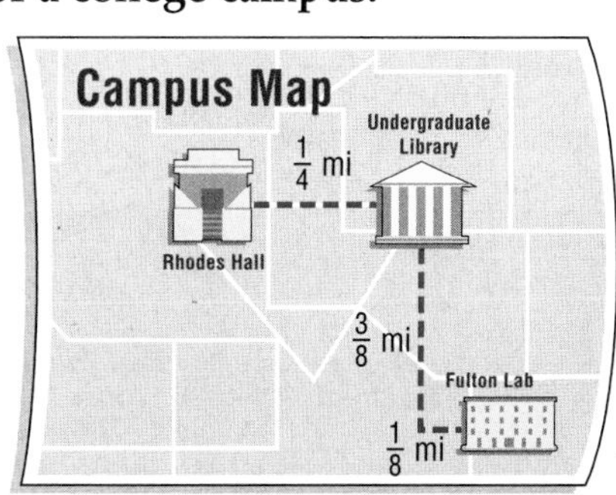

19. Kelly has her first class in Rhodes Hall and her second class in Fulton Lab. How far does she have to walk between her first and second classes?

20. She has 12 minutes between the end of her first class and the start of her second class. If she walks an average of 3 miles per hour, will she make it to her second class on time?

GEOGRAPHY For Exercises 21–23, use the map at the right that shows part of Minnesota and Wisconsin.

A coordinate grid has been superimposed on the map with the origin at St. Paul. The grid lines are 20 miles apart. Minneapolis is at $(-7, 3)$.

21. Estimate the coordinates for Duluth, St. Cloud, Eau Claire, and Rochester.

22. Find the distance between the following pairs of cities: Minneapolis and St. Cloud, St. Paul and Rochester, Minneapolis and Eau Claire, and Duluth and St. Cloud.

23. A radio station in St. Paul has a range of 75 miles. Which cities shown can receive the broadcast?

Find the possible values of a if the points with the given coordinates are the indicated distance apart.

24. $(4, 7), (a, 3); d = 5$

25. $(-4, a), (4, 2); d = 17$

26. $(5, a), (6, 1); d = \sqrt{10}$

27. $(a, 5), (-7, 3); d = \sqrt{29}$

28. $(6, -3), (-3, a); d = \sqrt{130}$

29. $(20, 5), (a, 9); d = \sqrt{340}$

30. GEOMETRY Triangle ABC has vertices $A(7, -4)$, $B(-1, 2)$, and $C(5, -6)$. Determine whether the triangle has three, two, or no sides that are equal in length.

31. GEOMETRY If the diagonals of a trapezoid have the same length, then the trapezoid is isosceles. Is trapezoid $ABCD$ with vertices $A(-2, 2)$, $B(10, 6)$, $C(9, 8)$, and $D(0, 5)$ isosceles? Explain.

32. GEOMETRY Triangle LMN has vertices $L(-4, -3)$, $M(2, 5)$, and $N(-13, 10)$. If the distance from point $P(x, -2)$ to L equals the distance from P to M, what is the value of x?

33. GEOMETRY Plot the points $Q(1, 7)$, $R(3, 1)$, $S(9, 3)$, and $T(7, d)$. Find the value of d so that each side of $QRST$ has the same length.

Find the distance between each pair of points with the given coordinates. Express in simplest radical form and as decimal approximations rounded to the nearest hundredth, if necessary.

EXTRA PRACTICE
See page 738, 753.
Math Online
Self-Check Quiz at algebra1.com

34. $(4, 2), \left(6, -\frac{2}{3}\right)$

35. $\left(5, \frac{1}{4}\right), (3, 4)$

36. $\left(\frac{4}{5}, -1\right), \left(2, -\frac{1}{2}\right)$

37. $\left(3, \frac{3}{7}\right), \left(4, -\frac{2}{7}\right)$

38. $\left(4\sqrt{5}, 7\right), \left(6\sqrt{5}, 1\right)$

39. $\left(5\sqrt{2}, 8\right), \left(7\sqrt{2}, 10\right)$

40. OPEN ENDED Plot two ordered pairs and find the distance between their graphs. Does it matter which ordered pair is first when using the Distance Formula? Explain.

41. REASONING Explain why the value calculated under the radical sign in the Distance Formula will never be negative.

42. REASONING Explain why there are two values for a in Example 3. Draw a diagram to support your answer.

43. CHALLENGE Plot $A(-4, 4)$, $B(-7, -3)$, and $C(4, 0)$, and connect them to form triangle ABC. Demonstrate two different ways to determine whether ABC is a right triangle.

44. *Writing in Math* Use the information on page 555 to explain how the Distance Formula can be used to find the distance between two cities. Explain how the Distance Formula is derived from the Pythagorean Theorem, and why the helicopter can or cannot make the trip without refueling.

STANDARDIZED TEST PRACTICE

45. Find the perimeter of a square $ABCD$ if two of the vertices are $A(3, 7)$ and $B(-3, 4)$.

A 12 units

B $12\sqrt{5}$ units

C $9\sqrt{5}$ units

D 45 units

46. REVIEW Helen is making a scale model of her room. She uses a $\frac{1}{16}$ scale, and the dimensions of her model are $w = 9$ inches and $\ell = 12$ inches. What are the actual dimensions of her room?

F $w = 3$ ft and $\ell = 4$ ft

G $w = 6\frac{3}{4}$ ft and $\ell = 9$ ft

H $w = 12$ ft and $\ell = 16$ ft

J $w = 144$ ft and $\ell = 192$ ft

Spiral Review

If c is the measure of the hypotenuse of a right triangle, find each missing measure. If necessary, round to the nearest hundredth. (Lesson 10-4)

47. $a = 7, b = 24, c = ?$

48. $b = 30, c = 34, a = ?$

49. $a = \sqrt{7}, c = \sqrt{16}, b = ?$

Solve each equation. Check your solution. (Lesson 10-3)

50. $\sqrt{p - 2} + 8 = p$

51. $\sqrt{r + 5} = r - 1$

52. $\sqrt{5t^2 + 29} = 2t + 3$

Solve each inequality. Then check your solution and graph it on a number line. (Lesson 6-1)

53. $8 \leq m - 1$

54. $3 > 10 + k$

55. $3x \leq 2x - 3$

56. $s + \frac{1}{6} \leq \frac{2}{3}$

57. TRAVEL Two trains leave the station at the same time going in opposite directions. The first train travels south at a speed of 60 miles per hour, and the second train travels north at a speed of 75 miles per hour. How many hours will it take for the trains to be 675 miles apart? (Lesson 2-9)

GET READY for the Next Lesson

PREREQUISITE SKILL Solve each proportion. (Lesson 2-6)

58. $\frac{x}{4} = \frac{3}{2}$

59. $\frac{20}{x} = \frac{-5}{2}$

60. $\frac{6}{9} = \frac{8}{x}$

61. $\frac{2}{3} = \frac{6}{x + 4}$

10-6 Similar Triangles

Main Ideas

- Determine whether two triangles are similar.
- Find the unknown measures of sides of two similar triangles.

New Vocabulary

similar triangles

GET READY for the Lesson

The image of an object being photographed is projected by the camera lens onto the film. The height of the image on the film can be related to the height of the object using similar triangles.

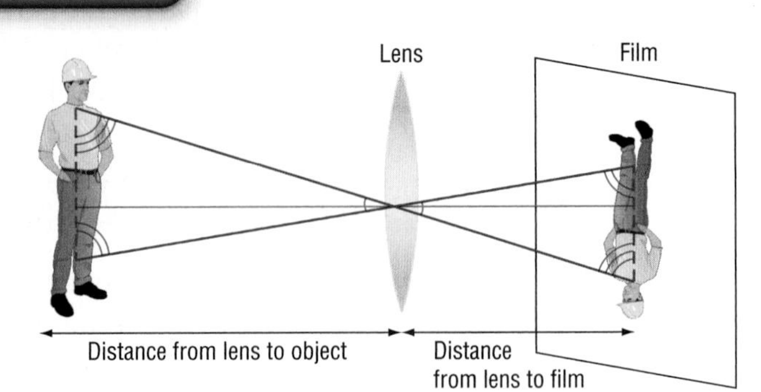

Similar Triangles **Similar triangles** have the same shape, but not necessarily the same size. There are two main tests for similarity.

- If the angles of one triangle and the corresponding angles of a second triangle have equal measures, then the triangles are similar.
- If the measures of the sides of two triangles form equal ratios, or are *proportional*, then the triangles are similar.

The triangles below are similar. The vertices of similar triangles are written in order to show the corresponding parts. So, $\triangle ABC \sim \triangle DEF$.

The symbol ~ is read *is similar to*.

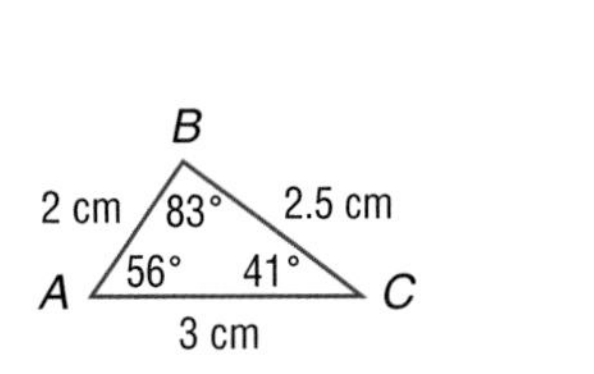

corresponding angles	corresponding sides
$\angle A$ and $\angle D$	$\overline{AB}$ and $\overline{DE} \rightarrow \frac{AB}{DE} = \frac{2}{4} = \frac{1}{2}$
$\angle B$ and $\angle E$	$\overline{BC}$ and $\overline{EF} \rightarrow \frac{BC}{EF} = \frac{2.5}{5} = \frac{1}{2}$
$\angle C$ and $\angle F$	$\overline{AC}$ and $\overline{DF} \rightarrow \frac{AC}{DF} = \frac{3}{6} = \frac{1}{2}$

KEY CONCEPT — Similar Triangles

Words If two triangles are similar, then the measures of their corresponding sides are proportional, and the measures of their corresponding angles are equal.

Symbols $\triangle$If $ABC \sim \triangle DEF$, then $\frac{AB}{DE} = \frac{BC}{EF} = \frac{AC}{DF}$.

Model

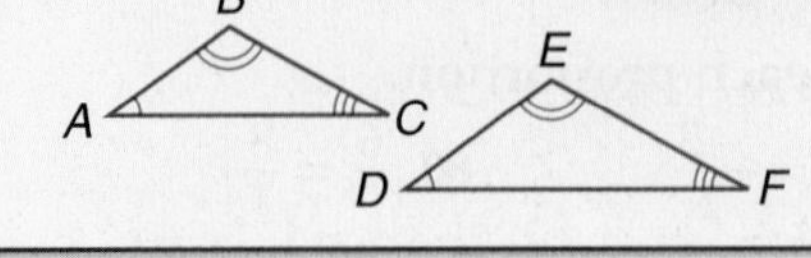

Concepts in Motion
Animation algebra1.com

EXAMPLE Determine Whether Two Triangles Are Similar

1 Determine whether the pair of triangles is similar. Justify your answer.

Remember that the sum of the measures of the angles in a triangle is 180°.

Concepts in Motion
BrainPOP®
algebra1.com

The measure of $\angle P$ is $180° - (51° + 51°)$ or $78°$.

In $\triangle MNO$, $\angle N$ and $\angle O$ have the same measure.

Let x = the measure of $\angle N$ and $\angle O$.

$$x + x + 78° = 180°$$
$$2x = 102°$$
$$x = 51°$$

So $\angle N = 51°$ and $\angle O = 51°$. Since the corresponding angles have equal measures, $\triangle MNO \sim \triangle PQR$.

1. Determine whether $\angle ABC$ with $m\angle A = 68$ and $m\angle B = m\angle C$ is similar to $\triangle DEF$ with $m\angle E = m\angle F = 56$. Justify your answer.

Find Unknown Measures Proportions can be used to find missing measures of the sides of similar triangles, when some of the measurements are known.

EXAMPLE Find Missing Measures

2 Find the missing measures for each pair of similar triangles.

a. Since the corresponding angles have equal measures, $\triangle TUV \sim \triangle WXY$. The lengths of the corresponding sides are proportional.

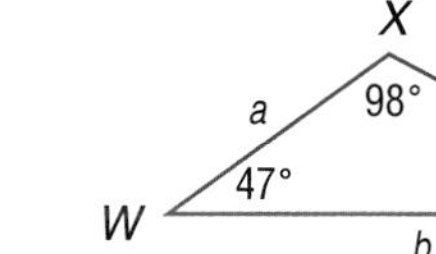

Corresponding Vertices

Always use the corresponding order of the vertices to write proportions for similar triangles.

$\frac{WX}{TU} = \frac{XY}{UV}$ Corresponding sides of similar triangles are proportional.

$\frac{a}{3} = \frac{16}{4}$ $WX = a, XY = 16, TU = 3, UV = 4$

$4a = 48$ Find the cross products.

$a = 12$ Divide each side by 4.

$\frac{WY}{TV} = \frac{XY}{UV}$ Corresponding sides of similar triangles are proportional.

$\frac{b}{6} = \frac{16}{4}$ $WY = b, XY = 16, TV = 6, UV = 4$

$4b = 96$ Find the cross products.

$b = 24$ Divide each side by 4.

The missing measures are 12 and 24.

b. $\triangle ABE \sim \triangle ACD$

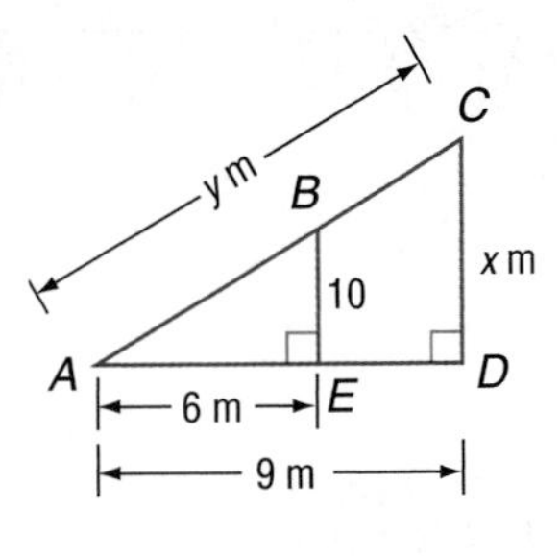

$\frac{BE}{CD} = \frac{AE}{AD}$	Corresponding sides of similar triangles are proportional.
$\frac{10}{x} = \frac{6}{9}$	$BE = 10$, $CD = x$, $AE = 6$, $AD = 9$
$90 = 6x$	Find the cross products.
$15 = x$	Divide each side by 6.
$15^2 + 9^2 = y^2$	Pythagorean Theorem
$\sqrt{306}$ or $17.49 = y$	Simplify.

CHECK Your Progress

2A.

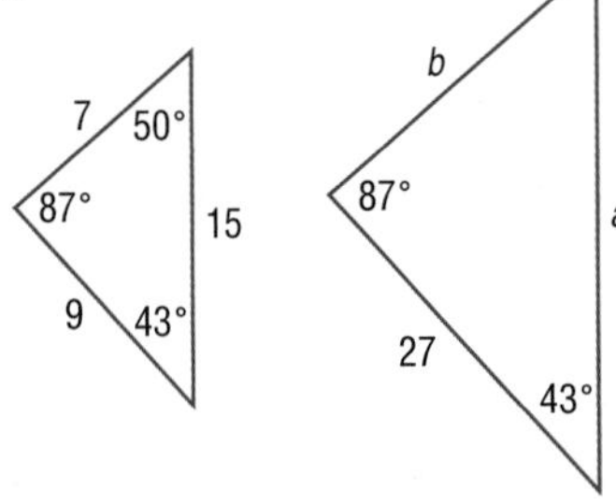

2B.

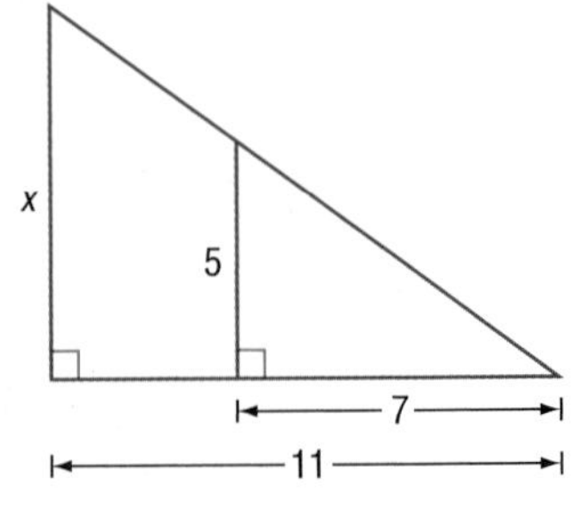

Online Personal Tutor at algebra1.com

Real-World EXAMPLE

The monument has a shape of an Egyptian *obelisk*. A pyramid made of solid aluminum caps the top of the monument.

Source: nps.gov

Real-World Link

3 SHADOWS Jenelle is standing near the Washington Monument in Washington, D.C. The shadow of the monument is 151.5 feet, and Jenelle's shadow is 1.5 feet. If Jenelle is 5.5 feet tall, how tall is the monument?

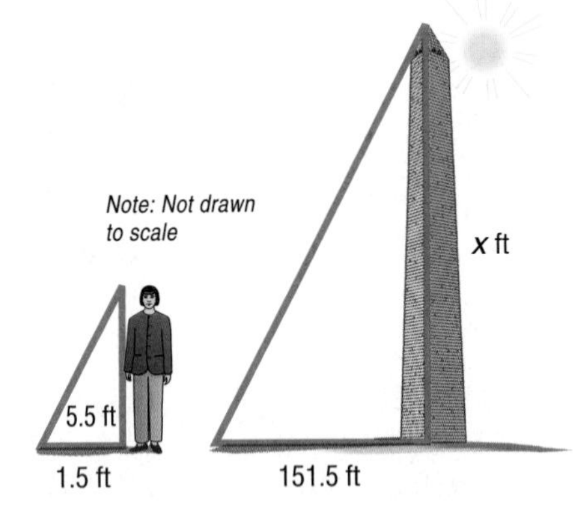

The shadows form similar triangles. Write a proportion that compares the heights of the objects and the lengths of their shadows.

Let x = the height of the monument.

Jenelle's shadow → $\frac{1.5}{151.5} = \frac{5.5}{x}$ ← Jenelle's height
monument's shadow → ← monument's height

$1.5x = 833.25$ Cross products

$x = 555.5$ Divide each side by 1.5.

The height of the monument is about 555.5 feet.

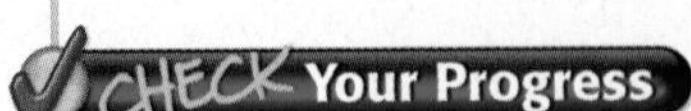

CHECK Your Progress

3. Jody is trying to follow the directions that explain how to pitch a triangular tent. The directions include a scale drawing where 1 inch = 4.5 feet. In the drawing, the tent is $1\frac{3}{4}$ inches tall. How tall should the actual tent be?

CHECK Your Understanding

Example 1 (p. 561)

Determine whether each pair of triangles is similar. Justify your answer.

1.

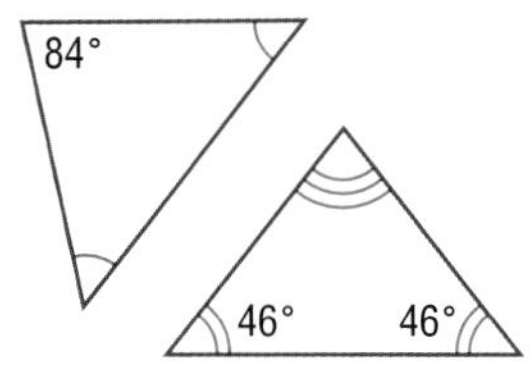

2.

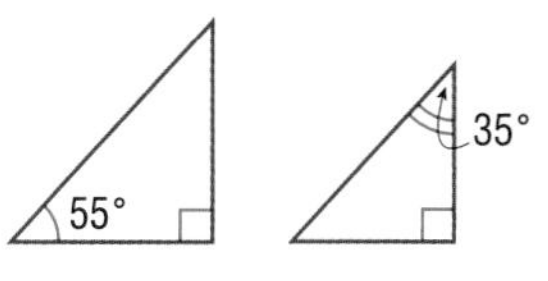

Example 2 (pp. 561–562)

For each set of measures given, find the measures of the missing sides if $\triangle ABC \sim \triangle DEF$.

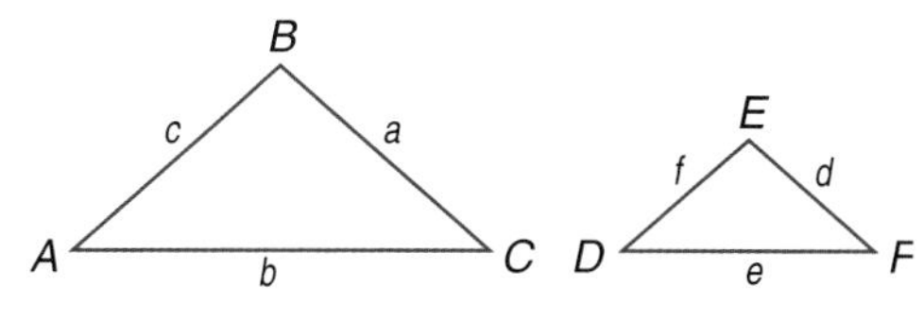

3. $c = 15, d = 7, e = 9, f = 5$

4. $a = 18, c = 9, e = 10, f = 6$

5. $a = 5, d = 7, f = 6, e = 5$

6. $a = 17, b = 15, c = 10, f = 6$

Example 3 (p. 562)

7. SHADOWS A 25-foot flagpole casts a shadow that is 10 feet long and the nearby building casts a shadow that is 26 feet long. How tall is the building?

Exercises

HOMEWORK HELP	
For Exercises	See Examples
8–13	1
14–21	2
22–24	3

Determine whether each pair of triangles is similar. Justify your answer.

8. **9.**

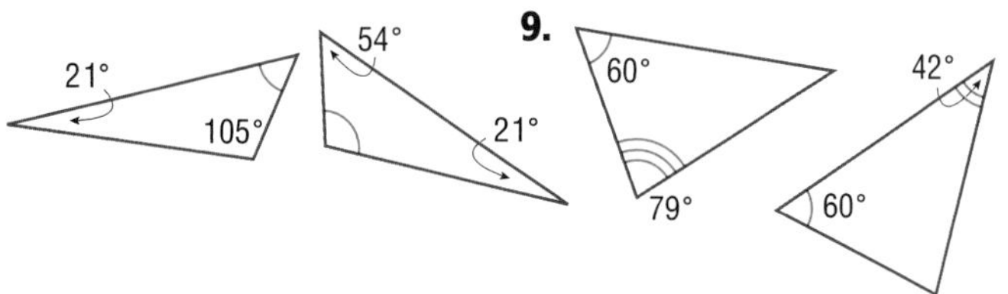

10.

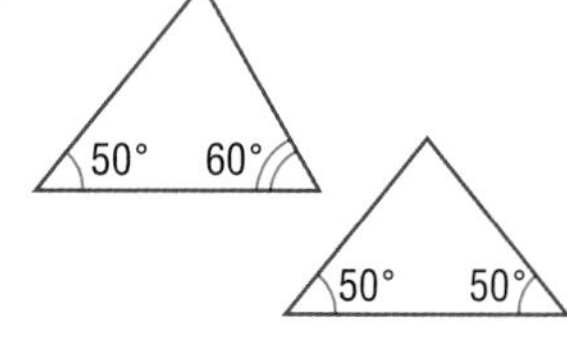

11.

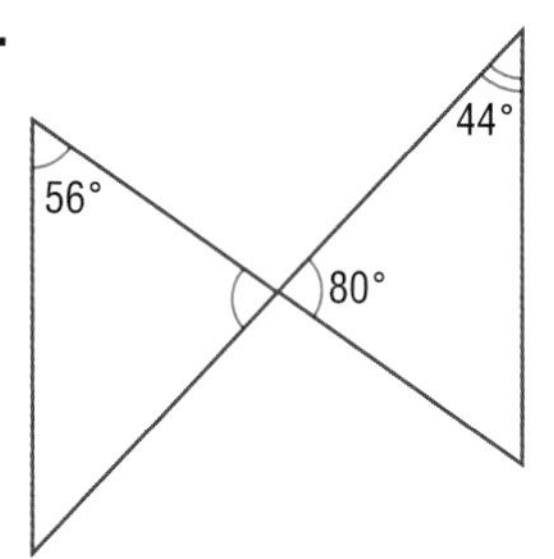

12.

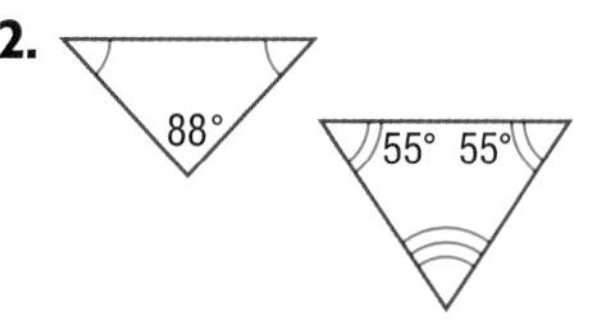

13.

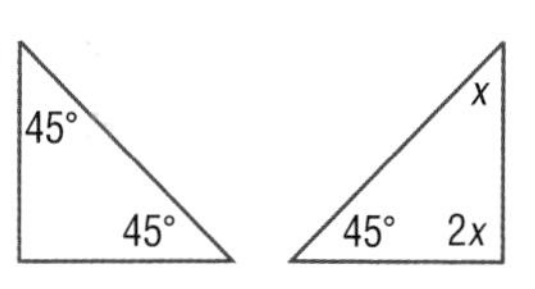

For each set of measures given, find the measures of the missing sides if $\triangle KLM \sim \triangle NOP$.

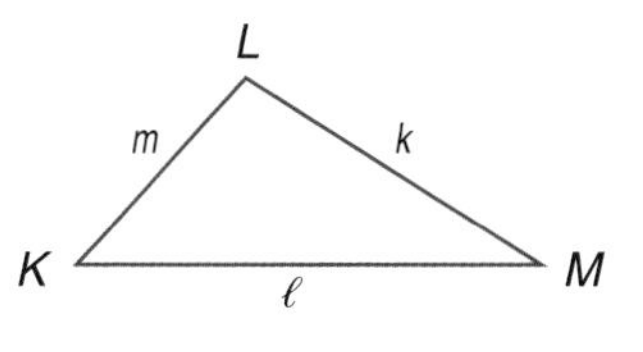

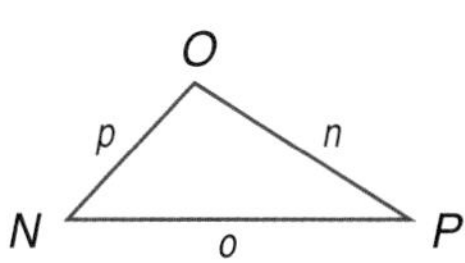

14. $k = 9, n = 6, o = 8, p = 4$

15. $k = 24, \ell = 30, m = 15, n = 16$

16. $m = 11, p = 6, n = 5, o = 4$

17. $k = 16, \ell = 13, m = 12, o = 7$

18. $n = 6, p = 2.5, \ell = 4, m = 1.25$

19. $p = 5, k = 10.5, \ell = 15, m = 7.5$

20. $n = 2.1, \ell = 4.5, p = 3.2, o = 3.4$

21. $m = 5, k = 12.6, o = 8.1, p = 2.5$

22. PHOTOGRAPHY Refer to the diagram of a camera at the beginning of the lesson. Suppose the image of a man who is 2 meters tall is 1.5 centimeters tall on film. If the film is 3 centimeters from the lens of the camera, how far is the man from the camera?

23. TOYS Diecast model cars use a scale of 1 inch : 2 feet of the real vehicle. The original vehicle has a window shaped like a right triangle. If the height of the window on the actual vehicle is 2.5 feet, what will the height of the window be on the model?

24. GOLF Jessica is playing miniature golf on a hole like the one shown at the right. She wants to putt her ball U so that it will bank at T and travel into the hole at R. Use similar triangles to find where Jessica's ball should strike the wall.

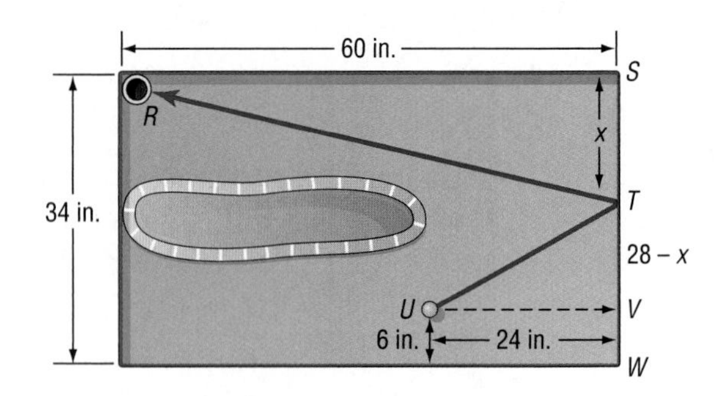

25. CRAFTS Melinda is working on a quilt pattern containing isosceles triangles whose sides measure 2 inches, 2 inches, and 2.5 inches. She has several square pieces of material that measure 4 inches on each side. From each square piece, how many triangles with the required dimensions can she cut?

MIRRORS For Exercises 26 and 27, use the diagram and the following information. Viho wanted to measure the height of a nearby building. He placed a mirror on the pavement at point P, 80 feet from the base of the building. He then backed away until he saw an image of the top of the building in the mirror.

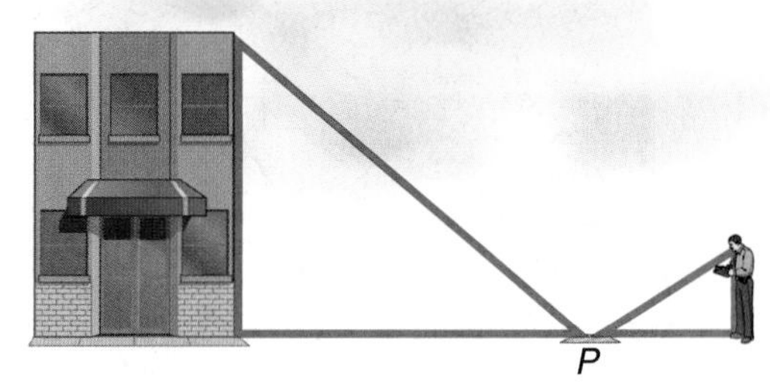

EXTRA PRACTICE
See page 738, 753.
Math Online
Self-Check Quiz at
algebra1.com

26. If Viho is 6 feet tall and he is standing 9 feet from the mirror, how tall is the building?

27. What assumptions did you make in solving the problem?

H.O.T. Problems

28. OPEN ENDED Draw and label a triangle ABC. Then draw and label a similar triangle MNO so that the area of $\triangle MNO$ is four times the area of $\triangle ABC$. Explain your strategy.

29. REASONING Determine whether the following statement is *sometimes*, *always*, or *never* true. Explain your reasoning.

If the measures of the sides of a triangle are multiplied by 3, then the measures of the angles of the enlarged triangle will have the same measures as the angles of the original triangle.

30. FIND THE ERROR Russell and Consuela are comparing the similar triangles below to determine their corresponding parts. Who is correct? Explain your reasoning.

Russell	Consuela
$m\angle X = m\angle T$	$m\angle X = m\angle V$
$m\angle Y = m\angle U$	$m\angle Y = m\angle U$
$m\angle Z = m\angle V$	$m\angle Z = m\angle T$
$\triangle XYZ, \triangle TUV$	$\triangle XYZ, \triangle VUT$

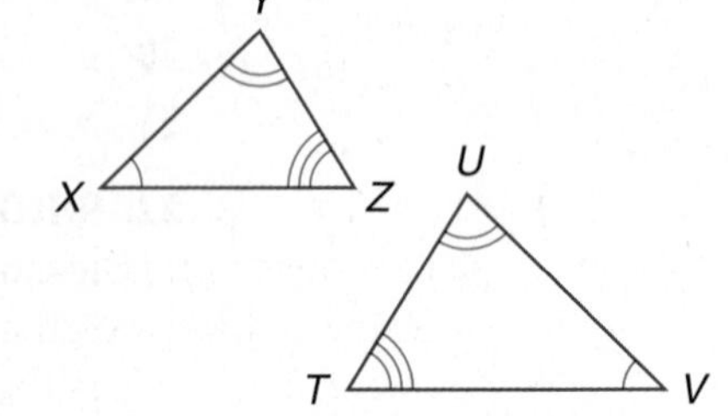

CRITICAL THINKING **For Exercises 31–33, use the following information.** The radius of one circle is twice the radius of another.

31. Are the circles similar? Explain your reasoning.

32. What is the ratio of their circumferences? Explain your reasoning.

33. What is the ratio of their areas? Explain your reasoning.

34. ***Writing in Math*** Use the information about photography on page 560 to explain how similar triangles are related to photography. Include an explanation of the effect of moving a camera with a zoom lens closer to the object being photographed and a description of what you could do to fit the entire image of a large object on the picture.

STANDARDIZED TEST PRACTICE

35. Which is a true statement about the figure?

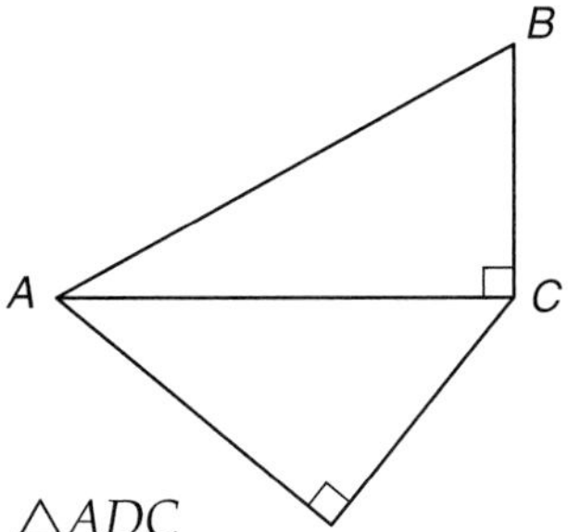

A $\triangle ABC \sim \triangle ADC$

B $\triangle ABC \sim \triangle ACD$

C $\triangle ABC \sim \triangle CAD$

D none of the above

36. **REVIEW** Kareem needs to know the length and width of his room but does not have a tape measure. He knows that the room is square and that it has an area of 121 square feet. What should Kareem do to find the dimensions of his room?

F Divide the area by the number of sides.

G Multiply the area by 2.

H Square the area.

J Take the square root of the area.

Spiral Review

Find the distance between each pair of points whose coordinates are given. Express answers in simplest radical form and as decimal approximations rounded to the nearest hundredth, if necessary. (Lesson 10-5)

37. $(1, 8), (-2, 4)$ **38.** $(4, 7), (3, 12)$ **39.** $\left(1, 5\sqrt{6}\right), \left(6, 7\sqrt{6}\right)$

The lengths of three sides of a triangle are given. Determine whether each triangle is a right triangle. (Lesson 10-4)

40. 25, 60, 65 **41.** 20, 25, 35 **42.** 49, 168, 175

Use elimination to solve each system of equations. (Lesson 5-3)

43. $2x + y = 4$
$x - y = 5$

44. $3x - 2y = -13$
$2x - 5y = -5$

45. $\frac{1}{3}x + \frac{1}{2}y = 8$
$\frac{1}{2}x - \frac{1}{4}y = 0$

46. **AVIATION** An airplane passing over Sacramento at an elevation of 37,000 feet begins its descent to land at Reno, 140 miles away. If the elevation of Reno is 4500 feet, what should the approximate slope of descent be? (*Hint*: 1 mi = 5280 ft) (Lesson 4-1)

READING MATH

The Language of Mathematics

The language of mathematics is a specific one, but it borrows from everyday language, scientific language, and world languages. To find a word's correct meaning, you will need to be aware of some confusing aspects of language.

Confusing Aspect	Words
Some words are used in English and in mathematics, but have distinct meanings.	factor, **leg**, prime, power, **rationalize**
Some words are used in English and in mathematics, but the mathematical meaning is more precise.	difference, even, **similar**, slope
Some words are used in science and in mathematics, but the meanings are different.	divide, **radical**, solution, variable
Some words are only used in mathematics.	decimal, **hypotenuse**, integer, quotient
Some words have more than one mathematical meaning.	base, **degree**, range, round, square
Sometimes several words come from the same root word.	polygon and polynomial, **radical** and **radicand**
Some mathematical words sound like English words.	**sum** and some, whole and hole, base and bass

Words in boldface are in this chapter.

Reading to Learn

1. How do the mathematical meanings of the following words compare to the everyday meanings?

a. factor

b. leg

c. rationalize

2. State two mathematical definitions for each word. Give an example for each definition.

a. degree

b. range

c. round

3. Each word below is shown with its root word and the root word's meaning. Find three additional words that come from the same root.

a. domain, from the root word *domus,* which means house

b. radical, from the root word *radix,* which means root

c. similar, from the root word *similis,* which means like

CHAPTER 10 Study Guide and Review

FOLDABLES™ Study Organizer — GET READY to Study

Be sure the following Key Concepts are noted in your Foldable.

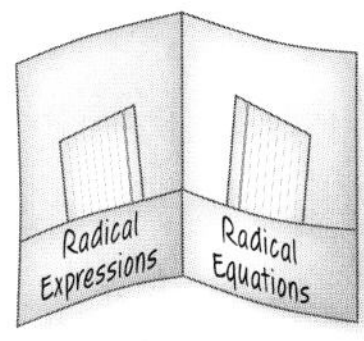

Key Concepts

Simplifying Radical Expressions

(Lesson 10-1)

- A radical expression is in simplest form when
- no radicands have perfect square factors other than 1,
- no radicands contain fractions,
- and no radicands appear in the denominator of a fraction.

Operations with Radical Expressions and Equations (Lessons 10-2 and 10-3)

- Radical expressions with like radicands can be added or subtracted.
- Use the FOIL Method to multiply radicand expressions.
- Solve radical equations by isolating the radical on one side of the equation. Square each side of the equation to eliminate the radical.

Pythagorean Theorem and Distance Formula (Lessons 10-4 and 10-5)

- If a and b are the measures of the legs of a right triangle and c is the measure of the hypotenuse, then $c^2 = a^2 + b^2$.
- If a and b are measures of the shorter sides of a triangle, c is the measure of the longest side, and $c^2 = a^2 + b^2$, then the triangle is a right triangle.
- The distance d between any two points with coordinates (x_1, y_1) and (x_2, y_2) is given by $d = \sqrt{(x_2 - x_1)^2 + (y_2 - y_1)^2}$.

Similar Triangles (Lesson 10-6)

- Similar Triangles have congruent corresponding angles and proportional corresponding sides. If $\triangle ABC \sim \triangle DEF$, then $\frac{AB}{DE} = \frac{BC}{EF} = \frac{AC}{DF}$.

Key Vocabulary

conjugate (p. 531)
converse (p. 551)
Distance Formula (p. 555)
extraneous solution (p. 542)
hypotenuse (p. 549)
leg (p. 549)
Pythagorean triple (p. 550)
radical equation (p. 541)
radical expression (p. 528)
radicand (p. 528)
rationalizing the denominator (p. 530)
similar triangles (p. 560)

Vocabulary Check

State whether each sentence is *true* or *false*. If *false*, replace the underlined word or number to make a true sentence.

1. The binomials $-3 + \sqrt{7}$ and $\underline{3 - \sqrt{7}}$ are conjugates.
2. In the expression $-4\sqrt{5}$, the radicand is $\underline{5}$.
3. The <u>longest</u> side of a right triangle is the hypotenuse.
4. After the first step in solving $\sqrt{3x + 19} = x + 3$, you would have $\underline{3x + 19 = x^2 + 9}$.
5. The two sides that form the right angle in a right triangle are called the <u>legs</u> of the triangle.
6. The expression $\frac{2x\sqrt{3x}}{\sqrt{6y}}$ is in simplest radical form.
7. A triangle with sides having measures of <u>25, 20, and 15</u> is a right triangle.
8. Two triangles are <u>similar</u> if the corresponding angles are congruent.

CHAPTER 10

Study Guide and Review

Lesson-by-Lesson Review

10–1 Simplifying Radical Expressions (pp. 528–534)

Simplify.

9. $\sqrt{\frac{60}{y^2}}$

10. $\sqrt{44a^2b^5}$

11. $(3 - 2\sqrt{12})^2$

12. $\frac{9}{3 + \sqrt{2}}$

13. $\frac{2\sqrt{7}}{3\sqrt{5} + 5\sqrt{3}}$

14. $\frac{\sqrt{3a^3b^4}}{\sqrt{8ab^{10}}}$

15. METEOROLOGY To estimate how long a thunderstorm will last, meteorologists use the formula $t = \sqrt{\frac{d^3}{216}}$, where t is time in hours and d is the diameter of the storm in miles. A storm is 10 miles in diameter. How long will it last?

Example 1 Simplify $\frac{3}{5 - \sqrt{2}}$.

$\frac{3}{5 - \sqrt{2}}$

$= \frac{3}{5 - \sqrt{2}} \cdot \frac{5 + \sqrt{2}}{5 + \sqrt{2}}$ Rationalize the denominator.

$= \frac{3(5) + 3\sqrt{2}}{5^2 - (\sqrt{2})^2}$ $(a - b)(a + b) = a^2 - b^2$

$= \frac{15 + 3\sqrt{2}}{25 - 2}$ $(\sqrt{2})^2 = 2$

$= \frac{15 + 3\sqrt{2}}{23}$ Simplify.

10–2 Operations with Radical Expressions (pp. 536–540)

Simplify each expression.

16. $2\sqrt{3} + 8\sqrt{5} - 3\sqrt{5} + 3\sqrt{3}$

17. $2\sqrt{6} - \sqrt{48}$

18. $4\sqrt{7k} - 7\sqrt{7k} + 2\sqrt{7k}$

19. $\sqrt{8} + \sqrt{\frac{1}{8}}$

Find each product.

20. $\sqrt{2}(3 + 3\sqrt{3})$

21. $(\sqrt{3} - \sqrt{2})(2\sqrt{2} + \sqrt{3})$

22. $(6\sqrt{5} + 2)(3\sqrt{2} + \sqrt{5})$

23. MOTION The velocity of a dropped object can be found using $v = \sqrt{2gd}$, where v is the velocity in feet per second, g is the acceleration due to gravity, and d is the distance in feet the object drops. Find the speed of a penny when it hits the ground, after being dropped off the Eiffel Tower. Use 32 feet per second squared for g and 984 feet for the height of the Eiffel Tower.

Example 2 $\sqrt{6} - \sqrt{54} + 3\sqrt{12} + 5\sqrt{3}$.

$\sqrt{6} - \sqrt{54} + 3\sqrt{12} + 5\sqrt{3}$

$= \sqrt{6} - \sqrt{3^2 \cdot 6} + 3\sqrt{2^2 \cdot 3} + 5\sqrt{3}$

$= \sqrt{6} - (\sqrt{3^2} \cdot \sqrt{6}) + 3(\sqrt{2^2} \cdot \sqrt{3}) + 5\sqrt{3}$

$= \sqrt{6} - 3\sqrt{6} + 3(2\sqrt{3}) + 5\sqrt{3}$

$= \sqrt{6} - 3\sqrt{6} + 6\sqrt{3} + 5\sqrt{3}$

$= -2\sqrt{6} + 11\sqrt{3}$

Example 3 Find $(2\sqrt{3} - \sqrt{5})(\sqrt{10} + 4\sqrt{6})$.

$(2\sqrt{3} - \sqrt{5})(\sqrt{10} + 4\sqrt{6})$

$= (2\sqrt{3})(\sqrt{10}) + (2\sqrt{3})(4\sqrt{6})$
$+ (-\sqrt{5})(\sqrt{10}) + (-\sqrt{5})(4\sqrt{6})$

$= 2\sqrt{30} + 8\sqrt{18} - \sqrt{50} - 4\sqrt{30}$

$= 2\sqrt{30} + 8\sqrt{3^2 \cdot 2} - \sqrt{5^2 \cdot 2} - 4\sqrt{30}$

$= -2\sqrt{30} + 19\sqrt{2}$

Mixed Problem Solving
For mixed problem-solving practice, see page 753.

10-3 Radical Equations (pp. 541–546)

Solve each equation. Check your solution.

24. $10 + 2\sqrt{b} = 0$

25. $\sqrt{a+4} = 6$

26. $\sqrt{7x-1} = 5$

27. $\sqrt{\frac{4a}{3}} - 2 = 0$

28. $\sqrt{x+4} = x - 8$

29. $\sqrt{3x-14} + x = 6$

30. FREE FALL Assuming no air resistance, the time t in seconds that it takes an object to fall h feet can be determined by $t = \frac{\sqrt{h}}{4}$. If a skydiver jumps from an airplane and free falls for 8 seconds before opening the parachute, how many feet does the skydiver fall?

Example 4 Solve $\sqrt{5 - 4x} - 6 = 7$.

$\sqrt{5-4x} - 6 = 7$	Original equation
$\sqrt{5-4x} = 13$	Add 6 to each side.
$5 - 4x = 169$	Square each side.
$-4x = 164$	Subtract 5 from each side.
$x = -41$	Divide each side by -4.

10-4 The Pythagorean Theorem (pp. 549–554)

If c is the measure of the hypotenuse of a right triangle, find each missing measure. If necessary, round answers to the nearest hundredth.

31. $a = 30, b = 16, c = ?$

32. $b = 4, c = 56, a = ?$

33. $a = 6, b = 10, c = ?$

34. $a = 10, c = 15, b = ?$

35. $a = 18, c = 30, b = ?$

36. $a = 1.2, b = 1.6, c = ?$

Determine whether the following side measures form right triangles.

37. 9, 16, 20

38. 20, 21, 29

39. 9, 40, 41

40. $18, \sqrt{24}, 30$

41. MOVING The door of Julio's apartment measures 7 feet high and 3 feet wide. Julio would like to buy a square table that is 7 feet on a side. If the table cannot go through the door sideways, will it fit diagonally? Explain.

Example 5 Find the length of the missing side.

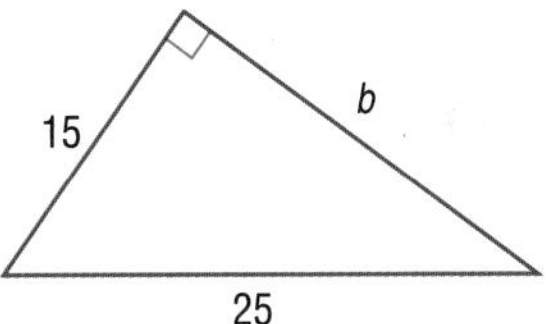

$c^2 = a^2 + b^2$	Pythagorean Theorem
$25^2 = 15^2 + b^2$	$c = 25$ and $a = 15$
$625 = 225 + b^2$	Evaluate squares.
$400 = b^2$	Subtract 225 from each side.
$20 = b$	Take the square root of each side.

Example 6 Determine whether the side measures, 6, 10, and 12, form a right triangle.

$c^2 = a^2 + b^2$	Pythagorean Theorem
$12^2 \stackrel{?}{=} 6^2 + 10^2$	$a = 6$, $b = 10$, and $c = 12$
$144 \stackrel{?}{=} 36 + 100$	Multiply.
$144 \neq 136$	Add. These side measures do *not* form a right triangle.

10-5 The Distance Formula (pp. 555–559)

Find the distance between each pair of points with the given coordinates. Express in simplest radical form and as decimal approximations rounded to the nearest hundredth if necessary.

42. (9, −2), (1, 13) **43.** (4, 2), (7, 9)

44. (4, 8), (−7, 12) **45.** (−2, 6), (5, 11)

Find the value of a if the points with the given coordinates are the indicated distance apart.

46. $(-3, 2), (1, a); d = 5$

47. $(5, -2), (a, -3); d = \sqrt{170}$

48. SAILING A boat leaves the harbor and sails 5 miles east and 3 miles north to an island. The next day, they travel to a fishing spot 10 miles south and 4 miles west of the harbor. How far is it from the fishing spot to the island?

Example 7 Find the distance between the points with coordinates (−5, 1) and (1, 5).

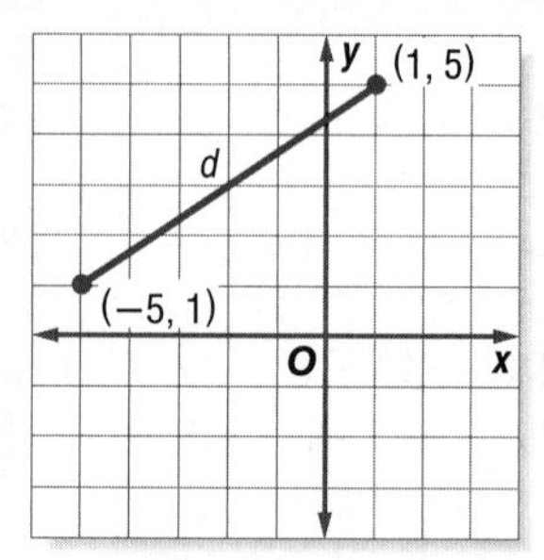

Use the Distance Formula.

$d = \sqrt{(x_2 - x_1)^2 + (y_2 - y_1)^2}$

$= \sqrt{(1 - (-5))^2 + (5 - 1)^2}$

$= \sqrt{6^2 + 4^2}$ Simplify.

$= \sqrt{36 + 16}$ Evaluate squares.

$= \sqrt{52}$ or about 7.21 units Simplify.

10-6 Similar Triangles (pp. 560–565)

For each set of measures given, find the measures of the remaining sides if $\triangle ABC \sim \triangle DEF$.

49. $c = 16, b = 12, a = 10, f = 9$

50. $a = 8, c = 10, b = 6, f = 12$

51. $c = 12, f = 9, a = 8, e = 11$

52. $b = 20, d = 7, f = 6, c = 15$

53. HOUSES Josh plans to make a model of his house in the scale 1 inch = 6 feet. If the height to the top of the roof on the house is 24 feet, what will the height of the model be?

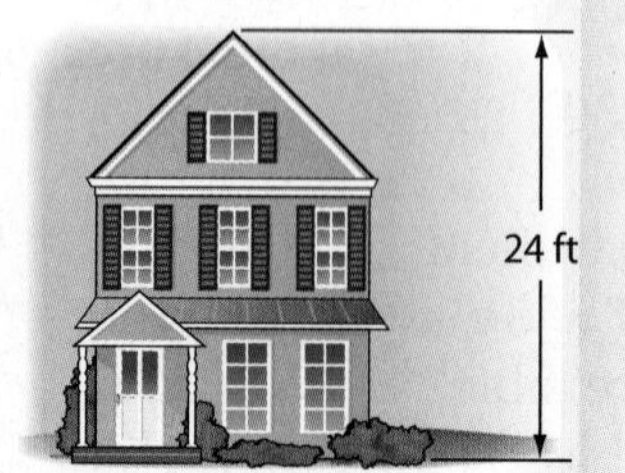

Example 8 Find the measure of side a if the two triangles are similar.

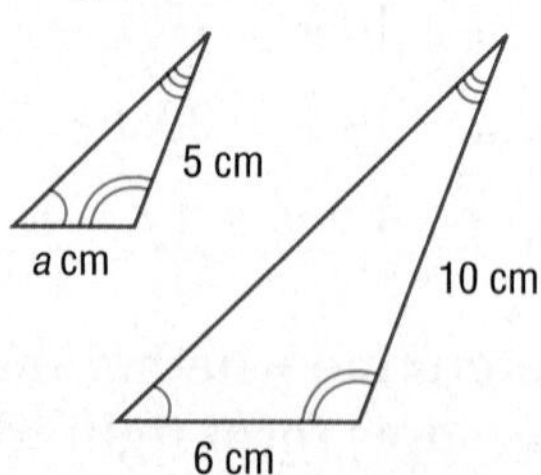

$\frac{10}{5} = \frac{6}{a}$ Corresponding sides of similar triangles are proportional.

$10a = 30$ Find the cross products.

$a = 3$ Divide each side by 10.

CHAPTER 10

Practice Test

Simplify.

1. $2\sqrt{27} - 4\sqrt{3}$
2. $\sqrt{6} + \sqrt{\frac{2}{3}}$
3. $\sqrt{6}(4 + \sqrt{12})$
4. $\sqrt{\frac{10}{3}} \cdot \sqrt{\frac{4}{30}}$
5. $(1 - \sqrt{3})(3 + \sqrt{2})$
6. $\sqrt{112x^4y^6}$

Solve each equation. Check your solution.

7. $\sqrt{10x} = 20$
8. $\sqrt{4s} + 1 = 11$
9. $\sqrt{4x + 1} = 5$
10. $x = \sqrt{-6x - 8}$
11. $x = \sqrt{5x + 14}$
12. $\sqrt{4x - 3} = 6 - x$

If c is the measure of the hypotenuse of a right triangle, find each missing measure. If necessary, round to the nearest hundredth.

13. $a = 8, b = 10, c = ?$
14. $a = 6\sqrt{2}, c = 12, b = ?$

15. **DISC GOLF** The sport of disc golf is similar to golf except that the players throw a disc into a basket instead of hitting a ball into a cup. Bob's first disc lands 10 feet short and 12 feet to the left of the basket. On his next throw, the disc lands 5 feet to the right and 2 feet beyond the basket. Assuming that the disc traveled in a straight line, how far did the disc travel on the second throw?

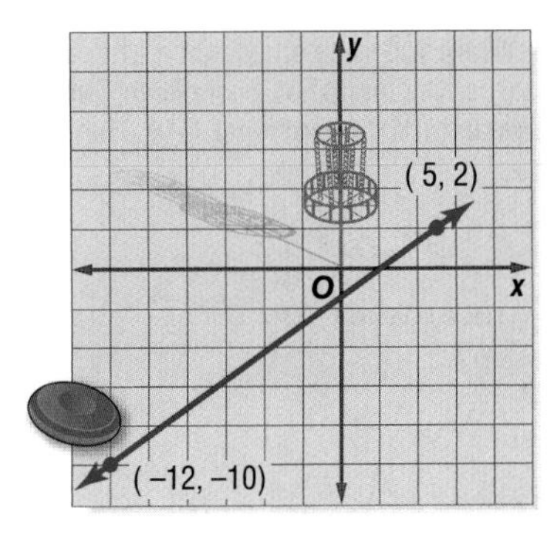

16. **SPORTS** A hiker leaves her camp in the morning. How far is she from camp after walking 9 miles due west and then 12 miles due north?

Find the distance between each pair of points with the given coordinates. Express in simplest radical form and as decimal approximations rounded to the nearest hundredth, if necessary.

17. (4, 7), (4, −2)
18. (−1, 1), (1, −5)
19. (−9, 2), (21, 7)

For each set of measures given, find the measures of the missing sides if $\triangle ABC \sim \triangle JKH$.

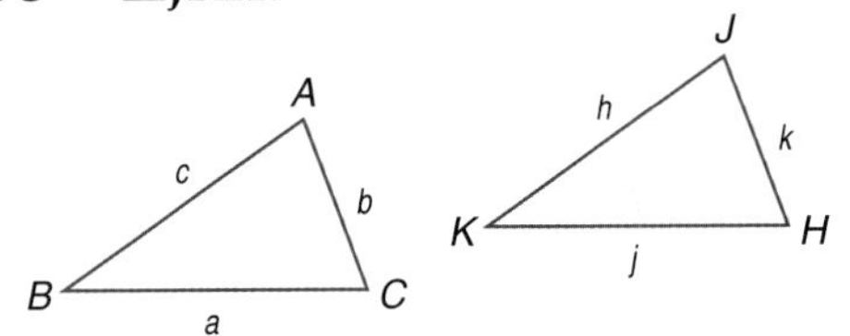

20. $c = 20, h = 15, k = 16, j = 12$
21. $c = 12, b = 13, a = 6, h = 10$
22. $k = 5, c = 6.5, b = 7.5, a = 4.5$
23. $h = 1\frac{1}{2}, c = 4\frac{1}{2}, k = 2\frac{1}{4}, a = 3$

24. **MULTIPLE CHOICE** Find the area of the rectangle.

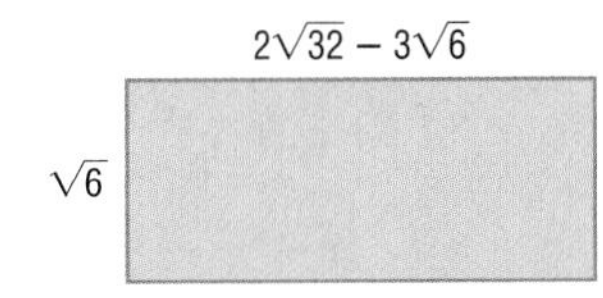

A $2\sqrt{32} - 18$ units^2

B $16\sqrt{2} - 4\sqrt{6}$ units^2

C $16\sqrt{3} - 18$ units^2

D $32\sqrt{3} - 18$ units^2

25. **SHADOWS** Suppose you are standing near the flag pole in front of your school and you want to know its height. The flag pole casts a 22-foot shadow. You cast a 3-foot shadow. If you are 5 feet tall, how tall is the flag pole?

CHAPTER 10

Standardized Test Practice

Cumulative, Chapters 1–10

Read each question. Then fill in the correct answer on the answer document provided by your teacher or on a sheet of paper.

1. Aliyah is creating two flower beds in her yard. She wants the rectangles to be similar. Using the dimensions given, find the approximate length of the side labeled x.

A 8 feet

B 15 feet

C 20 feet

D 50 feet

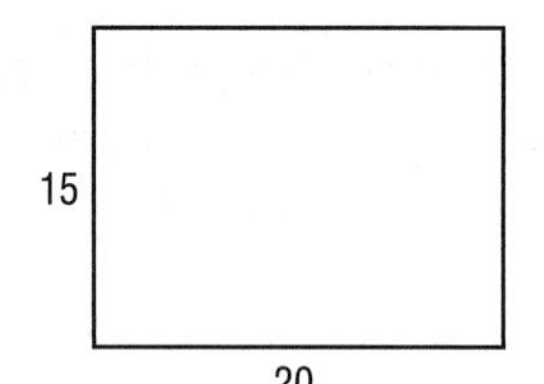

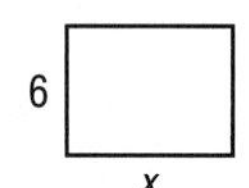

2. What is the area of the square in the diagram?

F 4.2 units^2

G 9 units^2

H 8.4 units^2

J 18 units^2

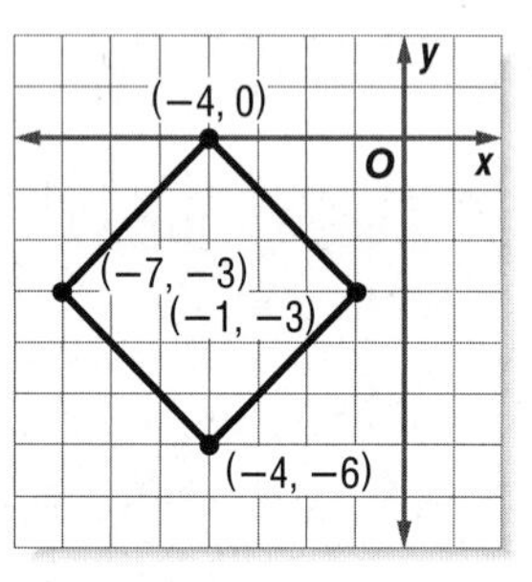

3. GRIDDABLE Carol must earn more than \$360 from selling boxes of oranges in order to go on a trip with the student council. If each box is sold for \$12.50, what is the least number of boxes that she must sell?

4. The number of jazz CDs that Vince owns j is 4 less than the number of pop CDs p that he owns. Which equation represents the number of pop CDs he owns?

A $j = p + 4$

B $p = j - 4$

C $p = j + 4$

D $p = 4j$

5. Matt, Nate, and Mark are walking to meet each other at a corner. At one point, Matt is at the corner, Nate is 60 feet from Matt, and Mark is 20 feet from Matt. About how far is Mark from Nate?

F 6 ft **G** 57 ft **H** 63 ft **J** 80 ft

6. Which linear function includes the points at (−4, 7) and (4, 5)?

A $f(x) = -4x + 6$

B $f(x) = -\frac{1}{4}x - \frac{9}{4}$

C $f(x) = \frac{1}{4}x - 6$

D $f(x) = -\frac{1}{4}x + 6$

TEST-TAKING TIP

QUESTION 6 When you write an equation, check that the given values make a true statement. For example, in Question 6, substitute the values of the coordinates (−4, 7) into your equation to check.

7. If $\triangle ABC$ is similar to $\triangle DEF$, what is the length of y?

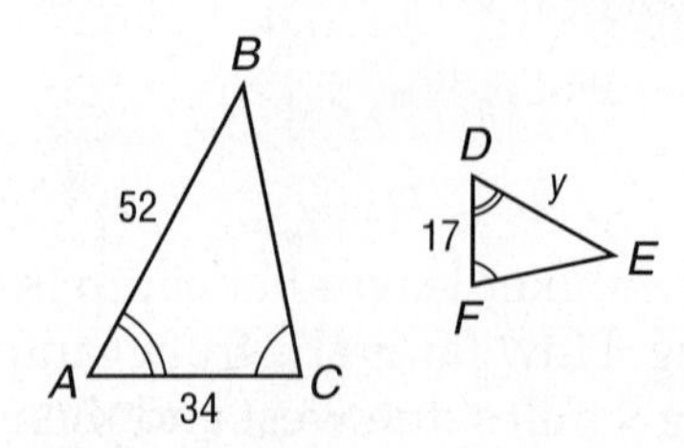

F 11.1 **G** 17 **H** 26 **J** 104

8. The perimeter p of a square may be found by using the formula $\left(\frac{1}{4}\right)p = \sqrt{a}$, where a is the area of the square. What is the perimeter of the square with an area of 64 square feet?

A 2

B 12

C 32

D 64

9. Ms. Milo's classroom is square, with sides that are 20 feet long. What is the approximate distance from one corner of the room to the other corner diagonally?

F 10 ft G 20 ft H 25 ft J 28 ft

10. Kayla wants to build a square cement patio in the corner of her yard as pictured below. What will be the area of the cement patio?

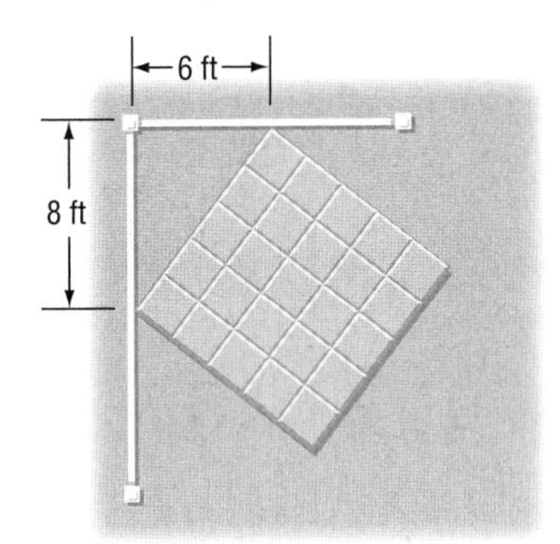

A 10 ft^2 C 64 ft^2

B 16 ft^2 D 100 ft^2

11. What property was used to simplify the expression $\frac{1}{4}(8x - 12) = 2x - 3$?

F Distributive Property

G Associative Property

H Commutative Property

J Not here

12. GRIDDABLE The coordinate grid below shows three squares.

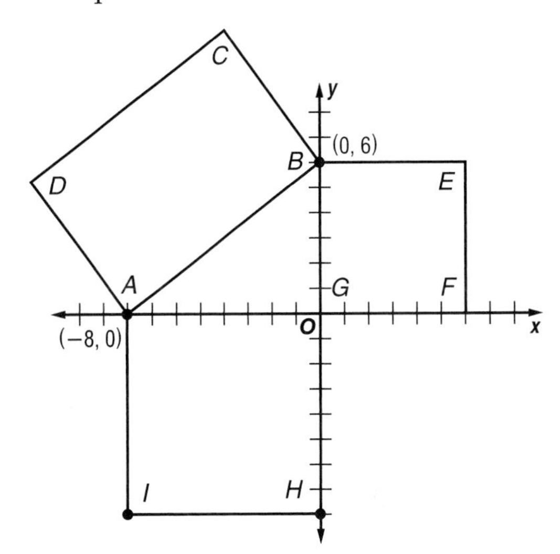

Which value best represents the area of square $ABCD$ in square units?

Pre-AP

Record your answers on a sheet of paper. Show your work.

13. Haley hikes 3 miles north, 7 miles east, and then 6 miles north again.

a. Draw a diagram showing the direction and distance of each segment of Haley's hike. Label Haley's starting point, her ending point, and the distance, in miles, of each segment of her hike.

b. To the nearest tenth of a mile, how far (in a straight line) is Haley from her starting point?

c. How did your diagram help you to find Haley's distance from her starting point?

d. Describe the direction and distance of Haley's return trip back to her starting position if she used the same trail.

NEED EXTRA HELP?													
If You Missed Question...	1	2	3	4	5	6	7	8	9	10	11	12	13
Go to Lesson or Page...	10-6	10-5	6-2	2-1	10-4	4-4	10-6	10-3	10-4	10-4	1-2	10-4	10-4

Rational Expressions and Equations

BIG Ideas

- Solve problems involving inverse variation.
- Simplify, add, subtract, multiply, and divide rational expressions.
- Divide polynomials.
- Solve rational equations.

Key Vocabulary

complex fraction (p. 621)
extraneous solutions (p. 629)
inverse variation (p. 577)
rational expression (p. 583)

Real-World Link

Marching Band Performing operations on rational expressions is an important part of working with equations. For example, the uniform manager for The Ohio State Marching Band can use rational expressions to determine the number of uniforms that can be repaired in a certain time given the number of tailors available and the time needed to repair each uniform.

Rational Expressions and Equations Make this Foldable to help you organize information about rational expressions and equations. Begin with a sheet of $8\frac{1}{2}''$ by $11''$ paper.

1. **Fold** in half lengthwise.

2. **Fold** the top to the bottom.

3. **Open.** Cut along the second fold to make two tabs.

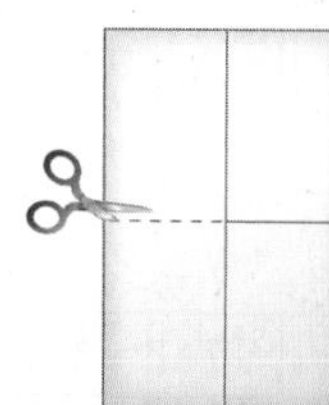

4. **Label** each tab as shown.

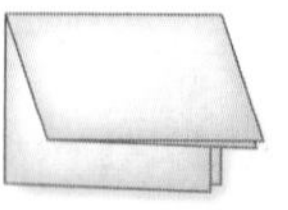

GET READY for Chapter 11

Diagnose Readiness You have two options for checking Prerequisite Skills.

Option 2

Math Online Take the Online Readiness Quiz at algebra1.com.

Option 1

Take the Quick Check below. Refer to the Quick Review for help.

QUICKCheck

Solve each proportion. (Lesson 2-6)

1. $\frac{y}{9} = -\frac{7}{16}$
2. $\frac{4}{x} = \frac{2}{10}$
3. $\frac{3}{15} = \frac{1}{n}$
4. $\frac{x}{8} = \frac{0.21}{2}$
5. $\frac{1.1}{0.6} = \frac{8.47}{n}$
6. $\frac{9}{8} = \frac{y}{6}$
7. $\frac{2.7}{3.6} = \frac{8.1}{a}$
8. $\frac{0.19}{2} = \frac{x}{24}$

Find the greatest common factor for each pair of monomials. (Lesson 8-1)

9. 30, 42
10. $60r^2, 45r^3$
11. **GAMES** There are 64 girls and 80 boys who attend an after-school program. For a game, the boys are going to split into groups and the girls are going to split into groups. The number of people in each group has to be the same. How large can the groups be?

Factor each polynomial. (Lessons 8-1 through 8-4)

12. $3c^2d - 6c^2d^2$
13. $6mn + 15m$
14. $x^2 + 11x + 24$
15. $x^2 + 4x - 45$
16. $2x^2 + x - 21$
17. $3x^2 - 12x + 9$
18. **AREA** The area of a rectangle can be represented by the expression $x^2 + 7x + 12$. What expressions represent the sides of the rectangle?

QUICKReview

EXAMPLE 1

Solve the proportion $\frac{4}{z} = \frac{13}{5}$.

$\frac{4}{z} = \frac{13}{5}$	Original equation
$4 \cdot 5 = 13 \cdot z$	Cross multiply.
$20 = 13z$	Simplify.
$\frac{20}{13} = \frac{13z}{13}$	Divide each side by 13.
$\frac{20}{13} = z$	Simplify.

EXAMPLE 2

Find the greatest common factor of 12 and 18.

3, 2, 2	Factors of 12
3, 3, 2	Factors of 18
$3 \cdot 2$	The product of the common factors

The greatest common factor of 12 and 18 is 6.

EXAMPLE 3

Factor $12x^2y^3 - 3xy$.

3, 2, 2, x, x, y, y, y	Factors of $12x^2y^3$
-1, 3, x, y	Factors of $-3xy$
$3 \cdot x \cdot y$	The product of the common factors

Factor out the common factors from each term of the expression.

$3xy(4xy^2) - 3xy(1)$	Rewrite the terms using the GCF.
$= 3xy(4xy^2 - 1)$	Distributive Property

Graphing Calculator Lab

Investigating Inverse Variation

You can use a data collection device to investigate the relationship between volume and pressure.

SET UP the Lab

- Connect a syringe to the gas pressure sensor. Then connect the data collection device to both the sensor and the calculator as shown.
- Start the data collection program and select the sensor.

ACTIVITY

Step 1 Open the valve between the atmosphere and the syringe. Set the inside ring of the syringe to 20 mL and close the valve. This ensures that the amount of air inside the syringe will be constant throughout the experiment.

Step 2 Press the plunger of the syringe down to the 5 mL mark. Wait for the pressure gauge to stop changing, then take the data reading. Enter 5 as the volume on the calculator. The pressure will be measured in atmospheres (atm).

Step 3 Repeat step 2, pressing the plunger down to 7.5 mL, 10.0 mL, 12.5 mL, 15.0 mL, 17.5 mL, and 20.0 mL. Record the volume as you take each data reading.

Step 4 After taking the last data reading, use STAT PLOT to create a line graph of the data.

ANALYZE THE RESULTS

1. Does the pressure vary directly as the volume? Explain.
2. As the volume changes from 10 to 20 mL, what happens to the pressure?
3. Predict what the pressure of the gas in the syringe would be if the volume was increased to 40 mL.
4. Add a column to the data table to find the product of the volume and the pressure for each data reading. What pattern do you observe?
5. **MAKE A CONJECTURE** The relationship between the pressure and volume of a gas is called Boyle's Law. Write an equation relating the volume v in milliliters and pressure p in atmospheres in your experiment.

Inverse Variation

Main Ideas

- Graph inverse variations.
- Solve problems involving inverse variation.

New Vocabulary

inverse variation
product rule

GET READY for the Lesson

The number of revolutions of the pedals made when riding a bicycle at a constant speed varies inversely as the gear ratio of the bicycle. In other words, as the gear ratio *decreases*, the revolutions per minute (rpm) *increase*. This is why shifting to a lower gear allows you to pedal with less difficulty when riding uphill.

Pedaling Rates to Maintain Speed of 10 mph

Gear Ratio	Rate
117.8	89.6
108.0	97.8
92.6	114.0
76.2	138.6
61.7	171.2
49.8	212.0
40.5	260.7

Study Tip

Look Back
To review **direct variation**, see Lesson 4-2.

Graph Inverse Variation Recall that some situations in which y increases as x increases are *direct variations*. If y varies directly as x, we can represent this relationship with an equation of the form $y = kx$, where $k \neq 0$. However, in the application above, as one value increases the other value decreases. When the product of two variables remains constant, the relationship forms an **inverse variation**. We say *y varies inversely as x* or *y is inversely proportional to x*. Recall that the constant k is called the *constant of variation*.

KEY CONCEPT — *Inverse Variation*

y varies inversely as x if there is some nonzero constant k such that $xy = k$.

Real-World EXAMPLE — Graph an Inverse Variation

1 **DRIVING The time t it takes to travel a certain distance varies inversely as the rate r at which you travel. The equation $rt = 250$ can be used to represent a person driving 250 miles. Complete the table and draw a graph of the relation.**

r (mph)	5	10	15	20	25	30	35	40	45	50
t (hours)										

Solve for t when $r = 5$.

$rt = 250$ Original equation

$5t = 250$ Replace r with 5.

$t = \frac{250}{5}$ Divide each side by 5.

$t = 50$ Simplify.

(continued on the next page)

Solve for t using the other values of r. Complete the table.

r (mph)	5	10	15	20	25	30	35	40	45	50
t (hours)	50	25	16.67	12.5	10	8.33	7.14	6.25	5.56	5

Next, graph the ordered pairs.

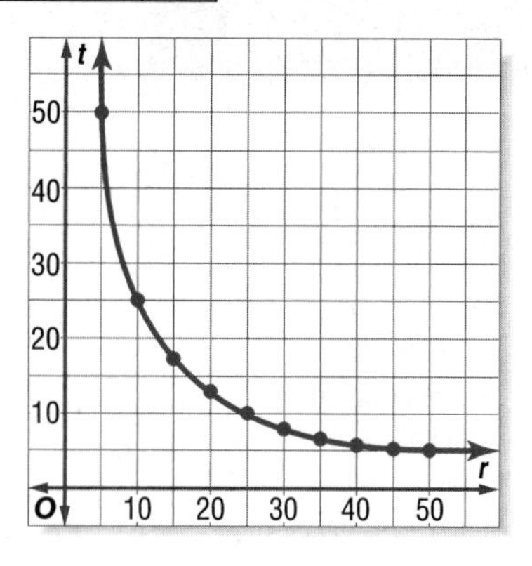

Because rate cannot be negative, it is only reasonable to use positive values for r.

The graph of an inverse variation is not a straight line like the graph of a direct variation. As the rate r increases, the time t that it takes to travel the same distance decreases.

CHECK Your Progress

1. Graph $64 = xy$ using a table.

Graphs of inverse variations can also be drawn using negative values of x.

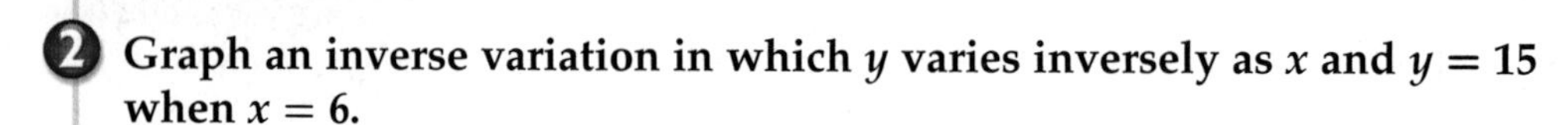

EXAMPLE Graph an Inverse Variation

2 Graph an inverse variation in which y varies inversely as x and $y = 15$ when $x = 6$.

Solve for k.

$xy = k$ Inverse variation equation

$(6)(15) = k$ $x = 6, y = 15$

$90 = k$ The constant of variation is 90.

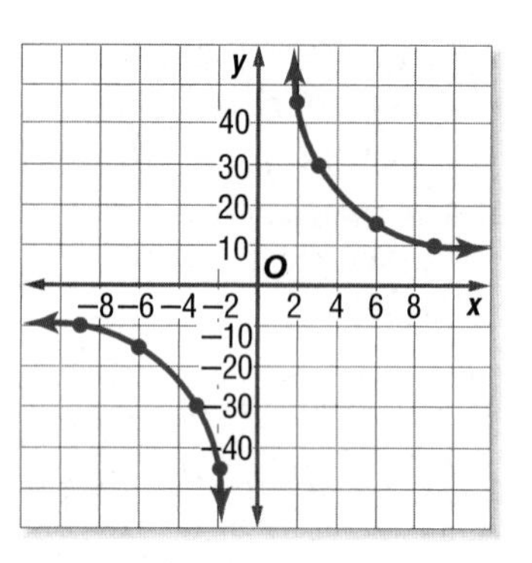

Choose values for x and y with a product of 90.

x	−9	−6	−3	−45	0	2	3	6	9
y	−10	−15	−30	−2	undefined	45	30	15	10

CHECK Your Progress

2. Graph an inverse variation equation in which y varies inversely as x when $y = 12$ and when $x = 4$.

Use Inverse Variation If (x_1, y_1) and (x_2, y_2) are solutions of an inverse variation, then $x_1, y_1 = k$ and $x_2y_2 = k$.

$x_1y_1 = k$ and $x_2y_2 = k$

$x_1y_1 = x_2y_2$ Substitute x_2y_2 for k.

Study Tip

Proportions

Notice that the proportion for inverse variations is different from the proportion for direct variation, $\frac{x_1}{x_2} = \frac{y_1}{y_2}$.

The equation $x_1y_1 = x_2y_2$ is called the **product rule** for inverse variations. You can use this equation to form a proportion.

$x_1y_1 = x_2y_2$ Product rule for inverse variations

$\frac{x_1y_1}{x_2y_1} = \frac{x_2y_2}{x_2y_1}$ Divide each side by x_2y_1.

$\frac{x_1}{x_2} = \frac{y_2}{y_1}$ Simplify.

EXAMPLE Solve for x or y

3 **If y varies inversely as x and $y = 4$ when $x = 7$, find x when $y = 14$.**

Let $x_1 = 7$, $y_1 = 4$, and $y_2 = 14$. Solve for x_2.

Method 1 Use the product rule.

$x_1y_1 = x_2y_2$ Product rule for inverse variations

$7 \cdot 4 = x_2 \cdot 14$ $x_1 = 7$, $y_1 = 4$, and $y_2 = 14$

$\frac{28}{14} = x_2$ Divide each side by 14.

$x_2 = 2$ Simplify.

Method 2 Use a proportion.

$\frac{x_1}{x_2} = \frac{y_2}{y_1}$ Proportion for inverse variations

$\frac{7}{x_2} = \frac{14}{4}$ $x_1 = 7$, $y_1 = 4$, and $y_2 = 14$

$28 = 14x_2$ Cross multiply.

$x_2 = 2$ Divide each side by 14.

CHECK Your Progress

3. If y varies inversely as x and $y = 4$ when $x = -8$, find y when $x = -4$.

Real-World EXAMPLE Use Inverse Variation

Study Tip

Levers

A lever is a bar with a pivot point called the *fulcrum*. For a lever to balance, the lesser weight must be positioned farther from the fulcrum.

4 **PHYSICAL SCIENCE When two people are balanced on a seesaw, their distances from the center of the seesaw are inversely proportional to their weights. If a 118-pound person sits 1.8 meters from the center of the seesaw, how far should a 125-pound person sit from the center to balance the seesaw?**

Let $w_1 = 118$, $d_1 = 1.8$, and $w_2 = 125$. Solve for d_2.

$w_1d_1 = w_2d_2$ Product rule for inverse variations

$118 \cdot 1.8 = 125 \cdot d_2$ Substitution

$\frac{212.4}{125} = d_2$ Divide each side by 125.

$1.7 \approx d_2$ Simplify.

To balance the seesaw, the second person should sit about 1.7 meters from the center.

CHECK Your Progress

4. **EARTH SCIENCE** As the temperature increases, the level of water in a river decreases. When the temperature was 90° Fahrenheit, the water level was 11 feet. If the temperature was 110° Fahrenheit, what was the level of water in the river?

Online **Personal Tutor at** algebra1.com

CHECK Your Understanding

Examples 1, 2 (pp. 577–578)

Graph each variation if y varies inversely as x.

1. $y = 24$ when $x = 8$

2. $y = -6$ when $x = -2$

Example 3 (p. 579)

Write an inverse variation equation that relates x and y. Assume that y varies inversely as x. Then solve.

3. If $y = 2.7$ when $x = 8.1$, find x when $y = 5.4$.

4. If $x = \frac{1}{2}$ when $y = 16$, find x when $y = 32$.

5. If $y = 12$ when $x = 6$, find y when $x = 8$.

6. If $y = -8$ when $x = -3$, find y when $x = 6$.

Example 4 (p. 579)

7. MUSIC When under equal tension, the frequency of a vibrating string from a piano varies inversely with the string length. If a string that is 420 millimeters in length vibrates at a frequency of 523 cycles a second, at what frequency will a 707-millimeter string vibrate?

Exercises

HOMEWORK HELP	
For Exercises	See Examples
8–13	1, 2
14, 15	4
16–23	3

Graph each variation if y varies inversely as x.

8. $y = 24$ when $x = -8$

9. $y = 3$ when $x = 4$

10. $y = 5$ when $x = 15$

11. $y = -4$ when $x = -12$

12. $y = 9$ when $x = 8$

13. $y = 2.4$ when $x = 8.1$

14. MUSIC The pitch of a musical note varies inversely as its wavelength. If the tone has a pitch of 440 vibrations per second and a wavelength of 2.4 feet, find the pitch of a tone that has a wavelength of 1.6 feet.

15. COMMUNITY SERVICE Students at Roosevelt High School are collecting canned goods for a local food pantry. They plan to distribute flyers to homes in the community asking for donations. Last year, 12 students were able to distribute 1000 flyers in nine hours. How long would it take if 15 students hand out 1000 flyers this year?

Write an inverse variation equation that relates x and y. Assume that y varies inversely as x. Then solve.

16. If $y = 8.5$ when $x = -1$, find x when $y = -1$.

17. If $y = 8$ when $x = 1.55$, find x when $y = -0.62$.

18. If $y = 6.4$ when $x = 4.4$, find x when $y = 3.2$.

19. If $y = 1.6$ when $x = 0.5$, find x when $y = 3.2$.

20. If $y = 12$ when $x = 5$, find y when $x = 3$.

21. If $y = 7$ when $x = -2$, find y when $x = 7$.

22. If $y = 4$ when $x = 4$, find y when $x = 7$.

23. If $y = -6$ when $x = -2$, find y when $x = 5$.

24. TRAVEL The Zalinski family can drive the 220 miles to their cabin in 4 hours at 55 miles per hour. Son Jeff claims that they could save half an hour if they drove 65 miles per hour, the speed limit. Is Jeff's claim true? Explain.

Write an inverse variation equation that relates x and y. Assume that y varies inversely as x. Then solve.

25. Find the value of y when $x = 7$ if $y = -7$ when $x = \frac{2}{3}$.

26. Find the value of y when $x = 32$ if $y = -16$ when $x = \frac{1}{2}$.

27. If $x = 6.1$ when $y = 4.4$, find x when $y = 3.2$.

28. If $x = 0.5$ when $y = 2.5$, find x when $y = 20$.

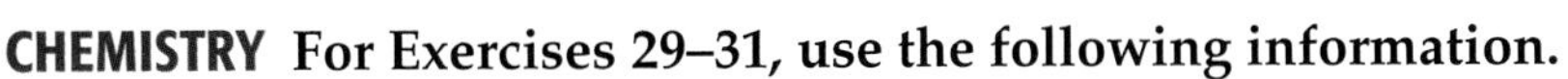

CHEMISTRY For Exercises 29–31, use the following information.
Boyle's Law states that the volume of a gas V varies inversely with applied pressure P.

29. Write an equation to show this relationship.

30. Pressure on 60 cubic meters of a gas is raised from 1 atmosphere to 3 atmospheres. What new volume does the gas occupy?

31. A helium-filled balloon has a volume of 22 cubic meters at sea level where the air pressure is 1 atmosphere. The balloon is released and rises to a point where the air pressure is 0.8 atmosphere. What is the volume of the balloon at this height?

Real-World Link

American sculptor Alexander Calder was the first artist to use mobiles as an art form.

Source: infoplease.com

32. GEOMETRY A rectangle is 36 inches wide and 20 inches long. How wide is a rectangle of equal area if its length is 90 inches?

33. ART Anna is designing a mobile to suspend from a gallery ceiling. A chain is attached eight inches from the end of a bar that is 20 inches long. On the shorter end of the bar is a sculpture weighing 36 kilograms. She plans to place another piece of artwork on the other end of the bar. How much should the second piece of art weigh if she wants the bar to be balanced?

Determine whether the data in each table represent an inverse variation or a direct variation. Explain.

34.

x	y
−6	3
−2	9
2	−9

35.

x	y
5	2.5
8	6
11	11.5

36.

x	y
−3	−7
−2	−10.5
4	5.25

H.O.T. Problems

37. OPEN ENDED Give a real-world situation or phenomena that can be modeled by an indirect variation equation. Use the correct terminology to describe your example and explain why this situation is an indirect variation.

38. REASONING Determine which situation is an example of inverse variation. Justify your answer.

a. Emily spends \$2 each day for snacks on her way home from school. The total amount she spends each week depends on the number of days school was in session.

b. A business donates \$200 to buy prizes for a school event. The number of prizes that can be purchased depends upon the price of each prize.

CHALLENGE For Exercises 39 and 40, assume that y varies inversely as x.

39. If the value of x is doubled, what happens to the value of y?

40. If the value of y is tripled, what happens to the value of x?

41. ***Writing in Math*** Use the data provided on page 577 to explain how the gears on a bicycle are related to inverse variation. Include an explanation of why the gear ratio affects the pedaling speed of the rider.

STANDARDIZED TEST PRACTICE

42. Which function *best* describes the graph?

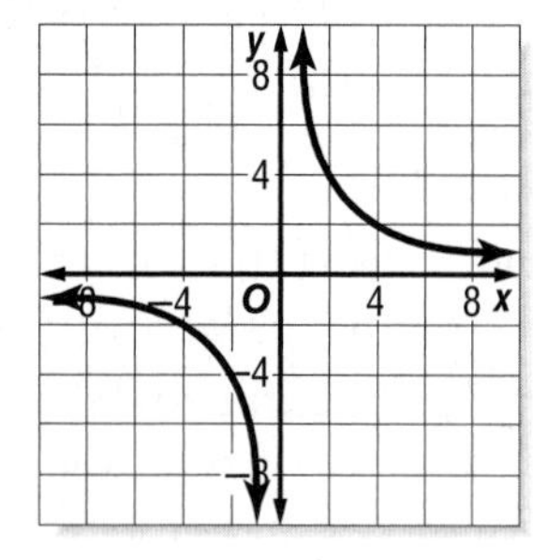

A $xy = 8$

B $xy = -8$

C $x + y = 8$

D $y = x + 8$

43. REVIEW A submarine is currently 200 feet below sea level. If the submarine begins to descend at a rate of 35 feet per minute, which equation could be used to determine t the time in minutes it will take the submarine to reach a depth of 2750 feet below sea level?

F $2750 = 200 + (-35t)$

G $-2750 = (-200 + 35)t$

H $2750 = (200 + 35)t$

J $-2750 = -200 + (-35t)$

Spiral Review

For each set of measures given, find the measures of the missing sides if $\triangle ABC \sim \triangle DEF$. (Lesson 10-6)

44. $a = 3, b = 10, c = 9, d = 12$

45. $b = 8, c = 4, d = 21, e = 28$

Find the possible values of a if the points with the given coordinates are the indicated distance apart. (Lesson 10-5)

46. $(3, 2), (a, 9); d = \sqrt{113}$

47. $(a, 6), (13, -6); d = 13$

48. $(-7, 1), (2, a); d = \sqrt{82}$

49. MUSIC Two musical notes played at the same time produce harmony. The closest harmony is produced by frequencies with the greatest GCF. A, C, and C sharp have frequencies of 220, 264, and 275, respectively. Which pair of these notes produce the closest harmony? (Lesson 8-1)

Solve each system of inequalities by graphing. (Lesson 6-8)

50. $y \le 3x - 5$
$y > -x + 1$

51. $y \ge 2x + 3$
$2y \ge -5x - 14$

52. $x + y \le 1$
$x - y \le -3$
$y \ge 0$

Solve each equation. (Lesson 2-5)

53. $7(2y - 7) = 5(4y + 1)$

54. $w(w + 2) = 2w(w - 3) + 16$

GET READY for the Next Lesson

PREREQUISITE SKILL Find the greatest common factor for each set of monomials. (Lesson 8-1)

55. 36, 15, 45

56. 48, 60, 84

57. 210, 330, 150

58. $17a, 34a^2$

59. $12xy^2, 18x^2y^3$

60. $12pr^2, 40p^4$

11-2 Rational Expressions

Main Ideas

- Identify values excluded from the domain of a rational expression.
- Simplify rational expressions.

New Vocabulary

rational expression
excluded values

GET READY for the Lesson

The intensity I of an image on a movie screen is inversely proportional to the square of the distance d between the projector and the screen. Recall from Lesson 11-1 that this can be represented by the equation $I = \frac{k}{d^2}$, where k is a constant.

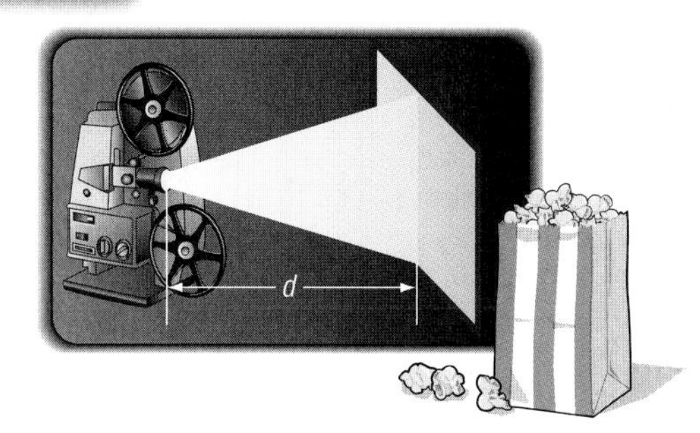

EXCLUDED VALUES OF RATIONAL EXPRESSIONS The expression $\frac{k}{d^2}$ is an example of a **rational expression**. A **rational expression** is an algebraic fraction whose numerator and denominator are polynomials.

Because a rational expression involves division, the denominator may not equal zero. Any values of a variable that result in a denominator of zero must be excluded from the domain of that variable. These are called **excluded values** of the rational expression.

EXAMPLE Excluded Values

State the excluded value for each rational expression.

a. $\frac{5m + 3}{m - 6}$

Exclude the values for which $m - 6 = 0$, because the denominator cannot equal 0.

$m - 6 = 0 \rightarrow m = 6$ Therefore, m cannot equal 6.

b. $\frac{x^2 - 5}{x^2 - 5x + 6}$

Exclude the values for which $x^2 - 5x + 6 = 0$.

$x^2 - 5x + 6 = 0$

$(x - 2)(x - 3) = 0$ Factor.

$x - 2 = 0$ or $x - 3 = 0$ Zero Product Property

$x = 2$ $x = 3$ Therefore, x cannot equal 2 or 3.

Study Tip

Look Back

You can review the Zero Product Property in Lesson 8-2.

CHECK Your Progress State the excluded values.

1A. $\frac{16x + 5}{3x}$

1B. $\frac{2x + 1}{3x^2 + 14x - 5}$

EXAMPLE Use Rational Expressions

2 HISTORY The ancient Egyptians probably used levers to help them maneuver the giant blocks they used to build the pyramids. The diagram shows how the devices may have worked using a 10-foot lever.

Cross-Curricular Project

Math Online You can use a rational expression to determine how an amusement park can finance a new roller coaster. Visit algebra1.com.

a. The mechanical advantage of a lever is $\frac{L_E}{L_R}$, where L_E is the length of the effort arm and L_R is the length of the resistance arm. Calculate the mechanical advantage of the lever the Egyptian worker is using.

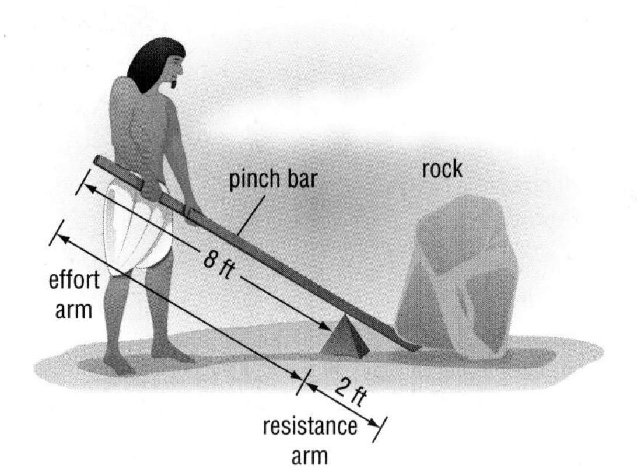

Let b represent the length of the bar and e represent the length of the effort arm. Then $b - e$ is the length of the resistance arm.

Use the expression for mechanical advantage to write an expression for the mechanical advantage in this situation.

$$\frac{L_E}{L_R} = \frac{e}{b - e} \quad L_E = e, L_R = b - e$$

$$= \frac{8}{10 - 8} \quad e = 8, b = 10$$

$$= 4 \quad \text{Simplify.}$$

The mechanical advantage is 4.

b. The force placed on the rock is the product of the mechanical advantage and the force applied to the end of the lever. If the Egyptian worker can apply a force of 180 pounds, what is the greatest weight he can lift with the lever?

Since the mechanical advantage is 4, the Egyptian worker can lift $4 \cdot 180$ or 720 pounds with this lever.

CHECK Your Progress

2. Kelli is going to lift a 535-pound rock using a 7-foot lever. If she places the fulcrum 2 feet from the rock, how much force will she have to use to lift the rock?

Simplify Rational Expressions Simplifying rational expressions is similar to simplifying fractions with numbers. To simplify a rational expression, you must eliminate any common factors in the numerator and denominator. To do this, use their greatest common factor (GCF). Remember that $\frac{ab}{ac} = \frac{a}{a} \cdot \frac{b}{c}$ and $\frac{a}{a} = 1$. So, $\frac{ab}{ac} = 1 \cdot \frac{b}{c}$ or $\frac{b}{c}$.

When a rational expression is in simplest form, the numerator and denominator have no common factors other than 1 or -1.

STANDARDIZED TEST EXAMPLE Expression Involving Monomials

3 **Which expression is equivalent to $\frac{(-3x^2)(4x^5)}{9x^6}$?**

A $\frac{4}{3}x$ B $\frac{4}{3x}$ C $\frac{-4}{3x}$ D $\frac{-4}{3}x$

Test-Taking Tip

Eliminate possibilities Sometimes you can eliminate some of the answer choices before solving the problem. For example, since there is only one negative factor in the expression in Example 3, the simplified expression must be negative. You can eliminate choices A and B.

Read the Test Item

The expression $\frac{(-3x^2)(4x^5)}{9x^6}$ represents the product of two monomials and the division of that product by another monomial.

Solve the Test Item

Step 1 Find the product in the numerator. $(-3x^2)(4x^5) = -12x^7$

Step 2 Find the GCF of the numerator and denominator. $\frac{(3x^6)(-4x)}{(3x^6)(3)}$

Step 3 Simplify. The correct answer is D. $\frac{(3x^6)(-4x)}{(3x^6)(3)}$ or $\frac{-4x}{3}$

CHECK Your Progress

3. Which expression is equivalent to $\frac{16c^2b^4}{8c^3b}$?

F $\frac{2b^3}{c}$ G $\frac{b^3}{2c}$ H $\frac{1}{2b^3c}$ J $2b^3c$

Online Personal Tutor at algebra1.com

You can use the same procedure to simplify a rational expression in which the numerator and denominator are polynomials. Determine the excluded values using the *original* expression rather than the simplified expression.

EXAMPLE Excluded Values

4 **Simplify $\frac{3x - 15}{x^2 - 7x + 10}$. State the excluded values of x.**

$\frac{3x - 15}{x^2 - 7x + 10} = \frac{3(x - 5)}{(x - 2)(x - 5)}$ Factor.

$= \frac{3\overset{1}{(x - 5)}}{(x - 2)\underset{1}{(x - 5)}}$ Divide the numerator and denominator by the GCF, $x - 5$.

$= \frac{3}{x - 2}$ Simplify.

Exclude the values for which $x^2 - 7x + 10$ equals 0.

$x^2 - 7x + 10 = 0$ The denominator cannot equal zero.

$(x - 5)(x - 2) = 0$ Factor.

$x = 5$ or $x = 2$ Zero Product Property Therefore, $x \neq 5$ and $x \neq 2$.

CHECK Your Progress

Simplify each expression. State the excluded values.

4A. $\frac{12x + 36}{x^2 - x - 12}$ **4B.** $\frac{x^2 - 2x - 35}{x^2 - 9x + 14}$

CHECK Your Understanding

Example 1 (p. 583)

State the excluded values for each rational expression.

1. $\dfrac{4a}{3 + a}$
2. $\dfrac{x^2 - 9}{2x + 6}$
3. $\dfrac{n + 5}{n^2 + n - 20}$

Examples 3, 4 (pp. 584–585)

Simplify each expression. State the excluded values of the variables.

4. $\dfrac{56x^2y}{70x^3y^2}$
5. $\dfrac{x^2 - 49}{x + 7}$
6. $\dfrac{x + 4}{x^2 + 8x + 16}$
7. $\dfrac{3x - 9}{x^2 - 7x + 12}$
8. $\dfrac{a^2 + 4a - 12}{a^2 + 2a - 8}$
9. $\dfrac{2x^2 - x - 21}{2x^2 - 15x + 28}$

Example 2 (p. 584)

AQUARIUMS For Exercises 10 and 11, use the following information.
Jenna has guppies in her aquarium. One week later, she adds four neon fish.

10. Define a variable. Then write an expression that represents the fraction of neon fish in the aquarium.

11. Suppose that two months later the guppy population doubles. Jenna still has four neons, and she buys 5 different tropical fish. Write an expression that shows the fraction of neons in the aquarium after the other fish have been added.

Exercises

HOMEWORK HELP	
For Exercises	See Examples
12–17	1
18–21	2
22–39	3, 4

State the excluded values for each rational expression.

12. $\dfrac{m + 3}{m - 2}$
13. $\dfrac{3b}{b + 5}$
14. $\dfrac{3n + 18}{n^2 - 36}$
15. $\dfrac{2x - 10}{x^2 - 25}$
16. $\dfrac{n^2 - 36}{n^2 + n - 30}$
17. $\dfrac{25 - x^2}{x^2 + 12x + 35}$

PHYSICAL SCIENCE For Exercises 18 and 19, use the following information.
To pry the lid off a crate, a crowbar that is 18.2 inches long is used as a lever. It is placed so that 1.5 inches of its length extends inward from the edge of the crate.

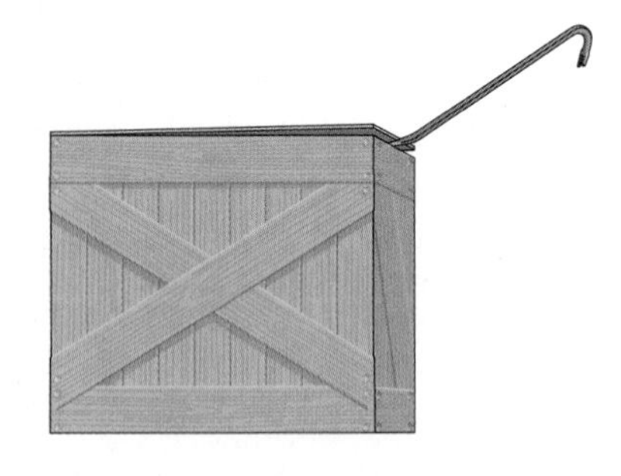

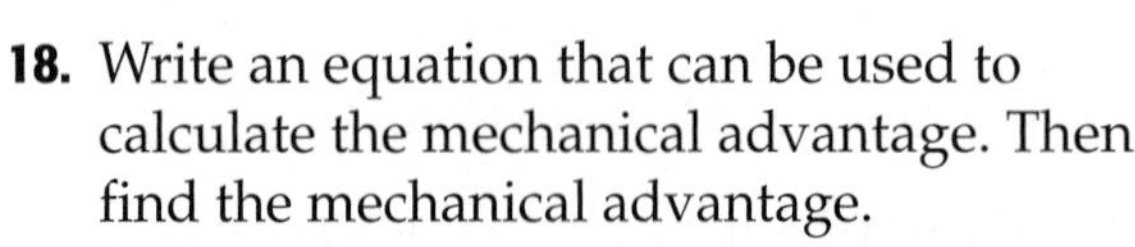

18. Write an equation that can be used to calculate the mechanical advantage. Then find the mechanical advantage.

19. If a force of 16 pounds is applied to the end of the crowbar, what is the force placed on the lid?

COOKING For Exercises 20 and 21, use the following information.
The formula $t = \dfrac{40(25 + 1.85a)}{50 - 1.85a}$ relates the time t in minutes that it takes to cook an average-size potato in an oven that is at an altitude of a thousands of feet.

20. What is the value of a for an altitude of 4500 feet?

21. Calculate the time it takes to cook a potato at an altitude of 3500 feet and at 7000 feet. How do your cooking times compare at these two altitudes?

Simplify each expression. State the excluded values of the variables.

22. $\dfrac{35yz^2}{14y^2z}$
23. $\dfrac{14a^3b^2}{42ab^3}$
24. $\dfrac{64qr^2s}{16q^2rs}$
25. $\dfrac{9x^2yz}{24xyz^2}$
26. $\dfrac{7a^3b^2}{21a^2b + 49ab^3}$
27. $\dfrac{3m^2n^3}{36mn^3 - 12m^2n^2}$
28. $\dfrac{x^2 + x - 20}{x + 5}$
29. $\dfrac{z^2 + 10z + 16}{z + 2}$
30. $\dfrac{4x + 8}{x^2 + 6x + 8}$
31. $\dfrac{2y - 4}{y^2 + 3y - 10}$
32. $\dfrac{m^2 - 36}{m^2 - 5m - 6}$
33. $\dfrac{a^2 - 9}{a^2 + 6a - 27}$
34. $\dfrac{x^2 + x - 2}{x^2 - 3x + 2}$
35. $\dfrac{b^2 + 2b - 8}{b^2 - 20b + 64}$
36. $\dfrac{x^2 - x - 20}{x^3 + 10x^2 + 24x}$
37. $\dfrac{n^2 - 8n + 12}{n^3 - 12n^2 + 36n}$
38. $\dfrac{4x^2 - 6x - 4}{2x^2 - 8x + 8}$
39. $\dfrac{3m^2 + 9m + 6}{4m^2 + 12m + 8}$

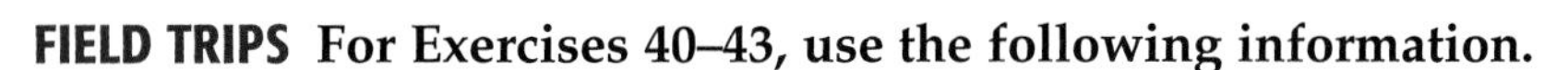

FIELD TRIPS For Exercises 40–43, use the following information.

Mrs. Hoffman's art class is taking a trip to the museum. A bus that can seat up to 56 people costs \$450 for the day, and group rate tickets at the museum cost \$4 each.

40. If there are no more than 56 students going on the field trip, write an expression for the total cost for n students to go to the museum.
41. Write a rational expression that could be used to calculate the cost of the trip per student.
42. How many students must attend in order to keep the cost under \$15 per student?
43. How would you change the expression for cost per student if the school were to cover the cost of two adult chaperones?

Real-World Career
Farmer
Farmers use mathematics to optimize usage of space, crop output, and profits while running their farms.

Math Online
For more information, go to algebra1.com.

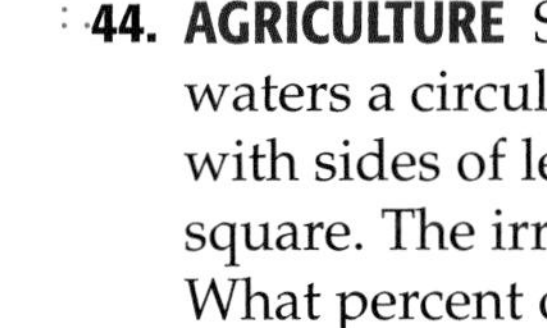

44. **AGRICULTURE** Some farmers use an irrigation system that waters a circular region in a field. Suppose a square field with sides of length $2x$ is irrigated from the center of the square. The irrigation system can reach a radius of x. What percent of the field is irrigated to the nearest whole percent?

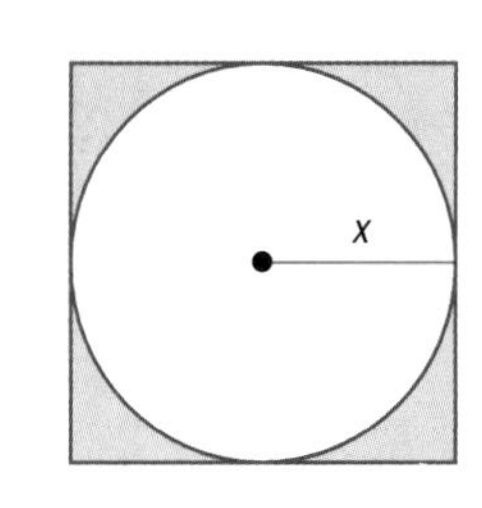

H.O.T. Problems

45. **OPEN ENDED** Write a rational expression involving one variable for which the excluded values are -4 and -7. Explain how you found the expression.

46. **REASONING** Explain why -2 may not be the only excluded value of a rational expression that simplifies to $\dfrac{x - 3}{x + 2}$.
47. **CHALLENGE** Two students graphed the following equations on their calculators.

$$y = \frac{x^2 - 16}{x - 4} \qquad y = x + 4$$

They were surprised to see that the graphs appeared to be identical. Explain how the graphs are different.

48. ***Writing in Math*** Use the information on page 583 to explain how rational expressions can be used in a movie theater. Include a description of how you determine the excluded values of the rational expression that is given.

STANDARDIZED TEST PRACTICE

49. The area of each wall in LaTisha's room is $x^2 + 3x + 2$ square feet. A gallon of paint will cover an area of $x^2 - 2x - 3$ square feet. Which expression gives the number of gallons of paint that LaTisha will need to buy to paint her room?

A $\frac{4x}{x-3}$

B $\frac{x+2}{x-3}$

C $\frac{4x+4}{x-3}$

D $\frac{4x+8}{x-3}$

50. REVIEW What is the volume of the triangular prism shown below?

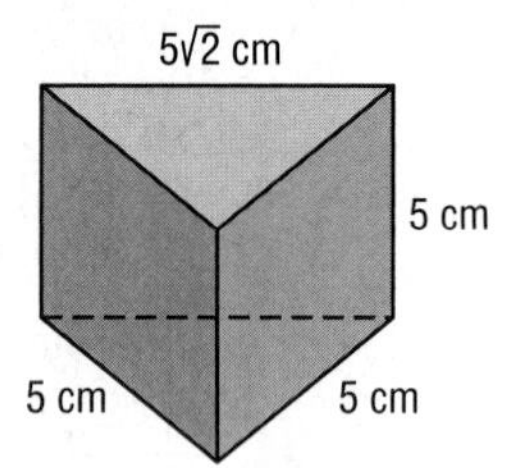

F 12.5 cm^3

G $25\sqrt{2} \text{ cm}^3$

H 62.5 cm^3

J $125\sqrt{2} \text{ cm}^3$

Spiral Review

Write an inverse variation equation that relates x and y. Assume that y varies inversely as x. Then solve. (Lesson 11-1)

51. If $y = 6$ when $x = 10$, find y when $x = -12$.

52. If $y = 16$ when $x = \frac{1}{2}$, find x when $y = 32$.

53. If $y = -9$ when $x = 6$, find x when $y = 3$.

54. If $y = -2.5$ when $x = 3$, find y when $x = -8$.

For each set of measures given, find the measures of the missing sides if $\triangle KLM \sim \triangle NOP$. (Lesson 10-6)

55. $k = 5, \ell = 3, m = 2, n = 10$

56. $\ell = 9, m = 3, n = 12, p = 4.5$

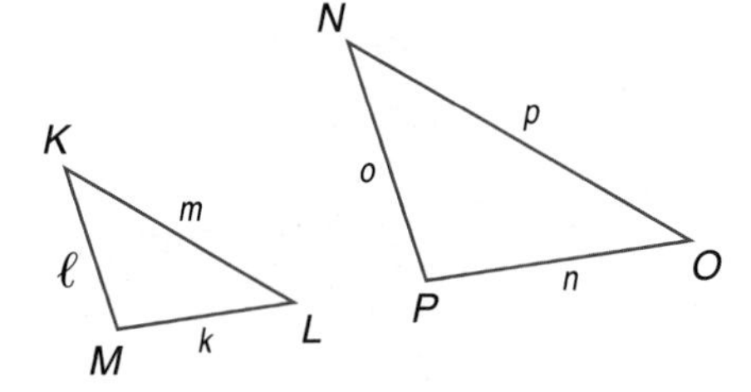

Solve each equation. Check your solution(s). (Lesson 10-3)

57. $\sqrt{a+3} = 2$

58. $\sqrt{2z+2} = z - 3$

59. $\sqrt{13 - 4p} - p = 8$

60. $\sqrt{3r^2 + 61} = 2r + 1$

61. GROCERIES The Ricardos drink about 5 gallons of milk every 2.5 weeks. At this rate, how much money will they spend on milk in a year if it costs \$3.58 a gallon? (Lesson 2-6)

GET READY for the Next Lesson

PREREQUISITE SKILL **Complete.**

62. 84 in. = ___ ft

63. 4.5 m = ___ cm

64. 4 h 15 min = ___ s

65. 18 mi = ___ ft

66. 3 days = ___ h

67. 220 mL = ___ L

EXTEND 11-2

Graphing Calculator Lab
Rational Expressions

When simplifying rational expressions, you can use a TI-83/84 Plus graphing calculator to support your answer. If the graphs of the original expression and the simplified expression coincide, they are equivalent.

ACTIVITY

Simplify $\frac{x^2 + 5x}{x^2 + 10x + 25}$.

Step 1 Factor the numerator and denominator.

$$\frac{x^2 + 5x}{x^2 + 10x + 25} = \frac{x(x + 5)}{(x + 5)(x + 5)} = \frac{x}{(x + 5)}$$

When $x = -5$, $x + 5 = 0$. Therefore, x cannot equal -5 because you cannot divide by zero.

Step 2 Graph the original expression.

- Set the calculator to **Dot** mode.
- Enter $\frac{x^2 + 5x}{x^2 + 10x + 25}$ as **Y1** and graph.

KEYSTROKES: MODE ▼ ▼ ▼ ▼ ▶

ENTER Y= (X,T,θ,n x^2 + 5 X,T,θ,n) ÷ (X,T,θ,n x^2 + 10 X,T,θ,n + 25) ZOOM 6

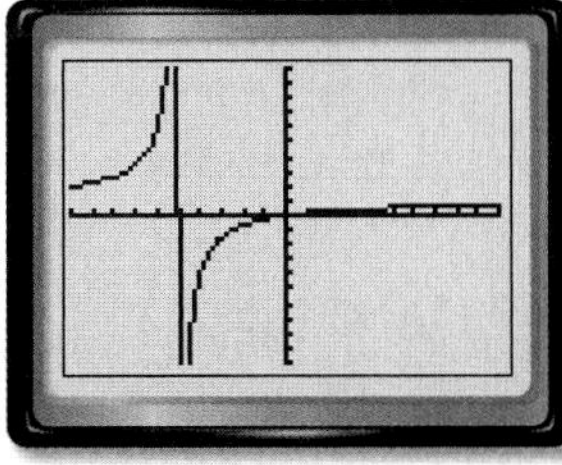

[−10, 10] scl: 1 by [−10, 10] scl: 1

Step 3 Graph the simplified expression.

- Enter $\frac{x}{(x + 5)}$ as **Y2** and graph.

KEYSTROKES: Y= ▼ X,T,θ,n ÷

(X,T,θ,n + 5) GRAPH

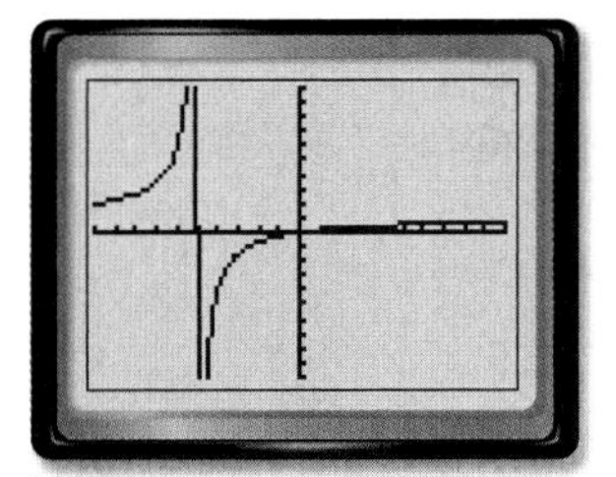

[−10, 10] scl: 1 by [−10, 10] scl: 1

Since the graphs overlap, the two expressions are equivalent.

Exercises

Simplify each expression. Then verify your answer graphically. Name the excluded values.

1. $\frac{3x + 6}{x^2 + 7x + 10}$

2. $\frac{2x + 8}{x^2 + 6x + 8}$

3. $\frac{5x^2 + 10x + 5}{3x^2 + 6x + 3}$

4. Simplify $\frac{2x - 9}{4x^2 - 18x}$ and answer each question using the **TABLE** menu.

a. How can you use the **TABLE** function to verify that the original expression and the simplified expression are equivalent?

b. How does the **TABLE** function show you that an x-value is excluded?

11-3 Multiplying Rational Expressions

Main Ideas

- Multiply rational expressions.
- Use dimensional analysis with multiplication.

GET READY for the Lesson

There are 25 lights around a patio. Each light is 40 watts, and the cost of electricity is 15 cents per kilowatt-hour. You can use the expression below to calculate the cost of using the lights for h hours.

$$25 \text{ lights} \cdot \frac{40 \text{ watts}}{\text{light}} \cdot \frac{1 \text{ kilowatt}}{1000 \text{ watts}} \cdot \frac{15 \text{ cents}}{1 \text{ kilowatt} \cdot \text{hour}} \cdot \frac{1 \text{ dollar}}{100 \text{ cents}} \cdot h \text{ hours}$$

MULTIPLY RATIONAL EXPRESSIONS The multiplication expression above is similar to the multiplication of rational expressions. Recall that to multiply fractions, you multiply numerators and multiply denominators. You can use the same method to multiply rational expressions.

Study Tip

Rational Expressions

From this point on, you may assume that no denominator of a rational expression in this text has a value of zero.

EXAMPLE Expressions Involving Monomials

1 Find $\frac{5ab^3}{8c^2} \cdot \frac{16c^3}{15a^2b}$.

Method 1 Divide by the greatest common factor after multiplying.

$$\frac{5ab^3}{8c^2} \cdot \frac{16c^3}{15a^2b} = \frac{80ab^3c^3}{120a^2bc^2}$$ ← Multiply the numerators. ← Multiply the denominators.

$$= \frac{\overset{1}{\cancel{40abc^2}}(2b^2c)}{\underset{1}{\cancel{40abc^2}}(3a)}$$ The GCF is $40abc^2$.

$$= \frac{2b^2c}{3a}$$ Simplify.

Method 2 Divide by the common factors before multiplying.

$$\frac{5ab^3}{8c^2} \cdot \frac{16c^3}{15a^2b} = \frac{\overset{1\,1\,b^2}{\cancel{5}\cancel{a}\cancel{b^3}}}{\underset{1\,1}{\cancel{8}\cancel{c^2}}} \cdot \frac{\overset{2\,c}{\cancel{16}\cancel{c^3}}}{\underset{3\,a\,1}{\cancel{15}\cancel{a^2}\cancel{b}}}$$ Divide by common factors 5, 8, a, b, and c^2.

$$= \frac{2b^2c}{3a}$$ Multiply.

CHECK Your Progress Find each product.

1A. $\frac{5c^3d}{c^4d} \cdot \frac{f^2d^3c}{10cf^4}$

1B. $\frac{16g^2h^3}{8gh^2} \cdot \frac{3g^2h}{32hj^2}$

Math Online Extra Examples at algebra1.com

Sometimes you must factor a quadratic expression before you can simplify a product of rational expressions.

EXAMPLE Expressions Involving Polynomials

2 **Find** $\frac{x-5}{x} \cdot \frac{x^2}{x^2-2x-15}$.

$$\frac{x-5}{x} \cdot \frac{x^2}{x^2-2x-15} = \frac{x-5}{x} \cdot \frac{x^2}{(x-5)(x+3)}$$ Factor the denominator.

$$= \frac{\overset{x}{\cancel{x^2}}\overset{1}{\cancel{(x-5)}}}{\underset{1}{\cancel{x}}\underset{1}{\cancel{(x-5)}}(x+3)}$$ The GCF is $x(x-5)$.

$$= \frac{x}{x+3}$$ Simplify.

CHECK Your Progress **Find each product.**

2A. $\frac{x+3}{x} \cdot \frac{5}{x^2+7x+12}$

2B. $\frac{y^2-3y-4}{y+5} \cdot \frac{y+5}{y^2-4y}$

DIMENSIONAL ANALYSIS When you multiply fractions that involve units of measure, you can divide by the units in the same way that you divide by variables. Recall that this process is called *dimensional analysis.*

Real-World EXAMPLE Dimensional Analysis

Real-World Link

American sprinter Thomas Burke won the 100-meter dash at the first modern Olympics in Athens, Greece, in 1896 in 12.0 seconds.

Source: olympics.org

3 **OLYMPICS In the 2004 Summer Olympics in Athens, Greece, Justin Gatlin of the United States won the gold medal for the 100-meter sprint. His winning time was 9.85 seconds. What was his speed in kilometers per hour? Round to the nearest hundredth.**

$$\frac{100\text{ m}}{9.85\text{ s}} \cdot \frac{1\text{ k}}{1000\text{ m}} \cdot \frac{60\text{ s}}{1\text{ min}} \cdot \frac{60\text{ min}}{1\text{ h}} = \frac{100\cancel{\text{ m}}}{9.85\cancel{\text{ s}}} \cdot \frac{1\text{ k}}{1000\cancel{\text{ m}}} \cdot \frac{60\cancel{\text{ s}}}{1\cancel{\text{ min}}} \cdot \frac{60\cancel{\text{ min}}}{1\text{ h}}$$

$$= \frac{\overset{1}{\cancel{100}} \cdot 1 \cdot 60 \cdot 60 \cdot \text{k}}{9.85 \cdot \underset{10}{\cancel{1000}} \cdot 1 \cdot 1\text{ h}}$$ Simplify.

$$= \frac{60 \cdot 60\text{ k}}{9.85 \cdot 10\text{ h}}$$ Multiply.

$$= \frac{3600\text{ k}}{98.5\text{ h}}$$ Multiply.

$$= \frac{36.54\text{ k}}{1\text{ h}}$$ Divide numerator and denominator by 98.5.

His speed was 36.54 kilometers per hour.

CHECK Your Progress

3. SPEED Todd is driving to his grandparents' house at 65 miles per hour. How fast is he going in feet per second?

Online Personal Tutor at algebra1.com

✓ CHECK Your Understanding

Find each product.

Example 1 (p. 590)

1. $\frac{64y^2}{5y} \cdot \frac{5y}{8y}$

2. $\frac{15s^2t^3}{12st} \cdot \frac{16st^2}{10s^3t^3}$

Example 2 (p. 591)

3. $\frac{m+4}{3m} \cdot \frac{4m^2}{(m+4)(m+5)}$

4. $\frac{x^2-4}{2} \cdot \frac{4}{x-2}$

5. $\frac{n^2-16}{n+4} \cdot \frac{n+2}{n^2+-8n+16}$

6. $\frac{x-5}{x^2-7x+10} \cdot \frac{x^2+x-6}{5}$

Example 3 (p. 591)

7. Find $\frac{24 \text{ feet}}{1 \text{ second}} \cdot \frac{60 \text{ seconds}}{1 \text{ minute}} \cdot \frac{60 \text{ minutes}}{1 \text{ hour}} \cdot \frac{1 \text{ mile}}{5280 \text{ feet}}$.

8. **SPACE** The Moon is about 240,000 miles from Earth. How many days would it take a spacecraft to reach the Moon if it travels at an average speed of 100 miles per minute?

Exercises

HOMEWORK HELP	
For Exercises	See Examples
9–12	1
13–22	2
23, 24	3

Find each product.

9. $\frac{8}{x^2} \cdot \frac{x^4}{4x}$

10. $\frac{10r^3}{6n^3} \cdot \frac{42n^2}{35r^3}$

11. $\frac{10y^3z^2}{6wx^3} \cdot \frac{12w^2x^2}{25y^2z^4}$

12. $\frac{3a^2b}{2gh} \cdot \frac{24g^2h}{15ab^2}$

13. $\frac{(x-8)}{(x+8)(x-3)} \cdot \frac{(x+4)(x-3)}{(x-8)}$

14. $\frac{(n-1)(n+1)}{(n+1)} \cdot \frac{(n-4)}{(n-1)(n+4)}$

15. $\frac{(z+4)(z+6)}{(z-6)(z+1)} \cdot \frac{(z+1)(z-5)}{(z+3)(z+4)}$

16. $\frac{(x-1)(x+7)}{(x-7)(x-4)} \cdot \frac{(x-4)(x+10)}{(x+1)(x+10)}$

17. $\frac{x^2-25}{9} \cdot \frac{x+5}{x-5}$

18. $\frac{y^2-4}{y^2-1} \cdot \frac{y+1}{y+2}$

19. $\frac{x+3}{x+4} \cdot \frac{x}{x^2+7x+12}$

20. $\frac{n}{n^2+8n+15} \cdot \frac{2n+10}{n^2}$

21. $\frac{b^2+12b+11}{b^2-9} \cdot \frac{b+9}{b^2+20b+99}$

22. $\frac{a^2-a-6}{a^2-16} \cdot \frac{a^2+7a+12}{a^2+4a+4}$

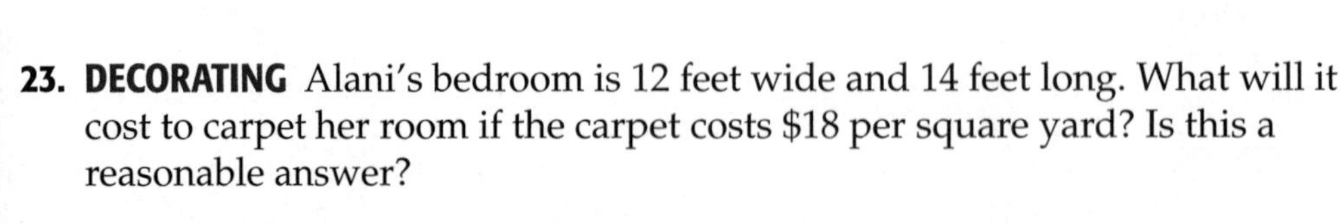

23. **DECORATING** Alani's bedroom is 12 feet wide and 14 feet long. What will it cost to carpet her room if the carpet costs $18 per square yard? Is this a reasonable answer?

Real-World Link

A system of floating exchange rates among international currencies was established in 1976. It was needed because the old system of basing a currency's value on gold had become obsolete.

Source: infoplease.com

24. **EXCHANGE RATES** While traveling in Canada, Johanna bought some gifts to bring home. She bought 2 T-shirts that cost a total of $21.95 (Canadian). If the exchange rate at the time was 1 U.S. dollar for 1.21 Canadian dollars, how much did Johanna spend in U.S. dollars?

25. **RESEARCH** Use the Internet or other sources to research exchange rates for the U.S. dollar against a foreign currency of your choosing over the last six months. What has been the average rate of exchange? What has been the overall trend of the exchange rate? What current events have affected the change in rates?

Find each product.

26. $\dfrac{2.54 \text{ centimeters}}{1 \text{ inch}} \cdot \dfrac{12 \text{ inches}}{1 \text{ foot}} \cdot \dfrac{3 \text{ feet}}{1 \text{ yard}}$

27. $\dfrac{60 \text{ kilometers}}{1 \text{ hour}} \cdot \dfrac{1000 \text{ meters}}{1 \text{ kilometer}} \cdot \dfrac{1 \text{ hour}}{60 \text{ minutes}} \cdot \dfrac{1 \text{ minute}}{60 \text{ seconds}}$

28. $\dfrac{32 \text{ feet}}{1 \text{ second}} \cdot \dfrac{60 \text{ seconds}}{1 \text{ minute}} \cdot \dfrac{60 \text{ minutes}}{1 \text{ hour}} \cdot \dfrac{1 \text{ mile}}{5280 \text{ feet}}$

29. $10 \text{ feet} \cdot 18 \text{ feet} \cdot 3 \text{ feet} \cdot \dfrac{1 \text{ yard}^3}{27 \text{ feet}^3}$

30. **CITY MAINTENANCE** Street sweepers can clean 3 miles of streets per hour. A city owns 2 street sweepers, and each sweeper can be used for three hours before it comes in for an hour to refuel. During an 18 hour shift, how many miles of street can be cleaned?

TRAINS For Exercises 31–33, use the following information.

Trying to get into a train yard one evening, all of the trains are backed up for 2 miles along a system of tracks. Assume that each car occupies an average of 75 feet of space on a track and that the train yard has 5 tracks.

31. Write and solve an expression that could be used to determine the number of train cars involved in the backup.

32. How many train cars are involved in the backup?

33. Suppose that there are 8 attendants doing safety checks on each car, and it takes each vehicle an average of 45 seconds for each check. Approximately how many hours will it take for all the vehicles in the backup to exit?

H.O.T. Problems

34. **OPEN ENDED** Write two rational expressions with a product of $\frac{2}{x}$.

35. **CHALLENGE** Identify the expressions that are equivalent to $\frac{x}{y}$. Explain why the expressions are equivalent.

a. $\dfrac{x+3}{y+3}$ **b.** $\dfrac{3-x}{3-y}$ **c.** $\dfrac{3x}{3y}$ **d.** $\dfrac{x^3}{y^3}$ **e.** $\dfrac{n^3x}{n^3y}$

36. **FIND THE ERROR** Amiri and Hoshi multiplied $\frac{x-3}{x+3}$ and $\frac{4x}{x^2-4x+3}$. Who is correct? Explain your reasoning.

Amiri

$$\frac{x-3}{x+3} \cdot \frac{4x}{x^2-4x+3} = \frac{(x-3)4x}{(x+3)(x-3)(x-1)} = \frac{4x}{(x+3)(x-1)}$$

Hoshi

$$\frac{x-3}{x+3} \cdot \frac{4x}{x^2-4x+3} = \frac{4x}{x^2-4x+3} = \frac{1}{x^2+3}$$

37. **CHALLENGE** Explain why $-\dfrac{x+6}{x-5}$ is not equivalent to $\dfrac{-x+6}{x-5}$. What property of mathematics was used to reach this conclusion?

38. ***Writing in Math*** Use the information provided on page 590 to explain how multiplying rational expressions can determine the cost of electricity. Include an expression that you could use to determine the cost of using 60-watt light bulbs instead of 40-watt bulbs.

STANDARDIZED TEST PRACTICE

39. In order to stay in a low-Earth orbit, an object must reach a speed of about 17,500 miles per hour. How fast is this in meters per second? (1609.34 meters ≈ 1 mile)

A 10.9 m/s

B 7823.18 m/s

C 469,390.8 m/s

D 2.8×10^7 m/s

40. REVIEW Stanley used toothpicks to make the shapes below. If x is a shape's order in the pattern (for the first shape $x = 1$, for the second shape $x = 2$, and so on), which expression can be used to find the number of toothpicks needed to make any shape in the pattern?

F $3x - 3$

G $4x$

H $3x + 1$

J $4x + 3$

Spiral Review

State the excluded values for each rational expression. (Lesson 11-2)

41. $\frac{s + 6}{s^2 - 36}$

42. $\frac{a^2 - 25}{a^2 + 3a - 10}$

43. $\frac{x + 3}{x^2 + 6x + 9}$

Write an inverse variation equation that relates x and y. Assume that y varies inversely as x. Then solve. (Lesson 11-1)

44. If $y = 9$ when $x = 8$, find x when $y = 6$.

45. If $y = 2.4$ when $x = 8.1$, find y when $x = 3.6$.

46. If $y = 24$ when $x = -8$, find y when $x = 4$.

47. If $y = 6.4$ when $x = 4.4$, find x when $y = 3.2$.

Solve each inequality. Then check your solution. (Lesson 6-2)

48. $\frac{g}{8} < \frac{7}{2}$

49. $3.5r \geq 7.35$

50. $\frac{9k}{4} > \frac{3}{5}$

Simplify. Assume that no denominator is equal to zero. (Lesson 7-2)

51. $\frac{-7^{12}}{7^9}$

52. $\frac{20p^6}{8p^8}$

53. $\frac{24a^3b^4c^7}{6a^6c^2}$

54. FINANCE The total amount of money Antonio earns mowing lawns and doing yard work varies directly with the number of days he works. At one point, he earned \$340 in 4 days. At this rate, how long will it take him to earn \$935? (Lesson 4-2)

GET READY for the Next Lesson

Factor each polynomial (Lessons 8-3 and 8-4)

55. $x^2 - 3x - 40$

56. $n^2 - 64$

57. $x^2 - 12x + 36$

58. $a^2 + 2a - 35$

59. $2x^2 - 5x - 3$

60. $3x^3 - 24x^2 + 36x$

11-4 Dividing Rational Expressions

Main Ideas

- Divide rational expressions.
- Use dimensional analysis with division.

GET READY for the Lesson

Most soft drinks come in aluminum cans. Although more cans are used today than in the 1970s, the demand for new aluminum has declined. This is due in large part to the great number of cans that are recycled. In recent years, approximately 63.9 billion cans were recycled annually. This represents $\frac{5}{8}$ of all cans produced.

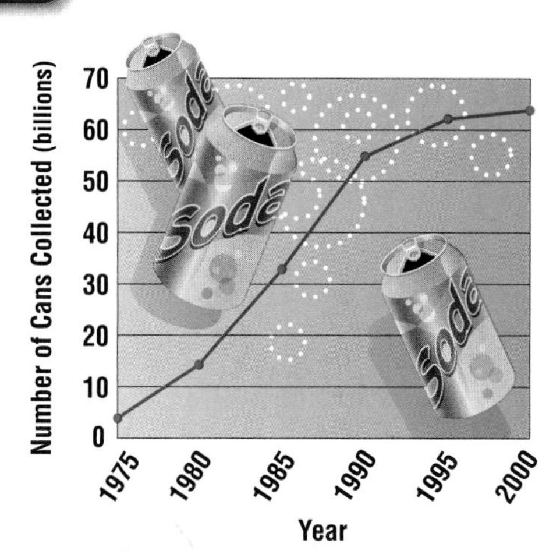

DIVIDE RATIONAL EXPRESSIONS Recall that to divide fractions, you multiply by the reciprocal of the divisor. You can use this same method to divide rational expressions.

EXAMPLE 1 Divide by Fractions

Find each quotient.

a. $\frac{5x^2}{7} \div \frac{10x^3}{21}$

$\frac{5x^2}{7} \div \frac{10x^3}{21} = \frac{5x^2}{7} \cdot \frac{21}{10x^3}$ Multiply by $\frac{21}{10x^3}$, the reciprocal of $\frac{10x^3}{21}$.

$= \frac{\overset{1}{\cancel{5x^2}}}{\underset{1}{\cancel{7}}} \cdot \frac{\overset{3}{\cancel{21}}}{\underset{2x}{\cancel{10x^3}}}$ Divide by common factors 5, 7, and x^2.

$= \frac{3}{2x}$ Simplify.

b. $\frac{n+1}{n+3} \div \frac{2n+2}{n+4}$

$\frac{n+1}{n+3} \div \frac{2n+2}{n+4} = \frac{n+1}{n+3} \cdot \frac{n+4}{2n+2}$ Multiply by $\frac{n+4}{2n+2}$, the reciprocal of $\frac{2n+2}{n+4}$.

$= \frac{n+1}{n+3} \cdot \frac{n+4}{2(n+1)}$ Factor $2n + 2$.

$= \frac{\overset{1}{\cancel{n+1}}}{n+3} \cdot \frac{n+4}{2\underset{1}{\cancel{(n+1)}}}$ The GCF is $n + 1$.

$= \frac{n+4}{2(n+3)}$ or $\frac{n+4}{2n+6}$ Simplify.

CHECK Your Progress Find each quotient.

1A. $\frac{15y^2}{4x} \div \frac{5y}{8x^3}$

1B. $\frac{27c^3d^2}{11d} \div \frac{2c^3}{9d^3e}$

1C. $\frac{b+4}{3b+2} \div \frac{3b+12}{b+1}$

1D. $\frac{6b-12}{3b+15} \div \frac{12b+18}{b+5}$

Sometimes you must factor a quadratic expression before you can simplify the quotient of rational expressions.

EXAMPLE Expression Involving Polynomials

2 **Find** $\frac{m^2+3m+2}{4} \div \frac{m+2}{m+1}$.

$\frac{m^2+3m+2}{4} \div \frac{m+2}{m+1} = \frac{m^2+3m+2}{4} \cdot \frac{m+1}{m+2}$ Multiply by the reciprocal, $\frac{m+1}{m+2}$.

$= \frac{(m+1)(m+2)}{4} \cdot \frac{m+1}{m+2}$ Factor $m^2 + 3m + 2$.

$= \frac{(m+1)\cancel{(m+2)}^{1}}{4} \cdot \frac{m+1}{\cancel{m+2}_{1}}$ The GCF is $m + 2$.

$= \frac{(m+1)^2}{4}$ Simplify.

CHECK Your Progress

Find each quotient.

2A. $\frac{p^2-4}{5p} \div \frac{p-2}{p+q}$

2B. $\frac{q^2+3q+2}{12} \div \frac{q+1}{q^2+4}$

Real-World Link

The first successful Mars probe was the Mariner 4, which arrived at Mars on July 14, 1965.

Source: NASA

DIMENSIONAL ANALYSIS You can divide rational expressions that involve units of measure by using dimensional analysis.

Real-World EXAMPLE

3 **SPACE In April, 2001, NASA launched the *Mars Odyssey* spacecraft. It took 200 days for the spacecraft to travel 466,000,000 miles from Earth to Mars. What was the average speed of the spacecraft in miles per hour? Round to the nearest mile per hour.**

$rt = d$ rate • time = distance

$r \cdot 200 \text{ days} = 466{,}000{,}000 \text{ mi}$ $t = 200$ days, $d = 466{,}000{,}000$

$r = \frac{466{,}000{,}000 \text{ mi}}{200 \text{ days}}$ Divide each side by 200 days.

$= \frac{466{,}000{,}000 \text{ miles}}{200 \cancel{\text{days}}} \cdot \frac{1 \cancel{\text{day}}}{24 \text{ hours}}$ Convert days to hours.

$= \frac{466{,}000{,}000 \text{ miles}}{4800 \text{ hours}}$ or about $\frac{97{,}083 \text{ miles}}{1 \text{ hour}}$

Thus, the spacecraft traveled at a rate of about 97,083 miles per hour.

3. On July 7, 2003, the rover *Opportunity* was launched. It landed on Mars on January 25, 2004. Assuming that *Opportunity* traveled the same distance as the *Mars Odyssey Spacecraft,* how fast did *Opportunity* travel? Round to the nearest mile per hour.

Online Personal Tutor at algebra1.com

CHECK Your Understanding

Find each quotient.

Example 1 (p. 595)

1. $\frac{10n^3}{7} \div \frac{5n^2}{21}$

2. $\frac{2a}{3} \div \frac{a^7}{b^3}$

3. $\frac{3m + 15}{m + 4} \div \frac{m + 5}{6m + 24}$

4. $\frac{3n^2 - 12}{n - 6} \div \frac{(n - 6)(n - 2)}{n + 4}$

Example 2 (p. 596)

5. $\frac{k + 3}{k^2 + 4k + 4} \div \frac{2k + 6}{k + 2}$

6. $\frac{2x + 4}{x^2 + 11x + 18} \div \frac{x + 1}{x^2 + 5x + 6}$

Example 3 (p. 596)

7. Express 85 kilometers per hour in meters per second.

8. Express 32 pounds per square foot as pounds per square inch.

9. COOKING Latisha was making candy using a two-quart pan. As she stirred the mixture, she noticed that the pan was about $\frac{2}{3}$ full. If each piece of candy has a volume of about $\frac{3}{4}$ ounce, approximately how many pieces of candy will Latisha make? (*Hint*: There are 32 ounces in a quart.)

Exercises

HOMEWORK HELP	
For Exercises	See Examples
10–15	1
16–21	2
28, 29	3

Find each quotient.

10. $\frac{a^2}{b^2} \div \frac{a}{b^3}$

11. $\frac{n^4}{p^2} \div \frac{n^2}{p^3}$

12. $\frac{10m^2}{7n^2} \div \frac{25m^4}{14n^3}$

13. $\frac{a^4bc^3}{g^2h^3} \div \frac{ab^2c^2}{g^3h^3}$

14. $\frac{3x + 12}{4x - 18} \div \frac{2x + 8}{x + 4}$

15. $\frac{4a - 8}{2a - 6} \div \frac{2a - 4}{a - 4}$

16. $\frac{x^2 + 2x + 1}{2} \div \frac{x + 1}{x - 1}$

17. $\frac{n^2 + 3n + 2}{4} \div \frac{n + 1}{n + 2}$

18. $\frac{a^2 + 8a + 16}{a^2 - 6a + 9} \div \frac{2a + 8}{3a - 9}$

19. $\frac{b + 2}{b^2 + 4b + 4} \div \frac{2b + 4}{b + 4}$

20. $\frac{x^2 + x - 2}{x^2 + 5x + 6} \div \frac{x^2 + 2x - 3}{x^2 + 7x + 12}$

21. $\frac{x^2 + 2x - 15}{x^2 - x - 30} \div \frac{x^2 - 3x - 18}{x^2 - 2x - 24}$

22. What is the quotient when $\frac{2x + 6}{x + 5}$ is divided by $\frac{2}{x + 5}$?

23. Find the quotient when $\frac{m - 8}{m + 7}$ is divided by $m^2 - 7m - 8$.

Complete.

24. $24 \text{ yd}^3 =$ ____ ft^3

25. $0.35 \text{ m}^3 =$ ____ cm^3

26. 330 ft/s = ____ mi/h

27. $1730 \text{ plants/km}^2 =$ ____ plants/m^2

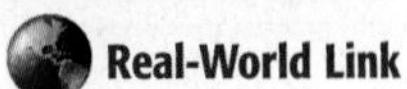
Real-World Link

The Ironman Championship Triathlon held in Hawaii consists of a 2.4-mile swim, a 112-mile bicycle ride, and a 26.2-mile run.

Source: www.infoplease.com

28. TRIATHLONS Sadie is training for an upcoming triathlon and plans to run the full length of the running section today. Jorge offered to ride his bicycle to help her maintain her pace. If Sadie wants to finish her run in about 4 hours and 28 minutes, how fast should Jorge ride in miles per hour?

29. VOLUNTEERING Tyrell is passing out orange drink from a 3.5-gallon cooler. If each cup of orange drink is 4.25 ounces, about how many cups can he hand out? (*Hint*: There are 128 ounces in a gallon.)

LANDSCAPING For Exercises 30 and 31, use the following information.

A landscaping supervisor needs to determine how many truckloads of dirt must be removed from a site before a brick patio can be completed. The truck bed has the shape shown at the right.

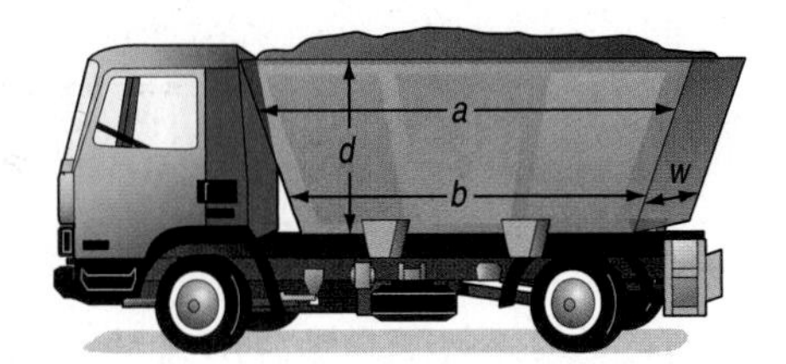

30. Write an equation involving units that represents the volume of the truck bed in cubic yards. Use the formula $V = \frac{d(a + b)}{2} \cdot w$ with $a = 10$ feet, $b = 17$ feet, $w = 4$ feet, and $d = 3.5$ feet.

31. The supervisor found that there are 45 cubic yards of dirt that must be removed from the site. Write an equation involving units that represents the number of truckloads that will be required to remove all of the dirt. How many trips will they have to take to remove all the dirt?

TRUCKS For Exercises 32 and 33, use the following information.

The speedometer of John's truck uses the revolutions of his tires to calculate the speed of the truck.

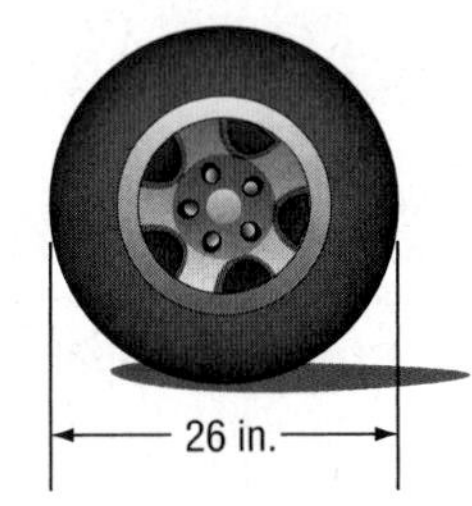

32. How many times per minute do the tires revolve when the truck is traveling at 55 miles per hour?

33. Suppose John buys tires with a diameter of 30 inches. When the speedometer reads 55 miles per hour, the tires would still revolve at the same rate as before. However, with the new tires, the truck travels a different distance in each revolution. Calculate the actual speed when the speedometer reads 55 miles per hour.

SCULPTURE For Exercises 34 and 35, use the following information.

A sculptor had a block of marble in the shape of a cube with sides x feet long. A piece that was $\frac{1}{2}$-foot thick was chiseled from the bottom of the block. Later, the sculptor removed a piece $\frac{3}{4}$-foot wide from the side of the marble block.

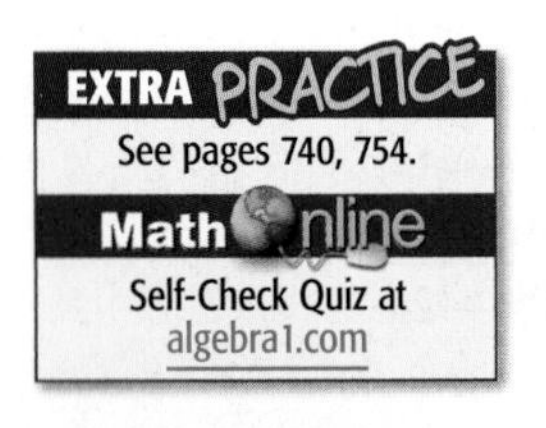

34. Write a rational expression that represents the volume of the block of marble.

35. If the remaining marble was cut into pieces weighing 85 pounds each, write an expression that represents the weight of the original block of marble.

H.O.T. Problems

36. OPEN ENDED Give an example of a real-world situation that could be modeled by the quotient of two rational expressions. Provide an example of this quotient.

37. REASONING Tell whether the following statement is *always, sometimes,* or *never* true. Explain your reasoning. *For a real number x, there is a reciprocal y.*

38. **CHALLENGE** Which expression is *not* equivalent to the reciprocal of $\frac{x^2 - 4y^2}{x + 2y}$? Justify your answer.

a. $\dfrac{1}{\frac{x^2 - 4y^2}{x + 2y}}$ b. $\dfrac{-1}{2y - x}$ c. $\dfrac{1}{x - 2y}$ d. $\dfrac{1}{x} - \dfrac{1}{2y}$

39. ***Writing in Math*** Use the information about soft drinks and aluminum on page 595 to explain how you can determine the number of aluminum soft drink cans made each year. Include a rational expression that will give the amount of new aluminum needed to produce x aluminum cans today when $\frac{5}{8}$ of the cans are recycled and 33 cans are produced from a pound of aluminum.

STANDARDIZED TEST PRACTICE

40. Which expression could be used to represent the width of the rectangle?

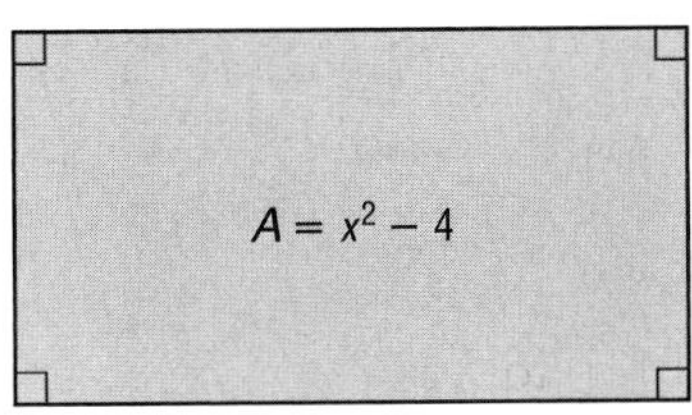

A $x - 2$ C $x + 2$

B $(x + 2)(x - 2)^2$ D $(x + 2)(x - 2)$

41. **REVIEW** What is the surface area of the regular square pyramid?

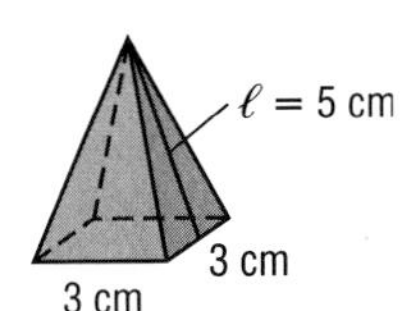

F 39 cm^2 H 83 cm^2

G 65 cm^2 J 117 cm^2

Spiral Review

Find each product. (Lesson 11-3)

42. $\dfrac{x - 5}{x^2 - 7x + 10} \cdot \dfrac{x - 2}{1}$

43. $\dfrac{x^2 + 3x - 10}{x^2 + 8x + 15} \cdot \dfrac{x^2 + 5x + 6}{x^2 + 4x + 4}$

44. $\dfrac{x + 4}{4y} \cdot \dfrac{16y}{x^2 + 7x + 12}$

Simplify each expression. (Lesson 11-2)

45. $\dfrac{c - 6}{c^2 - 12c + 36}$

46. $\dfrac{25 - x^2}{x^2 + x - 30}$

47. $\dfrac{a + 3}{a^2 + 4a + 3}$

48. $\dfrac{n^2 - 16}{n^2 - 8n + 16}$

49. **MANUFACTURING** Global Sporting Equipment sells tennis racket covers for \$2.35 each. It costs the company \$0.68 in materials and labor for each cover and \$1300 each month for equipment and building rental. Write an equation that gives the net profit the company makes, if x is the number of racket covers they can make in a month. (Lesson 4-4)

GET READY for the Next Lesson

PREREQUISITE SKILL Simplify. (Lesson 7-2)

50. $\dfrac{6x^2}{x^4}$

51. $\dfrac{5m^4}{25m}$

52. $\dfrac{b^6c^3}{b^3c^6}$

53. $\dfrac{12x^3y^2}{28x^4y}$

54. $\dfrac{7x^4z^2}{z^3}$

READING MATH

Rational Expressions

Several concepts need to be applied when reading rational expressions.

A fraction bar acts as a grouping symbol, where the entire numerator is divided by the entire denominator.

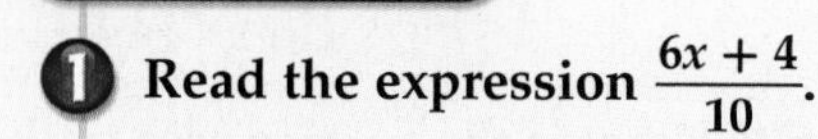

1 Read the expression $\frac{6x + 4}{10}$.

It is correct to read the expression as *the quantity six x plus four divided by ten.*

It is incorrect to read the expression as *six x divided by ten plus four, or six x plus four divided by ten.*

If a fraction consists of two or more terms divided by a one-term denominator, the denominator divides each term.

EXAMPLE

2 Simplify $\frac{6x + 4}{10}$.

It is correct to write $\frac{6x + 4}{10} = \frac{6x}{10} + \frac{4}{10}$.

$$= \frac{3x}{5} + \frac{2}{5} \text{ or } \frac{3x + 2}{5}$$

It is also correct to write $\frac{6x + 4}{10} = \frac{2(3x + 2)}{2 \cdot 5}$.

$$= \frac{\cancel{2}(3x + 2)}{\cancel{2} \cdot 5} \text{ or } \frac{3x + 2}{5}$$

It is incorrect to write $\frac{6x + 4}{10} = \frac{\overset{3x}{\cancel{6x}} + 4}{\underset{5}{\cancel{10}}} = \frac{3x + 4}{5}$.

Reading to Learn

Write the verbal translation of each rational expression.

1. $\frac{m + 2}{4}$
2. $\frac{3x}{x - 1}$
3. $\frac{a + 2}{a^2 + 8}$
4. $\frac{x^2 - 25}{x + 5}$
5. $\frac{x^2 - 3x + 18}{x - 2}$
6. $\frac{x^2 + 2x - 35}{x^2 - x - 20}$

Simplify each expression.

7. $\frac{3x + 6}{9}$
8. $\frac{2n - 12}{8}$
9. $\frac{5x^2 - 25x}{10x}$
10. $\frac{x + 3}{x^2 + 7x + 12}$
11. $\frac{x + y}{x^2 + 2xy + y^2}$
12. $\frac{x^2 - 16}{x^2 - 8x + 16}$

11-5 Dividing Polynomials

Main Ideas

- Divide a polynomial by a monomial.
- Divide a polynomial by a binomial.

GET READY for the Lesson

Suppose a partial bolt of fabric is used to make marching band flags. The original bolt was 36 yards long, and $7\frac{1}{2}$ yards of the fabric were used to make a banner for the band. Each flag requires $1\frac{1}{2}$ yards of fabric. The expression $\dfrac{36 \text{ yards} - 7\frac{1}{2} \text{ yards}}{1\frac{1}{2} \text{ yards}}$ can be used to represent the number of flags that can be made using the bolt of fabric.

Divide Polynomials By Monomials To divide a polynomial by a monomial, divide each term of the polynomial by the monomial.

EXAMPLE Divide Polynomials by Monomials

a. Find $(3r^2 - 15r) \div 3r$.

$$(3r^2 - 15r) \div 3r = \frac{3r^2 - 15r}{3r} \quad \text{Write as a rational expression.}$$

$$= \frac{3r^2}{3r} - \frac{15r}{3r} \quad \text{Divide each term by } 3r.$$

$$= \frac{\overset{r}{\cancel{3r^2}}}{\underset{1}{\cancel{3r}}} - \frac{\overset{5}{\cancel{15r}}}{\underset{1}{\cancel{3r}}} \quad \text{Simplify each term.}$$

$$= r - 5 \quad \text{Simplify.}$$

b. Find $(n^2 + 10n + 12) \div 5n$.

$$(n^2 + 10n + 12) \div 5n = \frac{n^2 + 10n + 12}{5n} \quad \text{Write as a rational expression.}$$

$$= \frac{n^2}{5n} + \frac{10n}{5n} + \frac{12}{5n} \quad \text{Divide each term by } 5n.$$

$$= \frac{\overset{n}{\cancel{n^2}}}{\underset{5}{\cancel{5n}}} + \frac{\overset{2}{\cancel{10n}}}{\underset{1}{\cancel{5n}}} + \frac{12}{5n} \quad \text{Simplify each term.}$$

$$= \frac{n}{5} + 2 + \frac{12}{5n} \quad \text{Simplify.}$$

Study Tip

Alternative Method

You could also solve Example 1a as $\frac{3r^2 - 15r}{3r} = \frac{3r(r-5)}{3r}$ or $r - 5$.

CHECK Your Progress Find each quotient.

1A. $(3q^3 - 6q) \div 3q$

1B. $(4t^5 - 5t^2 - 12) \div 2t^2$

Divide Polynomials by Binomials You can use algebra tiles to model some quotients of polynomials.

ALGEBRA LAB

Dividing Polynomials

Use algebra tiles to find $(x^2 + 3x + 2) \div (x + 1)$.

Step 1 Model the polynomial $x^2 + 3x + 2$.

Step 2 Place the x^2 tile at the corner of the product mat. Place one of the 1 tiles as shown to make a length of $x + 1$ because $x = 1$ is the divisor.

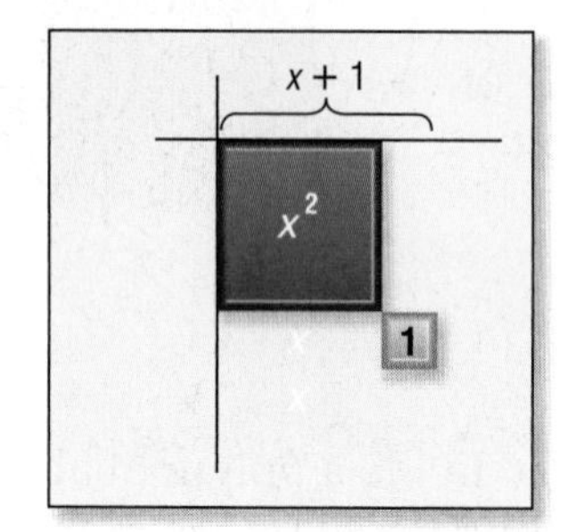

Step 3 Use the remaining tiles to make a rectangular array. Make sure the length of the rectangle, $x + 1$, does not change.

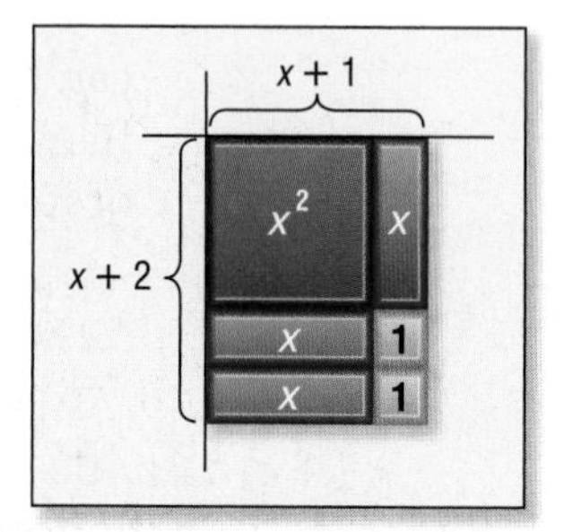

The width of the array, $x + 2$, is the quotient. This is because the dividend now fills the entire rectangular area.

MODEL AND ANALYZE

Use algebra tiles to find each quotient.

1. $(x^2 + 3x - 4) \div (x - 1)$

2. $(x^2 - 5x + 6) \div (x - 2)$

3. $(x^2 - 16) \div (x + 4)$

4. $(2x^2 - 4x - 6) \div (x - 3)$

5. Describe what happens when you try to model $(3x^2 - 4x + 3) \div (x + 2)$. What do you think the result means?

EXAMPLE Divide a Polynomial by a Binomial

2 **Find $(s^2 + 6s - 7) \div (s + 7)$.**

$(s^2 + 6s - 7) \div (s + 7) = \dfrac{s^2 + 6s - 7}{(s + 7)}$ Write as a rational expression.

$= \dfrac{(s + 7)(s - 1)}{(s + 7)}$ Factor the numerator.

$= \dfrac{\overset{1}{\cancel{(s + 7)}}(s - 1)}{\underset{1}{\cancel{(s + 7)}}}$ Divide by the GCF.

$= s - 1$ Simplify.

CHECK Your Progress Find each quotient.

2A. $(b^2 - 2b - 15) \div (3 + b)$

2B. $(x^2 + 3x - 28) \div (x + 7)$

If you cannot factor and divide by a common factor, you can use a long division process similar to the one you use to divide numbers.

EXAMPLE Long Division

3 **Find $(x^2 + 3x - 24) \div (x - 4)$.**

Step 1 Divide the first term of the dividend, x^2, by the first term of the divisor, x.

$$\begin{array}{rll} & x & x^2 \div x = x \\ x - 4 \overline{)} & x^2 + 3x - 24 & \\ & \underline{(-)\ x^2 - 4x} & \text{Multiply } x \text{ and } x - 4. \\ & 7x & \text{Subtract.} \end{array}$$

Step 2 Divide the first term of the partial dividend, $7x - 24$, by the first term of the divisor, x.

$$\begin{array}{rll} & x + 7 & 7x \div x = 7 \\ x - 4 \overline{)} & x^2 + 3x - 24 & \\ & \underline{(-)\ x^2 - 4x} & \\ & 7x - 24 & \text{Subtract and bring down the } -24. \\ & \underline{(-)\ 7x - 28} & \text{Multiply 7 and } x - 4. \\ & 4 & \text{Subtract.} \end{array}$$

So, $(x^2 + 3x - 24) \div (x - 4)$ is $x + 7$ with a remainder of 4. This answer can be written as $x + 7 + \frac{4}{x - 4}$.

Study Tip

Factors

In Example 3, since there is a nonzero remainder, $x - 4$ is not a factor of $x^2 + 3x - 24$.

CHECK Your Progress **Find each quotient.**

3A. $(y^2 + 3y + 12) \div (y + 3)$ **3B.** $(3x^2 + 9x - 15) \div (x + 5)$

Some dividends have missing terms. These are terms that have zero as their coefficient. In this situation, you must rewrite the dividend, including the missing term with a coefficient of zero.

EXAMPLE Polynomial with Missing Terms

4 **Find $(a^3 + 8a - 24) \div (a - 2)$.**

$$\begin{array}{rll} & a^2 + 2a + 12 & \\ a - 2 \overline{)} & a^3 + 0a^2 + 8a - 24 & \text{Insert an } a^2 \text{ term that has a coefficient of 0.} \\ & \underline{(-)\ a^3 - 2a^2} & \text{Multiply } a^2 \text{ and } a - 2. \\ & 2a^2 + 8a & \text{Subtract and bring down } 8a. \\ & \underline{(-)\ 2a^2 - 4a} & \text{Multiply } 2a \text{ and } a - 2. \\ & 12a - 24 & \text{Subtract and bring down 24.} \\ & \underline{(-)\ 12a - 24} & \text{Multiply 12 and } a - 2. \\ & 0 & \text{Subtract.} \end{array}$$

Therefore, $(a^3 + 8a - 24) \div (a - 2) = a^2 + 2a + 12$.

Concepts in Motion
Animation algebra1.com

CHECK Your Progress **Find each quotient.**

4A. $(c^4 + 2c^3 + 6c - 10) \div (c + 2)$ **4B.** $(6x^3 + 16x^2 - 60x + 39) \div (2x + 10)$

Online **Personal Tutor at** algebra1.com

CHECK Your Understanding

Find each quotient.

Example 1 (p. 601)

1. $(5q^2 + q) \div q$
2. $(4z^3 + 1) \div 2z$
3. $(4x^3 + 2x^2 - 5) \div 2x$
4. $\dfrac{14a^2b^2 + 35ab^2 + 2a^2}{7a^2b^2}$

Example 2 (p. 602)

5. $(n^2 + 7n + 12) \div (n + 3)$
6. $(r^2 + 12r + 36) \div (r + 9)$

Example 3 (p. 603)

7. $(2b^2 + 3b - 5) \div (2b - 1)$
8. $(x^2 + x + 12) \div (x - 3)$

Example 4 (p. 603)

9. $\dfrac{4m^3 + 5m - 21}{2m - 3}$
10. $\dfrac{2n^4 + 2n^2 - 4}{n^2 - 1}$

11. **ENVIRONMENT** The equation $C = \dfrac{120{,}000p}{1 - p}$ models the cost C in dollars for a manufacturer to reduce pollutants by p percent. How much will the company have to pay to remove 75% of the pollutants it emits?

Exercises

HOMEWORK HELP	
For Exercises	See Examples
12–17	1
18–21	2
22–25	3
26–29	4

Find each quotient.

12. $(9m^2 + 5m) \div 6m$
13. $(8k^2 - 6) \div 2k$
14. $(x^2 + 9x - 7) \div 3x$
15. $(a^2 + 7a - 28) \div 7a$
16. $\dfrac{9s^3t^2 - 15s^2t + 24t^3}{3s^2t^2}$
17. $\dfrac{12a^3b + 16ab^3 - 8ab}{4ab}$
18. $(x^2 + 9x + 20) \div (x + 5)$
19. $(x^2 + 6x - 16) \div (x - 2)$
20. $(n^2 - 2n - 35) \div (n + 5)$
21. $(s^2 + 11s + 18) \div (s + 9)$
22. $(z^2 - 2z - 30) \div (z + 7)$
23. $(a^2 + 4a - 22) \div (a - 3)$
24. $(3p^2 + 20p + 11) \div (p + 6)$
25. $(3x^3 + 8x^2 + x - 7) \div (x + 2)$
26. $(6x^3 - 9x^2 + 6) \div (2x - 3)$
27. $(9g^3 + 5g - 8) \div (3g - 2)$
28. Determine the quotient when $6n^3 + 5n^2 + 12$ is divided by $2n + 3$.
29. What is the quotient when $4t^3 + 17t^2 - 1$ is divided by $4t + 1$?

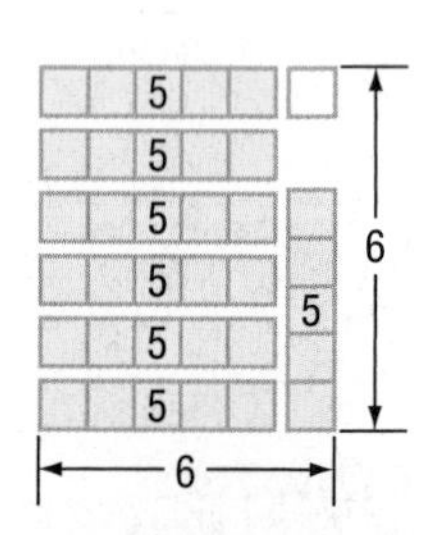

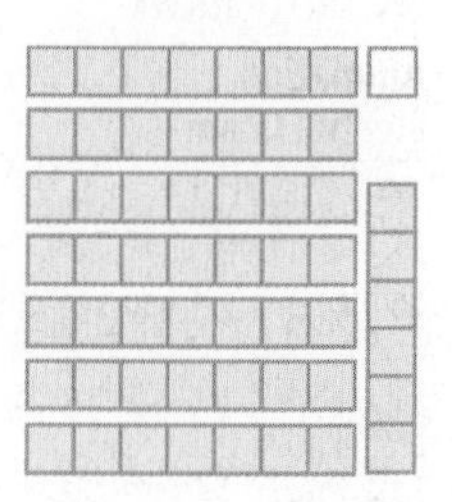

GEOMETRY **For Exercises 30–34, refer to the diagrams at the right.**

30. The first picture models $6^2 \div 7$. Notice that the square is divided into seven equal parts. What are the quotient and the remainder?
31. What division problem does the second picture model?
32. Draw diagrams for $3^2 \div 4$ and $2^2 \div 3$.
33. Do you observe a pattern in the previous exercises? Express this pattern algebraically.
34. Use long division to find $x^2 \div (x + 1)$. Does this result match your expression from the previous exercise?

Use long division to find the expression that represents the missing side.

35. $A = x^2 - 3x - 18$; sides: $x - 6$ and ?

36. $A = 4x^2 + 16x + 16$; sides: $2x + 4$ and ?

ROAD TRIP For Exercises 37–40, use the following information.
The Ski Club is taking two vans to Colorado. The first van has been on the road for 20 minutes, and the second van has been on the road for 35 minutes.

37. Write an expression for the amount of time that each van has spent on the road after an additional t minutes.

38. Write a ratio for the first van's time on the road to the second van's time on the road. Then use long division to rewrite this ratio as an expression.

39. Use the expression you wrote to find the ratio of the first van's time on the road to the second van's time on the road after 15 minutes, 60 minutes, 200 minutes, and 500 minutes.

40. As t increases, the ratio of the vans' times approaches 1. If t continues to increase, will this ratio ever be equal to 1?

Find each quotient.

41. $(21d^2 - 29d - 12) \div \left(\frac{7}{3}d - 4\right)$

42. $(x^2 - 4x + 4) \div (x + 3)$

43. $(3x^3 + 15x^2 - 12x) - 44 \div (x + 5)$

44. $\left(\frac{3}{2}x^2 - 8x - 32\right) \div (3x + 8)$

45. $\left(-5x^5 - \frac{5}{2}x^4 - 20x^3 + 5x\right) \div \left(5x + \frac{5}{2}\right)$

46. $(14y^5 + 21y^4 - 6y^3 - 9y^2 + 32y + 48) \div (2y + 3)$

47. GEOMETRY The volume of a prism with a triangular base is $10w^3 + 23w^2 + 5w - 2$. The height of the prism is $2w + 1$, and the height of the triangle is $5w - 1$. What is the measure of the base of the triangle? $\left(\textit{Hint}: V = \frac{1}{2}Bh\right)$

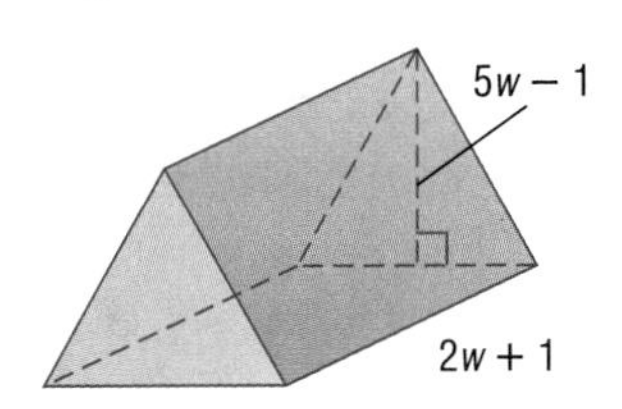

FUNCTIONS For Exercises 48–51, consider the function $f(x) = \frac{3x + 4}{x - 1}$.

48. Rewrite the function as a quotient plus a remainder. Then graph the quotient, ignoring the remainder.

49. Graph the original function using a graphing calculator.

50. How are the graphs of the function and quotient related?

51. What happens to the graph near the excluded value of x?

Since water boils at a lower temperature in high altitudes, more water is lost through evaporation when cooking. To counteract this, add about 20% more water to the recipe.

Source: crisco.com

BOILING POINT For Exercises 52–54, use the following information.
The temperature at which water boils decreases by approximately 0.9°F for every 500 feet you are above sea level. The boiling point of water at sea level is 212°F.

52. Write an equation that gives the temperature at which water boils for x every foot you are above sea level.

53. Mount Whitney, the tallest point in California, is 14,494 feet above sea level. At approximately what temperature does water boil on Mount Whitney?

54. Write an expression for the quotient of the boiling point of water at any height x and the boiling point of water at half that height.

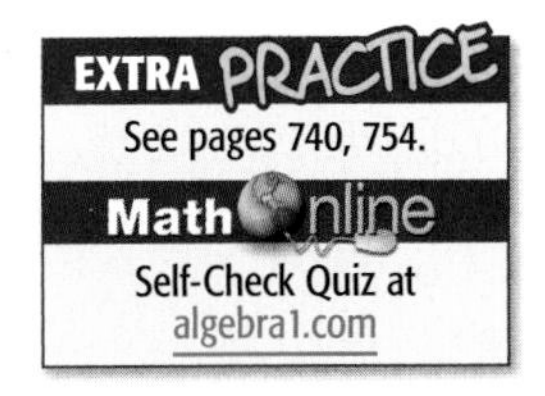

H.O.T. Problems

55. **OPEN ENDED** Write a third-degree polynomial that has a missing term. Rewrite the polynomial so that it can be divided by $x + 5$ using long division.

56. **Which One Doesn't Belong?** Select the divisor of $2x^2 - 9x + 9$ that does not belong with the other three. Explain your reasoning.

$x + 3$	$x - 2$	$2x - 3$	$2x + 3$

CHALLENGE Find the value of k in each situation.

57. k is an integer and $x + k$ is a factor of $x^2 + 7x + 12$.

58. When $x^2 + 7x + k$ is divided by $x + 2$, there is a remainder of 2.

59. $x + 7$ is a factor of $x^2 - 2x - k$.

60. ***Writing in Math*** Use the information about sewing on page 601 to describe how division can be used in sewing. Include a convincing argument to show that $\frac{a-b}{c} = \frac{a}{c} - \frac{b}{c}$.

STANDARDIZED TEST PRACTICE

61. Which expression represents the length of the rectangle?

$A = m^2 + 4m - 32$; width $m - 4$

A $m + 7$

B $m - 8$

C $m - 7$

D $m + 8$

62. **REVIEW** Paul and Lupe are building a shelter for their dog. The length of the shelter is 4.5 feet and the width is 2.5 feet. If each corner is a right angle, what is the length of each diagonal?

F 26.5 feet

G 20.25 feet

H 5.15 feet

J 3.25 feet

Spiral Review

Find each quotient. (Lesson 11-4)

63. $\frac{x^2 + 5x + 6}{x^2 - x - 12} \div \frac{x + 2}{x^2 + x - 20}$

64. $\frac{m^2 + m - 6}{m^2 + 8m + 15} \div \frac{m^2 - m - 2}{m^2 + 9m + 20}$

Find each product. (Lesson 11-3)

65. $\frac{b^2 + 19b + 84}{b - 3} \cdot \frac{b^2 - 9}{b^2 + 15b + 36}$

66. $\frac{z^2 + 16z + 39}{z^2 + 9z + 18} \cdot \frac{z + 5}{z^2 + 18z + 65}$

67. **BUSINESS** Jorge Martinez has budgeted \$150 to have business cards printed. A card printer charges \$11 to set up each job and an additional \$6 per box of 100 cards printed. What is the greatest number of cards Mr. Martinez can have printed? (Lesson 6-3)

GET READY for the Next Lesson

PREREQUISITE SKILL Find each sum. (Lesson 7-4)

68. $(6n^2 - 6n + 10m^3) + (5n - 6m^3)$

69. $(3x^2 + 4xy - 2y^2) + (x^2 + 9xy + 4y^2)$

70. $(a^3 - b^3) + (-3a^3 - 2a^2b + b^2 - 2b^3)$

71. $(2g^3 + 6h) + (-4g^2 - 8h)$

CHAPTER 11

Mid-Chapter Quiz

Lessons 11-1 through 11-5

Graph each variation if y varies inversely as x. (Lesson 11-1)

1. $y = 28$ when $x = 7$
2. $y = -6$ when $x = 9$
3. If y varies inversely as x and $y = 3$ when $x = 6$, find x when $y = -14$.
4. If y varies inversely as x and $y = -6$ when $x = 9$, find y when $x = 6$.

5. **DESIGN** The height of a rectangular tank varies inversely with the area of the base. If the tank has a height of 2 feet when the area of the base is 9 square feet, how tall will the tank be if the area of the base is 6 square feet? (Lesson 11-1)

State the excluded value(s). (Lesson 11-2)

6. $\frac{16x + 5}{3x}$
7. $\frac{12y + 4}{3y + 6}$
8. $\frac{x^2 + 1}{x^2 - 1}$
9. $\frac{15x}{3x^2 - x - 2}$

Simplify each expression. (Lesson 11-2)

10. $\frac{28a^2}{49ab}$
11. $\frac{y + 3y^2}{3y + 1}$
12. $\frac{b^2 - 3b - 4}{b^2 - 13b + 36}$
13. $\frac{3n^2 + 5n - 2}{3n^2 - 13n + 4}$

14. **LANDSCAPING** Kenyi is helping his parents landscape their yard and needs to move some large rocks. He plans to use a 6-foot bar as a lever. He positions the fulcrum 1 foot from the end of the bar touching the rock. If the rock weighs 200 pounds, how much force does he need to apply to the bar to lift the rock? (Lesson 11-2)

Find each product. (Lesson 11-3)

15. $\frac{3m^2}{2m} \cdot \frac{18m^2}{9m}$
16. $\frac{5a + 10}{10x^2} \cdot \frac{4x^3}{a^2 + 11a + 18}$
17. $\frac{4n + 8}{n^2 - 25} \cdot \frac{n - 5}{5n + 10}$
18. $\frac{x + 1}{3x^2 - 5x - 2} \cdot \frac{x - 2}{x^2 - 1}$
19. $\frac{x^2 - x - 6}{x^2 - 9} \cdot \frac{x^2 + 7x + 12}{x^2 + 4x + 4}$
20. $\frac{a^2 + 7a + 10}{a + 1} \cdot \frac{3a + 3}{a + 2}$

21. **TOYS** If a remote control car is advertised to travel at a speed of 44 feet per second, how fast can the car travel in miles per hour? (Lesson 11-3)

22. **MULTIPLE CHOICE** Which expression *best* represents the length of the rectangle? (Lesson 11-4)

$A = x^2 - 9$ (area of rectangle); width $\frac{x^2 - x - 12}{x - 4}$

A $x + 4$

B $x + 3$

C $x - 3$

D $x - 4$

Find each quotient. (Lessons 11-4 and 11-5)

23. $\frac{a}{a + 3} \div \frac{a + 11}{a + 3}$
24. $\frac{4z + 8}{z + 3} \div (z + 2)$
25. $\frac{b^2 - 9}{4b} \div (b - 3)$
26. $\frac{m^2 - 16}{5m} \div (m + 4)$
27. $\frac{(2x - 1)(x - 2)}{(x - 2)(x - 3)} \div \frac{(2x - 1)(x + 5)}{(x - 3)(x - 1)}$
28. $(9xy^2 - 15xy + 3) \div 3xy$
29. $(2x^2 - 7x - 16) \div (2x + 3)$
30. $\frac{y^2 - 19y + 9}{y - 4}$

31. **DECORATING** Anoki wants to put a decorative border 3 feet above the floor around his bedroom walls. If the border comes in 5-yard rolls, how many rolls of border should Anoki buy? (Lesson 11-5)

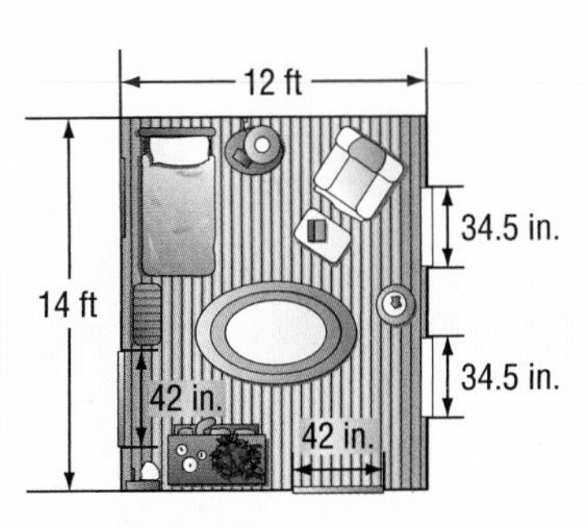

11-6 Rational Expressions with Like Denominators

Main Ideas

- Add rational expressions with like denominators.
- Subtract rational expressions with like denominators.

GET READY for the Lesson

The graph at the right shows the results of a survey that asked families how often they eat takeout. To determine what fraction of those surveyed eat takeout more than once a week, you can use addition. Remember that percents can be written as fractions with denominators of 100.

How Many Times a Week Families Eat Takeout

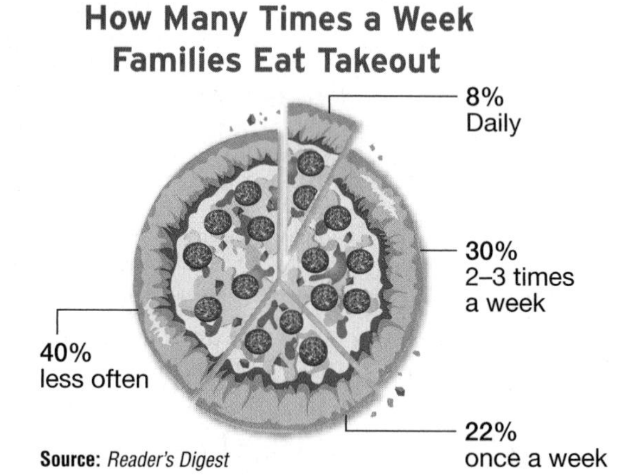

Source: *Reader's Digest*

2–3 times a week	plus	daily	equals	more than once a week.
$\frac{30}{100}$	$+$	$\frac{8}{100}$	$=$	$\frac{38}{100}$

Thus, $\frac{38}{100}$ or 38% eat takeout more than once a week.

Add Rational Expressions Recall that to add fractions with like denominators, you add the numerators and then write the sum over the common denominator. You can add rational expressions with like denominators in the same way. Answers should always be expressed in simplest form.

EXAMPLE Numbers in Denominator

1 **Find $\frac{3n}{12} + \frac{7n}{12}$.**

$\frac{3n}{12} + \frac{7n}{12} = \frac{3n + 7n}{12}$ The common denominator is 12.

$= \frac{10n}{12}$ Add the numerators.

$= \frac{\overset{5}{\cancel{10}}n}{\underset{6}{\cancel{12}}}$ or $\frac{5n}{6}$ Divide by the common factor, 2, and simplify.

CHECK Your Progress **Find each sum.**

1A. $\frac{8x}{6} + \frac{5x}{6}$

1B. $\frac{4x}{5dy} + \frac{7}{5dy}$

Sometimes the denominators of rational expressions are binomials. As long as each rational expression has exactly the same binomial as its denominator, the process of addition is the same.

EXAMPLE Binomials in Denominator

2 **Find $\frac{2x}{x+1} + \frac{2}{x+1}$.**

$\frac{2x}{x+1} + \frac{2}{x+1} = \frac{2x+2}{x+1}$ The common denominator is $x + 1$.

$= \frac{2(x+1)}{x+1}$ Factor the numerator.

$= \frac{2(\overset{1}{\cancel{x+1}})}{\underset{1}{\cancel{x+1}}}$ Divide by the common factor, $x + 1$.

$= \frac{2}{1}$ or 2 Simplify.

CHECK Your Progress **Find each sum.**

2A. $\frac{3y}{3+y} + \frac{y^2}{3+y}$ **2B.** $\frac{15x}{33x+9} + \frac{3}{33x+9}$

EXAMPLE Find a Perimeter

3 **GEOMETRY Find an expression for the perimeter of parallelogram *PQRS*.**

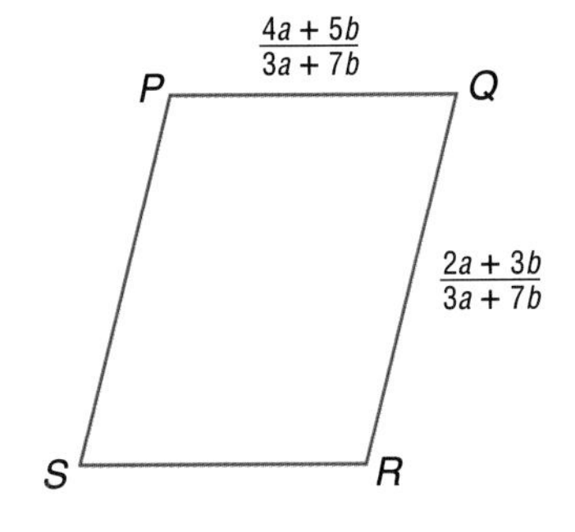

Remember that opposite sides of a parallelogram have the same length.

$P = 2\ell + 2w$ Perimeter formula

$= 2\left(\frac{4a+5b}{3a+7b}\right) + 2\left(\frac{2a+3b}{3a+7b}\right)$ $\ell = \frac{4a+5b}{3a+7b}$, $w = \frac{2a+3b}{3a+7b}$

$= \frac{2(4a+5b)+2(2a+3b)}{3a+7b}$ The common denominator is $3a + 7b$.

$= \frac{8a+10b+4a+6b}{3a+7b}$ Distributive Property

$= \frac{12a+16b}{3a+7b}$ Combine like terms.

$= \frac{4(3a+4b)}{3a+7b}$ Factor.

The perimeter can be represented by the expression $\frac{4(3a+4b)}{3a+7b}$.

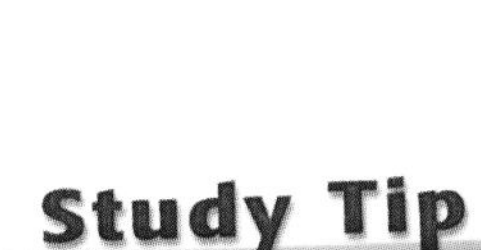

Study Tip

Common Misconceptions

You may be tempted to divide out common factors like the $3a$ in the final step of Example 3. But remember that every term of the numerator and the denominator must be multiplied or divided by a number for the fraction to remain equivalent.

CHECK Your Progress

Find an expression for the perimeter of each figure.

3A.

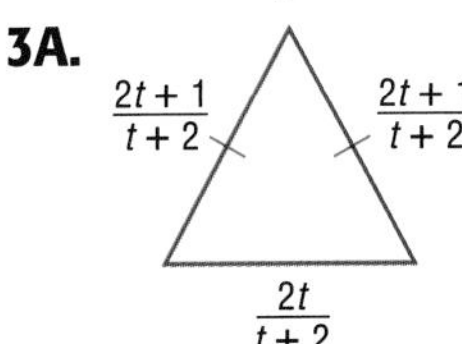

3B.

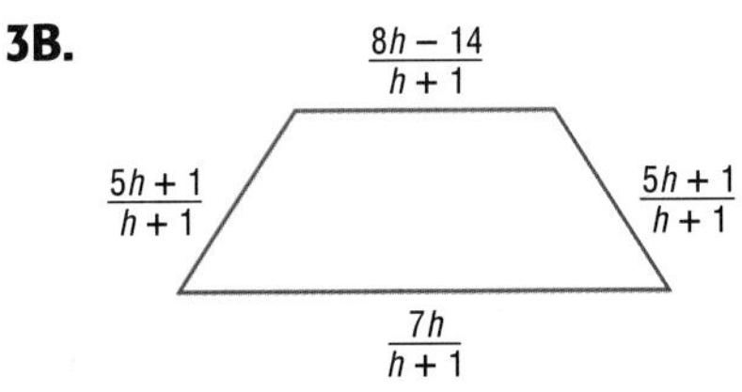

Online Personal Tutor at algebra1.com

Subtract Rational Expressions To subtract rational expressions with like denominators, subtract the numerators and write the difference over the common denominator. Recall that to subtract an expression, you add its additive inverse. As with addition, answers should always be expressed in simplest form.

Common Misconception

Adding the additive inverse will help you avoid the following error in the numerator.

$(3x + 4) - (x - 1) =$
$3x + 4 - x - 1.$

EXAMPLE Subtract Rational Expressions

4 **a.** Find $\frac{3x + 4}{x - 2} - \frac{x - 1}{x - 2}$.

$\frac{3x + 4}{x - 2} - \frac{x - 1}{x - 2} = \frac{(3x + 4) - (x - 1)}{x - 2}$ The common denominator is $x - 2$.

$= \frac{(3x + 4) + [-(x - 1)]}{x - 2}$ The additive inverse of $(x - 1)$ is $-(x - 1)$.

$= \frac{3x + 4 - x + 1}{x - 2}$ Distributive Property

$= \frac{2x + 5}{x - 2}$ Simplify.

b. Find $\frac{3m - 5}{m + 4} - \frac{4m + 2}{m + 4}$.

$\frac{3m - 5}{m + 4} - \frac{4m + 2}{m + 4} = \frac{(3m - 5) - (4m + 2)}{m + 4}$ The common denominator is $m + 4$.

$= \frac{(3m - 5) + [-(4m + 2)]}{m + 4}$ The additive inverse of $(4m + 2)$ is $-(4m + 2)$.

$= \frac{3m - 5 - 4m - 2}{m + 4}$ Distributive Property

$= \frac{-m - 7}{m + 4}$ Simplify.

CHECK Your Progress Find each difference.

4A. $\frac{2h + 4}{h + 1} - \frac{5 + h}{h + 1}$

4B. $\frac{17h + 4}{15h - 5} - \frac{2h - 6}{15h - 5}$

Sometimes you must express a denominator as its additive inverse to have like denominators.

EXAMPLE Inverse Denominators

5 Find $\frac{2m}{m - 9} + \frac{4m}{9 - m}$.

$\frac{2m}{m - 9} + \frac{4m}{9 - m} = \frac{2m}{m - 9} + \frac{4m}{-(m - 9)}$ Rewrite $9 - m$ as $-(m - 9)$.

$= \frac{2m}{m - 9} - \frac{4m}{m - 9}$ Rewrite so the denominators are the same.

$= \frac{2m - 4m}{m - 9}$ The common denominator is $m - 9$.

$= \frac{-2m}{m - 9}$ Subtract.

CHECK Your Progress Find each sum.

5A. $\frac{3n}{n - 4} + \frac{6n}{4 - n}$

5B. $\frac{t^2}{t - 3} + \frac{3}{3 - t}$

Math Online **Extra Examples at** algebra1.com

CHECK Your Understanding

Find each sum.

Example 1 (p. 608)

1. $\frac{a+2}{4} + \frac{a-2}{4}$
2. $\frac{12z}{7} + \frac{-5z}{7}$

Example 2 (p. 609)

3. $\frac{2-n}{n-1} + \frac{1}{n-1}$
4. $\frac{4t-1}{1-4t} + \frac{2t+3}{1-4t}$

Example 3 (p. 609)

5. Find an expression for the perimeter of the figure.

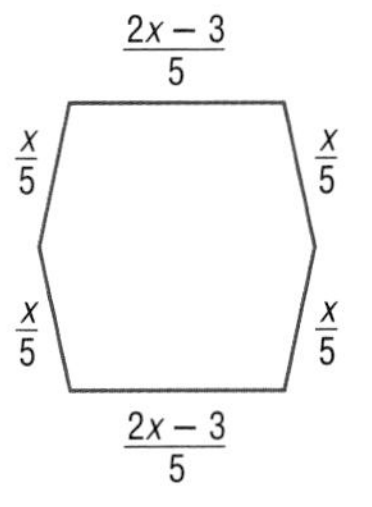

Find each difference.

Example 4 (p. 610)

6. $\frac{5a}{12} - \frac{7a}{12}$
7. $\frac{7}{n-3} - \frac{4}{n-3}$

Example 5 (p. 610)

8. $\frac{3m}{m-2} - \frac{6}{2-m}$
9. $\frac{x^2}{x-y} - \frac{y^2}{y-x}$

Exercises

HOMEWORK HELP	
For Exercises	See Examples
10, 11	1
12–17	2
18, 19	3
20–23	4
24, 25	5

Find each sum.

10. $\frac{m}{3} + \frac{2m}{3}$
11. $\frac{x+3}{5} + \frac{x+2}{5}$
12. $\frac{2y}{y+3} + \frac{6}{y+3}$
13. $\frac{3r}{r+5} + \frac{15}{r+5}$
14. $\frac{k-5}{k-1} + \frac{4}{k-1}$
15. $\frac{n-2}{n+3} + \frac{-1}{n+3}$
16. What is the sum of $\frac{12x-7}{3x-2}$ and $\frac{9x-5}{2-3x}$?
17. Find the sum of $\frac{11x-5}{2x+5}$ and $\frac{11x+12}{2x+5}$.

GEOMETRY For Exercises 18 and 19, use the following information.

Each figure has a perimeter of x units.

a. 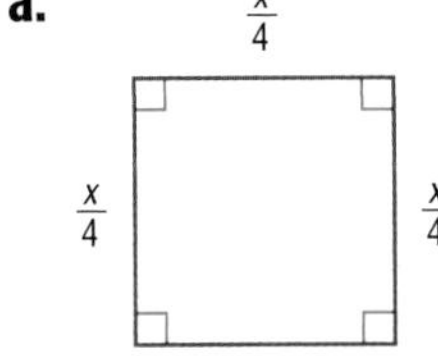

b. 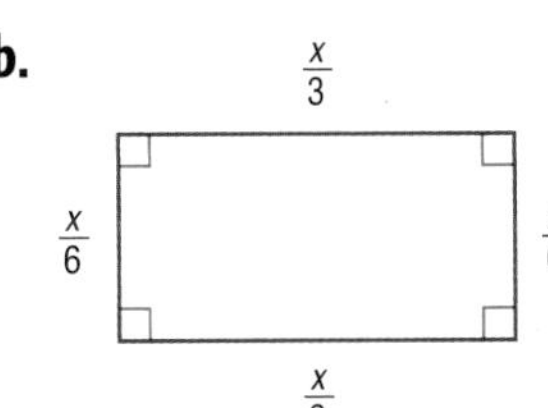

c. 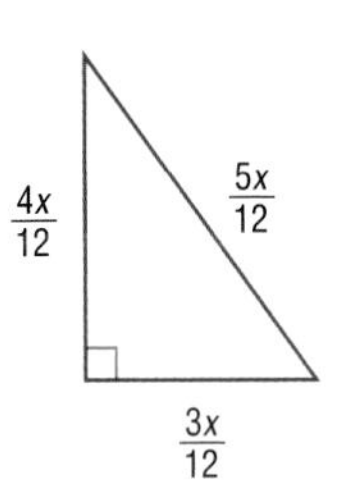

18. Find the ratio of the area of each figure to its perimeter.
19. Which figure has the greatest ratio?

Find each difference.

20. $\frac{5x}{7} - \frac{3x}{7}$
21. $\frac{x+4}{5} - \frac{x+2}{5}$
22. $\frac{5}{3x-5} - \frac{3x}{3x-5}$
23. $\frac{8}{3t-4} - \frac{6t}{3t-4}$
24. $\frac{2x}{x-2} - \frac{2x}{2-x}$
25. $\frac{5y}{y-3} - \frac{5y}{3-y}$

26. **POPULATION** The population of Aurora, Illinois, is 31,536 greater than the population of Springfield, Illinois. Write an expression for the fraction of the Aurora population that is under 19 years old if the population of Springfield is n.

Aurora Population	
Age	**Number of People**
Under 5 years	15,095
5 to 9 years	13,256
10 to 14 years	10,873
15 to 19 years	10,042

Source: U.S. Census Bureau

27. **CONSERVATION** The freshman class chose to plant spruce and pine trees at a wildlife sanctuary for a service project. Some students can plant 140 trees on Saturday, and others can plant 20 trees after school on Monday and again on Tuesday. Write an expression for the fraction of the trees that could be planted on these days if n represents the number of spruce trees and there are twice as many pine trees.

28. **SCHOOL** Most schools create daily attendance reports to keep track of their students. The school office manager knows that, of the 960 students, 45 are absent due to illness, 10 are excused for appointments, and both the wrestling team and the choir are at competitions. Though she doesn't know exactly how many people attended the competitions, she does know that there are twice as many people in the choir as there are on the wrestling team. Write an expression that gives the percentage of students who are absent.

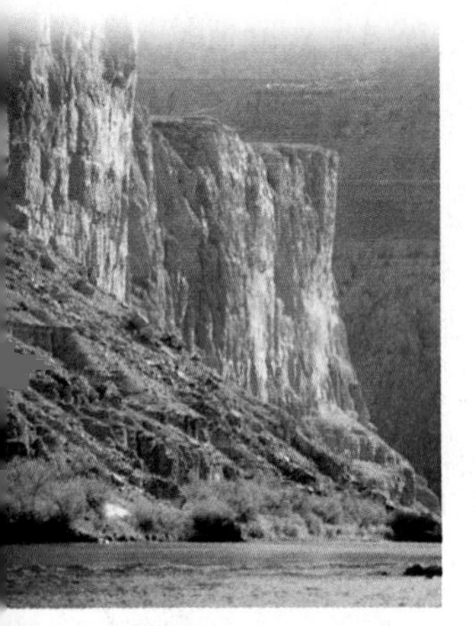

Find each sum or difference.

29. $\frac{4}{7m-2}+\frac{7m}{2-7m}$

30. $\frac{b-15}{2b+12}-\frac{-3b+8}{2b+12}$

31. $\frac{a+5}{6}-\frac{a+3}{6}$

32. $\frac{10a-12}{2a-6}-\frac{6a}{6-2a}$

33. $\frac{2}{x+7}-\frac{-5}{x+7}$

34. $\frac{15x}{5x+1}+\frac{-3}{-1-5x}$

35. **GEOMETRIC DESIGN** The Jerome Student Center has a square room that is 25 feet wide and 25 feet long. The walls are 10 feet high, and each wall is painted white with a red diagonal stripe as shown. What fraction of the walls are painted red?

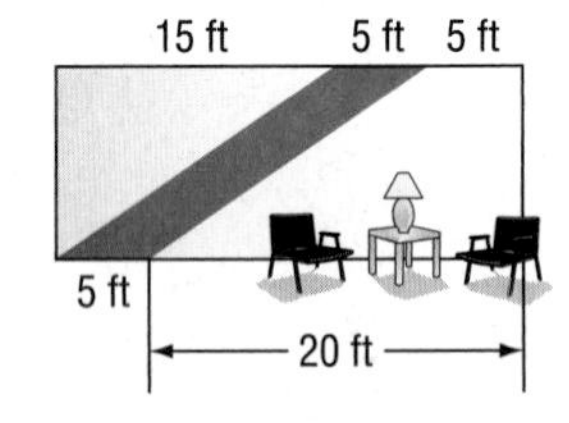

Real-World Link

Due to its popularity, the Grand Canyon is one of the most threatened natural areas in the United States.

Source: *The Wildlife Foundation*

HIKING For Exercises 36 and 37, use the following information.

A tour guide recommends that hikers carry a gallon of water on hikes to the bottom of the Grand Canyon. Water weighs 62.4 pounds per cubic foot, and one cubic foot of water contains 7.48 gallons.

EXTRA PRACTICE
See pages 740, 754.

Math Online
Self-Check Quiz at algebra1.com

36. Tanika plans to carry two 1-quart bottles and four 1-pint bottles for her hike. Write a rational expression for this amount of water written as a fraction of a cubic foot.

37. How much does this amount of water weigh?

38. **OPEN ENDED** Describe a real-life situation that could be expressed by adding two rational expressions that are fractions. Explain what the denominator and numerator represent in both expressions.

39. FIND THE ERROR Russell and Ginger are finding the difference of $\frac{7x+2}{4x-3}$ and $\frac{x-8}{3-4x}$. Who is correct? Explain your reasoning.

Russell

$$\frac{7x+2}{4x-3} - \frac{x-8}{3-4x} = \frac{7x+2}{4x-3} + \frac{x-8}{4x-3}$$
$$= \frac{7x+x+2-8}{4x-3}$$
$$= \frac{8x-6}{4x-3}$$
$$= \frac{2(4x-3)}{4x-3}$$
$$= 2$$

Ginger

$$\frac{7x+2}{4x-3} - \frac{x-8}{3-4x} = \frac{-2-7x}{3-4x} + \frac{x-8}{3-4x}$$
$$= \frac{-2-8-7x+x}{3-4x}$$
$$= \frac{-6-8x}{3-4x}$$
$$= \frac{-2(3-4x)}{3-4x}$$
$$= -2$$

40. CHALLENGE Which of the following rational expressions is *not* equivalent to the others?

a. $\frac{3}{2-x}$ **b.** $\frac{-3}{x-2}$ **c.** $-\frac{3}{2-x}$ **d.** $-\frac{3}{x-2}$

41. ***Writing in Math*** Use the chart on page 608 to explain how rational expressions can be used to interpret graphics. Include an explanation of how the numbers in the graphic relate to rational expressions and a description of how to add rational expressions with denominators that are additive inverses.

STANDARDIZED TEST PRACTICE

42. Which is an expression for the perimeter of the pentagon?

A $\frac{16r}{s+3r}$

B $\frac{16r}{2s+6r}$

C $\frac{32}{s+3r}$

D $\frac{32}{4s+12r}$

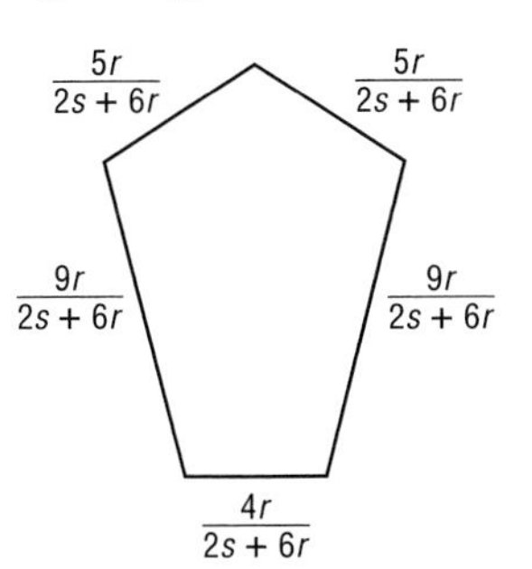

43. REVIEW Shelby sells cosmetics door-to-door. She makes $5 an hour and 17% commission on the total dollar value on whatever she sells. To the nearest dollar, how much money will she make if she sells $300 dollars worth of product and works 30 hours?

F $201 H $255

G $226 J $283

Spiral Review

Find each quotient. (Lessons 11-4 and 11-5)

44. $\frac{x^3-7x+6}{x-2}$

45. $\frac{56x^3+32x^2-63x-36}{7x+4}$

46. $\frac{b^2-9}{4b} \div (b-3)$

47. $\frac{x}{x+2} \div \frac{x^2}{x^2+5x+6}$

GET READY for the Next Lesson

PREREQUISITE SKILL **Find the least common multiple for each set of numbers.**

48. 4, 9, 12 **49.** 45, 10, 6 **50.** 16, 20, 25 **51.** 36, 48, 60

11-7 Rational Expressions with Unlike Denominators

Main Ideas

- Add rational expressions with unlike denominators.
- Subtract rational expressions with unlike denominators.

New Vocabulary

least common multiple (LCM)

least common denominator (LCD)

GET READY for the Lesson

The President of the United States is elected every four years, and senators are elected every six years. A certain senator is elected in 2004, the same year as a presidential election, and is reelected in subsequent elections. In what year is the senator's reelection the same year as a presidential election?

Add Rational Expressions The least number of years that will pass until the next election for both a specific senator and the President is the least common multiple of 4 and 6. The **least common multiple (LCM)** is the least number that is a multiple of two or more numbers. You can also find the LCM of a set of polynomials.

EXAMPLE LCMs of Polynomials

1 a. Find the LCM of $15m^2b^3$ and $18mb^2$.

Find the prime factors of each coefficient and variable expression.

$15m^2b^3 = 3 \cdot 5 \cdot m \cdot m \cdot b \cdot b \cdot b \qquad 18mb^2 = 2 \cdot 3 \cdot 3 \cdot m \cdot b \cdot b$

Use each prime factor the greatest number of times it appears in any of the factorizations.

$15m^2b^3 = 3 \cdot 5 \cdot m \cdot m \cdot b \cdot b \cdot b \qquad 18mb^2 = 2 \cdot 3 \cdot 3 \cdot m \cdot b \cdot b$

LCM $= 2 \cdot 3 \cdot 3 \cdot 5 \cdot m \cdot m \cdot b \cdot b \cdot b$ or $90m^2b^3$

b. Find the LCM of $x^2 + 8x + 15$ and $x^2 + x - 6$.

Express each polynomial in factored form.

$x^2 + 8x + 15 = (x + 3)(x + 5) \qquad x^2 + x - 6 = (x - 2)(x + 3)$

Use each factor the greatest number of times it appears.

LCM $= (x + 3)(x + 5)(x - 2)$

CHECK Your Progress

Find the LCM of each pair of polynomials.

1A. $28m^2n$ and $12m^2n^3p$ **1B.** $x^2 - 2x - 8$ and $x^2 - 5x - 14$

To add fractions with unlike denominators, you need to rename the fractions using the least common multiple (LCM) of the denominators, known as the **least common denominator (LCD)**. You can add rational expressions with unlike denominators in the same way.

KEY CONCEPT — *Add Rational Expressions*

Use the following steps to add rational expressions with unlike denominators.

Step 1 Find the LCD.

Step 2 Change each rational expression into an equivalent expression with the LCD as the denominator.

Step 3 Add rational expressions with like denominators.

Step 4 Simplify if necessary.

EXAMPLE Polynomial Denominators

a. Find $\frac{a+1}{a} + \frac{a-3}{3a}$.

Factor each denominator and find the LCD.

$a = a$

$3a = 3 \cdot a \qquad \text{LCD} = 3a$

Since the denominator of $\frac{a-3}{3a}$ is already $3a$, only $\frac{a+1}{a}$ needs to be renamed.

$$\frac{a+1}{a} + \frac{a-3}{3a} = \frac{3(a+1)}{3(a)} + \frac{a-3}{3a} \qquad \text{Multiply } \frac{a+1}{a} \text{ by } \frac{3}{3}.$$

$$= \frac{3a+3}{3a} + \frac{a-3}{3a} \qquad \text{Distributive Property}$$

$$= \frac{3a+3+a-3}{3a} \qquad \text{Add the numerators.}$$

$$= \frac{\overset{1}{\cancel{4a}}}{\underset{1}{\cancel{3a}}} \text{ or } \frac{4}{3} \qquad \text{Divide by the common factor, } a\text{, and simplify.}$$

b. Find $\frac{y-2}{y^2+4y+4} + \frac{y-2}{y+2}$.

$$\frac{y-2}{y^2+4y+4} + \frac{y-2}{y+2} = \frac{y-2}{(y+2)^2} + \frac{y-2}{y+2} \qquad \text{Factor the denominators.}$$

$$= \frac{y-2}{(y+2)^2} + \frac{y-2}{y+2} \cdot \frac{y+2}{y+2} \qquad \text{The LCD is } (y+2)^2.$$

$$= \frac{y-2}{(y+2)^2} + \frac{y^2-4}{(y+2)^2} \qquad (y-2)(y+2) = y^2-4$$

$$= \frac{y-2+y^2-4}{(y+2)^2} \qquad \text{Add the numerators.}$$

$$= \frac{y^2+y-6}{(y+2)^2} \text{ or } \frac{(y-2)(y+3)}{(y+2)^2} \qquad \text{Simplify.}$$

Study Tip

Checking Answers

You can check to see that you have simplified rational expressions involving variables by substituting a few values. If the results are different for the original expression and the simplified expression, check for an error in arithmetic.

Find each sum.

2A. $\frac{4d^2}{d} + \frac{d+2}{d^2}$

2B. $\frac{b+3}{b} + \frac{b-5}{b+1}$

Subtract Rational Expressions As with addition, to subtract rational expressions with unlike denominators, you must first rename the expressions using a common denominator.

EXAMPLE Polynomials in Denominators

3 **Find each difference.**

a. $\frac{4}{3a-6} - \frac{a}{a+2}$

$$\frac{4}{3a-6} - \frac{a}{a+2} = \frac{4}{3(a-2)} - \frac{a}{a+2}$$ Factor.

$$= \frac{4(a+2)}{3(a-2)(a+2)} - \frac{3a(a-2)}{3(a-2)(a+2)}$$ The LCD is $3(a-2)(a+2)$.

$$= \frac{4(a+2) - 3a(a-2)}{3(a-2)(a+2)}$$ Subtract the numerators.

$$= \frac{4a + 8 - 3a^2 + 6a}{3(a-2)(a+2)}$$ Multiply.

$$= \frac{-3a^2 + 10a + 8}{3(a-2)(a+2)}$$ Add like terms.

b. $\frac{h-2}{h^2+4h+4} - \frac{h-4}{h^2-4}$

$$\frac{h-2}{(h+2)^2} - \frac{h-4}{(h+2)(h-2)} = \frac{(h-2)}{(h+2)^2} \cdot \frac{(h-2)}{(h-2)} - \frac{(h-4)}{(h+2)(h-2)} \cdot \frac{(h+2)}{(h+2)}$$

$$= \frac{(h-2)(h-2)}{(h+2)^2(h-2)} - \frac{(h-4)(h+2)}{(h+2)^2(h-2)}$$ The LCD is $(h+2)^2(h-2)$.

$$= \frac{h^2 - 4h + 4}{(h+2)^2(h-2)} - \frac{h^2 - 2h - 8}{(h+2)^2(h-2)}$$ Multiply.

$$= \frac{(h^2 - 4h + 4) - (h^2 - 2h - 8)}{(h+2)^2(h-2)}$$ Subtract.

$$= \frac{h^2 - h^2 - 4h + 2h + 4 + 8}{(h+2)^2(h-2)}$$ Combine like terms.

$$= \frac{-2h + 12}{(h-2)(h+2)^2}$$ Simplify.

CHECK Your Progress

Find each sum.

3A. $\frac{5}{2\ell+2} + \frac{\ell}{\ell+5}$

3B. $\frac{k-3}{k^2+k-12} + \frac{k}{k^2-9}$

3C. $\frac{x}{x-3} - \frac{3}{x+2}$

3D. $\frac{m}{m^2-2m-8} - \frac{m+3}{m-4}$

Online **Personal Tutor at** algebra1.com

CHECK Your Understanding

Example 1 (p. 614) **Find the LCM for each pair of expressions.**

1. $5a^2, 7a$ **2.** $2x - 4, 3x - 6$ **3.** $n^2 + 3n - 4, (n - 1)^2$

4. MUSIC A music director wants to form a group of students to sing and dance at community events. Sometimes the music is 2-part, 3-part, or 4-part harmony, and she would like to have the same number of voices on each part. What is the least number of students that would allow for an even distribution on all these parts?

Example 2 (p. 615) **Find each sum.**

5. $\frac{6}{5x} + \frac{7}{10x^2}$ **6.** $\frac{3z}{6w^2} + \frac{z}{4w}$

7. $\frac{2y}{y^2 - 25} + \frac{y + 5}{y - 5}$ **8.** $\frac{a + 2}{a^2 + 4a + 3} + \frac{6}{a + 3}$

Example 3 (p. 616) **Find each difference.**

9. $\frac{a}{a - 4} - \frac{4}{a + 4}$ **10.** $\frac{b + 8}{b^2 - 16} - \frac{1}{b - 4}$

11. $\frac{2y}{y^2 + 7y + 12} - \frac{y + 2}{y + 4}$ **12.** $\frac{x}{x - 2} - \frac{3}{x^2 + 3x - 10}$

Exercises

HOMEWORK HELP

For Exercises	See Examples
13–18	1
19–30	2
31–39	3

Find the LCM for each pair of expressions.

13. a^2b, ab^3 **14.** $7xy, 21x^2y$ **15.** $x - 4, x + 2$

16. $2n - 5, n + 2$ **17.** $x^2 + 5x - 14, (x - 2)^2$ **18.** $p^2 - 5p - 6, p + 1$

Find each sum.

19. $\frac{3}{x^2} + \frac{5}{x}$ **20.** $\frac{2}{a^3} + \frac{7}{a^2}$ **21.** $\frac{7}{6a^2} + \frac{5}{3a}$

22. $\frac{3}{7m} + \frac{4}{5m^2}$ **23.** $\frac{3}{x + 5} + \frac{4}{x - 4}$ **24.** $\frac{n}{n + 4} + \frac{3}{n - 3}$

25. $\frac{7a}{a + 5} + \frac{a}{a - 2}$ **26.** $\frac{6x}{x - 3} + \frac{x}{x + 1}$ **27.** $\frac{-3}{5 - a} + \frac{5}{a^2 - 25}$

28. $\frac{18}{y^2 - 9} + \frac{-7}{3 - y}$ **29.** $\frac{x}{x^2 + 2x + 1} + \frac{1}{x + 1}$ **30.** $\frac{2x + 1}{(x - 1)^2} + \frac{x - 2}{x^2 + 3x - 4}$

Find each difference.

31. $\frac{7}{3x} - \frac{3}{6x^2}$ **32.** $\frac{5a}{7x} - \frac{3a}{21x^2}$ **33.** $\frac{x^2 - 1}{x + 1} - \frac{x^2 + 1}{x - 1}$

34. $\frac{m - 1}{m + 1} - \frac{4}{2m + 5}$ **35.** $\frac{2x}{x^2 - 5x} - \frac{-3x}{x - 5}$ **36.** $\frac{-3}{a - 6} - \frac{-6}{a^2 - 6a}$

37. $\frac{n}{5 - n} - \frac{3}{n^2 - 25}$ **38.** $\frac{3a + 2}{6 - 3a} - \frac{a + 2}{a^2 - 4}$ **39.** $\frac{k}{2k + 1} - \frac{2}{k + 2}$

40. PET CARE Kendra takes care of pets while their owners are out of town. One week she has three dogs that all eat the same kind of dog food. The first dog eats a bag of food every 12 days, the second dog eats a bag every 15 days, and the third dog eats a bag every 16 days. How many bags of food should Kendra buy for one week?

41. CHARITY Maya, Makalla, and Monya can walk one mile in 12, 15, and 20 minutes respectively. They plan to participate in a walk-a-thon to raise money for a local charity. Sponsors have agreed to pay $2.50 for each mile that is walked. What is the total number of miles the girls will walk in one hour? How much money will they raise?

42. COMPUTERS Computer owners need to follow a regular maintenance schedule to keep their computers running effectively and efficiently. The table shows several items that should be performed on a regular basis. If Jamie got his computer 53 weeks ago and he has been appropriately maintaining it, how many weeks will it be until he has to perform all of the items on the same week?

Maintenance	Frequency
back up files	every 3 weeks
scan files for viruses	every 3 weeks
check for operating system patches	every 8 weeks
update virus software	every 9 weeks

Find each sum or difference.

43. $\frac{4}{15x^2} - \frac{5}{3x}$

44. $\frac{x^2}{4x^2 - 9} + \frac{x}{(2x + 3)^2}$

45. $\frac{11x}{3y^2} - \frac{7x}{6y}$

46. $\frac{k}{k + 5} - \frac{3}{k - 3}$

47. $\frac{a^2}{a^2 - b^2} + \frac{a}{(a - b)^2}$

48. $\frac{x^2 + 4x - 5}{x^2 - 2x - 3} - \frac{2}{x + 1}$

49. $\frac{3x}{x^2 + 3x + 2} - \frac{3x - 6}{x^2 + 4x + 4}$

50. $\frac{5x}{3x^2 + 19x - 14} - \frac{1}{9x^2 - 12x + 4}$

EXTRA PRACTICE
See pages 741, 754.
Math Online
Self-Check Quiz at
algebra1.com

H.O.T. Problems

51. REASONING Describe how to find the LCD of two rational expressions with unlike denominators.

52. OPEN ENDED Write two rational expressions in which the LCD is twice the denominator of one of the expressions.

53. REASONING Explain how to rename rational expressions using their LCD.

54. CHALLENGE Janelle says that a shortcut for adding fractions with unlike denominators is to add the cross products for the numerator and write the denominator as the product of the denominators. For example, $\frac{2}{7} + \frac{5}{8} = \frac{2 \cdot 8 + 5 \cdot 7}{7 \cdot 8} = \frac{51}{56}$. Explain why the method will always work or provide a counterexample to show that it does not always work.

55. *Writing in Math* Use the information about elections on page 614 to explain how rational expressions can be used to describe elections. Include an explanation of how to determine the least common multiple of two or more rational expressions.

STANDARDIZED TEST PRACTICE

56. $\frac{x}{7x - 3} + \frac{x + 2}{15x + 30} =$

A $\frac{22x - 3}{210x - 90}$

B $\frac{22x - 3}{105x - 45}$

C $\frac{41x - 6}{210x - 90}$

D $\frac{41x - 6}{105x - 45}$

57. REVIEW A rectangle with length x and width $2x$ is inside a rectangle with length 12 and width 8. Which expression represents the area of the shaded region in terms of x?

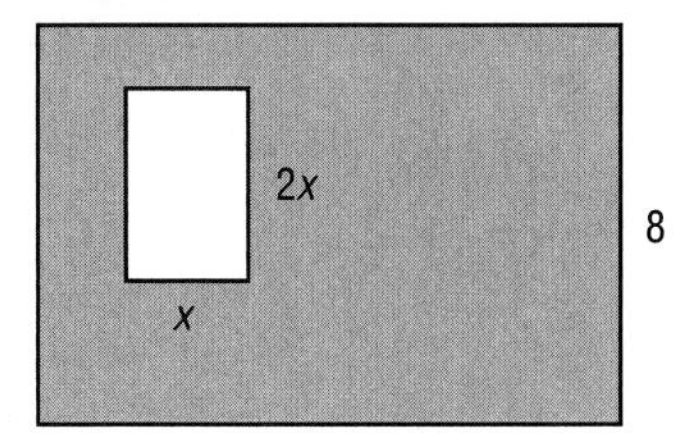

F $96 - x$

G $96 - 2x$

H $96 - x^2$

J $96 - 2x^2$

Spiral Review

Find each sum. (Lesson 11-6)

58. $\frac{3m}{2m + 1} + \frac{3}{2m + 1}$

59. $\frac{4x}{2x + 3} + \frac{5}{2x + 3}$

60. $\frac{2y}{y - 3} + \frac{5}{3 - y}$

Find each quotient. (Lesson 11-5)

61. $\frac{b^2 + 8b - 20}{b - 2}$

62. $\frac{t^3 - 19t + 9}{t - 4}$

63. $\frac{4m^2 + 8m - 19}{2m + 7}$

Determine the best method to solve each system of equations. Then solve the system. (Lesson 5-5)

64. $2x + 3y = 9$
$-x + 5y = 28$

65. $y = \frac{1}{4}x$
$-x + 3y = -6$

66. CURRENCY The table shows the exchange rate for the American dollar to the British pound over the course of 30 days. Create a graph to display this data. Determine whether the graph shows a positive or negative correlation and draw a line of fit. (Lesson 5-7)

Day	5	10	15	20	25	30
British Pound to 1 American Dollar	0.5532	0.5621	0.5715	0.5832	0.5681	0.5721

GET READY for the Next Lesson

PREREQUISITE SKILL Find each quotient. (Lesson 11-4)

67. $\frac{x}{2} \div \frac{3x}{5}$

68. $\frac{a^2}{5b} \div \frac{4a}{10b^2}$

69. $\frac{x + 7}{x} \div \frac{x + 7}{x + 3}$

70. $\frac{3n}{2n + 5} \div \frac{12n^2}{2n + 5}$

71. $\frac{3x}{x + 2} \div (x - 1)$

72. $\frac{x^2 + 7x + 12}{x + 6} \div (x + 3)$

11-8 Mixed Expressions and Complex Fractions

Main Ideas

- Simplify mixed expressions.
- Simplify complex fractions.

New Vocabulary

- mixed expression
- complex fraction

GET READY for the Lesson

Katelyn bought $2\frac{1}{2}$ pounds of chocolate chip cookie dough. If the average cookie requires $1\frac{1}{2}$ ounces of dough, the number of cookies that Katelyn can bake can be found by simplifying the expression $\frac{2\frac{1}{2} \text{ pounds}}{1\frac{1}{2} \text{ ounces}}$.

Simplify Mixed Expressions Recall that a number like $2\frac{1}{2}$ is a mixed number because it is the sum of an integer, 2, and a fraction, $\frac{1}{2}$. An expression like $3 + \frac{x+2}{x-3}$ is called a **mixed expression** because it contains the sum of a monomial, 3, and a rational expression, $\frac{x+2}{x+3}$. Changing mixed expressions to rational expressions is similar to changing mixed numbers to improper fractions.

EXAMPLE 1 Mixed Expression to Rational Expression

Simplify $3 + \frac{6}{x+3}$.

$3 + \frac{6}{x+3} = \frac{3(x+3)}{x+3} + \frac{6}{x+3}$ The LCD is $x + 3$.

$= \frac{3(x+3)+6}{x+3}$ Add the numerators.

$= \frac{3x+9+6}{x+3}$ or $\frac{3x+15}{x+3}$ Distributive Property

CHECK Your Progress **Simplify each expression.**

1A. $\frac{6y}{4y+8} + 5y$

1B. $15 - \frac{17x+5}{5x+10}$

Simplify Complex Fractions If a fraction has one or more fractions in the numerator or denominator, it is called a **complex fraction**. You simplify an algebraic complex fraction in the same way that you simplify a numerical complex fraction.

numerical complex fraction

$$\frac{\frac{8}{3}}{\frac{7}{5}} = \frac{8}{3} \div \frac{7}{5}$$
$$= \frac{8}{3} \cdot \frac{5}{7}$$
$$= \frac{40}{21}$$

algebraic complex fraction

$$\frac{\frac{a}{b}}{\frac{c}{d}} = \frac{a}{b} \div \frac{c}{d}$$
$$= \frac{a}{b} \cdot \frac{d}{c}$$
$$= \frac{ad}{bc}$$

KEY CONCEPT *Simplifying a Complex Fraction*

Any complex fraction $\frac{\frac{a}{b}}{\frac{c}{d}}$, where $b \neq 0$, $c \neq 0$, and $d \neq 0$, can be expressed as $\frac{ad}{bc}$.

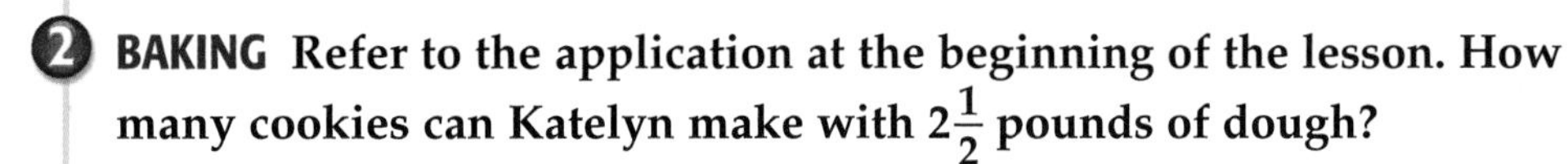

Complex Fraction Involving Numbers

2 BAKING Refer to the application at the beginning of the lesson. How many cookies can Katelyn make with $2\frac{1}{2}$ pounds of dough?

To find the total number of cookies, divide the amount of cookie dough by the amount of dough needed for each cookie.

$$\frac{2\frac{1}{2}\text{ pounds}}{1\frac{1}{2}\text{ ounces}} = \frac{2\frac{1}{2}\text{ pounds}}{1\frac{1}{2}\text{ ounces}} \cdot \frac{16\text{ ounces}}{1\text{ pound}}$$ Convert pounds to ounces. Divide by common units.

$$= 16 \cdot \frac{2\frac{1}{2}}{1\frac{1}{2}}$$ Simplify.

$$= \frac{\frac{16}{1} \cdot \frac{5}{2}}{\frac{3}{2}}$$ Express each term as an improper fraction.

$$= \frac{\frac{80}{2}}{\frac{3}{2}}$$ Multiply in the numerator.

$$= \frac{80 \cdot 2}{2 \cdot 3}$$ $\frac{\frac{a}{b}}{\frac{c}{d}} = \frac{ad}{bc}$

$$= \frac{160}{6} \text{ or } 26\frac{2}{3}$$ Simplify.

Study Tip

Order of Operations

Recall that when applying the order of operations, simplify the numerator and denominator of a complex fraction before proceeding with division.

CHECK Your Progress

2. The Centralville High School Cooking Club has $12\frac{1}{2}$ pounds of flour with which to make tortillas. If there are $3\frac{3}{4}$ cups of flour in a pound and it takes about $\frac{1}{3}$ cup of flour per tortilla, how many tortillas can they make?

Online Personal Tutor at algebra1.com

EXAMPLE Complex Fraction Involving Monomials

3 **Simplify** $\dfrac{\frac{x^2y^2}{a}}{\frac{x^2y}{a^3}}$.

$$\frac{\frac{x^2y^2}{a}}{\frac{x^2y}{a^3}} = \frac{x^2y^2}{a} \div \frac{x^2y}{a^3}$$ Rewrite as a division sentence.

$$= \frac{x^2y^2}{a} \cdot \frac{a^3}{x^2y}$$ Rewrite as multiplication by the reciprocal.

$$= \frac{\overset{1}{\cancel{x^2}}\overset{y}{\cancel{y^2}}}{\underset{1}{\cancel{a}}} \cdot \frac{\overset{a^2}{\cancel{a^3}}}{\underset{1}{\cancel{x^2}}\underset{1}{\cancel{y}}}$$ Divide by common factors x^2, y, and a.

$$= a^2y$$ Simplify.

CHECK Your Progress **Simplify each expression.**

3A. $\dfrac{\frac{g^3h}{b}}{\frac{gh^3i}{b^2}}$

3B. $\dfrac{\frac{-24m^3n^5}{p^2q}}{\frac{16pm^2}{n^4q}}$

EXAMPLE Complex Fraction Involving Polynomials

4 **Simplify** 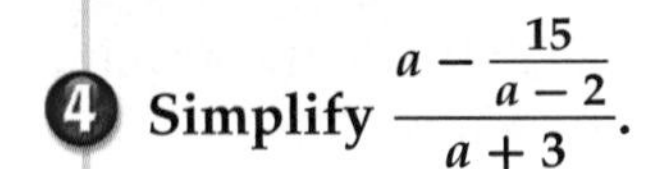 $\dfrac{a - \frac{15}{a-2}}{a+3}$.

The numerator contains a mixed expression. Rewrite it as a rational expression first.

$$\frac{a - \frac{15}{a-2}}{a+3} = \frac{\frac{a(a-2)}{a-2} - \frac{15}{a-2}}{a+3}$$ The LCD of the fractions in the numerator is $a - 2$.

$$= \frac{\frac{a^2 - 2a - 15}{a-2}}{a+3}$$ Simplify the numerator.

$$= \frac{\frac{(a+3)(a-5)}{a-2}}{a+3}$$ Factor.

$$= \frac{(a+3)(a-5)}{a-2} \div (a+3)$$ Rewrite as a division sentence.

$$= \frac{(a+3)(a-5)}{a-2} \cdot \frac{1}{a+3}$$ Multiply by the reciprocal of $a + 3$.

$$= \frac{\overset{1}{\cancel{(a+3)}}(a-5)}{a-2} \cdot \frac{1}{\underset{1}{\cancel{a+3}}}$$ Divide by the GCF, $a + 3$.

$$= \frac{a-5}{a-2}$$ Simplify.

CHECK Your Progress **Simplify each expression.**

4A. $\dfrac{\frac{b}{b+3} + 2}{b^2 - 2b - 8}$

4B. $\dfrac{1 + \frac{2c^2 - 6c - 10}{c+7}}{2c+1}$

Math Online **Extra Examples at algebra1.com**

CHECK Your Understanding

Example 1 (p. 620)

Write each mixed expression as a rational expression.

1. $3 + \frac{4}{x}$

2. $7 + \frac{5}{6y}$

3. $\frac{a-1}{3a} + 2a$

Example 2 (p. 622)

4. ENTERTAINMENT The student talent committee is arranging the performances for their holiday pageant. The first-act performances and their lengths are shown in the table. What is the average length of each performance?

Holiday Pageant Line-Up

Performance	Length (min)
A	7
B	$4\frac{1}{2}$
C	$6\frac{1}{2}$
D	$8\frac{1}{4}$
E	$10\frac{1}{5}$

Examples 3, 4 (p. 620)

Simplify each expression.

5. $\frac{3\frac{1}{2}}{4\frac{3}{4}}$

6. $\frac{\frac{x^3}{y^2}}{\frac{y^3}{x}}$

7. $\frac{\frac{x-y}{a+b}}{\frac{x^2-y^2}{a^2-b^2}}$

Exercises

HOMEWORK HELP

For Exercises	See Examples
8–19	1
20, 21	2
22–25	3
26–33	4

Write each mixed expression as a rational expression.

8. $8 + \frac{3}{n}$

9. $4 + \frac{5}{a}$

10. $2x + \frac{x}{y}$

11. $6z + \frac{2z}{w}$

12. $2m - \frac{4+m}{m}$

13. $3a - \frac{a+1}{2a}$

14. $b^2 + \frac{a-b}{a+b}$

15. $r^2 + \frac{r-4}{r+3}$

16. $5n^2 + \frac{n+3}{n^2-9}$

17. $3s^2 - \frac{s+1}{s^2-1}$

18. $(x-5) + \frac{x+2}{x-3}$

19. $(p+4) + \frac{p+1}{p-4}$

20. PARTIES The student council is planning a party for the school volunteers. There are five 66-ounce unopened bottles of soda left from a recent dance. When poured over ice, $5\frac{1}{2}$ ounces of soda fills a cup. How many servings of soda can they get from the bottles they have?

21. SCIENCE When air is pumped into a bicycle tire, the pressure P required varies inversely as the volume of the air V and is given by the equation $P = \frac{k}{V}$. If the pressure is 30 lb/in^2 when the volume is $1\frac{2}{3}$ cubic feet, find the pressure when the volume is $\frac{3}{4}$ cubic feet.

Simplify each expression.

22. $\frac{5\frac{3}{4}}{7\frac{2}{3}}$

23. $\frac{8\frac{2}{7}}{4\frac{4}{5}}$

24. $\frac{\frac{a}{b^3}}{\frac{a^2}{b}}$

Simplify each expression.

25. $\dfrac{\frac{n^3}{m^2}}{\frac{n^2}{m^2}}$

26. $\dfrac{\frac{x+4}{y-2}}{\frac{x^2}{y^2}}$

27. $\dfrac{\frac{s^3}{t^2}}{\frac{s+t}{s-t}}$

28. $\dfrac{\frac{y^2-1}{y^2+3y-4}}{y+1}$

29. $\dfrac{\frac{a^2-2q-3}{a^2-1}}{a-3}$

30. $\dfrac{\frac{n^2+2n}{n^2+9n+18}}{\frac{n^2-5n}{n^2+n-30}}$

31. $\dfrac{\frac{x^2+4x-21}{x^2-9x+18}}{\frac{x^2+3x-28}{x^2-10x+24}}$

32. $\dfrac{x-\frac{15}{x-2}}{x-\frac{20}{x-1}}$

33. $\dfrac{n+\frac{35}{n+12}}{n-\frac{63}{n-2}}$

SIRENS For Exercises 34 and 35, use the following information.
As an ambulance approaches, the siren sounds different than if it were sitting still. If the ambulance is moving toward you at v miles per hour and blowing its siren at a frequency of f, then you hear the siren as if it were blowing at a frequency of h. This can be defined by the equation $h = \dfrac{f}{1-\frac{v}{s}}$, where s is the speed of sound, approximately 760 miles per hour.

34. Simplify the complex fraction in the formula.

35. Suppose a siren blows at 45 cycles per minute and is moving toward you at 65 miles per hour. Find the frequency of the siren as you hear it.

36. **POPULATION** According to a recent census, Union City, New Jersey, was the most densely populated city in the U.S., and Anchorage City, Alaska, was one of the least. The population of Union City was 67,088, and the population of Anchorage City was 260,283. The land area of Union City is about 1.3 square miles, and the land area of Anchorage City is about 1,697.3 square miles. How many more people were there per square mile in Union City than in Anchorage City?

37. What is the product of $\frac{2b^2}{5c}$ and the quotient of $\frac{4b^3}{2c}$ and $\frac{7b^3}{8c^2}$?

H.O.T. Problems

38. **OPEN ENDED** Think of a real-world complex fraction and explain how you would simplify it.

39. **CHALLENGE** Which expressions are equivalent to 0?

a. $\dfrac{a}{1-\frac{3}{a}} + \dfrac{a}{\frac{3}{a}-1}$

b. $\dfrac{a-\frac{1}{3}}{b} - \dfrac{a+\frac{1}{3}}{b}$

c. $\dfrac{\frac{1}{2}+2a}{b-1} + \dfrac{2a+\frac{1}{2}}{1-b}$

40. **FIND THE ERROR** Bolton and Lian found the LCD of $\frac{4}{2x+1} - \frac{5}{x+1} + \frac{2}{x-1}$. Who is correct? Explain your reasoning.

Bolton	Lian
$\frac{4}{2x+1} - \frac{5}{x+1} + \frac{2}{x-1}$	$\frac{4}{2x+1} - \frac{5}{x+1} + \frac{2}{x-1}$
LCD: $(2x+1)(x+1)(x-1)$	LCD: $2(x+1)(x-1)$

41. ***Writing in Math*** Refer to the information on page 620. Explain how complex fractions are used in baking. Include an explanation of the process used to simplify a complex fraction.

STANDARDIZED TEST PRACTICE

42. The perimeter of hexagon $ABCDEF$ is 12. Which expression can be used to represent the measure of $\overline{BC}$?

A $\dfrac{6n-96}{n-8}$

B $\dfrac{9n-96}{n-8}$

C $\dfrac{3n-48}{2n-16}$

D $\dfrac{9n-96}{4n-32}$

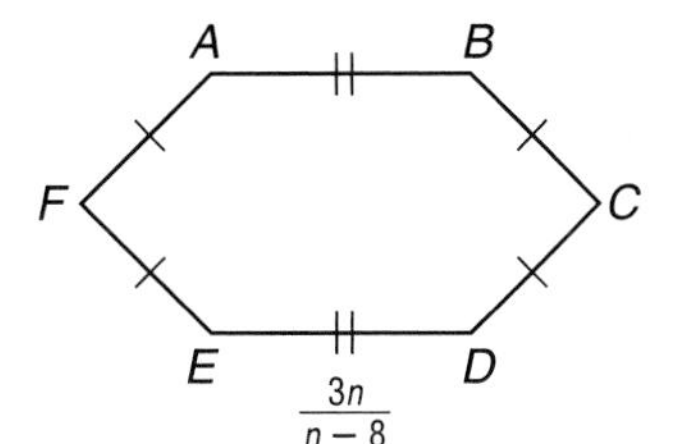

43. REVIEW Ms. Roberts is draining her cylindrical above-ground pool. The pool has a radius of 12 feet and a height of 4 feet. If water is pumped out at a constant rate of 5.5 gallons per minute, about how long will it take to drain the pool? (1 ft^3 = 7.5 gal)

F 43.9 min

G 3.8 h

H 35.2 h

J 41.1 h

Spiral Review

Find each sum. (Lesson 11-7)

44. $\dfrac{12x}{4y^2} + \dfrac{8}{6y}$

45. $\dfrac{a}{a-b} + \dfrac{b}{2b+3a}$

46. $\dfrac{a+3}{3a^2-10a-8} + \dfrac{2a}{a^2-8a+16}$

Find each difference. (Lesson 11-6)

47. $\dfrac{7}{x^2} - \dfrac{3}{x^2}$

48. $\dfrac{x}{(x-3)^2} - \dfrac{3}{(x-3)^2}$

49. $\dfrac{2}{t^2-t-2} - \dfrac{t}{t^2-t-2}$

FAMILIES For Exercises 50–52, refer to the graph. (Lesson 7-1)

50. Write each number in the graph using scientific notation.

51. How many times as great is the amount spent on food as the amount spent on clothing? Express your answer in scientific notation.

52. What percent of the total amount is spent on housing?

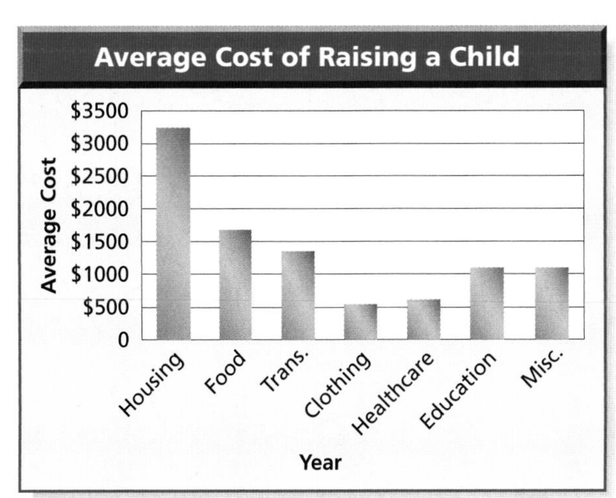

Source: University of Minnesota

GET READY for the Next Lesson

PREREQUISITE SKILL Solve each equation. (Lessons 2-2, 2-3, and 2-4)

53. $-12 = \dfrac{x}{4}$

54. $1.8 = g - 0.6$

55. $\dfrac{3}{4}n - 3 = 9$

56. $7x^2 = 28$

57. $3.2 = \dfrac{-8+n}{-7}$

58. $\dfrac{-3n-(-4)}{-6} = -9$

11-9 Rational Equations and Functions

Main Ideas

- Solve rational equations.
- Eliminate extraneous solutions.

New Vocabulary

rational equations
work problems
rate problems
extraneous solutions

GET READY for the Lesson

The Washington, D.C., Metrorail is one of the safest subway systems in the world, serving a population of more than 3.5 million. It is vital that a rail system of this size maintain a consistent schedule. Rational equations can be used to determine the exact positions of trains at any given time.

Washington Metropolitan Area Transit Authority

Train	Distance
Red Line	19.4 mi
Orange Line	24.14 mi
Blue Line	19.95 mi
Green Line	20.59 mi
Yellow Line	9.46 mi

Solve Rational Equations **Rational equations** are equations that contain rational expressions. You can use cross products to solve rational equations, but *only* when both sides of the equation are single fractions.

EXAMPLE Use Cross Products

Solve $\frac{12}{x+5} = \frac{4}{(x+2)}$.

$\frac{12}{x+5} = \frac{4}{(x+2)}$	Original equation
$12(x+2) = 4(x+5)$	Cross multiply.
$12x + 24 = 4x + 20$	Distributive Property
$8x = -4$	Add $-4x$ and -24 to each side.
$x = -\frac{4}{8}$ or $-\frac{1}{2}$	Divide each side by 8.

Study Tip

Proportions
If the rational equation is a proportion, you can cross multiply to solve.

CHECK Your Progress

Solve each equation.

1A. $\frac{7}{y-3} = \frac{3}{y+1}$

1B. $\frac{13}{f+10} = \frac{2}{7f}$

Another method you can use to solve rational equations is to multiply each side of the equation by the LCD of all of the fractions on both sides of the equation. This will eliminate all of the fractions. This method works for any rational equation.

EXAMPLE Use the LCD

Solve $\frac{n-2}{n} - \frac{n-3}{n-6} = \frac{1}{n}$.

$\frac{n-2}{n} - \frac{n-3}{n-6} = \frac{1}{n}$ Original equation

$$n(n-6)\left(\frac{n-2}{n}-\frac{n-3}{n-6}\right)=n(n-6)\left(\frac{1}{n}\right)$$ The LCD is $n(n-6)$.

$$\left(\frac{\overset{1}{\cancel{n}}(n-6)}{1}\cdot\frac{n-2}{\underset{1}{\cancel{n}}}\right)-\left(\frac{n\overset{1}{\cancel{(n-6)}}}{1}\cdot\frac{n-3}{\underset{1}{\cancel{n-6}}}\right)=\frac{\overset{1}{\cancel{n}}(n-6)}{1}\cdot\frac{1}{\underset{1}{\cancel{n}}}$$ Distributive Property

$$(n-6)(n-2)-n(n-3)=n-6$$ Simplify.

$$(n^2-8n+12)-(n^2-3n)=n-6$$ Multiply.

$$n^2-8n+12-n^2+3n=n-6$$ Subtract.

$$-5n+12=n-6$$ Simplify.

$$-6n=-18$$ Subtract.

$$n=3$$ Divide.

CHECK Your Progress **Solve each equation.**

2A. $\frac{2b-5}{b-2}-2=\frac{3}{b+2}$

2B. $1+\frac{1}{c+2}=\frac{28}{c^2+2c}$

Recall that to find the roots of a quadratic function, find the values of x when $y=0$. The roots of a rational function are found similarly.

Study Tip

Look Back

To review **solving quadratic equations by factoring,** see Lessons 9-3 through 9-6.

EXAMPLE Rational Functions

3 **Find the roots of $f(x)=\frac{x^2-x-12}{x-2}$.**

$$f(x)=\frac{x^2-x-12}{x-2}$$ Original function

$$0=\frac{x^2-x-12}{x-2}$$ $f(x)=0$

$$0=\frac{(x-4)(x+3)}{x-2}$$ Factor.

When $x=4$ and -3, the numerator becomes zero, so $f(x)=0$. Therefore, the roots of the function are 4 and -3.

CHECK Your Progress **Find the roots of each function.**

3A. $f(x)=\frac{x^2+3x-18}{x-3}$

3B. $f(x)=\frac{x^2+6x+8}{x^2+x-2}$

Rational equations can be used to solve **work problems**.

EXAMPLE Work Problem

4 **LAWN CARE It takes Abbey two hours to mow and trim Mr. Morely's lawn. When Jamal worked with her, the job took only 1 hour and 20 minutes. How long would it have taken Jamal to do the job himself?**

Explore Since it takes Abbey two hours to do the yard, she can finish $\frac{1}{2}$ the job in one hour. Thus, her rate of work is $\frac{1}{2}$ of the job per hour. The amount of work Jamal can do in one hour can be represented by $\frac{1}{t}$. To determine how long it takes Jamal to do the job alone, use the formula Abbey's portion of the job + Jamal's portion of the job = 1 completed yard.

(continued on the next page)

Plan The time that both of them worked was $1\frac{1}{3}$ hours. Each rate multiplied by this time results in the amount of work done by each person.

Solve

Abbey's portion	plus	Jamal's portion	equals	1 job.
$\frac{1}{2}\left(\frac{4}{3}\right)$	$+$	$\frac{1}{t}\left(\frac{4}{3}\right)$	$=$	1

$\frac{4}{6} + \frac{4}{3t} = 1$ Multiply.

$6t\left(\frac{4}{6} + \frac{4}{3t}\right) = 6t \cdot 1$ The LCD is $6t$.

$\overset{1}{\cancel{6t}}\left(\frac{4}{\underset{1}{\cancel{6}}}\right) + \overset{2}{\cancel{6t}}\left(\frac{4}{\underset{1}{\cancel{3t}}}\right) = 6t$ Distributive Property

$4t + 8 = 6t$ Simplify.

$8 = 2t$ Add $-4t$ to each side.

$4 = t$ Divide each side by 2.

Check The time that it would take Jamal to do the yard by himself is 4 hours. This seems reasonable because the combined efforts of the two took longer than half of Abbey's usual time.

Study Tip

Work Problems

When solving work problems, remember that each term should represent the portion of a job completed in one unit of time.

4. Lupe can paint a 60 square foot wall in 40 minutes. Working with her friend Steve, the two of them can paint the wall in 25 minutes. How long would it take Steve to do the job himself?

Online Personal Tutor at algebra1.com

Rational equations can also be used to solve **rate problems**.

EXAMPLE Rate Problem

5 **TRANSPORTATION Refer to the application at the beginning of the lesson. The Yellow Line runs between Huntington and Mt. Vernon Square. Suppose one train leaves Mt. Vernon Square at noon and arrives at Huntington 24 minutes later, and a second train leaves Huntington at noon and arrives at Mt. Vernon Square 28 minutes later. At what time do the two trains pass each other?**

Determine the rates of both trains. The total distance is 9.46 miles.

Train 1 $\frac{9.46 \text{ mi}}{24 \text{ min}}$ **Train 2** $\frac{9.46 \text{ mi}}{28 \text{ min}}$

Next, since both trains left at the same time, the time both have traveled when they pass will be the same. And since they started at opposite ends of the route, the sum of their distances is equal to the total route, 9.46 miles.

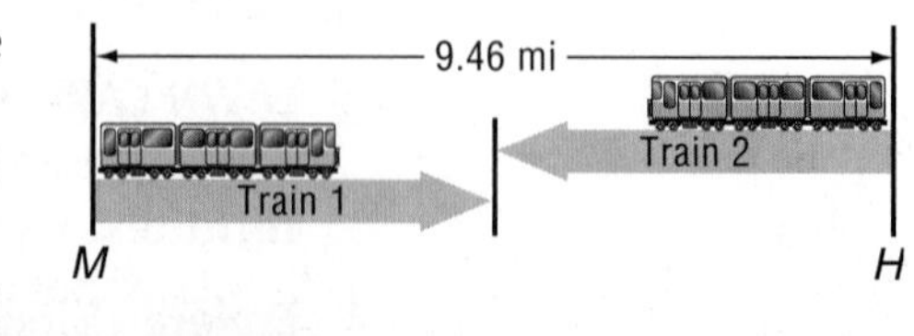

Study Tip

Rate Problems

You can solve rate problems, also called *uniform motion problems,* more easily if you first make a drawing.

	r	t	d
Train 1	$\frac{9.46 \text{ mi}}{24 \text{ min}}$	t min	$\frac{9.46t}{24}$ mi
Train 2	$\frac{9.46 \text{ mi}}{28 \text{ min}}$	t min	$\frac{9.46t}{28}$ mi

Concepts in Motion
Animation
algebra1.com

$$\frac{9.46t}{24} + \frac{9.46t}{28} = 9.46$$ The sum of the distances is 9.46.

$$168\left(\frac{9.46t}{24} + \frac{9.46t}{28}\right) = 168 \cdot 9.46$$ The LCD is 168.

$$\frac{\overset{7}{\cancel{168}}}{1} \cdot \frac{9.46t}{\underset{1}{\cancel{24}}} + \frac{\overset{6}{\cancel{168}}}{1} \cdot \frac{9.46t}{\underset{1}{\cancel{28}}} = 1589.28$$ Distributive Property

$$66.22t + 56.76t = 1589.28$$ Simplify.

$$122.98t = 1589.28$$ Add.

$$t = 12.92$$ Divide each side by 122.98.

The trains passed at about 12.92 or about 13 minutes after leaving their stations. This would be 12:13 P.M.

CHECK Your Progress

5. Debbie leaves the house walking at a rate of 3 miles per hour. After 10 minutes, her mother realizes that Debbie has forgotten her homework and leaves the house riding a bicycle at a rate of 10 miles per hour. How many minutes after Debbie initially left the house will her mother catch up to her?

Extraneous Solutions Multiplying each side of an equation by the LCD of two rational expressions can yield results that are not solutions to the original equation. Recall that such solutions are called **extraneous solutions**. Rational equations can have both valid solutions and extraneous solutions.

Vocabulary Link

Extraneous
Everyday Use: irrelevant, unimportant

Extraneous solution
Math Use: a result that is not a solution of the original equation

EXAMPLE Extraneous Solutions

6 **Solve $\frac{3x}{x-1} + \frac{6x-9}{x-1} = 6$.**

$$\frac{3x}{x-1} + \frac{6x-9}{x-1} = 6$$ Original equation

$$(x-1)\left(\frac{3x}{x-1} + \frac{6x-9}{x-1}\right) = (x-1)6$$ The LCD is $x - 1$.

$$\overset{1}{\cancel{(x-1)}}\left(\frac{3x}{\underset{1}{\cancel{x-1}}}\right) + \overset{1}{\cancel{(x-1)}}\left(\frac{6x-9}{\underset{1}{\cancel{x-1}}}\right) = (x-1)6$$ Distributive Property

$$3x + 6x - 9 = 6x - 6$$ Simplify.

$$9x - 9 = 6x - 6$$ Add like terms.

$$3x = 3$$ Add 9 to each side.

$$x = 1$$ Divide each side by 3.

Notice that 1 is an excluded value for x. If we substitute 1 for x in the original equation, we get undefined expressions. Since $x = 1$ is an extraneous solution, this equation has no solution.

CHECK Your Progress **Solve each equation.**

6A. $\frac{3x}{x-4} - \frac{4x+4}{x-4} = -1$ **6B.** $\frac{n^2-3n}{n^2-4} - \frac{10}{n^2-4} = 2$

Concepts in Motion
Interactive Lab
algebra1.com

CHECK Your Understanding

Examples 1, 2, 6 (pp. 626–627, 629)

Solve each equation. State any extraneous solutions.

1. $\frac{2}{x} = \frac{3}{x+1}$

2. $\frac{3x}{5} + \frac{3}{2} = \frac{7x}{10}$

3. $\frac{x+2}{x-2} - \frac{2}{x+2} = -\frac{7}{3}$

4. $\frac{n^2-n-6}{n^2-n} - \frac{n-5}{n-1} = \frac{n-3}{n^2-n}$

Example 3 (p. 627)

Find the roots of each function.

5. $f(x) = \frac{x^2-8x+15}{x^2+5x-6}$

6. $f(x) = \frac{x^2-x-6}{x^2+8x+12}$

Example 4 (pp. 627–628)

7. BASEBALL Omar has 32 hits in 128 times at bat. He wants his batting average to be .300. His current average is $\frac{32}{128}$ or .250. How many at bats does Omar need to reach his goal if he gets a hit in each of his next b at bats?

Example 5 (pp. 628–629)

8. LANDSCAPING Kumar is filling a 3.5-gallon bucket to water plants at a faucet that flows at a rate of 1.75 gallons a minute. If he were to add a hose that flows at a rate of 1.45 gallons per minute, how many minutes would it take him to fill the bucket? Round to the nearest tenth of a minute.

Exercises

HOMEWORK HELP	
For Exercises	See Examples
9–12	1
13–18	2
19–22	6
23–26	3
27, 28	4
29–32	5

Solve each equation. State any extraneous solutions.

9. $\frac{4}{a} = \frac{3}{a-2}$

10. $\frac{3}{x} = \frac{1}{x-2}$

11. $\frac{x-3}{x} = \frac{x-3}{x-6}$

12. $\frac{x}{x+1} = \frac{x-6}{x-1}$

13. $\frac{2n}{3} + \frac{1}{2} = \frac{2n-3}{6}$

14. $\frac{5}{4} + \frac{3y}{2} = \frac{7y}{6}$

15. $\frac{a-1}{a+1} - \frac{2a}{a-1} = -1$

16. $\frac{m-1}{m+1} - \frac{2m}{m-1} = -1$

17. $\frac{4x}{2x+3} - \frac{2x}{2x-3} = 1$

18. $\frac{a}{3a+6} - \frac{a}{5a+10} = \frac{2}{5}$

19. $\frac{2n}{n-1} + \frac{n-5}{n^2-1} = 1$

20. $\frac{5}{5-p} - \frac{p^2}{p-5} = -8$

21. $\frac{x^2-x-6}{x+2} + \frac{x^3+x^2}{x} = 3$

22. $\frac{x-\frac{6}{5}}{x} - \frac{x-10\frac{1}{5}}{x-5} = \frac{x+21}{x^2-5x}$

Find the roots of each function.

23. $f(x) = \frac{x^2-x-12}{x^2+2x-35}$

24. $f(x) = \frac{x^2+3x-4}{x^2+9x+20}$

25. $f(x) = \frac{x^3+x^2-6x}{x-1}$

26. $f(x) = \frac{x^3-4x^2-12x}{x+2}$

27. CAR WASH Ian and Nadya can each wash a car and clean its interior in about 2 hours, but Raul needs 3 hours to do the work. If the three work together, how long will it take to clean seven cars?

28. JOBS Ron works as a dishwasher and can wash 500 plates in two hours and 15 minutes. Occasionally, the busser, Chris, helps. Together they can finish 500 plates in 77 minutes. About how long would it take Chris to finish all of the plates by himself?

Real-World Link Started in 1851, the America's Cup is the oldest and most prestigious race in yachting. This race of just two boats involves 3 laps of an 18.55-nautical-mile course.

Source: americascup.com

SWIMMING POOLS **For Exercises 29 and 30, use the following information.**

The pool in Kara's backyard is cleaned and ready to be filled for the summer. It measures 15 feet long and 10 feet wide with an average depth of 4 feet.

29. If Kara's hose runs at a rate of 5 gallons per minute, how long will it take to fill the pool?

30. Kara's neighbor's hose runs at a rate of 9 gallons per minute. How long will it take to fill the pool using both hoses?

BOATING **For Exercises 31 and 32, use the following information.**

Jim and Mateo live across a lake from each other at a distance of about 3 miles. Jim can row his boat to Mateo's house in 1 hour and 20 minutes. Mateo can drive his power boat the same distance in a half hour.

31. If they leave their houses at the same time and head for each other, how long will it be before they meet?

32. How far from the nearest shore will they be when they meet?

33. QUIZZES Each week, Mandy's algebra teacher gives a 10-point quiz. After 5 weeks, Mandy has earned a total of 36 points for an average of 7.2 points per quiz. She would like to raise her average to 9 points. On how many quizzes must she score 10 points in order to reach her goal?

34. PAINTING Morgan can paint a standard-sized house in about 5 days. For his latest assignment, Morgan is going to hire two assistants. At what rate must these assistants work for Morgan to meet a deadline of two days?

GRAPHING CALCULATOR **For Exercises 35–37, use a graphing calculator.**

For each rational function, a) describe the shape of the graph, b) use factoring to simplify the function, and c) determine the roots of the function.

35. $f(x) = \frac{x^2 - x - 30}{x - 6}$ **36.** $f(x) = \frac{x^3 + x^2 - 2x}{x + 2}$ **37.** $f(x) = \frac{x^3 + 6x^2 + 12x}{x}$

AIRPLANES **For Exercises 38 and 39, use the following information.**

Headwinds push against a plane and reduce its total speed, while tailwinds push on a plane and increase its total speed. Let w = the speed of the wind, r = the speed set by the pilot, and s = the total speed.

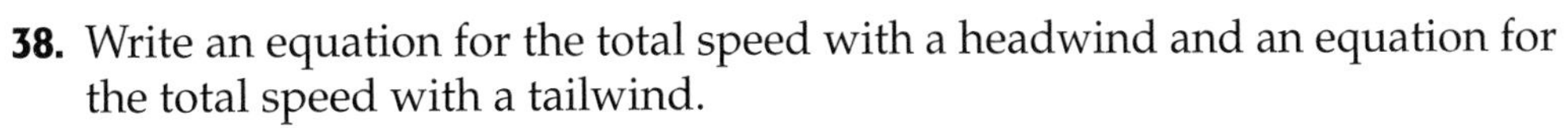

38. Write an equation for the total speed with a headwind and an equation for the total speed with a tailwind.

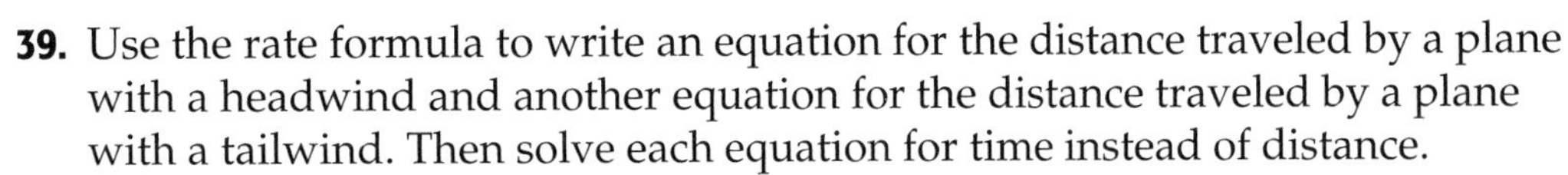

39. Use the rate formula to write an equation for the distance traveled by a plane with a headwind and another equation for the distance traveled by a plane with a tailwind. Then solve each equation for time instead of distance.

H.O.T. Problems

40. OPEN ENDED Write an expression that models a real-world situation where work is being done.

41. REASONING Find a counterexample for the following statement.

The solution of a rational equation can never be zero.

42. CHALLENGE Solve $\frac{\frac{x+3}{x-2} \cdot \frac{x^2+x-2}{x+5}}{x-1} + 2 = 0$.

43. ***Writing in Math*** Refer to the information on page 626. How are rational equations important in the operation of a subway system? Include in your answer an explanation of how rational equations can be used to approximate the time that trains will pass each other if they leave distant stations and head toward each other.

STANDARDIZED TEST PRACTICE

44. A group of band students went to a restaurant after the football game. They agreed to split the bill equally. When the bill arrived, three people discovered that they had forgotten their wallets. The others in the group agreed to make up the difference by paying an extra $2.70 each. If the total bill was $117, how many band students went to dinner?

A 13 **C** 10

B 11 **D** 9

45. REVIEW Lorenzo's math test scores are shown below.

85%, 92%, 95%

If he earns a 92% on the next test, then

F the median would decrease.

G the mean would decrease.

H the median would increase.

J the mean would increase.

Spiral Review

Simplify each expression. (Lesson 11-8)

46. $\dfrac{\dfrac{x^2+8x+15}{x^2+x-6}}{\dfrac{x^2+2x-15}{x^2-2x-3}}$

47. $\dfrac{\dfrac{a^2-6a+5}{a^2+13a+42}}{\dfrac{a^2-4a+3}{a^2+3a-18}}$

48. $\dfrac{x+2+\dfrac{2}{x+5}}{x+6+\dfrac{6}{x+1}}$

Find each difference. (Lesson 11-7)

49. $\dfrac{3}{2m-3} - \dfrac{m}{6-4m}$

50. $\dfrac{y}{y^2-2y+1} - \dfrac{1}{y-1}$

51. $\dfrac{a+2}{a^2-9} - \dfrac{2a}{6a^2-17a-3}$

52. CHEMISTRY One solution is 50% glycol, and another is 30% glycol. How much of each solution should be mixed to make a 100 gallon solution that is 45% glycol? (Lesson 2-9)

Cross-Curricular Project

Math and Science

Building the Best Roller Coaster It is time to complete your project. Use the information and data you have gathered about the building and financing of a roller coaster to prepare a portfolio or Web page. Be sure to include graphs and/or tables in the presentation.

Math Online Cross-Curricular Project at algebra1.com

CHAPTER 11 Study Guide and Review

FOLDABLES™ Study Organizer — GET READY to Study

Be sure the following Key Concepts are noted in your Foldable.

Rational Expressions | Rational Equations

Key Concepts

Inverse Variation (Lesson 11-1)

- You can use $\frac{x_1}{x_2} = \frac{y_1}{y_2}$ to solve problems involving inverse variation.

Rational Expressions (Lessons 11-2 to 11-4)

- Excluded values are values of a variable that result in a denominator of zero.
- Multiplying rational expressions is similar to multiplying rational numbers.
- Divide rational expressions by multiplying by the reciprocal of the divisor.

Dividing Polynomials (Lesson 11-5)

- To divide a polynomial by a monomial, divide each term of the polynomial by the monomial.

Rational Expressions (Lessons 11-6 and 11-7)

- Add (or subtract) rational expressions with like denominators by adding (or subtracting) the numerators and writing the sum (or difference) over the denominator.
- Rewrite rational expressions with unlike denominators using the least common denominator (LCD). Then add or subtract.

Complex Fractions (Lesson 11-8)

- Simplify complex fractions by writing them as division problems.

Solving Rational Equations (Lesson 11-9)

- Use cross product rule to solve rational equations with a single fraction on each side of the equals sign.

Key Vocabulary

Vocabulary Check

State whether each sentence is *true* or *false*. If *false*, replace the underlined word or number to make a true sentence.

1. A mixed expression is a fraction whose numerator and denominator are polynomials.
2. The complex fraction $\frac{\frac{4}{5}}{\frac{2}{3}}$ can be simplified as $\frac{6}{5}$.
3. The equation $\frac{x}{x-1} + \frac{2x-3}{x-1}$ has an extraneous 1.
4. The mixed expressions $6 - \frac{a-2}{a+3}$ can be rewritten as $\frac{5a+16}{a+3}$.
5. The least common multiple for $(x^2 - 144)$ and $(x + 12)$ is $(x + 12)$.
6. The equation $x_1y_1 = x_2y_2$ is called the product rule for inverse variations.
7. The excluded values for $\frac{4x}{x^2 - x - 12}$ are -3 and 4.
8. When the product of two values remains constant, the relationship forms an inverse variation.

Lesson-by-Lesson Review

11-1 Inverse Variation (pp. 577–582)

Write an inverse variation equation that relates x and y. Assume that y varies inversely as x. Then solve.

9. If $y = 28$ when $x = 42$, find y when $x = 56$.

10. If $y = 35$ when $x = 175$, find x when $y = 75$.

11. PHYSICS If a 135-pound person sits 5 feet from the center of a seesaw and a 108-pound person is on the other end, how far from the center should the 108-pound person sit to balance?

Example 1 **If y varies inversely as x and $y = 24$ when $x = 30$, find x when $y = 10$.**

Let $x_1 = 30$, $y_1 = 24$, and $y_2 = 10$. Solve for x_2. Use a proportion.

$\frac{x_1}{x_2} = \frac{y_2}{y_1}$ Proportion for inverse variations

$\frac{30}{x_2} = \frac{10}{24}$ $x_1 = 30$, $y_1 = 24$, and $y_2 = 10$

$720 = 10x$ Cross multiply.

$72 = x_2$ Divide each side by 10.

Thus, $x = 72$ when $y = 10$.

11-2 Operations with Radical Expressions (pp. 583–588)

Simplify each expression.

12. $\frac{3x^2y}{12xy^3z}$

13. $\frac{n^2 - 3n}{n - 3}$

14. $\frac{a^2 - 25}{a^2 + 3a - 10}$

15. $\frac{x^2 + 10x + 21}{x^3 + x^2 - 42x}$

16. $\frac{b^2 - 5b + 6}{b^2 - 13b + 36}$

17. $\frac{3x^3}{3x^3 + 6x^2}$

Example 2 **Simplify $\frac{x + 4}{x^2 + 12x + 32}$.**

$\frac{x + 4}{x^2 + 12x + 32} = \frac{\overset{1}{\cancel{x + 4}}}{\underset{1}{\cancel{(x + 4)}}(x + 8)}$ Factor.

$= \frac{1}{x + 8}$ Simplify.

11-3 Multiplying Rational Expressions (pp. 590–594)

Find each product.

18. $\frac{7b^2}{9} \cdot \frac{6a^2}{b}$

19. $\frac{5x^2y}{8ab} \cdot \frac{12a^2b}{25x}$

20. $(3x + 30) \cdot \frac{10}{x^2 - 100}$

21. $\frac{3a - 6}{a^2 - 9} \cdot \frac{a + 3}{a^2 - 2a}$

22. $\frac{b^2 + 19b + 84}{b - 3} \cdot \frac{b^2 - 9}{b^2 + 15b + 36}$

Example 3 **Find $\frac{6m^2n^4}{12} \cdot \frac{3m^3n^2}{mn}$.**

$\frac{6m^2n^4}{12} \cdot \frac{3m^3n^2}{mn}$

$= \frac{2mn^3}{4} \cdot \frac{m^2n}{1}$ Divide by GCF $3mn$.

$= \frac{2m^3n^4}{4}$ or $\frac{m^3n^4}{2}$ Multiply.

Mixed Problem Solving
For mixed problem-solving practice, see page 754.

11-4 Dividing Rational Expressions (pp. 595–599)

Find each quotient.

23. $\frac{p^3}{2q} \div \frac{p^3}{4q}$

24. $\frac{y^2}{y+4} \div \frac{3y}{y^2-16}$

25. $\frac{3x-12}{y+4} \div (y^2 - 6y + 8)$

26. $\frac{2m^2+7m-15}{m+5} \div \frac{9m^2-4}{3m+2}$

27. **PIZZA** On average, Americans eat 18 acres of pizza a day. If an average slice of pizza is about 5 square inches, how many pieces of pizza is this? (*Hint*: 43,560 square feet per acre)

Example 4 Find $\frac{y^2-16}{y^2-64} \div \frac{y+4}{y-8}$.

$\frac{y^2-16}{y^2-64} \div \frac{y+4}{y-8}$

$= \frac{y^2-16}{y^2-64} \cdot \frac{y-8}{y+4}$ Multiply by the reciprocal of $\frac{y+4}{y-8}$.

$= \frac{(y-4)(y+4)}{(y-8)(y+8)} \cdot \frac{y-8}{y+4}$ Factor.

$= \frac{(y-4)\overset{1}{\cancel{(y+4)}}}{\underset{1}{\cancel{(y-8)}}(y+8)} \cdot \frac{\overset{1}{\cancel{y-8}}}{\underset{1}{\cancel{y+4}}}$ Simplify.

$= \frac{y-4}{y+8}$

11-5 Dividing Polynomials (pp. 601–606)

Find each quotient.

28. $(4a^2b^2c^2 - 8a^3b^2c + 6abc^2) \div 2ab^2$

29. $(x^3 + 7x^2 + 10x - 6) \div (x + 3)$

30. $(48b^2 + 8b + 7) \div (12b - 1)$

31. $(4t^2 + 17t - 1) \div (4t + 1)$

32. **GEOMETRY** The volume of a prism with a triangular base is $x^3 + 6.5x^2 + 8.5x - 6$. If the height of the prism is $2x - 1$, what is the area of the triangular base?

Example 5 Find $(x^3 - 2x^2 - 22x + 21) \div (x - 3)$.

$$\begin{array}{r l}
x^2 + x - 19 & \\
x - 3\overline{)x^3 - 2x^2 - 22x + 21} & \\
\underline{x^3 - 3x^2} \quad\quad\quad\quad & \text{Multiply } x^2 \text{ and } x - 3. \\
x^2 - 22x \quad\quad & \text{Subtract.} \\
\underline{x^2 - 3x} \quad\quad & \text{Multiply } x \text{ and } x - 3. \\
-19x + 21 & \text{Subtract.} \\
\underline{-19x + 57} & \text{Multiply } -19 \text{ and } x - 3. \\
-36 & \text{Subtract.}
\end{array}$$

The quotient is $x^3 + x - 19 - \frac{36}{x-3}$.

11-6 Rational Expressions with Like Denominators (pp. 608–613)

Find each sum or difference.

33. $\frac{m+4}{5} + \frac{m-1}{5}$

34. $\frac{-5}{2n-5} + \frac{2n}{2n-5}$

35. $\frac{a^2}{a-b} + \frac{-b^2}{a-b}$

36. $\frac{7a}{b^2} - \frac{5a}{b^2}$

37. $\frac{2x}{x-3} - \frac{6}{x-3}$

38. $\frac{m^2}{m-n} - \frac{2mn-n^2}{m-n}$

Example 6 Find $\frac{n^2+10n}{n+5} + \frac{25}{n+5}$.

$\frac{n^2+10n}{n+5} + \frac{25}{n+5}$

$= \frac{n^2+10n+25}{n+5}$ Add the numerators.

$= \frac{(n+5)(n+5)}{(n+5)}$ Factor.

$= n + 5$ Simplify.

11-7 Rational Expressions with Unlike Denominators (pp. 614–619)

Find each sum or difference.

39. $\frac{2c}{3d^2} + \frac{3}{2cd}$

40. $\frac{r^2 + 21r}{r^2 - 9} + \frac{3r}{r + 3}$

41. $\frac{7}{3a} - \frac{3}{6a^2}$

42. $\frac{2x}{2x + 8} - \frac{4}{5x + 20}$

Example 7 Find $\frac{3}{y + 1} - \frac{y}{y + 3}$.

$$\frac{3}{y+1} - \frac{y}{y+3}$$
$$= \frac{y+3}{y+3} \cdot \frac{3}{y+1} - \frac{y}{y+3} \cdot \frac{y+1}{y+1}$$
$$= \frac{3y+9}{(y+3)(y+1)} - \frac{y^2+y}{(y+3)(y+1)}$$
$$= \frac{-y^2+2y+9}{(y+3)(y+1)}$$

11-8 Mixed Expressions and Complex Fractions (pp. 620–625)

Write each mixed expression as a rational expression.

43. $4 + \frac{x}{x - 2}$

44. $2 - \frac{x + 2}{x^2 - 4}$

Simplify each expression.

45. $\dfrac{x + \frac{35}{x + 2}}{x + \frac{42}{x + 13}}$

46. $\dfrac{y + 9 - \frac{6}{y + 4}}{y + 4 + \frac{2}{y + 1}}$

Example 8 Simplify $\dfrac{\frac{a^2b^4}{c}}{\frac{a^3b}{c^2}}$.

$\dfrac{\frac{a^2b^4}{c}}{\frac{a^3b}{c^2}} = \frac{a^2b^4}{c} \div \frac{a^3b}{c^2}$ Rewrite as a division sentence.

$= \frac{a^2b^4}{c} \cdot \frac{c^2}{a^3b}$ Multiply by the reciprocal.

$= \frac{b^3c}{a}$ Simplify.

11-9 Rational Equations and Functions (pp. 626–632)

Solve each equation. State any extraneous solutions.

47. $\frac{4x}{3} + \frac{7}{2} = \frac{7x}{12} - 14$

48. $\frac{11}{2x} - \frac{2}{3x} = \frac{1}{6}$

49. $\frac{3}{x^2 + 3x} + \frac{x + 2}{x + 3} = \frac{1}{x}$

50. $\frac{1}{n + 4} - \frac{1}{n - 1} = \frac{2}{n^2 + 3n - 4}$

51. JOBS Normally, it takes Jeffery 1 hour 45 minutes to mow and trim an average lawn. When Lupe worked with him, an average lawn only took an hour. How long would it take Lupe to mow and trim an average yard on her own?

Example 9 Solve $\frac{5n}{6} + \frac{1}{n - 2} = \frac{n + 1}{3(n - 2)}$.

$$\frac{5n}{6} + \frac{1}{n-2} = \frac{n+1}{3(n-2)}$$
$$6(n-2)\left(\frac{5n}{6} + \frac{1}{n-2}\right) = 6(n-2)\frac{n+1}{3(n-2)}$$
$$\frac{\overset{1}{\cancel{6}}(n-2)(5n)}{\underset{1}{\cancel{6}}} + \frac{6\overset{1}{\cancel{(n-2)}}}{\underset{1}{\cancel{n-2}}} = \frac{\overset{2}{\cancel{6}}\overset{1}{\cancel{(n-2)}}(n+1)}{\underset{1}{\cancel{3}}\underset{1}{\cancel{(n-2)}}}$$

$(n - 2)(5n) + 6 = 2(n + 1)$ Simplify.

$5n^2 - 12n + 4 = 0$ Simplify.

$(5n - 2)(n - 2) = 0$ Factor.

$n = \frac{2}{5}$ or $n = 2$

When you check the value 2, you get a zero in the denominator. So, 2 is an extraneous solution.

CHAPTER 11 Practice Test

WINTER **For Exercises 1 and 2 use the following information.**

An ice sculptor has a cube of ice. The length of each side of the cube is x inches. To begin a sculpture, he removes $\frac{3}{4}$ of an inch from the top. Then he removes $\frac{1}{2}$ inch from the width and length of the block.

1. Write an expression that represents the current volume of the block.
2. The sculptor scraps his idea and decides to divide the block into $x - 2$ blocks. What is the volume of the smaller blocks?

Solve each variation. Assume that y varies inversely as x.

3. If $y = 21$ when $x = 40$, find y when $x = 84$.
4. If $y = 22$ when $x = 4$, find x when $y = 16$.

5. **MULTIPLE CHOICE** Willie can type a 200 word essay in 6 hours. Myra can type the same essay in $4\frac{1}{2}$ hours. If they work together, how long will it take them to type the essay?

   A $2\frac{3}{5}$ hr C $1\frac{4}{7}$ hr

   B $2\frac{4}{7}$ hr D $1\frac{3}{5}$ hr

Simplify each rational expression. State the excluded values of the variables.

6. $\dfrac{5 - 2m}{6m - 15}$

7. $\dfrac{3 + x}{2x^2 + 5x - 3}$

8. $\dfrac{4c^2 + 12c + 9}{2c^2 - 11c - 21}$

9. $\dfrac{x + 4 + \frac{5}{x - 2}}{x + 6 + \frac{15}{x - 2}}$

10. $\dfrac{1 - \frac{9}{t}}{1 - \frac{81}{t^2}}$

11. $\dfrac{\frac{5}{6} + \frac{u}{t}}{\frac{2u}{t} - 3}$

Perform the indicated operations.

12. $\dfrac{2x}{x - 7} - \dfrac{14}{x - 7}$

13. $\dfrac{n + 3}{2n - 8} \cdot \dfrac{6n - 24}{2n + 1} \div (z - 3)$

14. $(10m^2 + 9m - 36) \div (2m - 3)$

15. $\dfrac{x^2 + 4x - 32}{x + 5} \cdot \dfrac{x - 3}{x^2 - 7x + 12}$

16. $\dfrac{4x^2 + 11x + 6}{x^2 - x - 6} \div \dfrac{x^2 + 8x + 16}{x^2 + x - 12}$

17. $(10z^4 + 5z^3 - z^2) \div 5z^3$

18. $\dfrac{y}{7y + 14} + \dfrac{6}{-3y + 6}$

19. $\dfrac{x + 5}{x + 2} + 6$

Solve each equation.

20. $\dfrac{2}{3t} + \dfrac{1}{2} = \dfrac{3}{4t}$

21. $\dfrac{2c}{c - 4} - 2 = \dfrac{4}{c + 5}$

22. **FINANCE** Barrington High School is raising money to build a house for Habitat for Humanity by doing lawn work for friends and neighbors. Scott can rake a lawn and bag the leaves in 5 hours, while Kalyn can do it in 3 hours. If Scott and Kalyn work together, how long will it take them to rake a lawn and bag the leaves?

23. **MULTIPLE CHOICE** Which expression can be used to represent the area of the triangle?

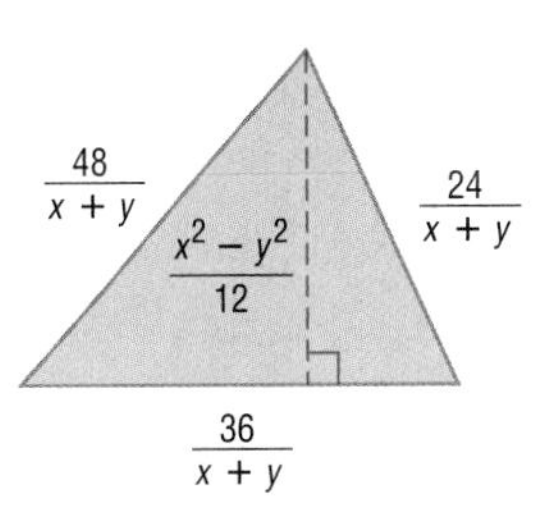

   F $2(x - y)$ H $4(x - y)$

   G $\frac{3}{2}(x - y)$ J $\dfrac{108}{x + y}$

CHAPTER 11

Standardized Test Practice

Cumulative Chapters 1–11

Read each question. Then fill in the correct answer on the answer document provided by your teacher or on a sheet of paper.

1. Which expression is equivalent to $\frac{(3y^5)(6y^4)}{9y^2}$?

A $2y^7$
B y^{11}
C $2y^{18}$
D y^{18}

2. What is the area of the polygon below?

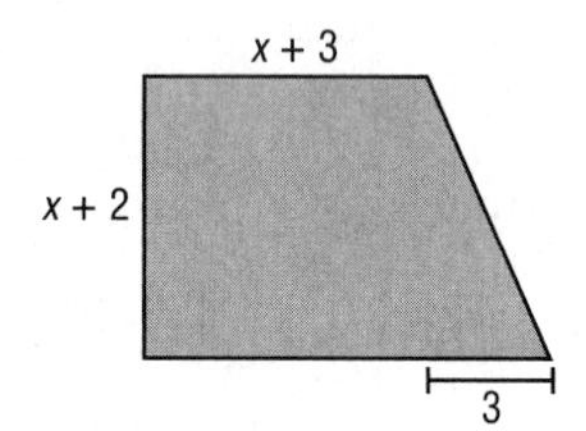

F $x^2 + 5x + 6$
G $x^2 + \frac{13}{2}x + 9$
H $x^2 + 8x + 12$
J $x^2 + \frac{13}{2}x + 12$

3. GRIDDABLE If $4x + y = 12$ and $x = -2$, then what value is y equal to?

4. What is the area of the polygon below?

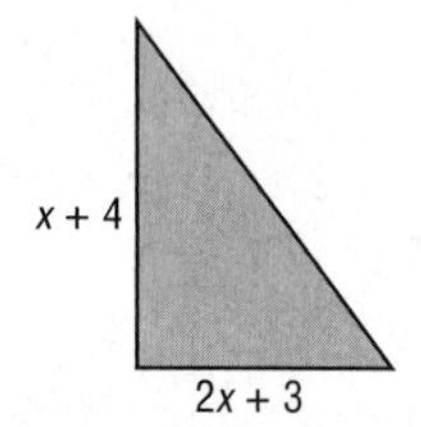

A $2x^2 + 11x + 12$
B $x^2 + \frac{11}{2}x + 6$
C $3x + 7$
D $x^2 + 6$

5. Which expression is equivalent to $\frac{16a^2b^3}{32ab^7}$?

F $2ab^4$
G $\frac{b^4}{2a}$
H $\frac{a}{2b^4}$
J $\frac{a^3}{2b^{10}}$

6. Which graph represents a system of equations with infinitely many solutions?

A

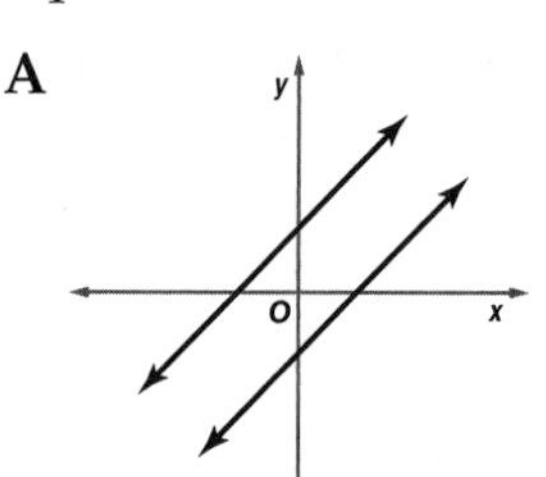

B

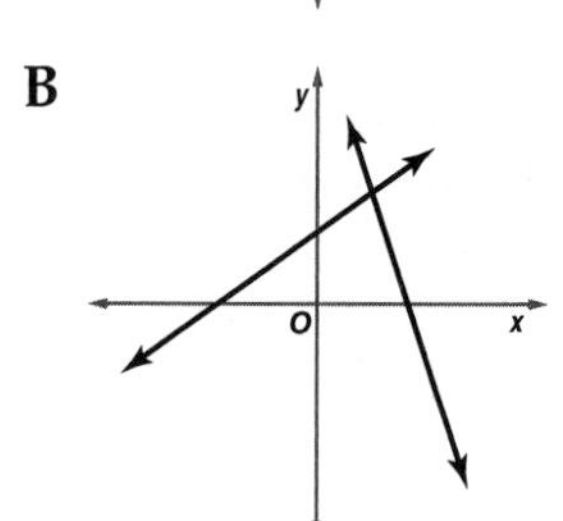

C

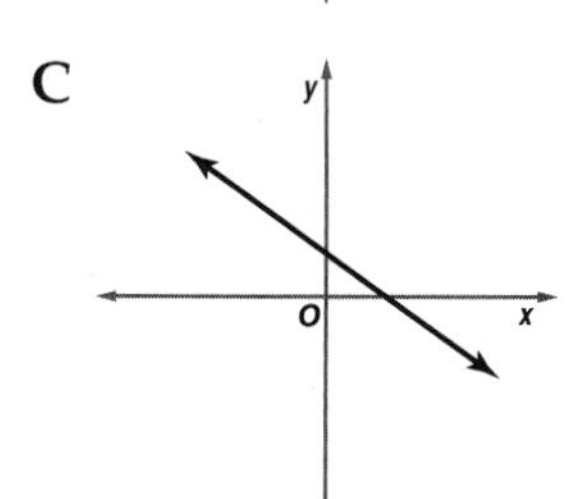

D

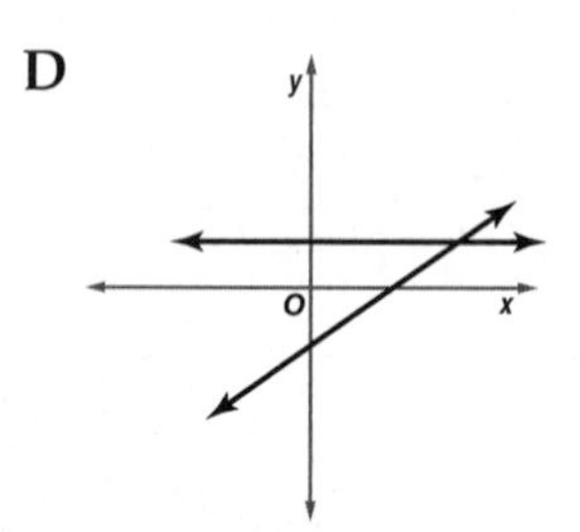

TEST-TAKING TIP

Question 6 If you don't know how to solve a problem, eliminate the answer choices you know are incorrect and then guess from the remaining choices. Even eliminating only one answer choice greatly increases your chance of guessing the correct answer.

7. Lauren sold T-shirts for 10 days in a row as a fundraiser. In those 10 days, she sold 120 T-shirts for an average of 12 T-shirts per day. For how many days must she sell 20 T-shirts to bring her average to 18 T-shirts per day?

F 10 **H** 30

G 20 **J** 40

8. The graph of the equation $y = x^2 - x - 6$ is shown below.

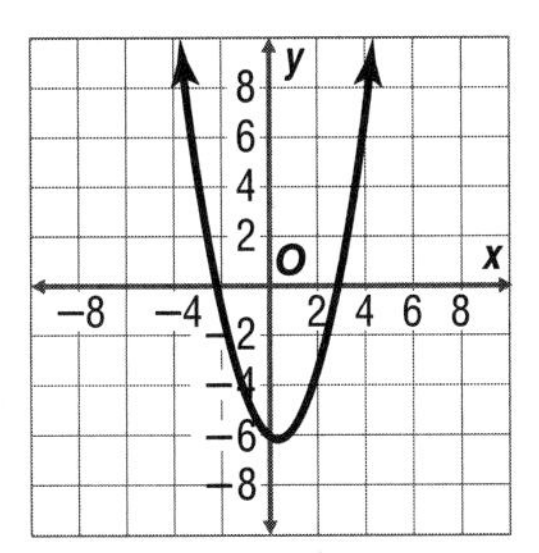

For what value or values of x is $y = 0$?

A $x = -2$ only

B $x = -3$ only

C $x = -2$ and $x = 3$

D $x = 2$ and $x = -3$

9. Sally wants to sod a corner of her yard as pictured below. How much sod will she need to cover the triangular area?

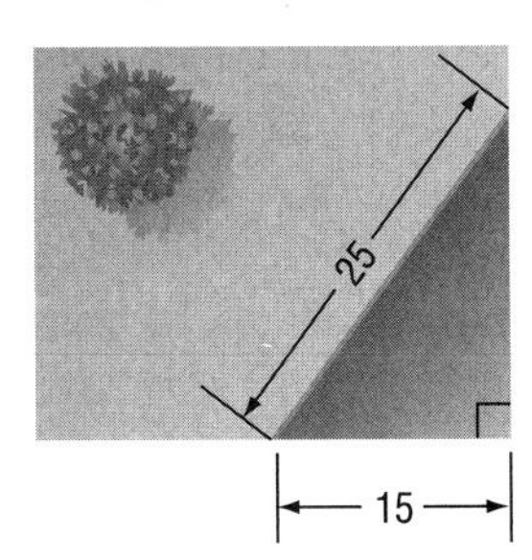

F 150 ft^2 **H** 187.5 ft^2

G 300 ft^2 **J** 375 ft^2

10. Which graph below represents the solution set for $h = -4$ and $h > 2$?

A −5 −4 −3 −2 −1 0 1 2 3 4 5

B −5 −4 −3 −2 −1 0 1 2 3 4 5

C −5 −4 −3 −2 −1 0 1 2 3 4 5

D −5 −4 −3 −2 −1 0 1 2 3 4 5

11. GRIDDABLE Peter drives to his grandmother's house every Sunday. He lives 120 miles from his grandmother and he drives this distance in 2.4 hours. At this rate, how long, in hours, would it take him to drive 300 miles?

Pre-AP

Record your answers on a sheet of paper. Show your work.

12. A 12-foot ladder is placed against the side of a building so that the bottom of the ladder is 6 feet from the base of the building.

a. Suppose the bottom of the ladder is moved closer to the base of the building. Does the height that the ladder reaches increase or decrease?

b. What conclusion can you make about the height the ladder reaches and the distance between the bottom of the ladder and the base of the building?

c. Does this relationship form an inverse variation? Explain your reasoning.

NEED EXTRA HELP?												
If You Missed Question...	1	2	3	4	5	6	7	8	9	10	11	12
Go to Lesson...	7-2	11-3	3-3	11-3	11-2	4-5	11-9	9-2	10-4	6-4	3-6	11-1

CHAPTER 12

Statistics and Probability

BIG Ideas

- Identify various sampling techniques.
- Count outcomes using the Fundamental Counting Principle.
- Determine probabilities.

Key Vocabulary

combination (p. 657)
compound event (p. 663)
permutation (p. 655)
sample (p. 642)

Real-World Link

U.S. Senate The United States Senate forms committees to focus on different issues. These committees are made up of senators from various states and political parties. You can use probability to find how many ways these committees can be formed.

FOLDABLES™ Study Organizer

Statistics and Probability Make this Foldable to help you organize what you learn about statistics and probability. Begin with a sheet of $8\frac{1}{2}$" × 11" paper.

1. **Fold** in half lengthwise.

2. **Fold** the top to the bottom twice.

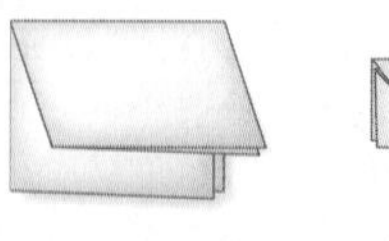

3. **Open.** Cut along the second fold to make four tabs.

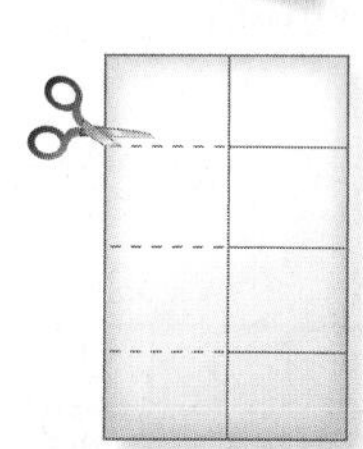

4. **Label** as shown.

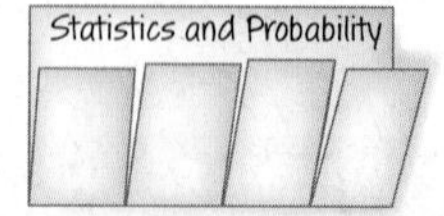

GET READY for Chapter 12

Diagnose Readiness You have two options for checking Prerequisite Skills.

Option 2

Math Online Take the Online Readiness Quiz at algebra1.com.

Option 1

Take the Quick Check below. Refer to the Quick Review for help.

QUICKCheck

Determine the probability of each event if you randomly select a cube from a bag containing 6 red cubes, 4 yellow cubes, 3 blue cubes, and 1 green cube. (Prerequisite Skill)

1. P(red) **2.** P(blue) **3.** P(not red)

4. GAMES Paul is going to roll a game cube with 3 sides painted red, two painted blue, and 1 painted green. What is the probability that a red side will land face up?

Find each product. (Prerequisite Skill)

5. $\frac{4}{5} \cdot \frac{3}{4}$ **6.** $\frac{5}{12} \cdot \frac{6}{11}$

7. $\frac{7}{20} \cdot \frac{4}{19}$ **8.** $\frac{4}{32} \cdot \frac{7}{32}$

9. $\frac{13}{52} \cdot \frac{4}{52}$ **10.** $\frac{56}{100} \cdot \frac{24}{100}$

Write each fraction as a percent. Round to the nearest tenth. (Prerequisite Skill)

11. $\frac{7}{8}$ **12.** $\frac{33}{80}$

13. $\frac{107}{125}$ **14.** $\frac{625}{1024}$

15. CONCERTS At a local concert, 585 of 2000 people were under the age of 18. What percentage of the audience were under 18? Round to the nearest tenth.

QUICKReview

EXAMPLE 1

Determine the probability of selecting a green cube if you randomly select a cube from a bag containing 6 red cubes, 4 yellow cubes, and 1 green cube.

There is 1 green cube and a total of 11 cubes in the bag.

$$\frac{1}{11} = \frac{\text{number of green cubes}}{\text{total number of cubes}}$$

The probability of selecting a green cube is $\frac{1}{11}$.

EXAMPLE 2

Find $\frac{5}{4} \cdot \frac{2}{3}$.

$\frac{5}{4} \cdot \frac{2}{3} = \frac{5 \cdot 2}{4 \cdot 3}$ Multiply both the numerators and the denominators.

$= \frac{10}{12}$ or $\frac{5}{6}$ Simplify.

EXAMPLE 3

Write the fraction $\frac{14}{17}$ as a decimal. Round to the nearest tenth.

$\frac{14}{17} = 0.823$ Simplify and round.

0.823×100 Multiply the decimal by 100.

$= 82.3$ Simplify.

$\frac{14}{17}$ written as a percent is 82.3%.

12-1 Sampling and Bias

Main Ideas

- Identify various sampling techniques.
- Recognize a biased sample.

New Vocabulary

sample
population
random sample
simple random sample
stratified random sample
systematic random sample
biased sample
convenience sample
voluntary response sample

GET READY for the Lesson

Manufacturing music CDs involves burning copies from a master. However, not every burn is successful. Because it is costly to check every CD, manufacturers monitor production by randomly checking CDs for defects.

Sampling Techniques A **sample** is some portion of a larger group, called the **population,** selected to represent that group. Sample data are often used to estimate a characteristic within an entire population, such as voting preferences prior to elections. A **random sample** of a population is selected so that it is representative of the entire population. The sample is chosen without any preference. There are several ways to pick a random sample.

KEY CONCEPT *Random Samples*

Type	Definition	Example
Simple Random Sample	A simple random sample is a sample that is as equally likely to be chosen as any other sample from the population.	The 26 students in a class are each assigned a different number from 1 to 26. Then three of the 26 numbers are picked at random.
Stratified Random Sample	In a stratified random sample, the population is first divided into similar, nonoverlapping groups. A sample is then selected from each group.	The students in a school are divided into freshmen, sophomores, juniors, and seniors. Then two students are randomly selected from each group of students.
Systematic Random Sample	In a systematic random sample, the items are selected according to a specified time or item interval.	Every 2 minutes, an item is pulled off the assembly line. or Every twentieth item is pulled off the assembly line.

EXAMPLE Classify a Random Sample

ECOLOGY Ten lakes in Minnesota are selected randomly. Then 2 liters of water are drawn from each of the ten lakes.

a. Identify the sample and suggest a population from which it was selected.

The sample is ten 2-liter containers of lake water, one from each of 10 lakes. The population is lake water from all of the lakes in Minnesota.

b. Classify the sample as *simple, stratified,* or *systematic.*

This is a simple random sample. Each of the ten lakes was equally likely to have been chosen from the list.

CHECK Your Progress

BARBECUE Refer to the information at the left. The cooks lined up randomly within their category, and every tenth cook in each category was selected.

1A. Identify the sample and a population from which it was selected.

1B. Classify the sample as *simple, stratified,* or *systematic.*

Online Personal Tutor at algebra1.com

Real-World Link

Each year, Meridian, Texas, hosts The National Championship Barbecue Cook-Off. In 2003, there were 190 cooks in the competition, and they competed in one of three categories: brisket, chicken, or pork spare ribs.

Source: bbq.htcomp.net

Biased Sample Random samples are unbiased. In a **biased sample,** one or more parts of a population are favored over others.

EXAMPLE Identify Sample as Biased or Unbiased

2 Identify each sample as *biased* or *unbiased.* Explain your reasoning.

a. MANUFACTURING Every 1000th bolt is pulled from the production line and measured for length.

The sample is chosen using a specified interval. This is an unbiased sample because it is a systematic random sample.

b. MUSIC Every tenth customer in line for a certain rock band's concert tickets is asked about his or her favorite rock band.

The sample is a biased sample because customers in line for concert tickets are more likely to name the band giving the concert as a favorite.

CHECK Your Progress

2. POLITICS A journalist visited a senior center and chose 10 individuals randomly to poll about various political topics.

Two popular forms of samples that are often biased include convenience samples and voluntary response samples.

KEY CONCEPT *Biased Samples*

Type	Definition	Example
Convenience Sample	A convenience sample includes members of a population who are easily accessed.	To check spoilage, a produce worker selects 10 apples from the top of the bin. The 10 apples are unlikely to represent all of the apples in the bin.
Voluntary Response Sample	A voluntary response sample involves only those who want to participate in the sampling.	A radio call-in show records that 75% of its 40 callers voiced negative opinions about a local football team. Those 40 callers are unlikely to represent the entire local population. Volunteer callers are more likely to have strong opinions and are typically more negative than the entire population.

EXAMPLE Identify and Classify a Biased Sample

3 BUSINESS The travel account records from 4 of the 20 departments in a corporation are to be reviewed. The accountant states that the first 4 departments to voluntarily submit their records will be reviewed.

a. Identify the sample and a population from which it was selected.

The sample is the travel account records from 4 departments in the corporation. The population is the travel account records from all 20 departments in the corporation.

b. Classify the sample as *convenience* or *voluntary response.*

Since the departments voluntarily submit their records, this is a voluntary response sample.

CHECK Your Progress

POLL A principal asks the students in her school to write down the name of a favorite teacher on an index card. She then tabulates the results from the first 20 and the last 20 cards received.

3A. Identify the sample and a population from which it was selected.

3B. Classify the sample as *convenience* or *voluntary response.*

EXAMPLE Identify the Sample

4 NEWS REPORTING Rafael needs to determine whether students in his school believe that an arts center should be added to the school. He polls 15 of his friends who sing in the chorale. Twelve of them think the school needs an arts center, so Rafael reports that 80% of the students surveyed support the project.

a. Identify the sample.

The sample is a group of students from the chorale.

b. Suggest a population from which the sample was selected.

The population for the survey is all of the students in the school.

c. State whether the sample is *unbiased* (random) or *biased*. If unbiased, classify it as *simple, stratified,* or *systematic*. If biased, classify it as *convenience* or *voluntary response.*

The sample was from the chorale. So the reported support is not likey to be representative of the student body. The sample is biased. Since Rafael polled only his friends, it is a convenience sample.

> **Study Tip**
>
> **Random Sample**
>
> A sample is *random* if every member of the population has an equal probability of being chosen for the sample.

CHECK Your Progress

ELECTIONS To estimate the leading candidate, a candidate's committee randomly sends a survey to the people on their mailing list. The returns indicate that their candidate is leading by a margin of 58% to 42%.

4A. Identify the sample.

4B. Suggest a population from which the sample was selected.

4C. State whether the sample is *unbiased* (random) or *biased*. If unbiased, classify it as *simple, stratified,* or *systematic*. If biased, classify it as *convenience* or *voluntary response.*

CHECK Your Understanding

Examples 1–4 (pp. 642–644)

Identify each sample, suggest a population from which it was selected, and state whether it is *unbiased* (random) or *biased*. If unbiased, classify the sample as *simple*, *stratified*, or *systematic*. If biased, classify as *convenience* or *voluntary response*.

1. **NEWSPAPERS** The local newspaper asks readers to write letters stating their preferred candidates for mayor.

2. **SCHOOL** A teacher needs a sample of work from four students in her first-period math class to display at the school open house. She selects the work of the first four students who raise their hands.

3. **BUSINESS** A hardware store wants to assess the strength of nails it sells. Store personnel select 25 boxes at random from among all of the boxes on the shelves. From each of the 25 boxes, they select one nail at random and subject it to a strength test.

4. **SCHOOL** A class advisor hears complaints about an incorrect spelling of the school name on pencils sold at the school store. The advisor goes to the store and asks Namid to gather a sample of pencils and look for spelling errors. Namid grabs the closest box of pencils and counts out 12 pencils from the top of the box. She checks the pencils, returns them to the box, and reports the results to the advisor.

Exercises

HOMEWORK HELP	
For Exercises	See Examples
5–18	1-4

Identify each sample, suggest a population from which it was selected, and state whether it is *unbiased* (random) or *biased*. If unbiased, classify the sample as *simple*, *stratified*, or *systematic*. If biased, classify as *convenience* or *voluntary response*.

5. **SCHOOL** Pieces of paper with the names of three sophomores are drawn from a hat containing identical pieces of paper with all sophomores' names.

6. **FOOD** Twenty shoppers outside a fast-food restaurant are asked to name their preferred cola between two choices.

7. **RECYCLING** An interviewer goes from house to house on weekdays between 9 A.M. and 4 P.M. to determine how many people recycle.

8. **POPULATION** Ten people from each of the 86 counties in a state are chosen at random and asked their opinion on a state issue.

9 **SCOOTERS** A scooter manufacturer is concerned about quality control. The manufacturer checks the first five scooters off the line in the morning and the last five off the line in the afternoon for defects.

10. **SCHOOL** To determine who will speak for her class at the school board meeting, Ms. Finchie used the numbers appearing next to her students' names in her grade book. She writes each of the numbers on an identical piece of paper and shuffles the pieces of papers in a box. Without seeing the contents of the box, one student draws 3 pieces of paper from the box. The students with these numbers will speak for the class.

Identify each sample, suggest a population from which it was selected, and state whether it is *unbiased* (random) or *biased*. If unbiased, classify the sample as *simple, stratified,* or *systematic*. If biased, classify as *convenience* or *voluntary response*.

Real-World Link

In 2002, Washington led the nation in cherry production by growing 194 million pounds of cherries.

Source: USDA

11. FARMING An 8-ounce jar was filled with corn from a storage silo by dipping the jar into the pile of corn. The corn in the jar was then analyzed for moisture content.

12. COURTS The gender makeup of district court judges in the United States is to be estimated from a sample. All judges are grouped geographically by federal reserve districts. Within each of the 11 federal reserve districts, all judges' names are assigned a distinct random number. In each district, the numbers are then listed in order. A number between 1 and 20 inclusive is selected at random, and the judge with that number is selected. Then every 20th name after the first selected number is also included in the sample.

13. TELEVISION A television station asks its viewers to share their opinions about a proposed golf course to be built just outside the city limits. Viewers can call one of two 800 numbers. One number represents a "yes" vote, and the other number represents a "no" vote.

14. GOVERNMENT To discuss leadership issues shared by all United States Senators, the President asks four of his closest colleagues in the Senate to meet with him.

15. FOOD To sample the quality of the Bing cherries throughout the produce department, the produce manager picks up a handful of cherries from the edge of one case and checks to see if these cherries are spoiled.

16. MANUFACTURING During the manufacture of high-definition televisions, units are checked for defects. Within the first 10 minutes of a work shift, a television is randomly chosen from the line of completed sets. For the rest of the shift, every 15th television on the line is checked for defects.

17. BUSINESS To get reaction about a benefits package, a company uses a computer program to randomly pick one person from each of its departments.

18. MOVIES A magazine is trying to determine the most popular actor of the year. It asks its readers to mail the name of their favorite actor to their office.

COLLEGE For Exercises 19 and 20, use the following information.
The graph at the right reveals that 56% of survey respondents did not have a formal financial plan for a child's college tuition.

19. Write a statement to describe what you do know about the sample.

20. What additional information would you like to have about the sample to determine whether the sample is biased?

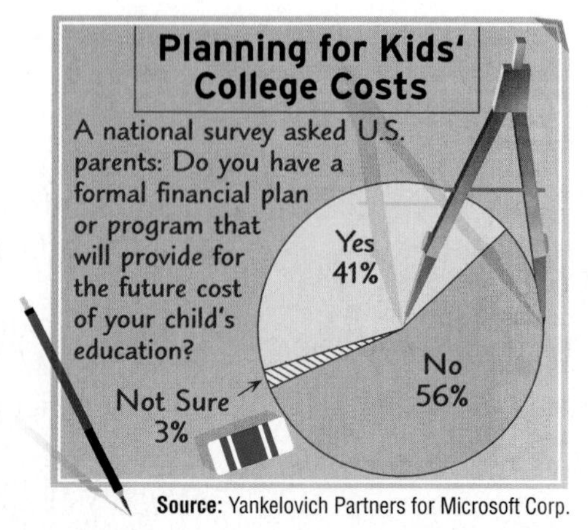

Source: Yankelovich Partners for Microsoft Corp.

DESIGN A SURVEY **For Exercises 21–23, describe an unbiased way to conduct each survey.**

21. **SCHOOL** Suppose you want to sample the opinion of the students in your school about a new dress code.

22. **ELECTIONS** Suppose you are running for mayor of your city and want to know if you are likely to be elected.

23. **PICK A TOPIC** Write a question you would like to conduct a survey to answer. Then describe an unbiased way to conduct your survey.

24. **FAMILY** Study the graph at the right. Describe the information that is revealed in the graph. What information is there about the type or size of the sample?

Source: National Pork Producers Council

25. **FARMING** Suppose you are a farmer and want to know if your tomato crop is ready to harvest. Describe an unbiased way to determine whether the crop is ready to harvest.

26. **MANUFACTURING** Suppose you want to know whether the infant car seats manufactured by your company meet the government standards for safety. Describe an unbiased way to determine whether the seats meet the standards.

H.O.T. Problems

27. **REASONING** Describe how the following three types of sampling techniques are similar and how they are different.
    - simple random sample
    - stratified random sample
    - systematic random sample

28. **REASONING** Explain the difference between a convenience sample and a voluntary response sample.

29. **OPEN ENDED** Give a real-world example of a biased sample.

30. **CHALLENGE** The following is a proposal for surveying a stratified random sample of the student body.

    Divide the student body according to those who are on the basketball team, those who are in the band, and those who are in the drama club. Then take a simple random sample from each of the three groups. Conduct the survey using this sample.

    Study the proposal. Describe its strengths and weaknesses. Is the sample a stratified random sample? Explain.

31. ***Writing in Math*** Refer to the information on page 642 to explain why sampling is important in manufacturing. Describe two different ways, one biased and one unbiased, to pick which CDs to check.

STANDARDIZED TEST PRACTICE

32. To predict the candidate who will win the seat in city council, which method would give the newspaper the *most* accurate result?

A Ask every fifth person that passes a reporter in the mall.

B Use a list of registered voters and call every 20th person.

C Publish a survey and ask readers to reply.

D Ask reporters at the newspaper.

33. REVIEW Which equation *best* represents the relationship between x and y?

x	y
−1	8
0	5
1	2
2	−1
3	−4
4	−7

F $y = 8 - 3x$

G $y = 5x - 3$

H $y = -3x + 5$

J $y = 3x + 5$

Spiral Review

Solve each equation. (Lesson 11-9)

34. $\frac{10}{3y} - \frac{5}{2y} = \frac{1}{4}$

35. $\frac{3}{r+4} - \frac{1}{r} = \frac{1}{r}$

36. $\frac{1}{4m} + \frac{2m}{m-3} = 2$

Simplify. (Lesson 11-8)

37. $\dfrac{2 + \frac{5}{x}}{\frac{x}{3} + \frac{5}{6}}$

38. $\dfrac{a + \frac{35}{a+12}}{a+7}$

39. $\dfrac{\frac{t^2 - 4}{t^2 + 5t + 6}}{t-2}$

40. GEOMETRY The sides of a triangle have measures of $4\sqrt{24}$ centimeters, $5\sqrt{6}$ centimeters, and $3\sqrt{54}$ centimeters. What is the perimeter of the triangle? Write in simplest form. (Lesson 10-2)

Solve each equation by using the Quadratic Formula. Approximate any irrational roots to the nearest tenth. (Lesson 9-4)

41. $x^2 - 6x - 40 = 0$

42. $6b^2 + 15 = -19b$

43. $2d^2 = 9d + 3$

Find each product. (Lesson 7-6)

44. $(y + 5)(y + 7)$

45. $(c - 3)(c - 7)$

46. $(x + 4)(x - 8)$

GET READY for the Next Lesson

PREREQUISITE SKILL **Find each product.**

47. $3 \cdot 2 \cdot 1$

48. $11 \cdot 10 \cdot 9$

49. $6 \cdot 5 \cdot 4 \cdot 3 \cdot 2 \cdot 1$

50. $8 \cdot 7 \cdot 6 \cdot 5$

51. $19 \cdot 18 \cdot 17$

52. $30 \cdot 29 \cdot 28 \cdot 27$

READING MATH

Survey Questions

Even though taking a random sample eliminates bias or favoritism in the choice of a sample, questions may be worded to influence people's thoughts in a desired direction. Two different surveys on Internet sales tax had different results.

Question 1

Should there be sales tax on purchases made on the Internet?

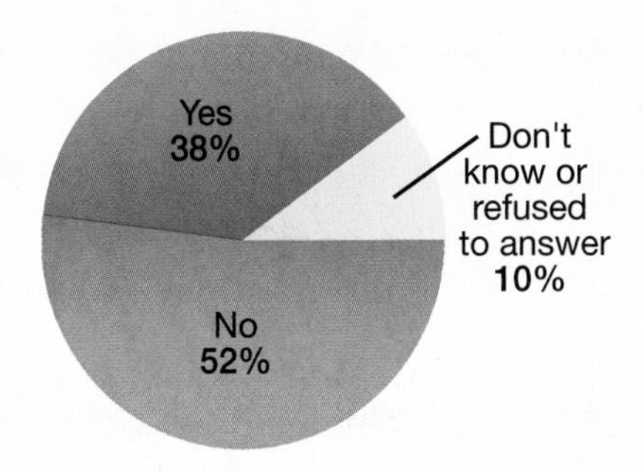

Question 2

Do you think people should or should not be required to pay the same sales tax for purchases made over the Internet as those bought at a local store?

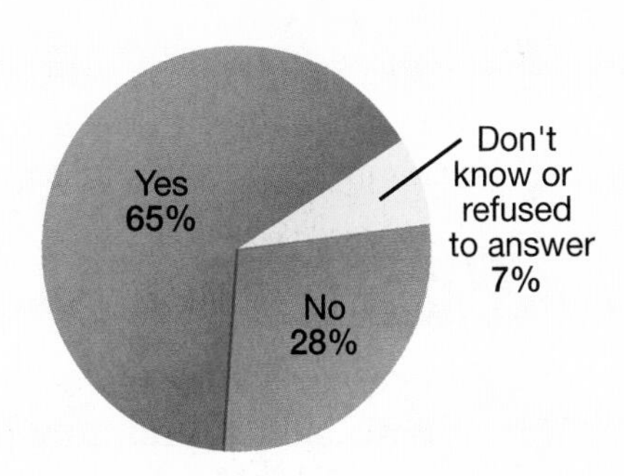

Notice the difference in Questions 1 and 2. Question 2 includes more information. Pointing out that customers pay sales tax for items bought at a local store may give the people answering the survey a reason to answer "yes." Asking the question in that way probably led people to answer the way they did.

Because they are random samples, the results of both of these surveys are accurate. However, the results could be used in a misleading way by someone with an interest in the issue. For example, an Internet retailer would prefer to state the results of Question 1. Be sure to think about survey questions carefully so the results can be interpreted correctly.

Reading to Learn

For Exercises 1–2, tell whether each question is likely to bias the results. Explain your reasoning.

1. On a survey on environmental issues:
   a. "Due to diminishing resources, should a law be made to require recycling?"
   b. "Should the government require citizens to participate in recycling efforts?"

2. On a survey on education:
   a. "Should schools fund extracurricular sports programs?"
   b. "The budget of the River Valley School District is short of funds. Should taxes be raised in order for the district to fund extracurricular sports programs?"

3. Suppose you want to determine whether to serve hamburgers or pizza at the class party.
   a. Write a survey question that would likely produce biased results.
   b. Write a survey question that would likely produce unbiased results.

12-2 Counting Outcomes

Main Ideas

- Count outcomes using a tree diagram.
- Count outcomes using the Fundamental Counting Principle.

New Vocabulary

tree diagram
sample space
event
Fundamental Counting Principle
factorial

GET READY for the Lesson

The Atlantic Coast Conference (ACC) football championship is decided by the number of conference wins. If there is a tie, the team with more nonconference wins is champion. If Florida State plays 3 nonconference games, a tree diagram can be used to show the different records they could have for those games.

Tree Diagrams One method used for counting the number of possible outcomes is to draw a **tree diagram.** The last column of a tree diagram shows all of the possible outcomes. The list of all possible outcomes is called the **sample space**, while any collection of one or more outcomes in the sample space is called an **event.**

EXAMPLE Tree Diagram

1 **A soccer team uses red jerseys for road games, white jerseys for home games, and gray jerseys for practice games. The team uses gray or black pants, and black or white shoes. Use a tree diagram to determine the number of possible uniforms.**

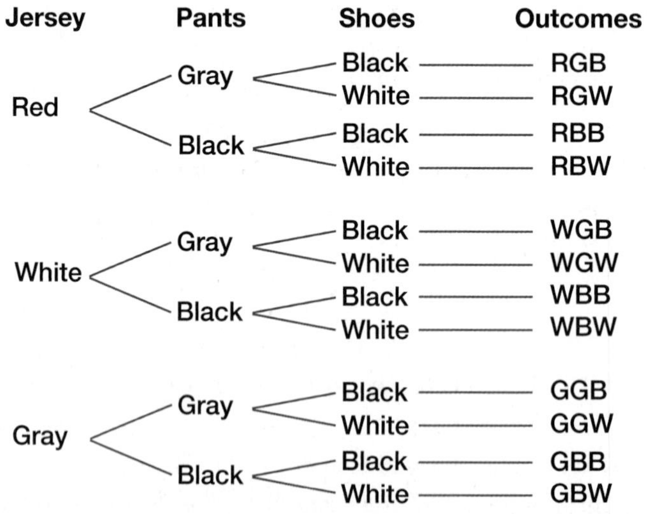

Jersey	Pants	Shoes	Outcomes
Red	Gray	Black	RGB
		White	RGW
	Black	Black	RBB
		White	RBW
White	Gray	Black	WGB
		White	WGW
	Black	Black	WBB
		White	WBW
Gray	Gray	Black	GGB
		White	GGW
	Black	Black	GBB
		White	GBW

The tree diagram shows that there are 12 possible uniforms.

CHECK Your Progress

1. At the cafeteria, you have several options for a sandwich. You can choose either white (W) or wheat (E) bread. You can choose turkey (T), ham (H), or roast beef (R). You can choose mustard (M) or mayonnaise (A). Use a tree diagram to determine the number of possibilities for your sandwich.

The Fundamental Counting Principle The number of possible uniforms in Example 1 can also be found by multiplying the number of choices for each item. If the team can choose from 3 different colored jerseys, 2 different colored pants, and 2 different colored pairs of shoes, there are $3 \cdot 2 \cdot 2$, or 12, possible uniforms. This example illustrates the **Fundamental Counting Principle**.

Study Tip

Fundamental Counting Principle

This rule for counting outcomes can be extended to any number of events.

KEY CONCEPT *Fundamental Counting Principle*

If an event M can occur in m ways and is followed by an event N that can occur in n ways, then the event M followed by event N can occur in $m \cdot n$ ways.

EXAMPLE Fundamental Counting Principle

2 **The Uptown Deli offers a lunch special in which you can choose from 10 different sandwiches, 12 different side dishes, and 7 different beverages. How many different lunch specials can you order?**

Multiply to find the number of lunch specials.

sandwich choices		side dish choices		beverage choices		number of specials
10	$\cdot$	12	$\cdot$	7	$=$	840

CHECK Your Progress

2. When ordering a certain car, there are 7 colors for the exterior, 8 colors for the interior, and 4 choices of interior fabric. How many different possibilities are there for color and fabric when ordering this car?

EXAMPLE Counting Arrangements

3 **Mackenzie is setting up a display of the ten most popular video games from the previous week. If she places the games side-by-side on a shelf, in how many different ways can she arrange them?**

Multiply the number of choices for each position.

- Mackenzie has ten games from which to choose for the first position.
- After choosing a game for the first position, there are nine games left from which to choose for the second position.
- There are now eight choices for the third position.
- This process continues until all positions have been filled.

The number of arrangements is

$n = 10 \cdot 9 \cdot 8 \cdot 7 \cdot 6 \cdot 5 \cdot 4 \cdot 3 \cdot 2 \cdot 1$ or 3,628,800.

There are 3,628,800 different ways to arrange the video games.

CHECK Your Progress

3. Student Council has a president, vice-president, treasurer, secretary, and two representatives from each of the four grades. For the school assembly they were all required to sit in a row up on the stage. In how many different ways can they arrange themselves?

Study Tip

Technology

You can use a TI 83/84 Plus graphing calculator to find 10! by pushing 10 [MATH], scroll to PRB, 4 [ENTER].

The expression $n = 10 \cdot 9 \cdot 8 \cdot 7 \cdot 6 \cdot 5 \cdot 4 \cdot 3 \cdot 2 \cdot 1$ used in Example 3 can be written as 10! using a **factorial**.

KEY CONCEPT *Fundamental Counting Principle*

Words The expression $n!$, read n factorial, where n is greater than zero, is the product of all positive integers beginning with n and counting backward to 1.

Symbols $n! = n \cdot (n - 1) \cdot (n - 2) \cdot \ldots \cdot 3 \cdot 2 \cdot 1$

Example $5! = 5 \cdot 4 \cdot 3 \cdot 2 \cdot 1$ or 120

By definition, $0! = 1$.

EXAMPLE Factorial

4 **Find the value of each expression.**

a. 6!

$6! = 6 \cdot 5 \cdot 4 \cdot 3 \cdot 2 \cdot 1$

$= 720$

b. 10!

$10! = 10 \cdot 9 \cdot 8 \cdot 7 \cdot 6 \cdot 5 \cdot 4 \cdot 3 \cdot 2 \cdot 1$

$= 3{,}628{,}800$

CHECK Your Progress

4A. 5!

4B. 8!

Real-World Link

In 2005, there were 658 roller coasters in the United States.

Type	Number
Wood	121
Steel	537
Inverted	45
Stand Up	9
Suspended	9
Bobsled	4

Source: Roller Coaster Database

EXAMPLE Use Factorials to Solve a Problem

5 **ROLLER COASTERS Zach and Kurt are going to an amusement park. They cannot decide in which order to ride the 12 roller coasters in the park.**

a. In how many different orders can they ride all of the roller coasters if they ride each once?

Use a factorial.

$12! = 12 \cdot 11 \cdot 10 \cdot 9 \cdot 8 \cdot 7 \cdot 6 \cdot 5 \cdot 4 \cdot 3 \cdot 2 \cdot 1$ Definition of factorial

$= 479{,}001{,}600$ Simplify.

b. If they only have time to ride 8 of the roller coasters, how many ways can they do this?

Use the Fundamental Counting Principle to count the sample space.

$s = 12 \cdot 11 \cdot 10 \cdot 9 \cdot 8 \cdot 7 \cdot 6 \cdot 5$ Fundamental Counting Principle

$= 19{,}958{,}400$ Simplify.

CHECK Your Progress

José needs to speak with six college representatives.

5A. In how many different orders can he speak to these people if he only speaks to each person once?

5B. He decides that he will not have time to talk to two of the people. In how many ways can he speak to the others?

 **Personal Tutor at** algebra1.com

Check Your Understanding

Examples 1–3 (pp. 650–651)

For Exercises 1–3, suppose the spinner at the right is spun three times.

1. Draw a tree diagram to show the sample space.
2. How many outcomes involve both green and blue?
3. How many outcomes are possible?

Example 4 (p. 652)

4. Find the value of 8!.

Example 5 (p. 652)

5. **SCHOOL** In a science class, each student must choose a lab project from a list of 15, write a paper on one of 6 topics, and give a presentation about one of 8 subjects. How many ways can students choose to do their assignments?

Exercises

HOMEWORK HELP

For Exercises	See Examples
6, 7, 16	1
8–10	2, 3
11–14	4
15, 17–20	5

Draw a tree diagram to show the sample space for each event. Determine the number of possible outcomes.

6. earning an A, B, or C in English, math, and science classes
7. buying a computer with a choice of a CD-ROM, a CD recorder, or a DVD drive, one of 2 monitors, and either a printer or a scanner

For Exercises 8–10, determine the possible number of outcomes.

8. Three dice, one red, one white, and one blue are rolled. How many outcomes are possible?
9. How many outfits are possible if you choose one each of 5 shirts, 3 pairs of pants, 3 pairs of shoes, and 4 jackets?
10. **TRAVEL** Suppose four different airlines fly from Seattle to Denver. Those same four airlines and two others fly from Denver to St. Louis. In how many ways can a traveler use these airlines to book a flight from Seattle to St. Louis?

Find the value of each expression.

11. 4!
12. 7!
13. 11!
14. 13!

COMMUNICATIONS For Exercises 15 and 16, use the following information.
A new 3-digit area code is needed to accommodate new telephone numbers.

15. If the first digit must be odd, the second digit must be a 0 or a 1, and the third digit can be anything, how many area codes are possible?
16. Draw a tree diagram to show the different area codes using 4 or 5 for the first digit, 0 or 1 for the second digit, and 7, 8, or 9 for the third digit.

SOCCER For Exercises 17–19, use the following information.
The Columbus Crew is playing FC Dallas in a best three out of five championship soccer series.

17. What are the possible outcomes of the series?
18. How many outcomes require only four games be played to determine the champion?
19. How many ways can FC Dallas win the championship?

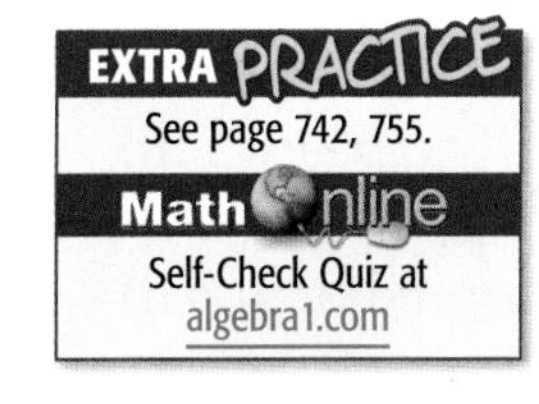

20. **GAMES** William has been dealt seven different cards in a game he is playing. How many different ways are there for him to play his cards if he is required to play one card at a time?

H.O.T. Problems

21. OPEN ENDED Give a real-world example of an event that has $7 \cdot 6$ or 42 outcomes.

22. CHALLENGE To get to and from school, Tucker can walk, ride his bike, or get a ride with a friend. Suppose that one week he walked 60% of the time, rode his bike 20% of the time, and rode with his friend 20% of the time, not necessarily in that order. How many outcomes represent this situation? Assume that he returns home the same way that he went to school.

23. ***Writing in Math*** Refer to the information on page 650 to explain how possible win/loss records can be determined in football. Demonstrate how to find the number of possible outcomes for a team's four home games.

STANDARDIZED TEST PRACTICE

24. A car manufacturer offers a sports car in 4 different models with 6 different option packages. Each model is available in 12 different colors. How many different possibilities are available for this car?

A 48   B 54   C 76   D 288

25. REVIEW Ko collects soup can labels to raise money for his school. He receives \$4 for every 75 labels that he collects. If Ko wants to raise \$96, how many labels does he need?

F 7200   G 1800   H 900   J 24

Spiral Review

PRINTING For Exercises 26–28, use the following information.
To determine the quality of calendars printed at a local shop, the last 10 calendars printed each day are examined. (Lesson 12-1)

26. Identify the sample.

27. Suggest a population from which it was selected.

28. State whether it is *unbiased* (random) or *biased*. If unbiased, classify the sample as *simple, stratified,* or *systematic*. If biased, classify as *convenience* or *voluntary response*.

Solve each equation. (Lesson 11-9)

29. $\frac{-4}{a+1} + \frac{3}{a} = 1$ **30.** $\frac{3}{x} + \frac{4x}{x-3} = 4$ **31.** $\frac{d+3}{d+5} + \frac{2}{d-9} = \frac{5}{2d+10}$

Find each sum. (Lesson 11-7)

32. $\frac{2x+1}{3x-1} + \frac{x+4}{x-2}$ **33.** $\frac{4n}{2n+6} + \frac{3}{n+3}$

GET READY for the Next Lesson

PREREQUISITE SKILL Colored marbles are placed in a bag and selected at random. There are 12 yellow, 15 red, 11 green, 16 blue, and 14 black marbles in the bag. Find each probability.

34. P(green)   **35.** P(black)   **36.** P(yellow or blue)

37. P(green or red)   **38.** P(purple)   **39.** P(yellow or red or green)

12-3 Permutations and Combinations

Main Ideas

- Determine probabilities uing permutations.
- Determine probabilities using combinations.

New Vocabulary

permutation
combination

GET READY for the Lesson

The United States Senate forms various committees by selecting senators from both political parties. The Senate Health, Education, Labor, and Pensions Committee of the 109th Congress was made up of 10 Republican senators and 9 Democratic senators. How many different ways could the committee have been selected? Assume that the members of the committee were selected in no particular order. This is an example of a combination.

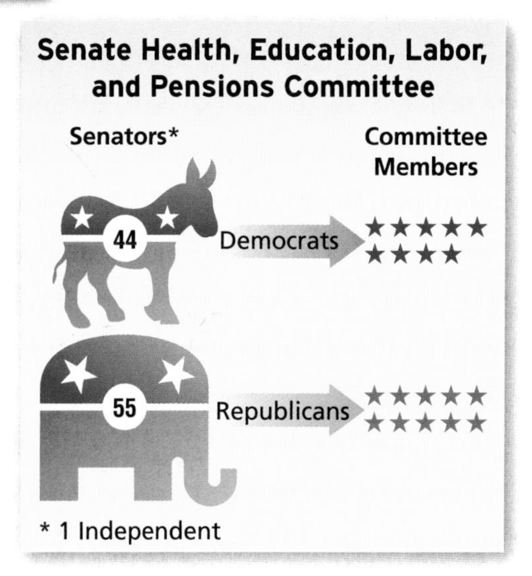

Permutations An arrangement or listing in which order or placement is important is called a **permutation**.

EXAMPLE Tree Diagram Permutation

1 **EMPLOYMENT The manager of a coffee shop needs to hire two employees, one to work at the counter and one to work at the drive-through window. Katie, Bob, and Alicia all applied for the jobs. How many possible ways can the manager place them?**

Use a tree diagram to show the possible arrangements.

Counter	Drive-Through	Outcomes
Katie (K)	Bob	KB
	Alicia	KA
Bob (B)	Katie	BK
	Alicia	BA
Alicia (A)	Katie	AK
	Bob	AB

There are 6 different ways for the 3 applicants to hold the 2 positions.

Study Tip

Common Misconception

When arranging two objects *A* and *B* using a permutation, the arrangement *AB* is different from the arrangement *BA*.

CHECK Your Progress

1. **MUSIC** At Rock City Music Store, customers can purchase CDs, cassettes, and downloads. They can choose from rock, jazz, hip-hop, and gospel. How many possible ways are there for a customer to buy music?

In Example 1, the positions are in a specific order, so each arrangement is unique. The symbol ${}_3P_2$ denotes the number of permutations when arranging 3 applicants in 2 positions. You can also use the Fundamental Counting Principle to determine the number of permutations.

$$\begin{aligned} {}_3P_2 &= \underset{\text{ways to choose first employee}}{3} \cdot \underset{\text{ways to choose second employee}}{2} \\ &= 3 \cdot 2 \cdot \frac{1}{1} && \tfrac{1}{1} = 1 \\ &= \frac{3 \cdot 2 \cdot 1}{2 \cdot 1} && \text{Multiply.} \\ &= \frac{3!}{2!} && 3 \cdot 2 \cdot 1 = 3!,\ 2 \cdot 1 = 2! \end{aligned}$$

In general, ${}_nP_r$ is used to denote the number of permutations of n objects taken r at a time.

KEY CONCEPT — *Permutation*

Word The number of permutations of n objects taken r at a time is the quotient of $n!$ and $(n - r)!$.

Symbols ${}_nP_r = \frac{n!}{(n - r)!}$

EXAMPLE Permutation and Probability

2 **A word processing program requires a user to enter a 7-digit registration code made up of the digits 1, 2, 4, 5, 6, 7, and 9. Each number has to be used, and no number can be used more than once.**

a. How many different registration codes are possible?

Since the order of the numbers in the code is important, this situation is a permutation of 7 digits taken 7 at a time.

${}_nP_r = \frac{n!}{(n - r)!}$ Definition of permutation

${}_7P_7 = \frac{7!}{(7 - 7)!}$ $n = 7, r = 7.$

${}_7P_7 = \frac{7 \cdot 6 \cdot 5 \cdot 4 \cdot 3 \cdot 2 \cdot 1}{1}$ or 5040 Since $0! = 1$, 5040 codes are possible.

Study Tip

Permutations
The number of permutations of n objects taken n at a time is $n!$. ${}_nP_n = n!$

b. What is the probability that the first three digits of the code are even?

Use the Fundamental Counting Principle to determine the number of ways for the first three digits to be even.

- There are three even digits and four odd digits.
- The number of choices for the first three digits, if they are even, is $3 \cdot 2 \cdot 1$.
- The number of choices for the remaining odd digits is $4 \cdot 3 \cdot 2 \cdot 1$.
- The number of favorable outcomes is $3 \cdot 2 \cdot 1 \cdot 4 \cdot 3 \cdot 2 \cdot 1$ or 144.

$$P(\text{first 3 digits even}) = \frac{144}{5040} \quad \begin{matrix} \leftarrow \text{number of favorable outcomes} \\ \leftarrow \text{number of possible outcomes} \end{matrix}$$

$$= \frac{1}{35} \quad \text{Simplify.}$$

The probability that the first three digits are even is $\frac{1}{35}$ or about 3%.

CHECK Your Progress

BICYCLES A combination bike lock requires a three-digit code made up of the digits 0, 1, 2, 3, 4, 5, 6, 7, 8, and 9. No number can be used more than once.

2A. How many different combinations are possible?

2B. What is the probability that all of the digits are odd numbers?

Combinations An arrangement or listing in which order is not important is called a **combination**. For example, if you are choosing 2 salad ingredients from a list of 10, the order in which you choose the ingredients does not matter.

KEY CONCEPT *Combination*

Words The number of combinations of n objects taken r at a time is the quotient of $n!$ and $(n - r)!r!$.

Symbols ${}_nC_r = \frac{n!}{(n - r)!r!}$

Study Tip

Common Misconception

Not all everyday uses of the word *combination* are descriptions of mathematical combinations. For example, the combination to a lock is described by a permutation.

EXAMPLE Combinations and Probability

3 SCHOOL A group of 7 seniors, 5 juniors, and 4 sophomores have volunteered to be peer tutors. Mr. DeLuca needs to choose 12 students out of the group.

a. How many ways can the 12 students be chosen?

The order in which the students are chosen does not matter, so we must find the number of combinations of 16 students taken 12 at a time.

$${}_nC_r = \frac{n!}{(n - r)!r!} \quad \text{Definition of combination}$$

$${}_{16}C_{12} = \frac{16!}{(16 - 12)!12!} \quad n = 16, r = 12$$

$$= \frac{16!}{4!12!} \quad 16 - 12 = 4$$

$$= \frac{16 \cdot 15 \cdot 14 \cdot 13 \cdot \overset{1}{\cancel{12!}}}{4! \cdot \underset{1}{\cancel{12!}}} \quad \text{Divided by the GCF, 12!.}$$

$$= \frac{43{,}680}{24} \text{ or } 1820 \quad \text{Simplify.}$$

There are 1820 ways to choose 12 students out of 16.

(continued on the next page)

b. If the students are chosen randomly, what is the probability that 4 seniors, 4 juniors, and 4 sophomores will be selected?

To find the probability, there are three questions to consider.

- How many ways can 4 seniors be chosen from 7?
- How many ways can 4 juniors be chosen from 5?
- How many ways can 4 sophomores be chosen from 4?

Using the Fundamental Counting Principle, the number of combinations with 4 students from each grade is the product of the three combinations.

ways to choose 4 seniors out of 7		ways to choose 4 juniors out of 5		ways to choose 4 sophomores out of 4
$({}_7C_4)$	$\cdot$	$({}_5C_4)$	$\cdot$	$({}_4C_4)$

Study Tip

Combinations
The number of combinations of n objects taken n at a time is 1. ${}_nC_n = 1$

$({}_7C_4)({}_5C_4)({}_4C_4) = \frac{7!}{(7-4)!4!} \cdot \frac{5!}{(5-4)!4!} \cdot \frac{4!}{(4-4)!4!}$ Definition of combination

$= \frac{7!}{3!4!} \cdot \frac{5!}{1!4!} \cdot \frac{4!}{0!4!}$ Simplify.

$= \frac{7 \cdot 6 \cdot 5}{3!} \cdot \frac{5}{1}$ Divide by the GCF, 4!.

$= 175$ Simplify.

Finally, there are 175 ways to choose this particular combination out of 1820 possible combinations.

$P(\text{4 seniors, 4 juniors, 4 sophomores}) = \frac{175}{1820}$ ← number of favorable outcomes / ← number of possible outcomes

$= \frac{5}{52}$ Simplify.

The probability that Mr. DeLuca will randomly select 4 seniors, 4 juniors, and 4 sophomores is $\frac{5}{52}$ or about 10%.

CHECK Your Progress

PARADE A group of 7 Army veterans, 5 Air Force veterans, 6 Navy veterans, and 4 Marine veterans have volunteered to march in the Memorial Day Parade.

3A. In how many ways can 12 veterans be chosen to march?

3B. If the 12 veterans are chosen randomly, what is the probability that 3 veterans from each branch of the military will be selected?

Online Personal Tutor at algebra1.com

CHECK Your Understanding

Example 1, 3 (pp. 655, 657)

Determine whether each situation involves a *permutation* or *combination*. Explain your reasoning.

1. choosing 6 books from a selection of 12 for summer reading
2. choosing digits for a personal identification number

Example 2 (pp. 656–657)

Evaluate each expression.

3. ${}_8P_5$

4. $({}_{10}P_5)({}_3P_2)$

For Exercises 5–7, use the following information.
The digits 0 through 9 are written on index cards. Three of the cards are randomly selected to form a three-digit code.

5. Does this situation represent a permutation or a combination? Explain.
6. How many different codes are possible?
7. What is the probability that all three digits will be odd?

Example 3
(pp. 657–658)

8. A diner offers a choice of two side items from the list with each entrée. How many ways can two items be selected?

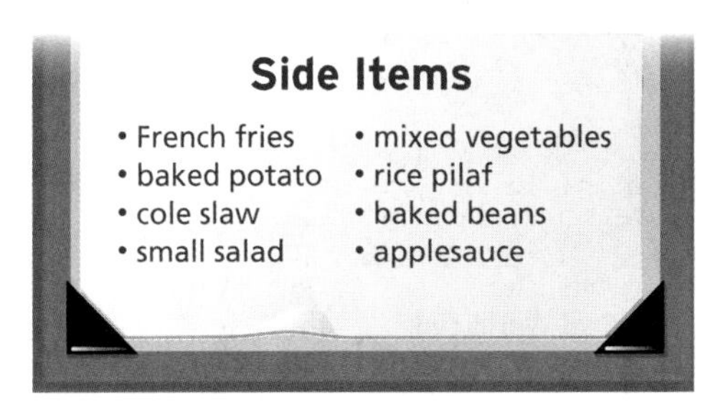

9. PROBABILITY 15 marbles out of 20 must be randomly selected. There are 7 red marbles, 8 purple marbles, and 5 green marbles from which to choose. What is the probability that 5 of each color is selected?

Evaluate each expression.

10. ${}_7C_5$ **11.** $({}_6C_2)({}_4C_3)$

Exercises

HOMEWORK HELP	
For Exercises	See Examples
12–19, 32, 36, 38, 43	1, 3
20–31, 33–35, 37, 39–40, 44–45	2, 3

Determine whether each situation involves a *permutation* or *combination*. Explain your reasoning.

12. team captains for the soccer team
13. three mannequins in a display window
14. a hand of 10 cards from a selection of 52
15. the batting order of the New York Yankees
16. first-place and runner-up winners for the table tennis tournament
17. a selection of 5 DVDs from a group of eight
18. selection of 2 candy bars from six equally-sized bars
19. the selection of 2 trombones, 3 clarinets, and 2 trumpets for a jazz combo

Evaluate each expression.

20. ${}_{12}P_3$ **21.** ${}_4P_1$ **22.** ${}_6C_6$ **23.** ${}_7C_3$
24. ${}_{15}C_3$ **25.** ${}_{20}C_8$ **26.** ${}_{15}P_3$ **27.** ${}_{16}P_5$
28. $({}_7P_7)({}_7P_1)$ **29.** $({}_{20}P_2)({}_{16}P_4)$ **30.** $({}_3C_2)({}_7C_4)$ **31.** $({}_8C_5)({}_5P_5)$

SCHOOL For Exercises 32–35, use the following information.
Mrs. Moyer's class has to choose 4 out of 12 people for an activity committee.

32. Does the selection involve a permutation or a combination? Explain.
33. How many different groups of students could be selected?
34. Suppose the students are selected for the positions of chairperson, activities planner, activity leader, and treasurer. How many different groups of students could be selected?
35. What is the probability that any one of the students is chosen to be the chairperson?

SOFTBALL **For Exercises 36 and 37, use the following information.**
The manager of a softball team needs to prepare a batting lineup using her nine starting players.

36. Is this situation a permutation or a combination?

37. How many different lineups can she make?

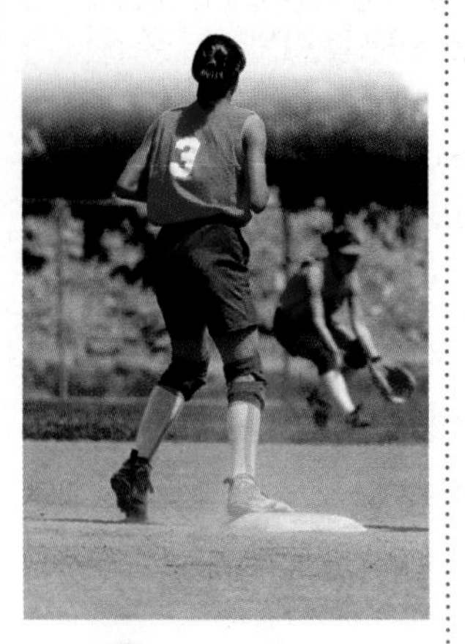

Real-World Link

The game of softball was developed in 1888 as an indoor sport for practicing baseball during the winter months.

Source: encyclopedia.com

GAMES **For Exercises 38–40, use the following information.**
For a certain game, each player rolls five dice at the same time.

38. Do the outcomes of rolling the five dice represent a permutation or a combination? Explain.

39. How many outcomes are possible?

40. What is the probability that all five dice show the same number on a single roll?

BUSINESS **For Exercises 41 and 42, use the following information.**
There are six positions available in the research department of a software company. Of the applicants, 15 are men and 10 are women.

41. In how many ways could 4 men and 2 women be chosen if each were equally qualified?

42. What is the probability that five women would be selected if the positions were randomly filled?

DINING **For Exercises 43–45, use the following information.**
For lunch in the school cafeteria, you can select one item from each category to get the daily combo.

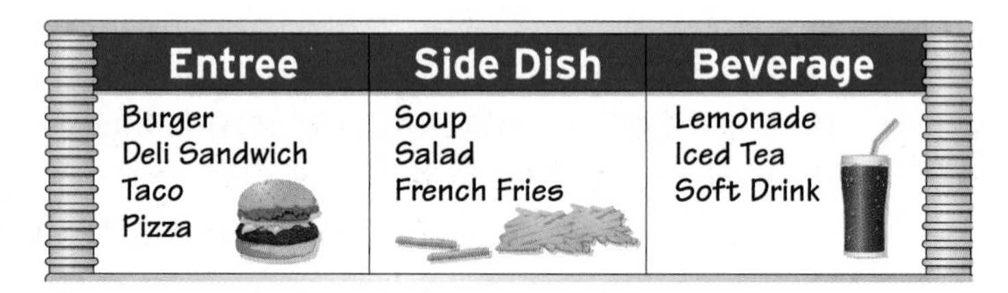

Entree	Side Dish	Beverage
Burger	Soup	Lemonade
Deli Sandwich	Salad	Iced Tea
Taco	French Fries	Soft Drink
Pizza		

43. Find the number of possible meal combinations.

44. If a side dish is chosen at random, what is the probability that a student will choose soup?

45. What is the probability that a student will randomly choose a sandwich and soup?

SWIMMING **For Exercises 46–48, use the following information.**
A swimming coach plans to pick four swimmers out of a group of 6 to form the 400-meter freestyle relay team.

46. How many different teams can he form?

47. The swimmers have been chosen for the relay team. The coach must now decide in which order the four swimmers should swim. He timed the swimmers in each possible order and chose the best time. How many relays did the four swimmers have to swim so that the coach could collect all the data necessary?

48. If Tomás is chosen to be on the team, what is the probability that he will swim in the third leg?

Cross-Curricular Project

Math Online You can use permutations and combinations to analyze data on U.S. amusement parks. Visit algebra1.com

49. BASKETBALL The coach had to select 5 out of 12 players on his basketball team to start the game. How many different groups of players could be selected to start the game?

SPORTS For Exercises 50 and 51, use the following information.
Central High School is competing against West High School at a track meet. Each team entered four girls to run the 1600-meter event. The top three finishers are awarded medals.

50. If there are only the runners from Central and West in this race, how many different ways can the runners place first, second, and third?

51. If all eight runners have an equal chance of placing, what is the probability that the first and second place finishers are from West and the third place finisher is from Central?

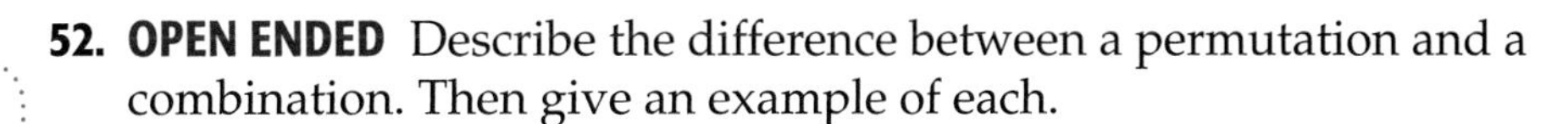

52. OPEN ENDED Describe the difference between a permutation and a combination. Then give an example of each.

53. Which One Doesn't Belong? Determine which situation does not belong. Explain your reasoning.

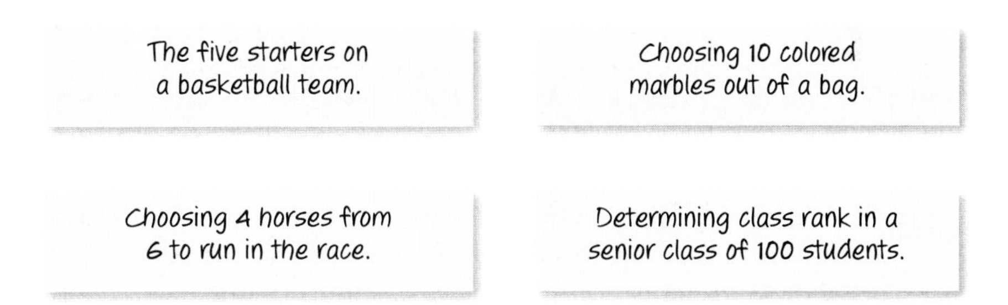

54. FIND THE ERROR Eric and Alisa are taking a trip to Washington, D.C., to visit the Lincoln Memorial, the Jefferson Memorial, the Washington Monument, the White House, the Capitol Building, the Supreme Court, and the Pentagon. Both are finding the number of ways they can choose to visit 5 of these 7 sites. Who is correct? Explain your reasoning.

Eric	Alisa
${}_7C_5 = \frac{7!}{2!}$ $= 2520$	${}_7C_5 = \frac{7!}{2!5!}$ $= 21$

CHALLENGE For Exercises 55 and 56, use the following information.
Larisa is trying to solve a word puzzle. She needs to arrange the letters H, P, S, T, A, E, O into a two-word arrangement.

55. How many different arrangements of the letters can she make?

56. If each arrangement has an equal chance of occurring, what is the probability that she will form the words "tap shoe" on her first try?

57. ***Writing in Math*** Refer to the information on page 655 to explain how combinations can be used to form Senate committees. Discuss why the formation of a Senate committee is a combination and not a permutation. Explain how this would change if the selection of the committee was based on seniority.

STANDARDIZED TEST PRACTICE

58. Julie remembered that the 4 digits of her locker combination were 4, 9, 15, and 22, but not their order. What is the maximum number of attempts Julie could have to make before her locker opens?

A 4

B 16

C 24

D 256

59. REVIEW Jimmy has $23 in a jar at home, and he is saving to buy a $175 video game system. If he can save $15 a week, which equation could be used to determine w, the number of weeks it will take Jimmy to buy the video game system?

F $23 = 175 + 15w$

G $23 = 15(w + 175)$

H $175 = 15w + 23$

J $175 = 23w + 15$

Spiral Review

60. The Sanchez family acts as a host family for a foreign exchange student during each school year. It is equally likely that they will host a girl or a boy. In how many different ways can they host boys and girls over the next four years? (Lesson 12-2)

61. MANUFACTURING Every 15 minutes, a CD player is taken off the assembly line and tested. State whether this sample is *unbiased* (random) or *biased*. If unbiased, classify the sample as *simple*, *stratified*, or *systematic*. If biased, classify as *convenience* or *voluntary response*. (Lesson 12-1)

Simplify each expression. (Lesson 11-2)

62. $\dfrac{x + 3}{x^2 + 6x + 9}$ **63.** $\dfrac{x^2 - 49}{x^2 - 2x - 35}$ **64.** $\dfrac{n^2 - n - 20}{n^2 + 9n + 20}$

Find the distance between each pair of points with the given coordinates. Express answers in simplest radical form and as decimal approximations rounded to the nearest hundredth if necessary. (Lesson 10-5)

65. (12, 20), (16, 34) **66.** (−18, 7), (2, 15) **67.** $(-2, 5), \left(-\frac{1}{2}, 3\right)$

Solve each equation by using the Quadratic Formula. Approximate irrational roots to the nearest hundredth. (Lesson 9-4)

68. $m^2 + 4m + 2 = 0$ **69.** $2s^2 + s - 15 = 0$ **70.** $2n^2 - n = 4$

GET READY for the Next Lesson

PREREQUISITE SKILL Find each sum or difference. (Pages 694–695)

71. $\frac{8}{52} + \frac{4}{52}$ **72.** $\frac{7}{32} + \frac{5}{8}$ **73.** $\frac{5}{15} + \frac{6}{15} - \frac{2}{15}$

74. $\frac{15}{24} + \frac{11}{24} - \frac{3}{4}$ **75.** $\frac{2}{3} + \frac{15}{36} - \frac{1}{4}$ **76.** $\frac{16}{25} + \frac{3}{10} - \frac{1}{4}$

12-4 Probability of Compound Events

Main Ideas

- Find the probability of two independent events or dependent events.
- Find the probability of two mutually exclusive or inclusive events.

New Vocabulary

simple event
compound event
independent events
dependent events
complements
mutually exclusive
inclusive

GET READY for the Lesson

The weather forecast for Saturday calls for rain in Chicago and Los Angeles. By using the probabilities for both cities, we can find other probabilities. What is the probability that it will rain in both cities? only in Chicago? Chicago or Los Angeles?

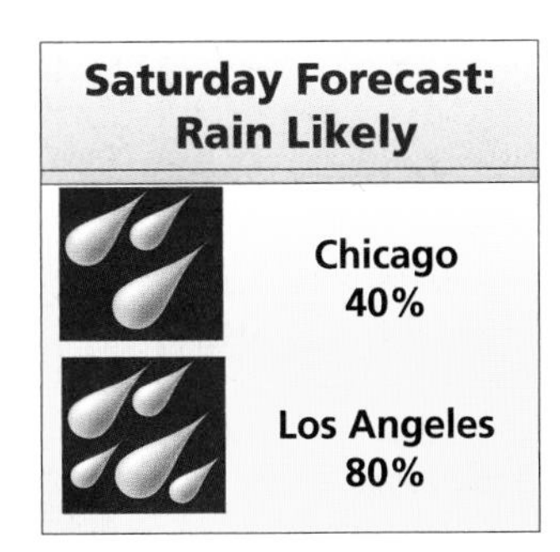

Independent and Dependent Events A single event, like rain in Los Angeles, is called a **simple event**. Suppose you wanted to determine the probability that it will rain in both Chicago and Los Angeles. This is an example of a **compound event**, which is made up of two or more simple events. The weather in Chicago does not affect the weather in Los Angeles. These two events are called **independent events** because the outcome of one event does not affect the outcome of the other.

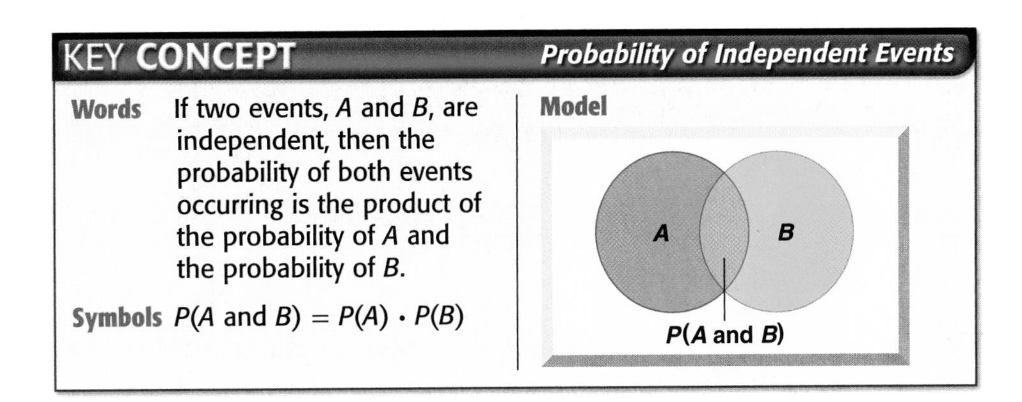

KEY CONCEPT *Probability of Independent Events*

Words If two events, A and B, are independent, then the probability of both events occurring is the product of the probability of A and the probability of B.

Symbols $P(A \text{ and } B) = P(A) \cdot P(B)$

Model

EXAMPLE Independent Events

1 Refer to the application above. Find the probability that it will rain in Chicago and Los Angeles.

$P(\text{Chicago and Los Angeles})$

$= P(\text{Chicago}) \cdot P(\text{Los Angeles})$ Probability of independent events

$= 0.4 \cdot 0.8$ $40\% = 0.4$ and $80\% = 0.8$

$= 0.32$ Multiply.

The probability that it will rain in Chicago and Los Angeles is 32%.

CHECK Your Progress

1. Find the probability of rain in Chicago and no rain in Los Angeles.

When the outcome of one event affects the outcome of another event, the events are **dependent events.** For example, drawing a marble from a bag, not returning it, then drawing a second marble are dependent events because the probability of drawing the second marble depends on what marble was drawn first.

Study Tip

More Than Two Dependent Events

Notice that the formula for the probability of dependent events can be applied to more than two events.

KEY CONCEPT *Probability of Dependent Events*

Words If two events, A and B, are dependent, then the probability of both events occurring is the product of the probability of A and the probability of B after A occurs.

Symbols $P(A \text{ and } B) = P(A) \cdot P(B \text{ following } A)$

EXAMPLE Dependent Events

2 **A bag contains 8 red marbles, 12 blue marbles, 9 yellow marbles, and 11 green marbles. Three marbles are randomly drawn from the bag one at a time and not replaced. Find each probability if the marbles are drawn in the order indicated.**

a. ***P*(red, blue, green)**

The selection of the first marble affects the selection of the next marble since there is one less marble from which to choose. So, the events are dependent.

First marble: $P(\text{red}) = \frac{8}{40} \text{ or } \frac{1}{5}$ ← $\frac{\text{number of red marbles}}{\text{total number of marbles}}$

Second marble: $P(\text{blue}) = \frac{12}{39} \text{ or } \frac{4}{13}$ ← $\frac{\text{number of blue marbles}}{\text{number of marbles remaining}}$

Third marble: $P(\text{green}) = \frac{11}{38}$ ← $\frac{\text{number of green marbles}}{\text{number of marbles remaining}}$

$P(\text{red, blue, green}) = P(\text{red}) \cdot P(\text{blue}) \cdot P(\text{green})$

$= \frac{1}{5} \cdot \frac{4}{13} \cdot \frac{11}{38}$ Substitution

$= \frac{44}{2470} \text{ or } \frac{22}{1235}$ Multiply.

b. ***P*(yellow, yellow, *not* green)**

Notice that after selecting a yellow marble, not only is there one fewer marble from which to choose, there is also one fewer yellow marble. Also, since the marble that is not green is selected after the first two marbles, there are 29 – 2 or 27 marbles that are not green.

$P(\text{yellow, yellow, } not \text{ green}) = P(\text{yellow}) \cdot P(\text{yellow}) \cdot P(\text{not green})$

$= \frac{9}{40} \cdot \frac{8}{39} \cdot \frac{27}{38}$

$= \frac{1944}{59{,}280} \text{ or } \frac{81}{2470}$

CHECK Your Progress

2A. P(red, green, green) **2B.** P(red, blue, *not* yellow)

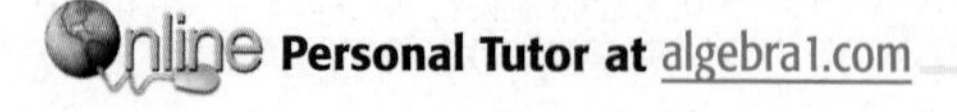

In part b of Example 2, the events for drawing a marble that is green and for drawing a marble that is *not* green are called **complements.** Consider the probabilities for drawing the third marble.

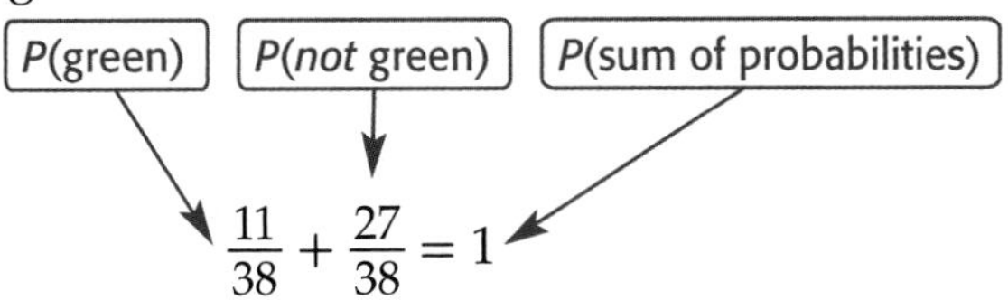

$$\frac{11}{38} + \frac{27}{38} = 1$$

This is always true for any two complementary events.

Vocabulary Link

Complement

Everyday Use Two people are said to complement each other if their strengths and weaknesses balance each other.

Math Use A complement is one of two parts that make up a whole.

Mutually Exclusive and Inclusive Events Events that cannot occur at the same time are called **mutually exclusive.** Suppose you want to find the probability of rolling a 2 *or* a 4 on a die. Since a die cannot show both a 2 and a 4 at the same time, the events are mutually exclusive.

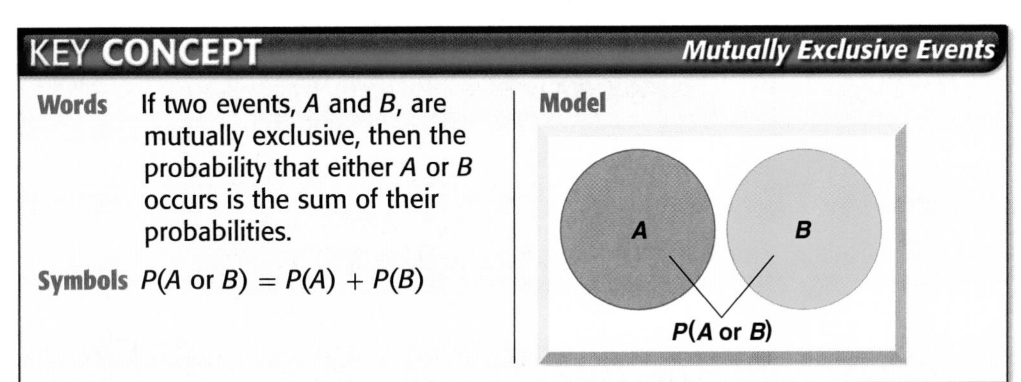

EXAMPLE Mutually Exclusive Events

3 During a magic trick, a magician randomly draws one card from a standard deck of cards. What is the probability that the card drawn is a heart or a diamond?

Since a card cannot be both a heart and a diamond, the events are mutually exclusive.

$P(\text{heart}) = \frac{13}{52}$ or $\frac{1}{4}$ ← $\frac{\text{number of hearts}}{\text{total number of cards}}$

$P(\text{diamond}) = \frac{13}{52}$ or $\frac{1}{4}$ ← $\frac{\text{number of diamonds}}{\text{total number of cards}}$

$P(\text{heart or diamond}) = P(\text{heart}) + P(\text{diamond})$ Definition of mutually exclusive events

$= \frac{1}{4} + \frac{1}{4}$ Substitution

$= \frac{2}{4}$ or $\frac{1}{2}$ Add.

The probability of drawing a heart or a diamond is $\frac{1}{2}$.

CHECK Your Progress

3. What is the probability that the card drawn is an ace or a face card?

Suppose you want to find the probability of randomly selecting an ace or a spade from a standard deck of cards. Since it is possible to draw a card that is both an ace and a spade, these events are not mutually exclusive. They are called **inclusive** events. The following formula allows you to find the probability of inclusive events.

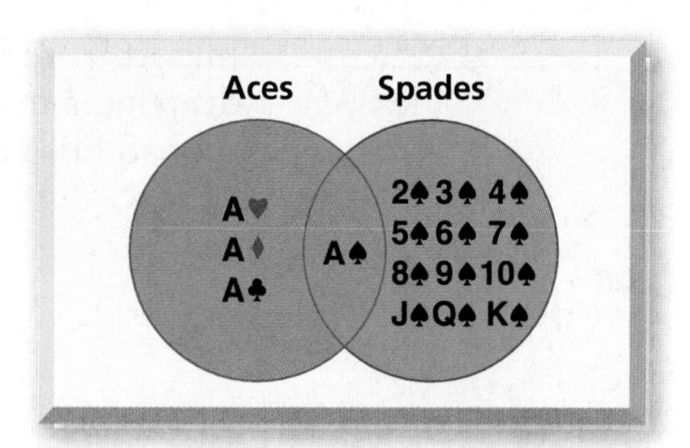

Study Tip

Inclusive Events

If the formula for mutually exclusive events was used to find the probability for inclusive events, events in the intersection would be counted twice.

KEY CONCEPT — *Mutually Inclusive Events*

Words If two events, A and B, are inclusive, then the probability that either A or B occurs is the sum of their probabilities decreased by the probability of both occurring.

Symbols $P(A \text{ or } B) = P(A) + P(B) - P(A \text{ and } B)$

Model

A B

P(A or B)

STANDARDIZED TEST EXAMPLE

4 **GAMES In the game of bingo, balls or tiles are numbered 1 through 75. These numbers correspond to columns on a bingo card, as shown in the table. A number is selected at random. What is the probability that it is a multiple of 4 or is in the O column?**

B	I	N	G	O
1–15	16–30	31–45	46–60	61–75

A $\frac{1}{5}$ **B** $\frac{2}{5}$ **C** $\frac{1}{2}$ **D** $\frac{4}{5}$

Since the numbers 64, 68, and 72 are multiples of 4 and they are also in the O column, these events are inclusive.

P(multiple of 4 or O column)

$$P(\text{multiple of 4}) + P(\text{O column}) - P(\text{multiple of 4 and O column})$$

$$= \frac{18}{75} + \frac{15}{75} - \frac{3}{75} \quad \text{Substitution}$$

$$= \frac{18 + 15 - 3}{75} \quad \text{LCD is 75.}$$

$$= \frac{30}{75} \text{ or } \frac{2}{5} \quad \text{Simplify.}$$

The correct choice is B.

CHECK Your Progress

4. Refer to the table above. What is the probability that a number selected is even or is in the N column?

F $\frac{1}{5}$ **G** $\frac{2}{5}$ **H** $\frac{1}{2}$ **J** $\frac{4}{5}$

Concepts in Motion
BrainPOP®
algebra1.com

CHECK Your Understanding

Example 1 (p. 663)

BUSINESS For Exercises 1–3, use the following information.
Mr. Salyer is a buyer for an electronics store. He received a shipment of 5 hair dryers in which one is defective. He randomly chose 3 of the hair dryers to test.

1. Determine whether choosing the hair dryers are independent or dependent events.

2. What is the probability that he selected the defective dryer?

3. Suppose the defective dryer is one of the three that Mr. Salyer tested. What is the probability that the last one tested was the defective one?

Examples 1, 2 (pp. 663–664)

A bin contains colored chips as shown in the table. Find each probability.

Color	Number
Blue	8
Red	5
Green	6
Yellow	2

4. drawing a red chip, replacing it, then drawing a green chip

5. choosing green, then blue, then red, replacing each chip after it is drawn

6. selecting two yellow chips without replacement

7. choosing green, then blue, then red without replacing each chip

Examples 3, 4 (pp. 665–666)

A student is selected at random from a group of 12 male and 12 female students. There are 3 male students and 3 female students from each of the 9th, 10th, 11th, and 12th grades. Find each probability.

8. P(9th or 12th grader)

9. P(male or female)

10. P(10th grader or female)

11. P(male or not 11th grader)

Example 4 (p. 666)

12. STANDARDIZED TEST PRACTICE At the basketball game, 50% of the fans cheered for the home team. In the same crowd, 20% of the fans were waving banners. What is the probability that a fan cheered for the home team and waved a banner?

A $\frac{1}{20}$ B $\frac{1}{10}$ C $\frac{1}{5}$ D $\frac{2}{5}$

Exercises

HOMEWORK HELP	
For Exercises	See Examples
13–19, 31–32	1
20–27	2
28–30	3
33–34	4

A die is rolled and a spinner like the one at the right is spun. Find each probability.

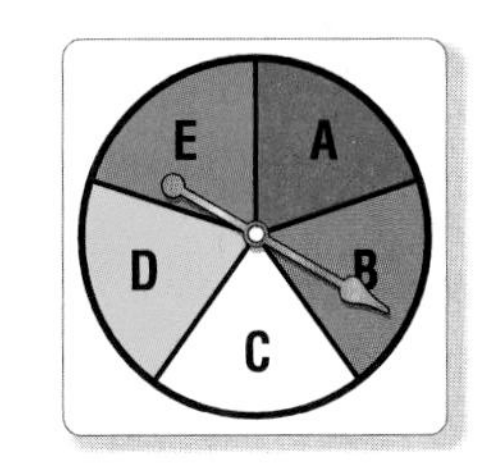

13. P(3 and D)

14. P(an odd number and a vowel)

15. P(a prime number and A)

16. P(2 and A, B, or C)

BIOLOGY For Exercises 17–19, use the diagram and following information.
Each person carries two types of genes for eye color. The gene for brown eyes (B) is dominant over the gene for blue eyes (b). That is, if a person has one gene for brown eyes and the other for blue, that person will have brown eyes. The Punnett square at the right shows the genes for two parents.

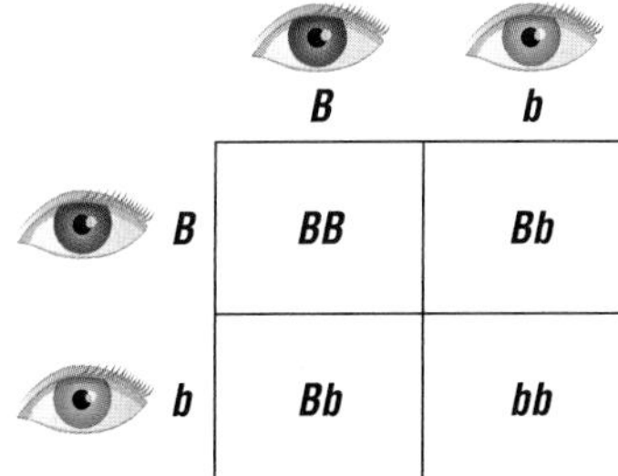

17. What is the probability that any child will have blue eyes?

18. What is the probability that the couple's two children both have brown eyes?

19. Find the probability that the first or the second child has blue eyes.

SAFETY For Exercises 20–23, use the following information.
A carbon monoxide detector system uses two sensors, A and B. If carbon monoxide is present, there is a 96% chance that sensor A will detect it, a 92% chance that sensor B will detect it, and a 90% chance that both sensors will detect it.

20. Draw a Venn diagram that illustrates this situation.

21. If carbon monoxide is present, what is the probability that it will be detected?

22. What is the probability that carbon monoxide would go undetected?

23. Do sensors A and B operate independently of each other? Explain.

A bag contains 2 red, 6 blue, 7 yellow, and 3 orange marbles. Once a marble is selected, it is not replaced. Find each probability.

24. P(2 orange)

25. P(blue, then red)

26. P(2 yellows in a row then orange)

27. P(blue, then yellow, then red)

ECONOMICS For Exercises 28–30, use the table below that compares the total number of hourly workers who earned the minimum wage of $5.15 with those making less than minimum wage.

Number of Hourly Workers (thousands), 2005			
Age (years)	Total	At $5.15	Below $5.15
16–24	16,174	272	750
25+	57,765	249	733

Source: U.S. Bureau of Labor Statistics

Real-World Link

The first federal minimum wage was set in 1938 at $0.25 per hour. That was the equivalent of $3.22 in 2005.

Source: U.S. Department of Labor

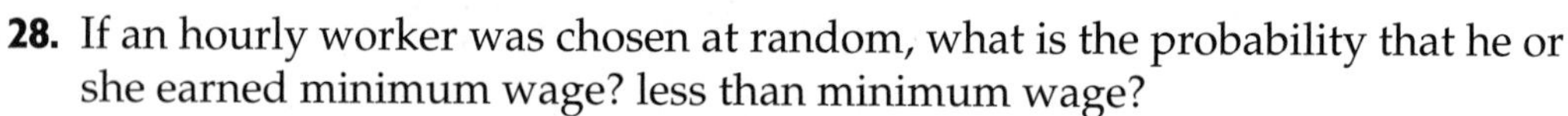

28. If an hourly worker was chosen at random, what is the probability that he or she earned minimum wage? less than minimum wage?

29. What is the probability that a randomly-chosen hourly worker earned less than or equal to minimum wage?

30. If you randomly chose an hourly worker from each age group, which would you expect to have earned no more than minimum wage? Explain.

Raffle tickets numbered 1 through 30 are placed in a box. Tickets for a second raffle numbered 21 to 48 are placed in another box. One ticket is randomly drawn from each box. Find each probability.

31. Both tickets are even.

32. Both tickets are greater than 20 and less than 30.

33. The first ticket is greater than 10, and the second ticket is less than 40 or odd.

34. The first ticket is greater than 12 or prime, and the second ticket is a multiple of 6 or a multiple of 4.

GEOMETRY For Exercises 35–37, use the figure and the following information.
Two of the six angles in the figure are chosen at random.

35. What is the probability of choosing an angle inside $\angle ABC$ or an obtuse angle?

36. What is the probability of selecting a straight angle or a right angle?

37. Find the probability of picking a 20° angle or a 130° angle.

A
30° 50°
C B

38. **RESEARCH** Use the Internet or other reference to investigate various blood types. Use this information to determine the probability of a child having blood type O if the father has blood type A(Ai) and the mother has blood type B(Bi).

A dart is thrown at a dartboard like the one at the right. If the dart can land anywhere on the board, find the probability that it lands in each of the following.

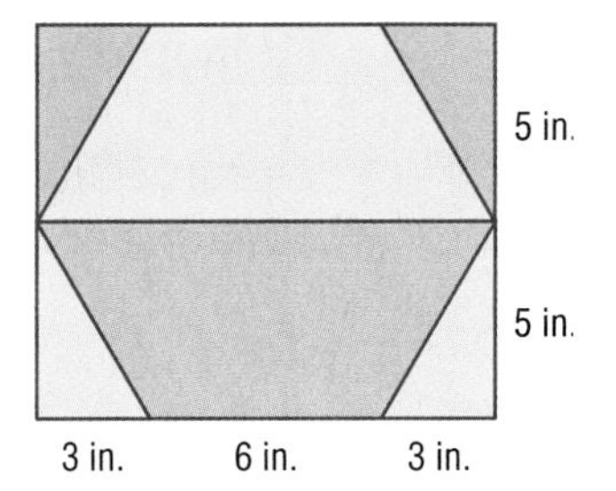

39. a triangle or a red region
40. a trapezoid or a blue region
41. a blue triangle or a red triangle
42. a square or a hexagon

H.O.T. Problems

43. **OPEN ENDED** Explain how dependent events are different from independent events. Give specific examples in your explanation.

44. **FIND THE ERROR** On the school debate team, 6 of the 14 girls are seniors, and 9 of the 20 boys are seniors. Chloe and Amber are both seniors on the team. Each girl calculated the probability that either a girl or a senior would randomly be selected to argue a position at a state debate. Who is correct? Explain your reasoning.

Chloe

P(girl or senior)

$= \frac{14}{34} + \frac{15}{34} - \frac{6}{34}$

$= \frac{23}{34}$

Amber

P(girl or senior)

$= \frac{6}{34} + \frac{15}{34} - \frac{14}{34}$

$= \frac{7}{34}$

45. **REASONING** Find a counterexample for the following statement.
If two events are independent, then the probability of both events occurring is less than 1.

CHALLENGE For Exercises 46–49, use the following information.
A sample of high school students were asked if they
A) drive a car to school,
B) are involved in after-school activities, or
C) have a part-time job.
The results are shown in the Venn diagram.

Event A
Event B
36
38
8
2
25
5
3
Event C
3

46. How many students were surveyed?
47. How many students said that they drive a car to school?
48. If a student is chosen at random, what is the probability that he or she does all three?
49. What is the probability that a randomly chosen student drives a car to school or is involved in after-school activities or has a part-time job?

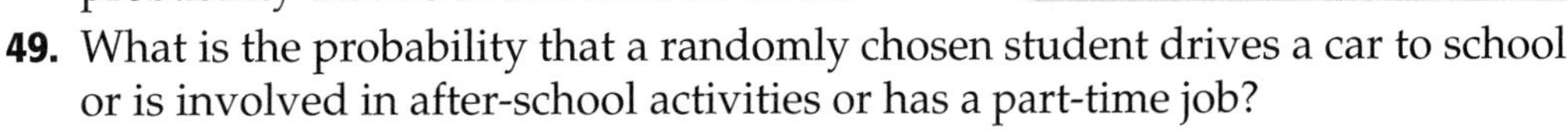

50. ***Writing in Math*** Refer to the information on page 663 to explain how probabilities are used by meteorologists. Illustrate how compound probabilities can be used to predict the weather.

STANDARDIZED TEST PRACTICE

51. A bag contains 8 red marbles, 5 blue marbles, 4 green marbles, and 7 yellow marbles. Five marbles are randomly drawn from the bag one at a time and not replaced. What is the probability that the first three marbles drawn are red?

A $\frac{1}{27}$ C $\frac{7}{253}$

B $\frac{28}{1771}$ D $\frac{7}{288}$

52. REVIEW The sum of a number x and -12 is 64. Which equation shows this relationship?

F $x - 12 = 64$

G $x + 12 = 64$

H $12x = 64$

J $64x = -12$

Spiral Review

CIVICS For Exercises 53 and 54, use the following information.
Stratford City Council wants to form a 3-person parks committee. Five people have applied to be on the committee. (Lesson 12-2)

53. How many committees are possible?

54. What is the probability of any one person being selected if each has an equal chance?

55. BUSINESS A real estate developer built a strip mall with seven different-sized stores. Ten businesses have shown interest in renting space in the mall. The developer must decide which business would be best suited for each store. How many different arrangements are possible? (Lesson 12-1)

Find each quotient. Assume that no denominator has a value of 0. (Lesson 11-4)

56. $\frac{s}{s+7} \div \frac{s-5}{s+7}$

57. $\frac{2m^2 + 7m - 15}{m + 2} \div \frac{2m - 3}{m^2 + 5m + 6}$

Simplify. (Lesson 10-1)

58. $\sqrt{45}$

59. $\sqrt{128}$

60. $\sqrt{40b^4}$

61. $\sqrt{120a^3b}$

62. $3\sqrt{7} \cdot 6\sqrt{2}$

63. $\sqrt{3}(\sqrt{3} + \sqrt{6})$

GET READY for the Next Lesson

PREREQUISITE SKILL Express each fraction as a decimal. Round to the nearest thousandth. (pp. 700–701)

64. $\frac{9}{24}$

65. $\frac{2}{15}$

66. $\frac{63}{128}$

67. $\frac{5}{52}$

68. $\frac{8}{36}$

69. $\frac{11}{38}$

70. $\frac{81}{2470}$

71. $\frac{18}{1235}$

CHAPTER 12

Mid-Chapter Quiz

Lessons 12-1 through 12-4

Identify each sample, suggest a population from which it was selected, and state whether it is *unbiased* (random) or *biased*. If unbiased, classify the sample as *simple*, *stratified*, or *systematic*. If biased, classify as *convenience* or *voluntary response*. (Lesson 12-1)

1. Every other household in a neighborhood is surveyed to determine how to improve the neighborhood park.
2. Every other household in a neighborhood is surveyed to determine the favorite candidate for the state's governor.

Find the number of outcomes for each event. (Lesson 12-2)

3. A die is rolled, and two coins are tossed.
4. A certain model of mountain bike comes in 5 sizes, 4 colors, with regular or off-road tires, and with a choice of 1 of 5 accessories.

5. **MULTIPLE CHOICE** There are seven teams in a league, but only four teams qualify for the post-season tournament. How many ways can the four spaces on the tournament bracket be filled by the teams in the league? (Lesson 12-2)

   A 210

   B 420

   C 840

   D 5040

Find each value. (Lesson 12-3)

6. ${}_{13}C_8$
7. ${}_9P_6$
8. $({}_5C_2)({}_7C_4)$
9. $({}_{10}P_5)({}_{13}P_8)$

10. **SCHOOL** The students in Ms. Kish's homeroom had to choose 4 out of the 7 people who were nominated to serve on the Student Council. How many different groups of students could be selected?

11. **FLOWERS** A vase holds 5 carnations, 6 roses, and 3 lilies. Eliza picks four at random to give to her grandmother. What is the probability of selecting two roses and two lilies? (Lesson 12-3)

12. **MULTIPLE CHOICE** In a standard 52-card deck, what is the probability of randomly drawing an ace or a club? (Lesson 12-4)

    F $\frac{1}{13}$

    G $\frac{4}{13}$

    H $\frac{1}{4}$

    J $\frac{17}{52}$

A ten-sided die, numbered 1 through 10, is rolled. Find each probability. (Lesson 12-4)

13. P(odd or greater than 4)
14. P(less than 3 or greater than 7)

15. **ACTIVITIES** There are 650 students in a high school. The Venn diagram shows the number of students involved in the band and at least one sport. What is the probability a student is randomly selected who participates in one of these extracurricular activities? (Lesson 12-4)

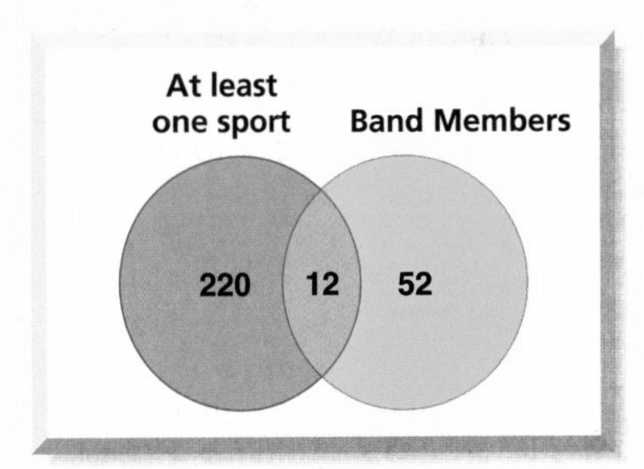

16. **MULTIPLE CHOICE** A teacher took a survey of his class of 28 students about their favorite foods. Thirteen students chose pizza, 7 chose ice cream, 5 chose steak, and 3 chose chicken. If the teacher randomly selects one person's favorite food for a party, what is the probability that he chooses ice cream or steak? (Lesson 12-4)

    A $\frac{2}{7}$

    B $\frac{3}{7}$

    C $\frac{4}{7}$

    D $\frac{5}{7}$

12-5 Probability Distributions

Main Ideas

- Use random variables to compute probability.
- Use probability distributions to solve real-world problems.

New Vocabulary

random variable

discrete random variable

probability distribution

GET READY for the Lesson

The owner of a pet store asked customers how many pets they owned. The results of this survey are shown in the table.

Number of Pets	Number of Customers
0	3
1	37
2	33
3	18
4	9

Random Variables and Probability A **random variable** is a variable with a value that is the numerical outcome of a random event. A **discrete random variable** has a finite number of possible outcomes. In the situation above, we can let the random variable X represent the number of pets owned. Thus, X can equal 0, 1, 2, 3, or 4.

EXAMPLE Random Variable

Refer to the application above.

a. Find the probability that a randomly chosen customer has 2 pets.

There is only one outcome in which there are 2 pets owned, and there are 100 survey results.

$$P(X = 2) = \frac{\text{2 pets owned}}{\text{customers surveyed}}$$

$$= \frac{33}{100}$$

The probability is $\frac{33}{100}$ or 33%.

b. Find the probability that a randomly chosen customer has at least 3 pets.

There are $18 + 9$ or 27 customers who own at least 3 pets.

$$P(X \geq 3) = \frac{27}{100}$$

The probability is $\frac{27}{100}$ or 27%.

Reading Math

Notation The notation $P(X = 2)$ means the same as $P(\text{2 pets})$, the probability of a customer having 2 pets.

CHECK Your Progress

GRADES On an algebra test, there are 7 students with As, 9 students with Bs, 11 students with Cs, 3 students with Ds, and 2 students with Fs.

1A. Find the probability that a randomly chosen student has a C.

1B. Find the probability that a randomly chosen student has at least a B.

Probability Distributions The probability of every possible value of the random variable X is called a **probability distribution.**

> **KEY CONCEPT** *Properties of Probability Distributions*
>
> 1. The probability of each value of X is greater than or equal to 0 and less than or equal to 1.
> 2. The sum of the probabilities of all the values of X is 1.

The probability distribution for a random variable can be given in a table or in a **probability histogram**. A probability distribution table and a probability histogram for the application at the beginning of the lesson are shown below.

Probability Distribution Table

X = Number of Pets	$P(X)$
0	0.03
1	0.37
2	0.33
3	0.18
4	0.09

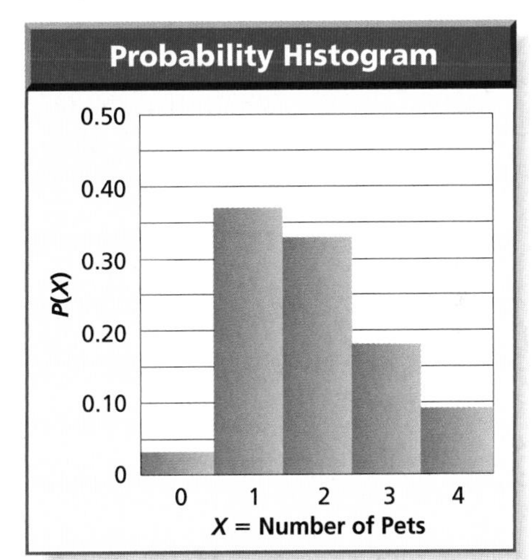

Real-World Link

In 1900, there were 8000 registered cars in the United States. By 2002, there were over 136 million. This is an increase of more than 1,699,900.

Source: *The World Almanac*

EXAMPLE Probability Distribution

2 **CARS The table shows the probability distribution of the number of vehicles per household for the Columbus, Ohio, area.**

Vehicles per Household Columbus, OH

X = Number of Vehicles	Probability
0	0.10
1	0.42
2	0.36
3+	0.12

Source: U.S. Census Bureau

a. Show that the distribution is valid.

Check to see that each property holds.

1. For each value of X, the probability is greater than or equal to 0 and less than or equal to 1.
2. $0.10 + 0.42 + 0.36 + 0.12 = 1$, so the probabilities add up to 1.

b. What is the probability that a household has fewer than 2 vehicles?

Recall that the probability of a compound event is the sum of the probabilities of each individual event.

The probability of a household having fewer than 2 vehicles is the sum of the probability of 0 vehicles and the probability of 1 vehicle.

$P(X < 2) = P(X = 0) + P(X = 1)$ Sum of individual probabilities

$= 0.10 + 0.42 \text{ or } 0.52$ $P(X = 0) = 0.10, P(X = 1) = 0.42$

(continued on the next page)

c. Make a probability histogram of the data.

Draw and label the vertical and horizontal axes. Remember to use equal intervals on each axis. Include a title.

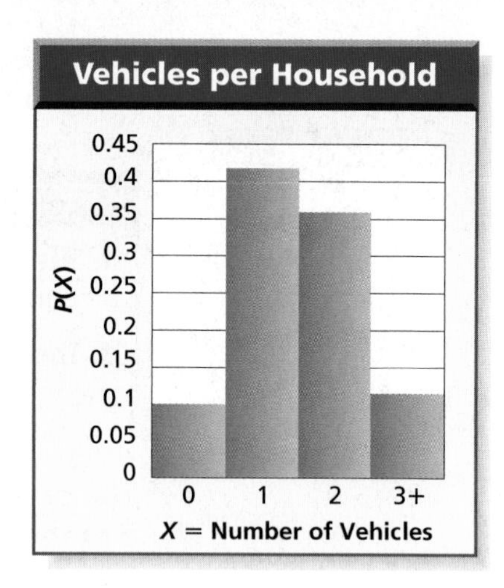

CHECK Your Progress

The table shows the probability distribution of adults who have played golf by age range.

2A. Show that the distribution is valid.

2B. What is the probability that an adult golfer is 35 years or older?

2C. Make a probability histogram of the data.

Golfers by Age	
X = Age	Probability
18–24	0.13
25–34	0.18
35–44	0.21
45–54	0.19
55–64	0.12
65+	0.17

Online **Personal Tutor at** algebra1.com

CHECK Your Understanding

Example 1 (p. 672)

For Exercises 1–3, use the table that shows the possible sums when rolling two dice and the number of ways each sum can be found.

Sum of Two Dice	2	3	4	5	6	7	8	9	10	11	12
Ways to Achieve Sum	1	2	3	4	5	6	5	4	3	2	1

1. Draw a table to show the sample space of all possible outcomes.
2. Find the probabilities for $X = 4$, $X = 5$, and $X = 6$.
3. What is the probability that the sum of two dice is greater than 6 on three separate rolls?

Example 2 (pp. 673–674)

GRADES For Exercises 4–6, use the table that shows a class's grade distribution, where A = 4.0, B = 3.0, C = 2.0, D = 1.0, and F = 0.

X = Grade	0	1.0	2.0	3.0	4.0
Probability	0.05	0.10	0.40	0.40	0.05

4. Show that the probability distribution is valid.
5. What is the probability that a student passes the course?
6. What is the probability that a student chosen at random from the class receives a grade of B or better?

Exercises

HOMEWORK HELP	
For Exercises	See Examples
7–8, 11, 15	1
9–10, 12–14, 16–17	2

For Exercises 7–10, the spinner shown is spun three times.

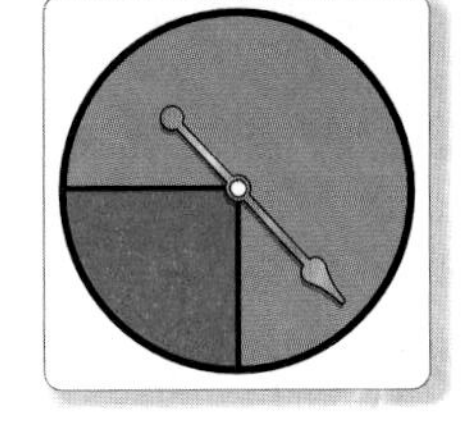

7. Write the sample space with all possible outcomes.
8. Find the probability distribution X, where X represents the number of times the spinner lands on blue for $X = 0$, $X = 1$, $X = 2$, and $X = 3$.
9. Make a probability histogram.
10. Do all possible outcomes have an equal chance of occurring? Explain.

SALES For Exercises 11–14, use the following information.
A music store manager takes an inventory of the top 10 CDs sold each week. After several weeks, the manager has enough information to estimate sales and make a probability distribution table.

Number of Top 10 CDs Sold Each Week	0–100	101–200	201–300	301–400	401–500
Probability	0.10	0.15	0.40	0.25	0.10

11. Define a random variable and list its values.
12. Show that this is a valid probability distribution.
13. In a given week, what is the probability that fewer than 400 CDs sell?
14. In a given week, what is the probability that more than 200 CDs sell?

EDUCATION For Exercises 15–17, use the table, which shows the education level of persons aged 25 and older in the United States.

X = Level of Education	Probability
Some High School	0.154
High School Graduate	0.320
Some College	0.172
Associate's Degree	0.082
Bachelor's Degree	0.179
Advanced Degree	0.093

15. If a person was randomly selected, what is the probability that he or she completed at most some college?
16. Make a probability histogram of the data.
17. Explain how you can find the probability that a randomly selected person has earned at least a bachelor's degree.

SPORTS For Exercises 18 and 19, use the graph that shows the sports most watched by women on TV.

18. Determine whether this is a valid probability distribution. Justify your answer.
19. Based on the graph, in a group of 35 women how many would you expect to say they watch figure skating?

Top 5 Sports Watched by Women

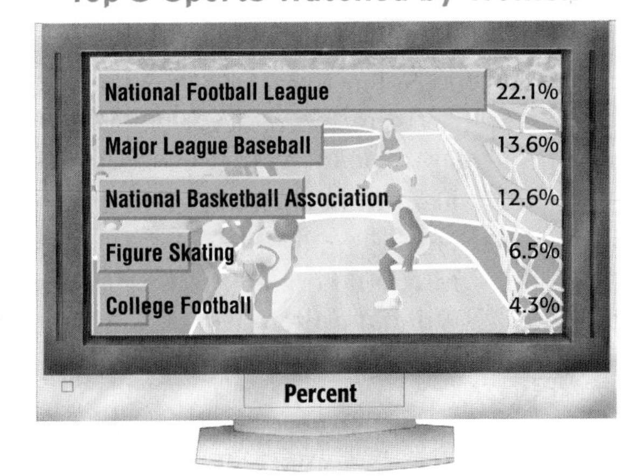

Source: ESPN Sports Poll

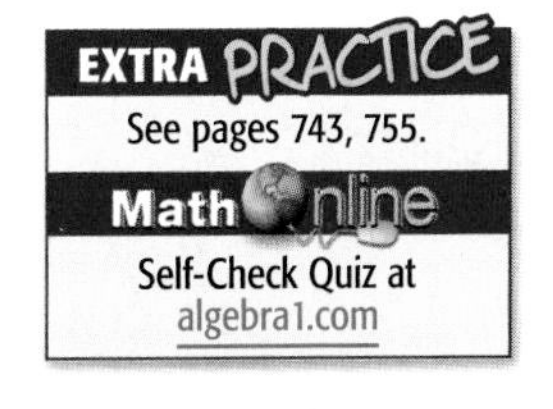

H.O.T. Problems

20. **OPEN ENDED** Describe real-life data that could be displayed in a probability histogram.

21. **CHALLENGE** Suppose a married couple keeps having children until they have a girl. Let the random variable X represent the number of children in their family. Assume that the probability of having a boy or a girl is each $\frac{1}{2}$.

   a. Calculate the probability distribution for $X = 1, 2, 3$, and 4.

   b. Find the probability that the couple will have more than 4 children.

22. ***Writing in Math*** Refer to the information on page 672 to explain how a pet store owner could use a probability distribution. How could the owner create a probability distribution and use it to establish a frequent buyer program?

STANDARDIZED TEST PRACTICE

23. The table shows the probability distribution for the number of heads when four coins are tossed. What is the probability that no more than two heads show on a random toss?

X = Number of Heads	Probability P(X)
0	0.0625
1	0.25
2	0.375
3	0.25
4	0.0625

A 0.3125

B 0.375

C 0.6875

D 0.875

24. **REVIEW** Mr. Perez works 40 hours a week at The Used Car Emporium. He earns $7 an hour and 10% commission on every car he sells. If his hourly wage is increased to $7.50 and his commission to 13%, how much money would he earn in a week if he sold $20,000 worth of cars?

F $1700

G $2200

H $2300

J $2900

Spiral Review

A card is drawn from a standard deck of 52 cards. Find each probability. (Lesson 12-4)

25. P(ace or 10)

26. P(3 or diamond)

27. P(odd number or spade)

Evaluate. (Lesson 12-3)

28. ${}_{10}C_7$

29. ${}_{12}C_5$

30. $({}_6P_3)({}_5P_3)$

SAVINGS For Exercises 31–32, use the following information.
Selena is investing her $900 tax refund in a certificate of deposit that matures in 4 years. The interest rate is 4.25% compounded quarterly. (Lesson 9-6)

31. Determine the balance in the account after 4 years.

32. Her friend Monique invests the same amount of money at the same interest rate, but her bank compounds interest monthly. Determine how much she will have after 4 years.

33. Which type of compounding appears more profitable? Explain.

GET READY for the Next Lesson

PREREQUISITE SKILL Write each fraction as a percent rounded to the nearest whole number. (pages 702–703)

34. $\frac{16}{80}$

35. $\frac{20}{52}$

36. $\frac{30}{114}$

37. $\frac{57}{120}$

38. $\frac{72}{340}$

39. $\frac{54}{162}$

12-6 Probability Simulations

Main Ideas

- Use theoretical and experimental probability to represent and solve problems involving uncertainty.
- Perform probability simulations to model real-world situations involving uncertainty.

New Vocabulary

theoretical probability
experimental probability
relative frequency
empirical study
simulation

GET READY for the Lesson

Researchers at a pharmaceutical company expect a new drug to work successfully in 70% of patients. To test the drug's effectiveness, the company performs three clinical studies of 100 volunteers who use the drug for six months. The results of the studies are shown in the table.

Study Of New Medication

Result	Study 1	Study 2	Study 3
Expected Success Rate	70%	70%	70%
Condition Improved	61%	74%	67%
No Improvement	39%	25%	33%
Condition Worsened	0%	1%	0%

Theoretical and Experimental Probability The probability we have used to describe events in previous lessons is theoretical probability. **Theoretical probabilities** are determined mathematically and describe what should happen. In the situation above, the expected success rate of 70% is a theoretical probability.

A second type of probability is **experimental probability,** which is determined using data from tests or experiments. Experimental probability is the ratio of the number of times an outcome occurred to the total number of events or trials. This ratio is also known as the **relative frequency.**

$$\text{experimental probability} = \frac{\text{frequency of an outcome}}{\text{total number of trials}}$$

EXAMPLE Experimental Probability

1 **MEDICAL RESEARCH Refer to the application at the beginning of the lesson. What is the experimental probability that the drug was successful for a patient in Study 1?**

$$\text{experimental probability} = \frac{61}{100} \quad \begin{matrix} \leftarrow \text{frequency of successes} \\ \leftarrow \text{total number of patients} \end{matrix}$$

The experimental probability of Study 1 is $\frac{61}{100}$ or 61%.

CHECK Your Progress

1. Trevor says he is able to make at least 63% of free throws he takes. To prove this, he decides to take 50 free throws, of which he made 33. Did his experimental probability support his assertion?

It is often useful to perform an experiment repeatedly, collect and combine the data, and analyze the results. This is known as an **empirical study**.

EXAMPLE Empirical Study

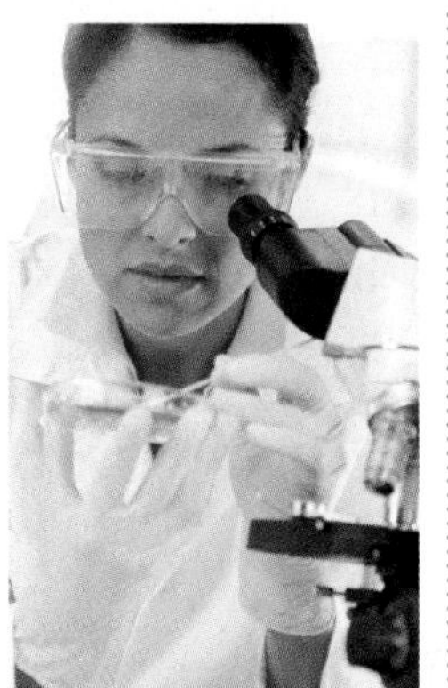

Real-World Career...

Medical Scientist

Many medical scientists conduct research to advance knowledge of living organisms, including viruses and bacteria.

Math Online

For more information, go to algebra1.com.

2 **Refer to the application at the beginning of the lesson. What is the experimental probability of success for all three studies?**

The number of successful outcomes of the three studies was 61 + 74 + 67 or 202 out of the 300 total patients.

$$\text{experimental probability} = \frac{202}{300} \text{ or } \frac{101}{150}$$

The experimental probability of the three studies was $\frac{101}{150}$ or about 67%.

CHECK Your Progress

2. Refer to Check Your Progress 1. Trevor decided to shoot 50 free throws two more times. He makes 29 of the first 50 free throws and 34 of the second 50. What is the experimental probability of all three tests?

Performing Simulations A **simulation** allows you to find an experimental probability by using objects to act out an event that would be difficult or impractical to perform.

Algebra Lab

Simulations

COLLECT THE DATA

- Roll a die 20 times. Record the value on the die after each roll.
- Determine the experimental probability distribution for X, the value on the die.
- Combine your results with the rest of the class to find the experimental probability distribution for X given the new number of trials. *(20 • the number of students in your class)*

ANALYZE THE DATA

1. Find the theoretical probability of rolling a 2.

2. Find the theoretical probability of rolling a 1 or a 6.

3. Find the theoretical probability of rolling a value less than 4.

4. Compare the experimental and theoretical probabilities. Which pair of probabilities was closer to each other: your individual probabilities or your class's probabilities?

5. Suppose each person rolls the die 50 times. Explain how this would affect the experimental probabilities for the class.

6. What can you conclude about the relationship between the number of experiments in a simulation and the experimental probability?

Reading Math

Law of Large Numbers The *Law of Large Numbers* states that as the number of trials increases, the experimental probability gets closer to the theoretical probability.

You can conduct simulations of the outcomes for many problems by using one or more objects such as dice, coins, marbles, or spinners. The objects you choose should have the same number of outcomes as the number of possible outcomes of the problem, and all outcomes should be equally likely.

EXAMPLE Simulation

3 **In one season, Malcolm made 75% of the field goals he attempted.**

a. What could be used to simulate his kicking a field goal? Explain.

You could use a spinner like the one at the right, where 75% of the spinner represents making a field goal.

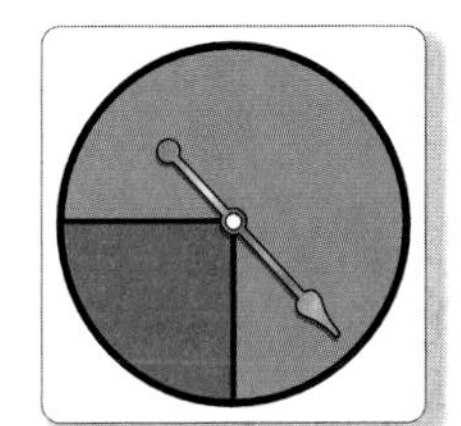

b. Describe a way to simulate his next 8 attempts.

Spin the spinner once to simulate a kick. Record the result, then repeat this 7 more times.

Study Tip

Alternate Simulation

You could also use a 6-sided die to simulate this, where 1, 2, and 3 represent field goals made and 4 is field goals missed. If you roll a 5 or a 6, ignore it and roll again.

CHECK Your Progress

In a trivia game, Becky answered an average of two out of three questions correctly.

3A. What could be used to simulate her correctly answering a question? Explain.

3B. Describe a way to simulate the next 12 questions.

EXAMPLE Theoretical and Experimental Probability

4 **DOGS Ali raises purebred dogs. One of her dogs had a litter of four puppies. What is the most likely mix of male and female puppies? Assume that $P(\text{male}) = P(\text{female}) = \frac{1}{2}$.**

a. What objects can be used to simulate the possible outcomes of the puppies?

Each puppy can be male or female, so there are $2 \cdot 2 \cdot 2 \cdot 2$ or 16 possible outcomes for the litter. Use a simulation that also has 2 outcomes for each of 4 events. One possible simulation would be to toss four coins, one for each puppy, with heads representing female and tails representing male.

b. Find the theoretical probability that there are two female and two male puppies.

There are 16 possible outcomes, and the number of combinations that have two female and two male puppies is ${}_4C_2$ or 6. So the theoretical probability is $\frac{6}{16}$ or $\frac{3}{8}$.

c. The results of a simulation Ali performed are shown in the table at the right. What is the experimental probability that there are three male puppies?

Outcomes	Frequency
4 female, 0 male	3
3 female, 1 male	13
2 female, 2 male	18
1 female, 3 male	12
0 female, 4 male	4

Ali performed 50 trials and 12 of those resulted in three males. So, the experimental probability is $\frac{12}{50}$ or 24%.

(continued on the next page)

d. How does the experimental probability compare to the theoretical probability of a litter with three males?

Theoretical probability:

$P(3 \text{ males}) = \frac{{}_4C_3}{16}$ $\qquad \frac{\text{combinations with 3 male puppies}}{\text{possible outcomes}}$

$= \frac{4}{16}$ or 25% $\qquad$ Simplify.

The experimental probability, 24%, is very close to the theoretical probability.

CHECK Your Progress

QUALITY CONTROL Brandon inspects automobile frames as they come through on the assembly line. On average, he finds a weld defect in one out of ten of the frames each day. He sends these back to correct the defect.

4A. What objects can be used to model the possible outcomes of the automobiles per hour?

4B. What is the theoretical probability that there is one automobile found with defects in a certain hour?

4C. The results of a simulation Brandon performed are shown in the table at the right. What is the experimental probability that there will be one defect found in a certain hour?

4D. How does the experimental probability compare to the theoretical probability of one defect found in a certain hour?

Defects	Frequency
0	14
1	3
2	2
3	1

Online **Personal Tutor at** algebra1.com

CHECK Your Understanding

Example 1 (p. 677)

GAMES Games at the fair require the majority of people who play to lose in order for game owners to make a profit. Therefore, new games need to be tested to make sure they have sufficient difficulty. The results of three test groups are listed in the table. The owners would like a maximum of 33% of players to win the game. There were 50 participants in each test group.

Result	Group 1	Group 2	Group 3
Winners	13	15	19
Losers	37	35	31

1. What is the experimental probability that the participant was a winner in the second group?

Example 2 (p. 678)

2. What is the experimental probability of winning for all three groups?

3. A baseball player has a batting average of .300. That is, he gets a hit 30% of the time he is at bat. What could be used to simulate the player taking a turn at bat?

For Exercises 4–6, roll a die 25 times and record your results.

Example 3 (p. 679)

4. Based on your results, what is the probability of rolling a 3?

5. Based on your results, what is the probability of rolling a 5 or an odd number?

6. Compare your results to the theoretical probabilities.

Example 4 (p. 679)

ASTRONOMY For Exercises 7–10, use the following information.
Enrique is writing a report about meteorites and wants to determine the probability that a meteor reaching Earth's surface hits land. He knows that 70% of Earth's surface is covered by water. He places 7 blue marbles and 3 brown marbles in a bag to represent hitting water $\left(\frac{7}{10}\right)$ and hitting land $\left(\frac{3}{10}\right)$. He draws a marble from the bag, records the color, and then replaces the marble. Enrique drew 56 blue and 19 brown marbles.

7. Did Enrique choose an appropriate simulation? Explain.

8. What is the theoretical probability that a meteorite reaching Earth's surface hits land?

9. Based on his results, what is the probability that a meteorite hits land?

10. Using the experimental probability, how many of the next 500 meteorites that strike Earth would you expect to hit land?

Exercises

HOMEWORK HELP	
For Exercises	See Examples
11–13	1, 2
14–17	3
18–29	4

GOVERNMENT For Exercises 11–13, use the following information.
The Lewiston School Board sent surveys to randomly selected households to determine needs for the school district. The results of the survey are shown.

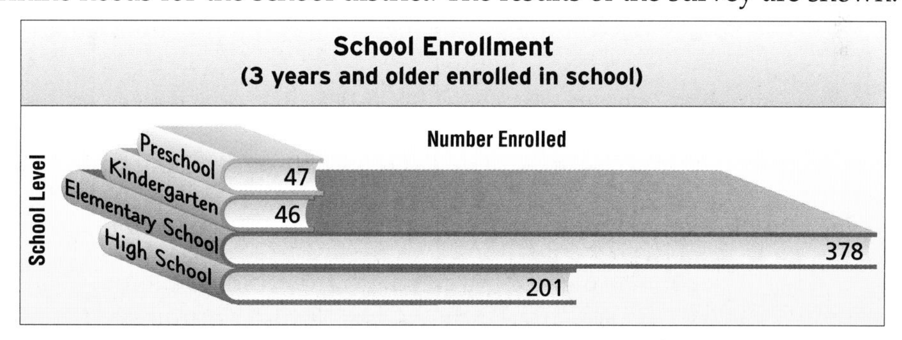

11. Find the experimental probability distribution for the number of people enrolled at each level.

12. Based on the survey, what is the probability that a student chosen at random is in elementary school or high school?

13. Suppose the school district is expecting school enrollment to increase by 1800 over the next 5 years due to new homes in the area. Of the new enrollment, how many will most likely be in kindergarten?

14. What could you use to simulate guessing on 15 true-false questions?

15. There are 12 cans of cola, 8 cans of diet cola, and 4 cans of root beer in a cooler. What could be used for a simulation to determine the probability of randomly picking any one type of soft drink?

For Exercises 16 and 17, use the following information.
A mall randomly gives each shopper one of 12 different gifts during a sale.

16. What could be used to perform a simulation of this situation? Explain.

17. How could you use this simulation to model the next 100 gifts handed out?

For Exercises 18 and 19, toss 3 coins, one at a time, 25 times and record your results. Find each probability based on your results.

18. What is the probability that any two coins will show heads?

19. What is the probability that the first and third coins show tails?

For Exercises 20–22, roll two dice 50 times and record the sums.

20. Based on your results, what is the probability that the sum is 8?
21. Based on your results, what is the probability that the sum is 7, or the sum is greater than 5?
22. If you roll the dice 25 more times, which sum would you expect to see about 10% of the time?

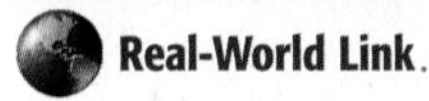

Real-World Link

Labrador retrievers are the most popular breed of dog in the United States.

Source: American Kennel Club

RESTAURANTS For Exercises 23–25, use the following information.
A family restaurant gives away a free toy with each child's meal. There are eight different toys that are randomly given. There is an equally likely chance of getting each toy each time.

23. What objects could be used to perform a simulation of this situation?
24. Conduct a simulation until you have one of each toy. Record your results.
25. Based on your results, how many meals must be purchased so that you get all 8 toys?

ANIMALS For Exercises 26–29, use the following information.
Refer to Example 4 on page 679. Suppose Ali's dog has a litter of 5 puppies.

26. List the possible outcomes of the genders of the puppies.
27. Perform a simulation and list your results in a table.
28. Based on your results, what is the probability that there are 3 females and two males in the litter?
29. What is the experimental probability that the litter has at least three males?

ENTERTAINMENT For Exercises 30–32, use the following information.
A CD changer contains 5 CDs with 14 songs each. When "Random" is selected, each CD is equally likely to be chosen as each song.

30. Use a graphing calculator to perform a simulation of randomly playing 20 songs from the 5 CDs. Record your answer.
KEYSTROKES: MATH ◄ 5 1 , 70 , 20) ENTER
31. Do the experimental probabilities for your simulation support the statement that each CD is equally likely to be chosen? Explain.
32. Based on your results, what is the probability that the first three songs played are on the third disc?

EXTRA PRACTICE
See page 743, 755.
Math Online
Self-Check Quiz at algebra1.com

H.O.T. Problems

33. **OPEN ENDED** Describe a real-life situation that could be represented by a simulation. What objects would you use for this experiment?

34. **CHALLENGE** The captain of a football team believes that the coin the referee uses for the opening coin toss gives an advantage to one team. The referee has players toss the coin 50 times each and record their results. Based on the results, do you think the coin is fair? Explain your reasoning.

Player	1	2	3	4	5	6
Heads	38	31	29	27	26	30
Tails	12	19	21	23	24	20

35. ***Writing in Math*** Refer to the information on page 677 to explain how simulations can be used in health care. Include an explanation of experimental probability and why more trials are better than fewer trials when considering experimental probability.

STANDARDIZED TEST PRACTICE

36. Ramón tossed two coins and rolled a die. What is the probability that he tossed two tails and rolled a 3?

A $\frac{1}{4}$ C $\frac{5}{12}$

B $\frac{1}{6}$ D $\frac{1}{24}$

37. REVIEW Miranda bought a DVD boxed set for $\frac{1}{3}$ off the original price and another 10% off the sale price. If the original cost of the DVD set was $44.99, what price did Miranda pay?

F $29.99 H $27.79

G $28.34 J $26.99

Spiral Review

For Exercises 38–40, use the probability distribution for the random variable X, the number of computers per household. (Lesson 12-5)

Computers per Household	
X = Number of Computers	$P(X)$
0	0.579
1	0.276
2	0.107
3+	0.038

Source: U.S. Dept. of Commerce

38. Show that the probability distribution is valid.

39. If a household is chosen at random, what is the probability that it has at least 2 computers?

40. Determine the probability of randomly selecting a household with no more than one computer.

For Exercises 41–43, use the following information.
A jar contains 18 nickels, 25 dimes, and 12 quarters. Three coins are randomly selected one at a time. Find each probability. (Lesson 12-4)

41. picking three dimes, replacing each after it is drawn

42. a nickel, then a quarter, then a dime without replacing the coins

43. 2 dimes and a quarter, without replacing the coins, if order does not matter

Determine whether the following side measures would form a right triangle. (Lesson 10-4)

44. 5, 7, 9

45. $3\sqrt{34}$, 9, 15

46. 36, 86.4, 93.6

Cross-Curricular Project

Algebra and Physical Science

Building the Best Roller Coaster It is time to complete your project. Use the information and data you have gathered about the building and financing of a roller coaster to prepare a portfolio or Web page. Be sure to include graphs, tables, and/or calculations in the presentation.

Math Online **Cross-Curricular Project at** algebra1.com

CHAPTER 12

Study Guide and Review

Download Vocabulary Review from algebra1.com

FOLDABLES™ Study Organizer GET READY to Study

Be sure the following Key Concepts are noted in your Foldable.

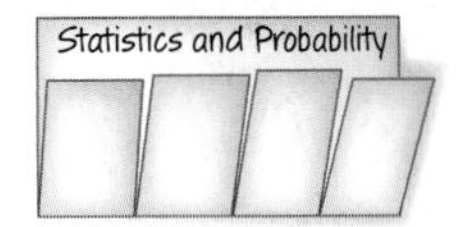

Key Concepts

Sampling and Bias (Lesson 12-1)

- Simple random sample, stratified random sample, and systematic random sample are types of unbiased, or random, samples.
- Convenience sample and voluntary response sample are types of biased samples.

Counting Outcomes, Permutations, and Combinations (Lessons 12-2 and 12-3)

- If an event *M* can occur m ways and is followed by an event *N* that can occur *n* ways, the event *M* followed by event *N* can occur $m \cdot n$ ways.
- In a permutation, the order of objects is important. ${}_nP_r = \frac{n!}{(n-r)!}$
- In a combination, the order of objects is not important. ${}_nC_r = \frac{n!}{(n-r)!r!}$

Probability of Compound Events (Lesson 12-4)

- For independent events, use $P(A \text{ and } B) = P(A) \cdot P(B)$.
- For dependent events, use $P(A \text{ and } B) = P(A) \cdot P(B \text{ following } A)$.
- For mutually exclusive events, use $P(A \text{ or } B) = P(A) + P(B)$.
- For inclusive events, use $P(A \text{ or } B) = P(A) + P(B) - P(A \text{ and } B)$.

Probability Distributions and Simulations (Lessons 12-5 and 12-6)

- For each value of X, $0 \leq P(X) \leq 1$. The sum of the probabilities of each value of X is 1.
- Theoretical probability describes expected outcomes, while experimental probability describes tested outcomes.
- Simulations are used to perform experiments that would be difficult or impossible to perform in real life.

Key Vocabulary

biased sample (p. 643)
combination (p. 657)
complements (p. 665)
compound event (p. 663)
convenience sample (p. 643)
dependent events (p. 664)
discrete random variable (p. 672)
empirical study (p. 678)
event (p. 650)
experimental probability (p. 677)
factorial (p. 652)
inclusive (p. 666)
independent events (p. 663)
population (p. 642)
random sample (p. 642)
sample (p. 642)
simple random sample (p. 642)
stratified random sample (p. 642)
systematic random sample (p. 642)
voluntary response sample (p. 643)

Vocabulary Check

Choose the word or term that best completes each sentence.

1. The arrangement in which order is important is called a (combination, permutation).
2. The notation 10! refers to a (prime factor, factorial).
3. Rolling one die and then another die are (dependent, independent) events.
4. The sum of probabilities of complements equals (0, 1).
5. Randomly drawing a coin from a bag and then drawing another coin are dependent events if the coins (are, are not) replaced.
6. Events that cannot occur at the same time are (mutually exclusive, inclusive).
7. The sum of the probabilities in a probability distribution equals (0, 1).
8. (Theoretical, Experimental) probabilities are precise and predictable.

Lesson-by-Lesson Review

12-1 Sampling and Bias (pp. 642–648)

Identify the sample, suggest a population from which it was selected, and state whether the sample is *unbiased* (random) or *biased*. If unbiased, classify the sample as *simple*, *stratified*, or *systematic*. If biased, classify as *convenience* or *voluntary response*.

9. **SCIENCE** A laboratory technician needs a sample of results of chemical reactions. She selects test tubes from the first 8 experiments performed on Tuesday.

10. **CANDY BARS** To ensure that all of the chocolate bars are the appropriate weight, every 50th bar on the conveyor belt in the candy factory is removed and weighed.

Example 1 GOVERNMENT To determine whether voters support a new trade agreement, 5 people from the list of registered voters in each state and in the District of Columbia are selected at random. Identify the sample, suggest a population from which it was selected, and state whether the sample is *unbiased* (random) or *biased*. If unbiased, classify the sample as *simple*, *stratified*, or *systematic*. If biased, classify as *convenience* or *voluntary response*.

Since $5 \times 51 = 255$, the sample is 255 registered voters in the United States.

The sample is unbiased. It is an example of a stratified random sample.

12-2 Counting Outcomes (pp. 650–654)

Determine the number of outcomes for each event.

11. **MOVIES** Samantha wants to watch 3 videos one rainy afternoon. She has a choice of 3 comedies, 4 dramas, and 3 musicals.

12. **BOOKS** Marquis buys 4 books, one from each category. He can choose from 12 mystery, 8 science fiction, 10 classics, and 5 biographies.

13. **SOCCER** The Jackson Jackals and the Westfield Tigers are going to play a best three-out-of-five games soccer tournament.

Example 2 When Jerri packs her lunch, she can choose to make a turkey (T) or roast beef (R) sandwich on French (F) or sourdough bread (S). She also can pack an apple (A) or an orange (O). Draw a tree diagram to show the number of different ways Jerri can select these items.

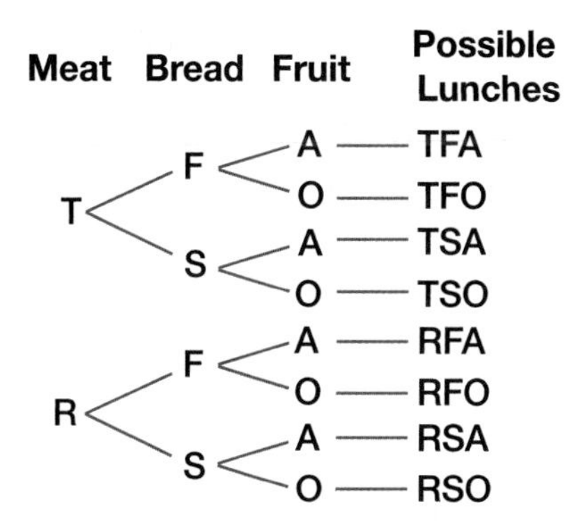

There are 8 different ways for Jerri to select these items.

12-3 Permutations and Combinations (pp. 653–662)

Evaluate each expression.

14. ${}_4P_2$ **15.** ${}_8C_3$

16. ${}_4C_4$ **17.** $({}_7C_1)({}_6C_3)$

18. $({}_7P_3)({}_7P_2)$ **19.** $({}_3C_2)({}_4P_1)$

CLASS PHOTO For Exercises 20 and 21, use the following information.

The French teacher at East High School wants to arrange the 7 students who joined the French club for a yearbook photo.

20. Does this situation involve a permutation or a combination?

21. How many different ways can the 7 students be arranged?

Example 3 **Find ${}_{12}C_8$.**

$${}_{12}C_8 = \frac{12!}{(12-8)!8!}$$
$$= \frac{12!}{4!8!}$$
$$= \frac{12 \cdot 11 \cdot 10 \cdot 9}{4!}$$
$$= 495$$

Example 4 **Find ${}_9P_4$.**

$${}_9P_4 = \frac{9!}{(9-4)!}$$
$$= \frac{9!}{5!}$$
$$= \frac{9 \cdot 8 \cdot 7 \cdot 6 \cdot 5 \cdot 4 \cdot 3 \cdot 2 \cdot 1}{5 \cdot 4 \cdot 3 \cdot 2 \cdot 1}$$
$$= 3024$$

12-4 Probability of Compound Events (pp. 663–670)

A bag of colored paper clips contains 30 red clips, 22 blue clips, and 22 green clips. Find each probability if three clips are drawn randomly from the bag and are not replaced.

22. P(blue, red, green) **23.** P(red, red, blue)

One card is randomly drawn from a standard deck of 52 cards. Find each probability.

24. P(heart or red) **25.** P(10 or spade)

26. BASEBALL Travis Hafner of the Cleveland Indians has a batting average of .391, which means he has gotten a hit 39.1% of the time. Victor Martinez bats directly after Hafner and has a batting average of .375. What are the chances that both men will get hits their first time up to bat?

Example 5 **A box contains 8 red chips, 6 blue chips, and 12 white chips. Three chips are randomly drawn from the box and not replaced. Find P(red, white, blue).**

First chip: $P(\text{red}) = \frac{8}{26}$ $\quad \frac{\text{red chips}}{\text{total chips}}$

Second chip: $P(\text{white}) = \frac{12}{25}$ $\quad \frac{\text{white chips}}{\text{chips remaining}}$

Third chip: $P(\text{blue}) = \frac{6}{24}$ $\quad \frac{\text{blue chips}}{\text{chips remaining}}$

P(red, white, blue)

$$= P(\text{red}) \cdot P(\text{white}) \cdot P(\text{blue})$$
$$= \frac{8}{26} \cdot \frac{12}{25} \cdot \frac{6}{24}$$
$$= \frac{576}{15600} \text{ or } \frac{12}{325}$$

Mixed Problem Solving
For mixed problem-solving practice, see page 755.

12-5 Probability Distributions (pp. 672–676)

ACTIVITIES The table shows the probability distribution for the number of extracurricular activities in which students at Boardwalk High School participate.

27. Show that the probability distribution is valid.

28. If a student is chosen at random, what is the probability that the student participates in 1 to 3 activities?

29. Make a probability histogram of the data.

Extracurricular Activities	
X = Number of Activities	Probability
0	0.04
1	0.12
2	0.37
3	0.30
4+	0.17

Example 6 A local cable provider asked its subscribers how many television sets they had in their homes. The results of their survey are shown in the probability distribution.

Televisions per Household	
X = Number of Televisions	Probability
1	0.18
2	0.36
3	0.34
4	0.08
5+	0.04

a. Show that the probability distribution is valid.

For each value of X, the probability is greater than or equal to 0 and less than or equal to 1. $0.18 + 0.36 + 0.34 + 0.08 + 0.04 = 1$, so the sum of the probabilities is 1.

b. If a household is selected at random, what is the probability that it has fewer than 4 televisions?

$$\begin{aligned} P(X < 4) &= P(X = 1) + P(X = 2) + P(X = 3) \\ &= 0.18 + 0.36 + 0.34 \\ &= 0.88 \end{aligned}$$

The probability that a randomly selected household has fewer than 4 televisions is 88%.

12-6 Probability Simulations (pp. 677–683)

BIOLOGY **While studying flower colors in biology class, students are given the Punnett square below. The Punnett square shows that red parent plant flowers (Rr) produce red flowers (RR and Rr) and pink flowers (rr).**

	R	r
R	RR	Rr
r	Rr	rr

30. If 5 flowers are produced, find the theoretical probability that there will be 4 red flowers and 1 pink flower.

31. Describe items that the students could use to simulate the colors of 5 flowers.

32. The results of a simulation of flowers are shown in the table. What is the experimental probability that there will be 3 red flowers and 2 pink flowers?

Outcomes	Frequency
5 red, 0 pink	15
4 red, 1 pink	30
3 red, 2 pink	23
2 red, 3 pink	7
1 red, 4 pink	4
0 red, 5 pink	1

Example 7 **A group of 3 coins are tossed.**

a. Find the theoretical probability that there will be 2 heads and 1 tail.

Each coin toss can be heads or tails, so there are $2 \cdot 2 \cdot 2$ or 8 possible outcomes. There are 3 possible combinations of 2 heads and one tail, HHT, HTH, or TTH. So, the theoretical probability is $\frac{3}{8}$.

b. The results of a simulation in which three coins are tossed ten times are shown in the table. What is the experimental probability that there will be 2 heads and 1 tail?

Outcomes	Frequency
3 heads, 0 tails	1
2 heads, 1 tail	4
1 head, 2 tails	3
0 heads, 3 tails	2

Of the 10 trials, 4 resulted in 2 heads and 1 tail, so the experimental probability is $\frac{4}{10}$ or 40%.

c. Compare the theoretical probability of 2 heads and 1 tail and the experimental probability of 2 heads and 1 tail.

The theoretical probability is $\frac{3}{8}$ or 37.5%, while the experimental probability is $\frac{4}{10}$ or 40%. The probabilities are close.

CHAPTER 12 Practice Test

Identify the sample, suggest a population from which it was selected, and state whether it is *unbiased* (random) or *biased*. If unbiased, classify the sample as *simple*, *stratified*, or *systematic*. If biased, classify as *convenience* or *voluntary response*.

1. **DOGS** A veterinarian needs a sample of dogs in her kennel to be tested for fleas. She selects the first five dogs who run from the pen.

2. **LIBRARIES** A librarian wants to sample book titles checked out on Wednesday. He randomly chooses a book checked out each hour that the library is open.

There are two roads from Ashville to Bakersville, four roads from Bakersville to Clifton, and two roads from Clifton to Derry.

3. Draw a tree diagram showing the possible routes from Ashville to Derry.
4. How many different routes are there from Ashville to Derry?

Determine whether each situation involves a *permutation* or a *combination*. Then determine the number of possible arrangements.

5. Six students in a class meet in a room that has nine chairs.
6. the top four finishers in a race with ten participants
7. A class has 15 girls and 19 boys. A committee is formed with two girls and two boys, each with a distinct responsibility.

A bag contains 4 red, 6 blue, 4 yellow, and 2 green marbles. Once a marble is selected, it is not replaced. Find each probability.

8. P(blue, green)
9. P(yellow, yellow)
10. P(red, blue, yellow)
11. P(blue, red, not green)

The spinner is spun, and a die is rolled. Find each probability.

12. P(yellow, 4)
13. P(red, even)
14. P(purple or white, not prime)
15. P(green, even or less than 5)

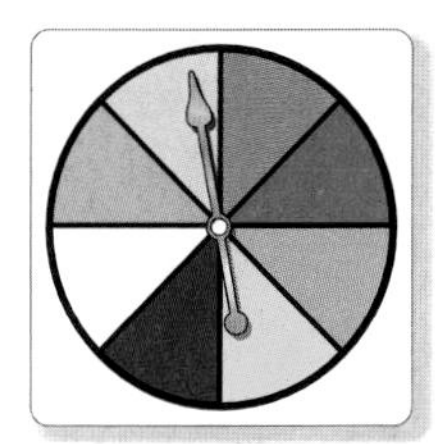

A magician randomly selects a card from a standard deck of 52 cards. Without replacing it, the magician has a member of the audience randomly select a card. Find each probability.

16. P(club, heart)
17. P(jack or queen)
18. P(black 7, diamond)
19. P(queen or red, jack of spades)
20. P(black 10, ace or heart)

The table shows the number of ways the coins can land heads up when four coins are tossed at the same time. Find each probability.

Four Coins Tossed	
Number of Heads	Possible Outcomes
0	1
1	4
2	6
3	4
4	1

21. P(no heads)
22. P(at least two heads)
23. P(two tails)

24. **MULTIPLE CHOICE** Two numbers a and b can be arranged in two different orders: a, b and b, a. In how many ways can three numbers be arranged?

    A 3 **B** 4 **C** 5 **D** 6

25. **MULTIPLE CHOICE** If a coin is tossed three times, what is the probability that the results will be heads exactly one time?

    F $\frac{2}{3}$ **H** $\frac{3}{8}$

    G $\frac{1}{5}$ **J** $\frac{1}{8}$

CHAPTER 12

Standards Practice

Cumulative, Chapters 1–12

Read each question. Then fill in the correct answer on the answer document provided by your teacher or on a sheet of paper.

1. The table shows the results of a survey given to 600 customers at a music store.

Favorite Music	Percent
Jazz	12
Pop	58
Classical	14
Other	16

Based on these data, which of the following statements is true?

A More than half of the customers' favorite music is classical or jazz.

B More customers' favorite is pop music than all other types of music.

C More customers' favorite music is something other than jazz, pop, or classical.

D The number of customers whose favorite music is pop is more than five times the number of customers whose favorite music is jazz.

2. Which equation describes a line that has a y-intercept of -3 and a slope of 6?

F $y = -3x + 6$

G $y = (-3 + x)6$

H $y = (-3x + 1)6$

J $y = 6x - 3$

TEST-TAKING TIP

Question 2 Know the slope-intercept form of linear equations, $y = mx + b$, and understand the definition of slope.

3. GRIDDABLE Hailey is driving her car at a rate of 65 miles per hour. What is her rate in miles per second? Round to the nearest hundredth.

4. The diagram shown below is a scale drawing of Sally's backyard. She is ordering a cover for her swimming pool that costs \$5.00 per square foot.

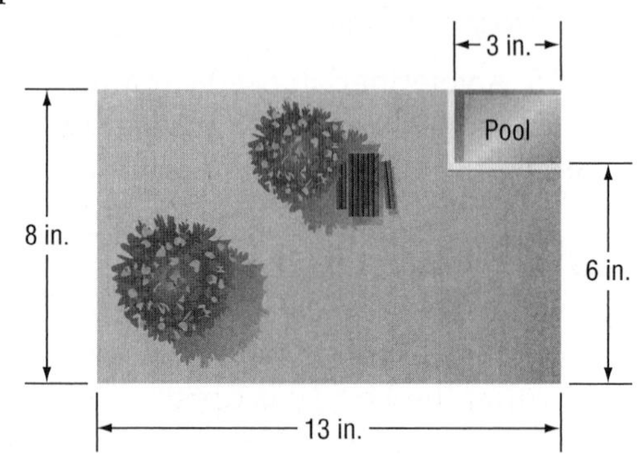

What information must be provided in order to find the total cost of the cover?

A The width of the swimming pool.

B The thickness of the cover.

C The scale of inches to feet on the drawing.

D The amount that Sally has budgeted for the cover.

5. At Marvin's Pizza Place, 30% of the customers order pepperoni pizza. Also, 65% of the customers order a cola to drink. What is the probability that a customer selected at random orders a pepperoni pizza and a cola?

F $\frac{95}{100}$ **H** $\frac{35}{100}$

G $\frac{1}{3}$ **J** $\frac{39}{200}$

6. When graphed, which function would appear to be shifted 3 units down from the graph of $f(x) = x^2 + 2$?

A $f(x) = x^2 + 5$

B $f(x) = x^2$

C $f(x) = x^2 - 3$

D $f(x) = x^2 - 1$

7. GRIDDABLE Laura's Pizza Shop has your choice of 5 meats, 3 cheeses, and 4 vegetables. How many different combinations are there if you choose 1 meat, 1 cheese, and 1 vegetable?

Math Online **Standardized Test Practice at** algebra1.com

8. Carlos rolled a 6 sided die 60 times. The results of his rolls are shown in the table.

Number	Frequency
1	12
2	9
3	14
4	10
5	7
6	8

What is the difference between the theoretical probability and the experimental probability for rolling a 5?

F 5%

G 12%

H 17%

J 30%

9. Miguel rolled a six-sided die 60 times. The results are shown in the table below.

Side	Number of Times Landed
1	12
2	10
3	8
4	15
5	7
6	8

Which number has the same experimental probability as theoretical probability?

A 1

B 2

C 3

D 4

10. Milla sold T-shirts for 10 days in a row for her band fund-raiser. In those 10 days, she sold 120 T-shirts for an average of 12 T-shirts per day. For how many days must she sell 20 T-shirts to bring her average to 18 T-shirts per day?

F 10

G 20

H 30

J 40

11. Maryn is filling up a cylindrical water can. How many times more water could she fit in the can if the radius was doubled?

A 2 times

B 3 times

C 4 times

D 8 times

Pre-AP

Record your answers on a sheet of paper. Show your work.

12. At WackyWorld Pizza, the Random Special is a random selection of two different toppings on a large cheese pizza. The available toppings are pepperoni, sausage, onion, mushrooms, and green peppers.

a. How many different Random Specials are possible? Show how you found your answer.

b. If you order the Random Special, what is the probability that it will have onions?

c. If you order the Random Special, what is the probability that it will have neither onion nor green peppers?

NEED EXTRA HELP?												
If You Missed Question...	1	2	3	4	5	6	7	8	9	10	11	12
Go to Lesson or Page...	12-6	4-3	11-4	2-4	12-4	9-1	12-4	12-6	12-6	11-9	708	12-2 and 12-3

Student Handbook

Built-In Workbooks

Reference

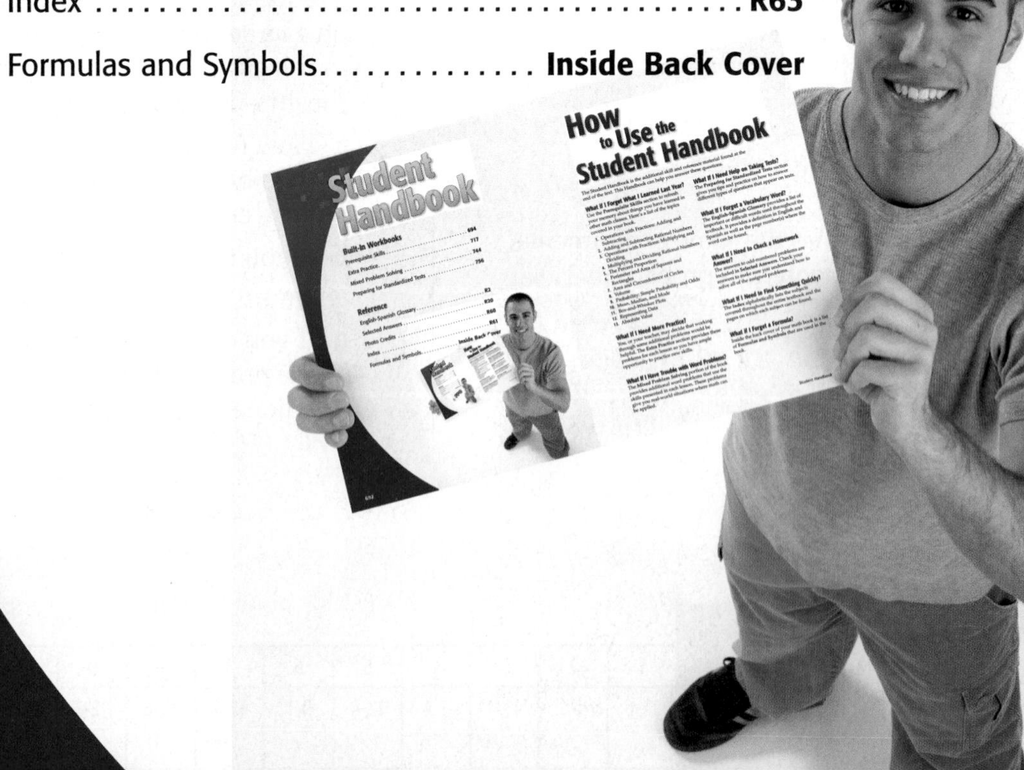

How to Use the Student Handbook

The Student Handbook is the additional skill and reference material found at the end of the text. This Handbook can help you answer these questions.

What if I Forget What I Learned Last Year?

Use the **Prerequisite Skills** section to refresh your memory about things you have learned in other math classes. Here's a list of the topics covered in your book.

1. Operations with Fractions: Adding and Subtracting
2. Adding and Subtracting Rational Numbers
3. Operations with Fractions: Multiplying and Dividing
4. Multiplying and Dividing Rational Numbers
5. The Percent Proportion
6. Perimeter and Area of Squares and Rectangles
7. Area and Circumference of Circles
8. Volume
9. Probability: Simple Probability and Odds
10. Mean, Median, and Mode
11. Box-and-Whisker Plots
12. Representing Data
13. Absolute Value

What If I Need More Practice?

You, or your teacher, may decide that working through some additional problems would be helpful. The **Extra Practice** section provides these problems for each lesson so you have ample opportunity to practice new skills.

What If I Have Trouble with Word Problems?

The **Mixed Problem Solving** portion of the book provides additional word problems that use the skills presented in each lesson. These problems give you real-world situations where math can be applied.

What If I Need Help on Taking Tests?

The **Preparing for Standardized Tests** section gives you tips and practice on how to answer different types of questions that appear on tests.

What If I Forget a Vocabulary Word?

The **English-Spanish Glossary** provides a list of important or difficult words used throughout the textbook. It provides a definition in English and Spanish as well as the page number(s) where the word can be found.

What If I Need to Check a Homework Answer?

The answers to odd-numbered problems are included in **Selected Answers**. Check your answers to make sure you understand how to solve all of the assigned problems.

What If I Need to Find Something Quickly?

The **Index** alphabetically lists the subjects covered throughout the entire textbook and the pages on which each subject can be found.

What if I Forget a Formula?

Inside the back cover of your math book is a list of **Formulas and Symbols** that are used in the book.

Prerequisite Skills

Prerequisite Skills

1 Operations with Fractions: Adding and Subtracting

To add or subtract fractions with the same denominator, add or subtract the numerators and write the sum or difference over the denominator.

EXAMPLE

1 Find each sum or difference.

a. $\frac{3}{5} + \frac{1}{5}$

$\frac{3}{5} + \frac{1}{5} = \frac{3+1}{5}$ The denominators are the same. Add the numerators

$= \frac{4}{5}$ Simplify.

b. $\frac{5}{9} - \frac{4}{9}$

$\frac{5}{9} - \frac{4}{9} = \frac{5-4}{9}$ The denominators are the same. Subtract the numerators.

$= \frac{1}{9}$ Simplify.

To write a fraction in simplest form, divide both the numerator and the denominator by their greatest common factor (GCF).

EXAMPLE

2 Write each fraction in simplest form.

a. $\frac{4}{16}$

$\frac{4}{16} = \frac{4 \div 4}{16 \div 4}$ Divide 4 and 16 by their GCF, 4.

$= \frac{1}{4}$ Simplify.

b. $\frac{24}{36}$

$\frac{24}{36} = \frac{24 \div 12}{36 \div 12}$ Divide 24 and 36 by their GCF, 12.

$= \frac{2}{3}$ Simplify.

EXAMPLE

3 Find each sum or difference. Write in simplest form.

a. $\frac{7}{16} - \frac{1}{16}$

$\frac{7}{16} - \frac{1}{16} = \frac{6}{16}$ The denominators are the same. Subtract the numerators.

$= \frac{3}{8}$ Simplify.

b. $\frac{5}{8} + \frac{7}{8}$

$\frac{5}{8} + \frac{7}{8} = \frac{12}{8}$ The denominators are the same. Add the numerators.

$= 1\frac{4}{8}$ or $1\frac{1}{2}$ Rename $\frac{12}{8}$ as a mixed number in simplest form.

To add or subtract fractions with unlike denominators, first find the least common denominator (LCD). Rename each fraction with the LCD, and then add or subtract. Simplify if necessary.

EXAMPLE

4 **Find each sum or difference. Write in simplest form.**

a. $\frac{2}{9} + \frac{1}{3}$

$\frac{2}{9} + \frac{1}{3} = \frac{2}{9} + \frac{3}{9}$ The LCD for 9 and 3 is 9. Rename $\frac{1}{3}$ as $\frac{3}{9}$.

$= \frac{5}{9}$ Add the numerators.

b. $\frac{1}{2} + \frac{2}{3}$

$\frac{1}{2} + \frac{2}{3} = \frac{3}{6} + \frac{4}{6}$ The LCD for 2 and 3 is 6. Rename $\frac{1}{2}$ as $\frac{3}{6}$ and $\frac{2}{3}$ as $\frac{4}{6}$.

$= \frac{7}{6}$ or $1\frac{1}{6}$ Simplify.

c. $\frac{3}{8} - \frac{1}{3}$

$\frac{3}{8} - \frac{1}{3} = \frac{9}{24} - \frac{8}{24}$ The LCD for 8 and 3 is 24. Rename $\frac{3}{8}$ as $\frac{9}{24}$ and $\frac{1}{3}$ as $\frac{8}{24}$.

$= \frac{1}{24}$ Simplify.

d. $\frac{7}{10} - \frac{2}{15}$

$\frac{7}{10} - \frac{2}{15} = \frac{21}{30} - \frac{4}{30}$ The LCD for 10 and 15 is 30. Rename $\frac{7}{10}$ as $\frac{21}{30}$ and $\frac{2}{15}$ as $\frac{4}{30}$.

$= \frac{17}{30}$ Simplify.

Exercises

Find each sum or difference.

1. $\frac{2}{5} + \frac{1}{5}$
2. $\frac{2}{7} - \frac{1}{7}$
3. $\frac{4}{3} + \frac{4}{3}$
4. $\frac{3}{9} + \frac{4}{9}$
5. $\frac{5}{16} - \frac{4}{16}$
6. $\frac{7}{2} - \frac{4}{2}$

Simplify.

7. $\frac{6}{9}$
8. $\frac{7}{14}$
9. $\frac{28}{40}$
10. $\frac{16}{100}$
11. $\frac{27}{99}$
12. $\frac{24}{180}$

Find each sum or difference. Write in simplest form.

13. $\frac{2}{9} + \frac{1}{9}$
14. $\frac{2}{15} + \frac{7}{15}$
15. $\frac{2}{3} + \frac{1}{3}$
16. $\frac{7}{8} - \frac{3}{8}$
17. $\frac{4}{9} - \frac{1}{9}$
18. $\frac{5}{4} - \frac{3}{4}$
19. $\frac{1}{2} + \frac{1}{4}$
20. $\frac{1}{2} - \frac{1}{3}$
21. $\frac{4}{3} + \frac{5}{9}$
22. $1\frac{1}{2} - \frac{3}{2}$
23. $\frac{1}{4} + \frac{1}{5}$
24. $\frac{2}{3} + \frac{1}{4}$
25. $\frac{3}{2} + \frac{1}{2}$
26. $\frac{8}{9} - \frac{2}{3}$
27. $\frac{3}{7} + \frac{5}{14}$
28. $\frac{13}{20} - \frac{2}{5}$
29. $1 - \frac{1}{19}$
30. $\frac{9}{10} - \frac{3}{5}$
31. $\frac{3}{4} - \frac{2}{3}$
32. $\frac{4}{15} + \frac{3}{4}$
33. $\frac{11}{12} - \frac{4}{15}$
34. $\frac{3}{11} + \frac{1}{8}$
35. $\frac{94}{100} - \frac{11}{25}$
36. $\frac{3}{25} + \frac{5}{6}$

2 Adding and Subtracting Rational Numbers

You can use a number line to add rational numbers.

EXAMPLE

1 Use a number line to find each sum.

a. $-3 + (-4)$

Step 1 Draw an arrow from 0 to -3.

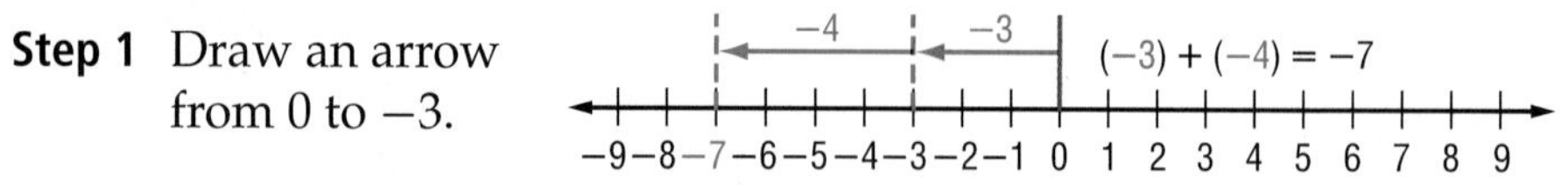

Step 2 Then draw a second arrow 4 units to the left to represent adding -4.

Step 3 The second arrow ends at the sum -7. So, $-3 + (-4) = -7$.

b. $2.5 + (-3.5)$

Step 1 Draw an arrow from 0 to 2.5.

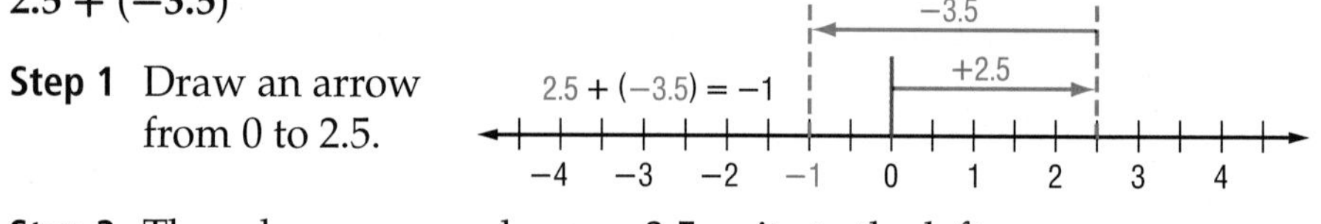

Step 2 Then draw a second arrow 3.5 units to the left.

Step 3 The second arrow ends at the sum -1. So, $2.5 + (-3.5) = -1$.

You can use absolute value to add rational numbers.

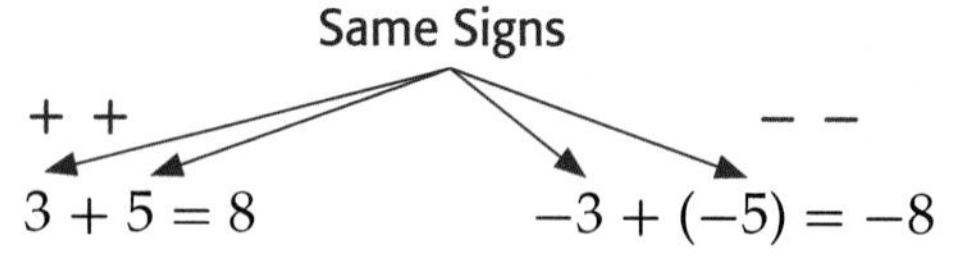

$3 + 5 = 8$

3 and 5 are positive, so the sum is positive.

$-3 + (-5) = -8$

-3 and -5 are negative, so the sum is negative.

Different Signs

+ −

$3 + (-5) = -2$

Since -5 has the greater absolute value, the sum is negative.

− +

$-3 + 5 = 2$

Since 5 has the greater absolute value, the sum is positive.

EXAMPLE

2 Find each sum.

a. $-11 + (-7)$

$$-11 + (-7) = -(|-11| + |-7|)$$ Both numbers are negative, so the sum is negative.

$$= -(11 + 7)$$

$$= -18$$

b. $\frac{7}{16} + \left(-\frac{3}{8}\right)$

$$\frac{7}{16} + \left(-\frac{3}{8}\right) = \frac{7}{16} + \left(-\frac{6}{16}\right)$$ The LCD is 16. Replace $-\frac{3}{8}$ with $-\frac{6}{16}$.

$$= +\left(\left|\frac{7}{16}\right| - \left|-\frac{6}{16}\right|\right)$$ Subtract the absolute values.

$$= +\left(\frac{7}{16} - \frac{6}{16}\right)$$ Since the number with the greater absolute value is $\frac{7}{16}$, the sum is positive.

$$= \frac{1}{16}$$

Every positive rational number can be paired with a negative rational number. These pairs are called **opposites**. A number and its opposite are **additive inverses** of each other. Additive inverses can be used when you subtract rational numbers.

EXAMPLE

3 Find each difference.

a. **18 − 23**

$18 - 23 = 18 + (-23)$	To subtract 23, add its inverse.
$= -(\lvert -23 \rvert - \lvert 18 \rvert)$	Subtract the absolute values.
$= -(23 - 18)$	The absolute value of −23 is greater, so the result is negative.
$= -5$	Simplify.

b. **−32.25 − (−42.5)**

$-32.25 - (-42.5) = -32.25 + 42.5$	To subtract −42.5, add its inverse.
$= \lvert 42.5 \rvert - \lvert -32.25 \rvert$	Subtract the absolute values.
$= 42.5 - 32.25$	The absolute value of 42.5 is greater, so the result is positive.
$= 10.25$	Simplify.

Exercises Find each sum.

1. $-8 + 13$
2. $-11 + 19$
3. $41 + (-63)$
4. $80 + (-102)$
5. $-77 + (-46)$
6. $-92 + (-64)$
7. $-1.6 + (-3.8)$
8. $-32.4 + (-4.5)$
9. $-38.9 + 24.2$
10. $-7.007 + 4.8$
11. $43.2 + (-57.9)$
12. $38.37 + (-61.1)$
13. $\frac{6}{7} + \frac{2}{3}$
14. $\frac{3}{18} + \frac{6}{17}$
15. $-\frac{4}{11} + \frac{3}{5}$
16. $-\frac{2}{5} + \frac{17}{20}$
17. $-\frac{4}{15} + \left(-\frac{9}{16}\right)$
18. $-\frac{16}{40} + \left(-\frac{13}{20}\right)$

Find each difference.

19. $-19 - 8$
20. $16 - (-23)$
21. $9 - (-24)$
22. $12 - 34$
23. $22 - 41$
24. $-9 - (-33)$
25. $-58 - (-42)$
26. $79.3 - (-14.1)$
27. $1.34 - (-0.458)$
28. $-9.16 - 10.17$
29. $67.1 - (-38.2)$
30. $72.5 - (-81.3)$
31. $-\frac{1}{6} - \frac{2}{3}$
32. $\frac{1}{2} - \frac{4}{5}$
33. $-\frac{7}{8} - \left(-\frac{3}{16}\right)$
34. $-\frac{1}{12} - \left(-\frac{3}{4}\right)$
35. $2\frac{1}{4} - 6\frac{1}{3}$
36. $5\frac{3}{10} - 1\frac{31}{50}$

Prerequisite Skills

3 Operations with Fractions: Multiplying and Dividing

To multiply fractions, multiply the numerators and multiply the denominators.

EXAMPLE

1 Find each product.

a. $\frac{2}{5} \cdot \frac{1}{3}$

$\frac{2}{5} \cdot \frac{1}{3} = \frac{2 \cdot 1}{5 \cdot 3}$ Multiply the numerators. Multiply the denominators.

$= \frac{2}{15}$ Simplify.

b. $\frac{7}{3} \cdot \frac{1}{11}$

$\frac{7}{3} \cdot \frac{1}{11} = \frac{7 \cdot 1}{3 \cdot 11}$ Multiply the numerators. Multiply the denominators.

$= \frac{7}{33}$ Simplify.

If the fractions have common factors in the numerators and denominators, you can simplify before you multiply by canceling.

EXAMPLE

2 Find each product. Simplify before multiplying.

a. $\frac{3}{4} \cdot \frac{4}{7}$

$\frac{3}{4} \cdot \frac{4}{7} = \frac{3}{\cancel{4}_1} \cdot \frac{\cancel{4}^1}{7}$ Divide by the GCF, 4.

$= \frac{3}{7}$ Multiply.

b. $\frac{4}{9} \cdot \frac{45}{49}$

$\frac{4}{9} \cdot \frac{45}{49} = \frac{4}{\cancel{9}_1} \cdot \frac{\cancel{45}^5}{49}$ Divide by the GCF, 9.

$= \frac{20}{49}$ Multiply.

Two numbers whose product is 1 are called **multiplicative inverses** or **reciprocals**.

EXAMPLE

3 Name the reciprocal of each number.

a. $\frac{3}{8}$

$\frac{3}{8} \cdot \frac{8}{3} = 1$ The product is 1.

The reciprocal of $\frac{3}{8}$ is $\frac{8}{3}$.

b. $\frac{1}{6}$

$\frac{1}{6} \cdot \frac{6}{1} = 1$ The product is 1.

The reciprocal of $\frac{1}{6}$ is 6.

c. $2\frac{4}{5}$

$2\frac{4}{5} = \frac{14}{5}$ Write $2\frac{4}{5}$ as an improper fraction.

$\frac{14}{5} \cdot \frac{5}{14} = 1$ The product is 1.

The reciprocal of $2\frac{4}{5}$ is $\frac{5}{14}$.

To divide one fraction by another fraction, multiply the dividend by the multiplicative inverse of the divisor.

EXAMPLE

4 Find each quotient.

a. $\frac{1}{3} \div \frac{1}{2}$

$\frac{1}{3} \div \frac{1}{2} = \frac{1}{3} \cdot \frac{2}{1}$ — Multiply $\frac{1}{3}$ by $\frac{2}{1}$, the reciprocal of $\frac{1}{2}$.

$= \frac{2}{3}$ — Simplify.

b. $\frac{3}{8} \div \frac{2}{3}$

$\frac{3}{8} \div \frac{2}{3} = \frac{3}{8} \div \frac{3}{2}$ — Multiply $\frac{3}{8}$ by $\frac{3}{2}$, the reciprocal of $\frac{2}{3}$.

$= \frac{9}{16}$ — Simplify.

c. $4 \div \frac{5}{6}$

$4 \div \frac{5}{6} = \frac{4}{1} \cdot \frac{6}{5}$ — Multiply 4 by $\frac{6}{5}$, the reciprocal of $\frac{5}{6}$.

$= \frac{24}{5}$ or $4\frac{4}{5}$ — Simplify.

d. $\frac{3}{4} \div 2\frac{1}{2}$

$\frac{3}{4} \div 2\frac{1}{2} = \frac{3}{4} \cdot \frac{2}{5}$ — Multiply $\frac{3}{4}$ by $\frac{2}{5}$, the reciprocal of $2\frac{1}{2}$.

$= \frac{6}{20}$ or $\frac{3}{10}$ — Simplify.

Exercises

Find each product.

1. $\frac{3}{4} \cdot \frac{1}{5}$
2. $\frac{2}{7} \cdot \frac{1}{3}$
3. $\frac{1}{5} \cdot \frac{3}{20}$
4. $\frac{2}{5} \cdot \frac{3}{7}$
5. $\frac{5}{2} \cdot \frac{1}{4}$
6. $\frac{7}{2} \cdot \frac{3}{2}$
7. $\frac{1}{3} \cdot \frac{2}{5}$
8. $\frac{2}{3} \cdot \frac{1}{11}$

Find each product. Simplify before multiplying if possible.

9. $\frac{2}{9} \cdot \frac{1}{2}$
10. $\frac{15}{2} \cdot \frac{7}{15}$
11. $\frac{3}{2} \cdot \frac{1}{3}$
12. $\frac{1}{3} \cdot \frac{6}{5}$
13. $\frac{9}{4} \cdot \frac{1}{18}$
14. $\frac{11}{3} \cdot \frac{9}{44}$
15. $\frac{2}{7} \cdot \frac{14}{3}$
16. $\frac{2}{11} \cdot \frac{110}{17}$
17. $\frac{1}{3} \cdot \frac{12}{19}$
18. $\frac{1}{3} \cdot \frac{15}{2}$
19. $\frac{30}{11} \cdot \frac{1}{3}$
20. $\frac{6}{5} \cdot \frac{10}{12}$

Name the reciprocal of each number.

21. $\frac{6}{7}$
22. $\frac{3}{2}$
23. $\frac{1}{22}$
24. $\frac{14}{23}$
25. $2\frac{3}{4}$
26. $5\frac{1}{3}$

Find each quotient.

27. $\frac{2}{3} \div \frac{1}{3}$
28. $\frac{16}{9} \div \frac{4}{9}$
29. $\frac{3}{2} \div \frac{1}{2}$
30. $\frac{3}{7} \div \frac{1}{5}$
31. $\frac{9}{10} \div \frac{3}{7}$
32. $\frac{1}{2} \div \frac{3}{5}$
33. $2\frac{1}{4} \div \frac{1}{2}$
34. $1\frac{1}{3} \div \frac{2}{3}$
35. $\frac{11}{12} \div 1\frac{2}{3}$
36. $\frac{3}{8} \div \frac{1}{4}$
37. $\frac{1}{3} \div 1\frac{1}{5}$
38. $\frac{3}{25} \div \frac{2}{15}$

Prerequisite Skills

4 Multiplying and Dividing Rational Numbers

The product of two numbers having the *same sign* is positive. The product of two numbers having *different signs* is negative.

EXAMPLE

1 Find each product.

a. **4(–5)**

$4(-5) = -20$ different signs → negative product

b. **(–12)(–14)**

$(-12)(-14)$ same signs → positive product

c. $\left(-\frac{3}{4}\right)\left(\frac{3}{8}\right)$

$\left(-\frac{3}{4}\right)\left(\frac{3}{8}\right) = -\frac{9}{32}$ different signs → negative product

You can evaluate expressions that contain rational numbers.

EXAMPLE

2 Evaluate $n^2\left(-\frac{5}{8}\right)$ if $n = -\frac{2}{5}$.

$n^2\left(-\frac{5}{8}\right) = \left(-\frac{2}{5}\right)^2\left(-\frac{5}{8}\right)$ Substitution

$= \left(\frac{4}{25}\right)\left(-\frac{5}{8}\right)$ $\left(-\frac{2}{5}\right)^2 = \left(-\frac{2}{5}\right)\left(-\frac{2}{5}\right)$ or $\frac{4}{25}$

$= -\frac{20}{200}$ or $-\frac{1}{10}$ different signs → negative product

The quotient of two numbers having the *same sign* is positive. The quotient of two numbers having *different signs* is negative.

EXAMPLE

3 Find each quotient.

a. $\frac{-51}{-3}$

$\frac{-51}{-3} = -51 \div (-3)$ Divide.

$= 17$ positive quotient

b. **245.66 ÷ (–14.2)**

$245.66 \div (-14.2) = -17.3$ Use a calculator.
different signs → negative quotient

c. $-\frac{2}{5} \div \frac{1}{4}$

$-\frac{2}{5} \div \frac{1}{4} = -\frac{2}{5} \cdot \frac{4}{1}$ Multiply by $\frac{4}{1}$, the reciprocal of $\frac{1}{4}$.

$= -\frac{8}{5}$ or $-1\frac{3}{5}$ different signs → negative quotient

You can use the Distributive Property to evaluate rational expressions.

EXAMPLE

4 **Evaluate if $\frac{ab}{c^2}$ if $a = -7.8$, $b = 5.2$, and $c = -3$. Round to the nearest hundredth.**

$\frac{ab}{c^2} = \frac{(-7.8)(5.2)}{(-3)^2}$ Replace a with -7.8, b with 5.2, and c with -3.

$= \frac{-40.56}{9}$ Find the numerator and denominator separately.

≈ -4.51 Use a calculator; different signs → negative quotient.

Exercises

Find each product.

1. $5(18)$
2. $8(22)$
3. $-12(15)$
4. $-24(8)$
5. $-47(-29)$
6. $-81(-48)$
7. $\left(\frac{4}{5}\right)\left(\frac{3}{8}\right)$
8. $\left(\frac{5}{12}\right)\left(\frac{4}{9}\right)$
9. $\left(-\frac{3}{5}\right)\left(\frac{5}{6}\right)$
10. $\left(-\frac{2}{5}\right)\left(\frac{6}{7}\right)$
11. $\left(-3\frac{1}{5}\right)\left(-7\frac{1}{2}\right)$
12. $\left(-1\frac{4}{5}\right)\left(-2\frac{1}{2}\right)$
13. $7.2(0.2)$
14. $6.5(0.13)$
15. $(-5.8)(2.3)$
16. $(-0.075)(6.4)$
17. $\frac{3}{5}(-5)(-2)$
18. $\frac{2}{11}(-11)(-4)$

Evaluate each expression if $a = -2.7$, $b = 3.9$, $c = 4.5$ and $d = -0.2$.

19. $-5c^2$
20. $-2b^2$
21. $-4ab$
22. $-5cd$
23. $ad - 8$
24. $ab - 3$
25. $d^2(b - 2a)$
26. $b^2(d - 3c)$

Find each quotient.

27. $-64 \div (-8)$
28. $-78 \div (-4)$
29. $-78 \div (-1.3)$
30. $108 \div (-0.9)$
31. $42.3 \div (-6)$
32. $68.4 \div (-12)$
33. $-23.94 \div 10.5$
34. $-60.97 \div 13.4$
35. $-32.25 \div (-2.5)$
36. $-98.44 \div (-4.6)$
37. $-\frac{1}{3} \div 4$
38. $-\frac{3}{4} \div 12$
39. $-7 \div \frac{3}{5}$
40. $-5 \div \frac{2}{7}$
41. $\frac{16}{36} \div \frac{24}{60}$
42. $-\frac{24}{56} \div \frac{31}{63}$
43. $\frac{14}{32} \div \left(-\frac{12}{15}\right)$
44. $\frac{80}{25} \div \left(-\frac{2}{3}\right)$

Evaluate each expression if $m = -8$, $n = 6.5$, $p = 3.2$, and $q = -5.4$. Round to the nearest hundredth.

45. $\frac{mn}{p}$
46. $\frac{np}{m}$
47. $mq \div np$
48. $pq \div mn$
49. $\frac{n + p}{m}$
50. $\frac{m + p}{q}$
51. $\frac{m - 2n}{-n + q}$
52. $\frac{p - 3q}{-q - m}$

5 The Percent Proportion

A **percent** is a ratio that compares a number to 100. To write a percent as a fraction, express the ratio as a fraction with a denominator of 100. Fractions should be stated in simplest form.

EXAMPLE

1 Express each percent as a fraction.

a. 25%

$25\% = \frac{25}{100}$ or $\frac{1}{4}$ Definition of percent

b. 107%

$107\% = \frac{107}{100}$ or $1\frac{7}{100}$ Definition of percent

c. 0.5%

$0.5\% = \frac{0.5}{100}$ Definition of percent

$= \frac{5}{1000}$ or $\frac{1}{200}$ Simplify.

In the **percent proportion**, the ratio of a part of something (part) to the whole (base) is equal to the percent written as a fraction.

part → $\frac{a}{b} = \frac{p}{100}$ ← percent
base →

Example: 10 (part) is 25% (percent) of 40 (base).

EXAMPLE

2 40% of 30 is what number?

$\frac{a}{b} = \frac{p}{100}$ The percent is 40, and the base is 30. Let a represent the part.

$\frac{a}{30} = \frac{40}{100}$ Replace b with 30 and p with 40.

$100a = 30(40)$ Find the cross products.

$100a = 1200$ Simplify.

$\frac{100a}{100} = \frac{1200}{100}$ Divide each side by 100.

$a = 12$ The part is 12. So, 40% of 30 is 12.

EXAMPLE

3 Kelsey took a survey of some of the students in her lunch period. 42 out of the 70 students Kelsey surveyed said their family had a pet. What percent of the students had pets?

$\frac{a}{b} = \frac{p}{100}$ The part is 42, and the base is 70. Let p represent the percent.

$\frac{42}{70} = \frac{p}{100}$ Replace a with 42 and b with 70.

$4200 = 70p$ Find the cross products.

$\frac{4200}{70} = \frac{70p}{70}$ Divide each side by 70.

$60 = p$ The percent is 60, so $\frac{60}{100}$ or 60% of the students had pets.

EXAMPLE

4 **67.5 is 75% of what number?**

$\frac{a}{b} = \frac{p}{100}$	The percent is 75, and the part is 67.5. Let b represent the base.
$\frac{67.5}{b} = \frac{75}{100}$	$75\% = \frac{75}{100}$, so $p = 75$. Replace a with 67.5 and p with 75.
$6750 = 75b$	Find the cross products.
$\frac{6750}{75} = \frac{75b}{75}$	Divide each side by 75.
$90 = b$	The base is 90, so 67.5 is 75% of 90.

Exercises

Express each percent as a fraction.

1. 5%
2. 60%
3. 11%
4. 120%
5. 78%
6. 2.5%
7. 0.9%
8. 0.4%
9. 1400%

Use the percent proportion to find each number.

10. 25 is what percent of 125?
11. 16 is what percent of 40?
12. 14 is 20% of what number?
13. 50% of what number is 80?
14. What number is 25% of 18?
15. Find 10% of 95.
16. What percent of 48 is 30?
17. What number is 150% of 32?
18. 5% of what number is 3.5?
19. 1 is what percent of 400?
20. Find 0.5% of 250.
21. 49 is 200% of what number?
22. 15 is what percent of 12?
23. 48 is what percent of 32?
24. Madeline usually makes 85% of her shots in basketball. If she shoots 20 shots, how many will she likely make?
25. Brian answered 36 items correctly on a 40-item test. What percent did he answer correctly?
26. José told his dad that he won 80% of the solitaire games he played yesterday. If he won 4 games, how many games did he play?
27. A glucose solution is prepared by dissolving 6 milliliters of glucose in 120 milliliters of solution. What is the percent of glucose in the solution?

HEALTH **For Exercises 28–30, use the following information.** The U.S. Food and Drug Administration requires food manufacturers to label their products with a nutritional label. The sample label shown at the right shows a portion of the information from a package of macaroni and cheese.

Nutrition Facts	
Serving Size 1 cup (228g)	
Servings per container 2	
Amount per serving	
Calories 250 Calories from Fat 110	
	%Daily value*
Total Fat 12g	**18%**
Saturated Fat 3g	**15%**
Cholesterol 30mg	**10%**
Sodium 470mg	**20%**
Total Carbohydrate 31g	**10%**
Dietary Fiber 0g	**0%**
Sugars 5g	
Protein 5g	
Vitamin A 4% • Vitamin C	2%
Calcium 20% • Iron	4%

28. The label states that a serving contains 3 grams of saturated fat, which is 15% of the daily value recommended for a 2000-Calorie diet. How many grams of saturated fat are recommended for a 2000-Calorie diet?
29. The 470 milligrams of sodium (salt) in the macaroni and cheese is 20% of the recommended daily value. What is the recommended daily value of sodium?
30. For a healthy diet, the National Research Council recommends that no more than 30 percent of total Calories come from fat. What percent of the Calories in a serving of this macaroni and cheese come from fat?

6 Perimeter and Area of Squares and Rectangles

Perimeter is the distance around a geometric figure. Perimeter is measured in linear units.

- To find the perimeter of a rectangle, multiply two times the sum of the length and width, or $2(\ell + w)$.
- To find the perimeter of a square, multiply four times the length of a side, or $4s$.

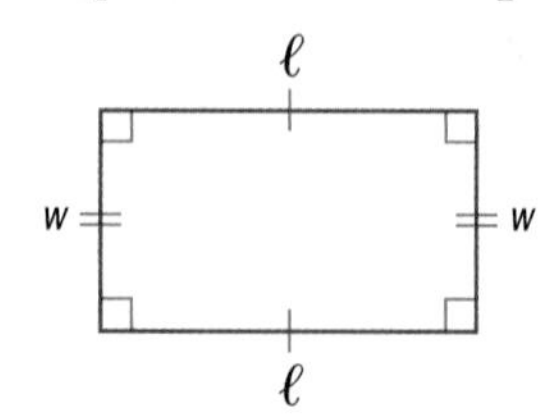

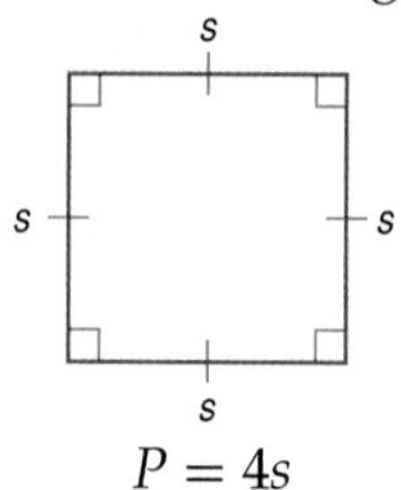

$P = 2(\ell + w)$ or $2\ell + 2w$ $\qquad$ $P = 4s$

Area is the number of square units needed to cover a surface. Area is measured in square units.

- To find the area of a rectangle, multiply the length times the width, or $\ell \cdot w$.
- To find the area of a square, find the square of the length of a side, or s^2.

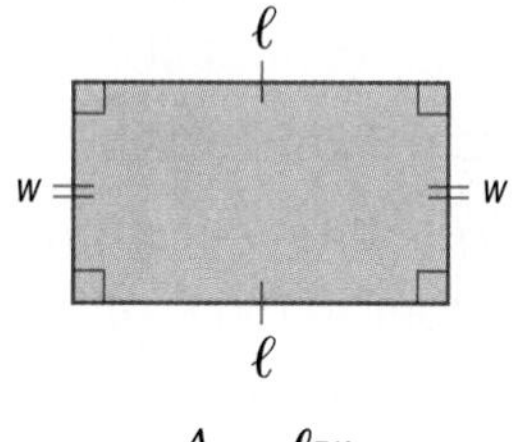

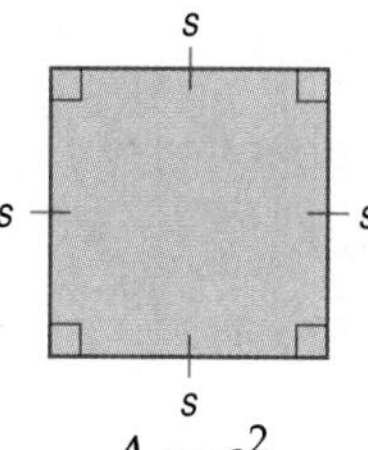

$A = \ell w$ $\qquad$ $A = s^2$

EXAMPLE

1 Find the perimeter and area of each rectangle.

a. **A rectangle has a length of 3 units and a width of 5 units.**

$P = 2(\ell + w)$	Perimeter formula
$= 2(3 + 5)$	Replace ℓ with 3 and w with 5.
$= 2(8)$	Add.
$= 16$	Multiply.
$A = \ell \cdot w$	Area formula
$= 3 \cdot 5$	Replace ℓ with 3 and w with 5.
$= 15$	Simplify.

The perimeter is 16 units, and the area is 15 square units.

b. **A rectangle has a length of 1 inch and a width of 10 inches.**

$P = 2(\ell + w)$	Perimeter formula
$= 2(1 + 10)$	Replace ℓ with 1 and w with 10.
$= 2(11)$	Add.
$= 22$	Multiply.
$A = \ell \cdot w$	Area formula
$= 1 \cdot 10$	Replace ℓ with 1 and w with 10.
$= 10$	Simplify.

The perimeter is 22 inches, and the area is 10 square inches.

EXAMPLE

2 Find the perimeter and area of each square.

a. A square has a side length of 8 feet.

$P = 4s$ Perimeter formula

$= 4(8)$ $s = 8$

$= 32$ Multiply.

$A = s^2$ Area formula

$= 8^2$ $s = 8$

$= 64$ $8^2 = 8 \cdot 8$ or 64

8 ft

The perimeter is 32 feet, and the area is 64 square feet.

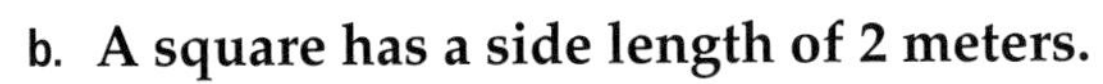

b. A square has a side length of 2 meters.

$P = 4s$ Perimeter formula

$= 4(2)$ $s = 2$

$= 8$ Multiply.

$A = s^2$ Area formula

$= 2^2$ $s = 2$

$= 4$ $2^2 = 2 \cdot 2$ or 4

2 m

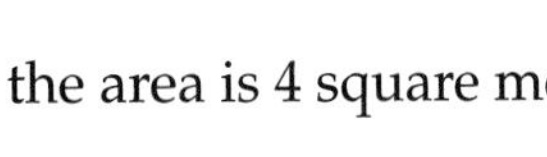

The perimeter is 8 meters, and the area is 4 square meters.

Exercises Find the perimeter and area of each figure.

1. 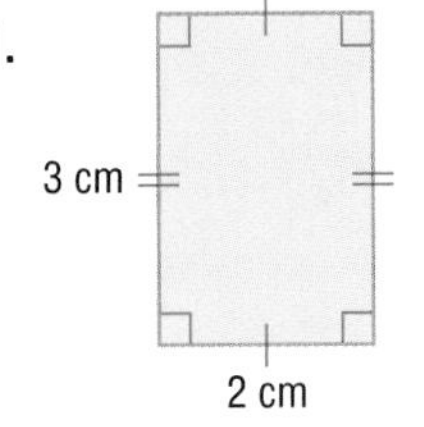

2. 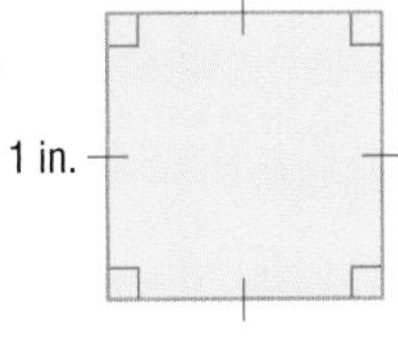

3. 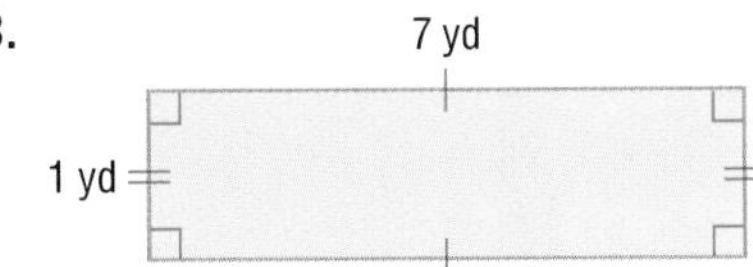

4. 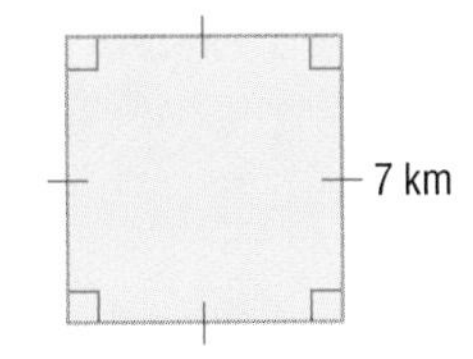

5. a rectangle with length 6 feet and width 4 feet
6. a rectangle with length 12 centimeters and width 9 centimeters
7. a square with length 3 meters
8. a square with length 15 inches
9. a rectangle with width $8\frac{1}{2}$ inches and length 11 inches
10. a rectangular room with width $12\frac{1}{4}$ feet and length $14\frac{1}{2}$ feet
11. a square with length 2.4 centimeters
12. a square garden with length 5.8 meters
13. **RECREATION** The Granville Parks and Recreation Department uses an empty city lot for a community vegetable garden. Each participant is allotted a space of 18 feet by 90 feet for a garden. What is the perimeter and area of each plot?

7 Area and Circumference of Circles

A **circle** is the set of all points in a plane that are the same distance from a given point.

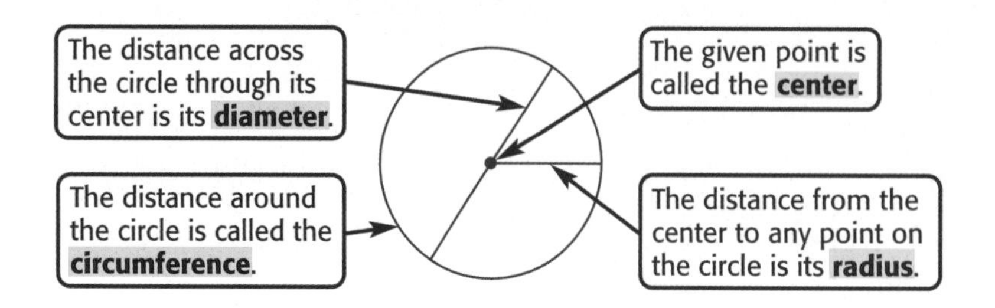

The formula for the circumference of a circle is $C = \pi d$ or $C = 2\pi r$.

EXAMPLE

1 Find the circumference of each circle to the nearest tenth.

a. The radius is 3 feet.

$C = 2\pi r$ Circumference formula

$= 2\pi(3)$ Replace r with 3.

$= 6\pi$ Simplify.

The exact circumference is 6π feet.

6 [π] [ENTER] 18.84955592

The circumference is about 18.8 feet.

b. The diameter is 24 centimeters.

$C = \pi d$ Circumference formula

$= \pi(24)$ Replace d with 24.

$= 24\pi$ Simplify.

$= 75.4$ Use a calculator to evaluate 24π.

The circumference is about 75.4 centimeters.

The formula for the area of a circle is $A = \pi r^2$.

EXAMPLE

2 Find the area of each circle to the nearest tenth.

a. The radius is 4 inches.

$A = \pi r^2$ Area formula

$= \pi(4)^2$ Replace r with 4.

$= 16\pi$ Simplify.

$= 50.3$ Use a calculator to evaluate 16π.

The area is about 50.3 square inches.

b. The diameter is 20 centimeters.

The radius is one half times the diameter, or 10 centimeters.

$A = \pi r^2$ Area formula

$= \pi(10)^2$ Replace r with 10.

$= 100\pi$ Simplify.

$= 314.2$ Use a calculator to evaluate 100π.

The area is about 314.2 square centimeters.

EXAMPLE

3 HISTORY Stonehenge is an ancient monument in Wiltshire, England. Historians are not sure who erected Stonehenge or why. It may have been used as a calendar. The giant stones of Stonehenge are arranged in a circle 30 meters in diameter. Find the circumference and the area of the circle.

$C = \pi d$	Write the formula.
$= \pi(30)$	Replace d with 30.
$= 30\pi$ or about 94.2	Simplify.

Find the radius to evaluate the formula for the area.

$A = \pi r^2$	Write the formula.
$= \pi(15)^2$	Replace r with $\frac{1}{2}$ (30) or 15.
$= 225\pi$ or about 706.9	Simplify.

The circumference of Stonehenge is about 94.2 meters, and the area is about 706.9 square meters.

Exercises

Find the circumference of each circle. Round to the nearest tenth.

1. 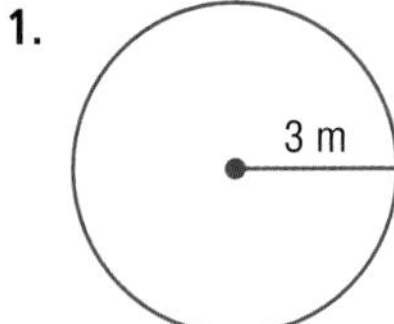

2. 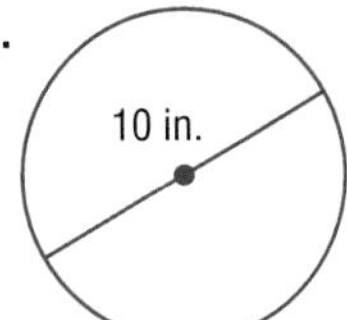

3. 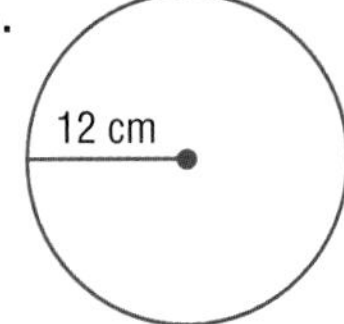

4. The radius is 1.5 kilometers.
5. The diameter is 1 yard.
6. The diameter is $5\frac{1}{4}$ feet.
7. The radius is $24\frac{1}{2}$ inches.

Find the area of each circle. Round to the nearest tenth.

8. 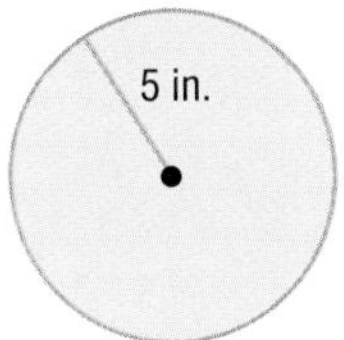

9. 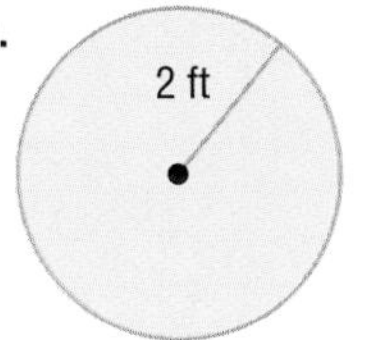

10. 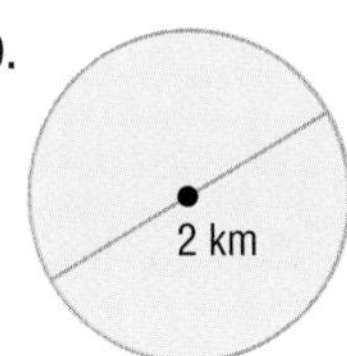

11. The diameter is 4 yards.
12. The radius is 1 meter.
13. The radius is 1.5 feet.
14. The diameter is 15 centimeters.
15. **GEOGRAPHY** Earth's circumference is approximately 25,000 miles. If you could dig a tunnel to the center of the Earth, how long would the tunnel be?
16. **CYCLING** The tire for a 10-speed bicycle has a diameter of 27 inches. Find the distance the bicycle will travel in 10 rotations of the tire.
17. **PUBLIC SAFETY** The Belleville City Council is considering installing a new tornado warning system. The sound emitted from the siren would be heard for a 2-mile radius. Find the area of the region that will benefit from the system.
18. **CITY PLANNING** The circular region inside the streets at DuPont Circle in Washington, D.C., is 250 feet across. How much area do the grass and sidewalk cover?

8 Volume

Volume is the measure of space occupied by a solid. Volume is measured in cubic units. The prism at the right has a volume of 12 cubic units.

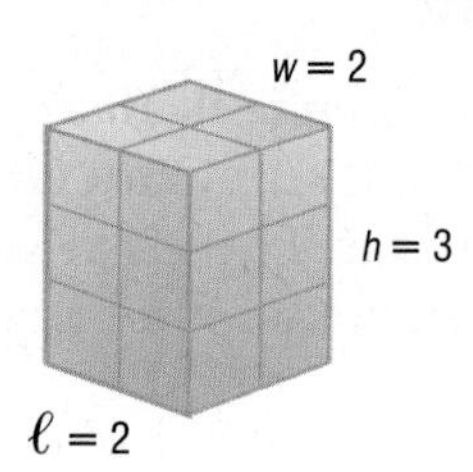

To find the volume of a rectangular prism, use the formula $V = \ell \cdot w \cdot h$. Stated in words, volume equals length times width times height.

EXAMPLE

Find the volume of the rectangular prism.

A rectangular prism has a height of 3 feet, width of 4 feet, and length of 2 feet.

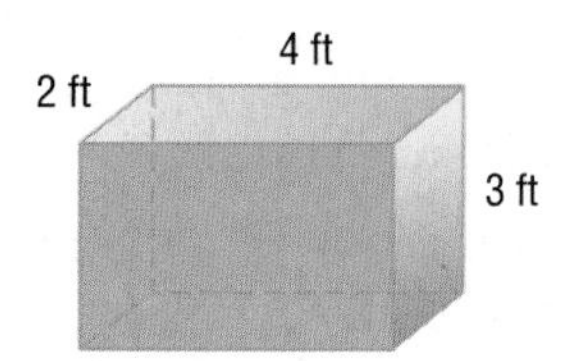

$V = \ell \cdot w \cdot h$ Volume formula

$= 2 \cdot 4 \cdot 3$ Replace ℓ with 2, w with 4, and h with 3.

$= 24$ Simplify.

The volume is 24 cubic feet.

Exercises **Find the volume of each rectangular prism given the length, width, and height.**

1. $\ell = 2$ in., $w = 5$ in., $h = \frac{1}{2}$ in.
2. $\ell = 12$ cm, $w = 3$ cm, $h = 2$ cm
3. $\ell = 6$ yd, $w = 2$ yd, $h = 1$ yd
4. $\ell = 100$ m, $w = 1$ m, $h = 10$ m

Find the volume of each rectangular prism.

5. 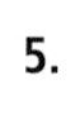

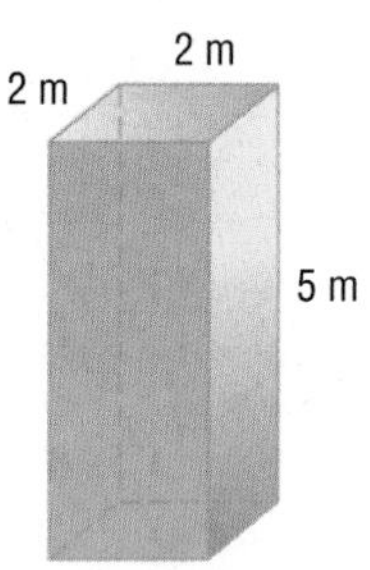

6. 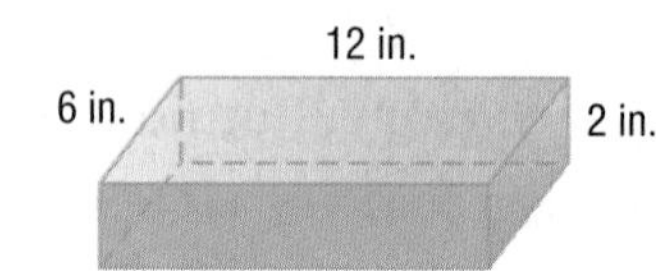

7. **AQUARIUMS** An aquarium is 8 feet long, 5 feet wide, and 5.5 feet deep. What is the volume of the tank?
8. **COOKING** What is the volume of a microwave oven that is 18 inches wide by 10 inches long with a depth of $11\frac{1}{2}$ inches?
9. **GEOMETRY** A cube measures 2 meters on a side. What is its volume?

FIREWOOD **For Exercises 10–12, use the following information.**
Firewood is usually sold by a measure known as a cord. A full cord may be a stack $8 \times 4 \times 4$ feet or a stack $8 \times 8 \times 2$ feet.

10. What is the volume of a full cord of firewood?
11. A "short cord" or "face cord" of wood is $8 \times 4 \times$ the length of the logs. What is the volume of a short cord of $2\frac{1}{2}$-foot logs?
12. If you have an area that is 12 feet long and 2 feet wide in which to store your firewood, how high will the stack be if it is a full cord of wood?

9 Probability and Odds

The **probability** of an event is the ratio of the number of favorable outcomes for the event to the total number of possible outcomes.

When there are n outcomes and the probability of each one is $\frac{1}{n}$, we say that the outcomes are **equally likely**. For example, when you roll a die, the 6 possible outcomes are equally likely because each outcome has a probability of $\frac{1}{6}$.

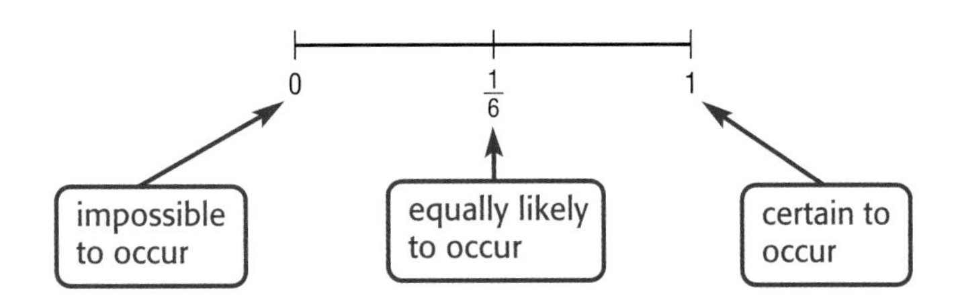

EXAMPLE

1 Find the probability of rolling an even number on a die.

There are six possible outcomes. Three of the outcomes are favorable. That is, three of the six outcomes are even numbers.

3 even numbers

Sample space: 1, 2, 3, 4, 5, 6 $\rightarrow \frac{3}{6}$

6 total possible outcomes

So, $P(\text{even number}) = \frac{3}{6}$ or $\frac{1}{2}$.

EXAMPLE

2 A bowl contains 5 red chips, 7 blue chips, 6 yellow chips, and 10 green chips. One chip is randomly drawn. Find each probability.

a. blue

There are 7 blue chips and 28 total chips.

$P(\text{blue chip}) = \frac{7}{28}$ ← number of favorable outcomes / ← number of possible outcomes

$= \frac{1}{4}$ The probability can be stated as $\frac{1}{4}$, 0.25, or 25%.

b. red or yellow

$P(\text{red or yellow}) = \frac{11}{28}$ ← number of favorable outcomes / ← number of possible outcomes

≈ 0.39 The probability can be stated as $\frac{11}{28}$, 0.39, or 39%.

c. not green

$P(\text{not green}) = \frac{18}{28}$ ← number of favorable outcomes / ← number of possible outcomes

$= \frac{9}{14}$ or about 0.64 The probability can be stated as $\frac{9}{14}$, about 0.64, or about 64%.

The **odds** of an event occurring is the ratio that compares the number of ways an event can occur (successes) to the number of ways it cannot occur (failures).

EXAMPLE

3 Find the odds of rolling a number less than 3.

There are 6 possible outcomes, 2 are successes and 4 are failures.

2 numbers less than 3

Sample space: 1, 2, 3, 4, 5, 6 $\quad \frac{2}{4}$ or $\frac{1}{2}$

4 numbers not less than 3

So, the odds of rolling a number less than tree are $\frac{1}{2}$ or 1:2.

Exercises

One coin is randomly selected from a jar containing 70 nickels, 100 dimes, 80 quarters, and 50 1-dollar coins. Find each probability.

1. P(quarter)
2. P(dime)
3. P(nickel or dollar)
4. P(quarter or nickel)
5. P(value less than \$1.00)
6. P(value greater than \$0.10)
7. P(value at least \$0.25)
8. P(value at most \$1.00)

Two dice are rolled and their sum is recorded. Find each probability.

9. P(sum less than 7)
10. P(sum less than 8)
11. P(sum is greater than 12)
12. P(sum is greater than 1)
13. P(sum is between 5 and 10)
14. P(sum is between 2 and 9)

One of the polygons is chosen at random. Find each probability.

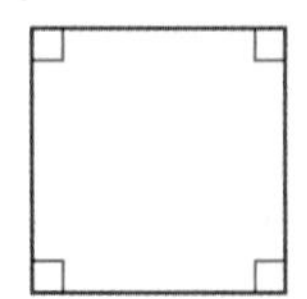 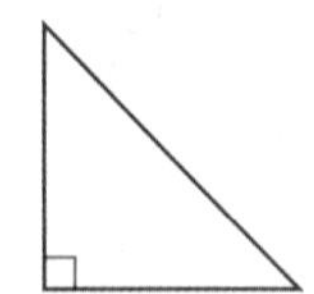  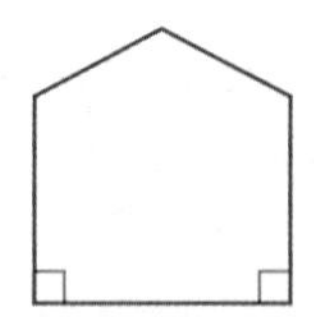 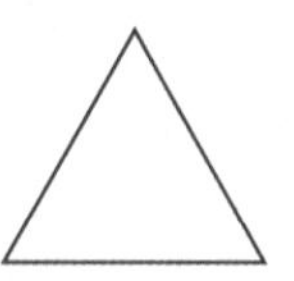 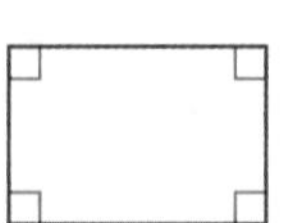

15. P(triangle)
16. P(pentagon)
17. P(not a triangle)
18. P(not a quadrilateral)
19. P(more than three sides)
20. P(more than one right angle)

Find the odds of each outcome if a computer randomly picks a letter in the name *The United States of America*.

21. the letter *a*
22. the letter *t*
23. a vowel
24. a consonant
25. an uppercase letter
26. a lowercase vowel

10 Mean, Median, and Mode

Measures of central tendency are numbers used to represent a set of data. Three types of measures of central tendency are mean, median, and mode.

The **mean** is the sum of the numbers in a set of data divided by the number of items.

EXAMPLE

1 Katherine is running a lemonade stand. She made \$3.50 on Tuesday, \$4.00 on Wednesday, \$5.00 on Thursday, and \$4.50 on Friday. What was her mean daily profit?

$$\text{mean} = \frac{\text{sum of daily profits}}{\text{number of days}}$$

$$= \frac{\$3.50 + \$4.00 + \$5.00 + \$4.50}{4}$$

$$= \frac{\$17.00}{4} \text{ or } \$4.25$$

Katherine's mean daily profit was \$4.25.

The **median** is the middle number in a set of data when the data are arranged in numerical order. If there are an even number of data, the median is the mean of the two middle numbers.

EXAMPLE

2 The table shows the number of hits Marcus made for his team. Find the median of the data.

Team Played	Number of Hits by Marcus
Badgers	3
Hornets	6
Bulldogs	5
Vikings	2
Rangers	3
Panthers	7

To find the median, order the numbers from least to greatest. The median is in the middle.

2, 3, 3, 5, 6, 7

$$\frac{3 + 5}{2} = 4$$

There is an even number of items. Find the mean of the middle two.

The median number of hits is 4.

The **mode** is the number or numbers that appear most often in a set of data. If no item appears most often, the set has no mode.

EXAMPLE

3 The table shows the heights in inches of the members of the 2005–2006 University of Dayton Men's Basketball team. What is the mode of the heights?

2005-2006 Dayton Flyers Men's Basketball Team				
74	78	79	80	78
72	81	83	76	78
76	75	77	79	72

Source: ESPN

The mode is the number that occurs most frequently. 78 occurs three times, 72, 76, and 79 each occur twice, and all the other heights occur once. The mode height is 78.

You can use measures of central tendency to solve problems.

EXAMPLE

4 On her first five history tests, Yoko received the following scores: 82, 96, 92, 83, and 91. What test score must Yoko earn on the sixth test so that her average (mean) for all six tests will be 90%?

$\text{mean} = \dfrac{\text{sum of the first five scores} + \text{sixth score}}{6}$ Write an equation.

$90 = \dfrac{82 + 96 + 92 + 83 + 91 + x}{6}$ Use x to represent the sixth score.

$90 = \dfrac{444 + x}{6}$ Simplify.

$540 = \dfrac{444 + x}{6}$ Multiply each side by 6.

$96 = x$ Subtract 444 from each side.

To have an average score of 90, Yoko must earn a 96 on the sixth test.

Exercises Find the mean, median, and mode for each set of data.

1. {1, 2, 3, 5, 5, 6, 13}
2. {3, 5, 8, 1, 4, 11, 3}
3. {52, 53, 53, 53, 55, 55, 57}
4. {8, 7, 5, 19}
5. {3, 11, 26, 4, 1}
6. {201, 201, 200, 199, 199}
7. {4, 5, 6, 7, 8}
8. {3, 7, 21, 23, 63, 27, 29, 95, 23}
9. **SCHOOL** The table shows the cost of some school supplies. Find the mean, median, and mode costs.

Cost of School Supplies	
Supply	Cost
Pencils	$0.50
Pens	$2.00
Paper	$2.00
Pocket Folder	$1.25
Calculator	$5.25
Notebook	$3.00
Erasers	$2.50
Markers	$3.50

10. **NUTRITION** The table shows the number of servings of fruits and vegetables that Cole eats one week. Find the mean, median, and mode.

Cole's Fruit and Vegetable Servings	
Day	Number of Servings
Monday	5
Tuesday	7
Wednesday	5
Thursday	4
Friday	3
Saturday	3
Sunday	8

11. **TELEVISION RATINGS** The ratings for the top television programs during one week are shown in the table at the right. Find the mean, median, and mode of the ratings. Round to the nearest hundredth.

Network Primetime Television Ratings	
Program	Rating
1	17.6
2	16.0
3	14.1
4	13.7
5	13.5
6	12.9
7	12.3
8	11.6
9	11.4
10	11.4

Source: Nielsen Media

12. **EDUCATION** Bill's scores on his first four science tests are 86, 90, 84, and 91. What test score must Bill earn on the fifth test so that his average (mean) will be exactly 88?
13. **BOWLING** Sue's average for 9 games of bowling is 108. What is the lowest score she can receive for the tenth game to have an average of 110?
14. **EDUCATION** Olivia has an average score of 92 on five French tests. If she earns a score of 96 on the sixth test,what will her new average score be?

11 Box-and-Whisker Plots

Data can be organized and displayed by dividing a set of data into four parts using the median and quartiles. This is a **box-and-whisker plot**.

EXAMPLE

1 Draw a box-and-whisker plot for these data.
14.03, 30.11, 16.03, 19.61, 18.15, 16.34, 20.43, 18.46, 22.24, 12.70, 8.25

Step 1 Order the data from least to greatest. Use this list to determine the quartiles.

8.25, 12.70, 14.03, 16.03, 16.34, 18.15, 18.46, 19.61, 20.43, 22.24, 30.11

↑ Q1 (14.03) ↑ Q2 (18.15) ↑ Q3 (20.43)

Determine the interquartile range.

$$\text{IQR} = 20.43 - 14.03 \text{ or } 6.4$$

Check to see if there are any outliers.

$$14.03 - 1.5(6.4) = 4.43 \qquad 20.43 + 1.5(6.4) = 30.03$$

Any numbers less than 4.43 or greater than 30.03 are outliers. The only outlier is 30.11.

Step 2 Draw a number line. Assign a scale to the number line and place bullets to represent the three quartile points, any outliers, the least number that is not an outlier, and the greatest number that is not an outlier.

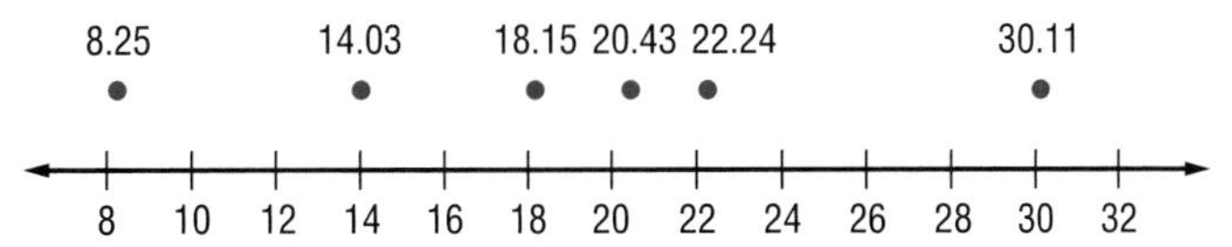

Step 3 Draw a box to designate the data between the upper and lower quartiles. Draw a vertical line through the point representing the median. Draw a line from the lower quartile to the least value that is not an outlier. Draw a line from the upper quartile to the greatest value that is not an outlier.

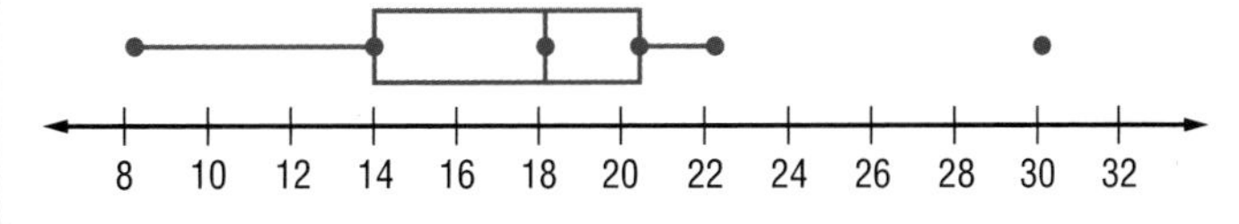

Exercises

For Exercises 1–3, use the box-and-whisker plot below.

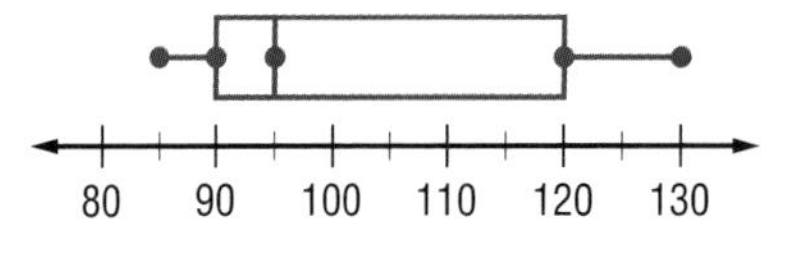

1. What is the median of the data?
2. What is the range of the data?
3. What is the interquartile range of the data?

Draw a box-and-whisker plot for each set of data.

4. 15, 8, 10, 1, 3, 2, 6, 5, 4, 27, 1
5. 20, 2, 12, 5, 4, 16, 17, 7, 6, 16, 5, 0, 5, 30
6. 4, 1, 1, 1, 10, 15, 4, 5, 27, 5, 14, 10, 6, 2, 2, 5, 8

⓬ Representing Data

Data can be displayed and organized by different methods. In a **frequency table**, you use tally marks to record and display the frequency of events. A **bar graph** compares different categories of data by showing each as a bar with a length that is related to the frequency.

EXAMPLE

1 **The frequency table shows the results of a survey of student's favorite sports. Make a bar graph to display the data.**

Sport	Tally	Frequency
Basketball	卌 卌 卌	15
Football	卌 卌 卌 卌 卌	25
Soccer	卌 卌 卌 |||	18
Baseball	卌 卌 卌 卌		21
Tennis	卌 卌 卌		16

Step 1 Draw a horizontal axis and a vertical axis. Label the axes as shown. Add a title.

Step 2 Draw a bar to represent each sport. The vertical scale is the number of students who chose each sport. The horizontal scale identifies the sport chosen.

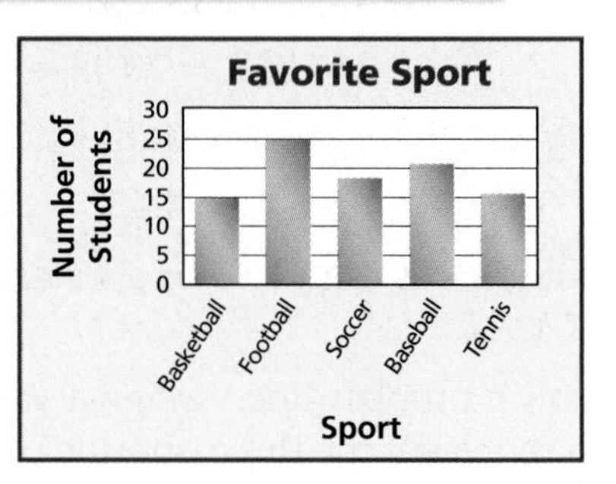

Another way to represent data is by using a *line graph*. A **line graph** usually shows how data changes over a period of time.

EXAMPLE

2 **Sales at the Marshall High School Store are shown in the table below. Make a line graph of the data.**

School Store Sales Amounts					
September	$670	December	$168	March	$412
October	$229	January	$290	April	$309
November	$300	February	$388	May	$198

Step 1 Draw a horizontal axis and a vertical axis and label them as shown. Include a title.

Step 2 Plot the points to represent the data.

Step 3 Draw a line connecting each pair of consecutive points.

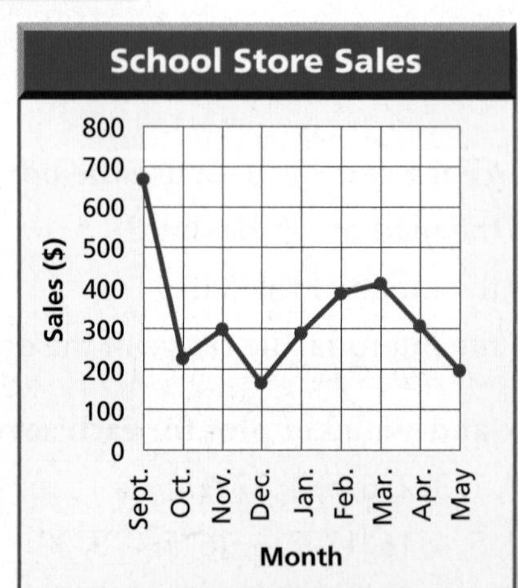

Another way to organize and display data is by using a **stem-and-leaf display**. In a stem-and-leaf display, the greatest common place value is used for the stems. The numbers in the next greatest place value are used to form the leaves.

EXAMPLE

3 ANIMALS The speeds (mph) of 20 of the fastest land animals are listed below. Use the data to make a stem-and-leaf display.

45	70	43	45	32	42	40	40	35	50
40	35	61	48	35	32	50	36	50	40

Source: *The World Almanac*

The greatest place value is tens. Thus, 32 miles per hour would have a stem of 3 and a leaf of 2.

Stem	Leaf
3	2 2 5 5 5 6
4	0 0 0 0 2 3 5 5 8
5	0 0 0
6	1
7	0

Key: *3|2 = 32*

Exercises

1. **SURVEYS** Alana surveyed several students to find the number of hours of sleep they typically get each night. The results are shown in the table. Make a bar graph of the data.

Hours of Sleep					
Alana	8	Kwam	7.5	Tomas	7.75
Nick	8.25	Kate	7.25	Sharla	8.5

2. **LAWN CARE** Marcus started a lawn care service. The chart shows how much money he made over the 15 weeks of summer break. Make a line graph of the data.

Lawn Care Profits ($)								
Week	1	2	3	4	5	6	7	8
Profit	25	40	45	50	75	85	95	95
Week	9	10	11	12	13	14	15	
Profit	125	140	135	150	165	165	175	

Use each set of data to make a stem-and-leaf display.

3. 6.5 6.3 6.9 7.1 7.3 5.9 6.0 7.0 7.2 6.6 7.1 5.8
4. 31 30 28 26 22 34 26 31 47 32 18 33 26 23 18 29
5. The frequency table below shows the ages of people attending a high school play. Make a bar graph to display the data.

Age	Tally	Frequency
under 20	卌卌卌卌卌卌卌卌卌 II	47
20–39	卌卌卌卌卌卌卌卌 III	43
40–59	卌卌卌卌卌卌 I	31
60 or over	卌 III	8

13 Absolute Value

On a number line, 4 is four units from zero in the positive direction, and −4 is four units from zero in the negative direction. This number line illustrates the meaning of **absolute value**.

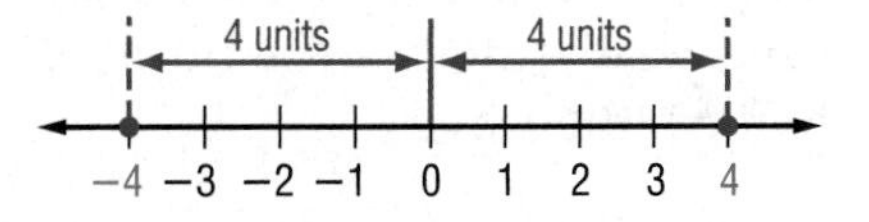

EXAMPLE Absolute Value of Rational Numbers

1 Find each absolute value.

a. $|-7|$

−7 is seven units from zero in the negative direction.

$|-7| = 7$

b. $\left|\frac{7}{9}\right|$

$\frac{7}{9}$ is seven-ninths unit from zero in the positive direction.

$\left|\frac{7}{9}\right| = \frac{7}{9}$

EXAMPLE Expressions with Absolute Value

2 Evaluate $15 - |x + 4|$ if $x = 8$.

$15 - \|x + 4\| = 15 - \|8 + 4\|$	Replace x with 8.
$= 15 - \|12\|$	$8 + 4 = 12$
$= 15 - 12$	$\|12\| = 12$
$= 3$	Simplify.

Exercises Find each absolute value.

1. $|-2|$
2. $|18|$
3. $|2.5|$
4. $\left|-\frac{5}{6}\right|$
5. $|-38|$
6. $|10|$
7. $|97|$
8. $|-61|$
9. $|-3.9|$
10. $|-6.8|$
11. $\left|-\frac{23}{56}\right|$
12. $\left|\frac{35}{80}\right|$

Evaluate each expression if $a = 6$, $b = \frac{2}{3}$, $c = \frac{5}{4}$, $x = 12$, $y = 3.2$, and $z = -5$.

13. $48 + |x - 5|$
14. $25 + |17 + x|$
15. $|17 - a| + 23$
16. $|43 - 4a| + 51$
17. $|z| + 13 - 4$
18. $28 - 13 + |z|$
19. $6.5 - |8.4 - y|$
20. $7.4 + |y - 2.6|$
21. $\frac{1}{6} + \left|b - \frac{7}{12}\right|$
22. $\left(b + \frac{1}{2}\right) - \left|-\frac{5}{6}\right|$
23. $|c - 1| + \frac{2}{5}$
24. $|-c| + \left(2 + \frac{1}{2}\right)$
25. $-a + |2x - a| + \frac{1}{2}$
26. $|y - 2z| - 3$
27. $3|3b - 8c| - 3$
28. $|2x - z| + 6b$
29. $-3|z| + 2(a + y)$
30. $-4|c - 3| + 2|z - a|$

Extra Practice

Lesson 1-1

(pages 6–9)

Write an algebraic expression for each verbal expression.

1. the sum of b and 21
2. the product of x and 7
3. the sum of 4 and 6 times a number z
4. the sum of 8 and -2 times n
5. one-half the cube of a number x
6. four-fifths the square of m

Evaluate each expression.

7. 2^4
8. 10^2
9. 7^3
10. 20^3
11. 3^6
12. 4^5

Write a verbal expression for each algebraic expression.

13. $2n$
14. 10^7
15. m^5
16. xy
17. $5n^2 - 6$
18. $9a^3 + 1$

Lesson 1-2

(pages 10–14)

Evaluate each expression.

1. $3 + 8 \div 2 - 5$
2. $4 + 7 \cdot 2 + 8$
3. $5(9 + 3) - 3 \cdot 4$
4. $9 - 3^2$
5. $(8 - 1) \cdot 3$
6. $4(5 - 3)^2$
7. $3(12 + 3) - 5 \cdot 9$
8. $5^3 + 6^3 - 5^2$
9. $16 \div 2 \cdot 5 \cdot 3 \div 6$
10. $7(5^3 + 3^2)$
11. $\dfrac{9 \cdot 4 + 2 \cdot 6}{6 \cdot 4}$
12. $25 - \frac{1}{3}(18 + 9)$

Evaluate each expression if $a = 2$, $b = 5$, $x = 4$, and $n = 10$.

13. $8a + b$
14. $48 + ab$
15. $a(6 - 3n)$
16. $bx + an$
17. $x^2 - 4n$
18. $3b + 16a - 9n$
19. $n^2 + 3(a + 4)$
20. $(2x)^2 + an - 5b$
21. $[a + 8(b - 2)]^2 \div 4$

Lesson 1-3

(pages 15–20)

Find the solution of each equation if the replacement sets are $x = \{0, 2, 4, 6, 8\}$ and $y = \{1, 3, 5, 7, 9\}$.

1. $x - 4 = 4$
2. $25 - y = 18$
3. $3x + 1 = 25$
4. $5y - 4 = 11$
5. $14 = \dfrac{96}{x} + 2$
6. $0 = \dfrac{y}{3} - 3$

Solve each equation.

7. $x = \dfrac{27 + 9}{2}$
8. $\dfrac{18 - 7}{13 - 2} = y$
9. $n = \dfrac{6(5) + 3}{2(4) + 3}$
10. $\dfrac{5(4) - 6}{2^2 + 3} = z$
11. $\dfrac{7^2 + 9(2 + 1)}{2(10) - 1} = t$
12. $a = \dfrac{3^3 + 5^2}{2(3 - 1)}$

Find the solution set for each inequality if the replacement sets are $x = \{4, 5, 6, 7, 8\}$ and $y = \{10, 12, 14, 16\}$.

13. $x + 2 > 7$
14. $x - 1 < 8$
15. $2x \leq 15$
16. $3y \geq 36$
17. $\dfrac{x}{3} < 2$
18. $\dfrac{5y}{4} \geq 20$

Lesson 1-4

(pages 21–25)

Find the value of n in each equation. Then name the property that is used.

1. $4 \cdot 3 = 4 \cdot n$
2. $\frac{5}{4} = n + 0$
3. $15 = 15 \cdot n$
4. $\frac{2}{3}n = 1$
5. $2.7 + 1.3 = n + 2.7$
6. $n\left(6^2 \cdot \frac{1}{36}\right) = 4$
7. $8n = 0$
8. $n = \frac{1}{9} \cdot 9$

Evaluate each expression. Name the property used in each step.

9. $\frac{2}{3}[15 \div (12 - 2)]$
10. $\frac{7}{4}\left[4 \cdot \left(\frac{1}{8} \cdot 8\right)\right]$
11. $[(18 \div 3) \cdot 0] \cdot 10$

Lesson 1-5

(pages 26–31)

Rewrite each expression using the Distributive Property. Then evaluate.

1. $5(2 + 9)$
2. $8(10 + 20)$
3. $6(y + 4)$
4. $9(3n + 5)$
5. $32\left(x - \frac{1}{8}\right)$
6. $c(7 - d)$

Use the Distributive Property to rewrite each expression. Then find the product.

7. $6 \cdot 55$
8. $15(108)$
9. $1689 \cdot 5$
10. 7×314
11. $36\left(5\frac{1}{4}\right)$
12. $\left(4\frac{1}{18}\right) \cdot 18$

Simplify each expression. If not possible, write *simplified*.

13. $13a + 5a$
14. $21x - 10x$
15. $8(3x + 7)$
16. $4m - 4n$
17. $3(5am - 4)$
18. $15x^2 + 7x^2$
19. $9y^2 + 13y^2 + 3$
20. $11a^2 - 11a^2 + 12a^2$
21. $6a + 7a + 12b + 8b$

Lesson 1-6

(pages 33–37)

Evaluate each expression using properties of numbers.

1. $23 + 8 + 37 + 12$
2. $19 + 46 + 81 + 54$
3. $10.25 + 2.5 + 3.75$
4. $22.5 + 17.6 + 44.5$
5. $2\frac{1}{3} + 6 + 3\frac{2}{3} + 4$
6. $5\frac{6}{7} + 15 + 4\frac{1}{7} + 25$
7. $6 \cdot 8 \cdot 5 \cdot 3$
8. $18 \cdot 5 \cdot 2 \cdot 5$
9. $0.25 \cdot 7 \cdot 8$
10. $90 \cdot 12 \cdot 0.5$
11. $5\frac{1}{3} \cdot 4 \cdot 6$
12. $4\frac{5}{6} \cdot 10 \cdot 12$

Simplify each expression.

13. $5a + 6b + 7a$
14. $8x + 4y + 9x$
15. $3a + 5b + 2c + 8b$
16. $\frac{2}{3}x^2 + 5x + x^2$
17. $(4p - 7q) + (5q - 8p)$
18. $8q + 5r - 7q - 6r$
19. $4(2x + y) + 5x$
20. $9r^5 + 2r^2 + r^5$
21. $12b^3 + 12 + 12b^3$
22. $7 + 3(uv - 6) + u$
23. $3(x + 2y) + 4(3x + y)$
24. $6.2(a + b) + 2.6(a + b) + 3a$

Lesson 1-7

(pages 39–44)

Identify the hypothesis and conclusion of each statement.

1. If an animal is a dog, then it barks.
2. If a figure is a pentagon, then it has five sides.
3. If $3x - 1 = 8$, then $x = 3$.
4. If 0.5 is the reciprocal of 2, then $0.5 \cdot 2 = 1$.

Identify the hypotheses and conclusion of each statement. Then write the statement in if-then form.

5. A square has four congruent sides.
6. $6a + 10 = 34$ when $a = 4$.
7. The video store is open every night.
8. The band will not practice on Thursday.

Find a counterexample for each conditional statement.

9. If the season is spring, then it does not snow.
10. If you live in Portland, then you live in Oregon.
11. If $2y + 4 = 10$, then $y < 3$.
12. If $a^2 > 0$, then $a > 0$.

Lesson 1-8

(pages 46–52)

Find each square root. If necessary, round to the nearest hundredth.

1. $\sqrt{121}$
2. $-\sqrt{36}$
3. $\sqrt{2.89}$
4. $-\sqrt{125}$
5. $\sqrt{\frac{81}{100}}$
6. $-\sqrt{\frac{36}{196}}$
7. $\pm\sqrt{9.61}$
8. $\pm\sqrt{\frac{7}{8}}$

Name the set or sets of numbers to which each real number belongs.

9. $-\sqrt{149}$
10. $\frac{5}{6}$
11. $\sqrt{\frac{8}{2}}$
12. $-\frac{66}{55}$
13. $\sqrt{225}$
14. $-\sqrt{\frac{3}{4}}$
15. $\frac{-1}{7}$
16. $\sqrt{0.0016}$

Replace each • with <, >, or = to make each sentence true.

17. $6.\overline{16} \bullet 6$
18. $3.88 \bullet \sqrt{15}$
19. $-\sqrt{529} \bullet -20$
20. $-\sqrt{0.25} \bullet -0.\overline{5}$
21. $\frac{1}{3} \bullet \frac{\sqrt{3}}{3}$
22. $\frac{1}{\sqrt{3}} \bullet \frac{\sqrt{3}}{3}$
23. $-\sqrt{\frac{1}{4}} \bullet -\frac{1}{4}$
24. $-\frac{1}{6} \bullet -\frac{1}{\sqrt{6}}$

Lesson 1-9

(pages 53–58)

Describe what is happening in each graph.

1. The graph shows the average monthly high temperatures for a city over a one-year period.

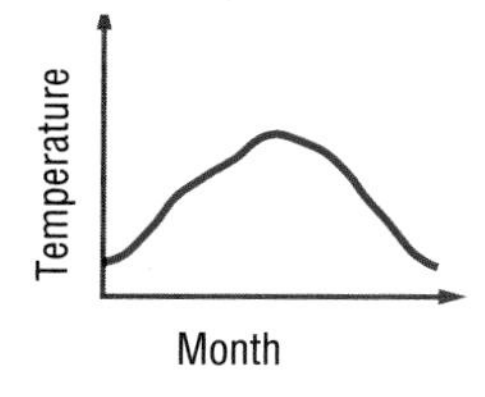

2. The graph shows the speed of a roller coaster car during a two- minute ride.

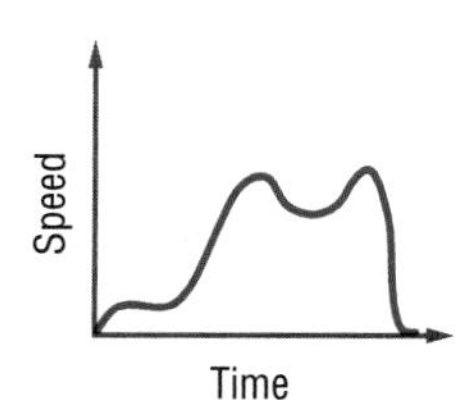

Extra Practice

Lesson 2-1

(pages 70–76)

Translate each sentence into an equation or formula.

1. A number z times 2 minus 6 is the same as m divided by 3.
2. The cube of a decreased by the square of b is equal to c.
3. Twenty-nine decreased by the product of x and y is the same as z.
4. The perimeter P of an isosceles triangle is the sum of twice the length of leg a and the length of the base b.
5. Thirty increased by the quotient of s and t is equal to v.
6. The area A of a rhombus is half the product of lengths of the diagonals a and b.

Translate each equation into a verbal sentence.

7. $0.5x + 3 = -10$
8. $\frac{n}{-6} = 2n + 1$
9. $18 - 5h = 13h$
10. $n^2 = 16$
11. $2x^2 + 3 = 21$
12. $\frac{m}{n} + 4 = 12$

Lesson 2-2

(pages 78–84)

Solve each equation. Check your solution.

1. $-2 + g = 7$
2. $9 + s = -5$
3. $-4 + y = -9$
4. $m + 6 = 2$
5. $t + (-4) = 10$
6. $v - 7 = -4$
7. $a - (-6) = -5$
8. $-2 - x = -8$
9. $d + (-44) = -61$
10. $b - (-26) = 41$
11. $p - 47 = 22$
12. $-63 - f = -82$
13. $c + 5.4 = -11.33$
14. $-6.11 + b = 14.321$
15. $-5 = y - 22.7$
16. $-5 - q = 1.19$
17. $n + (-4.361) = 59.78$
18. $t - (-46.1) = -3.673$
19. $\frac{7}{10} - a = \frac{1}{2}$
20. $f - \left(-\frac{1}{8}\right) = \frac{3}{10}$
21. $-4\frac{5}{12} = t - \left(-10\frac{1}{36}\right)$
22. $x + \frac{3}{8} = \frac{1}{4}$
23. $1\frac{7}{16} + s = \frac{9}{8}$
24. $17\frac{8}{9} = d + \left(-2\frac{5}{6}\right)$

Lesson 2-3

(pages 85–90)

Solve each equation. Check your solution.

1. $7p = 35$
2. $-3x = -24$
3. $2y = -3$
4. $62y = -2356$
5. $\frac{a}{-6} = -2$
6. $\frac{c}{-59} = -7$
7. $\frac{f}{14} = -63$
8. $84 = \frac{x}{97}$
9. $\frac{w}{5} = 3$
10. $\frac{q}{9} = -3$
11. $\frac{2}{5}x = \frac{4}{7}$
12. $\frac{z}{6} = -\frac{5}{12}$
13. $-\frac{5}{9}r = 7\frac{1}{2}$
14. $2\frac{1}{6}j = 5\frac{1}{5}$
15. $3 = 1\frac{7}{11}q$
16. $-1\frac{3}{4}p = -\frac{5}{8}$
17. $57k = 0.1824$
18. $0.0022b = 0.1958$
19. $5j = -32.15$
20. $\frac{w}{-2} = -2.48$
21. $\frac{z}{2.8} = -6.2$
22. $\frac{x}{-0.063} = 0.015$
23. $15\frac{3}{8} = -5p$
24. $-18\frac{1}{4} = 2.5x$

Lesson 2-4

(pages 92–97)

Solve each equation. Check your solution.

1. $2x - 5 = 3$
2. $4t + 5 = 37$
3. $7a + 6 = -36$
4. $47 = -8g + 7$
5. $-3c - 9 = -24$
6. $5k - 7 = -52$
7. $5s + 4s = -72$
8. $3x - 7 = 2$
9. $8 + 3x = 5$
10. $-3y + 7.569 = 24.069$
11. $7 - 9.1f = 137.585$
12. $6.5 = 2.4m - 4.9$
13. $\frac{n}{5} + 6 = -2$
14. $\frac{d}{4} - 8 = -5$
15. $-\frac{4}{13}y - 7 = 6$
16. $\frac{p+3}{10} = 4$
17. $\frac{h-7}{6} = 1$
18. $\frac{5f+1}{8} = -3$
19. $\frac{4n-8}{-2} = 12$
20. $\frac{-3t-4}{2} = 8$
21. $4.8a - 3 + 1.2a = 9$

Lesson 2-5

(pages 98–103)

Solve each equation. Check your solution.

1. $5x + 1 = 3x - 3$
2. $6 - 8n = 5n + 19$
3. $-3z + 5 = 2z + 5$
4. $\frac{2}{3}h + 5 = -4 - \frac{1}{3}h$
5. $\frac{1}{2}a - 4 = 3 - \frac{1}{4}a$
6. $6(y - 5) = 18 - 2y$
7. $-28 + p = 7(p - 10)$
8. $\frac{1}{3}(b - 9) = b + 9$
9. $-4x + 6 = 0.5(x + 30)$
10. $4(2y - 1) = -8(0.5 - y)$
11. $1.9s + 6 = 3.1 - s$
12. $2.85y - 7 = 12.85y - 2$
13. $2.9m + 1.7 = 3.5 + 2.3m$
14. $3(x + 1) - 5 = 3x - 2$
15. $\frac{x}{2} - \frac{1}{3} = \frac{x}{3} - \frac{1}{2}$
16. $\frac{6v-9}{3} = v$
17. $\frac{3t+1}{4} = \frac{3}{4}t - 5$
18. $0.4(x - 12) = 1.2(x - 4)$
19. $3y - \frac{4}{5} = \frac{1}{3}y$
20. $\frac{3}{4}x - 4 = 7 + \frac{1}{2}x$
21. $-0.2(1 - x) = 2(4 + 0.1x)$
22. $3.2(y + 1) = 2(1.4y - 3)$

Lesson 2-6

(pages 105–110)

Solve each proportion. If necessary, round to the nearest hundredth.

1. $\frac{4}{5} = \frac{x}{20}$
2. $\frac{b}{63} = \frac{3}{7}$
3. $\frac{y}{5} = \frac{3}{4}$
4. $\frac{7}{4} = \frac{3}{a}$
5. $\frac{t-5}{4} = \frac{3}{2}$
6. $\frac{x}{9} = \frac{0.24}{3}$
7. $\frac{n}{3} = \frac{n+4}{7}$
8. $\frac{12q}{-7} = \frac{30}{14}$
9. $\frac{1}{y-3} = \frac{3}{y-5}$
10. $\frac{x}{8.71} = \frac{4}{17.42}$
11. $\frac{a-3}{8} = \frac{3}{4}$
12. $\frac{6p-2}{7} = \frac{5p+7}{8}$
13. $\frac{2}{9} = \frac{k+3}{2}$
14. $\frac{5m-3}{4} = \frac{5m+3}{6}$
15. $\frac{w-5}{4} = \frac{w+3}{3}$
16. $\frac{96.8}{t} = \frac{12.1}{7}$
17. $\frac{r-1}{r+1} = \frac{3}{5}$
18. $\frac{4n+5}{5} = \frac{2n+7}{7}$

Lesson 2-7

(pages 111–115)

State whether each percent of change is a percent of increase or a percent of decrease. Then find each percent of change. Round to the nearest whole percent.

1. original: \$100
   new: \$67
2. original: 62 acres
   new: 98 acres
3. original: 322 people
   new: 289 people
4. original: 78 pennies
   new: 36 pennies
5. original: \$212
   new: \$230
6. original: 35 mph
   new: 65 mph

Find the final price of each item.

7. television: \$299
   discount: 20%
8. book: \$15.95
   sales tax: 7%
9. software: \$36.90
   sales tax: 6.25%
10. boots: \$49.99
    discount: 15%
    sales tax: 3.5%
11. jacket: \$65
    discount: 30%
    sales tax: 4%
12. backpack: \$28.95
    discount: 10%
    sales tax: 5%

Extra Practice

Lesson 2-8

(pages 117–121)

Solve each equation or formula for *x*.

1. $x + r = q$
2. $ax + 4 = 7$
3. $2bx - b = -5$
4. $\frac{x - c}{c + a} = a$
5. $\frac{x + y}{c} = d$
6. $\frac{ax + 1}{2} = b$
7. $d(x - 3) = 5$
8. $nx - a = bx + d$
9. $3x - r = r(-3 + x)$
10. $y = \frac{5}{9}(x - 32)$
11. $A = \frac{1}{2}h(x + y)$
12. $A = 2\pi r^2 + 2\pi rx$

Lesson 2-9

(pages 122–128)

1. **ADVERTISING** An advertisement for grape drink claims that the drink contains 10% grape juice. How much pure grape juice would have to be added to 5 quarts of the drink to obtain a mixture containing 40% grape juice?

2. **GRADES** In Ms. Pham's social studies class, a test is worth four times as much as homework. If a student has an average of 85% on tests and 95% on homework, what is the student's average?

3. **ENTERTAINMENT** At the Golden Oldies Theater, tickets for adults cost \$5.50 and tickets for children cost \$3.50. How many of each kind of ticket were purchased if 21 tickets were bought for \$83.50?

4. **FOOD** Wes is mixing peanuts and chocolate pieces. Peanuts sell for \$4.50 a pound and the chocolate sells for \$6.50 a pound. How many pounds of chocolate mixes with 5 pounds of peanuts to obtain a mixture that sells for \$5.25 a pound?

5. **TRAVEL** Missoula and Bozeman are 210 miles apart. Sheila leaves Missoula for Bozeman and averages 55 miles per hour. At the same time, Casey leaves Bozeman and averages 65 miles per hour as he drives to Missoula. When will they meet? How far will they be from Bozeman?

Lesson 3-1

(pages 143–148)

Express each relation as a table, a graph, and a mapping. Then determine the domain and range.

1. $\{(5, 2), (0, 0), (-9, -1)\}$
2. $\{(-4, 2), (-2, 0), (0, 2), (2, 4)\}$

Express the relation shown in each table, mapping, or graph as a set of ordered pairs. Then write the inverse of the relation.

3.

x	y
1	3
2	4
3	5
4	6
5	7

4.

x	y
−4	1
−2	3
0	1
2	3
4	1

5.

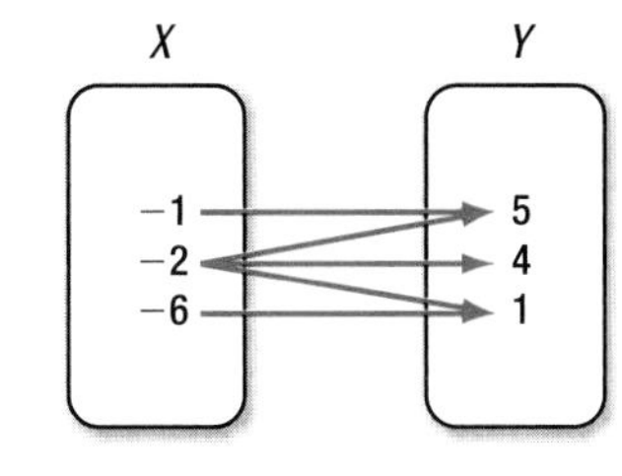

6.

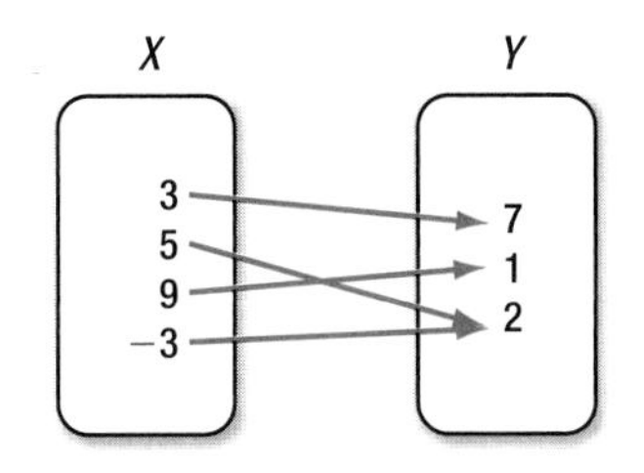

7.

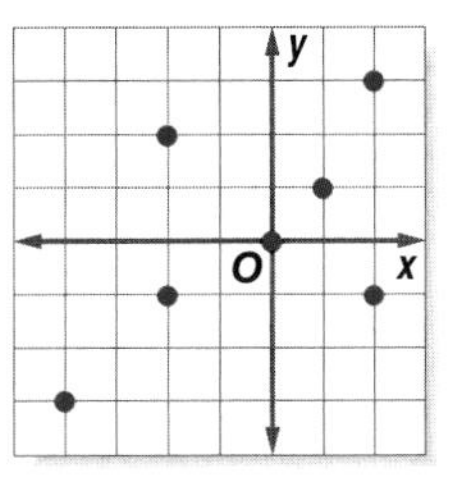

8.

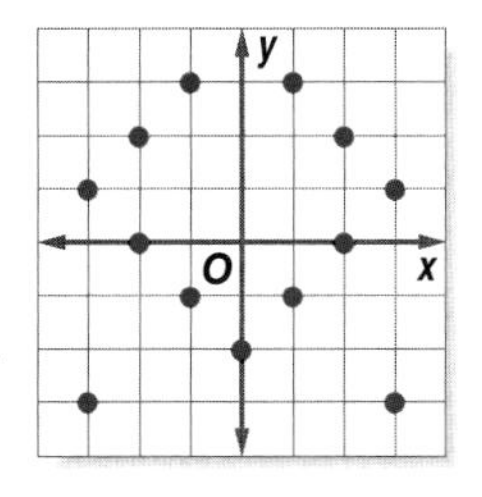

Lesson 3-2

(pages 149–154)

Determine whether each relation is a function.

1.

x	y
1	3
2	5
1	−7
2	9

2.

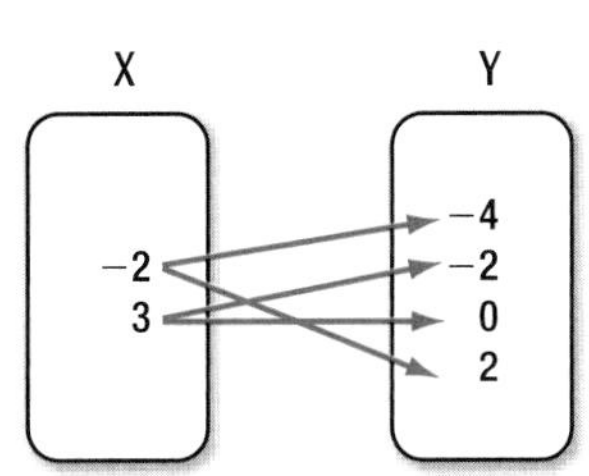

3.

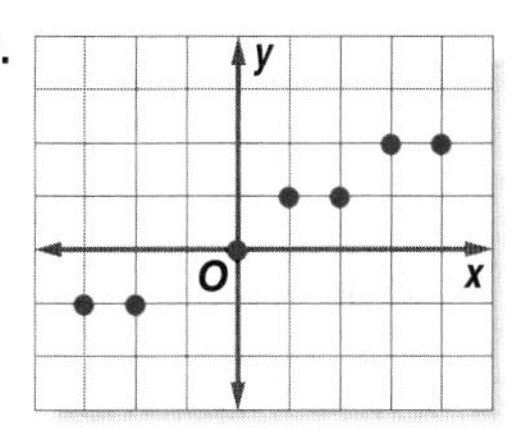

4. $x^2 + y = 11$
5. $y = 2$
6. $\{(-2, 4), (1, 3), (5, 2), (1, 4)\}$

If $f(x) = 2x + 5$ and $g(x) = 3x^2 - 1$, find each value.

7. $f(-4)$
8. $g(2)$
9. $f(3) - 5$
10. $g(a + 1)$

Lesson 3-3

(pages 156–162)

Determine whether each equation is a linear equation. If so, write the equation in standard form.

1. $3x = 2y$
2. $2x - 3 = y^2$
3. $4x = 2y + 8$
4. $5x - 7y = 2x - 7$
5. $2x + 5x = 7y + 2$
6. $\frac{1}{x} + \frac{5}{y} = -4$

Graph each equation.

7. $3x + y = 4$
8. $y = 3x + 1$
9. $3x - 2y = 12$
10. $2x - y = 6$
11. $2x - 3y = 8$
12. $y = -2$
13. $y = 5x - 7$
14. $x = 4$
15. $x + \frac{1}{3}y = 2$
16. $5x - 2y = 8$
17. $4.5x + 2.5y = 9$
18. $\frac{1}{2}x + 3y = 12$

Extra Practice

Lesson 3-4

(pages 166–171)

Determine whether each sequence is an arithmetic sequence. If it is, state the common difference.

1. $-2, -1, 0, 1, \ldots$
2. $3, 5, 8, 12, \ldots$
3. $2, 4, 8, 16, \ldots$
4. $-21, -16, -11, -6, \ldots$
5. $0, 0.25, 0.5, 0.75, \ldots$
6. $\frac{1}{3}, \frac{1}{9}, \frac{1}{27}, \frac{1}{81}, \ldots$

Find the next three terms of each arithmetic sequence.

7. $3, 13, 23, 33, \ldots$
8. $-4, -6, -8, -10, \ldots$
9. $-2, -1.4, -0.8, -0.2, \ldots$
10. $5, 13, 21, 29, \ldots$
11. $\frac{3}{4}, \frac{7}{8}, 1, \frac{9}{8}, \ldots$
12. $-\frac{1}{3}, -\frac{5}{6}, -\frac{4}{3}, -\frac{11}{6}, \ldots$

Find the *n*th term of each arithmetic sequence described.

13. $a_1 = 3, d = 6, n = 12$
14. $a_1 = -2, d = 4, n = 8$
15. $a_1 = -1, d = -3, n = 10$
16. $2\frac{1}{2}, 2\frac{1}{8}, 1\frac{3}{4}, 1\frac{3}{8}, \ldots$ for $n = 10$

Write an equation for the *n*th term of the arithmetic sequence. Then graph the first five terms in the sequence.

17. $-3, 1, 5, 9, \ldots$
18. $25, 40, 55, 70, \ldots$
19. $-9, -3, 3, 9, \ldots$
20. $-3.5, -2, -0.5, \ldots$

Lesson 3-5

(pages 172–176)

Find the next three terms in each sequence.

1. $12, 23, 34, 45, \ldots$
2. $39, 33, 27, 21, \ldots$
3. $6.0, 7.2, 8.4, 9.6, \ldots$
4. $15, 16, 18, 21, 25, 30, \ldots$
5. $w - 2, w - 4, w - 6, \ldots$
6. $13, 10, 11, 8, 9, 6, \ldots$

Write an equation in function notation for each relation.

7. 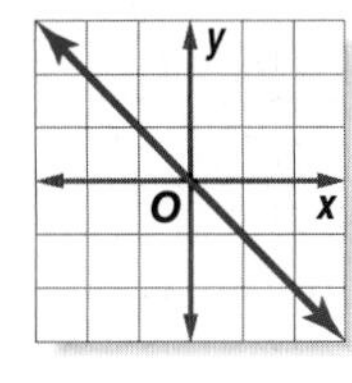

8. 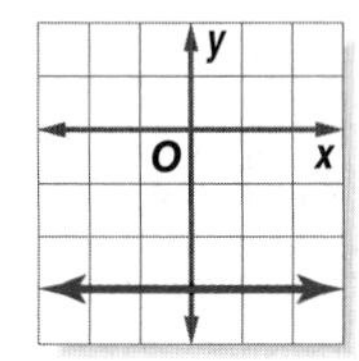

9. 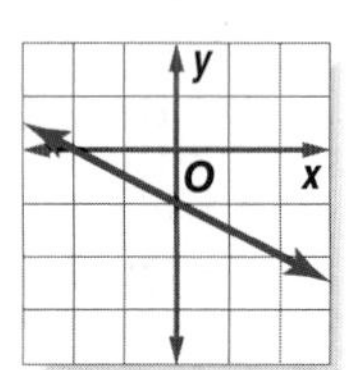

10. 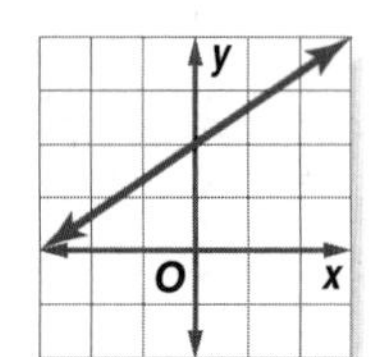

Lesson 4-1

(pages 187–195)

Find the slope of the line that passes through each pair of points.

1. 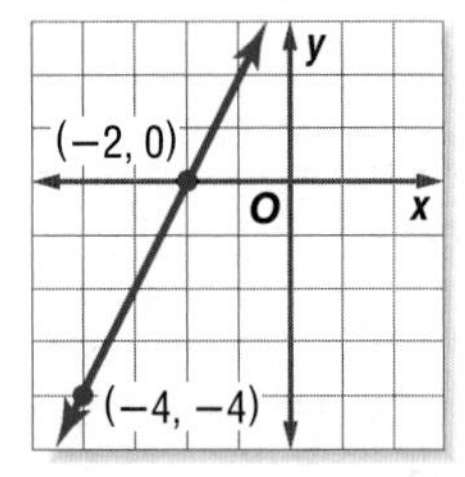

2. 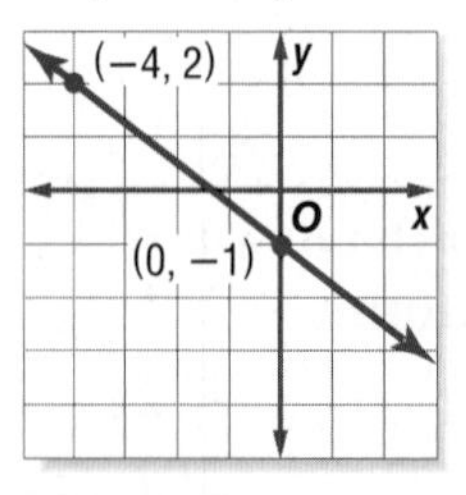

3. $(-2, 2), (3, -3)$
4. $(-2, -8), (1, 4)$
5. $(3, 4), (4, 6)$
6. $(-5, 4), (-1, 11)$
7. $(18, -4), (6, -10)$
8. $(-4, -6), (-4, -8)$
9. $(0, 0), (-1, 3)$
10. $(-8, 1), (2, 1)$

Find the value of *r* so the line that passes through each pair of points has the given slope.

11. $(-1, r), (1, -4), m = -5$
12. $(r, -2), (-7, -1), m = -\frac{1}{4}$

Lesson 4-2

(pages 196–202)

Name the constant of variation for each equation. Then determine the slope of the line that passes through each pair of points.

1. 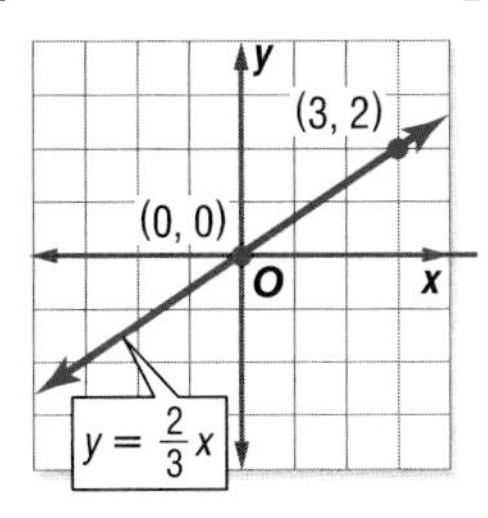

2. 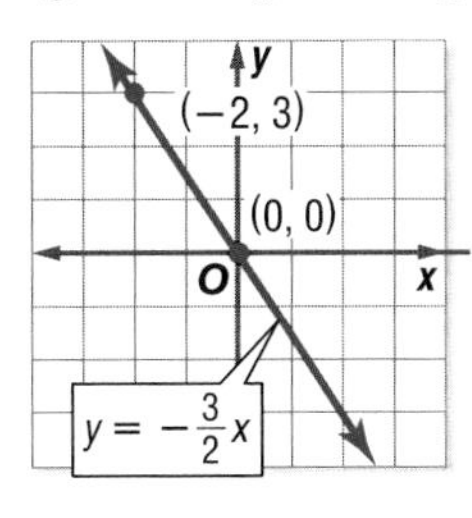

3. 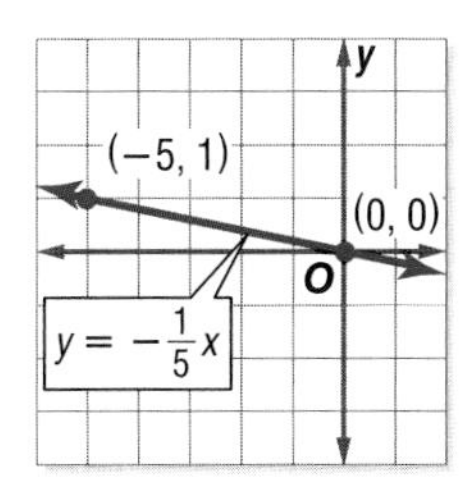

Graph each equation.

4. $y = 5x$
5. $y = -6x$
6. $y = -\frac{4}{3}x$

Suppose y varies directly as x. Write a direct variation equation that relates x and y. Then solve.

7. If $y = 45$ when $x = 9$, find y when $x = 7$.
8. If $y = -7$ when $x = -1$, find x when $y = -84$.

Lesson 4-3

(pages 204–209)

Write an equation, in slope-intercept form, of the line with the given slope and y-intercept.

1. m: 5, y-intercept: -15
2. m: -6, y-intercept: 3
3. m: 0.3, y-intercept: -2.6
4. m: $-\frac{4}{3}$, y-intercept: $\frac{5}{3}$
5. m: $-\frac{2}{5}$, y-intercept: 2
6. m: $\frac{7}{4}$, y-intercept: -2

Write an equation in slope-intercept form of the line shown in each graph.

7. 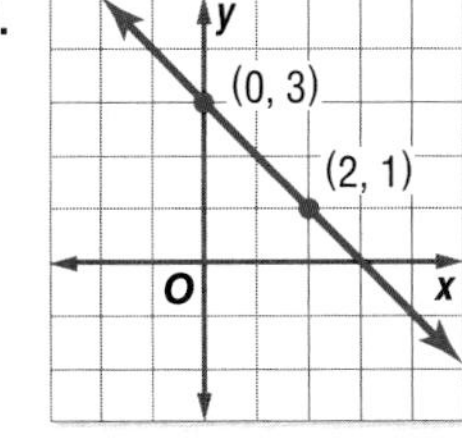

8. 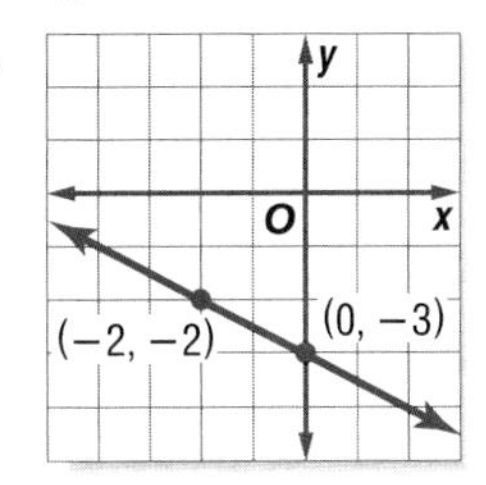

9. 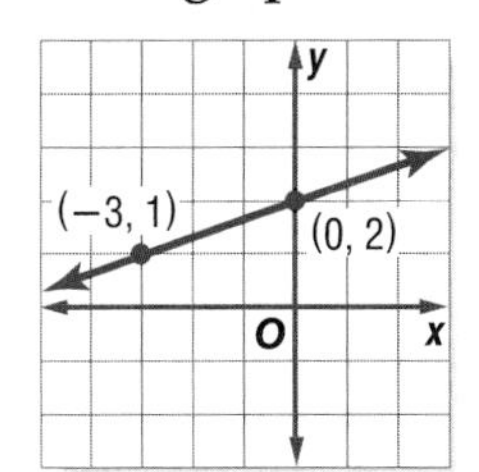

Graph each equation.

10. $y = 5x - 1$
11. $y = -2x + 3$
12. $3x - y = 6$

Lesson 4-4

(pages 213–218)

Write an equation of the line that passes through each point with the given slope.

1. $(0, 0)$; $m = -2$
2. $(-3, 2)$; $m = 4$
3. $(0, 5)$; $m = -1$
4. $(-2, 3)$; $m = -\frac{1}{4}$
5. $(1, -5)$; $m = \frac{2}{3}$
6. $\left(\frac{1}{2}, \frac{1}{4}\right)$; $m = 8$

Write an equation of the line that passes through each pair of points.

7. $(-1, 7), (8, -2)$
8. $(4, 0), (0, 5)$
9. $(8, -1), (7, -1)$
10. $(-2, 3), (1, 3)$
11. $(0, 0), (-4, 3)$
12. $\left(-\frac{1}{2}, \frac{1}{2}\right), \left(\frac{1}{4}, \frac{3}{4}\right)$

Lesson 4-5

(pages 220–225)

Write the point-slope form of an equation for a line that passes through each point with the given slope.

1. $(5, -2), m = 3$
2. $(0, 6), m = -2$
3. $(-3, 1), m = 0$
4. $(-2, -4), m = \frac{3}{4}$

Write each equation in standard form.

5. $y + 3 = 2(x - 4)$
6. $y + 3 = -\frac{1}{2}(x + 6)$
7. $y - 4 = -\frac{2}{3}(x - 5)$
8. $y + 2 = \frac{4}{3}(x - 6)$
9. $y - 1 = 1.5(x + 3)$
10. $y + 6 = -3.8(x - 2)$

Write each equation in slope-intercept form.

11. $y - 1 = -2(x + 5)$
12. $y + 3 = 4(x - 1)$
13. $y - 6 = -4(x - 2)$
14. $y + 1 = \frac{4}{5}(x + 5)$
15. $y - 2 = -\frac{3}{4}(x - 2)$
16. $y + \frac{1}{4} = \frac{2}{3}\left(x + \frac{1}{2}\right)$

Lesson 4-6

(pages 227–233)

Determine whether each graph shows a *positive correlation,* a *negative correlation,* or *no correlation.* If there is a correlation, describe its meaning in the situation.

1.

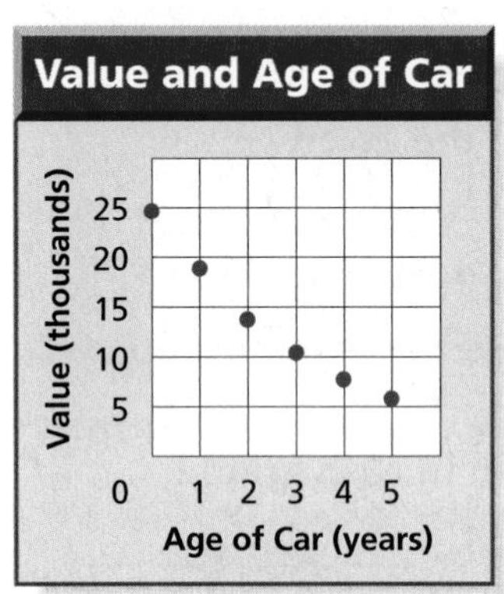

2.

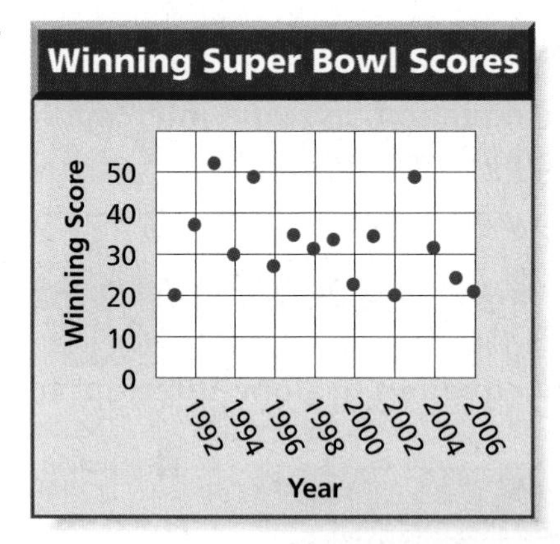

Use the scatter plot that shows the year and the number of TV-owning households in millions.

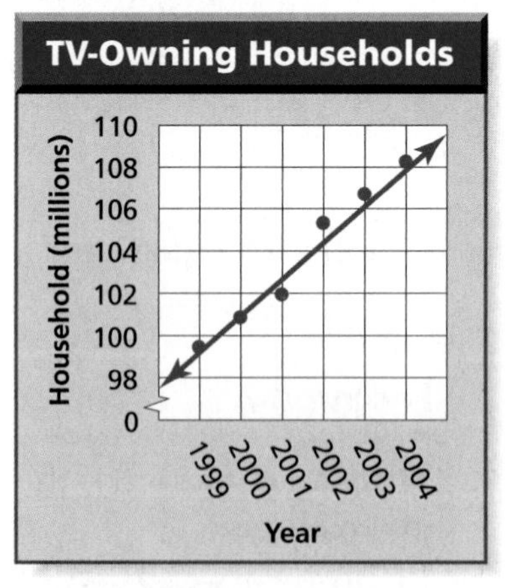

3. Describe the relationship that exists in the data.
4. Let x represent years since 1999. Use the points (1, 100.8) and (4, 106.7) to write the slope-intercept form of an equation for the line of fit shown in the scatter plot.
5. Predict the number of TV-owning households in 2015.

Lesson 4-7

(pages 236–251)

Write the slope-intercept form of an equation for the line that passes through the given point and is parallel to the graph of each equation. Then write an equation in slope-intercept form for the line that passes through the given point and is perpendicular to the graph of each equation.

1. $(1, 6), y = 4x - 2$
2. $(4, 6), y = 2x - 7$
3. $(-3, 0), y = \frac{2}{3}x + 1$
4. $(5, -2), y = -3x - 7$
5. $(0, 4), 3x + 8y = 4$
6. $(2, 3), x - 5y = 7$

Lesson 5-1

(pages 253–258)

Graph each system of equations. Then determine whether the system has *no* solution, *one* solution, or *infinitely many* solutions. If the system has one solution, name it.

1. $y = 3x$
   $4x + 2y = 30$
2. $x = -2y$
   $x + y = 1$
3. $y = x + 4$
   $3x + 2y = 18$
4. $x + y = 6$
   $x - y = 2$
5. $x + y = 6$
   $3x + 3y = 3$
6. $y = -3x$
   $4x + y = 2$
7. $2x + y = 8$
   $x - y = 4$
8. $\frac{1}{5}x - y = \frac{12}{5}$
   $3x - 5y = 6$
9. $x + 2y = 0$
   $y + 3 = -x$
10. $x + 2y = -9$
    $x - y = 6$
11. $x + \frac{1}{2}y = 3$
    $y = 3x - 4$
12. $\frac{2}{3}x + \frac{1}{2}y = 2$
    $4x + 3y = 12$

Lesson 5-2

(pages 260–265)

Use substitution to solve each system of equations. If the system does *not* have exactly one solution, state whether it has *no* solutions or *infinitely many* solutions.

1. $y = x$
   $5x = 12y$
2. $y = 7 - x$
   $2x - y = 8$
3. $x = 5 - y$
   $3y = 3x + 1$
4. $3x + y = 6$
   $y + 2 = x$
5. $x - 3y = 3$
   $2x + 9y = 11$
6. $3x = -18 + 2y$
   $x + 3y = 4$
7. $x + 2y = 10$
   $-x + y = 2$
8. $2x = 3 - y$
   $2y = 12 - x$
9. $6y - x = -36$
   $y = -3x$
10. $\frac{3}{4}x + \frac{1}{3}y = 1$
    $x - y = 10$
11. $x + 6y = 1$
    $3x - 10y = 31$
12. $3x - 2y = 12$
    $\frac{3}{2}x - y = 3$
13. $2x + 3y = 5$
    $4x - 9y = 9$
14. $x = 4 - 8y$
    $3x + 24y = 12$
15. $3x - 2y = -3$
    $25x + 10y = 215$

Lesson 5-3

(pages 266–270)

Use elimination to solve each system of equations.

1. $x + y = 7$
   $x - y = 9$
2. $2x - y = 32$
   $2x + y = 60$
3. $-y + x = 6$
   $y + x = 5$
4. $s + 2t = 6$
   $3s - 2t = 2$
5. $x = y - 7$
   $2x - 5y = -2$
6. $3x + 5y = -16$
   $3x - 2y = -2$
7. $x - y = 3$
   $x + y = 3$
8. $x + y = 8$
   $2x - y = 6$
9. $2s - 3t = -4$
   $s = 7 - 3t$
10. $-6x + 16y = -8$
    $6x - 42 = 16y$
11. $3x + 0.2y = 7$
    $3x = 0.4y + 4$
12. $9x + 2y = 26$
    $1.5x - 2y = 13$
13. $x = y$
    $x + y = 7$
14. $4x - \frac{1}{3}y = 8$
    $5x + \frac{1}{3}y = 6$
15. $2x - y = 3$
    $\frac{2}{3}x - y = -1$

Extra Practice

Lesson 5-4

(pages 272–278)

Use elimination to solve each system of equations.

1. $-3x + 2y = 10$
   $-2x - y = -5$
2. $2x + 5y = 13$
   $4x - 3y = -13$
3. $5x + 3y = 4$
   $-4x + 5y = -18$
4. $\frac{1}{3}x - y = -1$
   $\frac{1}{5}x - \frac{2}{5}y = -1$
5. $3x - 5y = 8$
   $4x - 7y = 10$
6. $x - 0.5y = 1$
   $0.4x + y = -2$
7. $x + 8y = 3$
   $4x - 2y = 7$
8. $4x - y = 4$
   $x + 2y = 3$
9. $3y - 8x = 9$
   $y - x = 2$
10. $x + 4y = 30$
    $2x - y = -6$
11. $3x - 2y = 0$
    $4x + 4y = 5$
12. $9x - 3y = 5$
    $x + y = 1$
13. $2x - 7y = 9$
    $-3x + 4y = 6$
14. $2x - 6y = -16$
    $5x + 7y = -18$
15. $6x - 3y = -9$
    $-8x + 2y = 4$

Lesson 5-5

(pages 280–284)

Determine the best method to solve each system of equations. Then solve the system.

1. $y = 2x + 1$
   $y = -3x + 1$
2. $y = 5x - 8$
   $y = 3x$
3. $x + 2y = -6$
   $x = y + 3$
4. $2x - 3y = 5$
   $y = -6x$
5. $x = -1$
   $y = 8$
6. $4x + y = 5$
   $-4x - 2y = 9$
7. $-7x + 3y = -4$
   $2x + 3y = 5$
8. $4x - y = 11$
   $x + 2y = 5$
9. $2y - x = -7$
   $x + 3y = 5$
10. $-13x + 8y = -6$
    $3x - 4y = 2$
11. $-x + 7y = 9$
    $-4x + 6y = -8$
12. $2x + 5y = 7$
    $5x - 2y = 13$
13. $12x - 3y = 7$
    $x = 2 + 13y$
14. $6x = 5$
    $9y - 2x = 7$
15. $17x + 8y = -4$
    $-8y - 2x = 9$

Lesson 6-1

(pages 294–299)

Solve each inequality. Check your solution, and then graph it on a number line.

1. $c + 9 \leq 3$
2. $d - (-3) < 13$
3. $z - 4 > 20$
4. $h - (-7) > -2$
5. $-11 > d - 4$
6. $2x > x - 3$
7. $2x - 3 \geq x$
8. $16 + w < -20$
9. $14p > 5 + 13p$
10. $-7 < 16 - z$
11. $1.1v - 1 > 2.1v - 3$
12. $\frac{1}{2}t + \frac{1}{4} \geq \frac{3}{2}t - \frac{2}{3}$
13. $9x < 8x - 2$
14. $-2 + 9n \leq 10n$
15. $a - 2.3 \geq -7.8$
16. $5z - 6 > 4z$

Define a variable, write an inequality, and solve each problem. Then check your solution.

17. The sum of a number and negative six is greater than 9.
18. Negative five times a number is less than the sum of negative six times the number and 12.

Lesson 6-2

(pages 301–307)

Solve each inequality. Check your solution.

1. $7b \geq -49$
2. $-5j < -60$
3. $\frac{w}{3} > -12$
4. $\frac{p}{5} < 8$
5. $-8f < 48$
6. $-0.25t \geq -10$
7. $\frac{g}{-8} < 4$
8. $-4.3x < -2.58$
9. $4c \geq -6$
10. $6 \leq 0.8n$
11. $\frac{2}{3}m \geq -22$
12. $-25 > -0.05a$
13. $-15a < -28$
14. $-\frac{7}{9}x < 42$
15. $0.375y \leq 32$
16. $-7y \geq 91$

Define a variable, write an inequality, and solve each problem. Then check your solution.

17. Negative one times a number is greater than -7.
18. Three fifths of a number is at least negative 10.
19. Seventy-five percent of a number is at most 100.

Lesson 6-3

(pages 308–313)

Solve each inequality. Check your solution.

1. $3y - 4 > -37$
2. $7s - 12 < 13$
3. $-5q + 9 > 24$
4. $-6v - 3 \geq -33$
5. $-2k + 12 < 30$
6. $-2x + 1 < 16 - x$
7. $15t - 4 > 11t - 16$
8. $13 - y \leq 29 + 2y$
9. $5q + 7 \leq 3(q + 1)$
10. $2(w + 4) \geq 7(w - 1)$
11. $-4t - 5 > 2t + 13$
12. $\left\{\frac{2t + 5}{3}\right\} < -9$
13. $\frac{z}{4} + 7 \geq -5$
14. $13r - 11 > 7r + 37$
15. $8c - (c - 5) > c + 17$
16. $-5(k + 4) \geq 3(k - 4)$
17. $9m + 7 < 2(4m - 1)$
18. $3(3y + 1) < 13y - 8$
19. $5x \leq 10(3x + 4)$
20. $3\left(a + \frac{2}{3}\right) \geq a - 1$

Lesson 6-4

(pages 315–320)

Solve each compound inequality. Then graph the solution set.

1. $2 + x < -5$ or $2 + x > 5$
2. $-4 + t > -5$ or $-4 + t < 7$
3. $3 \leq 2g + 7$ and $2g + 7 \leq 15$
4. $2v - 2 \leq 3v$ and $4v - 1 \geq 3v$
5. $3b - 4 \leq 7b + 12$ and $8b - 7 \leq 25$
6. $-9 < 2z + 7 < 10$
7. $5m - 8 \geq 10 - m$ or $5m + 11 < -9$
8. $12c - 4 \leq 5c + 10$ or $-4c - 1 \leq c + 24$
9. $2h - 2 \leq 3h \leq 4h - 1$
10. $3p + 6 < 8 - p$ and $5p + 8 \geq p + 6$
11. $2r + 8 > 16 - 2r$ and $7r + 21 < r - 9$
12. $-4j + 3 < j + 22$ and $j - 3 < 2j - 15$
13. $2(q - 4) \leq 3(q + 2)$ or $q - 8 \leq 4 - q$
14. $\frac{1}{2}w + 5 \geq w + 2 \geq \frac{1}{2}w + 9$
15. $n - (6 - n) > 10$ or $-3n - 1 > 20$
16. $-(2x + 5) \leq x + 5 \leq 2x - 9$

Extra Practice

Lesson 6-5

(pages 322–327)

Solve each open sentence. Then graph the solution set.

1. $|c - 5| = 4$
2. $|e + 3| = 7$
3. $|4 - g| = 6$
4. $|10 - k| = 8$
5. $|2j + 4| = 12$
6. $|2r - 6| = 10$
7. $|6 - 3w| = 8$
8. $|7 + 2x| = 14$
9. $|4z + 6| = 12$

Lesson 6-6

(pages 329–333)

Solve each open sentence. Then graph the solution set.

1. $|y - 9| < 19$
2. $|g + 6| > 8$
3. $|t - 5| \leq 3$
4. $|a + 5| \leq 0$
5. $|2m - 5| > 13$
6. $|14 - w| \geq 20$
7. $|3p + 5| \leq 23$
8. $|6b - 12| \leq 36$
9. $|25 - 3x| < 5$
10. $|4 - 5s| > 46$
11. $|4 - (1 - x)| \geq 10$
12. $\left|\frac{7 - 2b}{2}\right| \leq 3$

Lesson 6-7

(pages 334–339)

Determine which ordered pairs are part of the solution set for each inequality.

1. $x + y \geq 0$, $\{(0, 0), (1, -3), (2, 2), (3, -3)\}$
2. $2x + y \leq 8$, $\{(0, 0), (-1, -1), (3, -2), (8, 0)\}$

Graph each inequality.

3. $y \leq -2$
4. $x < 4$
5. $x + y < -2$
6. $x + y \geq -4$
7. $y > 4x - 1$
8. $3x + y > 1$

Lesson 6-8

(pages 341–345)

Solve each system of inequalities by graphing.

1. $x > 3$
$y < 6$
2. $y > 2$
$y > -x + 2$
3. $x \leq 2$
$y + 3 \geq 5$
4. $x + y \leq -1$
$2x + y \leq 2$
5. $y \geq 2x + 2$
$y \geq -x - 1$
6. $y \leq x + 3$
$y \geq x + 2$
7. $y - x \geq 0$
$y \leq 3$
$x \geq 0$
8. $y > 2x$
$x > -3$
$y < 4$
9. $y \leq x$
$x + y < 4$
$y \geq -3$

Lesson 7-1

(pages 358–364)

Determine whether each expression is a monomial. Write *yes* or *no*. Explain.

1. $n^2 - 3$
2. 53
3. $9a^2b^3$
4. $15 - x^2y$

Simplify.

5. $a^5(a)(a^7)$
6. $(r^3t^4)(r^4t^4)$
7. $(x^3y^4)(xy^3)$
8. $(bc^3)(b^4c^3)$
9. $(-3mn^2)(5m^3n^2)$
10. $[(3^3)^2]^2$
11. $(3s^3t^2)(-4s^3t^2)$
12. $x^3(x^4y^3)$
13. $(1.1g^2h^4)^3$
14. $-\frac{3}{4}a(a^2b^3c^4)$
15. $\left(\frac{1}{2}w^3\right)^2(w^4)^2$
16. $[(-2^3)^3]^2$

Lesson 7-2

(pages 366–373)

Simplify. Assume that no denominator is equal to zero.

1. $\frac{6^{10}}{6^7}$
2. $\frac{b^6c^5}{b^3c^2}$
3. $\frac{(-a)^4b^8}{a^4b^7}$
4. $\frac{(-x)^3y^3}{x^3y^6}$
5. $\frac{12ab^5}{4a^4b^3}$
6. $\frac{24x^5}{-8x^2}$
7. $\frac{-9h^2k^4}{18h^5j^3k^4}$
8. $\left(\frac{2a^2b^4}{3a^3b}\right)^2$
9. $a^5b^0a^{-7}$
10. $\frac{(-u^{-3}v^3)^2}{(u^3v)^{-3}}$
11. $\left(\frac{a^3}{b^2}\right)^{-3}$
12. $\left(\frac{2x}{y^{-3}}\right)^{-2}$
13. $\frac{(-r)s^5}{r^{-3}s^{-4}}$
14. $\frac{28a^{-4}b^0}{14a^3b^{-1}}$
15. $\frac{(j^2k^3m)^4}{(jk^4)^{-1}}$
16. $\left(\frac{-2x^4y}{4y^2}\right)^0$
17. $\left(\frac{-18x^0a^{-3}}{-6x^{-2}a^{-3}}\right)$
18. $\left(\frac{2a^3b^{-2}}{2^{-1}a^{-5}b^3}\right)^{-1}$
19. $\left(\frac{5n^{-1}m^2}{2nm^{-2}}\right)^0$
20. $\frac{(3ab^2c)^{-3}}{(2a^2bc^2)^2}$

Lesson 7-3

(pages 376–381)

State whether each expression is a polynomial. If the expression is a polynomial, identify it as a *monomial,* a *binomial,* or a *trinomial.*

1. $5x^2y + 3xy - 7$
2. 0
3. $\frac{5}{k} - k^2y$
4. $3a^2x - 5a$

Find the degree of each polynomial.

5. $a + 5c$
6. $14abcd - 6d^3$
7. $\frac{a^3}{4}$
8. 10
9. $-4h^5$
10. $\frac{x^2}{3} - \frac{x}{2} + \frac{1}{5}$
11. -6
12. $a^2b^3 - a^3b^2$

Arrange the terms of each polynomial so that the powers of *x* are in ascending order.

13. $2x^2 - 3x + 4x^3 - x^5$
14. $x^3 - x^2 + x - 1$
15. $2a + 3ax^2 - 4ax$
16. $-5bx^3 - 2bx + 4x^2 - b^3$
17. $x^8 + 2x^2 - x^6 + 1$
18. $cdx^2 - c^2d^2x + d^3$

Arrange the terms of each polynomial so that the powers of *x* are in descending order.

19. $5x^2 - 3x^3 + 7 + 2x$
20. $-6x + x^5 + 4x^3 - 20$
21. $5b + b^3x^2 + \frac{2}{3}bx$
22. $21p^2x + 3px^3 + p^4$
23. $3ax^2 - 6a^2x^3 + 7a^3 - 8x$
24. $\frac{1}{3}s^2x^3 + 4x^4 - \frac{2}{5}s^4x^2$

Lesson 7-4

(pages 384–388)

Find each sum or difference.

1. $(3a^2 + 5) + (4a^2 - 1)$
2. $(5x - 3) + (-2x + 1)$
3. $(6z + 2) - (9z + 3)$
4. $(-4n + 7) - (-7n - 8)$
5. $(-7t^2 + 4ts - 6s^2) + (-5t^2 - 12ts + 3s^2)$
6. $(6a^2 - 7ab - 4b^2) - (2a^2 + 5ab + 6b^2)$
7. $(4a^2 - 10b^2 + 7c^2) + (-5a^2 + 2c^2 + 2b)$
8. $(z^2 + 6z - 8) - (4z^2 - 7z - 5)$
9. $(4d + 3e - 8f) - (-3d + 10e - 5f + 6)$
10. $(7g + 8h - 9) + (-g - 3h - 6k)$
11. $(9x^2 - 11xy - 3y^2) - (x^2 - 16xy + 12y^2)$
12. $(-3m + 9mn - 5n) + (14m - 5mn - 2n)$
13. $(6 - 7y + 3y^2) + (3 - 5y - 2y^2) + (-12 - 8y + y^2)$
14. $(-7c^2 - 2c - 5) + (9c - 6) + (16c^2 + 3) + (-9c^2 - 7c + 7)$

Extra Practice

Lesson 7-5

(pages 390–395)

Find each product.

1. $-3(8x + 5)$
2. $3b(5b + 8)$
3. $1.1a(2a + 7)$
4. $\frac{1}{2}x(8x - 6)$
5. $7xy(5x^2 - y^2)$
6. $5y(y^2 - 3y + 6)$
7. $-ab(3b^2 + 4ab - 6a^2)$
8. $4m^2(9m^2n + mn - 5n^2)$
9. $4st^2(-4s^2t^3 + 7s^5 - 3st^3)$

Simplify.

10. $-3a(2a - 12) + 5a$
11. $6(12b^2 - 2b) + 7(-2 - 3b)$
12. $x(x - 6) + x(x - 2) + 2x$
13. $11(n - 3) + 2(n^2 + 22n)$
14. $-2x(x + 3) + 3(x + 3)$
15. $4m(n - 1) - 5n(n + 1)$

Solve each equation.

16. $-6(11 - 2x) = 7(-2 - 2x)$
17. $11(n - 3) + 5 = 2n + 44$
18. $a(a - 6) + 2a = 3 + a(a - 2)$
19. $q(2q + 3) + 20 = 2q(q - 3)$
20. $w(w + 12) = w(w + 14) + 12$
21. $x(x - 3) + 4x - 3 = 8x + x(3 + x)$
22. $-3(x + 5) + x(x - 1) = x(x + 2) - 3$
23. $n(n - 5) + n(n + 2) = 2n(n - 1) + 1.5$

Lesson 7-6

(pages 398–403)

Find each product.

1. $(d + 2)(d + 5)$
2. $(z + 7)(z - 4)$
3. $(m - 8)(m - 5)$
4. $(a + 2)(a - 19)$
5. $(c + 15)(c - 3)$
6. $(x + y)(x - 2y)$
7. $(2x - 5)(x + 6)$
8. $(7a - 4)(2a - 5)$
9. $(4x + y)(2x - 3y)$
10. $(7v + 3)(v + 4)$
11. $(7s - 8)(3s - 2)$
12. $(4g + 3h)(2g - 5h)$
13. $(4a + 3)(2a - 1)$
14. $(7y - 1)(2y - 3)$
15. $(2x + 3y)(4x + 2y)$
16. $(12r - 4s)(5r + 8s)$
17. $(-a + 1)(-3a - 2)$
18. $(2n - 4)(-3n - 2)$
19. $(x - 2)(x^2 + 2x + 4)$
20. $(3x + 5)(2x^2 - 5x + 11)$
21. $(4s + 5)(3s^2 + 8s - 9)$
22. $(5x - 2)(-5x^2 + 2x + 7)$
23. $(-n + 2)(-2n^2 + n - 1)$
24. $(x^2 - 7x + 4)(2x^2 - 3x - 6)$
25. $(x^2 + x + 1)(x^2 - x - 1)$
26. $(a^2 + 2a + 5)(a^2 - 3a - 7)$
27. $(5x^4 - 2x^2 + 1)(x^2 - 5x + 3)$

Lesson 7-7

(pages 404–409)

Find each product.

1. $(t + 7)^2$
2. $(w - 12)(w + 12)$
3. $(q - 4h)^2$
4. $(10x + 11y)(10x - 11y)$
5. $(4p + 3)^2$
6. $(2b - 4d)(2b + 4d)$
7. $(a + 2b)^2$
8. $(3x + y)^2$
9. $(6m + 2n)^2$
10. $(3m - 7d)^2$
11. $(5b - 6)(5b + 6)$
12. $(1 + x)^2$
13. $(5x - 9y)^2$
14. $(8a - 2b)(8a + 2b)$
15. $\left(\frac{1}{4}x + 4\right)^2$
16. $\left(\frac{1}{2}x - 10\right)\left(\frac{1}{2}x + 10\right)$
17. $\left(\frac{1}{3}n - m\right)\left(\frac{1}{3}n + m\right)$
18. $(a - 1)(a - 1)(a - 1)$
19. $(x + 2)(x - 2)(2x + 5)$
20. $(4x - 1)(4x + 1)(x - 4)$
21. $(x - 5)(x + 5)(x + 4)(x - 4)$
22. $(a + 1)(a + 1)(a - 1)(a - 1)$
23. $(n - 1)(n + 1)(n - 1)$
24. $(2c + 3)(2c + 3)(2c - 3)(2c - 3)$
25. $(4d + 5g)(4d + 5g)(4d - 5g)(4d - 5g)$

Lesson 8-1

(pages 420–423)

Factor each monomial completely.

1. $240mn$
2. $-64a^3b$
3. $-26xy^2$
4. $-231xy^2z$
5. $44rs^2t^3$
6. $-756m^2n^2$

Find the GCF of each set of monomials.

7. 16, 60
8. 15, 50
9. 45, 80
10. 29, 58
11. 55, 305
12. 126, 252
13. 128, 245
14. $7y^2, 14y^2$
15. $4xy, -6x$
16. $35t^2, 7t$
17. $16pq^2, 12p^2q, 4pq$
18. 5, 15, 10
19. $12mn, 10mn, 15mn$
20. $14xy, 12y, 20x$
21. $26jk^4, 16jk^3, 8j^2$

Extra Practice

Lesson 8-2

(pages 426–431)

Factor each polynomial.

1. $10a^2 + 40a$
2. $15wx - 35wx^2$
3. $27a^2b + 9b^3$
4. $11x + 44x^2y$
5. $16y^2 + 8y$
6. $14mn^2 + 2mn$
7. $25a^2b^2 + 30ab^3$
8. $2m^3n^2 - 16mn^2 + 8mn$
9. $2ax + 6xc + ba + 3bc$
10. $6mx - 4m + 3rx - 2r$
11. $3ax - 6bx + 8b - 4a$
12. $a^2 - 2ab + a - 2b$
13. $8ac - 2ad + 4bc - bd$
14. $2e^2g + 2fg + 4e^2h + 4fh$
15. $x^2 - xy - xy + y^2$

Solve each equation. Check your solutions.

16. $a(a - 9) = 0$
17. $d(d + 11) = 0$
18. $z(z - 2.5) = 0$
19. $(2y + 6)(y - 1) = 0$
20. $(4n - 7)(3n + 2)$
21. $(a - 1)(a + 1) = 0$
22. $10x^2 - 20x = 0$
23. $8b^2 - 12b = 0$
24. $14d^2 + 49d = 0$
25. $15a^2 = 60a$
26. $33x^2 = -22x$
27. $32x^2 = 16x$

Lesson 8-3

(pages 434–439)

Factor each trinomial.

1. $x^2 - 9x + 14$
2. $a^2 - 9a - 36$
3. $x^2 + 2x - 15$
4. $n^2 - 8n + 15$
5. $b^2 + 22b + 21$
6. $c^2 + 2c - 3$
7. $x^2 - 5x - 24$
8. $n^2 - 8n + 7$
9. $m^2 - 10m - 39$
10. $z^2 + 15z + 36$
11. $s^2 - 13st - 30t^2$
12. $y^2 + 2y - 35$
13. $r^2 + 3r - 40$
14. $x^2 + 5x - 6$
15. $x^2 - 4xy - 5y^2$
16. $r^2 + 16r + 63$
17. $v^2 + 24v - 52$
18. $k^2 - 27kj - 90j^2$

Solve each equation. Check your solutions.

19. $a^2 + 3a - 4 = 0$
20. $x^2 - 8x - 20 = 0$
21. $b^2 + 11b + 24 = 0$
22. $y^2 + y - 42 = 0$
23. $k^2 + 2k - 24 = 0$
24. $r^2 - 13r - 48 = 0$
25. $n^2 - 9n = -18$
26. $2z + z^2 = 35$
27. $-20x + 19 = -x^2$
28. $10 + a^2 = -7a$
29. $z^2 - 57 = 16z$
30. $x^2 = -14x - 33$
31. $22x - x^2 = 96$
32. $-144 = q^2 - 26q$
33. $x^2 + 84 = 20x$

Lesson 8-4

(pages 441–446)

Factor each trinomial, if possible. If the trinomial cannot be factored using integers, write *prime*.

1. $4a^2 + 4a - 63$
2. $3x^2 - 7x - 6$
3. $4r^2 - 25r + 6$
4. $2z^2 - 11z + 15$
5. $3a^2 - 2a - 21$
6. $4y^2 + 11y + 6$
7. $6n^2 + 7n - 3$
8. $5x^2 - 17x + 14$
9. $2n^2 - 11n + 13$
10. $5a^2 - 3a + 15$
11. $18v^2 + 24v + 12$
12. $4k^2 + 2k - 12$
13. $10x^2 - 20xy + 10y^2$
14. $12c^2 - 11cd - 5d^2$
15. $30n^2 - mn - m^2$

Solve each equation. Check your solutions.

16. $8t^2 + 32t + 24 = 0$
17. $6y^2 + 72y + 192 = 0$
18. $5x^2 + 3x - 2 = 0$
19. $9x^2 + 18x - 27 = 0$
20. $4x^2 - 4x - 4 = 4$
21. $12n^2 - 16n - 3 = 0$
22. $12x^2 - x - 35 = 0$
23. $18x^2 + 36x - 14 = 0$
24. $15a^2 + a - 2 = 0$
25. $14b^2 + 7b - 42 = 0$
26. $13r^2 + 21r - 10 = 0$
27. $35y^2 - 60y - 20 = 0$
28. $16x^2 - 4x - 6 = 0$
29. $28d^2 + 5d - 3 = 0$
30. $30x^2 - 9x - 3 = 0$

Lesson 8-5

(pages 447–452)

Factor each polynomial, if possible. If the polynomial cannot be factored, write *prime*.

1. $x^2 - 9$
2. $a^2 - 64$
3. $4x^2 - 9y^2$
4. $1 - 9z^2$
5. $16a^2 - 9b^2$
6. $8x^2 - 12y^2$
7. $a^2 - 4b^2$
8. $75r^2 - 48$
9. $x^2 - 36y^2$
10. $3a^2 - 16$
11. $9x^2 - 100y^2$
12. $49 - a^2b^2$
13. $5a^2 - 48$
14. $169 - 16t^2$
15. $8r^2 - 4$
16. $-45m^2 + 5$

Solve each equation by factoring. Check your solutions.

17. $4x^2 = 16$
18. $2x^2 = 50$
19. $9n^2 - 4 = 0$
20. $a^2 - \frac{25}{36} = 0$
21. $\frac{16}{9} - b^2 = 0$
22. $18 - \frac{1}{2}x^2 = 0$
23. $20 - 5g^2 = 0$
24. $16 - \frac{1}{4}p^2 = 0$
25. $\frac{1}{4}c^2 - \frac{4}{9} = 0$
26. $2q^3 - 2q = 0$
27. $3r^3 = 48r$
28. $100d - 4d^3 = 0$

Lesson 8-6

(pages 454–460)

Determine whether each trinomial is a perfect square trinomial. If so, factor it.

1. $x^2 + 12x + 36$
2. $n^2 - 13n + 36$
3. $a^2 + 4a + 4$
4. $x^2 - 10x - 100$
5. $2n^2 + 17n + 21$
6. $4a^2 - 20a + 25$

Factor each polynomial, if possible. If the polynomial cannot be factored, write *prime*.

7. $3x^2 - 75$
8. $4p^2 + 12pr + 9r^2$
9. $6a^2 + 72$
10. $s^2 + 30s + 225$
11. $24x^2 + 24x + 9$
12. $1 - 10z + 25z^2$
13. $28 - 63b^2$
14. $4c^2 + 2c - 7$

Solve each equation. Check your solutions.

5. $x^2 + 22x + 121 = 0$
16. $343d^2 = 7$
17. $(a - 7)^2 = 5$
$^2 + 10c + 36 = 11$
19. $16s^2 + 81 = 72s$
20. $9p^2 - 42p + 20 = -29$

Lesson 9-1

(pages 471–477)

Use a table of values to graph each function.

1. $y = x^2 + 6x + 8$
2. $y = -x^2 + 3x$
3. $y = -x^2$

Write the equation of the axis of symmetry, and find the coordinates of the vertex of the graph of each function. Identify the vertex as a maximum or minimum. Then graph the function.

4. $y = -x^2 + 2x - 3$
5. $y = 3x^2 + 24x + 80$
6. $y = x^2 - 4x - 4$
7. $y = 5x^2 - 20x + 37$
8. $y = 3x^2 + 6x + 3$
9. $y = 2x^2 + 12x$
10. $y = x^2 - 6x + 5$
11. $y = x^2 + 6x + 9$
12. $y = -x^2 + 16x - 15$
13. $y = 4x^2 - 1$
14. $y = -2x^2 - 2x + 4$
15. $y = 6x^2 - 12x - 4$
16. $y = -x^2 - 1$
17. $y = -x^2 + x + 1$
18. $y = -5x^2 - 3x + 2$

Lesson 9-2

(pages 480–485)

Solve each equation by graphing.

1. $a^2 - 25 = 0$
2. $n^2 - 8n = 0$
3. $d^2 + 36 = 0$
4. $b^2 - 18b + 81 = 0$
5. $x^2 + 3x + 27 = 0$
6. $-y^2 - 3y + 10 = 0$

Solve each equation by graphing. If integral roots cannot be found, estimate the roots by stating the consecutive integers between which the roots lie.

7. $x^2 + 2x - 3 = 0$
8. $-x^2 + 6x - 5 = 0$
9. $-a^2 - 2a + 3 = 0$
10. $2r^2 - 8r + 5 = 0$
11. $-3x^2 + 6x - 9 = 0$
12. $c^2 + c = 0$
13. $3t^2 + 2 = 0$
14. $-b^2 + 5b + 2 = 0$
15. $3x^2 + 7x = 1$
16. $x^2 + 5x - 24 = 0$
17. $8 - n^2 = 0$
18. $x^2 - 7x = 18$
19. $a^2 + 12a + 36 = 0$
20. $64 - x^2 = 0$
21. $-4x^2 + 2x = -1$
22. $5z^2 + 8z = 1$
23. $p = 27 - p^2$
24. $6w = -15 - 3w^2$

Lesson 9-3

(pages 486–491)

Solve each equation by taking the square root of each side. Round to the nearest tenth if necessary.

1. $x^2 - 4x + 4 = 9$
2. $t^2 - 6t + 9 = 16$
3. $b^2 + 10b + 25 = 11$
4. $a^2 - 22a + 121 = 3$
5. $x^2 + 2x + 1 = 81$
6. $t^2 - 36t + 324 = 85$

Find the value of c that makes each trinomial a perfect square.

7. $a^2 + 20a + c$
8. $x^2 + 10x + c$
9. $t^2 + 12t + c$
10. $y^2 - 9y + c$
11. $p^2 - 14p + c$
12. $b^2 + 13b + c$

Solve each equation by completing the square. Round to the nearest tenth if necessary.

13. $a^2 - 8a - 84 = 0$
14. $c^2 + 6 = -5c$
15. $p^2 - 8p + 5 = 0$
16. $2y^2 + 7y - 4 = 0$
17. $t^2 + 3t = 40$
18. $x^2 + 8x - 9 = 0$
19. $y^2 + 5y - 84 = 0$
20. $t^2 + 12t + 32 = 0$
21. $2x - 3x^2 = -8$
22. $2y^2 - y - 9 = 0$
23. $2z^2 - 5z - 4 = 0$
24. $8t^2 - 12t - 1 = 0$

Extra Practice

Lesson 9-4

(pages 493–499)

Solve each equation by using the Quadratic Formula. Round to the nearest tenth if necessary.

1. $x^2 - 8x - 4 = 0$
2. $x^2 + 7x - 8 = 0$
3. $x^2 - 5x + 6 = 0$
4. $y^2 - 7y - 8 = 0$
5. $m^2 - 2m = 35$
6. $4n^2 - 20n = 0$
7. $m^2 + 4m + 2 = 0$
8. $2t^2 - t - 15 = 0$
9. $5t^2 = 125$
10. $t^2 + 16 = 0$
11. $-4x^2 + 8x = -3$
12. $3k^2 + 2 = -8k$
13. $8t^2 + 10t + 3 = 0$
14. $3x^2 - \frac{5}{4}x - \frac{1}{2} = 0$
15. $-5b^2 + 3b - 1 = 0$
16. $n^2 - 3n + 1 = 0$
17. $2z^2 + 5z - 1 = 0$
18. $3h^2 = 27$

State the value of the discriminant for each equation. Then determine the number of real roots of the equation.

19. $3f^2 + 2f = 6$
20. $2x^2 = 0.7x + 0.3$
21. $3w^2 - 2w + 8 = 0$
22. $4r^2 - 12r + 9 = 0$
23. $x^2 - 5x = -9$
24. $25t^2 + 30t = -9$

Lesson 9-5

(pages 502–508)

Graph each function. State the y-intercept. Then use the graph to determine the approximate value of the given expression. Use a calculator to confirm the value.

1. $y = 7^x; 7^{1.5}$
2. $\left(\frac{1}{3}\right)^x; \left(\frac{1}{3}\right)^{5.6}$
3. $y = \left(\frac{3}{5}\right)^x; \left(\frac{3}{5}\right)^{-4.2}$

Graph each function. State the y-intercept.

4. $y = 3^x + 1$
5. $y = 2^x - 5$
6. $y = 2^{x+3}$
7. $y = 3^{x+1}$
8. $y = \left(\frac{2}{3}\right)^x$
9. $y = 5\left(\frac{2}{5}\right)^x$
10. $y = 5(3^x)$
11. $y = 4(5)^x$
12. $y = 2(5)^x + 1$
13. $y = \left(\frac{1}{2}\right)^{x+1}$
14. $y = \left(\frac{1}{8}\right)^x$
15. $y = \left(\frac{3}{4}\right)^x - 2$

Determine whether the data in each table display exponential behavior. Explain why or why not.

16.

x	−1	0	1	2
y	−5	−1	3	7

17.

x	1	2	3	4
y	25	125	625	3125

Lesson 9-6

(pages 510–514)

1. **MONEY** Marco deposited \$8500 in a 4-year certificate of deposit earning 7.25% compounded monthly. Write an equation for the amount of money Marco will have at the end of the four years. Then find the amount.
2. **TRANSPORTATION** Elise is buying a new car for \$21,500. The rate of depreciation on this type of car is 8% per year. Write an equation for the value of the car in 5 years. Then find the value of the car in 5 years.
3. **POPULATION** In 2000, the town of Belgrade had a population of 3422. For each of the next 8 years, the population increased by 4.9% per year. Find the projected population of Belgrade in 2008.

Lesson 10-1

(pages 528–534)

Simplify.

1. $\sqrt{50}$
2. $\sqrt{200}$
3. $\sqrt{162}$
4. $\sqrt{700}$
5. $\frac{\sqrt{3}}{\sqrt{5}}$
6. $\frac{\sqrt{72}}{\sqrt{6}}$
7. $\sqrt{\frac{8}{7}}$
8. $\sqrt{\frac{7}{32}}$
9. $\sqrt{\frac{5}{8}} \cdot \sqrt{\frac{2}{6}}$
10. $\sqrt{\frac{2}{3}} \cdot \sqrt{\frac{3}{2}}$
11. $\sqrt{\frac{2x}{30}}$
12. $\sqrt{\frac{50}{z^2}}$
13. $\sqrt{10} \cdot \sqrt{20}$
14. $\sqrt{7} \cdot \sqrt{3}$
15. $6\sqrt{2} \cdot \sqrt{3}$
16. $5\sqrt{6} \cdot 2\sqrt{3}$
17. $\sqrt{4x^4y^3}$
18. $\sqrt{200m^2y^3}$
19. $\sqrt{12ts^3}$
20. $\sqrt{175a^4b^6}$
21. $\sqrt{\frac{54}{g^2}}$
22. $\sqrt{99x^3y^7}$
23. $\frac{\sqrt{32c^5}}{9d^2}$
24. $\sqrt{\frac{27p^4}{3p^2}}$
25. $\frac{1}{3+\sqrt{5}}$
26. $\frac{2}{\sqrt{3}-5}$
27. $\frac{\sqrt{3}}{\sqrt{3}-5}$
28. $\frac{\sqrt{6}}{7-2\sqrt{3}}$

Lesson 10-2

(pages 536–540)

Simplify.

1. $3\sqrt{11} + 6\sqrt{11} - 2\sqrt{11}$
2. $6\sqrt{13} + 7\sqrt{13}$
3. $2\sqrt{12} + 5\sqrt{3}$
4. $9\sqrt{7} - 4\sqrt{2} + 3\sqrt{2}$
5. $3\sqrt{5} - 5\sqrt{3}$
6. $4\sqrt{8} - 3\sqrt{5}$
7. $2\sqrt{27} - 4\sqrt{12}$
8. $8\sqrt{32} + 4\sqrt{50}$
9. $\sqrt{45} + 6\sqrt{20}$
10. $2\sqrt{63} - 6\sqrt{28} + 8\sqrt{45}$
11. $14\sqrt{3t} + 8$
12. $7\sqrt{6x} - 12\sqrt{6x}$
13. $5\sqrt{7} - 3\sqrt{28}$
14. $7\sqrt{8} - \sqrt{18}$
15. $7\sqrt{98} + 5\sqrt{32} - 2\sqrt{75}$
16. $4\sqrt{6} + 3\sqrt{2} - 2\sqrt{5}$
17. $-3\sqrt{20} + 2\sqrt{45} - \sqrt{7}$
18. $4\sqrt{75} + 6\sqrt{27}$
19. $10\sqrt{\frac{1}{5}} - \sqrt{45} - 12\sqrt{\frac{5}{9}}$
20. $\sqrt{15} - \sqrt{\frac{3}{5}}$
21. $3\sqrt{\frac{1}{3}} - 9\sqrt{\frac{1}{12}} + \sqrt{243}$

Find each product.

22. $\sqrt{3}(\sqrt{5} + 2)$
23. $\sqrt{2}(\sqrt{2} + 3\sqrt{5})$
24. $(\sqrt{2} + 5)^2$
25. $(3 - \sqrt{7})(3 + \sqrt{7})$
26. $(\sqrt{2} + \sqrt{3})(\sqrt{3} + \sqrt{2})$
27. $(4\sqrt{7} + \sqrt{2})(\sqrt{3} - 3\sqrt{5})$

Lesson 10-3

(pages 541–546)

Solve each equation. Check your solution.

1. $\sqrt{5x} = 5$
2. $4\sqrt{7} = \sqrt{-m}$
3. $\sqrt{t} - 5 = 0$
4. $\sqrt{3b} + 2 = 0$
5. $\sqrt{x} - 3 = 6$
6. $5 - \sqrt{3x} = 1$
7. $2 + 3\sqrt{y} = 13$
8. $\sqrt{3g} = 6$
9. $\sqrt{a} - 2 = 0$
10. $\sqrt{2j} - 4 = 8$
11. $5 + \sqrt{x} = 9$
12. $\sqrt{5y + 4} = 7$
13. $7 + \sqrt{5c} = 9$
14. $2\sqrt{5t} = 10$
15. $\sqrt{44} = 2\sqrt{p}$
16. $4\sqrt{x - 5} = 15$
17. $4 - \sqrt{x - 3} = 9$
18. $\sqrt{10x^2 - 5} = 3x$
19. $\sqrt{2a^2 - 144} = a$
20. $\sqrt{3y + 1} = y - 3$
21. $\sqrt{2x^2 - 12} = x$
22. $\sqrt{b^2 + 16} + 2b = 5b$
23. $\sqrt{m + 2} + m = 4$
24. $\sqrt{3 - 2c} + 3 = 2c$

Lesson 10-4

(pages 549–554)

If c is the measure of the hypotenuse of a right triangle, find each missing measure. If necessary, round to the nearest hundredth.

1. $b = 20, c = 29, a = ?$
2. $a = 7, b = 24, c = ?$
3. $a = 2, b = 6, c = ?$
4. $b = 10, c = \sqrt{200}, a = ?$
5. $a = 3, c = 3\sqrt{2}, b = ?$
6. $a = 6, c = 14, b = ?$
7. $a = \sqrt{11}, c = \sqrt{47}, b = ?$
8. $a = \sqrt{13}, b = 6, c = ?$
9. $a = \sqrt{6}, b = 3, c = ?$
10. $b = \sqrt{75}, c = 10, a = ?$
11. $b = 9, c = \sqrt{130}, a = ?$
12. $a = 9, c = 15, b = ?$

Determine whether the following side measures form right triangles.

13. 14, 48, 50
14. 20, 30, 40
15. 21, 72, 75
16. $5, 12, \sqrt{119}$
17. 15, 39, 36
18. $10, 12, \sqrt{22}$
19. 2, 3, 4
20. $\sqrt{7}, 8, \sqrt{71}$

Lesson 10-5

(pages 555–559)

Find the distance between each pair of points whose coordinates are given. Express answers in simplest radical form and as decimal approximations rounded to the nearest hundredth if necessary.

1. $(4, 2), (-2, 10)$
2. $(-5, 1), (7, 6)$
3. $(4, -2), (1, 2)$
4. $(-2, 4), (4, -2)$
5. $(3, 1), (-2, -1)$
6. $(-2, 4), (7, -8)$
7. $(-5, 0), (-9, 6)$
8. $(5, -1), (5, 13)$
9. $(2, -3), (10, 8)$
10. $(-7, 5), (2, -7)$
11. $(-6, -2), (-5, 4)$
12. $(8, -10), (3, 2)$
13. $(4, -3), (7, -9)$
14. $(6, 3), (9, 7)$
15. $(10, 0), (9, 7)$
16. $(2, -1), (-3, 3)$
17. $(-5, 4), (3, -2)$
18. $(0, -9), (0, 7)$
19. $(-1, 7), (8, 4)$
20. $(-9, 2), (3, -3)$
21. $(3\sqrt{2}, 7), (5\sqrt{2}, 9)$
22. $(6, 3), (10, 0)$
23. $(3, 6), (5, -5)$
24. $(-4, 2), (5, 4)$

Find the possible values of a if the points with the given coordinates are the indicated distance apart.

25. $(0, 0), (a, 3); d = 5$
26. $(2, -1), (-6, a); d = 10$
27. $(1, 0), (a, 6); d = \sqrt{61}$
28. $(-2, a), (5, 10); d = \sqrt{85}$
29. $(15, a), (0, 4); d = \sqrt{274}$
30. $(3, 3), (a, 9); d = \sqrt{136}$

Lesson 10-6

(pages 560–565)

Determine whether each pair of triangles is similar. Justify your answer.

1.

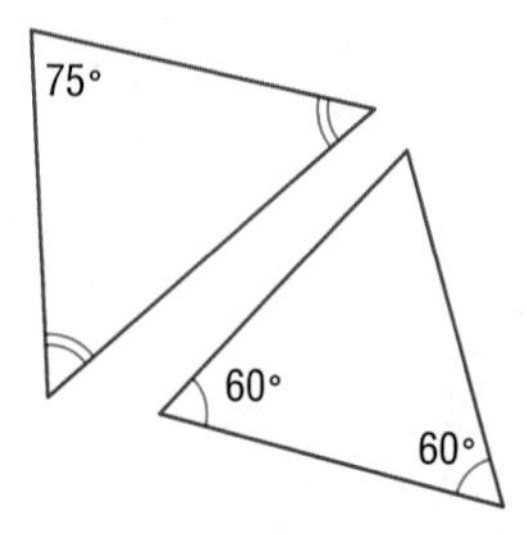

2.

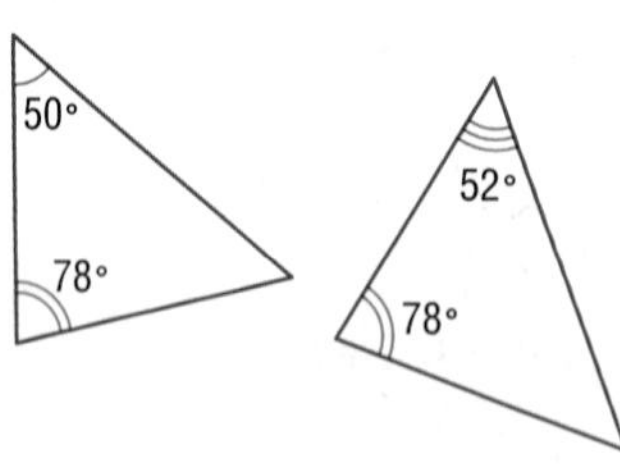

3.

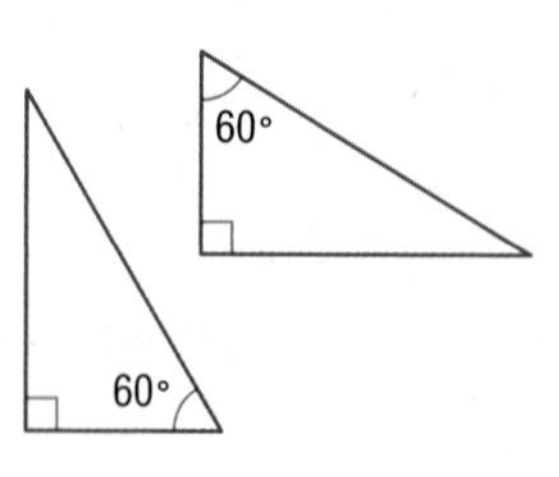

For each set of measures given, find the measures of the missing sides if $\Delta ABC \sim \Delta DEF$.

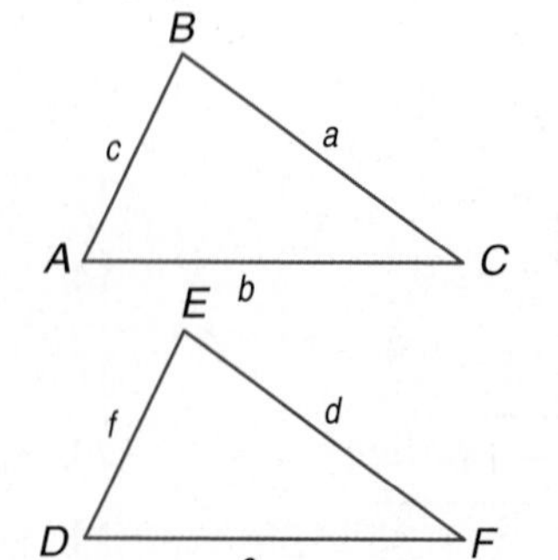

4. $a = 5, d = 10, b = 8, c = 7$
5. $a = 2, b = 3, c = 4, d = 3$
6. $a = 6, d = 4.5, e = 7, f = 7.5$
7. $a = 15, c = 20, b = 18, f = 10$
8. $f = 17.5, d = 8.5, e = 11, a = 1.7$

Lesson 11-1

(pages 577–582)

Graph each variation if y varies inversely as x.

1. $y = 10$ when $x = 7.5$
2. $y = -5$ when $x = 3$
3. $y = -6$ when $x = -2$
4. $y = 1$ when $x = -0.5$
5. $y = -2.5$ when $x = 3$
6. $y = -2$ when $x = -1$

Write an inverse variation equation that relates x and y. Assume that y varies inversely as x. Then solve.

7. If $y = 54$ when $x = 4$, find x when $y = 27$.
8. If $y = 18$ when $x = 6$, find x when $y = 12$.
9. If $y = 12$ when $x = 24$, find x when $y = 9$.
10. If $y = 8$ when $x = -8$, find y when $x = -16$.
11. If $y = 3$ when $x = -8$, find y when $x = 4$.
12. If $y = 27$ when $x = \frac{1}{3}$, find y when $x = \frac{3}{4}$.
13. If $y = -3$ when $x = -8$, find y when $x = 2$.
14. If $y = -3$ when $x = -3$, find x when $y = 4$.
15. If $y = -7.5$ when $x = 2.5$, find y when $x = -2.5$.
16. If $y = -0.4$ when $x = -3.2$, find x when $y = -0.2$.

Lesson 11-2

(pages 583–588)

State the excluded values for each rational expression.

1. $\frac{x}{x+1}$
2. $\frac{m}{n}$
3. $\frac{c-2}{c^2-4}$
4. $\frac{b^2-5b+6}{b^2-8b+15}$

Simplify each expression. State the excluded values of the variables.

5. $\frac{13a}{39a^2}$
6. $\frac{38x^2}{42xy}$
7. $\frac{p+5}{2(p+5)}$
8. $\frac{a+b}{a^2-b^2}$
9. $\frac{y+4}{y^2-16}$
10. $\frac{c^2-4}{c^2+4c+4}$
11. $\frac{a^2-a}{a-1}$
12. $\frac{x^2+4}{x^4-16}$
13. $\frac{r^3-r^2}{r-1}$
14. $\frac{4t^2-8}{4t-4}$
15. $\frac{6y^3-12y^2}{12y^2-18}$
16. $\frac{5x^2+10x+5}{3x^2+6x+3}$

Lesson 11-3

(pages 590–594)

Find each product.

1. $\frac{a^2b}{b^2c} \cdot \frac{c}{d}$
2. $\frac{6a^2n}{8n^2} \cdot \frac{12n}{9a}$
3. $\frac{2a^2d}{3bc} \cdot \frac{9b^2c}{16ad^2}$
4. $\frac{10n^3}{6x^3} \cdot \frac{12n^2x^4}{25n^2x^2}$
5. $\frac{6m^3n}{10a^2} \cdot \frac{4a^2m}{9n^3}$
6. $\frac{5n-5}{3} \cdot \frac{9}{n-1}$
7. $\frac{(a-5)(a+1)}{(a+1)(a+7)} \cdot \frac{(a+7)(a-6)}{(a+8)(a-5)}$
8. $\frac{x-1}{(x+2)(x-3)} \cdot \frac{x+2}{(x-3)(x-1)}$
9. $\frac{a^2}{a-b} \cdot \frac{3a-3b}{a}$
10. $\frac{2a+4b}{5} \cdot \frac{25}{6a+8b}$
11. $\frac{3}{x-y} \cdot \frac{x-y^2}{6}$
12. $\frac{x+5}{3x} \cdot \frac{12x^2}{x^2+7x+10}$
13. $\frac{4a+8}{a^2-25} \cdot \frac{a-5}{5a+10}$
14. $\frac{r^2}{r-s} \cdot \frac{r^2-s^2}{s^2}$
15. $\frac{x^2+10x+9}{x^2+11x+18} \cdot \frac{x^2+3x+2}{x^2+7x+6}$
16. $\frac{x^2-6x+5}{x^2+7x+12} \cdot \frac{x^2+14x+40}{x^2+5x-50}$

Lesson 11-4

(pages 595–599)

Find each quotient.

1. $\frac{5m^2n}{12a^2} \div \frac{30m^4}{18an}$
2. $\frac{25g^7h}{28t^3} \div \frac{5g^5h^2}{42s^2t^3}$
3. $\frac{6a+4b}{36} \div \frac{3a+2b}{45}$
4. $\frac{x^2y}{18z} \div \frac{2yz}{3x^2}$
5. $\frac{p^2}{14qr^3} \div \frac{2r^2p}{7q}$
6. $\frac{5d-f}{5d+f} \div (25d^2 - f^2)$
7. $\frac{t^2-2t-15}{t-5} \div \frac{t+3}{t+5}$
8. $\frac{5x+10}{x+2} \div (x+2)$
9. $\frac{3d}{2d^2-3d} \div \frac{9}{2d-3}$
10. $\frac{3v^2-27}{15v} \div \frac{v+3}{v^2}$
11. $\frac{3g^2+15g}{4} \div \frac{g+5}{g^2}$
12. $\frac{b^2-9}{4b} \div (b-3)$
13. $\frac{p^2}{y^2-4} \div \frac{p}{2-y}$
14. $\frac{k^2-81}{k^2-36} \div \frac{k-9}{k+6}$
15. $\frac{2a^3}{a+1} \div \frac{a^2}{a+1}$
16. $\frac{x^2-16}{16-x^2} \div \frac{7}{x}$
17. $\frac{y}{5} \div \frac{y^2-25}{5-y}$
18. $\frac{3m}{m+1} \div (m-2)$

Lesson 11-5

(pages 601–606)

Find each quotient.

1. $(2x^2 - 11x - 20) \div (2x + 3)$
2. $(a^2 + 10a + 21) \div (a + 3)$
3. $(m^2 + 4m - 5) \div (m + 5)$
4. $(x^2 - 2x - 35) \div (x - 7)$
5. $(c^2 + 6c - 27) \div (c + 9)$
6. $(y^2 - 6y - 25) \div (y + 7)$
7. $(3t^2 - 14t - 24) \div (3t + 4)$
8. $(2r^2 - 3r - 35) \div (2r + 7)$
9. $\frac{12n^2+36n+15}{6n+3}$
10. $\frac{10x^2+29x+21}{5x+7}$
11. $\frac{4t^3+17t^2-1}{4t+1}$
12. $\frac{2a^3+9a^2+5a-12}{a+3}$
13. $\frac{27c^2-24c+8}{9c-2}$
14. $\frac{4b^3+7b^2-2b+4}{b+2}$

Lesson 11-6

(pages 608–613)

Find each sum.

1. $\frac{4}{z} + \frac{3}{z}$
2. $\frac{a}{12} + \frac{2a}{12}$
3. $\frac{5}{2t} + \frac{-7}{2t}$
4. $\frac{y}{2} + \frac{y}{2}$
5. $\frac{b}{x} + \frac{2}{x}$
6. $\frac{y}{2} + \frac{y-6}{2}$
7. $\frac{x}{x+1} + \frac{1}{x+1}$
8. $\frac{2n}{2n-5} + \frac{5}{5-2n}$
9. $\frac{x-y}{2-y} + \frac{x+y}{y-2}$
10. $\frac{r^2}{r-s} + \frac{s^2}{r-s}$
11. $\frac{12n}{3n+2} + \frac{8}{3n+2}$
12. $\frac{6x}{x+y} + \frac{6y}{x+y}$

Find each difference.

13. $\frac{5x}{24} - \frac{3x}{24}$
14. $\frac{7p}{3} - \frac{8p}{3}$
15. $\frac{8k}{5m} - \frac{3k}{5m}$
16. $\frac{8}{m-2} - \frac{6}{m-2}$
17. $\frac{y}{b+6} - \frac{2y}{b+6}$
18. $\frac{a+2}{6} - \frac{a+3}{6}$
19. $\frac{2a}{2a+5} - \frac{5}{2a+5}$
20. $\frac{1}{4z+1} - \frac{(-4z)}{4z+1}$
21. $\frac{3a}{a-2} - \frac{3a}{a-2}$
22. $\frac{n}{n-1} - \frac{1}{1-n}$
23. $\frac{a}{a-7} - \frac{(-7)}{7-a}$
24. $\frac{2a}{6a-3} - \frac{(-1)}{3-6a}$

Lesson 11-7

(pages 614–619)

Find the LCM for each pair of expressions.

1. $27a^2bc, 36ab^2c^2$
2. $3m - 1, 6m - 2$
3. $x^2 + 2x + 1, x^2 - 2x - 3$

Find each sum.

4. $\frac{s}{3} + \frac{2s}{7}$
5. $\frac{5}{2a} + \frac{-3}{6a}$
6. $\frac{6}{5x} + \frac{7}{10x^2}$
7. $\frac{5}{xy} + \frac{6}{yz}$
8. $\frac{2}{t} + \frac{t+3}{s}$
9. $\frac{a}{a-b} + \frac{b}{2b+3a}$
10. $\frac{4a}{2a+6} + \frac{3}{a+3}$
11. $\frac{3t+2}{3t-2} + \frac{t+2}{t^2-4}$
12. $\frac{-3}{a-5} + \frac{-6}{a^2-5a}$

Find each difference.

13. $\frac{2n}{5} - \frac{3m}{4}$
14. $\frac{3z}{7w^2} - \frac{2z}{w}$
15. $\frac{s}{t^2} - \frac{r}{3t}$
16. $\frac{a}{a^2-4} - \frac{4}{a+2}$
17. $\frac{m}{m-n} - \frac{5}{m}$
18. $\frac{y+5}{y-5} - \frac{2y}{y^2-25}$
19. $\frac{t+10}{t^2-100} - \frac{1}{10-t}$
20. $\frac{2a-6}{a^2-3a-10} - \frac{3a+5}{a^2-4a-12}$

Lesson 11-8

(pages 620–625)

Write each mixed expression as a rational expression.

1. $4 + \frac{2}{x}$
2. $8 + \frac{5}{3t}$
3. $\frac{b+1}{2b+3b}$
4. $\frac{3z+z+2}{z}$
5. $\frac{2}{a-2} + a^2$
6. $\frac{3r^2+4}{2r+1}$

Simplify each expression.

7. $\frac{3\frac{1}{2}}{4\frac{3}{4}}$
8. $\frac{\frac{x^2}{y}}{\frac{y}{x^3}}$
9. $\frac{\frac{t^4}{u}}{\frac{t^3}{u^2}}$
10. $\frac{\frac{x-3}{x+1}}{\frac{x^2}{y^2}}$
11. $\frac{\frac{y}{3}+\frac{5}{6}}{2+\frac{5}{y}}$
12. $\frac{\frac{1}{x}+\frac{1}{y}}{\frac{1}{y}-\frac{1}{x}}$
13. $\frac{t-2}{\frac{t^2-4}{t^2+5t+6}}$
14. $\frac{a+\frac{2}{a+1}}{a-\frac{3}{a-2}}$

Lesson 11-9

(pages 626–632)

Solve each equation. State any extraneous solutions.

1. $\frac{k}{6} + \frac{2k}{3} = -\frac{5}{2}$
2. $\frac{2x}{7} + \frac{27}{10} = \frac{4x}{5}$
3. $\frac{18}{b} = \frac{3}{b} + 3$
4. $\frac{3}{5x} + \frac{7}{2x} = 1$
5. $\frac{2a-3}{6} = \frac{2a}{3} + \frac{1}{2}$
6. $\frac{3x+2}{x} + \frac{x+3}{x} = 5$
7. $\frac{2b-3}{7} - \frac{b}{2} = \frac{b+3}{14}$
8. $\frac{2y}{y-4} - \frac{3}{5} = 3$
9. $\frac{2t}{t+3} + \frac{3}{t} = 2$
10. $\frac{5x}{x+1} + \frac{1}{x} = 5$
11. $\frac{r-2}{r+2} - \frac{2r}{r+9} = 6$
12. $\frac{m}{m+1} + \frac{5}{m-1} = 1$
13. $\frac{2x}{x-3} - \frac{4x}{3-x} = 12$
14. $\frac{14}{b-6} = \frac{1}{2} + \frac{6}{b-8}$
15. $\frac{a}{4a+15} - 3 = -2$
16. $\frac{5x}{3x+10} + \frac{2x}{x+5} = 2$
17. $\frac{2a-3}{a-3} - 2 = \frac{12}{a+2}$
18. $\frac{z+3}{z-1} + \frac{z+1}{z-3} = 2$

Extra Practice

Lesson 12-1

(pages 642–648)

Identify each sample, suggest a population from which it was selected, and state whether it is *unbiased* (random) or *biased*. If unbiased, classify the sample as *simple, stratified,* or *systematic*. If biased, classify as *convenience* or *voluntary response*.

1. The sheriff has heard that many dogs in the county do not have licenses. He checks the licenses of the first ten dogs he encounters.
2. Every fifth car is selected from the assembly line. The cars are also identified by the day of the week during which they were produced.
3. A table is set up outside of a large department store. All people entering the store are given a survey about their preference of brand for blue jeans. As people leave the store, they can return the survey.
4. A community is considering building a new swimming pool. Every twentieth person on a list of residents is contacted for their opinion.

Lesson 12-2

(pages 650–654)

Draw a tree diagram to show the sample space for each event. Determine the number of possible outcomes.

1. choosing a dinner special at a restaurant offering lettuce salad or coleslaw; chicken, beef, or fish; and ice cream, pudding, or cookies
2. tossing a coin four times

Determine the number of possible outcomes.

3. A state's license plates feature two digits, a space, and two letters. Any digit or any letter can be used in the space.
4. At the Big Mountain Ski Resort, you can choose from three types of boots, four types of skis, and five types of poles.

Find the value of each expression.

5. $8!$
6. $1!$
7. $0!$
8. $5!$
9. $2!$
10. $9!$

Lesson 12-3

(pages 655–662)

Determine whether each situation involves a *permutation* or *combination*. Explain.

1. three topping flavors for a sundae from ten topping choices
2. selection and placement of four runners on a relay team from 8 runners
3. five rides to ride at an amusement park with twelve rides
4. first, second, and third place winners for a 10K race
5. a three-letter arrangement from eight different letters
6. selection of five digits from ten digits for a combination lock
7. selecting six items from twelve possible items to include in a custom gift basket

Evaluate each expression.

8. ${}_5P_2$
9. ${}_7P_7$
10. ${}_{10}C_2$
11. ${}_6C_5$
12. $({}_7P_3)({}_4P_2)$
13. $({}_8C_6)({}_7C_5)$
14. $({}_3C_2)({}_{10}P_{10})$
15. $({}_3P_2)({}_{10}C_{10})$

Extra Practice

Lesson 12-4

(pages 663–670)

A red die and a blue die are rolled. Find each probability.

1. P(red 1, blue 1)
2. P(red even, blue even)
3. P(red prime number, blue even)
4. P(red 6, blue greater than 4)
5. P(red greater than 2, blue greater than 3)

At a carnival game, toy ducks are selected from a pond to win prizes. Once a duck is selected, it is not replaced. The pond contains 8 red, 2 yellow, 1 gold, 4 blue, and 40 black ducks. Find each probability.

6. P(red, then gold)
7. P(2 black)
8. P(2 yellow)
9. P(black, then gold)
10. P(3 black, then red)
11. P(yellow, then blue, then gold)

Lesson 12-5

(pages 672–676)

Consider finding the product of the numbers shown rolling two dice and the number of ways each product can be found.

1. Draw a table to show the sample space of all possible outcomes.
2. Find the probability for $X = 9$, $X = 12$, and $X = 24$.
3. What is the probability that the product of two dice is greater than 15 on two separate rolls?

Use the table that shows a probability distribution for the number of customers that enter a particular store during a business day.

Number of Customers	0–500	501–1000	1001–1500	1501–2000	2000–2500
Probability	0.05	0.25	0.35	0.30	0.05

4. Define a random variable and list its values.
5. Show that this is a valid probability distribution.
6. What is the probability that fewer than 1001 customers enter in a day?

Lesson 12-6

(pages 677–683)

For Exercises 1–3, toss 4 coins, one at a time, 50 times and record your results.

1. Based on your results, what is the probability that any two coins show tails?
2. Based on your results, what is the probability that the first and fourth coins show heads?
3. What is the theoretical probability that all four coins show heads?

For Exercises 4–6, use the table that shows the results of a survey about household occupancy.

Number in Household	Number of Households
1	172
2	293
3	482
4	256
5 or more	148

4. Find the experimental probability distribution for the number of households of each size.
5. Based on the survey, what is the probability that a person chosen at random lives in a household with five or more people?
6. Based on the survey, what is the probability that a person chosen at random lives in a household with 1 or 2 people?

Mixed Problem Solving

Chapter 1 The Language and Tools of Algebra

(pp. 4–67)

GEOMETRY For Exercises 1 and 2, use the following information.

The surface area of a cone is given by $SA = \pi r^2 + \pi r\ell$, where r is the radius and ℓ is the slant height. (Lesson 1-1)

1. Write an expression that represents the surface area of the cone.
2. Suppose the radius and the slant height of a cone have the same measure r. Write an expression that represents the surface area of this cone.

SALES For Exercises 3 and 4, use the following information.

At the Farmer's Market, merchants can rent a small table for \$5.00 and a large table for \$8.50. One time, 25 small and 10 large tables were rented. Another time, 35 small and 12 large were rented. (Lesson 1-2)

3. Write an expression to show the total amount of money collected.
4. Evaluate the expression.

ENTERTAINMENT For Exercises 5–7, use the following information.

The Morrows are planning to go to a water park. The table shows the ticket prices. The family has 2 adults, 2 children, and a grandparent who wants to observe. They want to spend no more than \$55. (Lesson 1-3)

Admission Prices ($)		
Ticket	Full Day	Half Day
Adult	16.95	10.95
Child (6–18)	12.95	8.95
Observer	4.95	3.95

5. Write an inequality to show the cost for the family to go to the water park.
6. How much would it cost the Morrows to go for a full day? a half day?
7. Can the family go to the water park for a full day and stay within their budget?

RETAIL For Exercises 8–10, use the following information.

A department store is having a sale on children's clothing. The table shows the prices. (Lesson 1-4)

Shorts	T-Shirts	Tank Tops
\$7.99	\$8.99	\$6.99
\$5.99	\$4.99	\$2.99

8. Write three different expressions that represent 8 pairs of shorts and 8 tops.
9. Evaluate the three expressions in Exercise 8 to find the costs of the 16 items. What do you notice about all the total costs?
10. If you buy 8 shorts and 8 tops, you receive a discount of 15% on the purchase. Find the greatest and least amount of money you can spend on the 16 items at the sale.
11. **CRAFTS** Mandy makes baby blankets and stuffed rabbits to sell at craft fairs. She sells blankets for \$28 and rabbits for \$18. Write and evaluate an expression to find her total amount of sales if she sells 25 blankets and 25 rabbits. (Lesson 1-5)
12. **BASEBALL** Tickets to a baseball game cost \$18.95, \$12.95, or \$9.95. A hot dog and soda combo costs \$5.50. The Madison family is having a reunion. They buy 10 tickets in each price category and plan to buy 30 combos. What is the total cost for the tickets and meals? (Lesson 1-6)
13. **GEOMETRY** Two perpendicular lines meet to form four right angles. Write two different if-then statements for this definition. (Lesson 1-7)
14. **JOBS** Laurie mows lawns to earn extra money. She knows that she can mow at most 30 lawns in one week. She determines that she profits \$15 on each lawn she mows. Identify a reasonable domain and range for this situation and draw a graph. (Lesson 1-9)

GEOMETRY For Exercises 1–4, use the following information.

The lateral surface area L of a cylinder is two times π times the product of the radius r and the height h. (Lesson 2-1)

1. Write a formula for the lateral area of a cylinder.
2. Find the lateral area of a cylinder with a radius of 4.5 inches and a height of 7 inches. Use 3.14 for π and round the answer to the nearest tenth.
3. The total surface area T of a cylinder includes the area of the two bases of the cylinder, which are circles. The formula for the area of one circle is πr^2. Write a formula for the total surface area T of a cylinder.
4. Find the total surface area of the cylinder in Exercise 2. Round to the nearest tenth.

RIVERS For Exercises 5 and 6, use the following information.

The Congo River in Africa is 2900 miles long. That is 310 miles longer than the Niger River, which is also in Africa. **Source:** *The World Almanac* (Lesson 2-2)

5. Write an equation you could use to find the length of the Niger River.
6. What is the length of the Niger River?

ANIMALS For Exercises 7 and 8, use the following information.

The average length of a yellow-banded angelfish is 12 inches. This is 4.8 times as long as an average common goldfish.
Source: *Scholastic Records* (Lesson 2-3)

7. Write an equation you could use to find the length of the common goldfish.
8. What is the length of an average common goldfish?
9. **PETS** In 2003, there were 8949 Great Danes registered with the American Kennel Club. The number of registered Labrador Retrievers was 1750 more than sixteen times the number of registered Great Danes. How many registered Labrador Retrievers were there? **Source:** *The World Almanac* (Lesson 2-4)
10. **GEOMETRY** One angle of a triangle measures 10° more than the second. The measure of the third angle is twice the sum of the measures of the first two angles. Find the measure of each angle. (Lesson 2-5)
11. **POOLS** Tyler needs to add 1.5 pounds of a chemical to the water in his pool for each 5000 gallons of water. The pool holds 12,500 gallons. How much chemical should he add to the water? (Lesson 2-6)

SKIING For Exercises 12 and 13, use the following information.

Michael is registering for a ski camp in British Columbia, Canada. The cost of the camp is $1254, but the Canadian government imposes a general sales tax of 7%. (Lesson 2-7)

12. What is the total cost of the camp including tax?
13. As a U.S. citizen, Michael can apply for a refund of one-half of the tax. What is the amount of the refund he can receive?

FINANCE For Exercises 14 and 15, use the following information.

Allison is using a spreadsheet to solve a problem about investing. She is using the formula $I = Prt$, where I is the amount of interest earned, P is the amount of money invested, r is the rate of interest as a decimal, and t is the period of time the money is invested in years. (Lesson 2-8)

14. Allison needs to find the amount of money invested P for given amounts of interest, given rates, and given time. The formula needs to be solved for P to use in the spreadsheet. Solve the formula for P.
15. Allison uses these values in the formula in Exercise 17: $I = \$1848.75$, $r = 7.25\%$, $t = 6$ years. Find P.
16. **CHEMISTRY** Isaac had 40 gallons of a 15% iodine solution. How many gallons of a 40% iodine solution must he add to make a 20% iodine solution? (Lesson 2-9)

HEALTH For Exercises 1–3, use the following information.

The table shows suggested weights for adults for various heights in inches. (Lesson 3-1)

Height	Weight	Height	Weight
60	102	68	131
62	109	70	139
64	116	72	147
66	124	74	155

Source: *The World Almanac*

1. Graph the relation.
2. Do the data lie on a straight line? Explain.
3. Estimate a suggested weight for a person who is 78 inches tall. Explain your method.

SPORTS For Exercises 4–6, use the following information.

The table shows the winning times of the Olympic mens' 50-km walk for various years. The times are rounded to the nearest minute. (Lesson 3-2)

Year	Years Since 1980	Time
1980	0	229
1984	4	227
1988	8	218
1992	12	230
1996	16	224
2000	20	222

Source: ESPN

4. Graph the relation using columns 2 and 3.
5. Is the relation a function? Explain.
6. Predict a winning time for the 2008 games.

PLANETS For Exercises 7–9, use the following information.

An astronomical unit (AU) is used to express great distances in space. It is based upon the distance from Earth to the Sun. A formula for converting any distance d in miles to AU is $AU = \frac{d}{93{,}000{,}000}$. The table shows the average distances from the Sun of four planets in miles. (Lesson 3-3)

Planet	Distance from Sun
Mercury	36,000,000
Mars	141,650,000
Jupiter	483,750,000
Pluto	3,647,720,000

Source: *The World Almanac*

7. Find the number of AU for each planet rounded to the nearest thousandth.
8. How can you determine which planets are farther from the Sun than Earth?
9. Alpha Centauri is 270,000 AU from the Sun. How far is that in miles?

HOME DECOR For Exercises 10 and 11, use the following information.

Pam is having blinds installed at her home. The cost for installation for any number of blinds can be described by $c = 25 + 6.5x$. (Lesson 3-3)

10. Graph the equation.
11. If Pam has 8 blinds installed, what is the cost?

JEWELRY For Exercises 12 and 13, use the following information.

A necklace is made with beads placed in a circular pattern. The rows have the following numbers of beads: 1, 6, 11, 16, 21, 26, and 31. (Lesson 3-4)

12. Write a formula for the beads in each row.
13. If a larger necklace is made with 20 rows, find the number of beads in row 20.
14. **GEOMETRY** The table below shows the area of squares with sides of various lengths. (Lesson 3-5)

Side	Area	Side	Area
1	1	4	
2	4	5	
3	9	6	

Write the first 10 numbers that would appear in the area column. Describe the pattern.

FARMING **For Exercises 1–3, use the following information.**

The table shows wheat prices per bushel from 1940 through 2000. (Lesson 4-1)

Year	1940	1950	1960	1970
Price	\$0.67	\$2.00	\$1.74	\$1.33
Year	1980	1990	2000	
Price	\$3.91	\$2.61	\$2.62	

1. For which time period was the rate of change the greatest? the least?
2. Find the rate of change from 1940 to 1950.
3. Explain the meaning of the slope from 1980 to 1990.

SOUND **For Exercises 4 and 5, use the following information.**

The table shows the distance traveled by sound in water. (Lesson 4-2)

Time, x (seconds)	Distance, y (feet)
0	0
1	4820
2	9640
3	14,460
4	19,280

Source: New York Public Library

4. Write an equation that relates distance traveled to time.
5. Find the time for 72,300 feet.

POPULATION **For Exercises 6–8, use the following information.**

In 1990, the population of Wyoming was 453,588. Over the next decade, it increased by about 4019 per year. Source: *The World Almanac* (Lesson 4-3)

6. Assume the rate of change remains the same. Write a linear equation to find the population y of Wyoming at any time. Let x represent the number of years since 1990.
7. Graph the equation.
8. Estimate the population in 2015.

HEALTH **For Exercises 9 and 10, use the following information.**

A person with height of 60 inches should have a weight of 112 pounds and a person with height of 66 inches should have a weight of 136 pounds. Source: *The World Almanac* (Lesson 4-4)

9. Write a linear equation to estimate the weight of a person of any height.
10. Estimate the weight of a person who is 72 inches tall.

TRAVEL **For Exercises 11–13, use the following information.**

Between 1990 and 2000, the number of people taking cruises increased by about 300,000 each year. In 1990, about 3.6 million people took cruises. Source: *USA Today* (Lesson 4-5)

11. Write the point-slope form of an equation to find the total number of people taking a cruise y for any year x.
12. Write the equation in slope-intercept form.
13. Estimate the number of people who will take a cruise in 2010.

ADOPTION **For Exercises 14–16, use the following information.**

The table shows the number of children from Russia adopted by U.S. citizens. (Lesson 4-6)

Years Since 1996 x	Number of Children y
0	2454
1	3816
2	4491
3	4348
4	4269
5	4279
6	4939
7	5209

Source: *The World Almanac*

14. Draw a scatter plot and a line of fit.
15. Write the slope-intercept form of the equation for the line of fit.
16. Predict the number of children who will be adopted in 2015.

GEOMETRY **For Exercises 17 and 18, use the following information.**

A quadrilateral has sides with equations $y = -2x$, $2x + y = 6$, $y = \frac{1}{2}x + 6$, and $x - 2y = 9$. (Lesson 4-7)

17. Is the figure a rectangle?
18. Explain your reasoning.

Mixed Problem Solving

WORKING **For Exercises 1–3, use the following information.**

The table shows the percent of men and women 65 years and older that were working in the U.S. in the given years. (Lesson 5-1)

U.S. Workers over 65		
Year	Percent of Men	Percent of Women
2000	17.7	9.4
2003	18.6	10.6

Source: *The World Almanac*

1. Let the year 2000 be 0. Assume that the rate of change remains the same for years after 2003. Write an equation to represent the percent of working elderly men y in any year x.
2. Write an equation to represent the percent of working elderly women.
3. Assume the rate of increase or decrease in working men and women remains the same for years after 2003. Estimate when the percent of working men and women will be the same.

SPORTS **For Exercises 4–7, use the following information.**

The table shows the winning times for the men's and women's Triathlon World Championship for 1995 and 2000. (Lesson 5-2)

Year	Men's	Women's
2000	1:51:39	1:54:43
2005	1:49:31	1:58:03

Source: International Triathlon Union

4. The times in the table are in hours, minutes, and seconds. Rewrite the times in minutes rounded to the nearest minute.
5. Let the year 2000 be 0. Assume that the rate of change remains the same for years after 2000. Write an equation to represent the men's winning times y in any year x.
6. Write an equation to represent the women's winning times in any year.
7. If the trend continues, when would you expect the men's and women's winning times to be the same?

MONEY **For Exercises 8–10, use the following information.**

In 2004, the sum of the number of \$2 bills and \$50 bills in circulation was 1,857,573,945. The number of \$50 bills was 494,264,809 more than the number of \$2 bills. (Lesson 5-3)

8. Write a system of equations to represent this situation.
9. Find the number of each type of bill in circulation.
10. Find the amount of money that was in circulation in \$2 and \$50 bills.

SPORTS **For Exercises 11–14, use the following information.**

In the 2004 Summer Olympic Games, the total number of gold and silver medals won by the U.S. was 74. Gold medals are worth 3 points and silver medals are worth 2 points. The total points scored for gold and silver medals was 183. (Lesson 5-4) **Source:** *ESPN Almanac*

11. Write an equation for the sum of the number of gold and silver medals won by the U.S.
12. Write an equation for the sum of the points earned by the U.S. for gold and silver medals.
13. How many gold and silver medals did the U.S. win?
14. The total points scored by the U.S. was 212. Bronze medals are worth 1 point. How many bronze medals were won?

15. **GEOMETRY** Supplementary angles are two angles whose measures have the sum of 180 degrees. Angles X and Y are supplementary, and the measure of angle X is 24 degrees greater than the measure of angle Y. Write and solve a system of equations to find the measures of angles X and Y. (Lesson 5-5)

16. **CHEMISTRY** MX Labs needs to make 500 gallons of a 34% acid solution. The only solutions available are a 25% acid solution and a 50% acid solution. Write and solve a system of equations to find the number of gallons of each solution that should be mixed to make the 34% solution. (Lesson 5-5)

MONEY For Exercises 1 and 2, use the following information.

Scott's allowance for July is \$50. He wants to attend a concert that costs \$26. (Lesson 6-1)

1. Write and solve an inequality that shows how much money he can spend in July after buying a concert ticket.
2. He spends \$2.99 for lunch with his friends and \$12.49 for a CD. Write and solve an inequality that shows how much money he can spend after all of his purchases.

ANIMALS For Exercises 3 and 4, use the following information.

The world's heaviest flying bird is the great bustard. A male bustard can be up to 4 feet long and weigh up to 40 pounds. (Lesson 6-2)

3. Write inequalities to describe the ranges of lengths and weights of male bustards.
4. Male bustards are usually about four times as heavy as females. Write and solve an inequality that describes the range of weights of female bustards.

FOOD For Exercises 5–7, use this information.

Jennie wants to make at least \$75 selling caramel-coated apples at the County Fair. She plans to sell each apple for \$1.50. (Lesson 6-3)

5. Let a be the number of apples she makes and sells. Write an inequality to find the number of apples she needs to sell to reach her goal if it costs her \$0.30 per apple.
6. Solve the inequality.
7. Interpret the meaning of the solution.

RETAIL For Exercises 8–10, use the following information.

A sporting goods store is offering a \$15 coupon on any pair of shoes. (Lesson 6-4)

8. The most expensive pair of shoes is \$149.95 and the least expensive pair of shoes is \$24.95. What is the range of prices for customers who have the coupons?
9. You have a choice of buying a pair of shoes with a regular price of \$109.95 using the coupon or having a 15% discount on the price. Which option is best?
10. For what price of shoe is a 15% discount the same as \$15 off the regular price?

WEATHER For Exercises 11 and 12, use the following information.

The following are average normal monthly temperatures for Honolulu, Hawaii, in degrees Fahrenheit. (Lessons 6-5 and 6-6)

73, 73, 74, 76, 78, 79, 81, 81, 81, 80, 77, 74

11. What is the mean of the temperatures to the nearest whole degree?
12. Write an inequality to show the normal range of temperatures for Honolulu during the year.

QUILTING For Exercises 13–15, use the following information.

Ingrid is making a quilt in the shape of a rectangle. She wants the perimeter of the quilt to be no more than 318 inches. (Lesson 6-7)

13. Write an inequality for this situation.
14. Graph the inequality and name two different dimensions for the quilt.
15. What are the dimensions and area of the largest possible quilt Ingrid can make?

RADIO For Exercises 16–20, use the following information.

KSKY radio station is giving away tickets to an amusement park. Each child ticket costs \$15 and each adult ticket costs \$20. The station wants to spend no more than \$800 on tickets. They also want the number of child tickets to be greater than twice the number of adult tickets. (Lesson 6-8)

16. Write an inequality for the total cost of c child tickets and a adult tickets.
17. Write an inequality for the relationship between the number of child and adult tickets.
18. Write two inequalities that would assure you that the numbers of adult and child tickets would not be negative.
19. Graph the four inequalities to show possible numbers of tickets they can buy.
20. Give three possible combinations of child and adult tickets for the station to buy.

GEOMETRY For Exercises 1–3, use the following information.

If the side length of a cube is s, then the volume is presented by s^3 and the surface area is represented by $6s^2$. (Lesson 7-1)

1. Are the expressions for volume and surface area monomials? Explain.
2. If the side of a cube measures 3 feet, find the volume and surface area.
3. Find a side length s such that the volume and surface area have the same measure.
4. The volume of a cylinder can be found by multiplying the radius squared times the height times π, or $V = \pi r^2 h$. Suppose you have two cylinders. Each measure of the second is twice the measure of the first, so $V = \pi(2r)^2(2h)$. What is the ratio of the volume of the first cylinder to the second cylinder? (Lesson 7-2)

POPULATION For Exercises 5–7, use the following information.

The table shows the population density for Nevada for various years. (Lesson 7-3)

Year	Years Since 1930	People/Square Mile
1930	0	0.8
1960	30	2.6
1980	50	7.3
1990	60	10.9
2000	70	18.2

Source: *The World Almanac*

5. The population density d of Nevada from 1920 to 1990 can be modeled by $d = 0.005y^2 - 0.127y + 1$, where y represents the number of years since 1930. Identify the type of polynomial for $0.005y^2 - 0.127y + 1$.
6. What is the degree of this polynomial?
7. Predict the population density of Nevada for 2010. Explain your method.

RADIO For Exercises 8 and 9, use the following information.

From 1997 to 2000, the number of radio stations presenting primarily news and talk N and the total number of radio stations of all types R in the U.S. could be modeled by the following equations, where x is the number of years since 1997. (Lesson 7-4) Source: *The World Almanac*

$N = 25.7x + 1098.0 \qquad R = 74.2x + 10{,}246.3$

8. Find an equation that models the number of radio stations O that are *not* primarily news and talk for this time period.
9. If this trend continues, how many radio stations that are not news and talk will there be in the year 2015?

GEOMETRY For Exercises 10–12, use the following information.

The number of diagonals of a polygon can be found by using the formula $d = 0.5n(n - 3)$, where d is the number of diagonals and n is the number of sides of the polygon. (Lesson 7-5)

10. Use the Distributive Property to write the expression as a polynomial.
11. Find the number of diagonals for polygons with 3 through 10 sides.
12. Describe any patterns you see in the numbers you wrote in Exercise 11.

GEOMETRY For Exercises 13 and 14, use the following information.

A rectangular prism has dimensions of x, $x + 3$, and $2x + 5$. (Lesson 7-6)

13. Find the volume of the prism in terms of x.
14. Choose two values for x. How do the volumes compare?

MONEY For Exercises 15–17, use the following information.

Money invested in a certificate of deposit or CD collects interest once per year. Suppose you invest \$4000 in a 2-year CD. (Lesson 7-7)

15. If the interest rate is 5% per year, the expression $4000(1 + 0.05)^2$ can be evaluated to find the total amount of money you will have at the end of two years. Explain the numbers in this expression.
16. Find the amount of money at the end of two years.
17. Suppose you invest \$10,000 in a CD for 4 years at a rate of 6.25%. What is the total amount of money you will have at the end of 4 years?

FLOORING For Exercises 1 and 2, use the following information.

Eric is refinishing his dining room floor. The floor measures 10 feet by 12 feet. Flooring World offers a wood-like flooring in 1-foot by 1-foot squares, 2-foot by 2-foot squares, 3-foot by 3-foot squares, and 2-foot by 3-foot rectangular pieces. (Lesson 8-1)

1. Without cutting the pieces, which of the four types of flooring can Eric use in the dining room? Explain.
2. The price per piece of each type is shown in the table. If Eric wants to spend the least money, which should he choose? What will be the total cost of his choice?

Size	1 × 1	2 × 2	3 × 3	2 × 3
Price	\$3.75	\$15.00	\$32.00	\$21.00

FIREWORKS For Exercises 3–5, use the following information.

At a Fourth of July celebration, a rocket is launched with an initial velocity of 125 feet per second. The height h of the rocket in feet above sea level is modeled by the formula $h = 125t - 16t^2$, where t is the time in seconds after the rocket is launched. (Lesson 8-2)

3. What is the height of the rocket when it returns to the ground?
4. Let $h = 0$ in the equation $h = 125t - 16t^2$ and solve for t.
5. How many seconds will it take for the rocket to return to the ground?

FOOTBALL For Exercises 6–8, use the following information.

Some small high schools play six-man football as a team sport. The dimensions of the field are less than the dimensions of a standard football field. Including the end zones, the length of the field, in feet, is 60 feet more than twice the width. (Lesson 8-3)

6. Write an expression for the area.
7. If the area of the field is 36,000 square feet, what are the dimensions of the field?
8. What are the dimensions of the field in yards?

PHYSICAL SCIENCE For Exercises 9 and 10, use the following information.

Teril throws a ball upward while standing on the top of a 500-foot tall apartment building. Its height h, in feet, after t seconds is given by the equation $h = -16t^2 + 48t + 506$. (Lesson 8-4)

9. What do the values 48 and 506 in the equation represent?
10. The ball falls on a balcony that is 218 feet above the ground. How many seconds was the ball in the air?

DECKS For Exercises 11 and 12, use the following information.

Zelda is building a deck in her backyard. The plans for the deck show that it is to be 24 feet by 24 feet. Zelda wants to reduce one dimension by a number of feet and increase the other dimension by the same number of feet. (Lesson 8-5)

11. If the area of the reduced deck is 512 square feet, what are the dimensions of the deck?
12. Suppose Zelda wants to reduce the deck to one-half the area of the deck in the plans. Can she reduce each dimension by the same length and use dimensions that are whole numbers? Explain.

POOLS For Exercises 13–16, use the following information.

Susan wants to buy an above ground swimming pool for her yard. Model A is 42 inches deep and holds 1750 cubic feet of water. The length of the pool is 5 feet more than the width. (Lesson 8-6)

13. What is the area of water that is exposed to the air?
14. What are the dimensions of the pool?
15. A Model B pool holds twice as much water as Model A. What are some possible dimensions for this pool?
16. Model C has length and width that are both twice as long as Model A, but the height is the same. What is the ratio of the volume of Model A to Model C?

PHYSICAL SCIENCE For Exercises 1–4, use the following information.

A ball is released 6 feet above the ground and thrown vertically into the air. The equation $h = -16t^2 + 112t + 6$ gives the height of the ball if the initial velocity is 112 feet per second. (Lesson 9-1)

1. Write the equation of the axis of symmetry and find the coordinates of the vertex of the graph of the equation.
2. What is the ball's maximum height?
3. How many seconds after release does the ball reach its maximum height?
4. How many seconds is the ball in the air?

RIDES For Exercises 5–7, use this information.

A popular amusement park ride whisks riders to the top of a 250-foot tower and drops them at speeds exceeding 50 miles per hour. A function for the path of a rider is $h = -16t^2 + 250$, where h is the height and t is the time in seconds. (Lesson 9-2)

5. The ride stops the descent of the rider 40 feet above the ground. Write an equation that models the drop of the rider.
6. Solve the equation by graphing the related function. How many roots are there?
7. About how many seconds does it take to complete the ride?

PROJECTS For Exercises 8–10, use the following information.

Jude is making a poster for his science project. The poster board is 22 inches wide by 27 inches tall. He wants to cover two thirds of the area with text or pictures and leave a top margin 3 times as wide as the side margins and a bottom margin twice as wide as the side margins. (Lesson 9-3)

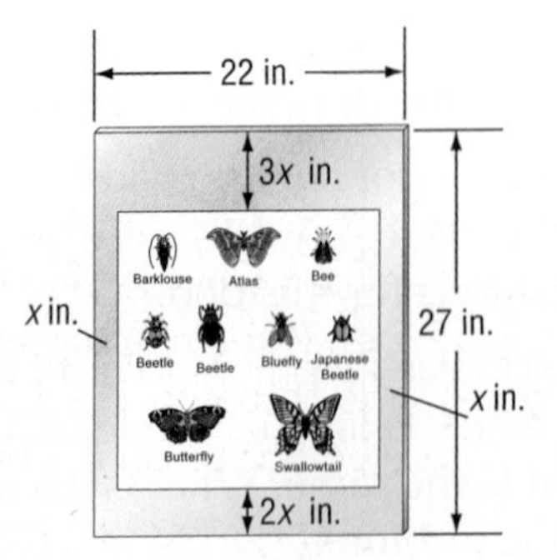

8. Write an equation for this situation.
9. Solve your equation for x by completing the square. Round to the nearest tenth.
10. What should be the widths of the margins?

TELEVISION For Exercises 11 and 12, use the following information.

The number of U.S. households with cable television has been on the rise. The percent of households with cable y can be approximated by the quadratic function $y = -0.11x^2 + 4.95x + 12.69$, where x stands for the number of years after 1977. (Lesson 9-4)

11. Use the Quadratic Formula to solve for x when $y = 30$. What do these values represent?
12. Is a quadratic function a good model for these data? Why or why not?

POPULATION For Exercises 13–15, use the following information.

The population of Asia from 1650 to 2000 can be estimated by the function $P(x) = 335(1.007)^x$, where x is the number of years since 1650 and the population is in millions of people. (Lesson 9-5)

13. Graph the function and name the y-intercept.
14. What does the y-intercept represent?
15. Use the function to approximate the number of people in Asia in 2050.

POPULATION For Exercises 16 and 17, use the following information.

The percent of the U.S. population P that is at least 65 years old can be approximated by $P = 3.86(1.013)^t$, where t represents the number of years since 1900. (Lesson 9-6)

16. To the nearest percent, how much of the population will be 65 years of age or older in the year 2010?
17. Predict the year in which people aged 65 years or older will represent 20% of the population if this trend continues. (*Hint:* Make a table.)

SATELLITES For Exercises 1–3, use the following information.

A satellite is launched into orbit 200 kilometers above Earth. The orbital velocity of a satellite is given by the formula

$v = \sqrt{\frac{Gm_E}{r}}$, where v is velocity in meters per second, G is a given constant, m_E is the mass of Earth, and r is the radius of the satellite's orbit. (Lesson 10-1)

1. The radius of Earth is 6,380,000 meters. What is the radius of the satellite's orbit in meters?
2. The mass of Earth is 5.97×10^{24} kilogram and the constant G is 6.67×10^{-11} N • m^2/kg^2 where N is in Newtons. Use the formula to find the orbital velocity of the satellite in meters per second.
3. The orbital period of the satellite can be found by using the formula $T = \frac{2\pi r}{v}$, where r is the radius of the orbit and v is the orbital velocity of the satellite in meters per second. Find the orbital period of the satellite in hours.

RIDES For Exercises 4–6, use the following information.

The designer of a roller coaster must consider the height of the hill and the velocity of the coaster as it travels over the hill. Certain hills give riders a feeling of weightlessness. The formula $d = \sqrt{\frac{2hv^2}{g}}$ allows designers to find the correct distance from the center of the hill that the coaster should begin its drop for maximum fun. (Lesson 10-2)

4. In the formula above, d is the distance from the center of the hill, h is the height of the hill, v is the velocity of the coaster at the top of the hill in meters per second, and g is a gravity constant of 9.8 meters per second squared. If a hill is 10 meters high and the velocity of the coaster is 10 m/s, find d.
5. Find d if the height of the hill is 10 meters but the velocity is 20 m/s. How does d compare to the value in Exercise 4?
6. Suppose you find the same formula in another book written as $d = 1.4\sqrt{\frac{hv^2}{g}}$. Will this produce the same value of d? Explain.
7. **PACKAGING** A cylindrical container of chocolate drink mix has a volume of about 162 in^3. The formula for volume of a cylinder is $V = \pi r^2 h$, where r is the radius and h is the height. The radius of the container can be found by using the formula $r = \sqrt{\frac{V}{\pi h}}$. If the height is 8.25 inches, find the radius of the container. (Lesson 10-3)

TOWN SQUARES For Exercises 8 and 9, use the following information.

Tiananmen Square in Beijing, China, is the largest town square in the world, covering 98 acres. **Source:** *The Guinness Book of World Records* (Lesson 10-4)

8. One square mile is 640 acres. Assuming that Tiananmen Square is a square, how many feet long is a side to the nearest foot?
9. To the nearest foot, what is the diagonal distance across Tiananmen Square?

PIZZA DELIVERY For Exercises 10 and 11, use the following information.

The Pizza Place delivers pizza to any location within a radius of 5 miles from the store for free. Tyrone drives 32 blocks north and then 45 blocks east to deliver a pizza. In this city, there are about 6 blocks per half mile. (Lesson 10-5)

10. Should there be a charge for the delivery? Explain.
11. Describe two delivery situations that would result in about 5 miles.
12. **BRIDGES** Truss bridges use triangles in their support beams. Mark plans to make a model of a truss bridge in the scale 1 inch = 12 feet. If the height of the triangles on the actual bridge is 40 feet, what will the height be on the model? (Lesson 10-6)

OPTOMETRY **For Exercises 1–3, use the following information.**

When a person does not have clear vision, an optometrist can prescribe lenses to correct the condition. The power P of a lens, in a unit called diopters, is equal to 1 divided by the focal length f, in meters, of the lens. The formula is $P = \frac{1}{f}$. (Lesson 11-1)

1. Graph the inverse variation $P = \frac{1}{f}$.
2. Find the powers of lenses with focal lengths +20 and −40 centimeters. (*Hint*: Change centimeters to meters.)
3. What do you notice about the powers in Exercise 2?

PHYSICS **For Exercises 4 and 5, use the following information.**

Some principles in physics, such as gravitational force between two objects, depend upon a relationship known as the inverse square law. This law states that two variables are related by the relationship $y = \frac{1}{x^2}$, where x is distance. (Lesson 11-2)

4. Make a table of values and graph $y = \frac{1}{x^2}$. Describe the shape of the graph.
5. If x represents distance, how does this affect the domain of the graph?

FERRIS WHEELS **For Exercises 6–8, use the following information.**

George Ferris built the first Ferris wheel for the World's Columbian Exposition in 1892. It had a diameter of 250 feet. (Lesson 11-3)

6. To find the speed traveled by a car located on the circumference of the wheel, you can find the circumference of a circle and divide by the time it takes for one rotation of the wheel. (Recall that $C = \pi d$.) Write a rational expression for the speed of a car rotating in time t.
7. Suppose the first Ferris wheel rotated once every 5 minutes. What was the speed of a car on the circumference in feet per minute?
8. Use dimensional analysis to find the speed of a car in miles per hour.

9. **MOTOR VEHICLES** In 1999, the U.S. produced 13,063,000 motor vehicles. This was 23.2% of the total motor vehicle production for the whole world. How many motor vehicles were produced worldwide in 1999? (Lesson 11-4)

10. **LIGHT** The speed of light is approximately 1.86×10^5 miles per second. The table shows the distances, in miles, of the planets from the Sun. Find the amount of time in minutes that it takes for light from the Sun to reach each planet. (Lesson 11-5)

Planet	Miles	Planet	Miles
Mercury	5.79×10^{10}	Jupiter	7.78×10^{11}
Venus	1.08×10^{11}	Saturn	1.43×10^{12}
Earth	1.496×10^{11}	Uranus	2.87×10^{12}
Mars	2.28×10^{11}	Pluto	4.50×10^{12}

11. **GEOGRAPHY** The land areas of all the continents, in thousands of square miles, are shown. What fraction of the land area of the world do North and South America occupy? (Lesson 11-6)

Continent	Area
North America	9400
South America	6900
Europe	3800
Asia	17,400
Africa	11,700
Oceania	3300
Antarctica	5400

Source: *The World Almanac*

12. **GARDENING** Celeste builds decorative gardens in her landscaping business. She uses either 35, 50, or 75 bricks per garden. What is the least number of bricks she should order that would allow her to build a whole number of each type of garden? (Lesson 11-7)

CRAFTS **For Exercises 13 and 14, use the following information.**

Ann makes tablecloths to sell at craft fairs. A small one takes one-half yard of fabric, a medium takes five-eighths yard, and a large takes one and one-quarter yard. (Lesson 11-8)

13. How many yards of fabric does she need to make one of each type of tablecloth?
14. One bolt contains 30 yards of fabric. Can she use the entire bolt by making an equal number of each type? Explain.

CAREERS For Exercises 1 and 2, use the following information.

The graph below shows the results of a survey of students who were asked their preferences for a future career. (Lesson 12-1)

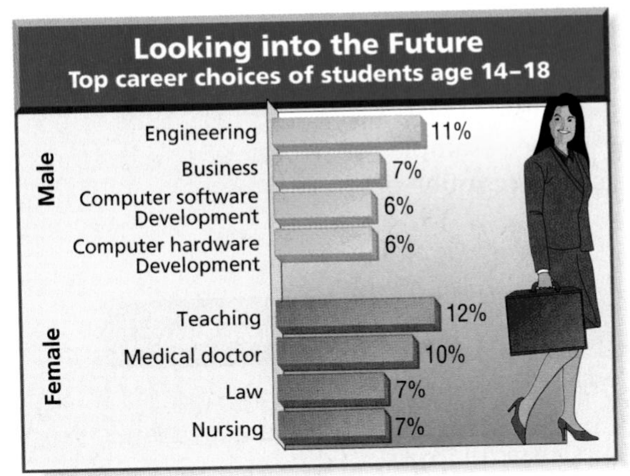

Source: *USA TODAY*

1. Write a statement to describe what you do know about the sample.
2. What additional information would you like to have about the sample to determine whether the sample is biased?

FLOWERS For Exercises 3–5, use this information.

A flower shop is making special floral arrangements for a holiday. The tables show the options available. (Lesson 12-2)

Vase	Deluxe	Standard	Economy
Cost	$12.00	$8.00	$5.00

Ribbon	Velvet	Satin
Cost	$3.00	$2.00

Flowers	Orchids	Roses	Daisies
Cost	$35.00	$20.00	$12.00

Card	Large	Small
Cost	$2.50	$1.75

3. How many floral arrangements are possible? Each arrangement has one vase, one ribbon, one type of flowers, and one card.
4. What is the cost of the most expensive arrangement? the least expensive?
5. What is the cost of each of the four most expensive arrangements?

GAMES For Exercises 6–8, use this information.

Melissa is playing a board game in which you make words to score points. There are 12 letters left in the box, and she must choose 4. She cannot see the letters. (Lesson 12-3)

6. Suppose the 12 letters are all different. In how many ways can she choose 4 of the 12?
7. She chooses A, T, R, and E. How many different arrangements of three letters can she make from her letters?
8. How many of the three-letter arrangements are words? List them.

DRIVING For Exercises 9 and 10, use the following information.

The table shows a probability distribution for various age categories of licensed drivers in the U.S. in 1998. (Lesson 12-5)

X = Age Category	Probability
under 20	0.053
20–34	0.284
35–49	0.323
50–64	0.198
65 and over	0.142

Source: *The World Almanac*

9. Is this a valid probability distribution? Justify your answer.
10. If a driver in the U.S. is randomly selected, what is the probability that the person is under 20 years old? 50 years old or over?

LOTTERIES For Exercises 11–13, use the following information.

In a certain state, lottery numbers are five-digit numbers. Each digit can be 1, 2, 3, 4, 5, or 6. Once a week, a random 5-digit number is chosen as the winning number. (Lesson 12-6)

11. How many five-digit numbers are possible? Explain how you calculated the number of possible outcomes.
12. Perform a simulation for winning the lottery. Describe the objects you used.
13. According to your experiment, if you buy one ticket, what is the experimental probability of winning the lottery?

Preparing for Standardized Tests

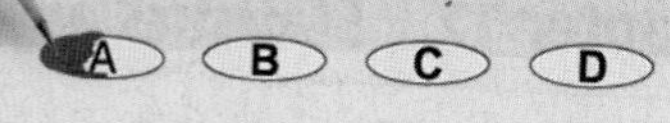

Becoming a Better Test-Taker

At some time in your life, you will probably have to take a standardized test. Sometimes this test may determine if you go on to the next grade or course, or even if you will graduate from high school. This section of your textbook is dedicated to making you a better test-taker.

TYPES OF TEST QUESTIONS In the following pages, you will see examples of four types of questions commonly seen on standardized tests. A description of each type of question is shown in the table below.

Type of Question	Description	See Pages
multiple choice	4 or 5 possible answer choices are given from which you choose the best answer.	757–760
gridded response	You solve the problem. Then you enter the answer in a special grid and shade in the corresponding circles.	761–764
short response	You solve the problem, showing your work and/or explaining your reasoning.	765–768
extended response	You solve a multi-part problem, showing your work and/or explaining your reasoning.	769–773

PRACTICE After being introduced to each type of question, you can practice that type of question. Each set of practice questions is divided into five sections that represent the concepts most commonly assessed on standardized tests.

- Number and Operations
- Algebra
- Geometry
- Measurement
- Data Analysis and Probability

USING A CALCULATOR On some tests, you are permitted to use a calculator. You should check with your teacher to determine if calculator use is permitted on the test you will be taking, and if so, what type of calculator can be used.

TEST-TAKING TIPS In addition to Test-Taking Tips like the one shown at the right, here are some additional thoughts that might help you.

- Get a good night's rest before the test. Cramming the night before does not improve your results.
- Budget your time when taking a test. Don't dwell on problems that you cannot solve. Just make sure to leave that question blank on your answer sheet.
- Watch for key words like NOT and EXCEPT. Also look for order words like LEAST, GREATEST, FIRST, and LAST.

TEST-TAKING TIP

If you are allowed to use a calculator, make sure you are familiar with how it works so that you won't waste time trying to figure out the calculator when taking the test.

Multiple-Choice Questions

Multiple-choice questions are the most common type of questions on standardized tests. These questions are sometimes called *selected-response questions.* You are asked to choose the best answer from four or five possible answers.

To record a multiple-choice answer, you may be asked to shade in a bubble that is a circle or an oval, or to just write the letter of your choice. Always make sure that your shading is dark enough and completely covers the bubble.

Incomplete Shading

Ⓐ Ⓑ Ⓒ Ⓓ

Too light shading

Ⓐ Ⓑ Ⓒ Ⓓ

Correct shading

Ⓐ Ⓑ Ⓒ Ⓓ

The answer to a multiple-choice question is usually not immediately obvious from the choices, but you may be able to eliminate some of the possibilities by using your knowledge of mathematics. Another answer choice might be that the correct answer is not given.

EXAMPLE

1 A storm signal flag is used to warn small craft of wind speeds that are greater than 38 miles per hour. The length of the square flag is always three times the length of the side of the black square. If y is the area of the black square and x is the length of the side of the flag, which equation describes the relationship between x and y?

STRATEGY

Elimination
You can eliminate any obvious wrong answers.

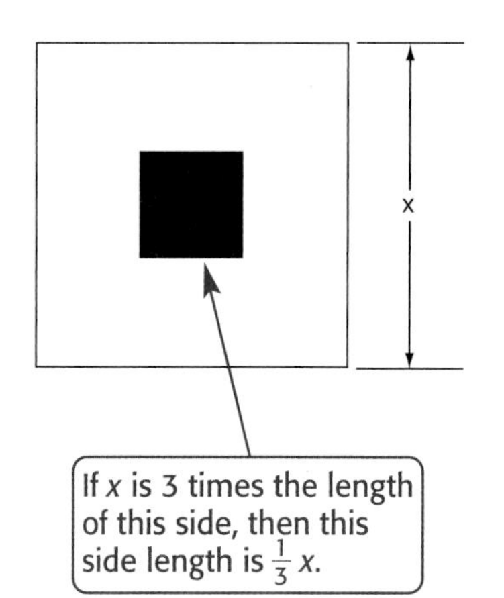

A $y = \frac{1}{3}x^2$

B $y = \frac{1}{9}x^2$

C $y = x^2 - 1$

D $y = 3x$

E $y = 9x$

For the area of a square, $A = s^2$. So, $A = x \cdot x$ or x^2.

The area of the black square is part of the area of the flag, which is x^2. Eliminate choices D and E because they do not include x^2.

$A = \left(\frac{1}{3}\right)x^2$ or $\frac{1}{9}x^2$ square units

So, $y = \frac{1}{9}x^2$. This is choice B.

Use some random numbers to check your choice.

Multiples of 3 make calculations easier.

Length of Flag (x)	Length of Black Square	Area of Black Square	Area $= \frac{1}{9}x^2$
12	4	16	$16 \stackrel{?}{=} \frac{1}{9}(12^2)$ ✓
27	9	81	$81 \stackrel{?}{=} \frac{1}{9}(27^2)$ ✓
60	20	400	$400 \stackrel{?}{=} \frac{1}{9}(60^2)$ ✓

Many multiple-choice questions are actually two- or three-step problems. If you do not read the question carefully, you may select a choice that is an intermediate step instead of the correct final answer.

EXAMPLE

2 Barrington can skateboard down a hill five times as fast as he can walk up the hill. If it takes 9 minutes to walk up the hill and skateboard back down, how many minutes does it take him to walk up the hill?

STRATEGY

Reread the Problem Read the problem carefully to find what the question is asking.

F 1.5 min G 4.5 min H 7.2 min J 7.5 min

Before involving any algebra, let's think about the problem using random numbers.

Skating is five times as fast as walking, so walking time equals 5 times the skate time. Use a table to find a pattern.

Use the pattern to find a general expression for walk time given any skate time.

Skate Time	Skate Time × 5 = Walk Time
6 min	$6 \cdot 5 = 30$ min
3 min	$3 \cdot 5 = 15$ min
2 min	$2 \cdot 5 = 10$ min
x min	$x \cdot 5 = 5x$ min

The problem states that the walk time and the skate time total 9 minutes.

Use the expression to write an equation for the problem.

$x + 5x = 9$ skate time + walk time = 9 minutes

$6x = 9$ Add like terms.

$x = 1.5$ Divide each side by 6.

Looking at the choices, you might think that choice F is the correct answer. But what does x represent, and what is the problem asking?

The problem asks for the time it takes to walk up the hill, but the value of x is the time it takes to skateboard. So, the actual answer is found using $5x$ or $5(1.5)$, which is 7.5 minutes.

The correct choice is J.

EXAMPLE

3 The Band Boosters are making ice cream to sell at an Open House. Each batch of ice cream calls for 5 cups of milk. They plan to make 20 batches. How many gallons of milk do they need?

STRATEGY

Units of Measure Make certain your answer reflects the correct unit of measure.

A 800 B 100 C 25 D 12.5 E 6.25

The Band Boosters need 5×20 or 100 cups of milk. However, choice B is not the correct answer. The question asks for *gallons* of milk.

4 cups = 1 quart and 4 quarts = 1 gallon, so 1 gallon = 4×4 or 16 cups.

$100 \text{ cups} \times \frac{1 \text{ gallon}}{16 \text{ cups}} = 6.25$ gallons, which is choice E.

Multiple Choice Practice

Choose the best answer.

Number and Operations

1. One mile on land is 5280 feet, while one nautical mile is 6076 feet. What is the ratio of the length of a nautical mile to the length of a land mile as a decimal rounded to the nearest hundredth?

A 0.87 **B** 1.01 **C** 1.15 **D** 5.68

2. The star Proxima Centauri is 24,792,500 million miles from Earth. The star Epsilon Eridani is 6.345 × 1013 miles from Earth. In scientific notation, how much farther from Earth is Epsilon Eridani than Proxima Centauri?

F 0.697×10^{14} mi **H** 6.097×10^{13} mi
G 3.866×10^{13} mi **J** 38.658×10^{12} mi

3. In 1976, the cost per gallon for regular unleaded gasoline was 61 cents. In 2002, the cost was \$1.29 per gallon. To the nearest percent, what was the percent of increase in the cost per gallon of gas from 1976 to 2002?

A 1% **B** 53% **C** 95% **D** 111%

4. The serial numbers on a particular model of personal data assistant (PDA) consist of two letters followed by five digits. How many serial numbers are possible if any letter of the alphabet and any digit 0–9 can be used in any position in the serial number?

F 676,000,000 **H** 6,760,000
G 67,600,000 **J** 676,000

Algebra

5. The graph shows the approximate relationship between the latitude of a location in the Northern Hemisphere and its distance in miles from the equator. If y represents the distance of a location from the equator and x represents the measure of latitude, which equation describes the relationship between x and y?

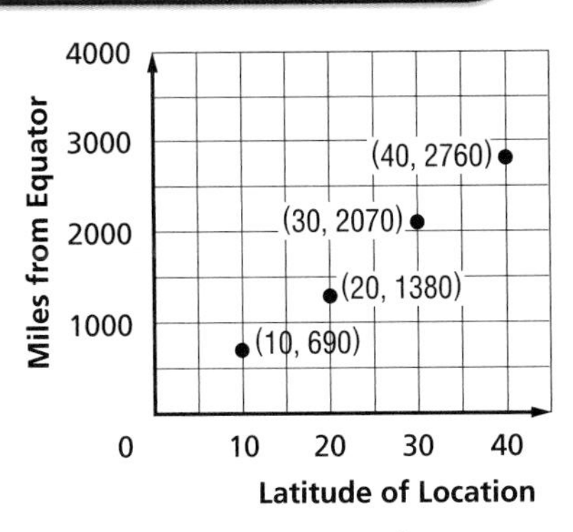

A $y = x + 69$ **C** $y = 69x$
B $y = x + 690$ **D** $y = 10x$

6. A particular prepaid phone card can be used from a pay phone. The charge is 30 cents to connect and then 4.5 cents per minute. If y is the total cost of a call in cents where x is the number of-minutes, which equation describes the relation between x and y?

F $y = 4.5x + 30$ **H** $y = 0.45x + .30$
G $y = 30x + 4.5$ **J** $y = 0.30x + 0.45$

7. Katie drove to the lake for a weekend outing. The lake is 100 miles from her home. On the trip back, she drove for an hour, stopped for lunch for an hour, and then finished the trip home. Which graph best represents her trip home and the distance from her home at various times?

A

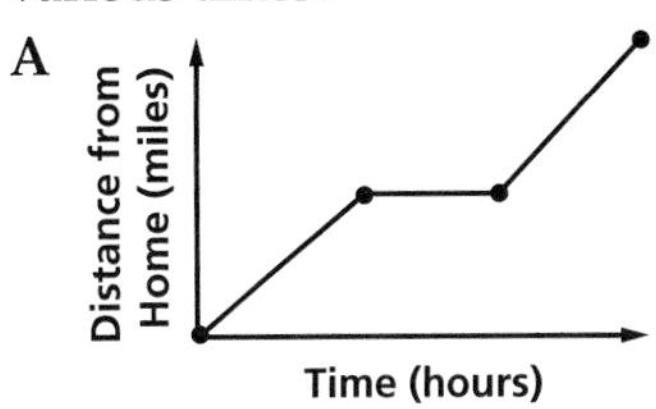

B

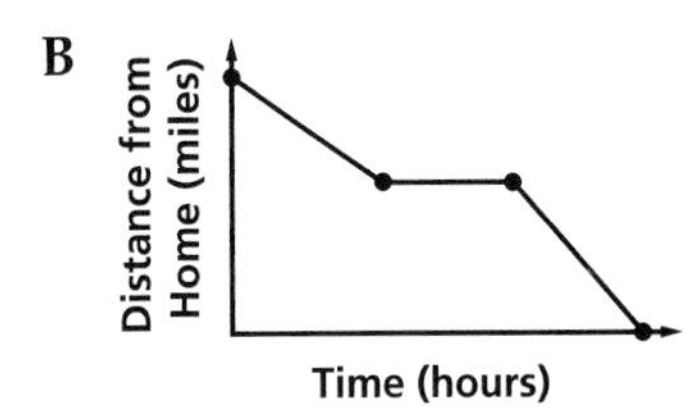

C

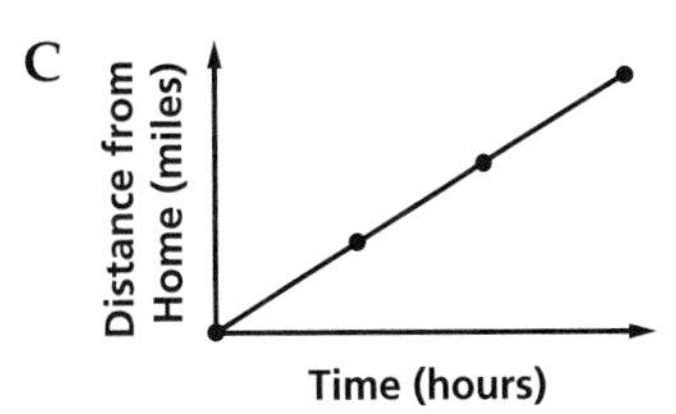

D

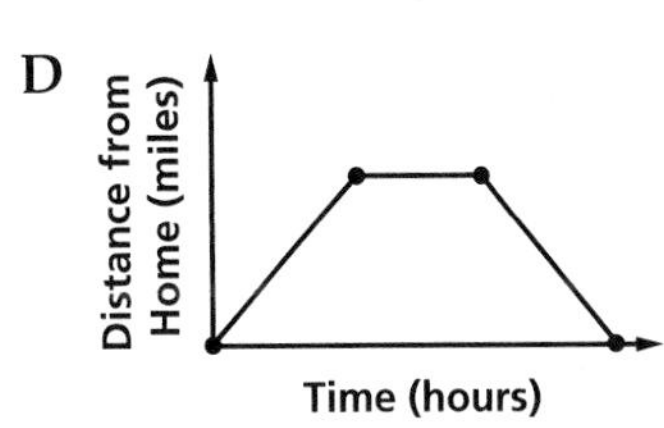

8. Temperature can be given in degrees Fahrenheit or degrees Celsius. The formula $F = \frac{9}{5}C + 32$ can be used to change any temperature given in degrees Celsius to degrees Fahrenheit. Solve the formula for C.

F $C = \frac{5}{9}(F - 32)$ **H** $C = \frac{5}{9}F - 32$
G $C = F + 32 - \frac{9}{5}$ **J** $C = \frac{9}{5}(F - 32)$

Geometry

9. Which of the following statements are true about the 4-inch quilt square?

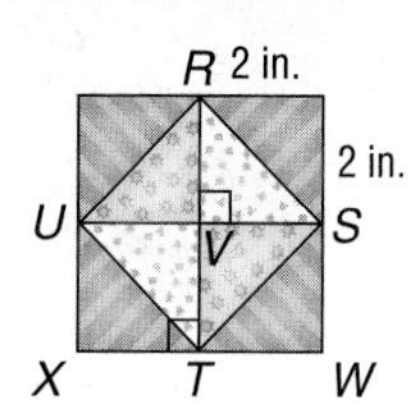

A $VSWT$ is a square.

B $UVTX \cong VSWT$

C Four right angles are formed at V.

D Only A and B are true.

E A, B, and C are true.

10. At the Daniels County Fair, the carnival rides are positioned as shown. What is the value of x?

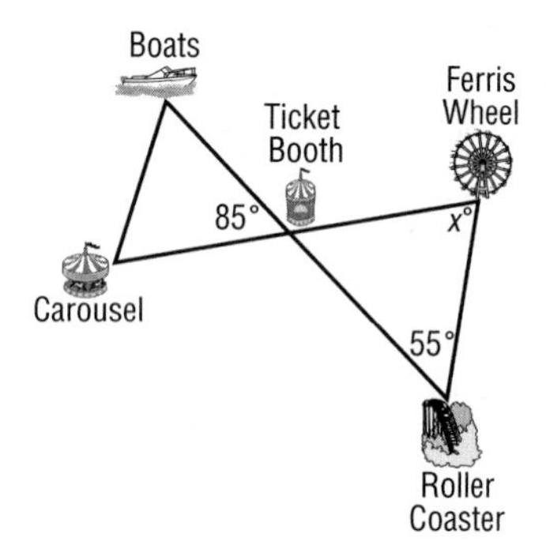

F 40

G 47.5

H 55

J 70

K 85

11. The diagram shows a map of the Clearwater Wilderness hiking area. To the nearest tenth of a mile, what is the distance from the Parking Lot to Bear Ridge using the most direct route?

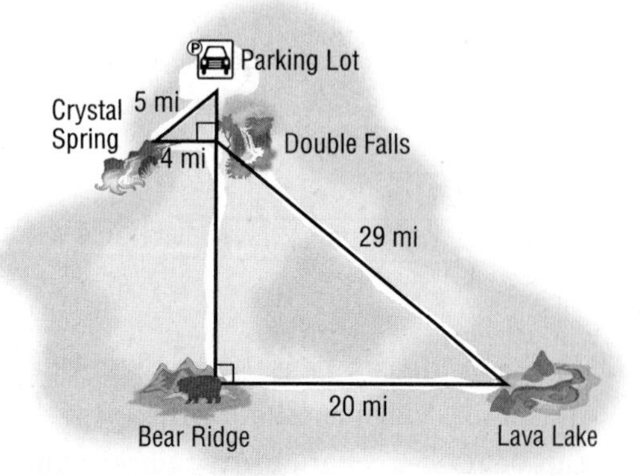

A 24 mi

B 25.5 mi

C 26 mi

D 30.4 mi

Measurement

12. Laura expects about 60 people to attend a party. She estimates that she will need one quart of punch for every two people. How many gallons of punch should she prepare?

F 7.5

G 15

H 30

J 34

13. Stone Mountain Manufacturers are designing two sizes of cylindrical cans below. What is the ratio of the volume of can A to the volume of can B?

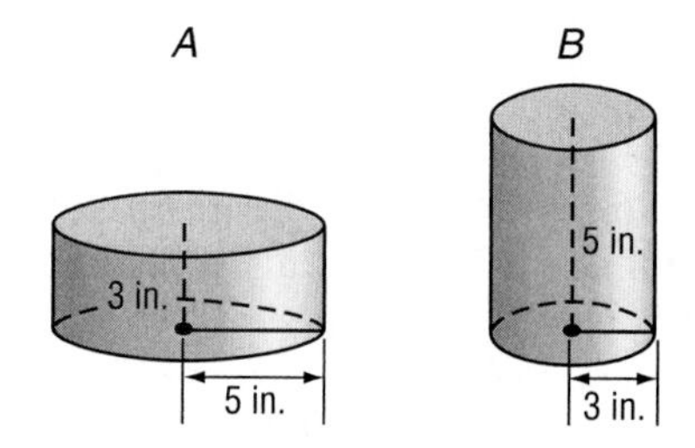

A 9 to 2

B 25 to 3

C 3 to 5

D 5 to 3

Data Analysis and Probability

14. The 2000 populations of the five least-populated U.S. states are shown in the table. Which statement is true about this set of data?

State	Population
Alaska	626,932
North Dakota	642,200
South Dakota	754,844
Vermont	608,827
Wyoming	493,782

Source: *U.S. Census Bureau*

F The mode of the data set is 642,200.

G The median of the data set is 626,932.

H The mean of the data set is 625,317.

J F and H are true.

K G and H are true.

Gridded-Response Questions

Gridded-response questions are other types of questions on standardized tests. These questions are sometimes called *student-produced responses* or *grid-ins,* because you must create the answer yourself, not just choose from four or five possible answers.

For gridded response, you must mark your answer on a grid printed on an answer sheet. The grid contains a row of four or five boxes at the top, two rows of ovals or circles with decimal and fraction symbols, and four or five columns of ovals, numbered 0–9. Since there is no negative symbol on the grid, answers are never negative. At the right is an example of a grid from an answer sheet.

How do you correctly fill in the grid?

EXAMPLE

1 Diego drove 174 miles to his grandmother's house. He made the drive in 3 hours without any stops. At this rate, how far in miles can Diego drive in 5 hours?

What value do you need to find?

You need to find the number of miles Diego can drive in 5-hours.

Write a proportion for the problem. Let m represent the number of miles.

$$\text{miles} \rightarrow \frac{174}{3} = \frac{m}{5} \leftarrow \text{miles} \quad (\text{hours} \rightarrow \text{denominators} \leftarrow \text{hours})$$

Solve the proportion.

$\frac{174}{3} = \frac{m}{5}$	Original proportion
$870 = 3m$	Find the cross products.
$290 = m$	Divide each side by 3.

How do you fill in the grid for the answer?

- Write your answer in the answer boxes.
- Write only one digit or symbol in each answer box.
- Do not write any digits or symbols outside the answer boxes.
- You may write your answer with the first digit in the left answer box, or with the last digit in the right answer box. You may leave blank any boxes you do not need on the right or the left side of your answer.
- Fill in only one bubble for every answer box that you have written in. Be sure not to fill in a bubble under a blank answer box.

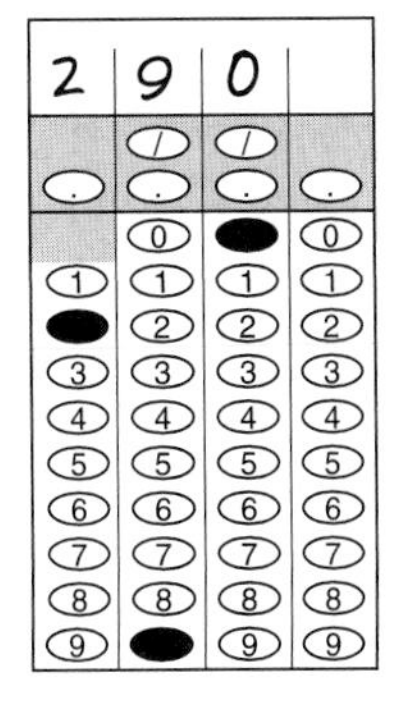

Many gridded response questions result in an answer that is a fraction or a decimal. These values can also be filled in on the grid.

How do you grid decimals and fractions?

EXAMPLE

2 What is the slope of the line that passes through (−2, 3) and (2, 4)?

STRATEGY

Decimals and Fractions
Fill in the grid with decimal and fraction answers.

Let $(-2, 3) = (x_1, y_1)$ and $(2, 4) = (x_2, y_2)$.

$m = \dfrac{y_2 - y_1}{x_2 - x_1}$ Slope formula

$= \dfrac{4 - 3}{2 - (-2)}$ or $\dfrac{1}{4}$ Substitute and simplify.

How do you grid the answer?

You can either grid the fraction $\frac{1}{4}$, or rewrite it as 0.25 and grid the decimal. Be sure to write the decimal point or fraction bar in the answer box. The following are acceptable answer responses that represent $\frac{1}{4}$ and 0.25.

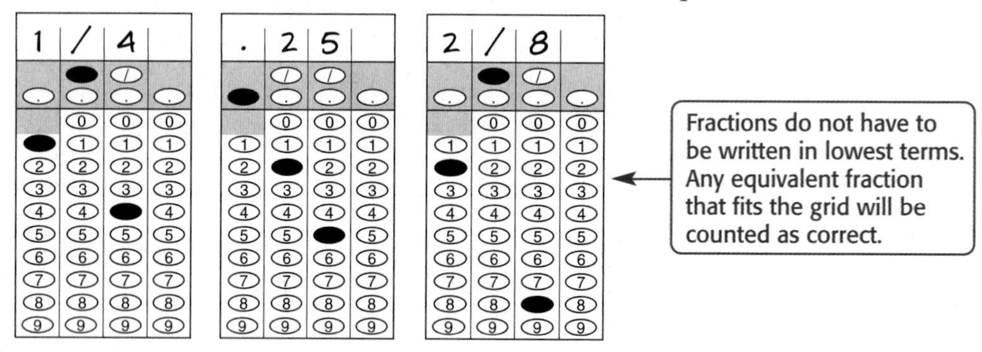

Some problems may result in an answer that is a mixed number. Before filling in the grid, change the mixed number to an equivalent improper fraction or decimal. For example, if the answer is $1\frac{1}{2}$, do not enter 1 1/2 as this will be interpreted as $\frac{11}{2}$. Instead, either enter 3/2 or 1.5.

How do you grid mixed numbers?

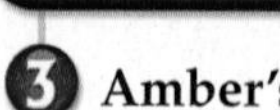

3 Amber's cookie recipe calls for $1\frac{1}{3}$ cups of coconut. If Amber plans to make 4 batches of cookies, how much coconut does she need?

Find the amount of coconut needed using a proportion.

coconut → batches →

$$\frac{1\frac{1}{3}}{1} = \frac{x}{4}$$

$$4\left(1\frac{1}{3}\right) = 1x$$

$$4\left(\frac{4}{3}\right) = x$$

$$\frac{16}{3} = x$$

Leave the answer as the improper fraction $\frac{16}{3}$, as you cannot correctly grid $5\frac{1}{3}$.

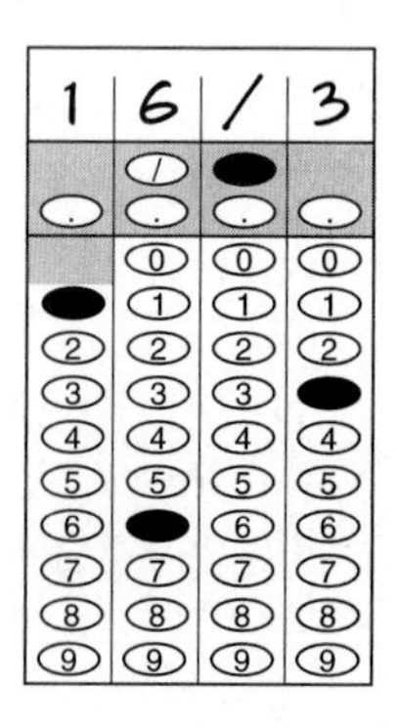

Gridded-Response Practice

Solve each problem. Then copy and complete a grid like the one shown on page 761.

Number and Operations

1. China has the most days of school per year for children with 251 days. If there are 365 days in a year, what percent of the days of the year do Chinese students spend in school?

2. Charles is building a deck and wants to buy some long boards that he can cut into various lengths without wasting any lumber. He would like to cut any board into all lengths of 24 inches, 48 inches, or 60 inches. In feet, what is the shortest length of boards that he can buy?

3. At a sale, an item was discounted 20%. After several weeks, the sale price was discounted an additional 25%. What was the total percent discount from the original price of the item?

4. The Andromeda Spiral galaxy is 2.2×10^6 light-years from Earth. The Ursa Minor dwarf is 2.5×10^5 light-years from Earth. How many times as far is the Andromeda Spiral as Ursa Minor dwarf from Earth?

5. Twenty students want to attend the World Language Convention. The school budget will only allow for four students to attend. In how many ways can four students be chosen from the twenty students to attend the convention?

Algebra

6. Find the y-intercept of the graph of the equation $3x + 4y - 5 = 0$.

7. Name the x-coordinate of the solution of the system of equations $2x - y = 7$ and $3x + 2y = 7$.

8. Solve $2b - 2(3b - 5) = 8(b - 7)$ for b.

9. Kersi read 36 pages of a novel in 2 hours. Find the number of hours it will take him to read the remaining 135 pages if he reads them at the same rate?

10. If $f(x) = x + 3$ and $g(x) = x^2 - 2x + 5$, find $6[f(g(1))]$.

11. The Donaldsons have a fish pond in their yard that measures 8 feet by 15 feet. They want to put a walkway around the pond that measures the same width on all sides of the pond, as shown in the diagram. They want the total area of the pond and walkway to be 294 square feet. What will be the width of the walkway in feet?

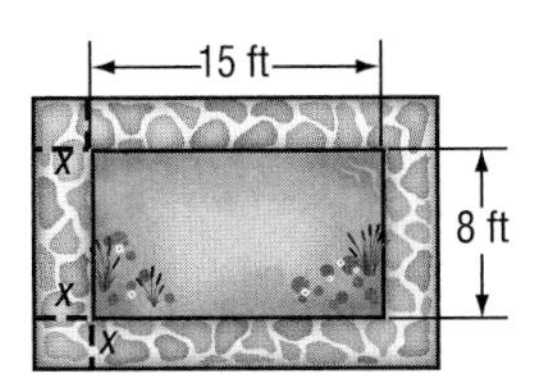

Geometry

12. $\triangle MNP$ is reflected over the x-axis. What is the x-coordinate of the image of point N?

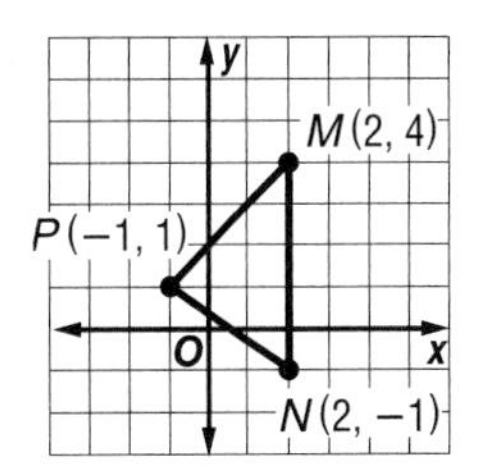

13. The pattern for the square tile shown in the diagram is to be enlarged so that it will measure 15 inches on a side. By what scale factor must the pattern be enlarged?

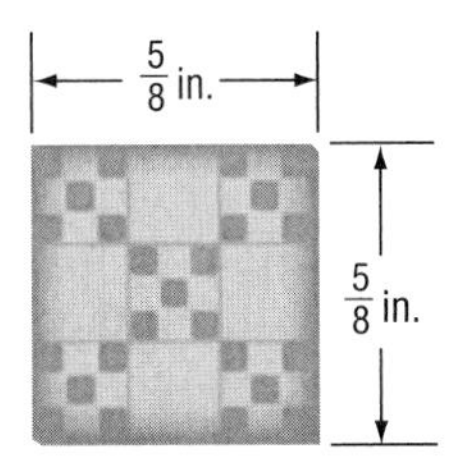

14. Find the measure of $\angle A$ to the nearest degree.

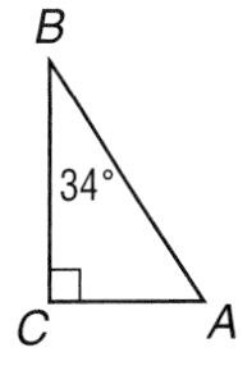

TEST-TAKING TIP

Question 14
Remember that the hypotenuse of a right triangle is always opposite the right angle.

15. Use the diagram for $\triangle ABO$ and $\triangle XBY$. Find the length of $\overline{BX}$.

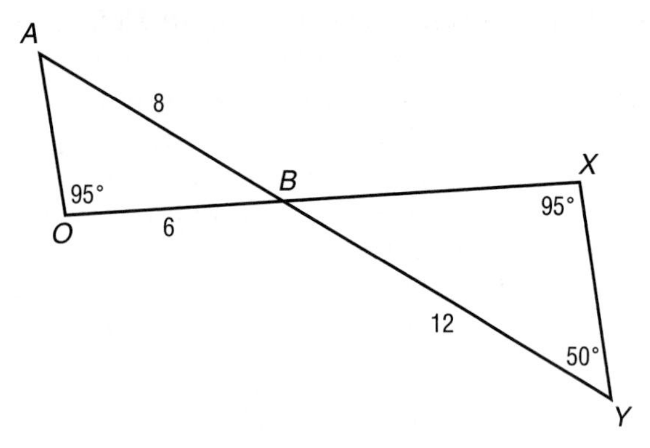

16. A triangle has a perimeter of 96 centimeters. The ratio of measures of its three sides is 6:8:10. Find the length of the longest side in centimeters.

17. The scale on a map of Texas is 0.75 inch = 5 miles. The distance on the map from San Antonio to Dallas is 8.25 inches. What is the actual distance from San Antonio to Dallas in miles?

Measurement

18. Pluto is 2756 million miles from the Sun. If light travels at 186,000 miles per second, how many minutes does it take for a particular ray of light to reach Pluto from the Sun? Round to the nearest minute.

19. Noah drove 342 miles and used 12 gallons of gas. At this same rate, how many gallons of gas will he use on his entire trip of 1140 miles?

20. The jumping surface of a trampoline is shaped like a circle with a diameter of 14 feet. Find the area of the jumping surface. Use 3.14 for π and round to the nearest square foot.

21. A cone is drilled out of a cylinder of wood. If the cone and cylinder have the same base and height, find the volume of the remaining wood. Use 3.14 for π. Round to the nearest cubic inch.

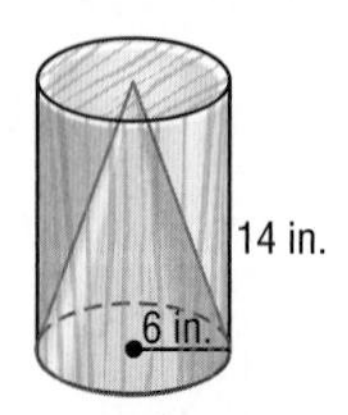

22. The front of a storage building is shaped like a trapezoid as shown. Find the area in square feet of the front of the storage building.

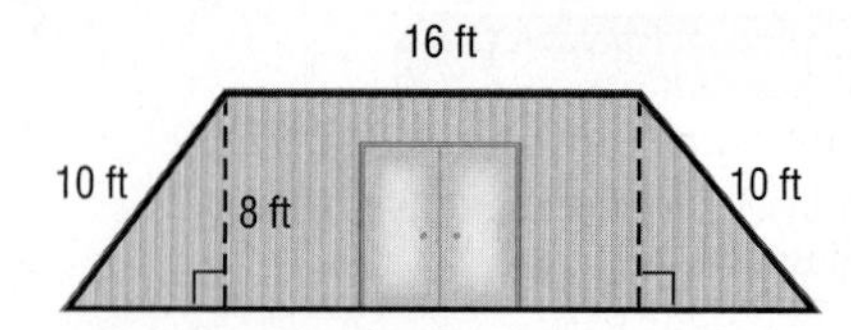

Data Analysis and Probability

23. The table shows the average size in acres of farms in the six states with the largest farms. Find the median of the farm data in acres.

Average Size of Farms for 2001	
State	**Acres per Farm**
Alaska	1586
Arizona	3644
Montana	2124
Nevada	2267
New Mexico	2933
Wyoming	3761
Vermont	608,827
Wyoming	493,782

24. The Lindley Park Pavilion is available to rent for parties. There is a fee to rent the pavilion and then a charge per hour. The graph shows the total amount you would pay to rent the pavilion for various numbers of hours. If a function is written to model the charge to rent the pavilion, where x is the number of hours and y is the total charge, what is the rate of change of the function?

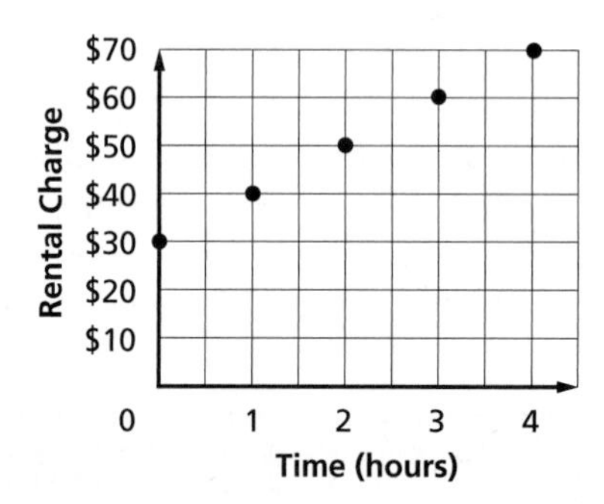

25. A particular game is played by rolling three tetrahedral (4-sided) dice. The faces of each die are numbered with the digits 1–4. How many outcomes are in the sample space for the event of rolling the three dice once?

26. In a carnival game, the blindfolded contestant draws two toy ducks from a pond without replacement. The pond contains 2 yellow ducks, 10 black ducks, 22 white ducks, and 8 red ducks. The best prize is won by drawing two yellow ducks. What is the probability of drawing two yellow ducks? Write your answer as a percent rounded to the nearest tenth of a percent.

Short-Response Questions

Short-response questions require you to provide a solution to the problem, as well as any method, explanation, and/or justification you used to arrive at the solution. These are sometimes called *constructed-response, open-response, open-ended, free-response,* or *student-produced questions.* The following is a sample rubric, or scoring guide, for scoring short-response questions.

Credit	Score	Criteria
Full	2	Full credit: The answer is correct and a full explanation is provided that shows each step in arriving at the final answer.
Partial	1	Partial credit: There are two different ways to receive partial credit. • The answer is correct, but the explanation provided is incomplete or incorrect. • The answer is incorrect, but the explanation and method of solving the problem is correct.
None	0	No credit: Either an answer is not provided or the answer does not make sense.

On some standardized tests, no credit is given for a correct answer if your work is not shown.

EXAMPLE

1 **Susana is painting two large rooms at her art studio. She has calculated that each room has 4000 square feet to be painted. It says on the can of paint that one gallon covers 300 square feet of smooth surface for one coat and that two coats should be applied for best results. What is the minimum number of 5-gallon cans of paint Susana needs to buy to apply two coats in the two rooms of her studio?**

Full Credit Solution

First find the total number of square feet to be painted.

$$4000 \times 2 = 8000 \text{ ft}^2$$

Since 1 gallon covers 300 ft^2, multiply 8000 ft^2 by the unit rate $\frac{1 \text{ gal}}{300 \text{ ft}^2}$.

$$8000 \text{ ft}^2 \times \frac{1 \text{ gal}}{300 \text{ ft}^2} = \frac{8000}{300} \text{ gal}$$

$$= 26\frac{2}{3} \text{ gal}$$

Each can of paint contains 5 gallons, so divide $26\frac{2}{3}$ gallons by 5 gallons.

$$26\frac{2}{3} \div 5 = \frac{80}{3} \div 5 = \frac{\cancel{80}^{16}}{3} \times \frac{1}{\cancel{5}_1} = \frac{16}{3} = 5\frac{1}{3}$$

Since Susana cannot buy a fraction of a can of paint, she needs to buy 6 cans of paint.

The steps, calculations, and reasoning are clearly stated.

The solution of the problem is clearly stated.

Partial Credit Solution

In this sample solution, the answer is correct; however there is no justification for any of the calculations.

There is no explanation of how $26\frac{2}{3}$ was obtained.

$$26\frac{2}{3} \div 5 = \frac{80}{3} \div 5$$

$$= \frac{\overset{16}{\cancel{80}}}{3} \times \frac{1}{\underset{1}{\cancel{5}}}$$

$$= \frac{16}{3}$$

$$= 5\frac{1}{3}$$

Susana will need to buy 6 cans of paint.

Partial Credit Solution

In this sample solution, the answer is incorrect. However, after the first statement, all of the calculations and reasoning are correct.

There are 4000 ft^2 to be painted and one gallon of paint covers 300 ft^2.

$$4000\text{ft}^2 \times \frac{1 \text{ gal}}{300 \text{ ft}^2} = \frac{4000}{300} \text{ gal}$$

$$= 13\frac{1}{3} \text{ gal}$$

The first step of doubling the square footage for painting the second room was left out.

Each can of paint contains five gallons. So 2 cans would contain 10 gallons, which is not enough. Three cans of paint would contain 15 gallons which is enough.

Therefore, Susana will need to buy 3 cans of paint.

No Credit Solution

The wrong operations are used, so the answer is incorrect. Also, there are no units of measure given with any of the calculations.

$300 \times 2 = 600$

$600 \div 5 = 120$

$4000 \div 120 = 33\frac{1}{3}$

Susana will need 34 cans of paint.

Short-Response Practice

Solve each problem. Show all your work.

Number and Operations

1. The world's slowest fish is the sea horse. The average speed of a sea horse is 0.001 mile per hour. What is the rate of speed of a sea horse in feet per minute?

2. Two buses arrive at the Central Avenue bus stop at 8 A.M. The route for the City Loop bus takes 35 minutes, while the route for the By-Pass bus takes 20 minutes. What is the next time that the two buses will both be at the Central Avenue bus-stop?

3. Toya's Clothing World purchased some denim jackets for $35. The jackets are marked up 40%. Later in the season, the jackets are discounted 25%. How much does the store lose or gain on the sale of one jacket at the discounted price?

4. A femtosecond is 10^{-15} second, and a millisecond is 10^{-3} second. How many times faster is a millisecond than a femtosecond?

5. Find the next three terms in the sequence.

$$1, 3, 9, 27, \ldots$$

Algebra

6. Find the slope of the graph of $5x - 2y + 1 = 0$.

7. Simplify $5 + x(1 - x) + 3x$. Write the result in the form $ax^2 + bx + c$.

8. Solve $17 - 3x \geq 23$.

9. The table shows what Gerardo charges in dollars for his consulting services for various numbers of hours. Write an equation that can be used to find the charge for any amount of time, where y is the total charge in dollars and x is the number of hours.

Hours	Charge	Hours	Charge
0	$25	2	$55
1	$40	3	$70

10. The population of Clark County, Nevada, was 1,375,765 in 2000 and 1,464,653 in 2001. Let x represent the years since 2000 and y represent the total population of Clark County. Suppose the county continues to increase at the same rate. Write an equation that represents the population of the county for any year after 2000.

Geometry

11. $\triangle ABC$ is dilated with scale factor 2.5. Find the coordinates of dilated $\triangle A'B'C'$.

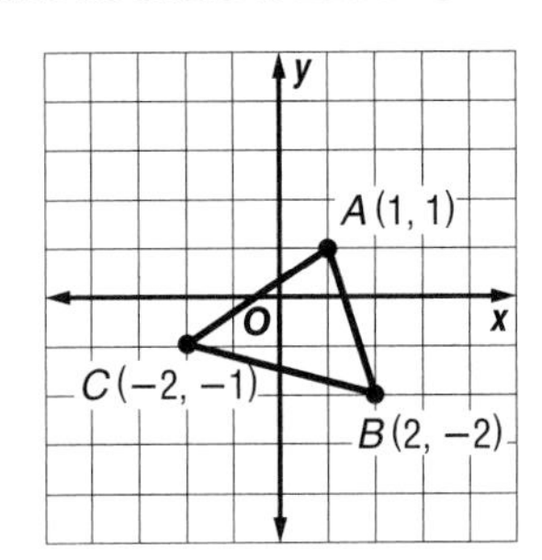

12. At a particular time in its flight, a plane is 10,000 feet above a lake. The distance from the lake to the airport is 5 miles. Find the distance in feet from the plane to the airport.

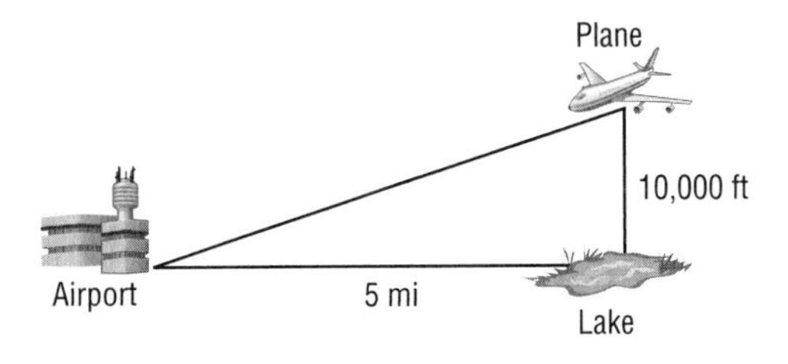

13. Refer to the diagram of the two similar triangles below. Find the value of a.

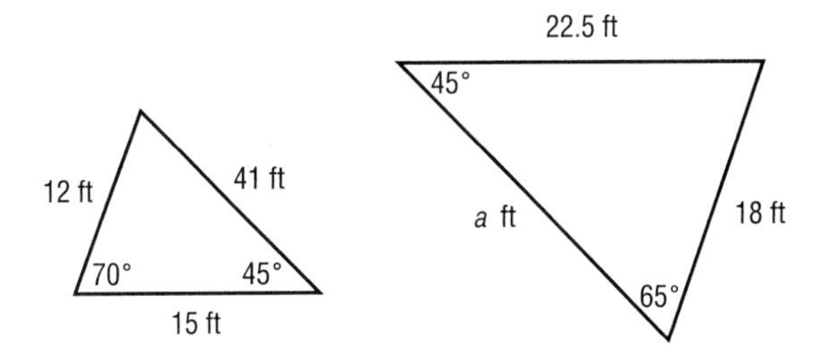

14. The vertices of two triangles are $P(3, 3)$, $Q(7, 3)$, $R(3, 10)$, and $S(-1, 4)$, $T(3, 4)$, $U(-1, 11)$. Which transformation moves $\triangle PQR$ to $\triangle STU$?

15. Aaron has a square garden in his front yard. The length of a side of the garden is 9 feet. Casey wants to plant 6 flowering bushes evenly spaced out along the diagonal. Approximately how many inches apart should he plant the bushes?

Measurement

16. One inch is equivalent to approximately 2.54 centimeters. Nikki is 61 inches tall. What is her height in centimeters?

17. During the holidays, Evan works at Cheese Haus. He packages gift baskets containing a variety of cheeses and sausages. During one four-hour shift, he packaged 20 baskets. At this rate, how many baskets will he package if he works 26 hours in one week?

18. Ms. Ortega built a box for her garden and placed a round barrel inside to be used for a fountain in the center of the box. The barrel touches the box at its sides as shown. She wants to put potting soil in the shaded corners of the box at a depth of 6 inches. How many cubic feet of soil will she need? Use 3.14 for π and round to the nearest tenth of a cubic foot.

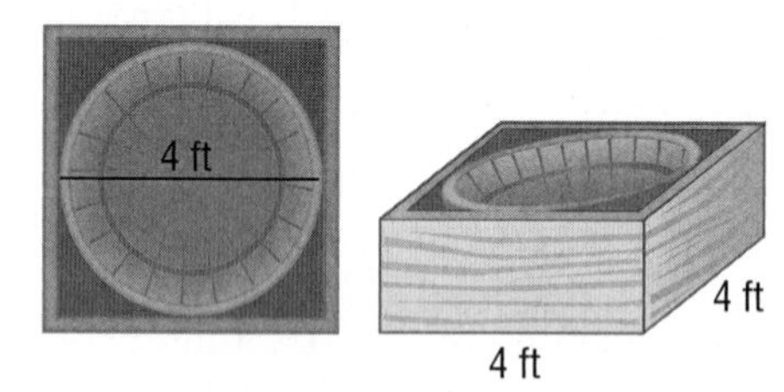

19. A line segment has its midpoint located at $(1, -5)$ and one endpoint at $(-2, -7)$. Find the length of the line to the nearest tenth.

TEST-TAKING TIP

Question 20
Most standardized tests will include any commonly used formulas at the front of the test booklet. Quickly review the list before you begin so that you know what formulas are available.

20. A child's portable swimming pool is 6 feet across and is filled to a depth of 8 inches. One gallon of water is 231 cubic inches. What is the volume of water in the pool in gallons? Use 3.14 as an approximation for π and round to the nearest gallon.

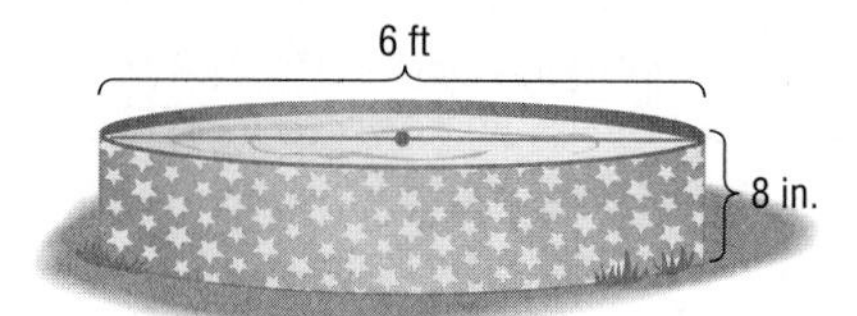

Data Analysis and Probability

21. The table shows the five lowest recorded temperatures on Earth. Find the mean of the temperatures.

Location	Temperature (°F)
Vostok, Antarctica	−138.6
Plateau Station, Antarctica	−129.2
Oymyakon, Russia	−96.0
Verkhoyansk, Russia	−90.0
Northice, Greenland	−87.0

22. Two six-sided dice are rolled. The sum of the numbers of dots on the faces of the two dice is recorded. What is the probability that the sum is 10?

23. The table shows the amount of a particular chemical that is needed to treat various sizes of swimming pools. Write the equation for a line to model the data. Let x represent the capacity of the pool in gallons and y represent the amount of the chemical in ounces.

Pool Capacity (gal)	Amount of Chemical (oz)
5000	15
10,000	30
15,000	45
20,000	60
25,000	75

24. Fifty balls are placed in a bin. They are labeled from 1 through 50. Two balls are drawn without replacement. What is the probability that both balls show an even number?

Extended-Response Questions

Extended-response questions are often called *open-ended* or *constructed-response questions.* Most extended-response questions have multiple parts. You must answer all parts correctly to receive full credit.

Extended-response questions are similar to short-response questions in that you must show all of your work in solving the problem, and a rubric is used to determine whether you receive full, partial, or no credit. The following is a sample rubric for scoring extended-response questions.

Credit	Score	Criteria
Full	4	A correct solution is given that is supported by well-developed, accurate explanations
Partial	3, 2, 1	A generally correct solution is given that may contain minor flaws in reasoning or computation or an incomplete solution. The more correct the solution, the greater the score.
None	0	An incorrect solution is given indicating no mathematical understanding of the concept, or no solution is given.

On some standardized tests, no credit is given for a correct answer if your work is not shown.

Make sure that when the problem says to *Show your work,* show every aspect of your solution including figures, sketches of graphing calculator screens, or reasoning behind computations.

EXAMPLE

The table shows the population density in the United States on April 1 in each decade of the 20th century.

a. Make a scatter plot of the data.

b. Alaska and Hawaii became states in the same year. Between what two census dates do you think this happened. Why did you choose those years?

c. Use the data and your graph to predict the population density in 2010. Explain your reasoning.

U.S. Population Density	
Year	**People Per Square Mile**
1910	31.0
1920	35.6
1930	41.2
1940	44.2
1950	50.7
1960	50.6
1970	57.4
1980	64.0
1990	70.3
2000	79.6

Source: U.S. Census Bureau

Full Credit Solution

Part a A complete scatter plot includes a title for the graph, appropriate scales and labels for the axes, and correctly graphed points.

- The student should determine that the year data should go on the x-axis while the people per square mile data should go on the y-axis.
- On the x-axis, each square should represent 10 years.
- The y-axis could start at 0, or it could show data starting at 30 with a broken line to indicate that some of the scale is missing.

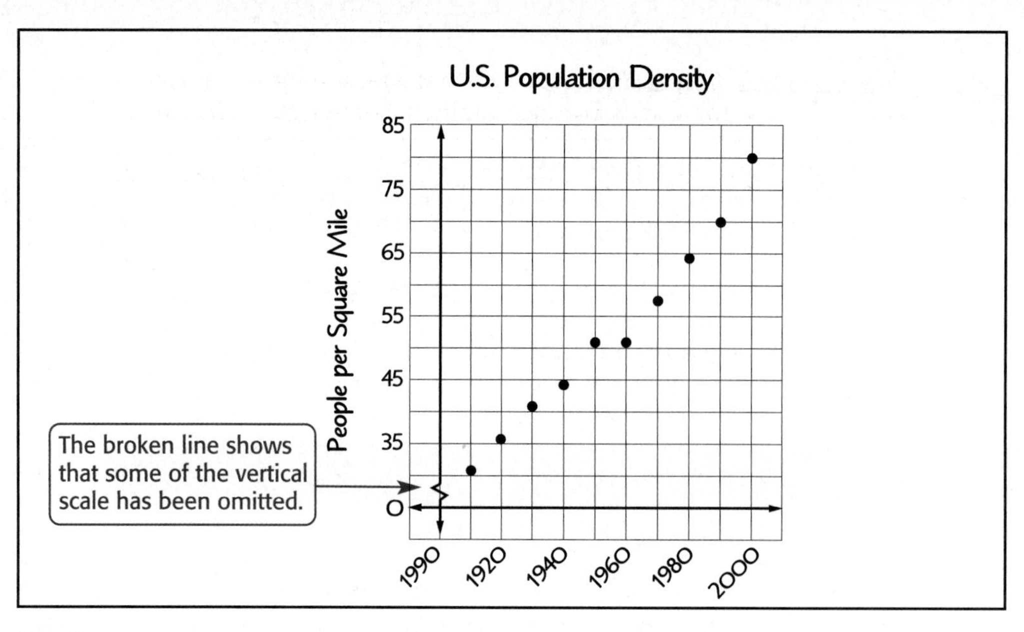

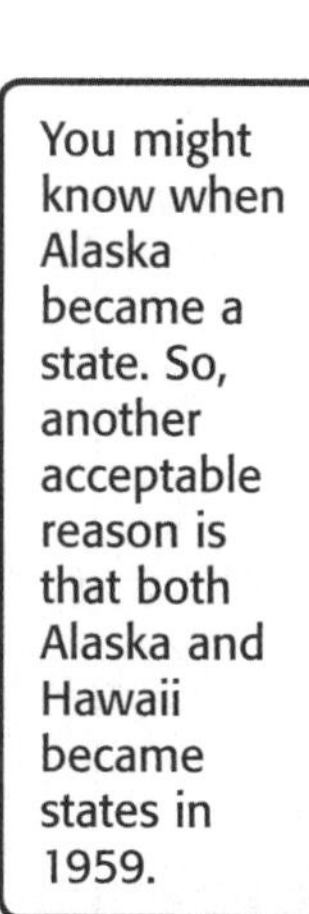

Part b

1950–1960, because when Alaska became a state it added little population but a lot of land, which made the people per square mile ratio less.

Part c

About 85.0, because the population per square mile would probably get larger so I connected the first point and the last point. The rate of change for each year was $\frac{79.6 - 31.0}{2000 - 1910}$ or about 0.54. I added 10×0.54, or 5.4 to 79.6 to get the next 10-year point.

Actually, any estimate from 84 to 86 might be acceptable. You could also use different points to find the equation for a line of best fit for the data, and then find the corresponding y value for $x = 2010$.

Partial Credit Solution

Part a This sample answer includes no labels for the graph or the axes and one of the-points is not graphed correctly.

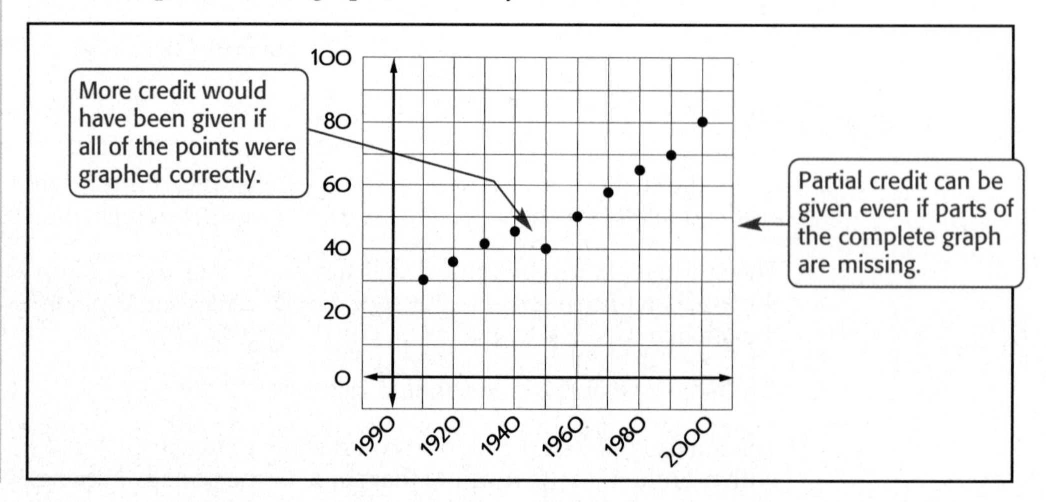

Part b Partial credit is given because the reasoning is correct, but the reasoning was based on the incorrect graph in Part a.

> 1940–1950, because when Alaska became a state it added little population but a lot of land, which made the people per square mile ratio less.

Part c Full credit is given for Part c.

> Suppose I draw a line of best fit through points (1910, 31.0) and (1990, 70.3). The slope would be $\frac{70.3 - 31.0}{1990 - 1910}$ about 0.49. Now use the slope and one of the points to find the y-intercept.
>
> $y = mx + b$
> $70.3 = 0.49(1990) + b$
> $-904.8 = b$
>
> So an equation of my line of best fit is $y = 0.49x - 904.8$.
>
> If $x = 2010$, then $y = 0.49(2010) - 904.8$ or about 80.1 people per square mile in the year 2010.

This sample answer might have received a score of 2 or 1. Had the student graphed all points correctly and gotten Part b correct, the score would probably have been a 3.

No Credit Solution

Part a

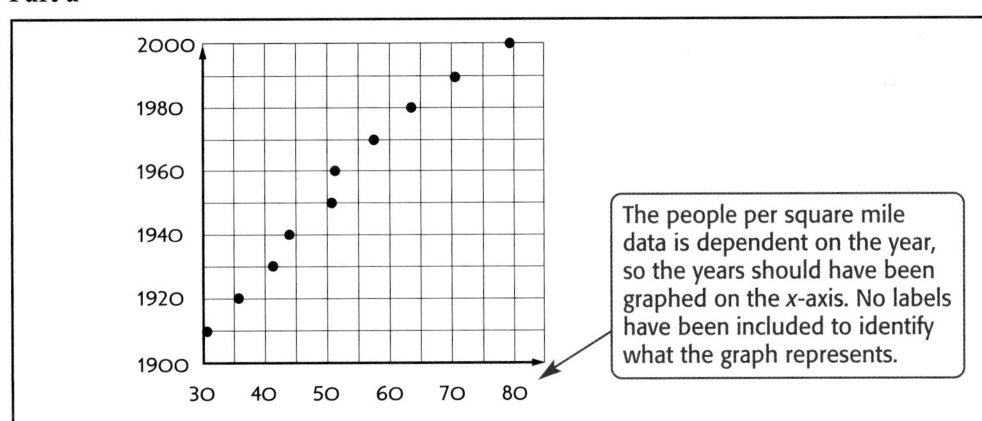

Part b

> I have no idea.

Part c

> 85, because it is the next grid line.

In this sample answer, the student does not understand how to represent data on a graph or how to interpret the data after the points are graphed.

Extended-Response Practice

Solve each problem. Show all your work.

Number and Operations

1. The table shows what one dollar in U.S. money was worth in five countries in 1990 and in 2006.

Money Equivalent to One U.S. Dollar		
Country	**1990 value**	**2006 Value**
Canada	1.16 Canadian dollars	1.13 Canadian dollars
European Union	0.83 euros	0.80 euros
Great Britain	0.62 pounds	0.54 pounds
India	16.96 rupees	45.67 rupees
Japan	146.25 yen	119.68 yen

a. For which country was the percent of increase or decrease in the number of units of currency that was equivalent to $1 the greatest from 1990 to 2006?

b. Suppose a U.S. citizen traveled to India in 1990 and in 2006. In which year would the traveler receive a better value for their money? Explain.

c. In 2006, what was the value of one euro in yen?

2. The table shows some data about the planets and the Sun. The radius is given in miles and the volume, mass, and gravity quantities are related to the volume and mass of Earth, which has a value of 1.

	Volume	Mass	Density	Radius	Gravity
Sun	1,304,000	332,950	0.26	434,474	28
Mercury	0.056	0.0553	0.98	1516	0.38
Venus	0.857	0.815	0.95	3760	0.91
Moon	0.0203	0.0123	0.61	1079	0.17
Mars	0.151	0.107	0.71	2106	0.38
Jupiter	1321	317.83	0.24	43,441	2.36
Saturn	764	95.16	0.12	36,184	0.92
Uranus	63	14.54	0.23	15,759	0.89
Neptune	58	17.15	0.30	15,301	1.12

a. Make and test a conjecture relating volume, mass, and density.

b. Describe the relationship between radius and gravity.

c. Can you be sure that the relationship in part **b** holds true for all planets? Explain.

Algebra

3. The graph shows the altitude of a glider during various times of his flight after being released from a tow plane.

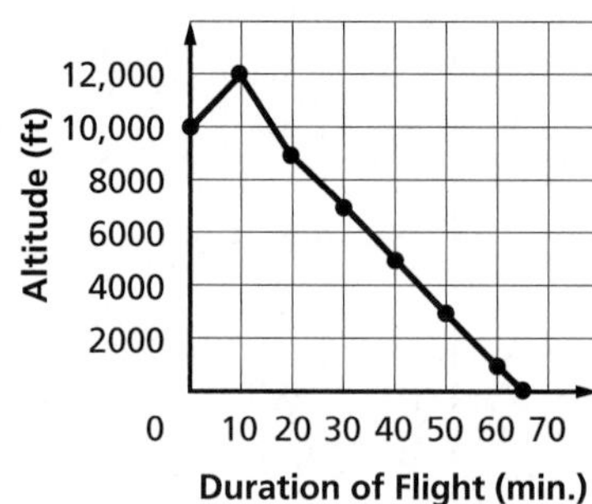

a. What point on the graph represents the moment the glider was released from the tow plane? Explain the meaning of this point in terms of altitude.

b. During which time period did the greatest rate of descent of the glider take place? Explain your reasoning.

c. How long did it take the glider to reach an altitude of 0 feet? Where is this point on the graph?

d. What is the equation of a line that represents the glider's altitude y as the time increased from 20 to 60 minutes?

e. Explain what the slope of the line in part **d** represents?

4. John has just received his learner's permit which allows him to practice driving with a licensed driver. His mother has agreed to take him driving every day for two weeks. On the first day, John will drive for 20 minutes. Each day after that, John's mother has agreed he can drive 15 minutes more than the day before.

a. Write a formula for the nth term of the sequence. Explain how you found the formula.

b. For how many minutes will John drive on the last day? Show how you found the number of minutes.

c. John's driver's education teacher requires that each student drive for 30 hours with an adult outside of class. Will John fulfill this requirement? Explain.

Geometry

5. Polygon *QUAD* is shown on a coordinate plane.

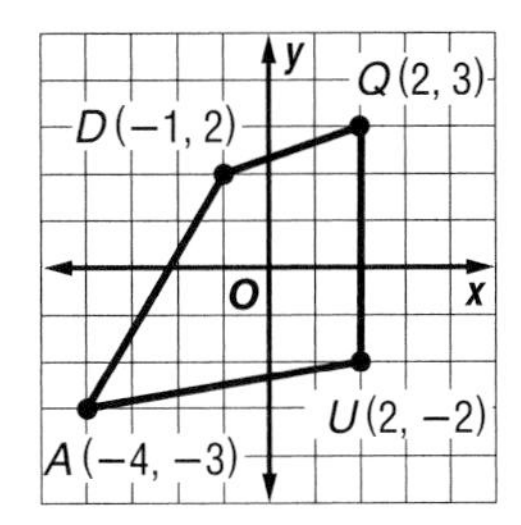

a. Find the coordinates of the vertices of $Q'U'A'D'$, which is the image of *QUAD* after a reflection over the y-axis. Explain.

b. Suppose point $M(a, b)$ is reflected over the y-axis. What will be the coordinates of the image M'?

c. Describe a reflection that will make *QUAD* look "upside down." What will be the coordinates of the vertices of the image?

6. The diagram shows a sphere with a radius of a and a cylinder with a radius and a height of a.

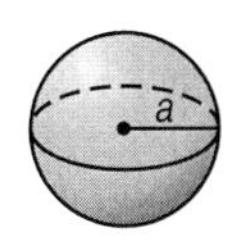

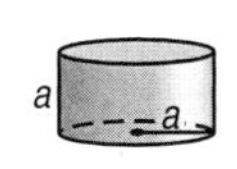

a. What is the ratio of the volume of the sphere to the volume of the cylinder?

b. What is the ratio of the surface area of the sphere to the surface area of the cylinder?

c. The ratio of the volume of another cylinder is 3 times the volume of the sphere shown. Give one possible set of measures for the radius and height of the cylinder in terms of a.

Measurement

7. Alexis is using a map of the province of Saskatchewan in Canada. The scale for the map shows that 2 centimeters on the map is 30 kilometers in actual distance.

a. The distance on the map between two cities measures 7 centimeters. What is the actual distance between the two cities in kilometers? Show how you found the distance.

TEST-TAKING TIP

Question 5

In a reflection over the x-axis, the x-coordinate remains the same, and the y-coordinate changes its sign. In a reflection over the y-axis, the y-coordinate remains the same, and the x-coordinate changes its sign.

b. Alexis is more familiar with distances in miles. The distance between two other cities is 8 centimeters. If one kilometer is about 0.62 mile, what is the distance in miles?

c. Alexis' entire trip measures 54 centimeters. If her car averages 25 miles per gallon of gasoline, how many gallons will she need to complete the trip? Round to the nearest gallon. Explain.

8. The diagram shows a pattern for a quilt square called Colorful Fan.

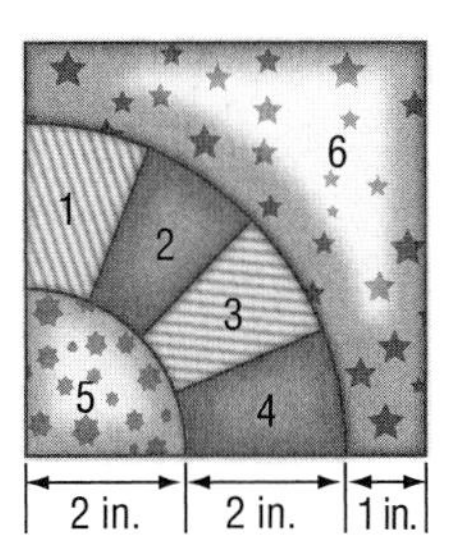

a. What is the area of region 6? Explain.

b. What is the area of region 1? Explain.

c. What is the ratio of the area of region 5 to the area of the entire square? Show how you found the ratio. Leave the ratio in terms of π.

Data Analysis and Probability

9. The table shows the Olympics winning times in the women's 1500-meter speed skating event. The times are to the nearest second.

Year	Time (s)	Year	Time (s)
1960	172	1984	124
1964	143	1988	121
1968	142	1992	126
1972	141	1994	122
1976	137	1998	118
1980	131	2002	114

a. Make a scatter plot of the data.

b. Use the data and your graph to predict the winning time in 2010.

10. There are 1320 ways for three students to win first, second, and third place during a debate.

a. How many students are on the debate team?

b. What is the probability that a student will come in one of the first three places if each student has an equal chance of succeeding?

c. If the teacher announces the third place winner, what is the probability that one particular other student will win first or second place?

Glossary/Glosario

Math Online A mathematics multilingual glossary is available at www.math.glencoe.com/multilingual_glossary. The glossary includes the following languages.

Arabic	Haitian Creole	Russian	Vietnamese
Bengali	Hmong	Spanish	
Cantonese	Korean	Tagalog	
English	Portuguese	Urdu	

Cómo usar el glosario en español:

1. Busca el término en inglés que desees encontrar.
2. El término en español, junto con la definición, se encuentran en la columna de la derecha.

English

Español

A

absolute value function (p. 324) A function written as $f(x) = |x|$, in which $f(x) \geq 0$ for all values of x.

función del valor absoluto Una función que se escribe $f(x) = |x|$, donde $f(x) \geq 0$, para todos los valores de x.

additive identity (p. 21) For any number a, $a + 0 = 0 + a = a$.

identidad de la adición Para cualquier número a, $a + 0 = 0 + a = a$.

algebraic expression (p. 6) An expression consisting of one or more numbers and variables along with one or more arithmetic operations.

expresión algebraica Una expresión que consiste en uno o más números y variables, junto con una o más operaciones aritméticas.

arithmetic sequence (p. 165) A numerical pattern that increases or decreases at a constant rate or value. The difference between successive terms of the sequence is constant.

sucesión aritmética Un patrón numérico que aumenta o disminuye a una tasa o valor constante. La diferencia entre términos consecutivos de la sucesión es siempre la misma.

axis of symmetry (p. 472) The vertical line containing the vertex of a parabola.

eje de simetría La recta vertical que pasa por el vértice de una parábola.

B

base (p. 6) In an expression of the form x^n, the base is x.

base En una expresión de la forma x^n, la base es x.

best-fit line (p. 228) The line that most closely approximates the data in a scatter plot.

recta de ajuste óptimo La recta que mejor aproxima los datos de una gráfica de dispersión.

biased sample (p. 643) A sample in which one or more parts of the population are favored over others.

muestra sesgada Muestra en que se favorece una o más partes de una población en vez de otras partes.

binomial (p. 376) The sum of two monomials.

binomio La suma de dos monomios.

boundary (p. 334) A line or curve that separates the coordinate plane into regions.

frontera Recta o curva que divide el plano de coordenadas en regiones.

C

coefficient (p. 29) The numerical factor of a term.

coeficiente Factor numérico de un término.

combination (p. 657) An arrangement or listing in which order is not important.

combinación Arreglo o lista en que el orden no es importante.

common difference (p. 165) The difference between the terms in a sequence.

diferencia común Diferencia entre términos consecutivos de una sucesión.

complements (p. 665) One of two parts of a probability making a whole.

complementos Una de dos partes de una probabilidad que forma un todo.

completing the square (p. 486) To add a constant term to a binomial of the form $x^2 + bx$ so that the resulting trinomial is a perfect square.

completar el cuadrado Adición de un término constante a un binomio de la forma $x^2 + bx$, para que el trinomio resultante sea un cuadrado perfecto.

complex fraction (p. 621) A fraction that has one or more fractions in the numerator or denominator.

fracción compleja Fracción con una o más fracciones en el numerador o denominador.

composite number (p. 420) A whole number, greater than 1, that has more than two factors.

número compuesto Número entero mayor que 1 que posee más de dos factores.

compound event (p. 663) Two or more simple events.

evento compuesto Dos o más eventos simples.

compound inequality (p. 315) Two or more inequalities that are connected by the words *and* or *or*.

desigualdad compuesta Dos o más desigualdades que están unidas por las palabras y u o.

compound interest (p. 511) A special application of exponential growth.

interés compuesto Aplicación especial de crecimiento exponencial.

conclusion (p. 39) The part of a conditional statement immediately following the word *then*.

conclusión Parte de un enunciado condicional que sigue inmediatamente a la palabra *entonces*.

conditional statements (p. 39) Statements written in the form *If A, then B*.

enunciados condicionales Enunciados de la forma *Si A, entonces B*.

conjugates (p. 531) Binomials of the form $a\sqrt{b} + c\sqrt{d}$ and $a\sqrt{b} - c\sqrt{d}$.

conjugados Binomios de la forma $a\sqrt{b} + c\sqrt{d}$ y $a\sqrt{b} - c\sqrt{d}$.

consecutive integers (p. 94) Integers in counting order.

enteros consecutivos Enteros en el orden de contar.

consistent (p. 253) A system of equations that has at least one ordered pair that satisfies both equations.

consistente Sistema de ecuaciones para el cual existe al menos un par ordenado que satisface ambas ecuaciones.

constant (p. 358) A monomial that is a real number.

constante Monomio que es un número real.

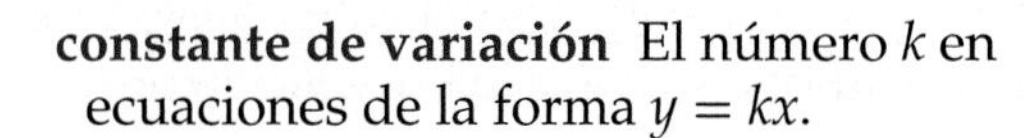

constant of variation (p. 196) The number k in equations of the form $y = kx$.

constante de variación El número k en ecuaciones de la forma $y = kx$.

continuous function (p. 55) A function that can be graphed with a line or a smooth curve.

función continua Función cuya gráfica puedes ser una recta o una curva suave.

convenience sample (p. 643) A sample that includes members of a population that are easily accessed.

muestra de conveniencia Muestra que incluye miembros de una población fácilmente accesibles.

converse (p. 551) The statement formed by exchanging the hypothesis and conclusion of a conditional statement.

recíproco Enunciado que se obtiene al inter cambiar la hipótesis y la conclusión de un enucnciado condicional dado.

coordinate (p. 48) The number that corresponds to a point on a number line.

coordenada Número que corresponde a un punto en una recta numérica.

coordinate plane (p. 53) The plane containing the x- and y-axes.

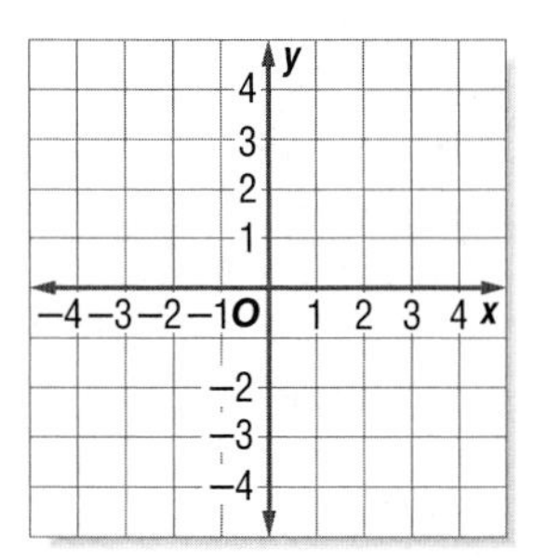

plano de coordenadas Plano que contiene los ejes x y y.

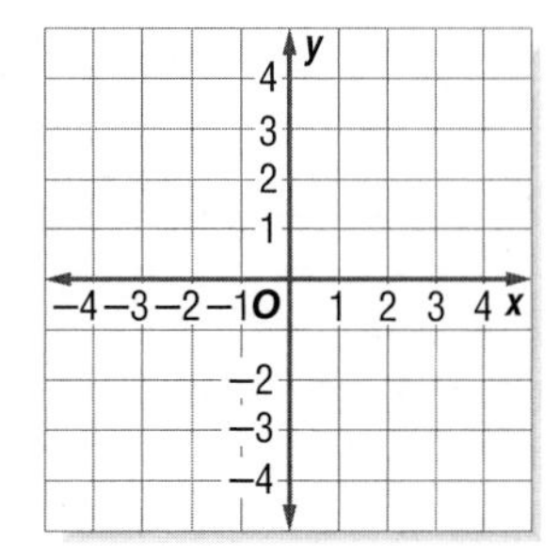

coordinate system (p. 53) The grid formed by the intersection of two number lines, the horizontal axis and the vertical axis.

sistema de coordenadas Cuadriculado formado por la intersección de dos rectas numéricas: los ejes x y y.

counterexample (p. 41) A specific case in which a statement is false.

contraejemplo Ejemplo específico de la falsedad de un enunciado.

D

deductive reasoning (p. 40) The process of using facts, rules, definitions, or properties to reach a valid conclusion.

razonamiento deductivo Proceso de usar hechos, reglas, definiciones o propiedades para sacar conclusiones válidas.

defining a variable (p. 71) Choosing a variable to represent one of the unspecified numbers in a problem and using it to write expressions for the other unspecified numbers in the problem.

definir una variable Consiste en escoger una variable para representar uno de los números desconocidos en un problema y luego usarla para escribir expresiones para otros números desconocidos en el problema.

degree of a monomial (p. 377) The sum of the exponents of all its variables.

grado de un monomio Suma de los exponentes de todas sus variables.

degree of a polynomial (p. 377) The greatest degree of any term in the polynomial.

grado de un polinomio El grado mayor de cualquier término del polinomio.

dependent (p. 253) A system of equations that has an infinite number of solutions.

dependiente Sistema de ecuaciones que posee un número infinito de soluciones.

dependent events (p. 664) Two or more events in which the outcome of one event affects the outcome of the other events.

eventos dependientes Dos o más eventos en que el resultado de un evento afecta el resultado de los otros eventos.

dependent variable (p. 54) The variable in a relation with a value that depends on the value of the independent variable.

variable dependiente La variable de una relación cuyo valor depende del valor de la variable independiente.

difference of squares (pp. 404, 448) Two perfect squares separated by a subtraction sign. $a^2 - b^2 = (a + b)(a - b)$ or $a^2 - b^2 = (a - b)(a + b)$.

diferencia de cuadrados Dos cuadrados perfectos separados por el signo de sustracción. $a2 - b2 = (a + b)(a - b)$ or $a2 - b2 = (a - b)(a + b)$.

dimensional analysis (p. 119) The process of carrying units throughout a computation.

análisis dimensional Proceso de tomar en cuenta las unidades de medida al hacer cálculos.

direct variation (p. 196) An equation of the form $y = kx$, where $k \neq 0$.

variación directa Una ecuación de la forma $y = kx$, donde $k \neq 0$.

discrete function (p. 55) A function of points that are not connected.

función discreta Función de puntos desconectados.

discrete random variable (p. 672) A variable with a value that is a finite number of possible outcomes.

variable aleatoria discreta Variable cuyo valor es un número finito de posibles resultados.

discriminant (p. 496) In the Quadratic Formula, the expression under the radical sign, $b^2 - 4ac$.

discriminante En la fórmula cuadrática, la expresión debajo del signo radical, $b^2 - 4ac$.

Distance Formula (p. 555) The distance d between any two points with coordinates (x_1, y_1) and (x_2, y_2) is given by the formula $d = \sqrt{(x_2 - x_1)^2 + (y_2 - y_1)^2}$.

Fórmula de la distancia La distancia d entre cualquier par de puntos con coordenadas (x_1, y_1) y (x_2, y_2) viene dada por la fórmula $d = \sqrt{(x_2 - x_1)^2 + (y_2 - y_1)^2}$.

domain (p. 55) The set of the first numbers of the ordered pairs in a relation.

dominio Conjunto de los primeros números de los pares ordenados de una relación.

E

element (p. 15) **1.** Each object or number in a set. (p. 715) **2.** Each entry in a matrix.

elemento **1.** Cada número u objeto de un conjunto. **2.** Cada entrada de una matriz.

elimination (p. 266) The use of addition or subtraction to eliminate one variable and solve a system of equations.

eliminación El uso de la adición o la sustracción para eliminar una variable y resolver así un sistema de ecuaciones.

empirical study (p. 678) Performing an experiment repeatedly, collecting and combining data, and analyzing the results.

estudio empírico Ejecución repetida de un experimento, recopilación y combinación de datos y análisis de resultados.

equation (p. 15) A mathematical sentence that contains an equals sign, =.

ecuación Enunciado matemático que contiene el signo de igualdad, =.

equivalent equations (p. 79) Equations that have the same solution.

ecuaciones equivalentes Ecuaciones que poseen la misma solución.

equivalent expressions (p. 29) Expressions that denote the same number.

expresiones equivalentes Expresiones que denotan el mismo número.

evaluate (p. 7) To find the value of an expression.

evaluar Calcular el valor de una expresión.

event (p. 650) Any collection of one or more outcomes in the sample space.

evento Cualquier colección de uno o más resultados de un espacio muestral.

excluded values (p. 583) Any values of a variable that result in a denominator of 0 must be excluded from the domain of that variable.

valores excluidos Cualquier valor de una variable cuyo resultado sea un denominador igual a cero, debe excluirse del dominio de dicha variable.

experimental probability (p. 677) What actually occurs when conducting a probability experiment, or the ratio of relative frequency to the total number of events or trials.

probabilidad experimental Lo que realmente sucede cuando se realiza un experimento probabilístico o la razón de la frecuencia relativa al número total de eventos o pruebas.

exponent (p. 6) In an expression of the form x^n, the exponent is n. It indicates the number of times x is used as a factor.

exponente En una expresión de la forma x^n, el exponente es n. Éste indica cuántas veces se usa x como factor.

exponential decay (p. 511) When an initial amount decreases by the same percent over a given period of time.

desintegración exponencial La cantidad inicial disminuye según el mismo porcentaje a lo largo de un período de tiempo dado.

exponential function (p. 502) A function that can be described by an equation of the form $y = a^x$, where $a > 0$ and $a \neq 1$.

función exponencial Función que puede describirse mediante una ecuación de la forma $y = a^x$, donde $a > 0$ y $a \neq 1$.

exponential growth (p. 510) When an initial amount increases by the same percent over a given period of time.

crecimiento exponencial La cantidad inicial aumenta según el mismo porcentaje a lo largo de un período de tiempo dado.

extraneous solutions (pp. 542, 629) Results that are not solutions to the original equation.

soluciones extrañas Resultados que no son soluciones de la ecuación original.

extremes (p. 106) In the ratio $\frac{a}{b} = \frac{c}{d}$, a and d are the extremes.

extremos En la razón $\frac{a}{b} = \frac{c}{d}$, a y d son los extremos.

F

factored form (p. 421) A monomial expressed as a product of prime numbers and variables in which no variable has an exponent greater than 1.

forma reducida Monomio escrito como el producto de números primos y variables y en el que ninguna variable tiene un exponente mayor que 1.

factorial (p. 652) The expression $n!$, read n *factorial*, where n is greater than zero, is the product of all positive integers beginning with n and counting backward to 1.

factorial La expresión $n!$, que se lee n factorial, donde n que es mayor que cero, es el producto de todos los números naturales, comenzando con n y contando hacia atrás hasta llegar al 1.

factoring (p. 426) To express a polynomial as the product of monomials and polynomials.

factorización La escritura de un polinomio como producto de monomios y polinomios.

factoring by grouping (p. 427) The use of the Distributive Property to factor some polynomials having four or more terms.

factorización por agrupamiento Uso de la Propiedad distributiva para factorizar polinomios que poseen cuatro o más términos.

factors (p. 6) In an algebraic expression, the quantities being multiplied are called factors.

factores En una expresión algebraica, los factores son las cantidades que se multiplican.

family of graphs (pp. 197, 478) Graphs and equations of graphs that have at least one characteristic in common.

familia de gráficas Gráficas y ecuaciones de gráficas que tienen al menos una característica común.

FOIL method (p. 399) To multiply two binomials, find the sum of the products of the *F*irst terms, the *O*uter terms, the *I*nner terms, and the *L*ast terms.

método FOIL Para multiplicar dos binomios, busca la suma de los productos de los primeros (*F*irst) términos, los términos exteriores (*O*uter), los términos interiores (*I*nner) y los últimos términos (*L*ast).

formula (p. 72) An equation that states a rule for the relationship between certain quantities.

fórmula Ecuación que establece una relación entre ciertas cantidades.

four-step problem-solving plan (p. 71)
Step 1 Explore the problem.
Step 2 Plan the solution.
Step 3 Solve the problem.
Step 4 Check the solution.

plan de cuatro pasos para resolver problemas
Paso 1 Explora el problema.
Paso 2 Planifica la solución.
Paso 3 Resuelve el problema.
Paso 4 Examina la solución.

function (pp. 53, 149) A relation in which each element of the domain is paired with exactly one element of the range.

función Una relación en que a cada elemento del dominio le corresponde un único elemento del rango.

function notation (p. 150) A way to name a function that is defined by an equation. In function notation, the equation $y = 3x - 8$ is written as $f(x) = 3x - 8$.

notación funcional Una manera de nombrar una función definida por una ecuación. En notación funcional, la ecuación $y = 3x - 8$ se escribe $f(x) = 3x - 8$.

Fundamental Counting Principle (p. 651) If an event M can occur in m ways and is followed by an event N that can occur in n ways, then the event M followed by the event N can occur in $m \times n$ ways.

Principio fundamental de contar Si un evento M puede ocurrir de m maneras y lo sigue un evento N que puede ocurrir de n maneras, entonces el evento M seguido del evento N puede ocurrir de $m \times n$ maneras.

G

general equation for exponential decay (p. 511) $y = C(1 - r)^t$, where y is the final amount, C is the initial amount, r is the rate of decay expressed as a decimal, and t is time.

ecuación general de desintegración exponencial $y = C(1 - r)^t$, donde y es la cantidad final, C es la cantidad inicial, r es la tasa de desintegración escrita como decimal y t es el tiempo.

general equation for exponential growth (p. 510) $y = C(1 + r)^t$, where y is the final amount, C is the initial amount, r is the rate of change expressed as a decimal, and t is time.

ecuación general de crecimiento exponencial $y = C(1 + r)^t$, donde y es la cantidad final, C es la cantidad inicial, r es la tasa de cambio del crecimiento escrita como decimal y t es el tiempo.

graph (pp. 48, 195) To draw, or plot, the points named by certain numbers or ordered pairs on a number line or coordinate plane.

graficar Marcar los puntos que denotan ciertos números en una recta numérica o ciertos pares ordenados en un plano de coordenadas.

greatest common factor (GCF) (p. 421) The product of the prime factors common to two or more integers.

máximo común divisor (MCD) El producto de los factores primos comunes a dos o más enteros.

H

half-plane (p. 334) The region of the graph of an inequality on one side of a boundary.

semiplano Región de la gráfica de una desigualdad en un lado de la frontera.

hypotenuse (p. 549) The side opposite the right angle in a right triangle.

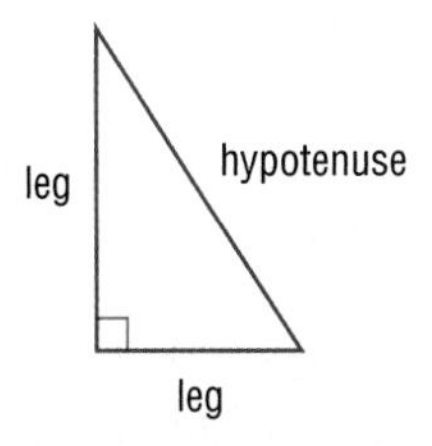

hipotenusa Lado opuesto al ángulo recto en un triángulo rectángulo.

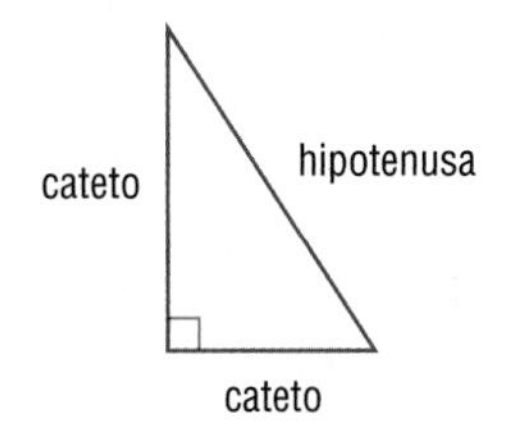

hypothesis (p. 39) The part of a conditional statement immediately following the word *if.*

hipótesis Parte de un enunciado condicional que sigue inmediatamente a la palabra *si.*

I

if-then statements (p. 39) Conditional statements in the form *If A, then B.*

enunciados si-entonces Enunciados condicionales de la forma *Si A, entonces B.*

inclusive (p. 666) Two events that can occur at the same time.

inclusivos Dos eventos que pueden ocurrir simultáneamente.

inconsistent (p. 253) A system of equations with no ordered pair that satisfy both equations.

inconsistente Un sistema de ecuaciones para el cual no existe par ordenado alguno que satisfaga ambas ecuaciones.

independent (p. 253) A system of equations with exactly one solution.

independiente Un sistema de ecuaciones que posee una única solución.

independent events (p. 663) Two or more events in which the outcome of one event does not affect the outcome of the other events.

eventos independientes El resultado de un evento no afecta el resultado del otro evento.

independent variable (p. 55) The variable in a function with a value that is subject to choice.

variable independiente La variable de una función sujeta a elección.

inductive reasoning (p. 172) A conclusion based on a pattern of examples.

razonamiento inductivo Conclusión basada en un patrón de ejemplos.

inequality (p. 16) An open sentence that contains the symbol <, ≤, >, or ≥.

desigualdad Enunciado abierto que contiene uno o más de los símbolos <, ≤, >, o ≥.

integers (p. 46) The set {..., −2, −1, 0, 1, 2, ...}.

enteros El conjunto {..., −2, −1, 0, 1, 2, ...}.

intersection (p. 315) The graph of a compound inequality containing *and;* the solution is the set of elements common to both inequalities.

intersección Gráfica de una desigualdad compuesta que contiene la palabra *y;* la solución es el conjunto de soluciones de ambas desigualdades.

inverse (p. 145) The inverse of any relation is obtained by switching the coordinates in each ordered pair.

inversa La inversa de una relación se halla intercambiando las coordenadas de cada par ordenado.

inverse variation (p. 577) An equation of the form $xy = k$, where $k \neq 0$.

variación inversa Ecuación de la forma $xy = k$, donde $k \neq 0$.

irrational numbers (p. 46) Numbers that cannot be expressed as terminating or repeating decimals.

números irracionales Números que no pueden escribirse como decimales terminales o periódicos.

L

least common denominator (LCD) (p. 615) The least common multiple of the denominators of two or more fractions.

mínimo denominador común (mcd) El mínimo común múltiplo de los denominadores de dos o más fracciones.

least common multiple (LCM) (p. 614) The least number that is a common multiple of two or more numbers.

mínimo común múltiplo (mcm) El número menor que es múltiplo común de dos o más números.

legs (p. 549) The sides of a right triangle that form the right angle.

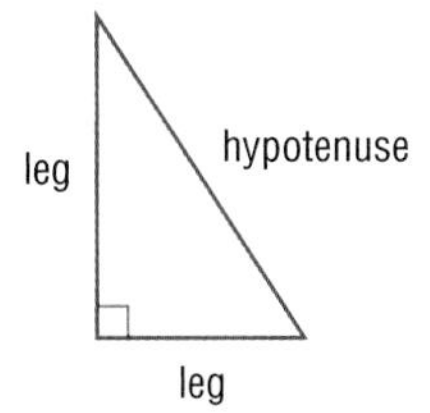

catetos Lados de un triángulo rectángulo que forman el ángulo recto del mismo.

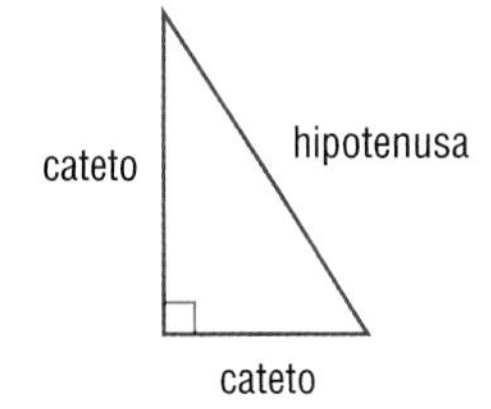

like terms (p. 29) Terms that contain the same variables, with corresponding variables having the same exponent.

términos semejantes Expresiones que tienen las mismas variables, con las variables correspondientes elevadas a los mismos exponentes.

linear equation (p. 155) An equation in the form $Ax + By = C$, with a graph that is a straight line.

ecuación lineal Ecuación de la forma $Ax + By = C$, cuya gráfica es una recta.

linear extrapolation (p. 216) The use of a linear equation to predict values that are outside the range of data.

extrapolación lineal Uso de una ecuación lineal para predecir valores fuera de la amplitud de los datos.

Glossary/Glosario

linear function (p. 155) A function with ordered pairs that satisfy a linear equation.

función lineal Función cuyos pares ordenados satisfacen una ecuación lineal.

linear interpolation (p. 230) The use of a linear equation to predict values that are inside of the data range.

interpolación lineal Uso de una ecuación lineal para predecir valores dentro de la amplitud de los datos.

line of fit (p. 228) A line that describes the trend of the data in a scatter plot.

recta de ajuste Recta que describe la tendencia de los datos en una gráfica de dispersión.

look for a pattern (p. 172) Find patterns in sequences to solve problems.

buscar un patrón Encontrar patrones en sucesiones para resolver problemas.

lower quartile (p. 713) Divides the lower half of the data into two equal parts.

cuartil inferior Éste divide en dos partes iguales la mitad inferior de un conjunto de datos.

M

mapping (p. 143) Illustrates how each element of the domain is paired with an element in the range.

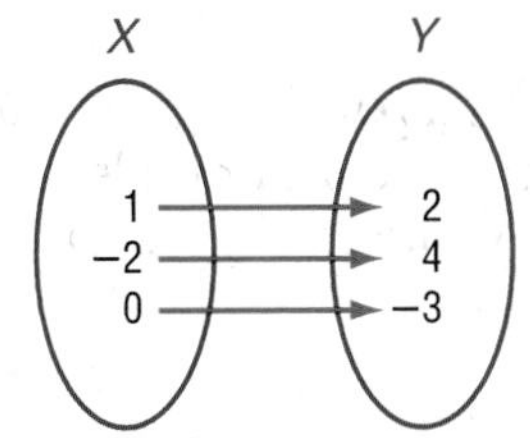

aplicaciones Ilustra la correspondencia entre cada elemento del dominio con un elemento del rango.

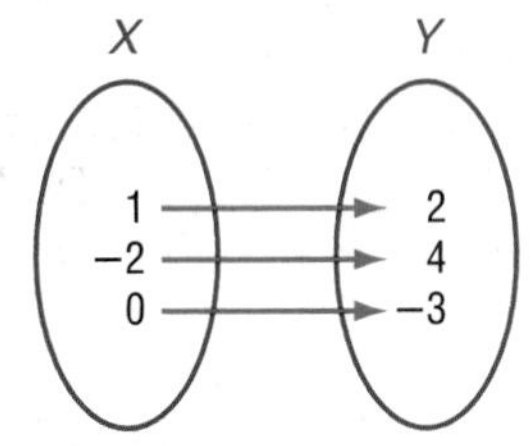

maximum (p. 472) The highest point on the graph of a curve.

máximo El punto más alto en la gráfica de una curva.

means (p. 106) The middle terms of the proportion.

medios Los términos centrales de una proporción.

minimum (p. 472) The lowest point on the graph of a curve.

mínimo El punto más bajo en la gráfica de una curva.

mixed expression (p. 620) An expression that contains the sum of a monomial and a rational expression.

expresión mixta Expresión que contiene la suma de un monomio y una expresión racional.

mixture problems (p. 122) Problems in which two or more parts are combined into a whole.

problemas de mezclas Problemas en que dos o más partes se combinan en un todo.

monomial (p. 358) A number, a variable, or a product of a number and one or more variables.

monomio Número, variable o producto de un número por una o más variables.

multiplicative identity (p. 21) For any number a, $a \cdot 1 = 1 \cdot a = a$.

identidad de la multiplicación Para cualquier número a, $a \cdot 1 = 1 \cdot a = a$.

multiplicative inverses (p. 21) Two numbers with a product of 1.

inversos multiplicativos Dos números cuyo producto es igual a 1.

multi-step equations (p. 92) Equations with more than one operation.

ecuaciones de varios pasos Ecuaciones con más de una operación.

mutually exclusive (p. 665) Events that cannot occur at the same time.

mutuamente exclusivos Eventos que no pueden ocurrir simultáneamente.

N

natural numbers (p. 46) The set {1, 2, 3, ...}.

números naturales El conjunto {1, 2, 3, ...}.

negative correlation (p. 227) In a scatter plot, as x increases, y decreases.

correlación negativa En una gráfica de dispersión, a medida que x aumenta, y disminuye.

negative number (p. 46) Any value less than zero.

número negativo Cualquier valor menor que cero.

O

open sentence (p. 15) A mathematical statement with one or more variables.

enunciado abierto Un enunciado matemático que contiene una o más variables.

ordered pair (p. 53) A set of numbers or coordinates used to locate any point on a coordinate plane, written in the form (x, y).

par ordenado Un par de números que se usa para ubicar cualquier punto de un plano de coordenadas y que se escribe en la forma (x, y).

order of operations (p. 10)

1. Evaluate expressions inside grouping symbols.
2. Evaluate all powers.
3. Do all multiplications and/or divisions from left to right.
4. Do all additions and/or subtractions from left to right.

orden de las operaciones

1. Evalúa las expresiones dentro de los símbolos de agrupamiento.
2. Evalúa todas las potencias.
3. Multiplica o divide de izquierda a derecha.
4. Suma o resta de izquierda a derecha.

origin (p. 53) The point where the two axes intersect at their zero points.

origen Punto donde se intersecan los dos ejes en sus puntos cero.

P

parabola (p. 471) The graph of a quadratic function.

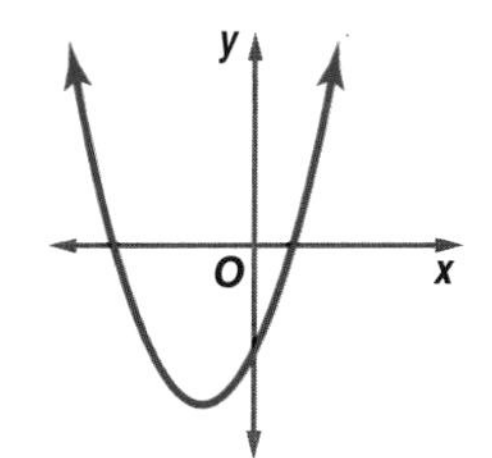

parábola La gráfica de una función cuadrática.

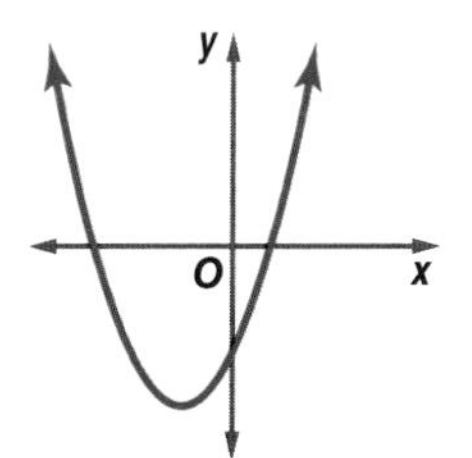

parallel lines (p. 236) Lines in the same plane that never intersect and have the same slope.

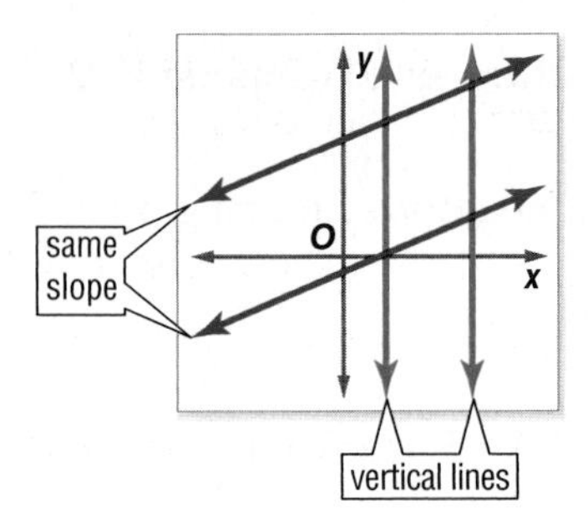

rectas paralelas Rectas en el mismo plano que no se intersecan jamás y que tienen pendientes iguales.

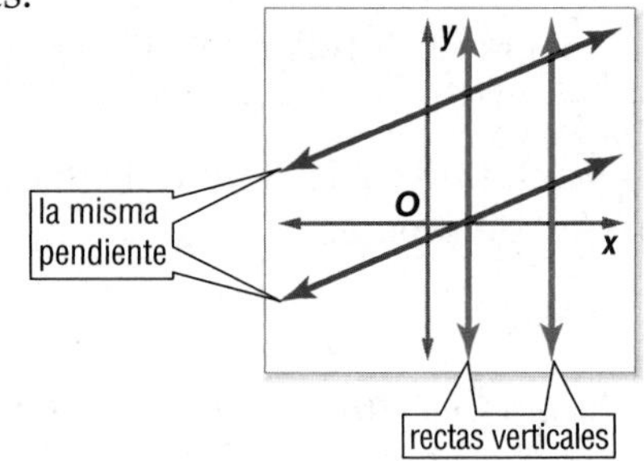

parent graph (p. 197) The simplest of the graphs in a family of graphs.

gráfica madre La gráfica más sencilla en una familia de gráficas.

percent of change (p. 111) When an increase or decrease is expressed as a percent.

porcentaje de cambio Cuando un aumento o disminución se escribe como un tanto por ciento.

percent of decrease (p. 111) The ratio of an amount of decrease to the previous amount, expressed as a percent.

porcentaje de disminución Razón de la cantidad de disminución a la cantidad original, escrita como un tanto por ciento.

percent of increase (p. 111) The ratio of an amount of increase to the previous amount, expressed as a percent.

porcentaje de aumento Razón de la cantidad de aumento a la cantidad original, escrita como un tanto por ciento.

perfect square (p. 46) A number with a square root that is a rational number.

cuadrado perfecto Número cuya raíz cuadrada es un número racional.

perfect square trinomial (p. 454) A trinomial that is the square of a binomial.
$(a + b)^2 = (a + b)(a + b) = a^2 + 2ab + b^2$ or
$(a - b)^2 = (a - b)(a - b) = a^2 - 2ab - b^2$

trinomio cuadrado perfecto Un trinomio que es el cuadrado de un binomio.
$(a + b)^2 = (a + b)(a + b) = a^2 + 2ab + b^2$ o
$(a - b)^2 = (a - b)(a - b) = a^2 - 2ab - b^2$

permutation (p. 655) An arrangement or listing in which order is important.

permutación Arreglo o lista en que el orden es importante.

piecewise function (p. 324) A function written using two or more expressions.

función por partes Función que se escribe usando dos o más expresiones.

point-slope form (p. 220) An equation of the form $y - y_1 = m(x - x_1)$, where m is the slope and (x_1, y_1) is a given point on a nonvertical line.

forma punto-pendiente Ecuación de la forma $y - y_1 = m(x - x_1)$, donde m es la pendiente y (x_1, y_1) es un punto dado de una recta no vertical.

polynomial (p. 376) A monomial or sum of monomials.

polinomio Un monomio o la suma de monomios.

population (p. 642) A large group of data usually represented by a sample.

población Grupo grande de datos, representado por lo general por una muestra.

positive correlation (p. 227) In a scatter plot, as x increases, y increases.

correlación positiva En una gráfica de dispersión, a medida que x aumenta, y aumenta.

positive number (p. 46) Any value that is greater than zero.

número positivos Cualquier valor mayor que cero.

power (p. 6) An expression of the form x^n, read *x to the nth power.*

prime factorization (p. 420) A whole number expressed as a product of factors that are all prime numbers.

prime number (p. 420) A whole number, greater than 1, with only factors that are 1 and itself.

prime polynomial (p. 443) A polynomial that cannot be written as a product of two polynomials with integral coefficients.

principal square root (p. 49) The nonnegative square root of a number.

probability distribution (p. 673) The probability of every possible value of the random variable x.

probability histogram (p. 673) A way to give the probability distribution for a random variable and obtain other data.

product (p. 6) In an algebraic expression, the result of quantities being multiplied is called the product.

product rule (p. 578) If (x_1, y_1) and (x_2, y_2) are solutions to an inverse variation, then $y_1x_1 = y_2x_2$.

proportion (p. 105) An equation of the form $\frac{a}{b} = \frac{c}{d}$ stating that two ratios are equivalent.

Pythagorean Theorem (p. 549) If a and b are the measures of the legs of a right triangle and c is the measure of the hypotenuse, then $c^2 = a^2 + b^2$.

Pythagorean triple (p. 550) Whole numbers that satisfy the Pythagorean Theorem.

potencia Una expresión de la forma x^n, se lee *x a la enésima potencia.*

factorización prima Número entero escrito como producto de factores primos.

número primo Número entero mayor que 1 cuyos únicos factores son 1 y sí mismo.

polinomio primo Polinomio que no puede escribirse como producto de dos polinomios con coeficientes enteros.

raíz cuadrada principal La raíz cuadrada no negativa de un número.

distribución de probabilidad Probabilidad de cada valor posible de una variable aleatoria x.

histograma probabilístico Una manera de exhibir la distribución de probabilidad de una variable aleatoria y obtener otros datos.

producto En una expresión algebraica, se llama producto al resultado de las cantidades que se multiplican.

regla del producto Si (x_1, y_1) y (x_2, y_2) son soluciones de una variación inversa, entonces $y_1x_1 = y_2x_2$.

proporción Ecuación de la forma $\frac{a}{b} = \frac{c}{d}$ que afirma la equivalencia de dos razones.

Teorema de Pitágoras Si a y b son las longitudes de los catetos de un triángulo rectángulo y si c es la longitud de la hipotenusa, entonces $c^2 = a^2 + b^2$.

Triple pitagórico Números enteros que satisfacen el Teorema de Pitágoras.

Q

quadratic equation (p. 480) An equation of the form $ax^2 + bx + c = 0$, where $a \neq 0$.

Quadratic Formula (p. 493) The solutions of a quadratic equation in the form $ax^2 + bx + c = 0$, where $a \neq 0$, are given by the formula $x = \frac{-b \pm \sqrt{b_2 - 4ac}}{2a}$.

quadratic function (p. 471) An equation of the form $y = ax^2 + bx + c$, where $a \neq 0$.

ecuación cuadrática Ecuación de la forma $ax^2 + bx + c = 0$, donde $a \neq 0$.

Fórmula cuadrática Las soluciones de una ecuación cuadrática de la forma $ax^2 + bx + c = 0$, donde $a \neq 0$, vienen dadas por la fórmula $x = \frac{-b \pm \sqrt{b^2 - 4ac}}{2a}$.

función cuadrática Función de la forma $y = ax^2 + bx + c$, donde $a \neq 0$.

R

radical equations (p. 541) Equations that contain radicals with variables in the radicand.

ecuaciones radicales Ecuaciones que contienen radicales con variables en el radicando.

radical expression (p. 528) An expression that contains a square root.

expresión radical Expresión que contiene una raíz cuadrada.

radical sign (p. 49) The symbol $\sqrt{\ }$, used to indicate a nonnegative square root.

signo radical El símbolo $\sqrt{\ }$, que se usa para indicar la raíz cuadrada no negativa.

radicand (p. 528) The expression that is under the radical sign.

radicando La expresión debajo del signo radical.

random sample (p. 642) A sample that is chosen without any preference, representative of the entire population.

muestra aleatoria Muestra tomada sin preferencia alguna y que es representativa de toda la población.

random variable (p. 672) A variable with a value that is the numerical outcome of a random event.

variable aleatoria Una variable cuyos valores son los resultados numéricos de un evento aleatorio.

range (p. 55) The set of second numbers of the ordered pairs in a relation.

rango Conjunto de los segundos números de los pares ordenados de una relación.

rate (p. 107) The ratio of two measurements having different units of measure.

tasa Razón de dos medidas que tienen distintas unidades de medida.

rate of change (p. 187) How a quantity is changing over time.

tasa de cambio Cómo cambia una cantidad con el tiempo.

rate problems (p. 628) Rational equations are used to solve problems involving transportal rates.

problemas de tasas Ecuaciones racionales que se usan para resolver problemas de tasas de transportación.

ratio (p. 105) A comparison of two numbers by division.

razón Comparación de dos números mediante división.

rational approximation (p. 50) A rational number that is close to, but not equal to, the value of an irrational number.

aproximación racional Número racional que está cercano, pero que no es igual, al valor de un número irracional.

rational equations (p. 626) Equations that contain rational expressions.

ecuaciones racionales Ecuaciones que contienen xpresiones racionales.

rational expression (p. 583) An algebraic fraction with a numerator and denominator that are polynomials.

expresión racional Fracción algebraica cuyo numerador y denominador son polinomios.

rationalizing the denominator (p. 530) A method used to eliminate radicals from the denominator of a fraction.

racionalizar el denominador Método que se usa para eliminar radicales del denominador de una fracción.

rational numbers (p. 46) The set of numbers expressed in the form of a fraction $\frac{a}{b}$, where a and b are integers and $b \neq 0$.

números racionales Conjunto de los números que pueden escribirse en forma de fracción $\frac{a}{b}$, donde a y b son enteros y $b \neq 0$.

real numbers (p. 46) The set of rational numbers and the set of irrational numbers together.

números reales El conjunto de los números racionales junto con el conjunto de los números irracionales.

reciprocal (p. 21) The multiplicative inverse of a number.

recíproco Inverso multiplicativo de un número.

relation (p. 55) A set of ordered pairs.

relación Conjunto de pares ordenados.

relative frequency (p. 677) The number of times an outcome occurred in a probability experiment.

frecuencia relativa Número de veces que aparece un resultado en un experimento probabilístico.

replacement set (p. 15) A set of numbers from which replacements for a variable may be chosen.

conjunto de sustitución Conjunto de números del cual se pueden escoger sustituciones para una variable.

roots (p. 428) The solutions of a quadratic equation.

raíces Las soluciones de una ecuación cuadrática.

S

sample (p. 642) Some portion of a larger group selected to represent that group.

muestra Porción de un grupo más grande que se escoge para representarlo.

sample space (p. 650) The list of all possible outcomes.

espacio muestral Lista de todos los resultados posibles.

scale (p. 108) A ratio or rate used when making a model of something that is too large or too small to be conveniently shown at actual size.

escala Razón o tasa que se usa al construir un modelo de algo que es demasiado grande o pequeño como para mostrarlo de tamaño natural.

scatter plot (p. 227) Two sets of data plotted as ordered pairs in a coordinate plane.

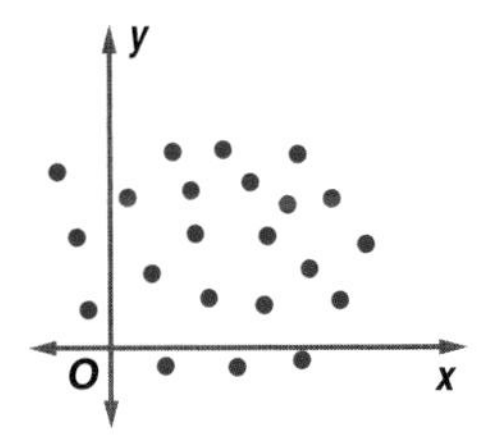

gráfica de dispersión Dos conjuntos de datos graficados como pares ordenados en un plano de coordenadas.

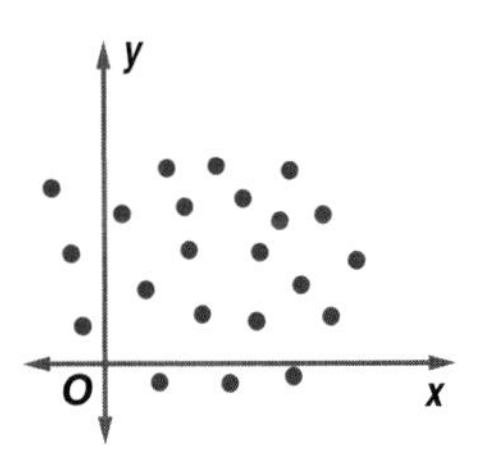

sequence (p. 165) A set of numbers in a specific order.

sucesión Conjunto de números en un orden específico.

set (p. 15) A collection of objects or numbers, often shown using braces { } and usually named by a capital letter.

conjunto Colección de objetos o números, que a menudo se exhiben usando paréntesis de corchete { } y que se identifican por lo general mediante una letra mayúscula .

Glossary/Glosario

set-builder notation (p. 295) A concise way of writing a solution set. For example, $\{t|t < 17\}$ represents the set of all numbers t such that t is less than 17.

notación de construcción de conjuntos Manera concisa de escribir un conjunto solución. Por ejemplo, $\{t|t < 17\}$ representa el conjunto de todos los números t que son menores o iguales que 17.

similar triangles (p. 560) Triangles having the same shape but not necessarily the same size.

semejantes Que tienen la misma forma, pero no necesariamente el mismo tamaño.

simple event (pp. 98, 663) A single event.

evento simple Un sólo evento.

simple random sample (p. 642) A sample that is as likely to be chosen as any other from the population.

muestra aleatoria simple Muestra de una población que tiene la misma probabilidad de escogerse que cualquier otra.

simplest form (p. 29) An expression is in simplest form when it is replaced by an equivalent expression having no like terms or parentheses.

forma reducida Una expresión está reducida cuando se puede sustituir por una expresión equivalente que no tiene ni términos semejantes ni paréntesis.

simulation (p. 678) Using an object to act out an event that would be difficult or impractical to perform.

simulación Uso de un objeto para representar un evento que pudiera ser difícil o poco práctico de ejecutar.

slope (p. 189) The ratio of the change in the y-coordinates (rise) to the corresponding change in the x-coordinates (run) as you move from one point to another along a line.

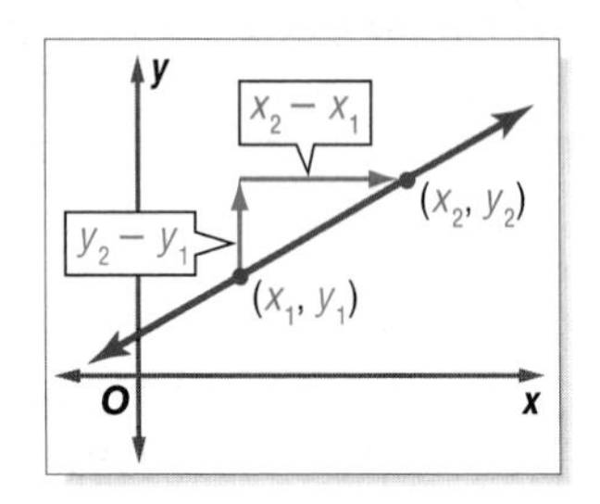

pendiente Razón del cambio en la coordenada y (elevación) al cambio correspondiente en la coordenada x (desplazamiento) a medida que uno se mueve de un punto a otro en una recta.

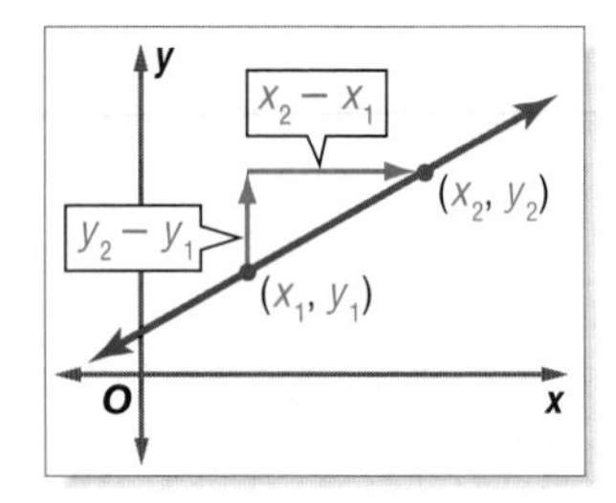

slope-intercept form (p. 204) An equation of the form $y = mx + b$, where m is the slope and b is the y-intercept.

forma pendiente-intersección Ecuación de la forma $y = mx + b$, donde m es la pendiente y b es la intersección y.

solution (pp. 15, 214) A replacement value for the variable in an open sentence.

solución Valor de sustitución de la variable en un enunciado abierto.

solution set (p. 15) The set of elements from the replacement set that make an open sentence true.

conjunto solución Conjunto de elementos del conjunto de sustitución que hacen verdadero un enunciado abierto.

solve an equation (p. 79) The process of finding all values of the variable that make the equation a true statement.

resolver una ecuación Proceso en que se hallan todos los valores de la variable que hacen verdadera la ecuación.

solving an open sentence (p. 15) Finding a replacement value for the variable that results in a true sentence or an ordered pair that results in a true statement when substituted into the equation.

resolver un enunciado abierto Hallar un valor de sustitución de la variable que resulte en un enunciado verdadero o un par ordenado que resulte en una proposición verdadera cuando se lo sustituye en la ecuación.

square root (p. 46) One of two equal factors of a number.

raíz cuadrada Uno de dos factores iguales de un número.

standard form (p. 155) The standard form of a linear equation is $Ax + By = C$, where $A \geq 0$, A and B are not both zero, and A, B, and C are integers with a greatest common factor of 1.

forma estándar La forma estándar de una ecuación lineal es $Ax + By = C$, donde $A \geq 0$, ni A ni B son ambos cero, y A, B, y C son enteros cuyo máximo común divisor es 1.

stratified random sample (p. 642) A sample in which the population is first divided into similar, nonoverlapping groups; a simple random sample is then selected from each group.

muestra aleatoria estratificada Muestra en que la población se divide en grupos similares que no se sobreponen; luego se selecciona una muestra aleatoria simple, de cada grupo.

substitution (p. 260) Use algebraic methods to find an exact solution of a system of equations.

sustitución Usa métodos algebraicos para hallar una solución exacta a un sistema de ecuaciones.

symmetry (p. 472) A geometric property of figures that can be folded and each half matches the other exactly.

simetría Propiedad geométrica de figuras que pueden plegarse de modo que cada mitad corresponde exactamente a la otra.

system of equations (p. 253) A set of equations with the same variables.

sistema de ecuaciones Conjunto de ecuaciones con las mismas variables.

system of inequalities (p. 341) A set of two or more inequalities with the same variables.

sistema de desigualdades Conjunto de dos o más desigualdades con las mismas variables.

systematic random sample (p. 642) A sample in which the items in the sample are selected according to a specified time or item interval.

muestra aleatoria sistemática Muestra en que los elementos de la muestra se escogen según un intervalo de tiempo o elemento específico.

T

term (p. 29) A number, a variable, or a product or quotient of numbers and variables.

término Número, variable o producto, o cociente de números y variables.

terms (p. 165) The numbers in a sequence.

términos Los números de una sucesión.

theoretical probability (p. 677) What should happen in a probability experiment.

probabilidad teórica Lo que debería ocurrir en un experimento probabilístico.

tree diagram (p. 650) A diagram used to show the total number of possible outcomes.

diagrama de árbol Diagrama que se usa para mostrar el número total de resultados posibles.

trinomials (p. 376) The sum of three monomials.

trinomios Suma de tres monomios.

U

uniform motion problems (p. 123) Problems in which an object moves at a certain speed, or rate.

problemas de movimiento uniforme Problemas en que el cuerpo se mueve a cierta velocidad o tasa.

union (p. 316) The graph of a compound inequality containing *or*; the solution is a solution of either inequality, not necessarily both.

unión Gráfica de una desigualdad compuesta que contiene la palabra o; la solución es el conjunto de soluciones de por lo menos una de las desigualdades, no necesariamente ambas.

variable (p. 6) Symbols used to represent unspecified numbers or values.

variable Símbolos que se usan para representar números o valores no especificados.

vertex (p. 472) The maximum or minimum point of a parabola.

vértice Punto máximo o mínimo de una parábola.

vertical line test (p. 150) If any vertical line passes through no more than one point of the graph of a relation, then the relation is a function.

prueba de la recta vertical Si cualquier recta vertical pasa por un sólo punto de la gráfica de una relación, entonces la relación es una función.

voluntary response sample (p. 643) A sample that involves only those who want to participate.

muestra de respuesta voluntaria Muestra que involucra sólo aquellos que quieren participar.

weighted average (p. 122) The sum of the product of the number of units and the value per unit divided by the sum of the number of units, represented by M.

promedio ponderado Suma del producto del número de unidades por el valor unitario dividida entre la suma del número de unidades y la cual se denota por M.

whole numbers (p. 46) The set {0, 1, 2, 3, ...}.

números enteros El conjunto {0, 1, 2, 3, ...}.

work problems (p. 627) Rational equations are used to solve problems involving work rates.

problemas de trabajo Las ecuaciones racionales se usan para resolver problemas de tasas de trabajo.

***x*-axis** (p. 53) The horizontal number line on a coordinate plane.

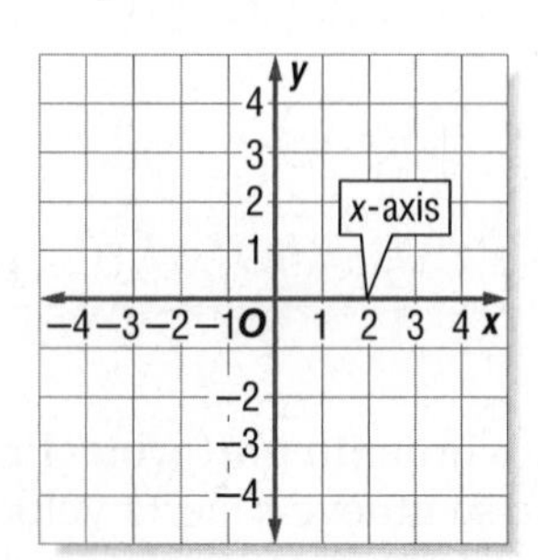

eje *x* Recta numérica horizontal que forma parte de un plano de coordenadas.

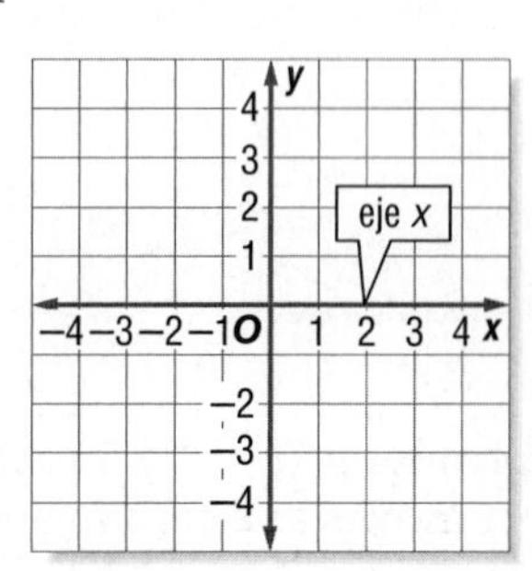

Glossary/Glosario

x-coordinate (p. 53) The first number in an ordered pair.

coordenada x El primer número de un par ordenado.

y-axis (p. 53) The vertical number line on a coordinate plane.

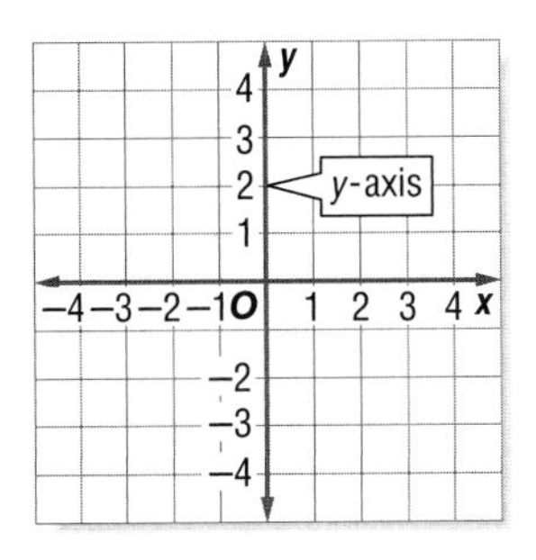

eje y Recta numérica vertical que forma parte de un plano de coordenadas.

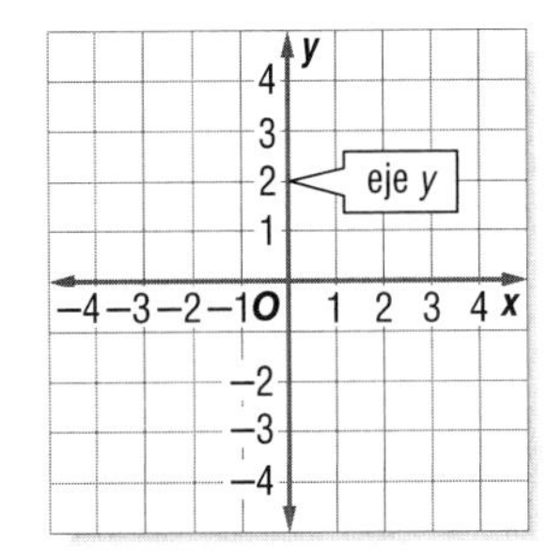

y-coordinate (p. 53) The second number in an ordered pair.

coordenada y El segundo número de un par ordenado.

zeros (p. 480) The roots, or x-intercepts, of a quadratic function.

ceros Las raíces o intersecciones x de una función cuadrática.

Glossary/Glosario

Selected Answers

Chapter 1 The Language and Tools of Algebra

Page 5 ***Chapter 1*** ***Get Ready***

1. 4 **3.** $\frac{3}{5}$ **5.** $\frac{13}{50}$ **7.** 17 **9.** $\frac{1}{6}$ **11.** 19.1 cm **13.** $135\frac{3}{4}$ ft **15.** 6.72 **17.** 1.8 **19.** 9 **21.** $\frac{5}{12}$

Pages 8–9 ***Lesson 1-1***

1. $n + 14$ **3.** $3n - 24$ **5.** $\frac{2}{5}j^2$ **7.** $20 - p$ **9.** 256 **11.** Sample answer: one half of n cubed **13.** $7 + x$ **15.** $5n$ **17.** $\frac{f}{10}$ **19.** $49 + 2x$ **21.** $k^2 - 11$ **23.** πr^2 **25.** 14 lb **27.** 1,000,000 **29.** 3375 **31.** Sample answer: one eighth of y **33.** Sample answer: w minus 24 **35.** Sample answer: r to the fourth power divided by 9 **37.** Sample answer: n cubed times p to the fifth power **39.** Sample answer: 12 times z squared divided by 5 **41.** $s + 12d$ **43.** Sometimes; the product is negative when a is a positive number, the product is positive when a is a negative value. The product is zero when a is zero. **45.** Sample answer: You can use the expression $4s$ to find the perimeter of a baseball diamond. The perimeter of a baseball diamond is four times the length of the sides and the sum of the four sides; $s + s + s + s$. **47.** G **49.** 6.76 **51.** 3.2 **53.** $\frac{1}{6}$

Pages 12–14 ***Lesson 1-2***

1. 23 **3.** 67 **5.** 26 **7.** $\frac{3}{15}$ or $\frac{1}{5}$ **9.** $\frac{11}{150}$ **11.** 20 **13.** $n(2n + 3)$; 44 cm^2 **15.** 14 **17.** 4 **19.** 14 **21.** 142 **23.** 36 **25.** 3 **27.** 1 **29.** 9 **31.** 149 **33.** $50(7.5) + 90(5)$; \$825 **35.** 4 **37.** 55 **39.** $\frac{3}{32}$ **41.** $s + 12c + 4b$ **43.** Chase; Leonora squared the incorrect quantity. **45.** Sample answer: using 1, 2, 3: $1 + 2 + 3 = 6$; $1 + 2 \cdot 3 = 7$; $1 \cdot 2 + 3 = 5$; $3 - 2 \cdot 1 = 1$; $(2 - 1) \cdot 3 = 3$ **47.** B **49.** $13p$ **51.** $20 + 2n$ **53.** $23{,}500 + 14m$ **55.** Sample answer: q squared minus 12 **57.** 0.425 **59.** 15.42 **61.** $3\frac{11}{35}$

Pages 18–20 ***Lesson 1-3***

1. 13 **3.** 12 **5.** 18 **7.** 27 **9.** 5 **11.** {10, 15, 20, 25} **13.** $39n + 10.95 \leq 102.50$; 2 sweaters **15.** 12 **17.** 17 **19.** 3 **21.** 5 **23.** 4 **25.** 78 **27.** 8 **29.** 5 **31.** 11 **33.** $3550 = 75x$; about 47 light bulbs **35.** {4, 5, 6, 7} **37.** {12, 14, 16} **39.** $3(41.99) + 26.99c \leq 300$, where c is the number of children. 6 children could go to the park with 3 adults. **41.** $\frac{1}{3}$ **43.** 2.4 **45.** {3.5, 4} **47.** 1000 Calories **49.** Sample answer: $x \geq 8$ represents the inequality because it includes the set of all whole numbers greater than or equal to 8. **51.** The solution set includes all numbers less than or equal to $\frac{1}{3}$. **53.** B **55.** D **57.** 53 **59.** 1004 **61.** Sample answer: n to the fifth power minus 8 **63.** Sample answer: 6 divided by the product of 5 and a **65.** $6 + \frac{1}{2}n$ **67.** $\frac{1}{4}n^3$ **69.** $\frac{4}{21}$ **71.** $\frac{2}{7}$ **73.** $\frac{11}{15}$

Pages 23–25 ***Lesson 1-4***

1. 0; Multiplicative Property of Zero
3. 6; Multiplicative Inverse
5. $6(12 - 48 \div 4)$
$= 6(12 - 12)$ Substitution
$= 6(0)$ Substitution
$= 0$ Multiplicative Property of Zero
7. $4(20) + 7$
$= 80 + 7$ Substitution
$= 87$ Substitution; 87 yr
9. 5; Multiplicative Identity
11. 0; Additive Identity **13.** $\frac{1}{2}$; Multiplicative Inverse
15. 1; Multiplicative Inverse **17.** $\frac{1}{2}$; Additive Inverse
19. $7 + (9 - 3^2)$
$= 7 + (9 - 9)$ Substitution
$= 7 + 0$ Substitution
$= 7$ Additive Identity
21. $[3 \div (2 \cdot 1)]\frac{2}{3}$
$= [3 \div 2]\frac{2}{3}$ Multiplicative Identity
$= \frac{3}{2} \cdot \frac{2}{3}$ Substitution
$= 1$ Multiplicative Inverse
23. $6 \cdot \frac{1}{6} + 5(12 \div 4 - 3)$
$= 6 \cdot \frac{1}{6} + 5(3 - 3)$ Substitution
$= 6 \cdot \frac{1}{6} + 5(0)$ Substitution
$= 6 \cdot \frac{1}{6} + 0$ Multiplicative Prop. of Zero
$= 1 + 0$ Multiplicative Inverse
$= 1$ Additive Identity
25. $2 \cdot \frac{22}{7} \cdot 14^2 + 2 \cdot \frac{22}{7} \cdot 14 \cdot 7$
$= 2 \cdot \frac{22}{7} \cdot 196 + 2 \cdot \frac{22}{7} \cdot 14 \cdot 7$ Substitution
$= \frac{44}{7} \cdot 196 + \frac{44}{7} \cdot 14 \cdot 7$ Substitution
$= 1232 + 616$ Substitution
$= 1848$ Substitution; 1848 in^2
27. 3; Multiplicative Identity
29. $3 + 5(4 - 2^2) - 1$
$= 3 + 5(4 - 4) - 1$ Substitution
$= 3 + 5(0) - 1$ Substitution
$= 3 + 0 - 1$ Multiplicative Property of Zero
$= 3 - 1$ Additive Identity
$= 2$ Substitution
31. $\left[\frac{5}{8}\left(1 + \frac{3}{5}\right)\right] \cdot 17$
$= \frac{5}{8}\left(\frac{8}{5}\right) \cdot 17$ Substitution
$= 1 \cdot 17$ Multiplicative Inverse
$= 17$ Multiplicative Identity
33. $25(2 - 0.3) + 80(2.5 - 1) + 40(10 - 6)$

35. Sometimes; sample answer: true: $x = 2, y = 1, z = 4, w = 3$; $2 \cdot 4 > 1 \cdot 3$; false: $x = 1, y = -1, z = -2, w = -3$; $1(-2) < (-1)(-3)$ **37.** Sample answer: $5 = 3 + 2$ and $3 + 2 = 4 + 1$ so $5 = 4 + 1$; $5 + 7 = 8 + 4$, and $8 + 4 = 12$, so $5 + 7 = 12$. **39.** You can use the Identity and Equality properties to see if data is the same. Answers should include the following: Reflexive: $r = r$, or Symmetric: $a = b$, so $b = a$; Southern California, week 1 $= a$, week 2 $= b$, week 3 $= c$. $a = b$ and $b = c$, so $a = c$. **41.** J **43.** {11, 12, 13} **45.** 2(213) – 59; $367 **47.** 80 **49.** 28

Pages 29–31 Lesson 1-5

1. 6(12) – 6(3); 54 **3.** 19(10) + 3(10); 220 **5.** 16(100 + 3); 1648 **7.** $8 + 2t$ **9.** $14m$ **11.** $9x$ **13.** 7(13) + 7(12); 175 **15.** 3(15) + 8(15); 165 **17.** 10(13) – 7(13); 39 **19.** 4(110,000 + 17,500); 510,000 **21.** 8(900 + 90); 7920 **23.** $\left(3 + \frac{1}{6}\right)48$; 152 **25.** $15 + 3n$ **27.** $-3x + 18$ **29.** $9b - 1$ **31.** $17a^2 + a$ **33.** $45x - 75$ **35.** $1956 **37.** $4p + 4q - 4r$ **39.** $30m^3 + 5n$ **41.** $7y^3 + y^4$ **43.** Sample answer: The numbers inside the parentheses are each multiplied by the number outside the parentheses, then the products are added. **45.** Courtney; she correctly used the Distributive Property. **47.** You can use the Distributive Property to calculate quickly by expressing any number as a sum or difference of a more convenient number. Answers should include the following: Both methods result in the correct answer. In one method you multiply then add, and in the other you add then multiply. **49.** J **51.** Multiplicative Identity **53.** Reflexive **55.** 15 **57.** 9 **59.** 168 cm^2

Pages 35–37 Lesson 1-6

1. Sample answer:
$14 + 18 + 26$
$= 14 + 26 + 18$ Commutative (+)
$= (14 + 26) + (18)$ Associative (+)
$= 40 + 18$ Add.
$= 58$ Add.

3. Sample answer:
$5 \cdot 3 \cdot 6 \cdot 4$
$= 5 \cdot 6 \cdot 3 \cdot 4$ Commutative (×)
$= (5 \cdot 6) \cdot (3 \cdot 4)$ Associative (×)
$= 30 \cdot 12$ Multiply.
$= 360$ Multiply.

5. $10x + 5y$ **7.** $14x + 6$ **9.** $xy + 10 + 2x$ **11.** 60 **13.** 16 **15.** 12 **17.** 440 **19.** 4.8 **21.** 60 **23.** 46.8 in^2 **25.** $5a + 2b$ **27.** $6x^2 + 5x$ **29.** $10x + 14$

31. $2(s + t) - s$
$= 2(s) + 2(t) - s$ Distributive
$= 2s + 2t - s$ Multiply.
$= 2t + 2s - s$ Commutative (+)
$= 2t + s(2 - 1)$ Distributive
$= 2t + s(1)$ Subtract.
$= 2t + s$ Multiplicative Identity

33. $6z^2 + (7 + z^2 + 6)$
$= 6z^2 + (z^2 + 7 + 6)$ Commutative (+)
$= (6z^2 + z^2) + (7 + 6)$ Associative (+)
$= z^2(6 + 1) + (7 + 6)$ Distributive
$= z^2(7) + 13$ Add.
$= 7z^2 + 13$ Multiply.

35. $5q$ **37.** $20m + 3n + 2mn$ **39.** $\frac{3}{4} + \frac{5}{3}s + \frac{4}{3}t$ **41.** $291 **43.** $149.35 **45.** $n \div 2 = 2 \div n$; This equation is true for $n = 2$ or $n = -2$ only because division is not commutative. The other three sentences illustrate the Commutative Property of Addition or Multiplication and are therefore true for all values of the variables. **47.** You can use the Commutative and Associative Properties to rearrange and group numbers for easier calculations. Answers should include: $d = (0.4 + 1.1) + (1.5 + 1.5) + (1.9 + 1.8 + 0.8)$. **49.** J **51.** simplified **53.** $15 + 6p$ **55.** $8d + 40 + 2f$ **57.** Sample answer: $50x \leq 180$; 3 loads **59.** 60 **61.** 13

Pages 42–44 Lesson 1-7

1. H: it is April; C: it might rain **3.** H: $34 - 3x = 16$; C: $x = 6$ **5.** H: a number is divisible by 10; C: it is divisible by 5; If a number is divisible by 10, then it is divisible by 5. **7.** The number is divisible by 2. **9.** No valid conclusion; the last digit is a 6. **11.** A book can have more than 384 pages. **13.** $x = 15$ **15.** H: both parents have red hair; C: their children have red hair **17.** H: $2n - 7 > 25$; C: $n > 16$ **19.** H: it is Monday; C: the trash is picked up; If it is Monday, then the trash is picked up. **21.** H: $x = 8$; C: $x^2 - 3x = 40$; If $x = 8$, then $x^2 - 3x = 40$. **23.** H: a triangle with all sides congruent; C: it is an equilateral triangle; If all the sides of a triangle are congruent, then it is an equilateral triangle. **25.** Ian will buy a DVD box set. **27.** The DVD box set cost $70 or more. **29.** A person born in North Carolina moved to California. **31.** $2 \cdot 3 = 6$ **33.** $4(15) - 8 = 52$ **35.** Sample answer:

P Q R

37. The perimeter is doubled. **41.** Sample answer: If a number is divisible by 2 and 3, then it must be a multiple of 6. **43.** No; sample answer: Let $a = 1$ and $b = 2$; then $1 * 2 = 1 + 2(2)$ or 5 and $2 * 1 = 2 + 2(1)$ or 4. **45.** Sample answer: You can use deductive reasoning to determine whether a hypothesis and its conclusion are both true or whether one or both are false. **47.** A **49.** A **51.** $23mn + 24$ **53.** $100d + 80d + 8d$; $(100 + 80 + 8)d$ **55.** 7; Reflexive **57.** $\frac{1}{4}$; Multiplicative Inverse **59.** 0; Multiplicative Property of Zero **61.** 64 **63.** 132.25

Pages 50–52 Lesson 1-8

1. integers, rationals **3.** irrationals **5.** No; Sample answer: $3 \div 4 = \frac{3}{4}$ **7.** Yes; Sample answer: $3.2 \div 1.5 = 2.1\overline{3}$

9. –5 –4 –3 –2 –1 0 1 2 3 4 5 6 7 8

11. –8 –7 –6 –5 –4 –3 –2 –1 0 1

13. 1.2 **15.** 19 **17.** > **19.** < **21.** $\sqrt{\frac{1}{30}}, 5\frac{4}{9}, \sqrt{30}, 13$ **23.** naturals, wholes, integers, rationals **25.** rationals **27.** irrationals **29.** No; Sample answer: $\sqrt{5} - \sqrt{5} = 0$ **31.** Yes; Sample answer: $3 \times 2 = 6$ **33.** No; Sample answer: $2\sqrt{3} \div \sqrt{3} = 2$ **35.** Yes; Sample answer: $\frac{3}{5} - \frac{1}{5} = \frac{2}{5}$ **37.** Yes; Sample answer: $-3 \times -8 = 24$

39. −2 −1 0 1 2 3 4 5 6

41. −10 −9

43. ± 0.8 **45.** -2.5 **47.** $\frac{5}{18}$ **49.** $<$ **51.** $<$ **53.** $\frac{5}{9}$, $\sqrt{\frac{16}{49}}$, 0.6 **55.** $-\sqrt{\frac{5}{8}}, -0.25, 0.\overline{14}, \sqrt{0.5}$ **57.** 6 **59.** 12
61. Side lengths: 1 unit, 2 units, 3 units, 4 units, 5 units; Perimeters: 4 units, 8 units, 12 units, 16 units, 20 units **63.** $4\sqrt{a}$ **65.** True; the average of $\sqrt{2}$ and $\sqrt{3}$ is a decimal number that does not terminate or repeat, so it is irrational. **69.** C **71.** $3 + 3 = 6$ **73.** $9x + 2y$ **75.** $4 + 80x + 32y$ **77.** $48

Pages 56–58 ***Lesson 1-9***

1. (2, 35); On day 2, the average temperature is about 35°F. **3.** independent: day; dependent: temperature **5.** Graph B
7.

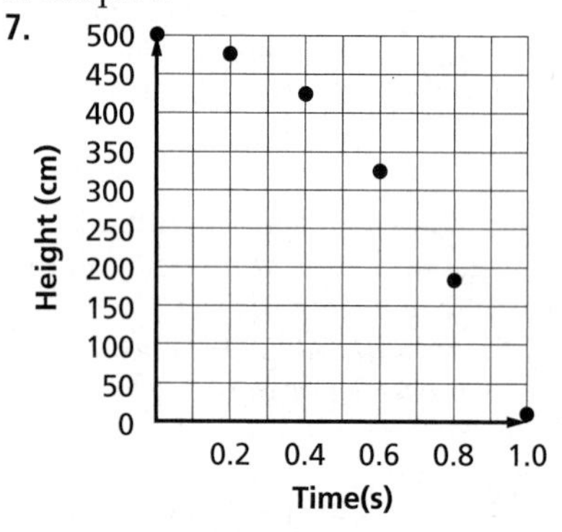

9. The function is discrete because the points are not connected with a line or a curve. **11.** (5, 25); The dog walker earns $25 for walking 5 dogs. **13.** (02, 3); In the year 2002, sales were about $3 million.
15. independent: year; dependent: sales **17.** Their altitude is increasing as they ascend. Then they go down a steep incline. They then make a longer climb up another hill. **19.** Graph B
21.

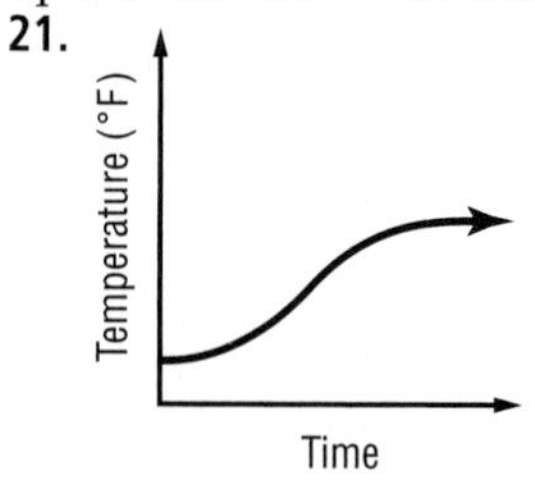

23. domain: {3, 4, 5, 6, 7}; range: {180, 360, 540, 720, 900} **25.** 1080; 1260; 1440 **29.** Sample answer: Real-world data can be recorded and visualized in a graph and by expressing an event as a function of another event. A graph gives you a visual representation of the situation which is easier to analyze and evaluate. During the first 24 hours, blood flow to the brain decreases to 50% at the moment of the injury and gradually increases to about 60%. Significant improvement occurs during the first two days.
31. F **33.** H: you use a computer; C: you can send email

35. $ab(a + b) = (ab)a + (ab)b$ Distributive
$= a(ab) + (ab)b$ Commutative (×)
$= (a \cdot a)b + a(b \cdot b)$ Associative (×)
$= a^2b + ab^2$ Substitution
37. $3x - 10$

Pages 60–64 ***Chapter 1*** ***Study Guide and Review***

1. true **3.** false; set **5.** false; real **7.** true **9.** $5y^2$
11. $2k - 8$ **13.** 32 **15.** six times the square of a number p **17.** $20b$ **19.** 3 **21.** 34 **23.** 96 **25.** 20
27. {7} **29.** {5} **31.** 13 **33.** 35 **35.** $p = 15(3) + 9(1)$; $p = \$54$ **37.** $\frac{1}{5}$; Multiplicative Inverse
39. $3(4 \div 4)^2 - \frac{1}{4}(4) = 3(1)^2 - \frac{1}{4}(4)$ Substitution
$= 3(1) - \frac{1}{4}(4)$ Substitution
$= 3 - \frac{1}{4}(4)$ Substitution
$= 3 - 1$ Multiplicative Inverse
$= 2$ Substitution
41. $\frac{1}{2} \cdot 2 + 2[2 \cdot 3 - 1] = \frac{1}{2} \cdot 2 + 2[6 - 1]$ Substitution
$= \frac{1}{2} \cdot 2 + 2(5)$ Substitution
$= \frac{1}{2} \cdot 2 + 10$ Substitution
$= 1 + 10$ Multiplicative Inverse
$= 11$ Substitution
43. $8(15) - 8(6) = 72$ **45.** $2w + 7v$ **47.** $9(550 + 225 + 110 + 150) = \$9{,}315$ **49.** $6 + 30x$ **51.** H: it is a school day; C: the day begins at 7:30 A.M.; If it is a school day, then the day begins at 7:30 A.M. **53.** H: lightning has struck twice; C: it has not done so in the same place; If lightning has struck twice, then it has not done so in the same place. **55.** irrationals
57.

−2 −1 0 1 2 3 4 5 6 7 8

59. $-\sqrt{34}, -\frac{47}{9}, \sqrt{27}$, and $5\frac{1}{5}$ **61.** (2, 60); This represents a score of 60 on the math test with 2 hours of study. **63.** Graph C

Chapter 2 Solving Linear Equations

Page 69 ***Chapter 2*** ***Get Ready***

1. $\frac{1}{2}t + 5$ **3.** $3a + b^2$ **5.** 15 **7.** 16 **9.** 63 **11.** 5
13. $40 **15.** 300% **17.** 51%

Pages 73–76 ***Lesson 2-1***

1. $2t - 8 = 70$ **3.** $\frac{1}{2}p = p - 3$ **5.** $1900 + 30w = 2500$; 20 weeks **7.** $C = 2\pi r$ **9.** Sample answer: 14 plus d equals 6 times d. **11.** Sample answer: The original cost of a suit is c. After a $25 discount, the suit costs $150. What is the original cost of the suit? **13.** Sample answer: $2r + 3s = 13$ **15.** Sample answer: $200 - 3n = 9$ **17.** Sample answer: $m^2 - n^3 = 16$ **19.** $0.46E = P$
21. $1912 + y = 1928$; 16 yr **23.** $A = bh$

25. $P = 2(a + b)$ **27.** Sample answer: d minus 14 equals 5. **29.** Sample answer: k squared plus 17 equals 53 minus j. **31.** Sample answer: $\frac{3}{4}$ of p plus $\frac{1}{2}$ equals p. **33.** Sample answer: Lindsey is 7 inches taller than Yolanda. If 2 times Yolanda's height plus Lindsey's height equals 193 inches, find Yolanda's height. **35.** $\frac{1}{2}(9 + n) = n - 3$ **37.** $V = \frac{1}{3}\pi r^2 h$ **39.** $V = \frac{4}{3}\pi r^3$ **41.** $4a + 15$ **43.** 9 min **45.** Sample answer: 4 times the quantity t minus s equals 5 times s plus 12.

47. Sample answer: $S = 3ah + \frac{a^2\sqrt{3}}{2}$; the area of the two triangular bases is $2\left(\frac{1}{2}\right)(a)\left(\frac{a\sqrt{3}}{2}\right)$, which simplifies to $\frac{a^2\sqrt{3}}{2}$. The area of the three rectangular sides is $3ah$. So, the total surface area S is the sum $3ah + \frac{a^2\sqrt{3}}{2}$.

49. Sample answer: Equations can be used to describe the relationships of the heights of various parts of a structure. The equation representing the Sears Tower is $1454 + a = 1707$. **51.** J **53.** 90 **55.** 9.49 **57.** $8d + 3$ **59.** $8a + 6b$ **61.** 396 **63.** 3.37 **65.** 1.65 **67.** $\frac{5}{6}$

Pages 81–84 ***Lesson 2-2***

1. 25 **3.** −3 **5.** −13 **7.** $\frac{3}{10}$ **9.** $\frac{5}{6}$ **11.** $n + 91 = 37$; −54 **13.** 23 **15.** −43 **17.** 22 **19.** −19 **21.** 73 **23.** $-1\frac{1}{9}$ **25.** $1\frac{1}{8}$ **27.** $n - 18 = 31$; 49 **29.** $n + (-16) = -21$; −5 **31.** Sample equation: $\ell + 10 = 34$; 24 mi **33.** −4.4 **35.** 7.7 **37.** 11.03 **39.** $n - \frac{1}{2} = -\frac{3}{4}$; $-\frac{1}{4}$ **41.** $\ell + 8.1 = 24.9$; 16.8 h **43.** −19 **45.** $14.6 + x = 14.7$; 0.1 million volumes **47.** $28.7 + 14.7 + 14.6 = x$; 58.0 million volumes **51a.** Sometimes; if $x = 0$, $x + x = x$ is true. **51b.** Always; any number plus 0 always equals the number. **53.** C **55.** $A = \pi r^2$ **57.** $<$ **59.** $=$ **61.** H: $y = 2$; C: $4y - 6 = 2$; If $y = 2$, then $4y - 6 = 2$. **63.** 18.20 **65.** $\frac{5}{12}$

Pages 88–90 ***Lesson 2-3***

1. −35 **3.** 15 **5.** −36 **7.** 10 m **9.** 5 **11.** 42 **13.** $\frac{1}{3} = -7n$; $-\frac{1}{21}$ **15.** −77 **17.** $5\frac{1}{3}$ **19.** −10 **21.** $-1\frac{3}{7}$ **23.** 18 **25.** 225 **27.** about 95 ft **29.** −5 **31.** −7 **33.** 14 **35.** $\frac{2}{5}n = -24$; −60 **37.** $12 = \frac{1}{5}n$; 60 **39.** $\frac{11}{15}$ **41.** 2.1 **43.** −3.5 **45.** $2\frac{1}{2}n = 1\frac{1}{5}$; $\frac{12}{15}$ **47.** 0.48 s **49.** $8x$ **51.** 53 g; 424 g

53. Sample answer: Both properties can be used to solve equations. The Multiplication Property of Equality says you can multiply each side of an equation by the same number. The Division Property of Equality says you can divide each side of an equation by the same number. Dividing each side of an equation by a number is the same as multiplying each side of the equation by the number's reciprocal. **55.** Camila; to find an equivalent equation with $1n$ on one side of the equation, you must divide each side by 8 or multiply each side by $\frac{1}{8}$. Casey incorrectly multiplied each side by 8. **57.** D **59.** 67 **61.** −72 **63.** $40 + 3n$ **65.** 0 **67.** $\frac{3}{19}$

Pages 95–97 ***Lesson 2-4***

1. −1 **3.** 56 **5.** 163 **7.** $12 - 2n = -34$; 23 **9.** $n + (n + 1) + (n + 2) = 42$; 13, 14, 15 **11.** −2 **13.** 5 **15.** 7 **17.** −15 **19.** 23 **21.** 34 **23.** $\frac{2}{3}n - 6 = -10$; −6 **25.** $n + (n + 2) + (n + 4) = 51$; 15, 17, 19 **27.** $n + (n + 1) + (n + 2) + (n + 3) = 94$; 22, 23, 24, 25 **29.** 450.5 mi **31.** 21,000 ft **33.** −42.72 **35.** −56 **37.** 1.5 **39.** 2 **45.** Never; let n and $n + 2$ be the even numbers and m, $m + 2$ be the odd numbers. Write and solve the equation $n + (n + 2) = m + (m + 2)$ gives the solution $n = m$. Thus, n must equal m, so n is not even or m is not odd. **47.** D **49.** A **51.** −120 **53.** $m + 9 = 56$; 47 models **55.** $13(100 + 1)$; 1313 **57.** $5m + \frac{n}{2}$ **59.** $3d$ **61.** $14t$ **63.** $-3f$

Pages 101–103 ***Lesson 2-5***

1. 4 **3.** 3 **5.** 3, 4 **7.** no solution **9.** all numbers **11.** 4 **13.** −3 **15.** −16 **17.** 1 **19.** −1 **21.** −3 **23.** 3 **25.** 120 **27.** 60 **29.** 17, 19 **31.** no solution **33.** all numbers **35.** 30 years **37.** −2 **39.** 4 **41.** 2.5 **43.** about 5.6 yr **45.** Sample answer: $3(x + 1) = x - 1$ **47a.** Incorrect; the 2 must be distributed over both g and 5; 6. **47b.** correct **49.** D **51.** 90 **53.** −2 **55.** (1, 8), (2, 12), (3, 4), (4, 2);

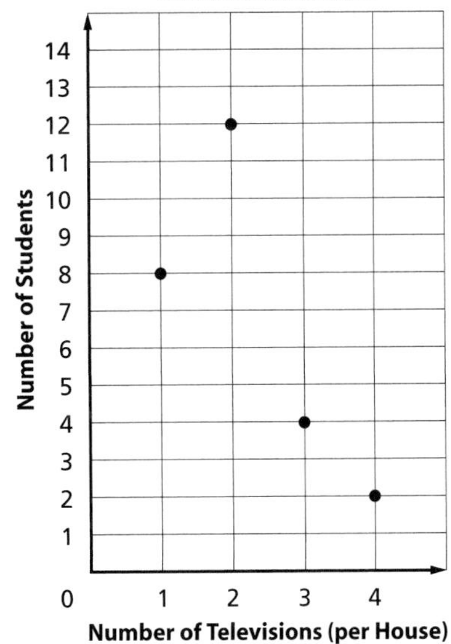

57. $3x^2 + x^2 + 7x$; $4x^2 + 7x$; Distributive Property and Substitution **59.** {1, 3, 5} **61.** $8\frac{1}{3}$ **63.** $\frac{4}{7}$ **65.** $\frac{1}{15}$

Pages 108–110 **Lesson 2-6**

1. yes **3.** no **5.** 15 **7.** about 14.6 gal **9.** yes **11.** no **13.** no **15.** 20 **17.** 18 **19.** $9\frac{1}{3}$ **21.** 1.32 **23.** 0.84 **25.** 2.28 **27.** 14 days **29.** $4\frac{1}{6}$ ft **31.** 19.33 **33.** 2.56 **35.** USA: $\frac{907}{2219}$; USSR/UT/Russia: $\frac{525}{1370}$; Germany: $\frac{388}{1230}$; GB: $\frac{189}{668}$; France: $\frac{199}{631}$; Italy: $\frac{189}{511}$; Sweden: $\frac{140}{476}$ **39.** $\frac{9}{19}$ **41.** C **43.** $1\frac{1}{11}$ **45.** $3\frac{3}{5}$ **47.** −125 **49.** Sample answer:

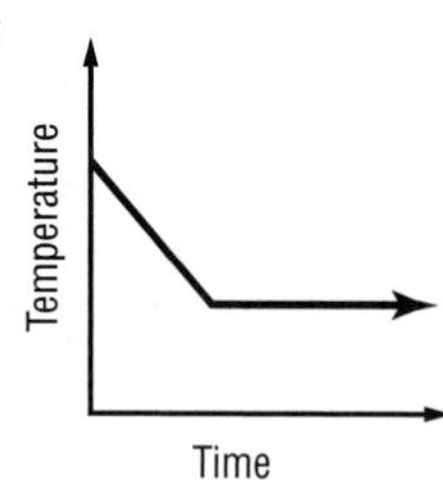

51. 4.3 **53.** 30% **55.** 40%

Pages 113–115 **Lesson 2-7**

1. dec.; 50% **3.** inc.; 14% **5.** 347 mi **7.** \$16.91 **9.** \$13.37 **11.** dec.; 28% **13.** inc.; 193% **15.** inc.; 14% **17.** dec.; 19% **19.** \$130,200 **21.** \$37.45 **23.** \$19.66 **25.** \$31.71 **27.** \$4.80 **29.** \$16.25 **31.** \$55.99 **33.** \$52.43 **35.** 47% **37.** China: about 1.52 billion people; India: about 1.53 billion people; United States: about 0.39 billion people; India **39.** always; x% of $y \Rightarrow \frac{x}{100} = \frac{P}{y}$ or $P = \frac{xy}{100}$; y% of $x \Rightarrow \frac{y}{100} = \frac{P}{x}$ or $P = \frac{xy}{100}$ **41.** Laura; Cory used the new number as the base instead of the original number. **43.** A **45.** 9 **47.** 24 **49.** −6 **51.** \$75,000 **53.** 20 **55.** −3 **57.** 11

Pages 119–121 **Lesson 2-8**

1. $x = \frac{b}{9}$ **3.** $y = 3c - a$ **5.** $h = \frac{2A}{b}$ **7.** 24 min **9.** $a = \frac{v - r}{t}$ **11.** $b = \frac{5 - t}{4}$ **13.** $m = \frac{5}{3}(b - a)$ **15.** $y = \frac{3c - 2}{b}$ **17.** $t = -\frac{r}{4}$ **19.** $a = \frac{2b + c}{5}$ **21.** $g = \frac{5 + m}{2 + h}$ **23.** 6 m **25.** 3 errors **27.** $t = \frac{-2 - p}{p}$ **29.** $t - 7 = r + 6$; $t = r + 13$ **31.** $\frac{5}{8}x = \frac{1}{2}y + 3$; $y = \frac{5}{4}x - 6$ **33.** 225 lb **35.** Sample answer: $A = \frac{5}{2}s^2$; the area of the square is s^2 and the area of the triangle is $\frac{1}{2}(3s)(s)$ or $\frac{3}{2}s^2$. So, the total area is $s^2 + \frac{3}{2}s^2$ or $\frac{5}{2}s^2$. **37.** Sample answer for a triangle: $A = \frac{1}{2}bh$; $b = \frac{2A}{h}$ **39.** B **41.** 10 g **43.** 3.75 **45.** $12 - 6t$ **47.** $-21a - 7b$

Pages 125–128 **Lesson 2-9**

1.

	Number of Pounds	Price per Pound	Total Price
Votive Wax	8	\$0.90	0.90(8)
Low Shrink	p	\$1.04	1.04p
Blend	$8 + p$	\$0.98	$0.98(8 + p)$

3. $10\frac{2}{3}$ lb

5.

	Quarts	Total Amount of Juice
20% Juice	$5 - n$	$0.20(5 - n)$
100% Juice	n	1.00n
50% Juice	5	0.50(5)

7. $1\frac{7}{8}$ qt; $3\frac{1}{8}$ qt **9.** 10 mph **11.** 2 h **13.** 87

15.

	Number of Ounces	Price per Ounce	Value
Gold	g	\$432	432g
Silver	$15 - g$	\$7.35	$7.35(15 - g)$
Alloy	15	\$177.21	177.21(15)

17. 6 oz **19.** 270 rolls of solid wrap, 210 rolls of print wrap **21.** $2.00b + 3.50(b - 36) = 281.00$ **23.** 38 doz **25.** 10 gal of cream, 25 gal of 2% milk **27.** 22.2 mph

29.

	r	t	$d = rt$
Eastbound Train	40	h	40h
Westbound Train	30	h	30h

31. $3\frac{1}{2}$ h **33.** $90t = 70t + 300$ **35.** No; it takes the sprinter $\frac{200}{8.2} \approx 24.39$ seconds to run 200 meters and it takes his opponent $\frac{200}{8} = 25$ seconds to run 200 meters. Since the sprinter lost 1 second at the start, his time would be 25.39, which is 0.39 second slower than his opponent's time. **37.** Sample answer: grade point average **39.** Sample answer: A weighted average is used to determine a skater's average. The score of the short program is added to twice the score of the long program. The sum is divided by 3. $\frac{4.9(1) + 5.2(2)}{1 + 2} = 5.1$ **41.** H **43.** $t = \frac{s - 4}{3}$ **45.** increase; 20% **47.** Sample answer: $80 + d = 115$; \$35

Pages 130–134 **Chapter 2** **Study Guide and Review**

1. true **3.** true **5.** true **7.** true **9.** $mn = 3(m + 8)$

11. Sample answer: The quotient of 56 and g equals seven minus three times g. **13.** Sample answer: Let s = the millions of sq. km of land surface on Earth; $\frac{1}{3}s = 50$; $s = 150$ million km^2 **15.** $-\frac{1}{15}$ **17.** −48 **19.** $x + (-71) = 29$; 100 **21.** 203 **23.** 3 **25.** $\frac{3}{8}n = 9$; 24 **27.** $25{,}000{,}000x = 45{,}000{,}000{,}000$; $x = 1800$ chopsticks **29.** −153 **31.** 136 **33.** $n + (n + 2) + (n + 4) = 39$; 11, 13, and 15 **35.** 5 **37.** 160 cans **39.** 18 **41.** dec.; 15% **43.** \$3.56 **45.** $k = \frac{4}{7}(s + g)$ or $k = \frac{4(s + g)}{7}$ or $k = \frac{4s + 4g}{7}$ **47.** $h = \frac{pq - z}{7}$ **49.** 27 lb Brand A, 18 lb Brand B **51.** $\frac{4}{25}$ mi/min or 0.16 mi/min

Chapter 3 Functions and Patterns

Page 141 ***Chapter 3*** ***Get Ready***

1. 6 **3.** −21 **5.** \$4.55 **7.** $y = 8 - x$ **9.** $y = -\frac{2}{3}x + 3$ **11.** $y = 3x + 1$

13.

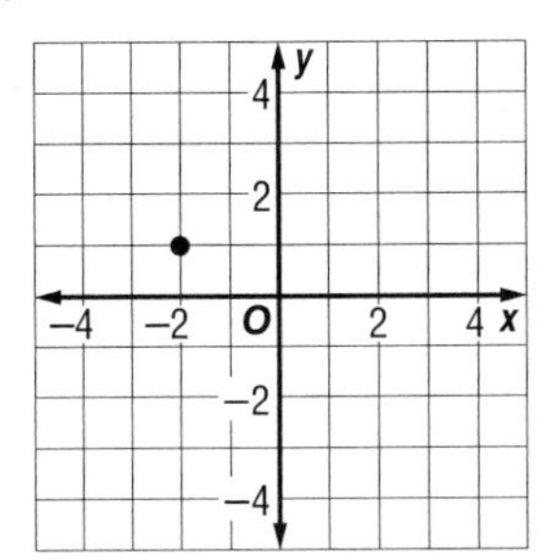

15.

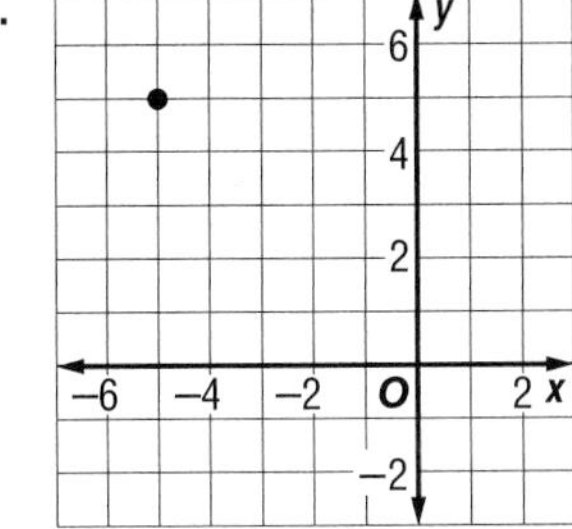

17.

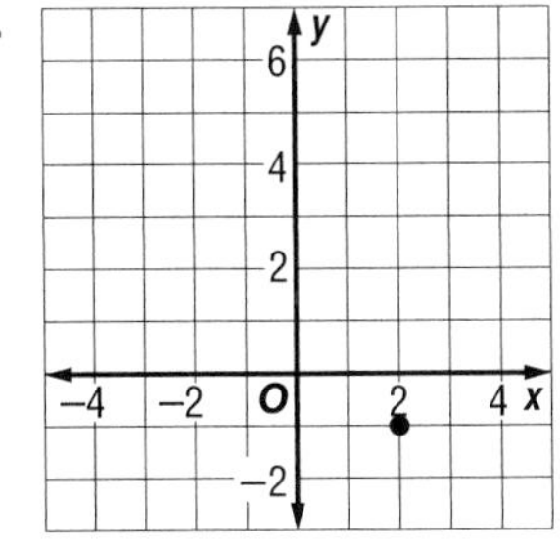

Pages 146–148 ***Lesson 3-1***

1.

x	y
5	−2
8	3
−7	1

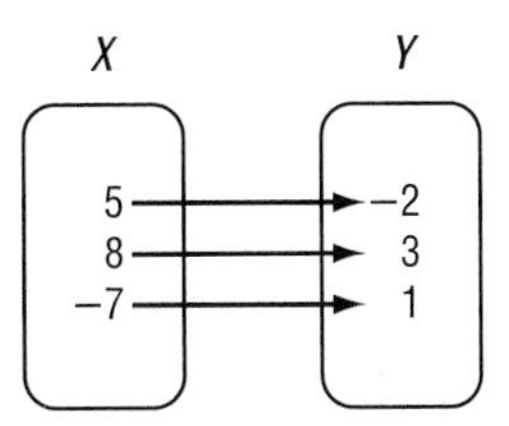

$D = \{-7, 5, 8\}$; $R = \{-2, 1, 3\}$

3. $D = \{0, 1000, 2000, 3000, 5000, 10{,}000\}$; $R = \{212.0, 210.2, 208.4, 206.5, 201.9, 193.7\}$

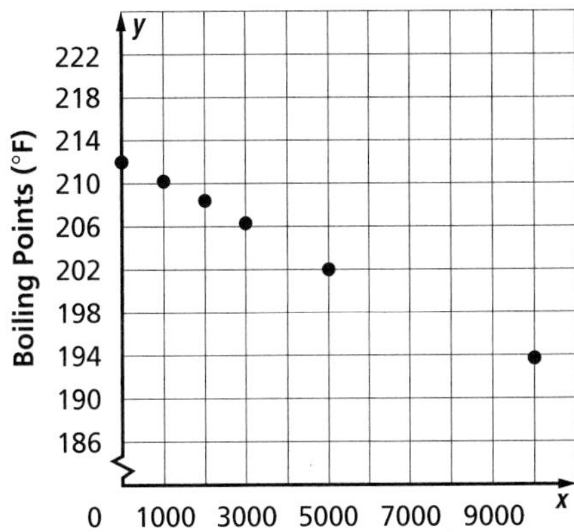

5. $\{(3, -2), (-6, 7), (4, 3), (-6, 5)\}$; $\{(-2, 3), (7, -6), (3, 4), (5, -6)\}$ **7.** $\{(-1, 2), (2, 4), (3, -3), (4, -1)\}$; $\{(2, -1), (4, 2), (-3, 3), (-1, 4)\}$

9.

x	y
3	8
3	7
2	−9
1	−9

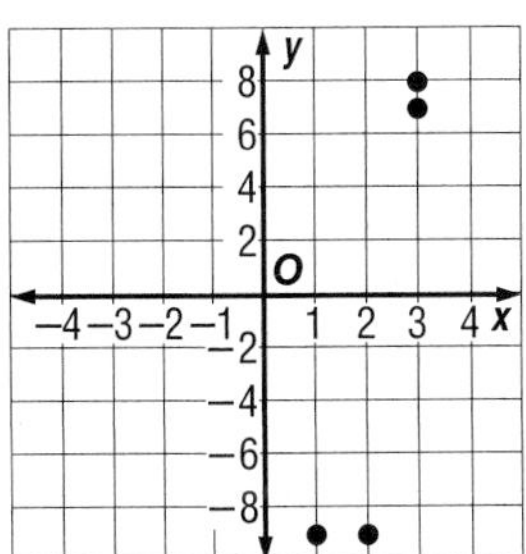

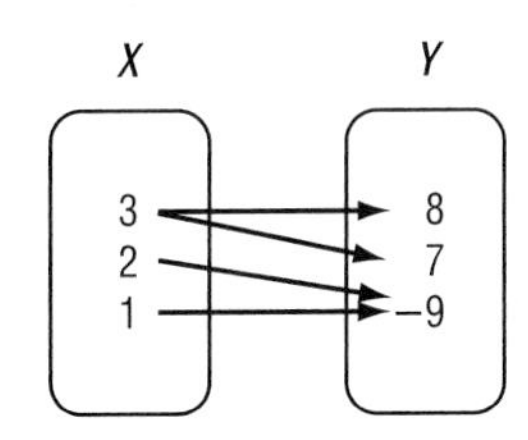

$D = \{1, 2, 3\}$; $R = \{-9, 7, 8\}$

11.

x	y
0	2
−5	1
0	6
−1	9

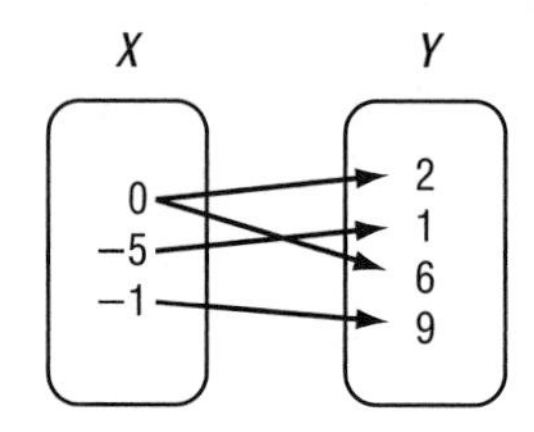

D ={−5, −1, 0}; R = {1, 2, 6, 9}

13.

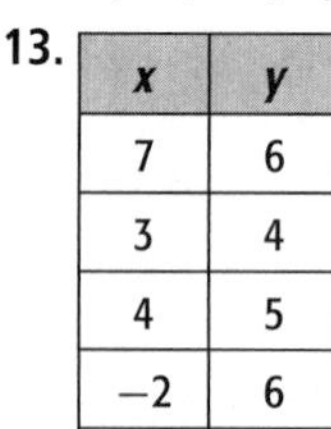

x	y
7	6
3	4
4	5
−2	6
−3	2

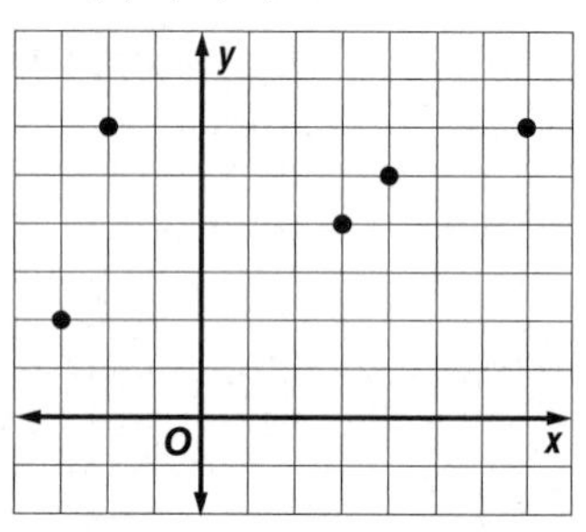

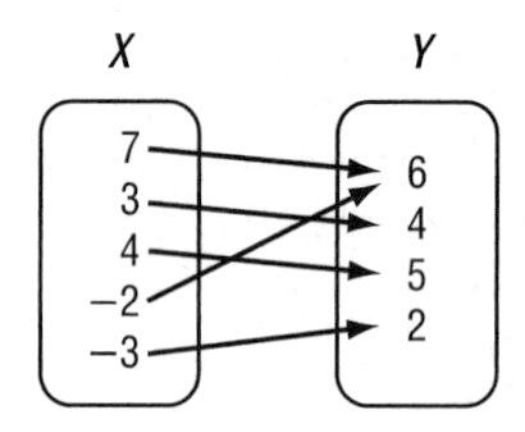

D ={−3, −2, 3, 4, 7}; R = {2, 4, 5, 6}
15. {1992, 1993, 1994, 1995, 1996, 1997, 1998, 1999, 2000, 2001, 2002} **17.** There are fewer students per computer in more recent years. So, the number of computers in schools has increased. **19.** 2007; 2014 **21.** Sample answer: 13.0; this means that the production of apples is projected to be 13.0 billion pounds in the year 2015.
23. {(0, 0), (4, 7), (8, 10.5), (12, 18), (16, 14.5)}; {(0, 0), (7, 4), (10.5, 8), (18, 12), (14.5, 16)} **25.** {(−3, 2), (−3, −8), (6, 5), (7, 4), (11, 4)}; {(2, −3), (−8, −3), (5, 6), (4, 7), (4, 11)} **27.** {(−3, −1), (−3, −3), (−3, −5), (0, 3), (2, 3), (4, 3)}; {(−1, −3), (−3, −3), (−5, −3), (3, 0), (3, 2), (3, 4)} **29.** {(1, 8), (3, 16), (4, 20), (7, 28)}; D = {1, 3, 4, 7}; R = {8, 16, 20, 28}; {(8, 1), (16, 3), (20, 4), (28, 7)}

31. D = {100, 105, 110, 115, 120, 125, 130}; R = {40, 42, 44, 46, 48, 50, 52}

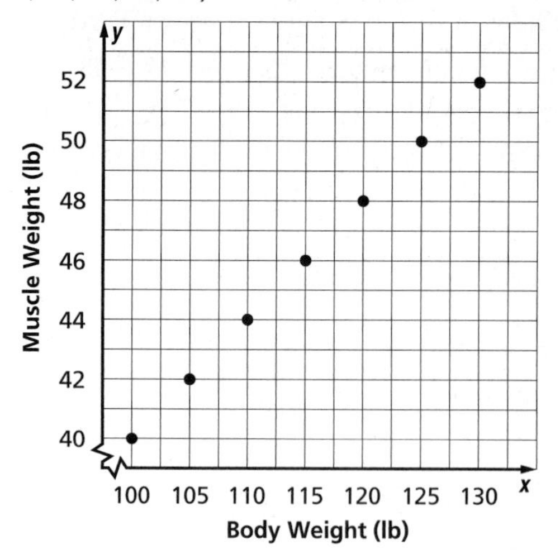

33.

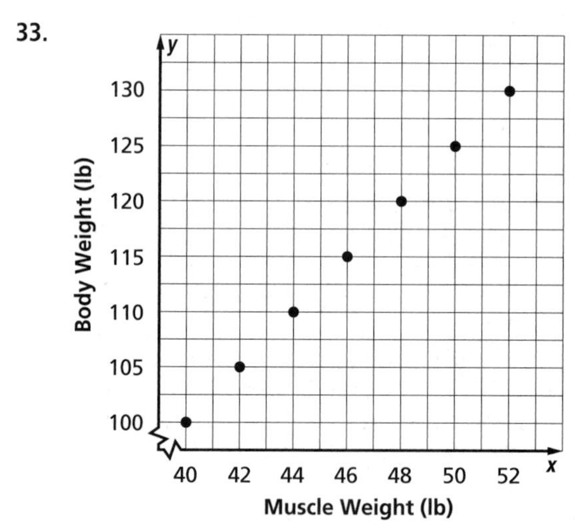

35. Sample answer: The number of movie tickets bought and the total cost of the tickets can be represented using a relation. The total cost depends on the number of tickets bought.
{(0, 0), (1, 9), (2, 18), (3, 27)}

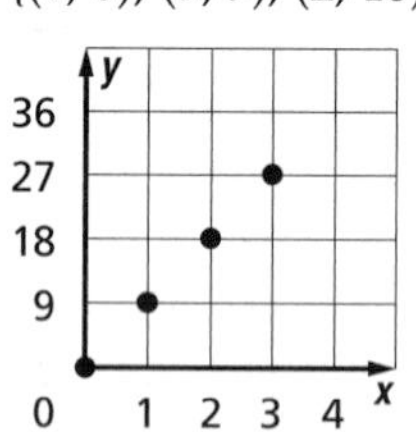

Number of Tickets	Total Cost
0	$0.00
1	$9.00
2	$18.00
3	$27.00

37. D **39.** 10 mL **41.** $h = \frac{6w - b}{3}$ **43.** 48 **45.** 7

Pages 152–154 ***Lesson 3-2***

1. yes **3.** no **5.** no **7.** 3 **9.** $4x + 15$ **11.** $t^2 - 3$
13. $f(x) = 0.25x$; $1.25, $3.00; It costs $1.25 to send 5 photos and $3.00 to send 12 photos. **15.** no
17. yes **19.** yes **21.** yes **23.** yes **25.** yes **27.** 16
29. 15 **31.** 16 **33.** $4c^2 - 4c$ **35.** $3k + 13$ **37.** $9r + 21$
39. $f(h) = 77 - 0.005h$; 76.5, 76, 72

41.

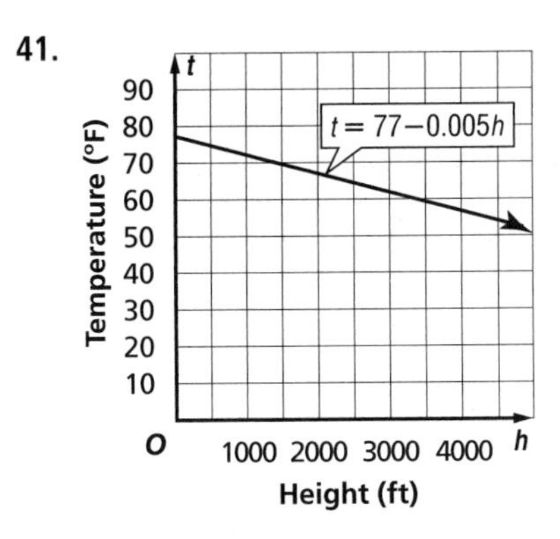

43.

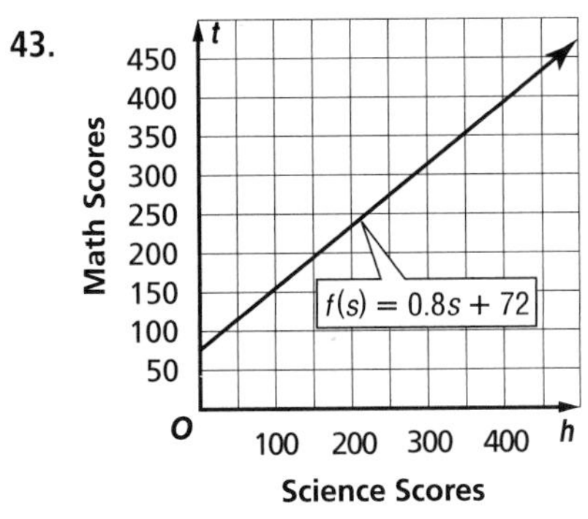

45. no **47.** yes **49.** No; one member of the domain is paired with two different members of the range. **51.** Sometimes; if each domain value is paired with a different range value, then the inverse is also a function. If two or more domain values are paired with the same range value, then the inverse is not a function. **53.** Sample answer: Functions can be used in meteorology to determine if there is a relationship between certain weather conditions. This can help to predict future weather patterns. As barometric pressure decreases, temperature increases. As barometric pressure increases, temperature decreases. The relation is not a function since there is more than one temperature for a given barometric pressure. However, there is still a pattern in the data and the two variables are related. **55.** F **57.** about 320 mph **59.** 21 **61.** −4

Pages 158–161 ***Lesson 3-3***

1. no **3.** yes; $3x - 2y = 25$ **5.** 12; −24; The x-intercept 12 means that after 12 seconds, the scuba diver is at a depth of 0 meters, or at the surface. The y-intercept −24 means that at time 0, the scuba diver is at a depth of −24 meters, or 24 meters below sea level.

7.

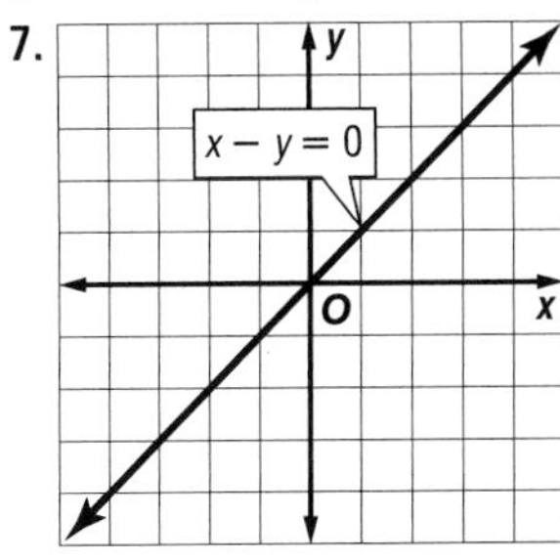

9.

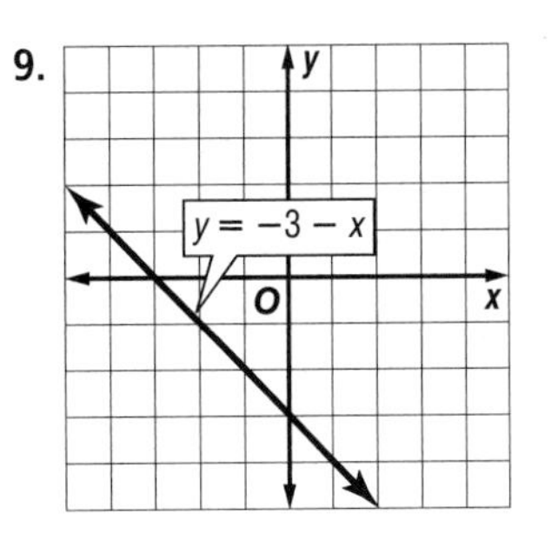

11.

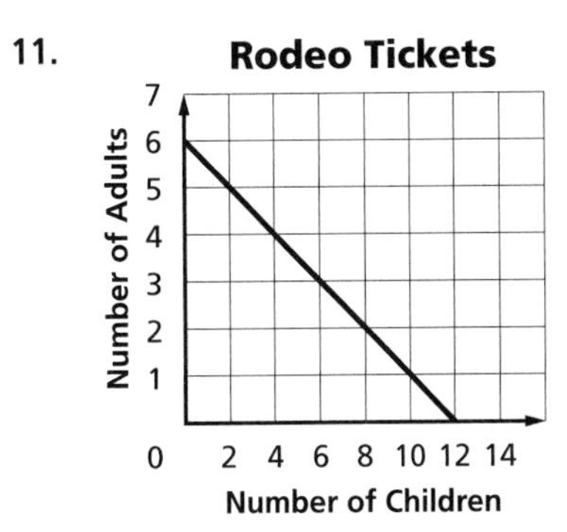

The x-intercept 12 means that 12 children could attend the rodeo if there are 0 adults attending. The y-intercept 6 means that 6 adults and 0 children could attend the rodeo. **13.** yes; $2x + y = 6$ **15.** yes; $y = -5$ **17.** no **19.** −2, 2, −2 **21.** 6, 20; The x-intercept represents the number of seconds that it takes the eagle to land. The y-intercept represents the initial height of the eagle. **23.** 8, 4; The x-intercept 8 means that it took Eva 8 minutes to get home. The y-intercept 4 means that Eva was initially 4 miles from home.

25.

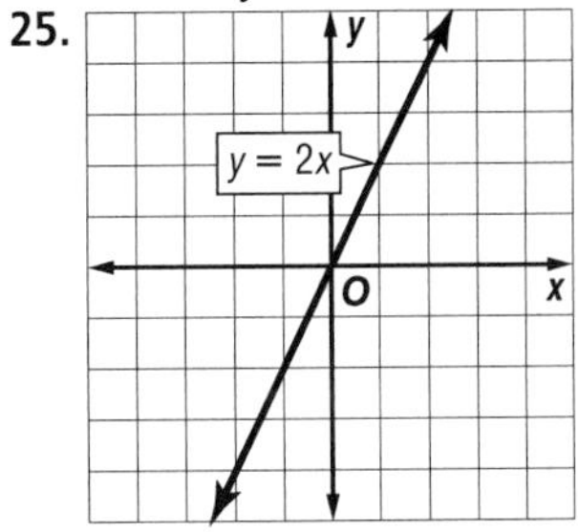

27.

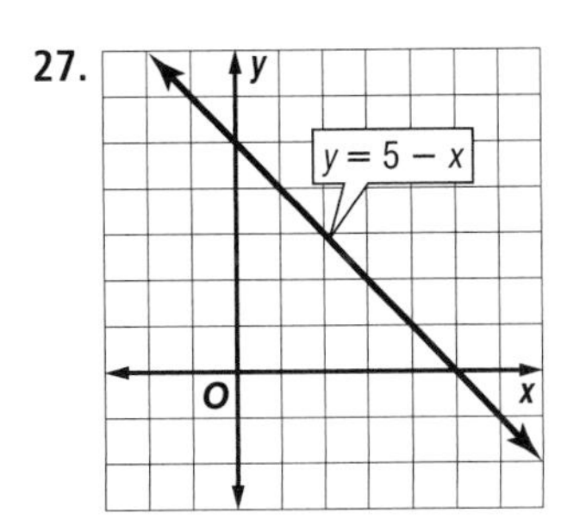

29.

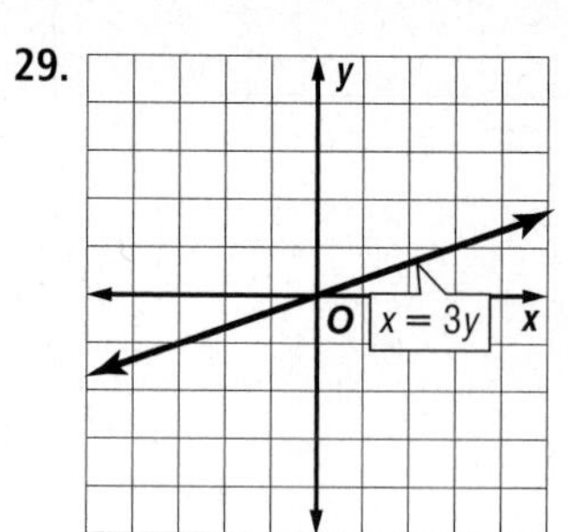

31.

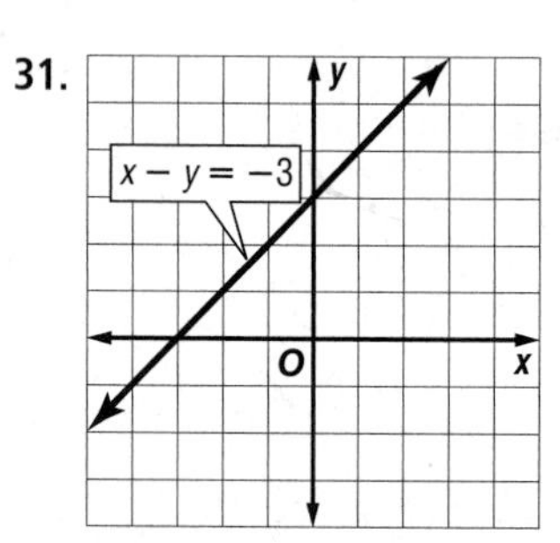

33.

t	d	t	d
0	0	10	2.10
2	0.42	12	2.52
4	0.84	14	2.94
6	1.26	16	3.36
8	1.68		

35. about 14 s **37.** 7.5, 15; No; the x-intercept 7.5 means that the length would be 7.5 inches if the width were 0. The y-intercept means that the width would be 7.5 inches if the length were 0. A rectangle cannot have only a length or only a width, so these values do not make sense in the context of the problem. **39.** no **41.** yes; $6m - 7n = -4$ **43.** no

45.

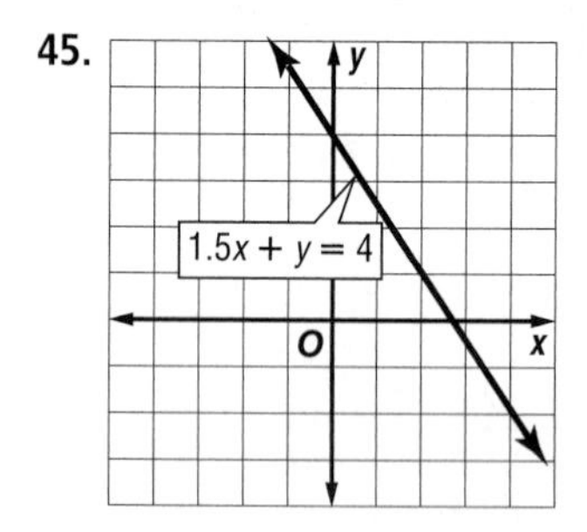

47.

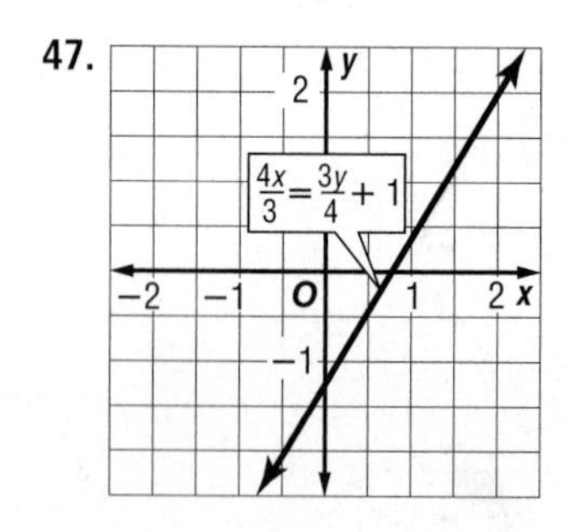

49.

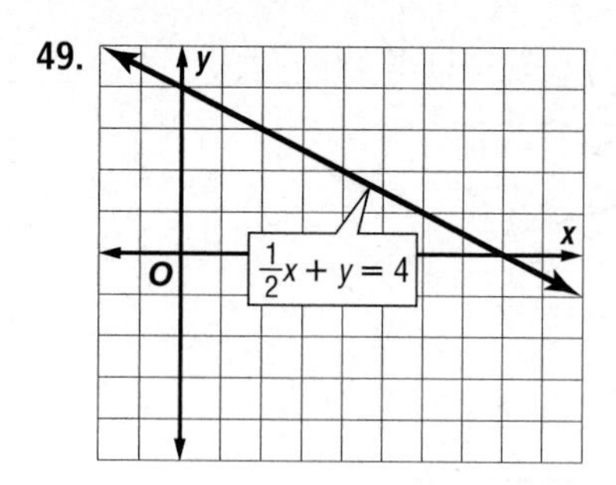

51. $\frac{7}{2}, -2$ **53.**

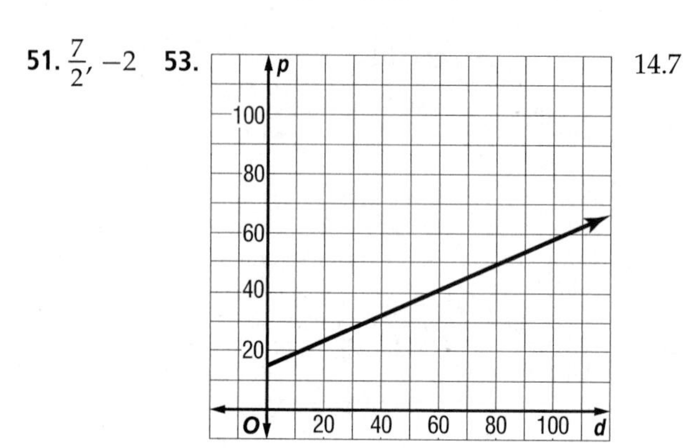

14.7

55. $2 \stackrel{?}{=} \frac{1}{2}(-4) + 4$

$2 \stackrel{?}{=} -2 + 4$

$2 = 2$

Since substituting the ordered pair in the equation yields a true equation, the point lies on the line.

57. Sample answer: $x = 5$ **59.** Substitute the values for x and y into the equation $2x - y = 8$. If the value of $2x - y$ is less than 8, then the point lies *above* the line. If the value of $2x - y$ is greater than 8, then the point lies *below* the line. If the value of $2x - y$ equals 8, then the point lies *on* the line. Sample answers: (1, 5) lies above the line, (5, 1) lies below the line, (6, 4) lies on the line. **61.** A **63.** D **65.** 4 **67.** $f(x) = 2.25x$; \$9.00, \$15.75; It costs \$9.00 to buy 4 energy bars and \$15.75 to buy 7 energy bars. **69.** −6

71. Sample answer:

Amount of Air in Balloon

Time

73. 3 **75.** −15

Pages 168–170 ***Lesson 3-4***

1. yes; −8 **3.** 35, 42, 49 **5.** 31 **7.** 45 **9.** the seventh week **11.** $a_n = 5.1n + 7$

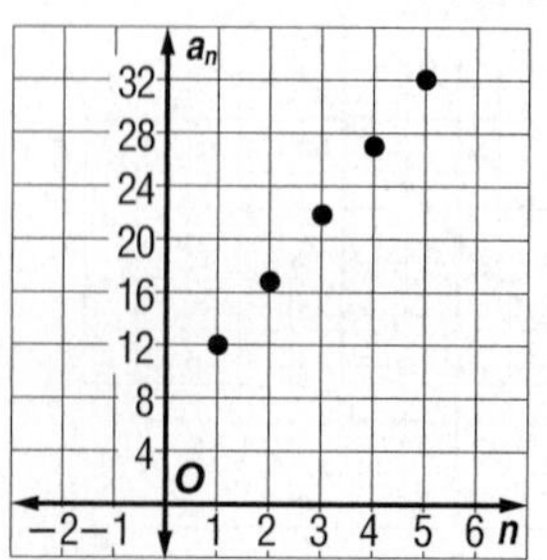

13. no **15.** yes; 0.5 **17.** 42, 48, 54 **19.** 5, 14, 23 **21.** 38 **23.** 1264 **25.** 25 **27.** 25 **29.** 28; yes, by 4 seats

31. $a_n = n + 7$

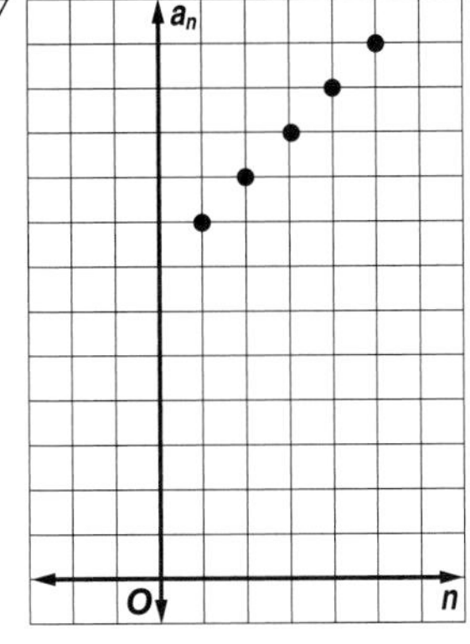

33. $a_n = 2n - 20$

35. $3\frac{7}{12}, 4\frac{1}{3}, 5\frac{1}{12}$ **37.** 9 **39.** −1 **41.** $a_n = 4n + 5$; 145 cm **43.** Sample answer: Yes; the rate of change is greater. **45.** $92,500 **47.** Yes; $4x + 5 - (2x + 5) = 2x$, $6x + 5 - (4x + 5) = 2x$, $8x + 5 - (6x + 5) = 2x$. The common difference is $2x$. **49.** Sample answer: The formula $a_t = 8.2t - 1.9$ represents the altitude a_t of the probe after t seconds. Replace t with 15 in the equation for a_t to find that the altitude of the probe after 15 seconds is 121.1 feet. **51.** J **53.** no **55.** yes; $2x - y = 3$ **57.** $2r + 3s = 13$

59., 61.

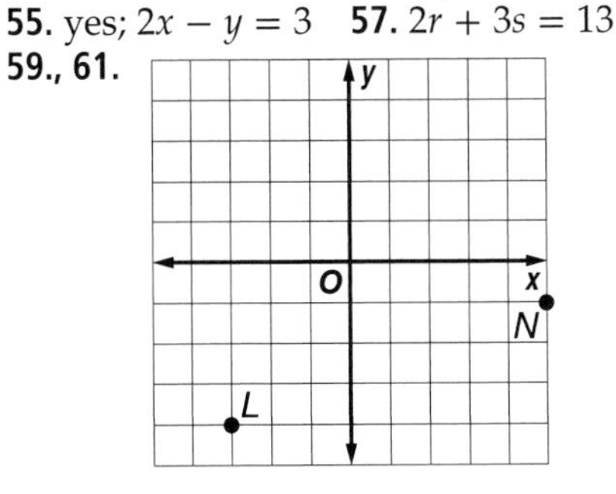

Pages 174–176 ***Lesson 3-5***

1. $f(x) = x$

3.

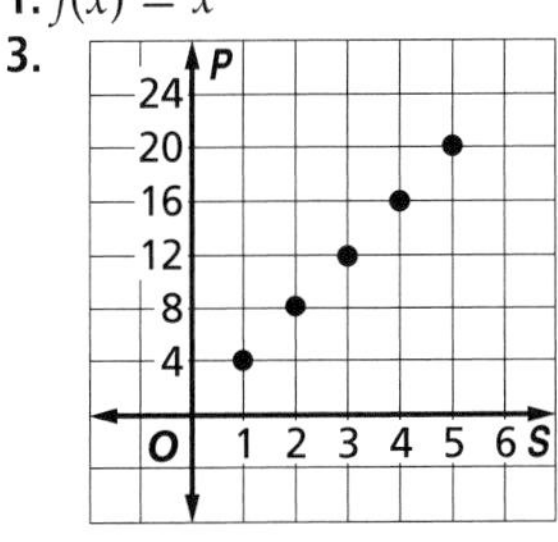

The length and the perimeter are proportional.

5.

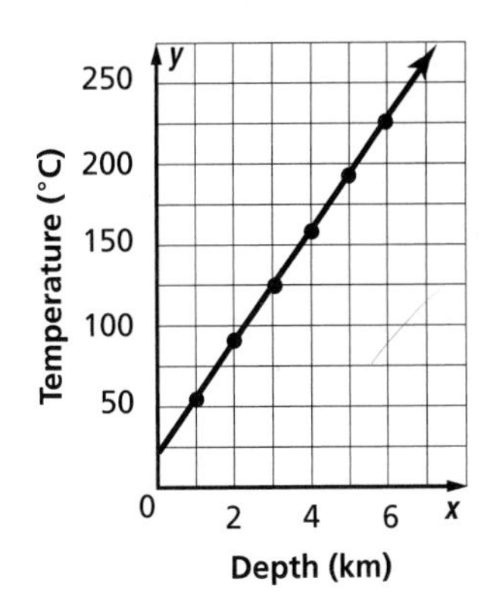

7. 370°C **9.** 10, 13, 11 **11.** 27, 35, 44 **13.** $4x + 1, 5x + 1, 6x + 1$ **15.** $f(x) = \frac{1}{2}x$ **17.** $f(x) = 6 - x$ **19.** 1, 1, 2, 3, 5, 8, 13, 21, 34, 55, 89, 144 **21.** $f(x) = -\frac{3}{2}x + 6$ **23.** $f(a) = -0.9a + 193$ **25.** Once you recognize a pattern, you can find a general rule that can be written as an algebraic expression. **27.** Sample answer: In scientific experiments you try to find a relationship or develop a formula from observing the results of your experiment. For every 11 cubic feet the volume of water increases, the volume of ice increases 12 cubic feet. **29.** J **31.** −1, 5, 11

33.

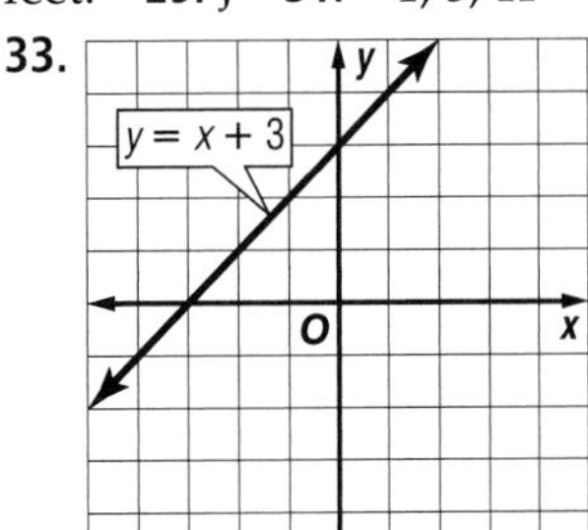

35.

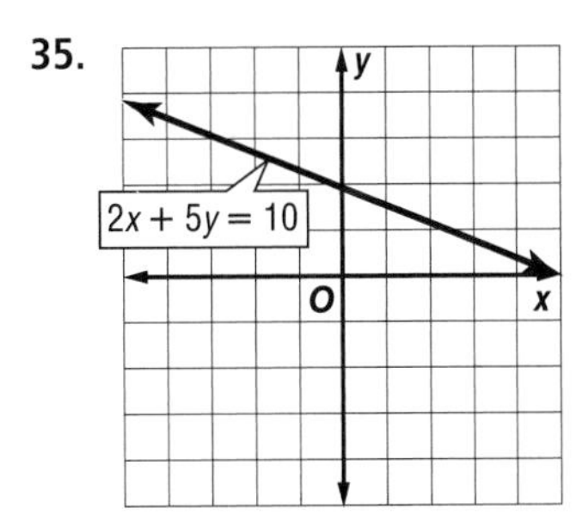

Pages 177–180 ***Chapter 3*** ***Study Guide and Review***

1. false; inverse **3.** true **5.** false; range **7.** true **9.** false; standard form

11.

x	y
−2	6
3	−2
3	0
4	6

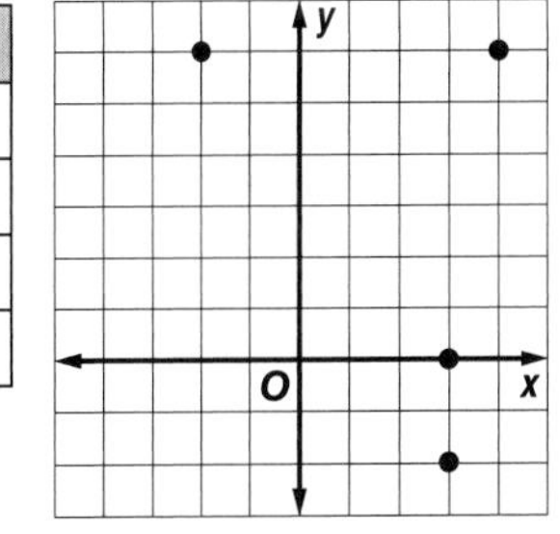

Selected Answers

11. (continued)

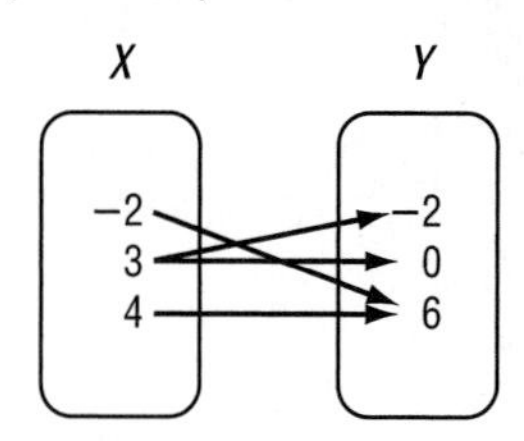

D = {−2, 3, 4}, R = {−2, 0, 6}
13. D = {45, 52, 55, 60, 80}; R = {72, 137, 118, 195, 300}
15. no **17.** 3 **19.** $a^2 - a + 1$ **21.** −8, 6, −8
23.

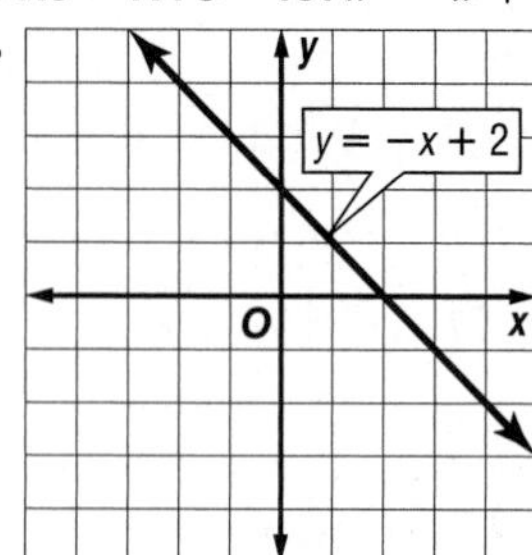

25.

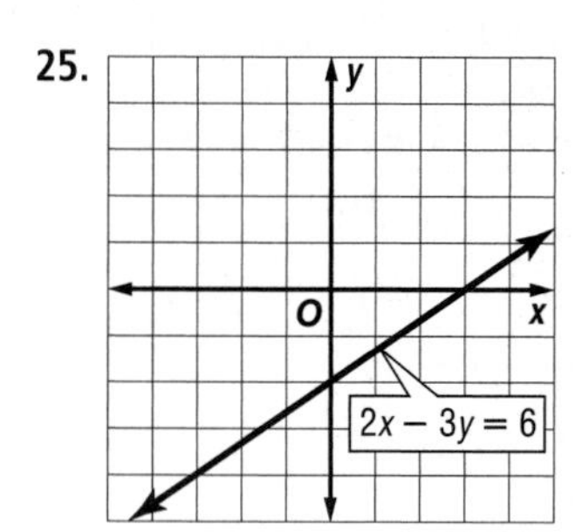

27. about 11 km

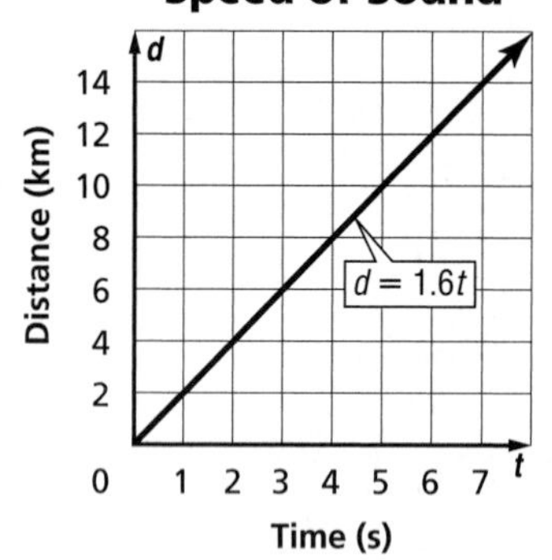

29. 0.6, 0.4, 0.2 **31.** 56 **33.** $a_n = 12 + 8.5n$; $182
35. $f(x) = -x - 1$ **37.** $f(x) = 1.25x$

Chapter 4 Analyzing Linear Equations

Page 185 ***Chapter 4*** ***Get Ready***

1. $\frac{1}{5}$ **3.** $-\frac{1}{4}$ **5.** $\frac{1}{3}$ **7.** 3 **9.** $-2\frac{3}{5}$ **11.** $-\frac{3}{4}$
13. $0.055 **15.** (−4, 1) **17.** (−1, −4) **19.** (−1, 3)

Pages 192–195 ***Lesson 4-1***

1. 4 **3.** 2.005; There was an average increase in ticket price of $2.005 per year. **5.** Sample answer: 1998–2000; Ticket prices show a sharp increase. **7.** $\frac{4}{5}$
9. $-\frac{3}{2}$ **11.** undefined **13.** 2 **15.** −6 **17.** $-\frac{1}{3}$
19. 12.75%; There was an average increase of 12.75% per year of teens who had cell phones. **21.** $-\frac{2}{5}$
23. 0 **25.** $\frac{3}{4}$ **27.** $-\frac{2}{3}$ **29.** $\frac{3}{8}$ **31.** undefined **33.** $\frac{3}{4}$
35. 6 **37.** undefined **39.** $\frac{3}{200}$ **41.** Sample answer: $\frac{1}{3}$
43. undefined **45.** $\frac{1}{3}$ **47.** 7 **49.** 12–14; steepest part of the graph **51.** '90–'95; '80–'85 **53.** a decline in enrollment
55.

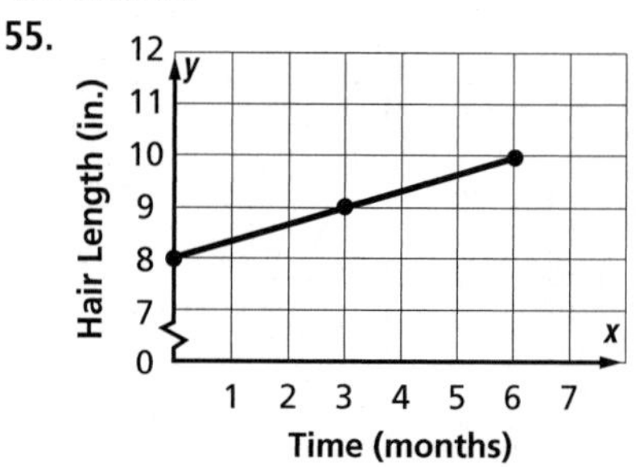

57. 12 ft 10 in. **59.** (−4, −5) is in Quadrant III and (4, 5) is in Quadrant I. The segment connecting them goes from lower left to upper right, which is a positive slope. **61.** No, they do not. Slope of $\overline{QR}$ is $\frac{4}{3}$ and slope of $\overline{RS}$ is $\frac{1}{3}$. If they lie on the same line, the slopes should be the same. **63.** Sample answer: Analysis of the slope of a roof might help to determine the materials of which it should be made and its functionality. To find the slope of the roof, find a vertical line that passes through the peak of the roof and a horizontal line that passes through the eave. Find the distances from the intersection of those two lines to the peak and to the eave. Use those measures as the rise and run to calculate the slope. A roof that is steeper than one with a rise of 6 and a run of 12 would be one with a rise greater than 6 and the same run. A roof with a steeper slope appears taller than one with a less steep slope. **65.** H **67.** $f(x) = 11 - x$
69. 3, 6, 9 **71.** −20, −11, −2 **73.** 9 **75.** $\frac{1}{8}$

Pages 200–202 ***Lesson 4-2***

1. $-\frac{1}{3}$; $-\frac{1}{3}$
3.

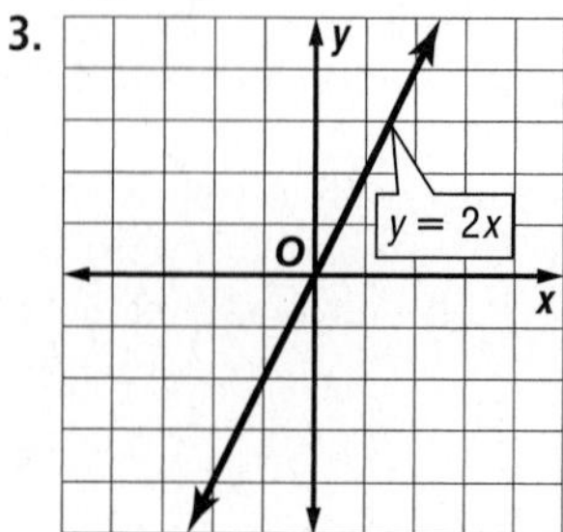

5. 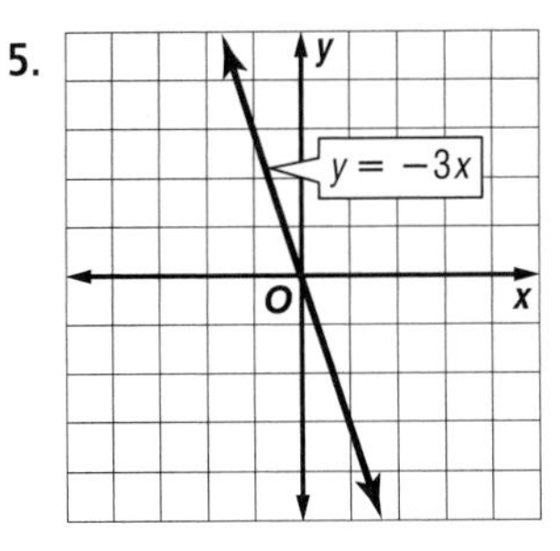

7. $y = \frac{9}{2}x$; 10 9. $y = 6x$ 11. $180 13. 4; 4

15. −1; −1 17. $-\frac{1}{4}$; $-\frac{1}{4}$

19. 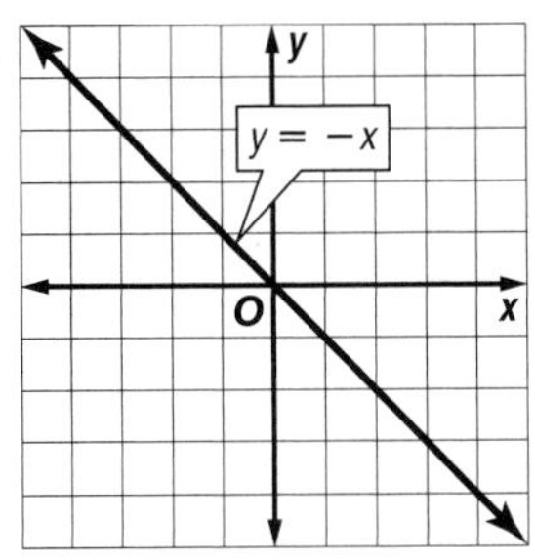

21. 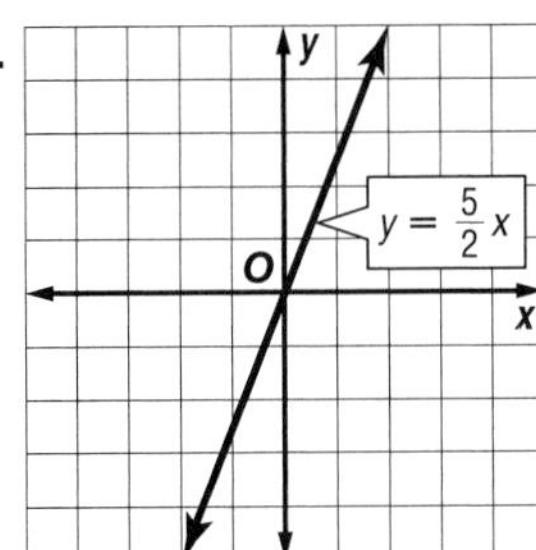

23. 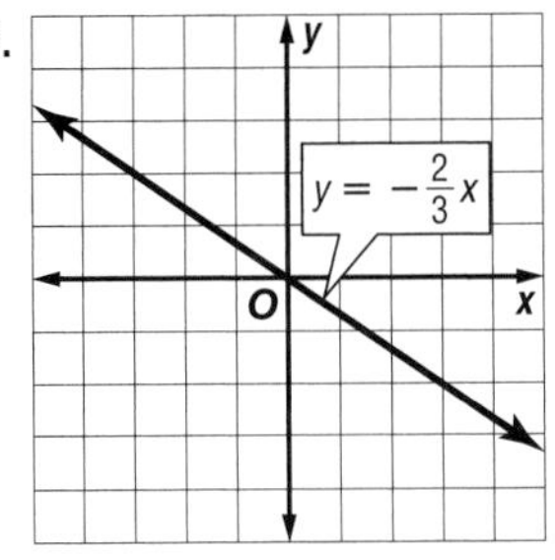

25. 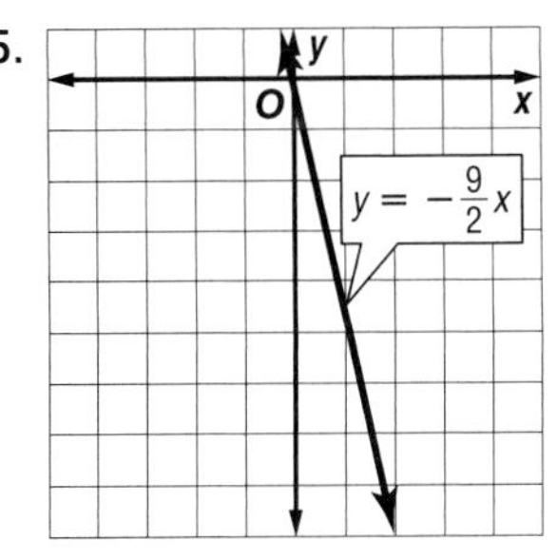

27. $y = -4x$; −5 29. $y = \frac{4}{5}x$; 26.25 31. 189 yd

33. 5 yr 4 mo 35. $y = -\frac{2}{3}x$; −4.4 37. $y = 9x$; $\frac{4}{3}$

39. 2 41. 3

43. $P = 4s$ 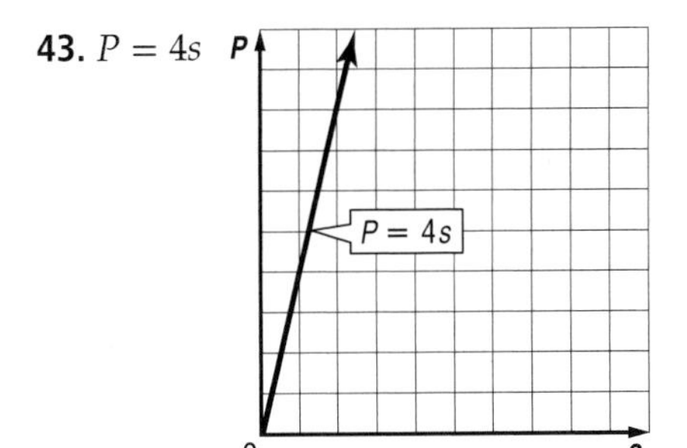

45. $C = 14.49p$ 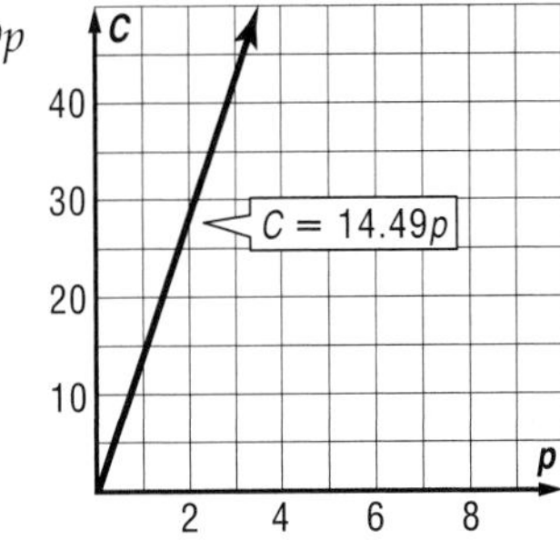

47. Sample answer: They all pass through (0, 0), but these have negative slopes. 49. Sample answer: Find the absolute value of k in each equation. The one with the greatest value of $|k|$ has the steeper slope.
51. Sample answer: $y = 0.50x$ represents the cost of x apples.

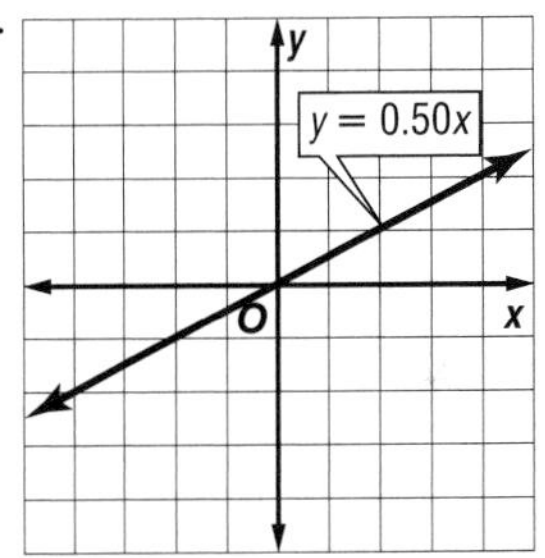

The rate of change, 0.50, is the cost per apple.
53. Sample answer: The slope of the equation that relates number of ringtones and total cost is the cost of each ringtone; $y = 1.5x$; The graph of this equation is less steep; the slope is less than the slope of the graph on page 196. 55. J 57. $f(x) = 4x + 1$ 59. $209.93
61. $y = 2x + 5$

Pages 207–209 ***Lesson 4-3***

1. $y = -3x + 1$ 3. $y = 2x - 1$

5. 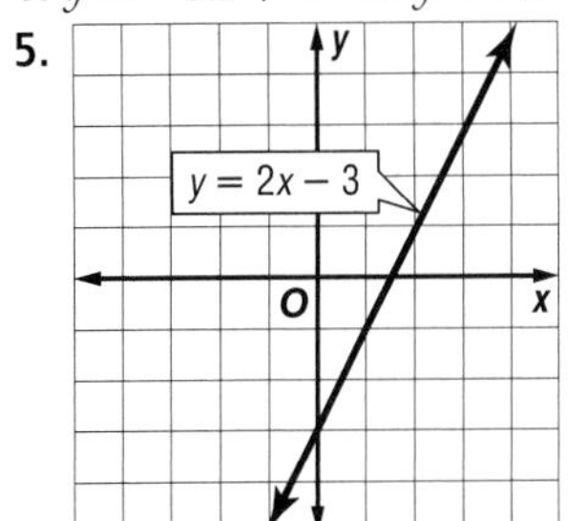

7.

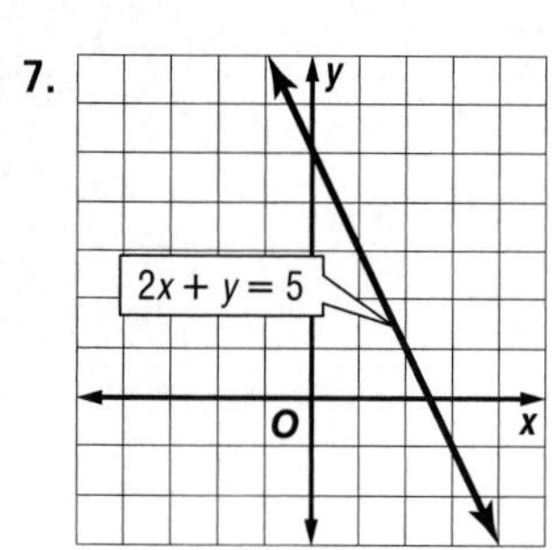

9. $T = 50 + 5w$

11. \$85 **13.** $y = 3x - 5$ **15.** $y = -\frac{3}{5}x + 12$ **17.** $y = -x$ **19.** $y = \frac{3}{2}x - 4$ **21.** $y = \frac{3}{2}x$ **23.** $y = 2$

25.

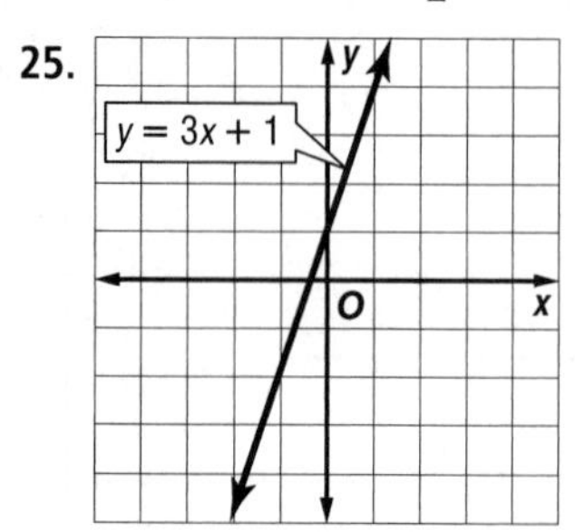

27.

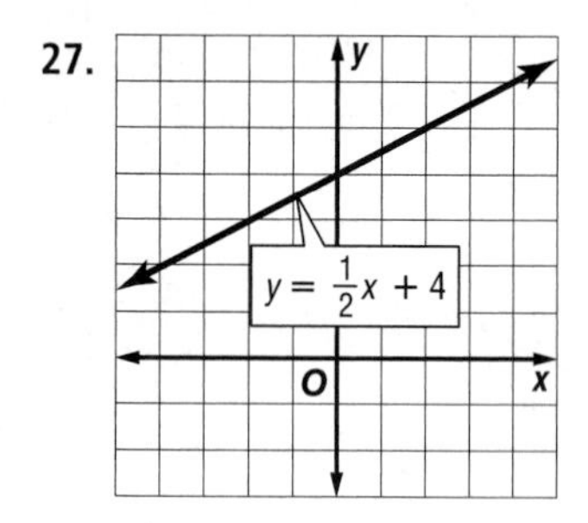

29.

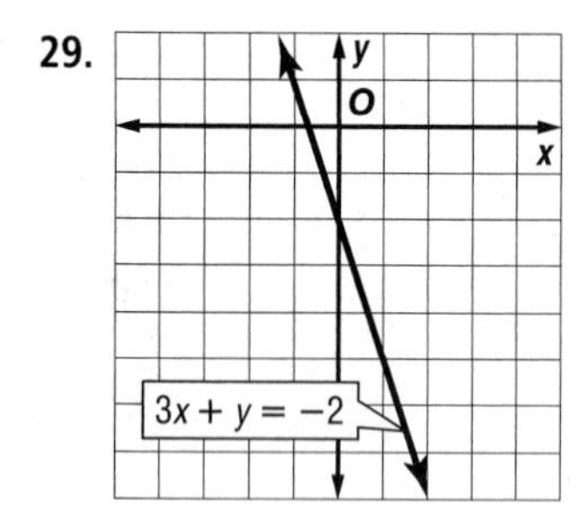

31.

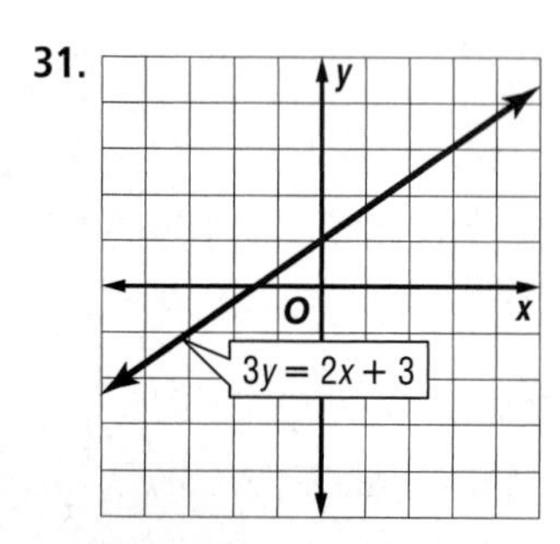

33. $y = 8t + 5$ **35.** $S = 23.4 + 1.2t$ **37.** $T = 258.15c + 2125$ **39.** $y = -x$ **41.** $y = 7$ **43.** $y = -5$ **45.** They all have a y-intercept of 3. **47.** Sample answer: The y-intercept is the flat fee in an equation that represents a price. If a mechanic charges \$25 plus \$40 per hour to work on your car, the graph representing this situation would have a y-intercept of \$25. **49.** G **51.** $y = \frac{3}{4}x$, $10\frac{2}{3}$ **53.** $-\frac{1}{4}$ **55.** 1932.2 mi **57.** $\frac{4}{5}$

Pages 216–218 ***Lesson 4-4***

1. $y = 2x - 10$ **3.** $y = -x + 2$ **5.** $y = -\frac{2}{3}x + 4$ **7.** A **9.** \$95 **11.** $y = -x + 3$ **13.** $y = -5x + 29$ **15.** $y = \frac{1}{2}x + \frac{1}{2}$ **17.** $y = -\frac{5}{3}x - 10$ **19.** $y = 2x - 8$ **21.** $y = 4x - 10$ **23.** $y = 5$ **25.** $y = \frac{1}{5}x + 6$ **27.** about 312 thousand or 312,000 **29.** 24.1 thousand or 24,100 **31.** $y = -\frac{1}{4}x + \frac{11}{16}$ **33.** $y = \frac{5}{3}x + 5$ **35.** (7, 0), (0, −2) **37.** Sample answer: Let y represent the quarts of water in a pitcher and let x represent the time in seconds.

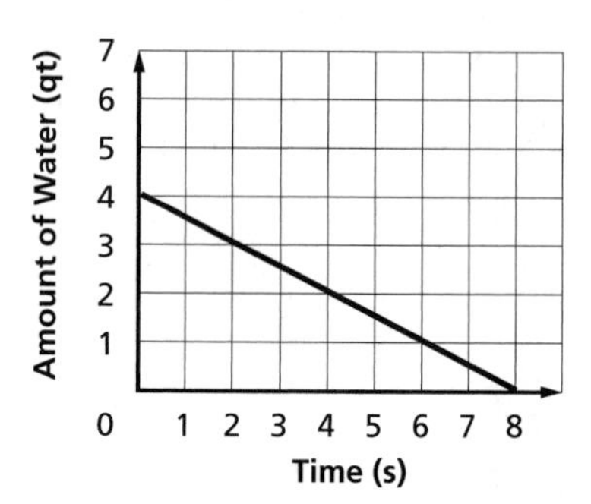

As the time increases by 1 second, the amount of water in the pitcher decreases by $\frac{1}{2}$ quart. An equation is $y = -\frac{1}{2}x + 4$. The slope $-\frac{1}{2}$ represents the rate at which the water is emptying from the pitcher, $\frac{1}{2}$ quart per second. The y-intercept 4 represents the initial amount of water in the pitcher, 4 quarts. **39.** Sample answer: Linear extrapolation is when you use a linear equation to predict values that are outside of the given points on the graph. You can use the slope-intercept form of the equation to find the y-value for any requested x-value. **41.** J

43.

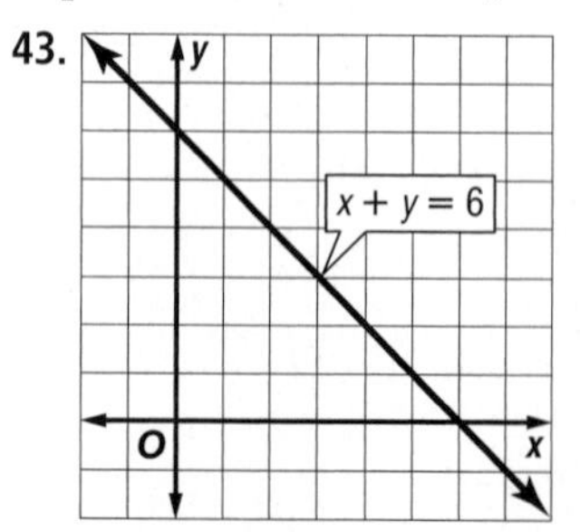

45. $V = 2.5b$ **47.** 45, 3; The x-intercept 45 means that it took Tara 45 minutes to walk home. The y-intercept 3 means that she was initially 3 miles from her home. **49.** D = {−2, 0, 5}; R = {−3, 1, 7} **51.** −7 **53.** −5

Pages 223–225 Lesson 4-5

1. $y - 3 = -2(x - 1)$ **3.** $y + 2 = 0$ **5.** $3x + 4y = 9$ **7.** $y = 2x - 10$ **9.** $y = x + 13$ **11.** $2x - y = -5$ **13.** $y + 3 = x + 4$ **15.** $y - 6 = 0$ **17.** $y + 3 = -\frac{5}{8}(x - 1)$ **19.** $y - 7 = 0$ **21.** $2x + y = -7$ **23.** $x - 2y = 12$ **25.** $2x + 5y = 26$ **27.** $17x - 10y = 3$ **29.** $y = 6x + 11$ **31.** $y = \frac{1}{2}x - 1$ **33.** $y = -\frac{1}{4}x - \frac{7}{2}$ **35.** $y = 9x + 5$ **37.** $y = 5x + 150$ **39.** $y - 35{,}170 = 410(x - 2001)$ **41.** 37,630 **43.** $5x - 2y = -11$ **45.** $y = x - 1$ **47.** $y = -3x - \frac{7}{4}$ **49.** $y + 6 = \frac{3}{2}(x - 1)$; $y = \frac{3}{2}x - \frac{15}{2}$; $3x - 2y = 15$ **51.** Sample answer: The total cost of going to the zoo is \$4 for parking plus \$9 per person; $y = 9x + 4$, $y - 5 = 9(x + 1)$. **53.** $f(x) = -60x + 720$ **55.** Write the definition of the slope using (x, y) as one point and (x_1, y_1) as the other. Then solve the equation so that the ys are on one side and the slope and xs are on the other. **57.** H **59.** $y = -1$ **61.** $y = 3x + 10$ **63.** 5 **65.** 90 **67.** $y = -\frac{1}{2}x + \frac{3}{2}$ **69.** $y = -5x + 11$ **71.** $y = 9$

Pages 230–233 Lesson 4-6

1. Positive; the longer you study, the better your test score.

3.

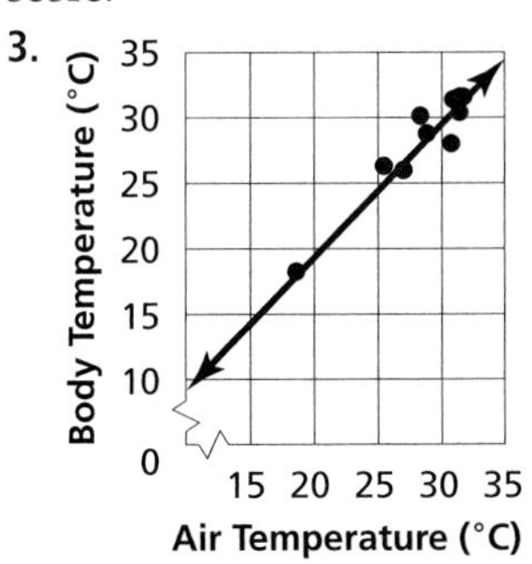

5. Sample answer: Using (31.2, 31.0) and (26.2, 25.6) $y = 1.08x - 2.696$. **7.** No; at this temperature, there would be no insects. **9.** no correlation **11.** Positive; the higher the sugar content, the more Calories. **13.** about 210 **15.** $y = -1200x + 12{,}000$ **17.** No; the equation would give a price of −\$48,000. In reality, this car would be an antique and would more than likely be valuable.

19.

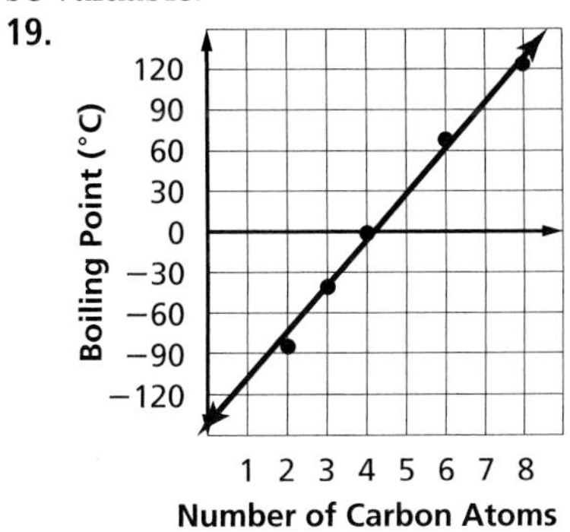

21. Sample answer: 32°C

23.

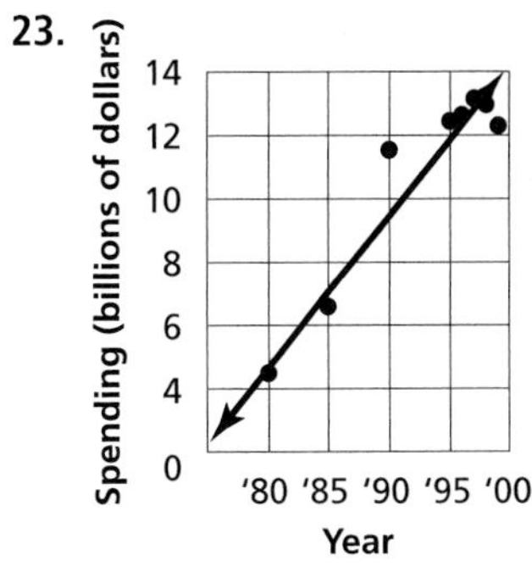

25. Sample answer: Using (0, 4.5) and (16, 12.7), $y = 0.5125x + 4.5$. **27.** Sample answer: The amount spent will probably not increase at a constant rate, so the linear equation that is useful in making predictions in the near future would not be useful for making predictions in the distant future. **33.** Linear extrapolation predicts values outside the range of the data set. Linear interpolation predicts values inside the range of the data. **35.** Sample answer: You can visualize a line to determine whether the data has a positive or negative correlation. The graph below shows the ages and heights of people. To predict a person's age given his or her height, write a linear equation for the line of fit. Then substitute the person's height and solve for the corresponding age. You can use the pattern in the scatter plot to make decisions.

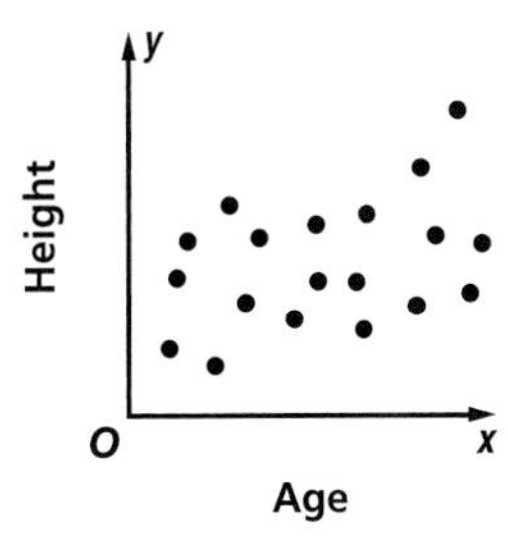

37. G **39.** $y - 3 = -2(x + 2)$ **41.** $C = 0.22m + 0.99$; \$3.63 **43.** $\frac{1}{10}$ **45.** $\frac{3}{2}$ **47.** $-\frac{4}{3}$

Pages 239–241 Lesson 4-7

1. $y = -2x - 1$ **3.** $y = 2x - 5$ **5.** Slope of $\overline{AC} = \frac{1 - 7}{-2 - 5}$ or $\frac{6}{7}$; slope of $\overline{BD} = \frac{-3 - 4}{3 - (-3)}$ or $-\frac{7}{6}$; the paths are perpendicular. **7.** $y = -\frac{5}{3}x + 8$ **9.** $y = -\frac{1}{6}x + \frac{1}{6}$ **11.** $y = 2x - 5$ **13.** $y = \frac{2}{3}x + 1$ **15.** $y = -\frac{1}{2}x + \frac{3}{2}$ **17.** $y = -\frac{1}{3}x + 6$ **19.** No; the slopes are $-\frac{1}{2}$ and $\frac{3}{2}$. **21.** $y = -\frac{1}{4}x + \frac{5}{4}$ **23.** $y = -2x - 1$ **25.** $y = -\frac{3}{2}x + 13$ **27.** $y = -\frac{1}{2}x + 2$ **29.** parallel

31. They are perpendicular because the slopes are 3 and $-\frac{1}{3}$. **33.** 0 **35.** Sample answer: If two equations have the same slope, then the lines are parallel. If the product of their slopes equals −1, then the lines are perpendicular. The graph of $y = \frac{3}{2}x$ is parallel to the

graph of $y = \frac{3}{2}x + 1$ because their slopes both equal $\frac{3}{2}$. The graph of $y = -\frac{2}{3}x$ is perpendicular to the graph of $y = \frac{3}{2}x + 1$ because the slopes are negative reciprocals of each other. **37.** G **39.** $y - 5 = -2(x - 3)$
41. $y + 3 = -\frac{1}{2}(x + 1)$ **43.** $c = \frac{7b + t}{6}$

Pages 242–246 ***Chapter 4*** ***Study Guide and Review***

1. true **3.** false; standard form **5.** false; parallel
7. false; 0 **9.** false; direct variation **11.** 3 **13.** $-\frac{1}{2}$
15. 0.08; an average decrease in cost of \$0.08 per year
17.

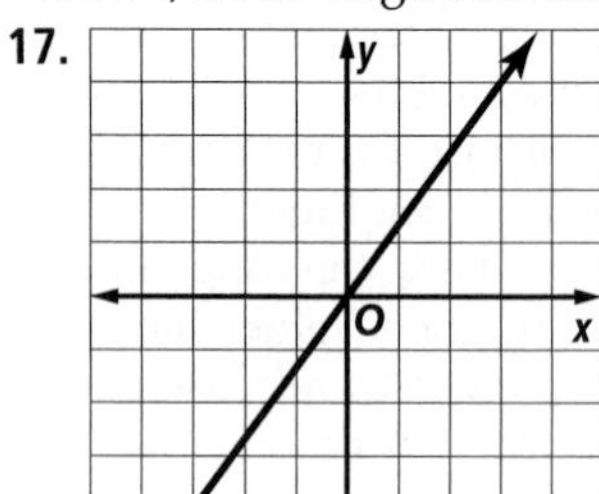

19. $y = 7.5x$; $y = 60$ **21.** $y = -x$; $y = -7$
23. $y = 3x + 2$ **25.** $y = 0x + 4$ or $y = 4$
27.

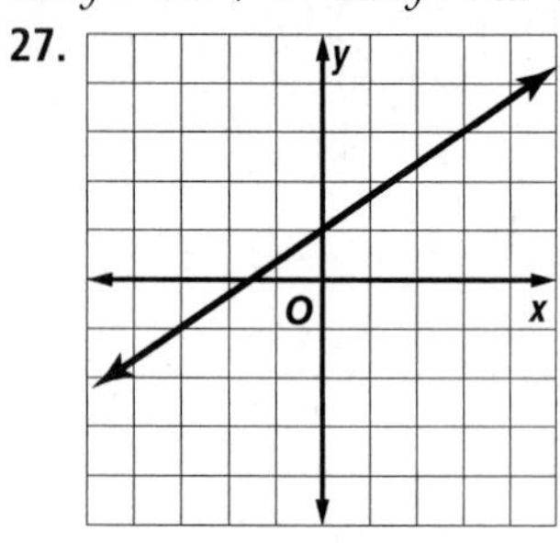

29.

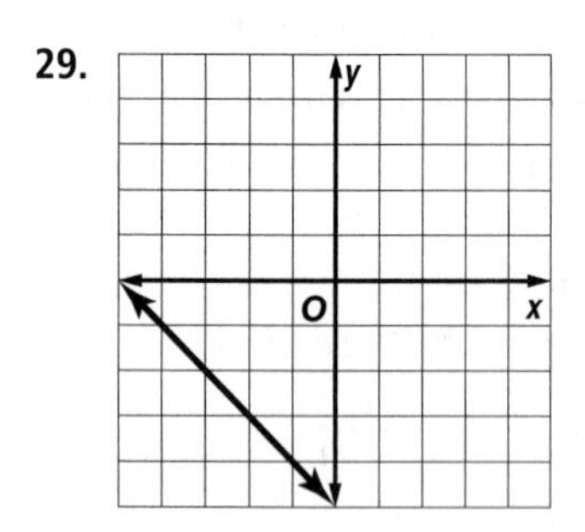

31. $y = 0.35 + 0.10x$ **33.** $y = -\frac{3}{5}x - \frac{3}{5}$
35. $y = -2x + 6$ **37.** $y = -5x + 25$
39. $y - 5 = -\frac{2}{3}(x + 3)$ **41.** $y + 3 = \frac{1}{2}(x - 5)$
43. $3x - 2y = 20$ **45.** $y = 2x + 3$
47. $y - 165 = 25(x - 8)$

49.

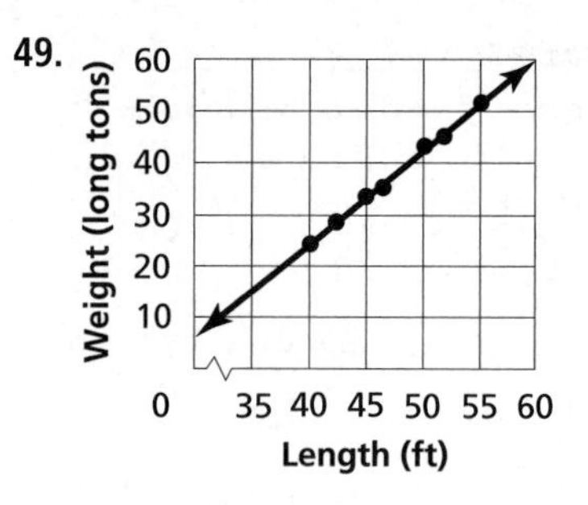

51. 38.9 long tons **53.** $y = 3x - 6$ **55.** $y = 5x - 15$
57. Yes, $\overline{AC} \perp \overline{AB}$.

Chapter 5 Solving Systems of Linear Equations

Page 251 ***Chapter 5*** ***Get Ready***

1.

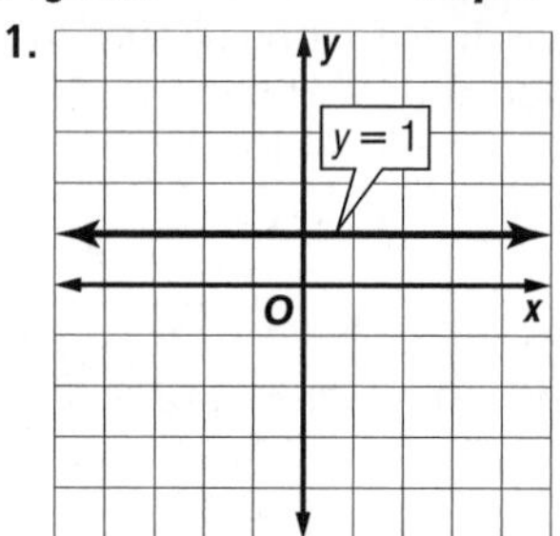

3.

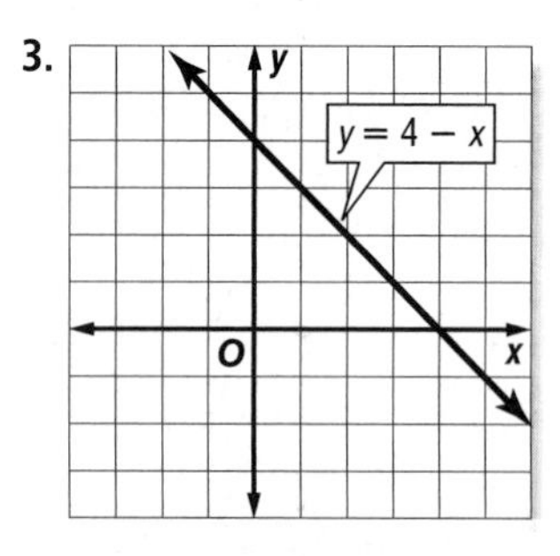

5.

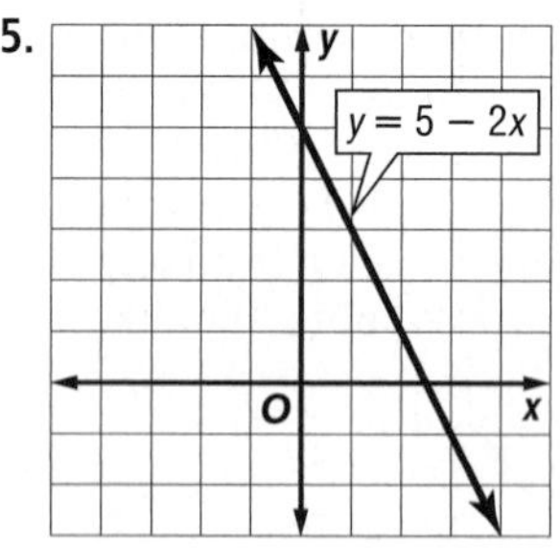

7.

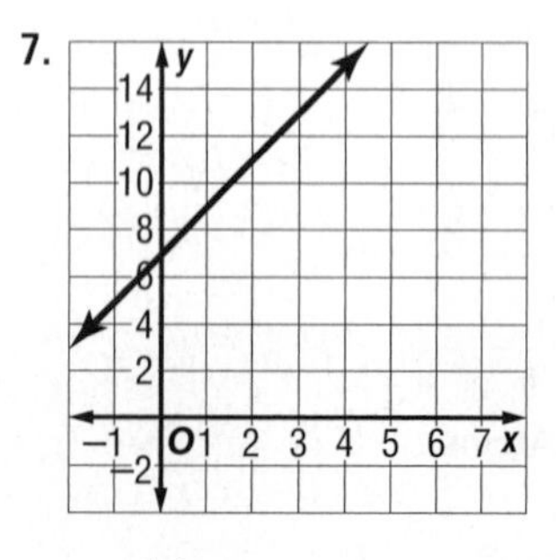

9. $a = \frac{16 - y}{8}$ **11.** $q = \frac{7m + n}{2m}$ **13.** $-6y$

Pages 255–258 **Lesson 5-1**

1. one **3.** infinitely many

5. one; (0, −4)

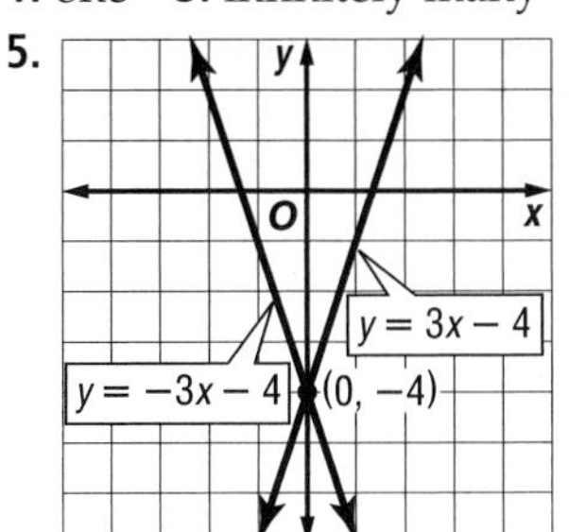

7. no solution

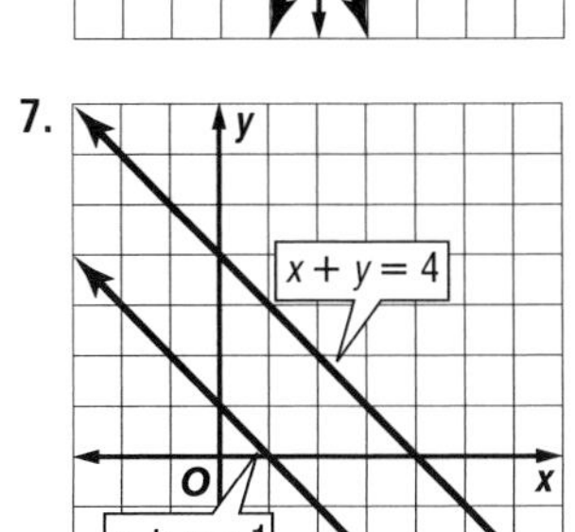
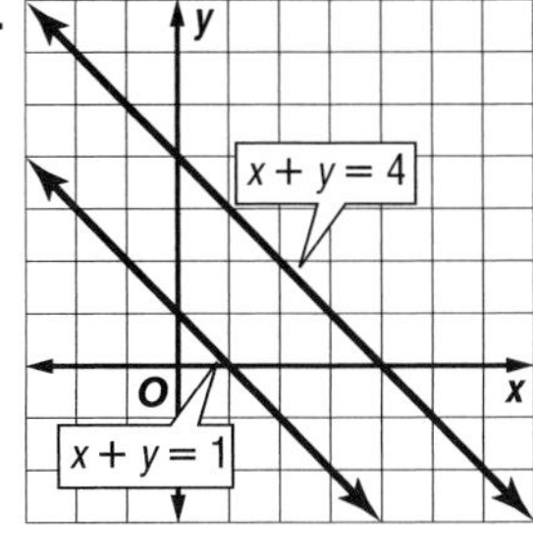

9. 13 m by 7 m **11.** one **13.** no solution **15.** no solution

17. one; (2, −2)

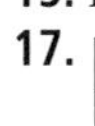
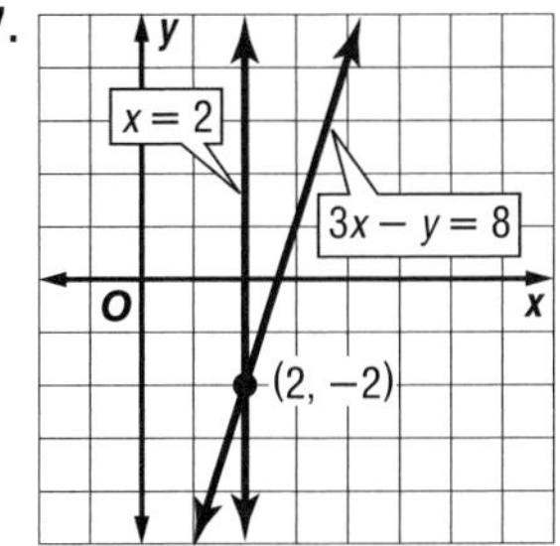

19. one; (2, −2)

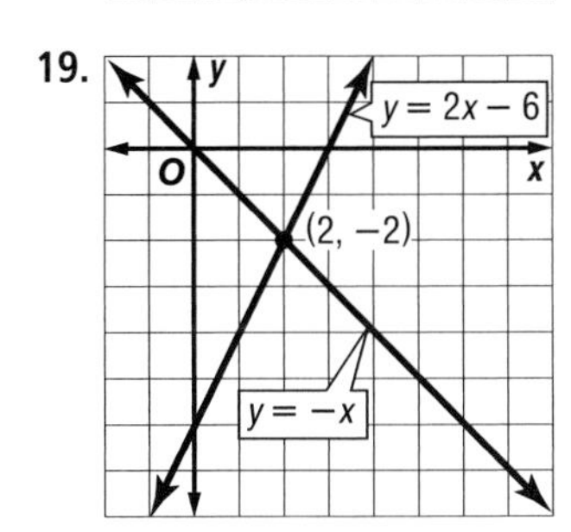

21. one; (2, 0)

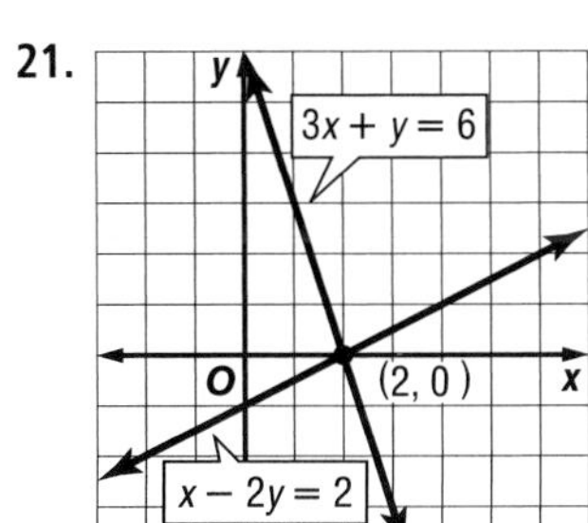

23. no solution

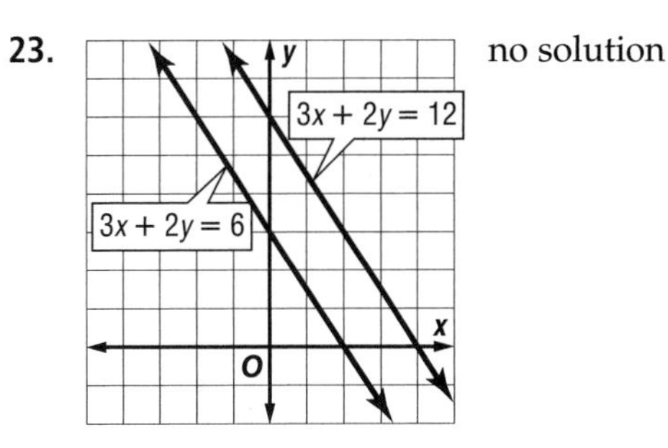

25. one; (−6, 8)

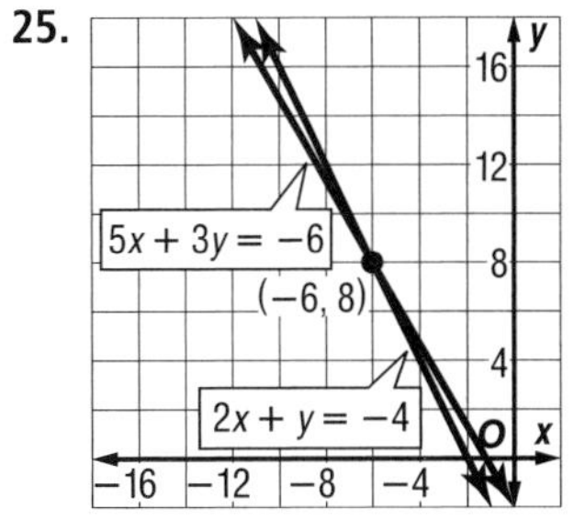

27. infinitely many

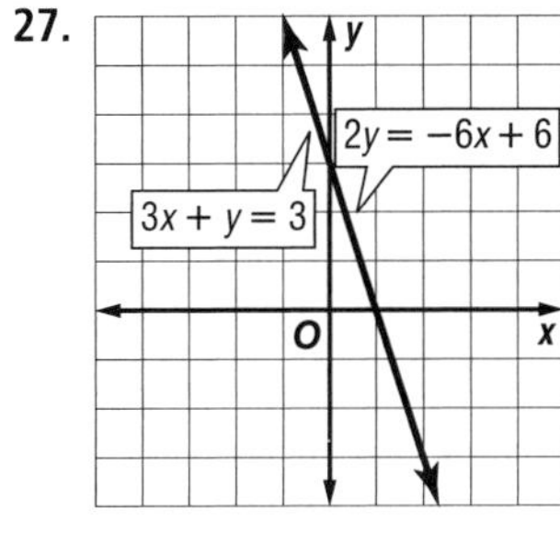

29. $40 **31.** 70 m; yes, 70 m is a reasonable height for balloons to fly.

33. infinitely many

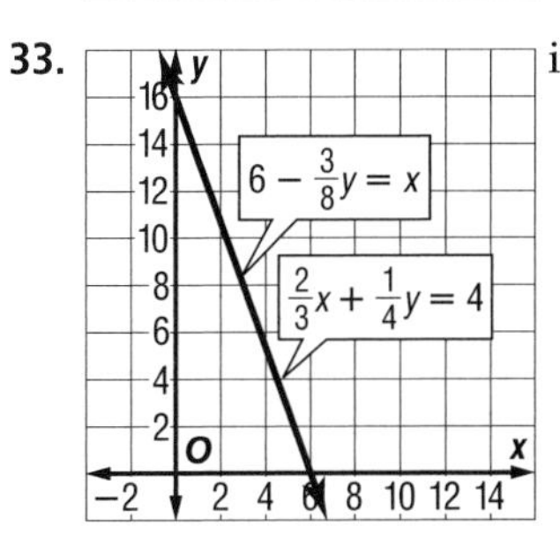

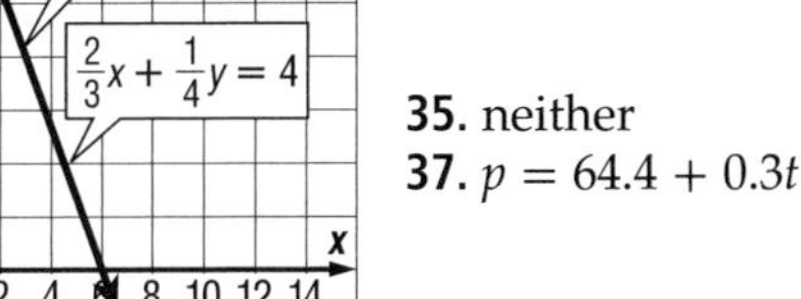

35. neither
37. $p = 64.4 + 0.3t$

39.

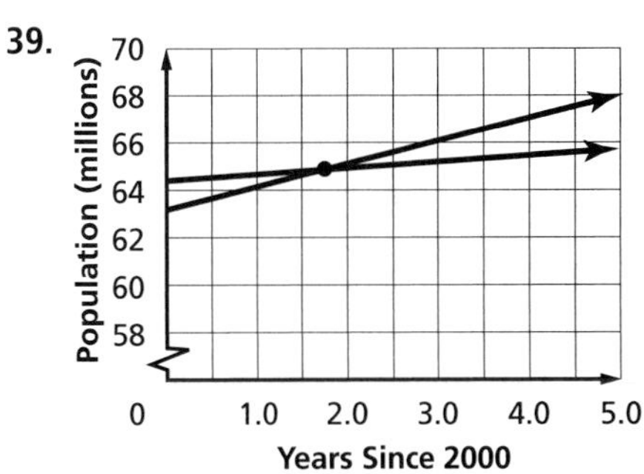

The solution (1.75, 65) means that 1.75 years after 2000, or in 2002, the population of the West and the Midwest were approximately equal, 65 million. **41.** Sample answer: $y = 5x + 3$, $y = -x - 3$, $2y = 10x - 6$
43. Graphs can show when the units sold of one item is greater than the units sold of the other item and when the units sold of the items are equal. The units sold of CD singles equaled the units sold of music videos in about 1.5 years, or between 2001 and 2002.

45. H **47.** $y = -2x + 3$ **49.** $y = 2x$ **51.** $q = \frac{7m - n}{10}$

Pages 263–265 Lesson 5-2

1. (5, –1) **3.** no solution **5.** (–2, 3) **7.** about 135.5 mi **9.** (16, 4) **11.** (2, 7) **13.** infinitely many **15.** (13, 30) **17.** (–4, 3) **19.** (2, 5) **21.** 320 gal of 25% acid, 180 gal of 50% acid **23.** (5, 2) **25.** $\left(2\frac{2}{3}, 4\frac{1}{3}\right)$ **27.** The second offer is better if she sells less than \$80,000. The first offer is better if she sells more than \$80,000. **29.** 23 yr **31.** Josh; $b = 3$ means that 3 bananas were bought. Solving the first equation for a gives $a = 4$. This means that 4 apples were bought.
33. Sample answer: $x + y = 3$
$3x + 2y = 6$
Discount movie tickets for one adult and one child cost \$3. The cost for 3 adults and two children is \$6. The solution (2, 1) means that an adult ticket costs \$2 and a child's ticket costs \$1. **35.** A **37.** C
39.

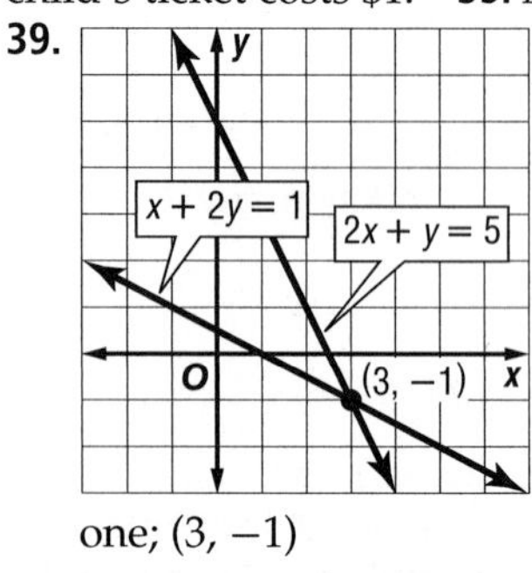

one; (3, –1)

41. Sample answer:

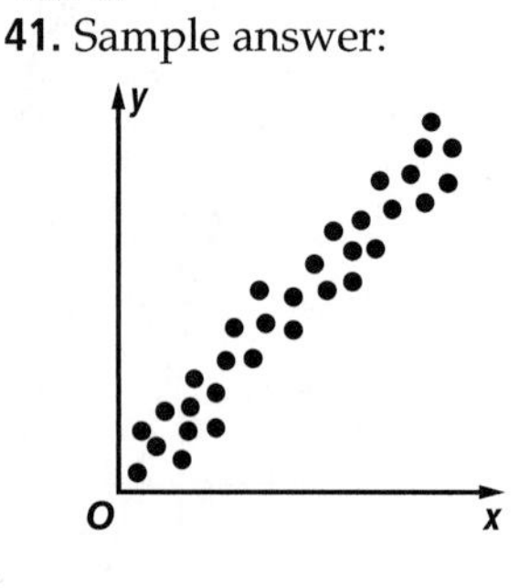

43. $-3a$ **45.** $-15g$ **47.** $3x + 3y$

Pages 268–270 Lesson 5-3

1. (–1, 3) **3.** (0, –5) **5.** 6, 18 **7.** (–1, –2) **9.** (5, 1) **11.** (–5, 20) **13.** (–5, 4) **15.** (2, –2) **17.** (4.5, 2) **19.** 36, 12 **21.** 6, 0 **23.** adult: \$16; student: \$9 **25.** $y = 0.0022x + 1.28$ **27.** Sample answer: The solution (23, 1.33) means that 23 years after 2002, or in 2025, the populations of China and India are predicted to be the same, 1.33 billion. **29.** (–1.4, 5630); About 1.4 years before 2004, or in 2002, the number of online catalogs and the number of print catalogs were both 5630. **31.** $A = 6$, $B = 3$; In each equation, replace x with 2 and y with 1 to get $2A + B = 15$ and $2A - B = 9$. Next, eliminate the B variables by adding the equations to get $4A = 24$. Divide each side by 4 to get $A = 6$. Now substitute 6 for A in either equation to get $B = 3$. **33.** B **35.** (2, 10) **37.** (–9, –7)
39.

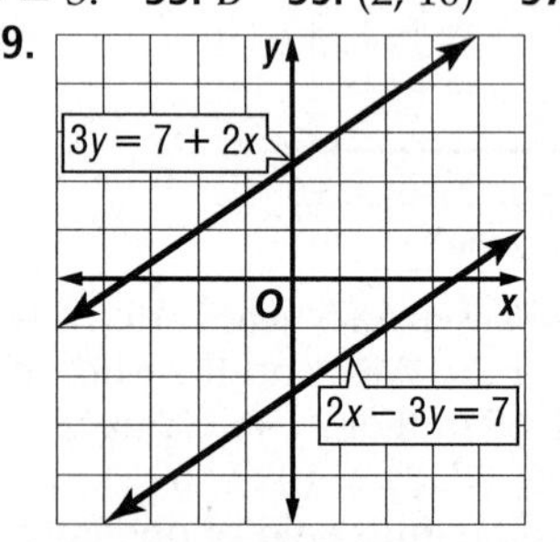

no solution

41. $\frac{12}{11}x$ or $1\frac{1}{11}x$ **43.** $12a - 30b$ **45.** $-20t + 10s$

Pages 275–278 Lesson 5-4

1. (2, –2) **3.** (5, –2) **5.** B **7.** (5, –2) **9.** (2, 1) **11.** (–4, –7) **13.** (–1, –2) **15.** (0, 1) **17.** 2, –5 **19.** (10, 12) **21.** (2, –8) **23.** 86 **25.** (1.5, 2); A batting token costs \$1.50 and a game of miniature golf costs \$2.00. **27.** (4, 1) **29.** If one of the variables cannot be eliminated by adding or subtracting the equations, you must multiply one or both of the equations by numbers so that a variable will be eliminated when the equations are added or subtracted. **31.** Sample answer: $3x + 2y = 5$, $5x - 10y = -6$ **33.** Sample answer: By having two equations that represent the time restraints, a manager can determine the best use of employee time. The following is a solution to the system of equations on top of page 272.

$$20c + 10b = 800 \rightarrow \quad 20c + 10b = 800$$
$$10c + 30b = 900 \rightarrow \; \underline{-20c - 60b = -1800}$$
$$-50b = -1000$$
$$\frac{-50b}{-50} = \frac{-1000}{-50}$$
$$b = 20$$
$$20c + 10b = 800$$
$$20c + 10(20) = 800$$
$$20c + 200 = 800$$
$$20c + 200 - 200 = 800 - 200$$
$$20c = 600$$
$$\frac{20c}{20} = \frac{600}{20}$$
$$c = 30$$

In order to make the most of the employee and oven time, the manager should make assignments to bake 30 batches of cookies and 20 loaves of bread.
35. J **37.** (6, 2) **39.** (11, 7) **41.** (–4, 4) **43.** 4; For every 4-feet increase in height, there is a 1-foot increase in horizontal distance. **45.** –8, 2, –8 **47.** 2

Pages 283–284 Lesson 5-5

1. elimination (×); (4, 1) **3.** substitution; (–3, –1) **5.** Sample answer: $3s + 5p = 233$ and $s = p + 11$; Denzell sold 25 pizzas and 36 subs. **7.** substitution; (2, 6) **9.** elimination (×) or substitution; (3, 1) **11.** elimination (–); no solution **13.** 880 books; If they sell this number, then their income and expenses both equal \$35,200. **15.** elimination (×); (–6, 2) **17.** Let x = the cost per pound of aluminum cans and let y = the cost per pound of newspaper; $9x + 9y = 3.77$ and $26x + 114y = 4.65$. **19.** Sample answer: $x + y = 12$ and $3x + 2y = 29$, where x represents the cost of a student ticket for the football game and y represents the cost of an adult ticket; substitution could be used to solve the system; (5, 7) means the cost of a student ticket is \$5 and the cost of an adult ticket is \$7.
21. Sample answer: You should always check that the answer makes sense in the context of the original problem. If it does not, you may have made an incorrect calculation. If (1, –7) was the solution, then it is probably incorrect since distance in this case cannot be a negative number. The solution should be recalculated. **23.** (0, –3) **25.** (–5, –1)

Pages 285–288 ***Chapter 5*** ***Study Guide and Review***

1. true **3.** true **5.** false; consistent and dependent **7.** false; inconsistent **9.** true

11. 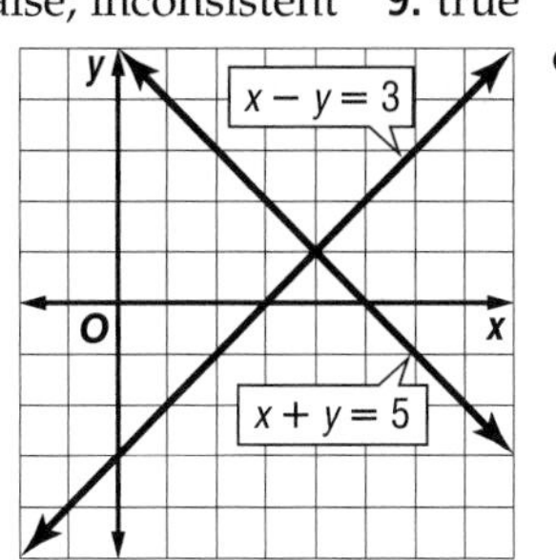

one; (4, 1)

13. 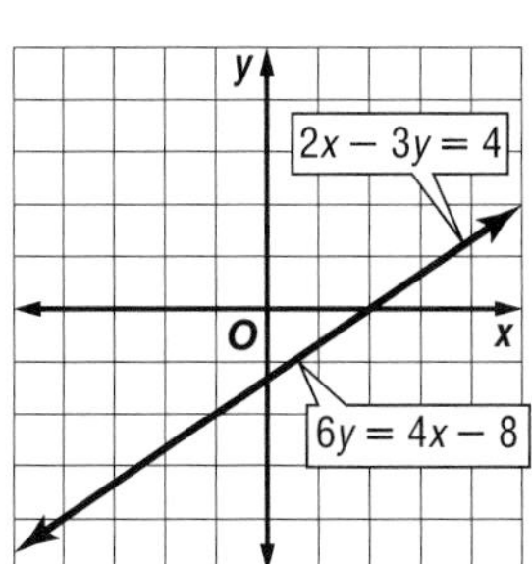

infinitely many

15. 15 mi; 3 h **17.** infinitely many **19.** (4, 0) **21.** (2, 2) **23.** (4, 1) **25.** Denver: 10th; Detroit: 19th **27.** (−9, −7) **29.** (0.6, 3) **31.** substitution; (1.6, 3.2) **33.** substitution or elimination (−); (0, 0) **35.** \$900 at 4% and \$600 at 6%

Chapter 6 Solving Linear Inequalities

Page 293 ***Chapter 6*** ***Get Ready***

1. −9 **3.** −19 **5.** 22 **7.** 4 **9.** 6, 7 **11.** 1.5 **13.** 15 **15.** 3

17. 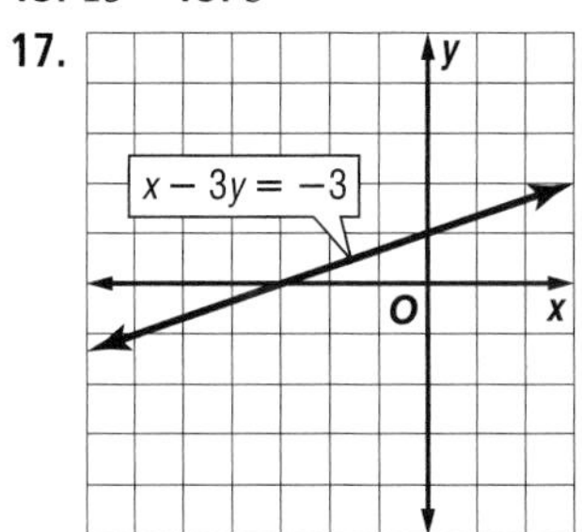

19. 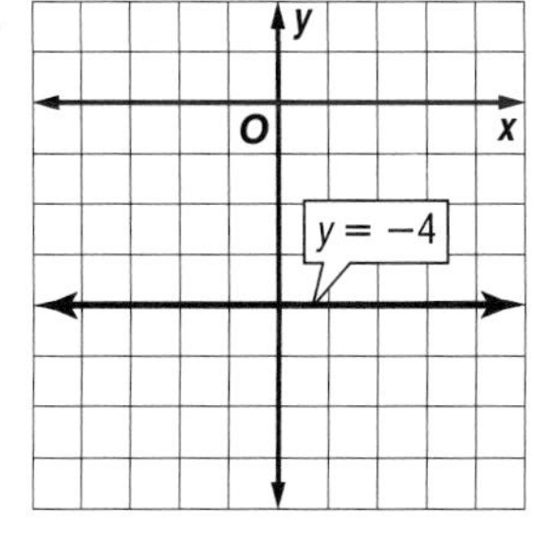

21. 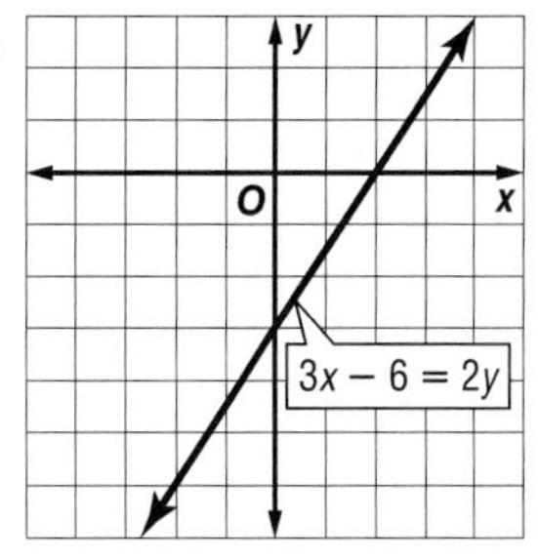

23. 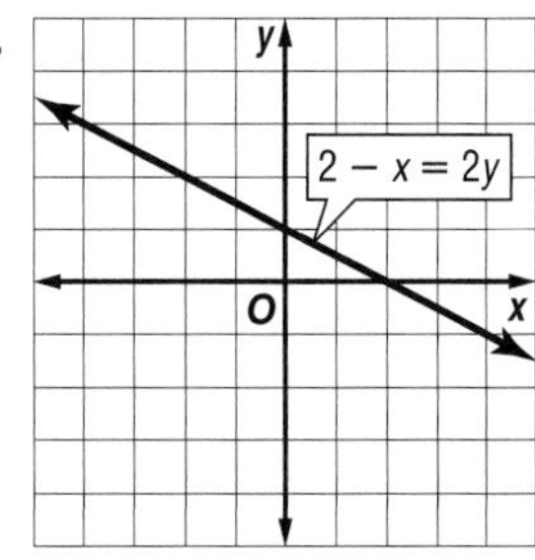

Pages 297–299 ***Lesson 6-1***

1. $\{t \mid t \geq 12\}$ 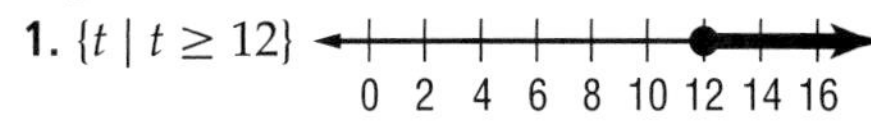

3. $\{a \mid a < -2\}$ 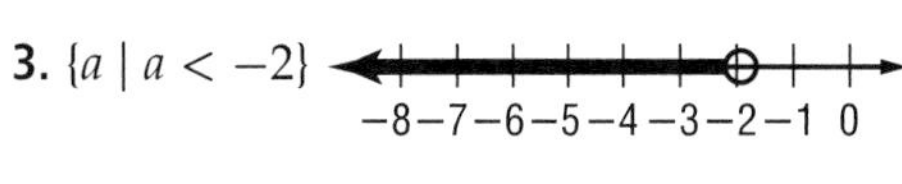

5. $\{n \mid n < 11\}$ 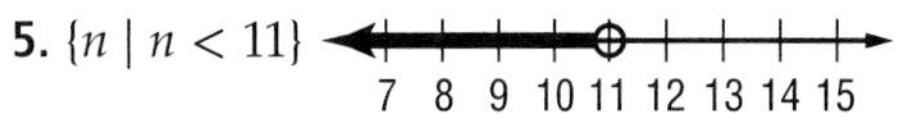

7. $\{r \mid r > 6\}$ 

9. Sample answer: Let n = the number, $n - 8 \leq 14$; $\{n \mid n \leq 22\}$. **11.** no more than 2043 lb

13. $\{d \mid d \leq 2\}$ 

15. $\{s \mid s > 4\}$ 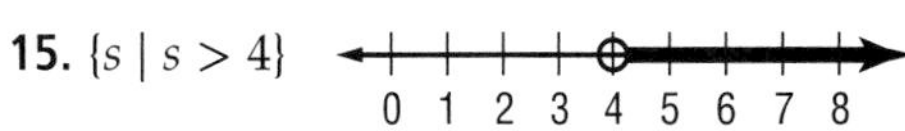

17. $\{r \mid r < -5\}$ 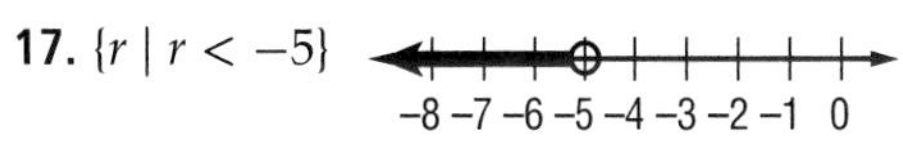

19. $\{q \mid q \leq 7\}$ 

21. $\{h \mid h < 30\}$

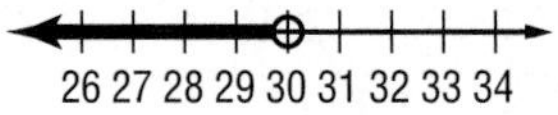

26 27 28 29 30 31 32 33 34

23. $\{c \mid c < -27\}$

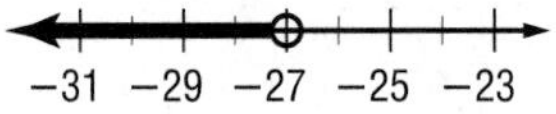

−31 −29 −27 −25 −23

25. $\{f \mid f < -3\}$

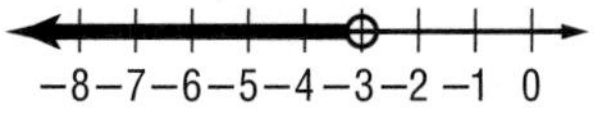

−8 −7 −6 −5 −4 −3 −2 −1 0

27. $\{w \mid w \geq 1\}$

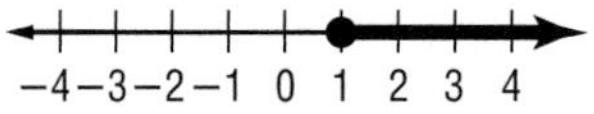

−4 −3 −2 −1 0 1 2 3 4

29. $\{a \mid a > -9\}$

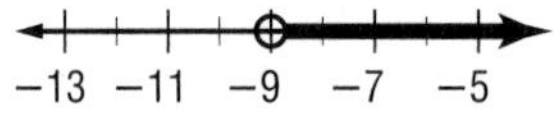

−13 −11 −9 −7 −5

31. Sample answer: Let n = the number, $n - 5 < 33$; $\{n \mid n < 38\}$. **33.** Sample answer: Let n = the number, $n + (-7) \leq 18$; $\{n \mid n \leq 25\}$. **35.** more than 5 million

37. $\{y \mid y > 5.6\}$

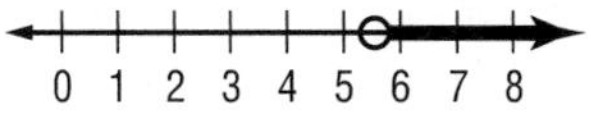

0 1 2 3 4 5 6 7 8

39. $\left\{a \,\middle|\, a > -\frac{1}{8}\right\}$

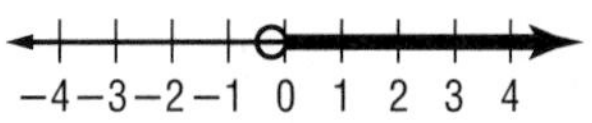

−4 −3 −2 −1 0 1 2 3 4

41. Sample answer: Let n = the number, $30 \leq n + (-8)$; $\{n \mid n \geq 38\}$. **43.** Sample answer: Let m = the amount of money in the account, $m - 1300 - 947 \geq 1500$; $\{m \mid m \geq 3747\}$. Mr. Hayashi must have at least \$3747 in his account. **45.** $12 < 4 + x$; more than 8 in. **47.** In both graphs, the line is darkened to the left. In the graph of $a < 4$, there is a circle at 4 to indicate that 4 is not included in the graph. In the graph of $a \leq 4$, there is a dot at 4 to indicate that 4 is included in the graph. **49.** Sample answers: $y + 1 < -2$, $y - 1 < -4$, $y + 3 < 0$ **51.** B **53.** A **55.** substitution; (16, 2) **57.** \$1.50; 3.75 **59.** 36 h **61.** 145 **63.** −21

Pages 305–307 ***Lesson 6-2***

1. $\{t \mid t < -108\}$ **3.** $\{r \mid r \geq 8\}$ **5.** Sample answer: Let n = the number, $-4n > 12$; $\{n \mid n < -3\}$. **7.** Sample answer: Let n = the number of DVDs sold, $15n > 5500$; $\{n \mid n \geq 366.6\}$, they sold at least 367 DVDs. **9.** $\{x \mid x > -5\}$ **11.** $\{s \mid s \leq 7\}$ **13.** $\{m \mid m < -68\}$ **15.** $\{r \mid r > 49\}$ **17.** $\{y \mid y \geq -24\}$ **19.** $\{x \mid x \geq 20\}$ **21.** Sample answer: Let n = the number, $7n > 28$; $\{n \mid n > 4\}$. **23.** Sample answer: Let n = the number, $24 \leq \frac{1}{3}n$; $\{n \mid n \geq 72\}$. **25.** $\{g \mid g \leq 24\}$ **27.** $\{d \mid d \leq -6\}$ **29.** $\{y \mid y > -16\}$ **31.** $\{s \mid s > -1\}$ **33.** Sample answer: Let b = the number of bags of mulch, $2.5b \geq 2000$; $\{b \mid b \geq 800\}$, at least 800 bags. **35.** $\{b \mid b \geq 13.5\}$ **37.** $\{w \mid w > -2.72\}$ **39.** $\left\{c \,\middle|\, c < -\frac{1}{10}\right\}$

41. $\{m \mid m \geq 3\}$

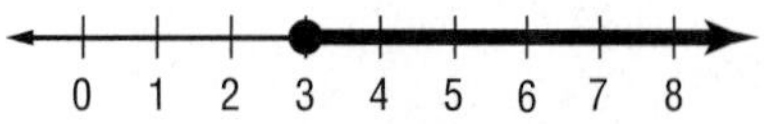

0 1 2 3 4 5 6 7 8

43a. 3.5 **43b.** −14 **43c.** −6 **45.** Sample answer: Let n = the number, $0.40n \leq 45$; $\{n \mid n \leq 112.5\}$. **47.** Sample answer: Let r = the radius of the flower garden, $2\pi r \leq 38$; $\{r \mid r \leq 6.04\}$, up to about 6 ft. **49.** Sample answer: Let v = the number of visits to the zoo, $128 < v(2 \cdot 19.50 + 2 \cdot 11.75)$; $\{v \mid v > 2.05\}$, at least 3 times. **51.** You could solve the inequality by multiplying each side by $-\frac{1}{7}$ or by dividing each side by −7. In either case, you must reverse the direction of the inequality symbol. **53a.** Sample answer: $2 > -3$, but $4 < 9$. **53b.** Sample answer: $-1 < 2$ and $-3 < -2$, but $3 > -4$. **55.** Sample answer: Inequalities can be used to compare the heights of cases of beverages. If x represents the number of cases of water and the cases must be no higher than 3 ft or 36 in., then $8x \leq 36$. To solve this inequality, divide each side by 8 and do not change the direction of the inequality. The solution is or $x \leq 4.5$. This means that the stack must be 4 cases high or fewer. **57.** J **59.** $\{g \mid g \leq -7\}$ **61.** 12 months, \$240 **63.** $y = -2$ **65.** 7 **67.** 12

Pages 311–313 ***Lesson 6-3***

1. $\{h \mid h \geq 7\}$ **3.** $\{y \mid y > -10.5\}$ **5.** $4n + 60 \leq 800$; $n \leq 185$; less than 185 lb **7.** $\{v \mid v \geq 0\}$ **9.** ∅ **11.** $\{a \mid a < 3\}$ **13.** $\{w \mid w > 56\}$ **15.** $\{f \mid f > -8\}$ **17.** $\{a \mid a < -4\}$ **19.** $\{k \mid k > 8\}$ **21.** Sample answer: Let n = the number, $\frac{2}{3}n + 8 > 12$; $\{n \mid n > 6\}$. **23.** Sample answer: Let n = the number, $10 \leq 4(2n + 3)$; $\left\{n \,\middle|\, n \geq -\frac{1}{4}\right\}$. **25.** Sample answer: Let w = the number of weeks, $1.25w > 90 - 75$; $\{w \mid w > 12\}$, it will take more than 12 weeks for the dog to reach a healthy weight. **27.** $\{a \mid a \leq 11\}$ **29.** $\{b \mid b$ is a real number.$\}$ **31.** $\{a \mid a \geq -9\}$

33.

$-5(k + 4) > 3(k - 4)$	Original inequality
$-5k - 20 > 3k - 12$	Distributive Property
$-5k - 20 + 5k > 3k - 12 + 5k$	Add $5k$ to each side.
$-20 > 8k - 12$	Simplify.
$-20 + 12 > 8k - 12 + 12$	Add 12 to each side.
$-8 > 8k$	Simplify.
$\frac{-8}{8} > \frac{8k}{8}$	Divide each side by 8.
$-1 > k$	Simplify.

35. at least 94 **37.** Sample answer: $7.5 + 1.25x + 0.15(7.5 + 1.25x) < 13$; 3 or fewer toppings **39.** $x < 0$ **41.** $-\frac{41}{3}$ **43.** 7, 9; 5, 7; 3, 5; 1, 3 **45.** $\{v \mid v \geq 4.5\}$ **47.** $3(n + 7) > 5n - 13$; $\{n \mid n < 17\}$ **49.** Sample answers: $2x + 5 < 2x + 3$; $2x + 5 > 2x + 3$ **51.** $4y + 9 > -3$; it is the only inequality that does not have a solution set of $\{y \mid y > 3\}$. **53.** C **55.** up to 416 mi

57. $\{t \mid t < 8\}$

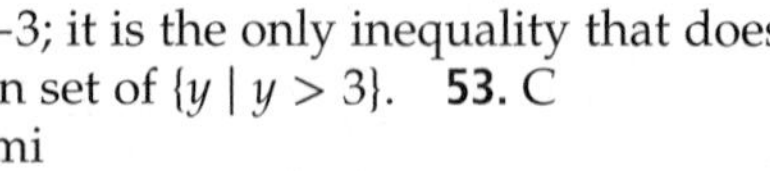

59. $2x - y = 5$ **61.** $y - 6 = 0$ **63.** Sample answer: '97–'99; A steeper segment means greater rate of change.

65.

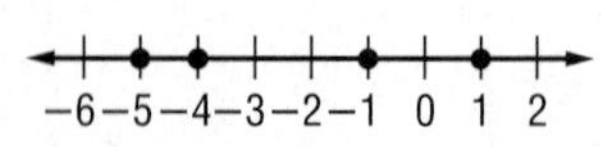

67.

Pages 317–320 **Lesson 6-4**

1. −3 −2 −1 0 1 2 3 4 5 6 7

3. $\{w \mid 3 < w < 8\}$ 0 1 2 3 4 5 6 7 8 9 10

5. $\{z \mid z < 4\}$ −2 −1 0 1 2 3 4 5 6 7 8

7. 11 psi $\le x \le$ 56 psi

9. −10 −9 −8 −7 −6 −5 −4 −3 −2 −1 0

11. −7 −6 −5 −4 −3 −2 −1 0 1 2 3

13. −5 −4 −3 −2 −1 0 1 2 3 4 5

15. $\{f \mid -13 \le f \le -5\}$ −14 −13 −12 −11 −10 −9 −8 −7 −6 −5 −4

17. $\{h \mid h < -1\}$ −5 −4 −3 −2 −1 0 1 2 3 4 5

19. $\{y \mid 3 < y < 6\}$ 0 1 2 3 4 5 6 7 8 9 10

21. $\{q \mid -1 < q < 6\}$ −3 −2 −1 0 1 2 3 4 5 6 7

23. $\{n \mid n \le 4\}$ 0 1 2 3 4 5 6 7 8 9 10

25. $t < 18$ or $t > 22$ **27.** $-7 < x < -3$
29. $x \le -7$ or $x \ge -6$ **31.** $x = 2$ or $x > 5$

33. ∅ −5 −4 −3 −2 −1 0 1 2 3 4 5

35. $\{b \mid b < -12$ or $b > -12\}$

37. between \$51 and \$110 inclusive **39.** Sample answer: Let n = the number, $-8 < 3n + 4 < 10$; $\{n \mid -4 < n < 2\}$. **41.** Sample answer: Let n = the number, $0 < \frac{1}{2}n \le 1$; $\{n \mid 0 < n \le 2\}$. **43.** $\{h \mid 15 \le h \le 50{,}000\}$; $\{h \mid 20 \le h \le 20{,}000\}$ **45.** Sample answer: troposphere: $a \le 10$, stratosphere: $10 < a \le 30$, mesosphere: $30 < a \le 50$, thermosphere: $50 < a \le 400$, exosphere: $a > 400$ **47.** Sample answer: $x < -2$ and $x > 3$ **49a.** $x \ge 5$ and $x \le 8$ **49b.** $x > 6$ or $x < 1$
51. A **53.** D **55.** $\{d \mid d \ge 5\}$ **57.** $\{t \mid t < 169\}$
59. 2.25 **61.** 5 **63.** 2 **65.** −1 **67.** −2

Pages 325–327 **Lesson 6-5**

1. $\{-13, 7\}$ −14 −12 −10 −8 −6 −4 −2 0 2 4 6 8

3. ∅ −5 −4 −3 −2 −1 0 1 2 3 4 5

5.

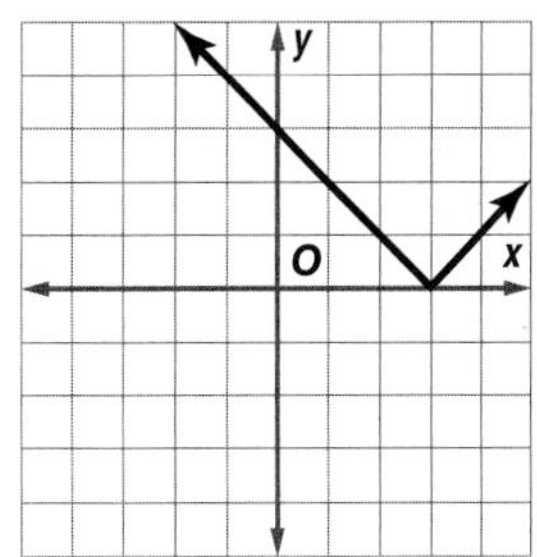

7. $\{-3, 13\}$

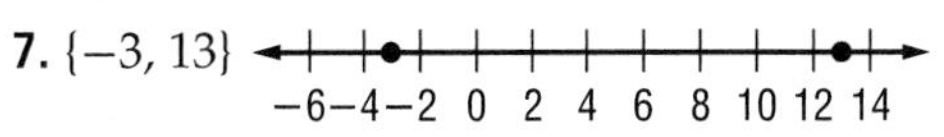

9. ∅ −5 −4 −3 −2 −1 0 1 2 3 4 5

11. $\{-0.8, 4\}$ −5 −4 −3 −2 −1 0 1 2 3 4 5

13. $\{-16, -4\}$ −16 −14 −12 −10 −8 −6 −4 −2 0 2 4

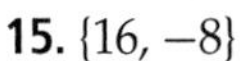
15. $\{16, -8\}$

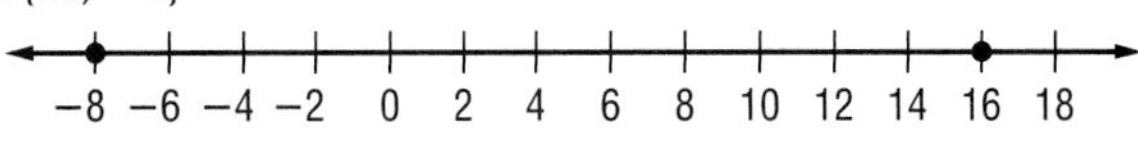

17. $\left\{-\frac{3}{2}, \frac{9}{2}\right\}$ −5 −4 −3 −2 −1 0 1 2 3 4 5

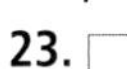
19. $|x| = 5$ **21.** $|x - 3| = 5$

23.

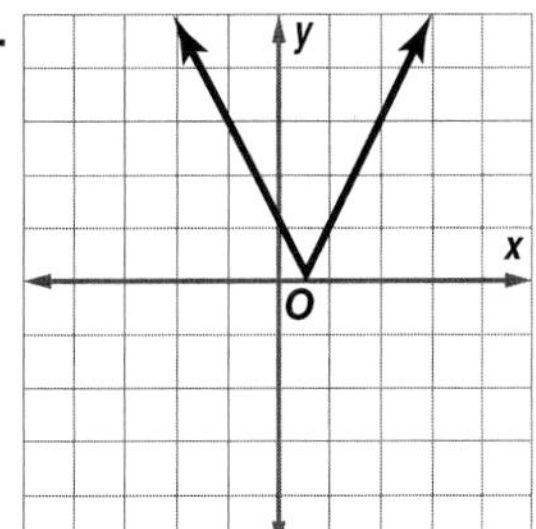

25.

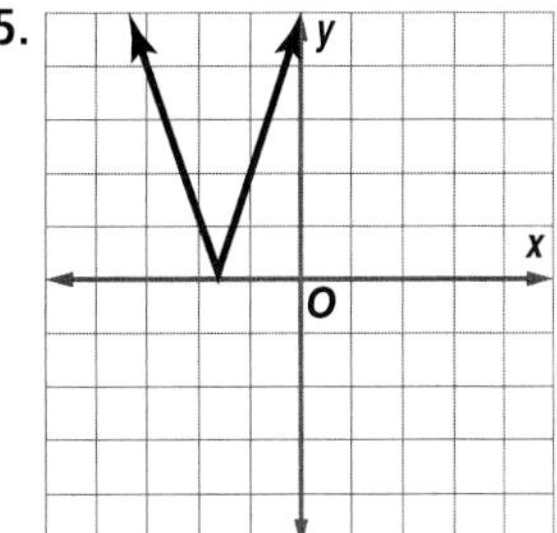

27.

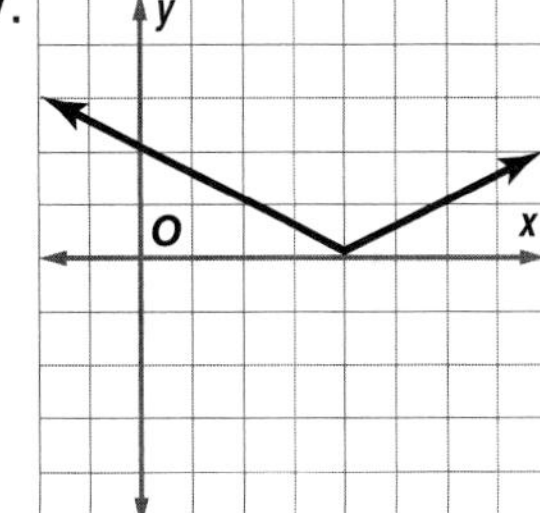

29.

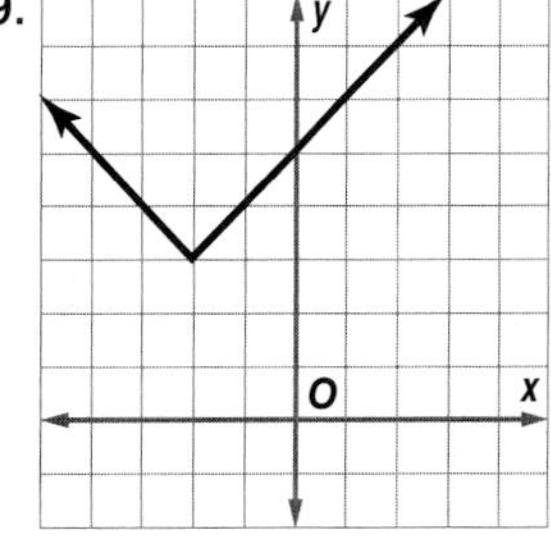

31. 5.5, −5.5

Selected Answers

33. D: all real numbers; $\{y|y \geq 4\}$ **35.** D: all real numbers; R: $\{y|y \geq 1\}$ **37.** 45–51%

39.

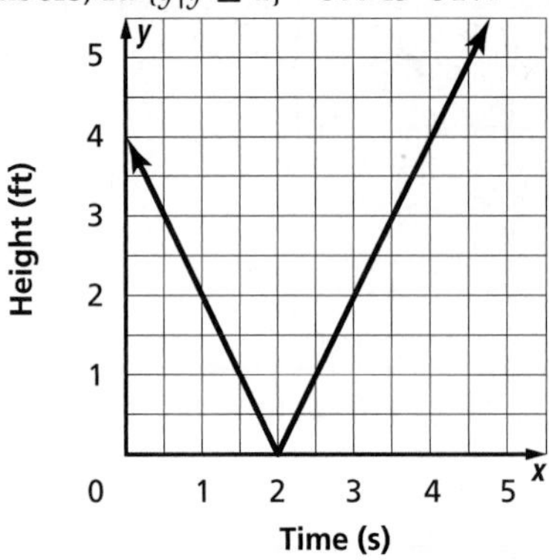

41. Sample answer: Let x = the time in minutes to run one mile. Then the time to run one mile is 10 ± 4. **43.** Sometimes; when c is a negative value, x is a positive value. **45.** Sometimes; when $c < 0$ and $0 < x < -2$, then the expression is less than 0. **47.** Leslie; you need to consider the case when the value inside the absolute value symbols is positive and the case when the value inside the absolute value symbols is negative. So $x + 3 = 2$ or $x + 3 = -2$.

49. C **51.** D **53.** $\{m|m > 5\}$ **55.** $\{w|w \leq 15\}$ **57.** $\frac{3}{2}$; 2

59. $\{(6, 0), (-3, 5), (2, -2), (-3, 3)\}$; $\{-3, 2, 6\}$; $\{-2, 0, 3, 5\}$; $\{(0, 6), (5, -3), (-2, 2), (3, -3)\}$

61. $\{(3, 4), (3, 2), (2, 9), (5, 4), (5, 8), (-7, 2)\}$; $\{-7, 2, 3, 5\}$; $\{2, 4, 8, 9\}$; $\{(4, 3), (2, 3), (9, 2), (4, 5), (8, 5), (2, -7)\}$

63. $x = \frac{3z + 2y}{e}$ **65.** $m \geq 2$ **67.** $z \leq 6$ **69.** $r \leq 7$

Pages 331–333 ***Lesson 6-6***

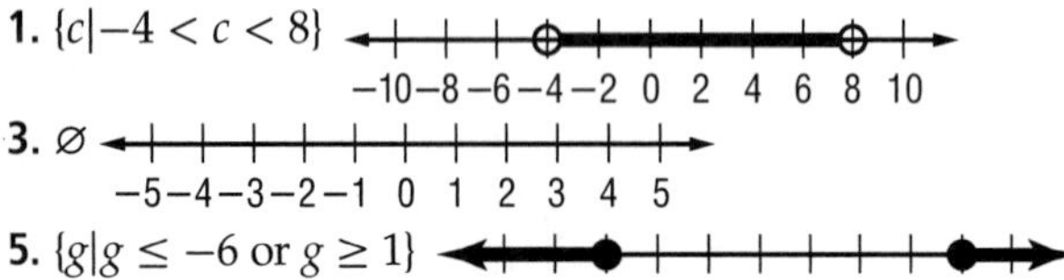

1. $\{c|-4 < c < 8\}$ −10 −8 −6 −4 −2 0 2 4 6 8 10

3. ∅ −5 −4 −3 −2 −1 0 1 2 3 4 5

5. $\{g|g \leq -6$ or $g \geq 1\}$ −8 −7 −6 −5 −4 −3 −2 −1 0 1 2

7. $\{d|1.499 \leq d \leq 1.501\}$

9. $\{t|-10 < t < -6\}$ −10 −9 −8 −7 −6 −5 −4 −3 −2 −1 0

11. $\{v|v < -4$ or $v > -2\}$

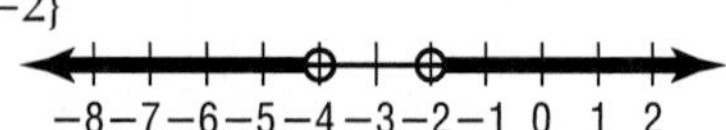

13. $\{a|a$ is a real number.$\}$

15. $\{n|-5 < n < 4\}$ −5 −4 −3 −2 −1 0 1 2 3 4 5

17. $\{p|2000 < p < 3000\}$ **19.** $|x| \leq 3$ **21.** $|x - 1| > 2$

23. $\left\{-8\frac{4}{5}, 8\right\}$ −10 −8 −6 −4 −2 0 2 4 6 8 10

25. $\{s|s$ is a real number.$\}$

27. $|p - 7.5| \leq 0.3$ **29.** $|s - 98| \leq 6$ **31.** $\{p \mid 28 \leq p \leq 32\}$ **33.** $|g - 52| \leq 5$ **35.** The solution of $|x - 2| > 6$ includes all values that are less than −4 or greater than 8. The solution of $|x - 2| < 6$ includes all values that are greater than −4 and less than 8.

37. $|x - 2| < 0.3$ **39.** C **41.** B

43. ∅ −5 −4 −3 −2 −1 0 1 2 3 4 5

45. $0 \leq x \leq 25$; $25.01 \leq x \leq 50$ **47.** $y = \frac{1}{2}x + \frac{3}{4}$

49.

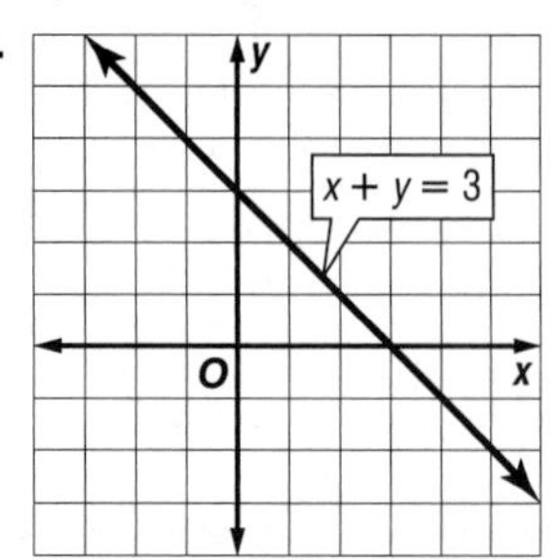

51.

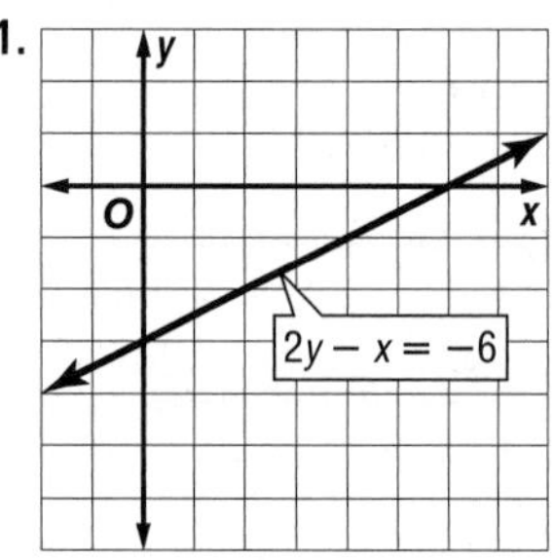

Pages 337–339 ***Lesson 6-7***

1.

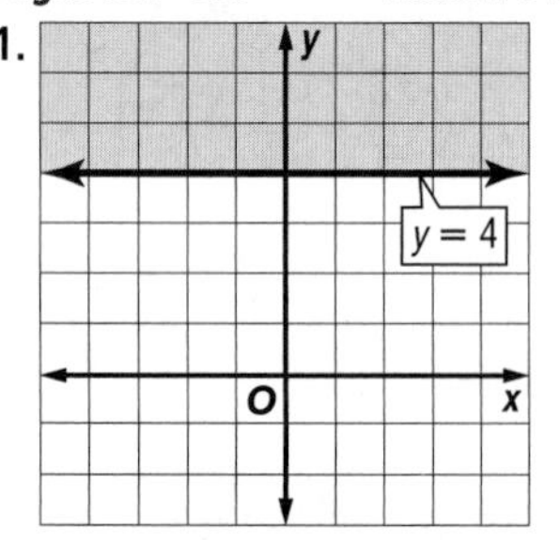

3. y = x + 3

5.

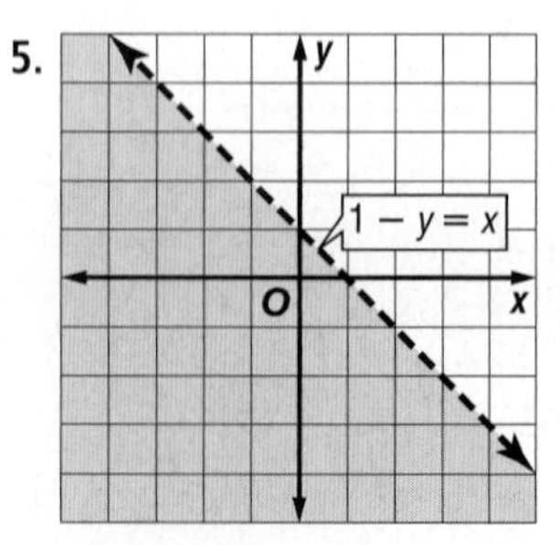

7. $12x + 3y \leq 60$

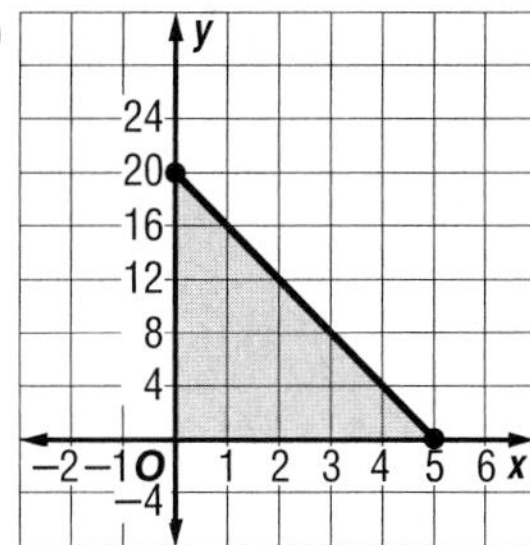

9.

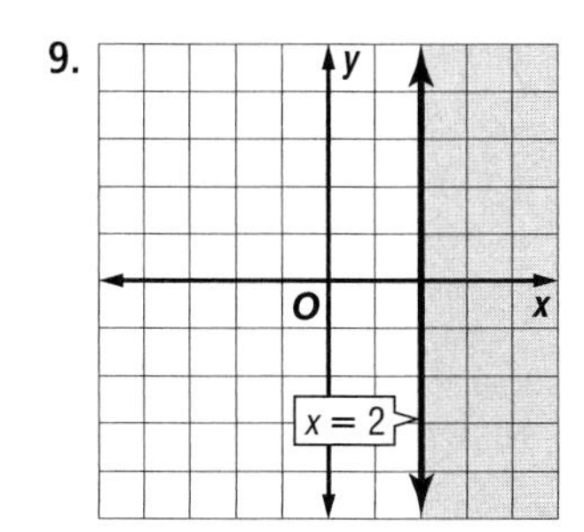

11.

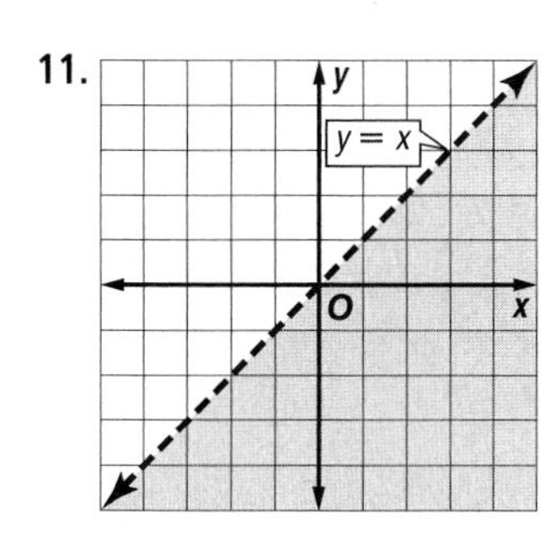

13.

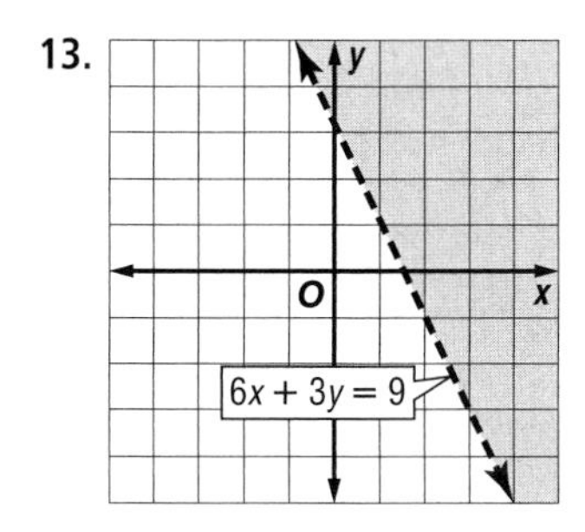

15.

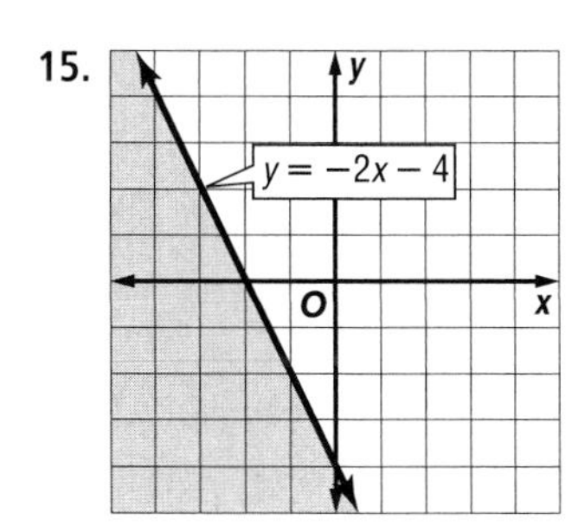

17.

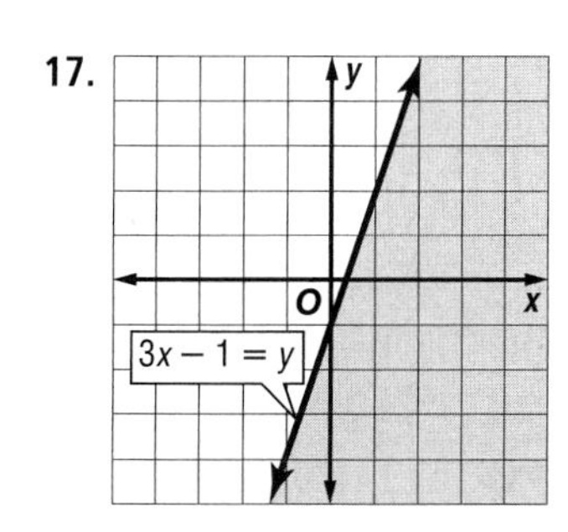

19. The solution set is limited to pairs of positive numbers. **21.** No, the weight will be greater than 4000 pounds. **23.** $\{(-1, 3), (-4, 5)\}$ **25.** $\{(-13, 10), (4, 4), (-6, -2)\}$ **27.** c **29.** d **31.** $\{(-3, 2)\}$

33.

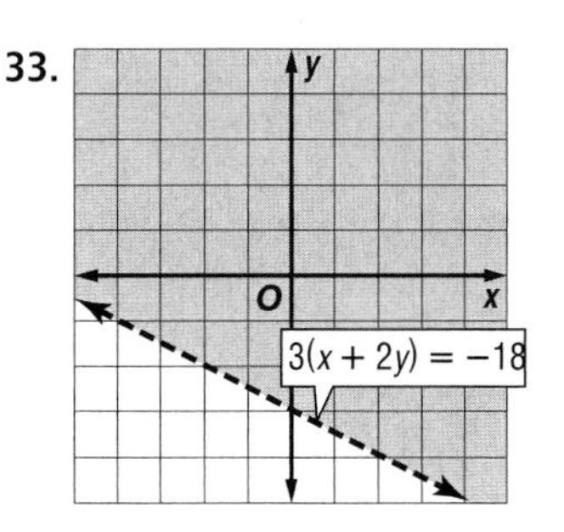

35. $8x + 4y \leq 96$ **37.** The graph of $y = x + 2$ is a line. The graph of $y < x + 2$ does not include the boundary $y = x + 2$, and it includes all ordered pairs in the half-plane that contains the origin. **39.** If the test point results in a true statement, shade the half-plane that contains the point. If the test point results in a false statement, shade the other half-plane. **41.** The amount of money spent in each category must be less than or equal to the budgeted amount. How much you spend on individual items can vary. The domain and range must be positive integers. Sample answers: Hannah could buy 5 cafeteria lunches and 3 restaurant lunches, 2 cafeteria lunches and 5 restaurant lunches, or 8 cafeteria lunches and 1 restaurant lunch. **43.** F
45. $\{d \mid 266 \leq d \leq 294\}$
47. $\{-2, 3\}$

−5 −4 −3 −2 −1 0 1 2 3 4 5

49. $y = -x + 4$ **51.** 31, 37 **53.** 48, 96
55. increase; 42%
57.

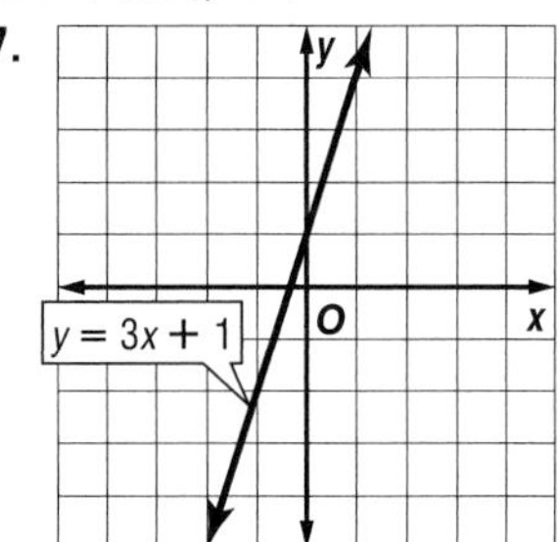

59.

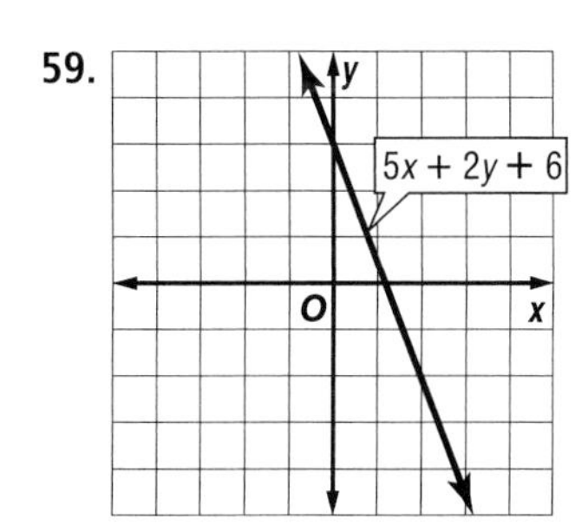

Selected Answers

Pages 343–345 Lesson 6-8

1.

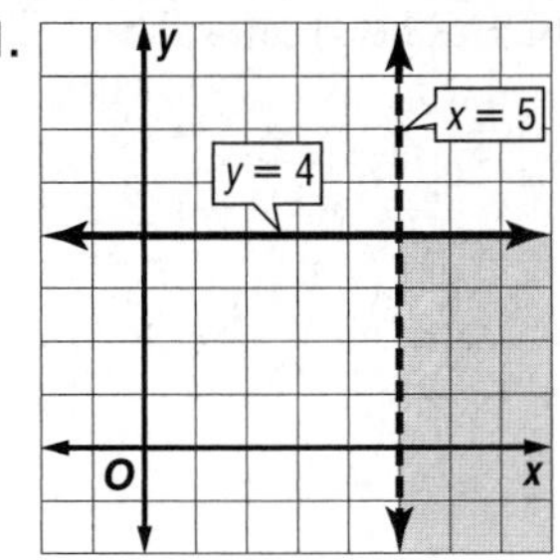

3.

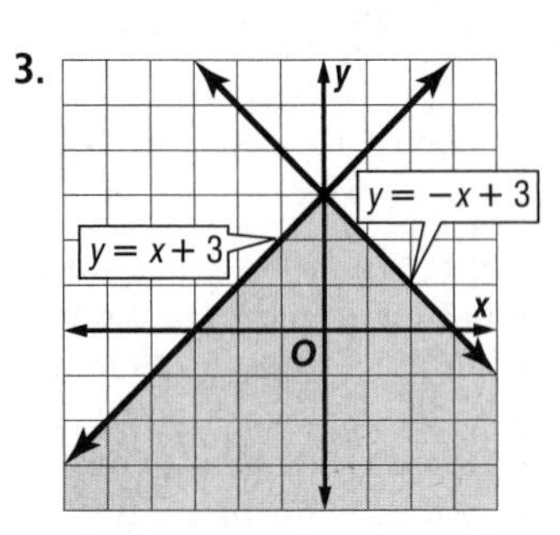

5. **Natasha's Daily Exercise**

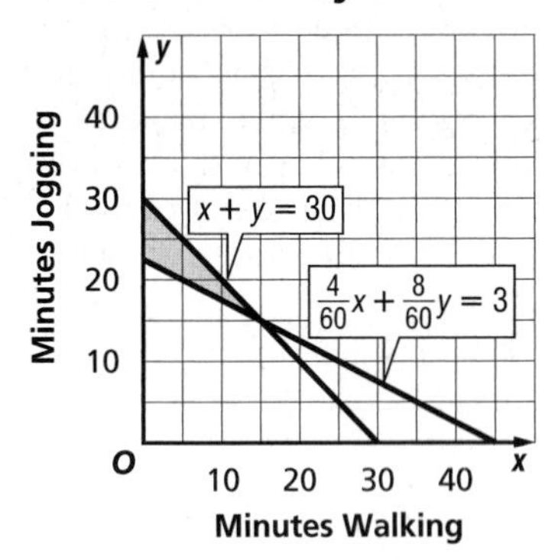

7.

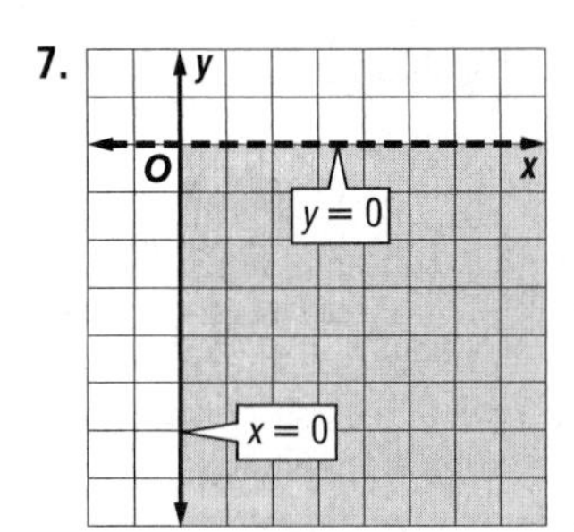

9.

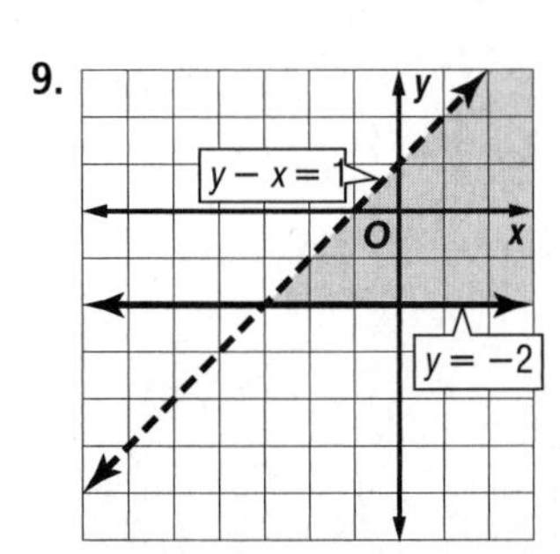

11.

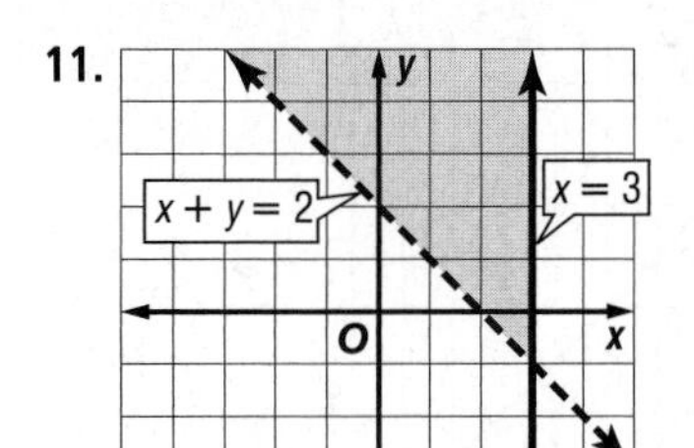

13.

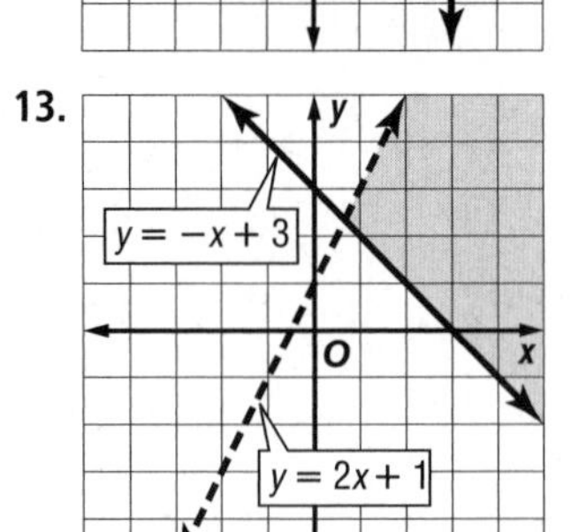

15.

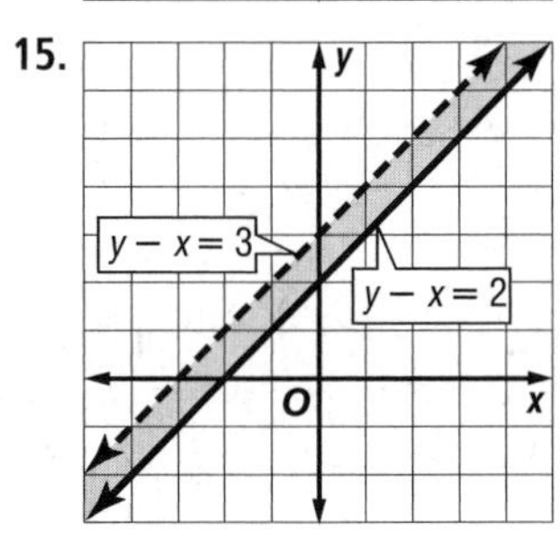

17.

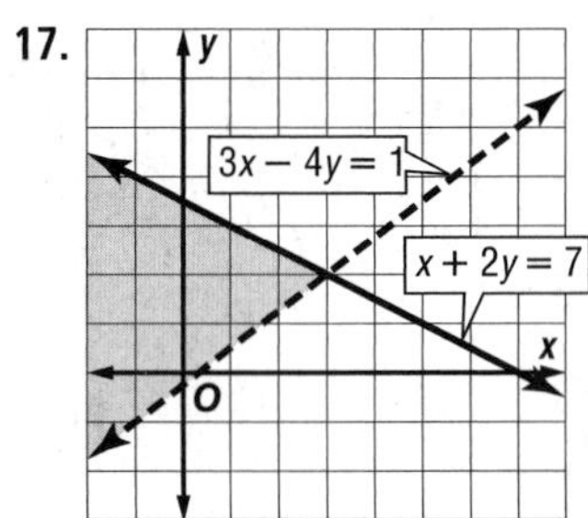

19. **Furniture Manufacturing**

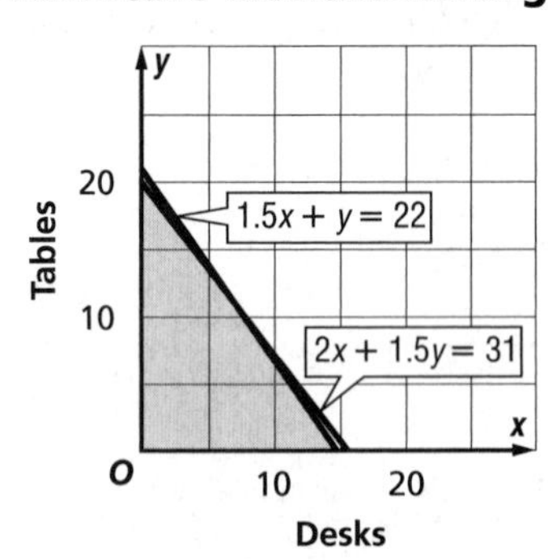

21. **Green Paint**

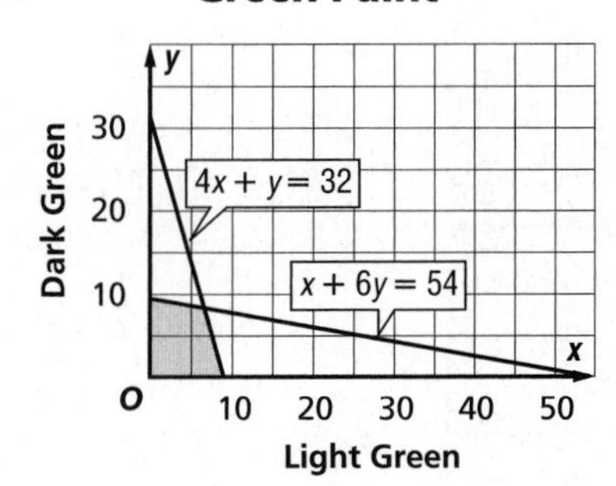

23.

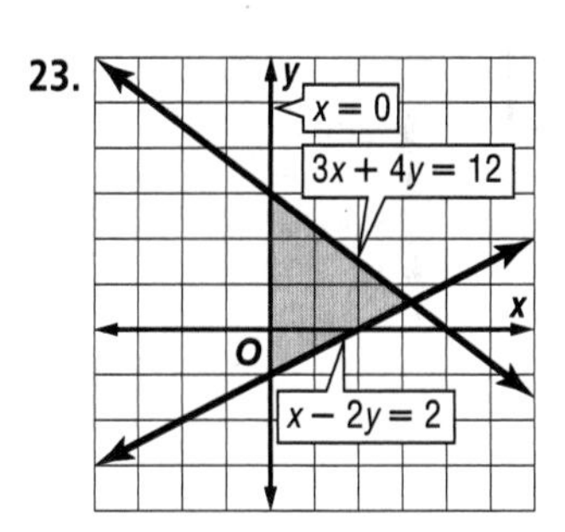

25.

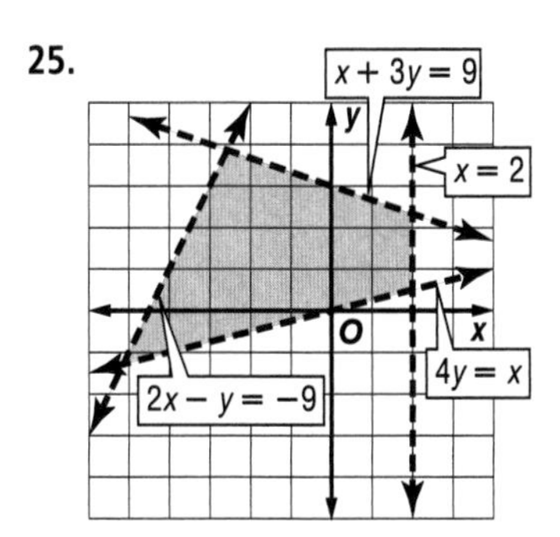

27. $y \leq \frac{2}{3}x + 2, y > -x - 3$ 29. Any point in the shaded region is a possible solution. For example, since (7, 8) is a point in the region, Mr. Hobson could plant corn for 7 days and soybeans for 8 days. In this case, he would use 15 days to plant 250(7) or 1750 acres of corn and 200(8) or 1600 acres of soybeans.

31.

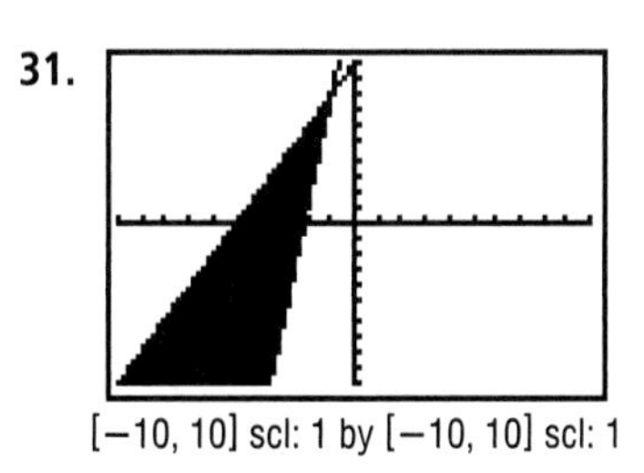

[−10, 10] scl: 1 by [−10, 10] scl: 1

33.

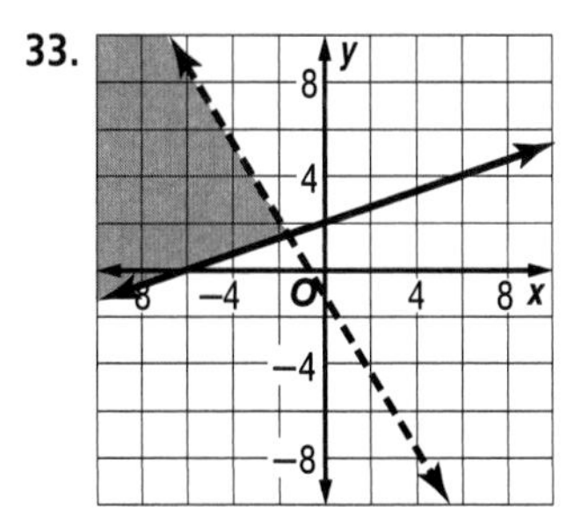

35. Sample answer: ∅

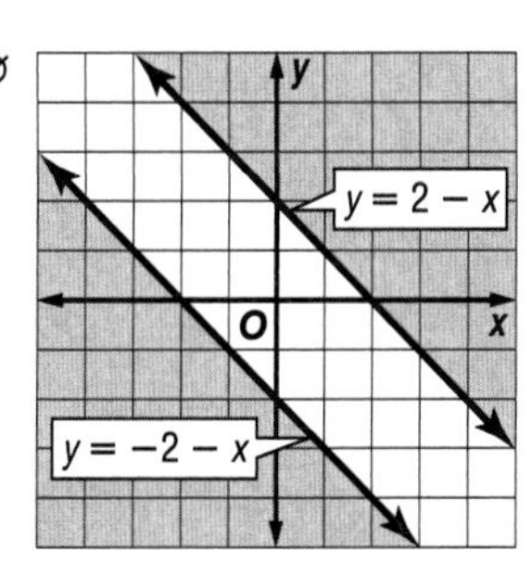

37. Jocelyn; the graph of $x + 2y \geq -2$ is the region representing $x + 2y = -2$ and the half-plane above it. 39. A

41. 43.

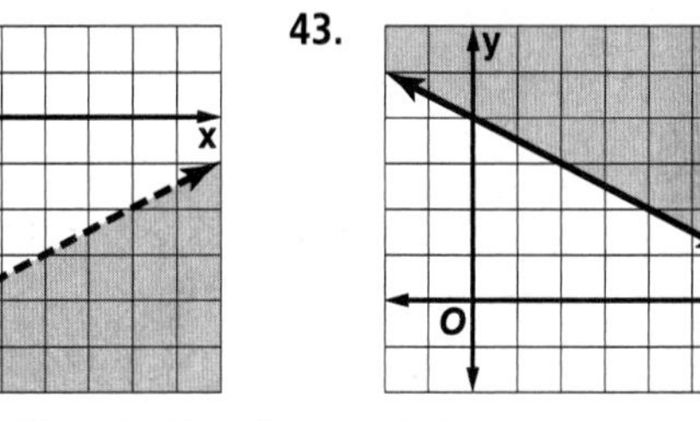

45. (2, −1) 47. (−2, 3) 49. yes 51. no

Pages 346–350 ***Chapter 6 Study Guide and Review***

1. true 3. false; greater than or equal to 5. false; intersection 7. false; or 9. true

11. $\{x \mid x < 25\}$

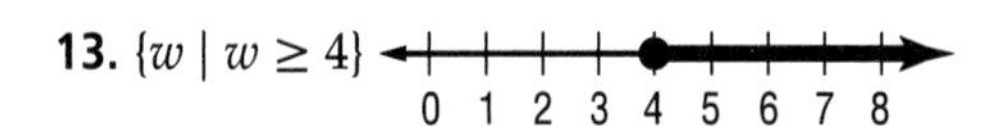

13. $\{w \mid w \geq 4\}$

0 1 2 3 4 5 6 7 8

15. Sample answer: Let n = the number, $16 < n + 31$; $\{n \mid n > -15\}$. 17. $\{v \mid v > 4\}$ 19. $\{m \mid m > -11\}$
21. Sample answer: Let n = the number, $0.8n \geq 24$; $\{n \mid n \geq 30\}$. 23. $\{y \mid y < 4\}$ 25. $\{x \mid x \geq -2\}$
27. Sample answer: Let n = the number, $\frac{2}{3}n - 27 \geq 9$; $\{n \mid n \geq 54\}$.

29.

−4 −3 −2 −1 0 1 2 3 4

31. $\{w \mid w$ is a real number$\}$.

33. $33 \leq m \leq 37$

31 32 33 34 35 36 37 38 39

35. $\{4, -1\}$

−5 −4 −3 −2 −1 0 1 2 3 4 5

37. $|c - 13.6| = 0.408$; 14.008 gal, 13.192 gal

39. ∅

−5 −4 −3 −2 −1 0 1 2 3 4 5

41. $\{x \mid x \leq -1$ or $x \geq 6\}$

−2 −1 0 1 2 3 4 5 6

43. $\{w \mid -19 \leq w \leq 3\}$

45.

y
O
x

47.

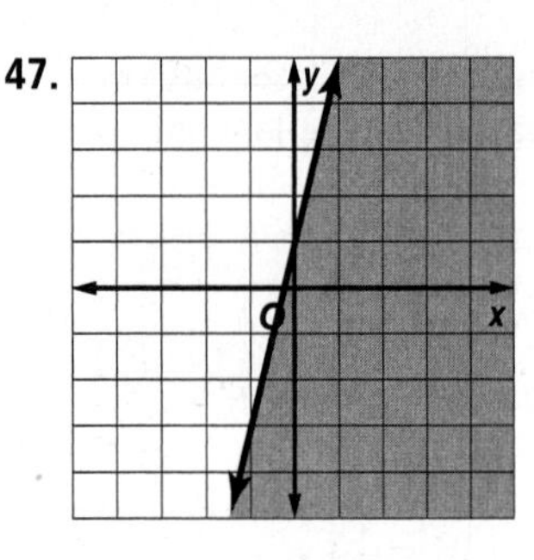

49. $95h + 0.08m \leq 500$; Yes, Brenda can afford to hire this moving company.

51.

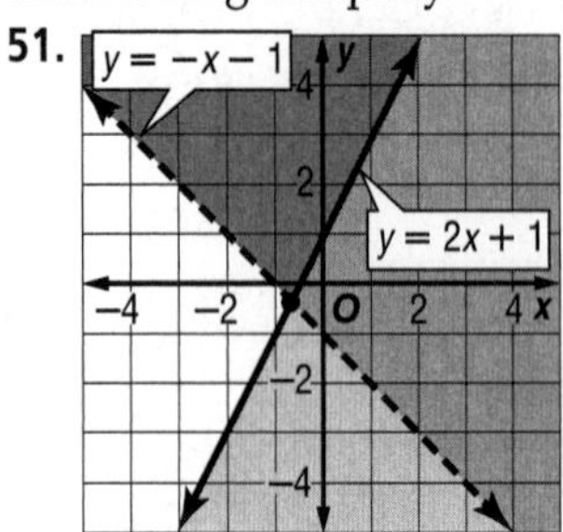

53.

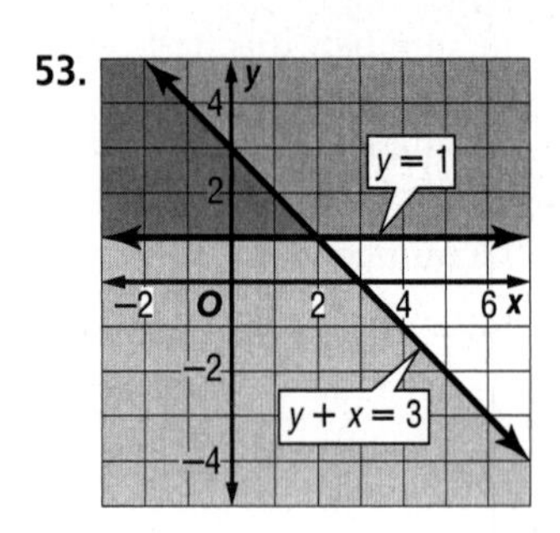

57. False; let $a = 2$. Then $(-a)^2 = (-2)^2 = 4$ and $-a^2 = -2^2 = -4$. **59.** False; let $a = 3$, $b = 4$, and $n = 2$. Then $(a + b)^n = (3 + 4)^2$ or 49 and $a^n + b^n = 3^2 + 4^2$ or 25. **61a.** no; $(5m)^2 = 25m^2$ **61b.** Yes; the power of a product is the product of the powers. **61c.** no; $(-3a)^2 = 9a^2$ **61d.** no; $2(c^7)^3 = 2c^{21}$ **63.** B

65.

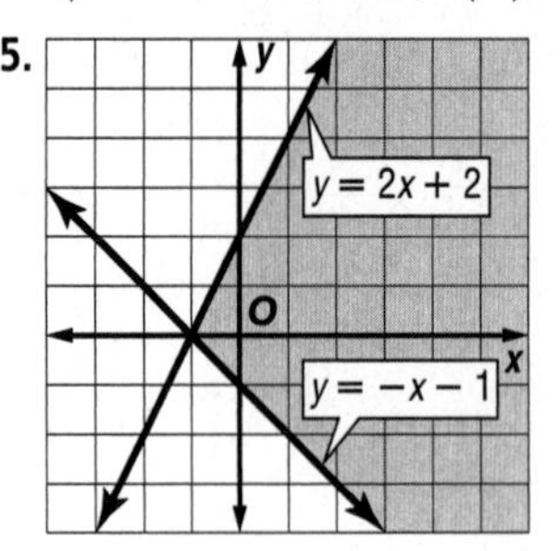

67.

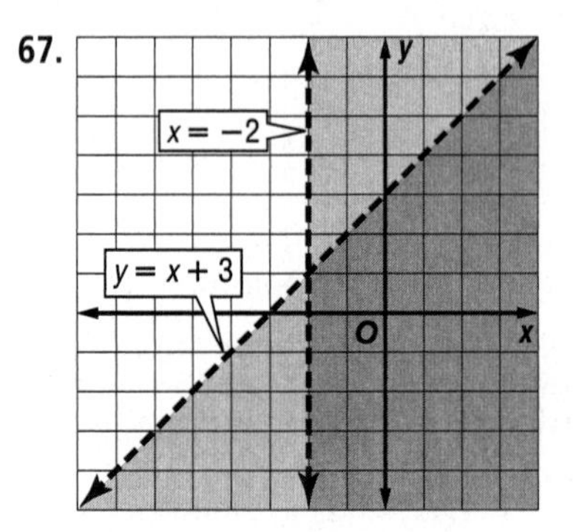

69. $\{(-4, 2), (-3, 0), (1, 4)\}$
71. $\{a \mid -2 < a < 3\}$

−4 −2 0 2 4 6

73. $\{m \mid m > 1\}$

−4 −2 0 2 4 6

75. $(-3, -4)$ **77.** $f(x) = -2x$ **79.** $\{(-5, 2), (-2, 3), (0, 5), (4, 9)\}$; $\{(2, -5), (3, -2), (5, 0), (9, 4)\}$ **81.** $\{(-3, 2), (0, 4), (2, -1), (-1, -3)\}$; $\{(2, -3), (4, 0), (-1, 2), (-3, -1)\}$ **83.** $\frac{1}{3}$ **85.** 2 **87.** $\frac{7}{18}$ **89.** $\frac{11}{8}$

Chapter 7 Polynomials

Page 357 ***Chapter 7*** ***Get Ready***

1. 2^5 **3.** 5^2 **5.** a^6 **7.** $\left(\frac{1}{2}\right)^5$ **9.** 9 **11.** 36 **13.** $\frac{16}{81}$ **15.** The probability of correctly guessing the outcome of a flipped penny six times in a row is $\frac{1}{64}$. **17.** 36π m² or about 113.04 **19.** 125 cm³

Pages 361–364 ***Lesson 7-1***

1. No; $5 - 7d$ shows subtraction, not multiplication. **3.** Yes; a single variable is a monomial. **5.** $36a^6b^4$ **7.** 3^8 or 6,561 **9.** $-8f^6g^3$ **11.** $6a^5b^6$ **13.** $800x^8y^{12}z^4$ **15.** Yes; $4x^3$ is the product of a number and three variables. **17.** No; $4n + 5m$ shows addition, not multiplication of variables. **19.** Yes; $\frac{1}{5}abc^{14}$ is the product of a number, $\frac{1}{5}$, and several variables. **21.** p^7q^5 **23.** $24j^8k^{13}$ **25.** $343b^9c^{18}$ **27.** 4^{12} or 16,777,216 **29.** a^4b^2 units² **31.** $-432c^2d^8$ **33.** $144a^8g^{14}$ **35.** $40b^{12}$ **37.** 10^5 *E* **39.** $30a^5b^7c^6$ **41.** $0.25x^6$ **43.** $\frac{-27}{64}c^3$ or $-0.421875c^3$ **45.** $-9x^3y^9$ **47.** $64k^9$ units³ **49.** $16\pi n^5$ units³ **51.** $(10^6)^4$ or 10^{24} **53.** The collision impact quadruples, since $2(2s)^2$ is $4(2s^2)$. **55.** $\frac{1}{4{,}194{,}304}$

Pages 370–373 ***Lesson 7-2***

1. 7^6 or 117,649 **3.** $\frac{q^4}{2p^5}$ **5.** $\frac{4a^4b^2}{c^6}$ **7.** $\frac{1}{y^4}$ **9.** 1 **11.** $\frac{g^8}{d^3c^5}$ **13.** D **15.** 3^6 or 729 **17.** y^2z^7 **19.** $\frac{81m^{28}}{256x^{20}y^{12}}$ **21.** 1 **23.** $\frac{1}{125}$ **25.** $\frac{8}{27}$ **27.** $4a^4c^4$ **29.** $3x^2y$ units **31.** $-\frac{1}{5a^5}$ **33.** $\frac{6k^{17}}{h^3}$ **35.** $\frac{19}{3z^{12}}$ **37.** $\frac{p}{q}$ **39.** $\frac{b^2}{5n^2z^3}$ **41.** 2^{-n} **43.** jet plane **45.** $\frac{1}{10^5} - \frac{1}{10^4}$ cm; $\frac{1}{100{,}000} - \frac{1}{10{,}000}$ cm **47.** 100 **49.** $\frac{a^3b^5}{ab^2} = a^3a^{-1}b^5b^{-2} = a^{3-1}b^{5-2} = a^2b^3$ **51.** 5^{6x-2} **53.** Since each number is obtained by dividing the previous number by 3, $3^1 = 3$ and $3^0 = 1$. **55.** You can compare pH levels by finding the ratio of one pH to another written in terms of the concentration c of hydrogen ions, $c = \left(\frac{1}{10}\right)^{\text{pH}}$. Sample answer: To compare a pH of 8 with a pH of 9 requires simplifying the quotient of powers, $\frac{\left(\frac{1}{10}\right)^8}{\left(\frac{1}{10}\right)^9} = 10$. Thus, a pH of 8 is ten times more acidic than a pH of 9. **57.** J **59.** $12x^8y^4$ **61.** $9c^2d^{10}$ **63.** $-108a^3b^9$ **65.** Sample answer: 3 oz of mozzarella, 4 oz of Swiss; 4 oz of mozzarella, 3 oz of Swiss; 5 oz of mozzarella, 3 oz of Swiss **67.** 37

Pages 378–381 Lesson 7-3

1. yes; binomial **3.** yes; trinomial **5.** 0 **7.** 5 **9.** $2a + 4x^2 - 7a^2x^3 - 2ax^5$ **11.** $x^3 + 3x^2y + 3xy^2 + y^3$ **13.** no **15.** yes; trinomial **17.** no **19.** $ab - 4x^2$ **21.** $\pi r^2 - r^2$ **23.** 1 **25.** 0 **27.** 4 **29.** 9 **31.** 6 **33.** 9 **35.** $7 + 9x^3 - 3x^5$ **37.** $4a + x^3 + 5a^2x^6$ **39.** $5y^4 + 2x^2 + 10x^3y^2 - 3x^9y$ **41.** $-5 - 8a^5x - a^2x^2 + 2ax^4$ **43.** $6x^2 + 2x - 1$ **45.** $x^2 - 2xb + b^2$ **47.** $4ax^3 + 9x^2 - 2a^2x + 3$ **49.** $4x^3y - x^2y^3 + 3xy^4 + y^4$ **51.** $t > 15$; For $t > 15$, the number of quadruplet births declines dramatically. **53.** about 92.15 in^3 **55.** Sample answer: −8 **57.** True; for the degree of a binomial to be zero, the highest degree of both terms would need to be zero. Then the terms would be like terms. With these like terms combined, the expression is not a binomial, but a monomial. Therefore, the degree of a binomial can never be zero. Only a monomial can have a degree of zero. **59.** A polynomial model of a set of data can be used to predict future trends in data. Answers should include the following:

t	H	Actual Data Values
0	70	70
1	78.75	79
2	89.5	90
3	97.75	97
4	105	103
5	118.75	115

The polynomial function models the data almost exactly for the first three values of t, and then closely for the next three values.
Someone might point to this model as evidence that the time people spend playing video games is on the rise. This model may assist video game manufacturers in predicting production needs. **61.** F **63.** $\frac{-5}{n^3}$ **65.** $-y^2m^{15}$ **67.** no **69.** no **71.** 66.7 mi **73.** ±1.8 **75.** $8n$ **77.** $a - 2b$

Pages 386–388 Lesson 7-4

1. $2p^2 + 6p$ **3.** $10cd - 3d + 4c - 6$ **5.** $11a^2 + 6a + 1$ **7.** $3ax^2 - 9x - 9a + 8a^2x$ **9.** $T = 2{,}829n + 224{,}395$ **11.** $4n^2 + 5$ **13.** $2a^2 - 6a + 8$ **15.** $5x + 2y + 3$ **17.** $10d^2 + 8$ **19.** $-8y^3 - 3y^2 - y + 17$ **21.** $-2x^2 + 8x + 8$ **23.** $4x + 2y$ **25.** $-4a + 6b - 5c$ **27.** $2y^2 - y + 5$ **29.** $D = 1.8n^2 - 42n + 860$ **31.** $60 - 2x$; $40 - 2x$; $80 - 2x$ **33.** 19 in. **35.** Sample answer: $6x^2 + 4x + 7$ and $4x^2 + 3x + 4$ **37.** $x + 1$ **39.** 4 **41.** A **43.** 5 **45.** 3 **47.** $\frac{7a^3b^3}{c}$ **49.** $1728n^{20}$ **51.** Sample answer: $y = 4x + 17$ **53.** There is a limit to how fast one can keyboard. **55.** $-2b - 18$ **57.** $27a + 45b - 21$ **59.** $-6a^2 + 15a - 21$

Pages 392–395 Lesson 7-5

1. $-15y^2 - 6y$ **3.** $8a^4x - 6ax^2 + 12x^3$ **5.** $5t^2 - 11t$ **7.** $20n^4 + 30n^3 - 14n^2 - 13n$ **9.** $10{,}000 - x$ **11.** $10,440 **13.** 20 **15.** 1 **17.** $2w^4 - 9w^3$ **19.** $-10y^3 - 35y^2$ **21.** $-3n^3p + 6np^2$ **23.** $30x^3 + 18x^4 - 66x^5$ **25.** $-3cd^3 - 2c^3d^3 + 4c^2d^2$ **27.** $-9x^3 + 2x^2$ **29.** $10n^4 + 5n^3 - n^2 + 44n$ **31.** $-2y^3 - 17y^2 + 24y - 6$ **33.** savings account: $1500; certificate of deposit: $4500 **35.** $1.50t + 1.25mt$ **37.** $-\frac{1}{3}$ **39.** 8 **41.** 1 **43.** $3x - 2x - 2 = x - 2$ **45.** $4a^5b - \frac{8}{3}a^3b^2 + 6a^2b^3$ **47.** $8p^4q^2 - 4p^2q^4 + 36p^5q^2 + 12p^2q^3$ **49.** $19x^3 - x^2 + 2x$ **51.** 7 **53.** $15p^2 + 8p + 6$ **55.** $2.20 **57.** $126 **59.** $x + 2$ **61.** Sample answer: $4x$ and $x^2 + 2x + 3$; $4x^3 + 8x^2 + 12x$ **63.** $2x + 1$ or $2x - 1$ **65.** The product of a monomial and a polynomial can be modeled using an area model. The area of the figure shown at the beginning of the lesson is the product of its length $2x$ and width $(x + 3)$. This product is $2x(x + 3)$, which when the Distributive Property is applied becomes $2x(x) + 2x(3)$ or $2x^2 + 6x$. This is the same result obtained when the areas of the algebra tiles are added together. **67.** J **69.** $-4y^2 + 5y + 3$ **71.** $7p^3 - 3p^2 - 2p - 7$ **73.** yes; binomial **75.** yes; monomial **77.** $9n + 4 \geq 7 - 13n$; $\left\{n \mid n \geq \frac{3}{22}\right\}$ **79.** $y = -2x - 3$ **81.** 7 **83.** no solution **85.** a^2 **87.** $-24y^4$ **89.** $-10n^3 + 40n^2 - 20n$

Pages 401–403 Lesson 7-6

1. $y^2 + 7y + 12$ **3.** $a^2 - 3a - 40$ **5.** $27p^2 - 21p + 2$ **7.** $\frac{6x^2 + 7x - 3}{2}$ or $3x^2 + \frac{7}{2}x - \frac{3}{2}$ **9.** $8x^4 + 24x^3 - 12x - 2$ **11.** $15m^4 + 13m^3 - 6m^2 + 11m - 63$ **13.** $n^2 + 13n + 42$ **15.** $a^2 - 8a + 15$ **17.** $p^2 - 8p - 20$ **19.** $3k^2 + 34k - 24$ **21.** $36g^2 + 51g + 18$ **23.** $18a^2 - 63a + 40$ **25.** $25m^2 - 60m + 36$ **27.** $49t^2 - 25$ **29.** $22a^2 + 21ab - 18b^2$ **31.** $6x^2 - \frac{17}{2}x + 3$ units^2 **33.** $9\pi x^2 + 24\pi x + 16\pi$ units^2 **35.** $a^3 - 11a^2 + 29a - 15$ **37.** $21k^3 + 34k^2 - 19k - 36$ **39.** $y^4 + y^3 - 38y^2 + 41y - 5$ **41.** $5t^2 - 34t + 56$ **43.** $63y^3 - 57y^2 - 36y$ units^3 **45.** $x - 2$, $x + 4$ **47.** bigger; 10 sq ft **49.** The three monomials that make up the trinomial are similar to the three digits that make up the 3-digit number. The single monomial is similar to a 1-digit number. With each procedure you perform 3 multiplications. The difference is that polynomial multiplication involves variables and the resulting product is often the sum of two or more monomials while numerical multiplication results in a single number. **51.** Sometimes; the product of $x + 1$ and $x^2 + 2x + 3$ is $x^3 + 3x^2 + 5x + 3$, which has 4 terms; the product of $y + 1$ and $x^3 + 2x^2 + 3x$ is $x^3y + 2x^2y + 3xy + x^3 + 2x^2 + 3x$, which has 6 terms. **53.** Multiplying binomials and two-digit numbers involve the use of the Distributive Property twice. Each procedure involves four multiplications and the addition of like terms. *(continued)*

$$\begin{aligned} 24 \times 36 &= (4 + 20)(6 + 30) \\ &= (4 + 20)6 + (4 + 20)30 \\ &= (24 + 120) + (120 + 600) \\ &= 144 + 720 \\ &= 864 \end{aligned}$$

The like terms in vertical two-digit multiplication are digits with the same place value. **55.** J **57.** $20a^3 + 2a^2 - 10a + 27$ **59.** -13 **61.** $2a + 1$ **63.** $y = \frac{ax - 2cz}{b}$ **65.** 23 **67.** 3.25 **69.** $49x^2$ **71.** $16y^4$ **73.** $9g^8$

Pages 407–409 ***Lesson 7-7***

1. $a^2 + 12a + 36$ **3.** $9x^2 + 54xy + 81y^2$ **5.** $x^4 - 12x^2y + 36y^2$ **7.** $0.5Bb + 0.5b^2$ **9.** $64x^2 - 25$ **11.** $16y^4 - 9z^2$ **13.** $y^2 + 8y + 16$ **15.** $a^2 - 10a + 25$ **17.** $49 - 56y + 16y^2$ **19.** 25% **21.** $c^2 - 4$ **23.** $81x^2 + 54x + 9$ **25.** $144p^2 - 9$ **27.** $m^2 + 14mn + 49n^2$ **29.** $9n^2 - 60np + 100p^2$ **31.** $16d^2 - 169$ **33.** $9a^4 - 6a^2b^2 + b^4$ **35.** $25x^8 - y^2$ **37.** $\frac{16}{25}x^2 + 16x + 100$ **39.** $p^4 - 25p^2 + 144$ **41.** $a^2 + 2a + 1$ **43.** 1 **45.** $(1189.66 - 3.14s^2 - 56.52s)$ ft^2

47.

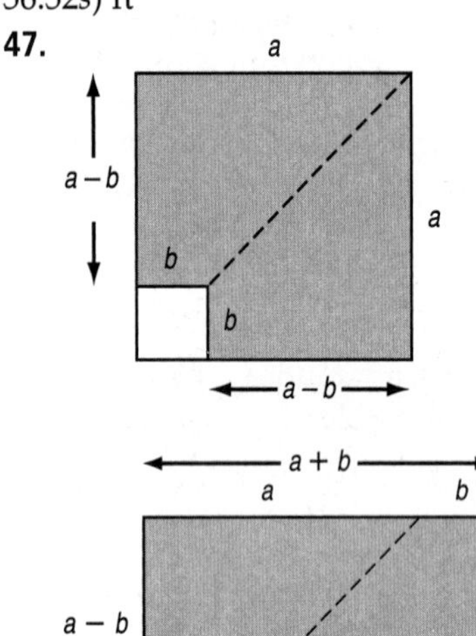

Area of rectangle $= (a - b)(a + b)$

or

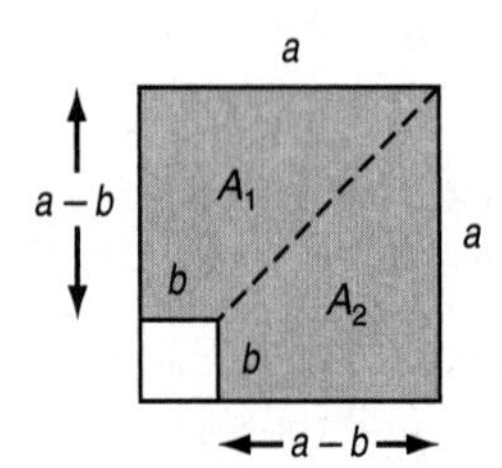

Area of a trapezoid $= \frac{1}{2}$(height)(base 1 + base 2)

$A_1 = \frac{1}{2}(a - b)(a + b)$

$A_2 = \frac{1}{2}(a - b)(a + b)$

Total area of shaded region

$= \left[\frac{1}{2}(a - b)(a + b)\right] + \left[\frac{1}{2}(a - b)(a + b)\right]$

$= (a - b)(a + b)$

49.

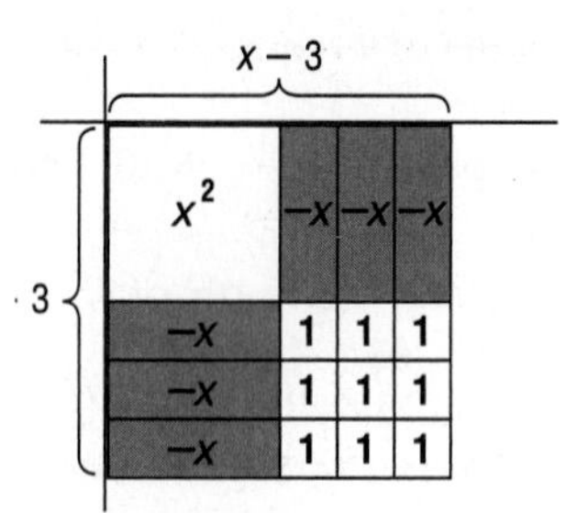

51a. $a^3 + 3a^2b + 3ab^2 + b^3$ **51b.** $x^3 + 6x^2 + 12x + 8$

51c.

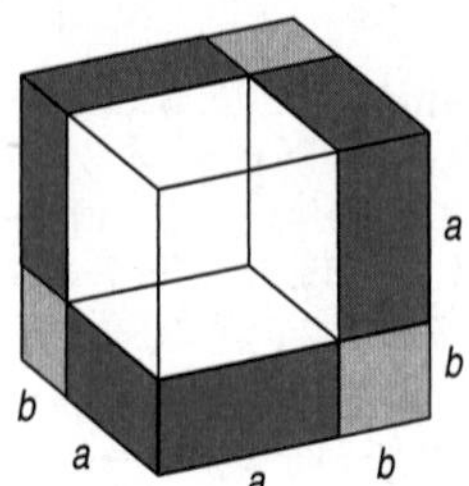

53. D **55.** $x^2 + 9x + 14$ **57.** $20y^2 - 29y + 6$ **59.** $3x^3 - 11x^2 + 14x - 8$ **61.** 4 **63.** 0 **65.** (0, 25) **67.** (5, –2) **69.** $y = -2.5x + 2$ **71.** 71 **73.** b

Pages 410–414 ***Chapter 7*** ***Study Guide and Review***

1. negative exponent **3.** Quotient of Powers **5.** trinomial **7.** polynomial **9.** Product of Powers **11.** y^7 **13.** $20a^5x^5$ **15.** $576x^5y^2$ **17.** $-\frac{1}{2}m^4n^8$ **19.** $16\pi x^6b^2$ **21.** $\frac{27b^3c^5}{64d^3}$ **23.** $\frac{27b}{14}$ **25.** $\frac{bx^3}{3ay^2}$ **27.** $\frac{1}{64a^6}$ **29.** $\frac{4n}{m^2}$ **31.** 2 **33.** 5 **35.** $3x^4 + x^2 - x - 5$ **37.** $2\pi x^2 + 12x + 5$ **39.** $4x^2 - 5xy + 6y^2$ **41.** $21m^4 - 10m - 1$ **43.** $-7p^2 - 2p + 25$ **45.** $4b^2 + 9b$ **47.** $61y^3 - 16y^2 + 167y - 18$ **49.** 2 **51.** $\$16.05x + \$27.52y + \$13.41z$ **53.** $4a^2 + 13a - 12$ **55.** $18x^2 - 0.125$ **57.** $12p^3 - 13p^2 + 11p - 6$ **59.** $x^2 - 36$ **61.** $64x^2 - 80x + 25$ **63.** $36a^2 - 60ab + 25b^2$ **65.** $0.25RR + 0.5Rr + 0.25rr$; 0.25

Chapter 8 Factoring

Page 419 ***Chapter 8*** ***Get Ready***

1. $12 - 3x$ **3.** $-7n^2 + 21n - 7$ **5.** $4(4e + 10p + 20t)$ **7.** $3n^2 + 11n - 20$ **9.** $-2x - 4xy + 96y^2$ **11.** $y^2 + 18y + 81$ **13.** $9m^2 + 30mn + 25n^2$

Pages 422–424 ***Lesson 8-1***

1. $2 \cdot 2 \cdot p \cdot p$ **3.** $-1 \cdot 2 \cdot 2 \cdot 5 \cdot 5 \cdot x \cdot x \cdot x \cdot y \cdot z \cdot z$ **5.** 5 **7.** $18y$ **9.** $4r$ **11.** $2 \cdot 3 \cdot 11 \cdot d \cdot d \cdot d \cdot d$ **13.** $-1 \cdot 7 \cdot 7 \cdot a \cdot a \cdot a \cdot b \cdot b$ **15.** $2 \cdot 2 \cdot 2 \cdot 2 \cdot 2 \cdot 5 \cdot p \cdot q \cdot q$ **17.** The minimum value is 40 mm; the factors of 96 whose sum when doubled is the least are 12 and 8. The maximum value is 194 mm; the factors of 96 whose sum when doubled is the greatest are 1 and 96. **19.** 9 **21.** 1 (relatively prime) **23.** $6d$ **25.** $5s$ **27.** $2mn$ **29.** base 1 cm, height 40 cm; base 2 cm, height 20 cm; base 4 cm, height 10 cm; base 5 cm, height 8 cm; base 8 cm, height 5 cm; base 10 cm, height 4 cm; base 20 cm, height 2 cm; base 40 cm, height 1 cm **31.** false; 2 **33.** Sample answer: $5x^2$ and $10x^3$; $5x^2 = 5 \cdot x \cdot x$ and $10x^3 = 2 \cdot 5 \cdot x \cdot x \cdot x$. The GCF is $5 \cdot x \cdot x$ or $5x^2$. **35.** D **37.** $4x^2 - 4x + 1$ **39.** $49p^4 + 56p^2 + 16$

41. $20h^2 + 52hk + 5k^2$ **43.** Sample answer: Let n = the number of video minutes, $10n + 50 \leq 500$; $n \leq 45$; up to 45 min. **45.** 0 **47.** $10x + 40$ **49.** $6g^2 - 8g$ **51.** $7(b + c)$

Pages 429–431 ***Lesson 8-2***

1. $9x(x + 4)$ **3.** $(5y + 4)(y - 3)$ **5.** 0, −5 **7.** $0, \frac{5}{3}$ **9.** $5(x + 6y)$ **11.** $2h(7g - 9)$ **13.** $x(15xy^2 + 25y + 1)$ **15.** $(x + 3)(x + 2)$ **17.** $(6x - 1)(3x - 5)$ **19.** $(2x - 3)(4a - 3)$ **21.** 0, 24 **23.** −4, 5 **25.** $\frac{3}{2}, \frac{8}{3}$ **27.** −4, 0 **29.** 3 s; yes **31.** $4xy^2z\,(3x + 10yz)$ **33.** $(9x^2 + 30xy + 25y^2)$ in^2 **35.** $\frac{1}{2}n(n - 3)$; 35 diagonals **37.** 63 games **39.** $3r^2(4 - \pi)$ **41.** Sample answer: $(x + 3)(x + 2) = 0$; set each factor equal to zero and solve each equation; $x + 3 = 0$, $x = -3$; $x + 2 = 0$, $x = -2$. The roots are −3 and −2. **43.** $(a^x - b^x)(a^y + b^y)$; Sample answer: Group terms with common factors, $(a^{x+y} + a^xb^y)$ and $(-a^yb^x - b^{x+y})$. Factor the GCF from each grouping to get $a^x(a^y + b^y) - b^x(a^y + b^y)$. Then use the Distributive Property to get $(a^x - b^x)(a^y + b^y)$. **45.** B **47.** a **49.** $3xy$ **51.** $4p^2 - 25q^2$ **53.** 37 shares **55.** $x^2 - 9x + 20$ **57.** $18a^2 - 6a - 4$ **59.** $8y^2 - 14y - 15$

Pages 437–439 ***Lesson 8-3***

1. $(x + 3)(x + 8)$ **3.** $(w - 3)(w + 16)$ **5.** $(y - 4)(y + 5)$ **7.** −1, −6 **9.** 1, 9 **11.** −13 and −12 or 12 and 13 **13.** $(c + 5)(c + 7)$ **15.** $(d - 5)(d - 2)$ **17.** $(g - 4)(g - 15)$ **19.** $(n - 6)(n + 9)$ **21.** $(z + 2)(z - 20)$ **23.** $(x - 2)(x + 15)$ **25.** −6, 2 **27.** 3, 16 **29.** −15, −1 **31.** −25, 2 **33.** $36 = \frac{n^2 - n}{2}$; 9 justices **35.** 120 m by 68 m **37.** $4x + 26$ **39.** −35, 15 **41.** In this trinomial, $b = 6$ and $c = 9$. This means that $m + n$ is positive and mn is positive. Only two positive numbers have both a positive sum and product. Therefore, negative factors of 9 need not be considered. **43.** Aleta; to use the Zero Product Property, one side of the equation must equal zero. **45.** −15, −9, 9, 15 **47.** 4, 6 **49.** C **51.** $-3, \frac{5}{2}$ **53.** $-\frac{9}{5}, 0$ **55.** $3pq^3$ **57.** at least 5 days **59.** $(a + 4)(3a + 2)$ **61.** $(2p + 7)(p - 3)$ **63.** $(2g - 3)(2g - 1)$

Pages 444–446 ***Lesson 8-4***

1. $(3a + 2)(a + 2)$ **3.** $2(p + 3)(p + 4)$ **5.** $3(2x - 1)(x + 3)$ **7.** $-3, -\frac{2}{3}$ **9.** $-\frac{5}{2}, \frac{4}{3}$ **11.** $(2x + 5)(x + 1)$ **13.** $(5d - 4)(d + 2)$ **15.** $(3g - 2)(3g - 2)$ **17.** $(x - 4)(2x + 5)$ **19.** prime **21.** $2(3r + 2)(r - 3)$ **23.** $-5, -\frac{2}{5}$ **25.** $-\frac{5}{4}, \frac{7}{3}$ **27.** $-\frac{2}{7}, 1$ **29.** $\frac{1}{3}, \frac{2}{5}$ **31.** 3 s **33.** $(12x + 20y)$ in.; The area of the square equals $(3x + 5y)(3x + 5y)$ in.^{2}, so the length of one side is $(3x + 5y)$ in. The perimeter is $4(3x + 5y)$ or $(12x + 20y)$ in. **35.** $-\frac{7}{3}, \frac{5}{2}$ **37.** $\frac{1}{2}, \frac{7}{2}$ **39.** 5 in. by 7 in. **41.** ±25, ±14, ±11, ±10 **43.** Sample answer: Find two numbers, m and n, that are the factors of ac and that add to b. **45.** G **47.** prime **49.** $-\frac{7}{5}, 4$ **51.** 0, 12

53. $1(5.5) + 8.09(5.5) = (1 + 8.09)(5.5)$ or $9.09(5.5)$ **55.** 6 **57.** 11 **59.** 15

Pages 450–452 ***Lesson 8-5***

1. $(n + 9)(n - 9)$ **3.** $2x^3(x + 7)(x - 7)$ **5.** prime **7.** $-\frac{5}{2}, \frac{5}{2}$ **9.** $-\frac{11}{7}, 0, \frac{11}{7}$ **11.** $(x + 7)(x - 7)$ **13.** prime **15.** $3(5 + 2p)(5 - 2p)$ **17.** $(12a + 7b)(12a - 7b)$ **19.** $(n + 2)(n - 2)(n + 5)$ **21.** $(z^2 + 4)(z - 2)(z + 2)$ **23.** $-\frac{6}{5}, \frac{6}{5}$ **25.** $-\frac{2}{3}, \frac{2}{3}$ **27.** $-\frac{2}{7}, \frac{2}{7}$ **29.** −18, 18 **31.** 2 in. **33.** $P = \frac{1}{2}d(v_1 + v_2)(v_1 - v_2)$ **35.** Sample answer: $x^2 - 25 = (x + 5)(x - 5)$ **37.** Manuel; $4x^2 + y^2$ is not the *difference* of squares. **39.** Sample answer: A maximum height would be 1 foot. To find the hang time of a student athlete who attains a maximum height of 1 foot, solve the equation $4t^2 - 1 = 0$. You can factor the left side using the difference of squares pattern since $4t^2$ is the square of $2t$ and 1 is the square of 1. Thus, the equation becomes $(2t + 1)(2t - 1) = 0$. Using the Zero Product Property, each factor can be set equal to zero, resulting in two solutions, $t = -\frac{1}{2}$ and $t = \frac{1}{2}$. Since time cannot be negative, the hang time is $\frac{1}{2}$ second. **41.** J **43.** $(2x - 1)(3x - 4)$ **45.** −16, −2 **47.** −2, 4 **49.** $x^2 + 2x + 1$ **51.** $9x^2 - 24x + 16$

Pages 458–460 ***Lesson 8-6***

1. yes; $(y + 4)^2$ **3.** $2(x^2 + 9)$ **5.** $(2x - 7)(4x + 5)$ **7.** −3 **9.** −1, 7 **11.** about 3.35 s **13.** no **15.** yes; $(5a - 12b)^2$ **17.** $4(a - 3b)(a + 3b)$ **19.** $2(5g + 2)^2$ **21.** $2(5n + 1)(2n + 3)$ **23.** $2(3y - 4)^2$ **25.** 5 **27.** $-\frac{2}{3}$ **29.** $-\frac{2}{5}$ **31.** $-3 \pm \sqrt{2}$ **33.** $6 \pm \sqrt{11}$ **35.** 144 ft **37.** yes; 2 s **39.** prime **41.** $2mn(m^2 - 4n)(2m + 3)$ **43.** $8x^2 - 22x + 14$ in^2 if $x > \frac{7}{4}$, $8x^2 - 34x + 35$ in^2 if $x < \frac{7}{4}$ **45.** Sample answer: $x^3 + 5x^2 - 4x - 20$; group terms with common factors and factor out the GCF from each grouping to get $(x^2 - 4)(x + 5)$. Then factor the perfect square trinomial to get $(x - 2)(x + 2)(x + 5)$. **47.** 4, −4 **49.** 100 **51.** C **53.** $\pm\frac{4}{3}$ **55.** $-3, \frac{1}{4}$ **57.** $\{r \mid r > -55\}$ **59.** $\{t \mid t > 19\}$ **61.** $y = 65x + 50$

Pages 461–464 ***Chapter 8*** ***Study Guide and Review***

1. false; composite **3.** false; sample answer: 64 **5.** false; $2^4 \cdot 3$ **7.** true **9.** true **11.** $2^2 \cdot 7 \cdot n \cdot n \cdot n$ **13.** $2 \cdot 3 \cdot 5^2 \cdot s \cdot t$ **15.** 5 **17.** $4ab$ **19.** $5n$ **21.** 12-in. square **23.** $(a - 4c)(a + b)$ **25.** $2a(13b + 9c + 16a)$ **27.** $(8m - 3n)(3a + 5b)$ **29.** $0, -\frac{7}{4}$ **31.** 0.75 s **33.** $(x - 12)(x + 3)$ **35.** $(r - 3)(r - 6)$ **37.** −6, 11 **39.** prime **41.** $(4b + 3)(3b + 2)$ **43.** $4, -\frac{5}{2}$ **45.** 2 s **47.** $2y(y - 8)(y + 8)$ **49.** $\left(\frac{1}{2}n - \frac{3}{4}r\right)\left(\frac{1}{2}n + \frac{3}{4}r\right)$ **51.** $-\frac{5}{3}, \frac{5}{3}$ **53.** $-\frac{5}{7}, \frac{5}{7}$ **55.** $(a + 9)^2$ **57.** $(2 - 7r)^2$ **59.** 0, 2 **61.** $\frac{9}{7}$ **63.** 1 in.

Chapter 9 Quadratic and Exponential Functions

Page 469 *Chapter 9* *Get Ready*

1.

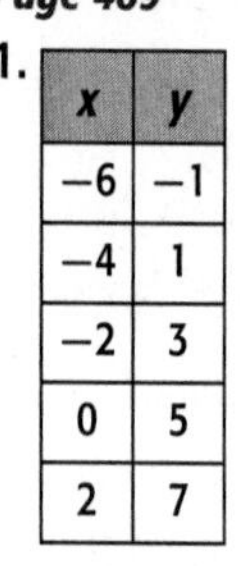

x	y
−6	−1
−4	1
−2	3
0	5
2	7

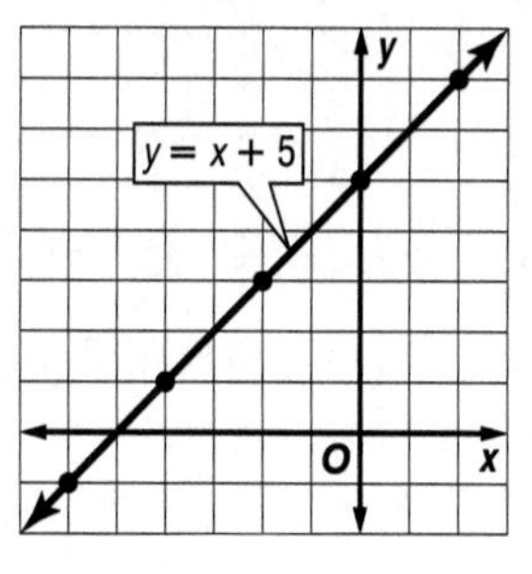

3.

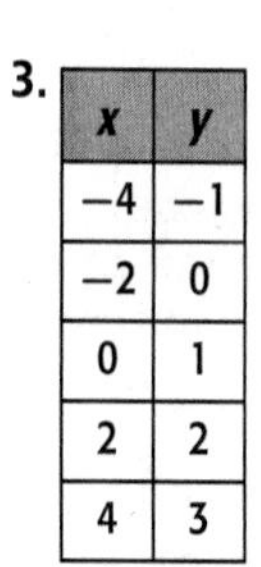

x	y
−4	−1
−2	0
0	1
2	2
4	3

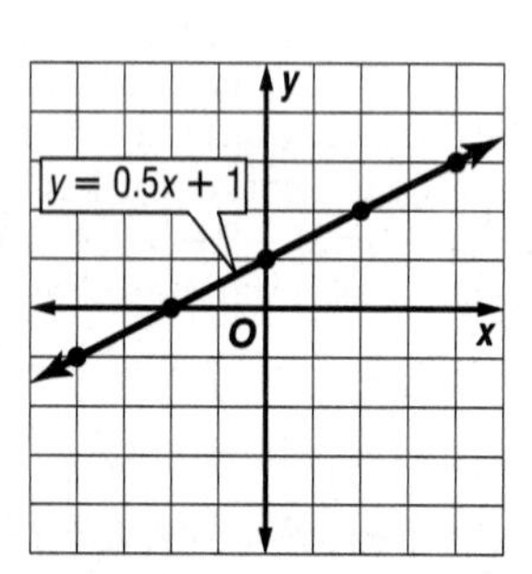

5.

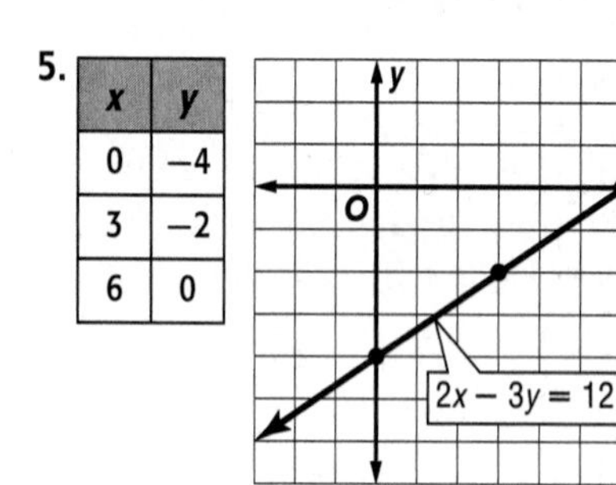

x	y
0	−4
3	−2
6	0

7. $T = 200 + 35m$

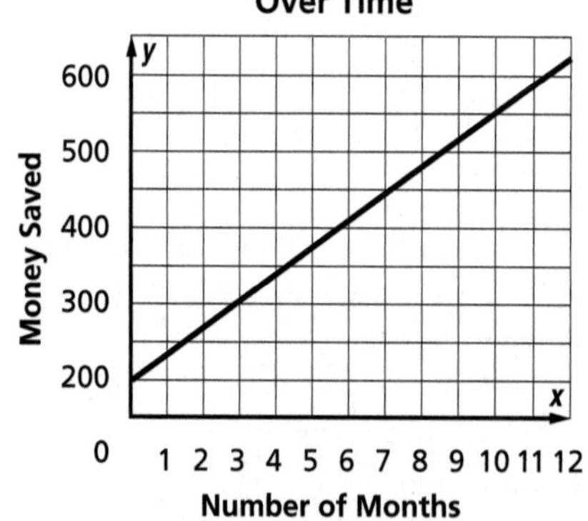

9. yes; $(a - 7)^2$ **11.** no **13.** no **15.** yes ; $(4s - 3)^2$
17. −16, −23, −30 **19.** 56, 64, 72

Pages 475–477 

1.

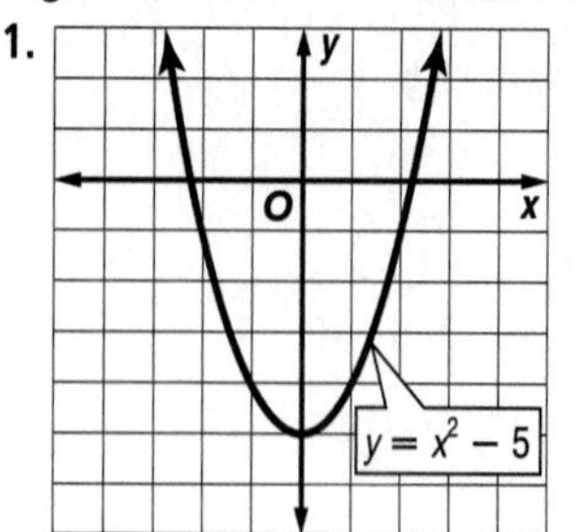

3.

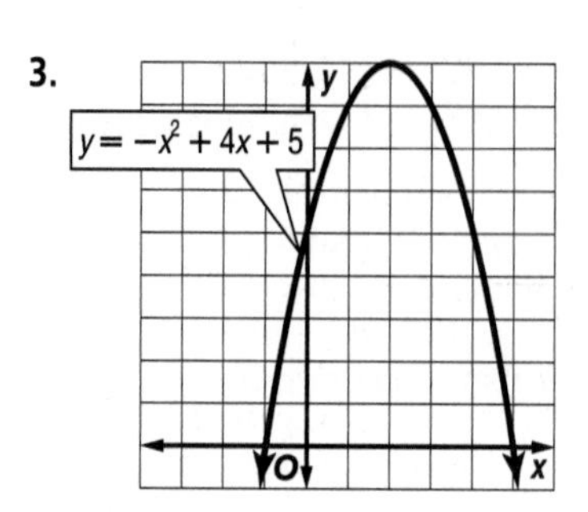

5. $x = -2$; $(-2, -13)$; min

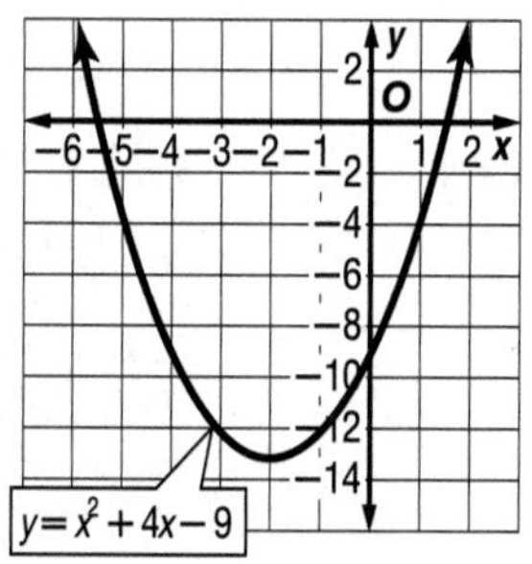

7. $x = 2$; $(2, 1)$; max

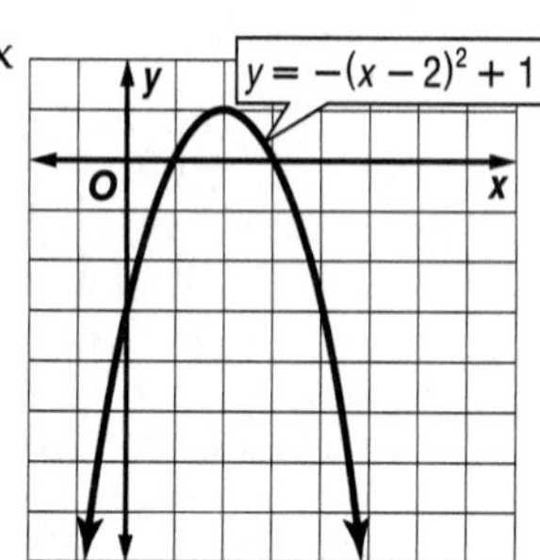

9. B **11.**

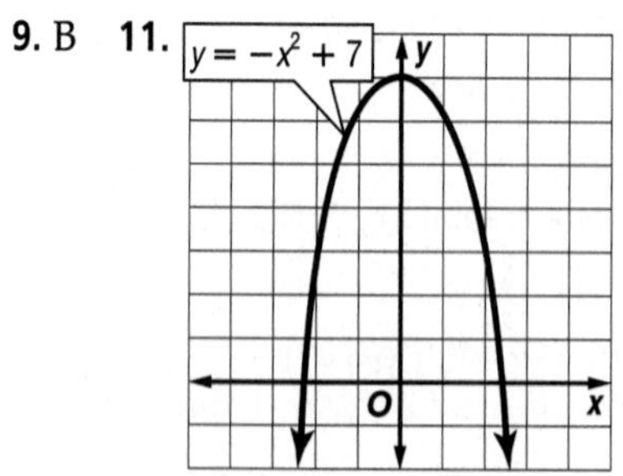

13.

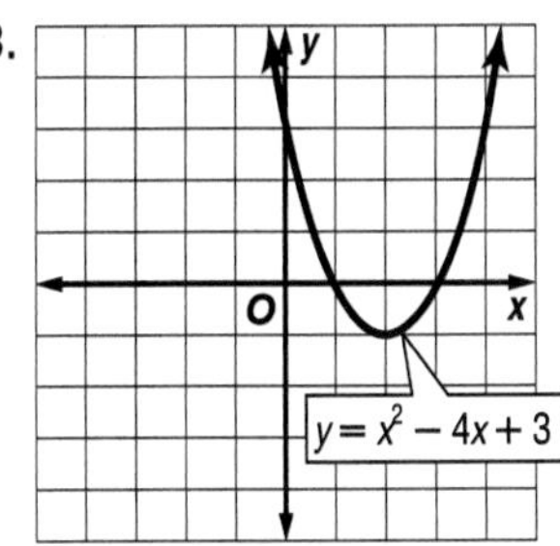

15.

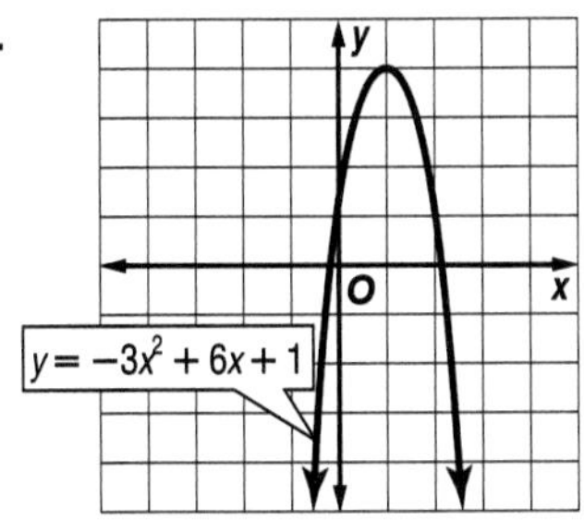

17. $x = 0$; $(0, 0)$; max

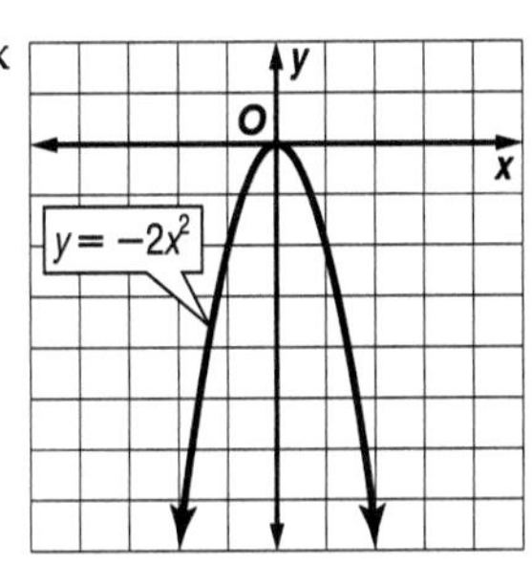

19. $x = 0$; $(0, 5)$; max

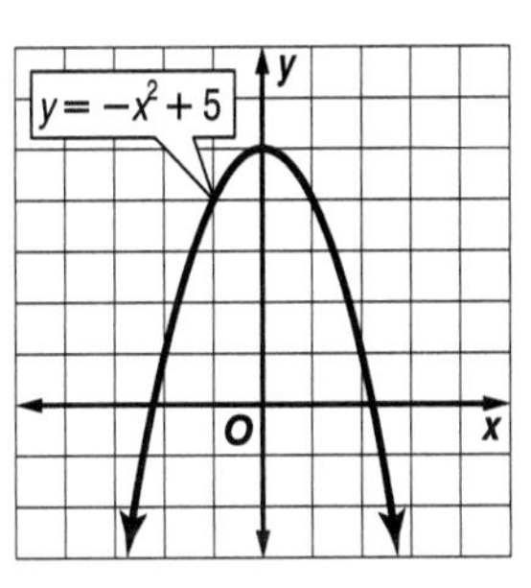

21. $x = -3$; $(-3, 24)$; max

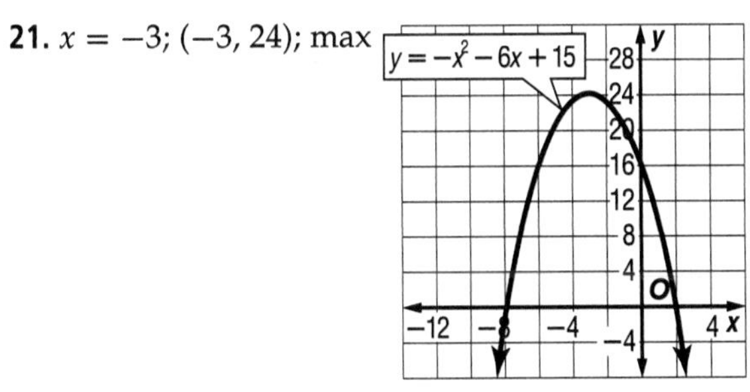

23. $x = 2$; $(2, 1)$; min

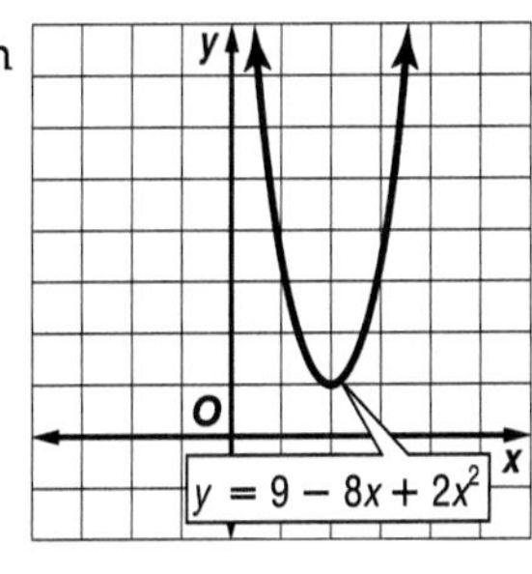

25. $x = \frac{5}{8}$ **27.** No; The height needs to be 20 ft to win a prize. **29.** 10 m; 100 m^2

31. $x = 5$; $(5, -2)$; min

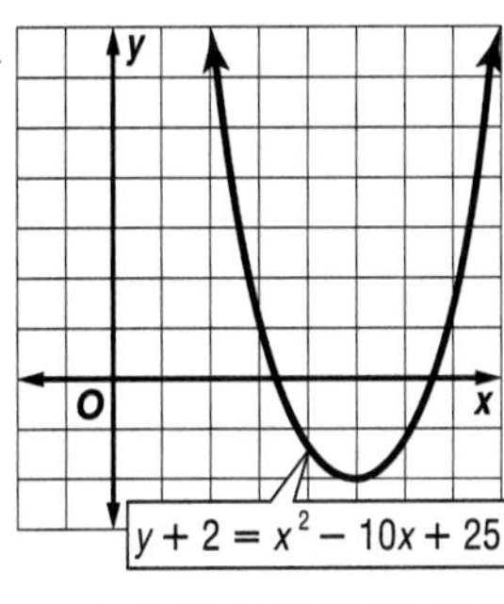

33. $x = -1$; $(-1, -1)$; min

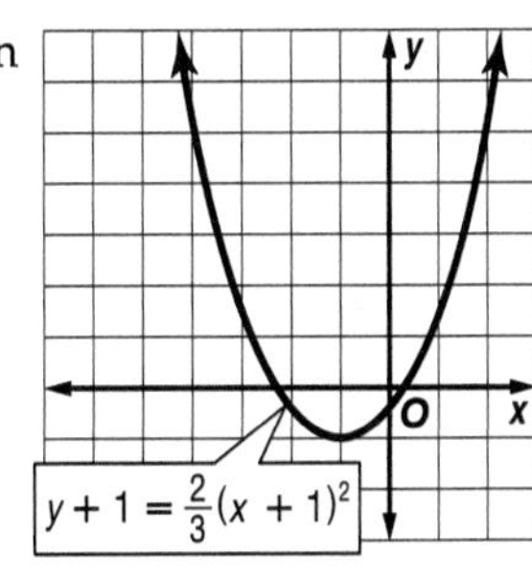

35. $x = -1$ **37.** 630 ft **39.** 74 ft

41. $t = 0$ and 5.625 s; The height of the ball is zero before it is kicked and again when the ball lands on the ground.

43.

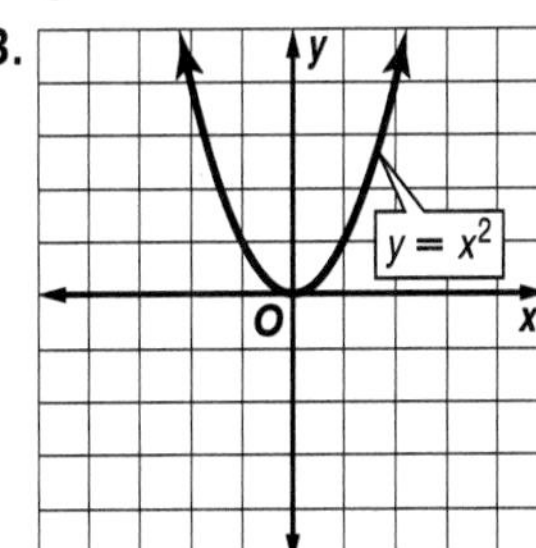

45. $\{y \mid y \geq -9\}$

47. domain: $x \leq -3$ and $x \geq 3$; range: $y \geq 0$

49. Sample answer: $y = 4x^2 + 3x + 5$; write the equation for the axis of symmetry of a parabola, $x = -\frac{b}{2a}$. From the equation, $b = 3$ and $2a = 8$, so $a = 4$. Substitute these values for a and b into the equation $y = ax^2 + bx + c$. **51.** B **53.** prime **55.** $(2m - 1)^2$ **57.** prime **59.** $13x + 20y$ **61.** $6p^2 - p - 18$ **63.** 8 **65.** −3.5

Pages 483–485 ***Lesson 9-2***

1.

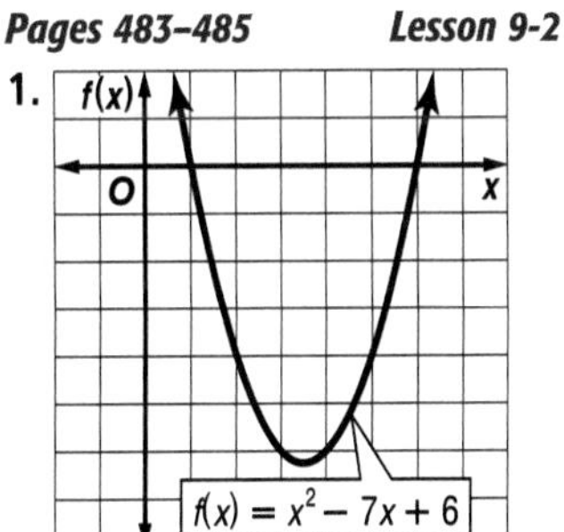

1, 6

3. 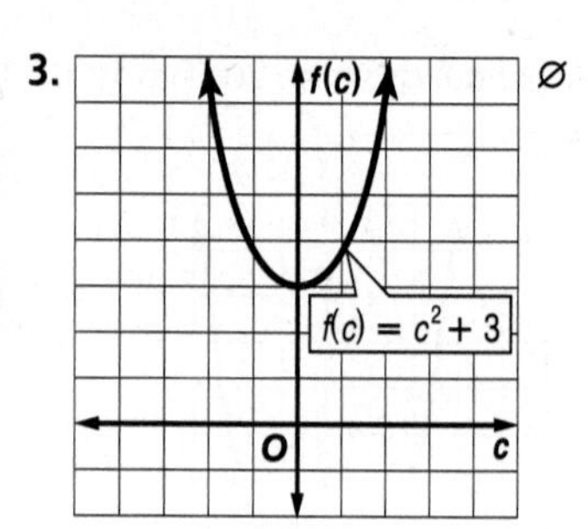

∅

17. 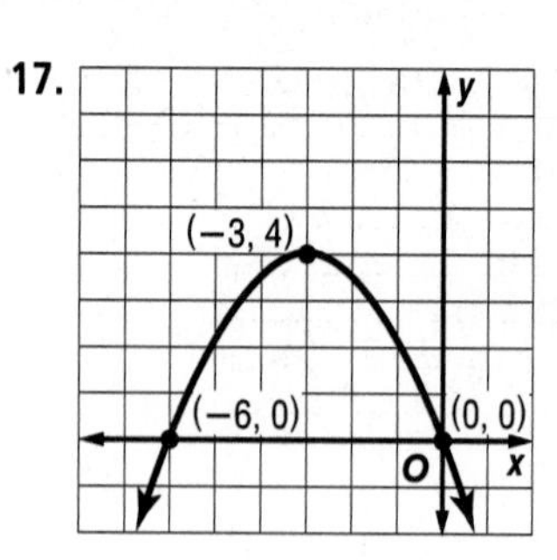

The graph is wider than the graph of $y = x^2$ and opens downward rather than upward. Also, the vertex is 3 units left and 4 units up from the vertex of the graph of $y = x^2$.

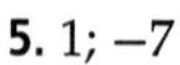
5. 1; −7

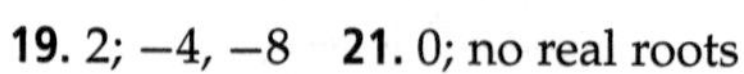
19. 2; −4, −8 21. 0; no real roots

7. 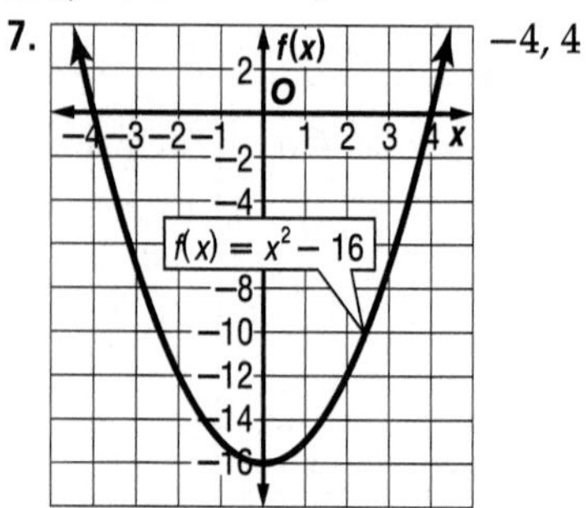

−4, 4

23. 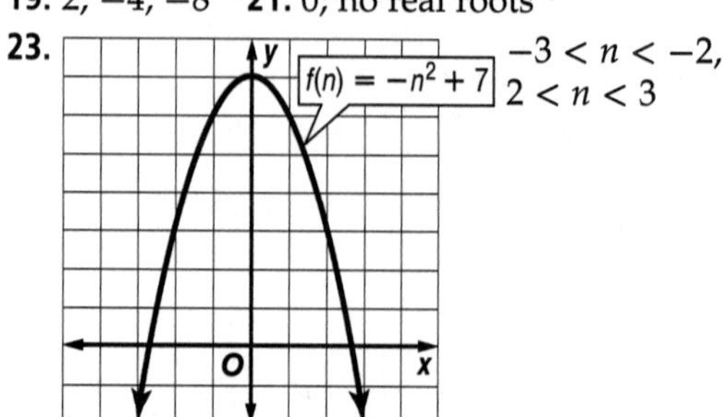

$-3 < n < -2$, $2 < n < 3$

9. −2, 6

11. 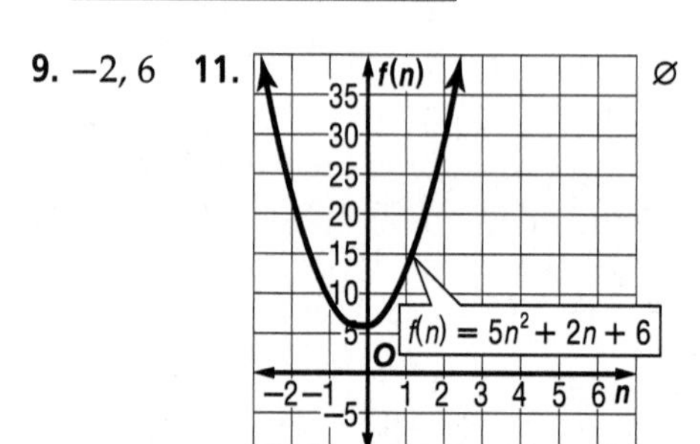

∅

25. 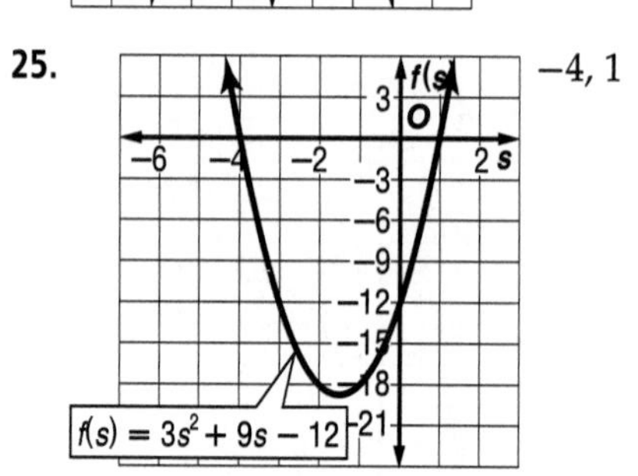

−4, 1

13. 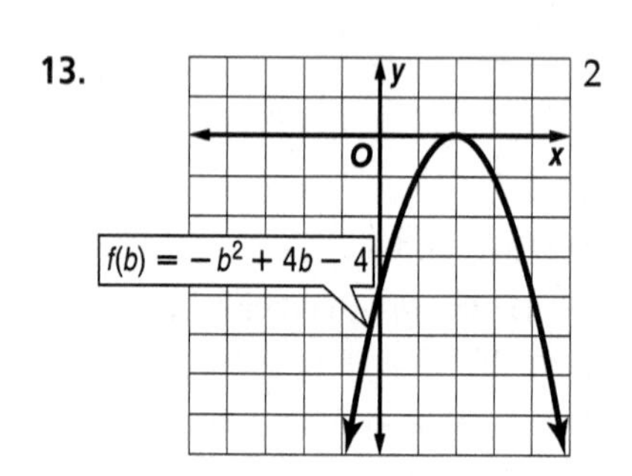

2

27. 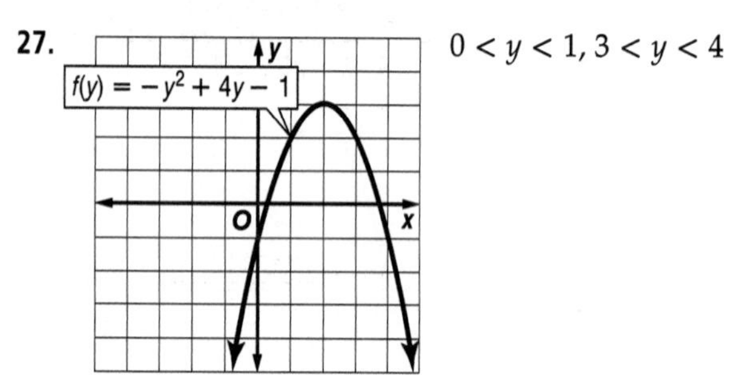

$0 < y < 1, 3 < y < 4$

15. 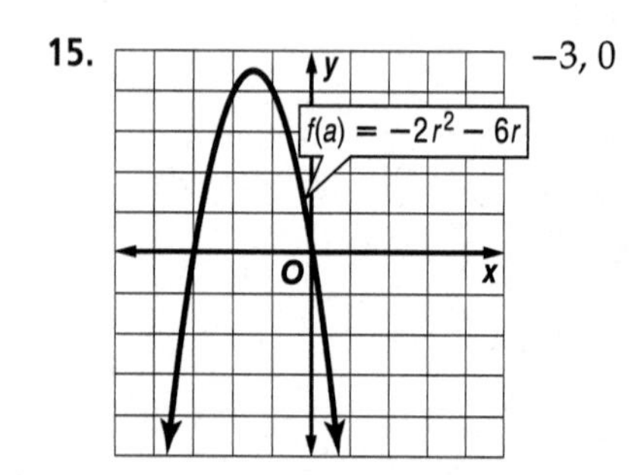

−3, 0

29. 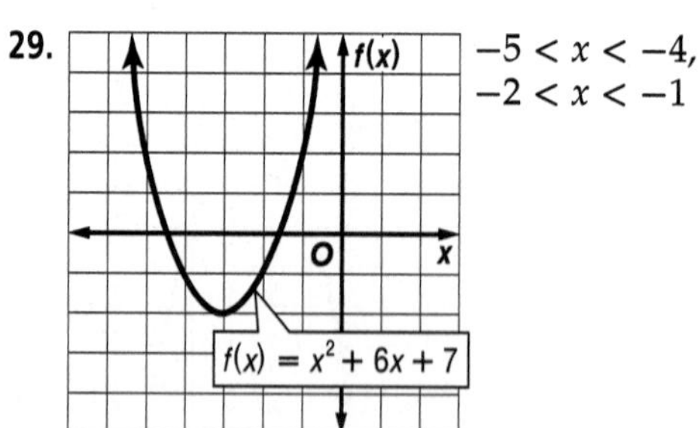

$-5 < x < -4$, $-2 < x < -1$

31. 4, 5

33. about 9 s **35.** 6 ft **37.** about 180 ft^2
39. $(500 - 2x)(400 - 2x) = 100{,}000$ or $4x^2 - 1800x + 100{,}000 = 0$; about 65 ft
41. Sample answer:

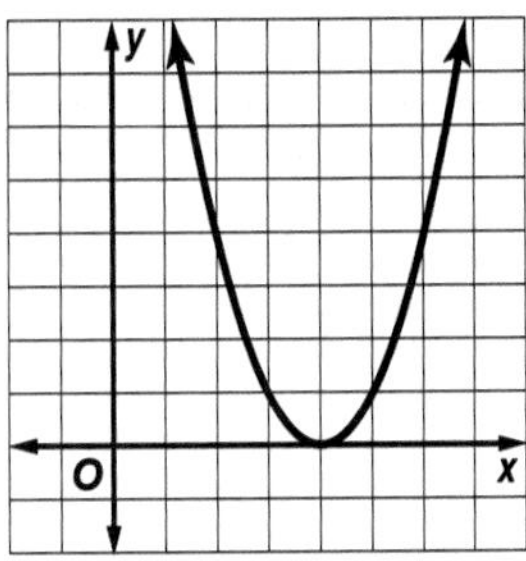

The only solution to the equation with the graph shown is 4.
43. Always; the shaded region of the graph includes y-values greater than 2.
45. D
47. $x = -3$; $(-3, 0)$; min

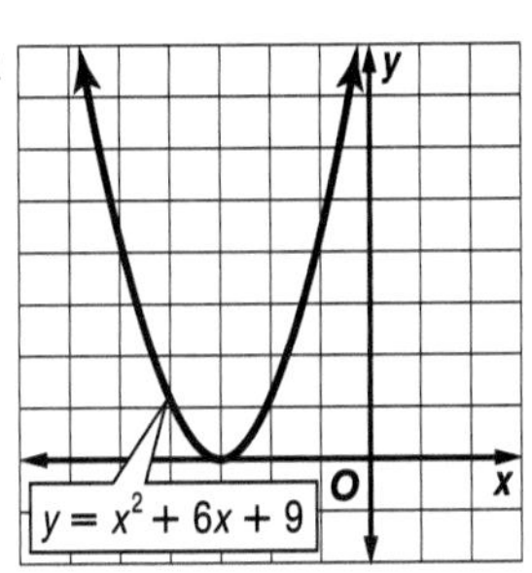

49. $x = 6$; $(6, -13)$; min

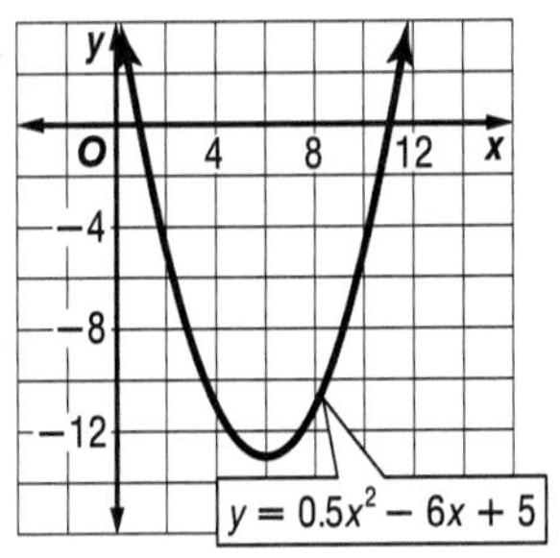

51. {5} **53.** $\frac{m^3}{3}$ **55.** $-\frac{m^5 y^4}{3}$ **57.** yes; $(a + 7)^2$ **59.** no

Pages 489–491 ***Lesson 9-3***
1. −2, 8 **3.** 36 **5.** −1, 7 **7.** −15, 1 **9.** −2.4, 0.1
11. 12 in. **13.** −6, 4 **15.** −12.2, −3.8 **17.** 25 **19.** $\frac{121}{4}$
21. −5, 2 **23.** −27, 7 **25.** −3, 7 **27.** 1, 7 **29.** 2.3
31. −0.1, 1.4 **33.** about 2017 **35.** 2.5 **37.** $\frac{1}{2}$, 3
39. −24, 24 **41.** $3 \pm \sqrt{9 - c}$
43. Sample answer:
$x^2 + 4x + 4$

x	1	1
x	1	1
x^2	x	x

45. There are no real solutions since completing the square results in $(x + 2)^2 = -8$ and the square of a number cannot be negative. **47.** Sample answer: Al-Khwarizmi used squares to geometrically represent quadratic equations. He represented x^2 by a square whose sides were each x units long. To this square, he added 4 rectangles with length x units long and width $\frac{8}{4}$ or 2 units long. This area represents 35.

To make this a square, four 4 × 4 squares must be added. To solve $x^2 + 8x = 35$ by completing the square, use the following steps.

$x^2 + 8x = 35$	
$x^2 + 8x + 16 = 35 + 16$	Since $\left(\frac{8}{2}\right)^2 = 16$, add 16 to each side.
$(x + 4)^2 = 51$	Factor $x^2 + 8x + 16$.
$x + 4 = \pm\sqrt{51}$	Take the square root of each side.
$x + 4 - 4 = \pm\sqrt{51} - 4$	Subtract 4 from each side.
$x = -4 \pm \sqrt{51}$	Simplify.
$x = -4 - \sqrt{51}$ or $x = -4 + \sqrt{51}$	Simplify.
$x \approx -11.14$ $\quad x \approx 3.14$	

The solutions are −11.14 and 3.14.

49. J **51.**

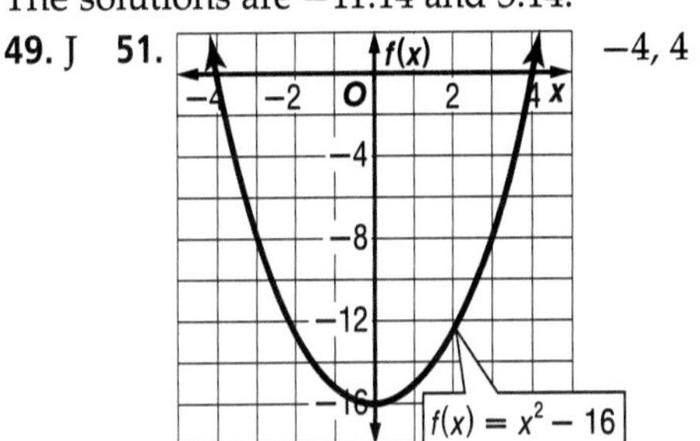

−4, 4

53. $A = (280 - 2x)x$ or $A = 280 - 2x^2$ **55.** ab^2
57. $-3 < x < 1$ **59.** (3, 6) **61.** (7, 2) **63.** 5 **65.** 9.4

Pages 497–499 ***Lesson 9-4***
1. −6, −1 **3.** −8.6, −1.4 **5.** about 18.8 cm by 18.8 cm
7. 0; 1 real root **9.** −10, −2 **11.** −0.8, 1 **13.** ∅ **15.** 5
17. −0.4, 3.9 **19.** −0.5, 0.6 **21.** 5 cm by 16 cm
23. 25; 2 real roots **25.** 0; 1 real root
27. $-\frac{61}{12}$, no real roots **29.** ∅ **31.** −0.3, 0.9 **33.** about 0.6 and 1.6 **35.** 2 **37.** 0 **39.** about 2.5 s **41.** Juanita; you must first write the equation in the form $ax^2 + bx + c = 0$ to determine the values of a, b, and c. Therefore, the value of c is −2, not 2. **43.** The function can be factored to $f(x) = (x - 4)^2$, so there is one real root at (4, 0). Using the discriminant to determine the number of roots involves more computation and potential for error. **45.** C **47.** 1, 7 **49.** −0.4, 12.4
51.

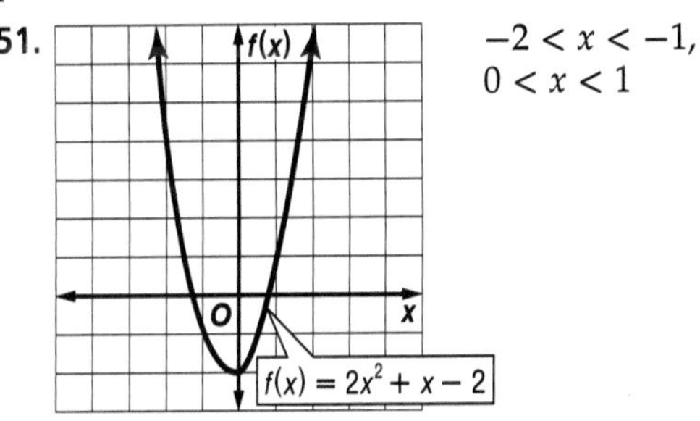

$-2 < x < -1$,
$0 < x < 1$

53. 12 cm **55.** $y^3(15x + y)$ **57.** $\{m \mid m > 5\}$
59. $\{k \mid k \leq -4\}$ **61.** $y = -7$ **63.** 16 **65.** 250

Pages 505–508 Lesson 9-5

1. 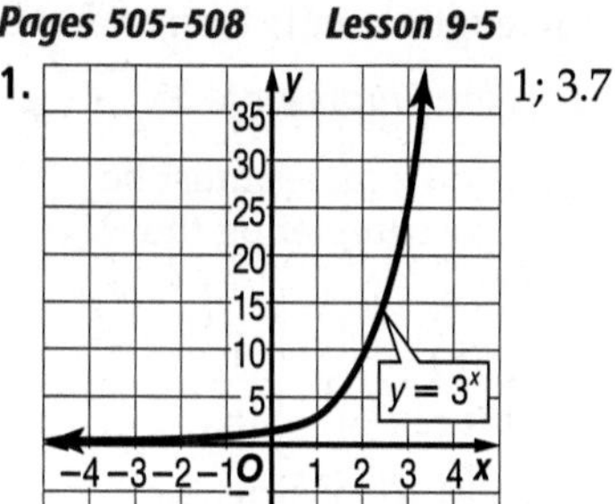

1; 3.7

3. 1; 5.8

5. 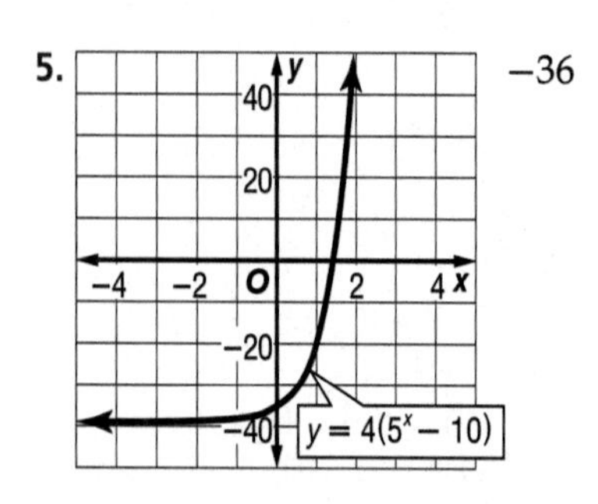

–36

7. Yes; the domain values are at regular intervals and the range values have a common factor 6.

9. 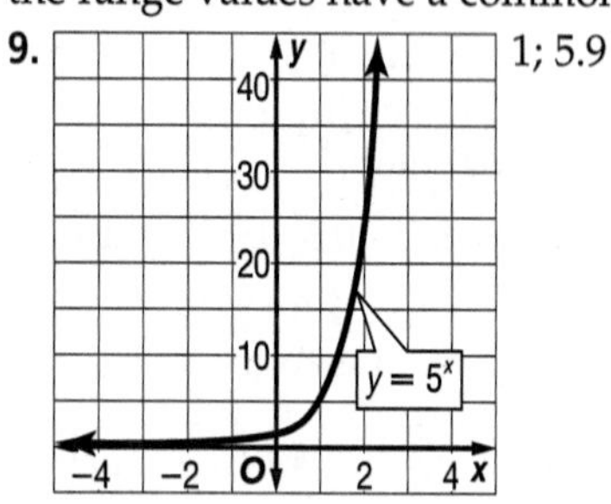

1; 5.9

11. 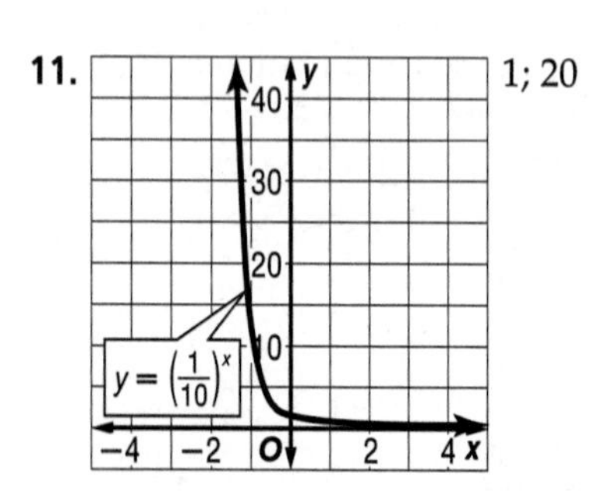

1; 20

13. 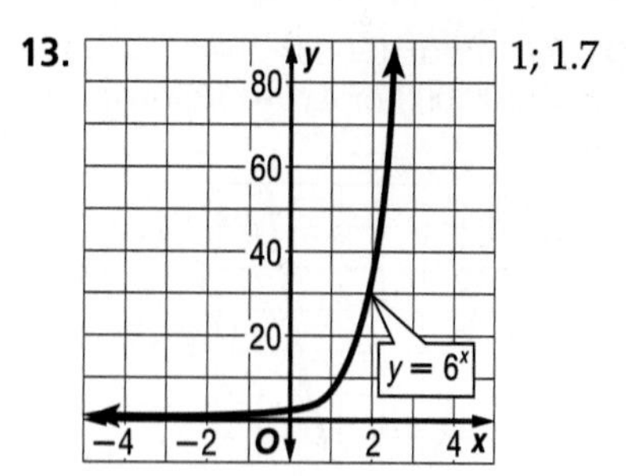

1; 1.7

15. 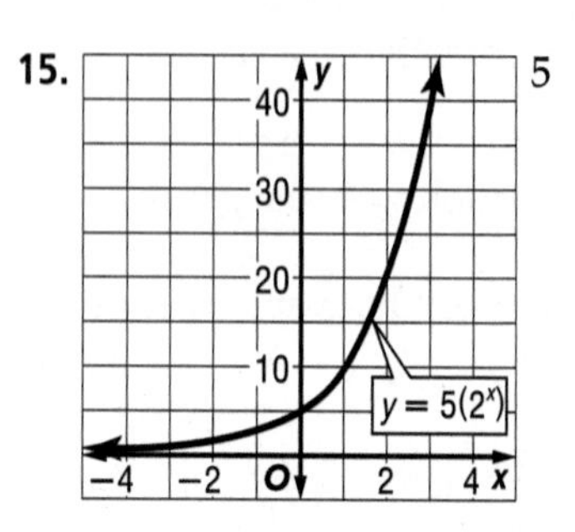

5

17. 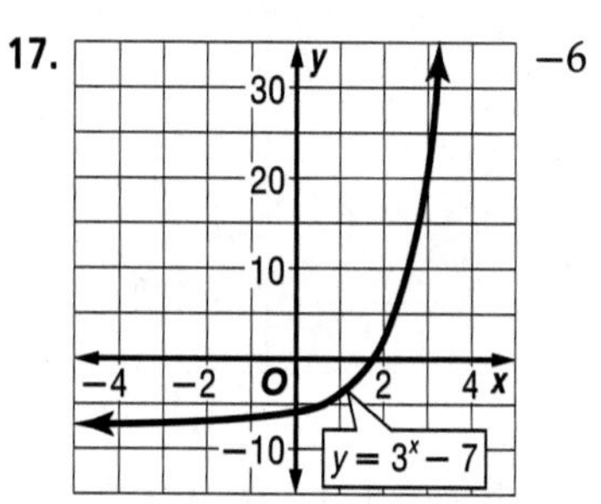

–6

19. about 312 **21.** about \$37.27 million; about \$41.74 million; about \$46.75 million **23.** No; the domain values are at regular intervals and the range values have a common difference 3. **25.** Yes; the domain values are at regular intervals and the range values have a common factor 0.75.

27. 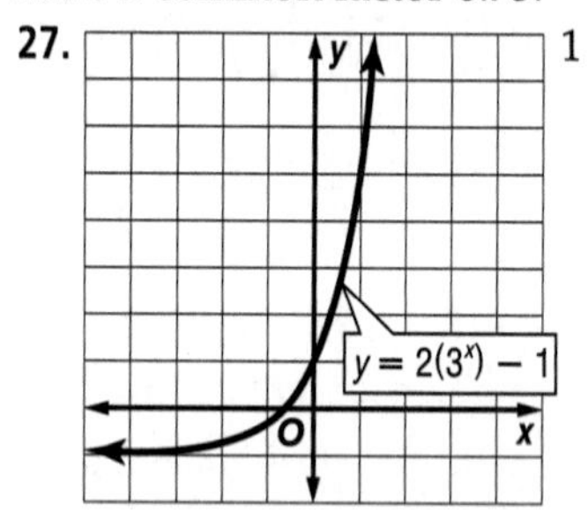

1

29. 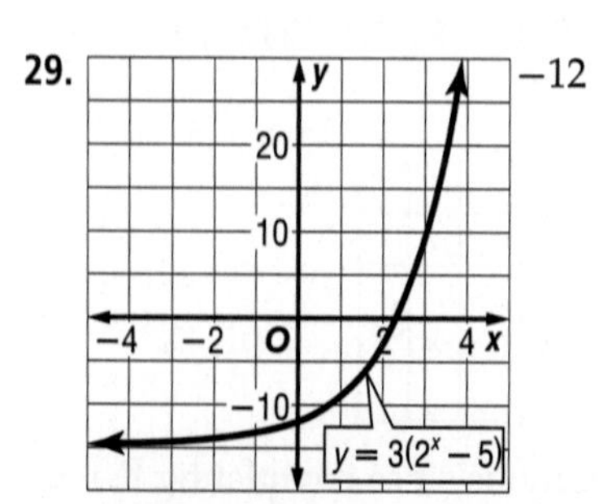

–12

31. quadratic **33.** quadratic **35.** linear
37. 27 schools

39.

Week	Distance (miles)
1	22
2	24.2
3	26.62
4	29.282

41. Never; the graph will never intersect the x-axis.
43. Hannah; the graph of $y = \left(\frac{1}{3}\right)^x$ decreases as x increases. **45.** a translation 2 units up
47. Sample answer: If the number of items on each level of a piece of art is a given number times the number of items on the previous level, an exponential function can be used to describe the situation. For the carving of the pliers, $y = 2^x$. For this situation, x is an integer between 0 and 8, inclusive. The values of y are 1, 2, 4, 8, 16, 32, 64, 128, and 256.

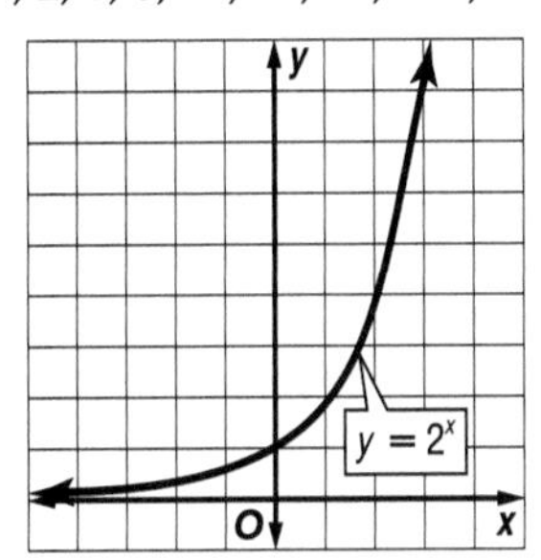

49. H **51.** −1.8, 0.3 **53.** 2, 5 **55.** −5.4, −0.6
57. prime **59.** $\{x \mid x > -5\}$ **61.** $\{y \mid y < -5\}$ **63.** 11.25
65. 144

Pages 512–514 ***Lesson 9-6***

1. $I = 221{,}000(1.089)^t$ **3.** about \$661.44
5. $W = 43.2(1.06)^t$ **7.** about 122,848,204 people
9. about \$2097.86 **11.** about 2,050,422 **13.** about 128 g
15. about 17,190 years ago **17.** Sample answer: Determine the amount of the investment if \$500 is invested at an interest rate of 7% compounded quarterly for 6 years. **19.** D
21. 1

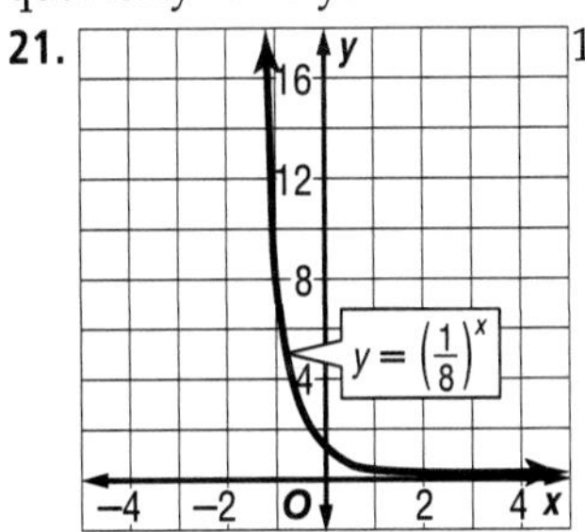

23. −20

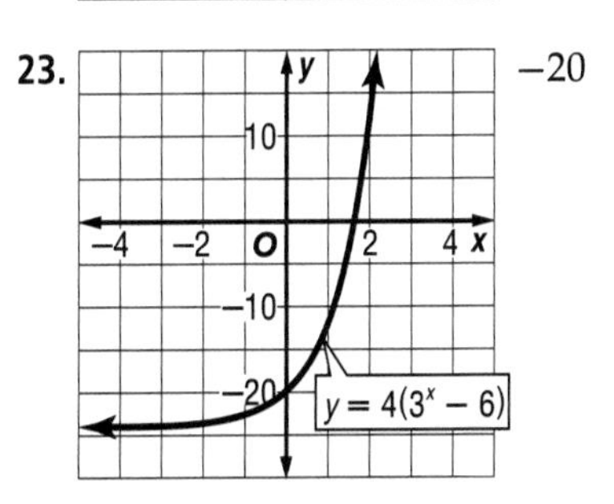

25. −0.6, 2.6 **27.** Yes; slope $= \frac{60}{250} = 0.24 < 0.33$.

Pages 517–520 ***Chapter 9*** ***Study Guide and Review***

1. true **3.** false; axis of symmetry **5.** false; exponential growth **7.** true **9.** true
11. $x = -1$; $(-1, -1)$; min

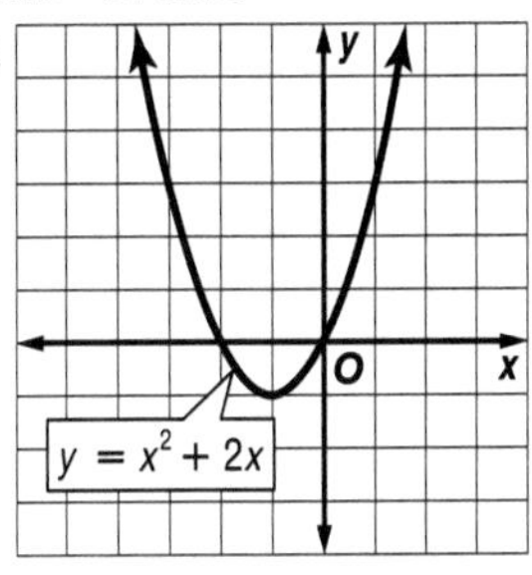

13. $x = 1\frac{1}{2}$; $\left(1\frac{1}{2}, -6\frac{1}{4}\right)$; min

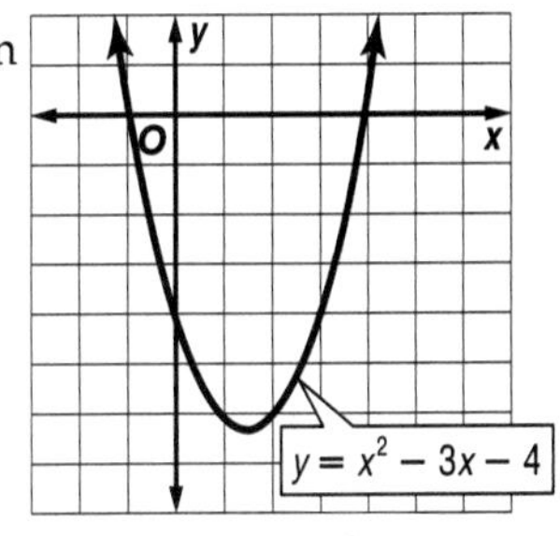

15. $x = 0$; $(0, 1)$; max

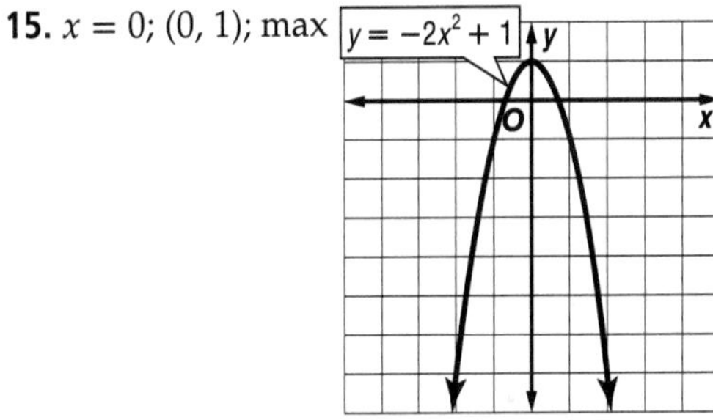

17. $x = 2$; $(2, 64)$ **19.** 64 ft
21. −3, 4

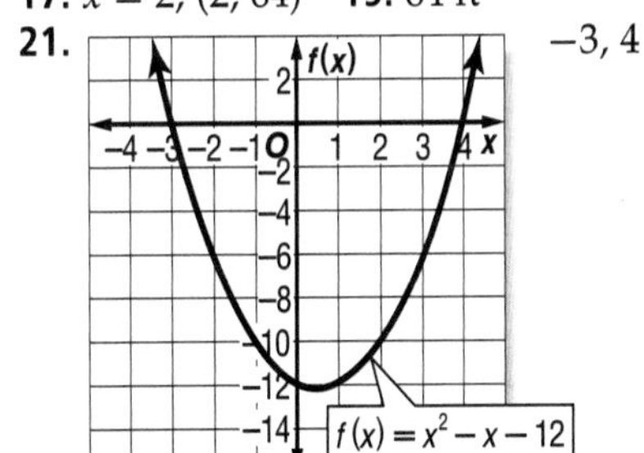

23. $-5 < x < -4, 0 < x < 1$

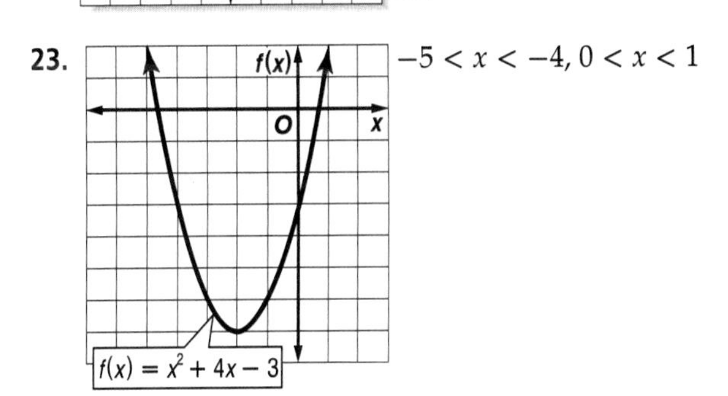

25. 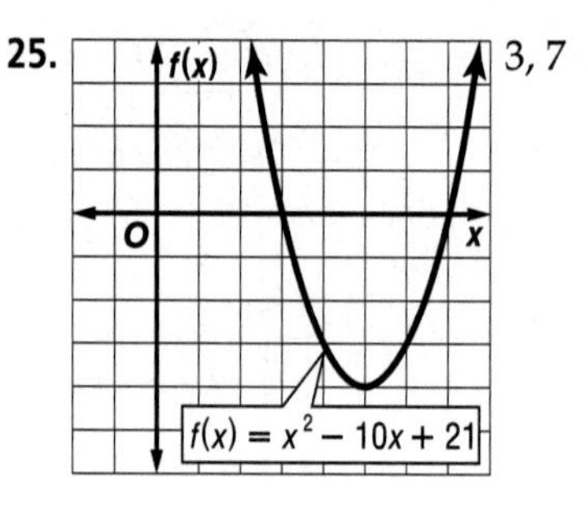

3, 7

27. −3 and 8 **29.** −4, 6 **31.** 2.3, 13.7 **33.** 6 cm **35.** −9, −1 **37.** −3.6, −0.4 **39.** −0.7, 0.5

41. 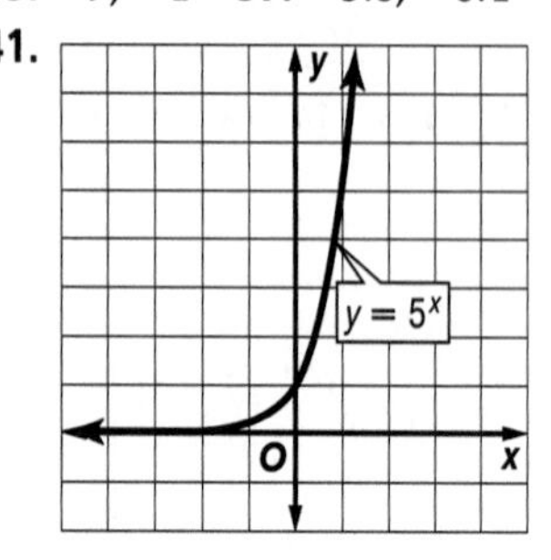

1; 3.1

43. 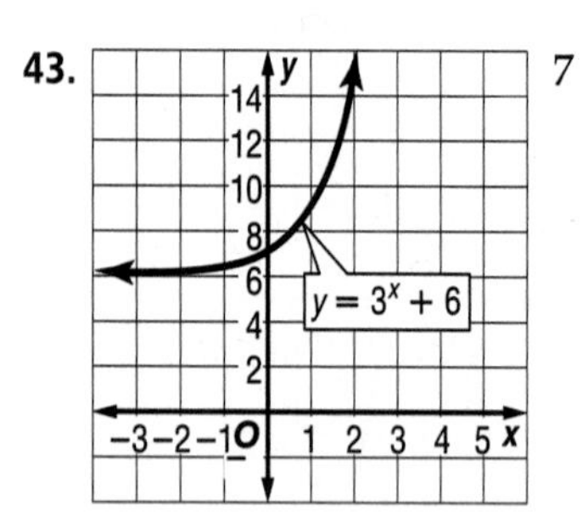

7

45. about 568 **47.** \$7705.48 **49.** \$1006.86 **51.** about \$570 billion

Chapter 10 Radical Expressions and Triangles

Page 527 ***Chapter 10*** ***Get Ready***

1. 5 **3.** 7.48 **5.** $16c$ **7.** $25m + 6n$ **9.** $\{-6, -4\}$ **11.** 8 cm **13.** no **15.** 11.375 feet

Pages 511–534 ***Lesson 10-1***

1. $2\sqrt{5}$ **3.** $8\sqrt{2}$ **5.** $3\sqrt{6}$ **7.** 28 ft² **9.** $2x^2|y^3|\sqrt{15x}$ **11.** $\frac{2\sqrt{6}}{3}$ **13.** $\frac{\sqrt{210}}{6}$ **15.** $\frac{8(3+\sqrt{2})}{7}$ **17.** $3\sqrt{2}$ **19.** $4\sqrt{5}$ **21.** $\sqrt{30}$ **23.** $84\sqrt{5}$ **25.** $2a^2\sqrt{10}$ **27.** $7x^3|y^3|\sqrt{3y}$ **29.** $\frac{\sqrt{6}}{3}$ **31.** $\frac{\sqrt{2t}}{4}$ **33.** $\frac{c^2\sqrt{5cd}}{2d^3}$ **35.** $\frac{54+9\sqrt{2}}{17}$ **37.** $2\sqrt{7} - 2\sqrt{2}$ **39.** $\frac{16+12\sqrt{3}}{-11}$ **41.** $3\sqrt{2d}$ **43.** $6\sqrt{55} \approx 44$ mph, $4\sqrt{165} \approx 51$ mph **45.** $s = \sqrt{A}$; $6\sqrt{2}$ in. **47.** $6\sqrt{5}$ m/s **49.** 2.4 km/s; The Moon has a much lower escape velocity than Earth. **51.** $x^2 + \frac{b}{a}x + \frac{b^2}{4a^2} = -\frac{c}{a} + \frac{b^2}{4a^2}$ **53.** 5; $\frac{1}{\sqrt{4(x-4)}} = \frac{1}{2}$ **55.** No; to solve the equation, take the square root of each side remembering that $a = \pm b$. Thus, $3x - 2 = 2x + 6$ yields $x = 8$; but $3x - 2 = -(2x + 6)$ yields $x = -\frac{4}{5}$ **57.** A lot of formulasand calculations that are used in space exploration contain radical expressions. To determine the escape velocity of a planet, you would need to know its mass and the radius. It would be very important to know the escape velocity of a planet before you landed on it so you would know if you had enough fuel and velocity to launch from it to get back into space. **59.** J **61.** 16, −32, 64 **63.** 144, 864, 5184 **65.** 0.08, 0.016, 0.0032 **67.** 84.9°C **69.** $(5x - 4)(7x - 3)$ **71.** $x^2 - x - 6$ **73.** $2t^2 - 11t - 6$ **75.** $15x^2 + 4xy - 3y^2$

Pages 538–540 ***Lesson 10-2***

1. $11\sqrt{3}$ **3.** $-\sqrt{5}$ **5.** $9\sqrt{5} + 5\sqrt{6}$ **7.** $4 + 4\sqrt{6}$ **9.** $P = 16 + 12\sqrt{6}$ ft; $A = 70 + 24\sqrt{6}$ ft² **11.** $11\sqrt{5}$ **13.** $-7\sqrt{15}$ **15.** $18\sqrt{x}$ **17.** $13\sqrt{3} + \sqrt{2}$ **19.** $5\sqrt{2} + 2\sqrt{3}$ **21.** $-\sqrt{7}$ **23.** $3\sqrt{2} + 10\sqrt{3}$ **25.** 4 **27.** $10\sqrt{3} + 16$ **29.** $19\sqrt{5}$ **31.** $20\sqrt{7} + 2\sqrt{5}$ in.; $52 - 16\sqrt{35}$ in² **33.** $15\sqrt{6}$ cm² **35.** $\frac{4\sqrt{10}}{5}$ **37.** $\frac{53\sqrt{7}}{7}$ **39.** Approximately 1000 feet; solve $\sqrt{\frac{3(1250)}{2}} - \sqrt{\frac{3h}{2}} = 4.57$; may guess and test, graphical, or analytical methods. **41.** No, each pipe would need to carry 500 gallons per minute, so the pipes would need at least a 6-inch radius. **43.** The velocity doubled because $\sqrt{4} = 2$. **45.** Sample answer: Let $x = 2$ and $y = 3$. $(\sqrt{2} + \sqrt{3})^2 = 2 + 2\sqrt{6} + 3$ or $5 + 2\sqrt{6}$ **47.** Sample answer: $a = 1$, $b = 1$; $(\sqrt{1+1})^2 = (\sqrt{1})^2 + (\sqrt{1})^2 = 1 + 1$ or 2 **49.** D **51.** $2\sqrt{10}$ **53.** $-14|x|y\sqrt{y}$ **55.** $\frac{5|c|\sqrt{2d}}{2}$ **57.** 4096 **59.** −5103 **61.** $\left\{\pm\frac{6}{11}\right\}$ **63.** $\{-4, -3, 4\}$ **65.** $x^2 - 4x + 4$ **67.** $x^2 + 12x + 36$ **69.** $4x^2 - 12x + 9$

Pages 543–546 ***Lesson 10-3***

1. 25 **3.** 7 **5.** 2 **7.** 3 **9.** 6 **11.** −12 **13.** −16 **15.** −63 **17.** 9 **19.** 29 **21.** 34 **23.** 2 **25.** 2, 3 **27.** 3 **29.** 10,569 lb **31.** 57 **33.** no solution **35.** $\frac{3}{25}$ **37.** 6 **39.** 2 **41.** 11 **43.** $r = \sqrt{\frac{A}{\pi}}$ **45.** $4\sqrt{3}$ m **47.** 9.8 m/s **49.** approximately 296 m/s **51.** 0.36 ft **53.** $\frac{24t^2}{\pi^2}$ **55.** 16°C **57.** 8 **59.** −2 **61.** 1.70 **63.** The solution may not satisfy the original equation. **65.** Alex; the square of $-\sqrt{x-5}$ is $x - 5$. **67.** You can determine the time it takes an object to fall from a given height using a radical equation. It would take a skydiver approximately 42 seconds to fall 10,000 feet. Using the equation, it would take 25 seconds. The time is different in the two calculations because air resistance slows the skydiver. **69.** H **71.** $20\sqrt{3}$ **73.** $8\sqrt{3}$ **75.** $3(\sqrt{10} - \sqrt{3})$ **77.** $6z^2 + 44z + 70$ **79.** $95° \le F \le 104°$ **81.** $2x + y = 15$ **83.** 3.9% increase **85.** 25 **87.** 10 **89.** $4\sqrt{13}$

Pages 551–554 ***Lesson 10-4***

1. 26 **3.** 8 **5.** 18.44 **7.** A **9.** yes; $10^2 + 24^2 = 26^2$ **11.** 30 **13.** 5 **15.** $\sqrt{45} \approx 6.71$ **17.** $\sqrt{27} \approx 5.20$ **19.** $\sqrt{124} \approx 11.14$ **21.** $\sqrt{40x} \approx 6.32x$ **23.** 11.40

25. 13.08 **27.** 20 **29.** 111.1 units^2 **31.** no; $6^2 + 12^2 \neq 18^2$ **33.** yes; $45^2 + 60^2 = 75^2$ **35.** yes; $4^2 + 7^2 = (\sqrt{65})^2$ **37.** 14.56 **39.** 212.6 ft **43.** 18 ft **45.** $4\sqrt{3}$ in. or about 6.93 in. **47.** 900 **49.** Compare the lengths of the sides. The hypotenuse is the longest side, which is always the side across from the right angle. **51.** B **53.** 144 **55.** 12 **57.** $-3\sqrt{z}$ **59.** 425 mph **61.** $f(x) = x - 2$ **63.** 10

Pages 557–559 ***Lesson 10-5***

1. 10 **3.** $3\sqrt{2}$ or 4.24 **5.** yes; $AB = \sqrt{68}$, $BC = \sqrt{85}$, $AC = \sqrt{85}$ **7.** 20.6 yd **9.** 2 or −14 **11.** 13 **13.** 17 **15.** $2\sqrt{13}$ or 7.21 **17.** $\sqrt{185}$ or 13.60 **19.** 0.53 mi **21.** Duluth, (44, 116); St. Cloud, (−46, 39); Eau Claire, (71, −8); Rochester, (27, −58) **23.** all cities except Duluth **25.** 17 or −13 **27.** −2 or −12 **29.** 2 or 38 **31.** $\sqrt{157} \neq \sqrt{101}$; The trapezoid is not isosceles. **33.** 9 **35.** $\frac{17}{4}$ or 4.25 **37.** $\frac{\sqrt{74}}{7}$ or 1.23 **39.** $2\sqrt{3}$ or 3.46 **41.** The values that are subtracted are squared before being added and the square of a negative number is always positive. The sum of two positive numbers is positive, so the distance will never be negative. **43.** Compare the slopes of the two potential legs to determine whether the slopes are negative reciprocals of each other. You can also compute the lengths of the three sides and determine whether the square of the longest side length is equal to the sum of the squares of the other two side lengths. Neither test holds true in this case because the triangle is not a right triangle. **45.** B **47.** 25 **49.** 3 **51.** 4

53. $\{m \mid m \geq 9\}$ 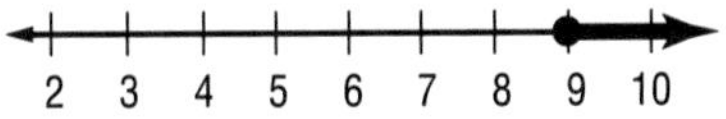

55. $\{x \mid x \leq -3\}$ 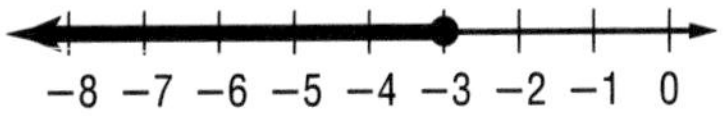

57. 5 hours **59.** −8 **61.** 5

Pages 563–565 ***Lesson 10-6***

1. No; the angle measures are not equal. **3.** $a = 21$, $b = 27$ **5.** $b = \frac{25}{7}$, $c = \frac{30}{7}$ **7.** 65 ft **9.** No; the angle measures are not equal. **11.** Yes; the angle measures are equal. **13.** Yes, the angle measures are equal. **15.** $o = 20$, $p = 10$ **17.** $n = \frac{112}{13}$, $p = \frac{84}{13}$ **19.** $n = 7$, $o = 10$ **21.** $\ell = 16.2$, $n = 6.3$ **23.** 1.25 in **25.** 8 **27.** Viho's eyes are 6 feet off the ground, Viho and the building each create right angles with the ground, and the two angles with the ground at P have equal measure. **29.** The angles will always be the same because the sides are proportional with a scale factor of three which means the triangles are similar and the angles are congruent. **31.** Yes; all circles are similar because they have the same shape. **33.** 4:1; The area of the first is πr^2 and the area of the other is $\pi(2r)^2 = 4\pi r^2$. **35.** D **37.** 5 **39.** 7 **41.** no **43.** (3, −2) **45.** (6, 12)

Pages 567–570 ***Chapter 10*** ***Study Guide and Review***

1. false, $-3 - \sqrt{7}$ **3.** true **5.** true **7.** true **9.** $\frac{2\sqrt{15}}{|y|}$ **11.** $57 - 24\sqrt{3}$ **13.** $\frac{3\sqrt{35} - 5\sqrt{21}}{-15}$ **15.** about 2 h **17.** $2\sqrt{6} - 4\sqrt{3}$ **19.** $\frac{9\sqrt{2}}{4}$ **21.** $\sqrt{6} - 1$ **23.** 250.95 ft/s **25.** 32 **27.** 3 **29.** 5 **31.** 34 **33.** $2\sqrt{34} \approx 11.66$ **35.** 24 **37.** no **39.** yes **41.** no **43.** $\sqrt{58}$ or 7.62 **45.** $\sqrt{74}$ or 8.60 **47.** 18 or −8 **49.** $d = \frac{45}{8}$, $e = \frac{27}{4}$ **51.** $b = \frac{44}{3}$, $d = 6$ **53.** 4 in.

Chapter 11 Rational Expressions and Equations

Page 575 ***Chapter 11*** ***Get Ready***

1. $-\frac{63}{16}$ **3.** 5 **5.** 4.62 **7.** 10.8 **9.** 6 **11.** 16 **13.** $3m(2n + 5)$ **15.** $(x - 5)(x + 9)$ **17.** $3(x - 3)(x - 1)$

Pages 580–582 ***Lesson 11-1***

1. 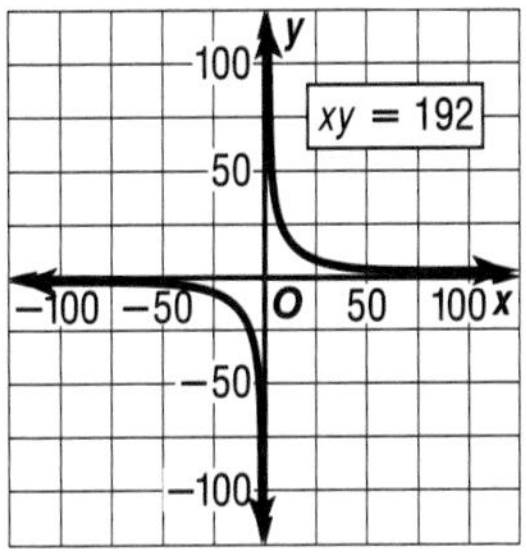

3. $xy = 21.87$; 4.05 **5.** $xy = 72$; 9 **7.** approximately 311 cycles per second

9. 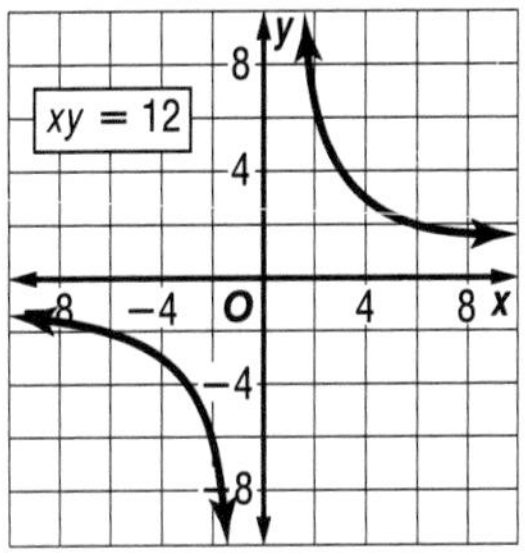

11. 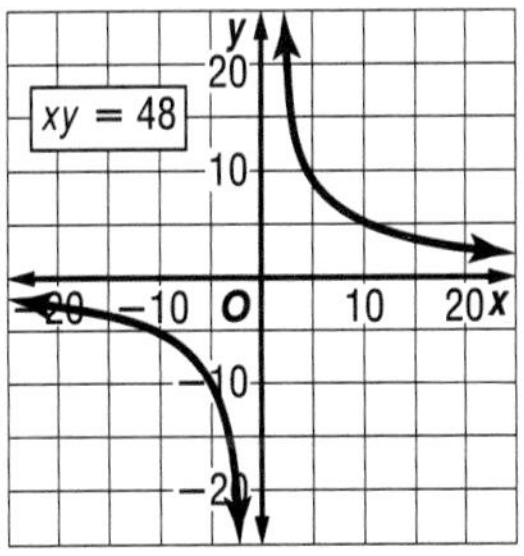

13. 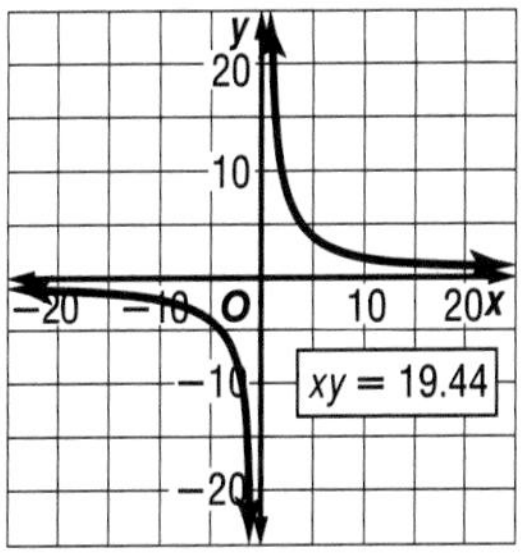

15. 7.2 h **17.** $xy = 12.4$; −20 **19.** $xy = 0.8$; 0.25

27. $xy = 26.84$; 8.3875 **29.** $PV = k$ or $P_1V_1 = P_2V_2$ **31.** 27.5 m^3 **33.** 24 kg **35.** neither **37.** Sample answer: Though it has been replaced by Einstein's theory of relativity, Newton's law of Gravitational Force is an example of an indirect variation which models real world situations. The gravitational force exerted on two objects is inversely proportional to the square of the distances between the two objects. The force exerted on the two objects, times the square of the distance between the two objects, is equal to the gravitational constant times the masses of the two objects. **39.** It is half. **41.** When the gear ratio is lower, the pedaling revolutions increase to keep a constant speed. Lower gears at a constant rate will cause a decrease in speed, while higher gears at a constant rate will cause an increase in speed. **43.** J **45.** $a = 6, f = 14$ **47.** $a = 8$ or 18 **49.** A and C sharp **51.**

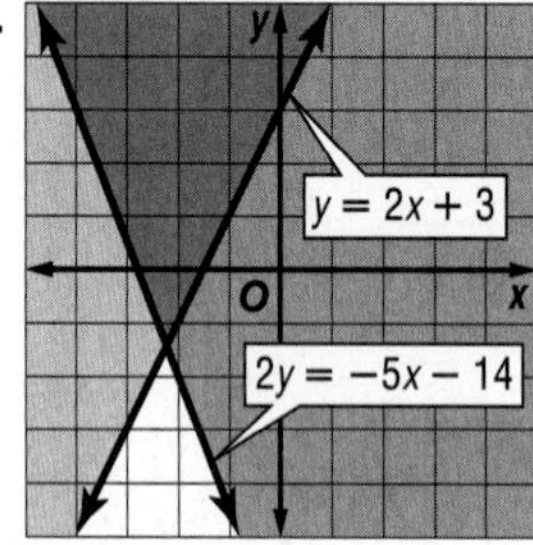

53. −9 **55.** 3 **57.** 30 **59.** $6xy^2$

Pages 586–588 **Lesson 11-2**

1. −3 **3.** −5, 4 **5.** $x - 7$; −7 **7.** $\frac{3}{x-4}$; 4, 3 **9.** $\frac{x+3}{x-4}$; $\frac{7}{2}$, 4 **11.** $\frac{4}{9+2g}$ **13.** −5 **15.** −5, 5 **17.** −5, −7 **19.** 178.13 lb **21.** The difference is 12 minutes. **23.** $\frac{a^2}{3b}$; 0, 0 **25.** $\frac{3x}{8z}$; 0, 0, 0 **27.** $\frac{mn}{12n - 4m}$; $m \neq 3n$, 0, 0 **29.** $z + 8$; −2 **31.** $\frac{2}{y+5}$; −5, 2 **33.** $\frac{a+3}{a+9}$; −9, 3 **35.** $\frac{(b+4)(b-2)}{(b-4)(b-16)}$; 4, 16 **37.** $\frac{n-2}{n(n-6)}$; 0, 6 **39.** $\frac{3}{4}$; −2, −1 **41.** $\frac{450 + 4n}{n}$ **43.** $\frac{450 + 4(n+2)}{n}$ **45.** Sample answer: $\frac{1}{(x+4)(x+7)}$; since the excluded values are −4 and −7, the denominator of the rational expression must contain the factors $(x + 4)$ and $(x + 7)$. **47.** Sample answer: The first graph has a hole at $x = 4$ because it is an excluded value of the equation. **49.** B **51.** $xy = 60$; −5 **53.** $xy = -54$; −18 **55.** $o = 6, p = 4$ **57.** 1 **59.** −3 **61.** \$372.32 **63.** 450 **65.** 95,040 **67.** 0.22

Pages 592–594 **Lesson 11-3**

1. $8y$ **3.** $\frac{4m}{3(m+5)}$ **5.** $\frac{n+2}{n-4}$ **7.** 16.36 mph **9.** $2x$ **11.** $\frac{4wy}{5xz^2}$ **13.** $\frac{x+4}{x+8}$ **15.** $\frac{(z+6)(z-5)}{(z-6)(z+3)}$ **17.** $\frac{(x+5)^2}{9}$ **19.** $\frac{x}{(x+4)^2}$ **21.** $\frac{b+1}{(b+3)(b-3)}$ **23.** \$336 if carpet is sold in square feet; it is not reasonable to assume that carpet is sold only in square feet. If the carpet is sold in square yards the price will be \$360. **27.** 16.67 m/s **29.** 20 yd^3 **31.** 5 tracks • 2 miles/1 track • 5280 feet/1 mile • 1 car/75 feet **33.** 1.1 h **35.** c and e; sample answer: the expressions each have a GCF which can be use to simplify the expressions. **37.** Sample answer: When the negative sign in front of the first expression is distributed, the numerator is $-x - 6$; Distributive Property. **39.** B **41.** −6, 6 **43.** −3 **45.** $xy = 19.44$; 5.4 **47.** $xy = 28.16$; 8.8 **49.** $\{r \mid r \geq 2.1\}$ **51.** -7^3 or −343 **53.** $\frac{4b^4c^5}{a^3}$ **55.** $(x + 5)(x - 8)$ **57.** $(x - 6)^2$ **59.** $(2x + 1)(x - 3)$

Pages 597–599 **Lesson 11-4**

1. $6n$ **3.** 18 **5.** $\frac{1}{2(k+2)}$ **7.** 23.61 m/s **9.** about 57 pieces **11.** n^2p **13.** $\frac{a^3cg}{b}$ **15.** $\frac{a-4}{a-3}$ **17.** $\frac{(n+2)^2}{4}$ **19.** $\frac{b+4}{2(b+2)^2}$ **21.** $\frac{(x-3)(x+4)}{(x-6)(x+3)}$ **23.** $\frac{1}{(m+7)(m+1)}$ **25.** 350,000 **27.** 0.00173 **29.** about 105.4 cups **31.** $n = 45\ yd^3 \div \frac{3.5\ ft(10\ ft + 17\ ft)}{2} \cdot 4ft \cdot \frac{1\ yd^3}{27\ ft^3}$; 7 loads **33.** 63.5 mph **35.** $(10 \cdot 85 \text{ pounds}) \div \frac{\left(x - \frac{1}{2}\right)\left(x - \frac{3}{4}\right)(x)}{x^3}$ **37.** Sometimes; sample answer: 0 has no reciprocal and it is a real number. **39.** Sample answer: Divide the number of cans recycled by $\frac{5}{8}$ to find the total number of cans produced. Answers should include $x = 63{,}900{,}000 \div \frac{5}{8} \cdot \frac{1 \text{ pound}}{33 \text{ cans}}$ **41.** F **43.** $\frac{x-2}{x+2}$ **45.** $\frac{1}{c-6}$ **47.** $\frac{1}{a+1}$ **49.** $y = 1.67x - 1300$ **51.** $\frac{m^3}{5}$ **53.** $\frac{3y}{7x}$

Pages 604–606 **Lesson 11-5**

1. $5q + 1$ **3.** $2x^2 + x - \frac{5}{2x}$ **5.** $n + 4$ **7.** $b + 2 - \frac{3}{2b-1}$ **9.** $2m^2 + 3m + 7$ **11.** \$360,000 **13.** $4k - \frac{3}{k}$ **15.** $\frac{a}{7} + 1 - \frac{4}{a}$ **17.** $3a^2 + 4b^2 - 2$ **19.** $x + 8$ **21.** $s + 2$ **23.** $a + 7 - \frac{1}{a-3}$ **25.** $3x^2 + 2x - 3 - \frac{1}{x+2}$ **27.** $3g^2 + 2g + 3 - \frac{2}{3g-2}$ **29.** $t^2 + 4t - 1$ **31.** $7^2 \div 8$ **33.** $x^2 \div (x + 1) = x - 1 + \frac{1}{x+1}$ **35.** $x + 3$ **37.** $t + 20$; $t + 35$ **39.** 0.7; 0.84; 0.946; 0.97 **41.** $9d + 3$ **43.** $3x^2 - 12 + \frac{16}{x} + 5$ **45.** $-x^4 - 4x^2 + 2x$ **47.** $2w + 4$

49.

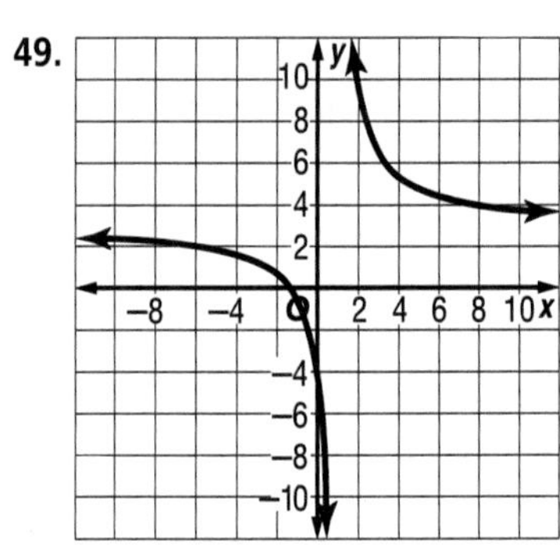

51. As x approaches 1 from the left y approaches negative infinity, and as x approaches y from the right y approaches positive infinity. **53.** 185.9° F **55.** Sample answer: $x^3 + 2x^2 + 8$; $x^3 + 2x^2 + 0x + 8$ **57.** 3, 4 **59.** 63 **61.** D **63.** $x + 5$ **65.** $b + 7$ **67.** 2300 cards **69.** $4x^2 + 13xy + 2y^2$ **71.** $2g^3 - 4g^2 - 2h$

Pages 611–613 ***Lesson 11-6***

1. $\frac{a}{2}$ **3.** $\frac{3-n}{n-1}$ **5.** $\frac{8x-6}{5}$ **7.** $\frac{3}{n-3}$ **9.** $\frac{x^2+y^2}{x-y}$ **11.** $\frac{2x+5}{5}$ **13.** 3 **15.** $\frac{n-3}{n+3}$ **17.** $\frac{22x+7}{2x+5}$ **19.** square **21.** $\frac{2}{5}$ **23.** -2 **25.** $\frac{10y}{y-3}$ **27.** $\frac{60}{n}$ **29.** $\frac{4-7m}{7m-2}$ **31.** $\frac{1}{3}$ **33.** $\frac{7}{x+7}$ **35.** $\frac{1}{5}$ **37.** 8.3 lb **39.** Russell; sample answer: Ginger factored incorrectly in the next-to-last step of her work. **41.** Sample answer: Since any rational number can be expressed as a fraction, values on graphics can be written as rational expressions for clarification. The numbers in the graphic are percents which can be written as rational expressions with a denominator of 100. To add rational expressions with inverse denominators, factor -1 out of either denominator so that it is like the other.

43. F **45.** $8x^2 - 9$ **47.** $\frac{x+3}{x}$ **49.** 90 **51.** 720

Pages 617–619 ***Lesson 11-7***

1. $35a^2$ **3.** $(n+4)(n-1)^2$ **5.** $\frac{12x+7}{10x^2}$ **7.** $\frac{y^2+12y+25}{(y-5)(y+5)}$ **9.** $\frac{a^2+16}{a^2-16}$ **11.** $\frac{-y^2-3y-6}{y^2+7y+12}$ **13.** a^2b^3 **15.** $(x-4)(x+2)$ **17.** $(x+7)(x-2)^2$ **19.** $\frac{3+5x}{x^2}$ **21.** $\frac{7+10a}{6a^2}$ **23.** $\frac{7x+8}{(x+5)(x-4)}$ **25.** $\frac{8a^2-9a}{(a+5)(a-2)}$ **27.** $\frac{3a+20}{(a-5)(a+5)}$ **29.** $\frac{2x+1}{(x+1)^2}$ **31.** $\frac{14x-3}{6x^2}$ **33.** $\frac{-2x}{x-1}$ **35.** $\frac{2+3x}{(x-5)}$ **37.** $\frac{-n^2-5n-3}{(n-5)(n+5)}$ **39.** $\frac{k^2-2k-2}{(2k+1)(k+2)}$ **41.** 12 miles; \$30 **43.** $\frac{4-25x}{15x^2}$ **45.** $\frac{22x-7xy}{6y^2}$ **47.** $\frac{a^3-a^2b+a^2+ab}{(a+b)(a-b)^2}$

49. $\frac{9x+6}{(x+1)(x+2)^2}$ **51.** Sample answer: To find the LCD, determine the least common multiple of all of the factors of the denominators. **53.** Sample answer: Multiply both the numerator and denominator by factors necessary to form the LCD. **55.** Sample answer: You can use rational expressions and their least common denominators to determine when elections will coincide. Use each factor of the denominators the greatest number of times it appears. **57.** J **59.** $\frac{4x+5}{2x+3}$ **61.** $b + 10$ **63.** $2m - 3 + \frac{2}{2m+7}$ **65.** (24, 6) **67.** $\frac{5}{6}$ **69.** $\frac{x+3}{x}$ **71.** $\frac{3x}{(x+2)(x-1)}$

Pages 623–625 ***Lesson 11-8***

1. $\frac{3x+4}{x}$ **3.** $\frac{6a^2+a-1}{3a}$ **5.** $\frac{14}{19}$ **7.** $\frac{a-b}{x+y}$ **9.** $\frac{4a+5}{a}$ **11.** $\frac{6wz+2z}{w}$ **13.** $\frac{6a^2-a-1}{2a}$ **15.** $\frac{r^3+3r^2+r-4}{r+3}$ **17.** $\frac{3s^3-3s^2-1}{s-1}$ **19.** $\frac{p^2+p-15}{p-4}$ **21.** $66\frac{2}{3}$ lb/in² **23.** $\frac{145}{84}$ **25.** n **27.** $\frac{s^3(s-t)}{t^2(s+t)}$ **29.** $\frac{1}{a-1}$ **31.** 1 **33.** $\frac{(n+5)(n-2)}{(n+12)(n-9)}$ **35.** 49.21 cycles/min **37.** $\frac{32b^2}{35}$ **39.** a and c **41.** Sample answer: Most measurements used in baking are fractions or mixed numbers, which are examples of rational expressions. Divide the expression in the numerator of a complex fraction by the expression in the denominator. **43.** J **45.** $\frac{3a^2+3ab-b^2}{(a-b)(2b+3a)}$ **47.** $\frac{4}{x^2}$ **49.** $-\frac{1}{t+1}$ **51.** 2.59×10^0 **53.** -48 **55.** 16 **57.** -14.4

Pages 630–632 ***Lesson 11-9***

1. 2 **3.** $-1, \frac{2}{5}$ **5.** 3, 5 **7.** 10 **9.** 8 **11.** 3 **13.** -3 **15.** 0 **17.** $\frac{1}{2}$ **19.** -4; extraneous 1 **21.** extraneous -2 and 0; no solution **23.** $-3, 4$ **25.** $0, -3, 2$ **27.** 5 h and 15 min **29.** 900 min or 15 h **31.** about 22 min **33.** 9 **35a.** line **35b.** $f(x) = \frac{(x+5)(x-6)}{x-6} = x + 5$ **35c.** -5 **37a.** parabola **37b.** $f(x) = x^2 + 6x + 12$ **37c.** no real roots **39.** $d = t(r-w)$, $d = t(r+w)$; $t = \frac{d}{r-w}$, $t = \frac{d}{r+w}$ **41.** Sample answer: $\frac{x}{4} = 0$ **43.** Sample answer: Rational equations are used in solving rate problems, so they can be used to determine traveling times, speeds, and distances related to subways. Since both trains leave at the same time, their traveling time is the same. The sum of the distances of both trains is equal to the total distance between the two stations. So, add the two expressions to represent the distance each train travels and solve for time. **45.** J **47.** $\frac{a-5}{a+7}$ **49.** $\frac{6+m}{2(2m-3)}$ **51.** $\frac{4a^2+7a+2}{(6a+1)(a-3)(a+3)}$

Pages 633–636 ***Chapter 11*** ***Study Guide and Review***

1. false; rational **3.** true **5.** false; $x^2 - 144$ **7.** true **9.** $xy = 1176$; 21 **11.** 6.25 ft or 6 feet 3 inches **13.** n **15.** $\frac{x+3}{x(x-6)}$ **17.** $\frac{x}{x+2}$ **19.** $\frac{3axy}{10}$ **21.** $\frac{3}{a^2-3a}$ **23.** $2p$

25. $\frac{3}{(x+4)(x-2)}$ **27.** 22,581,504 slices per day **29.** $x^2 + 2x - 2$ **31.** $t + 4 - \frac{5}{4t+1}$ **33.** $\frac{2m+3}{5}$ **35.** $a + b$ **37.** 2 **39.** $\frac{4c^2 + 9c}{6cd^2}$ **41.** $\frac{14a - 3}{6a^2}$ **43.** $\frac{5x - 8}{x - 2}$ **45.** $\frac{x^2 + 8x - 65}{x^2 + 8x + 12}$ **47.** -5 **49.** -1; extraneous 0 **51.** 1 hr and 40 minutes

Chapter 12 Statistics and Probability

Page 641 ***Chapter 12*** ***Get Ready***

1. $\frac{3}{7}$ **3.** $\frac{4}{7}$ **5.** $\frac{3}{5}$ **7.** $\frac{7}{95}$ **9.** $\frac{1}{52}$ **11.** 87.5% **13.** 85.6% **15.** 29.3%

Pages 645–648 ***Lesson 12-1***

1. a group of readers of a newspaper; all readers of the newspaper; biased; voluntary response **3.** 25 nails; all nails on the store shelves; unbiased; stratified **5.** 3 sophomores; all sophomores in the school; unbiased; simple **7.** people who are home between 9 A.M. and 4 P.M.; all people in the neighborhood; biased; convenience **9.** 10 scooters; all scooters manufactured on a particular production line during one day; biased; convenience **11.** an 8-oz jar of corn; all corn in the storage silo; biased; convenience **13.** a group of people who watch a television station; all people who watch the television station; biased; voluntary response **15.** a handful of Bing cherries; all Bing cherries in the produce department; biased; convenience **17.** a group of employees; all employees of the company; unbiased; stratified **19.** We know that the results are from a national survey conducted by Yankelovich Partners for Microsoft Corporation. **21.** Sample answer: Get a copy of the school's list of students and call every 10th person on the list. **25.** Sample answer: Randomly pick 5 rows from each field of tomatoes and then pick a tomato every 50 ft along each row. **27.** All three are unbiased samples. However, the methods for selecting each type of sample are different. In a simple random sample, a sample is as likely to be chosen as any other from the population. In a stratified random sample, the population is first divided into similar, nonoverlapping groups. Then a simple random sample is selected from each group. In a systematic random sample, the items are selected according to a specified time or item interval. **29.** Sample answer: Ask the members of the school's football team to name their favorite sport. **31.** Usually it is impossible for a company to test every item coming off its production lines. Therefore, testing a sample of these items is helpful in determining quality control. Sample Answers: A biased way to pick the CDs to be checked is to take the first 5 CDs coming off the production line in the morning. An unbiased way to pick the CDs to be checked is to take every 25th CD off the production line. **33.** H **35.** 8 **37.** $\frac{6}{x}$ **39.** $\frac{(t-2)^2}{t+3}$ **41.** -4, 10 **43.** -0.3, 4.8 **45.** $c^2 - 10c + 21$ **47.** 6 **49.** 720 **51.** 5814

Pages 653–654 ***Lesson 12-2***

1.

Spin 1	Spin 2	Spin 3	Outcomes
R	R	R	RRR
		B	RRB
		Y	RRY
		G	RRG
	B	R	RBR
		B	RBB
		Y	RBY
		G	RBG
	Y	R	RYR
		B	RYB
		Y	RYY
		G	RYG
	G	R	RGR
		B	RGB
		Y	RGY
		G	RGG
B	R	R	BRR
		B	BRB
		Y	BRY
		G	BRG
	B	R	BBR
		B	BBB
		Y	BBY
		G	BBG
	Y	R	BYR
		B	BYB
		Y	BYY
		G	BYG
	G	R	BGR
		B	BGB
		Y	BGY
		G	BGG
Y	R	R	YRR
		B	YRB
		Y	YRY
		G	YRG
	B	R	YBR
		B	YBB
		Y	YBY
		G	YBG
	Y	R	YYR
		B	YYB
		Y	YYY
		G	YYG
	G	R	YGR
		B	YGB
		Y	YGY
		G	YGG
G	R	R	GRR
		B	GRB
		Y	GRY
		G	GRG
	B	R	GBR
		B	GBB
		Y	GBY
		G	GBG
	Y	R	GYR
		B	GYB
		Y	GYY
		G	GYG
	G	R	GGR
		B	GGB
		Y	GGY
		G	GGG

3. 64 **5.** 720

7. 12;

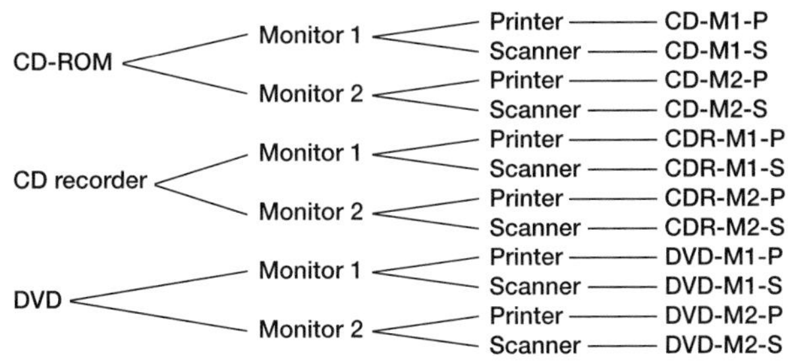

9. 180 **11.** 24 **13.** 39,916,800 **15.** 100 **17.** Columbus in three games : C-C-C; Columbus in four games: C-C-D-C, C-D-C-C, D-C-C-C; Columbus in 5 games: C-C-D-D-C, C-D-C-D-C, C-D-D-C-C, D-C-C-D-C, D-C-D-C-C, D-D-C-C-C; Dallas in three games: D-D-D; Dallas in four games: C-D-D-D, D-C-D-D, D-D-C-D; Dallas in five games: C-C-D-D-D, C-D-C-D-D, C-D-D-C-D, D-C-C-D-D, D-C-D-C-D, D-D-C-C-D **19.** 10 **21.** Sample answer: choosing 2 books in order from 7 books on a shelf **23.** Sample answer: You can make a chart showing all possible outcomes to help determine a football team's record. You can use a tree diagram or calculations to show 16 possible outcomes.
25. G **27.** all calendars printed in a day **29.** -3, 1 **31.** 1, 5.5 **33.** $\frac{2n+3}{n+3}$ **35.** $\frac{7}{34}$ **37.** $\frac{13}{34}$ **39.** $\frac{19}{34}$

Pages 658–662 ***Lesson 12-3***

1. Combination; order is not important. **3.** 6720 **5.** Permutation; order is important. **7.** $\frac{1}{12}$ **9.** $\frac{49}{646}$
11. 60 **13.** Permutation; order is important.
15. Permutation; order is important.
17. Combination; order is not important.
19. Combination; order is not important. **21.** 4
23. 35 **25.** 125,970 **27.** 524,160 **29.** 16,598,400
31. 6720 **33.** 495 **35.** $\frac{1}{12}$ **37.** 362,880 **39.** 7776
41. 61,425 **43.** 36 **45.** $\frac{1}{12}$ or about 8% **47.** 24
49. 792 **51.** $\frac{1}{7} \approx 14\%$ **53.** determining class rank in a senior class of 100 students **55.** 5040 **57.** Sample answer: Combinations can be used to determine how many different ways a committee can be formed by various members. Without seniority, order of selection is not important. Order is important when seniority is involved, so you need to find the number of permutations. **59.** H **61.** unbiased; systematic
63. $\frac{x+7}{x+5}$ **65.** $2\sqrt{53}$; 14.56 **67.** $2\frac{1}{2}$; 2.5 **69.** $\frac{5}{2}$, -3
71. $\frac{3}{13}$ **73.** $\frac{3}{5}$ **75.** $\frac{5}{6}$

Pages 667–670 ***Lesson 12-4***

1. independent **3.** $\frac{1}{3}$ **5.** $\frac{80}{3087}$ **7.** $\frac{4}{133}$ **9.** 1 **11.** $\frac{7}{8}$
13. $\frac{1}{30}$ **15.** $\frac{1}{10}$ **17.** $\frac{1}{4}$ **19.** $\frac{7}{16}$ **21.** 98% or 0.98
23. no; $P(A \text{ and } B) \neq P(A) \cdot P(B)$ **25.** $\frac{2}{51}$ **27.** $\frac{7}{408}$
29. ≈ 0.03 **31.** $\frac{1}{4}$ **33.** $\frac{23}{42}$ **35.** $\frac{5}{6}$ **37.** $\frac{1}{3}$ **39.** $\frac{5}{8}$ **41.** $\frac{1}{4}$
43. Sample answer: In a dependent event, an object is selected and not replaced. In an independent event, an object is selected and replaced. **45.** Sample answer: The probability of rolling a number less than or equal to six on a number cube and tossing heads or tails on a coin. **47.** 101 **49.** $\frac{39}{40}$ **51.** C **53.** 10 **55.** 604,800
57. $(m+5)(m+3)$ **59.** $8\sqrt{2}$ **61.** $2|a|\sqrt{30ab}$ **63.** $3 + 3\sqrt{2}$ **65.** 0.133 **67.** 0.096 **69.** 0.289 **71.** 0.015

Pages 674–676 ***Lesson 12-5***

1.

	1	2	3	4	5	6
1	2	3	4	5	6	7
2	3	4	5	6	7	8
3	4	5	6	7	8	9
4	5	6	7	8	9	10
5	6	7	8	9	10	11
6	7	8	9	10	11	12

3. $\frac{343}{1728}$ **5.** 0.95 **7.** RRR, RRB, RBR, RBB, BRR, BRB, BBR, BBB **9.**

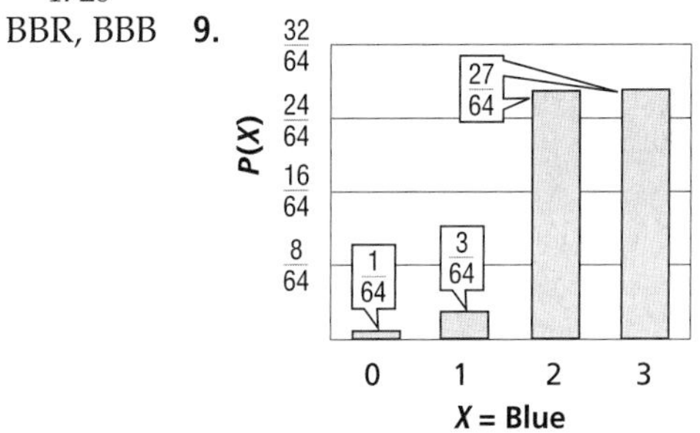

11. Let X = number of CDs; X = 100, 200, 300, 400, 500. **13.** 0.90 **15.** 0.646 **17.** Sample answer: Add the values for the bars representing bachelor's and advanced degrees. **19.** 3 **21a.** $P(X=1) = \frac{1}{2}$, $P(X=2) = \frac{1}{4}$, $P(X=3) = \frac{1}{8}$, $P(X=4) = \frac{1}{16}$ **21b.** $\frac{1}{16}$
23. C **25.** $\frac{2}{13}$ **27.** $\frac{25}{52}$ **29.** 792 **31.** $1065.82
33. Sample answer: Monthly; the interest earned is higher than quarterly. **35.** 38% **37.** 48% **39.** 33%

Pages 680–683 ***Lesson 12-6***

1. $\frac{3}{10}$ or 30% **3.** Sample answer: a spinner with 3 red sections of 10 equal sections, a spin on a red section simulates a hit **7.** Yes; 70% of the marbles in the bag represent water and 30% represent land. **9.** 0.25 or 25%
11. P(Preschool) = 0.060; P(Kindergarten) = 0.058; P(Elementary) = 0.480; P(High School) = 0.256; P(College) = 0.146 **13.** 123 **15.** Sample answer: a spinner divided into 3 sections, where $\frac{1}{2}$ represents cola, $\frac{1}{3}$ represents diet cola, and $\frac{1}{6}$ represents root beer
17. Sample answer: toss a coin and roll a number cube 100 times each **23.** Sample answer: 4 coins

33. Sample answer: a survey of 100 people voting in a two-person election where 50% of the people favor each candidate; 100 coin tosses **35.** Sample answer: Probability can be used to determine the likelihood that a medication or treatment will be successful. Experimental probability is determining probability based on trials or studies. To have the experimental more closely resemble the theoretical probability, the researches should perform more tials. **37.** J **39.** 0.145 **41.** $\frac{125}{1331}$ **43.** $\frac{80}{1749}$ **45.** yes

Pages 684–688 ***Chapter 12*** ***Study Guide and Review***

1. permutation **3.** independent **5.** are not **7.** 1 **9.** 8 test tubes with results of chemical reactions; the results of all chemical reactions performed; biased; convenience **11.** 720 **13.** 20 **15.** 56 **17.** 140 **19.** 12 **21.** 5040 **23.** $\frac{1595}{32,412}$ **25.** $\frac{4}{13}$ **27.** $0.04 + 0.12 + 0.37 + 0.30 + 0.17 = 1$ **29.** **Extracurricular Activities**

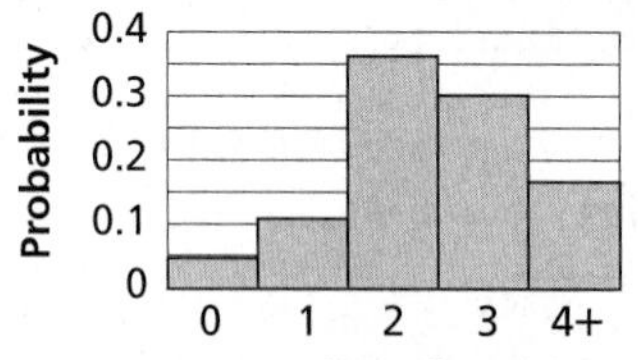

31. Sample answer: There are 6 possible outcomes. So, you could use a die.

Photo Credits

Cover (t) Courtesy Cedar Point, Sandusky, Ohio, USA; **cover (bkgd)** Created by Michael Trott with Mathematica. From "Graphica 1", Copyright © 1999 Wolfram Media, Inc.; **iv (tl)(tr)(bc)** David Dennison, **(bl)** Aaron Haupt; **v (tl)(tc)(tr)(bl)** David Dennison, **(bc)(br)** File Photo; **viii–ix** Beneluz Press/Index Stock Imagery/PictureQuest; **x–xi** John Warden/Index Stock Imagery; **xii–xiii** Gibson Stock Photography; **xiv–xv** Bill Brooks/Masterfile; **xvi–xvii** Peter Gridley/Getty Images; **xviii–xix** Lester Lefkowitz/CORBIS; **2–3** Beneluz Press/Index Stock Imagery/PictureQuest; **4** Jeremy Woodhouse/Masterfile; **12** BP/Taxi/Getty Images; **24** Troy Wayrynen/NewSport/CORBIS; **27** Mt. San Jacinto College, CA; **34** Gary Conner/Index Stock Imagery; **36** CORBIS; **38** Brand X/SuperStock; **40** Gail Mooney/Masterfile; **43** Diaphor Agency/Index Stock Imagery; **49** Aflo Foto Agency/Alamy Images; **58** E. John Thawley III/Alamy Images; **68** Icon SMI/CORBIS; **70** CORBIS; **71** Francisco Cruz/SuperStock; **74** The Everett Collection; **75** CORBIS; **81** Joseph Sohm/ChromoSohm/CORBIS; **82** H. Gousse/AP/Wide World Photos; **86** Bettmann/CORBIS; **88** Buddy Mays/CORBIS; **89** (t)Louis DeLuca/MLB Photos via Getty Images, (bl br)Mark Burnett; **92** Fritz Prenzel/Animals Animals; **96** Jess Stock/Getty Images; **102** Reuters/CORBIS; **108** Randy Wells/Getty Images; **112** HOF/NFL Photos; **114** Michael A. Dwyer/Stock Boston; **116** Alex Wilson/Getty Images; **118** Erika Nelson/World's Largest Things, Lucas, KS; **124** Keith Wood/Getty Images; **138–139** Stephanie Sinclair/Corbis; **140** Michael S. Yamashita/CORBIS; **144 147** CORBIS; **153** The Lawrence Journal-World, Nick Krug/AP/Wide World Photos; **160** Chris Harvey/Getty Images; **165** SuperStock; **169** Nikos Desyllas/SuperStock; **173** Reuters/CORBIS; **175** Gavriel Jecan/Getty Images; **184** NASA; **186** (t)The McGraw-Hill Companies, (b)PHOTOSPORT; **188** Tony Roberts/CORBIS; **189** Eric Sanford/Index Stock Imagery; **190** Michael Boys/CORBIS; **193** (l)Streeter Lecka/Getty Images, (c)CORBIS, (r)Laura Hinshaw/Index Stock Imagery; **199** John Warden/Index Stock Imagery; **201** age fotostock/SuperStock; **206** Carl Schneider/Getty Images; **208** David Young-Wolff/PhotoEdit; **210** Stockbyte/Getty Images; **215** Matt Brown/Icon SMI/CORBIS; **224** Schenectady Museum/Hall of Electrical History Foundation/CORBIS; **226** Bruce Burkhardt/CORBIS; **228** Aaron Haupt; **229** Gail Mooney/Masterfile; **232** NASA; **255** Lori Adamski Peek/Getty Images; **263** Kieran Doherty/REUTERS/Hulton Archive/Getty Images; **264** LWA-Stephen Welstead/CORBIS; **266** Alaska Stock LLC/Alamy Images; **269** Gibson Stock Photography; **274** Bruce Forster/Stone/Getty Images; **280** © Robert Holmes/CORBIS; **282** B2M Productions/Getty Images; **292** Buddy Mays/CORBIS; **296** Ilya Pitalev/ITAR-TASS/CORBIS; **297** Bruce Coleman, Inc./PictureQuest; **302** Galen Rowell/CORBIS; **306** Ticor Collection/San Diego Historical Society; **308** John Downer/Getty Images; **312** Min Roman/Masterfile; **315** Joe Schwartz; **316** Bruce Forster/Getty Images; **318** CORBIS; **319** Square Peg Productions/Getty Images; **329** LWA-Sharie Kennedy/CORBIS; **332** SuperStock; **336** Jose Luis Pelaez, Inc./CORBIS; **342** Andre Jenny/Alamy; **343** Bob Daemmrich; **348–349** Bill Brooks/Masterfile; **356** Ross Kinnaird/Allsport/Getty Images; **362** Duomo/CORBIS; **372** Jose Luis Pelaez, Inc./CORBIS; **374** Bob Daemmrich/PhotoEdit; **380** Paul Barton/CORBIS; **384** CORBIS; **385** Jim Cummins/Getty Images; **387** Robert D. Macklin; **391** Eric Kamp/PhotoTake NYC; **393** Kunio Owaki/CORBIS; **394** Tony Freeman/PhotoEdit; **402** D. Logan/H. Armstrong Roberts; **405** Mark Joseph/Getty Images; **408** Yves Herman/Reuters/CORBIS; **418** Tom Brakefield/Getty Images; **426** Pat Sullivan/AP/Wide World Photos; **430** Douglas Faulkner/Photo Researchers; **438** Steve Vidler/SuperStock; **447** Tim Shaffer/Reuters/CORBIS; **459** Esa Hiltula/Alamy Images; **468** Mark Gibson/Index Stock Imagery; **476** Robert Glusic/Getty Images; **483** Paul J. Sutton/Duomo/CORBIS; **488** Peter Gridley/Getty Images; **490** Stephen Simpson/Getty Images; **495** NASA/Science Source/Photo Researchers; **498** Zoom Agence/Allsport/Getty Images; **502** Don Gibson; **504** Gary Buss/Getty Images; **507** AFP/CORBIS; **511** Bob Krist/CORBIS; **524–525** David Wall/Alamy Images; **526** Joe McBride/Getty Images; **528** Joseph Drivas/Getty Images; **532** Bob Daemmrich/Stock Boston; **539** Alan Schein Photography/CORBIS; **541** Zefa Visual Media/IndexStock/PictureQuest; **544** NASA Dryden Flight Research Photo Collection; **545** Science Photo Library/Photo Researchers; **549** AP Photo/Tina Fineberg; **553** Paul L. Ruben; **556** Michael Zito/SportsChrome; **562** Andy Caulfield/Getty Images; **566** Phillip Spears/Getty Images; **574** Scott Boehm/Getty Images; **581** "Snake and the Cross", 1936. Alexander Calder. Private Collection/Art Resource, NY; **587** Ariel Skelley/CORBIS; **590** Sally Moskol/Index Stock Imagery; **591** Hulton Archive/Getty Images; **592** Bruce Fier/Getty Images; **596** NASA/Roger Ressmeyer/CORBIS; **598** Jay Dickman/CORBIS; **601** Larry Hamill; **605** K. Hackenberg/zefa/CORBIS; **612** Ken Samuelsen/Getty Images; **614** Joseph Sohm/Visions of America/CORBIS; **620** Buccina Studios/Getty Images; **631** Roger Ressmeyer/CORBIS; **640** Visions of America, LLC/Alamy Images; **643** Curtis R. Lantinga/Masterfile; **646** Michael S. Yamashita/CORBIS; **652** Lester Lefkowitz/CORBIS; **660** Inga Spence/Index Stock Imagery; **668** Tony Freeman/PhotoEdit; **673** Getty Images; **678** (t)CORBIS, (b)Geoff Butler; **682** O'Brien Productions/CORBIS; **684** Eclipse Studios.

Index

B

Index

Index

F

G

Index

Index

Index